TIME-DOMAIN METHODS FOR MICROWAVE STRUCTURES

IEEE Press
445 Hoes Lane, P.O. Box 1331
Piscataway, NJ 08855–1331

TIME-DOMAIN METHODS FOR MICROWAVE STRUCTURES

Analysis and Design

Edited by

Tatsuo Itoh
University of California at Los Angeles

Bijan Houshmand
*Jet Propulsion Laboratory,
California Institute of Technology*

IEEE PRESS

A Selected Reprint Volume

IEEE Microwave Theory and Techniques Society, *Sponsor*

The Institute of Electrical and Electronics Engineers, Inc., New York

This book may be purchased at a discount from the publisher
when ordered in bulk quantities. Contact:

IEEE Press Marketing
Attn: Special Sales
445 Hoes Lane, P.O. Box 1331
Piscataway, NJ 08855–1331
Fax: (732) 981-9334

For more information on the IEEE Press,
visit the IEEE home page: http://www.ieee.org/

Printed in the United States of America

10 9 8 7 6 5 4 3 2 1

ISBN 0-7803-1109-4

IEEE Order Number: PC4630

Library of Congress Cataloging-in-Publication Data

Time-domain methods for microwave structures / edited by Tatsuo Itoh, Bijan Houshmand.
p. cm.
"A selected reprint volume."
Includes bibliographical references and index.
ISBN 0-7803-1109-4 (alk. paper)
1. Microwave devices—Design and construction—Data processing. 2. Time-domain analysis. 3. Finite differences. I. Itoh, Tatsuo (date). II. Houshmand, Bijan (date).
TK7876.T5317 1997
621.381′33—dc21 97-35516
CIP

Contents

Chapter 3

Chapter 4

Chapter 5

Chapter 6

Preface

The purpose of this hybrid book is to provide a comprehensive collection of topics relating to the application of the Finite-Difference Time-Domain (FDTD) method to microwave structures. Important subjects such as numerical issues, geometry description, vector and parallel processing, methods to reduce the requirements for computational resources, and illustrative examples which demonstrate the range of application and level of numerical accuracy are covered. Each chapter consists of two parts: the first is a monograph on the specific subject written in a self-sustaining fashion; the second is a collection of selected reprints illustrating the state-of-the-art publication on the subject matter. It is our belief that this hybridization provides substantial utilities to the readers. Readers may study the fundamentals of the topics discussed in each chapter on one hand, and, on the other hand, they are exposed to the broad range of reference materials which by themselves have created milestones in the technology. Additionally, for busy scientists and engineers, the volume provides a quick reference. The FDTD papers have appeared in many different publications due to the popularity and variety of applications of the subject. As such, a search for the vast size of information is quite time-consuming for anyone; this volume was created to save such effort.

Chapter 1 begins with an overview of the FDTD algorithm. The Maxwell's equations in the Cartesian coordinate system are given, along with the corresponding finite-difference equations. The difference equations are the result of the center-difference approach which results in the leap-frog FDTD algorithm. A block diagram of the FDTD algorithm is also presented illustrating the sequence of operations for a time-domain simulation, including excitation, electromagnetic field updates, treatment of dielectric discontinuity and perfect electric conductors, and the absorbing boundary conditions.

Chapter 2 covers the fundamental theoretical and numerical aspects of the FDTD algorithm. The difference equations for all components of the electric and magnetic fields are derived from the time-domain Maxwell's equations. This chapter covers numerical stability, numerical dispersion, and the excitation issues for the FDTD algorithm. The absorbing boundary condition (ABC) is also discussed and detailed treatment of the Berenger's ABC is presented.

Chapter 3 describes the Conformal FDTD methods. The motivation for employing these methods is the added flexibility in the geometry representation. This chapter covers the local mesh refinement, curvilinear, and non-orthogonal grid. In addition, the finite element and finite volume formulation are also described. The numerical accuracy, stability, and computation cost for implementation of these formulations are also presented.

Chapter 4 covers the computation efficiency of the FDTD algorithm on various computer platforms ranging from RISC-based workstations to vector and parallel computers. This chapter first addresses the implementation of the FDTD algorithm on RISC-based workstations and vector processors. Then, the implementation of this algorithm on parallel computers is presented.

Chapter 5 addresses the issue of reduction in computational requirements for the FDTD algorithm. Computational requirements consist of memory and CPU time allocations. Signal processing algorithms are applied to the FDTD generated waveforms in order to efficiently extract parameters of interest such as S-parameters and resonant frequencies. The FDTD Diakoptics reduces the memory requirements by representing segments of the FDTD computation domain by their impulse responses. The formulation and numerical issues associated with these methodologies are presented.

Chapter 6 presents the application of the FDTD analysis technique to the simulation of the electromagnetic fields and the design of microwave integrated circuits. The use of the FDTD algorithm is demonstrated through a number of examples. The issues of excitation, absorbing boundary conditions, mesh refinements, and application of the signal processing methods are presented.

We would like to thank all the authors for their contributions, and all the reviewers for their thoughtful comments and suggestions.

Tatsuo Itoh
Department of Electrical Engineering,
University of California at Los Angeles

Bijan Houshmand
Jet Propulsion Laboratory,
California Institute of Technology

Introduction to FDTD Method for Planar Microwave Structures

BIJAN HOUSHMAND
JET PROPULSION LABORATORY
CALIFORNIA INSTITUTE OF TECHNOLOGY
4800 OAK GROVE DRIVE
PASADENA, CA 91109-8099

TATSUO ITOH, FELLOW, IEEE
DEPARTMENT OF ELECTRICAL ENGINEERING
UNIVERSITY OF CALIFORNIA AT LOS ANGELES
LOS ANGELES, CA 90024-1594

I. Introduction

Modeling of microwave structures has undergone rapid growth in the past decade. The driving force for this growth has been a combination of factors. These range from increasing the operating frequencies and integrating a complete system on a single module in which the proximity coupling becomes important, to the need for reducing the design cycle in which a design is expected to operate properly without further modifications after the initial fabrication. The accompanying growth in the available computation power has also provided a synergy for this growth. At this point a wealth of methodologies are available to address various modeling needs for microwave structures. These methodologies can be classified broadly as frequency- and time-domain methods. Each of these two classes can also be subdivided into more specialized categories such as integral and differential methods, or spectral and spatial methods. The pioneering work on the implementation of a numerical algorithm for the time solution of Maxwell's equations was presented by K. S. Yee in 1966 [1]. This method is currently known as the finite-difference time-domain (FDTD). References [2] and [3] provide a general comparison of the numerical properties of the FDTD and other popular numerical methods. The FDTD method is easy to implement and has been successfully used for simulating the performance of a broad class of microwave structures such as waveguides, multi-layer structures, discontinuities, transitions, filters, couplers, and antennas. Figure 1 illustrates the building blocks of the FDTD algorithm. Figures 2 and 3 show some of the recent applications of the FDTD to realistic microwave structures which include both active and non-linear devices. The details of the formulation, numerical issues, computer implementation, and range of applications of this method will be covered in this book.

II. FDTD Algorithm

The FDTD algorithm is derived directly from Maxwell's equations in the time domain. The components of the electric and magnetic fields in the Cartesian coordinates with the constitutive parameters (ε, μ), and an electric current vector source are given by:

$$-\mu \frac{\partial H_x}{\partial t} = \frac{\partial E_z}{\partial y} - \frac{\partial E_y}{\partial z} \tag{1}$$

$$-\mu \frac{\partial H_y}{\partial t} = \frac{\partial E_x}{\partial z} - \frac{\partial E_z}{\partial x} \tag{2}$$

$$-\mu \frac{H_z}{\partial t} = \frac{\partial E_y}{\partial x} - \frac{\partial E_x}{\partial y} \tag{3}$$

$$\varepsilon \frac{\partial E_x}{\partial t} = \frac{\partial H_z}{\partial y} - \frac{\partial H_y}{\partial z} - J_x \tag{4}$$

$$\varepsilon \frac{\partial E_y}{\partial t} = \frac{\partial H_x}{\partial z} - \frac{\partial H_z}{\partial y} - J_y \tag{5}$$

$$\varepsilon \frac{\partial E_z}{\partial t} = \frac{\partial H_y}{\partial x} - \frac{\partial H_x}{\partial y} - J_z \tag{6}$$

The flow graph in Fig. 1 shows the structure of the FDTD algorithm. The first step for the FDTD simulation of a microwave structure is the geometry and material description. The dielectric interfaces and location of the perfectly conducting surfaces are specified on the computational grid in this step. Chapter 3 is devoted to issues regarding the geometry description. The second step is the specification of the excitation signal. Issues regarding the signal excitation such as signal type and frequency

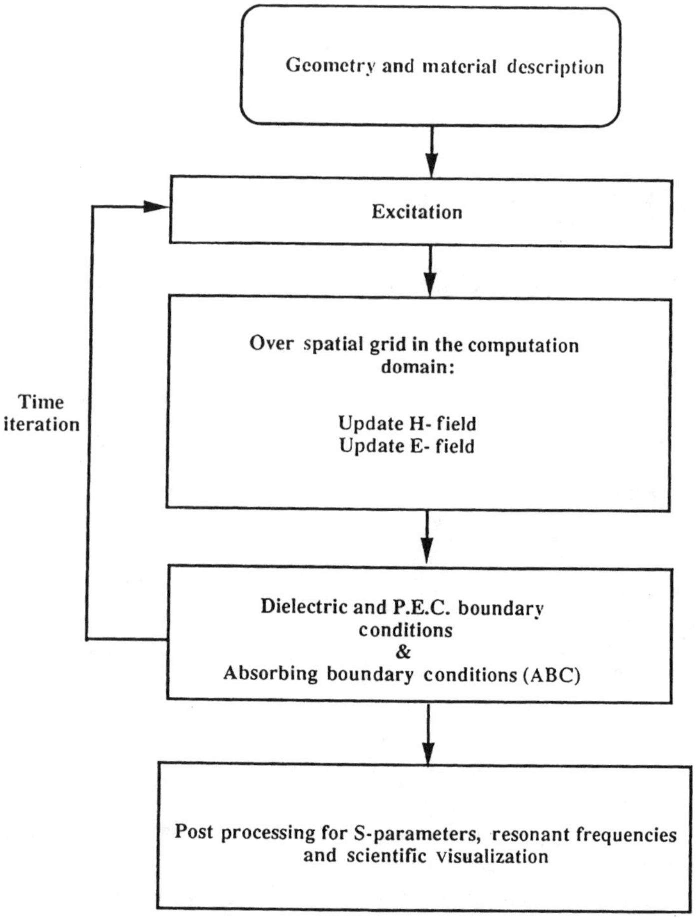

Fig. 1 Block diagram of the major building blocks for the FDTD algorithm.

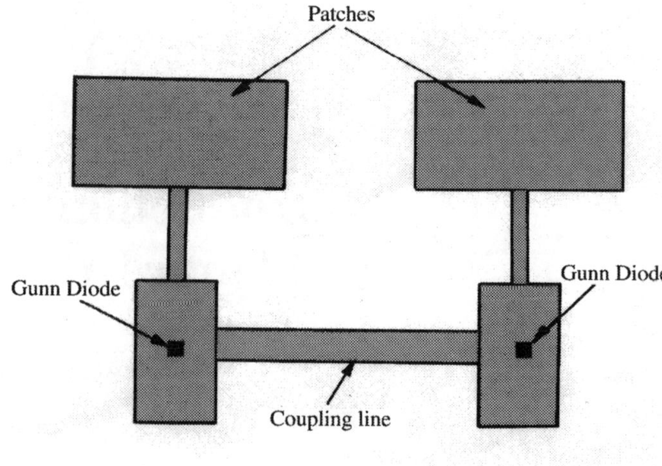

(a)

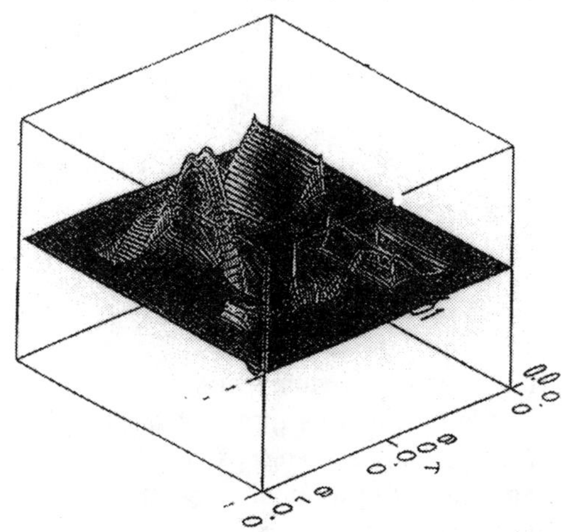

(b)

Fig. 2 Simulation of a two-element active antenna: (a) structure layout and (b) the steady field distribution perpendicular to the dielectric-air interface.

bandwidth are described in Chapter 2. Once the excitation signal is specified, the FDTD algorithm updates all the field components on the computational grid for each time step. The update equations are modified for the dielectric interfaces, and tangential electric field components are set to zero on the grid points corresponding to perfectly conducting layers. The tangential field components on the outer layers of the computational grid are updated using the absorbing boundary conditions. The field update equations are derived from Eq. (1) through (6). These equations are discretized with respect to the time and spatial variables. It can be shown that the central differencing scheme results in a second order accuracy for the field components with respect to the discretization variables. The resulting difference equations for the electric and magnetic field components are as follows:

$$H^{l+1/2}_{x\,i,j+1/2,k+1/2} = H^{l-1/2}_{x\,i,j+1/2,k+1/2} + \frac{\Delta t}{\mu \Delta z}(E^{l}_{y\,i,j,k} - E^{l}_{y\,i,j,k+1}) - \frac{\Delta t}{\mu \Delta y}(E^{l}_{z\,i,j,k} - E^{l}_{z\,i,j+1,k}) \tag{7}$$

$$H^{l+1/2}_{y\,i+1/2,j,k+1/2} = H^{l-1/2}_{y\,i+1/2,j,k+1/2} + \frac{\Delta t}{\mu \Delta z}(E^{l}_{x\,i,j,k} - E^{l}_{x\,i,j,k+1}) - \frac{\Delta t}{\mu \Delta x}(E^{l}_{z\,i,j,k} - E^{l}_{x\,i+1,j,k}) \tag{8}$$

$$H^{l+1/2}_{z\,i+1/2,j+1/2,k} = H^{l-1/2}_{z\,i+1/2,j+1/2,k} + \frac{\Delta t}{\mu \Delta y}(E^{l}_{x\,i,j,k} - E^{l}_{x\,i,j+1,k}) - \frac{\Delta t}{\mu \Delta x}(E^{l}_{y\,i,j,k} - E^{l}_{z\,i+1,j,k}) \tag{9}$$

$$E^{l+1}_{x\,i,j,k} = E^{l}_{x\,i,j,k} + \frac{\Delta t}{\varepsilon \Delta y}(H^{l+1/2}_{z\,i+1/2,j+1/2,k} - H^{l+1/2}_{z\,i+1/2,j-1/2,k}) - \frac{\Delta t}{\varepsilon \Delta z}(H^{l+1/2}_{y\,i+1/2,j,k+1/2} - H^{l+1/2}_{y\,i+1/2,j,k-1/2}) - \frac{\Delta t}{\varepsilon} J^{e+1/2}_{x_{i,j,k}} \tag{10}$$

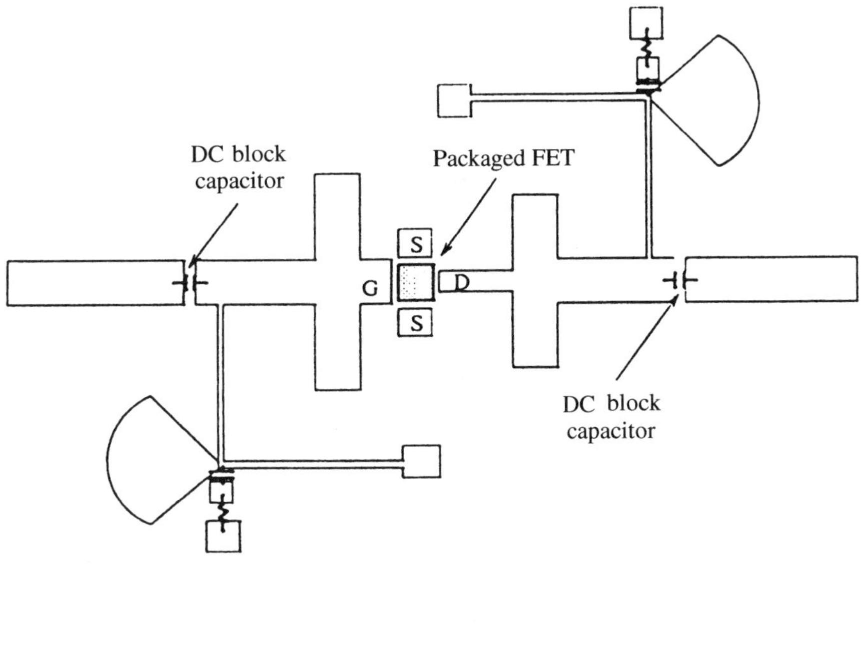

(a)

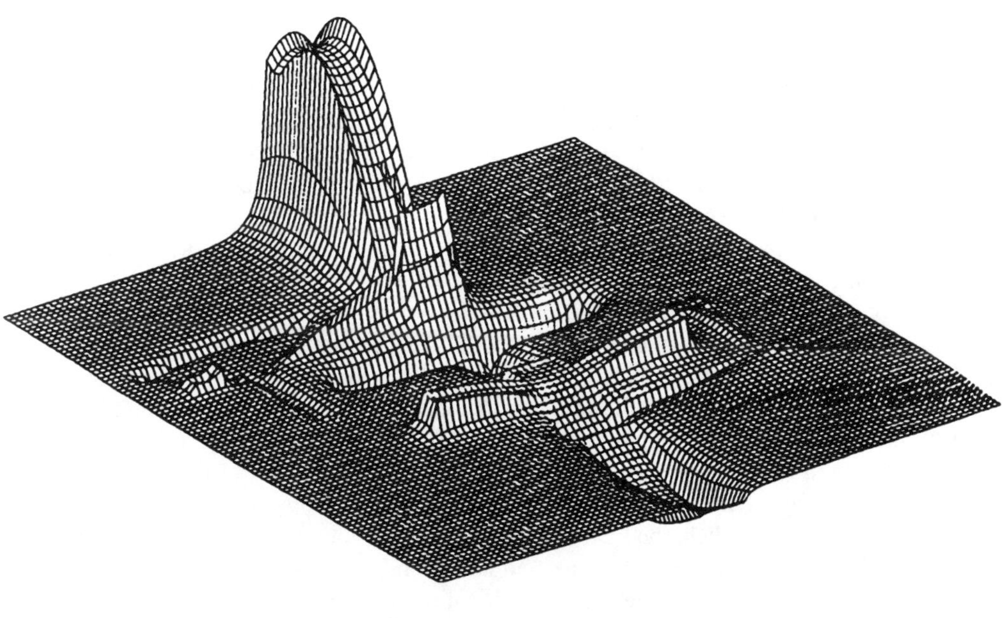

(b)

Fig. 3 Simulation of microwave amplifier: (a) structure layout and (b) the field distribution of the electric field component perpendicular to the dielectric-air interface at a time instant when the excitation pulse has traveled through the amplifier circuit.

$$E^{l+1}_{y\,i,j,k} = E^{l}_{y\,i,j,k} + \frac{\Delta t}{\varepsilon \Delta y}\left(H^{l+1/2}_{x\,i,j+1/2,k+1/2} - H^{l+1/2}_{x\,i,j+1/2,k-1/2}\right) - \frac{\Delta t}{\varepsilon \Delta x}\left(H^{l+1/2}_{z\,i+1/2,j+1/2,k} - H^{l+1/2}_{y\,i+1/2,j+1/2,k}\right) - \frac{\Delta t}{\varepsilon} J^{e+1/2}_{y_{i,j,k}} \quad (11)$$

$$E^{l+1}_{z\,i,j,k} = E^{l}_{z\,i,j,k} + \frac{\Delta t}{\varepsilon \Delta x}\left(H^{l+1/2}_{y\,i+1/2,j,k+1/2} - H^{l+1/2}_{y\,i-1/2,j,k+1/2}\right) - \frac{\Delta t}{\varepsilon \Delta y}\left(H^{l+1/2}_{x\,i,j+1/2,k+1/2} - H^{l+1/2}_{x\,i,j-1/2,k+1/2}\right) - \frac{\Delta t}{\varepsilon} J^{e+1/2}_{z_{i,j,k}} \quad (12)$$

Here, Δx, Δy, Δz are the spatial discretizations and the indices (i, j, k) correspond to the x, y, and z directions, respectively. The index (l) corresponds to the temporal discretization Δt. It is observed that the electric and magnetic fields are each displaced by a half step both in time and space indices. The relationship among the spatial and temporal discretizations is governed by the stability criteria, which is discussed in Chapter 2. The above update laws for the field components are modified for the nodes which are located in the boundary planes of the computational domain. The modified update laws, in general, simulate an outgoing wave and are called the absorbing boundary conditions (ABCs). The ABCs are also discussed in Chapter 2.

III. Future Trends and Growing Opportunities

As our understanding of the numerical issues related to the FDTD algorithms improves, the range of applications of this methodology also increases. The FDTD algorithms are being applied to radiation, scattering, and propagation problems where realistic material properties and geometry have been taken into account. For example, microwave structures including active elements have been analyzed by this method [4]. This is in contrast to the traditional approach where the passive structure is analyzed separately and the resulting S-parameters are combined with the active devices in a circuit simulator. This methodology will provide a more realistic account of wave-structure interaction with an increase in the operating frequencies and reduction in the structure dimensions. The FDTD has also been successfully applied to optical structures with nonlinear media [5,6]. Reference [5] presents first-time calculations of vector electromagnetic field solution of spatial optical soliton propagation and mutual deflection. Finally, the ability of the FDTD method for providing accurate computation of wave-material interaction can be utilized for study of wave interaction with biological materials. Reference [7] presents the application of the FDTD domain to the problem of cellular phone radiation interaction with the human body. FDTD is also applied to acoustic wave propagation as presented in Reference [8]. It is worth mentioning that the FDTD algorithms remain essentially the same for all the mentioned applications. As such, the FDTD is a truly general wave solver for Maxwell's equations, which is awaiting applications to a wide range of practical problems. It is hoped that the readers of this book might acquire a chance to look into such applications simulated herein.

IV. Selected Papers

The first paper is K. S. Yee's classic article which introduces the numerical method for the time-domain solutions of Maxwell's equations. The basic finite-difference equations for the field components are presented, and the locations of the electric and magnetic field components on a cell (currently known as the Yee's cell) are illustrated. The stability condition and relationship among the spatial and temporal discretizations are also presented. The author notes that the algorithm is computationally expensive for the computers available at the time. The method is applied to scattering from a perfectly conducting 2-D wedge.

The second paper is an overview of various numerical techniques for analysis and optimization of microwave structures. Both time-domain and frequency-domain methods are considered. These methods are broadly classified as differential and integral methods and further divided into time and frequency domains. The computational issues such as CPU time and memory requirements for time and frequency methods are discussed. The FDTD and mode matching methods are applied to the packaged microstrip via and the simulations results are compared with experimental measurements.

The third paper compares the field solutions for an aperture coupled multi-layer planar structure using the FDTD and the extended spectral domain methods. The metal thickness is taken into account for this circuit structure. The conventional spectral domain approach is modified to include the thickness effect. The computed S-parameters for the four-port network using both methods are presented.

The fourth paper presents the application of the FDTD method to microwave amplifiers. For this simulation the small signal model of the active device is included in the computation of the electromagnetic field components. The simulation includes the effects of passive matching network and the biasing structure. The simulation results are compared with measurements.

The fifth paper presents the application of the FDTD to optical structures. Field solution for spatial soliton propagation and mutual deflection in a 2-D homogeneous nonlinear dielectric medium is presented. The results are compared with the nonlinear Schrödinger equation models and numerical propagation models which generally make the paraxial approximation. It is noted that the FDTD algorithm provides a robust solution for this class of problems.

The sixth paper presents the formulation of a nonlinear FDTD Maxwell's equations solver. This method is used to model the interaction of an ultra short, optical pulsed Gaussian beam with a Kerr nonlinear material. Pulse-beam self-focusing, scattering of a pulsed-beam from a linear/nonlinear interface, and pulsed-beam propagation in nonlinear waveguides are presented. The formulation details are given for 2-D Maxwell's equations in Section II. Numerical results including the computational requirements are presented in Section III.

The seventh paper presents the application of the FDTD algorithm for modeling the interaction of biological material with electromagnetic waves. This method is applied to the study of radiation effects from a cellular phone on a human head. The 3-D field solutions are obtained by the FDTD algorithm, and the specific absorption rate (SAR) distributions are computed for an inhomogeneous model of a human head exposed to the electromagnetic waves radiated from a cellular phone.

The eighth paper presents the application of the FDTD method to the problem of the acoustic wave propagation. Acoustic equations are transformed into the finite difference equation similar to Maxwell's equations. The grid structure, grid dispersion, stability, and absorbing boundary conditions are discussed. This method is applied to sound reflection from a 3-D unflanged, open-ended circular cylinder. The FDTD-generated results are compared with the analytical solutions.

References

[1] K. S. Yee, "Numerical solution of initial boundary value problems involving Maxwell's equations in isotropic media," *IEEE Trans. Antennas Propagat.,* vol. AP-14, pp. 302–307, May 1966.

[2] Roberto Sorrentino, "Modeling and design of millimeter passive circuits: From 2 to 3D," 24th European Microwave Conference, vol. 1, pp. 48–61, 1994.

[3] Allan Tran, B. Houshmand, and T. Itoh, "Analysis of electromagnetic coupling through a thick aperture in multi-layer planar circuits using the extended spectral domain approach and finite difference time domain method," *IEEE Trans. Antennas Propagat.,* vol. 43, no. 9, pp. 921–926, 1995.

[4] C. N. Kuo, R. B. Wu, B. Houshmand, T. Itoh, "Modeling of microwave active devices using the FDTD analysis based on the voltage-source approach," *IEEE Microwave Guided Wave Lett.,* vol. 6, no. 5, pp. 199–201, May 1996.

[5] R. M. Joseph and A. Taflove, "Spatial soliton deflection mechanism indicated by FD-TD Maxwell's equations modeling," IEEE Photonics Technol. Lett., vol. 6, no. 10, October 1994.

[6] R. W. Ziolkowski and J. B. Judkins, "Application of the nonlinear finite difference time domain (NL-FDTD) method to pulse propagation in nonlinear media: Self-focusing and linear interfaces," *Radio Sci.,* vol. 28, pp. 901–911, 1993.

[7] H. Y. Chen and H. H. Wang, "Current and SAR induced in a human head model by the electromagnetic fields irradiated from a cellular phone," *IEEE Trans. Microwave Theory and Tech.,* vol. 42, no. 12, pp. 2249–2254, 1994.

[8] J. G. Maloney and Kathleen E. Cummings, "Adaptation of FDTD techniques to acoustic modeling," Proceedings of Applied Computational Electromagnetics at the Naval Postgraduate School, Monterey, CA, 1995, pp. 724–731.

Numerical Solution of Initial Boundary Value Problems Involving Maxwell's Equations in Isotropic Media

KANE S. YEE

Abstract—**Maxwell's equations are replaced by a set of finite difference equations. It is shown that if one chooses the field points appropriately, the set of finite difference equations is applicable for a boundary condition involving perfectly conducting surfaces. An example is given of the scattering of an electromagnetic pulse by a perfectly conducting cylinder.**

Introduction

SOLUTIONS to the time-dependent Maxwell's equations in general form are unknown except for a few special cases. The difficulty is due mainly to the imposition of the boundary conditions. We shall show in this paper how to obtain the solution numerically when the boundary condition is that appropriate for a perfect conductor. In theory, this numerical attack can be employed for the most general case. However, because of the limited memory capacity of present day computers, numerical solutions to a scattering problem for which the ratio of the characteristic linear dimension of the obstacle to the wavelength is large still seem to be impractical. We shall show by an example that in the case of two dimensions, numerical solutions are practical even when the characteristic length of the obstacle is moderately large compared to that of an incoming wave.

A set of finite difference equations for the system of partial differential equations will be introduced in the early part of this paper. We shall then show that with an appropriate choice of the points at which the various field components are to be evaluated, the set of finite difference equations can be solved and the solution will satisfy the boundary condition. The latter part of this paper will specialize in two-dimensional problems, and an example illustrating scattering of an incoming pulse by a perfectly conducting square will be presented.

Manuscript received August 24, 1965; revised January 28, 1966. This work was performed under the auspices of the U. S. Atomic Energy Commission.

The author is with the Lawrence Radiation Lab., University of California, Livermore, Calif.

Maxwell's Equation and the Equivalent Set of Finite Difference Equations

Maxwell's equations in an isotropic medium [1] are:[1]

$$\frac{\partial B}{\partial t} + \nabla \times E = 0, \tag{1a}$$

$$\frac{\partial D}{\partial t} - \nabla \times H = J, \tag{1b}$$

$$B = \mu H, \tag{1c}$$

$$D = \epsilon E, \tag{1d}$$

[1] In MKS system of units.

Reprinted from *IEEE Transactions on Antennas and Propagation*, pp. 302-307, May 1966.

where J, μ, and ϵ are assumed to be given functions of space and time.

In a rectangular coordinate system, (1a) and (1b) are equivalent to the following system of scalar equations:

$$-\frac{\partial B_x}{\partial t} = \frac{\partial E_z}{\partial y} - \frac{\partial E_y}{\partial z}, \tag{2a}$$

$$-\frac{\partial B_y}{\partial t} = \frac{\partial E_x}{\partial z} - \frac{\partial E_z}{\partial x}, \tag{2b}$$

$$\frac{\partial B_z}{\partial t} = \frac{\partial E_x}{\partial y} - \frac{\partial E_y}{\partial x}, \tag{2c}$$

$$\frac{\partial D_x}{\partial t} = \frac{\partial H_z}{\partial y} - \frac{\partial H_y}{\partial z} - J_x, \tag{2d}$$

$$\frac{\partial D_y}{\partial t} = \frac{\partial H_x}{\partial z} - \frac{\partial H_z}{\partial x} - J_y, \tag{2e}$$

$$\frac{\partial D_z}{\partial t} = \frac{\partial H_y}{\partial x} - \frac{\partial H_x}{\partial y} - J_z, \tag{2f}$$

where we have taken $\boldsymbol{A} = (A_x, A_y, A_z)$. We denote a grid point of the space as

$$(i, j, k) = (i\Delta x, j\Delta y, k\Delta z) \tag{3}$$

and for any function of space and time we put

$$F(i\Delta x, j\Delta y, k\Delta z, n\Delta t) = F^n(i, j, k). \tag{4}$$

A set of finite difference equations for (2a)–(2f) that will be found convenient for perfectly conducting boundary condition is as follows.

For (2a) we have

$$\frac{B_x^{n+1/2}(i, j+\frac{1}{2}, k+\frac{1}{2}) - B_x^{n-1/2}(i, j+\frac{1}{2}, k+\frac{1}{2})}{\Delta t} = \frac{E_y^n(i, j+\frac{1}{2}, k+1) - E_y^n(i, j+\frac{1}{2}, k)}{\Delta z} - \frac{E_z^n(i, j+1, k+\frac{1}{2}) - E_z^n(i, j, k+\frac{1}{2})}{\Delta y}. \tag{5}$$

The finite difference equations corresponding to (2b) and (2c), respectively, can be similarly constructed.

For (2d) we have

$$\frac{D_x^n(i+\frac{1}{2}, j, k) - D_x^{n-1}(i+\frac{1}{2}, j, k)}{\Delta t} = \frac{H_z^{n-1/2}(i+\frac{1}{2}, j+\frac{1}{2}, k) - H_z^{n-1/2}(i+\frac{1}{2}, j-\frac{1}{2}, k)}{\Delta y} - \frac{H_y^{n-1/2}(i+\frac{1}{2}, j, k+\frac{1}{2}) - H_y^{n-1/2}(i+\frac{1}{2}, j, k-\frac{1}{2})}{\Delta z} + J_x^{n-1/2}(i+\frac{1}{2}, j, k). \tag{6}$$

The equations corresponding to (2e) and (2f), respectively, can be similarly constructed.

The grid points for the E-field and the H-field are chosen so as to approximate the condition to be discussed below as accurately as possible. The various grid positions are shown in Fig. 1.

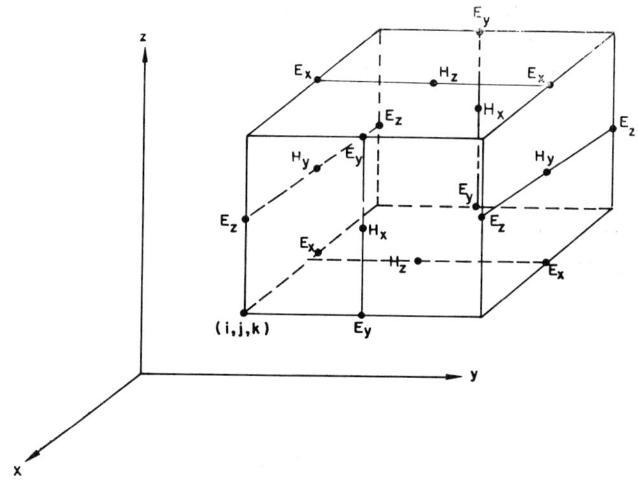

Fig. 1. Positions of various field components. The E-components are in the middle of the edges and the H-components are in the center of the faces.

Boundary Conditions

The boundary condition appropriate for a perfectly conducting surface is that the tangential components of the electric field vanish. This condition also implies that the normal component of the magnetic field vanishes on the surface. The conducting surface will therefore be approximated by a collection of surfaces of cubes, the sides of which are parallel to the coordinate axes. Plane surfaces perpendicular to the x-axis will be chosen so as to contain points where E_y and E_z are defined. Similarly, plane surfaces perpendicular to the other axes are chosen.

Grid Size and Stability Criterion

The space grid size must be such that over one increment the electromagnetic field does not change significantly. This means that, to have meaningful results, the linear dimension of the grid must be only a fraction of the wavelength. We shall choose $\Delta x = \Delta y = \Delta z$. For computational stability, it is necessary to satisfy a relation between the space increment and time increment Δt. When ϵ and μ are variables, a rigorous stability criterion is difficult to obtain. For constant ϵ and μ, computational stability requires that

$$\sqrt{(\Delta x)^2 + (\Delta y)^2 + (\Delta z)^2} > c\Delta t = \sqrt{\frac{1}{\epsilon\mu}}\,\Delta t, \tag{7}$$

where c is the velocity of light. If $c_{\max}$ is the maximum light velocity in the region concerned, we must choose

$$\sqrt{(\Delta x)^2 + (\Delta y)^2 + (\Delta z)^2} > c_{\max}\Delta t. \tag{8}$$

This requirement puts a restriction on Δt for the chosen Δx, Δy, and Δz.

Maxwell's Equations in Two Dimensions

To illustrate the method, we consider a scattering problem in two dimensions. We shall assume that the field components do not depend on the z coordinate of a point. Furthermore, we take ϵ and μ to be constants and $J \equiv 0$. The only source of our problem is then an "incident" wave. This incident wave will be "scattered" after it encounters the obstacle. The obstacle will be of a few "wavelengths" in its linear dimension. Further simplification can be obtained if we observe the fact that in cylindrical coordinates we can decompose any electromagnetic field into "transverse electric" and "transverse magnetic" fields if ϵ and μ are constants. The two modes of electromagnetic waves are characterized by

1) Transverse electric wave (TE)

$$H_x = H_y = 0, \qquad E_z = 0,$$
$$-\mu \frac{\partial H_z}{\partial t} = \frac{\partial E_y}{\partial x} - \frac{\partial E_x}{\partial y},$$
$$\epsilon \frac{\partial H_z}{\partial y} = \frac{\partial E_x}{\partial t}, \qquad -\frac{\partial H_z}{\partial x} = \epsilon \frac{\partial E_y}{\partial t}, \tag{9}$$

and

2) Transverse magnetic wave (TM)

$$E_x = E_y = 0, \qquad H_z = 0,$$
$$\epsilon \frac{\partial E_z}{\partial t} = \frac{\partial H_y}{\partial x} - \frac{\partial H_x}{\partial y}$$
$$\mu \frac{\partial H_x}{\partial t} = -\frac{\partial E_z}{\partial y}, \qquad \mu \frac{\partial H_y}{\partial t} = \frac{\partial E_z}{\partial x}. \tag{10}$$

Let C be a perfectly conducting boundary curve. We approximate it by a polygon whose sides are parallel to the coordinate axes. If the grid dimensions are small compared to the wavelength, we expect the approximation to yield meaningful results.

Letting

$$\tau = ct = \sqrt{\frac{1}{\mu\epsilon}}\, t \tag{11}$$

and

$$Z = \sqrt{\frac{\mu}{\epsilon}} = 376.7, \tag{12}$$

we can write the finite difference equations for the TE and TM waves.

TE waves:

$$H_z^{n+1/2}(i+\tfrac{1}{2}, j+\tfrac{1}{2}) = H_z^{n-1/2}(i+\tfrac{1}{2}, j+\tfrac{1}{2}) - \frac{1}{Z}\frac{\Delta\tau}{\Delta x}\left[E_y^n(i+1, j+\tfrac{1}{2}) - E_y^n(i, j+\tfrac{1}{2})\right] + \frac{1}{Z}\frac{\Delta\tau}{\Delta y}\, E_x^n(i+\tfrac{1}{2}, j+1) - E_x^n(i+\tfrac{1}{2}, j)] \tag{13a}$$

$$E_x^{n+1}(i+\tfrac{1}{2}, j) = E_x^n(i+\tfrac{1}{2}, j) + Z\frac{\Delta\tau}{\Delta y}\left[H_z^{n+1/2}(i+\tfrac{1}{2}, j+\tfrac{1}{2}) - H_z^{n+1/2}(i+\tfrac{1}{2}, j-\tfrac{1}{2})\right] \tag{13b}$$

$$E_y^{n+1}(i, j+\tfrac{1}{2}) = -Z\frac{\Delta\tau}{\Delta x}\left[H_z^{n+1/2}(i+\tfrac{1}{2}, j+\tfrac{1}{2}) - H_z^{n+1/2}(i-\tfrac{1}{2}, j+\tfrac{1}{2})\right]. \tag{13c}$$

TM waves:

$$E_z^{n+1}(i, j) = E_z^n(i, j) + Z\frac{\Delta\tau}{\Delta x}\left[H_y^{n+1/2}(i+\tfrac{1}{2}, j) - H_y^{n+1/2}(i-\tfrac{1}{2}, j)\right] - Z\frac{\Delta\tau}{\Delta y}\left[H_x^{n+1/2}(i, j+\tfrac{1}{2}) - H_x^{n+1/2}(i, j-\tfrac{1}{2})\right] \tag{14a}$$

$$H_x^{n+1/2}(i, j+\tfrac{1}{2}) = H_x^{n-1/2}(i, j+\tfrac{1}{2}) - \frac{1}{Z}\frac{\Delta\tau}{\Delta y}\left[E_z^n(i, j+1) - E_z^n(i, j)\right] \tag{14b}$$

$$H_y^{n+1/2}(i+\tfrac{1}{2}, j) = H_y^{n-1/2}(i+\tfrac{1}{2}, j) + \frac{1}{Z}\frac{\Delta\tau}{\Delta x}\left[E_z^n(i+1, j) - E_z^n(i, j)\right]. \tag{14c}$$

Numerical Computations for TM Waves

For further numerical discussion we shall limit ourselves to the TM waves. In this case we use the finite difference equations (14a)–(14c). The values for $E_z^0(i, j)$, $H_y^{1/2}(i+\frac{1}{2}, j)$, and $H_x^{1/2}(i, j-\frac{1}{2})$ are obtained from the incident wave.[2] Subsequent values are evaluated from the finite difference equations (14a)–(14c). The boundary condition is approximated by putting the boundary value of $E_z^n(i, j)$ equal to zero for any n.

To be specific, we shall consider the diffraction of an incident TM wave by a perfectly conducting square. The dimensions of the obstacle, as well as the profile of the incident wave, are shown in Fig. 2.

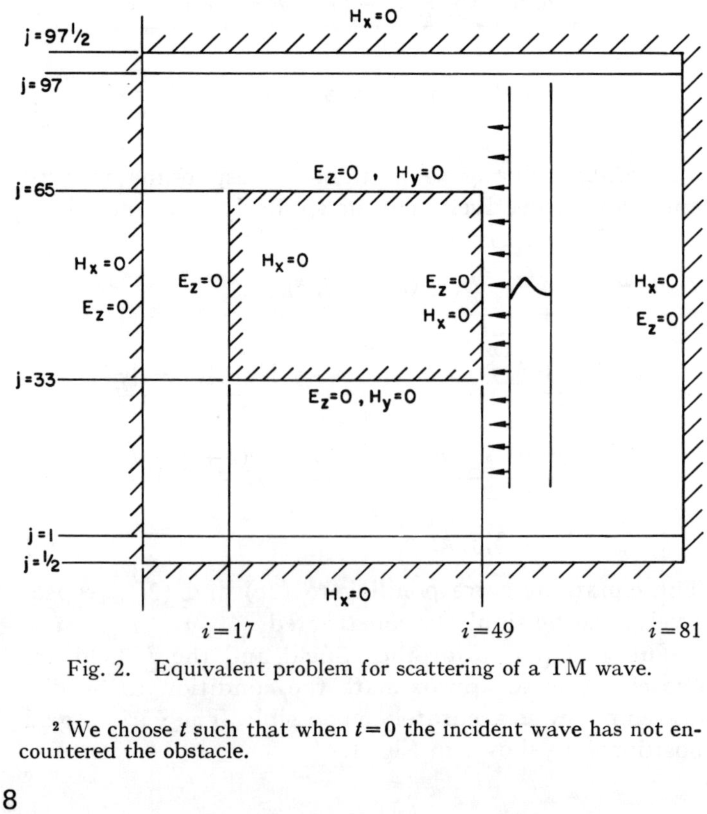

Fig. 2. Equivalent problem for scattering of a TM wave.

[2] We choose t such that when $t=0$ the incident wave has not encountered the obstacle.

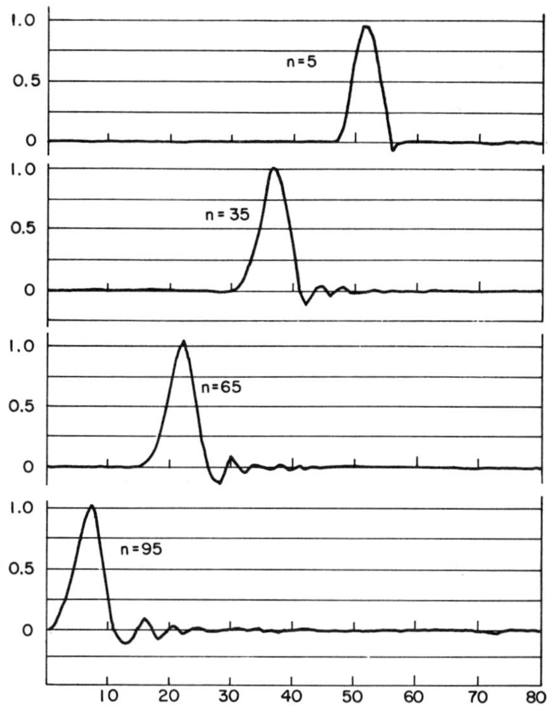

Fig. 3. Results of the calculation of E_z by means of (14a)–(14c) in the absence of the obstacle. The ordinate is in volts/meter and the abscissa is the number of horizontal increments. n is the number of time cycles.

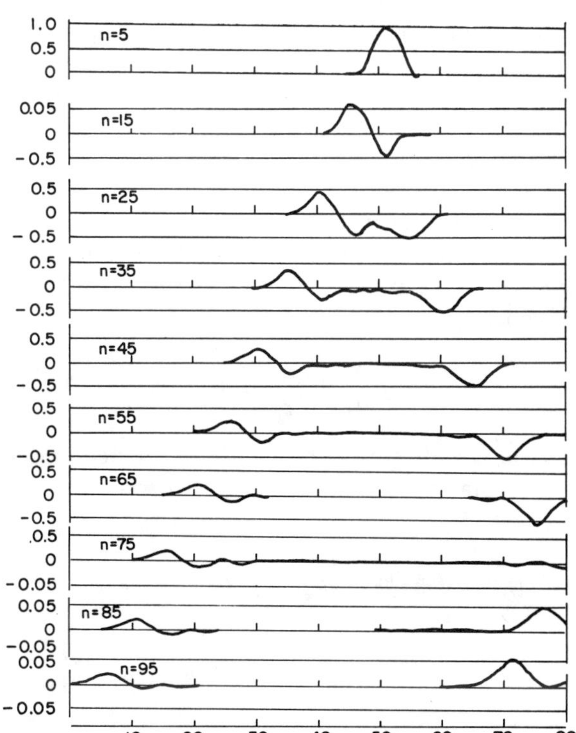

Fig. 4. E_z of the TM wave in the presence of the obstacle for various time cycles when $j=30$.

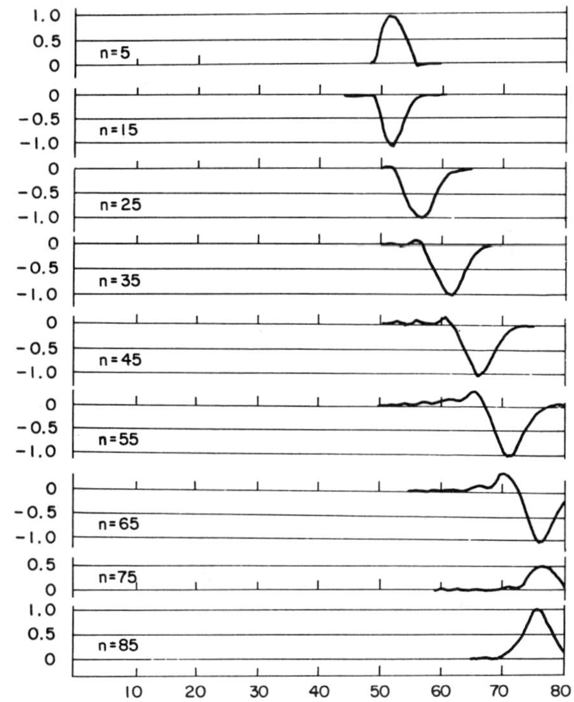

Fig. 5. E_z of the TM wave for various time cycles. $j=50$.

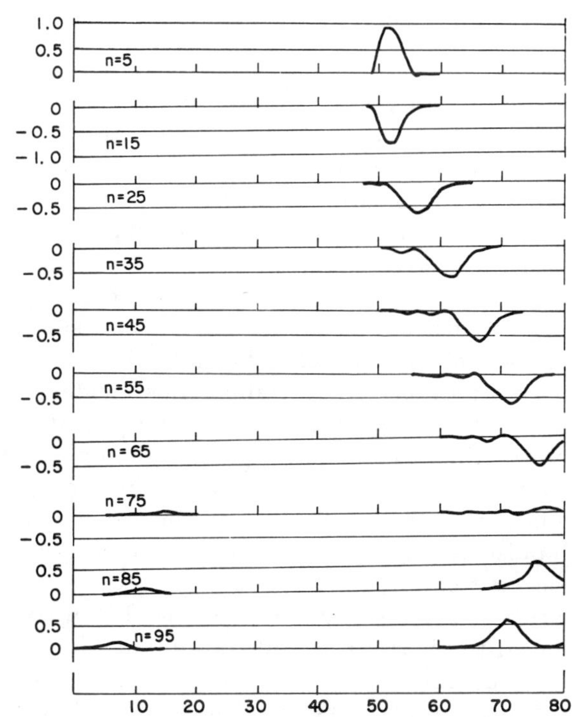

Fig. 6. E_z of the TM wave for various time cycles. $j=65$.

Let the incident wave be plane, with its profile being a half sine wave. The width of the incident wave is taken to be α units and the square has sides of length 4α units. Since the equations are linear, we can take $E_z = 1$ unit. The incident wave will have only an E_z component and an H_y component. We choose

$$\Delta x = \Delta y = \alpha/8 \tag{15a}$$

and

$$\Delta\tau = c\Delta t = \tfrac{1}{2}\Delta x = \alpha/16. \tag{15b}$$

A finite difference scheme over the whole x-y plane is impractical; we therefore have to limit the extent of our calculation region. We assume that at time $t = 0$, the left traveling plane wave is "near" the obstacle. For a *restricted* period of time, we can therefore replace the original problems by the equivalent problem shown in Fig. 2.

The input data are taken from the incident wave with

$$E_z(x, y, t) = \sin\left[\frac{(x - 50\alpha + ct)\pi}{8\alpha}\right]$$

$$0 \le x - 50\alpha + ct \le 8\alpha \tag{16a}$$

$$H_y(x, y, t) = \frac{1}{Z} E_z(x, y, t). \tag{16b}$$

From the differential equation satisfied by E_z we conclude that the results for the equivalent problem (see Fig. 2) should approximate those of the original problems, provided

$$0 \le n\Delta\tau \le 64\Delta\tau,$$

because the artificial boundary conditions will not affect our solution for this period of time.

For $n > 64$, however, only on certain points will the results of the equivalent problems approximate those of the original problems.

Numerical results are presented for the TM waves discussed above. To gain some idea of the accuracy of the finite difference equation, we have used the system (14a)–(14c) with the initial E_z being a half sine wave for the case of no obstacle. We note that the outer boundary conditions will not affect this incident wave as there is no H_x component in the incident wave. Ninety-five time cycles were run with the finite difference system (14a)–(14c), and the machine output is shown in Fig. 3. The oscillation and the widening of the initial pulse is due to the imperfection of the finite difference system.

Figure 4 shows the value of E_z of the TM wave as a function of the horizontal grid coordinate i for a fixed vertical grid coordinate $j = 30$. At the end of five time cycles, the wave just hits the obstacle. The line $j = 30$ does not meet the obstacle, but is "sufficiently" near the obstacle to be affected by a "partially reflected" wave. There is also a partially transmitted wave. The phase of the reflected wave is opposite that of the incident wave, as required by the boundary condition of the obstacle. There should also be a decrease in wave amplitude due to cylindrical divergence, but the calculation was not carried far enough to show this effect.

Figure 5 shows the value of E_z for the TM wave as a function of the horizontal grid coordinate i for $j = 50$. This line ($j = 50$) meets the obstacle, and hence we expect a reflected wave going to the right. These expectations are borne out in Fig. 5. After the reflected wave from the object meets the right boundary (see Fig. 2), the wave is reflected again. This effect is shown for the time cycles 75, 85, and 95.

Figure 6 is for $j = 65$. This line forms part of the boundary of the obstacle. Because of the required boundary condition, E_z is zero on the boundary point. To the right of the obstacle there is a "partially" reflected wave of about half the amplitude of a fully reflected wave. To the left of the obstacle there is a "transmitted" wave after 85 time cycles.

All these graphs were obtained by means of linear interpolation between the grid points. They have been redrawn for reproduction.

Comparison of the Computed Results With the Known Results on Diffraction of Pulses by a Wedge

There exist no exact results for the particular example we considered here. However, in the case when the obstacle is a wedge, Keller and Blank [2] and Friedlander [3] have solved the diffraction problem in closed forms. In addition, Keller [4] has also proposed a method to treat diffraction by polygonal cylinders. To carry out the method proposed by Keller [4], one would have to use some sort of finite difference scheme. The present difference scheme seems to be simpler to apply in practice. For a restricted period of time and on a restricted region, our results should be identical with those obtained from diffraction by a wedge. We present such a comparison along the points on the straight line coincident with the upper edge (i.e., $j = 65$).

Let the sides of a wedge coincide with the rays $\theta = 0$ and $\theta = \beta$. Let the physical space be $0 < r < \infty$, $0 < \theta < \beta$. Let this wedge be perfectly conducting. If the incident E_z is given by

$$E_z^{\text{inc}} = f\left\{t + \frac{r}{c}\cos(\theta - \theta_0)\right\},$$

where θ_0 is the direction of incidence, Friedlander [3] has shown that the solution to this diffraction problem is

$$\begin{aligned} E_z(r, \theta, t) &= \sigma(\theta - \theta_0) f\left\{t + \frac{r}{c}\cos(\theta - \theta_0)\right\} \\ &\quad - \sigma(\theta + \theta_0) f\left\{t + \frac{r}{c}\cos(\theta + \theta_0)^*\right\}^{3} \\ &\quad - \frac{1}{2\beta}\int_0^\infty \{Q(\theta - \theta_0, \xi) - Q(\theta + \theta_0, \xi)\} \\ &\quad \cdot f\left(t - \frac{r}{c}\cosh\xi\right) d\xi, \end{aligned} \tag{17}$$

[3] $(\theta+\theta_0)^* = (\theta+\theta_0) + 2m\beta$; where m is an integer so chosen that $-\beta < (\theta+\theta_0)^* \le \beta$.

where

$$\sigma(\psi) = 1 \quad (|\psi| < \pi);$$
$$\sigma(\psi) = 0 \quad (\pi < |\psi| < \beta);$$
$$\sigma(\psi + 2\beta) = \sigma(\psi)$$

and

$$k = \frac{\pi}{\beta}$$

$$Q(\psi, \xi) = -\frac{\sin k(\pi + \psi)}{\cosh k\xi - \cos k(\pi + \psi)} - \frac{\sin k(\pi - 4)}{\cosh k\xi - \cos k(\pi - 4)}.$$

At $t=0$, the incident wave hits the corner.

The discontinuities of the first two terms across the lines $\theta=\theta_0 \pm \pi$ and $\theta=-\theta_0 \pm \pi$ are compensated for by the contributions from the last integral. It can be shown that for

$$\theta = \theta_0 = \frac{\pi}{2}; \qquad \beta = \frac{3\pi}{2}$$

$$E_z\left(r, \frac{\pi}{2}, t\right) = f\left(t + \frac{r}{2}\right) - \frac{1}{2} f\left(t - \frac{r}{c}\right) - \frac{\sqrt{3}}{2\pi} \int_0^\infty \frac{f\left(t - \frac{r}{c} \cosh \xi\right)}{\cosh k\xi + \frac{1}{2}} d\xi. \quad (18)$$

For our incident wave we have[4]

[4] The origin of the wedge is taken to be the upper right-hand corner of the square. The zero time here differs from that of the numerical integration.

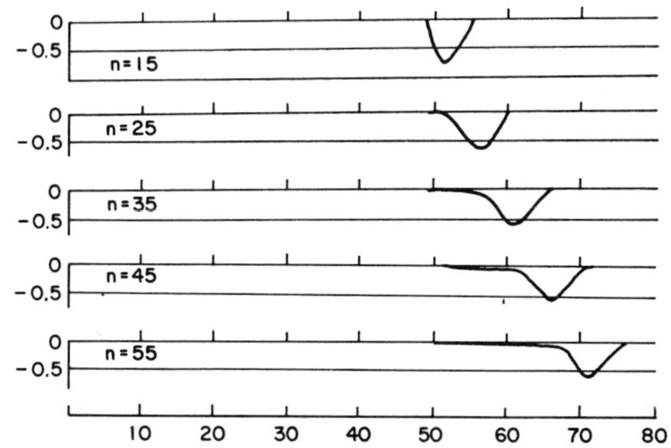

Fig. 7. Calculation of E_z for various sycles. These results are based on (18). The origin of the coordinate and of the time have been adjusted to agree with that used in the numerical calculation.

$$f(t) = \begin{cases} 0 < \dfrac{c\pi t}{\alpha} < 8 \\ 0 \quad \text{otherwise.} \end{cases}$$

Results of the computations based on (18) are shown in Fig. 7. The agreement with the graphs on Fig. 6 appears to be good, even for this coarse grid spacing.

Acknowledgment

The author wishes to thank Dr. C. E. Leith for helpful discussions and to express his gratitude to H. Barnett and W. P. Crowley for their assistance in the course of making the numerical calculations.

References

[1] J. Stratton, *Electromagnetic Theory*. New York: McGraw-Hill, 1941, p. 23.

[2] J. B. Keller and A. Blank, "Diffraction and reflection of pulses by wedges and corners," *Commun. Pure Appl. Math.*, vol. 4, pp. 75–94, June 1951.

[3] F. G. Friedlander, *Sound Pulses*. New York: Cambridge, 1958.

[4] J. B. Keller, *Electromagnetic Waves*. Madison, Wis.: Univ. of Wisconsin Press, 1961.

Modelling and Design of Millimetrewave Passive Circuits: From 2 to 3D

ROBERTO SORRENTINO
UNIVERSITY OF PERUGIA
ITALY

ABSTRACT—**With the advances of microwave and millimetre-wave circuit technology, fullwave electromagnetic approach has become necessary to model a number of phenomena (proximity effects, package interaction, crosstalk, etc.) dependent on complicated 3-D circuit configurations (multilevel circuits, vertical integration, multichip modules, finite metal thickness, etc.). This paper addresses a number of issues and reviews recent advances relevant to the modelling and optimisation of modern microwave circuits.**

1. Introduction

In recent years, the ever increasing needs for new applications in the area of communications and radars have spurred impressive advances in microwave technology exploiting higher and higher frequencies up to the millimetre- and submillimetre-wave ranges. New technologies and new configurations for microwave and millimetrewave circuits has been developed in a relatively short time. The dramatic evolution of millimitrewave circuit technology from waveguide to MIC's and MMIC's, has posed the necessity for new models and new design tools. Parallel to that, the advances in computer technology have made available extremely powerful analysis tools.

With the development of integrated circuit technology, the conventional waveguide technology has been gradually replaced in most applications by printed circuit configurations such as stripline and microstripline. In such transmission media the electromagnetic field is supported by longitudinal currents flowing in the strip conductor and is essentially confined to the thin dielectric substrate. As long as operating wavelengths were much larger of the cross sectional dimensions, conventional transmission line theory and quasi static approaches provided models accurate enough for the design of such circuits.

As higher frequency ranges were employed, more accurate models became necessary to account for dispersion and higher order mode effects. Fullwave techniques—typically the Spectral Domain Method—and phenomenological approaches—such as the Planar Circuit Approach—provided extremely useful models.

At millimeterwave frequencies, ohmic losses and tolerance problems in microstriplines and similar printed circuit transmission lines have stimulated the use of alternative configurations such as the quasi planar and E-plane circuits. In such configurations the electromagnetic field is not confined to the dielectric substrate but extends to and interacts with the waveguide housing. In the modelling of such transmission lines, two-dimensional models are totally inadequate. Still, because of the planarity of the conductors only two dimensional current densities are to be taken into account. In some cases, however, especially in the higher millimeterwave range, even the finite metallization thickness as well as waveguide grooves to hold the circuit can affect the circuit performance. Spectral domain techniques and mode-matching techniques were typically employed to provide useful design data for such circuits.

More recently, a new challenge to the microwave and millimeterwave designer has been posed by the advances in monolithic circuit and in general semiconductor-processing technology. Here, the complexity of the structures that support the electromagnetic (EM) field propagation is much higher than in conventional MIC's. Modern MMIC's configurations involve vertical integration of complex structures on multilayer substrates with signal transfer via electromagnetic coupling and/or via-hole connection. The third dimension can be used to create new revolutionary geometries to improve the circuit performances [1]. Current densities orthogonal to the substrate flow in through-hole connections and air bridges. High packing densities produce strong interaction between different circuit elements. The package interacts with the circuit so that it has to be considered as a part of the circuit itself.

It is apparent that only 3D electromagnetic tools can resolve the inherent complexity of such circuits. Such tools are needed

Reprinted with permission from *24th European Microwave Conference,* Roberto Sorrentino, "Modelling and Design of Millimeter Passive Circuits: From 2 to 3D," Vol. 1, pp. 48–61, 1994. © Nexus House.

not simply to characterize the various circuit elements, but also to analyse interaction phenomena between closely spaced elements, to characterise the package effects on circuit performance, to evaluate the behaviours of multichip modules.

This paper is aimed at reviewing the recent advances in the numerical modelling of microwave and millimeterwave passive circuits, enlightening the transition from 2- to 3-D models and trying to envisage the new possibilities in the CAD of microwave circuits that are being opened in the next future.

2. Overview: From 1 to 3D

Printed circuits are the basic technology for the realisation of most microwave and mm-wave passive components. It provides compact, low weight, low cost, easy to fabricate components. For many years, stripline and microstripline have been the most popular transmission line configurations. Parallel to the development of planar circuit technology, great effort has been devoted by a number of researchers in order to develop models and analysis methods for the characterisation of the propagation characteristics as well as discontinuities and components. A huge amount of work has been done based on numerous numerical and/or analytical techniques.

Initially, the design of microstrip and stripline circuits was based on simple transmission line models. Quasi static approaches were used to determine the propagation constants and characteristic impedances of the lines (see [2,3]). Lumped element models have been developed to characterise various types of discontinuities [4,5].

At the end of the 60's, with the use of increasingly higher frequencies, it was recognised that simple transmission line (thus 1-D) models were inadequate to represent the behaviour of microstrip circuits, particularly when low impedance levels (thus wider transmission lines) were involved. A second dimension was then included into the mathematical models for microstrip and stripline circuits, leading to the so-called planar circuit approach (PCA) [6–10]. The PCA can be regarded as a mathematical formalism to overcome the limitations inherent to the transmission line representation of microstrip circuits. Dispersion phenomena and fringe field effects are taken into account in a phenomenological manner by the use of effective dimensions and effective permittivities [9]. The PCA provided descriptions of microstrip components and discontinuities which are accurate enough for most practical applications, still permitting a relatively simple analytical treatment. At those times, three-dimensional full wave analyses would have required an excessive computer expenditure.

Full wave techniques for the analysis of printed circuit were being developed in the same years. The Spectral Domain Approach, pioneered in 1971 by Itoh and Mittra [11,12] soon became one of the most powerful and popular methods for computing the propagation characteristics not only of microstrips and striplines but of any printed circuit configuration, such as coplanar waveguide, coplanar lines, slotline, etc. [13]. First applied to eigenvalue problems, i.e. to determine the propagation characteristics of uniform guiding structures or the resonant modes of planar resonators, the SDA has later been extended to the solution of deterministic problems, such as the modelling of discontinuities [14,15].

The SDA has been an extremely useful and successful method that has permitted a number of practical design problems to be solved with modest numerical effort. In the standard version of SDA as applied to microstrip structures, the field problem is attacked by evaluating in the spectral domain the 2-D dyadic Green function for the electric field. After imposing the condition for zero tangential electric field on the metal strip(s), the problem is reduced to the solution of an algebraic system of equations by applying Galerkin technique.

SDA can be considered as a special case of Integral-Equation (IE) techniques. For an assessment on this subject see [16,17]. The basic steps in the solution process using IE techniques are as follows:

- EM field quantities are expressed in terms of an appropriate Green's function;
- an IE is set up by imposing the relevant boundary conditions;
- the IE is solved by the Method of Moments (MoM) [18,19].

Different formulations are possible depending on the choice of the dyadic Green's function, leading basically to the Electric Field Integral Equation (EFIE) and the Mixed Potential Integral Equation (MPIE). The former is commonly solved in the spectral domain and leads to the SDA. The MPIE on the contrary can be solved in the space domain. Although differing in the numerical details of the computer implementation, the two approaches are mathematically identical [17].

The possibility of incorporating edge singularities in the basis functions as well as of implementing various sub-domain basis functions provide excellent numerical efficiency and good flexibility in the analysis of planar circuits. SDA is a full wave approach, since all field components are taken into account, but should be considered as a 2.5-D method. Although 3-D effects, such as circuit package interaction, proximity effects, crosstalks, etc. can be predicted, finite metal thickness and vertical current densities are not taken into account. Additional efforts is required to incorporate such features into SDA codes [20,21].

Recent advances in the technology of microwave circuits involve not only the presence of several metallization layers, but also vertical metallization elements, such as via holes, air bridges, etc. Vertical integration of passive components is a possibility to achieve higher integration densities in MMIC's. Parallel to that, in view of commercial applications of microwaves, there is an increasing need to reduce the costs associated with the design and production of the circuits. This implies an increasing necessity of rigorous electromagnetic simulators for first pass design.

IE methods appear sometimes to be not flexible enough to deal with irregularly shaped 3-D structures. Differential techniques, such as Finite Element (FE) or Finite Difference (FD) methods, on the contrary can deal with arbitrarily shaped configurations, are relatively easy to implement (they don't require heavy analytical preprocessing) but require large memory storage and CPU time.

Thanks to the increased computing power available today, full 3-D EM simulators are now viable CAD tools. Although the complexity of practical circuits is still besides the present capabilities of EM simulators, the possibility of effective 3-D EM design of an entire MMIC can be foreseen as a standard routine in the next years. Combined differential and integral methods are, in the author's view, a very promising approach to fully exploit the potentials of both methods in terms of versatility and efficiency. In addition to that, there is a need not only for 3-D fullwave simulators, but also for efficient optimisation techniques. Specialists in EM field theory have seldom been interested in optimisation techniques, but as stand alone tools made available by specialists in circuit or system theory. On the basis of personal experience, it is believed, on the contrary, that a deeper understanding of optimisation techniques by the field theoretical people could lead to notable advances in the efficiency of numerical tools for EM simulators.

3. Electromagnetic Modelling

In 1984, a workshop on numerical techniques for microwave circuits was organised by T. Itoh and J. Mink at the 1984 IEEE MTT-S International Symposium, held in San Francisco. This memorable event was a very timely opportunity to assess the state of the art in the numerical techniques for microwave circuits. The interest stimulated was such that a special issue of the MTT Transaction was then devoted to the numerical techniques [22].

A great number of publications have since then been devoted to this topic (see for instance [23–25]). A detailed discussion on the characteristics of the various methods and on the respective advantages and disadvantages is outside the scope of this paper. The interested reader is addressed to the reference list. Here it is sufficient to outline some basic classifications of numerical methods and to identify the main parameters that address the choice among them.

Before going into that, let us identify the process of EM modelling of microwave circuits, by slightly modifying Miller's view [26].

Referring to Fig. 1, the EM modelling of microwave circuits can be viewed as consisting of the following steps:

- Description of the problem (geometry, electrical parameters, etc.) (INPUT).
- Excitation of the structure.
- Computation of the EM field in the structure by solution of Maxwell's equations.
- Extraction of terminal parameters (e.g. S-parameters (OUTPUT).

Our attention is focused here on the third step, i.e. the solution of Maxwell's equations. The Maxwell's equation solver can operate either in time domain (TD) or frequency domain (FD). Correspondingly, it can be defined as a [26]:

- field propagator: time-domain solver, where the initial value boundary value problem is updated in time;
- field solver: frequency domain solver requiring matrix inversion or system solution

The inverse or synthesis problem (Fig. 2) consists of finding (some) geometrical parameters that give a desired output (terminal description) from a specified input. The synthesis is usually performed in frequency domain using iterative algorithms. Essentially, the synthesis results from a large number of repeated analyses according to some search strategy (optimisation). Attempts to synthesise microwave structures directly in time domain, however, have been reported recently using a time reversal technique [27] or a hybrid frequency/time domain [28].

A variety of numerical techniques can be employed for building up a field solver or propagator. Among the most common are the following:

Integral Equation (IE) techniques, including the Spectral Domain Approach (SDA)

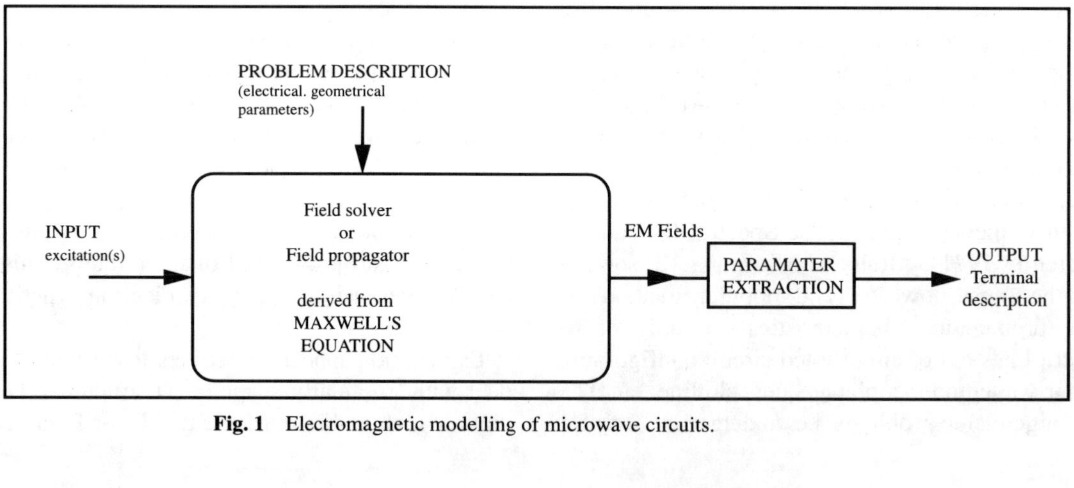

Fig. 1 Electromagnetic modelling of microwave circuits.

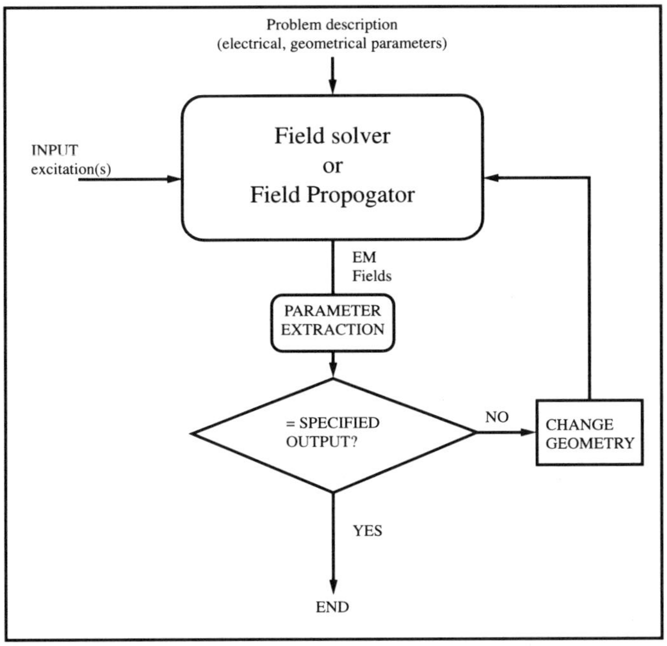

Fig. 2 The synthesis (inverse) problem.

Mode Matching (MM) Method
Finite element and Boundary Element Methods (FEM, BEM)
Finite Difference (FD) Method
TLM Method (and Spatial Network Method, SNM)

An attempt to classify numerical methods is presented in the next Section.

4. Classification of Numerical Methods

Numerical methods for the analysis of microwave structures belong basically to two categories corresponding to the numerical formulation of Maxwell's equations in differential or integral form. Strictly numerical methods, on the one hand, are based on the discretization of Maxwell's equations, (e.g. Finite Element, Finite Difference (in frequency or time domain)) or on the implementation of Huygens' principle (Transmission Line Matrix (TLM) method). Thanks to their high flexibility, they are well suited for problems with irregular geometries. On the other hand, methods requiring a high degree of mathematical preprocessing, such as the Spectral Domain Approach (SDA) or the Mode Matching (MM) method, are based on a continuous rather than discrete representation of the field quantities. They lead to very efficient computer codes, but are limited to problems with simple geometries.

Another important categorisation concerns the use of frequency or time domains. Dramatic advances in computing resources have recently made it possible the practical implementation of time domain techniques for the simulation and even the optimisation of microwave circuits. Not many years ago, because of their huge memory and CPU requirements, methods such as TLM or Finite Difference Time Domain (FDTD) were considered essentially as academic exercises. Instead, we are now facing an era where

Table I. Classification of Numerical Methods

	Frequency Domain	Time Domain
Differential Methods	FD	FDTD
	FDTLM	TLM
	FEM	
Integral Methods	MM	
	SDA	
	EFIE, MPIE	

such techniques are becoming increasingly popular and of practical utility.

Table I categorises the most popular numerical methods using the above criteria. Observe that some techniques originally developed and typically operating in one domain are currently being translated into the complementary domain (e.g. FDTLM [29]). Also, the Finite Element Method, conventionally applied in frequency domain, can be implemented in time domain. While differential methods have been used both in frequency and time domains, integral methods have been used in practice only in frequency domain.

The main objective of any numerical or analytical method for microwave and millimitrewave passive structures is to develop, with minimum effort (in terms of CPU time and memory space), numerical models having maximum accuracy, as well as maximum flexibility (i.e. capability to model a large variety of structures). Efficiency, accuracy and flexibility (besides the personal familiarity one may have with one specific method) are the main issues that determine the choice of a numerical method. The choice, however, is also strictly dependent on the problem at hand. No method can be thought as the best one, but, depending on the application, each can have advantages over the others. Such issues have been discussed many times in the literature (see, for instance, [23–25]). Rather than going into that, we would rather address here the issue of the complexity of microwave circuits.

5. Complexity of Microwave Circuits

The complexity of microwave circuits has constantly increased with the technological advances. We would like to identify here the complexity of the problem at hand, i.e. those parameters of a given structure that determine the amount of numerical and analytical effort required to compute its response. Defining the complexity of a circuit is not an easy task. The complexity itself is somewhat subjective, as the same problem has different degrees of complexity depending on the method employed and on the way the method is applied. The aim here is just to present some personal ideas on complexity, with no claim of completeness.

A first categorisation accounts for the number of dimensions of the problem, i.e., the number of dimensions comparable with the wavelength. Such a categorisation, introduced by Okoshi and Miyoshi [30], is illustrated in Fig. 3a [31].

Such a classification appears today to be too rough to quantify the numerical and analytical efforts associated with the modelling of microwave circuits. This figure was originally

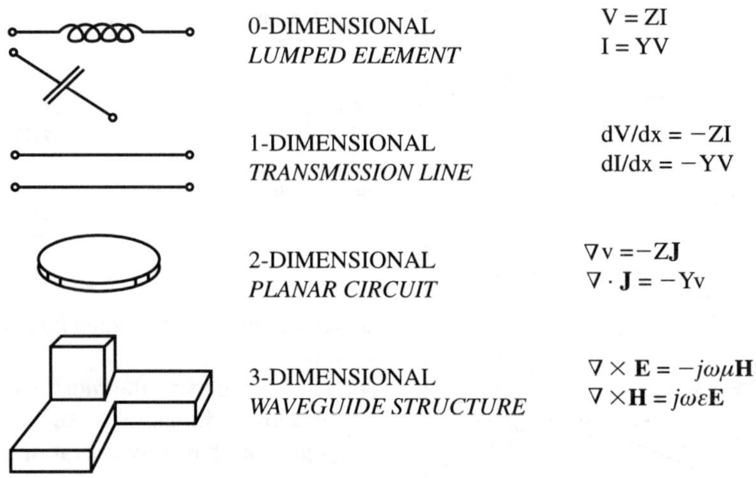

Fig. 3a Classification of microwave circuits (©1985 IEEE. Reprinted with permission from [31]).

used to introduce the concept of planar circuits, and, at the time, fullwave modelling was not yet a viable approach for 3-D structures. No attempt was made to identify and classify the complexity of 3-D boundary value problems.

The analytical/numerical effort required by the solution of a boundary value problem not simply depends on the number of dimensions involved, but also on the number of field variables necessary to describe the solution as well as on the frequency dependence of such a solution. The problem's complexity is related also to the analysis method, some methods being more suitable for certain rather than other classes of problems.

For example, the concept itself of planar circuit can involve different levels of complexity. While planar models were usually employed for the analysis of problems involving either TE or TM fields, Gwarek et al. [32] have recently shown that 2-D models can also be employed for a class of microwave circuits where fields are of the hybrid type, thus can be derived in terms of two, instead of one, scalar potentials.

Table II. Some Parameters Identifying the Complexity of a Microwave Circuit

	Parameter	Description
1	No. of dimensions	No. of space coordinates that fields depend on. Coordinates along which dependence is given are not considered.
2	No. of geometrical parameters	No. of dimensions necessary to define the geometry of the structure in a given reference system
3	No. of axes	No. of different axes of the feeding lines
4	No. of scalar potentials	No. of scalar potentials in terms of which the total EM field can be expressed
5	Modal dispersion	Frequency dependence of modal fields or eigensolutions. Related to 6.

Table II lists some parameters that can be used to quantify the problem's complexity. Correspondingly, various examples are quoted in Table III.

E- or H-plane discontinuities in rectangular waveguide (RW) are simple 2-D problems, as the dependence on one coordinate in given (assuming TE_{10} excitation). A E-plane dielectric obstacle creates an additional degree of complexity, because of the hybrid and dispersive nature of the fields, the problem still being a 2-D one (or vector 2-D [32]). A double step in both E- and H-planes makes the problem a 3-D one, where both TE and TM modes are excited. E- or H-plane T-junctions are still 2-D but are multiaxial problems. The magic T is a 3-D multiaxial structure exhibiting no modal dispersion (homogeneously filled).

Printed circuit discontinuities can be considered as 2.5-D as long as zero metal thickness can be assumed. The inclusion of current densities orthogonal to the substrate make the problem a 3-D one. This is the case of the air bridge or the via hole.

Complicated geometries, of course, increase the complexity of the structure. The number of geometrical dimensions necessary to describe the structure can be assumed to measure the complexity. For a given structure, such a parameter may depend on the coordinate system: a circle is easily described in a cylindrical system, but is difficult to describe in a rectangular one.

An additional issue, not included in Table II, that complicates significantly the modelling of microwave circuit is represented by loss, particularly conductor and radiation loss. While in conventional microwave circuit the perturbational approach to evaluate conductor loss was accurate enough, the use of miniaturised structure is such that skin depth is comparable with the metal thickness. More accurate evaluations of conductor loss are needed [33–37].

6. Analytically vs. Numerically Oriented Methods

We will now confine our discussion to two among the most popular methods, namely the Finite Difference Time Domain

Table III. Evaluation of the Complexity of Some Microwave Circuit Discontinuities

	Structure	Dimensions	Axes	Scalar Potentials	Modal dispersion
1	Homogeneous E- or H-plane RW step	2-D	1	1	No
2	E-plane dielectric obstacle	2-D	1	2	Yes
3	Double step in RW	3-D	1	2	No
4	E- or H- plane T-junction	2-D	2	1	No
5	Hybrid T	3-D	3	2	No
6	Step in stripline (zero thickness)	2.5-D	1	2	No
7	Step in microstripline (zero thickness)	2.5-D	1	2	Yes
8	Microstrip air bridge	3-D	1	2	Yes
9	Via hole through orthogonal strips	3-D	2	2	Yes

(FDTD) an the Mode Matching (MM) methods. They represent, on the one hand, time domain and frequency domain methods. On the other hand, they represent differential, or numerically oriented methods (based on discretized space representation), and integral or analytically oriented methods (using continuous space representation). Most of the conclusions drawn here in connection with FDTD and MM can be extended to other methods such as TLM and SDA.

The FDTD method has become very popular in recent years, because of its easy implementation and excellent flexibility [38–41]. It is also very suited for parallelization of the code [42]. Main problems arise from the large number of variables to be stored, high number of time iterations (especially for high-Q structures) and from imperfect absorbing boundary conditions when modelling open structures.

The MM method is one of the oldest and more popular analytical approaches to solve waveguide discontinuity problems [43,44]. In recent years it has been generalised so as to deal with 3-D problems. The 3-D MM consists of the segmentation of a 3D structure into simple volume elements (cells) where the EM fields are expanded in terms of modal series or in terms of Green's function. At the apertures connecting contiguous cells, the tangential fields are expanded into suitable 2-D basis functions (usually waveguide modes). The boundary conditions at the apertures are manipulated so as to obtain a linear system of equations in the expansion coefficient of the 2-D series. The method is described in some detail in [45,46]. Compared to other integral methods (SDA) it has the advantage that it naturally includes finite metallization thickness as well as vertical currents. Compared to differential methods like FDTD, it is less flexible, requires analytical preprocessing but less memory storage and CPU time.

Typical 3-D MMIC discontinuities are the via hole and the air bridge. Figure 3b shows the segmentation adopted for implementing the 3-D MM analysis of a microstrip via in a package. The via has been approximated with a rectangular rather than

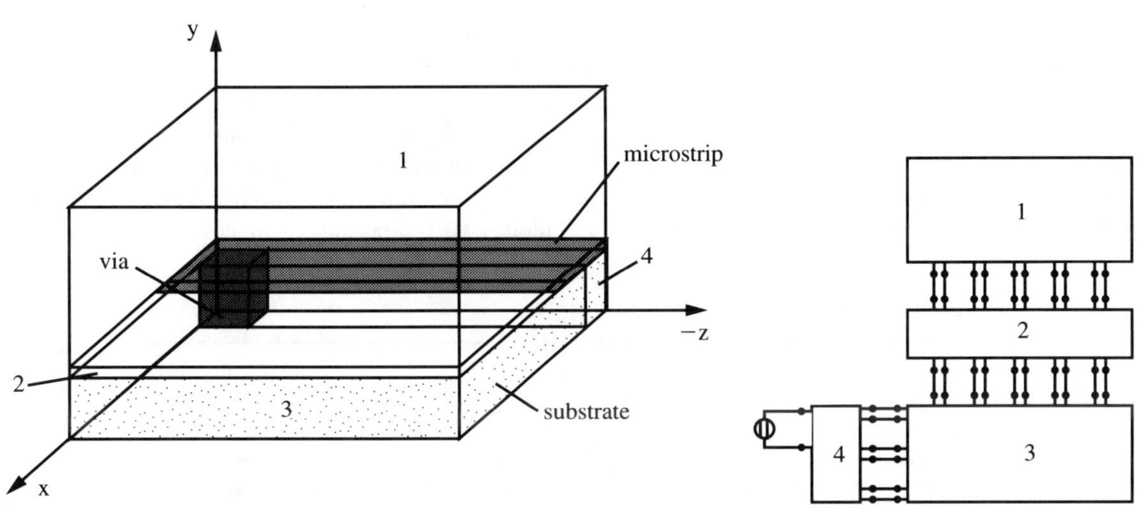

Fig. 3b The packaged microstrip via and the MM segmentation (© IEEE. Reprinted with permission from [45]).

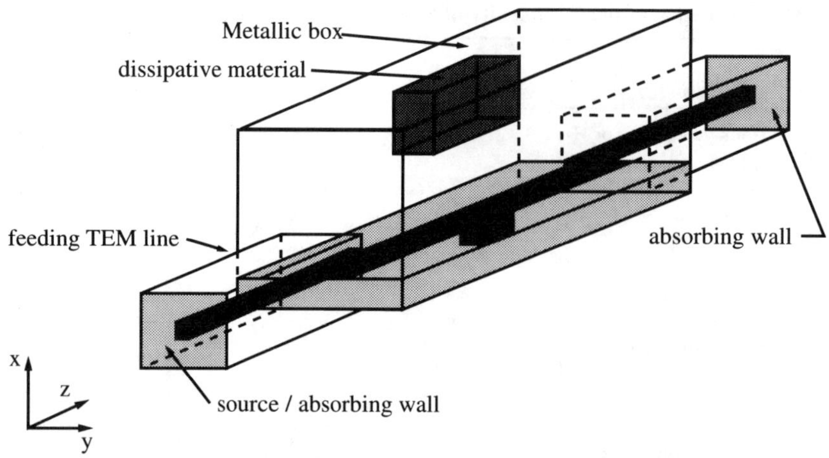

Fig. 4 The packaged microstrip via in the FDTD simulation (© IEEE. Reprinted with permission from [42]).

conical shape and the whole package has been reduced to one half for symmetry considerations. The structure has been partitioned into only 4 cells, each corresponding to a generalised multiport network. Excitation of the structure is made by impressing an electric field distribution (or a magnetic current density) at the end of region 4. The overall equivalent circuit is shown at the right of Fig. 3b.

Figure 4 shows the model adopted with the FDTD simulator. In this case, microstrip to coaxial connectors were included. This is not only for a model closer to reality, but also to improve the absorbing boundary conditions at feeding lines. A graded mesh was used to increase the space resolution close to the via without requiring excessive memory storage [47].

Figure 5 shows the computed behaviours of the via compared with measurements. A good agreement is observed between theories and experiment. Observe that package resonances are fully predicted by both methods. The 3-D MM analysis required about 10 minutes per frequency point on an IBM RISC 6000 workstation. The MM analysis involves the solution of a system of equation of the order of less than 500 variables, corresponding to the number of modes used in the field expansions.

Table IV describes the discretization used in FDTD. To run the FDTD code in a reasonable amount of time (hours), a DEC 12000 Massive Parallel computer was employed.

It is apparent that the FDTD code is much more computer intensive than the 3-D MM code. It must be observed, however, that the presence of the transitions could be taken into account only

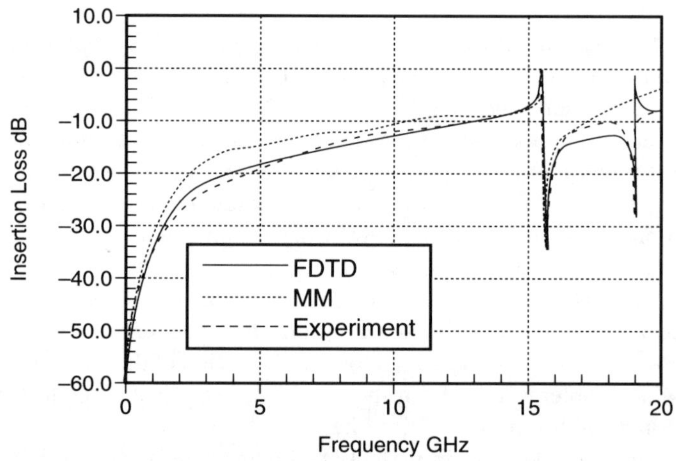

Fig. 5 Computed and measured behaviours of the via. © IEEE (Reprinted with permission from [42]).

with the FDTD code. In addition, the latter can be used to modify the shape of the microstrip package and to introduce pieces of damping materials to choke off the resonances [42]. With the MM code, only the parameters of the rectangular subregions can be changed. Any variations in the shape of the regions would require a new code to be built up. In addition, the FDTD code employed in [42] did not include any spectral estimation techniques for reducing the number of time iterations [48,49]. With such techniques, CPU time can be reduced substantially.

Table IV. CPU time per Δt versus mesh size

Nx * Ny * Nz	Mesh Dimension DEC 12000	DCPU Time mS per Δt	CPU Time mS per Δt RISC 6000-375
65 * 65 * 100	422500	1100	—
64 64 * 10	409600	630	3950
10 * 10 * 100	10000	480	80
51 * 55 * 105	294525	650	—

(© IEEE. Reprinted with permission from [42]).

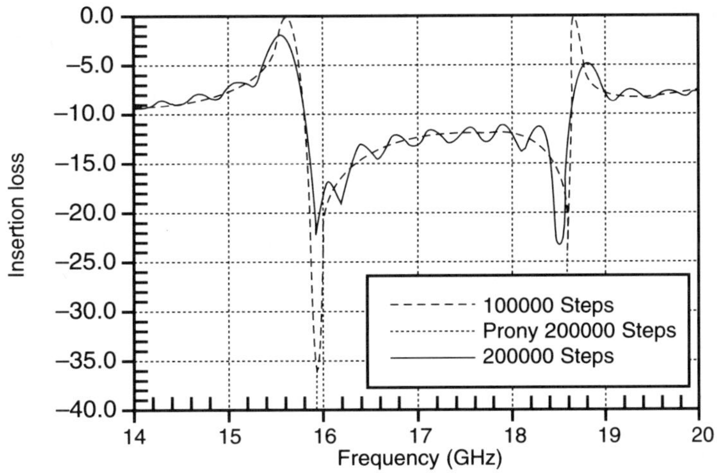

Fig. 6 FDTD simulations of the microstrip via with and without Prony's method.

The same structure of Fig. 5 has been simulated again after implementing Prony's method. Figure 6 shows the comparison between computed results. 20.000 time steps with Prony's algorithm give about the same results as 100.000 steps using regular FDTD.

Differential and integral techniques appear to be complementary to each other. While integral methods are numerically more efficient, differential methods are more flexible and easy to implement. As a consequence, integral methods are more suitable for routine analyses and optimisation of components of given geometries. Differential methods are especially suited for one shot analyses, when building up a new code, even if much more efficient, would require too much effort. Integral methods are extremely efficient for structures with simple shapes, while differential methods are particularly suited for complicated geometries.

It seems thus very reasonable to take advantage of both approaches by setting up hybrid methods where an integral method is used to model the "regular" part of the structure, while a differential method is used for the "irregular" one. First attempts in this direction appear to be extremely promising [50,51].

7. Optimisation with the Adjoint Network Method

The ultimate objective of numerical modelling is to set up an efficient design tool for microwave structures. Not only efficient analysis methods, but also optimisation algorithms are necessary. The most efficient optimisation techniques are based on the knowledge of the gradient of the objective function, i.e. its first derivatives with respect to all geometrical parameters. Because of the complexity of the full wave model of the component, the derivatives must be evaluated numerically, i.e. by finite difference approximation. For a component with M geometrical parameters, the computation of the objective function and its gradient requires $M + 1$ full wave analyses. Since the optimisation usually requires several hundreds iterations and the number of geometrical parameters is easily of the order of several tenths, the overall numerical effort can easily become exceedingly heavy.

This extremely lengthy procedure can be avoided by the use of the Adjoint Network Method (ANM) [5]. The ANM allows the gradient of the objective function to be computed with just one analysis of the (reciprocal) network, on the basis of the description of the constituent subnetworks. In this manner, the usual finite difference evaluation of the multidimensional gradient of the objective function can be avoided. The ANM has been known for a long time, but its applications have been limited to the area of conventional circuits. On the contrary, a distinct advantage of the MM technique is that an analytical description of the field variables is available. In this manner, using the microwave network formalism in connection with the segmentation of the structure into simple cells, the method can easily be transferred from circuit theory to microwave field theory applications.

A recent implementation of this concept has led to the direct optimisation of waveguide components using full wave EM simulators [52,53]. Practical examples have shown CPU accelerations of more than one order of magnitude in the CAD of waveguide filters and couplers. The efficiency of the approach increases with the number of the degrees of freedom, thus the geometrical complexity of the microwave structure. It is thus believed that this approach will be extremely powerful in the computer-aided design of complicated microwave circuits both conventional and MMIC's.

8. Conclusions

This paper has addressed a number of issues relevant to the modelling and optimisation of modern microwave circuits with reference to the structural complexity inherent to the most recent technologies (MMIC's, multichip modules, multilevel circuits). The complementarity of differential and integral methods in the modelling and optimisation of microwave circuits has been stressed along with the necessity of faster and more efficient optimisation algorithms.

References

[1] L. P. B. Katehi, "Novel transmission lines for the submillimeter-wave region," Proc. IEEE, vol. 80, n. 11, pp. 1771–1787, Nov. 1992.

[2] R. Mittra and T. Itoh, "Analysis of microstrip transmission lines," in L. Young and H. Sobol (Eds.), New York, pp. 67–141, Academic Press, 1974.

[3] K. C. Gupta et al., Microstrip Lines and Slotlines, Dedham, Mass, Artech House, 1979.

[4] H. M. Altschuler and A. A. Oliner, "Discontinuities in the center conductor of symmetric strip transmission lines," IRE Trans. Microwave Theory Tech., vol. MTT-8, pp. 328–339, May 1960.

[5] K. C. Gupta, Ramesh Garg, Rakesh Chadha, *Computer-aided design of microwave circuits,* Artech house, Dedham, Massachusetts, 1981.

[6] G. Biorci, S. Ridella, "On the theory of distributed three-layer n-port networks," Alta Frequenza, vol. 38, p. 615, 1969.

[7] P. P. Civalleri, S. Ridella, "Impedance and admittance matrices of distributed three-layer n-ports," IEEE Trans. on Circuit Theory, vol. CT-17, pp. 392–398, 1981.

[8] T. Okoshi, T. Miyoshi, "The planar circuit—A novel approach to microwave circuitry," Proc. IEEE Int. Conf. on Circuit and System Theory, Kyoto, Sept. 9–11, 1970.

[9] G. D'Inzeo, F. Giannini, C. M. Sodi, and R. Sorrentino, "Method of analysis and filtering properties of microwave planar networks," IEEE Trans. Microwave Theory Tech., vol. MTT-26, pp. 462–471, July 1978.

[10] G. Kompa, "S-matrix computation of microstrip discontinuities with a planar waveguide model," Arch. Elek. Ubertragung., vol. 30, pp. 58–64, Feb. 1976.

[11] T. Itoh and R. Mittra, "Dispersion characteristics of slot lines," Electron. Lett. Vol. 7, No. 13, pp. 364–365, July 1971.

[12] T. Itoh and R. Mittra, "Spectral domain approach for calculating the dispersion characteristics of microstrip lines," IEEE Trans. Microw. Theory Tech., vol. MTT-21, pp. 496–499, July 1973.

[13] R. H. Jansen, "The spectral-domain approach for microwave integrated circuits," IEEE Trans. Microwave Theory Tech., vol. MTT-33, pp. 1043–1056, Oct. 1985.

[14] R. H. Jansen, "Hybrid mode analysis of end effects of planar microwave and millimetrewave transmission lines," IEE Proc. Part H, vol. 128, pp. 77–86, Apr. 1981.

[15] Q. Zhang and T. Itoh, "Spectral domain analysis of scattering form E-plane circuit elements," IEEE Trans. Microwave Theory Tech., vol. MTT-35, pp. 138–150, Feb. 1987.

[16] J. Mosig, *Integral-Equation Techniques,* in T. Itoh (Ed.), "Numerical techniques for Microwave and Millimeter Wave Passive Structures," J. Wiley, New York, 1989.

[17] J. Mosig, *Integral-Equation Techniques for Three-Dimensional Microstrip Structures,* pp. 127–152.

[18] R. F. Harrington, *Field computation by Moment Methods,* McMillan, New York, 1968.

[19] J. J. H. Wang, *Generalized Moment Methods in Electromagnetism,* J. Wiley, New York, 1991.

[20] T. Kitazawa and R. Mittra, "Analysis of finline with finite metallization thickness," IEEE Trans. Microwave Theory Tech., vol. MTT-32, pp. 1484–1487, Nov. 1984.

[21] T. Becks and I. Wolff, "Full-wave analysis of 3D metallization structures using a spectral domain technique," IEEE Trans. on Microwave Theory Tech., vol. MTT-40, pp. 2219–2227, Dec. 1992.

[22] IEEE Trans. on Microwave Theory Tech., Special Issue on Numerical Methods, vol. MTT-33, Oct. 1985.

[23] T. Itoh (Ed.), *Numerical Techniques for Microwave and Millimeter Wave Passive Structures,* J. Wiley, New York 1989.

[24] R. Sorrentino (Ed.), *Numerical Methods for Passive Microwave and Millimeter Wave Structures,* IEEE Press, New York 1989.

[25] E. Yamashita (Ed), Analysis Methods for Electromagnetic Wave Problems, Artech House, Boston 1990.

[26] E. K. Miller, "A selective survey of computational electromagnetics," IEEE Trans. AP, vol. AP-36, No. 9, pp. 1281–1305, Sept. 1988.

[27] R. Sorrentino, Poman So, and W. J. R. Hoefer, "Numerical microwave synthesis by inversion of the TLM process," Proc. 21st Eu. Microwave Conf., Stuttgart, Sept. 1991, pp. 1273–1277.

[28] P. P. M. So, W. J. R. Hoefer, J. W. Bandler, R. M. Biernacki, and S. H. Chen, "Hybrid frequency/time domain field theory based CAD of microwave circuits," in Proc. 23rd Eu. Microwave Conf., Madrid (Spain) 1993, pp. 218–219.

[29] H. Jin and R. Vahldieck, "The frequency domain TLM method—A new concept," IEEE Trans. MTT, vol. MTT-40, pp. 2207–2218, Dec. 1992.

[30] T. Okoshi, T. Miyoshi, "The planar circuit—An approach to microwave integrated circuitry," IEEE Trans. Microwave Theory Tech., vol. MTT-20, pp. 245–252.

[31] R. Sorrentino, "Planar circuits, waveguide models, and segmentation method," (invited), IEEE Trans. Microwave Theory Tech., vol. MTT-33, pp. 1348–1353, Oct. 1985.

[32] W. K. Gwarek, T. Morawski, and C. Mroczkowski, "Application of the FD-TD method to the analysis of circuits described by the two-dimensional wave equation," IEEE Trans. MTT, vol. MTT-41, pp. 311–317, Feb. 1993.

[33] W. Heinrich, "The resistance of strip conductors embedded in a two-dimensional field distribution," 1993 IEEE MTT-S Int. Symp. Digest, Atlanta, June 1993, pp. 417–420.

[34] W. Heinrich, "Full-wave analysis of conductor losses on MMIC transmission lines," IEEE Trans. on Microwave Theory Tech., vol. MTT-38, pp. 1468–1472, Oct. 1990.

[35] T. E. van Deventer, P. B. Katehi, and A. A. Cangellaris, "High frequency conductor and dielectric losses in shielded microstrip," 1989 IEEE MTT-S Int. Symp. Digest, pp. 919–922.

[36] P. Waldow and I. Wolff, "The skin-effect at high frequencies," IEEE Trans. on Microwave Theory Tech., vol MTT-33, pp. 1076–1082, Oct. 1985.

[37] F. Alessandri, G. Baini, G. D'Inzeo, and R. Sorrentino, "Conductor loss computation in multiconductor MIC's by transverse resonance technique and modified perturbational method," IEEE Microwave and Guided Wave Letters, vol. 2, N. 6, pp. 250–252, June 1992.

[38] K. S. Yee, "Numerical solution of initial boundary value problems involving Maxwell's equations in isotropic media," IEEE Trans. AP, vol. AP-14, pp. 302–307, May 1966.

[39] A. Taflove and M. E. Brodwin, "Numerical Solution of Steady-State Electromagnetic Scattering Problems Using the Time-Dependent Maxwell's Equations," IEEE trans. MTT, vol. MTT-23, n. 8, pp. 623–630, August 1975.

[40] G. Mur, "Absorbing boundary conditions for the finite difference approximation of the time domain electromagnetic field equations," IEEE Trans. Electromagn. Compat., vol. 23, n. 4, pp. 377–382, Nov. 1981.

[41] Zhang X., J. Fang, K. K. Mei, Y. Liu, "Calculations of the dispersive characteristics of microstrips by the time-domain finite-difference method," IEEE Trans. Microwave Theory Techn., vol. MTT-36, No. 2, pp. 263–267, Feb. 1988.

[42] P. Mezzanotte, M. Mongiardo, L. Roselli, R. Sorrentino, W. Heinrich, "Analysis of Packaged Microwave Integrated Circuits by FDTD," IEEE Trans. Microwave Theory Tech., vol. MTT-42, Sept. 1994 (to appear).

[43] J. R. Whinnery and H. W. Jamieson, "Equivalent circuits for discontinuities in transmission lines," Proc. I.R.E., vol. 32, pp. 98–116, Feb. 1944.

[44] A. Wexler, "Solution of waveguide discontinuities by modal analysis," IEEE Trans. Microwave Theory Tech., vol. MTT-15, pp. 508–517, Sept. 1967.

[45] R. Sorrentino, F. Alessandri, M. Mongiardo, G. Avitabile, L. Roselli, "Full-wave modeling of via-hole grounds in microstrip by three-dimensional mode-matching technique," IEEE Trans. Microwave Theory Tech., vol. MTT-40, n. 12, pp. 2228–2234, Dec. 1992.

[46] F. Alessandri, G. Baini, M. Mongiardo and R. Sorrentino, "A 3-D Mode Matching Technique for the Efficient Analysis of Coplanar MMIC Discontinuities with Finite Metallization Thickness, IEEE Trans. Microwave Theory Tech., vol. MTT-41, Sept. 1993.

[47] D. H. Choi and W. J. R. Hoefer, "A graded mesh FD-TD algorithm for eigenvalue problems," Proc. of 17th Eu. M. C., pp. 413–417, Rome, 1987.

[48] W. L. Ko, R. Mittra, "A combination of FD-TD and Prony's methods for analyzing microwave integrated circuits," IEEE MTT, Vol. 39, No. 12, Dec. 1991, pp. 2176–2181.

[49] W. Kuempel, I. Wolff, "Digital Signal Processing of time domain field simulation results using the system identification Method," IEEE MTT-S Digest 1992, pp. 793–796.

[50] M. Mongiardo, R. Sorrentino, "Efficient and Versatile Analysis of Microwave Structures by Combined Mode Matching and Finite Difference Methods," IEEE Microwave and Guided Wave Letters, Vol. 3, n. 8, Aug. 1993, pp. 241–243.

[51] J. M. Reiter and F. Arndt, "A full-wave boundary contour mode-matching method (BCMM) for the rigorous CAD of single and cascaded optimized H-plane and E-plane bends," 1994 IEEE MTT-S Int. Microw. Symp. Digest, San Diego, 1994, pp. 1021–1024.

[52] F. Alessandri, M. Mongiardo, and R. Sorrentino, "New efficient fullwave optimization of microwave circuits by the adjoint network method," IEEE MGWL, vol. 3, No. 11, Nov. 1993.

[53] F. Alessandri, M. Dionigi, R. Sorrentino, M. Mongiardo, "A fullwave CAD tool of waveguide components using a high speed direct optimizer," 1994 IEEE MTT-S International Symposium Digest, San Diego, May 1994, pp. 1539–1542.

Analysis of Electromagnetic Coupling Through a Thick Aperture in Multilayer Planar Circuits Using the Extended Spectral Domain Approach and Finite Difference Time-Domain Method

Allen M. Tran, Bijan Houshmand, *Member, IEEE,* and Tatsuo Itoh, *Fellow, IEEE*

***Abstract*—Electromagnetic coupling plays a very important role in the design and development of integrated antenna arrays and microwave circuits. With the availability of accurate computer-aided design tools, proper characterization and analysis, this coupling phenomenon could be used to simplify the packaging and fabrication process, to reduce manufacturing cost, and also to add flexibility to the design of multilayer structures in many cases. An accurate and efficient spectral domain approach (SDA) is proposed as a means to study the coupling effect through an arbitrary aperture in a thick common ground plane of multilayer planar circuits. In addition, the finite difference time domain (FDTD) method is present as an alternate approach to this type of problem and also is used as a validation tool for the spectral domain method because of its simplicity, generality, and accuracy. The proposed methods are applied to different aperture coupled structures for illustration.**

I. Introduction

INTERESTS and demands for monolithic microwave and millimeter-wave integrated circuits and antennas have grown significantly during recent years. Research activities in this area are no longer in curiosity phase, but rather in the full scale development stage. Applications for integrated antennas range from navigation and telecommunications to military electronic warfare systems. The increased enthusiasm for these types of structures is attributed to their inherently attractive features such as reliability, reproducibility, direct integration with microwave circuitry, low cost, small size, light weight, low profile, and broadband and multifunction capabilities as well as tolerance to heat and radiation. Advances in this area also imply the necessity of accurate computer-aided design (CAD) methods and procedures. In many cases, empirical design is too expensive or not even feasible due to the complexity of the structures or funding limitation. Therefore, the accurate modeling of active and passive components in these structures is very crucial for the design process. Even in the situations where several design iterations are allowed and money is no object, it is much more efficient and much less time consuming to use an accurate CAD tool for the design. With a good analytical/numerical model, design feasibility and trade-offs can be studied on computers to arrive at the best possible design before committing to building any expensive prototype hardware. Furthermore, unexpected phenomena can also be predicted and prevented ahead of time to avoid costly redesigning efforts.

Manuscript received June 1, 1994; revised March 29, 1995. This work was supported in part by Hughes Aircraft Company through the California MICRO program.

A. M. Tran is with Hughes Aircraft Company, Radar and Communications Sector, El Segundo, CA 90245 USA.

B. Houshmand and T. Itoh are with the Department of Electrical Engineering, University of California, Los Angeles, CA 90024 USA.

IEEE Log Number 9413645.

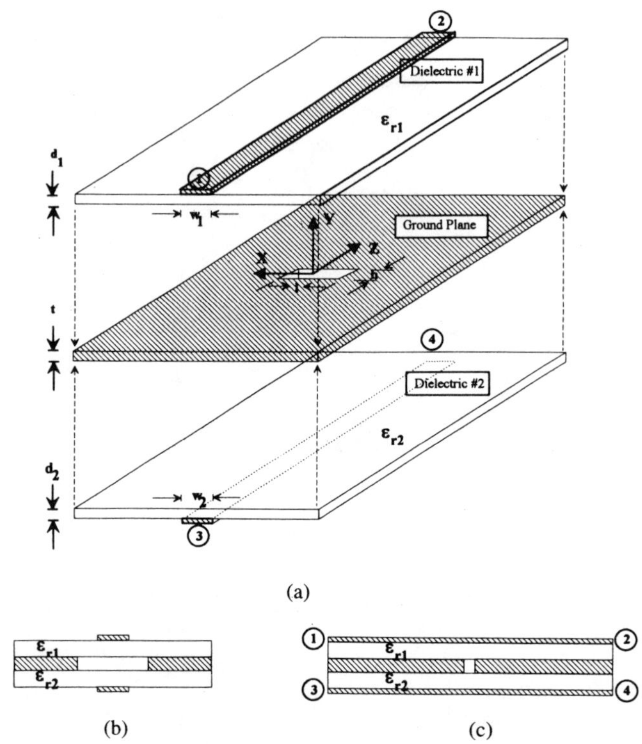

Fig. 1. (a) three-dimensional view, (b) end view, and (c) side view of typical aperture-coupled microstrip lines.

This paper proposes two different methods of modeling microstrip lines coupled through an arbitrarily shaped aperture residing in a thick common ground plane as depicted in Fig. 1. This multilayer structure is of interest in antenna design where the radiators and feed network are integrated to achieve a compact, low-profile planar antenna. A typical use of this coupling mechanism is to guide power from a beam-forming network on one substrate layer to subarrays of radiators on another substrate layer. One can visualize this by imagining

Reprinted from *IEEE Transactions on Antennas and Propagation,* pp. 921-926, 1995.

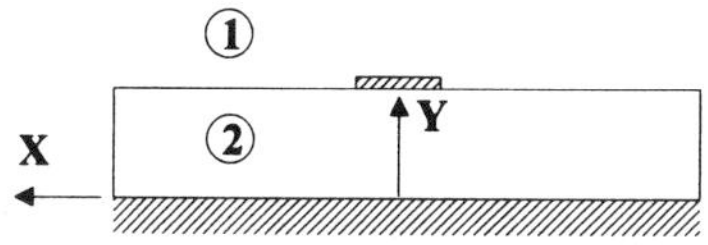

Fig. 2. Uniform microstrip line.

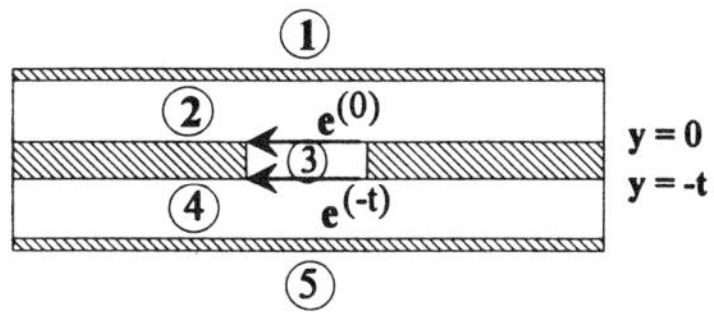

Fig. 3. Side view of the coupling junction.

ports 1 and 2 of the structure in Fig. 1 connected to subarrays of patch radiators, or any other types of printed radiators, and ports 3 and 4 to T/R modules of a feed network. In this case, the common ground plane is not only there for electrical purpose, but also to provide structural integrity for the antenna; therefore, a certain minimum thickness is specified. For the situations in which the antenna is required to operate in severe temperature environment, the common ground plane can also be used as a cold plate. Analyses of similar structures have been reported by Herscovici and Pozar [1] and Wakabayashi and Itoh [2] using reciprocity. In those analyses, however, the rectangular coupling slot was assumed to be narrow so that the longitudinal electric field component can be neglected. In addition, the thickness of the slot was not considered in those methods. In one of the proposed approaches, an extended version of the spectral domain method is utilized in conjunction with triangular subdomain basis functions to analyze the thick coupling aperture of arbitrary size and shape. The extension is, in a sense, removing the typical zero conductor thickness assumption of the original spectral domain formulation. The scattering parameters of the microstrip lines at the coupling junction are deterministically obtained in the spectral domain by a procedure similar to that described in [3]. The finite difference time domain (FDTD) method is present as an alternate approach and also a validation tool for the proposed spectral domain approach (SDA) because of its simplicity, flexibility, and good accuracy. Surface wave and radiation effects are automatically included in the formulation of SDA Green's function and also accounted for in FDTD by means of absorbing boundary conditions. Analytical procedures and numerical results for coupled microstrip lines via a thick aperture will be presented and compared between the extended SDA and FDTD to validate their accuracy and also to illustrate the thickness effect of the aperture region.

II. Analytical Formulation of the Extended Spectral Domain Approach

Assume that an incident wave is launched from $z = -\infty$ into port 1 (or any of the four ports) of the structure shown in Fig. 1. The scattering parameters at the four ports of a pair of coupled microstrip lines can be obtained through three main steps: eigenvalue formulation of the microstrip lines, coupling aperture field expansion, and determination of the junction scattering parameters using the method of moments (MoM). First, the top and bottom microstrip lines in Fig. 1 must be characterized without the presence of the coupling aperture. The thickness of the microstrip lines is neglected in this analysis because of its insignificant effect as discussed in [4]–[6]. Standard spectral domain method [7] or the immittance approach [8] can be utilized to compute the propagation constants β_n and current distributions $J_{x,\,z}$ of an infinitely long and uniform microstrip line depicted in Fig. 2.

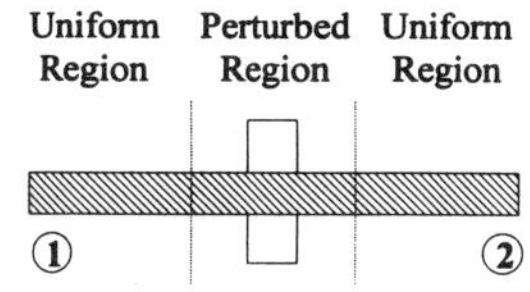

Fig. 4. Top view of the coupling junction.

As the result, the following matrix equation is obtained in the spectral domain at the microstrip-dielectric interface of the structure in Fig. 2

$$\begin{bmatrix} \tilde{Z}_{xx}(\alpha,\,\beta) & \tilde{Z}_{xz}(\alpha,\,\beta) \\ \tilde{Z}_{zx}(\alpha,\,\beta) & \tilde{Z}_{zz}(\alpha,\,\beta) \end{bmatrix} \cdot \begin{bmatrix} \tilde{J}_x(\alpha,\,\beta) \\ \tilde{J}_z(\alpha,\,\beta) \end{bmatrix} = \begin{bmatrix} \tilde{E}_x(\alpha,\,\beta) \\ \tilde{E}_z(\alpha,\,\beta) \end{bmatrix}. \quad (1)$$

It should be noted that (1) is the Fourier transform of the electric field integral equation (EFIE) in the conventional space domain analysis. To obtain the unknown current distributions, the first step is to expand them in terms of known basis functions $f_{x,\,z}(x)$

$$\tilde{J}_x(\alpha) = j \sum_{m=1}^{M} a_{x_m} \cdot \tilde{f}_{x_m}(\alpha)$$
$$\tilde{J}_z(\alpha) = \sum_{n=1}^{N} a_{z_n} \cdot \tilde{f}_{z_n}(\alpha) \quad (2)$$

where a_{x_m}, a_{z_n} are unknown coefficients to be determined

Next, eliminate the unknown fields on the left-hand side of (1) by applying Galerkin's method in the spectral domain and utilizing Parseval's relationship as described in [7] to arrive at

$$j \sum_{m=1}^{M} a_{x_m} \int_{-\infty}^{\infty} \tilde{f}^*_{x_p}(\alpha) \cdot \tilde{Z}_{xx}(\alpha,\,\beta) \cdot \tilde{f}_{x_m}(\alpha)\, d\alpha + \sum_{n=1}^{N} a_{z_n} \int_{-\infty}^{\infty} \tilde{f}^*_{x_p}(\alpha) \cdot \tilde{Z}_{xz}(\alpha,\,\beta) \cdot \tilde{f}_{z_n}(\alpha)\, d\alpha = 0$$
$$j \sum_{m=1}^{M} a_{x_m} \int_{-\infty}^{\infty} \tilde{f}^*_{z_q}(\alpha) \cdot \tilde{Z}_{zx}(\alpha,\,\beta) \cdot \tilde{f}_{x_m}(\alpha)\, d\alpha + \sum_{n=1}^{N} a_{z_n} \int_{-\infty}^{\infty} \tilde{f}^*_{z_q}(\alpha) \cdot \tilde{Z}_{zz}(\alpha,\,\beta) \cdot \tilde{f}_{z_n}(\alpha)\, d\alpha = 0$$
$$\text{where} \quad p = 1, \cdots, M \quad \text{and} \quad q = 1, \cdots, N. \quad (3)$$

Determinant equation (3) can be solved to obtain the propagation constant β and the unknown coefficients a_{x_m}, a_{z_n} of the current distributions. These parameters will be stored and

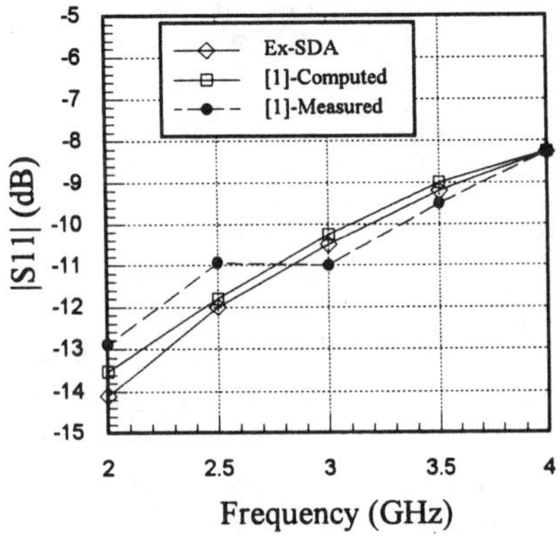

(a)

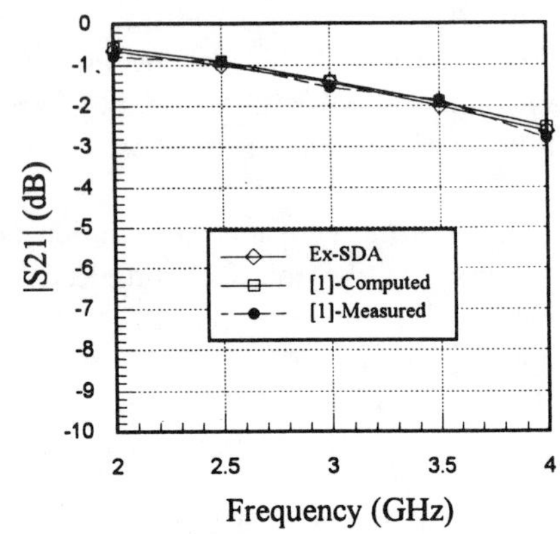

(b)

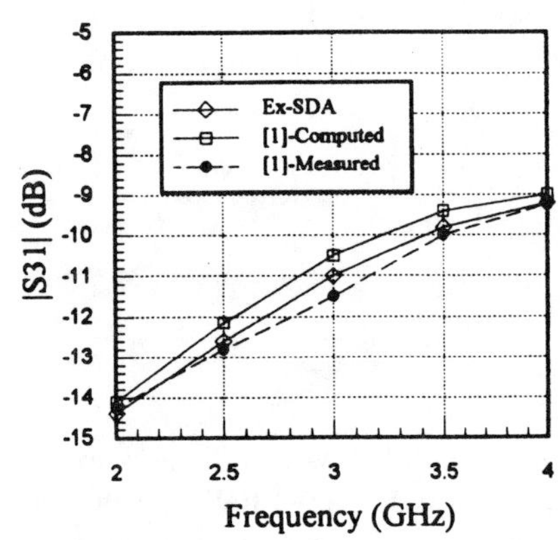

(c)

Fig. 5. S-parameters for structure in Fig. 1 with $\varepsilon_{r1} = \varepsilon_{r2} = 2.2$, $w_1 = w_2 = 2.54$ mm, $d_1 = d_2 = .762$ mm, $l = 15$ mm, $h = 1.0$ mm, aperture thickness $t = 0.0$.

used in the final step of the analysis to express the incident, reflected, through, and coupled current distributions.

The second part of this analysis is to extend the spectral domain method to handle the thick common ground plane. The fields in the thick aperture region (region 3 in Fig. 3) are expressed as Fourier series representations similar to the procedures in [9] and [10]. This procedure can also be viewed as mode expansion in an enclosed region

$$E_{x,z}^{(3)}(x, y, z) = \sum_{m=-\infty}^{\infty} \sum_{n=-\infty}^{\infty} \tilde{E}_{x,z}^{(3)}(y, \alpha_m, \beta_n) \cdot e^{-j\alpha_m x} \cdot e^{-j\beta_n z}$$

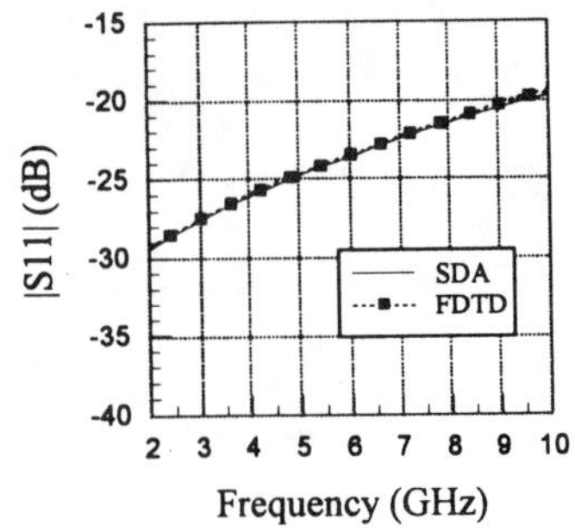

(a)

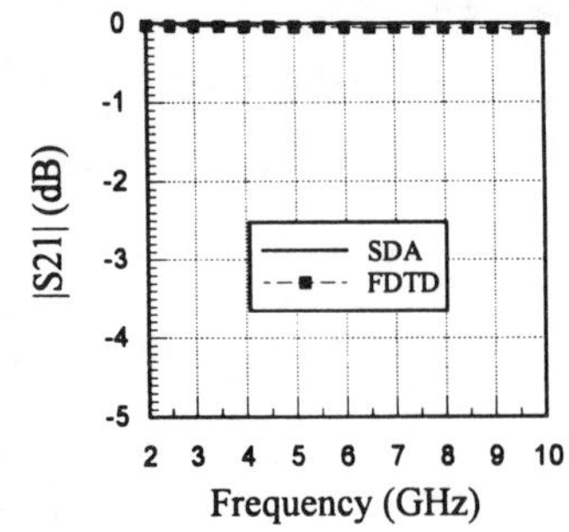

(b)

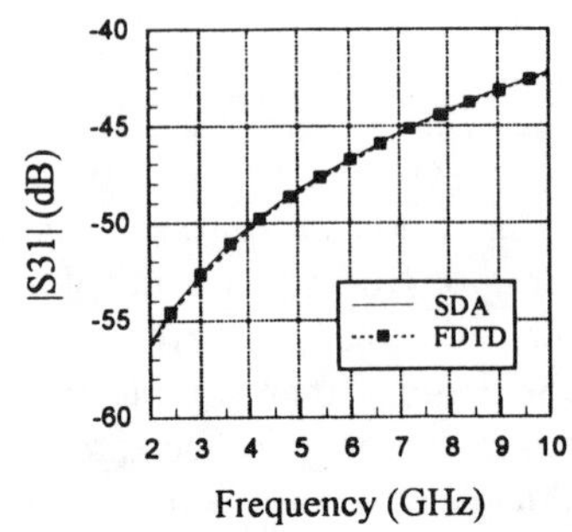

(c)

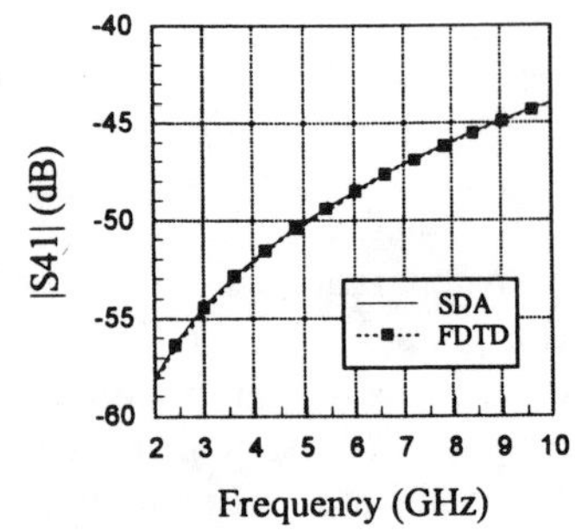

(d)

Fig. 6. S-parameters for circular coupling aperture with $\varepsilon_{r1} = \varepsilon_{r2} = 2.2$, $w_1 = w_2 = 2.54$ mm, $d_1 = d_2 = 0.762$ mm, $r = 0.50$ mm (radius of aperture), aperture thickness $t = 0.0$.

$$H_{x,z}^{(3)}(x, y, z) = \sum_{m=-\infty}^{\infty} \sum_{n=-\infty}^{\infty} \tilde{H}_{x,z}^{(3)}(y, \alpha_m, \beta_n) \cdot e^{-j\alpha_m x} \cdot e^{-j\beta_n z}. \quad (4)$$

Assume unknown aperture fields (or magnetic currents) over the coupling aperture at the interfaces $y = 0$ and $y = -t$

$$\vec{e}^{(0,-t)}(x, z) = \hat{x} \cdot e_x^{(0,-t)} + \hat{z} \cdot e_z^{(0,-t)}. \quad (5)$$

Expand the unknown aperture field components in terms of triangular subdomain basis functions [11] in the spatial domain and transform them into the spectral domain using the technique discussed in [12]

$$e_{x,z}^{(0,-t)}(x, z) = \sum_{n=1}^{N} c_{x,z_n}^{(0,-t)} \cdot \Delta_{x,z_n}(x, z). \quad (6)$$

Finally, a deterministic spectral domain procedure described in [3] and [10] is applied to express the current distributions on the microstrip lines in the vicinity of the coupling junction as the superposition of the known currents on the infinitely long, uniform microstrip lines and the unknown perturbed currents. The designation of the regions can be viewed from the top of the coupling structure as shown in Fig. 4.

In the uniform region of each port, the current distributions can be expressed as

$$J_{x,z}^{(uniform,1)}(x, z) = \sum_{m=1}^{M} a_{x,z_m} \cdot f_{x,z_m}(x) \cdot (e^{-j\beta_1 z} \pm \Gamma_1 \cdot e^{j\beta_1 z})$$

$$J_{x,\,z}^{(uniform,\,2)}(x,\,z) = \sum_{m=1}^{M} a_{x,\,z_m} \cdot f_{x,\,z_m}(x) \cdot T_2 \cdot e^{-j\beta_1 z}$$

$$J_{x,\,z}^{(uniform,\,3)}(x,\,z) = \sum_{m=1}^{M} a_{x,\,z_m} \cdot f_{x,\,z_m}(x) \cdot T_3 \cdot e^{+j\beta_2 z}$$

$$J_{x,\,z}^{(uniform,\,4)}(x,\,z) = \sum_{m=1}^{M} a_{x,\,z_m} \cdot f_{x,\,z_m}(x) \cdot T_4 \cdot e^{-j\beta_2 z} \quad (7)$$

where Γ_1, T_2, T_3, and T_4 are unknown complex scattering coefficients.

In the corresponding perturbed regions, the currents on the microstrip lines can be expressed as

$$J_{x,\,z}^{(perturbed,\,1)}(x,\,z) = J_{x,\,z}^{(uniform,\,1)}(x,\,z) + \sum_{p=1}^{P}\sum_{q=1}^{Q} b_{x,\,z_{pq}}^{(1)} \cdot u_{x,\,z_{pq}}(x) \cdot v_{x,\,z_{pq}}(z)$$

$$J_{x,\,z}^{(perturbed,\,2)}(x,\,z) = J_{x,\,z}^{(uniform,\,2)}(x,\,z) + \sum_{p=1}^{P}\sum_{q=1}^{Q} b_{x,\,z_{pq}}^{(2)} \cdot u_{x,\,z_{pq}}(x) \cdot v_{x,\,z_{pq}}(z)$$

$$J_{x,\,z}^{(perturbed,\,3)}(x,\,z) = J_{x,\,z}^{(uniform,\,3)}(x,\,z) + \sum_{p=1}^{P}\sum_{q=1}^{Q} b_{x,\,z_{pq}}^{(3)} \cdot u_{x,\,z_{pq}}(x) \cdot v_{x,\,z_{pq}}(z)$$

$$J_{x,\,z}^{(perturbed,\,4)}(x,\,z) = J_{x,\,z}^{(uniform,\,4)}(x,\,z) + \sum_{p=1}^{P}\sum_{q=1}^{Q} b_{x,\,z_{pq}}^{(4)} \cdot u_{x,\,z_{pq}}(x) \cdot v_{x,\,z_{pq}}(z). \quad (8)$$

At this point, the E- and H-fields of all five regions are known in terms of the unknown microstrip currents and aperture fields. The continuity conditions of tangential E- and H-fields are applied in the spectral domain at the interfaces $y = -(d+t)$, $-t$, 0, d to generate a set of equations which are the Fourier transforms of the mixed potential integral equations (MPIE). To obtain the microstrip currents and coupling aperture fields, MoM is applied in the spectral domain to generate a system of complex linear equations. This set of equations is very similar to that described in (3), except the right-hand side is no longer zero due to source excitation and the left-hand side contains both electric and magnetic terms. The reflection, through, and coupled coefficients are unknowns and can be solved for by matrix inversion

$$[\tilde{K}] \cdot \begin{bmatrix} c_{x,\,z_n}^{(0,\,-t)} \\ b_{x,\,z_{pq}}^{(m)} \\ \Gamma_1 \\ T_2,\,T_3,\,T_4 \end{bmatrix} = [\tilde{S}] \quad (9)$$

where $m = 1, 2, 3, 4$.

III. SDA Basis Functions

In the uniform microstrip line regions, full domain basis functions $f_x(x)$ and $f_z(x)$ are used to satisfy edge conditions and also to provide fast convergence. These functions are Maxwellian type functions defined in (10) below. In the perturbed regions, however, pulse and rooftop subsectional

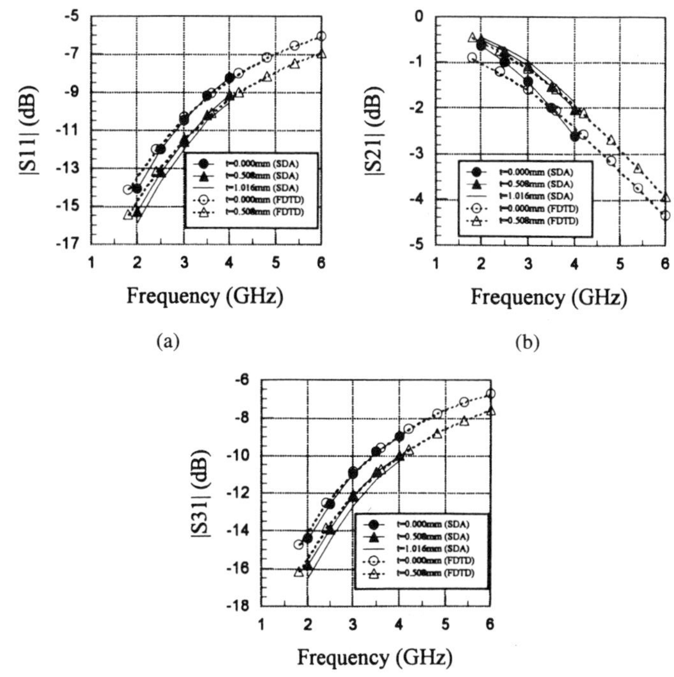

(a) (b) (c)

Fig. 7. S-parameters for structure in Fig. 1 with $\varepsilon_{r1} = \varepsilon_{r2} = 2.2$, $w_1 = w_2 = 2.54$ mm, $d_1 = d_2 = .762$ mm, $l = 15.0$ mm, $h = 1.0$ mm, aperture thickness $t = 0.0, 0.508, 1.016$ mm.

basis functions $u_{x,\,z}(x)$ and $v_{x,\,z}(z)$, discussed in [6], are used to accommodate possible high variations in the current distributions and also to add flexibility to the structure configuration. Triangular subdomain basis functions are used in the coupling aperture region to allow for arbitrary shape

$$f_{x_m}(x) = \frac{\sin(2m\pi x/w)}{\sqrt{1-(2x/w)^2}} \quad (10a)$$

$$f_{z_n}(x) = \frac{\cos[2(n-1)\pi x/w]}{\sqrt{1-(2x/w)^2}}. \quad (10b)$$

IV. FDTD

The aperture coupled microstrip line configuration is also analyzed by the application of the FDTD method as an alternate method and a comparison for the computed results by the extended SDA approach. The FDTD method used for this computation is based on the Yee's algorithm [13], where the time domain Maxwell's equations are solved on a cubical grid. This method is particularly applicable to the planar guiding structure where the geometry description is compatible to the grid structure. As a result, layers of dielectric materials and metallization can conveniently be handled by imposing the appropriate boundary conditions.

The time-domain simulation is carried out by launching an excitation signal at port 1 of the structure in Fig. 1. This signal can be expressed as

$$E(t,\,z_o) = e^{-[(t-t_o)/T]^2} \quad (11)$$

where z_o is the location of the excitation plane, t_o is the time delay, and T determines the frequency band width of the signal. For our simulation, T is 15 ps and $t_o = 2T$.

The excitation source is polarized in the vertical plane, perpendicular to the microstrip line at port 1. The S-parameters can be extracted from the time signals by the use of Fourier transformation. For example, S_{31} at frequency f is computed by

$$S_{31}(f) = \frac{V_3(f)}{V_1(f)} \tag{12}$$

where $V_n(f)$ is the Fourier transform of the observed signal at port n and related to the E-field as shown below

$$V_n(f) = -\int_0^d E(f, y, z = z_n) \cdot dy \tag{13}$$

where $E(f, y, z)$ is the observed electric field with the same polarization as the input signal and z_n is the observation location. The integration is carried out from the ground plane to the microstrip line.

The computational domain is terminated by the absorbing boundaries. For this structure, the Mur's first-order absorbing boundary condition is used successfully due to the propagation characteristic of the microstrip lines. For structures where the propagation field is more complex, such as fields close to a structure discontinuity, higher-order ABC's can be employed.

For the structure under consideration, the spatial discretization size is 0.25 mm for each unit cube and temporal discretization is 0.5 ps. It should be noted that the choice of spatial discretization is dictated by the geometry, while the temporal discretization is chosen such that the stability of the FDTD algorithm is ensured.

V. Results

Computer programs have been developed for both the extended SDA and FDTD to compute the scattering parameters of a pair of microstrip lines coupled through an arbitrary aperture residing in a thick common ground plane. The code for the extended SDA has been validated with published results [1] for a narrow rectangular slot with zero thickness in Fig. 5 and also against the FDTD code for a circular hole in the common ground, also of zero thickness, in Fig. 6. Fig. 7 provides results for the same rectangular aperture that is used for Fig. 5, but for different aperture thicknesses. Data in Fig. 7 was produced by the extended SDA and FDTD for comparison and validation and also to illustrate the effect of the metallization thickness. Numerical results indicate good agreement between the two proposed methods for all thicknesses. Fig. 7 also indicates that as the thickness of the aperture becomes larger, the coupling slot acts more like a cutoff waveguide which prevents energy from propagating through; therefore, the coupling coefficients decrease with aperture thickness. A similar type of phenomenon has been observed before in rectangular waveguides coupled through a thick rectangular slot. It has also been noticed that the magnitudes of the backward coupling coefficient ($|S_{31}|$) and the forward coupling coefficient ($|S_{41}|$) are significantly different for aperture with large width-to-length ratio as in the case of Fig. 6. This phenomenon is due to the existence of a considerable longitudinal field component in the x-direction of the coupling aperture which makes the coupler slightly directional. The results for the extended SDA were obtained on a 486DX2/66 MHz PC with a typical run time of 98 s per frequency point. The FDTD results were computed on a CRAY C-90 with a typical CPU time of 1.2 s for a computational box of 80 × 30 × 140 cells and 1500 time steps.

VI. Conclusions

The extended SDA and FDTD method have been applied successfully to the analysis of microstrip lines coupled through thick apertures. Numerical results show that the finite metallization thickness of the common ground plane affects the coupling characteristics of the structure significantly. Comparisons between the extended SDA and FDTD indicate good agreement between the two proposed methods and also imply results produced by these two methods are valid and accurate. Both the extended SDA and FDTD are quite flexible; therefore, they can also be applied to other types of discontinuities in printed antennas and circuits. Surface wave and radiation phenomena are included in the Green's function formulation of the SDA and also accounted for in FDTD by absorbing boundaries. Basis functions in the SDA method are chosen such that the edge conditions are properly satisfied, and also, a fast convergence rate is achieved.

References

[1] N. Herscovici and D. M. Pozar, "Full-wave analysis of aperture-coupled microstrip lines," *IEEE Trans. Microwave Theory Tech.*, vol. 39, pp. 1108–1114, July 1991.

[2] T. Wakabayashi and T. Itoh, "Three-dimensionally coupled microstrip lines via a rotated slot in a common ground plane," *1992 IEEE MTT-S Int. Microwave Symp. Dig.*, May 1992, pp. 1436–1440.

[3] J. McLean, H. Ling, and T. Itoh, "Full wave modeling of electrically wide microstrip open end discontinuities via a deterministic spectral domain method," *1990 IEEE MTT-S Int. Microwave Symp. Dig.*, May 1990, pp. 1155–1158.

[4] T. Kitazawa and Y. Hayashi, "Quasistatic characteristics of a coplanar waveguide with thick metal coating," in *Proc. Inst. Elect. Eng., Microwaves Antennas Propagat.*, Feb. 1986, vol. 133, no. 1, pp. 18–20.

[5] T. Kitazawa, "Metallization thickness effect of striplines with anisotropic media: Quasistatic and hybrid-mode analysis," *IEEE Trans. Microwave Theory Tech.*, vol. 37, pp. 769–775, Apr. 1989.

[6] T. Kitazawa, Y. Hayashi, and M. Suzuki, "A coplanar waveguide with thick metal coating," *IEEE Trans. Microwave Theory Tech.*, vol. MTT-24, pp. 604–608, Sep. 1976.

[7] T. Itoh, *Numerical Techniques for Microwave and Millimeter—Wave Passive Structures.* New York: Wiley, 1989.

[8] ——, "Spectral domain immitance approach for dispersion characteristics of generalized printed transmission lines," *IEEE Trans. Microwave Theory Tech.*, vol. MTT-28, no. 7, pp. 733–736, July 1980.

[9] T. Kitazawa and T. Itoh, "Propagation characteristics of coplanar-type transmission lines with lossy media," *IEEE Trans. Microwave Theory Tech.*, vol. 39, pp. 1694–1700, Oct. 1991.

[10] A. M. Tran and T. Itoh, "Full-wave modeling of coplanar waveguide discontinuities with finite conductor thickness," *IEEE Trans. Microwave Theory Tech.*, vol. 41, pp. 1611–1615, Sep. 1993.

[11] S. M. Rao, D. R. Wilton, and A. W. Glisson, "Electromagnetic scattering by surfaces of arbitrary shape," *IEEE Trans. Antennas Propagat.*, vol. AP-30, no. 3, pp. 409–418, May 1982.

[12] B. Housmand, W. C. Chew, and S. W. Lee, "Fourier transform of a linear distribution with triangular support and its applications in electromagnetics," *IEEE Trans. Antennas Propagat.*, vol. 39, no. 2, pp. 252–254, Feb. 1991.

[13] K. Yee, "Numerical solution of initial boundary value problems involving Maxwell's equations in isotropic media," *IEEE Trans. Antennas Propagat.*, vol. AP-14, pp. 302–307, 1966.

Modeling of Microwave Active Devices Using the FDTD Analysis Based on the Voltage-Source Approach

Chien-Nan Kuo, Ruey-Beei Wu, Bijan Houshmand, and Tatsuo Itoh, *Fellow, IEEE*

***Abstract*— This letter describes a voltage-source-based formulation of the extended finite-difference time-domain algorithm for the purpose of modeling microwave devices. The device-wave interaction is fully characterized by replacing the lumped devices with equivalent voltage sources in the device region, which in turn generate electromagnetic fields according to Faraday's law. This formulation is applied to the analysis of a typical microwave amplifier, which includes a three-terminal active MESFET device. Simulation results are in good agreement with measured data.**

I. Introduction

With the advancement of computers, the finite-difference time-domain (FDTD) method has become a very popular tool for analysis of various electromagnetic problems, including microwave circuits. As the size of microwave circuits and spacing between circuit elements become smaller, the coupling between the closely spaced elements adds more complicated effects to circuit performances. Proper modeling of the lumped passive and/or active devices and electromagnetic waves is critical in circuit simulation. Extensive investigations have been presented in literatures aiming at the extension of the FDTD method to include lumped microwave devices in the full-wave analysis. The FDTD method has been extended to incorporate circuit quantities of a two-terminal device into the algorithm [1]–[3]. By recognizing that the equivalent circuit model consists of two-terminal lumped elements, three-terminal active devices can be modeled by incorporating these elements into several FDTD cells [4], [5]. This approach, however, may become quite cumbersome in the modeling of active devices with complicated equivalent circuit models. A more general approach substitutes the device with equivalent current sources and consequently allows direct access to all SPICE models in the FDTD calculation [6]. This approach has been found to be advantageous in the simulation of practical three-dimensional circuits such as amplifiers [7], [8].

Manuscript received November 16, 1995. This work was supported by Hughes Micro, JSEP contract F49620-92-C-0055, and Army Research Office contract DAAH04-93-G-0068.

C.-N. Kuo and T. Itoh are with the Electrical Engineering Department, University of California at Los Angeles, CA 90095 USA.

R.-B. Wu is with the Electrical Engineering Department, National Taiwan University, Taipei, Taiwan.

B. Houshmand is with the Jet Propulsion Laboratory, Pasadena, CA 91109 USA.

Publisher Item Identifier S 1051-8207(96)03431-9.

In the previous approach, the lumped devices are treated as current sources that interact with electromagnetic fields in accordance with Ampere's law. This letter presents a dual approach that employs voltage sources to represent the lumped devices and generate electromagnetic fields according to Faraday's law. The formulation is established to construct an equivalent circuit of the FDTD meshes as seen from the device. The approach is then applied to simulate practical three-dimensional microstrip amplifier circuits.

II. Formulation

An active device in a microwave circuit is typically very small in size compared to a wavelength, and it can be modeled by its equivalent lumped circuit with a very high degree of accuracy. In order to include the equivalent circuit model into the FDTD analysis, the active device can be replaced by equivalent voltage sources in the active region if the voltage sources satisfy the voltage-current relationship at the input/output ports and the scattering properties of the active device.

Consider an active device in a microstrip circuit. Fig. 1 shows the placement of the voltage sources at one port of the active device; each source is aligned to the FDTD grid edges beneath and perpendicular to the microstrip line with each ends connected to the microstrip line or a grounded via. Those vias, modeled as perfect conductors, provide a voltage reference to the voltage sources. Assume that the voltage sources lie in an active sheet parallel to the x-y plane. Without loss of generality, consider a voltage source with its center denoted by $(i, j+\frac{1}{2}, k)$, where two FDTD meshes contribute to the current flowing through the voltage source. For the right mesh, the integral form of Faraday's law can be expressed in terms of circuit quantities as

$$-L_1 \frac{dI_{\text{mesh},1}}{dt} = V_{\text{loop},1} + V_{\text{dev}} \tag{1}$$

where L_1 is the space inductance of FDTD cells, $V_{\text{loop},1}$ is the equivalent voltage obtained from the the loop integration of the E-field along the cell boundary other than the edge of the voltage source, and $I_{\text{mesh},1}$ is the loop current that flows along the right-hand side of the active sheet through the cell width Δx. A similar formulation can be derived for the left mesh to obtain a circuit equation for the loop current $I_{\text{mesh},2}$ which flows along the left-hand side of the active sheet. The device current I_{dev} flowing into the voltage source equals the

Reprinted from *IEEE Microwave and Guided Wave Letters*, pp. 199-201, May 1996.

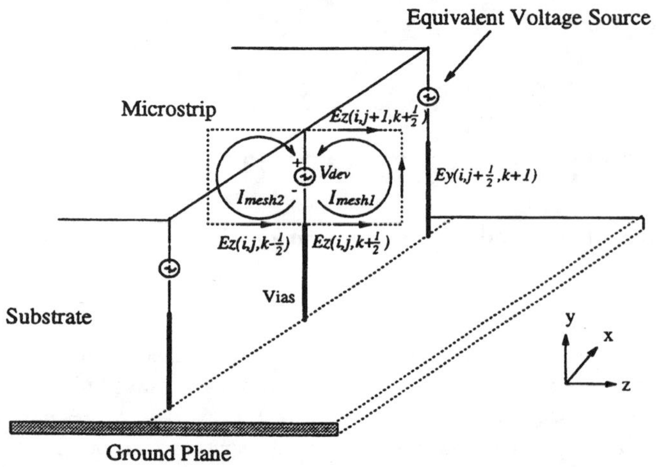

Fig. 1. The replacement of the active device by equivalent voltage sources on the FDTD grid edges.

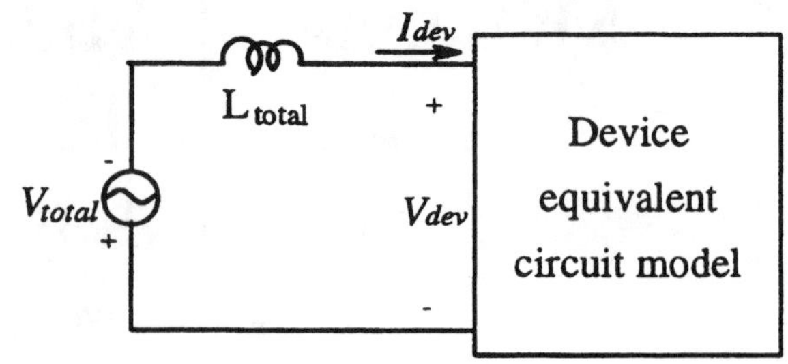

Fig. 2. The equivalent circuit of the configuration in Fig. 1, as seen from the device.

sum of the two loop current components on both sides of the active sheet.

In the case that there are N voltage sources placed across the entire width of the microstrip line, the device current I_{dev} is the sum of all the loop currents. The governing equation is derived as

$$-L_{\text{total}}\frac{dI_{\text{dev}}}{dt} = V_{\text{total}} + V_{\text{dev}} \tag{2}$$

where

$$V_{\text{total}} = \sum_{i}^{2N} \frac{V_{\text{loop},i}}{L_i/L_{\text{total}}} \tag{3}$$

$$L_{\text{total}} = 1 \Big/ \sum_{i}^{2N} \frac{1}{L_i} \tag{4}$$

and L_i is the space inductance of each mesh. The equivalent circuit of the governing equation is shown in Fig. 2. The voltage source/inductor pair is the equivalent circuit as seen from the device. Through this equivalent circuit, the field quantities in the FDTD algorithm are combined with the circuit quantities in the device model to describe the interaction between the device and electromagnetic fields.

The extended FDTD time-marching scheme can be described as follows:

Step 0) The electric fields E^{n-1}, magnetic fields $H^{n-\frac{1}{2}}$, device voltage $V_{\text{dev}}^{n-\frac{1}{2}}$, and device current $I_{\text{dev}}^{n-\frac{1}{2}}$ are given, where the superscript denotes the time step.

Step 1) The update of electric fields E^n is formed from E^{n-1} and $H^{n-\frac{1}{2}}$ by employing Ampere's law. Note that the electric fields on the active sheet have yet to be determined since Ampere's law is not applicable in the device region.

Step 2) The voltage source V_{total}^n is calculated from nearby electric fields by (3). The FDTD meshes as seen from the device satisfy (2). Combining (2) and the device circuit model, the values of the device voltage and current at the next time step, $V_{\text{dev}}^{n+\frac{1}{2}}$ and $I_{\text{dev}}^{n+\frac{1}{2}}$, can be solved by using typical circuit approaches, e.g., the backward difference scheme for the state equation [9], with the initial values of $V_{\text{dev}}^{n-\frac{1}{2}}$ and $I_{\text{dev}}^{n-\frac{1}{2}}$, respectively. The electric fields E^n on the active sheet are then determined by the device voltage V_{dev}^n, which is approximated by the average of $V_{\text{dev}}^{n+\frac{1}{2}}$ and $V_{\text{dev}}^{n-\frac{1}{2}}$.

Step 3) The magnetic fields $H^{n+\frac{1}{2}}$ are updated from $H^{n-\frac{1}{2}}$ and E^n by employing Faraday's law.

It is worth noting that Steps 1) and 3) in the updating algorithm for the electric and magnetic fields are exactly the same as those in the conventional FDTD algorithm.

III. Applications

The model is applied to analyze an actual microwave amplifier as in [8], which includes dc biasing circuits, impedance-matching circuits, and a three-terminal GaAs MESFET, NE72084. The layout of the amplifier and the equivalent lumped circuit model of the active device are depicted in Figs. 1 and 2(b) of [8].

The FDTD simulation is performed with uniform grids of space steps $\Delta x = 13$ mil, $\Delta y = 7.5$ mil, and $\Delta z = 10$ mil. The entire computation domain is divided into a grid of 90 (width) × 25 (height) × 230 (length) cells. Higdon's second-order absorbing boundary condition is used as in [10]. The radial stubs in the dc biasing circuits is modeled by the staircase approximation. The lumped resistors and capacitors in biasing circuits and matching circuits are treated as distributed elements and incorporated into the FDTD algorithm as in [1]. The dc power supply is modeled as an RF short circuit. The active region occupies eight cells in the longitudinal direction for the modeling of the packaged active device. For this two-port device, two sets of voltage sources are placed at the gate and drain ports. The numbers of voltage sources at the gate and drain ports are eight and four, respectively.

The S-parameters of the amplifier are calculated by taking the Fourier transform of the observed time response. Calculations show that the gain at 6 GHz is 9.33 dB and the matching point of the $|S_{11}|$ curve is 5.75 GHz. Measured data and the results of using current-source approach as in [8] are plotted in Fig. 3 for comparison. In Fig. 3(a), measured data smaller than -2 dB are truncated because of the choice of the scale in experiment. The matching point of measured data is at 5.6 GHz, and the gain is 9.23 dB at 6 GHz. The results of the voltage-source approach and the current-source approach are very similar, since both approaches are dual of each other. The

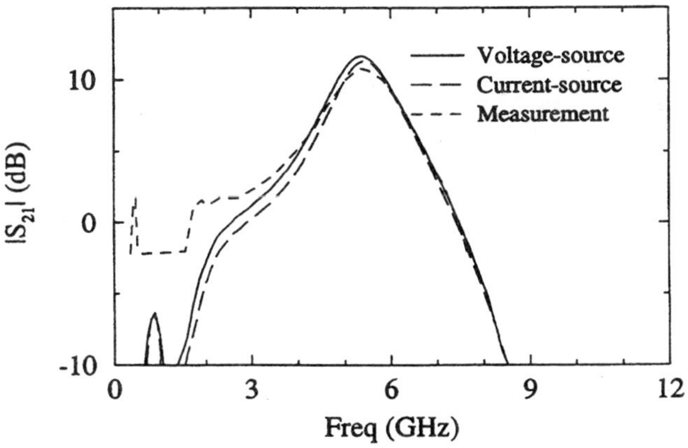

(a)

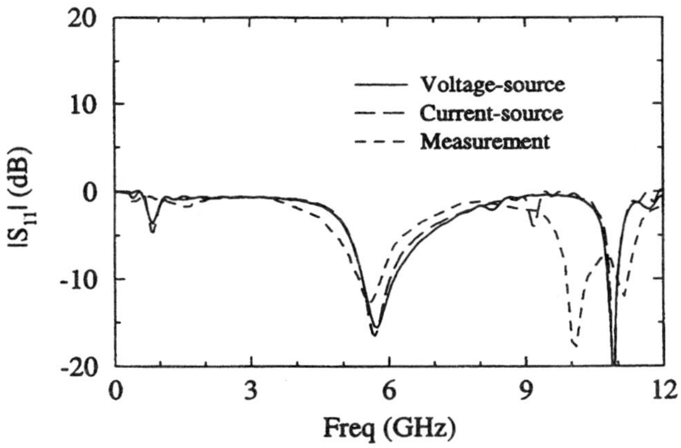

(b)

Fig. 3. The calculated results of two different approaches and measured S-parameters. (a) $|S_{21}|$. (b) $|S_{11}|$.

matching point is shifted if a different number of equivalet voltage sources is employed. The trend is the same as that in [11], which uses equivalent current sources. The spike in the $|S_{21}|$ curve at about 1 GHz is related to low-frequency oscillation. It has been found that the modeling of the dc power supply strongly affects the magnitude of the spike. In any case, FDTD simulations can predict those out-of-band dips near 1 and 11 GHz in the $|S_{11}|$ curve.

IV. Conclusion

A voltage-source approach is proposed for the extended FDTD analysis of microwave circuits, which may include three-terminal active devices. In this approach, the active device, substituted by equivalent voltage sources in the active region, can be viewed as a black box. The approach is then applied to the analysis of an actual microwave amplifier circuit. Simulation results show good agreement with measured data. In general, the voltage-source approach can be used for multi-port active devices. Utilizing equivalent voltage sources to model the device-wave interaction, the extended FDTD method can analyze the entire microwave circuit as a whole.

References

[1] W. Sui, D. A. Christensen, and C. H. Durney, "Extending the two-dimensional FD-TD method to hybrid electromagnetic systems with active and passive lumped elements," *IEEE Trans. Microwave Theory Tech.*, vol. 40, pp. 724–730, Apr. 1992.

[2] B. Toland, B. Houshmand, and T. Itoh, "Modeling of nonlinear active regions with the FDTD method," *IEEE Microwave Guided Wave Lett.*, vol. 3, no. 9, pp. 333–335, Sept. 1993.

[3] B. Toland, J. Lin, B. Houshmand, and T. Itoh, "FDTD analysis of an active antenna," *IEEE Microwave Guided Wave Lett.*, vol. 3, no. 11, pp. 423–425, Nov. 1993.

[4] Y.-S. Tsuei, A. C. Cangellaris, and J. L. Prince, "Rigorous electromagnetic modeling of chip-to-package (first-level) internonnections," *IEEE Trans. Components, Hybrids, and Manufacturing Tech.*, vol. 16, pp. 876–883, Dec. 1993.

[5] M. Piket-May, A. Taflove, and J. Baron, "FD-TD modeling of digital signal propagation in 3-D circuits with passive and active loads," *IEEE Trans. Microwave Theory Tech.*, vol. 42, no. 8, pp. 1514–1523, Aug. 1994.

[6] V. A. Thomas, M. E. Jones, M. Piket-May, A. Taflove, and E. Harrigan, "The use of SPICE lumped circuits as sub-grid models for FDTD analysis," *IEEE Microwave Guided Wave Lett.*, vol. 4, no. 5, pp. 141–143, May 1994.

[7] C.-N. Kuo, S. T. Chew, B. Houshmand, and T. Itoh, "FDTD simulation of a microwave amplifier," in *IEEE MTT-S Int. Microwave Symp.*, Orlando, FL, May 1995.

[8] C.-N. Kuo, V. A. Thomas, S. T. Chew, B. Houshmand, and T. Itoh, "Small signal analysis of active circuits using FDTD algorithm," *IEEE Microwave Guided Wave Lett.*, vol. 5, no. 7, pp. 216–218, July 1995.

[9] C. A. Desoer and E. S. Kuh, *Basic Circuit Theory*. New York: McGraw-Hill, 1969, chs. 12, 16.

[10] R. L. Higdon, "Absorbing boundary conditions for difference approximations to the multi-dimensional wave equation," *Math. Comput.*, vol. 47, no. 176, pp. 437–459, Oct. 1986.

[11] C.-N. Kuo, B. Houshmand, and T. Itoh, "FDTD analysis of active circuits with equivalent current source approach," in *IEEE AP-S Int. Symp.*, Newport Beach, CA, June 1995.

Spatial Soliton Deflection Mechanism Indicated by FD-TD Maxwell's Equations Modeling

Rose M. Joseph and Allen Taflove, *Fellow, IEEE*

***Abstract*— We present first-time calculations from the time-domain vector Maxwell's equations of spatial optical soliton propagation and mutual deflection, including carrier waves, in a 2-D homogeneous Kerr-type nonlinear dielectric. The nonlinear Schrödinger equation predicts that two co-propagating, in-phase spatial solitons remain bound to each other, executing a periodic separation. This disagrees with our new extensively tested finite-difference time-domain (FD-TD) solution of Maxwell's equations. FD-TD shows that co-propagating in-phase spatial solitons become unbound, i.e. diverge to arbitrarily large separations, if the ratio of soliton beamwidth to wavelength is order 1 or less. Not relying upon paraxial approximations or analogies to temporal soliton interactions, FD-TD appears to be a robust means of obtaining detailed models of the interaction of sub-picosecond pulsed light beams in nonlinear media directly in the space-time domain.**

I. Introduction

SPATIAL solitons have been observed in planar waveguides where self-focusing provides confinement of the beam in one transverse dimension and the refractive index profile of the waveguide provides confinement in the other transverse dimension [1]. The behavior of spatial solitons in Kerr-type nonlinear materials has also been predicted by nonlinear Schrödinger equation (NLSE) models and numerical beam propagation models [2] that generally make the paraxial approximation.

In principle, the behavior of electromagnetic fields in nonlinear dielectrics can be determined by solving Maxwell's equations subject to the assumption that the electric polarization has a nonlinear relation to the electric field. However, recent work using the finite-difference time-domain (FD-TD) methods [3]–[7], the nonlinear Maxwell's equations have not been solved directly. In this letter, new results are presented for interactions of spatial solitons in Kerr-type media using the FD-TD Maxwell's solver. It has been found for an important case that the FD-TD simulation of spatial solitons does not behave as described by NLSE. Further, FD-TD modeling allows simulation of the dynamics of pulsed spatial solitons and associated switching devices.

Manuscript received March 9, 1994; revised June 6, 1994. This work was supported in part by NSF Grant ECS-9218494, ONR Contract N00014-93-0133, and Cray Research, Inc.

R. M. Joseph and A. Taflove are with the Department of Electrical Engineering and Computer Science, McCormick School of Engineering, Northwestern University, Evanston, Illinois 60208-3118 USA.

IEEE Log Number 9404823.

II. 2-D Algorithm for Instantaneous Nonlinearity

Consider a 2-D transverse magnetic (TM) problem. Maxwell's equations for the electric and magnetic field intensities, E_z, H_x, and H_y are given by:

$$\frac{\partial \mu_0 H_x}{\partial t} = -\frac{\partial E_z}{\partial y}, \qquad \frac{\partial \mu_0 H_y}{\partial t} = \frac{\partial E_z}{\partial x},$$
$$\frac{\partial D_z}{\partial t} = \frac{\partial H_y}{\partial x} - \frac{\partial H_x}{\partial y} \tag{1}$$

Kerr nonresonant virtual electronic transitions in silica occur at time scales on the order of about 1 fs or less. In this letter, the FD-TD model for Kerr-type materials assumes instantaneous nonlinear response. The nonlinearity is modeled in the relation $D_z = \varepsilon_o \varepsilon E_z$, where

$$\varepsilon = n^2 = \left(n_o + n_2|E_z|^2\right)^2 \cong n_o^2 + 2n_o n_2 |E_z|^2 \tag{2}$$

In (2), n_o is unitless and n_2 has units of m^2/V^2. Equations (1) are first solved for H_x, H_y, and D_z by using the standard second-order accurate FD-TD scheme [8]. Then, the latest value of E_z can be obtained by iteration, using the new value of D_z and the old value of E_z:

$$E_z = \frac{D_z}{n_o^2 + 2n_o n_2 |E_z|^2} \tag{3}$$

Knowledge of E_z then permits another updating of H_x, H_y, and D_z, and the process repeats cyclically until time-stepping is completed. This procedure comprises the complete solution method for the space-time behaviour of the 2-D optical electromagnetic field.

III. Spatial Soliton Results

The modeling capabilities of this algorithm are demonstrated by 2-D calculations of propagating and mutually attracting and deflecting optical spatial solitons. The calculations are for a propagating sinusoidal beam that is switched on at $t = 0$ in Type-RN Corning glass with $n_o = 2.46$ and $n_2 = 1.25 \times 10^{-18}$ m^2/W [9]. The beam carrier frequency is 2.31×10^{14} Hz ($\lambda = 1.3$ μm); the initial peak electric field intensity is 6.87×10^9 V/m; and the fields have an initial hyperbolic secant transverse distribution with an intensity beamwidth (FWHM) of 0.65 μm. The computational domain is 95×31 μm.

The first Maxwell's equations calculation (Fig. 1(a)) selects the amplitude of the beam to balance its spreading and self-focusing. This provides a spatial soliton that retains its

Reprinted from *IEEE Photonics Technology Letters,* pp. 1251-1254, October 1994.

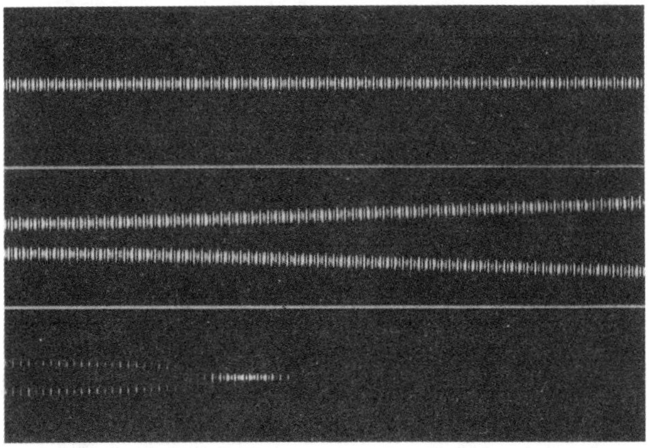

Fig. 1. FD-TD Maxwell's equations calculations of spatial solitons in Type RN Corning glass. Each beam has a 1.3-μm wavelength, an initial intensity beamwidth (FWHM) of 0.65 μm, and an initial peak electric field of 6870 V/μm. (a) Single spatial soliton; (b) repulsion of co-propagating spatialsolitons (initial center-to-center separation = 1.05 μm, relative carrier phase = π); (c) single coalescence and subsequent divergence for co-propagating spatial solitons (initial separation = 1.05 μm, relative carrier phase = 0).

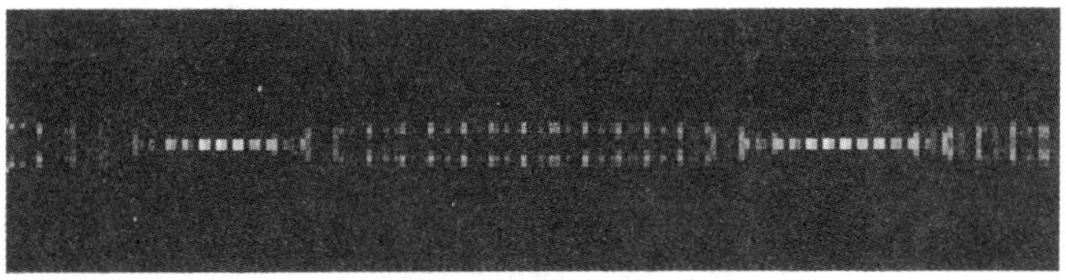

Fig. 2. FD-TD simulations showing restoration of the beam re-coalescence behavior after doubling the intensity beamwidth and separation parameters ofthe simulated beams, keeping the wavelength constant.

TABLE I
PROGRESSIVE AGREEMENT OF FD-TD AND NLSE RESULTS FOR PERIODICITY OF CO-PROPAGATING IN-PHASE SPATIAL SOLITIONS AS $B_I\lambda_d$ INCREASES

B_I, FWHM (μm)	B_I/λ_d	z_p (μm) NLSE	z_p (μm) FD-TD	Difference
0.65	1.22	9	∞	∞%
1.3	2.46	34	47	38%
2.6	4.9	135	153	13%

transverse electric field and magnetic field distributions. The second Maxwell's equations calculation (Fig. 1(b)) simulates the parallel co-propagation of two equal-amplitude spatial solitons separated by 1.05 μm center-to-center, where the solitons have a carrier phase difference of π radians. This computation provides the beam-to-beam repulsion expected from NLSE [10], [11].

The third Maxwell's equations calculation (Fig. 1(c)) simulates the parallel co-propagation of two equal-amplitude, in-phase spatial solitons (carrier phase difference of 0). NLSE predicts that the two in-phase solitons remain bound, executing a periodic relative motion [12] if the two beams have the appropriate amplitudes and spacing. Aitchison, *et al.* [13] indicate that two in-phase fundamental solitons with an input amplitude distribution of

$$A(z) = \frac{1}{kw}\left(\frac{n_o}{n_2}\right)^{1/2}\left[\text{sech}\left(\frac{x-x_o}{w}\right) + \text{sech}\left(\frac{x+x_o}{w}\right)\right] \quad (4)$$

oscillate with a period of

$$z_p = \frac{2z_o \sinh(2x_o/w)\cosh(x_o/w)}{2x_o/w + \sinh(2x_o/w)} \quad (5)$$

based on the NLSE theory of Desem and Chu [14]. Here, w is the characteristic width of the hyperbolic secant; $x_o = 1.42$ w; $2x_o$ is the center-to-center separation of the two beams; and $z_o = \pi^2 n_o w^2/\lambda$ is the usual soliton period. For the choice of parameters used in the FD-TD Maxwell's equations simulations, the predicted repetition period is $z_p = 9$ μm. However, as shown in Fig. 1(c), the FD-TD calculations show only a single beam coalescence and then subsequent beam divergence to arbitrarily large separations, yielding an effective $z_p = \infty$.

It was desired to understand why the nonlinear FD-TD Maxwell's equations model did not agree with the NLSE prediction in this case. The first possibility considered was that the FD-TD simulation was flawed because of inadequate grid resolution and/or inadequate decoupling of the beam interaction region from the weakly reflecting outer grid boundaries. In a series of exploratory modeling runs to address these issues, the space-time resolution of the FD-TD grid was progressively refined and the grid enlarged. These changes gave results identical to those of the original FD-TD model. Therefore, the original FD-TD model was concluded to be numerically converged and sufficiently free of the outer boundary artifact to yield plausible results.

The second possibility considered was that the ratio of intensity beamwidth, B_I to dielectric wavelength λ_d, was below the limit of applicability of NLSE. Because it is known that additional terms in the NLSE are required to model higher-order effects for temporal solitons, it was reasoned that basic NLSE modeling of co-propagating spatial solitons would be more physically meaningful if the two beams were widened relative to the optical wavelength while maintaining the same ratio of beamwidth to beam separation. This would reduce higher-order diffraction effects, hopefully bringing the test case into the region of validity for the simple NLSE model.

To test this possibility, two new FD-TD simulations were conducted where B_I and separation parameters of the simulated beams were each doubled keeping the dielectric wavelength, λ_d, constant. After the first doubling, the FD-TD-predicted spatial solitons began to qualitatively show the re-coalescence behavior predicted by NLSE, but with a 38% longer period of re-coalescence than the NLSE value. This FD-TD simulation is shown in Fig. 2. After the second doubling of beamwidth and beam separation, the FD-TD and NLSE predictions for z_p showed much better agreement, differing by only 13%. Results for these numerical experiments are shown in Table I.

It was concluded that there is a strong likelihood that co-propagating, in-phase optically narrow beams have only a single coalescence and then indefinite separation. The FD-TD model appears to properly predict the behavior of beams in nonlinear media both in the regime where the standard NLSE model breaks down ($B_I/\lambda_d < 1$) and the the regime

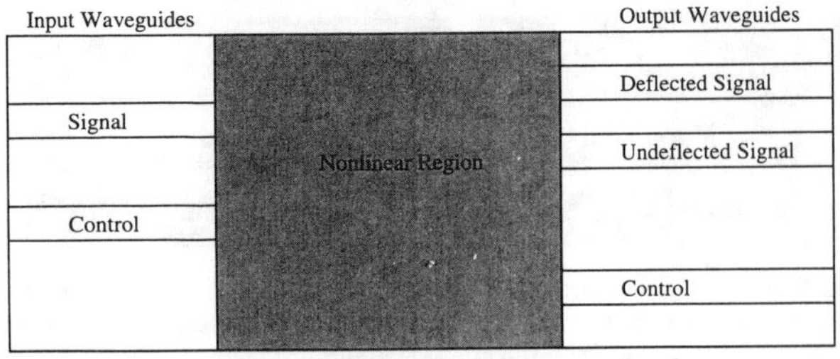

Fig. 3. Proposed all-optical switch based upon the single-time spatial soliton coalescence behavior indicated by FD-TD modeling. Pulsed optical signal and control beams are fed in at the left, interact in the Kerr nonlinear medium, and then couple into receptor waveguides. In the absence of the control, the signal propagates with zero deflection. In the presence of the control, and depending upon its carrier phase relative to the signal, there is either a single coalescence and then deflection to a collecting waveguide, or deflection without coalescence.

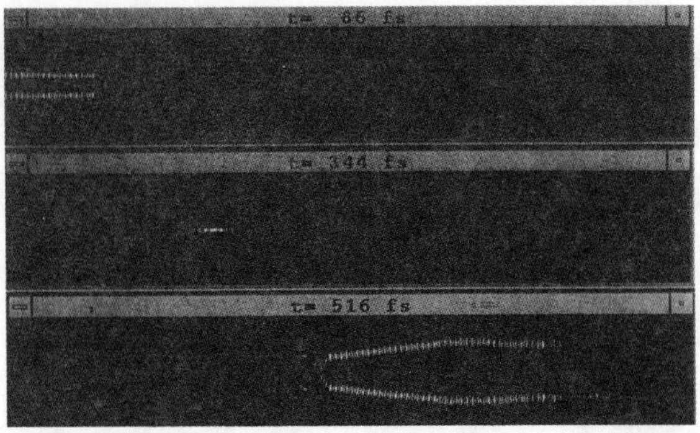

Fig. 4. Snapshots of the FD-TD-computed electric fields of 100-fs pulsed signal and control spatial solitons at the simulation times of 86 fs, 344 fs, and 516 fs for zero carrier phase between the pulses. Single-time spatial soliton coalescence behavior is indicated by FD-TD modeling for ultra-short pulses as well as continuous beams.

where the standard NLSE model is valid ($B_I/\lambda_d \gg 1$). The paraxial approximation inherent to NLSE, according to Lax *et al.* [15], accounts only for zeroth-order diffraction effects. Since the FD-TD model implements the fundamental Maxwell's curl equations, it makes no assumption about a preferred scattering direction. It naturally accounts for energy transport in arbitrary directions and should be exact for the computed optical electromagnetic fields up to the limit set by the grid resolution and Nyquist sampling theory.

IV. Pulsed Spatial Soliton Switch

The single-time spatial soliton coalescence behavior indicated by the FD-TD modeling studies discussed above provides the basis for the all-optical switch proposed in Fig. 3. This pulsed spatial soliton switch consists of a Kerr-type nonlinear interaction region (Corning glass Type-RN) with a pair of input and output waveguides on each side. Optical signal and control pulses are fed in at the left edge, interact in the nonlinear medium, and then couple into receptor waveguides. In the absence of the control pulse, the signal pulse propagates with zero deflection. In the presence of the control pulse, and depending upon its carrier phase relative to the signal pulse, there is either a single coalescence and then deflection to a collecting waveguide, or deflection without coalescence. (FD-TD studies have shown that optical pulses as short as 70 fs have the same coalescence/deflection behavior as continuous beams.) Note that the device of Fig. 3 differs from the all-optical spatial-soliton switch of Shi and Chi [10] which did not take advantage of the single-coalescence/single-divergence phenomenon, used continuous-wave excitation, and assumed a nonphysically high nonlinear coefficient.

Fig. 4 shows snapshots of the FD-TD-computed electric fields of 100-fs pulsed signal and control spatial solitons at the simulation times of 86 fs, 344 fs, and 516 fs for zero carrier phase between the pulses. The deflection of the signal pulse varies smoothly but not monotonically as the relative carrier phase is changed from 0 to π radians. The range of beam deflection complicates somewhat the design of a receptor waveguide that can efficiently capture the deflected beam for arbitrary phase differences between signal and control pulses. Yet, the principle appears proven and potentially useful.

V. Conclusion

This letter presented FD-TD Maxwell's equations calculations of spatial optical soliton propagation and mutual deflection in a 2–D homogeneous nonlinear dielectric medium. The FD-TD results show that co-propagating, in-phase optically narrow spatial solitons undergo only a single beam coalescence before diverging to arbitrarily large separations. This phenomenon provides a possible mechanism for constructing femtosecond all-optical switches spanning less than 100 mm in length in an existing type of Corning glass.

Acknowledgment

The authors acknowledge the contributions to the development of nonlinear Maxwell's equations algorithms by Dr. Peter Goorjian of NASA-Ames Research Center.

References

[1] J. S. Aitchison, *et al.*, "Observation of spatial optical solitons in a nonlinear glass waveguide," *Opt. Lett.,* vol. 15, pp. 471–473, 1990.

[2] V. E. Zakharov and A. B. Shabat, "Exact theory of two-dimensional self-focusing and one-dimensional self-modulation of waves in nonlinear media," *Soviet Phys. JETP*, vol. 34, pp. 62–69, 1972.

[3] P. M. Goorjian and A. Taflove, "Direct time integration of Maxwell's equations in nonlinear dispersive media for propagation and scattering of femtosecond electromagnetic solitons," *Opt. Lett.,* vol. 17, pp. 180–182, Feb. 1992.

[4] P. M. Goorjian, A. Taflove, R. M. Joseph, and S. C. Hagness, "Computational modeling of femtosecond optical solitons from Maxwell's equations," *IEEE J. Quantum Electron.,* vol. 28, (Special Issue Ultrafast Opti. and Electron.), pp. 2416–2422, Oct. 1992.

[5] R. M. Joseph, P. M. Goorjian, and A. Taflove, "Direct time integration of Maxwell's equations in 2-D dielectric waveguides for propagation and scattering of femtosecond electromagnetic solitons," *Op t. Lett.,* vol. 18, pp. 491–493, Apr. 1993.

[6] R. W. Ziolkowski and J. B. Judkins, "Applications of the nonlinear finite difference time domain (NL-FDTD) method to pulse propagation in nonlinear media: Self-focusing and linear interfaces," *Radio Science*, vol. 28, pp. 901–911, 1993.

[7] ——, "Full-wave vector Maxwell equation modeling of self-focusing of ultra-short optical pulses in a nonlinear Kerr medium exhibiting a finite response time," *Wave Motion,* vol. 10, pp. 186–198, 1993.

[8] A. Taflove, "Review of the formulation and application of the finite-difference time-domain method for numerical modeling of electromagnetic wave interactions with arbitrary structures," *Wave Motion*, vol. 10, pp. 547–582, Dec. 1988.

[9] M. A. Newhouse, "Glasses for nonlinear optics," in *Mater. Resear. Soc. Symp. Proc.*, vol. 244, 1991.

[10] T.-T. Shi and S. Chi, "Nonlinear photonic switching by using the spatial soliton collision," *Opt. Lett.,* vol. 15, pp. 1123–1125, 1990.
[11] J. Bian and A. K. Chan, *Proc. Integr. Photon. Res. Top. Mtg.,* (New Orleans, LA), Apr. 13–16, 1992, p. 240–241.
[12] J. P. Gordon, "Interaction forces among solitons in optical fibers," *Opt. Lett.,* vol. 8, pp. 596–598, 1983.
[13] J. S. Aitchison, *et al.*, "Experimental observation of spatial soliton interactions," *Opt. Lett.,* vol. 16, pp. 15–17, 1991.
[14] C Desem and P. L. Chu, "Reducing soliton interaction in single-mode opitcal fibers," *IEE Proc.*, vol. 134, pp. 145–151, 1987.
[15] M. Lax, W. H. Louisell, and W. B. McKnight, "From Maxwell to paraxial wave optics," *Phys. Rev. A.*, vol. 11, p. 1365, 1975.

Applications of the Nonlinear Finite Difference Time Domain (NL-FDTD) Method to Pulse Propagation in Nonlinear Media: Self-Focusing and Linear-Nonlinear Interfaces

RICHARD W. ZIOLKOWSKI AND JUSTIN B. JUDKINS
ELECTROMAGNETICS LABORATORY, DEPARTMENT OF ELECTRICAL AND COMPUTER ENGINEERING
THE UNIVERSITY OF ARIZONA, TUCSON

(Received January 6, 1993; revised April 12, 1993; accepted April 14, 1993)

In an effort to meet an ever increasing demand for more accurate and realistic integrated photonics simulations, we have developed a multidimensional, nonlinear finite difference time domain (NL-FDTD) Maxwell's equations solver. The NL-FDTD approach and its application to the modeling of the interaction of an ultrashort, optical pulsed Gaussian beam with a Kerr nonlinear material will be described. Typical examples from our studies of pulsed-beam self-focusing, the scattering of a pulsed-beam from a linear-nonlinear interface, and pulsed-beam propagation in nonlinear waveguides will be discussed.

1. INTRODUCTION

With the continuing and heightened interest in linear and nonlinear semiconductor and optically integrated devices, more accurate and realistic numerical simulations of these devices and systems are in demand. Such calculations provide a test-bed in which one can investigate new basic and engineering concepts, materials, and device configurations before they are fabricated. This encourages multiple concept and design iterations that result in enhanced performances and system integrations of those devices. They also provide a framework in which one can interpret complex experimental results and suggest further diagnostics or alternate protocols. Thus the time from device conceptualization to fabrication and testing could be enormously improved with numerical simulations that incorporate more realistic models of the linear and nonlinear material responses and the actual device geometries.

To date, most of the modeling of pulse propagation in and scattering from complex linear and nonlinear media has been accomplished with one-dimensional, scalar models. These models have become quite sophisticated; they have predicted and explained many of the nonlinear as well as linear effects in present devices and systems. Unfortunately, they cannot be used to explain many observed phenomena, and expectations are that they are not adequately modeling linear and nonlinear phenomena that could lead to new effects and devices. It is felt that vector and higher-dimensional properties of Maxwell's equations that are not currently included in existing scalar models, in addition to more detailed materials models, may significantly impact the scientific and engineering results. Moreover, because they are limited to simpler geometries, current modeling capabilties are not adequate for linear/nonlinear optical-component engineering design studies.

In this paper we describe numerically obtained, multidimensional, full-wave, vector Maxwell's equations solutions to problems describing the interaction of ultrashort, pulsed beams with a nonlinear Kerr material having a finite response time. These numerical solutions

Reprinted with permission from *Radio Science*, R. W. Ziolkowski and J. B. Judkins, "Application of the Nonlinear Finite Difference Time Domain (NL-FDTD) Method to Pulse Propagation in Nonlinear Media: Self Focusing and Linear Interfaces," Vol. 28, pp. 901-911, 1993. © American Geophysical Union.

have been obtained [*Ziolkowski and Judkins*, 1992*a*, *b*, *c*, 1993*a*, *b*, *c*] in two space dimensions and time with a nonlinear finite difference time domain (NL-FDTD) method which combines a generalization of a standard, FDTD, full-wave, vector, linear Maxwell's equations solver with a currently used phenomenological time relaxation (Debye) model of a nonlinear Kerr material. This NL-FDTD approach has been used to obtain numerical solutions in two space dimensions and time for nonlinear self-focusing in bulk, thermal Kerr media, for normal and oblique incidence linear-nonlinear interface problems, and for the propagation of pulses in nonlinear waveguiding strutures. Although these basic geometries are straightforward, the NL-FDTD approach can readily handle more complex, realistic structures.

The NL-FDTD method is beginning to resolve several very basic physics and engineering issues concerning the behavior of the full electromagnetic field during its interaction with a nonlinear medium. The various examples of transverse electric(TE) and transverse magnetic (TM) nonlinear optics problems described by *Ziolkowski and Judkins* [1992*a*, *b*, *c*, 1993*a*, *b*, *c*] highlight the differences between the scalar and the vector approaches, the effects of polarization, and the effects of the finite response time of the medium. Applying the NL-FDTD approach to self-focusing problems in bulk media, (1) we have shown the existence of back reflected power from the nonlinear self-focus when the medium is responding nearly instantaneously to the applied optical field; (2) we have discovered optical vortices are formed in the trailing wakefield behind the nonlinear self-focus; and (3) we have identified that the longitudinal field component plays a significant role in limiting the self-focusing process. Applying the NL-FDTD approach to both the TM and TE nonlinear interface problems, (1) we have characterized the performance on an optical diode (linear-nonlinear interface switch) to single-cycle pulsed Gaussian beams including the appearance of a nonlinear Goos-Hänchen effect, the stimulation of stable surface modes, and the effects of a finite response time of the Kerr material; (2) we have shown definitively that the linear-nonlinear interface does not act like an optical diode for a tightly focused, single-cycle pulsed Gaussian beam; and (3) we have characterized the performance of some basic linear-nonlinear slab waveguides as optical threshold devices.

In all of these analyses we have identified the role of the longitudinal field component (which is not taken into account in the scalar models), and the resulting transverse power flows in the associated scattering-coupling processes.

The Debye model for the Kerr nonlinearity is a standard choice and has been used to investigate finite response effects in Kerr media by several groups [e.g., *Mitchell and Moloney*, 1990, *Hayata et al.*, 1990, *Hayata et al.*, 1992]. Nonetheless, we have been extended this NL-FDTD model recently [*Ziolkowski and Judkins*, 1993*b*] to materials described by a Lorentz linear dispersion model and a Raman nonlinear model. Thus the NL-FDTD model can now deal with librational effects [*Reintjes*, 1984] as well as many other known nonlinear behaviors.

There have been a number of groups dealing with the numerical modeling of optical wave propagation in nonlinear materials using the full-wave, vector, time-independent Maxwell's equations by *Miyagi and S. Nishida* [1974, 1975] and *Pohl* [1970] and using the vector paraxial equations by *Pohl* [1972], *Sodha et al.* [1974], *Hayata and Koshiba* [1988], and *Hayata et al.* [1990]. These efforts have provided, for instance, the modal fields present in nonlinear waveguides and the resulting propagation behavior of beams in those guides. In contrast, the NL-FDTD approach is time dependent and accounts for the complete time evolution of the system as a pulse propagates in a Kerr medium having a finite response time with no envelope approximations. In particular, it provides a complete picture of the pulse behavior during the nonlinear self-focusing process and the scattering from a linear-nonlinear interface. Note that, because of the nonlinearities, such a pulse solution cannot be obtained from any sequence of single frequency, time-independent results; it can only be obtained from a direct time integration of Maxwell's equations. Thus, these time-independent and time-dependent modeling approaches yield additional and complementary information.

Related modeling of optical pulse propagation in nonlinear media has also been reported [*Goorjian and Taflove*, 1991, 1992*a*, 1992*b*; *Goorjian et al.*, 1992*a*; *Goorjian et al.*, 1992*b*; *Goorjian et al.*, 1993; and *Jamid and Al-Bader*, 1993]. The work by Goorjian and his coworkers (to date) has emphasized modeling soliton propagation effects; they have recovered one dimensional solitons (one space dimension and time) and solitons in two-dimensional TE guiding strutures. One-dimensional nonlinear soliton propagation has also been modeled with a FDTD approach by *Hile and Kath* [1993]. They have shown that the one-dimensional model recovers known nonlinear Schrödinger equation results. Nonlinear guided-wave structures are also being modeled now by several groups [e.g., *Ziolkowski and Judkins*, 1992*b*, *c*, 1993*b*; *Goorjian et al.*, 1992*a*; *Goorjian et al.*, 1993; and *Jamid and Al-Bader*, 1993]. The interest in these nonlinear guided wave structures stems from their potential applications to integrated photonics devices and circuits. Because of the versatility of the NL-FDTD approach, all of these groups hope to be able to simulate the behavior of more complicated nonlinear guided wave structures and devices in the near future.

2. NL-FDTD APPROACH

The NL-FDTD method discussed by *Ziolkowski and Judkins* [1992*a*, *b*, *c*, 1993*a*, *c*] solves numerically Maxwell's equations

$$\frac{\partial}{\partial t}\,[\mu_0\,\vec{H}] = -\nabla\times\vec{E} \tag{1}$$

$$\frac{\partial}{\partial t}\,[\epsilon_L\,\vec{E}] = \nabla\times\vec{H} - \frac{\partial}{\partial t}\,\vec{P}^{\mathrm{NL}}\,, \tag{2}$$

where the nonlinear polarization term $\vec{P}^{\mathrm{NL}} = \epsilon_0\,\chi^{\mathrm{NL}}(\vec{r},t,|\vec{E}|^2)\vec{E}$ is specified by solving simultaneously a Debye model for the third-order, nonlinear susceptibility χ^{NL} of the Kerr medium:

$$\frac{\partial}{\partial t}\,\chi^{\mathrm{NL}} + \frac{1}{\tau}\,\chi^{\mathrm{NL}} = \frac{1}{\tau}\,\epsilon_2\,|\vec{E}|^2\,. \tag{3}$$

The nonlinear susceptibility χ^{NL} is incorporated most simply in the FDTD approach by introducing the effective permittivity and conductivity of the Kerr medium

$$\epsilon_{\mathrm{eff}} = \epsilon_L + \epsilon_0\,\chi^{\mathrm{NL}} \tag{4}$$

$$\sigma_{\mathrm{eff}} = \epsilon_0\,\frac{\partial}{\partial t}\,\chi^{\mathrm{NL}}\,, \tag{5}$$

where ϵ_L is the linear permittivity, and by rewriting Maxwell's equations in the form

$$\frac{\partial}{\partial t}\,\vec{H} = -\frac{1}{\mu_0}\,\nabla\times\vec{E} \tag{6}$$

$$\frac{\partial}{\partial t}\,\vec{E} = \frac{1}{\epsilon_{\mathrm{eff}}}\,\nabla\times\vec{H} - \frac{\sigma_{\mathrm{eff}}}{\epsilon_{\mathrm{eff}}}\,\vec{E}\,. \tag{7}$$

This approach models the medium as having a finite response time τ. If T represents the pulse width, then by setting $T \gg \tau$, one obtains an instantaneous response model: $\chi^{\mathrm{NL}} \approx \epsilon_2\,|\vec{E}|^2$, that is, the medium follows the pulse. On the other hand, if $T \ll \tau$, then the finite response time effects are maximal, and the medium's response significantly lags the pulse. The NL-FDTD approach can treat both extremes. Moreover, the divergence equation associated with this system includes the nonlinear source term: $\nabla\cdot[\epsilon_L\,\vec{E}] = -\nabla\cdot\vec{P}^{\mathrm{NL}}$, which in the TM case provides the mechanism that couples the longitudinal to the transverse electric field components.

Because of the quadratic nature of the nonlinearity in (3), the nonlinear susceptibility, hence, permittivity must be strictly positive. However, the nonlinear conductivity, which from (5) is obtained as the time derivative of the nonlinear susceptibility, can be both positive and negative. This represents both loss and gain, respectively, in the medium. In the absence of any other dispersion mechanism, one would then expect some erosion of the pulse amplitude in the pulse's leading half as it propagates in this Kerr medium. On the other hand, the conductivity changes its sign along the trailing half of the pulse. This causes growth in the pulse amplitude and a shocklike structure to form along the trailing portion of the pulse. The sharpness and intensity of this shocklike structure depends on the finite response time; the structure will be more peaked the more instantaneous the medium's response is.

In two space dimensions and time with the coordinates (x, z, t) and with the choice of a TM_z-polarized wave, the NL-FDTD method solves for the complete time history of each of the components (E_x, E_z, H_y). The equations for a TE_z-polarized wave can be obtained by reciprocity: $\vec{E} \to \vec{H}$ and $\vec{H} \to -\vec{E}$, and they lead to the NL-FDTD solution of the components (E_y, H_x, H_z). Whereas the nonlinear source term strongly couples the transverse and longitudinal electric field components in the TM case, the corresponding magnetic field components in the TE case are driven by the transverse electric field component which exhibits the nonlinear growth. Additionally, when the linear-nonlinear interface problem is treated, Maxwell's equations naturally provide the boundary conditions appropriate for this lossy dielectric interface. Thus the linear- nonlinear interface problem can be handled without imposing any additional constraints on the fields. Moreover, more complex structures can be added to the simulation with little difficulty, giving the NL-FDTD approach a great deal of flexibility, particularly in comparison to the scalar models.

Note that the model defined by (1)-(7) ignores any linear dispersion effects and have taken the linear permittivity to be a constant $\epsilon_L = \epsilon_0$. This physically means that it is appropriate only for propagation distances shorter than the dispersion length of the material. As noted above, we have incorporated Lorentz linear dispersion and Raman nonlinearity models into the NL-FDTD approach [*Ziolkowski and Judkins*, 1993b]. We had investigated the nuances of a number of techniques introduced recently for modeling dispersive effects in the linear FDTD method by *Luebbers et al.* [1990], *Kashiwa and Fukai* [1990], *Lee et al.* [1991], and *Joseph et al.* [1991] and developed a stencil set that allows simultaneous solution of these models with Maxwell's equations. In particular, we are now solving in a self-consistent manner the system of equations:

$$\frac{\partial}{\partial t}\left[\mu_0 \vec{H}\right] = -\nabla \times \vec{E} \tag{1'}$$

$$\frac{\partial}{\partial t}\left[\epsilon_L \vec{E}\right] = \nabla \times \vec{H} - \frac{\partial}{\partial t}\vec{P} \tag{2'}$$

$$\frac{\partial^2}{\partial t^2}\vec{P}^L + \Gamma_L \frac{\partial}{\partial t}\vec{P}^L + \omega_L^2 \vec{P}^L = \epsilon_0 \chi_0 \omega_L^2 \vec{E} \quad \text{Lorentz model} \tag{8}$$

$$\frac{\partial^2}{\partial t^2}\chi^{\rm NL} + \Gamma_R \frac{\partial}{\partial t}\chi^{\rm NL} + \omega_R^2 \chi^{\rm NL} = \epsilon_R \omega_R^2 |\vec{E}|^2 \quad \text{Raman model}, \tag{9}$$

where $\vec{P} = \vec{P}^L + \vec{P}^{\rm NL}$ and $\vec{P}^{\rm NL} = \epsilon_0 \chi^{\rm NL} \vec{E}$. Details concerning this extended NL-FDTD model will be given in several manuscripts currently under preparation. Goorjian and his coworkers have also developed a similar capability to model pulse propagation under the influence of linear and nonlinear dispersive, linear and nonlinear diffractive, and time retardation effects in the medium.

Because of the versatility of the FDTD approach, we have been able to "turn on" the dispersion effects to analyze their impact on the self-focusing and the linear-nonlinear interface reflection-transmission processes. The results we describe below have been reaffirmed by the more complex NL-FDTD model defined by (1'), (2'), (8), and (9).

The NL-FDTD results to be reported below were obtained by carefully designing and testing the numerical grid, material parameters, and the algorithm based upon (1)-(7) to insure stability, accuracy, and efficiency. The basic stencils of the NL-FDTD algorithm in both the TE and the TM polarization cases are shown in Figure 1. These standard stencils represent two staggered grids: one for the electric field components and one for the magnetic field components. These are the standard choices associated with the two-space dimensional linear FDTD algorithm. The discrete versions of the TE and the TM forms of equations (3), (6), and (7) are centered in space and time on this numerical grid. In the TM case, the electric field components E_x and E_z

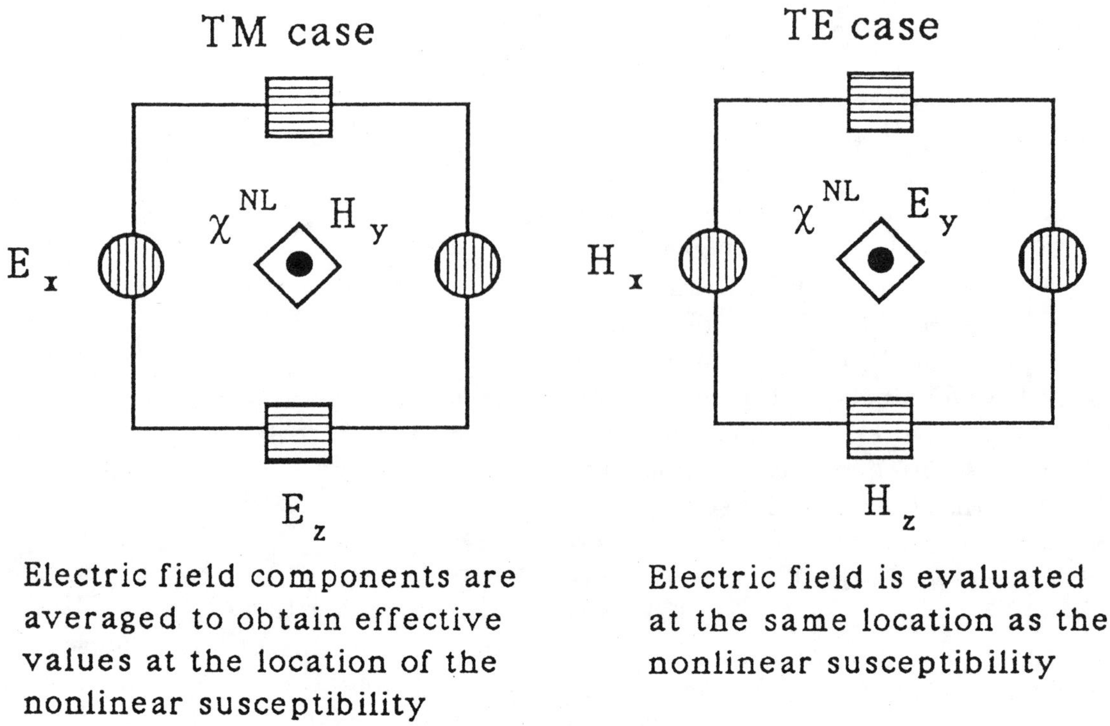

Fig. 1. The NL-FDTD TE and TM unit cell stencils.

are assigned to the edges and the magnetic field component H_y and the nonlinear susceptibility χ^{NL} to the center of the unit FDTD square cell. The nonlinearities in the Debye model (3) are obtained in the TM case by averaging the edge values in a unit FDTD cell to form effective E_x and E_z values at the center of that cell. The permittivities and conductivities are averaged across neighboring cells when necessary. For instance, the linear permittivity must be averaged across any interface in a linear medium. In addition, since the nonlinear susceptibility resides at the center of each cell, while the TM electric field components exist along its edges, the susceptibility must be averaged across every cell boundary, whenever the nonlinearity is present. The TE case follows immediately by reciprocity.

3. NL-FDTD RESULTS

We will specifically present NL-FDTD results obtained for the scattering of a pulsed Gaussian beam normally incident on a linear-nonlinear interface. This problem combines both the linear-nonlinear interface scattering and the nonlinear self-focusing effects. Interest in the linear-nonlinear interface problem is stimulated by the need to assess the potential of this geometry for an all-optical switch. If the pulse amplitude is below the critical value for the medium, the beam senses no interface and passes through unscathed. If the pulse amplitude is above the critical value for the medium, the beam experiences a strong reflection from the interface; and the transmitted beam experiences self-focusing.

In all of the interface problems we have considered, it has been assumed that the interface was in the far field of the source [*Ziolkowski and Judkins*, 1992*d*]. We thus used a single bipolar pulse excitation for the single-cycle cases. This initial pulse was given by the function

$$F(t) = x\,(1-x^2)^3\,H(1-|x|) \qquad ; \quad x = \frac{2t}{T} - 1\,, \tag{10}$$

where $H(x)$ is Heaviside's function. A total pulse width $T = 20.0\,fs$ corresponds to an effective wavelength of 4.0 μm. This initial driving function has both first and second time derivatives continuous at its endpoints, thus reducing the numerical noise initially generated in the grid. Note that the NL-FDTD approach can handle these single-cycle cases as readily as multiple-cycle cases which include pulses having an intrinsic carrier wave. Since most current optical systems deal directly with a carrier-wave type signal, the NL-FDTD approach can simulate the propagation and scattering effects associated with those systems. It can also simulate the behaviors of single-cycle optical devices and systems. The single cycle results represent behaviors that should be observed with optical systems currently under investigation and those being developed for future studies by the international optics community. However, in either case the full pulse is modeled with the NL-FDTD approach rather than only the envelope of the carrier wave, as it is with the nonlinear Schrödinger equation models. This enhanced modeling capability of the NL-FDTD approach allows one to model and distinguish the effects that arise in both the single- and multple-cycle cases.

A variety of TE and TM cases have been considered to establish the effects of the medium response time. This is accomplished simply by varying τ with respect to the input pulse width T. In particular, we have run cases with $T = 0.2\tau$ to $T = 20.0\,\tau$. These choices provided access to the nonlinear phenomena associated with mediums exhibiting either finite or instantaneous response times. Several grid sizes have been explored to assess the numerical stability and accuracy of the NL-FDTD approach. We found that an average spatial resolution of $\Delta z \leq \lambda/100$ was needed instead of the standard $\lambda/10$ rule-of-thumb for the wavelengths of interest, particularly in the instantaneous medium response cases. The signals steepened so quickly that this enhanced resolution was necessary to maintain the second-order accuracy of the FDTD approach. The grid size used in our linear-nonlinear interface model below was 1500×2000, where $\Delta x = \Delta z = 0.020\ \mu$m or $30.0\ \mu\text{m} \times 40.0\ \mu$m. This discretization provided an effective spatial resolution of $\Delta z = \lambda/160$ for the finite time pulse (4) that we considered. A Courant stability condition (the time step must be chosen for a two-dimensional problem with $\Delta x = \Delta z$ so that $\Delta t \leq \Delta z/\sqrt{2}c$) of $\Delta t/(\Delta z/c) = 0.38$ was selected; this means we chose $\Delta t = 0.045$ fs and $\Delta t = 0.018$ fs, respectively. We found that the algorithm did not model the physics well when it was run at the linear equation Courant limit, but did for values $\Delta t/(\Delta z/c) \leq 0.50$. The nonlinear version of the Courant limit is not known at this time. With four unknowns and additional overhead, the corresponding total memory requirement was approximately 10.0 MWords. The time requirement is proportional to the number of unknowns and the number of time steps. This 10.0 MWord problem took approximately 510 CONVEX/240 *cpu* minutes to compute 7550 time steps, approximately 0.5 μs per unknown. In terms of the problem parameters, this corresponds to the pulse propagating 40 μm $\sim$ 7 cT with a resolution of 240 cells over the spatial pulse width cT = 6.0 μm. The run time would be approximately a factor of 10 less using a single processor of a CRAY YMP/8-32.

The NL-FDTD results for an above-threshold, normally incident, pulsed TM Gaussian beam, linear-nonlinear interface results are shown in Figures 2a-2d. In each part the field structure in a windowed region of the overall simulation space is represented by a contour plot of the total electric field intensity $I = |E_x|^2 + |E_z|^2$. The incident beam field is shown in Figure 2*a*. It represents a Gaussian tapered (space), bipolar (time) pulse with an initial transverse waist $w_0 = 10.0\ \mu$m and a total initial pulse width $T = 20.0$ fs. For the results presented below the nonlinear medium parameters were set to the nearly instantaneous-regime value $\tau = 0.2\,T$, i.e., $T = 5.0\tau$, and $\epsilon_2 = 2.0 \times 10^{-18}\ (\text{m/V})^2$, and we set the input electric field amplitude to $E_0 = 9.33 \times 10^8 (\text{V/m}) = 2.8\ E_{\text{crit}}$ in the Kerr medium.

The interaction of the TM pulsed beam at the linear-nonlinear interface is shown in Figure 2*b* for a time just after the beam scatters from the interface which is located at $z = 0$. The reflected and transmitted pulsed beam fields are apparent. The transmitted pulsed beam begins to self-focus as it propagates into the Kerr medium as shown

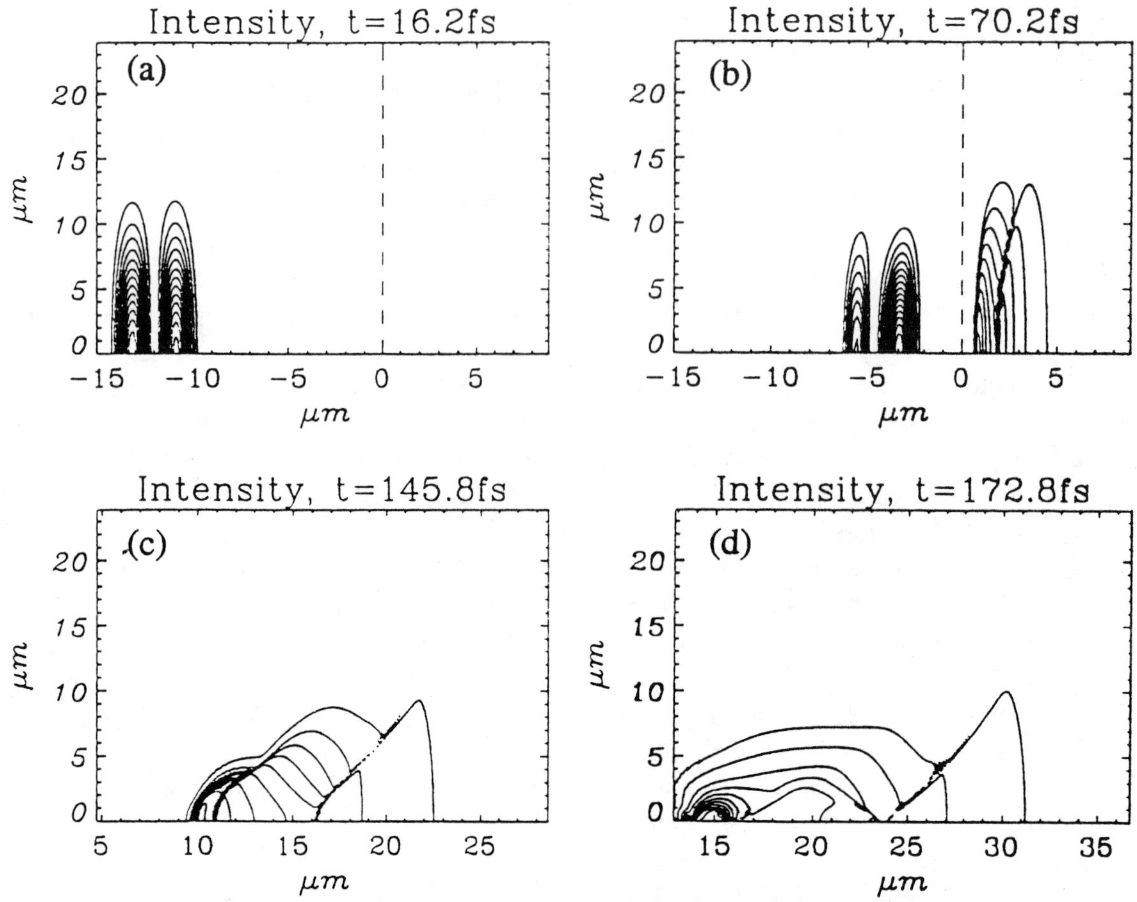

Fig. 2. Contour plots of the total electric field intensity $|E_x|^2 + |E_z|^2$ for the $T = 5\tau$ case. (a) The initial field, (b) The pulsed beam has just interacted with the linear-nonlinear interface, (c) The transmitted beam is self-focusing, and (d) The self-focusing region.

in Figure 2*c*. The intensity pattern in the focus region is shown in Figure 2*d*. Figures 2*c* and 2*d* recover the well-known horn pattern in which the front (large z) portion corresponds to the linear diffraction region, and the rear (smaller z) portion incorporates the nonlinear effects.

We have found that the growth of the longitudinal electric field component causes transverse power flows and reflections from this self-focusing region [*Ziolkowski and Judkins*, 1992*a*, *b*, *c*, 1993*a*, *b*, *c*]. Self-focusing is restrained by these nonlongitudinal power flow mechanisms since they channel power away from the focus. By studying the Poynting's vector $\vec{S} = E_x H_y \hat{z} - E_z H_y \hat{x}$ in the instantaneous response cases, we have demonstrated the existence of nonlinear optical vortices in the wake field of the focal region [*Ziolkowski and Judkins*, 1993c]. Negative longitudinal power flow occurs on the horn boundary, in the region around the focus, and in the wake field of the focus. Pure reflection occurs immediately behind the region of the maximum field intensity. Part of the negative flow in the focal region feeds the focus (power must be channeled into the focal region to produce the growth in the field components there), and part is channeled into the vortices. Positive longitudinal power flow occurs everywhere else and is particularly strong immediately within the focal region, where the nonlinear field growth occurs.

These vortices can be described in terms of turbulence in the power flow of the electromagnetic field in the self-focus region. The front portion of the incident beam generates a tapered waveguide which focuses the rest of the pulse. When the nonlinearity is nearly instantaneous and the field levels are large, the rate of the taper becomes very large. Thus the number of modes accessible to the beam increases. As the taper converts some of the field structure into these higher-order modes, nulls in the field components can develop. The presence of convective power flow is immediate and leads to the formation of the vortices. This behavior can occur whenever the power flow encounters a discontinuity in the guided-wave structure [*Ziolkowski and Grant*, 1986]. Vortices in nonlinear media have also been identified by several other groups [e.g., *Coullet et al.*, 1989, *Brambilla et al.*, 1991, *Akhmanov et al.*, 1992, and *McDonald et al.*, 1992]. These vortices are observed in the field patterns derived from the scalar nonlinear Schrödinger equation in planes transverse to the direction of propagation. In contrast, the vortices observed in the self-focus region are a direct result of the transverse power flow, which is intimately connected to the longitudinal field components, and are found in the planes containing the propagation axis.

Typical self-focusing results for the normal incidence problem are summarized in Figure 3 in which the waist of the energy of the pulsed beam is plotted as a function of its location along the direction of propagation. For low intensity, the beam diffracts as though it were in free space; for high intensity, some portion of the beam reflects, and the transmitted beam self-focuses in the Kerr medium. The actual reflection coefficient is substantially below the value anticipated from an equivalent monochromatic beam. This is due primarily to the fact that the pulse does not cause the medium to respond instantly; hence, much of the energy penetrates into the medium before the boundary is sufficiently reflective. The transmitted beam behaves as predicted from previously reported self-focusing beam propagation simulations.

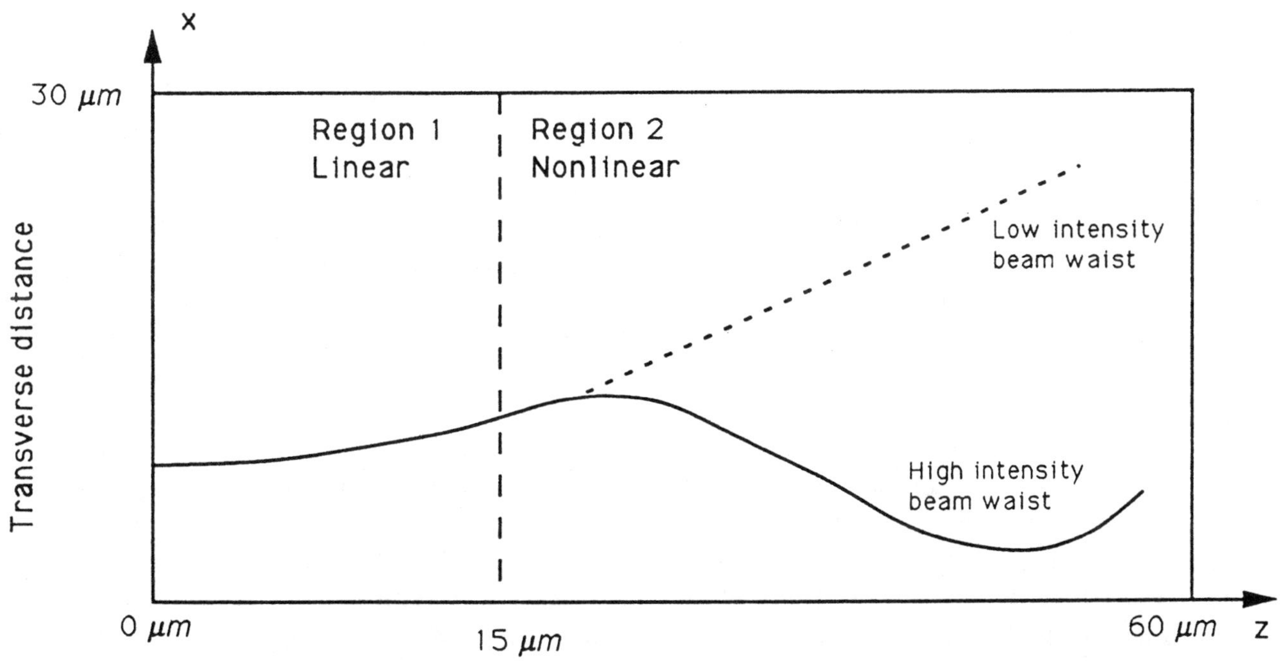

Fig. 3. The normal incidence problem is divided into two regions: the incident region which is linear and the transmission region which is nonlinear. The linear index is continuous across the interface so that a low-intensity beam will propagate unchanged. As the intensity increases, the pulse experiences some reflection and the transmitted pulse will self-focus if the initial transmitted field amplitude is above threshold.

4. CONCLUSIONS

Although the current NL-FDTD models are two-dimensional, we believe that the observed transverse power flow mechanisms, which remove power from the focal region, limit the self-focusing process and will prevent catastrophic focusing in three dimensions. This, in contradiction to the scalar nonlinear Schrödinger equation models, means the pulsed beam will not be focused to a point by the self-focusing process. Proof and comparisons with standard paraxial scalar models in three space dimensions await adequate computing resources.

We have completed our modeling of the scattering of an obliquely incident, focused, pulsed Gaussian beam from a linear-nonlinear interface. Our NL-FDTD calculations indicate that the anticipated switching properties of this optical diode switch are unrealizable with a tightly focused beam. These results recover the conclusions given theoretically by *Tomlinson et al.* [1982] and experimentally by *Smith et al.* [1981]. There is simply too much energy leakage into the medium from the initial pulsed beam components having wave vectors beyond critical and from the nonlinear coupling to the medium. Related nonlinear waveguide simulations, however, have recovered expected solitarylike wave emissions from a linear waveguide channel into a background nonlinear substrate. This linear-nonlinear interface class of problems has potential applications to nonlinear guided-wave couplers.

Our future efforts will include several directions. We will use the NL-FDTD model to study the interplay between the dispersive and the nonlinear effects of the medium in the presence of intense, applied optical pulsed beams. We will begin to include microscopic materials models in the NL-FDTD algorithm to study quantum effects associated with these systems. We will also be applying the NL-FDTD model to realistic device confirgurations; this will require modeling more complex propagation and scattering geometries.

Acknowledgments. This work was supported in part by the Lawrence Livermore National Laboratory under the auspices of the U.S. Department of Energy under contract W-7405-ENG-48.

REFERENCES

Akhmanov, S. A., M. A. Vorontsov, V. Yu. Ivanov, A. V. Larichev, and N. I. Zheleznykh, Controlling transverse-wave interactions in nonlinear optics: generation and interaction of spatiotemporal structures, *J. Opt. Soc. Am. B Opt. Phys.*, *9*(1), 78-90, 1992.

Brambilla, M., F. Battipede, L. A. Lugiato, V. Penna, F. Prati, C. Tamm, and C. O. Weiss, Transverse laser patterns, I, Phase singularity crystals, *Phys. Rev. A Gen. Phys.*, *43*(9), 5090-5113, 1991.

Coullet, P., L. Gil, and F. Rocca, Optical vortices, *Opt. Commun.*, *73*(5), 403-408, 1989.

Goorjian, P. M., and A. Taflove, FD-TD computational modeling of nonlinear electromagnetic phenomena using a nonlinear convolution approach, in *Proceedings of the North American URSI Radio Science Meeting*, p. 190, URSI, London, Ontario, June 1991.

Goorjian, P. M., and A. Taflove, Direct time integration of Maxwell's equations in nonlinear dispersive media for propagation and scattering of femtosecond electromagnetic solitons, *Opt. Lett.*, *17*(3), 180-182, 1992*a*.

Goorjian, P. M., and A. Taflove, Direct time integration of Maxwell's equations in nonlinear dispersive media for propagation and scattering of femtosecond electromagnetic solitons, in *Proceedings of the Integrated Photonics Research (IPR) Meeting*, pp. 380-383, Optical Society of America, New Orleans, LA April 13-15, 1992*b*.

Goorjian, P. M., A. C. Woo, and A. Taflove, Computational modeling of Maxwell's equations in nonlinear dispersive media for propagation and scattering of femtosecond electromagnetic solitons, in *Proceedings of the URSI Radio Science Meeting*, p. 63, URSI, Chicago, IL , June 1992*a*.

Goorjian, P. M., A. Taflove, R. M. Joseph, and S. C. Hagness, Computational modeling of femtosecond optical solitons from Maxwell's equations, *IEEE J. Quantum Electron.*, *QE-28*(10), 2416-2422, 1992*b*.

Goorjian, P. M., R. M. Joseph, and A. Taflove, Direct time integration of Maxwell's equations in 2-D dielectric waveguides for propagation and scattering of femtosecond electromagnetic solitons, in *Proceedings of the Integrated Photonics Research (IPR) Meeting*, pp. 124-127, Optical Society of America, Palm Springs, CA, March 22-24, 1993.

Hayata, K., and M. Koshiba, Full vectorial analysis of nonlinear-optical waveguides, *J. Opt. Soc. Am. B Opt. Phys.*, *5*(12), 2494-2501, 1988.

Hayata, K., A. Misawa, and M. Koshiba, Spatial polarization instabilities due to transverse effects in nonlinear guided-wave systems, *J. Opt. Soc. B Opt. Phys.*, *7*(7), 1268-1280, 1990.

Hayata, K., K. Kobayashi, and M. Koshiba, Intrapulse dynamics of stationary optical fields in nonlinear channel waveguides, *Opt. Lett.*, *17*(21), 1494-1496, 1992.

Hile, C. V., and W. L. Kath, Numerical and asymptotic solution of Maxwell's equations for nonlinear optical pulse propagation, in *Proceedings of the Integrated Photonics Research (IPR) Meeting*, pp. 308-311, Optical Society of America, Palm Springs, CA, March 22-24, 1993.

Jamid, H. A., and S. J. Al-Bader, Finite-difference time-domain approach to nonlinear guided waves, *Electronics Lett.*, *29*(1), 83-84, 1993.

Joseph, R. M., S. C. Hagness, and A. Taflove, Direct time integration of Maxwell's equations in linear dispersive media with absorption for scattering and propagation of femtosecond electromagnetic pulses, *Opt. Lett.*, *16*(18), 1412-1414, 1991.

Kashiwa, T., and I. Fukai, A treatment by the FD-TD method of the dispersive characteristics associated with electronic polarization, *Microwave and Opt. Tech. Lett.*, *3*(6), 203-205, 1990.

Lee, C. F., R. T. Shin and J. A. Kong, Finite difference method for electromagnetic scattering problems, in *PIER4: Progress in Electromagnetics Research*, Chap. 11, edited by J. A. Kong, Elsevier, New York, 1991.

Luebbers, R., F. P. Hunsberger, K. S. Kunz, R. B. Standler, and M. Schneider, A frequency dependent finite difference time domain formulation for transient propagation in plasmas, *IEEE Trans. Electromagn. Compat.*, *EMC-32*, 222-227, 1990.

McDonald, G. S., K. S. Syed, and W. J. Firth, Optical vortices in beam propagation through a self-defocusing medium, *Opt. Commun.*, *94*, 469-476, 1992.

Mitchell, C. S., and J. V. Moloney, Propagation and stability of optical pulses in nonlinear planar structures with instantaneous and finite response times, *IEEE J. Quantum Electron.*, *26*(12), 2115-2129, 1990.

Miyagi, M., and S. Nishida, TM-Type soliton in nonlinear self-focusing media, *Proc. IEEE*, *62*, 1284-1285, 1974.

Miyagi, M., and S. Nishida, TM waves in nonlinear self-focusing media, *Radio Sci.*, *10*, 833-838, 1975.

Pohl, D., Vectorial theory of self-trapped light beams, *Opt. Commun.*, *2*(7), 305-308, 1970.

Pohl, D., Self-focusing of TE_{01} and TM_{01} light beams: influence of longitudinal field components, *Phys. Rev. A Gen. Phys.*, *5*(4), 1906-1909, 1972.

Reintjes, J. F., *Nonlinear Optical Parameteric Processes in Liquids and Gases*, p. 329, Academic, San Diego, CA, 1984.

Sodha, M. S., V. P. Nayyar, and V. K. Tripathi, Asymmetric focusing of a laser beam in TEM_{01} doughnut mode in a nonlinear dielectric, *J. Opt. Soc. Am.*, *64*(7), 941-943, 1974.

Smith, P. W., W. J. Tomlinson, P. J. Maloney, and J.-P. Hermann, Experimental studies of a nonlinear interface, *IEEE J. Quantum Electron.*, *QE-17*(3), 340-348, 1981.

Tomlinson, W. J., J. P. Gordon, P. W. Smith, and A. E. Kaplan, Reflection of a gaussian beam at a nonlinear interface, *Appl. Opt.*, *21*(11), 2041-2051, 1982.

Ziolkowski, R. W., and J. B. Grant, Vortex formation near an iris in a rectangular waveguide, *IEEE Trans. Antennas and Propagat.*, *MTT-34*, 1164-1182, 1986.

Ziolkowski, R. W., and J. B. Judkins, Optical nonlinear wakefield vortices: results from full-wave vector maxwell equation dimulations in two dpatial Dimensions and Time, in *Proceedings of the Integrated Photonics Research (IPR) Meeting, Post-Deadline Papers*, pp. 50-51, Optical Society of America, New Orleans, LA, April 13-15, 1992*a*.

Ziolkowski, R. W., and J. B. Judkins, Full-wave, vector Maxwell equation modeling of self-focusing of ultrashort optical pulses in a nonlinear Kerr medium, in *Proceedings of the URSI Radio Science Meeting*, p. 65, URSI, Chicago, IL , June, 1992*b*.

Ziolkowski, R. W., and J. B. Judkins, Applications of discrete methods to pulse propagation in nonlinear media: self-focusing and linear-nonlinear interfaces", in *Proceedings of the URSI International EM Theory Symposium*, pp. 284-286, URSI, Sydney, Australia, August, 1992*c*.

Ziolkowski, R. W., and J. B. Judkins, Propagation characteristics of ultra-wide bandwidth pulsed gaussian beams, *J. Opt. Soc. Am. A Opt. Image Sci.*, *9*(11), 2021-2030, 1992*d*.

Ziolkowski, R. W., and J. B. Judkins, Full-wave vector Maxwell equation modeling of the self-focusing of ultrashort optical pulses in a nonlinear Kerr medium exhibiting a finite response time, *J. Opt. Soc. Am. B Opt. Phys.*, *10*(2), 186-198, 1993*a*.

Ziolkowski, R. W., and J. B. Judkins, Linear-nonlinear interfaces: results from full-wave, vector Maxwell equation NL-FDTD simulations, *Proceedings of the Integrated Photonics Research (IPR) Meeting*, pp. 128-131, Optical Society of America, Palm Springs, CA, March 22-24, 1993*b*.

Ziolkowski, R. W., and J. B. Judkins, Self-limiting effects and optical nonlinear wakefield vortices: results from full-wave vector Maxwell equation simulations of nonlinear self-focusing in two spatial dimensions and time, *Opt. Lett.*, in press, 1993*c*.

R. W. Ziolkowski and J. B. Judkins, Electromagnetics Laboratory, Department of Electrical and Computer Engineering, The University of Arizona, Tucson, AZ 85721.

Current and SAR Induced in a Human Head Model by the Electromagnetic Fields Irradiated from a Cellular Phone

Hsing-Yi Chen, *Member, IEEE*, and Hou-Hwa Wang

Abstract—**The near field finite-difference time-domain method was used to calculate the current and specific absorption rate (SAR) distributions in an inhomogeneous model of a human head exposed to the electromagnetic waves irradiated from a cellular phone. The human head was simulated by a model of 57 263 block cells with inhomogeneous dielectric constant and conductivity. The cellular phone was modeled by an equivalent dipole antenna with an equivalent resistor of 120 ohms located at the center gap between the two arms of this dipole antenna. The transmitted power of the cellular phone was assumed to be 0.6 watts at a frequency of 835 MHz. For the head near the dipole antenna in the range of 1.0 ~ 2.5 cm, the maximum currents and SAR's induced in the head were found in the ranges of 356 ~ 551 mA and 1.23 ~ 2.63 W/kg, respectively. It was also found that the maximum SAR induced in the head was below the IEEE's upper safety limit of 1.6 W/kg for the head to keep a distance from the dipole antenna by longer than 2.0 cm.**

I. Introduction

RECENTLY, there has been an increasing public concern about the health risk of electromagnetic (EM) waves emitted from cellular phones. Do cellular phones really cause brain tumors? Before answering this question, the most important thing for researchers may be to conduct innovative studies associated with analytical and numerical quantification of the biological effects caused by EM waves and to further study their potential hazards.

Cellular phones send their signals using very small surges of high-frequency EM waves, or microwaves, favored over most over-the-air telecommunication systems. Commercial systems of cellular phones used in the USA are in the frequency range of 820 ~ 850 MHz with 0.6 W power. In this study, the FDTD method [1] was used to calculate the currents and (SAR's) induced inside a human head due to the EM fields radiated from a cellular phone. To simplify calculations and make reasonable assumptions, the other parts of the human body, except the head, were excluded for the simulations. The human head was simulated by a model of 57 263 block cells with inhomogeneous dielectric constant and conductivity [2], [3]. The cell size used in the model of the human head was taken to be 0.5 cm. The microwave transmitter of the cellular phone was modeled by a wire dipole antenna fed with a sinusoidally driven voltage source. The driven voltage source was then transferred into a sinusoidally varying electric field with uniform amplitude and phase over the gap between the two arms of this wire dipole antenna. A magnetic field, 1/377 of the electric field, was similarly taken for the antenna gap. The frequency used for the simulation was chosen to be 835 MHz. Obtained results of electric currents and specific absorption ratio (SAR) distributions inside the human head were presented in figures. Finally, remarks on health effects, based on the calculated data, other scientific sources, and public health community reports [3], [4] were made in the conclusions.

Manuscript received July 6, 1993; revised January 21, 1994. This work was supported in part by the National Science Council of the Republic of China, under Grant NSC 83-0404-E-155-002.

The authors are with the Department of Electrical Engineering, Yuan-Ze Institute of Technology, Nei-Li, Taoyuan Shian, Taiwan 32026.

IEEE Log Number 9405384.

II. The FDTD Method

The basic finite-difference time-domain (FDTD) method was first proposed by [1] and later developed by many researchers [5]–[18]. The FDTD method is one of the most powerful techniques in computations involving electromagnetic waves. In the FDTD method, the coupled Maxwell's equations in the differential form are solved for various points of the scatters as well as its surroundings in a time-stepped manner until converged solutions are obtained.

The Maxwell's time-dependent curl equations may be written as follows:

$$\nabla \times \overline{E} = -\mu \frac{\partial \overline{H}}{\partial t} \tag{1}$$

$$\nabla \times \overline{H} = \sigma \overline{E} + \varepsilon \frac{\partial \overline{E}}{\partial t}. \tag{2}$$

The components of EM fields $\bar{E}$ and $\bar{H}$ are positioned at half-step interval around a unit cell. The parameter σ, μ and ε are the conductivity, permeability, and permittivity of the human body, respectively.

Using the MKS system of units, the following system of scalar equations is equivalent to Maxwell's equations in the rectangular coordinate system (x, y, z)

$$\frac{\partial E_x}{\partial t} = \frac{1}{\varepsilon}\left(\frac{\partial H_z}{\partial y} - \frac{\partial H_y}{\partial z} - \sigma E_x\right) \tag{3}$$

$$\frac{\partial E_y}{\partial t} = \frac{1}{\varepsilon}\left(\frac{\partial H_x}{\partial z} - \frac{\partial H_z}{\partial x} - \sigma E_y\right) \tag{4}$$

$$\frac{\partial E_z}{\partial t} = \frac{1}{\varepsilon}\left(\frac{\partial H_y}{\partial x} - \frac{\partial H_x}{\partial y} - \sigma E_z\right) \tag{5}$$

Reprinted from *IEEE Transactions on Microwave Theory and Techniques*, pp. 2249-2254, 1994.

$$\frac{\partial H_x}{\partial t} = \frac{1}{\mu}\left(\frac{\partial E_y}{\partial z} - \frac{\partial E_z}{\partial y}\right) \tag{6}$$

$$\frac{\partial H_y}{\partial t} = \frac{1}{\mu}\left(\frac{\partial E_z}{\partial x} - \frac{\partial E_x}{\partial z}\right) \tag{7}$$

$$\frac{\partial H_z}{\partial t} = \frac{1}{\mu}\left(\frac{\partial E_x}{\partial y} - \frac{\partial E_y}{\partial x}\right). \tag{8}$$

Following Yee's notation, we assume that the grid point of the space is defined as $(i,\ j,\ k)$ with coordinates $(i\delta_x,\ j\delta_y,\ k\delta_z)$, where $\delta = \delta_x = \delta_y = \delta_z$ is the cubic cell size, and δ_t is the time increment. Making the substitution

$$\tilde{E} = \left(\frac{\varepsilon_0}{\mu_0}\right)^{1/2} E. \tag{9}$$

A set of finite-difference equations is obtained as follows:

$$\begin{aligned}&\tilde{E}_x^{n+1}(i+1/2, j, k)\\ &\quad= CA_x(i+1/2, j, k)\tilde{E}_x^n(i+1/2, j, k)\\ &\qquad+ CB_x(i+1/2, j, k)\Big[H_z^{n+1/2}(i+1/2, j+1/2, k)\\ &\qquad- H_z^{n+1/2}(i+1/2, j-1/2, k)\\ &\qquad+ H_y^{n+1/2}(i+1/2, j, k-1/2)\\ &\qquad- H_y^{n+1/2}(i+1/2, j, k+1/2)\Big]\end{aligned} \tag{10}$$

$$\begin{aligned}&\tilde{E}_y^{n+1}(i, j+1/2, k)\\ &\quad= CA_y(i, j+1/2, k)\tilde{E}_y^n(i, j+1/2, k)\\ &\qquad+ CB_y(i, j+1/2, k)\Big[H_x^{n+1/2}(i, j+1/2, k+1/2)\\ &\qquad- H_x^{n+1/2}(i, j+1/2, k-1/2)\\ &\qquad+ H_z^{n+1/2}(i-1/2, j+1/2, k)\\ &\qquad- H_z^{n+1/2}(i+1/2, j+1/2, k)\Big]\end{aligned} \tag{11}$$

$$\begin{aligned}&\tilde{E}_z^{n+1}(i, j, k+1/2)\\ &\quad= CA_z(i, j, k+1/2)\tilde{E}_z^n(i, j, k+1/2)\\ &\qquad+ CB_z(i, j, k+1/2)\Big[H_y^{n+1/2}(i+1/2, j, k+1/2)\\ &\qquad- H_y^{n+1/2}(i-1/2, j, k+1/2)\\ &\qquad+ H_x^{n+1/2}(i, j-1/2, k+1/2)\\ &\qquad- H_x^{n+1/2}(i, j+1/2, k+1/2)\Big]\end{aligned} \tag{12}$$

$$\begin{aligned}&H_x^{n+1/2}(i, j+1/2, k+1/2)\\ &\quad= H_x^{n-1/2}(i, j+1/2, k+1/2)\\ &\qquad+ RC\Big[\tilde{E}_y(i, j+1/2, k+1)\\ &\qquad- \tilde{E}_y^n(i, j+1/2, k) + \tilde{E}_z^n(i, j, k+1/2)\\ &\qquad- \tilde{E}_z^n(i, j+1, k+1/2)\Big]\end{aligned} \tag{13}$$

$$\begin{aligned}&H_y^{n+1/2}(i+1/2, j, k+1/2)\\ &\quad= H_y^{n-1/2}(i+1/2, j, k+1/2) + RC\\ &\qquad\times\Big[\tilde{E}_z^n(i+1, j, k+1/2) - \tilde{E}_z^n(i, j, k+1/2)\\ &\qquad+ \tilde{E}_x^n(i+1/2, j, k)\\ &\qquad- \tilde{E}_x^n(i+1/2, j, k+1)\Big]\end{aligned} \tag{14}$$

$$\begin{aligned}&H_z^{n+1/2}(i+1/2, j+1/2, k)\\ &\quad= H_z^{n-1/2}(i+1/2, j+1/2, k) + RC\\ &\qquad\times\Big[\tilde{E}_x^n(i+1/2, j+1, k) - \tilde{E}_x^n(i+1/2, j, k)\\ &\qquad+ \tilde{E}_y^n(i, j+1/2, k) - \tilde{E}_y^n(i+1, j+1/2, k)\Big]\end{aligned} \tag{15}$$

where

$$CA_x(i+1/2, j, k) = \frac{2\varepsilon_x(i+1/2, j, k) - \sigma_x(i+1/2, j, k)\delta_t}{2\varepsilon_x(i+1/2, j, k) + \sigma_x(i+1/2, j, k)\delta_t} \tag{16}$$

$$CA_y(i, j+1/2, k) = \frac{2\varepsilon_y(i, j+1/2, k) - \sigma_y(i, j+1/2, k)\delta_t}{2\varepsilon_y(i, j+1/2, k) + \sigma_y(i, j+1/2, k)\delta_t} \tag{17}$$

$$CA_z(i, j, k+1/2) = \frac{2\varepsilon_z(i, j, k+1/2) - \sigma_z(i, j, k+1/2)\delta_t}{2\varepsilon_z(i, j, k+1/2) + \sigma_z(i, j, k+1/2)\delta_t} \tag{18}$$

$$CB_x(i+1/2, j, k) = \frac{2\varepsilon_0 RC}{2\varepsilon_x(i+1/2, j, k) + \sigma_x(i+1/2, j, k)\delta_t} \tag{19}$$

$$CB_y(i, j+1/2, k) = \frac{2\varepsilon_0 RC}{2\varepsilon_y(i, j+1/2, k) + \sigma_y(i, j+1/2, k)\delta_t} \tag{20}$$

$$CB_z(i, j, k+1/2) = \frac{2\varepsilon_0 RC}{2\varepsilon_z(i, j, k+1/2) + \sigma_z(i, j, k+1/2)\delta_t} \tag{21}$$

$$RC = \frac{\delta_t}{\delta(\mu_0\varepsilon_0)^{1/2}} = 0.5. \tag{22}$$

In the above equations $\varepsilon_0 = 8.854 \times 10^{-12}$ farads/meter is the permittivity of free space, $\mu_0 = 4\pi \times 10^{-7}$ henrys/meter is the permeability of free-space, $\varepsilon_x, \varepsilon_y, \varepsilon_z$ are the permittivities, $\sigma_x, \sigma_y, \sigma_z$ are the conductivities of the media at respective locations.

An important problem encountered in solving the time-domain electromagnetic-field equation, by the FDTD method, is the absorbing boundary conditions. Absorbing boundary conditions have been described in [5], [10], [11]. In our formulation, the second-order approximation of absorbing boundary conditions [11] is used for the near-field irradiation problems. For near-field applications, the FDTD computation space has only one region, i.e., there is no scattering region, as shown in Fig. 1. The external absorbing boundaries are placed at a distance of $4\sim 7\delta$ on all sides of the scattering object, where δ is the cubic cell size.

In our near-field simulations, the scattering object was the human head modeled by 57 263 cubic cells of each cell size 0.5 cm. The radiation source of the cellular phone

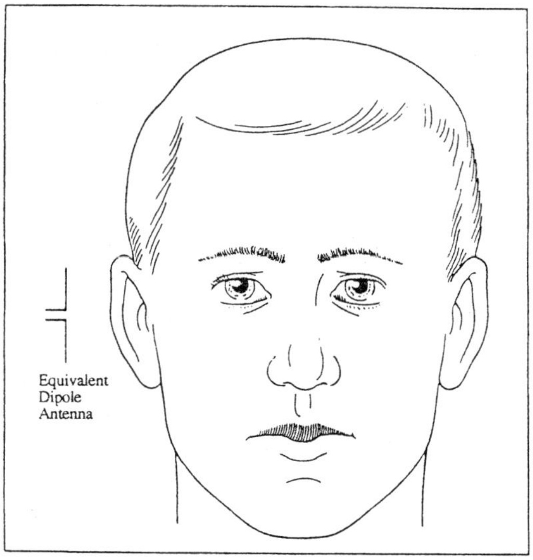

Fig. 1. Geometry of the cellular phone modeled by an equivalent dipole antenna and a human head.

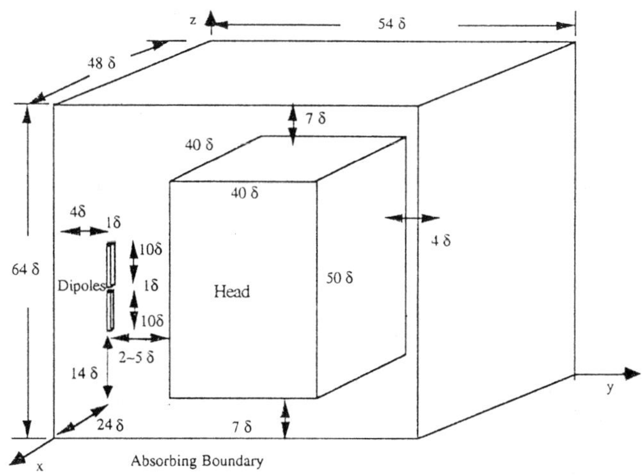

Fig. 2. Geometry of the equivalent dipole antenna and the human head used for the simulations by the near-field FDTD method. A sinusoidal time varying voltage source $V = 12 \sin(\omega t)$ volts was located in the center gap of the dipole antenna. A 120 ohms resistor was assumed at the center gap of the dipole antenna. The human head was constructed with 57 263 cubic cells and divided into 50 layers. The cell size δ was taken to be 0.5 cm.

was modeled by an equivalent dipole antenna. After having obtained the induced electric field by the FDTD method, the local specific absorption rate (SAR) at each cell in the head was calculated as follows:

$$\mathrm{SAR}\,(i,j,k) = \frac{\sigma(i,j,k)[E_{\max}(i,j,k) - E_{\min}(i,j,k)]^2}{8\rho(i,j,k)} \quad (\mathrm{W/kg}) \tag{23}$$

where (i,j,k) is the grid point of the space of the head in Cartesian coordinates. $E_{\max}$ and $E_{\min}$ are the positive and negative electric field peak values, respectively. σ and ρ are the conductivity and mass density, respectively.

The total SAR at the kth cross-section in the head model is calculated as follows:

$$I(k) = \sum_{i=1}^{n}\sum_{j=1}^{m} \mathrm{ABS}[\sigma(i,j,k) + j\omega\varepsilon(i,j,k)] \times \frac{[E_{\max}(i,j,k) - E_{\min}(i,j,k)]\delta^2}{2\sqrt{2}} \quad (\mathrm{Amp}). \tag{24}$$

The total induced currents through the kth layer of the head model are calculated as follows: where ε is the dielectric constant of the tissue of human head, δ is the cubic cell size, and ABS means "the absolute value".

III. Numerical Results

Fig. 2 shows the geometry of a human head model and an equivalent dipole antenna under the simulations by the near-field FDTD method. In the ANSI/IEEE C95.1–1992 standard [18], the parameters of concern with respect to human health are the rate at which a user absorbs electromagnetic energy, called the SAR, and the incident power density of the radiation. In order to find the power density in the air near the cellular phone, a set of equivalent resistors R_a of 40 ~ 200 ohms were assumed at the center gap of the equivalent dipole antenna.

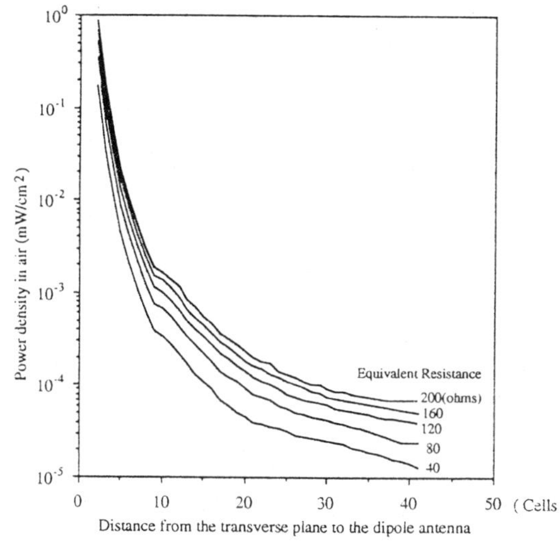

Fig. 3. The power density in the air irradiated from an equivalent dipole antenna with equivalent resistors of 40 to 200 ohms. One cell size was taken to be 0.5 cm.

The transmitted power of the cellular phone was assumed to be $P = 0.6$ watts. Following the equation $P = V^2/(2R_a)$, the voltage across the center gap of the equivalent antenna was calculated as a function of R_a, where V was the maximum value of voltage. A metallic material with $\varepsilon_r = 1.0$ and $\sigma = 3.72 \times 10^7$ S/m was employed to simulate the equivalent dipole antenna. The total length of the equivalent dipole antenna was assumed to be 10.5 cm. Fig. 3 shows the obtained results of incident power density vs. the distance between the antenna and the transverse plane for a set of equivalent antenna resistors of 40 ~ 200 ohms. From Fig. 3, it was found that the maximum power density of 0.86 mW/cm^2 was on the transverse plane located at a distance of 0.5 cm (one cell) from the antenna for the equivalent antenna resistor of 200 ohms. The maximum power densities on the transverse planes near the antenna for equivalent antenna resistors under

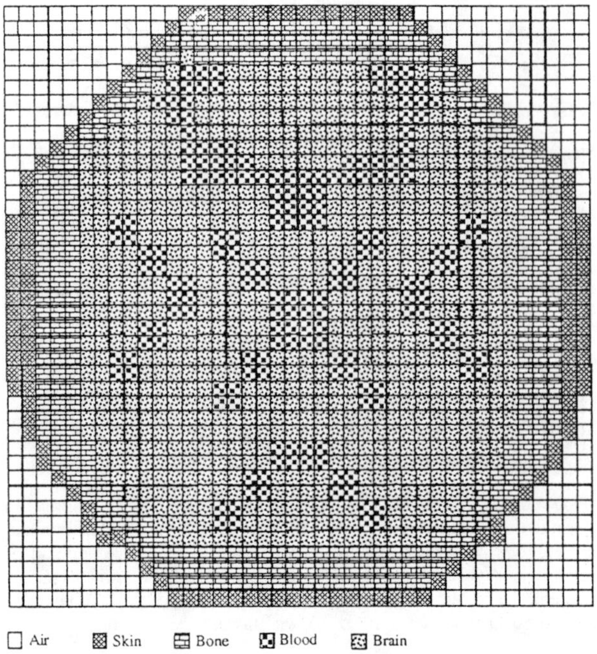

Fig. 4. The cubic structure at layer 35 counted from the bottom of the human head model.

TABLE I
TISSUE TYPES USED IN THE HUMAN HEAD MODEL AND THEIR MASS DESITIES ρ (1000 Kg/m^3), RELATIVE DIELECTRIC CONSTANTS ε_r, AND CONDUCTIVITIES σ (S/m) AT 835 MHz.

Tissue Type	ρ	σ	ε_r
air	0.0012	0.0	1.0
muscle	1.05	1.14	53
bone	1.20	0.04	5.4
blood	1.00	1.26	59
brain	1.05	1.46	51
skin	1.00	0.63	23
eye	1.00	1.04	59

120 ohms were below the maximum allowable power density of 0.57 mW/cm^2 at the frequencies used by cellular phones for whole-body exposure under the uncontrolled environment [18]. In the following study, the equivalent antenna resistor was chosen to be 120 ohms so that the incident power densities were under the allowable value of 0.57 mW/cm^2. To simplify calculations and make reasonable assumptions, the other parts of the human body, except the head, were excluded for the simulations. The human head was separated by a set of distance data of $1.0 \sim 2.5$ cm measured from the equivalent dipole antenna as shown in Fig. 2. The human head model was constructed with 57 263 cubic cells. The total number of layers used in the human head model was 50. The 35th layer counted from the bottom of the head, with each cell size $\delta =$ 0.5 cm is shown in Fig. 4. Both the dielectric constant ϵ and conductivity σ of the human body at 835 MHz were obtained from literatures [2], [3] as shown in Table I. The calculations were made with an initial sinusoidal time varying electric field $E_z = (V/\delta)\sin(\omega t)$ located at the center gap of the antenna. A magnetic field, 1/377 of the electric field, was similarly taken for the center gap of the equivalent dipole antenna. The SAR's induced at layers of 10, 20, 25, 30, and 40 counted from the

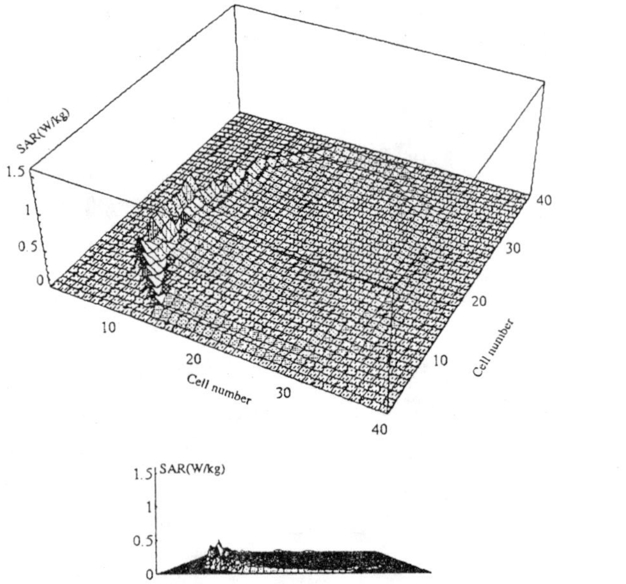

Fig. 5. SAR distributions in the 10th layer counted from the bottom of the head. The distance between the antenna and the head was assumed to be 2.0 cm.

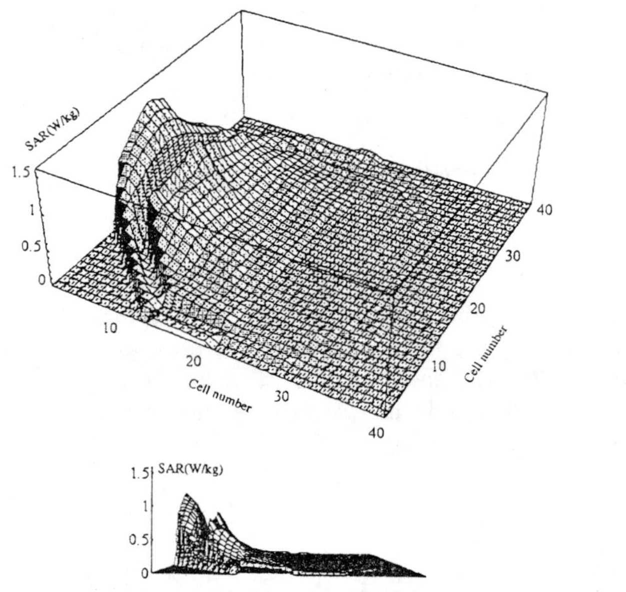

Fig. 6. SAR distributions in the 20th layer counted from the bottom of the head. The distance between the antenna and the head was assumed to be 2.0 cm.

bottom of the head for a distance of 2.0 cm measured from the antenna are shown in Figs. 5–9. From Figs. 5–9, it was found that the SAR values decreased very quickly as the unit cell moved away from the antenna. The maximum SAR was found in a cell at the left hand side of the 25th layer counted from the bottom of the head near to the antenna. Figs. 10 and 11 show the maximum SAR's and currents induced in the head for a set of distance data in the range of $1.0 \sim 2.5$ cm. From Figs. 10 and 11, it was found that the maximum currents and maximum SAR's induced in the head were in the ranges of $356 \sim 551$ mA and $1.228 \sim 2.627$ W/kg for the distance in the range of $1.0 \sim 2.5$ cm, respectively. From Fig. 10, for the

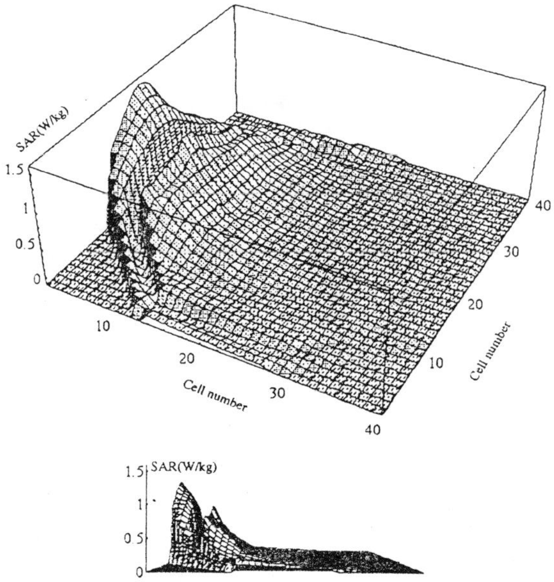

Fig. 7. SAR distributions in the 25th layer counted from the bottom of the head. The distance between the antenna and the head was assumed to be 2.0 cm.

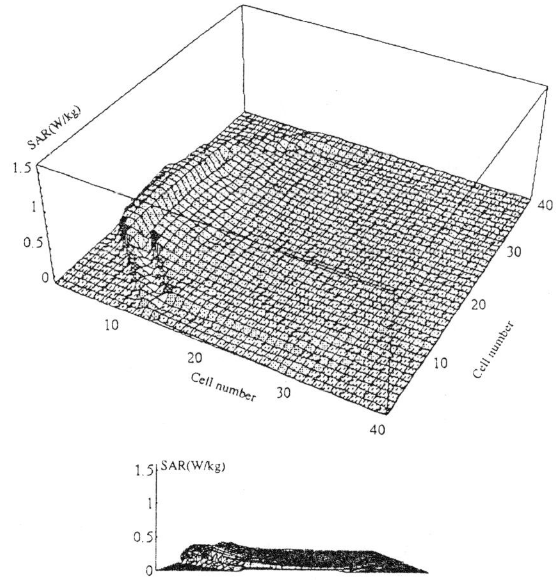

Fig. 9. SAR distributions in the 40th layer counted from the bottom of the head. The distance between the antenna and the head was assumed to be 2.0 cm.

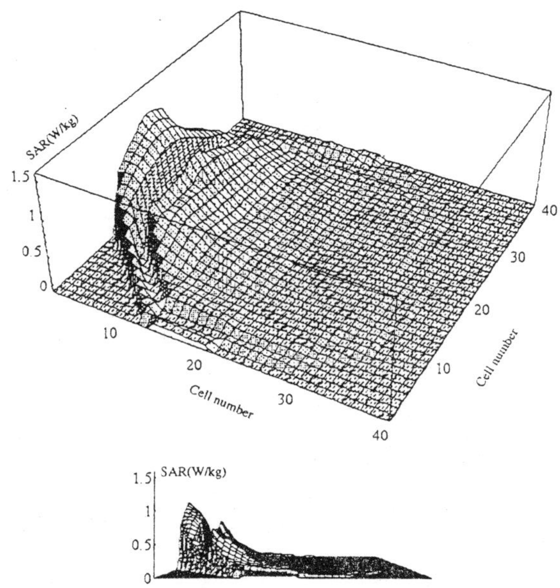

Fig. 8. SAR distributions in the 30th layer counted from the bottom of the head. The distance between the antenna and the head was assumed to be 2.0 cm.

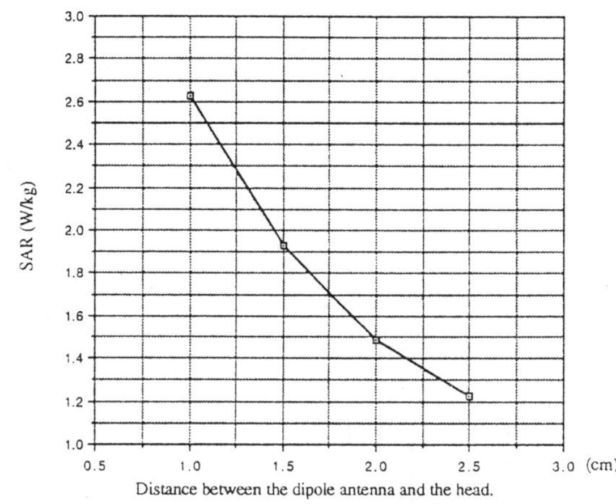

Fig. 10. Maximum SAR's induced in the head for the case of an equivalent resistance $R_a = 120$ ohms used in the dipole antenna.

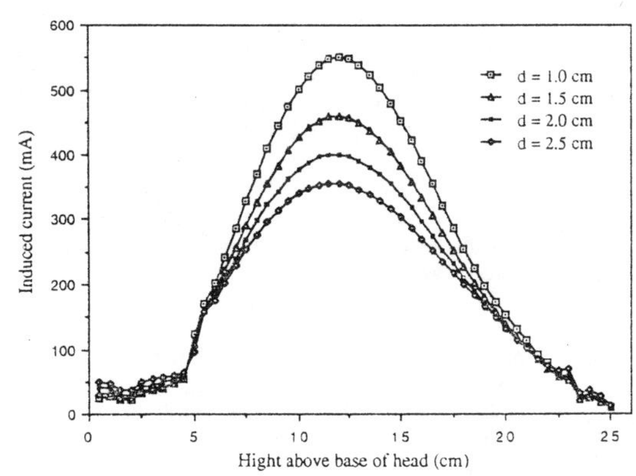

Fig. 11. Current distributions in the head for the case of an equivalent resistance $R_a = 120$ ohms used in the dipole antenna. d is the distance between the dipole antenna and the head.

distance between the dipole antenna and the head being longer than 2.0 cm, the maximum SAR is below the protection value of 1.6 W/kg for peak SAR delivered to any one gram of tissue assumed in the ANSI/IEEE protection guide [19].

IV. Conclusion

The current and SAR distributions in a model of an inhomogeneous human head composed of 57 263 cubical cells of dimension 0.5 cm were studied by using the near-field FDTD method under the irradiation of a cellular phone. Maximum currents of $356 \sim 551$ mA induced in an inhomogeneous human head were calculated for the distance between the dipole antenna and the head in the range of $1.0 \sim 2.5$ cm, while the maximum SAR's induced in the head were found in the range of 1.228 ~ 2.627 W/kg. The maximum SAR of 1.48 W/kg induced in the head was found for the distance of 2 cm, which was below the protection value of 1.6 W/kg for peak SAR delivered to any one gram of tissue assumed in the ANSI/IEEE protection guide [19]. It should be noted that the larger values of SAR's were obtained in the skin tissue. However, one skin cell of cubical shape has 0.125 cm^3 volume (0.5×0.5×0.5) and weighs 0.125 g. According to the standard, it means the average maximum SAR over top 8 values will be below the maximum SAR (1.48 W/kg) because the standard applies only to SAR's averaged over one gram of cubical shape. In practical use, the distance between the cellular phone and the human head is longer than 2.0 cm. If the distance between the cellular phone and the head is kept in the range of 2.0 to 3.0 cm, the SAR values induced in the human head exposed to the electromagnetic waves irradiated from a cellular phone should be below the protection value of 1.6 W/kg allowed by the ANSI/IEEE code. There are recent reports that the dielectric properties of bone may be closer to those of muscle at microwave frequencies than previously thought [20]. Because the relative dielectric constant of muscle used in this study is about 10 times of that used in bone. From the concept of boundary conditions, the obtained internal electric fields induced inside the head will decrease if the dielectric constant of bone is replaced by a higher value. So, the maximum SAR's will decrease due to the higher dielectric constant used in bone. Indeed, the maximum SAR of 1.48 decreased to 1.31 if the dielectric constant of bone was replaced by that of muscle under the same conditions. Distribution of absorbed energy is likely to be important in determining biological effects due to the EM radiation. Absorption of EM energy varies not only with the frequency, but also rather drastically with polarization. Vertically polarized EM waves were used for the simulations in this paper. Horizontally polarized EM waves may be used for the simulations, but the results will be smaller than those obtained by the vertically polarized EM waves [3].

References

[1] K. S. Yee, "Numerical solution of initial boundary value problems involving Maxwell's equations in isotropic media," *IEEE Trans. Antennas Propagat.*, vol. AP-14, no. 5, pp. 302–307, 1966.

[2] C. C. Johnson and A. W. Guy, "Nonionizing electromagnetic wave effects in biological materials and system," *Proc. IEEE*, vol. 60, pp. 692–718, 1972.

[3] C. H. Durney, H. Massoudi, and M. F. Iskander, *Radiofrequency Radiation Dosimetry Handbook, USAFAM-TR-85-73*, Prepared for USAF School of Aerospace Medicine, Aerospace Medical Division (AFSC), Books Air Force Base, TX, 1986.

[4] J. P. Reilly, *Electrical Stimulation and Electropathology*, Cambridge, UK: Cambridge University Press, 1992.

[5] C. D. Taylor, D. H. Lam, and T. H. Shumpert, "Electromagnetic pulse scattering in time-varying inhomogeneous media," *IEEE Trans. Antennas Propagat.*, vol. AP-17, pp. 585–589, 1969.

[6] A. Taflove, "Application of the finite-difference time-domain method to sinusoidal steady-state electromagnetic penetration problems," *IEEE Trans. Electromag . Compat.*, vol. EMC-22, no. 3, pp. 191–202, 1980.

[7] K. S. Kunz and K. M. Lee, "A Three-dimensional finite-difference solution of the external response of an aircraft to a complex transient em environment Part 1: The method and its implementation," *IEEE Trans. Electromag. Compat.*, vol. EMC-20, pp. 328–332, 1978.

[8] R. Holland, "THREDE: A free-field EMP coupling and scattering code," *IEEE Trans. Nucl. Sci.*, vol. NS-24, pp. 2416–2421, 1971.

[9] K. Umashanker and A. Taflove, "A novel method to analyzer electromagnetic scattering of complex objects," *IEEE Trans. Electromag. Compat.*, vol. EMC-24, pp. 397–405, 1982.

[10] A. Taflove and M. E. Brodwin, "Computation of electromagnetic fields and induced temperatures within a model of the microwave irradiated human eye," *IEEE Trans. Microwave Theory Tech.*, vol. MTT-23, pp 888–896, 1975.

[11] G. Mur, "Absorbing boundary conditions for the finite-difference approximation of the time-domain electromagnetic-field equations," *IEEE Trans. Electromag. Compat.*, vol. EMC-23, no. 4, pp. 377–382, 1981.

[12] R. J. Luebbers and J. Beggs, " FDTD calculation of wide-band antenna gain and efficiency," *IEEE Trans. Antennas Propagat.*, vol. 40, pp. 1403–1407, Nov. 1992.

[13] J. Y. Chen and O. P. Gandhi, "Currents induced in an anatomically based model of a human for exposure to vertically polarized electromagnetic pulses," *IEEE Trans. Microwave Theory Tech.*, vol. 39, pp. 31–39, Jan. 1991.

[14] J. G. Maloney, G. S. Smith, and W. R. Scott, "Accurate computation of the radiation from simple antennas using the finite-difference time-domain method," *IEEE Trans. Antennas Propagat.*, vol. 38, pp. 1059–1068, July 1990.

[15] P. H. Harms, J. F. Lee, and R. Mittra, "Study of the nonorthogonal FDTD method versus the conventional FDTD Technique for computing resonant frequencies of cylindrical cavities," *IEEE Trans. Antennas Propagat.*, vol. 40, pp. 741–746, Apr. 1992.

[16] C. M. Furse, S. P. Mathur, and O. P. Gandhi, "Improvements to the finite-difference time-domain method for calculating the radar cross section of a perfectly conducting target," *IEEE Trans. Microwave Theory Techniq.*, vol. 38, pp. 919–927, July 1990.

[17] P. J. Dimbylow, "Finite difference time–domain calculation of absorbed power in the ankle for 10–100 MHz plane wave exposure," *IEEE, Trans. Biomed. Eng.*, vol. 38, pp. 423–428, May 1991.

[18] J. A. Shaw, C. H. Durney, and D. A. Christensen, "Computer-aided design of two-dimensional electric-type hyperthermia applicators using the finite-difference time-domain method," *IEEE, Trans. Biomed. Eng.*, vol. 38,pp. 861–870, Sept. 1991.

[19] *Safety levels with respect to human exposure to radio-frequency electromagnetic fields, 3 kHz to 300 GHz*, ANSI Standard C95.1–1992.

[20] C. Gabriel, "Dielectric properties of biological material," *NATO Conf.*, Rome, Italy, May 18, 1993.

Adaptation of FDTD Techniques to Acoustic Modeling

James G. Maloney[1] and **Kathleen E. Cummings**[2]

[1]Signature Technology Lab, Georgia Tech Research Institute
[2]School of Electrical and Computer Engineering
Georgia Institute of Technology
Atlanta, Georgia 30332

I. INTRODUCTION

The Finite-Difference Time-Domain (FDTD) method has been used with great success to model various electromagnetics configurations, ranging from electromagnetic scattering (RCS) to driven antenna problems. The method has been extended over the years to allow frequency dispersive materials, surface impedance boundary conditions, sub-cell models of electrically thin material, etc. The quality of the absorbing boundary conditions used to truncate the computational domain has also continued to improve.

In this presentation, we will discuss the adaptation of a large part of the FDTD body of knowledge to the problem of acoustics. At a basic level, there are a number of parallels between Maxwell's equations and the coupled acoustic equations derived from the equations of continuity and momentum (assuming negligible viscosity and small perturbations from rest). It is natural to adapt the FDTD models which have found such great success in EM to the modeling of a variety of acoustics problems. This paper will present the many issues involved in the development of such an acoustic model including: discretization of the pressure and velocity equation (using Yee-like cells), derivation of the grid dispersion and stability relations, adaptation of absorbing boundary conditions, modeling of lossy walls, excitation of the model, etc. The well-known acoustics problem of the open-ended pipe was used to validate the model. The comparison of the numerical results and the analytical solution was excellent. As time permits, the application of the developed methodology to a variety of acoustics problems will also be presented.

This adaption of a popular EM modeling technique to another field, acoustics, is a prime example of the new multi-disciplinary nature of scientific research. This presentation will demonstrate how EM researchers can extend their techniques to a large variety of problems.

II. DISCRETIZATION OF ACOUSTIC EQUATIONS

If one assumes small perturbations from rest and negligible viscocity then the acoustic equations reduce to a form that is very similar to Maxwell's equations,[1]

Maxwell's equations	Acoustic equations	
$\epsilon_o \frac{\partial \vec{\mathcal{E}}}{\partial t} = \nabla \times \vec{\mathcal{H}}$	$\delta_o \frac{\partial \vec{u}}{\partial t} = -\nabla p$	(1)
$\mu_o \frac{\partial \vec{\mathcal{H}}}{\partial t} = \nabla \times \vec{\mathcal{E}}$	$\frac{\partial p}{\partial t} = -\delta_o c^2 \nabla \cdot \vec{u}$	(2)

[1]This small perturbation from rest and negligible viscocity assumption is in general valid for sound levels below 110 dB [1]. For reference, speech is from +30 to +80 dB, 120 dB is discomfort, and 140 dB is the pain threshold [2].

Reprinted with permission from *Proceedings of Applied Computational Electromagnetics at the Naval Postgraduate School,* Monterey, CA, J. G. Maloney and Kathleen E. Cummings, "Adaptation of FDTD Techniques to Acoustic Modeling," pp. 724-731, 1995. © ACES.

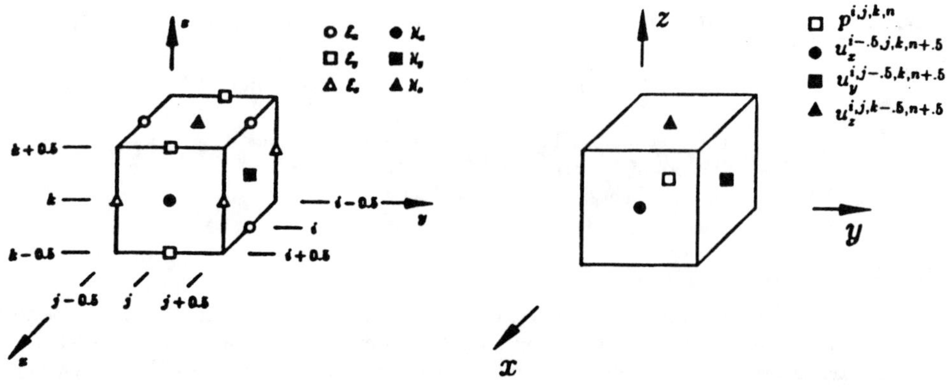

Figure 1: (a) Yee cell for Electromagnetics, (b) Yee-like cell for Acoustics. Both cells are displayed for the rectangular coordinates case.

where $\vec{u}$ is the gas particle velocity, comprised of velocity components in each coordinate direction, p is the deviation from ambient pressure, c is the speed of sound in the medium, and δ_o is the density of the gas at rest. [3]. In rectangular coordinates, these coupled acoustic equations can be expanded to yield

$$\delta_0 \frac{\partial \vec{u}}{\partial t} = -\frac{\partial p}{\partial x}\hat{x} - \frac{\partial p}{\partial y}\hat{y} - \frac{\partial p}{\partial z}\hat{z} \quad \text{and} \quad \frac{\partial p}{\partial t} = -\delta_0 c^2 \left[\frac{\partial u_x}{\partial x} + \frac{\partial u_y}{\partial y} + \frac{\partial u_z}{\partial z}\right]. \tag{3}$$

In FDTD modeling of Maxwell's equations, the solution space is discretized using the Yee cell, see Figure 1(a). The vector components of the electric and magnetic field are distributed around the unit-cell to allow the differential operators to be approximated by centered finite differences which yields a second order accurate formulation. In a similar fashion, we distribute the scalar pressure and 3 vector components of the velocity around an acoustic Yee-like unit-cell, see Figure 1(b). In rectangular coordinates, the discretized update equations thus become

$$u_x^{i-.5,j,k,n+.5} = u_x^{i-.5,j,k,n-.5} - \left(\frac{c\Delta t}{\Delta x}\right)\left(\frac{1}{c\delta_0}\right)\left(p^{i,j,k,n} - p^{i-1,j,k,n}\right), \tag{4}$$

$$u_y^{i,j-.5,k,n+.5} = u_y^{i,j-.5,k,n-.5} - \left(\frac{c\Delta t}{\Delta y}\right)\left(\frac{1}{c\delta_0}\right)\left(p^{i,j,k,n} - p^{i,j-1,k,n}\right), \tag{5}$$

$$u_z^{i,j,k-.5,n+.5} = u_z^{i,j,k-.5,n-.5} - \left(\frac{c\Delta t}{\Delta z}\right)\left(\frac{1}{c\delta_0}\right)\left(p^{i,j,k,n} - p^{i,j,k-1,n}\right), \tag{6}$$

and

$$\begin{aligned} p^{i,j,k,n+1} = p^{i,j,k,n} &- \left(\frac{c\Delta t}{\Delta x}\right)(c\delta_0)\left(u_x^{i+.5,j,k,n+.5} - u_x^{i-.5,j,k,n+.5}\right) \\ &- \left(\frac{c\Delta t}{\Delta y}\right)(c\delta_0)\left(u_y^{i,j+.5,k,n+.5} - u_y^{i,j-.5,k,n+.5}\right) \\ &- \left(\frac{c\Delta t}{\Delta z}\right)(c\delta_0)\left(u_z^{i,j,k+.5,n+.5} - u_z^{i,j,k-.5,n+.5}\right), \end{aligned} \tag{7}$$

where the notation $p(x, y, z, t) = p(i\Delta x, j\Delta y, k\Delta z, n\Delta t) = p^{i,j,k,n}$ is utilized.

These equations are updated in time using a leap-frog scheme. First, the u's at time level $n + .5$ are computed from p's at time level n and previous u's at time level $n - .5$. Then, the p's at time level $n + 1$ are computed from u's at time level $n + .5$ and previous p's at time level n. This process repeats until the temporal simulation is complete.

II.1 GRID DISPERSION

With any discretized solution to wave type equations, one should be concerned about grid dispersion. Grid dispersion is the fact that waves in the numerical grid travel at speeds slightly different than in real space. Futhermore, this grid dispersion tends to be frequency and angle of propagation dependent. Therefore, it is difficult to calibrate for, and the best strategy is to try to minimize the amount of grid dispersion.

To solve for the dispersion relation, we assume a general plane wave is propagating throughout the grid, i.e.,

$$p, u_x, u_y, u_z \propto e^{j(\omega t - k_x x - k_y y - k_z z)} \tag{8}$$

where k_x, k_y, and k_z are the x, y, and z components of the numerical grid k. Using (8) to eliminate half of the field components in (4)-(7) yields

$$\left[1 - e^{-j\omega\Delta t}\right] u_x^{i-.5,j,k,n+.5} = -\left(\frac{c\Delta t}{\Delta x}\right)\left(\frac{1}{c\delta_o}\right)\left[1 - e^{-jk_x\Delta x}\right] p^{i,j,k,n}, \tag{9}$$

$$\left[1 - e^{-j\omega\Delta t}\right] u_y^{i,j-.5,k,n+.5} = -\left(\frac{c\Delta t}{\Delta y}\right)\left(\frac{1}{c\delta_o}\right)\left[1 - e^{-jk_y\Delta y}\right] p^{i,j,k,n}, \tag{10}$$

$$\left[1 - e^{-j\omega\Delta t}\right] u_z^{i,j,k-.5,n+.5} = -\left(\frac{c\Delta t}{\Delta z}\right)\left(\frac{1}{c\delta_o}\right)\left[1 - e^{-jk_z\Delta z}\right] p^{i,j,k,n}, \tag{11}$$

and

$$\begin{aligned}\left[e^{j\omega\Delta t} - 1\right] p^{i,j,k,n} = -(c\delta_o)\Bigg(& \left(\frac{c\Delta t}{\Delta x}\right)\left[e^{jk_x\Delta x} - 1\right] u_x^{i-.5,j,k,n+.5} \\ & + \left(\frac{c\Delta t}{\Delta y}\right)\left[e^{jk_y\Delta y} - 1\right] u_y^{i,j-.5,k,n+.5} \\ & + \left(\frac{c\Delta t}{\Delta z}\right)\left[e^{jk_z\Delta z} - 1\right] u_z^{i,j,k-.5,n+.5}\Bigg)\end{aligned} \tag{12}$$

Then, substituting (9)-(11) into (12) yields after some algebraic manipulation

$$\left[\sin^2(\omega\Delta t) - \left(\frac{c\Delta t}{\Delta x}\right)^2 \sin^2\left(\frac{k_x\Delta x}{2}\right) - \left(\frac{c\Delta t}{\Delta y}\right)^2 \sin^2\left(\frac{k_y\Delta x}{2}\right) - \left(\frac{c\Delta t}{\Delta z}\right)^2 \sin^2\left(\frac{k_z\Delta z}{2}\right)\right] p^{i,j,k,n} = 0. \tag{13}$$

In general $p^{i,j,k,n}$ is not zero and thus

$$\sin^2(\omega\Delta t) = \left(\frac{c\Delta t}{\Delta x}\right)^2 \sin^2\left(\frac{k_x\Delta x}{2}\right) + \left(\frac{c\Delta t}{\Delta y}\right)^2 \sin^2\left(\frac{k_y\Delta x}{2}\right) + \left(\frac{c\Delta t}{\Delta z}\right)^2 \sin^2\left(\frac{k_z\Delta z}{2}\right). \tag{14}$$

This is the dispersion relation for the discretized solution; it is identical to the dispersion relation for the FDTD technique applied to Maxwell's equations. For a uniform grid (Δs=Δx=Δy=Δz), it is commonly re-written in the form

$$\sin\left(\frac{\pi}{2}\frac{\Delta s}{\lambda_o}\right) = \left(\frac{c\Delta t}{\Delta s}\right)\sqrt{\sin^2\left(\pi\cos\alpha\frac{\Delta s}{\lambda_n}\right) + \sin^2\left(\pi\cos\beta\frac{\Delta s}{\lambda_n}\right) + \sin^2\left(\pi\cos\gamma\frac{\Delta s}{\lambda_n}\right)} \tag{15}$$

where λ_o is the wavelength in real space, λ_n is the numerical wavelength, and $\cos\alpha$, $\cos\beta$, $\cos\gamma$ are the direction cosines of the propagating plane wave. Then the phase error per cell is

$$\Theta_{\text{err}} = 360^\circ\left(\frac{\Delta s}{\lambda_n} - \frac{\Delta s}{\lambda_o}\right). \tag{16}$$

II.2 STABILITY

Another important criteria is the stability of the algorithm. The stability relation is derived in a similiar fashion. Again, we assume a general plane wave is propagating throughout the grid, i.e.,

$$p, u_x, u_y, u_z \propto \xi^n e^{-j(k_x x + k_y y + k_z z)} \tag{17}$$

but now we have not constrained ourselves only to time-harmonic plane waves. Thus, we have

$$\left[1-\xi^{-1}\right] u_x^{i-.5,j,k,n+.5} = -\left(\frac{c\Delta t}{\Delta x}\right)\left(\frac{1}{c\delta_o}\right)\left[1-e^{-jk_x\Delta x}\right] p^{i,j,k,n}, \tag{18}$$

$$\left[1-\xi^{-1}\right] u_y^{i,j-.5,k,n+.5} = -\left(\frac{c\Delta t}{\Delta y}\right)\left(\frac{1}{c\delta_o}\right)\left[1-e^{-jk_y\Delta y}\right] p^{i,j,k,n}, \tag{19}$$

$$\left[1-\xi^{-1}\right] u_z^{i,j,k-.5,n+.5} = -\left(\frac{c\Delta t}{\Delta z}\right)\left(\frac{1}{c\delta_o}\right)\left[1-e^{-jk_z\Delta z}\right] p^{i,j,k,n}, \tag{20}$$

and

$$\begin{aligned}[\xi-1]\,p^{i,j,k,n} = -(c\delta_o)\Bigg(& \left(\frac{c\Delta t}{\Delta x}\right)\left[e^{jk_x\Delta x}-1\right] u_x^{i-.5,j,k,n+.5} \\ + & \left(\frac{c\Delta t}{\Delta y}\right)\left[e^{jk_y\Delta y}-1\right] u_y^{i,j-.5,k,n+.5} \\ + & \left(\frac{c\Delta t}{\Delta z}\right)\left[e^{jk_z\Delta z}-1\right] u_z^{i,j,k-.5,n+.5}\Bigg)\end{aligned} \tag{21}$$

Again, substituting (18)-(20) into (21) yields after some manipulation

$$\xi^2 - 2A\xi + 1 = 0 \tag{22}$$

where

$$A = 1 - 2\left(\frac{c\Delta t}{\Delta x}\right)^2\sin^2\left(\frac{k_x\Delta x}{2}\right) - 2\left(\frac{c\Delta t}{\Delta y}\right)^2\sin^2\left(\frac{k_y\Delta x}{2}\right) - 2\left(\frac{c\Delta t}{\Delta z}\right)^2\sin^2\left(\frac{k_z\Delta z}{2}\right). \tag{23}$$

The solutions of (22) are

$$\xi = A \pm \sqrt{A^2 - 1}. \tag{24}$$

A growing (unstable) solution will occur if $|\xi| > 1$ This only occurs when $|A| > 1$; thus,

$$(c\Delta t) > \frac{1}{\sqrt{\left(\frac{1}{\Delta x}\right)^2 + \left(\frac{1}{\Delta y}\right)^2 + \left(\frac{1}{\Delta z}\right)^2}} \quad \textbf{unstable}. \tag{25}$$

However, when $|A| \leq 1$, then $|\xi| = 1$ and the grid is stable; thus,

$$(c\Delta t) \leq \frac{1}{\sqrt{\left(\frac{1}{\Delta x}\right)^2 + \left(\frac{1}{\Delta y}\right)^2 + \left(\frac{1}{\Delta z}\right)^2}} \quad \text{is } \textbf{stable}. \tag{26}$$

This stability criteria is the same as for the FDTD solution of Maxwell's equations.

III. ADAPTATION OF PML FOR ACOUSTICS

The most important recent advance in FDTD modeling is the introduction of the perfectly matched layer (PML) absorbing boundary condition [4]. Basically, the PML ABC is a region of fictitious lossy material that surrounds the computational space. This fictitious material has the desired property that waves at any angle and any frequency are not reflected upon entering the lossy region. Then the loss is adjusted to sufficiently attenuate the energy that entered the PML region. This material is called fictitious because fields within the PML layer are artificially split and obey a special set of equations that can not be cast in the form of the usual equations with some effective material properties.

The special equations, cast for use in acoustics, are

$$\frac{\partial u_x}{\partial t} + \sigma_x u_x = -\left(\frac{1}{\delta_o}\right)\frac{\partial}{\partial x}(p_x + p_y + p_z), \tag{27}$$

$$\frac{\partial u_y}{\partial t} + \sigma_y u_y = -\left(\frac{1}{\delta_o}\right)\frac{\partial}{\partial y}(p_x + p_y + p_z), \tag{28}$$

$$\frac{\partial u_z}{\partial t} + \sigma_z u_z = -\left(\frac{1}{\delta_o}\right)\frac{\partial}{\partial z}(p_x + p_y + p_z) \tag{29}$$

and

$$\frac{\partial p_x}{\partial t} + \sigma_x^* p_x = -\left(c^2\delta_o\right)\frac{\partial u_x}{\partial x}, \tag{30}$$

$$\frac{\partial p_y}{\partial t} + \sigma_y^* p_y = -\left(c^2\delta_o\right)\frac{\partial u_y}{\partial y}, \tag{31}$$

$$\frac{\partial p_z}{\partial t} + \sigma_z^* p_z = -\left(c^2\delta_o\right)\frac{\partial u_z}{\partial z} \tag{32}$$

where the scalar pressure p has been broken into 3 fictitious components p_x, p_y, p_z and loss terms σ_x, σ_y, σ_z, σ_x^*, σ_y^*, σ_z^* have been introduced.

For example, if this material is placed at the positive x end of the solution space, then set $\sigma_x=\sigma_x^*$, $\sigma_y=\sigma_y^*=0$, $\sigma_z=\sigma_z^*=0$. The wave will not be reflected upon encountering the PML region,

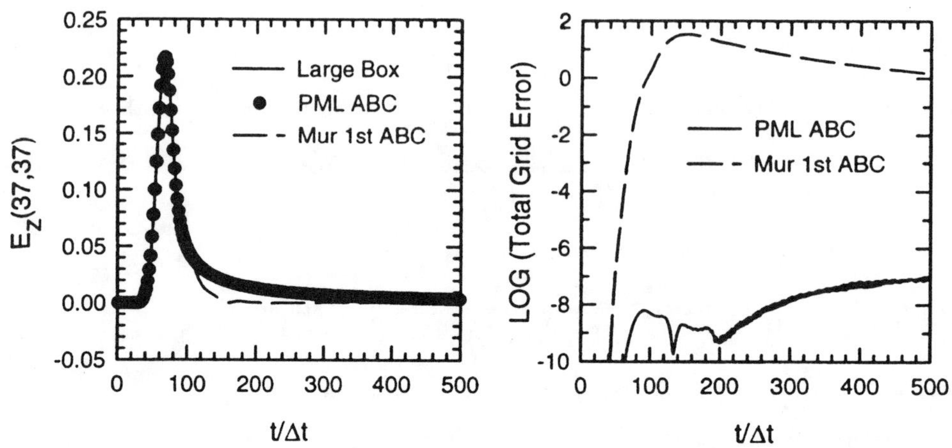

Figure 2: (a) Comparison of Ez(37,37) for PML and Mur 1st order type ABC, (b) Comparison of total energy error in the solution space.

and the wave will be attenuated as it propagates in the PML region. A similiar treatment is placed on the other faces of the solution space. The edges and corners however require some special attention, and the reader is refered to [4] for the details.

In order to verify the accuracy of the PML ABC, consider a simple 2D problem in which the computational space is defined to be 50 by 50 cells, with the pressure in the middle set equal to a gaussian pulse. Figure 2(a) compares the exact results (solid line) computed using a 500 by 500 box, the PML-surrounded box results (solid dots), and the simple Mur 1st order type ABC (Dashed line). Clearly, the PML-surrounded box solution and the exact solution are in good agreement, while the simple ABC is in error. The total energy error in the computational grid is compared in Figure 2(b). The PML results (solid line) are nearly 10 orders of magnitude better than the simple Mur 1st order type ABC (dashed line).

IV. VERIFICATION

The traditional method of verifying the accuracy of a numerical model is to select a problem with a well-known theoretical solution and to compare the results of using the numerical model to the theoretically expected results. For this paper, a three-dimensional unflanged, open-ended circular cylinder was modelled using the previously described acoustics formulation of FDTD. The model was excited with a Gaussian pulse. The reflection coefficient was calculated from the numerical solution of the propagation of the Gaussian pulse in the cylinder over time and compared to the theoretically derived reflection coefficent [5]. The comparison of the experimentally calculated and the theoretically expected values $|R|$ and l/a versus ka (where k is the wave number, $2\pi/\lambda$, and a is the radius of the cylinder) are shown in Figure 3. As can be seen, the agreement between the FDTD and the theoretical results is excellent. The accuracy of this finite-difference time-domain model is therefore verified.

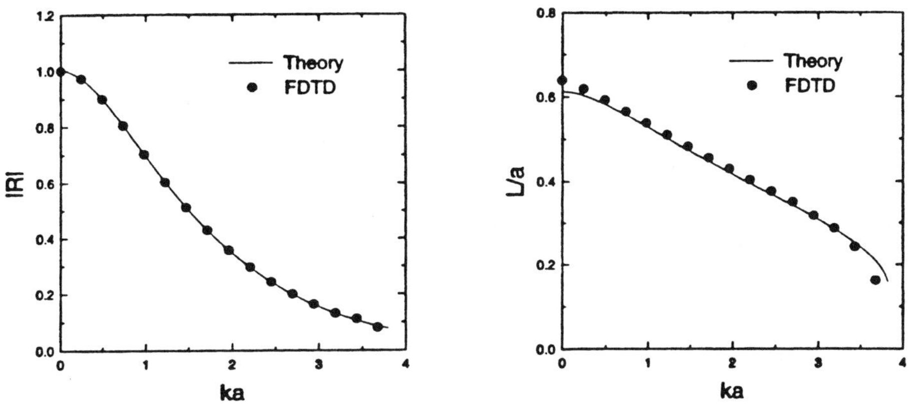

Figure 3: Comparison of $|R|$ and l/a, finite-difference time-domain solution and theoretical solution.

V. APPLICATION TO SPEECH MODELING

One important application of this new acoustics FDTD modelling technique is the problem of modelling speech production. Most current methods of modelling speech production involve either directly analyzing the acoustic speech waveform, which assumes a simplified linear model, or solving the acoustic wave equation in a series of cylindrical tubes, which assumes a simplified geometry. Although these classes of models have been successful for speech coding, recognition, synthesis, and analysis for many years, recent advances have proceeded at a very slow pace. Many researchers believe that the simplified models of speech production have been pushed as far as they will go, and that, in order to continue to advance the field of digital speech processing, improved models of speech production must be developed. This FDTD model of acoustics can provide a mechanism for accurately modeling acoustic wave propagation in the speech production process.

We are currently applying the FDTD model of acoustics discussed above to the problem of accurately modeling speech production. In this new model, acoustic wave propagation is solved for directly, in a geometry that accurately represents the vocal tract.

We believe that the FDTD approach has several distinct advantages over both traditional speech modeling techniques and finite element and boundary element models. First, a FDTD solution provides full knowledge of the acoustic flow at every point in the vocal tract for every time. Second, the finite-difference time-domain method makes no assumption of linearity between the input and the output (note that this is not true for frequency-domain finite-difference methods). Third, a FDTD model allows many parameters such as the geometry of the vocal tract and the excitation source to be altered easily. Finally, finite-different solutions are highly parallelizable, allowing for the efficient use of computing resources.

Several experiments have been performed using FDTD models of the vocal tract during vowel production. This has been accomplished in two dimensions using published X-ray data. Recently, MRI data of the vocal tract during vowel production has been obtained and is being used to develop a three-dimensional model of speech production. Results from these experiments will be presented at the conference, as time permits.

VI. CONCLUSIONS

This paper has presented an adaptation of a propular EM modeling technique, FDTD, to the field of acoustics. The parallels between the EM and acoustic coupled equations were discussed, and the discretization of a three-dimensional acoustic cell and of the coupled acoustic equations were presented. The dispersion relation and stability criteria were derived for the acoustic FDTD model and were shown to be parallel to the FDTD formulation for EM. The perfectly matched layer (PML) ABC that has recently been introduced for EM was also adapted to this FDTD acoustic model, and it was shown that the PML solution is 10 orders of magnitude better than a simple Mur 1st order type ABC. The accuracy of the acoustic FDTD model was verified by modeling a well-known acoustics problem, the unflanged, open-ended circular cylinder, and comparing the numerical and analytic solutions for the magnitude and phase of the reflection coefficient. Finally, the application of this FDTD acoustic model to an important problem, modeling speech production, was discussed briefly.

ACKNOWLEDGEMENT

This material is based upon work supported under National Science Foundation Grant # CDA-9211129 and a grant from the American Speech-Language-Hearing Association.

References

[1] L. L. Beranek, *Acoustics.* Woodbury, NY: Acoustical Society of America, American Institute of Physics, 1993.

[2] L. L. Beranek, *Acoustical Measurements.* Woodbury, NY: Acoustical Society of America, American Institute of Physics, 1988.

[3] J. J. Bowman, T. B. A. Senior, and P. L. E. Uslenghi, Electromagnetic and Acoustical Scattering by Simple Shapes. New York: Hemisphere Pub., 1987.

[4] J. P. Berenger, "A Perfectly Matched Layer for the Absorption of electromagnetic waves," *J. Computational Physics*, Oct. 1994.

[5] H. Levine and J. Schwinger "On the Radiation of Sound from an Unflanged Circular Pipe," *Physical Review*, Vol 3, No 4, pp 383-406, 1948.

Numerical Issues Regarding Finite-Difference Time-Domain Modeling of Microwave Structures

ALLEN TAFLOVE, FELLOW, IEEE
DEPARTMENT OF ELECTRICAL AND COMPUTER ENGINEERING
MCCORMICK SCHOOL OF ENGINEERING AND APPLIED SCIENCE
NORTHWESTERN UNIVERSITY
2145 SHERIDAN ROAD, EVANSTON, IL 60208

I. Introduction

Accurate numerical modeling of full-vector electromagnetic (EM) wave interactions in arbitrary microwave structures is difficult. Typical microwave structures of engineering interest have three-dimensional (3-D) shapes, apertures, cavities, materials, and embedded passive and active circuit devices that produce complex fields that cannot easily be resolved into finite sets of modes. Proper numerical modeling of such fields requires volumetric sampling at sub-wavelength resolution to avoid aliasing of magnitude and phase information. The goal is to provide a self-consistent model of the mutual coupling of the electrically small cells making up the microwave structure.

This paper summarizes a number of key technical issues and the corresponding research literature regarding a powerful approach for numerical modeling of microwave structures: the finite-difference time-domain (FDTD) solution of Maxwell's curl equations. FDTD is very simple in concept. However, it is remarkably robust, providing highly accurate modeling predictions for a wide variety of free-space and guided-wave electromagnetic interaction problems. It has become a viable alternative to frequency-domain method of moments (MM) and finite-element (FE) technology because of its dimensionally reduced computational burdens and ability to directly simulate the dynamics of wave propagation.

Topics discussed include: (1) general characteristics of space-grid time-domain models; (2) basics of the Yee algorithm for Maxwell's equations; (3) fundamental aspects of numerical stability; (4) numerical dispersion of the Yee algorithm for both second-order and optimized fourth-order accurate spatial central differences; (5) incident wave source conditions, including the zoned-grid formulation and the resistive voltage source; (6) absorbing boundary conditions; (7) unstructured gridding; and (8) digital signal processing and spectrum estimation techniques.

II. General Characteristics of Space-Grid Time-Domain Models

FDTD and related finite-volume time-domain (FVTD) approaches are direct solution methods for Maxwell's time-dependent curl equations. Such methods employ no potentials. Rather, they are based upon volumetric sampling of the unknown $\vec{E}$ and $\vec{H}$ near-field distributions within and surrounding the structure of interest and over a period of time. The sampling in space is at sub-wavelength resolution set by the user to properly sample, in the Nyquist sense, the highest near-field spatial frequencies thought to be important in the physics of the problem. Typically, 10 to 20 samples per wavelength are needed. The sampling in time is selected to ensure numerical stability.

Numerical Wave Analogs

FDTD and FVTD are marching-in-time procedures that simulate the continuous actual electromagnetic waves in a finite region by sampled-data numerical analogs propagating in a computer data space. Time-stepping continues as the numerical wave analogs propagate in the space grid to causally connect the physics of the modeled region. For simulations where the modeled region must extend to infinity, an absorbing boundary condition (ABC) is employed at the outer grid truncation planes that ideally permits all outgoing wave analogs to exit the region with negligible reflection. Phenomena such as induction of surface currents, scattering and multiple scattering, aperture penetration, and cavity excitation are modeled time-step by time-step by the action of the curl equations analog. Self-consistency of these modeled phenomena is generally assured if their spatial and temporal variations are well resolved by the space and time sampling process. In fact, the goal is to provide a self-consistent model of the mutual coupling of all electrically small volume cells in the structure and its near field, even if the structure spans tens of wavelengths in 3-D and there are hundreds of millions of space cells.

Time-stepping is continued until the desired late-time pulse response is observed at the field points of interest. For linear wave interaction problems, the sinusoidal response at these field points can be obtained over a wide band of frequencies by discrete Fourier transformation of the computed field-vs.-time waveforms at these points. Prolonged "ringing" of the computed field waveforms due to a high quality factor (Q) or large electrical size of the structure being modeled requires a combination of extending the computational window in time and extrapolation of the windowed data before Fourier transformation.

The primary FDTD and FVTD algorithms used today are fully explicit, second-order accurate grid-based solvers employing highly vectorizable and concurrent schemes for time-marching the six vector components of the electromagnetic near field at each of the volume cells. The explicit nature of the solvers is maintained by either leapfrog or predictor-corrector time integration schemes. Present methods differ primarily in how the space grid is set up. In fact, gridding methods can be categorized according to the degree of structure or regularity in the mesh cells, as follows.

Almost-Completely Structured Grids. In this case, the grid is organized so that its space cells are congruent wherever possible. The most basic example of such a mesh is the pioneering work of Yee [1], who employed a uniform Cartesian mesh having rectangular unit cells along with staircasing to approximate the surface of a structural feature that does not lie along a grid coordinate axis. Later work showed that it is possible to modify the size and shape of the grid cells immediately adjacent to the structural feature in question to conformally fit its surface [2,3]. This is computationally efficient because there are relatively few modified space cells requiring special treatment to specify the presence of a non-conforming structural feature. In fact, the number of modified cells is proportional to the surface area of the structure, and becomes progressively smaller relative to the number of regular mesh cells filling the structure volume as the structure size increases. As a result, the computer memory and running time needed to implement a fully conformal model can approximate that required for a stepped-surface model. A disadvantage is that special geometry-generation software must be constructed.

Surface-Fitted Grids. Here, the space grid is globally distorted to fit the shape of the structure of interest. The mesh can be divided into multiple zones to accommodate a set of distinct surface features [4]. The major advantage of this approach is that well-developed mesh generation software of this type is available. The major disadvantage is that, relative to the baseline Yee mesh, there is substantial added computer burden due to memory allocations for the position and stretching factors of each space cell. There are also extra computer arithmetic operations to implement Maxwell's equations at each cell and to enforce field continuity at the interfaces of adjacent cells. Another disadvantage is the possible presence of numerical dissipation in the time-stepping algorithm used for such meshes. This can limit the range of electrical size of the structure being modeled due to numerical wave attenuation artifacts across the structure.

Completely Unstructured Grids. Here, the space containing the structure of interest is completely filled with a collection of grid cells of varying sizes and shapes, but conforming to the structure surface [5]. As for the case of surface-fitted meshes, geometry generation software is available and capable of modeling complicated 3-D shapes possibly having volumetric inhomogeneities. The major disadvantage is that it is relatively difficult to map the unstructured mesh computations onto the architecture of either parallel vector computers such as the Cray C-90 or massively parallel machines such as the Intel Paragon. The structure-specific irregularity of the mesh mandates a robust pre-processing algorithm that optimally assigns specific mesh cells to specific processors.

At present, the choice of computational algorithm and mesh is not straightforward. Clearly, there are important tradeoff decisions to be made. For the next several years, we can expect considerable progress in this area as various groups develop their favored approaches and perform validations.

Predictive Dynamic Range

For computational modeling of microwave structures using space-grid time-domain codes, it is useful to consider the concept of predictive dynamic range. Let the power density of the primary (or exciting) wave in the space grid be P_0 W/m^2. Further, let the minimum observable power density of a secondary (reflected) wave be $P_s W/m^2$, where "minimum observable" means that the accuracy of the field computation degrades due to numerical artifacts to poorer than n dB (some desired figure of merit) at lower levels than P_s. Then, we can define the predictive dynamic range as $10 \log(P_0/P_s)$ decibels. This definition is well suited for space-grid time-domain codes for two reasons: (1) it squares nicely with the concept of a measurement noise floor, which is intuitive to most electrical engineers, and (2) it succinctly quantifies the reality that the desired numerical wave analogs propagating in the space grid exist in an additive noise environment due to nonphysical propagating wave analogs caused by the imperfect ABC, similar to unwanted reflections from the walls of an anechoic chamber.

A less intuitive but equally important source of error that affects the computational dynamic range is the gradual but progressive accumulation of phase (delay) error by the numerical wave analogs due to the finite precision of the mathematical operations simulating Maxwell's equations. This error depends upon both the frequency content of the propagating wave analog and its direction of propagation within the grid. Its effect mounts as the electrical size and Q of the structure increase, a situation that requires the numerical waves to be tracked over greater propagation distances and times, and thereby allows an increased phase error accumulation.

Over the past ten years, researchers have accumulated solid evidence for a predictive dynamic range in the order of 40 to 50dB for space-grid time-domain codes when used to model a wide variety of microwave structures of moderate electrical size (smaller than a few wavelengths). This value is reasonable if one considers the additive noise due to the imperfect ABC to be the primary limiting factor for such structures, since every analytical ABC in use yielded effective reflection coefficients in the range of about 0.3% to 3% (–50dB to –30dB). As will be discussed later, recent developments provide the possibility of virtually eliminating the error due to the ABC, thereby reducing the computational noise floor to the level set by the phase error accumulation.

III. Basics of the Yee Algorithm for Maxwell's Equations

Maxwell's Time-Dependent Curl Equations

Consider a source-free region of space that is linear, isotropic, non-dispersive, and possibly lossy. Then, using the MKS system of units, Maxwell's curl equations are given by:

$$\frac{\partial \vec{H}}{\partial t} = -\frac{1}{\mu}\nabla \times \vec{E} - \frac{\rho'}{\mu}\vec{H} \qquad \frac{\partial \vec{E}}{\partial t} = \frac{1}{\varepsilon}\nabla \times \vec{H} - \frac{\sigma}{\varepsilon}\vec{E} \tag{1a, 1b}$$

where μ is the magnetic permeability in H/m, ρ' is an equivalent magnetic loss (resistivity) in Ω/m, ε is the electric permittivity in F/m, σ is the electric conductivity in S/m, $\vec{E}$ is the electric field in V/m, $\vec{D} = \varepsilon\vec{E}$ is the electric flux density in C/m^2, $\vec{H}$ is the magnetic field in A/m, and $\vec{B} = \mu\vec{H}$ is the magnetic flux density in W/m^2 (tesla). We now write out the vector components of the curl operators in (1) to yield the following system of six coupled scalar equations equivalent to Maxwell's curl equations in the 3-D rectangular coordinate system (x,y,z):

$$\frac{\partial H_x}{\partial t} = \frac{1}{\mu}\left(\frac{\partial E_y}{\partial z} - \frac{\partial E_z}{\partial y} - \rho' H_x\right) \tag{2a}$$

$$\frac{\partial E_x}{\partial t} = \frac{1}{\varepsilon}\left(\frac{\partial H_z}{\partial y} - \frac{\partial H_y}{\partial z} - \sigma E_x\right)$$

$$\frac{\partial H_y}{\partial t} = \frac{1}{\mu}\left(\frac{\partial E_z}{\partial x} - \frac{\partial E_x}{\partial z} - \rho' H_y\right) \tag{2b}$$

$$\frac{\partial E_y}{\partial t} = \frac{1}{\varepsilon}\left(\frac{\partial H_x}{\partial z} - \frac{\partial H_z}{\partial x} - \sigma E_y\right)$$

$$\frac{\partial H_z}{\partial t} = \frac{1}{\mu}\left(\frac{\partial E_x}{\partial y} - \frac{\partial E_y}{\partial x} - \rho' H_z\right) \tag{2c}$$

$$\frac{\partial E_z}{\partial t} = \frac{1}{\varepsilon}\left(\frac{\partial H_y}{\partial x} - \frac{\partial H_x}{\partial y} - \sigma E_z\right)$$

Basic Concepts Underlying Yee's Algorithm

The system of six coupled partial differential equations of (2) forms the basis of the FDTD numerical algorithm for electromagnetic wave interactions with general 3-D objects, first presented in 1966 by K. S. Yee [1]. The Yee algorithm persists in having great usefulness since its fundamental basis is so robust. Namely:

1. It solves for *both* electric and magnetic fields in time and space using the coupled Maxwell's curl equations rather than solving for either field alone with a wave equation. This is analogous to the combined-field integral equation formulation of the method of moments, wherein both $\vec{E}$ and $\vec{H}$ boundary conditions are enforced on the surface of a material structure. Using both and $\vec{E}$ and $\vec{H}$ information, the solution is more robust than using either alone, that is, accurate for a wider class of structures. Both electric and magnetic material properties can be modeled in a straightforward manner. Further, features unique to each field such as tangential $\vec{H}$ singularities near edges and corners, azimuthal $\vec{H}$ singularities near thin wires, and radial $\vec{E}$ singularities near points, edges, and thin wires can be individually modeled.
2. The Yee algorithm centers its $\vec{E}$ and $\vec{H}$ components in 3-D space, so that every $\vec{E}$ component is surrounded by four circulating $\vec{H}$ components, and every $\vec{H}$ component is surrounded by four circulating $\vec{E}$ components. The resulting finite-difference expressions for the space derivatives used in the curl operators are central in nature and second-order accurate. Further, the Yee arrangement provides a beautifully simple picture of 3-D space being filled by an interlinked array of Faraday's law and Ampere's law contours. It is possible to identify Yee $\vec{E}$ components associated with displacement flux linking $\vec{H}$ loops, and $\vec{H}$ components associated with magnetic flux linking $\vec{E}$ loops. In effect, the Yee algorithm simulates both the pointwise, differential form *and* the macroscopic, integral form of Maxwell's equations. The latter is very useful in specifying field boundary conditions and singularities.
3. The Yee algorithm centers its $\vec{E}$ and $\vec{H}$ components in time in what is commonly termed a leapfrog arrangement. All of the $\vec{E}$ computations in the 3-D space of interest are completed for a particular time point using $\vec{H}$ data stored in the computer memory. Then, all of the $\vec{H}$ computations in the 3-D space are completed using the $\vec{E}$ data just computed and stored in memory. The cycle can begin again with the re-computation of $\vec{E}$ based upon the newly obtained $\vec{H}$. This process continues until time-stepping is concluded. This leapfrog time-stepping process is fully explicit, thereby completely avoiding the problems involved with simultaneous equations and matrix inversion. The resulting finite-difference expressions for the time derivatives used in the curl equations are central in nature, second-order accurate, and non-dissipative.

Notation and Finite-Difference Equations

Following Yee, we denote any $\vec{E}$ or $\vec{H}$ component, f, evaluated at space point $(i, j, k) = (i\Delta x, j\Delta y, k\Delta z)$ in the space lattice and at time point $t_n = n\Delta t$ as $f(i\Delta x, j\Delta y, k\Delta z, n\Delta t) \equiv f^{\,n}_{i,k,l}$. Here,

Δx, Δy, Δz, and Δt are assumed to be uniform over the space grid and observation time interval, and i, j, k, and n are integers. Yee's expression for the first space derivative of f in the x direction, evaluated at the fixed time, t_n, is given by:

$$\left.\frac{\partial f}{\partial x}\right|_{(i\Delta x,\, j\Delta y,\, k\Delta z,\, n\Delta t)} = \frac{f^n_{i+1/2,j,k} - f^n_{i-1/2,j,k}}{\Delta x} + O[(\Delta x)^2] \tag{3a}$$

This permits interleaving the $\vec{E}$ and $\vec{H}$ components in the space lattice at intervals of $\Delta x/2$. For example, the difference of two adjacent $\vec{E}$ components, separated by Δx and located $\pm\, \Delta x/2$ on either side of an $\vec{H}$ component, would be used to provide a numerical approximation for $\partial E/\partial x$ to permit stepping the $\vec{H}$ component in time. Yee's expression for the first time derivative of f, evaluated at the fixed space point (i, j, k), follows by analogy:

$$\left.\frac{\partial f}{\partial t}\right|_{(i\Delta x,\, j\Delta y,\, k\Delta z,\, n\Delta t)} = \frac{f^{n+1/2}_{i,j,k} - f^{n-1/2}_{i,j,k}}{\Delta t} + O[(\Delta t)^2] \tag{3b}$$

We now apply the above ideas and notation to achieve a numerical approximation of Maxwell's curl equations in 3-D. For example, consider the E_z part of (2c). Substituting for the time and space derivatives, we have:

$$\frac{E_z|^{n+1}_{i,j,k} - E_z|^n_{i,j,k}}{\Delta t} = \frac{1}{\varepsilon_{i,j,k}}\left(\frac{H_y|^{n+1/2}_{i+1/2,j,k} - H_y|^{n+1/2}_{i-1/2,j,k}}{\Delta x} - \frac{H_x|^{n+1/2}_{i,j+1/2,k} - H_x|^{n+1/2}_{i,j-1/2,k}}{\Delta y} - \sigma_{i,j,k} E_z|^{n+1/2}_{i,j,k}\right) \tag{4}$$

Note that all field quantities on the right-hand side are evaluated at time step $n + ½$, including the electric field term, E_z, appearing due to the electric conductivity, σ. Since E_z at time step $n + ½$ is not assumed to be stored in the computer's memory (only the previous values of E_z at time step n are assumed to be in memory), we need some way to estimate this term. A very good way is to use what is called a "semi-implicit" approximation wherein E_z at time step $n + ½$ is assumed to be the arithmetic average of the stored value of E_z at time step n and the yet-to-be computed new value of E_z at time step $n + 1$:

$$E_z|^{n+1/2}_{i,j,k} = \frac{E_z|^{n+1}_{i,j,k} + E_z|^n_{i,j,k}}{2} \tag{5}$$

Substituting into (4), collecting terms, and dividing both sides by $1 + \sigma_{i,j,k}\Delta t/2\varepsilon_{i,j,k}$ yields:

$$E_z|^{n+1}_{i,j,k} = \left(\frac{1 - \dfrac{\sigma_{i,j,k}\Delta t}{2\varepsilon_{i,j,k}}}{1 + \dfrac{\sigma_{i,j,k}\Delta t}{2\varepsilon_{i,j,k}}}\right) E_z|^n_{i,j,k} + \left(\frac{\dfrac{\Delta t}{\varepsilon_{i,j,k}}}{1 + \dfrac{\sigma_{i,j,k}\Delta t}{2\varepsilon_{i,j,k}}}\right) \cdot \left(\frac{H_y|^{n+1/2}_{i+1/2,j,k} - H_y|^{n+1/2}_{i-1/2,j,k}}{\Delta x} - \frac{H_y|^{n+1/2}_{i,j+1/2,k} - H_y|^{n+1/2}_{i,j-1/2,k}}{\Delta y}\right) \tag{6}$$

This is an explicit expression for $E_z|^{n+1}_{i,j,k}$ that is numerically stable for positive values of σ. By analogy, we can derive finite-difference expressions based on Yee's algorithm for the other two electric field components. For the magnetic field, a similar procedure results in explicit updating equations having the following form:

$$H_x|^{n+1/2}_{i,j,k} = \left(\frac{1 - \dfrac{\rho'_{i,j,k}\Delta t}{2\mu_{i,j,k}}}{1 + \dfrac{\rho'_{i,j,k}\Delta t}{2\mu_{i,j,k}}}\right) H_x|^{n-1/2}_{i,j,k} + \left(\frac{\dfrac{\Delta t}{\mu_{i,j,k}}}{1 + \dfrac{\rho'_{i,j,k}\Delta t}{2\mu_{i,j,k}}}\right) \cdot \left(\frac{E_y|^n_{i,j,k+1/2} - E_y|^n_{i,j,k-1/2}}{\Delta z} - \frac{E_z|^n_{i,j+1/2,k} - E_z|^n_{i,j-1/2,k}}{\Delta y}\right) \tag{7}$$

With a Maxwell's equations system appearing as in (6) and (7), the new value of a field vector component at any space lattice point depends only on its previous value and the previous values of the components of the other field vector at adjacent points. Therefore, at any given time step, the computation of a field vector can proceed either one point at a time or, if p parallel processors are employed concurrently, p points at a time.

To implement the above system for a region having a continuous variation of material properties with spatial position, it is desirable to define and store updating coefficients for each field vector component. For the electric field algorithm coefficients at point (i, j, k):

$$C_a|_{i,j,k} = \frac{1 - \dfrac{\sigma_{i,j,k}\Delta t}{2\varepsilon_{i,j,k}}}{1 + \dfrac{\sigma_{i,j,k}\Delta t}{2\varepsilon_{i,j,k}}} \qquad C_{b_1}|_{i,j,k} = \frac{\dfrac{\Delta t}{\varepsilon_{i,j,k}\Delta_1}}{1 + \dfrac{\sigma_{i,j,k}\Delta t}{2\varepsilon_{i,j,k}}} \qquad C_{b_2}|_{i,j,k} = \frac{\dfrac{\Delta t}{\varepsilon_{i,j,k}\Delta_2}}{1 + \dfrac{\sigma_{i,j,k}\Delta t}{2\varepsilon_{i,j,k}}} \tag{8}$$

For the magnetic field algorithm coefficients at point (i, j, k):

$$D_a|_{i,j,k} = \frac{1 - \dfrac{\rho'_{i,j,k}\Delta t}{2\mu_{i,j,k}}}{1 + \dfrac{\rho'_{i,j,k}\Delta t}{2\mu_{i,j,k}}} \qquad D_{b_1}|_{i,j,k} = \frac{\dfrac{\Delta t}{\mu_{i,j,k}\Delta_1}}{1 + \dfrac{\rho'_{i,j,k}\Delta t}{2\mu_{i,j,k}}} \qquad D_{b_2}|_{i,j,k} = \frac{\dfrac{\Delta t}{\mu_{i,j,k}\Delta_2}}{1 + \dfrac{\rho'_{i,j,k}\Delta t}{2\mu_{i,j,k}}} \tag{9}$$

In (8) and (9), Δ_1 and Δ_2 denote the two possible lattice space increments used for each field component calculation. For a cubic lattice, $\Delta x = \Delta y = \Delta z = \Delta$, and thus, $\Delta_1 = \Delta_2 = \Delta$. For this case, $C_{b_1} = C_{b_2} = C_b$ and $D_{b_1} = D_{b_2} = D_b$, reducing the storage requirement to two coefficients per field vector component. For this case, the approximate total computer storage needed is $18N$, where N is the number of space cells in the

FDTD lattice. The finite-difference system of (6) and (7) can now be rewritten more simply as:

$$E_z\big|_{i,j,k}^{n+1} = C_a\big|_{i,j,k} E_z\big|_{i,j,k}^{n} + C_b\big|_{i,j,k} \cdot (H_y\big|_{i+1/2,j,k}^{n+1/2} - H_y\big|_{i-1/2,j,k}^{n+1/2} - H_x\big|_{i,j+1/2,k}^{n+1/2} + H_x\big|_{i,j-1/2,k}^{n+1/2}) \tag{10}$$

$$H_x\big|_{i,j,k}^{n+1/2} = D_a\big|_{i,j,k} H_x\big|_{i,j,k}^{n-1/2} + D_b\big|_{i,j,k} \cdot (E_y\big|_{i,j,k+1/2}^{n} - E_y\big|_{i,j,k-1/2}^{n} - E_z\big|_{i,j+1/2,k}^{n} + E_z\big|_{i,j-1/2,k}^{n}) \tag{11}$$

For a space region with a finite number of media, the computer storage requirement can be further reduced by defining an integer array, MEDIA(i, j, k), for each field component. This array stores an integer pointer, m, at each grid location of a field component, enabling the proper algorithm coefficients to be extracted. Now, the approximate total computer storage needed is reduced to $12N$. The reduction in computer storage comes at some cost, however, since additional computer instructions must be executed at each field point to obtain m from the MEDIA array and then extract the $C(m)$ or $D(m)$ algorithm coefficients. On a vectorizing computer like the Cray, these additional instructions can inhibit the flow of vectorized computations. However, this effect can be mitigated through careful programming.

Taking advantage of the integer nature of the MEDIA arrays, even further reduction in computer storage can be achieved by packing the MEDIA pointer integers in a separate bit-packed array. For example, a 64-bit word can be divided into 16 four-bit words, each of which can be used to specify up to $2^4 = 16$ distinct media. Thus, if there are no more than 16 possible media to be modeled in the space lattice, the overall computer storage for the MEDIA arrays can be reduced by this form of word-packing by a factor of 15/16 (94%). Efficient vectorized routines for word-packing/unpacking of this type are available on the Cray, and have appeared in at least one widely used FDTD code (Lawrence Livermore National Laboratory's "TSAR") to minimize storage requirements for the MEDIA pointers and yet not unduly sacrifice speed.

Many electromagnetic wave interaction problems involve non-permeable media and can be implemented on a uniform cubic FDTD space lattice. For such problems, the finite difference system exemplified by (10) and (11) can be further simplified by defining the proportional electric field vector, $\hat{\vec{E}} = (\Delta t/\mu_0\Delta)\vec{E}$, where μ_0 is the vacuum permeability and $\Delta = \Delta x = \Delta y = \Delta z$ is the space increment of the cubic lattice. Assuming that $\hat{E}_x$, $\hat{E}_y$, and $\hat{E}_z$ are stored in the computer memory, the simplified system is exemplified by:

$$H_x\big|_{i,j,k}^{n+1/2} = H_x\big|_{i,j,k}^{n-1/2} + \hat{E}_y\big|_{i,j,k+1/2}^{n} - \hat{E}_y\big|_{i,j,k-1/2}^{n} + \hat{E}_z\big|_{i,j-1/2,k}^{n} - \hat{E}_z\big|_{i,j+1/2,k}^{n} \tag{12}$$

$$m = \text{MEDIA}_{EX}\big|_{i,j,k} \tag{13}$$

$$\hat{E}_x\big|_{i,j,k}^{n+1} = C_a(m)\hat{E}_x\big|_{i,j,k}^{n} + \hat{C}_b(m) \cdot (H_z\big|_{i,j+1/2,k}^{n+1/2} - H_z\big|_{i,j-1/2}^{n+1/2} + H_y\big|_{i,j,k-1/2}^{n+1/2} - H_y\big|_{i,j,k+1/2}^{n+1/2})$$

where we have defined in (13) a scaled coefficient, $\hat{C}_b(m) = (\Delta t/\mu_0\Delta)C_b(m)$. This simplified finite-difference system eliminates the three multiplications previously needed to compute the magnetic field vector components, and requires storage of MEDIA integer arrays only for the electric field components.

IV. Numerical Stability

The numerical algorithms defined by the finite-difference systems reviewed in Section III require that the time increment, Δt, have a specific bound relative to the lattice space increments, Δx, Δy, and Δz. This bound is necessary to avoid numerical instability, an undesirable possibility with explicit differential equation solvers that can cause the computed results to spuriously increase without limit as time-marching continues.

For simplicity, we assume that the FDTD modeling space is filled with a homogeneous lossless material. Following the procedure outlined in [6], the full 3-D FDTD algorithm corresponding to (6) and (7) is divided into separate time and space eigenvalue problems. That is, we assume the existence of 3-D Fourier spatial mode spectra for the $\vec{E}$ and $\vec{H}$ component data in the grid. The complete spectrum of eigenvalues for these spatial modes due to the Yee numerical space differentiation process is determined, and compared to the stable spectrum of eigenvalues indicated by the Yee numerical time-differentiation process (i.e., having the magnitude of the growth factor at each time step less than or equal to 1). By requiring that the complete spectrum of spatial eigenvalues be contained within the stable range, we are assured that all possible numerical wave modes in the grid are stable, i.e., they cannot grow spuriously without limit. This places an upper bound on the size of the time step relative to the grid space discretizations:

$$\Delta t \le \frac{1}{v_{\max}\sqrt{\dfrac{1}{(\Delta x)^2} + \dfrac{1}{(\Delta y)^2} + \dfrac{1}{(\Delta z)^2}}} \tag{14}$$

where $v_{\max}$ is the maximum electromagnetic wave phase velocity within the media being modeled. Note that the corresponding numerical stability criterion set forth in Eqs. 7 and 8 of Yee's original paper [1] is incorrect. If the grid is cubic ($\Delta x = \Delta y = \Delta z = \Delta$), (14) simplifies to $\Delta t \le \Delta/v_{\max}\sqrt{3}$. On the other hand, if we let one of the space increments (e.g., Δz) become very large compared to the other two, the stability condition reduces to:

$$\Delta t \le \frac{1}{v_{\max}\sqrt{\dfrac{1}{(\Delta x)^2} + \dfrac{1}{(\Delta y)^2}}} = \frac{\Delta}{v_{\max}\sqrt{2}} \text{ for } \Delta x = \Delta y = \Delta \tag{15}$$

In fact, this is precisely the numerical stability condition for the 2-D TM or TE cases.

V. Numerical Dispersion

The numerical algorithms for Maxwell's curl equations defined by the finite-difference systems reviewed in Section III cause dispersion of the simulated wave modes in the computational lattice. That is, the phase velocity of numerical wave modes in the FDTD grid can differ from the vacuum speed of light, c, in fact varying with the modal wavelength, the direction of propagation in the grid, and the grid discretization. A useful way to view this phenomenon is that the FDTD algorithm effectively embeds the electromagnetic wave interaction structure of interest in a tenuous "numerical aether" having a permittivity very close to vacuum, but not quite. This "aether" causes propagating waves to accumulate delay or phase errors that can lead to non-physical results such as broadening and ringing of single-pulse waveforms, imprecise cancellation of multiple scattered waves, spurious anisotropy, and pseudo-refraction. Numerical dispersion is a factor in FDTD modeling that must be accounted for to understand the operation of the algorithm and its accuracy limits, especially for electrically large structures.

Basic Yee Algorithm

The procedure for the numerical dispersion analysis involves assuming for $\vec{E}$ and $\vec{H}$ the space-time dependence of a plane monochromatic wave (in a homogeneous medium) of arbitrary wavevector $\vec{k}$ and angular frequency ω, and then substituting into the full 3-D FDTD algorithm corresponding to (6) and (7). After algebraic manipulation, an equation is obtained that relates the *numerical* wavevector components, $\tilde{k}_x$, $\tilde{k}_y$, and $\tilde{k}_z$, the wave frequency, ω, and the grid space and time increments, Δx, Δy, Δz, and Δt[7]:

$$\left(\frac{1}{c\Delta t}\right)^2 \sin^2(\omega\Delta t/2) = \frac{1}{\Delta x^2}\sin^2(\tilde{k}_x\Delta x/2) \tag{16}$$

$$+ \frac{1}{\Delta y^2}\sin^2(\tilde{k}_y\Delta y/2) + \frac{1}{\Delta z^2}\sin^2(\tilde{k}_z\Delta z/2)$$

where $c = 1/\sqrt{\mu\varepsilon}$ is the speed of light in the homogeneous material being modeled. This equation, the numerical dispersion relation, can then be solved for the numerical phase velocity of the wave propagating in the FDTD mesh for the desired grid discretization, wavevector, and wave frequency. In this manner, we can determine deviations from the ideal case which is characterized by the following analytical dispersion relation:

$$\frac{\omega^2}{c^2} = k_x^2 = k_y^2 + k_z^2 \tag{17}$$

Although at first glance (16) bears little resemblance to the ideal case of (17), we can easily show that the two dispersion relations are identical in the limit as Δt, Δx, Δy, and Δz all go to zero. Qualitatively, this suggests that numerical dispersion can be reduced to any degree that is desired if we only use fine-enough FDTD gridding.

To quantitatively assess the dependence of numerical dispersion upon FDTD grid discretization, we shall take as an example the 2-D ($\Delta z = \infty$), assuming for simplicity square unit cells ($\Delta x = \Delta y = \Delta$) and wave propagation at an angle α with respect to the positive x-axis ($\tilde{k}_x = \tilde{k}\cos\alpha$; $\tilde{k}_y = \tilde{k}\sin\alpha$). Then, numerical dispersion relation (16) simplifies to

$$\left(\frac{\Delta}{c\Delta t}\right)^2 \sin^2\left(\frac{\omega\Delta t}{2}\right) = \sin^2\left(\frac{\tilde{k}\Delta\cos\alpha}{2}\right) + \sin^2\left(\frac{\tilde{k}\Delta\sin\alpha}{2}\right) \tag{18}$$

Equation (18) can be conveniently solved for the numerical wavevector $\tilde{k}$ at any wave propagation angle α by applying Newton's method. For this case, the numerical phase velocity $\tilde{v}_p$ is given by $\tilde{v}_p/c = 2\pi/\tilde{k}_{\text{final}}$, where $\tilde{k}_{\text{final}}$ is the final result of the Newton's method iterations.

Reference [7] provided graphs that illustrated the variation of the numerical phase velocity with wave propagation angle in a 2-D FDTD grid. Three different grid resolutions of the propagating wave were examined: coarse ($\Delta = \lambda_0/5$); coarse-normal ($\Delta = \lambda_0/10$); and fine-normal ($\Delta = \lambda_0/20$). For each resolution, the time-step relation $c\Delta t = \Delta/2$ was maintained. This relation is commonly used in 2-D and 3-D FDTD codes to satisfy the numerical stability criterion (Section IV) with ample safety margin. It was shown that the numerical phase velocity is always less than c, the free-space speed of light, reaching a maximum at 45° (oblique incidence) and a minimum at 0° and 90° (incidence along either Cartesian grid axis) for all grid resolutions. This represents a numerical phase velocity anisotropy that is inherent in the Yee algorithm. Further, the phase velocity error relative to c (and thus, the velocity anisotropy) was found to diminish by approximately a 4:1 factor each time that the grid cell size was halved. For example, the worst-case phase velocity error at $\Delta = \lambda_0/10$ is -1.3%, but is only -0.3% at $\Delta = \lambda_0/20$. For the latter case, a sinusoidal numerical wave analog traveling over a $10\lambda_0$ distance (200 grid cells) would develop a lagging phase error of about 11°. This error is cumulative, i.e., it increases linearly with wave propagation distance. It represents a limitation of the classic Yee algorithm that is troublesome for electrically large or high-Q microwave structures. Such structures may require a fourth-order accurate FDTD algorithm to reduce phase errors to an acceptable degree, as discussed below.

It was also found that the numerical phase velocity at each angle of incidence in the grid diminishes as the propagating wave is more coarsely resolved, eventually reaching a sharp threshold where the numerical phase velocity goes to zero and the wave can no longer propagate in the FDTD grid. This represents a numerical low pass filtering effect that is inherent in the Yee algorithm, wherein the wavelength of propagating numerical modes has a lower bound of 2 to 3 space cells, depending upon the propagation direction.

As a result, FDTD modeling of pulses having large bandwidths results in progressive pulse distortion as the higher spatial frequency components propagate more slowly than the lower spatial frequency components, and the very high spatial frequency components with wavelengths less than 2 to 3 cells are rejected. This numerical dispersion causes pulse broadening and leaves a residue of high-frequency ringing on the trailing edge due to the relatively slowly propagating high-frequency components. The pulse distortion can be bounded by obtaining the Fourier spatial frequency spectrum of the desired pulse and selecting a grid cell size so that the principal spectral components are resolved with at least 10 cells per wavelength. This would limit the spread of numerical phase velocities of the principal spectral components to less than 1%, regardless of the wave propagation angle in the grid.

In addition to numerical phase velocity errors, anisotropy, and pulse distortion effects, numerical dispersion can lead to spurious refraction of propagating numerical modes if the grid cell size is a function of position in the grid. Such variable gridding also varies the local grid resolution of the propagating numerical wave modes, and thereby perturbs the modal phase velocity distribution. This leads to weak non-physical reflection and refraction of grid modes at interfaces between grid regions having different cell sizes, even if these interfaces are located in free-space regions (just as physical waves undergo reflection and refraction at interfaces of dielectric media having different indices of refraction). The degree of these non-physical effects is dependent upon the magnitude and abruptness of the change of the grid wave-mode velocity distribution, and can be estimated using conventional theory for physical wave refraction at dielectric interfaces.

We have noted that, in the limit of infinitesimal Δt and Δ, dispersion relation (16) reduces to the ideal dispersion relation (17). This also occurs if Δt, Δ, and the direction of wave propagation are suitably chosen. For example, in a 3-D cubic lattice, reduction to the ideal dispersion case can be demonstrated for a numerical plane wave propagating along a lattice diagonal ($\tilde{k}_x = \tilde{k}_y = \tilde{k}_z = \tilde{k}/\sqrt{3}$) if $\Delta t = \Delta/c\sqrt{3}$ (exactly the limit set by numerical stability). Similarly, in a 2-D square grid, the ideal dispersion case results for a numerical plane wave propagating along a grid diagonal ($\tilde{k}_x = \tilde{k}_y = \tilde{k}/\sqrt{2}$) if $\Delta t = \Delta/c\sqrt{2}$ (again the limit set by numerical stability). Finally, in one dimension, the ideal dispersion case ("magic time step") is obtained for $\Delta t = \Delta/c$ (again the limit set by numerical stability). These reductions to ideal dispersion have little practical value for 2-D and 3-D FDTD models, occurring uniquely for diagonal wave propagation and not some more general set of wave directions. However, the reduction to the ideal dispersion case in one dimension is very interesting since it implies that the Yee algorithm (based upon finite-difference *approximations*) can *exactly* solve the continuous, one-dimensional wave equation.

Dispersion-Optimized Yee Algorithm with Fourth-Order Accurate Spatial Central Differences

It is possible to substantially reduce the magnitude of numerical phase dispersion artifacts for the Yee algorithm by incorporating a fourth-order accurate central-difference scheme for the space derivatives of the curl operator. For purposes of illustration, consider the usual Yee time-stepping relation for the E_z component within a zero-loss, uniform Cartesian mesh comprising cubic unit cells. All remaining field components follow by analogy. Blindly applying the classical fourth-order accurate space-derivative analog to $\vec{H}$ on the Yee field template results in:

$$E_z\Big|_{i,j,k}^{n+1} = E_z\Big|_{i,j,k}^{n} + \frac{\Delta t}{24\varepsilon_{i,j,k}\Delta} \tag{19}$$

$$\begin{bmatrix} (-H_y|_{i+3/2,j,k}^{n+1/2} + 27H_y|_{i+1/2,j,k}^{n+1/2} - 27H_y|_{i-1/2,j,k}^{n+1/2} + H_y|_{i-3/2,j,k}^{n+1/2}) \\ -\ (-H_x|_{i,j+3/2,k}^{n+1/2} + 27H_x|_{i,j+1/2,k}^{n+1/2} - 27H_x|_{i,j-1/2,k}^{n+1/2} + H_x|_{i,j-3/2,k}^{n+1/2}) \end{bmatrix}$$

However, (19) can be optimized by slightly "tweaking" the coefficients in the space-derivative analog to zero-out numerical phase velocity errors in particular desired wave propagation directions in the grid [8]. First write (19) in a generalized form that separates the contribution of the original Yee second-order space-differencing and what amounts to be a correction term due to the fourth-order space-differencing:

$$E_z\Big|_{i,j,k}^{n+1} = E_z\Big|_{i,j,k,}^{n} + \frac{\Delta t}{\varepsilon_{i,j,k}\Delta} \tag{20}$$

$$\begin{Bmatrix} [A(H_y|_{i+1/2,j,k}^{n+1/2} - H_y|_{i-1/2,j,k}^{n+1/2}) + B(H_y|_{i+3/2,j,k}^{n+1/2} - H_y|_{i-3/2,j,k}^{n+1/2})] \\ -\ [A(H_x|_{i,j+1/2,k}^{n+1/2} - H_x|_{i,j-1/2,k}^{n+1/2}) + B(H_x|_{i,j+3/2,k}^{n+1/2} - H_x|_{i,j-3/2,k}^{n+1/2})] \end{Bmatrix}$$

We again assume for $\vec{E}$ and $\vec{H}$ the space-time dependence of a plane monochromatic wave of arbitrary wavevector $\vec{k}$ and angular frequency ω, and substitute into the full 3-D FDTD algorithm corresponding to (20). For waves propagating in the x-y plane, it can be shown that the numerical phase velocity error is eliminated at the propagation angles 0°, 45°, 90°, 135°, 180°, 225°, 270°, and 315° by having A and B satisfy the following simultaneous equations:

$$A\sin(\pi/R) + B\sin(3\pi/R) = 2\sin(\pi/2R) \tag{21a}$$

$$A\sin(\pi/R\sqrt{2}) + B\sin(3\pi/R\sqrt{2}) = \sqrt{2}\sin(\pi/2R) \tag{21b}$$

where $R = \lambda_0/\Delta$ and $c\Delta t = \Delta/2$.

Reference [8] considered an example of this optimization for a coarse grid of $R = 5$. Here, (21) yields $A = 1.13880$ and $B = -0.05398$. This compares to $A = 1.125$ and $B = -0.041\overline{6}$ and indicated by application of the standard fourth-order space differencing of (19). It was found that the dispersion-optimized fourth-order algorithm on the coarse grid has numerical phase velocity errors only about 1/100th those of the second-order algorithm on a fine grid of $R = 20$. Further, the optimized algorithm zeroes out the phase velocity error at twice as many propagation angles. As a result, optimized fourth-order space differencing permits FDTD modeling of microwave structures that are at least 100 times the electrical size of those permitted by second-order differencing while maintaining an absolute criterion for the phase error accumulation of propagating numerical waves within the structure.

There is one principal disadvantage of using fourth-order space differencing in FDTD. Although straightforward to set up in homogeneous material regions, the larger stencil on the Yee mesh needed to calculate fourth-order differences is troublesome when dealing with material discontinuities due to the presence of conductors or dielectric inhomogeneities. In fact, special boundary conditions must be derived and programmed for such discontinuities. This substantially complicates Yee's simple original second-order formulation and associated computer programming, and is the primary reason why most working FDTD codes today do *not* take advantage of fourth-order spatial differencing.

VI. Incident Wave Source Conditions

Pulsed Electric Field Source

The simplest way to excite an incident wave in an FDTD model of a microwave waveguide or stripline is to specify a pulsed electric field distribution in a transverse cross section of the microwave structure designated as the source plane. In certain cases, the transverse distribution of the desired propagating mode is known analytically, allowing the pulsed numerical wave that is launched to immediately represent the desired mode without the presence of undesired propagating modes and below cut-off reactive fields. In other cases where the transverse distribution of the desired mode can only be approximated, the numerical field source specifies only a subset of the complete transverse distribution. Here, in addition to the desired mode, there is the possibility of generating undesired numerical waves that represent multi-moding and reactive fields. A substantial buffer length of waveguiding structure may be required to permit the undesired numerical waves to decay. An example of an approximate waveguide modal source is the pulsing of a collinear array of electric field components in the FDTD mesh to simulate a probe extending from the waveguide wall to dead-end at a point halfway to the opposing wall. An example of an approximate stripline source is the pulsing of a similar collinear array of electric fields extending from a ground plane to the metal trace of the stripline.

The principal concern with this type of source condition is that its effective impedance is zero. As a result, numerical waves reflected from a properly modeled discontinuity further down the waveguiding system will nonphysically re-reflect from the source condition if it is still operating. This non-physical reflection is total or partial depending upon whether the source condition completely or partially specifies the transverse source plane. A way to eliminate this problem is to make sure that, before the reflecting numerical waves reach the source plane, the pulsed source waveform has been smoothly transitioned to zero and a program flag has been set so that execution diverts around the source statements. In effect, the source plane has been removed and replaced by standard free-space Yee cells by the time that the reflecting numerical waves reach its position. The reflecting numerical waves can then freely propagate through what was the source plane to reach an absorbing boundary condition (ABC) representing the extension of the waveguiding system to infinity. This procedure requires elongation of the simulated waveguiding system between the source plane and the first reflective discontinuity to permit the complete evolution and return to zero of the pulsed source waveform before arrival of the initial reflection at the source plane.

Total-Field/Reflected-Field Formulation

The total-field/reflected-field formulation (the application of the total-field/scattered-field method of [9] to guided-wave systems) is based on the linearity of Maxwell's equations. The finite-difference approximations of the Yee algorithm can be applied with equal validity to the incident-field vector components, the reflected-field vector components, and the total-field vector components (the sum of incident plus reflected). FDTD codes can utilize this property to zone the numerical space lattice modeling a microwave structure into two distinct regions separated by a non-physical virtual surface which serves to connect the fields in each region.

Region 1, the primary zone of the microwave structure, is denoted as the total-field region since it is assumed that the Yee algorithm operates on total-field components. The interacting structures of interest are embedded within this region. Region 2, the secondary zone, is denoted as the reflected-field region since it is assumed that the Yee algorithm operates only on reflected-field components. This implies that there is no incident wave in Region 2. A connecting condition is provided at the interface of Regions 1 and 2 to assure consistency of the numerical space derivative operations across the interface and simultaneously generate an arbitrary incident wave in Region 1 having a user-specified transverse distribution and time waveform. This connecting condition almost completely confines the numerical incident wave to Region 1. Yet, it is transparent to outgoing numerical reflected wave modes which are free to enter Region 2 and subsequently disappear from the grid due to the action of the ABC.

Unlike the simple pulsed electric field source discussed above, there is no need to transition the total-field/reflected-field source to zero and remove it from the grid, and there is no need to elongate the simulated waveguide system to causally isolate the source from adjacent waveguide discontinuities that generate reflections. Here, the source can be continuously operated in close proximity to adjacent discontinuities. The only disadvantage is a slight increase in the complexity of programming the source, which is now discussed.

Consider the problem of consistency at the interface of Regions 1 and 2. When a required Yee spatial difference is taken across the interface plane, the field to be used in the difference expression on the Region-1 side of the interface is assumed to be stored as a total field; whereas the field to be used on the Region-2 side of the interface is assumed to be stored as a reflected field. It would be inconsistent to perform an arithmetic difference between a reflected-field value and a total-field value. This problem can be solved by using the known values of the incident field vectors, $\vec{E}_{\text{inc}}$ and $\vec{H}_{\text{inc}}$. To illustrate this solution, assume that we wish to generate a $+x$-directed incident waveguide mode having an E_z component. Let the interface between Regions 1 and 2 be the y-z plane at grid coordinate i_s containing total-field E_z components. Applying the Yee algorithm (initially without worrying about consistency), the time-stepping relation for a typical E_z in this plane might be written as:

$$E_{z,\text{tot}}\Big|_{i_s,j,k}^{n+1} = \underbrace{E_{z,\text{tot}}\Big|_{i_s,j,k}^{n}}_{\substack{\text{assumed stored in}\\ \text{computer memory}}} + \frac{\Delta t}{\varepsilon_0 \Delta} \underbrace{\begin{bmatrix} \left(H_{y,\text{tot}}\Big|_{i_s+1/2,j,k}^{n+1/2} - H_{y,\text{refl}}\Big|_{i_s-1/2,j,k}^{n+1/2}\right) \\ -\left(H_{x,\text{tot}}\Big|_{i_s,j+1/2,k}^{n+1/2} - H_{x,\text{tot}}\Big|_{i_s,j-1/2,k}^{n+1/2}\right) \end{bmatrix}}_{\text{assumed stored in computer memory}} \tag{22}$$

Note that the H_y component at $i_s + ½$ (in Region 1 just to the right of E_z) is assumed to be stored in computer memory as a total-field quantity, while the H_y component at $i_s - ½$ (in Region 2 just to the left of E_z) is assumed to be stored in computer memory as a reflected-field quantity. As it stands, (22) is incorrect since we are subtracting unlike magnetic field quantities in the bracket. For consistency here, we must subtract total H_y from total H_y to advance total E_z in time. Conveniently, (22) can be made correct simply by adding another term which is a function of the assumed-known incident wave:

$$E_{z,\mathrm{tot}}\Big|_{i_s,j,k}^{n+1} = \underbrace{E_{z,\mathrm{tot}}\Big|_{i_s,j,k}^{n}}_{\substack{\text{assumed stored in}\\ \text{computer memory}}} + \underbrace{\frac{\Delta t}{\varepsilon_0\Delta}\begin{bmatrix}\left(H_{y,\mathrm{tot}}\Big|_{i_s+1/2,j,k}^{n+1/2} - H_{y,\mathrm{refl}}\Big|_{i_s-1/2,j,k}^{n+1/2}\right)\\ -\left(H_{x,\mathrm{tot}}\Big|_{i_s,j+1/2,k}^{n+1/2} - H_{x,\mathrm{tot}}\Big|_{i_s,j-1/2,k}^{n+1/2}\right)\end{bmatrix}}_{\text{assumed stored in computer memory}} - \frac{\Delta t}{\varepsilon_0\Delta}\underbrace{H_{y,\mathrm{inc}}\Big|_{i_s-1/2,j,k}^{n+1/2}}_{\text{assumed known}} \tag{23}$$

since $H_{y,\mathrm{tot}}\big|_{i_s,j,k}^{n+1/2} = H_{y,\mathrm{refl}}\big|_{i_s-1/2,j,k}^{n+1/2} + H_{y,\mathrm{inc}}\big|_{i_s-1/2,j,k}^{n+1/2}$, which was desired. This can be done as a correction (implemented only at grid plane i_s) after uniformly applying (22) in a single Fortran DO loop to update the reflected-field and total-field E_z everywhere in the grid. The correction strategy avoids needlessly breaking up the DO loop for updating E_z, and permits maximum speed on a vectorizing machine like the Cray.

We are not quite finished with the Region 1/Region 2 interface. Consider applying the Yee algorithm to update the H_y component at grid point $i_s - ½$ (in Region 2 just to the left of the interface):

$$H_{y,\mathrm{refl}}\Big|_{i_s-1/2,j,k}^{n+1/2} = \underbrace{H_{y,\mathrm{refl}}\Big|_{i_s-1/2,j,k}^{n-1/2}}_{\substack{\text{assumed stored in}\\ \text{computer memory}}} + \underbrace{\frac{\Delta t}{\varepsilon_0\Delta}\begin{bmatrix}\left(E_{z,\mathrm{tot}}\Big|_{i_s,j,k}^{n} - E_{z,\mathrm{refl}}\Big|_{i_s-1,j,k}^{n}\right)\\ -\left(E_{x,\mathrm{refl}}\Big|_{i_s-1/2,j,k+1/2}^{n} - E_{x,\mathrm{refl}}\Big|_{i_s-1/2,j,k-1/2}^{n}\right)\end{bmatrix}}_{\text{assumed stored in computer memory}} \tag{24}$$

Note that the E_z component at i_s (in Region 1 just to the right of the H_y component of interest) is assumed to be stored in computer memory as a total-field quantity, while the E_z component at $i_s - 1$ (in Region 2 just to the left of the H_y component of interest) is assumed to be stored in memory as a reflected-field quantity. As it stands, (24) is incorrect since we are subtracting unlike electric field quantities in the bracket. For consistency here, we must subtract reflected E_z from reflected E_z to advance reflected H_y in time. As before, (24) can be made correct simply by adding another term which is a function of the assumed-known incident wave:

$$H_{y,\mathrm{refl}}\Big|_{i_s-1/2,j,k}^{n+1/2} = \underbrace{H_{y,\mathrm{refl}}\Big|_{i_s-1/2,j,k}^{n-1/2}}_{\substack{\text{assumed stored in}\\ \text{computer memory}}} + \underbrace{\frac{\Delta t}{\varepsilon_0\Delta}\begin{bmatrix}\left(E_{z,\mathrm{tot}}\Big|_{i_s,j,k}^{n} - E_{z,\mathrm{refl}}\Big|_{i_s-1,j,k}^{n}\right)\\ -\left(E_{x,\mathrm{refl}}\Big|_{i_s-1/2,j,k+1/2}^{n} - E_{x,\mathrm{refl}}\Big|_{i_s-1/2,j,k-1/2}^{n}\right)\end{bmatrix}}_{\text{assumed stored in computer memory}} - \frac{\Delta t}{\mu_0\Delta}\underbrace{E_{z,\mathrm{inc}}\Big|_{i_s,j,k}^{n}}_{\text{assumed known}} \tag{25}$$

since $E_{z,\mathrm{tot}}\big|_{i_s,j,k}^{n} - E_{z,\mathrm{inc}}\big|_{i_s,j,k}^{n} = E_{z,\mathrm{refl}}\big|_{i_s,j,k}^{n}$, which was desired. As before, this can be done as a correction (implemented only at grid plane $i_s - ½$) after uniformly applying (24) in a single Fortran DO loop to update the scattered-field and total-field H_y's everywhere in the grid.

The combination of (23) and (25) fulfills the consistency requirement at the Region 1/Region 2 interface and properly generates the E_z and H_y components of the desired incident waveguide mode propagating in the $+x$-direction in Region 1. As needed, analogous consistency conditions at grid planes i_s and $i_s - ½$ are applied to components E_y and H_z, respectively, to complete the generation of the desired incident waveguide mode.

Resistive Source and Load Conditions

When using FDTD to model microstrips or waveguides, it is very useful to be able to simulate the commonly used excitation by a metal probe connected to a voltage source having a finite source resistance (usually matched to the characteristic impedance of the line). This is a much more realistic situation than the zero-impedance electric field source condition discussed in Section VI.

A highly effective resistive voltage source can be obtained using Maxwell's curl $\vec{H}$ equation, which is suitable for time-stepping the electric field. The basis of this formulation, reported in [10] and [11], showed that lumped circuit elements can be accounted for in Maxwell's equations by adding a lumped electric current density term, J_L, to the conduction and displacement current densities. Eq. (1b) now becomes:

$$\frac{\partial \vec{E}}{\partial t} = \frac{1}{\varepsilon}\nabla \times \vec{H} - \frac{\sigma}{\varepsilon}\vec{E} - \frac{1}{\varepsilon}\vec{J}_L \tag{26}$$

For example, let the lumped element be z-oriented in the grid in free space at $E_z|_{i,j,k}$. Here, the assumed positive direction of the lumped element current, I_L, is $+z$, so that the local current density, J_L, is given by $J_L = I_L/\Delta x\Delta y$. For general circuit components including resistors, capacitors, inductors, and non-linear devices, I_L is respectively a scalar multiple, time-derivative, time-integral, or non-linear function of the electric potential, $V = E_z|_{i,j,k}\Delta z$, developed across the lumped element. Finally, we shall write the Yee finite-difference analog to the curl of $\vec{H}$ as observed at $E_z|_{i,j,k}$ in shorthand notation as:

$$\nabla \times H_{i,j,k}^{n+1/2} = \frac{H_y\Big|_{i+1/2,j,k}^{n+1/2} - H_y\Big|_{i-1/2,j,k}^{n+1/2}}{\Delta x} + \frac{H_x\Big|_{i,j-1/2,k}^{n+1/2} - H_x\Big|_{i,j+1/2,k}^{n+1/2}}{\Delta y} \tag{27}$$

Then, the following time-stepping relation specifies the presence of the lumped circuit element in the FDTD grid, where the lumped current is evaluated at the mid time step, $n + \frac{1}{2}$ [14]:

$$E_z\big|_{i,j,k}^{n+1} = E_z\big|_{i,j,k}^{n} + \frac{\Delta t}{\varepsilon_0}\nabla \times H_{i,j,k}^{n+1/2} - \frac{\Delta t}{\varepsilon_0 \Delta x \Delta y} I_L^{n+1/2} \tag{28}$$

Since the lumped current is a function of the electric field at the circuit element, this requires averaging the values of $E_z\big|_{i,j,k}^{n+1}$ and $E_z\big|_{i,j,k}^{n}$, an operation that yields a numerically stable semi-implicit time-stepping algorithm. Generalization of all of this to x and y orientations of a lumped element is straightforward by permuting the coordinate subscripts of the field quantities.

Now, assuming a z-directed resistor located in free-space at the field component $E_z\big|_{i,j,k}$, the voltage-current characteristic that describes its behavior in a semi-implicit manner appropriate for stable operation of the FDTD field solver is:

$$I_z\big|_{i,j,k}^{n+1/2} = \frac{\Delta z}{2R}\left(E_z\big|_{i,j,k}^{n+1} + E_z\big|_{i,j,k}^{n}\right), \qquad J_L = \frac{I_z\big|_{i,j,k}^{n+1/2}}{\Delta x \Delta y} \tag{29a}$$

where R is the value of the resistance and Δx, Δy, and Δz are the grid increments in the x, y, and z directions. The corresponding time-stepping relation for $E_z\big|_{i,j,k}$ is [14]:

$$E_z\big|_{i,j,k}^{n+1} = \left(\frac{1 - \dfrac{\Delta t \Delta z}{2R\varepsilon_0 \Delta x \Delta y}}{1 + \dfrac{\Delta t \Delta z}{2R\varepsilon_0 \Delta x \Delta y}}\right) E_z\big|_{i,j,k}^{n} + \left(\frac{\dfrac{\Delta t}{\varepsilon_0}}{1 + \dfrac{\Delta t \Delta z}{2R\varepsilon_0 \Delta x \Delta y}}\right) \nabla \times H_{i,j,k}^{n+1/2} \tag{29b}$$

It is important to note that implementing (29b) at two adjacent electric field components between a microstrip signal line and a ground plane simulates the presence of two parallel resistors terminating the line. The effective load resistance is thereby halved.

Again assuming a z-directed lumped element located in free-space at $E_z\big|_{i,j,k}$, the voltage-current characteristic that models a resistive voltage source in a semi-implicit manner is:

$$I_z\big|_{i,j,k}^{n+1/2} = \frac{\Delta z}{2R_S}\left(E_z\big|_{i,j,k}^{n+1} + E_z\big|_{i,j,k}^{n}\right) + \frac{V_S^{n+1/2}}{R_S}, \qquad J_L = \frac{I_z\big|_{i,j,k}^{n+1/2}}{\Delta x \Delta y} \tag{30a}$$

where $V_S^{n+1/2}$ is the source voltage and R_s is the internal source resistance. The corresponding time-stepping relation for $E_z\big|_{i,j,k}$ is [14]:

$$E_z\big|_{i,j,k}^{n+1} = \left(\frac{1 - \dfrac{\Delta t \Delta z}{2R_S\varepsilon_0 \Delta x \Delta y}}{1 + \dfrac{\Delta t \Delta z}{2R_S\varepsilon_0 \Delta x \Delta y}}\right) E_z\big|_{i,j,k}^{n} + \left(\frac{\dfrac{\Delta t}{\varepsilon_0}}{1 + \dfrac{\Delta t \Delta z}{2R_S\varepsilon_0 \Delta x \Delta y}}\right) \nabla \times H_{i,j,k}^{n+1/2} + \left(\frac{\dfrac{\Delta t}{R_S\varepsilon_0 \Delta x \Delta y}}{1 + \dfrac{\Delta t \Delta z}{2R_S\varepsilon_0 \Delta x \Delta y}}\right) V_S^{n+1/2} \tag{30b}$$

Equation (30b) can be applied to any E_z component where a pointwise resistive source is desired. For a stripline, this would be an E_z linking the ground plane to a signal trace. For a waveguide, this would be an E_z linking the waveguide wall to a metal probe extending halfway across the narrow dimension of the guide and then dead-ending in free space. None of the restrictions noted in Section VI would apply for the resistive source. The resistive source is the most realistic means to use FDTD to simulate the physical connection of common RF/microwave power sources to microwave structures. A cautionary note is that any cutoff or higher-order modes generated by the actual physical connection will also appear in the FDTD model.

VII. Absorbing Boundary Counditions

Background

FDTD modeling of guided-wave microwave and optical structures often requires the simulation of such structures extending beyond the outer grid boundaries. An outstanding problem here is the implementation of an accurate absorbing boundary condition (ABC) that permits such a simulation. The key problem is that the propagation in a waveguide can be multimodal and dispersive, and the ABC must absorb energy having widely varying spatial distributions and propagation velocities.

Typical FDTD ABCs are categorized into two classes: (1) analytical and (2) numerical analogs of material absorbers. Analytical ABCs perform best for narrowband energy propagation where the group velocity is well defined. Much literature has been developed for free-space analytical ABCs of first-order, second-order, and third-order accuracy including one-way wave equations [12,13], error-canceling superabsorbers [14], outgoing wave annihilators [15,16], and extrapolation in space and time [17]. Less attention has been paid to numerical analogs of material absorber layers [18,19]. However, as discussed below, analogs of non-physical material absorbers show promise for constructing ABCs that can deal more effectively with wideband dispersive propagation.

When applied to terminate dispersive guided wave structures, the analytical ABCs have been specialized to account for group velocity variations of waveguide modes with frequency. For example, [20] reported the use of a composite ABC operator consisting of N-multiplicative first-order linear differential terms. This operator exactly annihilates a propagating pulsed mode at all points along a transverse plane of the waveguide at a set of N-desired frequencies, $\{f_i\}$, within the pulse spectrum, corresponding to a set of N-group velocities, $\{v_g(f_i)\}$. However, the annihilation is incomplete (i.e., reflection is present) for spectral components at frequencies falling between $\{f_i\}$. The number, N, of frequencies at which the annihilation is exact, and thus the bandwidth of the resulting ABC, can be increased by adding multiplicative terms to the operator.

A variation of this approach was reported in [21] which approximated the exact v_g vs. f variation with either a linear or a second-order Padé expression. Reference [22] derived a rigorous analytical termination by Laplace transformation of the exact v_g versus f variation. However, this algorithm is global in time, requiring the evaluation of a convolution integral for each mode. Overall, the methods of [20,21,22] require knowledge of the propagating mode or its velocity distribution. None provide the ability to simulate a matched absorber over the entire propagation regime of a waveguide with little or no *a priori* knowledge.

The Perfectly Matched Layer Absorber

A new approach employing a numerical analog of a non-physical material absorber has been shown to yield a robust multi-modal dispersive ABC for FDTD simulation of guided-wave systems above cutoff. This approach is based on the perfectly matched layer (PML) concept [23]. The new approach has the advantages of being local in time and space, general, and extremely accurate over a wide range of group velocities. It requires no knowledge of the modal distribution or dispersive nature of the propagating field.

The PML ABC provides a means to terminate FDTD grids essentially without reflection using a non-physical lossy medium adjacent to the outer grid boundary. In a PML medium matched to free space, waves propagate with the vacuum speed of light and the wave impedance is that of vacuum independent of frequency and angle of incidence. Here, the electromagnetic field components are split into subcomponents, and the field update equations are modified to account for the decomposition. This decomposition introduces an additional degree of freedom in specifying loss and impedance matching. Reference [23] reported that PML achieves local reflections of outgoing cylindrical waves in a 2-D grid as low as 1/3,000th those of standard second-order analytical ABCs. This was confirmed in [24], which also reported the global error energy in the grid declining by 10^{12}. Reference [24] also extended PML to 3-D unbounded grids and obtained data indicating local and global error reductions comparable to those in 2-D.

Consider the use of PML to terminate general 2-D waveguiding structures for modes having the field components E_x, E_y, and H_z. Following the notation of [23], there are four rather than the usual three coupled field equations because H_z is split into two subcomponents: H_{zx} and H_{zy}.

$$\varepsilon_d \frac{\partial E_x}{\partial t} + \sigma_y E_x = \frac{\partial (H_{zx} + H_{zy})}{\partial y}, \qquad \varepsilon_d \frac{\partial E_y}{\partial t} + \sigma_x E_y = -\frac{\partial (H_{zx} + H_{zy})}{\partial x} \tag{31}$$

$$\mu_d \frac{\partial H_{zx}}{\partial t} + \sigma_x^* H_{zx} = -\frac{\partial E_y}{\partial x}, \qquad \mu_d \frac{\partial H_{zy}}{\partial t} + \sigma_y^* H_{zy} = \frac{\partial E_x}{\partial y} \tag{32}$$

Electric loss, σ, and magnetic loss, σ^*, can be assigned to the electric and magnetic field components as indicated to cause exponential decay of propagating fields in the PML region. If

$$\frac{\sigma_x}{\varepsilon_d} = \frac{\sigma_x^*}{\mu_d}, \qquad \frac{\sigma_y}{\varepsilon_d} = \frac{\sigma_y^*}{\mu_d} \tag{33}$$

is enforced, where ε_d and μ_d are the PML permittivity and permeability, this decay can take place in a frequency-independent manner without impacting the wave impedance.

Now, consider as a specific subset of (31) and (32) the case of a 2-D dielectric-filled (ε_d, μ_d) waveguide structure propagating one or more TM modes in the $+x$ direction. At $x_{\max}$, the waveguide is loaded with PML that has σ_x and σ_x^* matched according to (33) along with $\sigma_y = \sigma_y^* = 0$ to permit reflectionless transmission across the vacuum-PML interface. Assume as in [26] that the loss in the PML region increases quadratically with depth, d, from zero at $d = 0$, the interface between the waveguide dielectric and the PML, to $\sigma_{\max}$ at $d = \delta$, the conducting wall backing the PML. Then, $\sigma_{\max}$ can be chosen to bound the apparent reflection coefficient

$$R = e^{-\frac{2\sigma_{max}\delta}{3\varepsilon_d c}} \tag{34}$$

to some desired low level, say 10^{-4}. Following [23], exponential time-stepping is used [8]. The following is a sample exponentially differenced time advance:

$$E_y\big|_{i,j+1/2}^{n+1} = e^{-\sigma\Delta t/\varepsilon_d} E_y\big|_{i,j+1/2}^{n} + \frac{1}{\sigma\Delta x}\left(e^{-\sigma\Delta t/\varepsilon_d} - 1\right)\left(H_z\big|_{i+1/2,j+1/2}^{n+1/2} - H_z\big|_{i-1/2,j+1/2}^{n+1/2}\right) \tag{35}$$

Sample Results

Reference [8] reported the use of PML to obtain a robust dispersive ABC for FDTD simulation of guided-wave systems. An air-filled 2-D PEC parallel-plate waveguide with a wall separation of 40mm ($f_{\text{cutoff}} = 3.75$GHz) was considered. The excitation consisted of an 83.3-ps Gaussian pulse modulating a 7.5-GHz carrier, which launched a $+x$-directed TM_1 mode towards the PML termination. The incident pulse contained significant energy below cutoff, and the group velocity of the pulse spectral components ranged from zero at cutoff to $\approx 0.98c$ well above cutoff. The PML was selected to be either 8 or 32 grid cells thick with optimized reflectivity and loss grading. The reflected wave was observed at the midpoint of the air-PML interface, and was allowed to evolve over many thousands of time steps to properly model the slowly propagating spectral components near f_{cutoff}. For the 8-cell PML, reflections between -60 and -100dB were noted in the frequency range 4 to 20GHz. For the 32-cell PML, the reflections dropped below -100dB in the range 6 to 18GHz. This example demonstrated the ability of the PML ABC to absorb very wideband propagating energy having a strongly dispersive group velocity.

Reference [25] reported the PML ABC used to terminate the FDTD model of a 2-D asymmetric dielectric slab optical waveguide. This consisted of a 1.5-μm film of permittivity $\varepsilon_r = 10.63$ sandwiched between an infinite substrate of $\varepsilon_r = 9.61$ and an infinite region of air. A 17-fs Gaussian pulse modulating a 200-THz carrier at the left edge of the three-layer system was used to launch three distinct $+x$-directed modes having widely varying frequency-dependent propagation factors. The system was terminated by extending each dielectric layer into its matching 16-cell thick PML region. Results for the composite reflection coefficient at the PML interface (representing the total retrodirected energy in all three regions) showed that the PML ABC exhibited reflections below -80dB across the entire spectrum of the incident field. This demonstrated the absorptive capability of the PML ABC for dispersive multi-modal propagation.

In summary, the PML ABC permits highly accurate ultrawideband terminations of PEC and dielectric waveguides in FDTD grids. It is local in time and space and can provide broadband reflection coefficients below -80dB for dispersive and multimodal energy. This yields a highly effective combination of broadband absorption and robustness.

VIII. Nonorthogonal Unstructured Grids[1]

Unstructured grids are composed of arrays of general fitted polyhedral cells. While every cell shares a common face and edge, the vertices of the grid can no longer be listed in a regular sense. These grids can conform to highly complex shapes while maintaining good cell aspect ratios and global uniformity. A general method to formulate an explicit time-marching solution of Maxwell's equations using unstructured grids was introduced in [25]. (See also [5].) This was founded upon deriving an explicit operator based on Maxwell's equations in integral form and evaluating the discrete surface and line integrals. A key need was to introduce a scheme which projects the face normal components of the fields onto the dual grid edges. It was shown that insufficient accuracy of the projection method could cause the numerical solution to become unstable after a very large number of time steps. This problem was overcome, and a numerically stable, second-order accurate projection of the normal fields onto the grid edges was reported. This was later incorporated into other algorithms for microstrip line applications in [27,28].

The class of algorithms presented in [5,26,27,28] provides an efficient and accurate means for treating 3-D electromagnetic wave problems based on unstructured grids, assuming that a second-order accurate solver is used. This class is referred to here as the generalized Yee algorithm, since analysis reveals that it is truly a straightforward extension of Yee's classic space-time meshing approach.

[1]This section was adapted from [8], Chapter 11 (Sections 11.6, 7), written by Stephen Gedney and Faisa Lansing.

Generalized Yee Algorithm

The generalized Yee algorithm is based on a direct 3-D solution of Maxwell's equations in integral form which extends Yee's concept of interleaved electric and magnetic fields to a dual lattice formed by a primary and secondary grid. The primary grid is composed of general space-filling polyhedra, while the secondary (or dual grid) is composed of the closed polyhedra whose edges connect the centroids of adjacent primary cells, thereby penetrating shared faces. Here an electric field vector is defined along each edge of a primary grid cell, and a magnetic field vector is located along each secondary grid cell edge. Faraday's and Ampere's laws are implemented by discretizing the surface and line integrals over each primary and secondary grid cell face, respectively. To this end, the fields are assumed to be constant over each face and along each edge. The time derivative is then approximated using a central difference, yielding

$$\sum_{j=1}^{N_{e_i}} \vec{p}_j \cdot \vec{E}^n_{i,j} = -A_i\left(\frac{\vec{B}^{n+1/2}_i - \vec{B}^{n-1/2}_i}{\Delta t}\right) \cdot \hat{n}_p \tag{36}$$

$$\sum_{j=1}^{N_{e_i}} \vec{s}_j \cdot \vec{H}^{n+1/2}_{i,j} = A_i\left(\frac{\vec{D}^{n+1}_i - \vec{D}^n_i}{\Delta t}\right) \cdot \hat{n}_s + \sigma_{\text{ave}} A_i\left(\frac{\vec{D}^{n+1}_i + \vec{D}^n_i}{2\varepsilon_{\text{ave}}}\right) \cdot \hat{n}_s + A_i \vec{J}^{n+1/2}_i \cdot \hat{n}_s \tag{37}$$

where the superscripts indicate the time index; N_{e_i} is the number of edges bounding the ith face of the primary grid in (36) or the secondary grid in (37); A_i is the area of the ith face; $\vec{p}_j$ and $\vec{s}_j$ are the length vectors of the jth edge bounding the primary and secondary grid faces, respectively; and ε_{ave} and σ_{ave} are average material properties. It can be shown that, by performing a closed surface integral over each cell, Gauss's laws are also satisfied based on this formulation.

Upon solving for the latest values of B and D in (36) and (37), respectively, the following explicit time-stepping algorithm is obtained:

$$\vec{B}^{n+1/2}_i \cdot \hat{n}_p = \vec{B}^{n-1/2}_i \cdot \hat{n}_p - \frac{\Delta t}{A_i} \sum_{j=1}^{N_{e_i}} \vec{p}_j \cdot \vec{E}^n_{i,j} \tag{38}$$

$$\vec{D}^{n+1}_i \cdot \hat{n}_s = \left(\frac{\varepsilon_{\text{ave}} - \sigma_{\text{ave}}\Delta t/2}{\varepsilon_{\text{ave}} + \sigma_{\text{ave}}\Delta t/2}\right) \vec{D}^n_i \cdot \hat{n}_s + \left(\frac{\varepsilon_{\text{ave}}\Delta t}{\varepsilon_{\text{ave}} + \sigma_{\text{ave}}\Delta t/2}\right)\left(\frac{1}{A_i}\sum_{j=1}^{N_{e_i}} \vec{s}_j \cdot \vec{H}^{n+1/2}_{i,j} - \vec{J}^{n+1/2}_i \cdot \hat{n}_s\right) \tag{39}$$

By specifying the electric field along the edges of the primary grid cell at time $t = n\Delta t$ and the magnetic field normal to each face at time $t = (n - ½)\Delta t$, the normal magnetic flux density at a time $t = (n + ½)\Delta t$ can be updated using (38). The magnetic field $\vec{H} = \vec{B}/\mu$ can then be used to update the displacement flux density in (39).

If the fields are discretized over a regular orthogonal grid, (38) and (39) are equivalent to the traditional Yee algorithm. This holds true because, in this case, $\hat{n}_p \equiv \hat{s}$ and $\hat{n}_s \equiv \hat{p}$. However, if the grid is unstructured, the edge vector is *not necessarily parallel* to the normal vector, and the normal field must be projected onto the complementary grid edge. Since the normal field alone cannot uniquely describe the edge component, a secondary expression must be introduced.

The projection of the magnetic field onto the secondary grid edge requires an interpolation of the local magnetic flux densities normal to adjacent faces. This is similar to what is required for a nonorthogonal structured grid projection. However, due to the unstructured nature of the grid, the adjacent faces are randomly oriented and random in number. Therefore, a more general projection scheme is required. It must be chosen such that: (1) the flux projected onto the edges has zero divergence in a charge-free medium, and (2) the time-stepping algorithm maintains numerical stability. The scheme of [5,26], and a slightly modified scheme in [27,28], each meet these criteria.

Planar Generalized Yee Algorithm

The memory requirements of the generalized Yee algorithm can be greatly reduced by exploiting symmetries in the model. Consider now geometries that have planar symmetry, namely 3-D geometries that can be uniquely described by a projection onto a 2-D plane. For example, many microwave printed circuits have planar symmetry. It was shown in [29] that by exploiting this symmetry, the entire 3-D circuit geometry can then be described uniquely by a 2-D grid. Subsequently, the grid used to analyze the 3-D problem can be described by an unstructured 2-D grid in a transverse plane and as a regular grid in the third dimension. Thus, only the 2-D grid needs to be stored. Further, it is found that the sparse time-stepping matrices need only be stored for a single layer of grid cells. This greatly relaxes the memory requirements of the algorithm to the extent that it is actually as memory-efficient as basic FDTD.

The grid of the planar generalized Yee (PGY) algorithm is described as being regular along the vertical direction (assumed to be the z-direction) and having horizontal cross sections described by a 2-D unstructured grid. Conceptually, this grid can be generated by extruding a 2-D unstructured grid in the vertical direction and segmenting it at discrete heights. A secondary grid is staggered within this primary grid such that its vertices lie at the centroids of the primary grid cells, and the edges of the secondary grid connect the centroids by passing through the faces of the primary grid. The electric and magnetic fields are then decomposed into orthogonal components:

$$\vec{E} = \vec{E}_t + \hat{z}E_z; \qquad \vec{H} = \vec{H}_t + \hat{z}H_z \tag{40}$$

Subsequently, the transverse electric and magnetic fields are mapped onto the horizontal edges of the primary and secondary grids, respectively. Likewise, the vertical electric and magnetic fields are mapped onto the vertical edges of the primary and secondary grids, respectively. The vector fields are assumed to be constant along their respective edge lengths, as well as over the dual face through which they pass.

Based on the above discretization, Faraday's and Ampere's laws are approximated by choosing the surfaces of integration to be the faces of the secondary and primary grids, respectively. The time derivative is then approximated using a central-difference expression, which is accurate to the second-order if the fields are staggered in time. This leads to the following explicit time-stepping expressions [29]:

$$H_{z_j}\Big|_k^{n+1} = H_{z_j}\Big|_k^n - \frac{\Delta t}{\mu_j(k)A_p}\sum_{i=1}^{N_{p_j}} E_{t_i}\Big|_k^{n+1/2}\ell_i^p \tag{41}$$

$$B_{t_i}\Big|_{k+1/2}^{n+1} = B_{t_i}\Big|_{k+1/2}^n - \frac{c\Delta t}{\ell_i^p\Delta z}\begin{bmatrix}(E_{z_m}\big|_{k+1/2}^{n+1/2} - E_{z_{m+1}}\big|_{k+1/2}^{n+1/2})\Delta z\\ +(E_{t_i}\big|_{k+1}^{n+1/2} - E_{t_i}\big|_k^{n+1/2})\ell_i^p\end{bmatrix} \tag{42}$$

$$\begin{aligned}E_{z_m}\Big|_{k+1/2}^{n+3/2} = {} & \left(\frac{\varepsilon_m\big|_{k+1/2} - \sigma_m\big|_{k+1/2}\Delta t/2}{\varepsilon_m\big|_{k+1/2} + \sigma_m\big|_{k+1/2}\Delta t/2}\right)E_{z_m}\Big|_{k+1/2}^{n+1/2}\\ & - \left(\frac{\Delta t/A_s}{\varepsilon_m\big|_{k+1/2} + \sigma_m\big|_{k+1/2}\Delta t/2}\right)\sum_{i=1}^{N_{s_j}} H_{t_i}\Big|_{k+1/2}^{n+1}\ell_i^s\end{aligned} \tag{43}$$

$$\begin{aligned}D_{t_i}\Big|_k^{n+3/2} = {} & \left(\frac{\varepsilon_i\big|_{k+1/2} - \sigma_i\big|_{k+1/2}\Delta t/2}{\varepsilon_i\big|_{k+1/2} + \sigma_i\big|_{k+1/2}\Delta t/2}\right)D_{t_i}\Big|_k^{n+1/2}\\ & - \left(\frac{\varepsilon_i\big|_{k+1/2}\Delta t/(\ell_i^s\Delta z)}{\varepsilon_i\big|_{k+1/2} + \sigma_i\big|_{k+1/2}\Delta t/2}\right)\\ & \cdot\begin{bmatrix}(H_{z_j}\big|_k^{n+1} - H_{z_{j+1}}\big|_k^{n+1})\Delta z\\ +(H_{t_i}\big|_{k-1/2}^{n+1} - H_{t_i}\big|_{k+1/2}^{n+1})\ell_i^s\end{bmatrix}\end{aligned} \tag{44}$$

where D_t and B_t are the flux densities in the transverse plane; k is the index along z; A_p and A_s are the areas of the primary and secondary grid faces, respectively; N_{p_i} and N_{s_j} are the number of edges bounding the ith primary and the jth secondary grid faces, respectively; and ℓ_i^p, ℓ_i^s are the lengths of the primary and secondary grid edges, respectively. The material parameters ε, μ, and σ are assumed to be piecewise homogeneous in both the z-direction as well as in the transverse direction and can be computed using the methods discussed earlier.

It is recognized that the flux densities updated in (42) and (44) are normal to the faces. However, the corresponding field intensities on the dual edges passing through these faces are not necessarily normal to the faces. As a result, the flux densities must be projected onto the edges before the dual fields can be updated and an auxiliary operator must be introduced to perform the projection. The projection operators discussed earlier are also used here. The principal difference is that due to the orthogonality of the vertical and transverse fields, only fields within the transverse plane are needed for the interpolation. This results in a simplified form, discussed in [29].

The emergence of the above techniques to efficiently generalize the Yee algorithm to unstructured meshes will have a large impact upon the modeling of microwave structures. Generalized Yee algorithms permit FDTD to provide surface-conforming (rather than staircased) simulations of complex structure shapes in a manner similar to that afforded by the finite-element method (FEM). This will markedly relax FDTD grid resolution requirements. Further, the generalization of FDTD to unstructured meshes retains the highly desirable *no-matrix* attribute of the original Yee algorithm, permitting much larger numbers of field unknowns to be solved by FDTD than any frequency-domain FEM. Here, the present advantage in problem size for FDTD is a decisive 1,000:1, i.e., hundred of millions of field components can be solved in a single problem using FDTD versus hundreds of thousands for FEM. And, the generalization of FDTD retains its dynamic-field solution method which allows a single modeling run to calculate pulse propagation and scattering, ultrawideband phasor domain characteristics via the discrete Fourier transform of the impulsive results, and the response of nonlinear materials and devices.

IX. Digital Signal Processing and Spectrum Estimation

Typical FDTD models of high-speed circuit structures use a grid resolution dictated by the fine dimensions of the circuit board layers, vias, and similar components. This resolution is almost always much finer than needed to resolve the smallest spectral wavelength propagating in the circuit. As a result, with the time step Δt bound to Δ by numerical stability considerations, it is common to run FDTD simulations for tens of thousands of time steps in order to fully evolve the impulse responses needed for calculating impedances, *S*-parameters, and resonant frequencies of passive structures in electronic circuits operating at UHF and microwave frequencies.

A brute-force way to mitigate the burden of a lengthy FDTD run is to simply truncate it before the impulse response has fully evolved. However, this has the effect of viewing the true impulse response through a rectangular window of duration $T = N_{max}\Delta t$. In the frequency domain, this windowing is translated into the convolution of the true spectrum with the function $\sin(f)/f$. This convolution widens the peaks in the spectral response, causes other distortions, and can mask weak spectral signatures. Now, the distortion can be reduced and the spectral resolution increased by lengthening the window duration. Unfortunately, this causes FDTD simulation times for certain classes of problems (particularly resonators) to be so lengthy as to be virtually prohibitive.

Recently, there has been significant progress in addressing this problem. The most fruitful approach has been to apply contemporary analysis techniques from the discipline of digital signal processing and spectrum estimation. The strategy is to extrapolate the electromagnetic field time waveform by 10:1 or more beyond the actual FDTD time-stepping window, allowing a good estimate of the complete system response with a 90% or greater reduction in the FDTD computation time. In effect, a low-computational-burden extrapolation process at a *limited number* of observation points in the grid replaces a computationally intensive FDTD process involving *every point* in the grid. Then the reflected and transmitted field spectra and the associated impedances and *S*-parameters at the observation points can be efficiently obtained by FFT of the extrapolated waveforms or, in certain cases, from the coefficients of the extrapolation process.

A number of digital signal processing and spectrum estimation techniques have appeared in the recent FDTD literature. While good results have been obtained with each of these methods for appropriate classes of modeling problems, there are emerging tradeoff considerations as FDTD practitioners become more familiar with the state of modern signal-processing theory. A goal of this section is to illuminate the tradeoffs as currently understood.

Prony's Method

Prony's method has been used to extrapolate transmission line matrix (TLM) and FDTD waveforms for microwave circuits by a number of investigators [30–33]. Because it is a technique for modeling sampled data as a linear combination of complex exponentials, it is particularly suitable for calculating the resonant frequency and Q of a resonating structure, since the impulse response of such a structure is characterized by a superposition of decaying exponentials. While Prony's method is not a spectral estimation technique in the usual sense, it is closely related to the least-squares linear prediction methods used for parametric spectral estimation.

Reference [32] reported accurate results using Prony's method for the resonant frequencies and Q factors of the three lowest TE_{0n} modes of a cylindrical dielectric resonator. A window of over 131,000 time steps was needed to obtain converged results for the Q of the TE_{02} mode when using an FFT applied to the windowed FDTD data. However, using Prony's method of order $p = 10$ and a decimation factor of 50, only 3,000 iterations were required, a reduction of about 98%. In fact, reductions can be *even greater* in the case of structures having even higher Q or with very closely spaced resonant frequencies.

Prony's method shows good resolution with relatively short data sequences and presents no windowing problem. The main difficulty with this approach lies in its determination of the user-selected order of the model. If the order is less than the number of actual modes excited in the structure, the spectral resolution is poor. However, if the order is selected to be too high, spurious (non-physical) modes appear.

Autoregressive Models

The most popular time series modeling approach used in modern spectral estimation is the class of linear predictors or autoregressive (AR) models. This is because an accurate estimate of the AR parameters can be derived by solving a set of linear equations. In contrast to Prony's method, which uses a sum of deterministic exponential functions to fit the data, the AR approach constructs a random model to fit a statistical database to the second-order.

There exists many published approaches to evaluating the coefficients of the autoregression. In fact, a primary categorization of AR techniques is the means used to calculate these parameters. Three specific recent applications of AR to the extrapolation of FDTD electromagnetic field records can be categorized according to this criterion: (1) the covariance method, (2) the forward-backward method, and (3) the nonlinear predictor. The choice of approach is very important, since these three methods have substantially different operational characteristics.

Covariance Method. A key issue that arises when using the covariance-based AR model is choosing the order of the model. The use of a low-order AR model of this type causes the extrapolated waveform to attenuate quickly in a non-physical manner. To compensate, a high-order model is required. However, a high-order model can cause divergence problems in some cases because of statistical instabilities introduced by the large order. These competing effects cause the results of the covariance-based AR model to be sensitive to the model order, somewhat diminishing its usefulness. Using a covariance-based AR method, [34] reported the extrapolation to over 30,000 time steps of FDTD field records spanning only 2,000 to 3,500 time steps. A decimation factor of 10 was used, and the order of the AR model was 50. Very good accuracy was observed for the extrapolated time waveform for an edge-coupled strip line bandpass filter. A second example discussed in [34] involved a stripline double-stub structure whose lengthy propagation delays along the stubs required extrapolating the FDTD data to 100,000 time steps to achieve a converged transmission coefficient. This was accomplished using a 91st-order AR model working from an FDTD data base covering iterations 3,100 to 8,900, and decimated by a factor of 25. Although very good agreement was observed with benchmark frequency-domain data, the results of the AR model were found to be very sensitive to the order of the model.

Forward-Backward Method. The results of a covariance-based AR model can be sensitive to the order of the model because its weights often have low accuracy. The low accuracy results, in turn, because the ensemble calculation that is rigorously required to find the covariance is instead substituted in these methods by using the law of large numbers and by approximating the covariance with inexact functions of the known time signal. The forward-backward method is a useful alternative approach based on an unconstrained least-squares estimation of the AR parameters. It avoids the problems of the covariance-based AR models by working directly with the time-domain data rather than calculating the covariance functions of the data. Forward-backward was used for the first time in [35] to extrapolate FDTD-computed field records. Results were given for a test problem involving a rectangular PEC plate embedded in a high-permittivity half space and excited by a sinusoidal point source. Both an autocorrelation-based AR model and the forward-backward model were trained on the same FDTD field waveform record, data points from 201 to 800 time steps, and then used to generate an extrapolated sequence extending over 5,000 time steps. The forward-backward method exhibited two key advantages: (1) it provided more accurate spectra, and (2) its order was much lower, 5 versus 50. It was concluded that the forward-backward AR model provides a very good approach to FDTD signal extrapolation and spectrum estimation both in terms of accuracy and speed.

Nonlinear Predictor. In addition to the use of covariance-based AR linear prediction to extrapolate FDTD waveform records (discussed earlier), [34] reported the application of a non-linear predictor having as its basic form an exponential autoregressive (EXPAR) model. EXPAR was applied to the large double-stub stripline problem modeled previously using a 91st-order covariance-based linear AR model, and similarly used to extrapolate an 8,900-step FDTD data record to 100,000 time steps. The non-linear EXPAR model provided excellent accuracy for S_{21} over a wide dynamic range upon Fourier transformation of the 100,000-step record. Moreover, the EXPAR results were less sensitive to the model order than were the results of the covariance-based approach. However, the order of the EXPAR model still exceeded 80. A necessary line of inquiry involves applying the forward-backward linear model to problems of this size and complexity to establish whether its characteristics can reduce the required model order.

System Identification

The system identification (SI) method is an alternative to convolution for characterizing a discrete-time linear system. This approach specifies the time-domain output/input relation as a recursive difference equation. Because SI has feedback, it is termed an infinite-duration impulse response (IIR) filter. It is known that an IIR filter can provide a less complex and lower-order realization than a convolutional finite-duration impulse response (FIR) filter for low-pass systems. Therefore, SI should be more efficient for processing FDTD data than convolution since FDTD is inherently a low-pass system (its spectrum of space/time Fourier modes is bounded by the Nyquist limit set by the grid resolution). A SI model was applied in [36] to calculate the resonant frequencies of a PEC rectangular cavity partially filled with a dielectric slab. The cavity was excited at its center plane by imposing a pulsed TE_{10} mode field distribution. The time waveform of this excitation was used as the input signal to the SI model. Here the parameter search began with all set to zero, and the values were updated at each FDTD time step (no decimation). Upon convergence of the parameters, the resonant frequencies could be derived with no Fourier transform of the output signal. Instead, the Z-transform of the recursion relation could be evaluated, thereby obtaining the resonant frequencies directly from the poles of the SI model.

Overall, SI requires minimal additional computational cost. It is a rapidly converging recursive processing of the input and output values of the FDTD simulation, and can be performed concurrently with the usual FDTD time-stepping at any number of observation points in the grid. In addition to permitting an extrapolation of the local field values at the observation points, the frequency spectrum is immediately available without needing to apply a Fourier transform. Further, the input/output nature of the SI formulation permits an efficient natural treatment of diakoptic splitting of the computation space without needing to use convolutions [36].

X. Reprints of Key Papers

Fourteen reprints of key papers in the development of the FDTD method for modeling microwave structures are included in this chapter. We now briefly discuss the significance of these papers.

The Yee [1], Taflove-Brodwin [6], and Mur [13] papers established the basis of the FDTD method with respect to the basic gridding and algorithm, its numerical stability and suitability for sinusoidal steady-state simulations, and the first exposition of a stable second-order accurate ABC. Holland's paper [37] was the first to extend the Yee algorithm to generalized nonorthogonal coordinates, presaging current work in unstructured meshing. Choi and Hoefer [38] pioneered the application of FDTD to waveguide cavity problems, including the postprocessing of the data to obtain resonant frequencies. Zhang et al. [39] and Sheen et al. [40] pioneered the application of FDTD to microstrips and microstrip circuits, including the postprocessing of the computed data to obtain line impedances, propagation constants, and *S*-parameters. Maloney and Smith [41] pioneered the application of FDTD to antennas (especially ultrawideband antennas), demonstrated the possibility of achieving highly accurate FDTD modeling results for antenna input impedance, and showed how FDTD visualization of an antenna's transient radiation characteristics could provide much design insight.

Maloney and Smith [42] helped to originate FDTD modeling of frequency-dependent surface impedance, such as that due to the skin effect. Luebbers and Hunsberger [43] pioneered the efficient recursive convolution formulation for FDTD modeling of practical linear dispersive materials having arbitrary combinations of Debye and Lorentz relaxations. Piket-May et al. [11] helped to originate 3-D FDTD models of lumped resistors, capacitors, inductors, diodes, and transistors in digital and microwave circuits; they also demonstrated that it is possible to use FDTD to simulate real-world circuit modules of great complexity, thereby gaining significant design insight.

Berenger [23] pioneered the PML ABC, a complete theoretical departure from any earlier approach, achieving in the process tens of decibels improved performance. Chen et al. [34] reported among the best results for digital signal postprocessing of FDTD data to extrapolate windowed impulse responses and obtain *S*-parameters for electrically large and high-*Q* structures. Finally, Madsen [5] reported his pioneering work on numerically stable and efficient algorithms for unstructured meshes involving a generalization of the Yee algorithm.

XI. Summary and Conclusions

This paper summarized a number of key technical issues regarding FDTD numerical modeling of microwave structures. There has been substantial recent progress in FDTD techniques, particularly in the areas of absorbing boundary conditions, unstructured meshes, and digital signal processing and spectrum estimation techniques. These technical advances have greatly improved the ability of FDTD methods to model microwave and optical circuits of realistic complexity. At the cutting edge, FDTD modeling capabilities challenge those of the frequency-domain FEM in meshing complexity, and overwhelm the frequency-domain FEM in problem size, modeling bandwidth, and treatment of non-linearities. Combined with rapid advances in personal computers and workstations, these technical advances have resulted in an exponential increase in FDTD research and modeling applications over the past ten years [44].

References

[1] K. S. Yee, "Numerical solution of initial boundary value problems involving Maxwell's equations in isotropic media," *IEEE Trans. Antennas Propagat.,* vol. AP-14, pp. 302–307, May 1966.

[2] A. Taflove, K. R. Umashankar, B. Beker, F. A. Harfoush, and K. S. Yee, "Detailed FD-TD analysis of electromagnetic fields penetrating narrow slots and lapped joints in thick conducting screens," *IEEE Trans. Antennas Propagat.,* vol. 36, pp. 247–257, Feb. 1988.

[3] T. G. Jurgens, A. Taflove, K. R. Umashankar, and T. G. Moore, "Finite-difference time-domain modeling of curved surfaces," *IEEE Trans. Antennas Propagat.,* vol. 40, pp. 357–366, April 1992.

[4] V. Shankar, A. H. Mohammadian, and W. F. Hall, "A time-domain finite-volume treatment for the Maxwell equations," *Electromagn.,* vol. 10, pp. 127–145, Jan.-June 1990.

[5] N. K. Madsen, "Divergence preserving discrete surface integral methods for Maxwell's curl equations using non-orthogonal unstructured grids," *J. Computat. Phys.,* vol. 119, pp. 34–45, 1995.

[6] A. Taflove and M. E. Brodwin, "Numerical solution of steady-state electromagnetic scattering problems using the time-dependent Maxwell's equations," *IEEE Trans. Microwave Theory Tech.,* vol. MTT-23, pp. 623–630, Aug. 1975.

[7] A. Taflove, "Review of the formulation and applications of the finite-difference time-domain method for numerical modeling of electromagnetic wave interactions with arbitrary structures," *Wave Motion,* vol. 10, pp. 547–582, Dec. 1988.

[8] A. Taflove, *Computational Electrodynamics: The Finite-Difference Time-Domain Method,* Boston, MA: Artech House, 1995.

[9] K. R. Umashankar and A. Taflove, "A novel method to analyze electromagnetic scattering of complex objects," *IEEE Trans. Electromagn. Compat.,* vol. EMC-24, pp. 397–405, Nov. 1982.

[10] W. Sui, D. A. Christensen, and C. H. Durney, "Extending the two-dimensional FD-TD method to hybrid electromagnetic systems with active and passive lumped elements," *IEEE Trans. Microwave Theory Tech.,* vol. MTT-40, pp. 724–730, April 1992.

[11] M. J. Piket-May, A. Taflove, and J. Baron, "FDTD modeling of digital signal propagation in three-dimensional circuits with passive and active loads," *IEEE Trans. Microwave Theory Tech.,* vol. MTT-42, pp. 1514–1523, Aug. 1994.

[12] B. Engquist and A. Majda, "Absorbing boundary conditions for the numerical simulation of waves," *Math. of Comput.,* vol. 31, pp. 629–651, 1977.

[13] G. Mur, "Absorbing boundary conditions for the finite-difference approximation of the time-domain electromagnetic field equations," *IEEE Trans. Electromagn. Compat.,* vol. EMC-23, pp. 377–382, Nov. 1981.

[14] K. K. Mei and J. Fang, "Superabsorption—a method to improve absorbing boundary conditions," *IEEE Trans. Antennas Propagat.,* vol. AP-40, pp. 1001–1010, Sept. 1992.

[15] A. Bayliss and E. Turkel, "Radiation boundary conditions for wave-like equations," *Commun. Pure Appl. Math.,* vol. 33, pp. 707–725, 1980.

[16] A. Bayliss, M. Gunzburger, and E. Turkel, "Boundary conditions for the numerical solution of elliptic equations in exterior regions," *SIAM J. Appl. Math.,* vol. 42, pp. 430–451, 1982.

[17] Z. P. Liao, H. L. Wong, B. P. Yang, and Y. F. Yuan, "A transmitting boundary for transient wave analysis," *Scientia Sinica (series A),* vol. 27, pp. 1063–1076, Oct. 1984.

[18] R. Holland and J. W. Williams, *IEEE Trans. Nucl. Sci.,* vol. NS-30, p. 4583, 1983.

[19] C. M. Rappaport and L. Bahrmasel, "An absorbing boundary condition based on anechoic absorber for EM scattering computation," *J. Electromagn. Waves Appl.,* vol. 6, pp. 1621–1634, Dec. 1992.

[20] Z.-Q. Bi, K.-L. Wu, C. Wu, and J. Litva, "A dispersive boundary condition for microstrip component analysis using the FD-TD method,"

IEEE Trans. Microwave Theory Tech., vol. MTT-40, pp. 774–777, April 1992.

[21] J. C. Olivier and D. A. McNamara, "Analysis of multiport discontinuities in waveguide using a pulsed FDTD approach," *IEEE Trans. Microwave Theory Tech.,* vol. MTT-42, pp. 2229–2238, Dec. 1994.

[22] F. Moglie, T. Rozzi, P. Marcozzi, and A. Schiavoni, "A new termination condition for the application of FDTD techniques to discontinuity problems in close homogeneous waveguide," *IEEE Microwave Guided Wave Lett.,* vol. 2, pp. 475–477, Dec. 1992.

[23] J. P. Berenger, "A perfectly matched layer for the absorption of electromagnetic waves," *J. Computat. Phys.,* vol. 114, pp. 185–200, 1994.

[24] D. S. Katz, E. T. Thiele, and A. Taflove, "Validation and extension to three dimensions of the Berenger PML absorbing boundary condition for FD-TD meshes," *IEEE Microwave Guided Wave Lett.,* vol. 4, pp. 268–270, Aug. 1994.

[25] C. E. Reuter, R. M. Joseph, E. T. Thiele, D. S. Katz, and A. Taflove, "Ultrawideband absorbing boundary condition for termination of waveguiding structures in FD-TD simulations," *IEEE Microwave Guided Wave Lett.,* vol. 4, pp. 344–346, Oct. 1994.

[26] N. Madsen, "Divergence preserving discrete surface integral methods for Maxwell's equations using nonorthogonal unstructured grids," *Technical Report* UCRL-JC-109787, Lawrence Livermore National Laboratory, Feb. 1992.

[27] S. Gedney and F. Lansing, "Full wave analysis of printed microstrip devices using a generalized Yee algorithm," in *IEEE Antennas Propagat. Symp. Digest,* Ann Arbor, MI, 1993.

[28] S. Gedney, F. Lansing, and D. Rascoe, "Full wave analysis of microwave monolithic circuit devices using a generalized Yee algorithm based on unstructured grids," *New Technology Report* no. 19186/8768, Pasadena, CA; Jet Propulsion Laboratory, Sept. 1994.

[29] S. Gedney, F. Lansing, and D. Rascoe, "Full-wave analysis of microwave monolithic circuit devices using a generalized Yee algorithm based on an unstructured grid," *IEEE Trans. Microwave Theory Tech.,* vol. 44, pp. 1393–1400, Aug. 1996.

[30] J. L. Dubard, D. Pompei, J. LeRoux, and A. Papiernik, "Characterization of microstrip antennas using the TLM simulation associated with a Prony-Pisrenko method," *Int. J. Num. Mod.,* vol. 3, pp. 269–285, 1990.

[31] W. L. Ko and R. Mittra, "A combination of FD-TD and Prony's methods for analyzing microwave integrated circuits," *IEEE Trans. Microwave Theory Tech.,* vol. MTT-39, pp. 2176–2181, Dec. 1991.

[32] J. A. Pereda, L. A. Vielva, A. Vegas, and A. Prieto, "Computation of resonant frequencies and quality factors of open dielectric resonators by a combination of the finite difference time domain (FDTD) and Prony's methods," *IEEE Microwave Guided Wave Lett.,* vol. 2, pp. 431–433, Nov. 1992.

[33] K. Naishadham and X. P. Lin, "Application of spectral domain Prony's method to the FD-TD analysis of planar microstrip circuits," *IEEE Trans. Microwave Theory Tech.,* vol. MTT-42, pp. 2391–2398, Dec. 1994.

[34] J. Chen, C. Wu, T. K. Lo, K. L. Wu, and J. Litva, "Using linear and nonlinear predictors to improve the computational efficiency of the FD-TD algorithm," *IEEE Trans. Microwave Theory Tech.,* vol. MTT-42, pp. 1992–1997, Oct. 1994.

[35] V. Jandhyala, E. Michielssen, and R. Mittra, "FDTD signal extrapolation using the forward-backward autoregressive (AR) model," *IEEE Microwave Guided Wave Lett.,* vol. 4, pp. 163–165, June 1994.

[36] T. W. Huang, B. Houshmand, and T. Itoh, "Fast sequential FDTD diakoptics method using the system identification technique," *IEEE Microwave Guided Wave Lett.,* vol. 3, pp. 378–380, Oct. 1993.

[37] R. Holland, "Finite-difference solution of Maxwell's equations in generalized nonorthogonal coordinates," *IEEE Trans. Nuclear Science,* vol. NS-30, pp. 4589–4591, Dec. 1983.

[38] D. H. Choi and W. J. R. Hoefer, "The finite-difference time-domain method and its application to eigenvalue problems," *IEEE Trans. Microwave Theory Tech.,* vol. MTT-34, pp. 1464–1470, Dec. 1986.

[39] X. Zhang, J. Fang, K. K. Mei, and Y. Liu, "Calculation of the dispersive characteristics of microstrips by the time-domain finite-difference method," *IEEE Trans. Microwave Theory Tech.,* vol. MTT-36, pp. 263–267, Feb. 1988.

[40] D. M. Sheen, S. A. Ali, M. D. Abouzahra, and J. A. Kong, "Application of the three-dimensional finite-difference time-domain method to the analysis of planar microstrip circuits," *IEEE Trans. Microwave Theory Tech.,* vol. MTT-38, pp. 849–857, July 1990.

[41] J. G. Maloney, G. S. Smith, and W. R. Scott, Jr., "Accurate computation of the radiation from simple antennas using the finite-difference time-domain method," *IEEE Trans. Antennas Propagat.,* vol. AP-38, pp. 1059–1068, July 1990.

[42] J. G. Maloney and G. S. Smith, "The use of surface impedance concepts in the finite-difference time-domain method," *IEEE Trans. Antennas Propagat.,* vol. AP-40, pp. 38–48, Jan. 1992.

[43] R. J. Luebbers and F. Hunsberger, "FDTD for the *N*th-order dispersive media," *IEEE Trans. Antennas Propagat.,* vol. AP-40, pp. 1297–1301, Nov. 1992.

[44] K.L. Shlager and J.B. Schneider, "A selective survey of the finite-difference time-domain literature," *IEEE Antennas Propagat. Mag.,* vol. 37, pp. 39–57, Aug. 1995.

Numerical Solution of Steady-State Electromagnetic Scattering Problems Using the Time-Dependent Maxwell's Equations

ALLEN TAFLOVE AND MORRIS E. BRODWIN, SENIOR MEMBER, IEEE

Abstract—**A numerical method is described for the solution of the electromagnetic fields within an arbitrary dielectric scatterer of the order of one wavelength in diameter. The method treats the irradiation of the scatterer as an initial value problem. At $t = 0$, a plane-wave source of frequency f is assumed to be turned on. The diffraction of waves from this source is modeled by repeatedly solving a finite-difference analog of the time-dependent Maxwell's equations. Time stepping is continued until sinusoidal steady-state field values are observed at all points within the scatterer. The envelope of the standing wave is taken as the steady-state scattered field. As an example of this method, the computed results for a dielectric cylinder scatterer are presented. An error of less than ± 10 percent in locating and evaluating the standing-wave peaks within the cylinder is achieved for a program execution time of 1 min. The extension of this method to the solution of the fields within three-dimensional dielectric scatterers is outlined.**

I. INTRODUCTION

THE accurate determination of the electromagnetic fields within an arbitrary, inhomogeneous, dielectric scatterer is both an important theoretical problem and a practical objective of workers investigating the effects of microwaves upon living tissue. Exact analytical solutions are obtained only for simple scatterers like the sphere and the circular cylinder, which may be solved using separation of variables. For complicated scatterers like most body organs, we must resort to some numerical method if an accurate model is to be examined.

Manuscript received June 18, 1974; revised February 18, 1975.
The authors are with the Department of Electrical Engineering, Technological Institute, Northwestern University, Evanston, Ill. 60201.

The computer techniques relevant to this problem that have appeared in the literature may be called, as a class, frequency-domain methods. These methods are based upon the assumption of an exp $(j2\pi ft)$ time dependence in the fundamental Maxwell's equations. In general, methods of this type derive a set of linear equations for either field variables or field expansion coefficients, and then solve the linear system with a suitable matrix-inversion scheme.

Wu and Tsai [1] solve two-dimensional scattering by an arbitrary dielectric cylinder. They develop a coupled integral equation pair for the electric field and its normal derivative at the surface of the scatterer. They then derive a corresponding set of linear equations for the surface fields using the moment method of Harrington [2]. Solution of this set of equations allows computation of the interior fields using Huygens' integrals. This method allows the very accurate solution of a homogeneous dielectric cylinder, about one free-space wavelength in circumference, by inverting an 80-by-80 matrix.

Reprinted from *IEEE Transactions on Microwave Theory and Techniques*, pp. 623-630, August 1975.

McDonald and Wexler [3] solve a two-dimensional radiating antenna with dielectric obstacles. They employ a finite-element solution of the Helmholtz equation within a restricted region, and use an integral equation constraint along the contour of this region to take into account the unbounded exterior. Solution is obtained by inverting a 60-by-60 matrix.

Wilton and Mittra [4] solve two-dimensional scattering by an arbitrary dielectric cylinder. They expand the fields inside and outside of the scatterer in terms of free-space-type wave functions, wherever valid, and in terms of the analytic continuation of these wave functions, wherever required. The unknown set of coefficients is determined by enforcing the field boundary conditions at a number of points along the surface of the scatterer. Sufficient points are selected to represent the shape of the scatterer.

These three methods may be extended to more complicated scattering problems. Their accuracy is excellent when a sufficiently large set of linear equations is solved. However, each method may have two problems when very complicated inhomogeneous scatterers like body organs are considered. First, programming a complex scatterer requires the (possibly lengthy) derivation of a set of linear equations appropriate only for that scatterer. Second, solution of such a problem with high accuracy may require such a large, dense matrix to be inverted that the available fast, direct access computer storage is exhausted. As a comparison with the 80-by-80 matrix of Wu and Tsai, the maximum size dense matrix solvable using direct access storage, on the Northwestern University CDC 6400, is 400 by 400 [5].

The numerical method discussed in this paper is a time-domain approach, which treats the irradiation of the scatterer as an initial value problem. At $t = 0$, a plane-wave source of frequency f is assumed to be turned on. The propagation of waves from this source is simulated by solving a finite-difference analog of the time-dependent Maxwell's equations on a lattice of points, including the scatterer. Time stepping is continued until the sinusoidal steady state is achieved at each point. The field envelope, or maximum absolute value, during the final half-wave cycle of time stepping is taken as the magnitude of the phasor of the steady-state field.

This method has two advantages relative to frequency-domain approaches. First, and most important, it is simple to implement for complicated scatterers, because arbitrary dielectric parameters may be assigned to each lattice point. Second, its memory requirement is not prohibitive for many scatterers of interest. For example, the Northwestern computer can process a 125-by-250-point grid for two-dimensional problems, or a 20-by-20-by-40-point lattice for three-dimensional problems, without resorting to noncore storage. This is sufficient, using symmetry, to process a 12-wavelength-diam cylinder or a 2-wavelength-diam sphere.

This method has two disadvantages relative to frequency-domain approaches. First, its accuracy is only about ±10 percent, which is at least one order of magnitude worse than that of the other methods. Second, the required program execution time may be excessive for some computer budgets. Typically, a two-dimensional problem solved on a 25-by-50-point grid, requires 1 min (at a rate of about 10 s/100 time steps). A three-dimensional problem solved on a 20-by-20-by-40-point lattice, requires 30 min (at a rate of about 5 min/100 time steps).

The important elements of the initial-value-problem approach to scattering problems are discussed below.

II. THE YEE ALGORITHM [6]

Using the MKS system of units, and assuming that the dielectric parameters μ, ϵ, and σ are independent of time, the following system of scalar equations is equivalent to Maxwell's equations in the rectangular coordinate system (x,y,z):

$$\frac{\partial H_x}{\partial t} = \frac{1}{\mu}\left(\frac{\partial E_y}{\partial z} - \frac{\partial E_z}{\partial y}\right) \tag{1a}$$

$$\frac{\partial H_y}{\partial t} = \frac{1}{\mu}\left(\frac{\partial E_z}{\partial x} - \frac{\partial E_x}{\partial z}\right) \tag{1b}$$

$$\frac{\partial H_z}{\partial t} = \frac{1}{\mu}\left(\frac{\partial E_x}{\partial y} - \frac{\partial E_y}{\partial x}\right) \tag{1c}$$

$$\frac{\partial E_x}{\partial t} = \frac{1}{\epsilon}\left(\frac{\partial H_z}{\partial y} - \frac{\partial H_y}{\partial z} - \sigma E_x\right) \tag{1d}$$

$$\frac{\partial E_y}{\partial t} = \frac{1}{\epsilon}\left(\frac{\partial H_x}{\partial z} - \frac{\partial H_z}{\partial x} - \sigma E_y\right) \tag{1e}$$

$$\frac{\partial E_z}{\partial t} = \frac{1}{\epsilon}\left(\frac{\partial H_y}{\partial x} - \frac{\partial H_x}{\partial y} - \sigma E_z\right). \tag{1f}$$

Yee introduces a set of finite-difference equations for the system of (1a)–(1f). Following Yee's notation, we denote a space lattice point as

$$(i,j,k) = (i\delta,j\delta,k\delta) \tag{2}$$

and any function of space and time as

$$F^n(i,j,k) = F(i\delta,j\delta,k\delta,n\delta t) \tag{3}$$

where $\delta = \delta x = \delta y = \delta z$ is the space increment, and δt is the time increment. Yee uses finite difference expressions for the space and time derivatives that are both simply programmed and second-order accurate in δ and in δt, respectively,

$$\frac{\partial F^n(i,j,k)}{\partial x} = \frac{F^n(i+\frac{1}{2},j,k) - F^n(i-\frac{1}{2},j,k)}{\delta} + 0(\delta^2) \tag{4}$$

$$\frac{\partial F^n(i,j,k)}{\partial t} = \frac{F^{n+1/2}(i,j,k) - F^{n-1/2}(i,j,k)}{\delta t} + 0(\delta t^2). \tag{5}$$

To achieve the accuracy of (4), and to realize all of the space derivatives of (1a)–(1f), Yee positions the components of $\bar{E}$ and $\bar{H}$ about a unit cell of the lattice as shown in Fig. 1. To achieve the accuracy of (5), he evaluates $\bar{E}$ and $\bar{H}$ at alternate half-time steps. The result of these assumptions is the following system of finite-difference equations for the system of (1a)–(1f):

$$\begin{aligned}H_x^{n+1/2}&\left(i,j+\tfrac{1}{2},k+\tfrac{1}{2}\right)\\ &= H_x^{n-1/2}\left(i,j+\tfrac{1}{2},k+\tfrac{1}{2}\right) + \frac{\delta t}{\mu\left(i,j+\tfrac{1}{2},k+\tfrac{1}{2}\right)\delta}\\ &\cdot\begin{bmatrix}E_y^n\left(i,j+\tfrac{1}{2},k+1\right) - E_y^n\left(i,j+\tfrac{1}{2},k\right) + \\ E_z^n\left(i,j,k+\tfrac{1}{2}\right) - E_z^n\left(i,j+1,k+\tfrac{1}{2}\right)\end{bmatrix}\end{aligned} \tag{6a}$$

$$\begin{aligned}H_y^{n+1/2}&\left(i+\tfrac{1}{2},j,k+\tfrac{1}{2}\right)\\ &= H_y^{n-1/2}\left(i+\tfrac{1}{2},j,k+\tfrac{1}{2}\right) + \frac{\delta t}{\mu\left(i+\tfrac{1}{2},j,k+\tfrac{1}{2}\right)\delta}\\ &\cdot\begin{bmatrix}E_z^n\left(i+1,j,k+\tfrac{1}{2}\right) - E_z^n\left(i,j,k+\tfrac{1}{2}\right) + \\ E_x^n\left(i+\tfrac{1}{2},j,k\right) - E_x^n\left(i+\tfrac{1}{2},j,k+1\right)\end{bmatrix}\end{aligned} \tag{6b}$$

$$\begin{aligned}H_z^{n+1/2}&\left(i+\tfrac{1}{2},j+\tfrac{1}{2},k\right)\\ &= H_z^{n-1/2}\left(i+\tfrac{1}{2},j+\tfrac{1}{2},k\right) + \frac{\delta t}{\mu\left(i+\tfrac{1}{2},j+\tfrac{1}{2},k\right)\delta}\\ &\cdot\begin{bmatrix}E_x^n\left(i+\tfrac{1}{2},j+1,k\right) - E_x^n\left(i+\tfrac{1}{2},j,k\right) + \\ E_y^n\left(i,j+\tfrac{1}{2},k\right) - E_y^n\left(i+1,j+\tfrac{1}{2},k\right)\end{bmatrix}\end{aligned} \tag{6c}$$

$$\begin{aligned}E_x^{n+1}&\left(i+\tfrac{1}{2},j,k\right)\\ &= \left[1 - \frac{\sigma\left(i+\tfrac{1}{2},j,k\right)\delta t}{\epsilon\left(i+\tfrac{1}{2},j,k\right)}\right]E_x^n\left(i+\tfrac{1}{2},j,k\right) + \frac{\delta t}{\epsilon\left(i+\tfrac{1}{2},j,k\right)\delta}\\ &\cdot\begin{bmatrix}H_z^{n+1/2}\left(i+\tfrac{1}{2},j+\tfrac{1}{2},k\right) - H_z^{n+1/2}\left(i+\tfrac{1}{2},j-\tfrac{1}{2},k\right) + \\ H_y^{n+1/2}\left(i+\tfrac{1}{2},j,k-\tfrac{1}{2}\right) - H_y^{n+1/2}\left(i+\tfrac{1}{2},j,k+\tfrac{1}{2}\right)\end{bmatrix}\end{aligned} \tag{6d}$$

$$\begin{aligned}E_y^{n+1}&\left(i,j+\tfrac{1}{2},k\right)\\ &= \left[1 - \frac{\sigma\left(i,j+\tfrac{1}{2},k\right)\delta t}{\epsilon\left(i,j+\tfrac{1}{2},k\right)}\right]E_y^n\left(i,j+\tfrac{1}{2},k\right) + \frac{\delta t}{\epsilon\left(i,j+\tfrac{1}{2},k\right)\delta}\\ &\cdot\begin{bmatrix}H_x^{n+1/2}\left(i,j+\tfrac{1}{2},k+\tfrac{1}{2}\right) - H_x^{n+1/2}\left(i,j+\tfrac{1}{2},k-\tfrac{1}{2}\right) + \\ H_z^{n+1/2}\left(i-\tfrac{1}{2},j+\tfrac{1}{2},k\right) - H_z^{n+1/2}\left(i+\tfrac{1}{2},j+\tfrac{1}{2},k\right)\end{bmatrix}\end{aligned} \tag{6e}$$

$$\begin{aligned}E_z^{n+1}&\left(i,j,k+\tfrac{1}{2}\right)\\ &= \left[1 - \frac{\sigma\left(i,j,k+\tfrac{1}{2}\right)\delta t}{\epsilon\left(i,j,k+\tfrac{1}{2}\right)}\right]E_z^n\left(i,j,k+\tfrac{1}{2}\right) + \frac{\delta t}{\epsilon\left(i,j,k+\tfrac{1}{2}\right)\delta}\\ &\cdot\begin{bmatrix}H_y^{n+1/2}\left(i+\tfrac{1}{2},j,k+\tfrac{1}{2}\right) - H_y^{n+1/2}\left(i-\tfrac{1}{2},j,k+\tfrac{1}{2}\right) + \\ H_x^{n+1/2}\left(i,j-\tfrac{1}{2},k+\tfrac{1}{2}\right) - H_x^{n+1/2}\left(i,j+\tfrac{1}{2},k+\tfrac{1}{2}\right)\end{bmatrix}.\end{aligned} \tag{6f}$$

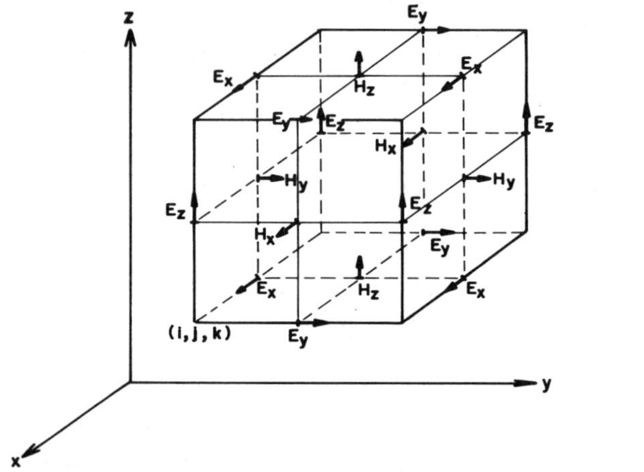

Fig. 1. Positions of the field components about a unit cell of the Yee lattice.

With the system of (6a)–(6f), the new value of a field vector component at any lattice point depends only on its previous value and on the previous values of the components of the other field vector at adjacent points. Therefore, at any given time step, the computation of a field vector may proceed one point at a time. Computer storage must be provided for 11 quantities at each unit cell of the lattice: the 6 field vector components, the values of ϵ and σ, and the maximum $|E_x|$, $|E_y|$, and $|E_z|$ achieved during the final half-wave cycle of time stepping.

The choice of δ and δt is motivated by the reasons of accuracy and stability, respectively. To ensure the accuracy of the computed results, δ must be taken as a small fraction of either the minimum wavelength expected in the model or the minimum scatterer dimension. Thus the field cannot change significantly over one space increment, and the cubic lattice approximation to the smooth scatterer surface cannot be too coarse. To ensure the stability of the time-stepping algorithm of (6a)–(6f), δt is chosen to satisfy

$$v_{\max}\delta t \le \left(\frac{1}{\delta x^2} + \frac{1}{\delta y^2} + \frac{1}{\delta z^2}\right)^{-1/2} \tag{7}$$

where $v_{\max}$ is the maximum wave phase velocity expected within the model. The corresponding stability criterion set forth by Yee in (7) and (8) of his paper is incorrect. The derivation of (7) is outlined in the Appendix.

III. THE LATTICE TRUNCATION CONDITIONS

A basic problem with any finite-difference solution of Maxwell's equations is the treatment of the field vector components at the lattice truncation. Because of limited computer storage, the lattice cannot cover a sufficiently large portion of space so that the scattered wave at the lattice truncation might be closely approximated. Necessarily, the lattice must terminate close to the scatterer in a region where the nature of the scattered wave is unknown.

Proper truncation of the lattice requires that any outgoing wave disappear at the lattice boundary without reflection during the continuous time stepping of the algorithm. Improper truncation results in error for all time steps after the boundary wave reflections return to the vicinity of the scatterer. This is exemplified by Yee's fixing of the field components at each truncation point at zero, for all time steps. This "hard" truncation condition does not take into account the values of the fields of any possible outgoing wave and causes boundary reflections in a way analogous to reflection at the surface of a conductor.

One way to achieve a "soft" lattice truncation is to assign linearly increasing values of σ near the lattice boundary so that absorption of outgoing waves is achieved. However, for a small value of reflection, the distance over which absorption takes place must be of the order of one wavelength. For many problems, this is an intolerable assignment of computer storage to points not in the scatterer. In addition, any incident wave propagating parallel to such a lattice boundary must suffer distortion due to the variation of phase velocity along the wavefront. Therefore, this type of truncation is not suitable.

A more desirable soft lattice truncation relates in a simple way the values of the field components at the truncation points to field component values at points one or more δ within the boundary. This is illustrated using the one-dimensional lattice of Fig. 2, for the time-step relation $c\delta t = \delta$

$$E_z^n(M) = E_z^{n-1}(M-1). \tag{8a}$$

Equation (8a) simulates the free-space propagation of the magnitude of E_z from the point $M-1$ to the truncation point M in one time step (the numerical propagation delay implied by the time-step relation). This is an exact truncation condition for the lattice of Fig. 2, in that all possible $+y$-directed waves are absorbed at M without reflection. If we wish to simulate the truncation of this lattice in an infinite dielectric half-space of refractive index m, (8a) is modified to

$$E_z^n(M) = E_z^{n-m}(M-1). \tag{8b}$$

Unfortunately, no simple, exact soft truncation condition analogous to (8a), (8b) is apparent for either the two- or three-dimensional space lattices. This is because any particular outgoing wave cannot be assumed to be plane and normally incident on one lattice boundary plane. At any truncation point, the local angle of incidence of this wave, relative to the boundary, is unknown. Further, several different waves having varying local angles of incidence may arrive at the same time. No simple, single truncation condition can account for all of these possibilities. Therefore, we can arrive at only an approximate truncation condition that reduces the effective lattice boundary reflection coefficient to an acceptable level.

An example of a set of simple, approximate, soft lattice truncation conditions is illustrated using the two-dimensional lattice of Fig. 3, for the time-step relation $c\delta t = \frac{1}{2}\delta$

$$E_z^n(i,0) = (E_z^{n-2}(i-1,1) + E_z^{n-2}(i,1) + E_z^{n-2}(i+1,1))/3 \tag{9a}$$

$$E_z^n(i,49) = (E_z^{n-2}(i-1,48) + E_z^{n-2}(i,48) + E_z^{n-2}(i+1,48))/3 \tag{9b}$$

$$H_y^n(\tfrac{1}{2},j) = (H_y^{n-2}(1\tfrac{1}{2},j-1) + H_y^{n-2}(1\tfrac{1}{2},j) + H_y^{n-2}(1\tfrac{1}{2},j+1))/3 \tag{9c}$$

$$H_y^n(50\tfrac{1}{2},j) = (H_y^{n-2}(49\tfrac{1}{2},j-1) + H_y^{n-2}(49\tfrac{1}{2},j) + H_y^{n-2}(49\tfrac{1}{2},j+1))/3. \tag{9d}$$

Equations (9a)–(9d) allow the field value at any lattice truncation point to rise to approach the field value of any outgoing wave, thus lowering the effective reflection coefficient at the boundary. This is done by simulating the propagation of an outgoing wave from the lattice plane adjacent to the truncation, to the lattice plane at the truncation, in a number of time steps corresponding to the propagation delay. The averaging process is used to take into account all possible local angles of incidence of the outgoing wave at the lattice boundary, and possible multiple incidences.

The effectiveness of (9a)–(9d) in reducing lattice boundary reflections is illustrated in Fig. 4, which plots the computed propagation of an outgoing cylindrical wave in the grid of Fig. 3. The outgoing wave is a Gaussian pulse originating from an excitation at grid point 25,24. The pulse is approximately 200 ps long, and has a maximum $|E_z| = 1000$. δ is set at 0.3 cm, and δt is set at 5 ps. In the figure, contours of constant E_z are superimposed on the left half of the grid (the right half is omitted because of symmetry) at intervals of 20 time steps. The pulse

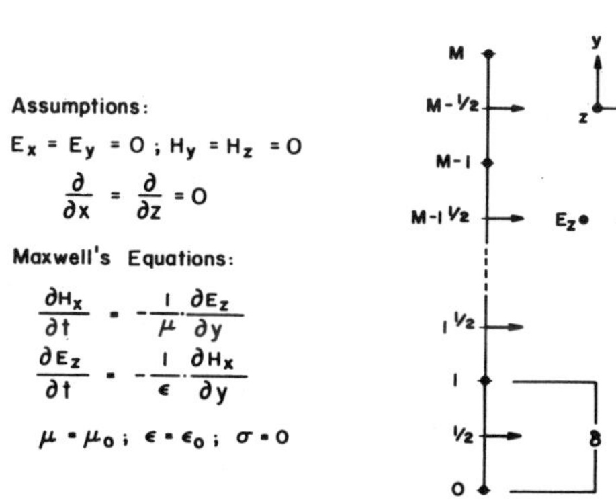

Fig. 2. One-dimensional lattice, illustrating a soft lattice truncation at M.

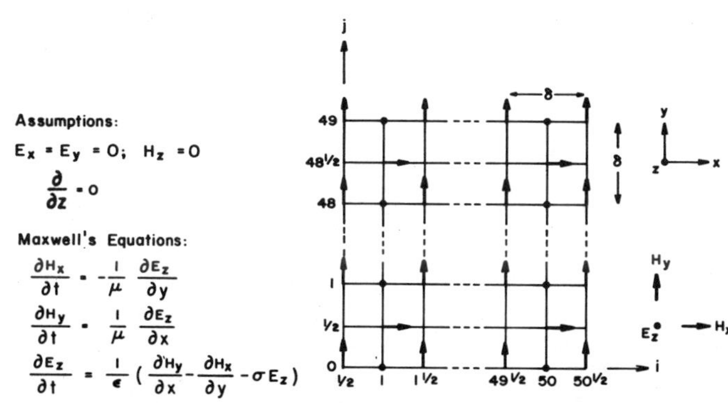

Fig. 3. Two-dimensional lattice, illustrating a set of approximate soft lattice truncation conditions.

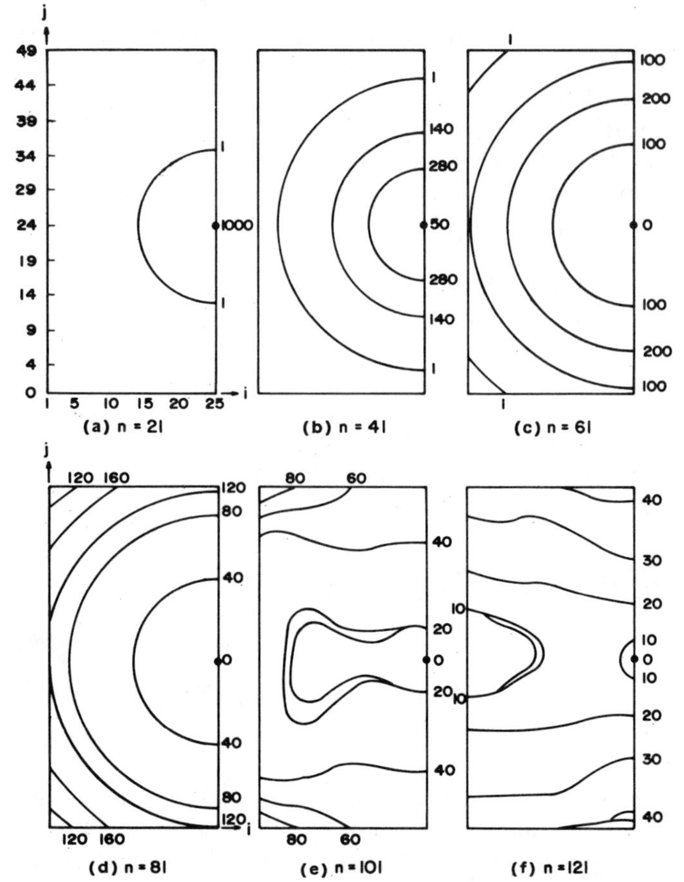

Fig. 4. Propagation of an outgoing cylindrical wave in the grid of Fig. 3.

appears to propagate off the edges of the grid, as desired, with only slight distortion of the nominal cylindrical shape. The residual field magnitude in Fig. 4(e)(f) is about one order of magnitude less than the peak outgoing fields at the grid boundaries in Fig. 4(c) (d). Further, the residual field continues to propagate off the grid. This example shows that (9a)–(9d) reduce grid-boundary reflection to the point where a first-order correct solution of continuous irradiation scattering problems is feasible.

Truncation conditions (9a)–(9d) are useful for an assumed y-directed incident plane wave, with field components E_z and H_x for scattering problems. First, (9a) and (9b) reduce to an exact truncation, similar to (8a), for such a wave. Second, (9c) and (9d) have no effect on the propagation of such a wave, which lacks an H_y component. Therefore, this lattice truncation effectively makes the lattice boundary invisible to a y-directed incident plane wave.

IV. THE PLANE-WAVE-SOURCE CONDITION

We now consider the simulation of a continuous, sinusoidal, incident plane wave for use in scattering problems. The simplest approach is to vary the electric field at all points along one endface of the lattice in a sinusoidal manner. This lattice plane would then radiate the desired plane wave. However, such a specification of field values at a lattice boundary plane, without consideration of the values of the fields of any possible outgoing scattered waves, represents a hard lattice truncation condition and would cause undesired reflections.

An example of a soft plane-wave-source condition is illustrated using the two-dimensional lattice of Fig. 3

$$E_z{}^n(i,2) \leftarrow 1000 \sin\,(2\pi f n\delta t) + E_z{}^n(i,2). \qquad (10)$$

Equation (10) is a modification of the Maxwell's equations algorithm for all points on the grid line $j = 2$. At each point on this line, the computer first calculates $E_z{}^n$ in the normal manner of the algorithm, and stores the value in memory. Then the value of the sinusoid is calculated and added to the stored value of $E_z{}^n$. Finally, this modified value of $E_z{}^n$ is stored in memory, In effect, (10) simulates the linear superposition of a y-directed plane wave and the ambient field along the grid line $j = 2$; (10) generates the desired sinusoidal, incident plane wave. But most importantly, (10) permits any scattered, outgoing wave to propagate right through the plane-wave source at $j = 2$ without reflection and reach the soft grid truncation at $j = 0$ to be absorbed. This condition simulates a plane wave originating at infinity, and a scattered wave returning to infinity, without permitting any interaction between the two waves except at the scatterer.

V. THE SYMMETRY CONDITIONS

An important savings of computer memory and program execution time results if even symmetry of the dielectric scatterer about one or two lattice planes can be assumed. An example of the programming of even scatterer symmetry is illustrated using the two-dimensional lattice of Fig. 3. The scatterer is assumed to be evenly symmetric about the grid line $i = 25\frac{1}{2}$

$$\mu,\epsilon,\sigma(25\tfrac{1}{2} + I,j) = \mu,\epsilon,\sigma(25\tfrac{1}{2} - I,j). \qquad (11a)$$

The incident radiation is assumed to be a $+y$-directed plane wave. The incident field components E_z and H_x are uniform in the x–z plane and thus naturally have even symmetry about $i = 25\frac{1}{2}$. Therefore, the E_z and H_x components of the total field must also possess even symmetry about $i = 25\frac{1}{2}$

$$E_z{}^n,H_x{}^n(25\tfrac{1}{2} + I,j) = E_z{}^n,H_x{}^n(25\tfrac{1}{2} - I,j). \qquad (11b)$$

Using the symmetry, the grid of Fig. 3 may be truncated at the line $i = 26$. The required truncation condition

$$E_z{}^n(26,j) = E_z{}^n(25,j) \qquad (12)$$

allows calculation of the complete set of field components, with full specification of the assumed symmetry of the problem.

VI. RESULTS OF TWO-DIMENSIONAL SCATTERING PROBLEMS

In this section, we shall present the computed results for the internal electric field of a uniform, circular, dielectric cylinder scatterer. The cylinder is assumed to be infinite in the z direction. The incident radiation is assumed to be a $+y$-directed TM wave of frequency 2.5 GHz. Because there is no variation of either scatterer geometry or incident fields in the z direction, this problem may be treated as the two-dimensional scattering of the incident wave, with only E_z-, H_x-, and H_y-fields present. Thus the two-dimensional grid of Fig. 3 is used.

The geometry of the scatterer relative to the grid is illustrated in Fig. 5. The cylinder axis is chosen as the line $25\frac{1}{2},24\frac{1}{2},k$, allowing symmetry condition (12) to be used to truncate the grid at $i = 26$. Soft grid truncation conditions (9a), (9b), and (9c) are used to tuncate the grid at $j = 0$, $j = 49$, and $i = \frac{1}{2}$, respectively. Soft plane-wave-source condition (10) is used to generate the incident wave at $j = 2$. The grid coordinates internal to the cylinder, determined by

$$((i - 25\tfrac{1}{2})^2 + (j - 24\tfrac{1}{2})^2)^{1/2} \leq 20 \tag{13}$$

are assigned the dielectric parameters ϵ_d, μ_0, and σ_d. All grid points outside of the cylinder are assigned the parameters of free space. Equation (13) leads to a stepped-edge approximation of the circular boundary of the scatterer. The program is started by setting all field components of the grid equal to zero. The plane-wave source is activated at $n = 1$, the first time step of the algorithm, and left on during the entire run. The program is time stepped to $n_{\max}$ chosen large enough so that the sinusoidal steady state is achieved.

The first cylinder scatterer program has the following parameters: $\epsilon_d = 4\epsilon_0$, $\sigma_d = 0$, $\delta = 0.3$ cm $= \lambda_d/20$, $\delta t = 5$ ps $= \delta/2c$. The choice of δ implies that the cylinder has a radius equal to λ_d. The choice of δt implies that one wave cycle requires 80 time steps of the algorithm. The computed results of this program are detailed in Fig. 6(a), which graphs the envelope of $E_z^n(25,j)$, and in Fig. 6(b), which graphs the envelope of $E_z^n(15,j)$ for $460 \leq n \leq 500 = n_{\max}$. The exact solution, calculated using the summed-series technique of Jones [7], is plotted with each computed solution for comparison. The computed solution locates the positions of the peaks and nulls of the envelope of E_z^n

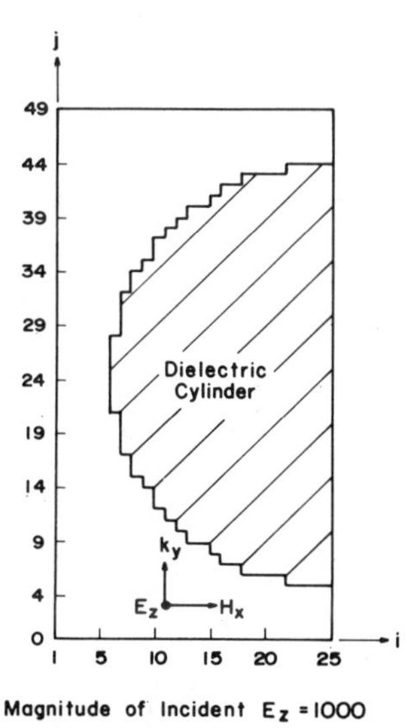

Fig. 5. Geometry of the cylindrical dielectric scatterer relative to the grid of Fig. 3.

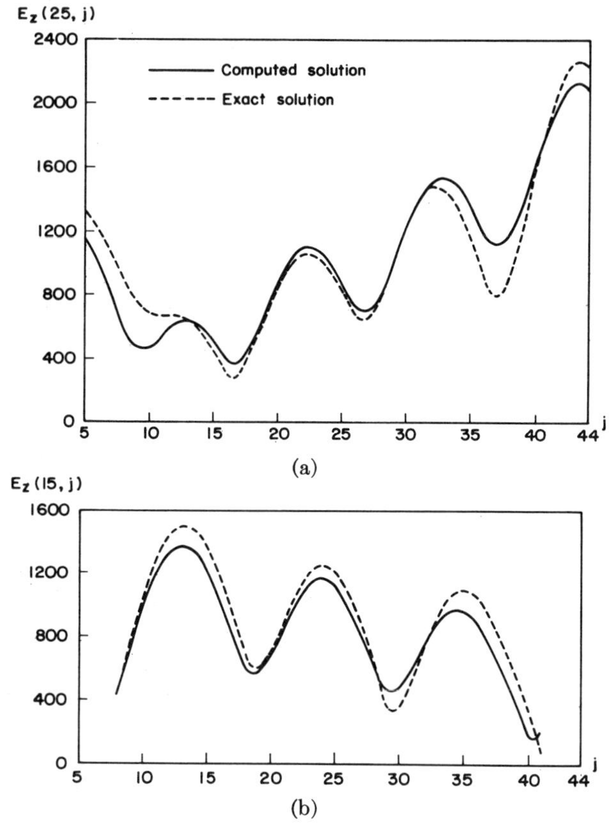

Fig. 6. Results of the first cylinder scatterer program.

with a maximum error of $\pm\delta$, or about ± 3 percent of the diameter of the scatterer. The computed solution gives the magnitude of the envelope at each peak with a maximum error of ± 10 percent. The execution time is 50 s using the CDC 6400.[1]

The second cylinder scatterer program has the following parameters: $\epsilon_d = 47\epsilon_0$, $\sigma_d = 2.2$ mho/m, $\delta = 0.6$ mm $\approx \lambda_d/28$, $\delta t = 1$ ps $= \delta/2c$. The choice of δ implies that the cylinder has a radius equal to $\frac{2}{3}\lambda_d$. The choice of δt implies that one wave cycle requires 400 time steps of the algorithm. The dielectric parameters are chosen to simulate human tissue with high water content. The radius of the scatterer is chosen to equal that of the eyeball. The computed results of this program are detailed in Fig. 7(a), which graphs the envelope of $E_z^n(25,j)$, and in Fig. 7(b), which graphs the envelope of $E_z^n(15,j)$ for $400 \leq n \leq 600$. The exact solution is plotted with each computed solution for comparison. The computed solution locates the positions of the peaks and nulls of the envelope with a maximum error of $\pm 3\delta$, or about ± 8 percent of the diameter of the scatterer. The computed solution gives the magnitude of the central peak of the envelope with an error of ± 5 percent. The execution time is 60 s.

There are two main sources of error in the results of Figs. 6 and 7. The first is the imperfection of the soft grid truncation of (9a)–(9c). The second is the stepped-edge ap-

[1] The listing of the 134-card Fortran IV source deck is available from the authors.

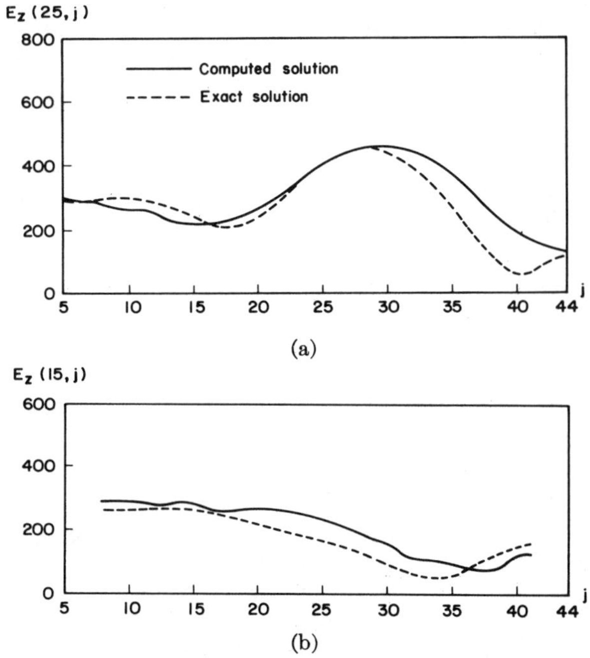

Fig. 7. Results of the second cylinder scatterer program.

proximation of the boundary of the scatterer. These error sources decrease linearly with δ, at most. Overall, the solutions of Figs. 6 and 7 may be described as better than first-order accurate. Comparable accuracy should be obtained with arbitrary two-dimensional scatterers of the general size of the cylinders examined. The only program modification required is a specification of the dielectric parameters of the arbitrary scatterer at each grid point.

VII. EXTENSION TO THREE-DIMENSIONAL SCATTERERS

The solution of three-dimensional dielectric scattering problems with this technique requires the full Yee algorithm of (6a)–(6f), in conjunction with the corresponding lattice. A sufficiently accurate lattice truncation condition, similar to (9a)–(9d), is formulated. For this case, care must be taken in setting the truncation condition to avoid algorithm instability. This may be seen by considering the application of (7) to a cubic lattice in two or three dimensions

$$\delta t \leq \frac{\delta}{c\sqrt{2}} = 0.707\delta/c \qquad \text{(2 dimensions)} \tag{14a}$$

$$\delta t \leq \frac{\delta}{c\sqrt{3}} = 0.577\delta/c \qquad \text{(3 dimensions).} \tag{14b}$$

If, for convenience in programming the truncation condition, the relation $\delta t = 0.5\delta/c$ is used for both lattices, the three-dimensional lattice algorithm is closer to instability. Therefore, any perturbation of the basic algorithm with a soft lattice truncation is more likely to lead to instability in the three-dimensional case. This has been borne out by preliminary efforts in the programming of this case.

APPENDIX

DERIVATION OF THE STABILITY CRITERION

For convenience, we consider a normalized region of space with $\mu = 1$, $\epsilon = 1$, $\sigma = 0$, and $c = 1$. Letting $j = (-1)^{1/2}$, we rewrite Maxwell's equations as

$$j\,\nabla \times (\bar{H} + j\bar{E}) = \frac{\partial}{\partial t}(\bar{H} + j\bar{E}) \tag{15a}$$

or more simply as

$$j\,\nabla \times \bar{V} = \partial\bar{V}/\partial t \quad \text{where} \quad \bar{V} = \bar{H} + j\bar{E}. \tag{15b}$$

The stability of a particular numerical representation of (15b) can be examined simply by considering the following pair of eigenvalue problems:

$$\left.\frac{\partial}{\partial t}\right|_{\text{numerical}} \bar{V} = \lambda\bar{V} \tag{16a}$$

$$j\,\nabla\,|_{\text{numerical}} \times \bar{V} = \lambda\bar{V}. \tag{16b}$$

Using the numerical time derivative of (5), (16a) yields

$$\frac{\bar{V}^{n+1/2} - \bar{V}^{n-1/2}}{\delta t} = \lambda\bar{V}^{n}. \tag{17}$$

Defining a solution growth factor $q = \bar{V}^{n+1/2}/\bar{V}^{n}$, and substituting into (17), we solve for q

$$q = \lambda\delta t/2 \pm (1 + (\lambda\delta t/2)^2)^{1/2}. \tag{18}$$

Algorithm stability requires that $|q| \leq 1$ for all possible spatial modes in the lattice. For this to occur

$$\mathrm{Re}\,\lambda = 0 \quad |\mathrm{Im}\,\lambda| \leq 2/\delta t. \tag{19}$$

We now let

$$\bar{V}(l,m,n) = \bar{V}_0 \exp\left[j(k_x l\delta x + k_y m\delta y + k_z n\delta z)\right] \tag{20}$$

represent an arbitrary lattice spatial mode. Using the numerical space derivative formulation of (4), (16b) yields

$$-2\left[\frac{\sin(\frac{1}{2}k_x\delta x)}{\delta x}, \frac{\sin(\frac{1}{2}k_y\delta y)}{\delta y}, \frac{\sin(\frac{1}{2}k_z\delta z)}{\delta z}\right] \times \bar{V}(l,m,n) = \lambda\bar{V}(l,m,n). \tag{21}$$

After performing the cross product and writing the x, y, and z component equations, the resulting system is solved for λ^2

$$\lambda^2 = -4\left(\frac{\sin^2(\frac{1}{2}k_x\delta x)}{\delta x^2} + \frac{\sin^2(\frac{1}{2}k_y\delta y)}{\delta y^2} + \frac{\sin^2(\frac{1}{2}k_z\delta z)}{\delta z^2}\right). \tag{22}$$

For all possible k_x, k_y, and k_z

$$\mathrm{Re}\,\lambda = 0 \quad |\mathrm{Im}\,\lambda| \leq 2\left(\frac{1}{\delta x^2} + \frac{1}{\delta y^2} + \frac{1}{\delta z^2}\right)^{1/2}. \tag{23}$$

To satisfy stability condition (19) for the arbitrary lattice spatial mode, we set

$$2\left(\frac{1}{\delta x^2}+\frac{1}{\delta y^2}+\frac{1}{\delta z^2}\right)^{1/2} \le \frac{2}{\delta t}. \tag{24}$$

The algorithm stability condition follows immediately from (24). In an inhomogeneous region of space, it is difficult to determine a spectrum of λ analogous to (23) for all possible lattice spatial modes. For absolute algorithm stability, (7) suffices because it represents a "worst case" choice of δt.

ACKNOWLEDGMENT

The authors wish to thank Dr. G. O. Roberts of the Engineering Sciences Department of Northwestern University for his useful comments on algorithm stability.

REFERENCES

[1] T. K. Wu and L. L. Tsai, "Numerical analysis of electromagnetic fields in biological tissues," *Proc. IEEE* (Lett.), vol. 62, pp. 1167–1168, Aug. 1974.

[2] R. F. Harrington, *Field Computation by Moment Methods.* New York: Macmillan, 1968, ch. 3.

[3] B. H. McDonald and A. Wexler, "Finite-element solution of unbounded field problems," *IEEE Trans. Microwave Theory Tech.* (*1972 Symp. Issue*), vol. MTT-20, pp. 841–847, Dec. 1972.

[4] D. R. Wilton and R. Mittra, "A new numerical approach to the calculation of electromagnetic scattering properties of two-dimensional bodies of arbitrary cross section," *IEEE Trans. Antennas Propagat.*, vol. AP-20, pp. 310–317, May 1972.

[5] Vogelback Computing Center, Northwestern Univ., Evanston, Ill., Library no. NUCC288, Subroutine LUECS.

[6] K. S. Yee, "Numerical solution of initial boundary value problems involving Maxwell's equations in isotropic media," *IEEE Trans. Antennas Propagat.*, vol. AP-14, pp. 302–307, May 1966.

[7] D. S. Jones, *The Theory of Electromagnetism.* New York: Macmillan, 1964, pp. 450–452.

Absorbing Boundary Conditions for the Finite-Difference Approximation of the Time-Domain Electromagnetic-Field Equations

GERRIT MUR

Abstract—**When time-domain electromagnetic-field equations are solved using finite-difference techniques in unbounded space, there must be a method limiting the domain in which the field is computed. This is achieved by truncating the mesh and using absorbing boundary conditions at its artificial boundaries to simulate the unbounded surroundings. This paper presents highly absorbing boundary conditions for electromagnetic-field equations that can be used for both two- and three-dimensional configurations. Numerical results are given that clearly exhibit the accuracy and limits of applicability of highly absorbing boundary conditions. A simplified, but equally accurate, absorbing condition is derived for two-dimensional time-domain electromagnetic-field problems.**

Key Words—**Electromagnetic-field equations, time domain, finite-difference approximation, absorbing boundary conditions.**

I. INTRODUCTION

THE THREE-DIMENSIONAL finite-difference formulation [1] of time-domain electromagnetic-field problems is a convenient tool for solving scattering problems. The main advantages of such a method are that it can be easily applied to (infinitely) conducting obstacles and/or to dielectric and magnetic obstacles which can be either homogeneous or inhomogeneous. The obstacles can be of arbitrary shape. Furthermore, finite-difference techniques provide a very efficient way of solving Maxwell's equations [2]. In a finite-difference method, a space-time mesh is introduced and Maxwell's equations are replaced by a system of finite-difference equations on the mesh. The difficulty encountered when trying to solve field problems in this way arises from the fact that scattering problems are usually *open* problems, i.e., the domain in which the field has to be computed is unbounded. Since no computer can store an unlimited amount of data, a method has to be used for limiting the domain in which the field is computed. This is done by using a mesh of limited size, but one large enough to fully contain the obstacle, and by using a boundary condition on the outer surface of the mesh such that the unbounded surrounding is modeled as accurately as possible. Boundary conditions of this type are called absorbing boundary conditions. For finite-difference approximations of Maxwell's equations, absorbing boundary conditions have been described by Taylor *et al.* [3], who use a simple extrapolation method, and by Taflove and Brodwin [4], who simulate the outgoing waves and use an averaging process in an attempt to account for all possible angles of propagation of the outgoing waves. An alternative method, suggested by Taflove [5], is to introduce losses in the region that surrounds the structure to be modeled, thus absorbing both the outgoing waves and

Manuscript received March 19, 1981; revised July 9, 1981. This work was supported in part by a fellowship from the Royal Society.

The author was with the Department of Electrical and Electronic Engineering, University of Nottingham, Nottingham, England. He is now with the Laboratory of Electromagnetic Research, Department of Electrical Engineering, Delft University of Technology, P.O. Box 5031 2600 GA Delft, The Netherlands. (015) 786294

Reprinted from *IEEE Transactions on Electromagnetic Compatibility*, pp. 377-382, November 1981.

the waves that are reflected by the boundaries of the mesh. Numerical experiments, however, have shown that, in order to obtain accurate results, a relatively thick conducting layer is required which makes this way of absorbing the outgoing waves inefficient. Merewether [6] and Kunz and Lee [7] use the radiation condition at large distances from the center of the scatterer to obtain an absorbing boundary condition.

The absorbing boundary conditions mentioned above, although useful as first approximations, have the disadvantage of causing considerable reflections when the fields near the boundary of the mesh do not propagate in a specific direction (either the direction normal to the boundary of the mesh or the radial direction from the center of the obstacle). Furthermore, no general theory for improving upon these approximations is available. In this paper, a potentially superior method to those given above is derived for both the two- and three-dimensional electromagnetic-field equations; it is similar to the scalar derivation of Engquist and Majda [8]. The first approximation in the sense of this method appears to be comparable to those given above. The second and higher approximations are less subject to reflection problems, especially for fields grazing the outer boundary, making them highly absorbing boundary conditions. For two-dimensional fields, a simplified, but equally accurate, absorbing boundary condition will be derived. Numerical results will be presented to elucidate the usefulness of highly absorbing boundary conditions.

II. FINITE-DIFFERENCE APPROXIMATION OF MAXWELL'S EQUATIONS

Since, in this paper, we do not concern ourselves with obstacles or inhomogeneities but with the homogeneous region that surrounds them, we can, without loss of generality, confine our attention to Maxwell's equations for a vacuum region. In rectangular Cartesian coordinates we have

$$\mu_0 \partial_t \boldsymbol{H} = -\nabla \times \boldsymbol{E} \tag{1a}$$

$$\epsilon_0 \partial_t \boldsymbol{E} = \nabla \times \boldsymbol{H} \tag{1b}$$

where ϵ_0 and μ_0 denote the permittivity and the permeability of a vacuum, respectively. We now introduce a finite-difference approximation of (1) and, following Yee's notation [1], we denote a mesh point as

$$(i, j, k) = (i\delta, j\delta, k\delta)$$

where $\delta = \delta x = \delta y = \delta z$ is the space increment, and any function of space and time as

$$F^n(i, j, k) = F(i\delta, j\delta, k\delta, n\delta t)$$

where δt is the time increment. By positioning the field components of $\boldsymbol{E}$ and $\boldsymbol{H}$ on the mesh in the way that is depicted in Fig. 1 and evaluating $\boldsymbol{E}$ and $\boldsymbol{H}$ at alternate half-time steps, Yee obtained finite-difference expressions that have a local trucation error of the second order in all increments. The finite-difference approximation for the x-component of (1a) and

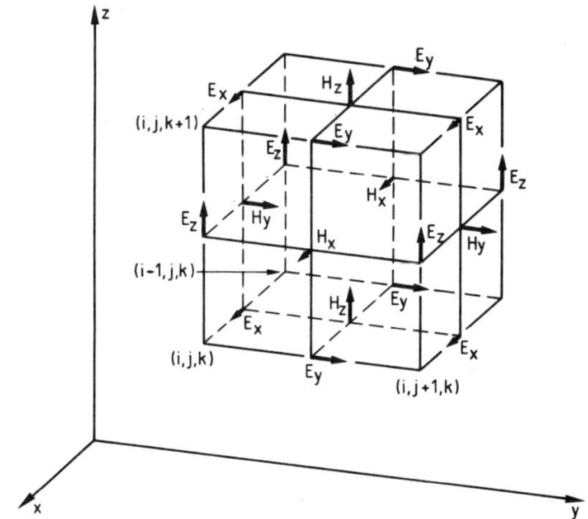

Fig. 1. The positions of the field components in Yee's mesh.

(1b) now reads

$$\begin{aligned} H_x^{n+1/2}&(i, j+1/2, k+1/2) \\ &= H_x^{n-1/2}(i, j+1/2, k+1/2) - (\delta t/\mu_0\delta) \\ &\quad \cdot (E_z^n(i, j+1, k+1/2) - E_z^n(i, j, k+1/2) \\ &\quad - E_y^n(i, j+1/2, k+1) + E_y^n(i, j+1/2, k)) \end{aligned} \tag{2a}$$

$$\begin{aligned} E_x^{n+1}&(i+1/2, j, k) \\ &= E_x^n(i+1/2, j, k) + (\delta t/\epsilon_0\delta) \\ &\quad \cdot (H_z^{n+1/2}(i+1/2, j+1/2, k) \\ &\quad - H_z^{n+1/2}(i+1/2, j-1/2, k) \\ &\quad - H_y^{n+1/2}(i+1/2, j, k+1/2) \\ &\quad + H_y^{n+1/2}(i+1/2, j, k-1/2)). \end{aligned} \tag{2b}$$

Similar finite-difference approximations for the y- and z-components of (1) can easily be derived. The condition for stability of (2) is

$$\delta t \leqslant \delta/(c_0\sqrt{3}) \tag{3}$$

where $c_0 = (\epsilon_0\mu_0)^{-1/2}$ denotes the speed of light in vacuo.

For a two-dimensional problem, we use only a single plane in the mesh and consequently (2) is considerably simplified. The condition for stability then reads

$$\delta t \leqslant \delta/(c_0\sqrt{2}). \tag{4}$$

In each of the coordinate directions, the mesh is truncated by enclosing it between two planes that are normal to the relevant coordinate axis and pass through mesh points (i, j, k), i, j, k integer. We then observe from Fig. 1 that all components of the electric-field vector $\boldsymbol{E}$ occurring in (2) applied to a particular point in the boundary of the mesh are tangential to this boundary while the relevant components of the magnetic-field vector $\boldsymbol{H}$ are normal to it. The latter field components can be evaluated by using the relevant finite-difference equation. The E-field components, however, cannot be evaluated in this way since this would require H-field components that are outside the mesh. In the next section, we shall describe absorbing boundary conditions that can be used for computing these E-field components. For absorbing boundary conditions, it is assumed that the fields near the boundary are outgoing waves and, therefore, they can be applied to only scattered fields, i.e., those that find their origin somewhere near the center of the mesh. Consequently, incident fields, if present on the mesh, should be subtracted from the total field near the boundary of the mesh. Finally, we note that, upon elimination of $\boldsymbol{H}$ from (2), we obtain

$$(\partial_x{}^2 + \partial_y{}^2 + \partial_z{}^2 - c_0{}^{-2}\partial_t{}^2)\,\boldsymbol{E} = \boldsymbol{0} \tag{5}$$

i.e., each component of the electric field independently satisfies the three-dimensional wave equation.

III. ABSORBING BOUNDARY CONDITIONS

In the previous section, we saw that absorbing boundary conditions for Maxwell's equations on the mesh depicted in Fig. 1 require absorbing boundary conditions for the three components of the electric field only. We also saw that each of these field components satisfies the three-dimensional scalar wave equation

$$(\partial_x{}^2 + \partial_y{}^2 + \partial_z{}^2 - c_0{}^{-2}\partial_t{}^2)W = 0. \tag{6}$$

In this section, we will present the necessary boundary conditions by using the method that is described in detail by Engquist and Majda [8]. We shall assume, without loss of generality, that the mesh is located in the region $0 \leqslant x$, and give boundary conditions for the plane $x = 0$.

A space-time plane-wave constituent traveling in the direction of decreasing x, with inverse velocity components s_x, s_y and s_z such that $s_x{}^2 + s_y{}^2 + s_z{}^2 = c_0{}^{-2}$, can be written as

$$W = \mathrm{Re}\,(\psi(t + (c_0{}^{-2} - s_y{}^2 - s_z{}^2)^{1/2}\,x + s_y y + s_z z)) \tag{7}$$

with $\mathrm{Re}(c_0{}^{-2} - s_y{}^2 - s_z{}^2)^{1/2} \geqslant 0$. For this outgoing wave, the first-order boundary condition

$$(\partial_x - c_0{}^{-1}(1 - (c_0 s_y)^2 - (c_0 s_z)^2)^{1/2}\partial_t)W|_{x=0} = 0 \tag{8}$$

would, for fixed values of s_y and s_z, determine a W on the outer surface that is consistent with an outgoing wave, i.e., it is absorbed. Since we do not know the angle of incidence of the wave approaching $x = 0$, an approximation in (8) is made. Writing

$$(1 - (c_0 s_y)^2 - (c_0 s_z)^2)^{1/2} = 1 + 0((c_0 s_y)^2 + (c_0 s_z)^2) \tag{9}$$

we obtain as a first approximation

$$(\partial_x - c_0{}^{-1}\partial_t)W|_{x=0} = 0. \tag{10}$$

Use of the next approximation to the square root

$$(1 - (c_0 s_y)^2 - (c_0 s_z)^2)^{1/2} = 1 - \tfrac{1}{2}((c_0 s_y)^2 + (c_0 s_z)^2) + 0(((c_0 s_y)^2 + (c_0 s_z)^2)^2) \tag{11}$$

yields the second approximation of the boundary condition

$$(c_0{}^{-1}\partial_{xt}{}^2 - c_0{}^{-2}\partial_t{}^2 + \tfrac{1}{2}(\partial_y{}^2 + \partial_z{}^2))W|_{x=0} = 0. \tag{12}$$

From (10) and (12), approximations of W can be determined on the outer boundary. Engquist and Majda, who arrived at the same boundary conditions using a different method, prove that these boundary conditions give well-posed initial-boundary-value problems. They also give a third approximation that they claim to be stable. Numerical experiments, in a two-dimensional configuration, with a difference approximation of the latter boundary condition, however, showed instabilities for values of $c_0\delta t/\delta$ near the maximum that is given by (4).

For two-dimensional electromagnetic-field problems, it is possible to simplify the second approximation. Assuming that the fields do not depend on z and are E-polarized, i.e., $\boldsymbol{E} = E_z \boldsymbol{i}_z$ and $\boldsymbol{H} = H_x \boldsymbol{i}_x + H_y \boldsymbol{i}_y$, (12) applies to E_z only. Now, in this case, we have from (1a)

$$\mu_0 \partial_t H_x = -\partial_y E_z. \tag{13}$$

Substituting (13) in (11), ($\partial_z \equiv 0$), with $W = E_z$, integrating with respect to t and using $E_z = 0$ for $t < 0$, we obtain

$$(\partial_x E_z - c_0{}^{-1}\partial_t E_z - (c_0\mu_0/2)\partial_y H_x)_{x=0} = 0 \tag{14}$$

which boundary condition is much simpler than (12) but still of the same order of approximation. We observe that the first two terms of (14) reduce to the first approximation. A similar boundary condition for the case of H-polarization can easily be derived.

IV. FINITE-DIFFERENCE APPROXIMATIONS OF ABSORBING BOUNDARY CONDITIONS

In this section, we present the finite-difference approximation of the absorbing boundary conditions from Section III. These approximations have a local truncation error of the second order in all increments. As we have seen above, we need absorbing boundary conditions for the E-field components that are tangential to the boundary of interest. Therefore, boundary conditions for the plane $x = 0$ are expressed in terms of E_y and E_z. We shall give the discretized form of the boundary condition for E_z at this boundary, the boundary conditions for E_y and the boundary conditions on the other planes following easily from the one for E_z. The finite-difference approximation of (10) was derived using centered differences in both the space and the time increments, it has a local truncation error of the second order in all increments. We prefer to present the actual formulas in a form that is directly manageable for the computer program. The first approximation (10) for E_z is discretized as follows:

$$E_z^{n+1}(0,j,k+1/2 = E_z^{n}(1,j,k+1/2) + \frac{c_0\delta t-\delta}{c_0\delta t+\delta}(E_z^{n+1}(1,j,k+1/2) - E_z^{n}(0,j,k+1/2)). \quad (15)$$

The second approximation (12) for E_z at the boundary $x = 0$ is discretized as

$$\begin{aligned} &E_z^{n+1}(0,j,k+1/2) \\ &= -E_z^{n-1}(1,j,k+1/2) + \frac{c_0\delta t-\delta}{c_0\delta t+\delta}(E_z^{n+1}(1,j,k+1/2) \\ &+ E_z^{n-1}(0,j,k+1/2)) + \frac{2\delta}{c_0\delta t+\delta}(E_z^{n}(0,j,k+1/2) \\ &+ E_z^{n}(1,j,k+1/2)) + \frac{(c_0\delta t)^2}{2\delta(c_0\delta t+\delta)} \\ &\cdot (E_z^{n}(0,j+1,k+1/2) - 2E_z^{n}(0,j,k+1/2) \\ &+ E_z^{n}(0,j-1,k+1/2) + E_z^{n}(1,j+1,k+1/2) \\ &- 2E_z^{n}(1,j,k+1/2) + E_z^{n}(1,j-1,k+1/2) \\ &+ E_z^{n}(0,j,k+3/2) - 2E_z^{n}(0,j,k+1/2) \\ &+ E_z^{n}(0,j,k-1/2) + E_z^{n}(1,j,k+3/2) \\ &- 2E_z^{n}(1,j,k+1/2) + E_z^{n}(1,j,k-1/2)). \quad (16) \end{aligned}$$

Finally, the second approximation (14) for the two-dimensional problem is discretized as

$$\begin{aligned} E_z^{n+1}(0,j) &= E_z^{n}(1,j) + \frac{c_0\delta t-\delta}{c_0\delta t+\delta}(E_z^{n+1}(1,j) \\ &- E_z^{n}(0,j)) - \frac{\mu_0 c_0}{2(c_0\delta t+\delta)} \\ &(H_x^{n+1/2}(0,j+1/2) - H_x^{n+1/2}(0,j-1/2) \\ &+ H_x^{n+1/2}(1,j+1/2) - H_x^{n+1/2}(1,j-1/2)) \quad (17) \end{aligned}$$

where we have deleted the z-dependence of the fields from our notation since the value of z is the same in all terms. Centered differences were used for deriving (16) and (17) and these finite-difference approximations also have a local truncation error of the second order in all increments. We note that the discretization of the first approximation (15) is identical to the discretization in [8]. The discretization of the second approximation (16), however, differs from the one in [8]. Ours has the advantage that it can be used closer to the vertices of the mesh that it requires less storage.

V. NUMERICAL RESULTS

In this section, we present numerical results that show the efficiency of highly absorbing boundary conditions. Since we need a relatively large-size mesh to exhibit clearly the properties of absorbing boundary conditions, we shall present results for two-dimensional configurations only; three-dimensional meshes of considerable size would give rise to storage requirements that were too large for the computer that was available to the author at the time this research was carried out. The insight gained by studying a two-dimensional configuration can easily be applied to three-dimensional configurations since the structure of (16) is the same for two- and three-dimensional problems. (The two-dimensional form of (16) is obtained by deleting from it the differentiations along the axis of cylindricity.) We present results for fields that do not depend on z and are E-polarized (i.e., $E = E_z i_z$). A square two-dimensional mesh is used ($0 \leqslant i \leqslant 34$, $0 \leqslant j \leqslant 34$) with absorbing boundary conditions on all four sides. We have a monochromatic isotropic point source of wavelength λ that is switched on at $t = 0$ and we use $\delta = \delta_x = \delta_y = 2c_0\delta t = 0.1\lambda$. This point source is modeled by adding a term representing a current

$$I_z(t) = C\sin(2\pi c_0 t/\lambda)\epsilon(t) \quad (18)$$

at $r_s = (x_s, y_s) = (i\delta, j\delta)$ to the relevant finite-difference equation. In (18), $\epsilon(t)$ denotes the Heaviside unit-step function. The results are given after 141 steps.

An isotropic point source has a circular radiation pattern and we investigate how well this pattern is maintained on the truncated mesh. Fig. 2 gives a contour plot of the radiation pattern on the mesh for a point source that is located at node $(i, j) = (5, 5)$, which location was chosen to show, as clearly as possible, the efficiency of highly absorbing boundary conditions. The numbers in the contour plot are proportional to the local electric-field strength.

The contour plot is given for both the first and the second approximation. The first approximation turns out to yield relatively poor results, especially at large distances from the source. With the second approximation, however, an almost circular pattern is obtained with only slight deformations at

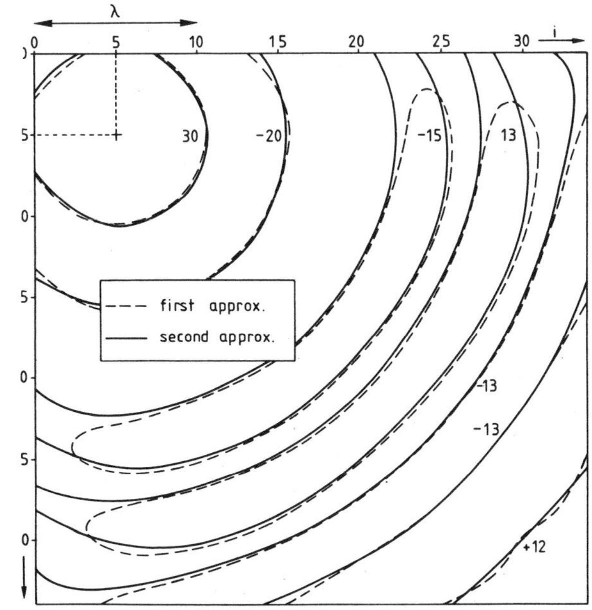

Fig. 2. Contour plot of the radiation pattern, after 141 time steps, of an isotropic source located at node (5, 5) of a 35*35-node mesh (arbitrary units).

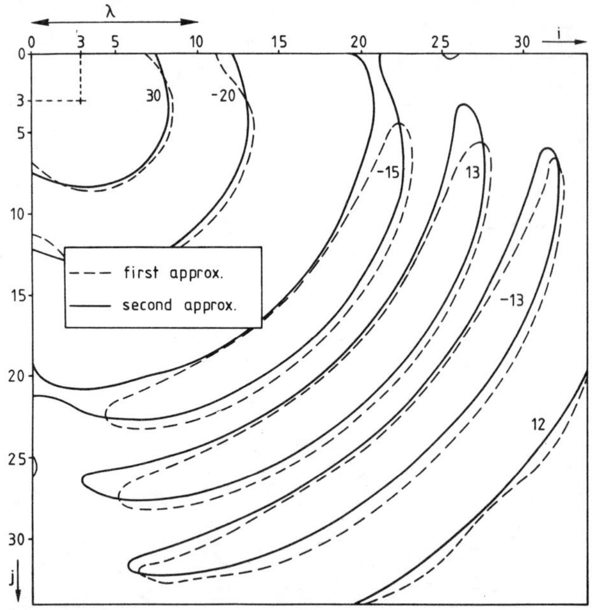

Fig. 3. Contour plot of the radiation pattern, after 141 time steps, of an isotropic source located at node (3, 3) of a 35*35-node mesh (arbitrary units).

points near the boundaries and far away from the source. These errors are caused by the fact that waves with grazing incidence on a boundary are not well absorbed, but partly reflected. For comparison, Fig. 3 gives the contour plot of the field of the same source that is now located closer to the boundary of the mesh, at node $(i, j) = (3, 3)$. A comparison with Fig. 2 is possible since the same contour lines are given. From Fig. 3, we observe that, near the boundary of the mesh, the second approximation is not accurate enough for sources this close to the boundary. It can be concluded that, on a mesh of the given dimensions, accurate results are obtained when the second approximation is used and when the source of the scattered field is located at about five or more nodes from the boundary of the mesh. With the first approximation, less accurate results are obtained and the source of the scattered field should be relatively far away from the boundary. Since (16) is easy to implement, it is advantageous to use the second approximation.

As to the simplified second approximation for two-dimensional fields (17), we note that it gives exactly the same results as the more complicated one that follows from (16).

For the point source under consideration, the exact radiated field is the solution of

$$(\partial_x{}^2 + \partial_y{}^2 - c_0{}^{-2}\partial_t{}^2)E_z = \mu_0 \partial_t I_z(t)\delta(r - r_s) \tag{19}$$

that is rotationally symmetric around r_s and that consists of outgoing waves. This solution is easily obtained [10] as

$$E_z(r, t) = -(\mu_0/2\pi) \int_0^\infty \partial_t I_z(t - (|r - r_s|^2 + \zeta^2)^{1/2}/c_0) \cdot (|r - r_s|^2 + \zeta^2)^{-1/2}\, d\zeta. \tag{20}$$

In Fig. 4, this exact solution is plotted as a function of the distance from the point source and at the same instant in time as the results that are given in Fig. 2. For comparison we have,

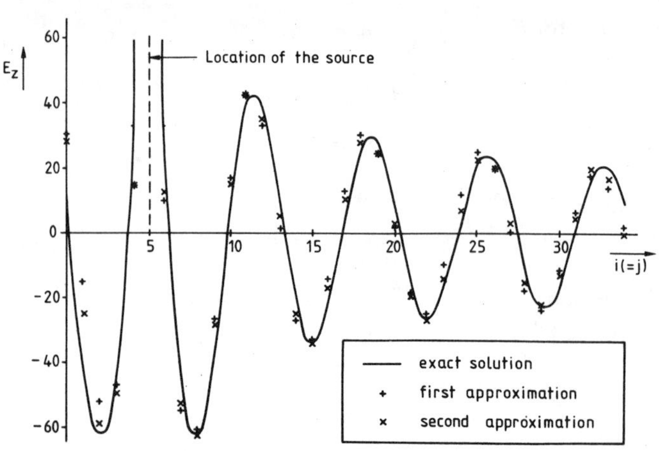

Fig. 4. A comparison of the results on the diagonal of the mesh of Fig. 2 with the exact solution.

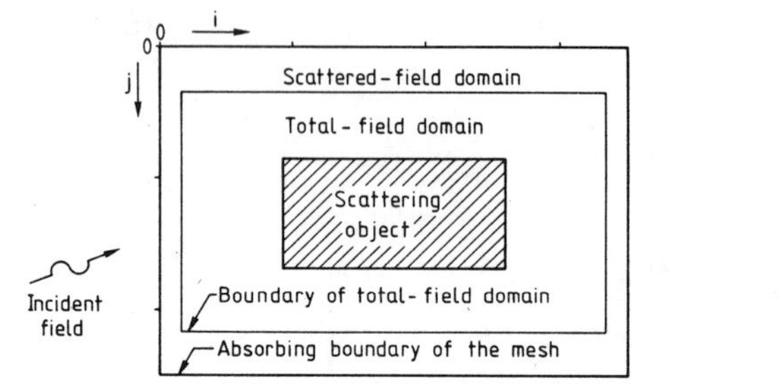

Fig. 5. The domain occupied by the mesh with the obstacle, the domain where the total field is computed, and the domain where the scattered field is computed (two-dimensional configuration).

in Fig. 4, also given the finite-difference results that are obtained on the diagonal of the mesh of Fig. 2 that passes through the point source. It turns out that, considering the fact that we have a relatively coarse mesh ($\delta = \lambda/10$), we have obtained accurate results.

VI. THE INCIDENT FIELD

In the present paper, we have used a point source to generate outgoing waves at the boundaries of the mesh. In practice, the outgoing waves are not generated by a source in the mesh but represent the scattered field $\boldsymbol{E}^s$ that is caused by the presence of an obstacle that scatters the incident field $\boldsymbol{E}^i$ (the incident field is assumed to be known.) If maximum accuracy is desired, it is advantageous to compute the total field $\boldsymbol{E} = \boldsymbol{E}^s + \boldsymbol{E}^i$ rather than the scattered field [5], [9], especially in those regions where the scattered field almost cancels the incident field (behind or in the interior of scattering objects). Absorbing boundary conditions, however, cannot be applied to the total field.

To overcome this difficulty, a boundary can be introduced that is located close to the absorbing boundaries of the mesh (see Fig. 5.) Inside this boundary the total field is computed, outside it the scattered field is computed, thus allowing the application of absorbing boundary conditions. When (2) is evaluated for the total (scattered) field at a node inside (outside) this boundary, but adjacent to it, the total field will often be required from nodes that are located outside (inside) it. For the relevant nodes, (2) is modified. When computing the total field inside the boundary, the incident field is added to the scattered field that is obtained from nodes outside the boundary and, similarly, when computing the scattered field outside the boundary, the incident field is subtracted from the total field obtained from inside the boundary. Since the incident field is known exactly at any place or instant in time the above procedure does not involve any approximations, moreover, it does not require any storage nor does it cause any spurious wave reflection.

ACKNOWLEDGMENT

The author wants to thank A. T. de Hoop for his suggestions concerning the work presented in this paper.

REFERENCES

[1] K. S. Yee, "Numerical solution of initial boundary value problems involving Maxwell's equations in isotropic media," *IEEE Trans. Antennas Propagat.*, vol. AP-14, pp. 302–307, May 1966.

[2] C. L. Longmire, "State of the art in IEMP and SGEMP calculations," *IEEE Trans. Nucl. Sci.*, vol. NS-22, pp. 2340–2344, Dec. 1975.

[3] C. D. Taylor, D.-H. Lam, and T. H. Shumpert, "Electromagnetic pulse scattering in time-varying inhomogeneous media," *IEEE Trans. Antennas Propagat.*, vol. AP-17, pp. 585–589, Sept. 1969.

[4] A. Taflove and M. E. Brodwin, "Numerical solution of steady-state electromagnetic scattering problems using the time dependent Maxwell's equations," *IEEE Trans. Microwave Theory Tech.*, vol. MTT-23, pp. 623–630, Aug. 1975.

[5] A. Taflove, "Application of the finite-difference time-domain method to sinusoidal steady-state electromagnetic penetration problems," *IEEE Trans. Electromagn. Compat.*, vol. EMC-22, pp. 191–202, 1980.

[6] D. E. Merewether, "Transient currents induced on a metallic body of revolution by an electromagnetic pulse," *IEEE Trans. Electromagn. Compat.*, vol. EMC-13, pp. 41–44, May 1971.

[7] K. S. Kunz and K.-M. Lee, "A three-dimensional finite-difference solution of the external response of an aircraft to a complex transient EM environment: Part I—The method and its implementation," *IEEE Trans. Electromagn. Compat.*, vol. EMC-20, pp. 328–333, May 1978.

[8] B. Engquist and A. Majda, "Absorbing boundary conditions for the numerical simulation of waves," *Math. Comp.*, vol. 31, pp. 629–651, July 1977.

[9] R. Holland, L. Simpson, and K. S. Kunz, "Finite-difference analysis of EMP coupling to lossy dielectric structures," *IEEE Trans. Electromagn. Compat.*, vol. EMC-22, pp. 203–209, 1980.

[10] J. D. Achenbach, *Wave Propagation in Elastic Solids*. Amsterdam, The Netherlands: North-Holland, 1973, ch. 3, p. 94.

Finite-Difference Solution of Maxwell's Equations in Generalized Nonorthogonal Coordinates

RICHARD HOLLAND
ELECTRO MAGNETIC APPLICATIONS, INC.
1025 HERMOSA DRIVE SE (87108)
P.O. BOX 8482 ALBUQUERQUE, NEW MEXICO 87198

ABSTRACT

This article describes a time-domain finite-difference algorithm for solving Maxwell's equations in generalized nonorthogonal coordinates. We believe this approach would be most useful for applications where a uniform, uncurved, but oblique, meshing scheme could be applied in lieu of staircasing. This algorithm is similar to conventional leapfrog-differencing schemes, but now an additional leapfrogging between covariant and contravariant field representations becomes necessary.

INTRODUCTION

Many real-world EMP coupling problems are characterized by geometries with oblique angles. In the past, finite-difference codes have all been written in orthogonal coordinates, and oblique surfaces represented by staircasing or interpolation.

In this context, a quotation from Stratton [1] is probably noteworthy: "Oblique systems might well be of the greatest practical importance; but they lead, unfortunately, to partial differential equations which cannot be mastered by present-day analysis." In the 42 years since this statement was written, largely due to the advent of modern computers, its validity has vanished. Time-domain finite-difference codes can be written in oblique systems nearly as easily as in orthogonal systems.

Representing an oblique surface by a staircase requires much finer meshing than representation in a matching oblique coordinate system. To this end, we present Maxwell's equations in generalized oblique coordinates and show how they may be differenced.

SUMMARY OF DIFFERENTIAL EQUATIONS

An oblique coordinate system (u^1, u^2, u^3) may be characterized by its unitary vectors $\underline{a}_i$ (see Figure 1) where

$$d\underline{r} = \frac{\partial \underline{r}}{\partial u^i} du^i = \underline{a}_i \, du^i \tag{1}$$

(repeated indices indicate summation). Equivalently, it may be characterized by its reciprocal unitary vectors (see Figure 2) $\underline{a}^i$

$$\underline{a}^i = \underline{a}_j \times \underline{a}_k / \sqrt{g} \tag{2}$$

where g is the determinant of the metric,

$$g_{ij} = \frac{\partial x^k}{\partial u^i} \frac{\partial x^k}{\partial u^j} \tag{3}$$

and the x^i are cartesian coordinates. The $\underline{a}^i$ point along coordinate lines and the $\underline{a}_i$ are normals to surfaces of constant u^i. For orthogonal coordinates, the $\underline{a}^i$ and $\underline{a}_i$ become colinear. Furthermore, $\underline{a}_i$ and $\underline{a}^i$ are sometimes normalized to unit vectors

$$\underline{i}_i = \frac{\underline{a}_i}{\sqrt{\underline{a}_i \cdot \underline{a}_i}} \qquad \underline{i}^i = \frac{\underline{a}^i}{\sqrt{\underline{a}^i \cdot \underline{a}^i}} \tag{4}$$

The $\underline{a}_i$ and $\underline{a}^i$ have the properties

$$\underline{a}^i \cdot \underline{a}_j = \delta_{ij} \qquad \underline{a}_i \cdot \underline{a}_j = g_{ij} \qquad \underline{a}^i \cdot \underline{a}^j = g^{ij} \tag{5}$$

where g^{ij} is the inverse of g_{ij}.

In general, a vector may be represented by its covariant components

$$\underline{E} = e_i \, \underline{a}^i \tag{6}$$

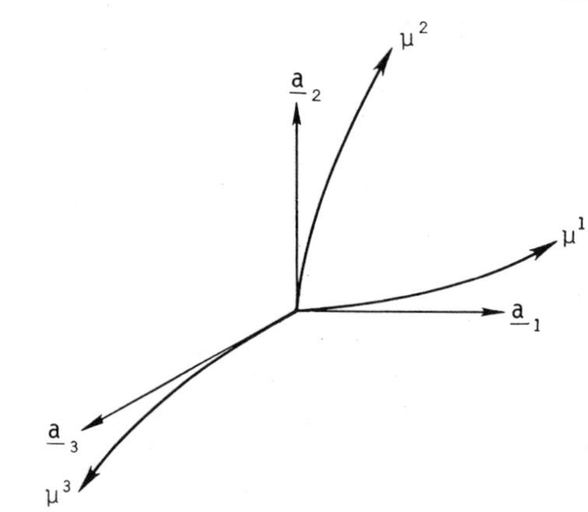

Figure 1. Unitary Vectors for an Oblique Coordinate System

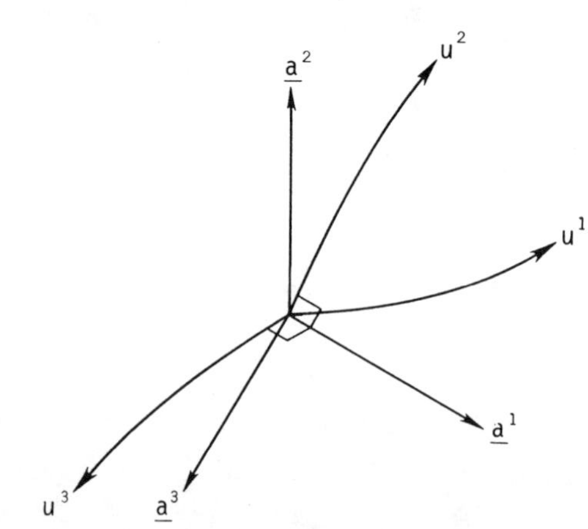

Figure 2. Reciprocal Unitary Vectors for an Oblique Coordinate System

Reprinted from *IEEE Transactions on Nuclear Science*, pp. 4589-4591, December 1983.

or by its contravariant components

$$\underline{E} = e^i \, \underline{a}_i \tag{7}$$

These components may have strange dimensions. The associated components having the usual dimensions are defined by

$$\underline{E} = E_i \, \underline{i}^i = E^i \, \underline{i}_i \tag{8}$$

where

$$E^1 = \sqrt{g_{11}} \, e^1 \qquad E_1 = \sqrt{g^{11}} \, e_1 \tag{9}$$

etc. The covariant and contravariant components are related by

$$e_i = g_{ij} \, e^j \qquad e^i = g^{ij} \, e_j \tag{10}$$

Thus, the physically real E^i and E_j are related by

$$E^i = G^{ij} \, E_j \qquad E_i = G_{ij} \, E^j \tag{11}$$

where

$$G^{ij} = \sqrt{\frac{g_{ii}}{g^{jj}}} \, g^{ij} \qquad G_{ij} = \sqrt{\frac{g_{jj}}{g^{ii}}} \, g_{ij} \quad \text{(no summation)} \tag{12}$$

(The G^{ij} and G_{ij} are inverse matrices, but are not tensors.)

In this formalism, the first component of the curl $\underline{E}$ equation becomes, in the presence of magnetic dissipation but no magnetic current,

$$-\mu \frac{\partial H^1}{\partial t} - \sigma^* H^1 = \sqrt{\frac{g_{11}}{g}} \left(\frac{\partial\left(E_3/\sqrt{g^{33}}\right)}{\partial u^2} - \frac{\partial\left(E_2/\sqrt{g^{22}}\right)}{\partial u^3} \right) \tag{13}$$

The first component of the curl $\underline{H}$ equation becomes

$$\varepsilon \frac{\partial E^1}{\partial t} + \sigma E^1 + J^1 = \sqrt{\frac{g_{11}}{g}} \left(\frac{\partial\left(H_3/\sqrt{g^{33}}\right)}{\partial u^2} - \frac{\partial\left(H_2/\sqrt{g^{22}}\right)}{\partial u^3} \right) \tag{14}$$

The other component equations are obtained by straightforward index permutation.

This system of equations may be advanced by the usual finite-difference leapfrog scheme [2,3] save that the contravariant-like H^i as obtained from the finite-difference equivalent of Eq. (13) must be converted by Eq. (11) to the covariant-like H_i before they are input to the finite-difference equivalent of Eq. (14). The E^i must undergo analogous conversion. This alternation between covariant-like and contravariant-like field components is the primary complication introduced by the use of oblique coordinates. One may view this alternation as a generalized leapfrogging in a fifth dimension.

EVALUATION OF FIELDS AND SURFACE CURRENT DENSITIES

The surface current density on a conductor face normal to $\underline{a}^1 = \underline{a}_2 \times \underline{a}_3/\sqrt{g}$ is

$$K^2 = -H_3 \sqrt{\frac{g_{22}}{g \, g^{11} \, g^{33}}} \qquad K^3 = H_2 \sqrt{\frac{g_{33}}{g \, g^{11} \, g^{22}}} \tag{15}$$

Currents on other surfaces may be analogously determined. Note that in an orthogonal system, the radicals in Eq. (15) all reduce to unity.

The surface magnetic current density on such a surface is

$$K^{*2} = E_3 \sqrt{\frac{g_{22}}{g \, g^{11} \, g^{33}}} \qquad K^{*3} = -E_2 \sqrt{\frac{g_{33}}{g \, g^{11} \, g^{22}}} \tag{16}$$

Since this magnetic current is usually nonexistent, the generalized coordinate E-field boundary condition on a conducting surface normal to $\underline{a}^1$ is

$$E_2 = E_3 = 0 \tag{17}$$

The magnitude of E in an oblique system is given from

$$|E|^2 = e^i e_i = \sum_{i=1}^{3} \frac{E^i \, E_i}{\sqrt{g_{ii} \, g^{ii}}} \tag{18}$$

In orthogonal systems, the radicals in Eq. (18) also reduce to unit and $|E|^2$ regains its familiar form as

$$|E|^2 = \sum_{i=1}^{3} E^i E_i \tag{19}$$

THE FINITE-DIFFERENCE EQUATIONS

Figure 3 illustrates a typical cell in a nonorthogonal system, and shows where the six field components are best evaluated. Differencing Eq. (13) in a leapfrog manner consistent with Figure 3 yields the new H^1 at the (I,J,K) mesh point in terms of the old E_i's:

(Equation 20 shown on following page)

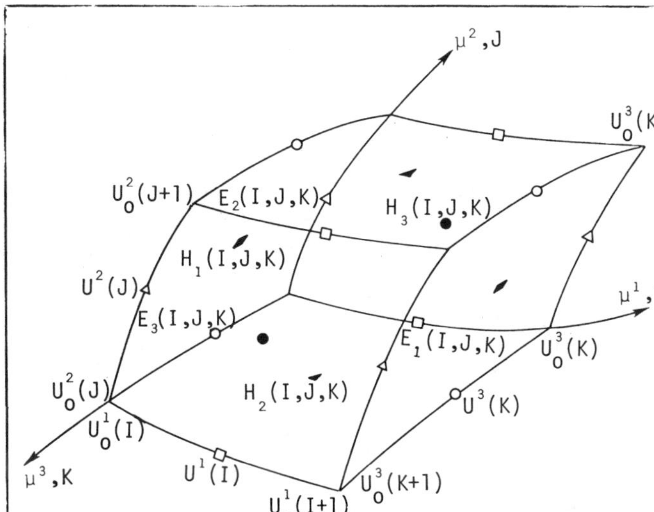

Figure 3. Field-Evaluation Points on a Finite-Difference Cell in a Generalized Non-Orthogonal Coordinate System

$$H^1(I,J,K)^{n+1} = \left.\frac{\mu(I,J,K)/\Delta t - \sigma^*(I,J,K)/2}{\mu(I,J,K)/\Delta t + \sigma^*(I,J,K)/2}\right|_{H_1} \cdot H^1(I,J,K)^n - \left.\frac{\sqrt{g_{11}/g}\ (I,J,K)}{\mu(I,J,K)/\Delta t + \sigma^*(I,J,K)/2}\right|_{H_1}$$

$$\cdot \left(\left.\frac{E_3/\sqrt{g^{33}}\ (I,J+1,K)^{n+\frac{1}{2}} - E_3/\sqrt{g^{33}}\ (I,J,K)^{n+\frac{1}{2}}}{U_0^2(J+1) - U_0^2(J)}\right|_{E_3} - \left.\frac{E_2/\sqrt{g^{22}}\ (I,J,K+1)^{n+\frac{1}{2}} - E_2/\sqrt{g^{22}}\ (I,J,K)^{n+\frac{1}{2}}}{U_0^3(K+1) - U_0^3(K)}\right|_{E_2} \right) \quad (20)$$

Here $|_{H_1}$ means μ and σ^* and $\sqrt{g_{11}/g}$ are evaluated at the H_1 mesh points, and $|_{E_i}$ means $\sqrt{g^{ii}}$ is evaluated at the E_i mesh point. $H^1(I,J,K)^{n+1}$ is $H^1(I,J,K)$ evaluated at $n\Delta t$, etc. The other two new contravariant-like H-field components are evaluated by index permutation of Eq. (20).

Obtaining the finite-difference covariant-like H-field components requires an interpolation as well as a linear combination of the contravariant-like output of Eq. (20) and its permutations. This is the case because the permutations of Eq. (20) yield H^2 and H^3 at different points on the unit cell (see Figure 3). The correct formula is

$$H_1(I,J,K)^{n+1} = G_{11}(I,J,K)\big|_{H_1} H^1(I,J,K)^{n+1} + \frac{G_{12}(I,J,K)|_{H_1}}{4} \cdot [H^2(I,J,K) + H^2(I,J+1,K) + H^2(I-1,J,K)$$

$$+ H^2(I-1,J+1,K)]^{n+1} + \frac{G_{13}(I,J,K)|_{H_1}}{4} \cdot [H^3(I,J,K) + H^3(I,J,K+1) + H^3(I-1,J,K) - H^3(I-1,J,K+1)]^{n+1} \quad (21)$$

The other contravariant-like H-field components are obtained by index permutation of Eq. (21).

Differencing Eq. (14) in a manner consistent with Figure 3 yields the new E^1's in terms of the old H_i's:

$$E^1(I,J,K)^{n+\frac{1}{2}} = \left.\frac{\varepsilon(I,J,K)/\Delta t - \sigma(I,J,K)/2}{\varepsilon(I,J,K)/\Delta t + \sigma(I,J,K)/2}\right|_{E_1} \cdot E^1(I,J,K)^{n-\frac{1}{2}} + \left.\frac{\sqrt{g_{11}/g}\ (I,J,K)}{\varepsilon(I,J,K)/\Delta t + \sigma(I,J,K)/2}\right|_{E_1} \cdot \left(- J^1(I,J,K)^n\big|_{E_1}\right.$$

$$\left. + \left.\frac{H_3/\sqrt{g^{33}}\ (I,J,K)^n - H_3/\sqrt{g^{33}}\ (I,J-1,K)^n}{U^2(J) - U^2(J-1)}\right|_{H_3} - \left.\frac{H_2/\sqrt{g^{22}}\ (I,J,K)^n - H_2/\sqrt{g^{22}}\ (I,J,K-1)^n}{U^3(K) - U^3(K-1)}\right|_{H_2} \right) \quad (22)$$

The other two new contravariant-like E-field components are obtained by index permutation of Eq. (22).

The finite-difference covariant-like E-field components are obtained in analogy with Eq. (21)

$$E_1(I,J,K)^{n+\frac{1}{2}} = G_{11}(I,J,K)\big|_{E_1} E^1(I,J,K)^{n+\frac{1}{2}} + \frac{G_{12}(I,J,K)|_{E_1}}{4} \cdot [E^2(I,J,K) + E^2(I+1,J,K) + E^2(I,J-1,K)$$

$$+ E^2(I+1,J-1,K)]^{n+\frac{1}{2}} + \frac{G_{13}(I,J,K)|_{E_1}}{4} \cdot [E^3(I,J,K) + E^3(I+1,J,K) + E^3(I,J,K-1) + E^3(I+1,J,K-1)]^{n+\frac{1}{2}} \quad (23)$$

The other contravariant-like E-field components are, of course, obtained by index permutation.

The use of nonorthogonal coordinates will be by far easiest in systems which are oblique, but not curvilinear. In this case, all the metrics in the curl equations [like Eqs. (13) and (14)] can be taken outside the differential operator. Also, the metrics and G's of Eqs. (20) through (23) will be independent of (I,J,K), and each must be calculated at only a single point in space and time for the entire problem.

ACKNOWLEDGEMENT

The oblique-coordinate notation used here appears in Reference [1], pp. 38-47. The development of a system of equations which can be treated by proved leapfrog finite-difference techniques [beginning with Eq. (11)] has not been found elsewhere.

REFERENCES

1. Stratton, J.A., Electromagnetic Theory, McGraw-Hill, New York, 1941, pp. 38-47.
2. Yee, K.S., IEEE Trans. Ant. & Prop., Vol. AP-14, May 1966, pp. 302-307.
3. Longmire, C.L., IEEE Trans. Nucl. Sci., Vol. NS-22, December 1975, pp. 2340-2344.

The Finite-Difference–Time-Domain Method and its Application to Eigenvalue Problems

DOK HEE CHOI AND WOLFGANG J. R. HOEFER, SENIOR MEMBER, IEEE

Abstract —This paper describes the application of the finite-difference method in the time domain to the solution of three-dimensional (3-D) eigenvalue problems. Maxwell's equations are discretized in space and time, and steady-state solutions are then obtained via Fourier transform. While achieving the same accuracy and versatility as the TLM method, the finite-difference–time-domain (FD–TD) method requires less than half the CPU time and memory under identical simulation conditions. Other advantages over the TLM method include the absence of dielectric boundary errors in the treatment of 3-D inhomogeneous planar structures, such as microstrip. Some numerical results, including dispersion curves of a microstrip on anisotropic substrate, are presented.

I. Introduction

THE TLM METHOD has been successfully applied to various microwave circuit problems for more than ten years. The special advantages of the TLM technique over other numerical methods are well illustrated by Johns and Beurle [1] in their original paper on the method. Since then, several improvements have been made to this technique by various authors in order to enhance the accuracy of the solution and economize CPU time and memory space [2]–[5]. Mariki [7] has extended the TLM method to analyze anisotropic media, and Saguet and Pic [4] as well as Al-Mukhtar and Sitch [5] have employed a graded mesh to make the algorithm faster and more efficient.

Although the graded mesh algorithm reduces memory space requirements, it demands far more iterations than the original method for equal frequency resolution [5]. This is especially obvious in the case of three-dimensional (3-D) simulations where the grade ratio N requires an additional N^2-1 iterations. These requirements of large computer resources may critically limit the applicability of the TLM technique.

Thus, Saguet [16] has proposed a simplified node which reduces the number of variables to be processed and stored at each node by one third. However, this modification increases the velocity error. A further reduction in computational expenditure has been proposed by the authors [8]; instead of the original vector solution, we obtain a scalar potential solution using a scalar 3-D network. However, the scalar approach is limited to problems which lead to uncoupled modal solutions, i.e., TE and TM or LSE and LSM fields.

The major reason for the large CPU memory demand of this technique resides in the basic 3-D TLM concept. In order to represent each electromagnetic component, each 3-D unit cell requires 26 real memory spaces, 12 for pulse storage and 14 for additional network parameters. Therefore, each operation on each node involves a large number of variables, requiring considerable computer CPU space and time. Furthermore, experience has shown that the number of iterations increases with the complexity of the structure under study. For example, the accurate analysis of a finline requires easily over 1000 iterations. Given these massive requirements, we have searched for an alternative numerical technique that possesses the advantages of the TLM approach but needs fewer computer resources. As a result, a new algorithm is proposed based on both the finite-difference–time-domain (FD–TD) and TLM methods.

The FD–TD method was first formulated by Yee [6], and has been applied extensively to scattering and coupling problems with open boundaries [9]–[15], i.e., to the solution of deterministic problems. We noted the similarity between this method and the TLM method, which has been widely used in the numerical solution of the electromagnetic eigenvalue problems in the time domain. Since the TLM method is based on the computation of the impulse response of a large mesh of transmission lines, much unwanted information is usually generated.

We have therefore developed a novel procedure which increases the numerical efficiency of the time-domain approach without sacrificing its advantages. The method differs from the classical FD–TD method in the assignment of initial field values and the application of the Fourier transform to the time-domain solution. In the following, we will describe this method and its application to some typical microwave problems.

II. Yee's Algorithm

Maxwell's equations have been expressed in finite-difference form by Yee [6] to solve two-dimensional (2-D) wave scattering problems. Subsequently, 3-D scattering problems have been solved by Taflove and other workers

Manuscript received March 22, 1986; revised July 8, 1986. This work was supported in part by the National Science and Engineering Research Council of Canada.

The authors are with the Department of Electrical Engineering, University of Ottawa, Ottawa, Ontario, Canada K1N 6N5.

IEEE Log number 8610560.

Reprinted from *IEEE Transactions on Microwave Theory and Techniques,* pp. 1464-1470, December 1986.

[9]–[15]. We will adopt Yee's original algorithm for the three-dimensional Maxwell's equations. Also, we will extend the concept further to include anisotropic media.

In a rectangular coordinate system, the source-free Maxwell's equations can be written as first-order hyperbolic equations

$$\begin{aligned} d\overline{E}/dt &= |A_x|\, d\overline{H}/dx + |A_y|\, d\overline{H}/dy + |A_z|\, d\overline{H}/dz \\ d\overline{H}/dt &= |B_x|\, d\overline{E}/dx + |B_y|\, d\overline{E}/dy + |B_z|\, d\overline{E}/dz \end{aligned} \tag{1}$$

where $\overline{E} = (E_x, E_y, E_z)^t$, $\overline{H} = (H_x, H_y, H_z)^t$, and

$$|A_x| = \begin{bmatrix} 0 & 0 & 0 \\ 0 & 0 & -\epsilon_{yy}^{-1} \\ 0 & \epsilon_{zz}^{-1} & 0 \end{bmatrix}, \quad |A_y| = \begin{bmatrix} 0 & 0 & \epsilon_{xx}^{-1} \\ 0 & 0 & 0 \\ -\epsilon_{zz}^{-1} & 0 & 0 \end{bmatrix}$$

$$|A_z| = \begin{bmatrix} 0 & -\epsilon_{xx}^{-1} & 0 \\ \epsilon_{yy}^{-1} & 0 & 0 \\ 0 & 0 & 0 \end{bmatrix}.$$

Here, ϵ_{xx}, ϵ_{yy}, and ϵ_{zz} are the diagonal elements of the permittivity tensor. All B_x, B_y, and B_z are expressed by replacing ϵ_{xx}, ϵ_{yy}, and ϵ_{zz} by $-\mu_{xx}$, $-\mu_{yy}$, and $-\mu_{zz}$ components, which are the diagonal elements of the permeability tensor. Then each of these scalar equations can be expressed in finite-difference form. Following Yee's nomenclature, any function of space and time is discretized

$$F^n(i, j, k) = F(i\Delta x, j\Delta y, k\Delta z, n\Delta t)$$

where $\Delta x = \Delta y = \Delta z = \Delta l$ is the space increment and Δt is the time increment. By positioning the components of $\overline{E}$ and $\overline{H}$ on the mesh as depicted in Fig. 1 and evaluating $\overline{E}$ and $\overline{H}$ at alternate half time steps, we obtain the components of Maxwell's equations.

$$\begin{aligned} &H_x^{n+1/2}(i, j+1/2, k+1/2) \\ &\quad = H_x^{n-1/2}(i, j+1/2, k+1/2) \\ &\qquad + s/\mu_{xx}(i, j+1/2, k+1/2)\big[E_y^n(i, j+1/2, k+1) \\ &\qquad - E_y^n(i, j+1/2, k) + E_z^n(i, j, k+1/2) \\ &\qquad - E_z^n(i, j+1, k+1/2)\big] \end{aligned} \tag{2a}$$

$$\begin{aligned} &H_y^{n+1/2}(i+1/2, j, k+1/2) \\ &\quad = H_y^{n-1/2}(i+1/2, j, k+1/2) \\ &\qquad + s/\mu_{yy}(i+1/2, j, k+1/2)\big[E_z^n(i+1, j, k+1/2) \\ &\qquad - E_z^n(i, j, k+1/2) + E_x^n(i+1/2, j, k) \\ &\qquad - E_x^n(i+1/2, j, k+1)\big] \end{aligned} \tag{2b}$$

$$\begin{aligned} &H_z^{n+1/2}(i+1/2, j+1/2, k) \\ &\quad = H_z^{n-1/2}(i+1/2, j+1/2, k) \\ &\qquad + s/\mu_{zz}(i+1/2, j+1/2, k)\big[E_x^n(i+1/2, j+1, k) \\ &\qquad - E_x^n(i+1/2, j, k) + E_y^n(i, j+1/2, k) \\ &\qquad - E_y^n(i+1, j+1/2, k)\big] \end{aligned} \tag{2c}$$

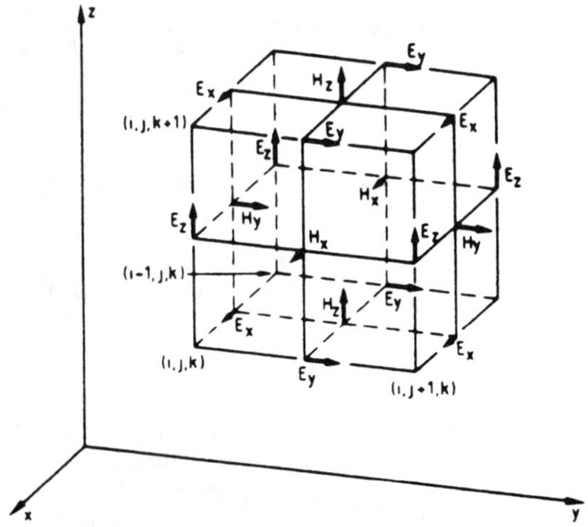

Fig. 1. The position of the field components in Yee's mesh.

$$\begin{aligned} &E_x^{n+1}(i+1/2, j, k) \\ &\quad = E_x^n(i+1/2, j, k) + s/\epsilon_{xx}(i+1/2, j, k) \\ &\qquad \cdot\big[H_z^{n+1/2}(i+1/2, j+1/2, k) \\ &\qquad - H_z^{n+1/2}(i+1/2, j-1/2, k) \\ &\qquad + H_y^{n+1/2}(i+1/2, j, k-1/2) \\ &\qquad - H_y^{n+1/2}(i+1/2, j, k+1/2)\big] \end{aligned} \tag{2d}$$

$$\begin{aligned} &E_y^{n+1}(i, j+1/2, k) \\ &\quad = E_y^n(i, j+1/2, k) + s/\epsilon_{yy}(i, j+1/2, k) \\ &\qquad \cdot\big[H^{n+1/2}(i, j+1/2, k+1/2) \\ &\qquad - H_x^{n+1/2n}(i, j+1/2, k-1/2) \\ &\qquad + H_z^{n+1/2}(i-1/2, j+1/2, k) \\ &\qquad - H_z^{n+1/2}(i+1/2, j+1/2, k)\big] \end{aligned} \tag{2e}$$

$$\begin{aligned} &E_z^{n+1}(i, j, k+1/2) \\ &\quad = E_z^n(i, j, k+1/2) + s/\epsilon_{zz}(i, j, k+1/2) \\ &\qquad \cdot\big[H_y^{n+1/2}(i+1/2, j, k+1/2) \\ &\qquad - H_y^{n+1/2}(i-1/2, j, k+1/2) \\ &\qquad + H_x^{n+1/2}(i, j-1/2, k+1/2) \\ &\qquad - H_x^{n+1/2}(i, j+1/2, k+1/2)\big] \end{aligned} \tag{2f}$$

where the stability factor $s = c\Delta t/\Delta l$, and c is the velocity of light. In these expressions, E and H are normalized such that the characteristic impedance of space is unity. The condition for stability of (1) in free space is [10]

$$s \leqslant 1/\sqrt{3}. \tag{3}$$

III. Boundary Conditions

So far, a space–time mesh has been introduced and Maxwell's equations have been replaced by a system of finite-difference equations. Difficulties arise when the domain in which the field must be computed is unbounded. Since no computer can store an unlimited amount of data, a special technique must be used to limit the domain in which the numerical computation is made, by introducing so-called absorbing or soft boundary conditions. These conditions have been described by Taylor *et al.* [9], who use a simple extrapolation method, and by Taflove and Brodwin [10], who simulate the outgoing waves and use an averaging process in an attempt to account for all possible angles of propagation of the outgoing waves. Kunz and Lee [12] use the radiation condition at a large distance from the center of the scatterer to obtain an absorbing boundary condition. Mur [13] employs a second-order radiation condition to improve the accuracy of the results. Although these schemes have been used in scattering problems in the past, no ideal reflection-free boundary condition has been proposed so far.

However, in the formulation of eigenvalue problems, only "hard boundaries"—usually represented by conducting walls—occur. At these boundaries, the tangential electric and the normal magnetic field components are maintained at zero. For example, on a perfectly conducting wall in the plane $i=1$ (see Fig. 1)

$$\text{for all } n \begin{cases} E_y^n(1, j+1/2, k) = 0 \\ E_z^n(1, j, k+1/2) = 0 \\ H_x^n(1, j+1/2, k+1/2) = 0 \end{cases}$$

(the third condition is implicit in the previous two, but its implementation reduces numerical errors).

IV. Initial Values

In most scattering problems, an impulsive or sinusoidal plane wave is injected at the beginning of the computation. However, in eigenvalue problems, the direction of the propagation vector is usually not known and depends on the space coordinates and on the eigenvalue that is to be found. In these cases, the logical choice is an isotropic pulse that propagates in the radial direction. The spatial pulse envelope should be wide enough with respect to the mesh size not to accumulate numerical errors due to overshoot and ringing as it propagates through the space lattice.

A better way to start the computation is to estimate the field distribution of the desired mode in the structure first and then choose the initial value accordingly. The experienced researcher usually has a good idea of the approximate modal field distributions in a structure and is therefore able to make an educated guess of the steady-state field pattern for a particular eigenmode. This procedure is equivalent to the excitation of a TLM mesh with a weighted impulse distribution, and is somewhat similar to the way in which one chooses appropriate basis functions in the spectral-domain approach.

V. Outline of the Numerical Procedure

The application of the FD–TD method will be discussed using a rectangular resonator as an example. A continuous medium is replaced with a 3-D uniform mesh. To solve the system of equations (2) in this mesh, initial values must be assigned first as described in the previous section. For a rectangular-type resonator, a simple sinusoidal function is an appropriate choice for the dominant mode eigenvalue. As n increases, the discrete time functions for $\bar{E}$ and $\bar{H}$ fields evolve towards the steady state which is characteristic of the desired mode in the geometry. In this way, the evolution of all six field components is obtained simultaneously at discrete time points $n\,\Delta t$. The final steady-state field distribution may be calculated by taking the time average of the time-domain solution at each mesh point. Thus, the steady-state solution is given by

$$F(i_0, j_0, k_0) = \sum_n |F^n(i_0, j_0, k_0)|/N. \tag{4}$$

This simple procedure to obtain the final field distribution is another advantage over the TLM method, which requires two simulations for finding the fields of a given mode.

In eigenvalue problems, the steady-state solution is a time-harmonic function, from which the eigenvalues can

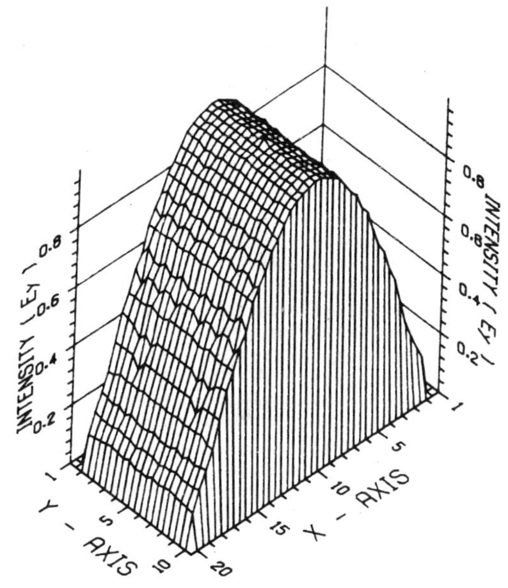

(a)

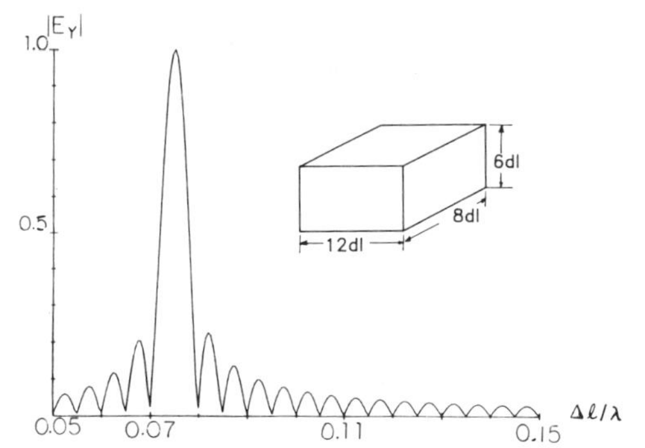

(b)

Fig. 2. (a) Field distribution of an empty rectangular resonator obtained with the FD–TD method. (b) Output spectrum obtained with both the FD–TD and TLM methods under identical conditions.

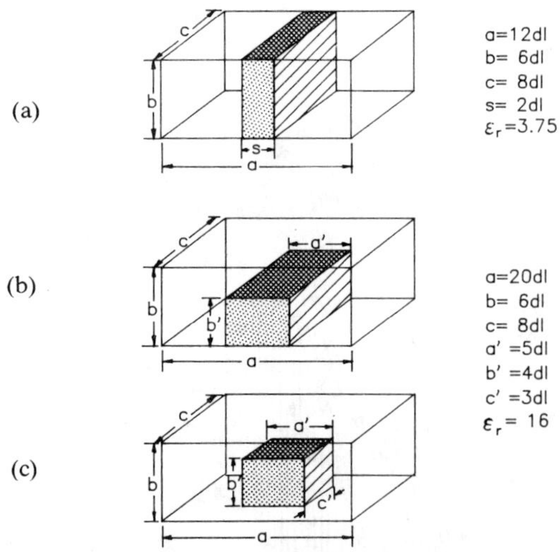

Fig. 3. Three-dimensional inhomogeneous resonators analyzed in this study.

TABLE I
COMPARISON OF RESULTS FOR THE NORMALIZED RESONANT FREQUENCIES $\Delta l/\lambda$ OBTAINED WITH THE TLM AND FD–TD METHODS UNDER IDENTICAL SIMULATION CONDITIONS FOR THE GEOMETRIES IN FIG. 3

Fig.3	Mode	s/a	TRM	TLM (CPU time)	FD–TD (CPU time)
a)	LSE	1/6	0.05220	0.0516 (144)	0.0517 (51)
		1/3	0.0445	0.0440 (145)	0.0442 (51)
b)	Hybrid			0.0278 (357)	0.0278 (117)
c)	Hybrid			0.0405 (357)	0.0405 (117)

be extracted by discrete Fourier transform, as in the TLM method

$$S(f) = \sum_n F^n(i_0, j_0, k_0) \exp(-j2\pi snf). \tag{5}$$

Both the stability factor s and the number of iterations n strongly affect the spectral response.

In order to test this algorithm for validity, it has been applied to a simple rectangular cavity with sides $12\,\Delta l \times 6\Delta l \times 8\Delta l$. We have assumed a dominant TE_{101} mode in the initial value assignment. The time-domain solution is given in Fig. 2(a). Discontinuous field figures are due to the numerical error caused by the finite-difference form of (2). Fig. 2(b) compares the frequency responses obtained with the FD–TD and TLM methods under identical conditions. The responses are not distinguishable. Five hundred iterations have been used. The peak of the solution is located at $\Delta l/\lambda = 0.0750$ in both methods. The exact analytical solution is 0.07511. Even though a small number of meshes is used in this algorithm when compared with the scattering problems in [9]–[13], it is noted that the accuracy in the solution of the eigenvalue problems is better than that of the scattering problems by one order of magnitude.

VI. NUMERICAL RESULTS

We have applied this technique to most of the examples described in the TLM literature and obtained practically identical results. The method requires less than one-half of

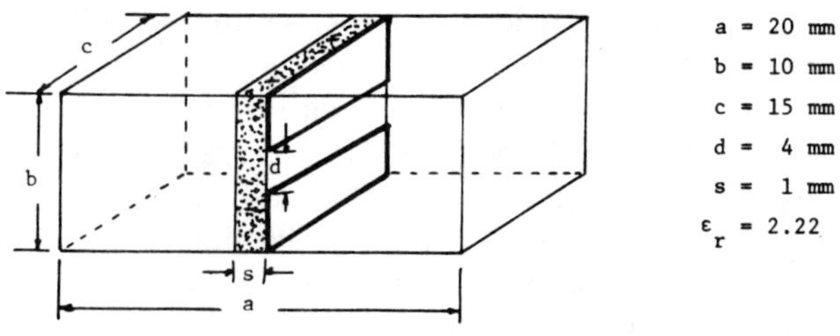

Fig. 4. Finline cavity.

TABLE II
RESONANT FREQUENCIES OF THE UNILATERAL FINLINE CAVITY IN FIG. 4, OBTAINED WITH VARIOUS METHODS

	Saguet [16]		This method	
	S.D.A.	Variable Mesh TLM	TDFD	T.L.M.
Resonant Frequency (GHz)	10.77	10.14	10.74	10.74
Number of Iterations		1000	600	600
CPU Time(s)			170	380

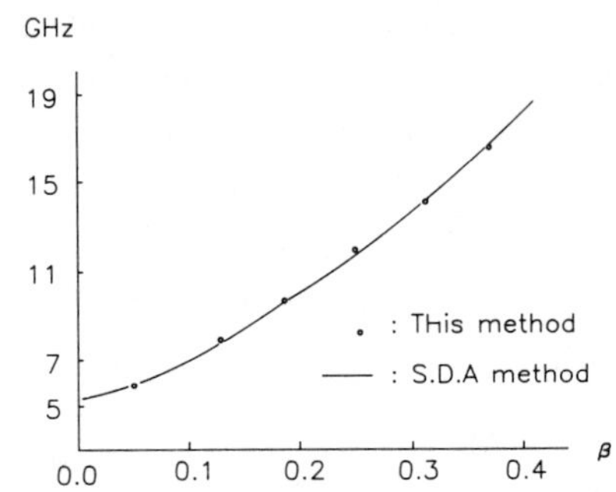

Fig. 5. Dispersion diagram of unilateral finline with the cross-sectional geometry given in Fig. 4.

the CPU time spent by the equivalent TLM program under identical conditions, including the initial excitation distribution. Furthermore, while the TLM procedure requires 22 real memory stores per 3-D node in an isotropic dielectric, the FD–TD method requires only seven real memory stores per node. Fig. 3(a), (b), and (c) shows structures for which solutions have been computed with this method. The dominant resonant frequencies of these structures are given and compared with the TLM solution in Table I. The inhomogeneous rectangular cavity of Fig. 3 (b) and (c) illustrates the capability of this algorithm to solve hybrid field problems. The number of nodes chosen in each problem is the same as that employed in the TLM solution.

Furthermore, we have computed the resonant frequency of a finline resonator (Fig. 4.) treated previously by Saguet [16]. Results are compared in Table II, which includes a value obtained with the spectral-domain method by Saguet. Fig. 5 shows the dispersion characteristics of a finline with the same cross section, as obtained with our method. Results calculated with our spectral-domain program are also shown in the same figure. In order to compare convergence of both time-domain methods, solutions obtained after every fifth iteration are drawn in Fig. 6. The results show virtually identical convergence.

To show the versatility of this method, the characteristics of a microstrip resonator on anisotropic substrate are

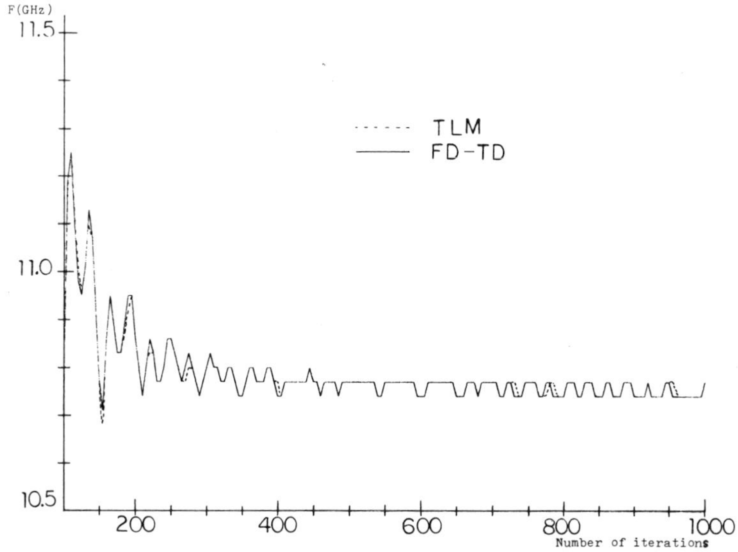

Fig. 6. Stability and convergence of the TLM and FD–TD methods as a function of the number of iterations. The solution of the finline problem in Fig. 4 is represented.

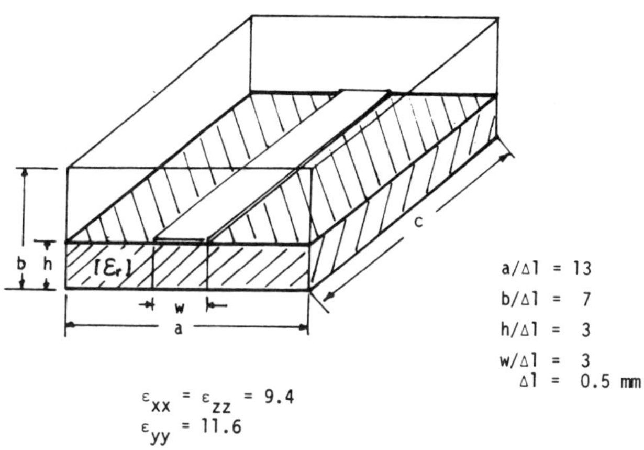

Fig. 7. Microstrip cavity on anisotropic substrate.

TABLE III
DOMINANT RESONANT FREQUENCIES OBTAINED BY BOTH THE TLM AND THE FD–TD METHODS

h \ c/Δl		6	8	10	15	20
TLM (GHz)	3Δl	15.72	12.24	10.14	7.14	5.46
	4Δl	14.22	11.16	9.18	6.36	4.98
	Average	14.97	11.70	9.66	6.75	5.22
FD - TD (GHz)		14.64	11.52	9.54	6.78	5.28

computed in the last example, shown in Fig. 7. Several different resonant frequencies obtained by changing the length c are tabulated in Table III. It is well known [17] that the TLM simulation of 3-D inhomogeneous planar structures involves dielectric interface ambiguity. The best way to resolve this error is to employ two dielectric substrate thicknesses differing by one Δl. In our case, $3\Delta l$ and $4\Delta l$ are used. The final result is obtained by taking the average of the solutions obtained for these two values. In order to illustrate this process, frequency spectra obtained with the TLM method for the two cases where c is equal to $10\Delta l$ are shown in Fig. 8. The solution obtained with the FD–TD method is also drawn in the same figure. As expected, the latter solution is located exactly between the two values obtained with the TLM method. This clearly illustrates the accuracy and convenience of the FD–TD method in such situations. Fig. 9 shows the dispersion characteristics of the microstrip which has the same cross section as that in Fig. 7. Again, both methods give very similar results except at higher frequencies, where the discretization errors associated with both methods become more pronounced, and their differences are more visible.

VII. CONCLUSIONS

The proposed new application of the FD–TD method to 3-D eigenvalue problems gives practically the same results

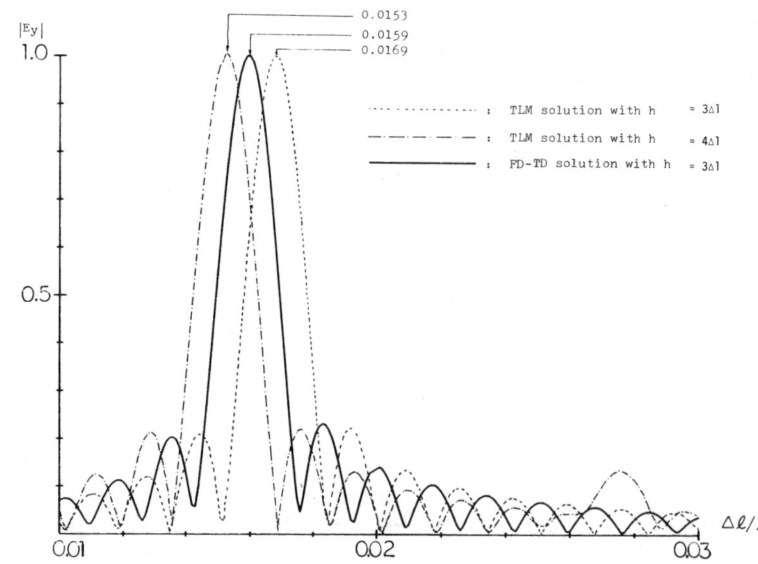

Fig. 8. Frequency spectra for the microstrip problem in Fig. 7, showing the effect of dielectric interface error in the TLM solution.

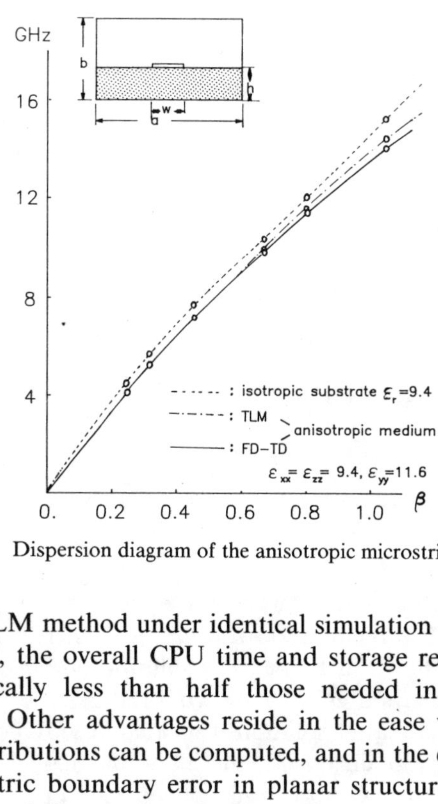

Fig. 9. Dispersion diagram of the anisotropic microstrip in Fig. 7.

as the TLM method under identical simulation conditions. However, the overall CPU time and storage requirements are typically less than half those needed in the TLM solution. Other advantages reside in the ease with which field distributions can be computed, and in the elimination of dielectric boundary error in planar structures. Further efforts are being made to implement losses, variable mesh size, and nonlinarity in the FD–TD procedure.

References

[1] P. B. Johns and R. L. Beurle, "Numerical solution of two-dimensional scattering problems using a transmission line matrix," *Proc. Inst. Elec. Eng.*, vol. 118, no. 9, pp. 1203–1208, Sept. 1971.

[2] S. Akhtarzad and P. B. Johns, "Solution of Maxwell's equations in three space dimensions and time by the TLM method of numerical analysis," *Proc. Inst. Elec. Eng.*, vol. 122, no. 12, pp. 1344–1348, Dec. 1975.

[3] Y. C. Shih and W. J. R. Hoefer, "The accuracy of TLM analysis of finned rectangular waveguides," *IEEE Trans. Microwave Theory Tech.*, vol. MTT-28, pp. 743–746, July 1980.

[4] P. Saguet and E. Pic, "Le maillage rectangulaire et le changement de maille dans la methode TLM en deux dimensions," *Electron. Lett.*, vol. 17, no. 7, pp. 277–279, Apr. 1981.

[5] D. A. Al-Mukhtar and J. E. Sitch, "Transmission line matrix method with irregularly graded space," *Proc. Inst. Elec. Eng.*, vol. 128, no. 6, pp. 299–305, Dec. 1981.

[6] K. S. Yee, "Numerical solution of initial boundary value problems involving Maxwell's equations in isotropic media," *IEEE Trans. Antennas Propagat.*, vol. AP-14, pp. 302–307, May 1966.

[7] G. E. Mariki, "Analysis of microstrip lines on inhomogeneous anisotropic substrates by the TLM numerical technique," Ph.D. thesis, Univ. California, Los Angeles, CA, June 1978.

[8] D. Choi and W. J. R. Hoefer, "The simulation of three dimensional wave propagation by a scalar TLM model," *IEEE MTT-S Symp. Dig.*, 1984, pp. 70–71.

[9] C. D. Taylor, D. H. Lam, and T. H. Shumpert, "Electromagnetic pulse scattering in time varying inhomogeneous media," *IEEE Trans. Antennas Propagat.*, vol. AP-17, pp. 586–589, Sept. 1969.

[10] A. Taflove and M. E. Brodwin, "Numerical solution of steady state electromagnetic scattering problems using the time dependent Maxwell's equations," *IEEE Trans. Microwave Theory Tech.*, vol. MTT-23, pp. 623–630, Aug. 1975.

[11] R. Holland, "Threde: A free-field EMP coupling and scattering code," *IEEE Trans. Nucl. Sci.*, vol. NS-24, pp. 2416–2421, Dec. 1977.

[12] K. S. Kunz and K. M. Lee, "A three-dimensional finite-difference solution of the external response of an aircraft to a complex transient EM environment: Part 1—The method and its implementation," *IEEE Trans. Electromagn. Compat.*, vol. EMC-20, pp. 328–33, May 1978.

[13] G. Mur, "Absorbing conditions for the finite-difference approximation of the time-domain electromagnetic-field equations," *IEEE Trans. Electromagn. Compat.*, vol. EMC-23, Nov. 1981.

[14] A. Taflove and K. R. Umashankar, "A hybrid moment method/finite difference time domain approach to electromagnetic coupling and aperture penetration into complex geometries," *IEEE Trans. Antennas Propagat.*, vol. AP-30, pp. 617–627, July 1982.

[15] K. R. Umashankar and A. Taflove, "A novel method to analyze electromagnetic scattering of complex objects," *IEEE Trans. Electromagn. Compat.*, vol. EMC-24, pp. 397–405, Nov. 1982.

[16] P. Saguet, "Le maillage parallelépipédique et le changement de maille dans la méthode TLM en trois dimensions," *Electron. Lett.* vol. 20, no. 5, pp. 222–24, Mar. 15, 1984.

[17] W. J. R. Hoefer, "The transmission-line matrix method—Theory and applications," *IEEE Trans. Microwave Theory Tech.*, vol. MTT–33, no. 10, pp. 882–893, Oct. 1985.

Calculations of the Dispersive Characteristics of Microstrips by the Time-Domain Finite Difference Method

XIAOLEI ZHANG, JIAYUAN FANG, KENNETH K. MEI, FELLOW, IEEE, AND YAOWU LIU

Abstract —The dispersive characteristics of microstrips have been investigated by many authors [1]–[4] using various numerical and empirical methods. Those results showed a lack of agreement with each other, and the true dispersive characteristics of microstrips still need to be identified. In this paper, a direct time-domain finite difference method is used to recharacterize the microstrip. Maxwell's equations are discretized both in time and space and a Gaussian pulse is used to excite the microstrip. The frequency-domain design data are obtained from the Fourier transform of the calculated time-domain field values. Since this method is completely independent of all the above-mentioned investigations, the new results can be considered as an impartial verification of the published results.

The comparison of the time-domain results and those from the frequency-domain methods has shown the integrity of the time-domain computations. Since this method is very general and can be applied to model many other microwave components, its success in the microstrip problem is an important step toward its general application.

I. Introduction

The application of the time-domain method to model microstrip problems is not new [7]–[11]. However, it is more frequently used to obtain qualitative results that graphically illustrate the field propagation and the transport of power than to obtain design data. It is interesting to note that when the time-domain method is used to obtain frequency-domain data, the investigators often go back to sinusoidal excitations [7]–[10]. This compromises the advantage of the time-domain approach, which is believed to be capable of producing wide-band frequency-domain results with a single computation process.

In the process of this investigation, it is found that the Fourier transform of the time-domain results are very sensitive to numerical errors, notably those resulting from the imperfect treatment of the absorbing boundary conditions used to truncate the numerical computations of that of an open structure. Thus, even though the time-domain results may be reasonably accurate, the frequency-domain results obtained from their Fourier transform may not be acceptable as useful data. That may explain, at least in part, why many investigators mysteriously reverted to sinusoidal signals while it seems obvious that a pulse in the time domain could have been used to advantage, and others turned to solve only the closed structures [11].

To improve the absorbing boundary conditions, we have used an open-circuit, short-circuit technique to cancel the reflections from the terminations of the finite difference meshes. The improvements in the resulting computations enabled us to get reliable frequency-domain data, i.e., the effective dielectric constant and the characteristic impedance, through Fourier transforms, and to compare with various numerical results of the microstrip, presently available in the open literature.

Manuscript received April 15, 1987; revised September 21, 1987. This work was supported in part by the Joint Services Electronics Program under Contract F49620-84-C-0057, the Office of Naval Research under Contract N00014-86-K-0420, and the National Science Foundation under Grant ECS-8311340.

The authors are with the Department of Electrical Engineering and Computer Sciences and the Electronics Research Laboratory, University of California, Berkeley, CA 94720.

IEEE Log Number 8717986.

II. Finite Difference Method and Absorbing Boundary Conditions

The finite difference method was first introduced by K. S. Yee [12] to solve electromagnetic scattering problems. In using this method, Maxwell's equations are discretized both in time and space. Knowing the initial, boundary and excitation conditions, the fields on the nodal points of the space–time mesh can be calculated in a leapfrog time marching manner. For more details of the method, [11]–[15] could be consulted.

Since the finite difference equations cannot be applied to the boundary nodes, an absorbing type of boundary condition must be used at the mesh boundary where an open space is to be simulated. There are several publications discussing the absorbing boundary conditions [16]–[17]. Most of the methods are quite good to terminate the finite difference equations, and they have been applied to the scattering problems [13]–[15]. But, they are not good enough for the microstrip problems, since the Fourier transform of the time-domain results is very sensitive to the imperfect boundary treatment. Minute errors in the time domain may produce fairly large errors in the frequency domain. To minimize the reflection error, we have chosen to use the open- and short-circuit boundary conditions [16]. In this method, the problem is solved twice for each absorbing boundary, once with vanishing tangential electric field (termination with electrically conducting wall) and once with vanishing tangential magnetic field (termination with magnetically conducting wall). The results of the two solutions are averaged to give the desired solution. The rationale of this method is quite simple. Because the reflection coefficient of a short-circuit boundary is -1, while that of an open circuit is $+1$, the

Reprinted from *IEEE Transactions on Microwave Theory and Techniques*, pp. 263-267, February 1988.

sum of the two results should have the reflected fields completely canceled out. However, caution must be exercised for computations of long duration, because the multiple reflections may not be canceled out. So, it is necessary to consider the size of the finite difference mesh domain to accommodate the anticipated duration of computations.

Using the open-circuit, short-circuit boundary conditions, we are able to reduce the reflection error to an acceptable level such that the corresponding frequency domain results are smooth and accurate.

III. Numerical Results

Using the time-domain method discussed above, a typical microstrip on a gallium arsenide substrate has been investigated. Unlike other investigations, which look for the propagation velocity of the field only [1], [3], the time-domain approach actually finds the space–time distributions of the fields everywhere within the finite difference space–time mesh. In many cases this provides clear pictures and illuminating details of the field variations. At the same time, the frequency-domain design data can also be easily obtained through Fourier transform of the time-domain fields.

A. Parameters Used in Computations

The microstrip structure used in our calculations is shown in Fig. 1. (Because of the symmetry of the problem, only half of the structure is considered.) The parameters used for most of our computations are as follows:

thickness of the substrate:	$H = 0.1$ mm
width of the metal strip:	1) $W = 0.075$ mm ($W/H = 0.75$)
	2) $W = 0.15$ mm ($W/H = 1.5$)
dielectric constant of the substrate:	$\varepsilon_r = 13.0$
thickness of the metal strip:	0.

To accommodate the structural details of the microstrip, we have used the mesh parameters as follows:

space interval	$dh = 0.1\ \text{mm}/16 = 6.25 \cdot 10^{-3}$ mm (for $W/H = 0.75$)
	$dh = 0.1\ \text{mm}/8 = 1.25 \cdot 10^{-2}$ mm (for $W/H = 1.5$)
	$n_1 = 30,\ n_2 = 55,\ n_3 = 160$
time step	$dt = k \cdot dh/c$ sec.

where c is the velocity of light in air and k is the constant restricted by stability criterion [13] ($k = 0.515$ in our calculation). The most often used values for r_1, r_2, r_3 are $r_1 = r_2 = r_3 = 1.0$. But sometimes other choices of r_1, r_2, r_3 (with $r_1, r_2, r_3 \geqslant 1$) are used to ease the problem with a particular W/H.

A Gaussian pulse excitation is used at the front surface, uniform under the strip (in plane "abcd" of Fig. 1) with only E_x component:

$$E_x(t) = \exp\left[-\frac{(t - t_0)^2}{T^2} \right] \tag{1}$$

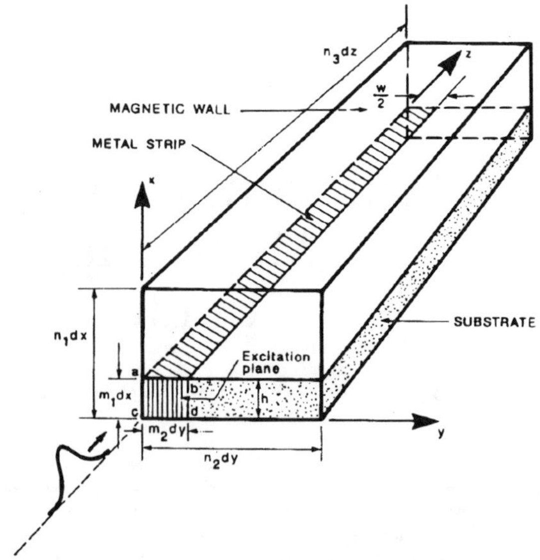

Fig. 1. Microstrip structure.

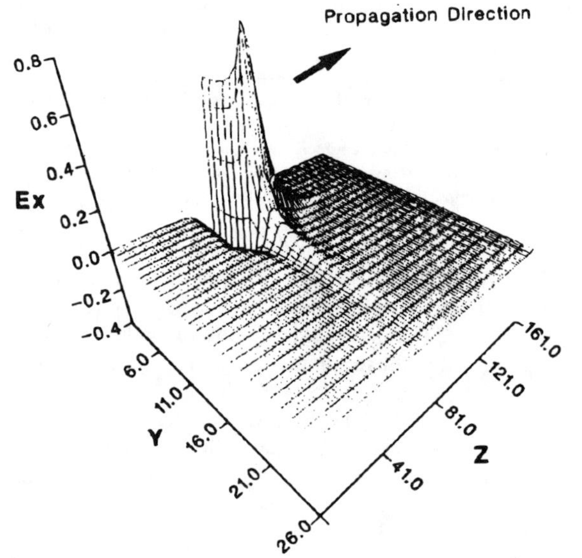

Fig. 2. Gaussian pulse propagation in microstrip (E_x component just underneath the air dielectric interface).

where $t_0 = 140dt$ and $T = 140dt$; elsewhere on the front surface, we set $E_x = E_y = 0$. The pulse width in space is about $20dh$, which is wide enough to obtain good resolution. The frequency spectrum of the pulse is from dc to about 700 GHz.

B. Time-Domain Fields

Fig. 2 shows the propagation of a Gaussian pulse along the microstrip. This is a plot of the space distributions of the vertical electric field on a surface just underneath the microstrip. Fig. 3 shows the time variations of vertical electric field at different positions along the propagation direction. The dispersive properties of the microstrip are

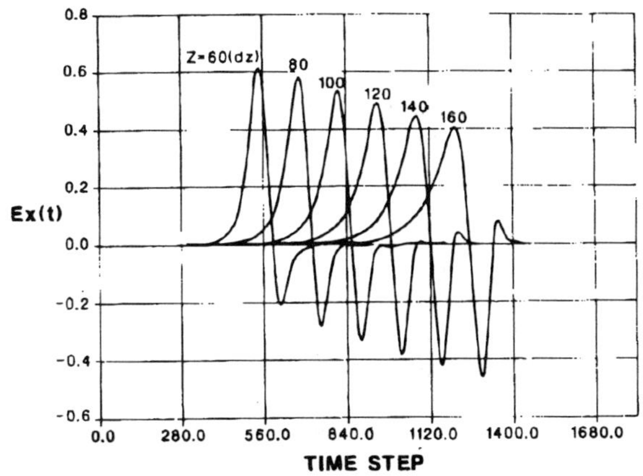

Fig. 3. Time variation of E_x at different positions along the direction of propagation.

quite obvious from the distortion of the pulse as it travels away from the feeding point. Indeed, the result (Fig. 3) is very similar to that calculated by Veghte and Balanis [6], who reconstructed the time-domain waveforms using the frequency-domain empirical formulas of effective dielectric constant.

C. Effective Dielectric Constant

The effective dielectric constant $\varepsilon_{reff}(\omega)$ can be used to account for the dispersive characteristics of the microstrip. It is calculated as follows:

Take the Fourier transforms of $E_x(t)$ at two different positions (underneath the center of the strip), with a separation of L, along the propagation direction:

$$E_x(\omega, z=0) = \int_{-\infty}^{\infty} E_x(t, z=0) e^{-j\omega t}\, dt \tag{2}$$

$$E_x(\omega, z=L) = \int_{-\infty}^{\infty} E_x(t, z=L) e^{-j\omega t}\, dt. \tag{3}$$

Taking the ratio of (2) and (3), we can get the transfer function of this section of microstrip, which is

$$e^{-\gamma(\omega)L} = \frac{Ex(\omega, z=L)}{Ex(\omega, z=0)} \tag{4}$$

where

$$\gamma(\omega) = \alpha(\omega) + j\beta(\omega). \tag{5}$$

The constant $\varepsilon_{reff}(\omega)$ is defined through $\beta(\omega)$ as

$$\beta(\omega) = \omega\sqrt{u_0 \varepsilon_0 \varepsilon_{reff}(\omega)} \tag{6}$$

or

$$\varepsilon_{reff}(\omega) = \frac{\beta^2(\omega)}{\omega^2 \varepsilon_0 \mu_0}. \tag{7}$$

The theoretical study of the effective dielectric constant has been done by many authors. Fig. 4 shows a set of results of different investigators which were gathered together by Kuester and Chang [1]. These results were obtained by using different numerical techniques and it can be observed that they do not agree with one another very well. Our result is plotted on the same figure.

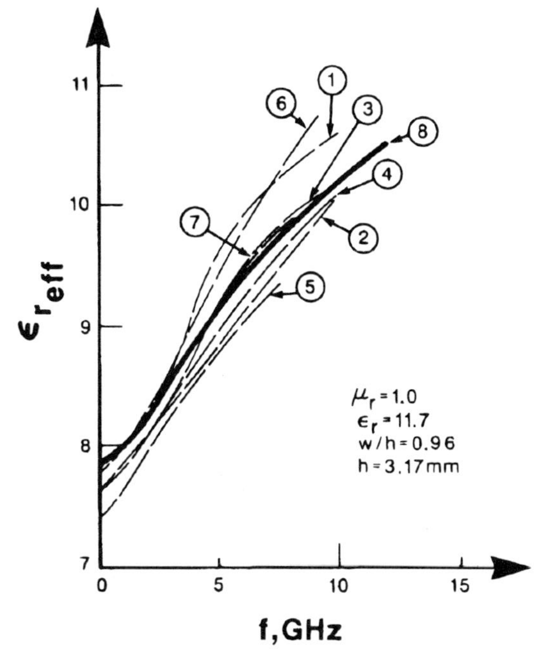

Fig. 4. Comparison of effective dielectric constant ε_{reff} as computed by different authors [1) Farrar and Adam, 2) Itoh and Mittra, 3) Van de Capelle and Luypaert, 4) Denlinger, 5) Schmitt and Sarges ($\varepsilon_r = 11.2$), 6) Chang and Kuester, 7) Pregla and Kowalaki, 8) time-domain result in this paper]. All of the results except the last one are from Kuester and Chang's paper [1].

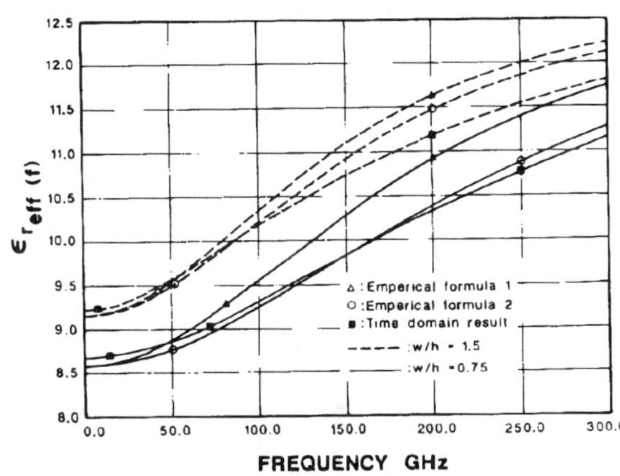

Fig. 5. Effective dielectric constant ($\varepsilon_r = 13.0$). Comparison of the results from time-domain method with two empirical formulas: 1) Edward and Owen [5], 2) Pramanick and Bhartia [6], [4].

Fig. 5 shows the comparisons of our result with the empirical formulas of Edward and Owen [5] and that of Pramanick and Bhartia ([4], [6]). For $W/H < 1$, our result is closer to that of Pramanick and Bhartia than to Edward and Owen's. For $W/H > 1$, both empirical formulas are off from our result in the frequency range shown. It is noticed that for both $W/H < 1$ and $W/H > 1$, the time-domain results provide fairly accurate results at dc (the dc value

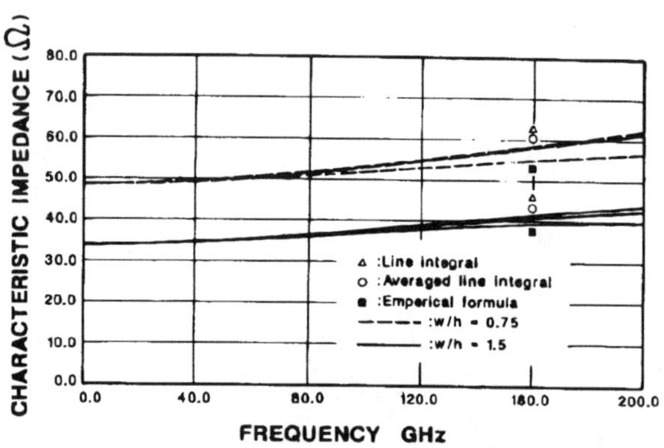

Fig. 6. Characteristic impedance ($\epsilon_r = 13.0$). Comparison of the results from time-domain method with empirical formula (Bianco *et al.* [5]).

used in these empirical formulas has 1 to 2 percent error, as claimed in [5]), and that the accuracy of the finite difference method, unlike the empirical formulas, is not affected by the ratio of W/H. In view of our agreement with Pramanick and Bhartia at $W/H<1$, it is not an unrealistic assumption to say that the time-domain results are more credible than those of the empirical formulas in the range $W/H>1$.

D. Characteristic Impedance

The variation of the characteristic impedance with frequency is obtained through the ratio of $V(\omega)/I(\omega)$. Here $I(\omega)$ is the Fourier transform of the current defined as the loop integral of the magnetic field around the metal strip. For $V(\omega)$, two kinds of definitions are used, one is the line integral of vertical electric field under the center of the strip, the other is the averaged line integral of vertical electric field under the whole strip.

Fig. 6 shows the comparison of our results with the result from the empirical formulas of Bianco *et al.* [5]. For a fairly large frequency range starting from dc, it turns out that the results of center line integral and averaged line integral are very close to each other, indicating that the voltage uniqueness is well satisfied in that frequency range.

IV. Conclusions

It has been demonstrated that the time-domain finite difference method is a credible numerical method capable of producing frequency-domain design data through Fourier transform. It has been a frequent practice to use a sinusoidal source in the time-domain approach to obtain frequency-domain results. Our experience leads us to believe that investigators use monochromatic sources because the Fourier transform of a time-domain result is very sensitive to time-domain errors. But the monochromatic approach is expensive, especially at low frequencies. This is because, using the explicit time-domain finite difference method to model the microstrip problems, will lead to an "oversampled" time mesh. That is, we are using a much smaller time step than that required by the Nyquist criterion. This situation is brought about by the Courant condition requiring $c\cdot dt < dh/\sqrt{3}$, (dh = smallest dimension of the mesh), and dh is kept small by the requirement of representing the small dimension of the microstrip. The above sequence of conditions limits the size of dt in the computation and results in a seemingly paradoxical situation in that it requires a longer time to calculate the response of a broad pulse than that of a sharp pulse. Consequently, it is much more economic to use a sharp Gaussian pulse source to obtain wide-band frequency results than to do it one frequency at a time. Indeed, it takes much more time to caluclate the response of a single sinusoidal wave than to calculate that of a sharp Gaussian pulse.

Many of the theoretical analyses of the microstrip are solely for microstrips, but the time-domain finite difference method is very general. With minor modifications this method has already been used by the authors to investigate many interesting microstrip-related microwave components and the results are to be published. This investigation paves the way for accurate modeling of passive microwave components at high frequencies and in mm-MMIC's.

References

[1] E. F. Kuester and D. C. Chang, "An appraisal of methods for computation of the dispersion characteristics of open microstrip," *IEEE Trans. Microwave Theory Tech.*, vol. MTT-27, pp. 691–694, July 1979.

[2] T. Itoh and R. Mittra, "Spectral-domain approach for calculating the dispersion characteristics of microstrip lines," *IEEE Trans. Microwave Theory Tech.*, vol. MTT-21, pp. 496–499, July 1973.

[3] M. Kobayashi and F. Ando, "Dispersion characteristics of open microstrip lines," *IEEE Trans. Microwave Theory Tech.*, vol. MTT-35, pp. 101–105, Feb. 1987.

[4] P. Pramanick and P. Bhartia, "An accurate description of dispersion in microstrip," *Microwave J.*, pp. 89–96, Dec. 1983.

[5] K. C. Gupta, R. Grag, and R. Chadha, *Computer-Aided Design of Microwave Circuits*, Dedham, MA: Artech House, 1981.

[6] R. L. Veghte and C. A. Balanis, "Dispersion of transient signals in microstrip transmission lines," *IEEE Trans. Microwave Theory Tech.*, vol. MTT-34, pp. 1427–1436, 1986.

[7] N. Yoshida and I. Fukai, "Transient analysis of a stripline having a corner in three-dimensional space," *IEEE Trans. Microwave Theory Tech.*, vol. MTT-32, pp. 491–498, May 1984.

[8] S. Koike, N. Yoshida, and I. Fukai, "Transient analysis of microstrip side-coupled filter in three-dimensional space," *Trans. IECE Japan*, vol. E69-B, pp. 1199–1205, Nov. 1986.

[9] S. Koike, N. Yoshida, and I. Fukai, "Transient analysis of microstrip gap in three-dimensional space," *IEEE Trans. Microwave Theory Tech.*, vol. MTT-33, pp. 726–730, Aug. 1985.

[10] S. Koike, N. Yoshida, and I. Fukai, "Transient analysis of coupling between crossing lines in three-dimensional space," *IEEE Trans. Microwave Theory Tech.*, vol. MTT-35, pp. 67–71, Jan. 1987.

[11] D. H. Choi and W. J. R. Hoefer, "The finite-difference time-domain method and its application to eigenvalue problems," *IEEE Trans. Microwave Theory Tech.*, vol. MTT-34, pp. 1464–1470, Dec. 1986..

[12] K. S. Yee, "Numerical solution of initial boundary value problems involving Maxwell's equations in isotropic media," *IEEE Trans. Antennas Propagat.*, vol. AP-14, pp. 302–307, May 1966.

[13] A. Taflove and M. E. Brodwin, "Numerical solution of steady-state electromangetic scattering problems using the time-dependent Maxwell's equations," *IEEE Trans. Microwave Theory Tech.*, vol. MTT-23, pp. 623–630, 1975.

[14] K. K. Mei, A. C. Cangellaris, and D. J. Angelakos, "Conformal time domain finite difference method," *Radio Sci.*, vol. 19, pp. 1145–1147, Sept.-Oct. 1984.

[15] A. C. Cangellaris, C. C. Lin, and K. K. Mei, "Point-matched time domain finite element methods for electromagnetic radiation and scatterings," University of California, Berkeley, Electronics Research Laboratory Memorandum No. UCB/ERL M85/25, Apr. 1985 (submitted to *IEEE Trans. Antennas Propagat.*).

[16] W. D. Smith, "A nonreflecting plane boundary for wave propagation problems," *J. Comp. Phys.*, vol. 15, pp. 492–503, 1974.

[17] G. Mur, "Absorbing boundary conditions for finite-difference approximation of time-domain electromagnetic field equations," *IEEE Trans. Electromagn. Compat.*, vol. EMC-23, pp. 1073–1077, 1981.

Application of the Three-Dimensional Finite-Difference Time-Domain Method to the Analysis of Planar Microstrip Circuits

DAVID M. SHEEN, SAMI M. ALI, SENIOR MEMBER, IEEE, MOHAMED D. ABOUZAHRA, SENIOR MEMBER, IEEE, AND JIN AU KONG, FELLOW, IEEE

Abstract —**A direct three-dimensional finite-difference time-domain (FDTD) method is applied to the full-wave analysis of various microstrip structures. The method is shown to be an efficient tool for modeling complicated microstrip circuit components as well as microstrip antennas. From the time-domain results, the input impedance of a line-fed rectangular patch antenna and the frequency-dependent scattering parameters of a low-pass filter and a branch line coupler are calculated. These circuits are fabricated and the measurements are compared with the FDTD results and shown to be in good agreement.**

I. Introduction

FREQUENCY-domain analytical work with complicated microstrip circuits has generally been done using planar circuit concepts in which the substrate is assumed to be thin enough that propagation can be considered in two dimensions by surrounding the microstrip with magnetic walls [1]–[6]. Fringing fields are accounted for by using either static or dynamic effective dimensions and permittivities. Limitations of these methods are that fringing, coupling, and radiation must all be handled empirically since they are not allowed for in the model. Also, the accuracy is questionable when the substrate becomes thick relative to the width of the microstrip. To fully account for these effects, it is necessary to use a full-wave solution.

Full-wave frequency-domain methods have been used to solve some of the simpler discontinuity problems [7], [8]. However, these methods are difficult to apply to a typical printed microstrip circuit.

Modeling of microstrip circuits has also been performed using Bergeron's method [9], [10]. This method is a modification of the transmission line matrix (TLM) method, and has limitations similar to the finite-difference time-domain (FDTD) method due to the discrete modeling of space and time [11], [12]. A unique problem with this method is that the dielectric interface and the perfectly conducting strip are misaligned by half a space step [12].

The FDTD method has been used extensively for the solution of two- and three-dimensional scattering problems [13]–[17]. Recently, FDTD methods have been used to effectively calculate the frequency-dependent characteristics of microstrip discontinuities [18]–[21]. Analysis of the fundamental discontinuities is of great importance since more complicated circuits can be realized by interconnecting microstrip lines with these discontinuities and using transmission line and network theory. Some circuits, however, such as patch antennas, may not be realized in this way. Additionally, if the discontinuities are too close to each other the use of network concepts will not be accurate due to the interaction of evanescent waves. To accurately analyze these types of structures it is necessary to simulate the entire structure in one computation. The FDTD method shows great promise in its flexibility in handling a variety of circuit configurations. An additional benefit of the time-domain analysis is that a broad-band pulse may be used as the excitation and the frequency-domain parameters may be calculated over the entire frequency range of interest by Fourier transform of the transient results.

In this paper, the frequency-dependent scattering parameters have been calculated for several printed microstrip circuits, specifically a line-fed rectangular patch antenna, a low-pass filter, and a rectangular branch line coupler. These circuits represent resonant microstrip structures on an open substrate; hence, radiation effects can be significant, especially for the microstrip antenna. Calculated results are presented and compared with experimental measurements.

The FDTD method has been chosen over the other discrete methods (TLM or Bergeron's) because it is extremely efficient, its implementation is quite straightforward, and it may be derived directly from Maxwell's equations. Many of the techniques used to implement this

Manuscript received September 29, 1989; revised March 21, 1990. This work was supported by NSF Grant 8620029-ECS, the Joint Services Electronics Program (Contract DAAL03-89-C-0001), RADC Contract F19628-88-K-0013, ARO Contract DAAL03-88-J-0057, ONR Contract N00014-89-J-1019, and the Department of the Air Force.

D. M. Sheen, S. M. Ali, and J. A. Kong are with the Department of Electrical Engineering and Computer Science and the Research Laboratory of Electronics, Massachusetts Institute of Technology, Cambridge, MA 02139.

M. D. Abouzahra is with MIT Lincoln Laboratory, Lexington, MA 02173.

IEEE Log Number 9036153.

Reprinted from *IEEE Transactions on Microwave Theory and Techniques,* pp. 849-857, July 1990.

method have been demonstrated previously [18]–[20]; however, simplification of the method has been achieved by using a simpler absorbing boundary condition [22]. This simpler absorbing boundary condition yields good results for the broad class of microstrip circuits considered by this paper. Additionally, the source treatment has been enhanced to reduce the source effects documented in [18]–[20].

II. Problem Formulation

The FDTD method is formulated by discretizing Maxwell's curl equations over a finite volume and approximating the derivatives with centered difference approximations. Conducting surfaces are treated by setting tangential electric field components to 0. The walls of the mesh, however, require special treatment to prevent reflections from the mesh termination.

A. Governing Equations

Formulation of the FDTD method begins by considering the differential form of Maxwell's two curl equations which govern the propagation of fields in the structures. For simplicity, the media are assumed to be piecewise uniform, isotropic, and homogeneous. The structure is assumed to be lossless (i.e., no volume currents or finite conductivity). With these assumptions, Maxwell's curl equations may be written as

$$\mu \frac{\partial \boldsymbol{H}}{\partial t} = -\nabla \times \boldsymbol{E} \tag{1}$$

$$\epsilon \frac{\partial \boldsymbol{E}}{\partial t} = \nabla \times \boldsymbol{H}. \tag{2}$$

In order to find an approximate solution to this set of equations, the problem is discretized over a finite three-dimensional computational domain with appropriate boundary conditions enforced on the source, conductors, and mesh walls.

B. Finite-Difference Equations

To obtain discrete approximations to these continuous partial differential equations the centered difference approximation is used on both the time and space first-order partial differentiations. For convenience, the six field locations are considered to be interleaved in space as shown in Fig. 1, which is a drawing of the FDTD unit cell [13]. The entire computational domain is obtained by stacking these rectangular cubes into a larger rectangular volume. The $\hat{x}$, $\hat{y}$, and $\hat{z}$ dimensions of the unit cell are Δx, Δy, and Δz, respectively. The advantages of this field arrangement are that centered differences are realized in the calculation of each field component and that continuity of tangential field components is automatically satisfied. Because there are only six unique field components within the unit cell, the six field components touching the shaded upper eighth of the unit cell in Fig. 1 are considered to be a unit node with subscript indices i, j, and k corresponding to the node numbers in the $\hat{x}$, $\hat{y}$, and $\hat{z}$ directions. This notation implicitly assumes the $\pm 1/2$ space indices and thus simplifies the notation, rendering the formulas directly implementable on the computer. The time steps are indicated with the superscript n. Using this field component arrangement, the above notation, and the centered difference approximation, the explicit finite difference approximations to (1) and (2) are

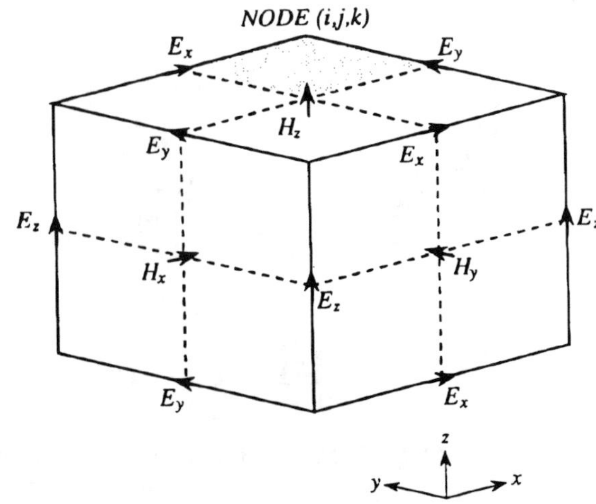

Fig. 1. Field component placement in the FDTD unit cell.

$$H^{n+1/2}_{x\,i,j,k} = H^{n-1/2}_{x\,i,j,k} + \frac{\Delta t}{\mu \Delta z}\left(E^n_{y\,i,j,k} - E^n_{y\,i,j,k-1}\right) - \frac{\Delta t}{\mu \Delta y}\left(E^n_{z\,i,j,k} - E^n_{z\,i,j-1,k}\right) \tag{3}$$

$$H^{n+1/2}_{y\,i,j,k} = H^{n-1/2}_{y\,i,j,k} + \frac{\Delta t}{\mu \Delta x}\left(E^n_{z\,i,j,k} - E^n_{z\,i-1,j,k}\right) - \frac{\Delta t}{\mu \Delta z}\left(E^n_{x\,i,j,k} - E^n_{x\,i,j,k-1}\right) \tag{4}$$

$$H^{n-1/2}_{z\,i,j,k} = H^{n-1/2}_{z\,i,j,k} + \frac{\Delta t}{\mu \Delta y}\left(E^n_{x\,i,j,k} - E^n_{x\,i,j-1,k}\right) - \frac{\Delta t}{\mu \Delta x}\left(E^n_{y\,i,j,k} - E^n_{y\,i-1,j,k}\right) \tag{5}$$

$$E^{n+1}_{x\,i,j,k} = E^n_{x\,i,j,k} + \frac{\Delta t}{\epsilon \Delta y}\left(H^{n-1/2}_{z\,i,j+1,k} - H^{n+1/2}_{z\,i,j,k}\right) - \frac{\Delta t}{\epsilon \Delta z}\left(H^{n+1/2}_{y\,i,j,k+1} - H^{n+1/2}_{y\,i,j,k}\right) \tag{6}$$

$$E^{n+1}_{y\,i,j,k} = E^n_{y\,i,j,k} + \frac{\Delta t}{\epsilon \Delta z}\left(H^{n+1/2}_{x\,i,j,k+1} - H^{n+1/2}_{x\,i,j,k}\right) - \frac{\Delta t}{\epsilon \Delta x}\left(H^{n+1/2}_{z\,i+1,j,k} - H^{n+1/2}_{z\,i,j,k}\right) \tag{7}$$

$$E^{n+1}_{z\,i,j,k} = E^n_{z\,i,j,k} + \frac{\Delta t}{\epsilon \Delta x}\left(H^{n+1/2}_{y\,i+1,j,k} - H^{n+1/2}_{y\,i,j,k}\right) - \frac{\Delta t}{\epsilon \Delta y}\left(H^{n+1/2}_{x\,i,j+1,k} - H^{n+1/2}_{x\,i,j,k}\right). \tag{8}$$

The half time steps indicate that E and H are alternately calculated in order to achieve centered differences for the time derivatives. In these equations, the permittivity and the permeability are set to the appropriate values depending on the location of each field component. For the electric field components on the dielectric–air interface the average of the two permittivities, $(\epsilon_0 + \epsilon_1)/2$, is used. The validity of this treatment is explained in [20].

Due to the use of centered differences in these approximations, the error is second order in both the space and time steps; i.e., if Δx, Δy, Δz, and Δt are proportional to Δl, then the global error is $O(\Delta l^2)$. The maximum time step that may be used is limited by the stability restriction of the finite difference equations,

$$\Delta t \leqslant \frac{1}{v_{\max}} \left(\frac{1}{\Delta x^2} + \frac{1}{\Delta y^2} + \frac{1}{\Delta z^2} \right)^{-1/2} \tag{9}$$

where $v_{\max}$ is the maximum velocity of light in the computational volume. Typically, $v_{\max}$ will be the velocity of light in free space unless the entire volume is filled with dielectric. These equations will allow the approximate solution of $E(r,t)$ and $H(r,t)$ in the volume of the computational domain or mesh; however, special consideration is required for the source, the conductors, and the mesh walls.

C. Source Considerations

The volume in which the microstrip circuit simulation is to be performed is shown schematically in Fig. 2. At $t = 0$ the fields are assumed to be identically 0 throughout the computational domain. A Gaussian pulse is desirable as the excitation because its frequency spectrum is also Gaussian and will therefore provide frequency-domain information from dc to the desired cutoff frequency by adjusting the width of the pulse.

In order to simulate a voltage source excitation it is necessary to impose the vertical electric field, E_z, in a rectangular region underneath port 1 as shown in Fig. 2. The remaining electric field components on the source plane must be specified or calculated. In [18]–[20] an electric wall source is used; i.e., the remaining electric field components on the source wall of the mesh are set to 0. An unwanted side effect of this type of excitation is that a sharp magnetic field is induced tangential to the source wall. This results in some distortion of the launched pulse. Specifically, the pulse is reduced in magnitude due to the energy stored in the induced magnetic field and a negative tail to the pulse is immediately evident. An alternative excitation scheme is to simulate a magnetic wall at the source plane. The source plane consists only of E_x and E_z components, with the tangential magnetic field components offset $\pm \Delta y/2$. If the magnetic wall is enforced by setting the tangential magnetic field components to zero just behind the source plane, then significant distortion of the pulse still occurs. If the magnetic wall is enforced directly on the source plane by using image theory (i.e., $H_{\tan}$ outside the magnetic wall is equal to $-H_{\tan}$ inside the magnetic wall), then the remaining electric field components on the source plane may be readily calculated using the finite-difference equations. Using this excitation, only a minimal amount of source distortion is apparent. The launched wave has nearly unit amplitude and is Gaussian in time and in the $\hat{y}$ direction:

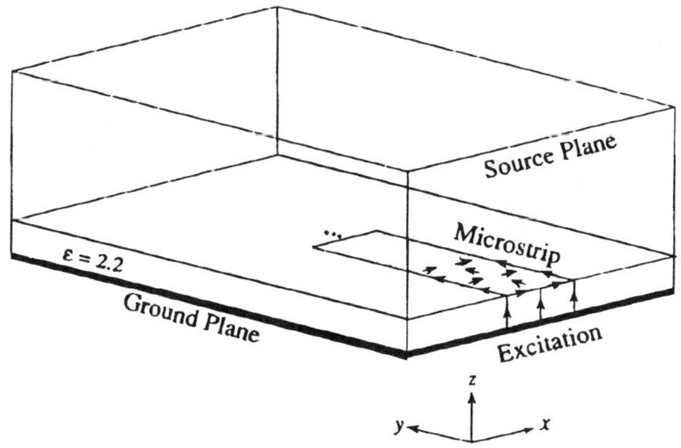

Fig. 2. Computational domain.

$$E_z = f_s(t) = e^{-(t-t_0)^2/T^2}. \tag{10}$$

It is assumed that excitation specified in this way will result in the fundamental mode only propagating down the microstrip in the frequency range of interest.

The finite-difference formulas are not perfect in their representation of the propagation of electromagnetic waves. One effect of this is numerical dispersion; i.e., the velocity of propagation is slightly frequency dependent even for uniform plane waves. In order to minimize the effects of numerical dispersion and truncation errors, the width of the Gaussian pulse is chosen for at least 20 points per wavelength at the highest frequency represented significantly in the pulse.

D. Conductor Treatment

The circuits considered in this paper have a conducting ground plane and a single dielectric substrate with metallization on top of this substrate in the ordinary microstrip configuration. These electric conductors are assumed to be perfectly conducting and have zero thickness and are treated by setting the electric field components that lie on the conductors to zero. The edge of the conductor should be modeled with electric field components tangential to the edge lying exactly on the edge of the microstrip as shown in Fig. 2.

E. Absorbing Boundary Treatment

Due to the finite capabilities of the computers used to implement the finite-difference equations, the mesh must be limited in the $\hat{x}$, $\hat{y}$, and $\hat{z}$ directions. The difference equations cannot be used to evaluate the field components tangential to the outer boundaries since they would require the values of field components outside of the mesh. One of the six mesh boundaries is a ground plane and its tangential electric field values are forced to be 0.

The tangential electric field components on the other five mesh walls must be specified in such a way that outgoing waves are not reflected using the absorbing boundary condition [22], [23]. For the structures considered in this paper, the pulses on the microstrip lines will be normally incident to the mesh walls. This leads to a simple approximate continuous absorbing boundary condition, which is that the tangential fields on the outer boundaries will obey the one-dimensional wave equation in the direction normal to the mesh wall. For the $\hat{y}$ normal wall the one-dimensional wave equation may be written

$$\left(\frac{\partial}{\partial y} - \frac{1}{v}\frac{\partial}{\partial t}\right)\boldsymbol{E}_{\tan} = 0. \tag{11}$$

This equation is Mur's first approximate absorbing boundary condition and it may be easily discretized using only field components on or just inside the mesh wall, yielding an explicit finite difference equation [22],

$$E_0^{n+1} = E_1^n + \frac{v\Delta t - \Delta y}{v\Delta t + \Delta y}\left(E_1^{n+1} - E_0^n\right) \tag{12}$$

where E_0 represents the tangential electric field components on the mesh wall and E_1 represents the tangential electric field components one node inside of the mesh wall. Similar expressions are immediately obtained for the other absorbing boundaries by using the corresponding normal directions for each wall. It should be noted that the normal incidence assumption is not valid for the fringing fields which are propagating tangential to the walls; therefore the sidewalls should be far enough away that the fringing fields are negligible at the walls. Additionally, radiation will not be exactly normal to the mesh walls. Second-order absorbing boundary conditions [22], [23] which account for oblique incidence will not work on the mesh walls where the microstrip is incident because these absorbing boundary conditions are derived in uniform space.

The results presented show that the first-order absorbing boundary treatment is sufficiently accurate and particularly well suited to the microstrip geometry. It is possible to obtain more accurate normal incidence absorbing boundary conditions [24], which have been used in the FDTD calculation of microstrip discontinuities [20]. Due to the more dynamic, resonant behavior of the circuits considered, Mur's first approximate absorbing boundary condition is used and this allows for accurate simulation of the various microstrip circuits.

F. Time Marching Solution

The finite difference equations, (3)–(8), are used with the above boundary and source conditions to simulate the propagation of a broad-band Gaussian pulse on the microstrip structure. The essential aspects of the time-domain algorithm are as follows:

- Initially (at $t = n = 0$) all fields are 0.
- The following are repeated until the response is ≈ 0:
  - Gaussian excitation is imposed on port 1.
  - $\boldsymbol{H}^{n+1/2}$ is calculated from FD equations.
  - $\boldsymbol{E}^{n+1}$ is calculated from FD equations.
  - Tangential $\boldsymbol{E}$ is set to 0 on conductors.
  - Save desired field quantities.
  - $n \to n+1$.
- Compute scattering matrix coefficients from time-domain results.

One additional consideration is that the reflections from the circuit will be reflected again by the source wall. To eliminate this, the circuit is placed a sufficient distance from the source, and after the Gaussian pulse has been fully launched, the absorbing boundary condition is switched on at the source wall.

G. Frequency-Dependent Parameters

In addition to the transient results obtained naturally by the FDTD method, the frequency-dependent scattering matrix coefficients are easily calculated.

$$[V]^r = [S][V]^i \tag{13}$$

where $[V]^r$ and $[V]^i$ are the reflected and incident voltage vectors, respectively, and $[S]$ is the scattering matrix. To accomplish this, the vertical electric field underneath the center of each microstrip port is recorded at every time step. As in [20], it is assumed that this field value is proportional to the voltage (which could be easily obtained by numerically integrating the vertical electric field) when considering propagation of the fundamental mode. To obtain the scattering parameter $S_{11}(\omega)$, the incident and reflected waveforms must be known. The FDTD simulation calculates the sum of incident and reflected waveforms. To obtain the incident waveform, the calculation is performed using only the port 1 microstrip line, which will now be of infinite extent (i.e., from source to far absorbing wall), and the incident waveform is recorded. This incident waveform may now be subtracted from the incident plus reflected waveform to yield the reflected waveform for port 1. The other ports will register only transmitted waveforms and will not need this computation. The scattering parameters, S_{jk}, may then be obtained by simple Fourier transform of these transient waveforms as

$$S_{jk}(\omega) = \frac{\mathcal{F}\mathcal{T}\{V_j(t)\}}{\mathcal{F}\mathcal{T}\{V_k(t)\}}. \tag{14}$$

Note that the reference planes are chosen with enough distance from the circuit discontinuities to eliminate evanescent waves. These distances are included in the definition of the circuit so that no phase correction is performed for the scattering coefficients. For all of the circuits considered, the only unique coefficients are in

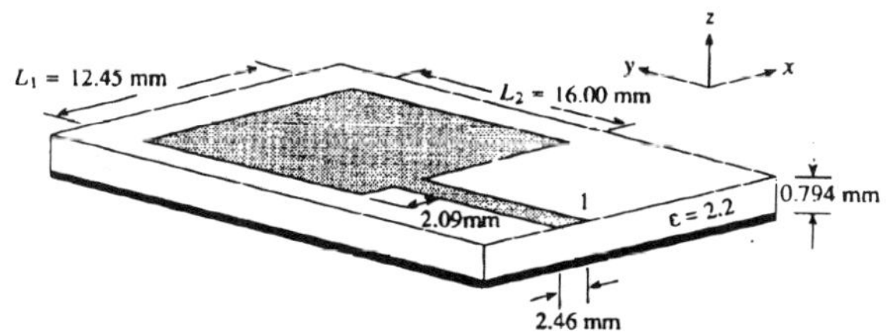

Fig. 3. Line-fed rectangular microstrip antenna detail.

the first column of the scattering matrix (i.e., $S_{11}(\omega), S_{21}(\omega), S_{31}(\omega), \cdots$).

III. Numerical Results

Numerical results have been computed for three configurations, a line-fed rectangular patch antenna, a low-pass filter, and a branch line coupler. These circuits have dimensions on the order of 1 cm, and the frequency range of interest is from dc to 20 GHz. The operating regions of all of these circuits are less than 10 GHz; however, the accuracy of the computed results at higher frequencies is examined. These circuits were constructed on Duroid substrates with $\epsilon = 2.2$ and thickness of 1/32 inch (0.794 mm). Scattering matrix coefficients were measured using an HP 8510 network analyzer, which is calibrated to 18 GHz, but provides measurement to 20 GHz.

A. Line-Fed Rectangular Microstrip Antenna

The actual dimensions of the microstrip antenna analyzed are shown in Fig. 3. The operating resonance approximately corresponds to the frequency where $L_1 =$ 1.245 mm $= \lambda/2$. Simulation of this circuit involves the straightforward application of the finite-difference equations, source, and boundary conditions. To model the thickness of the substrate correctly, Δz is chosen so that three nodes exactly match the thickness. An additional 13 nodes in the $\hat{z}$ direction are used to model the free space above the substrate. In order to correctly model the dimensions of the antenna, Δx and Δy have been chosen so that an integral number of nodes will exactly fit the rectangular patch. Unfortunately, this means the port width and placement will be off by a fraction of the space step. The sizes of the space steps are carefully chosen to minimize the effect of this error.

The space steps used are $\Delta x = 0.389$ mm, $\Delta y = 0.400$ mm, and $\Delta z = 0.265$ mm, and the total mesh dimensions are $60 \times 100 \times 16$ in the $\hat{x}$, $\hat{y}$, and $\hat{z}$ directions respectively. The rectangular antenna patch is thus $32\,\Delta x \times 40\,\Delta y$. The length of the microstrip line from the source plane to the edge of the antenna is $50\,\Delta y$, and the reference plane for port 1 is $10\,\Delta y$ from the edge of the patch. The microstrip line width is modeled as $6\,\Delta x$.

The time step used is $\Delta t = 0.441$ ps. The Gaussian half-width is $T = 15$ ps, and the time delay t_0 is set to be $3T$ so the Gaussian will start at approximately 0. The simulation is performed for 8000 time steps, somewhat longer than for other circuits, due to the highly resonant behavior of the antenna. The computation time for this circuit is approximately 12 h on a VAXstation 3500 workstation.

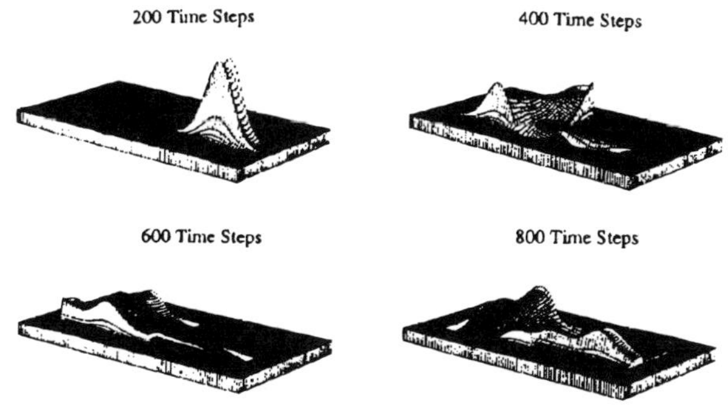

Fig. 4. Rectangular microstrip antenna distribution of $E_z(x, y, t)$ just underneath the dielectric interface at 200, 400, 600, and 800 time steps.

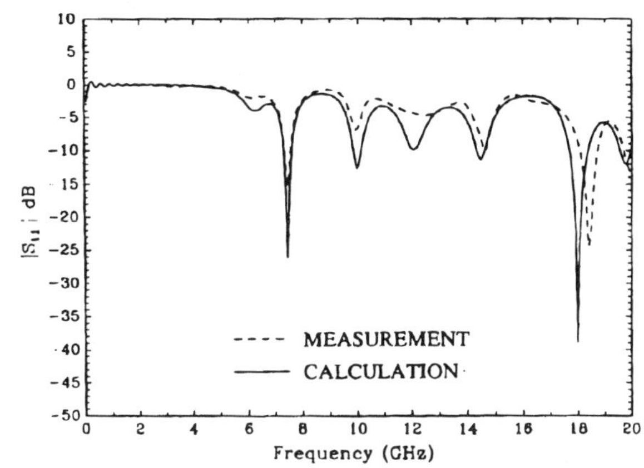

Fig. 5. Return loss of the rectangular antenna.

The spatial distribution of $E_z(x, y, t)$ just beneath the microstrip at 200, 400, 600, and 800 time steps is shown in Fig. 4, where the source Gaussian pulse and subsequent propagation on the antenna are observed. Notice that the absorbing boundary condition for the source has been implemented several nodes from the source plane to eliminate any undesirable effects of switching from source to absorbing boundary conditions. Three-dimensional properties of the propagation are observed including enhancement of the field near the edges of the microstrip.

The scattering coefficient results, shown in Fig. 5, show good agreement with the measured data. The operating resonance at 7.5 GHz is almost exactly shown by both theory and measurement. This result is a significant advancement over planar circuit techniques, which without empirical treatment will allow only $|S_{11}| = 1.0$ because they do not allow for radiation. Also, the resonance frequency calculated using planar circuit concepts will be sensitive to errors in the effective dimensions of the patch. Additional resonances are also in good agcement with experiment, except for the highest resonance near 18 GHz, which is somewhat shifted.

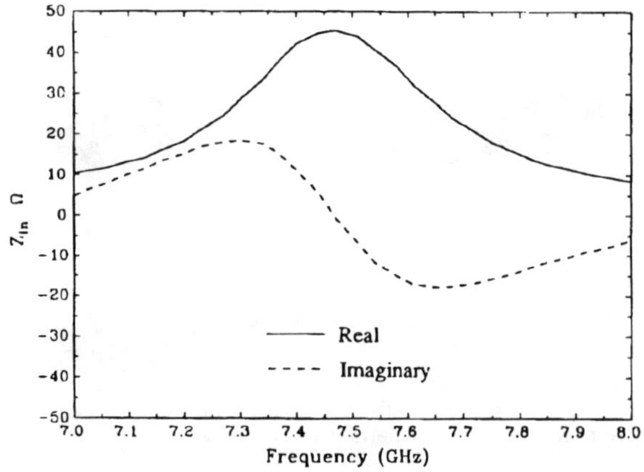

Fig. 6. Input impedance of the rectangular antenna near the operating resonance of 7.5 GHz as calculated from the FDTD results.

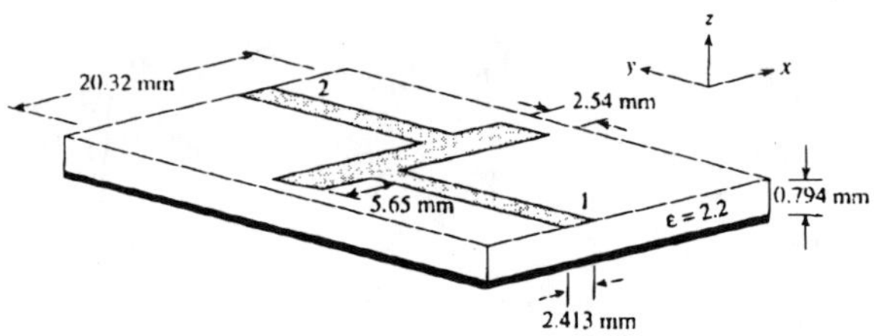

Fig. 7. Low-pass filter detail.

Input impedance for the antenna may be calculated from the $S_{11}(\omega)$ calculation by transforming the reference plane to the edge of the microstrip antenna,

$$Z_{in} = Z_0 \frac{1 + S_{11} e^{j2kL}}{1 - S_{11} e^{j2kL}} \tag{15}$$

where k is the wavenumber on the microstrip, L is the length from the edge of the antenna to the reference plane ($10\Delta y$), and Z_0 is the characteristic impedance of the microstrip line. For simplicity of the Z_{in} calculation, the microstrip is assumed to have a constant characteristic impedance of 50 Ω, and an effective permittivity of 1.9 is used to calculate the wavenumber. Results for the input impedance calculation near the operating resonance of 7.5 GHz are shown in Fig. 6.

B. Microstrip Low-Pass Filter

The low-pass filter analyzed is one designed according to the criteria established by [2] and is shown in Fig. 7. As in the rectangular antenna, Δx, Δy, and Δz are carefully chosen to fit the dimensions of the circuit. The space steps Δx and Δy are chosen to exactly match the dimensions of the rectangular patch; however, the locations and widths of the ports will be modeled with some error.

The space steps used are $\Delta x = 0.4064$ mm, $\Delta y = 0.4233$ mm, and $\Delta z = 0.265$ mm, and the total mesh dimensions are $80 \times 100 \times 16$ in the $\hat{x}$, $\hat{y}$, and $\hat{z}$ directions respectively. The long rectangular patch is thus $50\Delta x \times 6\Delta y$. The distance from the source plane to the edge of the long patch is $50\Delta y$, and the reference planes for ports 1 and 2 are $10\Delta y$ from the edges of the patch. The strip widths of ports 1 and 2 are modeled as $6\Delta x$.

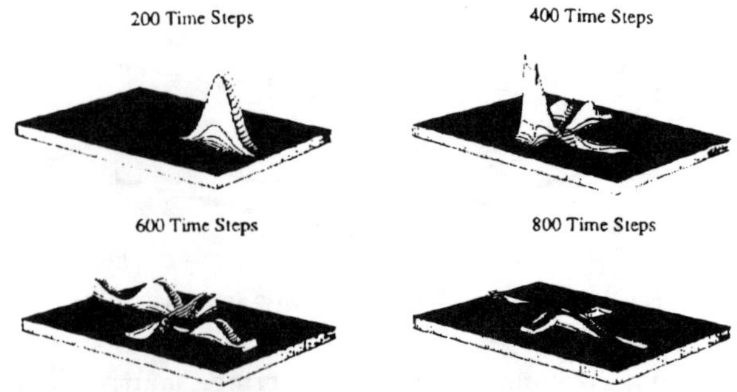

Fig. 8. Low-pass filter distribution of $E_z(x,y,t)$ just underneath the dielectric interface at 200, 400, 600, and 800 time steps.

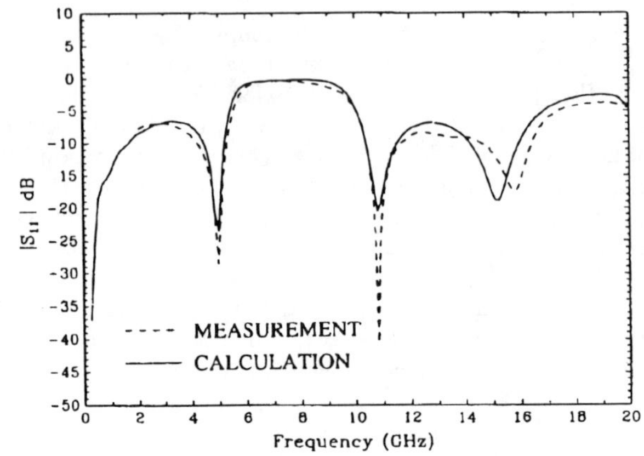

Fig. 9. Return loss of the low-pass filter.

The time step used is $\Delta t = 0.441$ ps. The Gaussian half-width is $T = 15$ ps and the time delay t_0 is set to be $3T$. The simulation is performed for 4000 time steps to allow the response on both ports to become nearly 0. The computation time for this circuit is approximately 8 h on a VAXstation 3500 workstation.

The spatial distribution of $E_z(x,y,t)$ just beneath the microstrip at 200, 400, 600, and 800 time steps is shown in Fig. 8. The scattering coefficient results, shown in Figs. 9 and 10, again show good agreement in the location of the response nulls. The desired low-pass filter performance is seen in the sharp S_{21} roll-off beginning at approximately 5 GHz. There is again some shift near the high end of the frequency range. In the S_{21} results, the stopband for the calculated curve is somewhat narrower than the measured results. Some experimentation with planar circuit techniques has led to the conclusion that this narrowing is caused predominantly by the slight misplacement of the ports inherent in the choice of Δx and Δy.

C. Microstrip Branch Line Coupler

The branch line coupler, shown in Fig. 11, is used to divide power equally between ports 3 and 4 from ports 1 or 2. This occurs at the frequency where the center-to-center distance between the four lines is a quarter wave-

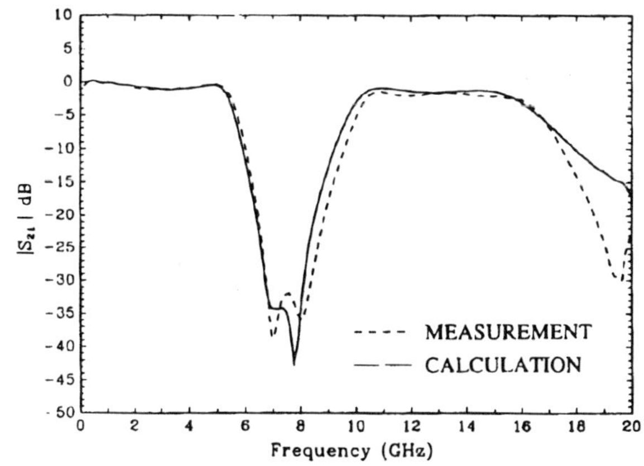

Fig. 10. Insertion loss of the low-pass filter.

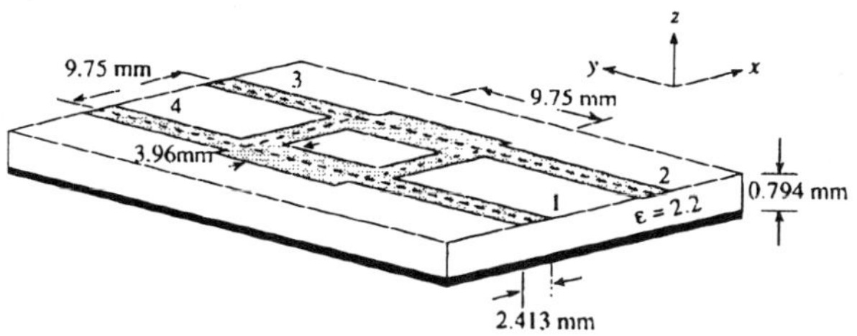

Fig. 11. Branch line coupler detail.

length. Also, the phase difference between ports 3 and 4 is 90° at this frequency. To model this circuit, Δx, Δy, and Δz are chosen to match the dimensions of the circuit as effectively as possible. The space step, Δz, is chosen to match the substrate thickness exactly. The space steps, Δx and Δy, are chosen to match the center-to-center distance (9.75 mm) exactly. Again, small errors in the other $\hat{x}$ and $\hat{y}$ dimensions occur.

The space steps used are $\Delta x = 0.406$ mm, $\Delta y = 0.406$ mm, and $\Delta z = 0.265$ mm and the total mesh dimensions are $60 \times 100 \times 16$ in the $\hat{x}$, $\hat{y}$, and $\hat{z}$ directions, respectively. The center-to-center distances are $24\,\Delta x$ and $24\,\Delta y$. The distance from the source plane to the edge of the coupler is $50\,\Delta y$, and the reference planes for ports 1 through 4 are $10\,\Delta y$ from the edges of the coupler. The strip widths of ports 1 through 4 are modeled as $6\,\Delta x$. The wide strips in the coupler are modeled as $10\,\Delta x$ wide.

The time step used is $\Delta t = 0.441$ ps. The Gaussian half-width is $T = 15$ ps and the time delay t_0 is set to be $3T$. The simulation is performed for 4000 time steps to allow the response on all four ports to become nearly zero. The computation time for this circuit is approximately 6 h on a VAXstation 3500 workstation.

The spatial distribution of $E_z(x, y, t)$ just beneath the microstrip at 200, 400, 600, and 800 time steps is shown in Fig. 12. The scattering coefficient results, shown in Fig. 13, again show good agreement in the location of the response nulls and crossover point. The desired branch line coupler performance is seen in the sharp S_{11} and S_{21} nulls which occur at approximately the same point (6.5 GHz) as the crossover in S_{31} and S_{41}. At this crossover point S_{31} and S_{41} are both approximately -3 dB, indicating that the power is being evenly divided between ports 3 and 4. The nulls in S_{11} and S_{21} at the operating point indicate that little power is being transmitted by ports 1 and 2. The phase difference between S_{31} and S_{41} is verified to be approximately 90° at the operating point (≈ 6.5 GHz) in both the calculated and measured coefficients, as shown in Fig. 14. Some shift in the location of the nulls is again observed and is likely due to the slight modeling errors in the widths of the lines.

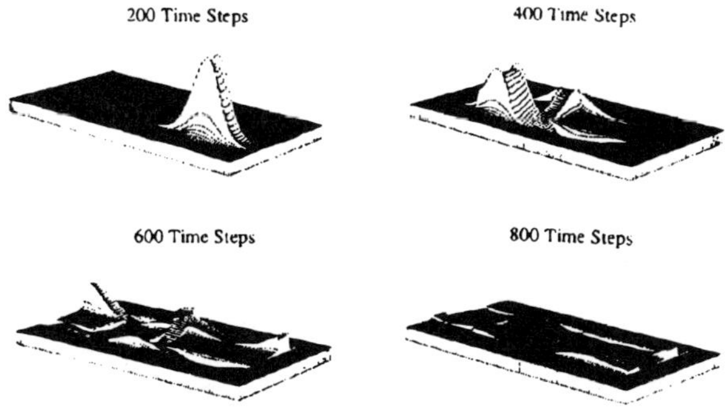

Fig. 12. Branch line coupler distribution of $E_z(x, y, t)$ just underneath the dielectric interface at 200, 400, 600, and 800 time steps.

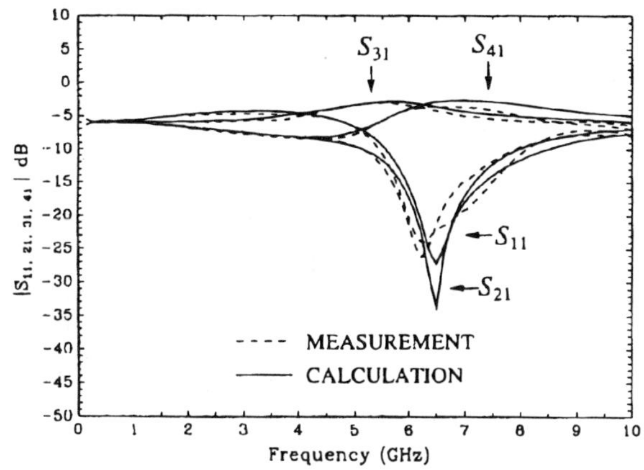

Fig. 13. Scattering parameters of the branch line coupler.

D. Error Discussion

In general, agreement between measured and calculated results has been good; however, there are several reasons to explain the small discrepancies in the results. The modeling error occurs primarily in the inability to match all of the circuit dimensions. The space steps, Δx, Δy, and Δz, may be freely chosen; this allows exact matching of only one circuit dimension (or two edges) in each of the $\hat{x}$, $\hat{y}$, and $\hat{z}$ directions. Another small source of error is the exclusion of dielectric and conductor loss in the FDTD calculation. This causes the calculated S parameters to be shifted up in amplitude from the measured data at the higher frequencies. Measurement errors occur because of the microstrip-to-coaxial transitions, which are

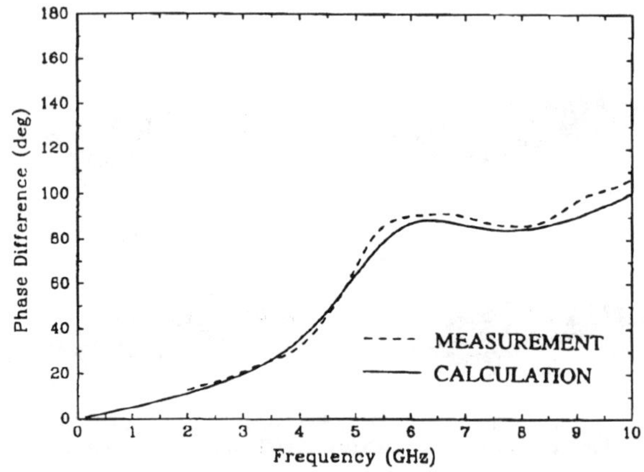

Fig. 14. Phase difference of port 3 relative to port 4 of the branch line coupler.

not de-embedded in the measurement, and because the network analyzer and its connectors are rated only to 18 GHz.

IV. Conclusions

The finite-difference time-domain method has been used to perform time-domain simulations of pulse propagation in several printed microstrip circuits. In addition to the transient results, frequency-dependent scattering parameters and the input impedance of a line-fed rectangular patch antenna have been calculated by Fourier transform of the time-domain results. These results have been verified by comparison with measured data. The versatility of the FDTD method allows easy calculation of many complicated microstrip structures. With the computational power of computers increasing rapidly, this method is very promising for the computer-aided design of many types of microstrip circuit components.

References

[1] G. D'Inzeo, F. Giannini, C. Sodi, and R. Sorrentino, "Method of analysis and filtering properties of microwave planar networks," *IEEE Trans. Microwave Theory Tech.*, vol. MTT-26, pp. 462–471, July 1978.

[2] G. D'Inzeo, F. Giannini, and R. Sorrentino, "Novel microwave integrated low-pass filters," *Electron. Lett.*, vol. 15, no. 9, pp. 258–260, Apr. 26, 1979.

[3] K. C. Gupta, R. Garg, and R. Chadha, *Computer-Aided Design of Microwave Circuits.* Dedham, MA: Artech House, 1981.

[4] T. Okoshi, *Planar Circuits for Microwaves and Lightwaves.* Berlin: Springer-Verlag, 1985.

[5] W. K. Gwarek, "Analysis of arbitrarily shaped two-dimensional microwave circuits by finite difference time domain method," *IEEE Trans. Microwave Theory Tech.*, vol. 36, pp. 738–744, Apr. 1988.

[6] K. C. Gupta and M. D. Abouzahra, "Planar circuit analysis," in *Numerical Techniques for Microwave and Millimeter-Wave Passive Structures*, T. Itoh, Ed. New York: Wiley, 1989, pp. 214–333.

[7] P. B. Katehi and N. C. Alexopoulos, "Frequency dependent characteristics of microstrip discontinuities in millimeter wave integrated circuits," *IEEE Trans. Microwave Theory Tech.*, vol. MTT-33, pp. 1029–1035, Oct. 1985.

[8] R. W. Jackson, "Full-wave, finite element analysis of irregular microstrip discontinuities," *IEEE Trans. Microwave Theory Tech.*, vol. 37, pp. 81–89, Jan. 1989.

[9] S. Koike, N. Yoshida, and I. Fukai, "Transient analysis of microstrip line on anisotropic substrate in three-dimensional space," *IEEE Trans. Microwave Theory Tech.*, vol. 36, pp. 34–43, Jan. 1988.

[10] T. Shibata, T. Hayashi, and T. Kimura, "Analysis of microstrip circuits using three-dimensional full-wave electromagnetic field analysis in the time domain," *IEEE Trans. Microwave Theory Tech.*, vol. 36, pp. 1064–1070, June 1988.

[11] S. Akhtarzad and P. B. Johns, "Solution of Maxwell's equations in three space dimensions and time by the T. L. M. method of numerical analysis," *Proc. Inst. Elec. Eng.*, vol. 122, no. 12, pp. 1344–1348, Dec. 1975.

[12] W. J. R. Hoefer, "The transmission-line matrix method—Theory and applications," *IEEE Trans. Microwave Theory Tech.*, vol. MTT-33, pp. 882–893, Oct. 1985.

[13] K. S. Yee, "Numerical solution of initial boundary value problems involving Maxwell's equations in isotropic media," *IEEE Trans. Antennas Propagat.*, vol. AP-14, pp. 302–307, May 1966.

[14] A. Taflove and M. E. Brodwin, "Numerical solution of steady state electromagnetic scattering problems using the time dependent Maxwell's equations," *IEEE Trans. Microwave Theory Tech.*, vol. MTT-23, pp. 623–630, Aug. 1975.

[15] A. Taflove, "Application of the finite difference time domain method to sinusoidal steady state electromagnetic penetration problems," *IEEE Trans. Electromagn. Compat.* vol. EMC-22, pp. 191–202, Aug. 1980.

[16] A. Taflove and K. R. Umashankar, "The finite difference time domain (FD-TD) method for electromagnetic scattering and interaction problems," *J. Electromagn. Waves and Appl.*, vol. 1, no. 3, pp. 243–267, 1987.

[17] A. C. Cangellaris, C. C. Lin, and K. K. Mei, "Point matched time domain finite element methods for electromagnetic radiation and scattering," *IEEE Trans. Antennas Propagat.*, vol. AP-35, pp. 1160–1173, Oct. 1987.

[18] X. Zhang, J. Fang, K. K. Mei, and Y. Liu, "Calculations of the dispersive characteristics of microstrips by the time-domain finite difference method," *IEEE Trans. Microwave Theory Tech.*, vol. 36, pp. 263–267, Feb. 1988.

[19] X. Zhang and K. K. Mei, "Time domain finite difference approach for the calculation of microstrip open-circuit end effect," *IEEE Trans. Microwave Theory Tech.*, in *IEEE MTT-S Int. Microwave Symp. Dig.*, 1988, pp. 363–366.

[20] X. Zhang and K. K. Mei, "Time domain finite difference approach to the calculation of the frequency dependent characteristics of microstrip discontinuities," *IEEE Trans. Microwave Theory Tech.*, vol. 36, pp. 1775–1787, Dec. 1988.

[21] D. H. Choi and W. J. R. Hoefer, "The finite-difference time-domain method and its application to eigenvalue problems," *IEEE Trans. Microwave Theory Tech.*, vol. MTT-34, pp. 1464–1470, Dec. 1986.

[22] G. Mur, "Absorbing boundary conditions for the finite difference approximation of the time domain electromagnetic field equations," *IEEE Trans. Electromagn. Compat.*, vol. EMC-23, pp. 377–382, Nov. 1981.

[23] B. Enquist and A. Majda, "Absorbing boundary conditions for the numerical simulation of waves," *Math. Comput.*, vol. 31, no. 139, pp. 629–651, July 1977.

[24] K. K. Mei and J. Fang, "A super-absorption boundary algorithm for one dimensional wave equations," submitted to *J. Comput. Phys.*

Accurate Computation of the Radiation from Simple Antennas Using the Finite-Difference Time-Domain Method

JAMES G. MALONEY, STUDENT MEMBER, IEEE, GLENN S. SMITH, FELLOW, IEEE, AND WAYMOND R. SCOTT, JR., MEMBER, IEEE

Abstract—**Two antennas are considered, a cylindrical monopole and a conical monopole. Both are driven through an image plane from a coaxial transmission line. Each of these antennas corresponds to a well-posed theoretical electromagnetic boundary value problem and a realizable experimental model. These antennas are analyzed by a straightforward application of the finite-difference time-domain (FD-TD) method. The computed results for these antennas, for both the time domain and the frequency domain, are shown to be in excellent agreement with accurate experimental measurements. The graphical displays presented for the transient near-zone and far-zone radiation from these antennas provide physical insight into the radiation process.**

I. Introduction

SIMPLE RADIATORS such as the cylindrical monopole/dipole, biconical monopole/dipole, circular loop and open-ended waveguides have received considerable attention in the literature. These radiators are generally the first ones discussed in a course on antenna analysis. They are analyzed using Maxwell's equations, and the theoretical results, such as input impedance, field patterns, etc., are compared with experimental measurements. When good agreement is obtained for the theory and experiment, one is presented with a convincing argument for the validity and usefulness of the theory.

The theoretical analysis of these radiators is generally carried out in two steps: the formulation of a theoretical model which corresponds, as closely as possible, to the actual antenna, and the analysis of the model using a particular mathematical technique. The theoretical model used for the antenna usually involves approximations introduced to simplify the analysis. For example, for the cylindrical dipole antenna an idealized source is often used, the so-called "delta-function generator" [1]. This source does not correspond to any realizable experimental model. Often the equations involved in the analysis of the antenna are also approximate. For example, for the cylindrical dipole antenna an approximate integral equation (thin-wire approximation) is often substituted for the exact integral equation. Approximations, like those mentioned above, lead to discrepancies between theoretical and experimental results, and it is often difficult to quantitatively ascertain the effects of the different approximations.

Accurate models and accurate theoretical analyses have been used with these simple radiators, but their complexity often puts them beyond the grasp of one first undertaking antenna analysis. For example, the cylindrical monopole can be modeled with the geometry shown in Fig. 1, which corresponds very closely to the arrangement used in experimental measurements. A rigorous mathematical analysis for this geometry due to Morris involves a coupled pair of singular integral equations whose numerical solution is a formidable task [2], [3].

The time-dependent Maxwell's differential equations can be represented by a set of difference equations that can be solved numerically using a digital computer. This procedure is often referred to as the finite-difference time-domain (FD-TD or TD-FD) method. This approach to solving electromagnetic problems is straightforward and easily understood by one first undertaking antenna analysis. In addition, the method is easily adapted to complex geometries, so simplified theoretical models are not required. The FD-TD method has been applied to problems of scattering and interaction, such as scattering from spheres, plates, etc., and the penetration of a field into the interior of cylindrical shields [4]-[9]. However, there has been little application of the method to antenna problems (driven).

In this paper the FD-TD method is used to analyze two radiators, a cylindrical monopole and a conical monopole, both antennas are driven through an image plane from a coaxial line. There are three objectives: first, to describe the formulation of theoretical models for these antennas that correspond to realizable experimental configurations; second, to apply the FD-TD method to these models and show that the theoretical results (time-domain and frequency-domain) are in excellent agreement with accurate experimental measurements; and third, to use the FD-TD results to describe the transient radiation from these antennas. The graphical displays presented

Manuscript received January 19, 1989; revised August 28, 1989. This work was supported in part by the Joint Services Electronics Program under Contracts DAAG29-84-K-0024 and DAAL03-87-K-0059.

The authors are with the School of Electrical Engineering, Georgia Institute of Technology, Atlanta, GA 30332.

IEEE Log Number 9035917.

Reprinted from *IEEE Transactions on Antennas and Propagation*, pp. 1059-1068, July 1990.

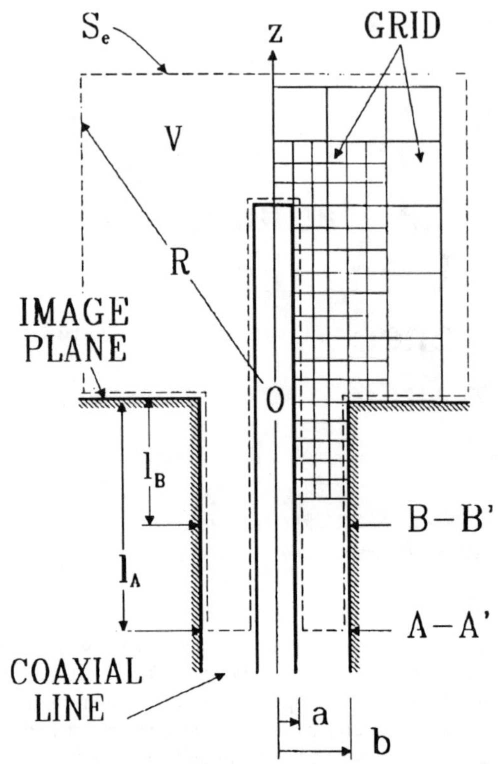

Fig. 1. Geometry for the electromagnetic boundary value problem —monopole antenna fed through an image plane from a coaxial transmission line.

for the transient radiation provide physical insight into the radiation process.

II. Formulation of the Antenna Problem

The theoretical analysis for an antenna begins with the formulation of a well-posed electromagnetic boundary value problem; ideally, one that closely corresponds to an actual antenna/experimental model. Fig. 1 shows the geometry for a monopole antenna fed through an image plane from a coaxial transmission line. This geometry will be used to illustrate the boundary value problem; other simple radiators can be handled in a similar manner.

The volume of free space V in which the electromagnetic field is to be determined surrounds the antenna and extends into the coaxial line to the depth $z = -l_A$. The boundary surface of the region is indicated by the dashed line in Fig. 1. All conductors are assumed perfect, and the field is to be determined within V for times $0 < t \leq t_0$.

To obtain a unique solution to Maxwell's equations within V for times $0 < t \leq t_0$, we must specify $\vec{\mathcal{E}}$ and $\vec{\mathcal{H}}$ within V at time $t = 0$. In addition, $\hat{n} \times \vec{\mathcal{E}}$ or $\hat{n} \times \vec{\mathcal{H}}$ must be specified on the boundary surface of V for all times $0 < t \leq t_0$ [10].

We will assume that the electromagnetic field ($\vec{\mathcal{E}}$ and $\vec{\mathcal{H}}$) is zero within V at time $t = 0$. On the cross section of the coaxial line at $A-A'$ ($z = -l_A$), the tangential component of the incident electric field ($-\hat{z} \times \vec{\mathcal{E}}^i$) is specified for times $0 < t \leq t_0$. Now this will be the only electric field at this cross section if we choose $2l_A > ct_0$, because the field reflected from the end of the line will not reach this cross section during the observation time. Since we have perfect conductors, $\hat{n} \times \vec{\mathcal{E}} = 0$ on the surface of the coaxial line, antenna and ground plane. The minimum distance to the surface S_e that surrounds the antenna above the image plane is $R_{\min}$. When $R_{\min} > ct_0 - l_A$, the electromagnetic field ($\hat{n} \times \vec{\mathcal{E}}$ and $\hat{n} \times \vec{\mathcal{H}}$) will be zero on the surface S_e for all observation times. A review of the above specifications shows that all the requirements have been met for obtaining a unique solution to Maxwell's equations within the volume V for times $0 < t \leq t_0$.

The dimensions of the coaxial transmission line, a and b, are chosen so that only a transverse electromagnetic (TEM) mode can propagate within the line for the signals of interest. Thus, on the cross section $A-A'$ the incident electric field is

$$\vec{\mathcal{E}}^i(t) = \frac{\mathcal{V}^i(t)}{\ln(b/a)r}\hat{r}. \tag{1}$$

The radiators considered represent two-dimensional electromagnetic problems. For example, the radiator in Fig. 1 is rotationally symmetric and is excited by a rotationally symmetric source. Therefore, the electromagnetic field is independent of the cylindrical coordinate ϕ, and Maxwell's equations can be expressed as two independent sets: one that involves only the components $\mathcal{E}_\phi$, $\mathcal{H}_r$, $\mathcal{H}_z$, the transverse electric (TE) field; and one that involves only the components $\mathcal{E}_r$, $\mathcal{E}_z$, $\mathcal{H}_\phi$, the transverse magnetic (TM) field. Since the excitation for the antenna in Fig. 1 is a TEM mode, which has only the field components $\mathcal{E}_r$, $\mathcal{H}_\phi$, only the rotationally symmetric TM modes are excited. The relevant Maxwell's equations are then

$$\frac{\partial \mathcal{E}_r}{\partial z} - \frac{\partial \mathcal{E}_z}{\partial r} = -\mu_0 \frac{\partial \mathcal{H}_\phi}{\partial t}, \tag{2a}$$

$$-\frac{\partial \mathcal{H}_\phi}{\partial z} = \epsilon_0 \frac{\partial \mathcal{E}_r}{\partial t}, \tag{2b}$$

$$\frac{1}{r}\frac{\partial (r\mathcal{H}_\phi)}{\partial r} = \epsilon_0 \frac{\partial \mathcal{E}_z}{\partial t}. \tag{2c}$$

In the FD–TD formulation both space and time are discretized. For the spatial increments Δr and Δz, and the time increment Δt the notation is

$$\mathcal{E}_z(r, z, t) = \mathcal{E}_z(i\Delta r, j\Delta z, n\Delta t) = \mathcal{E}_z^n(i, j).$$

For Yee's method of discretization, components of $\vec{\mathcal{E}}$ and $\vec{\mathcal{H}}$ are evaluated at interleaved spatial grid points and interleaved time steps [4]. The spatial grid points for the cylindrical system (r, ϕ, z) and the field components evaluated at these points are shown in Fig. 2. The Maxwell's equations (2) after discretization are

$$\begin{aligned}\mathcal{H}_\phi^{n+0.5}(i, j) = \mathcal{H}_\phi^{n-0.5}(i, j) &+ \frac{\Delta t}{\mu_0 \Delta r}[\mathcal{E}_z^n(i+0.5, j) \\ &- \mathcal{E}_z^n(i-0.5, j)] - \frac{\Delta t}{\mu_0 \Delta z}[\mathcal{E}_r^n(i, j+0.5) \\ &- \mathcal{E}_r^n(i, j-0.5)],\end{aligned} \tag{3a}$$

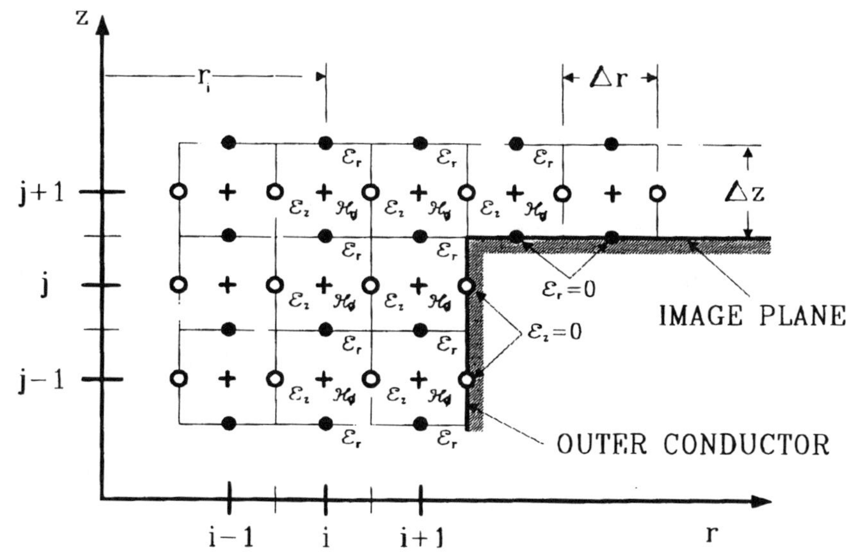

Fig. 2. Spatial grid and field components for the two-dimensional problem with cylindrical symmetry.

$$\mathcal{E}_r^{n+1}(i, j-0.5) = \mathcal{E}_r^n(i, j-0.5) - \frac{\Delta t}{\epsilon_0 \Delta z} \cdot [\mathcal{H}_\phi^{n+0.5}(i, j) - \mathcal{H}_\phi^{n+0.5}(i, j-1)], \quad (3b)$$

$$\mathcal{E}_z^{n+1}(i+0.5, j) = \mathcal{E}_z^n(i+0.5, j) + \frac{\Delta t}{\epsilon_0 \Delta r} \frac{1}{r_{i+0.5}} \cdot [r_{i+1}\mathcal{H}_\phi^{n+0.5}(i+1, j) - r_i \mathcal{H}_\phi^{n+0.5}(i, j)]. \quad (3c)$$

Note that the grid in Fig. 2 is arranged so that the electric field component tangential to the surface of a perfect conductor is evaluated at the surface. With the tangential electric field $\vec{\mathcal{E}}_r$ specified at the boundary $A - A'$, these equations (3a)-(3c) are used in a time-stepping procedure to determine the electromagnetic field in the volume V for times $0 < t \leq t_0$.

In a practical implementation an absorbing boundary condition is used at the surface S_e; this allows the observation period to be extended beyond $t = t_0$. Various **approximately** absorbing boundary conditions have been proposed; the one used here is due to Merewether [5]. In this approach the field near the boundary is assumed to have the functional form $f(t - R/c)/R$, and the tangential component of the electric field on the surface S_e is computed from local values within V by interpolation. The TEM field within the coaxial line has a known functional form $f(t - z/c)$; thus, an **exactly** absorbing boundary condition can be constructed within the line. In this procedure the incident field is additively injected at a plane $z = -l$, and the absorbing boundary condition, placed at $z = -(l + \Delta z)$, exactly absorbs the field of a TEM mode propagating in the $-z$ direction. This allows the cross section at which the incident field is specified to be moved closer to the ground plane; namely, in Fig. 1 $B - B'$ $(z = -l_B)$ is used instead of $A - A'$ $(z = -l_A)$.[1] This reduces both the time required for observation and the size of the grid.

[1]The plane $B - B'$ must be chosen so that all the evanescent TM modes present at $z = 0$ are insignificant at $z = -l_B$, typically $l_B = 3(b - a)$.

The spatial and temporal increments (Δr, Δz, and Δt) are chosen to satisfy the "domain of dependence condition" or "Courant-Friedrichs-Lewy condition" [11]:

$$c\Delta t \leq \sqrt{\frac{\Delta r^2 \Delta z^2}{\Delta r^2 + \Delta z^2}}. \quad (4)$$

In this work two spatial grid spacings are used: a fine spacing ($\Delta r_1 \approx \Delta z_1$) within the coaxial line and close to the antenna where the field is varying rapidly with spatial position, and a course grid ($\Delta r_2 = (3\text{–}5)\Delta r_1$, $\Delta z_2 = (3\text{–}5)\Delta z_1$) in the remainder of the space. The use of the dual grid reduces computer storage. Note, when (4) is satisfied for the fine grid it is automatically satisfied for the course grid. In the examples that follow

$$c\Delta t = \frac{\min(\Delta r_1, \Delta z_1)}{2}, \quad (5)$$

and the increments Δr_1, Δz_1 are chosen small enough to resolve the spatial variation of the field.

III. Comparison of Theoretical and Experimental Results

In this section the FD-TD method is used to analyze two antennas: a cylindrical monopole and a conical monopole, both are driven through an image plane from a coaxial transmission line. The theoretical models for each of these radiators fit the formulation presented in Section II and correspond closely to realizable experimental configurations.

A. Cylindrical Monopole

The cylindrical monopole antenna shown in Fig. 3(a) has been studied extensively both theoretically and experimentally; it is one of the canonical problems in antenna analysis. Various approximations have been used to make the theoretical analysis tractable, such as assuming a TEM field in the coaxial aperture at the image plane ($z = 0$) [12], [13]. An

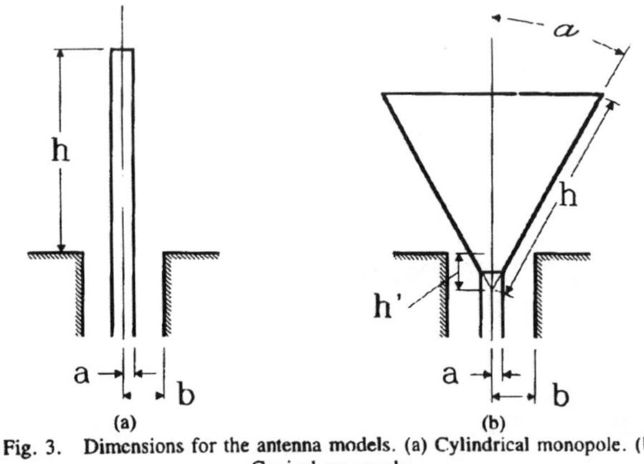

Fig. 3. Dimensions for the antenna models. (a) Cylindrical monopole. (b) Conical monopole.

accurate analysis of the correct boundary value problem, the one that corresponds to the experimental model, was carried out by Morris in 1980 and verified by accurate experimental measurements [2], [3]. Here a straightforward application of the FD–TD method to this antenna will be shown to produce results of comparable accuracy.

The parameters that describe the monopole antenna are the height h and the radii of the conductors of the coaxial line a and b. For the results shown in Figs. 4–7 $b/a = 2.30$, which corresponds to a characteristic impedance of 50 Ω for the coaxial line. Two grid spacings were used, as in Fig. 1: $\Delta r_1 = (b - a)/4$, $\Delta z_1 = h/203$; and $\Delta r_2 = 3\Delta r_1$, $\Delta z_2 = 3\Delta z_1$.

The cylindrical monopole was first studied for a Gaussian pulse excitation; the incident electric field in the transmission line at reference plane $A - A'$ being (1) with

$$\mathcal{V}^i(t) = V_0 \exp(-t^2/2\tau_p^2). \tag{6}$$

The antenna is characterized by the time $\tau_a = h/c$; this is the time required for light to travel its length. For the results shown in Figs. 4–7 the ratio of the characteristic time for the Gaussian pulse τ_p to the characteristic time for the antenna τ_a is $\tau_p/\tau_a = 8.04 \times 10^{-2}$.[2]

Fig. 4 shows the surface charge density on the coaxial line and antenna as a function of the normalized position z/h and the normalized time t/τ_a. This is the surface charge density on the inner conductor of the coaxial line for $-1.0 \leq z/h < 0.0$ and on the antenna for $0.0 \leq z/h \leq 1.0$. This bounce diagram clearly shows that the pulse travels up the coaxial line until it reaches the aperture ($z/h \approx 0.0$, $t/\tau_a \approx 1.0$; A in Fig. 4). At this point a portion of the pulse is reflected back into the line, and the remainder emerges on the antenna. The pulse is next reflected at the end of the antenna ($z/h \approx 1.0$, $t/\tau_a \approx 2.0$; B); then, it travels down to the aperture ($z/h \approx 0.0$, $t/\tau_a \approx 3.0$; C) where it is partially reflected and enters the coaxial line. This process is then repeated.

[2]The pulse duration was chosen short enough to resolve the reflections from different points on the antenna. However, the pulse duration is not so short that it excites propagating TM modes (other than the TEM mode) in the coaxial line. At the cutoff frequency for the lowest mode, the TM_{01}, the spectrum of the pulse is reduced by the factor $\approx 3 \times 10^{-32}$ from its maximum.

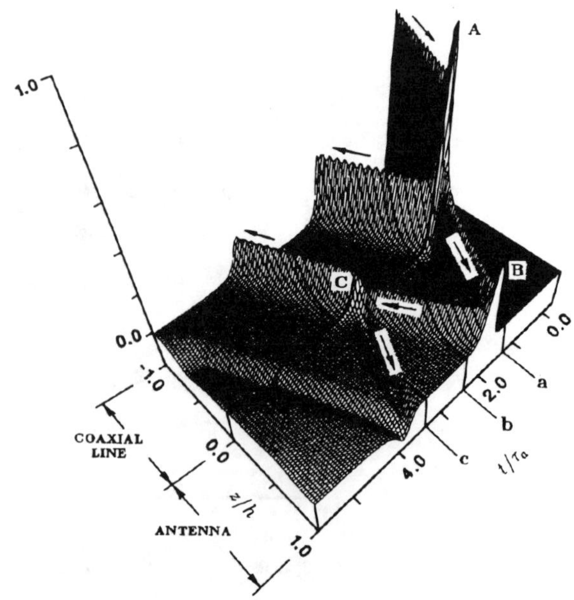

Fig. 4. Normalized surface charge density on the cylindrical monopole antenna as a function of the normalized position z/h and the normalized time t/τ_a: $b/a = 2.30$, $h/a = 65.8$, and $\tau_p/\tau_a = 8.04 \times 10^{-2}$.

The peaks at the points A and B are the result of the singularities in the charge density that occur at the sharp edges of the structure. The concentration of charge at A is produced by the singularity at the edge of the outer conductor of the coaxial line, while the spike at B is the singularity at the edge on the top of the antenna.

The electric field surrounding the monopole antenna is displayed on a gray scale in Fig. 5 for three times. Graphs of the surface charge density on the antenna are below each plot; these correspond to the slices marked a, b, and c in Fig. 4. The spacing between the conductors of the coaxial line is expanded in these plots to clarify the presentation.

In Fig. 5(a) the pulse has been reflected at the aperture and is traveling out along the antenna. The spherical wavefront centered on the aperture (W_1 in Fig. 5(a)) is attached to packets of charge on the antenna and image plane. A second wavefront (W_2) is produced when this pulse is reflected at the end of the antenna. This spherical wavefront, centered on the end of the antenna, is clearly shown in Fig. 5(b). The pulse, after reflection from the end, travels down the antenna, eventually being partially reflected at the aperture and entering the coaxial line. A third spherical wavefront (W_3), centered at the aperture, is produced on this reflection, Fig. 5(c). The effect the image plane has on the wavefront W_2 is easily understood when the dipole configuration of the antenna is viewed, Fig. 6. The wavefront W_{2R}, the reflection of wavefront W_2 at the image plane, is seen to be the same as the wavefront radiated from the opposite end of the dipole.

For times beyond those shown in Fig. 5, the pulse travels along the antenna alternately being reflected at the end and coaxial aperture and decreasing in amplitude. Eventually there is no charge on the antenna, and the radiation ceases. At each reflection a wavefront similar to one in Fig. 5 is produced;

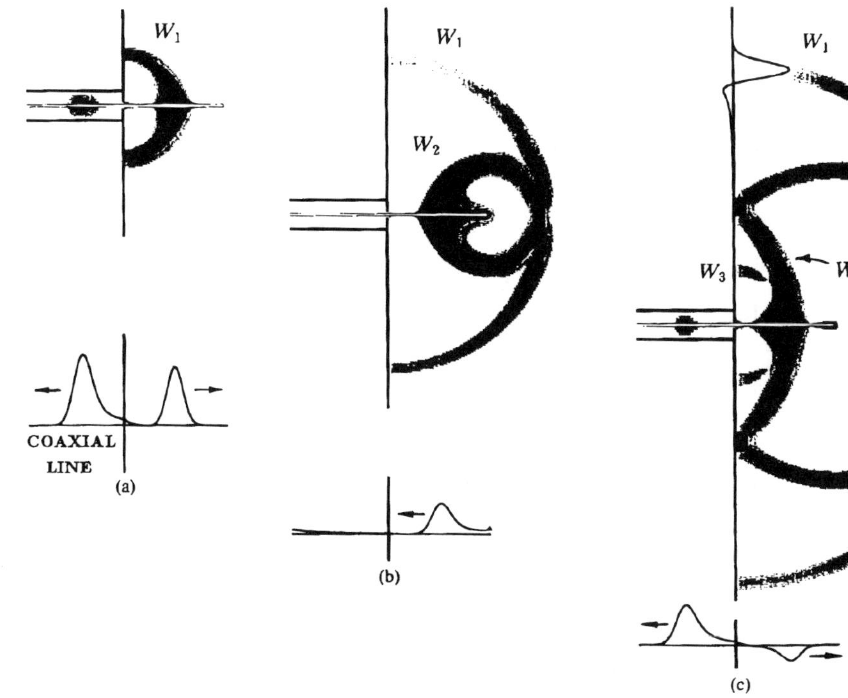

Fig. 5. Radiation of a Gaussian pulse from a cylindrical monopole antenna. Gray scale plots show the magnitude of the electric field, the line drawings show the surface charge density on the antenna. $b/a = 2.30$, $h/a = 65.8$, and $\tau_p/\tau_a = 8.04 \times 10^{-2}$.

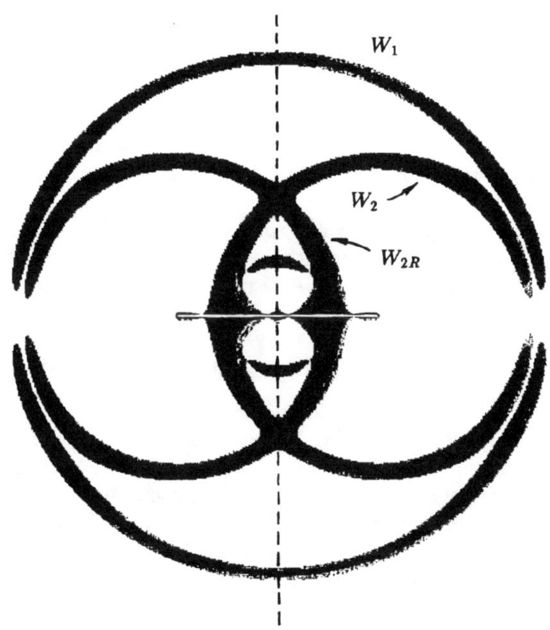

Fig. 6. Dipole configuration corresponding to Fig. 5(c).

for example, the next waveform W_4, which is similar to W_2, is produced when the pulse traveling up the antenna is again reflected at the end.

To complete the description of the radiation from this antenna, the electric field in the far zone will be determined. This is accomplished by using the electromagnetic form of Huygens' principle to continue the FD-TD results to the far zone. In particular, the electric field $\bar{\mathcal{E}}^r$ in the far zone is

$$\bar{\mathcal{E}}^r(\bar{r}, t) = \frac{\mu_0}{4\pi r} \oiint_S \{\hat{r} \times \hat{r} \times \frac{\partial}{\partial t'}[\hat{n} \times \bar{\mathcal{H}}(\bar{r}', t')] - \frac{1}{\zeta_0}\hat{r} \times \frac{\partial}{\partial t'}[\hat{n} \times \bar{\mathcal{E}}(\bar{r}', t')]\} \, dS|_{t'=t-|\bar{r}-\bar{r}'|/c}, \quad (7)$$

where $\bar{r}$ locates the point in the far zone, $\bar{r}'$ locates the point on the surface S surrounding the antenna, and the outward normal to this surface is $\hat{n}$.[3]

The far-zone electric field $\mathcal{E}_\theta^r$ for the cylindrical monopole antenna is shown in Fig. 7. The surface used in (7) for these calculations was the cylindrical boundary separating the fine and coarse grids in Fig. 1. Each trace in this figure shows the electric field at a fixed polar angle θ as a function of the normalized time t/τ_a. On these graphs the origin for the time, $\tau/\tau_a = 0$, and the amplitude for the field were selected to clarify the presentation. Notice that wavefronts from the same point on the antenna are always separated by a time interval which is a multiple of $2\tau_a$, the round-trip transit time for the pulse on the antenna. For example, wavefronts W_1 and W_3, which are centered on the drive point, are separated by the time $2\tau_a$, as are wavefronts W_2 and W_4, which are centered on the end. However, the relative position of the wavefront pairs, such as W_1, W_3 and W_2, W_4, changes with the viewing angle θ. For example, at $\theta = 90°$ the wavefronts W_1, W_2,

[3] The frequency domain counterpart to this formula is presented in the literature [14], [15]. Equation (7) is obtained by specializing this formula to the far zone and then taking the inverse Fourier transform.

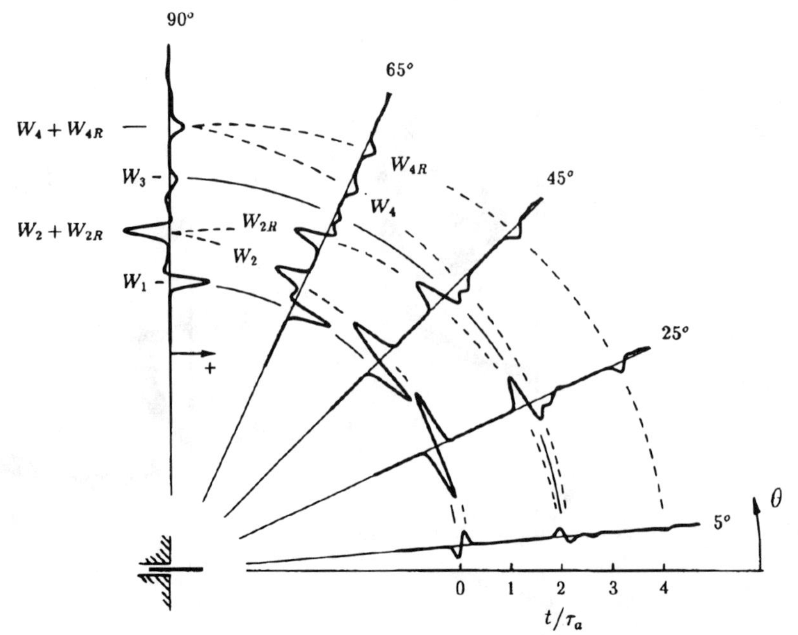

Fig. 7. Radiation of a Gaussian pulse from a cylindrical monopole antenna. Each trace shows the far-zone electric field $\mathcal{E}_\theta^r$ at a fixed polar angle θ as a function of the normalized time t/τ_a. $b/a = 2.30$, $h/a - 65.8$, and $\tau_p/\tau_a = 8.04 \times 10^{-2}$.

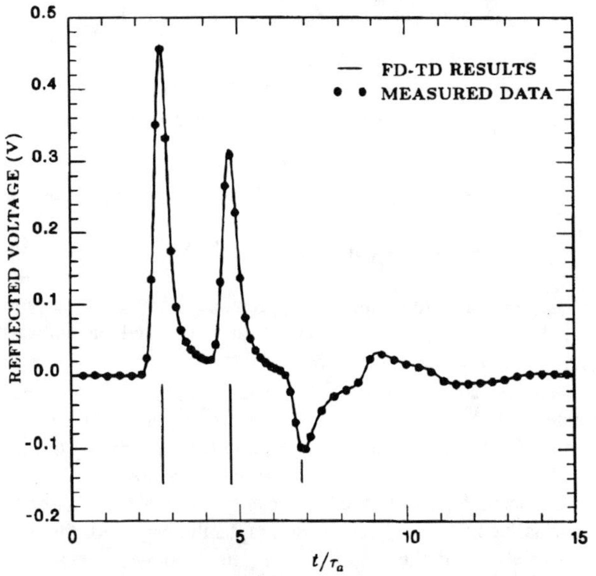

Fig. 8. Comparison of the theoretical (FD-TD) and the measured reflected voltages in the coaxial line for a cylindrical monopole antenna excited by a 1 V Gaussian pulse: $b/a = 2.30$, $h/a = 32.8$, and $\tau_p/\tau_a = 1.61 \times 10^{-1}$.

W_3 and W_4 are all separated, but at $\theta = 5°$ W_1 and W_2 are superposed as are W_3 and W_4. Of course, the far zone field off the end of the antenna is zero ($\theta = 0°$).

An experimental model was constructed for the cylindrical monopole antenna with the dimensions $b/a = 2.30$, $h/a = 32.8$. This model was mounted on an aluminum image plane (120 × 155 cm) that was surrounded by absorbing material to reduce reflections. The input reflection coefficient for this antenna was measured at a number of frequencies (50 MHz $\leq f \leq$ 18 GHz) using a Hewlett Packard Model 8409C Automated Network Analyzer. These data were then used with a fast Fourier transform to obtain the response of the antenna to a Gaussian pulse (6).

In Fig. 8 the reflected voltage measured in the coaxial line (dots) is compared with results computed by the FD-TD method (solid line) for a Gaussian pulse with $\tau_p/\tau_a = 1.61 \times 10^{-1}$. This is the reflected voltage at the plane $B - B'$ in Fig. 1 where only the TEM mode is present. The agreement between the two sets of data is excellent. Note that the peaks in the reflected voltage are spaced by about the round-trip transit time for the pulse on the antenna, $t/\tau_a = 2.0$.

As a second test, the FD-TD method was used to compute the response of a monopole excited by the signal

$$\mathcal{V}_i(t) = V_0 r(t) \sin(\omega t). \tag{8}$$

Here $r(t)$ is a ramp function which goes from 0 at time $t = 0$ to 1.0 at time $t = \tau_r$; τ_r is chosen to be several cycles of the sinusoid. The long-time response to this signal is approximately monochromatic and can be compared with frequency domain measurements.[4] The steady state sinusoidal voltage at the reference plane $B - B'$ in Fig. 1 is used to determine the reflection coefficient for the antenna, and this, in turn, determines the input admittance.

The geometry chosen corresponds to an antenna measured by Cooper: $b/a = 3.00$; $a/\lambda = 7.02 \times 10^{-3}$ [16]. The end of this antenna is hemispherical; it was modeled by six dis-

[4]The monochromatic case could be obtained by Fourier transforming a pulse response. This was not done, because the objective of this paper is to illustrate the direct use of the FD-TD method. For a single frequency the use of the Fourier transform with the FD-TD pulse response may also be less efficient, since the higher frequencies necessary in the pulse would require finer discretization in time and space.

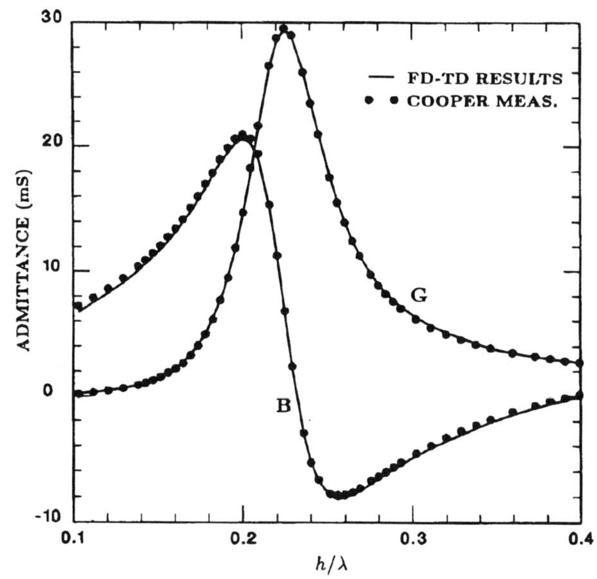

Fig. 9. Comparison of the theoretical (FD-TD) and the measured input admittances, $Y = G + jB$, for a cylindrical monopole antenna with a hemispherical end: $b/a = 3.00$, $a/\lambda = 7.02 \times 10^{-3}$.

crete steps in the FD-TD grid. In Fig. 9 the FD-TD results (solid line) are compared with Cooper's data (dots) for the input admittance measured for a range of lengths surrounding the first resonance of the antenna, $0.1 \leq h/\lambda \leq 0.4$. The agreement is very good, particularly when one considers that Cooper estimates the error bound on his data to be 5%.

B. Conical Monopole

The theoretical model usually used for the conical monopole antenna is a vertical, solid cone over an image plane. This structure is analyzed by expanding the electromagnetic field in spherical models [17], [18]. At the drive point of the antenna, the vertex of the cone, the input impedance is determined from the outwardly propagating and reflected TEM spherical waves. This theoretical model is only an approximation to the usual experimental model, a cone fed through an image plane by a coaxial transmission line as in Fig. 3(b). Thus, this modal analysis cannot produce results that are in extremely good agreement with experiment.

Here we use the FD-TD method to study the electromagnetic characteristics of the conical monopole antenna in Fig. 3(b). This antenna is described by the following parameters: b/a, h/a, h'/h and the half angle of the cone α. The antenna is characterized by a time $\tau_a = h/c$; this is the time required for light to travel over the slant height of the cone.

The grid spacings for the FD-TD method were essentially the same as those used with the cylindrical monopole. The side of the cone was handled by setting $\mathcal{E}_r$ and $\mathcal{E}_z$ equal to zero at all grid points inside the cone.

Fig. 10 shows the response of a conical monopole to a Gaussian pulse; the electric field in the transmission line being (1) with $\mathcal{V}^i(t)$ given by (6). The ratio of characteristic times is $\tau_p/\tau_a = 3.44 \times 10^{-2}$, and the other parameters for the antenna are $b/a = 2.30$, $h/a = 1.54 \times 10^2$, $h'/h = 1.34 \times 10^{-2}$ and $\alpha = 30°$. The electric field in the space surrounding the antenna is displayed on a gray scale, and the surface charge density on the antenna is shown below each plot. Again the spacing between the conductors of the coaxial line is expanded to clarify the presentation.

In Fig. 10(a) the Gaussian pulse has been reflected at the aperture and has moved out onto the antenna as the spherical wavefront W_1. When this pulse hits the top edge of the cone, Fig. 10(b), a second toroidal wavefront W_2, centered on the edge, is produced. Note that wavefront W_1 is attached to the packets of charge on the top of the cone and on the image plane, whereas waveform W_2 is attached to the packets of charge on the top and side of the cone.[5]

The charge packet on the side of the cone moves to the vertex, Fig. 10(c), where it eventually enters the coaxial line and is partially reflected to produce wavefront W_3, which is centered on the aperture, Fig. 10(d). The charge packet on the top of the cone, Figs. 10(b) and 10(c), travels across the top of the cone and produces a wavefront W_4 when it hits the opposite edge of the top, Fig. 10(d).

The far-zone electric field $\mathcal{E}_\theta^r$ for the conical monopole antenna is shown in Fig. 11. As for the cylindrical monopole, each trace is the electric field $\mathcal{E}_\theta^r$ at a fixed polar angle θ as a function of the normalized time t/τ_a. Notice that there are now two contributions from the wavefront W_1, marked W_1^A and W_1^B in Figs. 10 and 11. This is a result of the wavefront crossing itself on the top of the cone ($\theta = 0°$), as clearly seen in Figs. 10(c) and 10(d).

An experimental model was constructed for the conical monopole antenna with the dimensions $b/a = 2.30$, $h/a =$

[5]The division of the total field into specific wavefronts is somewhat arbitrary, since two or more of the wavefronts often overlap.

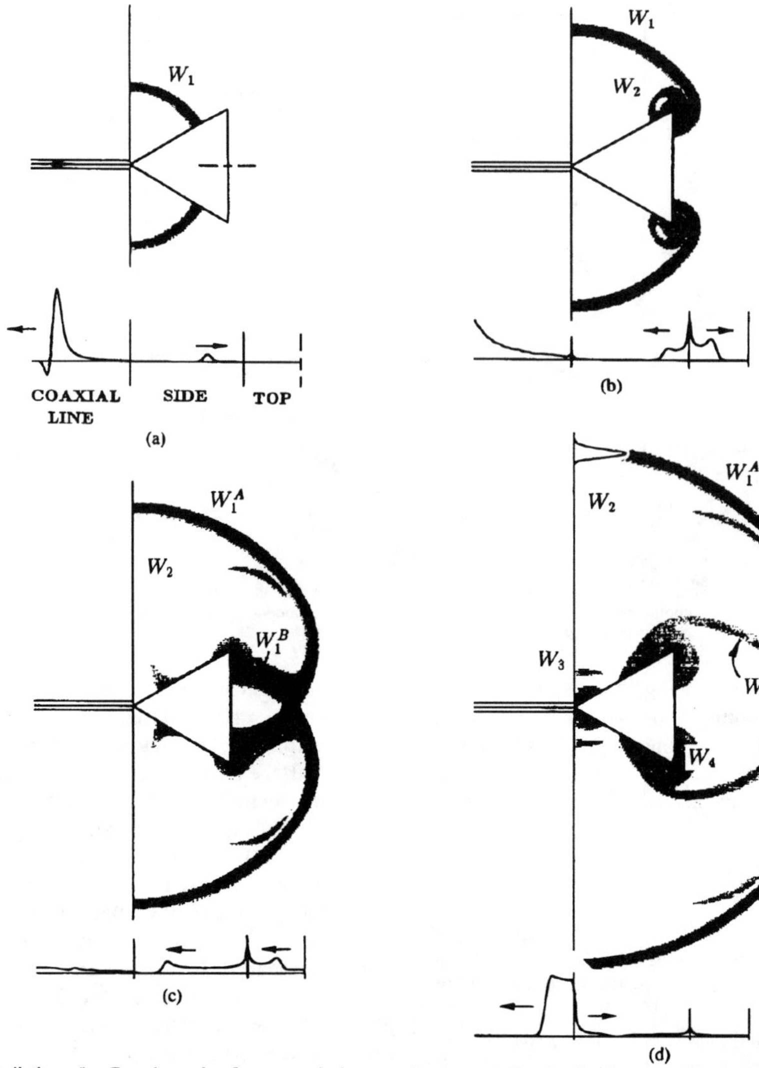

Fig. 10. Radiation of a Gaussian pulse from a conical monopole antenna. Gray scale plots show the magnitude of the electric field, the line drawings show the surface charge density on the antenna. $b/a = 2.30$, $h/a = 1.54 \times 10^2$, $h'/h = 1.34 \times 10^{-2}$, $\alpha = 30°$, and $\tau_p/\tau_a = 3.44 \times 10^{-2}$. Charge density is scaled in (b) and (c) by a factor of 15, and in (d) by a factor of 5.

23.1, $h'/h = 8.63 \times 10^{-2}$, and $\alpha = 30°$. The reflected voltage at the input of this antenna was measured in the same manner as for the cylindrical monopole. In Fig. 12 the measured reflected voltage (dots) is compared with results computed by the FD–TD method (solid line) for a Gaussian pulse with $\tau_p/\tau_a = 2.29 \times 10^{-1}$. The agreement between the two sets of data is excellent. The reflections due to the drive point and the near and far edges of the top are evident in this figure.

Note the small negative spike at the beginning of the waveform in Fig. 12. This is caused by the cone partially filling the coaxial aperture at $z = 0$, as in Fig. 3(b). This spike occurs whenever $h'/a > 1/\tan(\alpha)$. It is impressive that the FD–TD method handles this fine detail correctly.

IV. Summary and Conclusion

Two antennas were considered, a cylindrical monopole and a conical monopole, both are driven through an image plane from a coaxial transmission line. Each of these corresponds to a well-posed theoretical electromagnetic boundary value problem and a realizable experimental model. These antennas are structures of different complexity. The cylindrical monopole is a solid perfectly conducting structure with only right angle corners. The boundary surfaces of this antenna are at grid points where a tangential component of the electric field is set to zero. The conical monopole, however, involves corners of arbitrary angle (α) and a boundary surface (side) that does not pass through a specific set of grid points.

These antennas were analyzed by a straightforward application of the finite-difference time-domain method. The computed results are in excellent agreement with accurate experimental measurements. A survey of the literature for these antennas shows that no better results (in better agreement with measurement) have been obtained with analytical techniques or more involved numerical techniques.

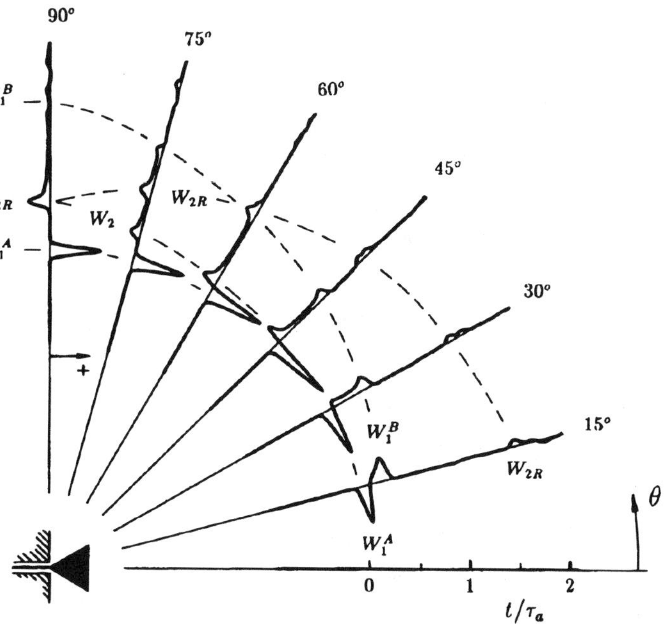

Fig. 11. Radiation of a Gaussian pulse from a conical monopole antenna. Each trace shows the far-zone electric field $\mathcal{E}_\theta^r$ at a fixed polar θ as a function of the normalized time t/τ_a. $b/a = 2.30$, $h/a = 1.54 \times 10^2$, $h'/h = 1.34 \times 10^{-2}$, $\alpha = 30°$, and $\tau_p/\tau_a = 3.44 \times 10^{-2}$.

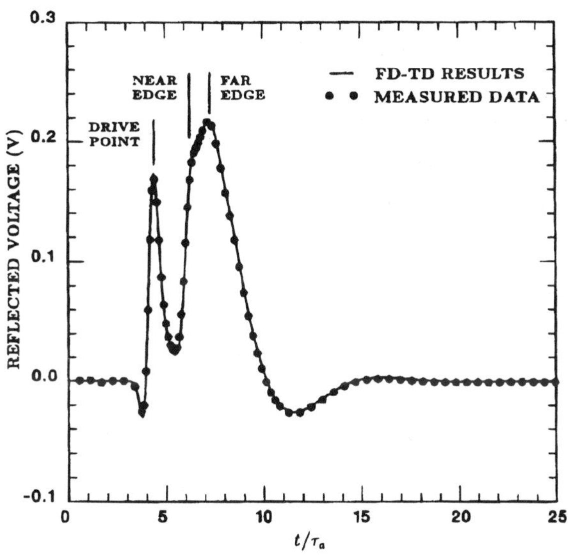

Fig. 12. Comparison of the theoretical (FD-TD) and the measured reflected voltages in the coaxial line for a conical monopole antenna excited by a 1 V Gaussian pulse: $b/a = 2.30$, $h/a = 23.1$, $h'/h = 8.63 \times 10^{-2}$, $\alpha = 30°$, and $\tau_p/\tau_a = 2.29 \times 10^{-1}$.

The graphical displays for the electromagnetic behavior of these antennas (the response to a pulse excitation) are presented in a manner that increases physical insight into the radiation process and is useful for instruction.

Acknowledgment

The authors would like to thank one of the reviewers for suggesting the inclusion of the graphs for the far-zone field, Figs. 7 and 11. These graphs enhance the description of the transient radiation.

References

[1] T. T. Wu and R. W. P. King, "Driving point and input admittance of linear antennas," *J. Appl. Phys.*, vol. 30, pp. 74–76, Jan. 1959.

[2] M. E. Morris, "Analysis of a finite length tubular monopole antenna driven by a coaxial line," Sandia Nat. Labs., Albuquerque, NM, Tech. Rep. SAND79-1557, May 1980.

[3] ——, "Precision admittance measurements of coaxially-driven monopole antennas—An experimental study of junction effects," Sandia Nat. Labs., Albuquerque, NM, Tech. Rep. SAND79-2047, May 1980.

[4] K. S. Yee, "Numerical solution of initial value problems involving Maxwell's equations in isotropic media," *IEEE Trans. Antennas Propagat.*, vol. AP-14, pp. 302–307, May 1966.

[5] D. E. Merewether, "Transient currents induced on a metallic body of revolution by an electromagnetic pulse," *IEEE Trans. Electromagn. Compat.*, vol. EMC-13, pp. 41–44, May 1971.

[6] A. Taflove and M. E. Brodwin, "Numerical solution of steady-state electromagnetic scattering problems using the time-dependent Maxwell's equations," *IEEE Trans. Microwave Theory Tech.*, vol. MTT-23, pp. 623–630, Aug. 1975.

[7] A. Taflove, "Application of the finite-difference time-domain method to sinusoidal steady-state electromagnetic-penetration problems," *IEEE Trans. Electromagn. Compat.*, vol. EMC-22, pp. 191–202, Aug. 1980.

[8] R. Holland, L. Simpson, and K. S. Kunz, "Finite-difference analysis of EMP coupling to lossy dielectric structures," *IEEE Trans. Electromagn. Compat.*, vol. EMC-22, pp. 203–209, Aug. 1980.

[9] A. Taflove and K. R. Umashankar, "Finite-difference time-domain (FD-TD) modeling of electromagnetic wave scattering and interaction problems," *IEEE Trans. Antennas Propagat. Soc. Newsletter*, pp. 5–20, Apr. 1988.

[10] J. A. Stratton, *Electromagnetic Theory*. New York: McGraw-Hill, 1941.

[11] E. Isaacson and H. B. Keller, *Analysis of Numerical Methods*. New York: Wiley, 1967.

[12] D. C. Chang, "On the electrically thick monopole Part I—Theoretical solution," *IEEE Trans. Antennas Propagat.*, vol. AP-16, pp. 58–64, Jan. 1968.

[13] C.-Y. Ting, "Theoretical study of a cylindrical antenna with a hemispherical cap," *IEEE Trans. Antennas Propagat.*, vol. AP-17, pp. 716–721, Nov. 1969.

[14] W. Franz, "Zur Formulierung des Huygensschen Prinzips," *Z. Naturforsch*, vol. 3a, pp. 500–506, 1948.

[15] C.-T. Tai, "Kirchoff theory: Scalar, vector, or dyadic," *IEEE Trans. Antennas Propagat.*, vol. AP-20, pp. 114–115, Jan. 1972.

[16] L. J. Cooper, "Monopole antennas on electrically thick conducting cylinders," Ph.D. dissertation, Harvard Univ., Cambridge, MA, Mar. 1975.

[17] S. A. Schelkunoff, *Advanced Antenna Theory*. New York: Wiley, 1952.

[18] R. W. P. King, *The Theory of Linear Antenna*. Cambridge, MA: Harvard Univ. Press, 1956, ch. VIII.

The Use of Surface Impedance Concepts in the Finite-Difference Time-Domain Method

James G. Maloney, *Student Member, IEEE,* and Glenn S. Smith, *Fellow, IEEE*

Abstract—**Surface impedance concepts are introduced into the finite-difference time-domain (FDTD) method. Lossy conductors are replaced by surface impedance boundary conditions (SIBC) reducing the solution space and producing significant computational savings. Specifically, a surface impedance boundary condition (SIBC) is developed to replace a lossy dielectric half-space. An efficient implementation of this FDTD-SIBC based on the recursive properties of convolution with exponentials is presented. Finally, three problems are studied to illustrate the accuracy of the FDTD-SIBC formulation: a plane wave incident on a lossy dielectric half-space, a line current over a lossy dielectric half-space, and wave propagation in a parallel-plate waveguide with lossy walls.**

I. Introduction

The finite-difference time-domain (FDTD) technique was first used to solve electromagnetics problems in 1966 [1]. The technique has since been successfully applied to a large number of scattering and interaction problems [2]. A higher order finite-difference time-domain algorithm has been proposed, and the problem of grid truncation has been extensively studied [3]. Recently, the technique has been successfully applied to driven antenna problems [4], and has been extended to allow the inclusion of dispersive dielectrics [5]. In this paper, surface impedance concepts will be introduced into the FDTD method, and the computational savings gained by using surface impedance concepts for problems containing lossy dielectric regions will be shown to be quite significant [6].

Surface impedance concepts have been used to simplify electromagnetic problems since the early 1940's [7]. One common example is the skin effect approximation used with highly conducting bodies [8]. The skin effect approximation provides a surface impedance boundary condition (SIBC) suitable for eliminating the conducting body from the solution space. The application of this skin effect SIBC into the FDTD method is the basis of this work.

In Section II, the computational savings gained by using surface impedance concepts in the FDTD technique will be discussed, and in Section III an SIBC for a lossy dielectric half-space will be derived. An efficient implementation of this SIBC based on an exponential approximation will be presented in Section IV, and finally, in Section V, various problems will be studied to demonstrate the applicability of this work.

II. Computational Savings

The two-region problem in Fig. 1 illustrates the computational savings gained by using surface impedance concepts in the FDTD method. The solution space consists of a region of free space ($\epsilon_1 = \epsilon_0$, $\mu_1 = \mu_0$, $\sigma_1 = 0$) and a region of lossy dielectric ($\epsilon_2 = \epsilon_{2r}\epsilon_0$, $\mu_2 = \mu_0$, $\sigma_2 \neq 0$) with the loss tangent $p_2 = \sigma_2/\omega\epsilon_2 > 1$ at the frequencies of interest.

First the computational requirements for the nonreduced, **original problem** will be examined, Fig. 2(a). Both the spatial grid increment Δs and the time increment Δt need to be considered. In general the spatial increment must be chosen small enough to resolve the field. The spatial variation of the field is greatest inside the lossy dielectric region, where it decays exponentially with the characteristic length δ, the skin depth,

$$\delta = \sqrt{\frac{2}{\omega\mu_2\sigma_2}} = \frac{\lambda_0}{\sqrt{2\pi^2 p_2 \epsilon_{2r}}}. \tag{1}$$

The spatial increment is chosen to be some fraction of the skin depth:

$$\Delta s = \frac{\delta}{N_s} = \frac{\lambda_0}{N_s\sqrt{2\pi^2 p_2 \epsilon_{2r}}} \tag{2}$$

where N_s is typically on the order of 8 to 16. The same spatial increment is used in both the conductor and free-space regions. The time increment is then constrained by the Courant stability condition in free space,[1] which is usually taken to be

$$c\Delta t = \frac{\Delta s}{2}; \tag{3}$$

therefore, the time increment Δt is

$$\Delta t = \frac{\delta}{2cN_s} = \frac{(\lambda_0/c)}{2N_s\sqrt{2\pi^2 p_2 \epsilon_{2r}}}. \tag{4}$$

The reduced problem shown in Fig. 2(b) results when

Manuscript received November 21, 1990; revised June 10, 1991. This work was supported in part by the Joint Services Electronics Program under Contracts DAAL03-87-K-0059, DAAL-03-90-C-0004, and DAAL-03-89-G-0078.

The authors are with the School of Electrical Engineering, Georgia Institute of Technology, Atlanta, GA 30332.

IEEE Log Number 9105720.

[1] The time increment must be chosen such that the Courant stability condition $v\Delta t/\Delta s \leq 1/\sqrt{D}$, where D is the number of dimensions in the solution space, is satisfied in every region of the solution space. Since the velocity in free space is larger than the velocity in the lossy dielectric region, the time increment determined for the free-space region is sufficient to satisfy the Courant condition in both regions.

Reprinted from *IEEE Transactions on Antennas and Propagation,* pp. 34-48, January 1992.

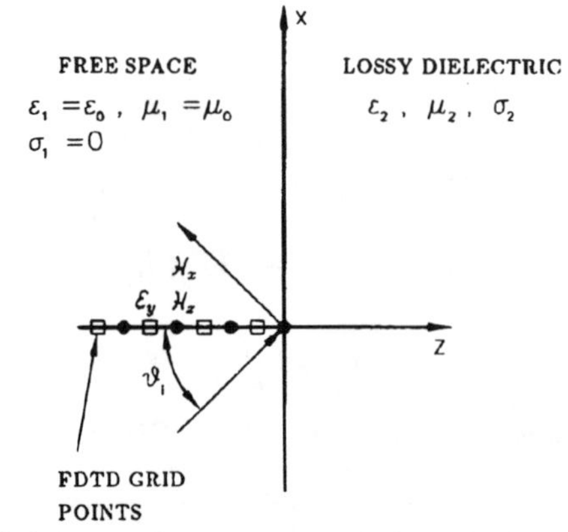

Fig. 1. Typical two-region electromagnetic problem: plane wave incident on a lossy dielectric half-space from free space at the angle ϑ_i.

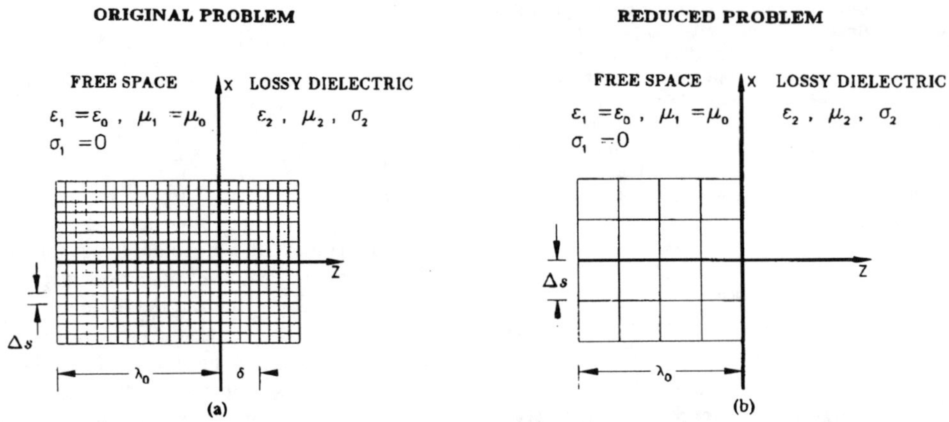

Fig. 2. Comparison of computational requirements. (a) Original, nonreduced problem. (b) Reduced problem.

surface impedance concepts are used, and the lossy dielectric region is replaced by a SIBC. With only the free-space region present in the solution space, the spatial and time increments can be much larger. The spatial variation of the field is now characterized by the wavelength in free space, λ_0. The spatial increment is chosen to be some fraction of this wavelength,

$$\Delta s = \frac{\lambda_0}{N_s} \tag{5}$$

where again N_s is typically on the order of 8 to 16. From (3) the time increment is

$$\Delta t = \frac{(\lambda_0/c)}{2N_s}. \tag{6}$$

The computational savings gained by replacing the original problem by the reduced problem can be seen by comparing (2) with (5) and (4) with (6). In the reduced problem, both the time and spatial increments are larger by the factor $\sqrt{2\pi^2 p_2 \epsilon_{2r}}$. Fig. 2 schematically shows the spatial grid savings. Clearly, many more spatial grid cells are needed for the original, nonreduced grid Fig. 2(a) than for the reduced grid Fig. 2(b).

Later in this paper, it will be shown that surface impedance concepts are in very good agreement with theory for loss tangents $p_2 \geq 10$. For this lower limit on the loss tangent, the time and spatial increments for the reduced problem are larger by a factor of approximately 15 than those for the original problem. For two- or three-dimensional problems, this can be a tremendous savings.

III. Time-Domain Surface Impedance Formulation

The surface impedance is inherently a frequency-domain concept. When it is transformed into the time-domain for use in methods such as FDTD, it is replaced by a convolution integral. The development of this integral is the subject of this section. Our discussion will be confined to the two-region problem shown in Fig. 1: a plane boundary separating free space from a region of lossy dielectric with loss tangent $p_2 = \sigma_2/\omega\epsilon_2 > 1$ for all frequencies of interest.

Consider a time-harmonic plane wave incident at angle ϑ_i on the lossy dielectric half-space of Fig. 1. The reflected and transmitted components of the field are easily obtained, and the following relationship can be defined for the tangential components of the field at the interface ($z = 0$):

$$\vec{E}_{\tan}(\omega) = Z(\omega)\left[\hat{n} \times \vec{H}_{\tan}(\omega)\right] \tag{7}$$

where $Z(\omega)$ is the surface impedance. When the inequality

$$|\epsilon_{2r} - j\sigma_2/\omega\epsilon_o| \gg \sin^2\vartheta_i \tag{8}$$

is satisfied, the surface impedance is simply

$$Z(\omega) \approx \frac{\eta_0}{\sqrt{\epsilon_{2r} - j\sigma_2/\omega\epsilon_0}} \tag{9}$$

for any orientation of the incident plane wave. Here η_0 is the wave impedance of free space. For future use it will be convenient to have the surface impedance (9) expressed in terms of the Laplace transform variable $s = j\omega$,

$$Z(s) = \frac{\eta_0}{\sqrt{\epsilon_{2r}}}\left(\frac{\sqrt{s}}{\sqrt{s + \sigma_2/\epsilon_2}}\right). \tag{10}$$

A time-harmonic expression such as (7) is not suitable for implementation in a time-domain technique such as FDTD; however, the use of the inverse Laplace transform allows (7) to be written in a suitable form in the time domain

$$\vec{\mathscr{E}}_{\tan}(t) = \mathscr{Z}(t)*[\hat{n} \times \vec{\mathscr{H}}_{\tan}(t)], \tag{11}$$

where the asterisk denotes convolution with respect to t. Now the time-domain surface impedance $\mathscr{Z}(t)$, the Laplace transform of (10), can be evaluated by using the transform pairs [9]

$$\frac{\sqrt{s+2a}}{\sqrt{s}} - 1 \Leftrightarrow ae^{-at}[I_1(at) + I_0(at)]U(t) \tag{12}$$

$$1 \Leftrightarrow \delta(t) \tag{13}$$

and by applying the functional Laplace transform relationship

$$f(s-a) \Leftrightarrow e^{at}\mathscr{F}(t) \tag{14}$$

to (12) twice. The resulting time-domain surface impedance is

$$\mathscr{Z}(t) = \frac{\eta_0}{\sqrt{\epsilon_{2r}}}\{ae^{at}[I_1(at) + I_0(at)]U(t) + \delta(t)\} \tag{15}$$

where

$$a \equiv -\sigma_2/2\epsilon_2. \tag{16}$$

Here I_0 and I_1 are the modified Bessel functions of the first kind of order zero and one, respectively, $U(t)$ is the Heaviside unit-step function, and $\delta(t)$ is the unit-impulse function. Substitution of (15) into (11) yields

$$\vec{\mathscr{E}}_{\tan}(t) = \frac{\eta_0}{\sqrt{\epsilon_{2r}}}\left\{[\hat{n} \times \vec{\mathscr{H}}_{\tan}(t)] + \int_0^t ae^{a\tau}[I_1(a\tau) + I_0(a\tau)][\hat{n} \times \vec{\mathscr{H}}_{\tan}(t-\tau)]\,d\tau\right\}. \tag{17}$$

This equation is the time-domain expression that is the basis for the rest of this paper. Notice, when the limit $\sigma \to \infty (a \to -\infty)$ is taken, (17) becomes the boundary condition for a perfect conductor: $\vec{\mathscr{E}}_{\tan} = 0$.

IV. Efficient Implementation

The convolution integral in (17) involves the tangential component of the magnetic field at all times previous to t. The direct implementation of this convolution integral as a convolution sum will be presented next. Since the computational requirements are quite large, an efficient implementation based on fitting a series of exponential functions to the impulse response, $\mathscr{Z}(t)$ (15), will be presented. The usual FDTD notation of

$$\mathscr{E}(x, y, z, t) = \mathscr{E}(I\Delta x, J\Delta y, K\Delta z, N\Delta t) = \mathscr{E}(I, J, K, N) = \mathscr{E}^N(I, J, K) \tag{18}$$

will be used, and when no confusion results, unnecessary arguments will be suppressed.

The discretization of (17) as a convolution sum is

$$\vec{\mathscr{E}}_{\tan}(N) = \frac{\eta_0}{\sqrt{\epsilon_{2r}}}\left\{[\hat{n} \times \vec{\mathscr{H}}_{\tan}(N)] + \sum_{N'=0}^{N-1} F_0(N')[\hat{n} \times \vec{\mathscr{H}}_{\tan}(N-N')]\right\} \tag{19}$$

where the function F_0, a part of the discretized impulse response, is

$$F_0(N) \equiv \int_\phi^{N+1}[1 - |\gamma - N|](a\Delta t)e^{(a\Delta t)\gamma} \cdot [I_1(a\Delta t\gamma) + I_0(a\Delta t\gamma)]\,d\gamma \tag{20}$$

where the lower limit for this integral is

$$\phi = \begin{cases} 0, & N = 0 \\ N-1, & N > 0 \end{cases}. \tag{21}$$

In obtaining this expression, the integral was discretized by assuming that $[\hat{n} \times \vec{\mathscr{H}}_{\tan}(t)]$ is piecewise linear. If it was assumed to be piecewise constant, then the $[1 - |\gamma - N|]$ term would not appear, and the lower limit for the integral in (20) would be simply $\phi = N$.

As previously mentioned, the computational requirements of (19) are quite large; however, by approximating F_0 by a series of exponential functions (Q in number),

$$F_0(N) \approx C_1 \underbrace{e^{A_1 N}}_{(\mu_1)^N} + C_2 \underbrace{e^{A_2 N}}_{(\mu_2)^N} + \cdots + C_Q \underbrace{e^{A_Q N}}_{(\mu_Q)^N} \tag{22}$$

the computational requirements can be made quite reasonable. The details of using Prony's method to determine the μ and the C in (22) are discussed in Appendix I. The convolution sum in (19) with the approximation (22) is

$$\sum_{N'=0}^{\infty} F_0(N')[\hat{n} \times \vec{\mathscr{H}}_{\tan}(N-N')] = \sum_{K=1}^{Q}\sum_{N'=0}^{\infty} C_K(\mu_K)^{N'}[\hat{n} \times \vec{\mathscr{H}}_{\tan}(N-N')] \equiv \sum_{K=1}^{Q} \vec{G}_K(N). \tag{23}$$

It can be easily shown that the terms $\vec{G}_K(N)$ can be recursively obtained using

$$\vec{G}_K(N) = C_K\left[\hat{n} \times \vec{\mathscr{H}}_{\tan}(N)\right] + \mu_K \vec{G}_K(N-1), \quad K = 1,2,3,\cdots,Q; \tag{24}$$

therefore, an efficient implementation of (17) is

$$\vec{\mathscr{E}}_{\tan}(N) = \frac{\eta_0}{\sqrt{\epsilon_{2r}}}\left\{\left[\hat{n} \times \vec{\mathscr{H}}_{\tan}(N)\right] + \sum_{K=1}^{Q} \vec{G}_K(N)\right\}. \tag{25}$$

The efficiency of this approach is evident on comparing (19) with (25). In contrast to (19) which requires the storage of many previous values of the tangential magnetic field, $\approx N_{\max}$, (25) only requires the storage of a small number, Q, of running sum variables. Furthermore, at each time step, (19) requires the many multiplications of the stored magnetic field values with the impulse response, $F_0(I)$, while (25) only requires two multiplies per running sum variable.

At first glance, the preprocessing time required per value of loss tangent, that is, the computation of the impulse response (20) and its exponential approximation (22), seems to be quite large. However, for two- or three-dimensional problems the computational savings outlined in Section II are much greater than this required amount of preprocessing. For a typical two-dimensional problem run on a Cyber 990, the preprocessing time was 15 CPU s per value of loss, and the SIBC-FDTD run-time was 200 CPU s. Without the use of surface impedance concepts, the run time assuming a loss tangent of 10 would have been approximately 675 000 CPU s.

V. Verification

In this section three different problems will be studied to illustrate the applicability of surface impedance concepts and to quantify the accuracy of the efficient implementation of the SIBC for use in the FDTD technique.

A. TE Plane Wave Incident on a Lossy Dielectric Half-Space

The first problem to be discussed is shown schematically in Fig. 1: a homogeneous TE plane wave is incident at the angle ϑ_i on a lossy dielectric half-space. The exact reflection coefficient for the TE case is

$$\Gamma_{\text{exact}}^{\text{TE}} = \frac{\cos\vartheta_i - \sqrt{\epsilon_{2r}(1 - jp_2) - \sin^2\vartheta_i}}{\cos\vartheta_i + \sqrt{\epsilon_{2r}(1 - jp_2) - \sin^2\vartheta_i}}. \tag{26}$$

Both the full summation implementation (19) and the efficient implementation (25) of the SIBC will be used to eliminate the lossy dielectric half-space in Fig. 1 from the solution space, and the results will be compared with the exact results. The reduced 2-D FDTD code, described in Appendix II, will be used to solve for the fields in the free-space region.

A difficulty is encountered in incorporating the SIBC, either (19) or (25), into the FDTD method. Equations (19) and (25) require the values of the tangential magnetic field, $\mathscr{H}_x$, to be at the same location in space and time as the values of the tangential electric field, $\mathscr{E}_y$. That is, both are evaluated at a point on the boundary at the same time. The FDTD method is based on the interleaved grid shown in Fig. 3; $\mathscr{H}_x$ and $\mathscr{E}_y$ are not co-located in space or in time. For this work, the $\mathscr{E}_y$ at the boundary were updated by simply assuming that the $\mathscr{H}_x$ components $\Delta s/2$ in front of the boundary and $\Delta t/2$ earlier in time were a good approximation of the required $\mathscr{H}_x$ components.

Initially, the incident field is assumed to be a time-harmonic signal with period τ_p;

$$\mathscr{E}_y^{\text{inc}}(z=0,t) = r(t)\sin(2\pi t/\tau_p) \tag{27}$$

where $r(t)$ is a ramp function which goes from zero to one over a few periods of the sinusoid. The sinusoidal amplitude will be sampled M times per period; that is,

$$\Delta t = \frac{\tau_p}{M}. \tag{28}$$

The parameter $(a\Delta t)$ that appears in the impulse response F_0 (20) is then

$$a\Delta t = \frac{-p_2\pi}{M}. \tag{29}$$

Fig. 4 compares the exponential approximation to $F_0(I)$ (22) (dots) with the exact value (20) (solid line) for loss tangents $p_2 = 1$, 10^2, and 10^4; $M = 128$ for all results. The excellent agreement demonstrates the accuracy of the exponential approximation.

Fig. 5 shows the reflection coefficient Γ, magnitude and phase, as a function of the angle of incidence ϑ_i for the particular case: $p_2 = 3$, $M = 128$. Three results are shown: the exact reflection coefficient calculated from the Fresnel formula (26) (solid line), the reflection coefficient determined analytically using the SIBC (dashed line),

$$\Gamma_{\text{SIBC}} = \frac{\cos\vartheta_i - \sqrt{\epsilon_{2r}(1 - jp_2)}}{\cos\vartheta_i + \sqrt{\epsilon_{2r}(1 - jp_2)}} \tag{30}$$

and the reflection coefficient determined using the efficient implementation of the FDTD-SIBC (25) (dots). For this low value of loss, $p_2 = 3$, the analytical results, SIBC-Analytical, differs slightly from the exact result. This is to be expected, since the SIBC is an approximation valid for large values of loss tangent. The efficient implementation of the FDTD-SIBC (25) is in excellent agreement with the analytical results for the SIBC. Furthermore, the agreement is good for angles of incidence varying from normal to grazing. To quantify the level of agreement between the SIBC-analytical, the SIBC-FDTD and the exact result, an error measure is defined as

$$\text{maximum relative error} = \max_{\vartheta_i}\left|\frac{\Gamma - \Gamma_{\text{exact}}}{\Gamma_{\text{exact}}}\right|. \tag{31}$$

For the results of Fig. 5, the maximum relative error of the SIBC-analytical is approximately 0.07 and that of the SIBC-FDTD is approximately 0.08.

In Fig. 6, the maximum relative error is displayed for

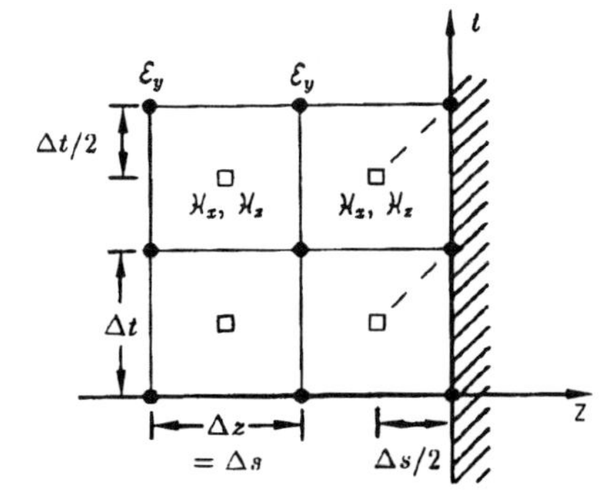

Fig. 3. Grid for the reduced two-dimensional FDTD problem showing node locations for SIBC implementation.

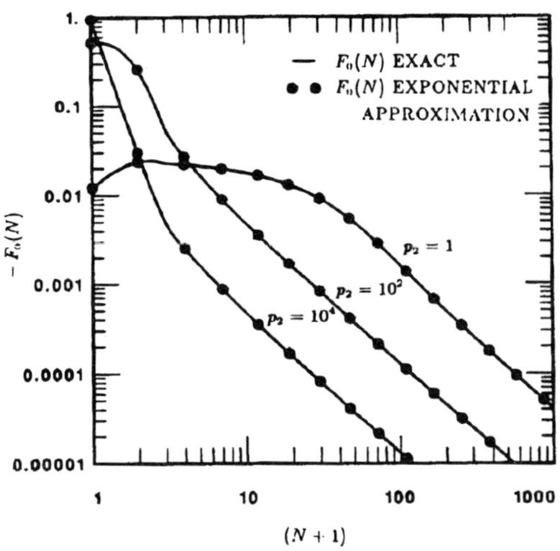

Fig. 4. Comparison of the discretized impulse response $F_0(N)$ with its exponential approximation for values of loss tangent $p_2 = 1.0, 10^2, 10^4$, $\epsilon_{2r} = 1.0$, and $M = 128$. The exponential approximations is for $Q = 20$.

loss tangents in the range $p_2 = 10^{-1}$ to $p_2 = 10^6$. As expected, the exponential formulation (25) (dots) agrees well with the full summation formulation (19) (solid line). For low values of loss, both of these formulations agree with the SIBC-analytical results (dashed line). For large values of loss, the error committed by using the $\mathscr{H}_x$ components in front of the boundary and at an earlier time causes the SIBC-FDTD results to deviate from the SIBC-analytical results. The error is seen to be reduced by a factor of four when the discretization is changed from $M = 32$ to $M = 128$, indicating that the maximum relative error is $O(\Delta s)$. Notice that the errors for large loss tangents are so small as to have no practical significance, even when $M = 32$.

The results presented so far are for an incident plane wave with harmonic-time dependence (27). However, the SIBC-FDTD can be easily applied with an incident plane wave of general time dependence as will be illustrated for the case of a **differentiated** Gaussian pulse:

$$\mathscr{E}_y^{\text{inc}}(z=0,t) = -(t/\tau_p)\exp\left[0.5 - 0.5(t/\tau_p)^2\right] \quad (32)$$

where the amplitude has been normalized to unity and τ_p is now the time from the zero crossing to the peak of the pulse.

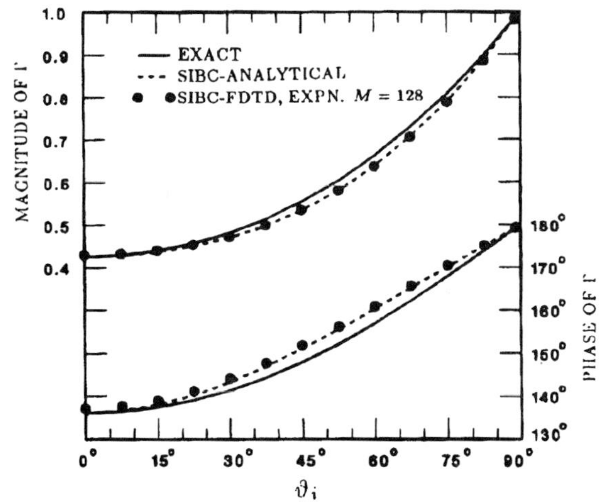

Fig. 5. Comparison of the exact reflection coefficient, both magnitude and phase, with the SIBC-analytical and the SIBC-FDTD results for $p_2 = 3.0$, $\epsilon_{2r} = 1.0$, and $M = 128$.

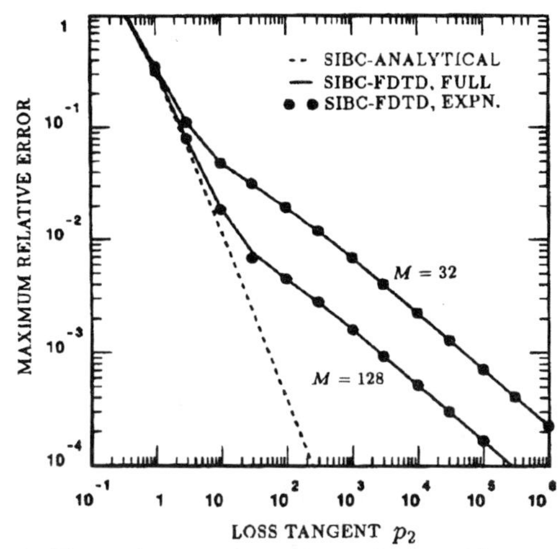

Fig. 6. The maximum relative error as a function of loss tangent. The behavior as a function of M is also shown. $\epsilon_{2r} = 1.0$.

Again the pulse will be sampled M times per τ_p, and the quantity $a\Delta t$ can be shown to be

$$a\Delta t = \frac{-p_{2p}}{2M} \quad (33)$$

where p_{2p} is the loss tangent at the peak frequency of (32), $\omega_p = 1/\tau_p$.

Fig. 7 compares the exact and the SIBC-FDTD results for the particular case of $\vartheta_i = 45°$, $M = 32$, $\epsilon_{2r} = 1$ with loss tangents $p_{2p} = 0.1$, 1.0, and 10.0. Both the reflected and total field at the boundary are shown, because the error is generally more evident in the smaller of these quantities. For the low-loss case ($p_{2p} = 0.1$), the smaller quantity is the reflected field, and for the high-loss case ($p_{2p} = 10.0$), the smaller quantity is the total field. The SIBC-FDTD (dots) are in excellent agreement with the exact results (solid line) when $p_{2p} = 10$, Figs. 7(e) and 7(f). The agreement de-

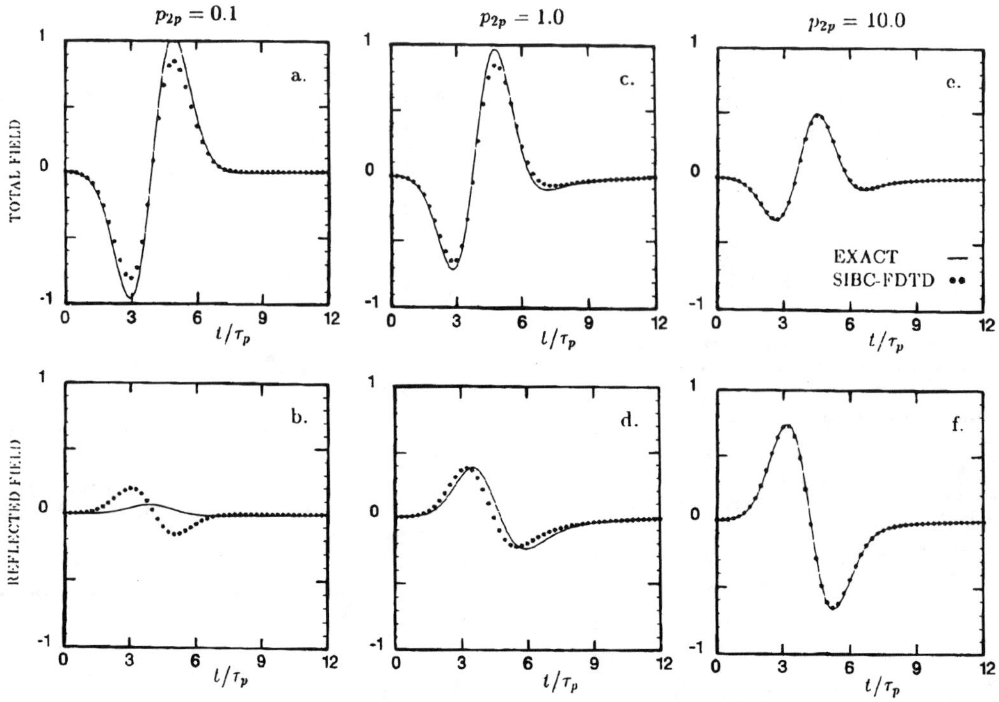

Fig. 7. Plane wave incident upon a lossy dielectric half-space. The pulse is a differentiated Gaussian. $\vartheta_i = 45°$, $\epsilon_{2r} = 1.0$, $M = 32$, and $p_{2p} = 0.1, 1.0, 10.0$.

creases slightly with decreasing loss tangent, $p_{2p} = 1.0$, Figs. 7(c) and 7(d). Finally, when $p_{2p} = 0.1$, the results for the reflected fields are very dissimilar, Fig. 7(b). Notice that the disagreement for the reflected fields, Fig. 7(b), is not evident in the total fields, Fig. 7(a). Of course, the failure of the SIBC for $p_{2p} = 0.1$ is to be expected, since the SIBC is an approximation valid for large values of loss tangent.

In summary, the results of this section, Figs. 5–7, show that the SIBC–FDTD is very suitable for problems involving homogeneous plane waves and planar regions of lossy dielectric, $p_2 > 1$.

B. Line Current Over Lossy Half-Space

In this section, the problem of an infinite line current $I(t)$ over a lossy half-space, Fig. 8, will be discussed. The temporal variation of the current $I(t)$ will be the differentiated Gaussian pulse (32). The electric field on the interface at a distance x from under the current will be considered. This problem can be described by the two dimensionless parameters $(k_p d)$ and (x/d), where $k_p = \omega_p / c$ is the wavenumber at the peak of the spectrum of the current $I(t)$.

An exact solution to this problem can be obtained using the plane wave spectrum [8]. The electric field incident upon the lossy dielectric half-space is determined to be

$$\vec{E}_{\tan}^{\text{inc}}(\omega, x) = \frac{\hat{y}}{d}\left(\frac{-c\mu_0 k_0 d}{2\pi}\right) I(\omega) \int_0^{\infty} \frac{1}{\sqrt{1-\xi^2}} \cdot \exp\left(-jk_0 d\sqrt{1-\xi^2}\right) \cos\left(k_0 d \frac{x}{d} \xi\right) d\xi. \quad (34)$$

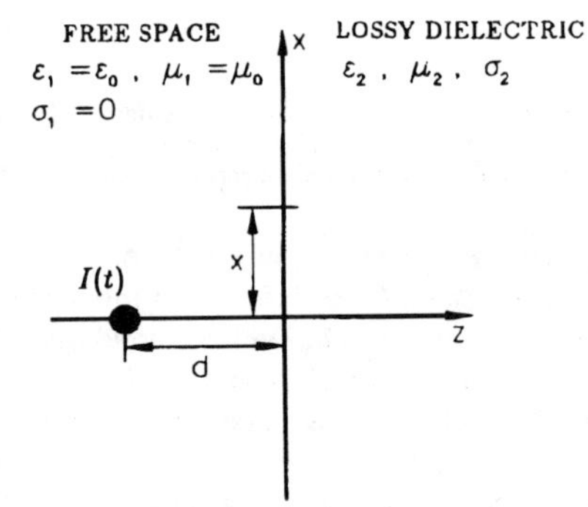

Fig. 8. Geometry for line current over lossy half-space.

The reflected electric field at the interface is then

$$\vec{E}_{\tan}^{\text{refl}}(\omega, x) = \frac{\hat{y}}{d}\left(\frac{-c\mu_0 k_0 d}{2\pi}\right) I(\omega) \int_0^{\infty} \frac{\sqrt{1-\xi^2} - \sqrt{\epsilon - \xi^2}}{\sqrt{1-\xi^2} + \sqrt{\epsilon - \xi^2}} \frac{1}{\sqrt{1-\xi^2}} \cdot \exp\left(-jk_0 d\sqrt{1-\xi^2}\right) \cos\left(k_0 d \frac{x}{d} \xi\right) d\xi \quad (35)$$

making the total electric field at the interface

$$\vec{E}_{\tan}^{\text{total}}(\omega, x) = \frac{\hat{y}}{d}\left(\frac{-c\mu_0 k_0 d}{2\pi}\right) I(\omega) \int_0^{\infty} \cdot \frac{2}{\sqrt{1-\xi^2}+\sqrt{\epsilon-\xi^2}} \cdot \exp\left(-jk_0 d\sqrt{1-\xi^2}\right)\cos\left(k_0 d\frac{x}{d}\xi\right) d\xi. \tag{36}$$

Here $I(\omega)$ is the spectrum of the current $I(t)$. These relations are used along with the fast Fourier transform to compute the exact results for this problem.

As the line current $I(t)$ is moved closer to the interface (decreasing $k_p d$), the inhomogeneous plane waves in the spectrum (34) become significant at the interface. Thus, with this problem we will see how the SIBC-FDTD handles inhomogeneous plane waves.

A standard total field code was used for the FDTD results. To obtain the reflected field, the FDTD code was first run without a material half-space and the incident field recorded. The reflected field was then found by subtracting the incident field from the total field. In all calculations, the efficient implementation (25) was used for the SIBC.

Once again, both the total field and the reflected field must be studied, because the error is generally more evident in the smaller quantity. For low-loss cases, the reflected field is the small quantity; whereas, for high-loss cases, the total field is the small quantity.

Figs. 9 and 10 compare the exact (solid line) and the SIBC-FDTD (dots) time-dependent waveforms at the position $x/d = 1$. The current $I(t)$ is the differentiated Gaussian pulse (32) sampled 32 times per τ_p (i.e., $M = 32$). For this illustration two values of loss tangent, $p_{2p} = 3.0, 30.0$, and two heights for the line current, $k_p d = 0.25, 1.0$, are used. The loss tangent $p_{2p} = 3.0$ will be considered first. When the height is $k_p d = 1.0$, Figs. 9(a) and 9(b), the SIBC-FDTD and the exact results are in good agreement. However, when the height is lowered to $k_p d = 0.25$, the reflected fields, Fig. 9(d), disagree. For the loss tangent $p_{2p} = 30.0$, the SIBC-FDTD and the exact results are in good agreement for both heights, Fig. 10.

Here we see a basic limitation of the SIBC, not associated with the FDTD implementation [8]. As the line current is lowered toward the interface, the inhomogeneous waves in the spectrum (34) become significant at the interface. The field along the interface then varies on a scale that can be much shorter than the free space wavelength λ_0. For the SIBC to be accurate, we must have [8]

$$\left(\frac{d}{\delta_p}\right)^2 > 1 \tag{37}$$

where δ_p is the skin depth at the frequency ω_p. This condition can be rearranged to give

$$(k_p d)^2 > \frac{2}{p_{2p}}. \tag{38}$$

Now we can see that for $p_{2p} = 30.0$, $(k_p d)^2$ must be greater than 0.067, which is valid for both heights in Fig. 10; however, for $p_{2p} = 3.0$, $(k_p d)^2$ must be greater than 0.667 which is valid only for $k_p d = 1.0$, Figs. 9(a) and 9(b), but not for $k_p d = 0.25$, Figs. 9(c) and 9(d).

C. Lossy Parallel-Plate Waveguide

As a final application of the SIBC-FDTD, we will consider time-harmonic excitation of the lossy parallel-plate waveguide shown in Fig. 11. The excitation is a TEM mode in the lossless waveguide on the left. This is a TM electromagnetic problem, only the field components $\mathscr{E}_x$, $\mathscr{E}_z$, and $\mathscr{H}_y$ are involved. The propagation through the lossy waveguide on the right will be studied.

The exact attenuation and propagation constants, α and β, for the lossy waveguide at the frequency ω_n can be found by solving the transcendental equation [11]

$$\tan\left(\frac{\omega_n}{2}\sqrt{\gamma_n^2+(2\pi)^2}\right) - \frac{1}{j\epsilon_{cr}(1-jp)} \cdot \sqrt{\frac{\gamma_n^2+(2\pi)^2\epsilon_{cr}(1-jp)}{\gamma_n^2+(2\pi)^2}} = 0 \tag{39}$$

for the quantity $\gamma_n \equiv (\alpha + j\beta)\lambda_0$ in terms of the quantities

$\omega_n = \omega a/\pi c$	normalized frequency
ϵ_{cr}	relative permittivity of the lossy conductor
p	loss tangent of the conductor at the frequency ω.

The direct use of an SIBC leads to a different result. The attenuation and propagation constants for the SIBC-analytical can be found by solving the transcendental equation

$$\tan\left(\frac{\omega_n}{2}\sqrt{\gamma_n^2+(2\pi)^2}\right) - j\frac{2\pi}{\sqrt{\epsilon_{cr}(1-jp)}\sqrt{\gamma_n^2+(2\pi)^2}} = 0. \tag{40}$$

These results are compared with those computed using a standard two-dimensional FDTD total field code with the lossy walls of the waveguide replaced by the efficient implementation of the SIBC-FDTD (25). The interior of the waveguide is discretized by a square grid with four cells from the center line to the top conductor, $N_x = 4$. When the incident field is a time-harmonic signal (27), the number of time steps per period, M, is

$$M = \frac{4N_x}{\omega_n} \tag{41}$$

and

$$a\Delta t = \frac{-p\pi}{M}. \tag{42}$$

The normalized attenuation $\alpha\lambda_0$ and normalized phase shift $(\beta - \beta_0)\lambda_0$ in the parallel-plate waveguide are shown in Fig. 12 as a function of the loss tangent p for the normalized frequency $\omega_n = 0.1$. Results for the exact (39) (solid line), the SIBC-analytical (40) (dashed line), and the SIBC-FDTD (dots) are displayed. Once again, the SIBC-FDTD results

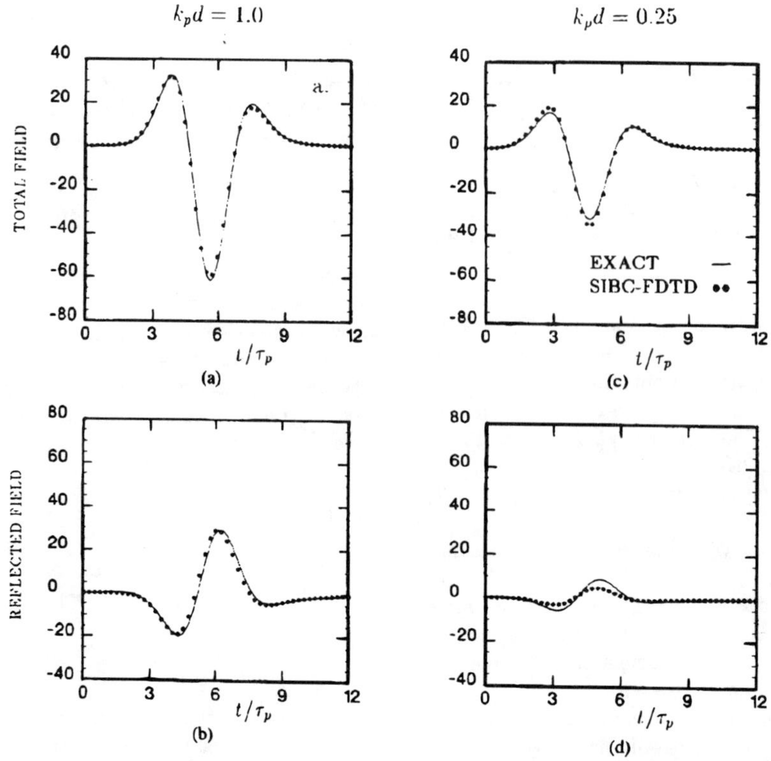

Fig. 9. Electric field along the interface at $x/d = 1$ for a line current $I(t)$ over a lossy half-space. The current is a differentiated Gaussian pulse, $M = 32$. $p_{2p} = 3.0$, $\epsilon_{2r} = 1.0$, and $k_p d = 1.0, 0.25$.

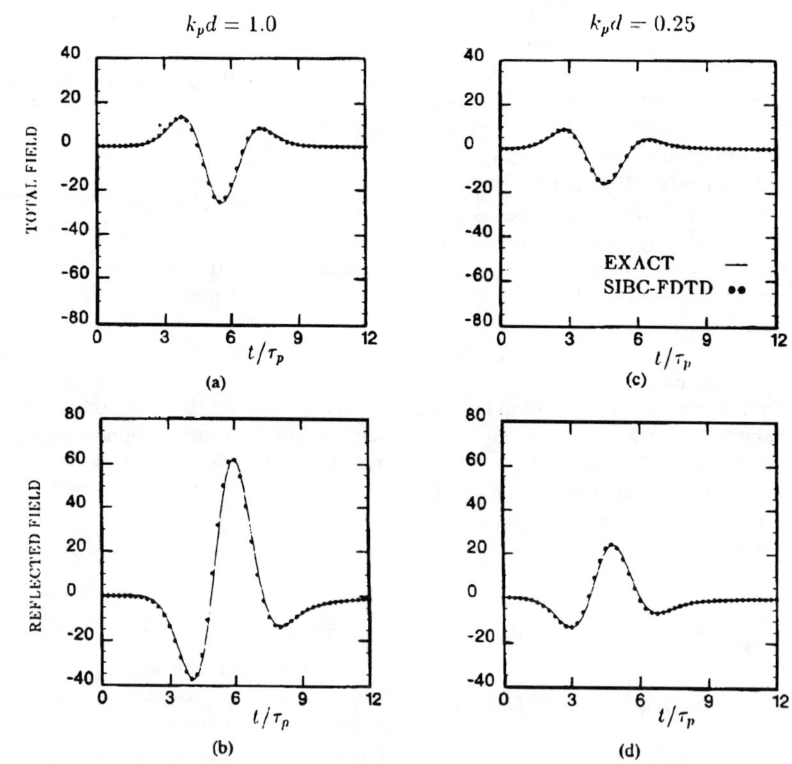

Fig. 10. Electric field along the interface at $x/d = 1$ for a line current $I(t)$ over a lossy half-space. The current is a differentiated Gaussian pulse, $M = 32$. $p_{2p} = 30.0$, $\epsilon_{2r} = 1.0$, and $k_p d = 1.0, 0.25$.

agree well with the SIBC-analytical results. The disagreement for lower values of loss tangent, $p < 10$, is seen to be a basic limitation of the SIBC, not associated with the FDTD implementation.

VI. Conclusion

In this paper the implementation of surface impedance concepts into the FDTD method has been discussed. An efficient algorithm for computer implementation based on exponential approximations has been presented. The implementation has been studied for both homogeneous and inhomogeneous plane wave incidence as well as for angles of incidence varying from normal to grazing. In fact, the parallel-plate waveguide results show that the implementation can handle propagation along the interface. Furthermore, results for fields with both harmonic-time dependence and general-time dependence (differentiated Gaussian pulse) have been presented. Overall, the FDTD results compare quite well with various exact theoretical results.

Appendix I

In this Appendix Prony's method of fitting an exponential approximation to data will be reviewed, and the minor modifications that were introduced for this research will be presented.

Prony's method fits an exponential approximation of the form

$$f(N) \approx C_1 \underbrace{e^{A_1 N}}_{(\mu_1)^N} + C_2 \underbrace{e^{A_2 N}}_{(\mu_2)^N} + \cdots + C_Q \underbrace{e^{A_Q N}}_{(\mu_Q)^N} \quad (43)$$

to a function $f(x)$ by sampling at $N_{\max}$ equally spaced points, $N = 0, 1, 2, \cdots, N_{\max} - 1$. This is a two-step procedure [10]. The first step is to find the μ. The set of equations

$$\begin{array}{ccccccccc} f_Q & + & f_{Q-1}\alpha_1 & + & f_{Q-2}\alpha_2 & + & \cdots & + & f_0\alpha_Q = 0 \\ f_{Q+1} & + & f_Q\alpha_1 & + & f_{Q-1}\alpha_2 & + & \cdots & + & f_1\alpha_Q = 0 \\ \vdots & & \vdots & & \vdots & & \ddots & & \vdots \\ f_{N_{\max}-1} & + & f_{N_{\max}-2}\alpha_1 & + & f_{N_{\max}-3}\alpha_2 & + & \cdots & + & f_Q\alpha_Q = 0 \end{array} \quad (44)$$

is solved using a least-square error algorithm for the α. The μ are then found as the roots of the polynomial

$$\mu^Q + \alpha_1 \mu^{Q-1} + \alpha_2 \mu^{Q-2} + \cdots + \alpha_{Q-1}\mu + \alpha_Q = 0. \quad (45)$$

The second step is to find the C. The set of equations

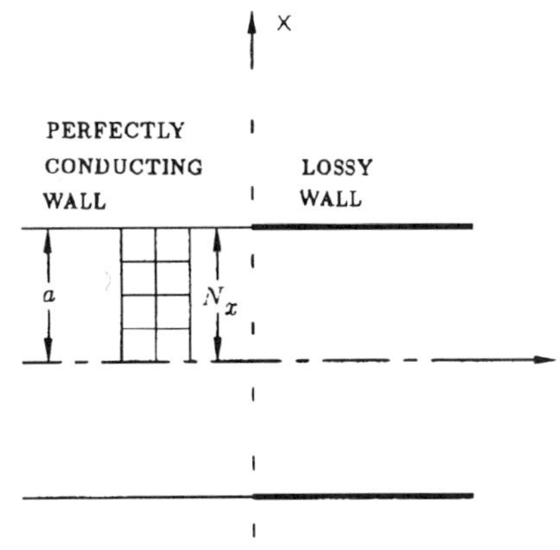

Fig. 11. Parallel-plate waveguide transition going from a perfectly conducting waveguide to one with lossy walls.

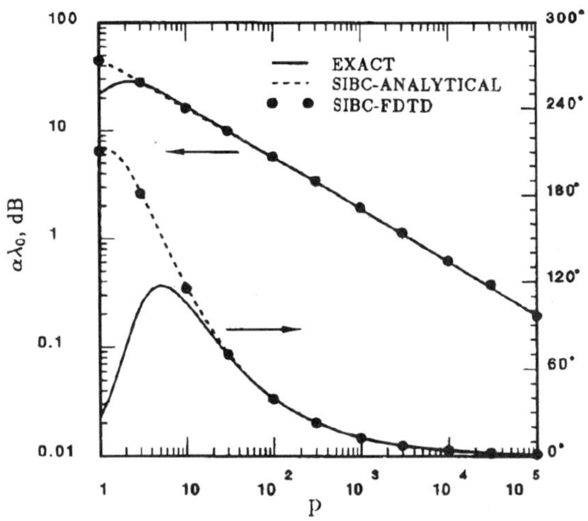

Fig. 12. Comparison of the exact, the SIBC-analytical, and the SIBC-FDTD results for $\alpha\lambda_0$ and $(\beta - \beta_0)\lambda_0$ for loss tangents varying from 1 to 10^5, $\epsilon_{2r} = 1.0$. $N_x = 4$, $\omega_n = 0.1$.

Several practical problems arise in the use of Prony's method. First, the number of exponentials, Q, used to fit the data must be chosen. Second, some of the μ determined by (45) are occasionally complex. For this research, the number of exponentials, Q, was overspecified; $Q = 20$. The resultant μ found in (45) were then examined. Any complex pairs of μ were replaced by their magnitude only. Finally, the list of μ was sorted and any duplicates were removed. The C were then determined using this reduced set of modified μ.

$$\begin{array}{ccccccccc} C_1 & + & C_2 & + & \cdots & + & C_Q & = & f_0 \\ \mu_1 C_1 & + & \mu_2 C_2 & + & \cdots & + & \mu_Q C_Q & = & f_1 \\ (\mu_1)^2 C_1 & + & (\mu_2)^2 C_2 & + & \cdots & + & (\mu_Q)^2 C_Q & = & f_2 \\ \vdots & & \vdots & & \ddots & & \vdots & & \vdots \\ (\mu_1)^{N_{\max}-1} C_1 & + & (\mu_2)^{N_{\max}-1} C_2 & + & \cdots & + & (\mu_Q)^{N_{\max}-1} C_Q & = & f_{N_{\max}-1} \end{array} \quad (46)$$

is solved for the C using a least-square error algorithm.

Appendix II

In this Appendix, the reduction of a two-dimensional problem to a one-dimensional problem and the subsequent formulation of the FDTD algorithm for the reduced problem will be presented.

In scattering problems, the incident field, $\vec{\mathscr{E}}^i$ and $\vec{\mathscr{H}}^i$, is known and the relevant set of Maxwell equations for the reflected field, $\vec{\mathscr{E}}^r$ and $\vec{\mathscr{H}}^r$, is

$$\nabla \times \vec{\mathscr{E}}^r + \mu \frac{\partial \vec{\mathscr{H}}^r}{\partial t} = 0 \tag{47}$$

$$\nabla \times \vec{\mathscr{H}}^r - \epsilon \frac{\partial \vec{\mathscr{E}}^r}{\partial t} = 0. \tag{48}$$

It is important to realize that any boundary conditions imposed by the scatterer are on the total fields, $\vec{\mathscr{E}} = \vec{\mathscr{E}}^i + \vec{\mathscr{E}}^r$, etc.; for example, at a perfect electric conductor

$$\vec{\mathscr{E}}_{\tan} = 0 \tag{49}$$

which in the scattering problem formulation would be

$$\vec{\mathscr{E}}^r_{\tan} = -\vec{\mathscr{E}}^i_{\tan}. \tag{50}$$

Hence, the various boundaries of the scatterer couple the known incident field to the reflected field.

The two-dimensional problem of a TE plane wave incident on a dielectric half is shown in Fig. 1. The unit vectors in the direction of propagation for the incident and reflected waves are

$$\hat{k}_i = -\sin\vartheta_i \hat{x} + \cos\vartheta_i \hat{z} \tag{51}$$

and

$$\hat{k}_r = -\sin\vartheta_i \hat{x} - \cos\vartheta_i \hat{z}. \tag{52}$$

The assumption that the reflected field, $\vec{\mathscr{E}}^r$ and $\vec{\mathscr{H}}^r$, only contains waves propagating in the $\hat{k}_r$ direction leads to a reduction in the dimensionality of the problem. The reflected field has the form of a traveling wave:

$$\vec{\mathscr{E}}^r \propto F\left(t - \frac{\vec{r}\cdot\hat{k}_r}{c}\right) \tag{53}$$

which in rectangular coordinates is

$$\vec{\mathscr{E}}^r \propto F\left(t + \frac{\sin\vartheta_i}{c}x + \frac{\cos\vartheta_i}{c}z\right). \tag{54}$$

From this expression it is evident that

$$\frac{\partial \vec{\mathscr{E}}^r}{\partial x} = \tan\vartheta_i \frac{\partial \vec{\mathscr{E}}^r}{\partial z} \tag{55}$$

and similarly for the magnetic field,

$$\frac{\partial \vec{\mathscr{H}}^r}{\partial x} = \tan\vartheta_i \frac{\partial \vec{\mathscr{H}}^r}{\partial z}. \tag{56}$$

The electric and magnetic fields in rectangular coordinates for these TE waves are simply

$$\vec{\mathscr{E}} = \mathscr{E}_y \hat{y}, \tag{57}$$

$$\vec{\mathscr{H}} = \mathscr{H}_x \hat{x} + \mathscr{H}_z \hat{z} \tag{58}$$

which reduces the Maxwell equations (47) and (48) to

$$-\frac{\partial \mathscr{E}_y}{\partial z} = -\mu \frac{\partial \mathscr{H}_x}{\partial t} \tag{59}$$

$$\frac{\partial \mathscr{E}_y}{\partial x} = -\mu \frac{\partial \mathscr{H}_z}{\partial t} \tag{60}$$

$$\frac{\partial \mathscr{H}_x}{\partial z} - \frac{\partial \mathscr{H}_z}{\partial x} = \epsilon \frac{\partial \mathscr{E}_y}{\partial t}. \tag{61}$$

Now (55) and (56) are used to eliminate the x dependence, yielding

$$-\frac{\partial \mathscr{E}_y}{\partial z} = -\mu \frac{\partial \mathscr{H}_x}{\partial t} \tag{62}$$

$$\tan\vartheta_i \frac{\partial \mathscr{E}_y}{\partial z} = -\mu \frac{\partial \mathscr{H}_z}{\partial t} \tag{63}$$

$$\frac{\partial \mathscr{H}_x}{\partial z} - \tan\vartheta_i \frac{\partial \mathscr{H}_z}{\partial z} = \epsilon \frac{\partial \mathscr{E}_y}{\partial t}. \tag{64}$$

These reduced equations are now discretized by replacing all the derivatives with central difference approximations. If the $\mathscr{E}$ and $\mathscr{H}$ are interleaved in time, then this discretization is a second-order accurate leap frog scheme. The grid is shown in Fig. 1, and the update equations are

$$\begin{aligned}
\mathscr{H}_x^{n+1/2}(K) &= \mathscr{H}_x^{n-1/2}(K) + \left(\frac{\Delta t}{\mu \Delta z}\right)[\mathscr{E}_y^n(K) - \mathscr{E}_y^n(K-1)] \\
\mathscr{H}_z^{n+1/2}(K) &= \mathscr{H}_z^{n-1/2}(K) - \tan\vartheta_i\left(\frac{\Delta t}{\mu \Delta z}\right)[\mathscr{E}_y^n(K) - \mathscr{E}_y^n(K-1)] \\
\mathscr{E}_y^{n+1}(K) &= \mathscr{E}_y^n(K) + \left(\frac{\Delta t}{\epsilon \Delta z}\right)[\mathscr{H}_x^{n+1/2}(K+1) - \mathscr{H}_x^{n+1/2}(K)] \\
&\quad - \tan\vartheta_i\left(\frac{\Delta t}{\epsilon \Delta z}\right)[\mathscr{H}_z^{n+1/2}(K+1) - \mathscr{H}_z^{n+1/2}(K)].
\end{aligned} \tag{65}$$

This set of equations is similar to the normal FDTD update equations. The major differences are that the $\mathscr{H}_x$ and $\mathscr{H}_z$ components are sampled at the same location and that the equations have a ϑ_i dependence.

The Courant stability condition for the typical FDTD algorithm is

$$\frac{v\Delta t}{\Delta s} \leq \frac{1}{\sqrt{D}} \tag{66}$$

where v is the maximum phase velocity, and D is the number of dimensions. A plane wave traveling at the speed of light c at an angle ϑ_i to the normal has a phase velocity of

$$v = \frac{c}{\cos\vartheta_i} \tag{67}$$

along the normal (z axis). Thus, using the conventional value $v\Delta t/\Delta z = 1/2$ yields

$$\Delta t = \frac{\Delta z \cos \vartheta_i}{2c}, \quad \text{or} \quad \frac{c\Delta t}{\Delta z} = \frac{\cos \vartheta_i}{2}. \tag{68}$$

Substitution of this relation into (65) yields equations which are suitable for computer implementation.

Acknowledgment

J. G. Maloney would like to thank the Joint Services Electronics Program (JSEP) and the DuPont Corporation for support received during the period of this research through a JSEP Graduate Fellowship and a DuPont Graduate Fellowship.

References

[1] K. S. Yee, "Numerical solution of initial boundary value problems involving Maxwell's equations in isotropic media," *IEEE Trans. Antennas Propagat.*, vol. AP-14, pp. 302-307, May 1966.

[2] A. Taflove, "Review of the formulation and applications of the finite-difference time-domain method for numerical modeling of electromagnetic wave interactions with arbitrary structures," *Wave Motion*, vol. 10, pp. 547-582, 1988.

[3] J. Fang, "Time domain finite difference computation for Maxwell's equations," Ph.D. dissertation, Univ. California, Berkeley, CA, Nov. 1989.

[4] J. G. Maloney, G. S. Smith, and W. R. Scott, Jr., "Accurate computation of the radiation from simple antennas using the finite-difference time-domain method," *IEEE Trans. Antennas Propagat.*, vol. 38, pp. 1059-1068, July 1990.

[5] R. Luebbers, F. P. Hunsberger, K. S. Kunz, R. B. Standler and M. Schneider, "A frequency-dependent finite-difference time-domain formulation for dispersive materials," *IEEE Trans. Electromagn. Compat.*, vol. 32, pp. 222-227, Aug. 1990.

[6] J. G. Maloney and G. S. Smith, "Implementation of surface impedance concepts in the finite-difference time-domain (FD-TD) technique," in *Proc. 1990 IEEE Antennas Propagat. Soc. Int. Symp.*, Dallas, TX, May 1990, vol. 4, pp. 1628-1631.

[7] M. A. Leontovich, "Approximate boundary conditions for the electromagnetic field on the surface of a good conductor," *Investigations on Radiowave Propagation*, Pt. II., 5-12, Printing house of the Academy of Sciences, Moscow. Translation: V. A. Fock, "Diffraction, refraction, and reflection of radio waves," Appendix, Air Force Cambridge Res. Center TN-57-102, 1957.

[8] G. S. Smith, "On the skin effect approximation," *Am. J. Phys.*, vol. 58, no. 10, pp. 996-1002, Oct. 1990.

[9] R. V. Churchill, *Operational Mathematics*. New York: McGraw-Hill, 1972, pp. 458-466.

[10] F. B. Hildebrand, *Introduction to Numerical Analysis*. New York: Dover, 1974, pp. 457-462.

[11] S. Ramo, J. R. Whinnery, and T. Van Duzer, *Fields and Waves in Communication Electronics*. New York: Wiley, 1965, pp. 379-382.

FDTD for Nth-Order Dispersive Media

Raymond J. Luebbers, *Senior Member, IEEE*, and Forrest Hunsberger

Abstract—**Previously, a method for applying the finite-difference time domain (FDTD) method to dispersive media with complex permittivity described by a function with a single first-order pole was presented. This method involved the recursive evaluation of a discrete convolution, and was therefore relatively efficient. In this paper, the recursive convolution approach is extended to media with dispersions described by multiple second-order poles. The significant change from the first-order implementation is that the single backstore variable for each second-order pole is complex. The approach is demonstrated for a pulsed plane wave incident on a medium with a complex permittivity described by two second-order poles, and excellent agreement is obtained with the exact solution.**

I. Introduction

The finite-difference time-domain (FDTD) method [1] is typically implemented with constant values of permittivity and conductivity. While this is adequate for narrow-band calculations or for transient calculations involving perfect conductors or low-loss dielectrics, it does not allow accurate transient calculations to be made for materials with significant frequency dependence in their constitutive parameters. In [2], a method for extending FDTD to media involving a first-order Debye dispersion relation for the complex permittivity was presented and applied to water. The improvement over FDTD results with constant constitutive parameters was significant. A similar approach to that given in [2] was applied in [3], and in [4], the method was applied to three-dimensional FDTD calculations of electromagnetic absorption in biological bodies. In [5], the method was extended to cold isotropic plasma. The complex permittivity for the plasma included regions with a negative real part of the complex permittivity, so that conventional FDTD could not be applied at all, yet quite accurate results were obtained.

Using this approach, the inclusion of the dispersive effects involves the calculation of a discrete convolution of the electric field with the time-domain susceptibility function for the material. While a straightforward evaluation of the convolution would be computationally burdensome, the convolution can be evaluated recursively since the time-domain susceptibilities for the first-order poles have an exponential time behavior.

An alternative approach involves adding an auxiliary difference equation to the FDTD formulation that relates the electric flux density D to the electric field E, thus substituting a differential equation for the convolution integral. Results have been reported using this method for media with single first- and second-order poles [6], [7]. However, it appears that the discrete convolution approach requires fewer backstored variables, and is simpler to implement than the auxiliary difference equation method.

In this paper, the discrete convolution method will be extended to media with multiple second-order Lorentz poles describing the complex permittivity. This generality allows much better modeling of materials which have strong frequency dependence in their constitutive parameters, such as biological materials, artificial dielectrics, and optical materials.

II. Discrete Convolution

In this section, we briefly review the discrete convolution approach as described in [2] and [5]. The notation will be changed slightly to accommodate multiple poles, and the conductivity term will be included to allow for zero-frequency conduction current.

In the time domain, the Maxwell curl equations are therefore

$$\nabla \times H = \frac{\partial D}{\partial t} + \sigma E \tag{1}$$

$$\nabla \times E = -\frac{\partial B}{\partial t}. \tag{2}$$

We will assume $B = \mu H$; however, dispersive magnetic materials can be included by a straightforward extension of this approach. For a linear dispersive medium, the relationship between D and E is given by

$$D(t) = \epsilon_\infty \epsilon_0 E(t) + \epsilon_0 \int_0^t E(t-\tau)\chi(\tau)\, d\tau \tag{3}$$

where ϵ_0 is the permittivity of free space, $\chi(\tau)$ is the electric susceptibility, and ϵ_∞ is the finite frequency permittivity. Using Yee notation with $t = n\,\Delta t$, we obtain from (3)

$$D(t) \approx D(n\,\Delta t) = D^n = \epsilon_\infty \epsilon_0 E^n + \epsilon_0 \int_0^{n\,\Delta t} E(n\,\Delta t - \tau)\chi(\tau)\, d\tau. \tag{4}$$

If we make the approximation that all field quantities are constant over each time interval Δt, and assume that all fields are zero for $t < 0$, then the integration becomes in part a summation so that (4) is equivalent to

$$D^n = \epsilon_\infty \epsilon_0 E^n + \epsilon_0 \sum_{m=0}^{n-1} E^{n-m} \int_{m\,\Delta t}^{(m+1)\Delta t} \chi(\tau)\, d\tau. \tag{5}$$

To simplify the following, we consider only a one-dimensional case, but the extension to three dimensions is

Manuscript received January 15, 1992.

R. J. Luebbers is with the Department of Electrical and Computer Engineering, Pennsylvania State University, University Park, PA 16802.

F. Hunsberger was with the Department of Electrical and Computer Engineering, Pennsylvania State University, University Park, PA 16802. He is now with M.I.T. Lincoln Laboratories, Lexington, MA 02173.

IEEE Log Number 9202722.

Reprinted from *IEEE Transactions on Antennas and Propagation*, pp. 1297-1301, November 1992.

straightforward. For this one-dimensional case, we include only the D_y, E_y, and H_z field components and assume propagation in the $\pm x$ directions. Then with $x = i\,\Delta x$, we have from (1)

$$\frac{D_y^{n+1}(i) - D_y^n(i)}{\Delta t} = -\frac{H_z^{n+1/2}(i+\frac{1}{2}) - H_z^{n+1/2}(i-\frac{1}{2})}{\Delta x} - \sigma E_y^{n+1}(i). \tag{6}$$

We need to eliminate the D_y terms from (6) in order to be able to solve for E_y^{n+1}. Using (5), we find that

$$D_y^{n+1}(i) - D_y^n(i) = (\epsilon_0\epsilon_\infty + \epsilon_0\chi^0)E_y^{n+1}(i) - \epsilon_0\epsilon_\infty E_y^n(i) - \epsilon_0\sum_{m=0}^{n-1} E_y^{n-m}(i)(\chi^m - \chi^{m+1}) \tag{7}$$

where

$$\chi^m = \int_{m\,\Delta t}^{(m+1)\,\Delta t}\chi(\tau)\,d\tau \tag{8}$$

and where it is assumed that the constitutive parameter values correspond to the electric field at spatial location $x = i\,\Delta x$. To simplify the final update equation, we also define

$$\Delta\chi^m = \chi^m - \chi^{m+1}. \tag{9}$$

Now substituting (7)–(9) into (6) and solving for E_y^{n+1}, we obtain

$$E_y^{n+1}(i) = \frac{\epsilon_\infty}{\frac{\sigma\,\Delta t}{\epsilon_0} + \epsilon_\infty + \chi^0}E_y^n(i) + \frac{1}{\frac{\sigma\,\Delta t}{\epsilon_0} + \epsilon_\infty + \chi^0}\sum_{m=0}^{n-1} E_y^{n-m}(i)\,\Delta\chi^m - \frac{\Delta t}{\left(\frac{\sigma\,\Delta t}{\epsilon_0} + \epsilon_\infty + \chi^0\right)\epsilon_0\,\Delta x}\cdot\left[H_z^{n+1/2}(i+\tfrac{1}{2}) - H_z^{n+1/2}(i-\tfrac{1}{2})\right]. \tag{10}$$

The FDTD update equation for the magnetic field is unchanged from that normally used [1], [2].

III. Recursive Convolution for Nth-Order Dispersion

In [2] and [5], it was shown that the convolution summation in (10) for a complex permittivity with a first-order pole (Debye) could be evaluated recursively. This was possible since the time-domain susceptibility function $\chi(\tau)$ was an exponential function of time. In this section, a recursive convolution for a dispersive medium which is described with multiple second-order (Lorentz) poles will be developed.

For a medium with P natural frequencies, the complex permittivity can be described as [8]

$$\epsilon(\omega) = \epsilon_\infty + (\epsilon_s - \epsilon_\infty)\sum_{p=1}^{P}\frac{G_p\,\omega_p^2}{\omega_p^2 + 2j\omega\delta_p - \omega^2} \tag{11}$$

with the condition that

$$\sum_{p=1}^{P} G_p = 1$$

where ϵ_s is the static permittivity, ω_p is the pth resonant frequency, δ_p is the pth damping coefficient, and a time convention of $e^{+j\omega t}$ is assumed. We obtain a time-domain susceptibility function χ_p for each term in the summation in (11) by considering the Fourier transform [9]

$$\chi_p(\tau) = \gamma_p e^{-\alpha_p\tau}\sin(\beta_p\tau)U(\tau) \Leftrightarrow \frac{\gamma_p\beta_p}{(\alpha_p^2 + \beta_p^2) + 2j\omega\alpha_p - \omega^2} \tag{12}$$

where $U(\tau)$ is the unit step function. Comparing (11) and (12), we see that (12) will provide the correct expression for χ_p provided that

$$\alpha_p = \delta_p$$
$$\beta_p = \sqrt{\omega_p^2 - \delta_p^2} \tag{13}$$
$$\gamma_p = \frac{\omega_p^2 G_p(\epsilon_s - \epsilon_\infty)}{\beta_p}.$$

The time dependence of χ_p in (12) is not in a form for which the corresponding discrete convolution can be updated recursively. However, if we define a complex *time-domain* susceptibility

$$\hat{\chi}_p(\tau) = -j\gamma_p e^{-(\alpha_p + j\beta_p)\tau}U(\tau) \tag{14}$$

so that

$$\chi_p(\tau) = \mathrm{Re}\left[\hat{\chi}_p(\tau)\right] \tag{15}$$

where the caret denotes complex quantities and $\mathrm{Re}[\circ]$ is the real operator, a recursive convolution can be implemented. Applying (8) and (9) to (14) readily produces

$$\hat{\chi}_p^m = \frac{-j\gamma_p}{\alpha_p - j\beta_p}e^{(-\alpha_p + j\beta_p)m\,\Delta t}\left[1 - e^{(-\alpha_p + j\beta_p)\Delta t}\right] \tag{16}$$

$$\Delta\hat{\chi}_p^m = \frac{-j\gamma_p}{\alpha_p - j\beta_p}e^{(-\alpha_p + j\beta_p)m\,\Delta t}\left[1 - e^{(-\alpha_p + j\beta_p)\Delta t}\right]^2 \tag{17}$$

with the important feature of (17) being the relationship

$$\Delta\chi_p^{m+1} = e^{(-\alpha_p + j\beta_p)\Delta t}\,\Delta\hat{\chi}_p^m \tag{18}$$

which allows the recursive evaluation of the convolution summation. If we now define

$$\psi_p^n = \sum_{m=0}^{n-1} E^{n-m}\,\Delta\chi_p^m = \mathrm{Re}\left[\hat{\psi}_p^n\right] \tag{19}$$

then the complex accumulation variable $\hat{\psi}_p^n$ for the pth Lorentz pole can be updated recursively [2], [5] as

$$\hat{\psi}_p^n = E^n \Delta \hat{\chi}_p^0 + e^{(-\alpha_p + j\beta_p)\Delta t}\hat{\psi}_p^{n-1}. \quad (20)$$

Finally, for the P poles, we obtain the quantities needed to update (10) as

$$\chi^0 = \sum_{p=1}^{P} \mathrm{Re}\left[\hat{\chi}_p^0\right] \quad (21)$$

and

$$\sum_{m=0}^{n-1} E_y^{n-m} \Delta \chi^m = \sum_{p=1}^{P} \mathrm{Re}\left[\hat{\psi}_p^n\right] \quad (22)$$

with each term in the summation of (22) updated recursively using (20).

While this derivation has included only Lorentz poles, any combination of first- and second-order poles could be included in the summation process. Application of this method to single but more complicated second-order poles that occur in the FDTD calculation of propagation in magnetized plasmas [10], [11] and ferrites [11] has been demonstrated, and the anisotropy of these materials has also been included.

Before demonstrating the effectiveness of this approach, a brief comparison with the differential equation method will be given. In [7], it is stated that for order-M dispersion, M previous values of D and $M-1$ additional previous values of E must be stored. Thus, for the following example which involves an order-4 dispersive material, the differential method would require storage of seven additional real variables, while the recursive convolution requires storage of two complex variables. These two complex variables correspond to four real variables in the computer storage required so that there is very nearly a 2:1 savings in storage with the recursive convolution method for this case. In addition, the differential method would require updating a fourth-order difference equation in two variables at each time step, while the update equation for the complex recursive accumulator variables given in (20) is comparatively simple.

IV. Demonstration

In order to demonstrate the effectiveness of this approach, the transient fields for a pulsed plane wave normally incident on the boundary between vacuum and a dispersive medium with two second-order Lorentz poles will be calculated. Referring to (11), $P = 2$ and the constants which determine the complex permittivity for the demonstration medium are given in Table I.

The conductivity σ is taken as zero. The frequency-domain complex relative permittivity for this medium is shown in Fig. 1. The corresponding time domain susceptibility from (12) is shown in Fig. 2.

For the one-dimensional FDTD calculation, 1500 spatial cells were used, with the vacuum–medium boundary at cell 500. A first-order Mur [12] absorbing boundary was located at cell 1, while the 1000 cell region containing the dispersive medium provided adequate time delay so that pulses reflecting from the boundary at cell 1500 did not return to the interface before the transients dissipated. The FDTD parameters were $\Delta x = 37.5\ \mu$m and $\Delta t = 0.125$ ps $= \Delta x / c$, where c is the speed of light in vacuum.

The excitation is a Gaussian pulse plane wave. The pulse initially has a spatial width of 256 cells between the 0.001 amplitude points and contains energy at frequencies somewhat beyond 80 GHz.

Spatial plots of electric field versus position for the pulse as it reflects from the interface and propagates in the dispersive medium are shown in Figs. 3–6 after increasing numbers of time steps. The extremely dispersive nature of the medium is clearly indicated.

The accuracy of the method is indicated by computing the reflection coefficient versus frequency from the tran-

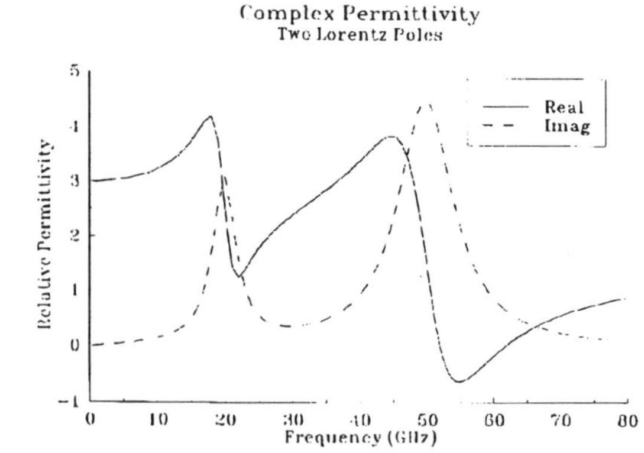

Fig. 1. Complex relative permittivity for the fourth-order dispersive medium with parameters given in Table I.

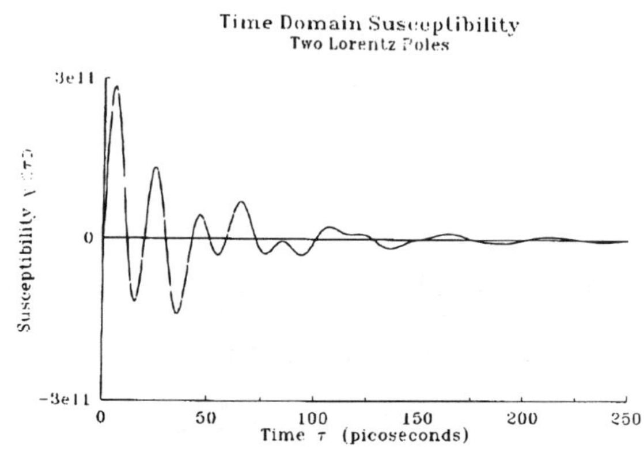

Fig. 2. Time domain susceptibility for the fourth-order dispersive media with parameters given in Table I.

TABLE I
Constants Used for Fourth-Order Dispersive Medium

$\epsilon_s = 3.0$	$\epsilon_\infty = 1.5$
$\omega_1 = 2\pi \cdot 20 \times 10^9$	$\omega_2 = 2\pi \cdot 50 \times 10^9$
$\delta_1 = 0.1\,\omega_1$	$\delta_2 = 0.1\,\omega_2$
$G_1 = 0.4$	$G_2 = 0.6$

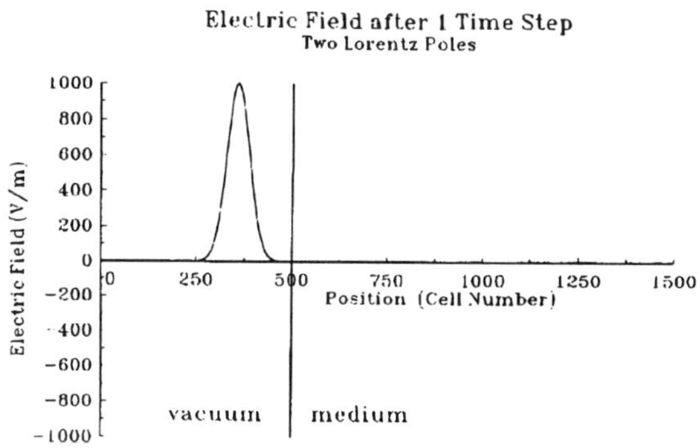

Fig. 3. Electric field versus position for a pulse plane wave incident from vacuum on the fourth-order dispersive medium after one FDTD time step.

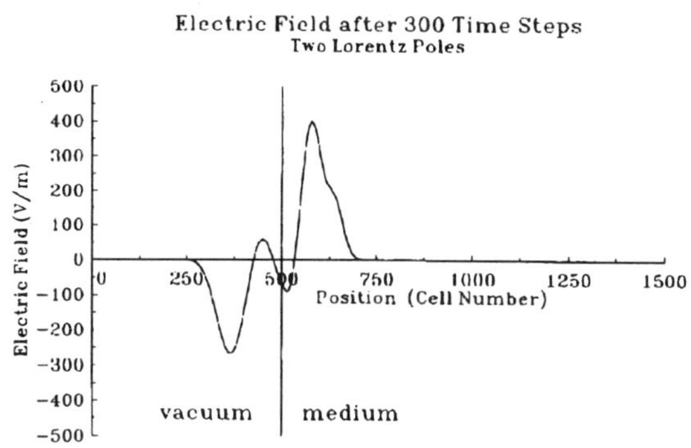

Fig. 4. Electric field versus position for a pulsed plane wave incident from vacuum on the fourth-order dispersive medium after 300 FDTD time steps.

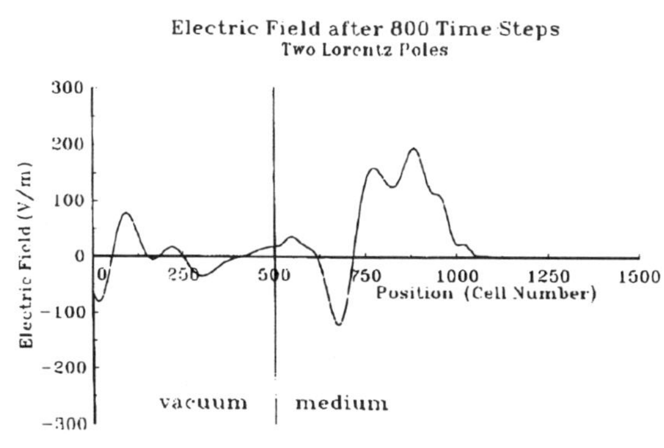

Fig. 5. Electric field versus position for a pulsed plane wave incident from vacuum on the fourth-order dispersive medium after 800 FDTD time steps.

sient fields. The FDTD computations were continued until the transients at the vacuum–medium interface were dissipated. This required approximately 2000 time steps or 250 ps, which then compared with Fig. 2, corresponds to the duration of the susceptibility response. Computation of the reflection coefficient involves taking the Fourier transform of the reflected field versus time at cell 499 and

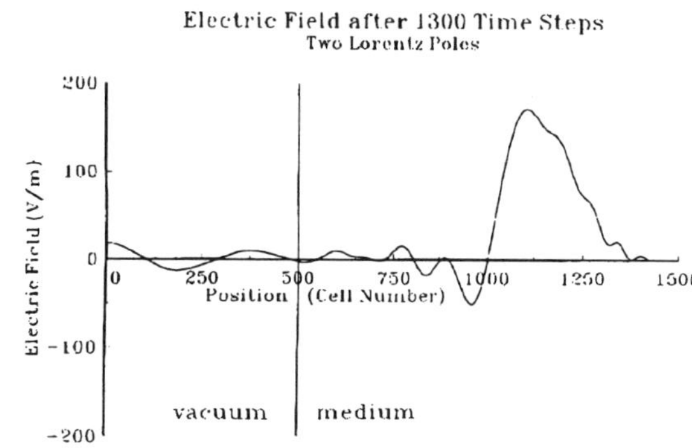

Fig. 6. Electric field versus position for a pulsed plane wave incident from vacuum on the fourth-order dispersive medium after 1300 FDTD time steps.

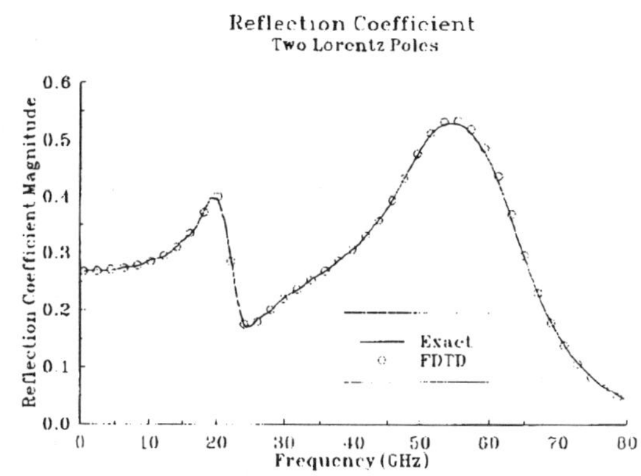

Fig. 7. Reflection coefficient magnitude computed from the transient FDTD fields for fourth-order dispersive medium compared with exact solution.

dividing by the Fourier transform of the incident pulse as it passes through that cell. The reflected field is obtained by subtracting the incident field (with the material region replaced with vacuum) from the total field (total fields are shown in Figs. 3–6) in cell 499 at each time step. The resulting reflection coefficient magnitude and phase are shown in Figs. 7 and 8, and display excellent agreement with the exact solution.

V. Conclusions

In this paper, the recursive convolution approach to applying FDTD to frequency dispersive materials has been extended to materials with an arbitrary number of second-order Lorentz poles in the complex frequency domain permittivity function. The computations for each second-order pole are similar to those given previously for first-order poles. For a first-order pole, one real backstore is required, while for a second-order pole, this backstored variable becomes complex. Results were given for reflection from a dispersive material with two second-order poles. The recursive convolution approach appears to require less computer memory than the differential equation methods which have appeared in the literature, and also appears simpler to implement.

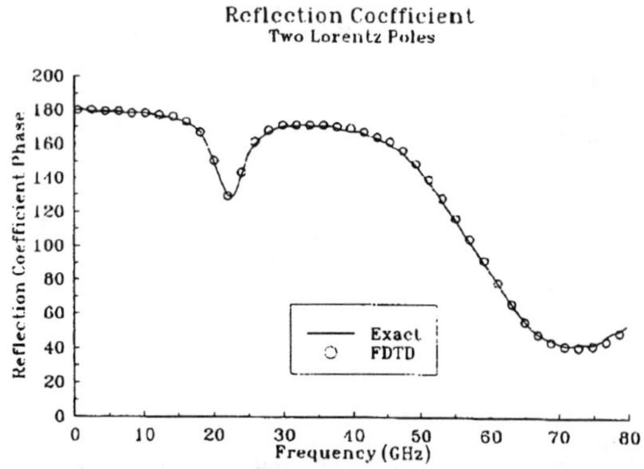

Fig. 8. Reflection coefficient phase computed from the transient FDTD fields for fourth-order dispersive medium compared with exact solution.

REFERENCES

[1] K. S. Yee, "Numerical solution of initial boundary value problems involving Maxwell's equations in isotropic media," *IEEE Trans. Antennas Propagat.*, vol. AP-14, May 1966.

[2] R. Luebbers, F. Hunsberger, K. Kunz, and R. Standler, "A frequency-dependent finite difference time domain formulation for dispersive materials," *IEEE Trans. Electromagn. Compat.*, vol. 32, Aug. 1990.

[3] M. Bui, S. Stuchly, and G. Costache, "Propagation of transients in dispersive dielectric media," *IEEE Trans. Microwave Theory Tech.*, vol. 39, July 1991.

[4] D. Sullivan, "A frequency-dependent FDTD method for biological applications," *IEEE Trans. Microwave Theory Tech.*, to be published.

[5] R. Luebbers, F. Hunsberger, and K. Kunz, "A frequency-dependent time domain formulation for transient propagation in plasma," *IEEE Trans. Antennas Propagat*, Jan. 1991.

[6] T. Kashiwa and I. Fukai, "A treatment of the dispersive characteristics associated with electronic polarization," *Microwave Opt. Technol. Lett.*, vol. 3, June 1990.

[7] R. Joseph, S. Hagness, and A. Taflove, "Direct time integration of Maxwell's equations in linear dispersive media with absorption for scattering and propagation of femtosecond electromagnetic pulses," *Opt. Lett.*, vol. 16, Sept. 15, 1991.

[8] C. A. Balanis, *Advanced Engineering Electromagnetics*. New York: Wiley, 1989.

[9] A. Papoulos, *The Fourier Integral and Its Application*. New York: McGraw-Hill, 1962.

[10] F. Hunsberger, R. Luebbers, and K. Kunz, "Finite-difference time-domain analysis of gyrotropic media I: Magnetized plasma," *IEEE Trans. Antennas Propagat.*, accepted for publication, Nov. 1991.

[11] F. Hunsberger, "Extension of the finite-difference time-domain method to gyrotropic media," Ph.D. dissertation, Dept. Elect. Comput. Eng., Pennsylvania State Univer., University Park, May 1991.

[12] G. Mur, "Absorbing boundary conditions for finite-difference approximation of the time-domain electromagnetic-field equations," *IEEE Trans. Electromagn. Compat.*, vol. EMC-23, Nov. 1981.

FD-TD Modeling of Digital Signal Propagation in 3-D Circuits With Passive and Active Loads

Melinda Piket-May, *Member, IEEE*, Allen Taflove, *Fellow, IEEE*, and John Baron, *Student Member, IEEE*

***Abstract*—Most existing computer-aided circuit design tools are limited when digital clock speeds exceed several hundred MHz. These tools may not deal effectively with the physics of UHF and microwave electromagnetic wave energy transport along metal surfaces such as ground planes or in the air away from metal paths that are common at or above this frequency range. In this paper, we discuss full-wave modeling of electronic circuits in three dimensions using the finite-difference time-domain (FD-TD) solution of Maxwell's equations. Parameters such as stripline complex line impedance, propagation constant, capacitance per unit length and inductance per unit length can be easily computed as a function of frequency. We also discuss FD-TD Maxwell's equations computational modeling of lumped-circuit loads and sources in 3-D, including resistors and resistive voltage sources, capacitors, inductors, diodes, and transistors. We believe that this approach will be useful in simulating the large-signal behavior of very high-speed nonlinear analog and digital devices in the context of the full-wave time-dependent electromagnetic field.**

I. Introduction

In the past, digital logic designers have not seriously considered the electromagnetic details of their systems. Clock speeds have been low enough and logic levels high enough that electromagnetic problems have been minor. But as voltage levels drop below one volt and clock speeds increase to 100 MHz and above, key electromagnetics issues must be addressed.

For example, most existing computer-aided circuit design tools (primarily SPICE) are limited at digital clock speeds exceeding 300 MHz. These tools do not deal with the physics of UHF/microwave electromagnetic wave energy transport along metal surfaces like ground planes, or in the air away from metal paths, that are common above this frequency. Effectively, electronic digital systems develop substantial analog wave effects when clock rates are high enough, with full-wave electromagnetic field phenomena markedly affecting the propagation, cross-talk, and radiation of electronic digital pulses in typical structures such as multilayer circuit boards and multichip modules. The general theme of this paper is that a synthesis of electromagnetic fields and waves and electronic circuit devices will eventually become necessary to design very high-speed digital systems.

The specific research summarized in this paper addresses numerical modeling of 3-D digital interconnects with passive and active loads directly from Maxwell's equations. We utilize the finite-difference time-domain (FD-TD) method, a simple, robust numerical technique for direct time integration of Maxwell's equations that has become the means to model a wide variety of electromagnetic wave problems [1]. Using the basic algorithm introduced by Yee [2], FD-TD implements the space derivatives of the curl operators via finite differences in regular interleaved (dual) Cartesian space meshes for the electric and magnetic fields. Simple leapfrog time integration is employed.

The literature indicates that 3-D FD-TD modeling of metallic digital interconnects has commenced. Liang *et al.* [3] modeled picosecond pulse propagation along co-planar waveguides above a gallium arsenide substrate. Shibata and Sano [4] modeled propagation along metal-insulator-semiconductor lines. Lam *et al.* [5] modeled digital signal propagation and radiation for VLSI packaging, and Maeda *et al.* [6] modeled digital pulse propagation through vias in a three-layer circuit board.

Work is also being done in FD-TD modeling of passive and active devices that can either source or terminate metallic transmission lines. Two groups reported incorporation of self-consistent models of charge transport into a 3-D FD-TD solver to model picosecond pulse generation by an optically excited gallium arsenide device: Sano and Shibata [7] with a drift-diffusion charge-transport model, and El-Ghazaly [8] with a Monte Carlo model. A third group Sui *et al.* [9], developed lumped-circuit models of typical passive and active circuit elements (resistors, inductors, capacitors, diodes, and transistors) incorporated into a 2-D FD-TD solver for exploring electromagnetic modeling of nanosecond-regime circuits.

The work of this paper follows directly from [6] and [9]. In Section III, we illustrate the power of FD-TD modeling to deduce a classical electromagnetic compatibility problem, ground loop current flow, in what may be the most complex 3-D passive circuit model so far, a module consisting of a stack of multilayer circuit boards penetrated by scores of via pins. Using a uniform space resolution of 0.004 inch, each layer, via, and pin of every circuit board was modeled. 60-million vector field unknowns were solved per modeling run, a factor of perhaps 100 times larger than the capacity of the largest SPICE or finite-element CAD tool available.

Manuscript teceived January 18, 1993; revised October 11, 1993. This work was supported in part by Cray Research, Inc., Eagan, Minnesota, and in part by Los Alamos National Laboratory, Los Alamos, NM, under a Cooperative Research and Development Agreement.

M. Piket-May is with the Department of Electrical and Computer Engineering, University of Colorado, Boulder, Colorado.

A. Taflove is with the Department of Electrical Engineering and Computer Science, McCormick School of Engineering, Northwestern University, Evanston, IL 60208-3118 USA.

J. Baron is with the Department of Electrical Engineering, Stanford University, Stanford, CA 94305-4055 USA.

IEEE Log Number 9402931.

Reprinted from *IEEE Transactions on Microwave Theory and Techniques*, pp. 1514-1523, August 1994.

In Section IV, we improve and extend the work of [9] to 3-D,[1] introducing a semi-implicit algorithm that permits the large-signal behavior of nonlinear circuit devices to be simulated in a numerically stable manner in the context of the FD-TD Maxwell's equations solver. We report FD-TD electromagnetic wave models of the resistor, resistive voltage source, capacitor, inductor, diode, and bipolar junction transistor.

II. FD-TD Algorithm and Extension

A. Basic FD-TD Formulation

The derivation of the basic FD-TD equations used to solve Maxwell's equations has been documented extensively [1], [2], [10] so it will not be repeated. In all of the cases to be discussed, a uniform, 3-D Cartesian Yee mesh is used with a unit cell of dimensions $\Delta x, \Delta y, and \Delta z$. Numerical stability is maintained by bounding the time step [11].

Consider Maxwell's curl $\overline{H}$equation, suitable for time-stepping the electric field

$$\nabla \times \overline{H} = \overline{J}_c + \frac{\partial \overline{D}}{\partial t} \tag{1a}$$

where the electric conduction current is $\overline{J}_c = \sigma \overline{E}$ and the electric displacement is $\overline{D} = \varepsilon \overline{E}$. Using central differencing with the standard Yee subscript/superscript notation for space and time coordinates, respectively, (1a) becomes

$$E_z\Big|_{i,j,k}^{n+1} = \left(\frac{1 - \frac{\sigma_{i,j,k}\Delta t}{2\epsilon_{i,j,k}}}{1 + \frac{\sigma_{i,j,k}\Delta t}{2\epsilon_{i,j,k}}} \right) E_z\Big|_{i,j,k}^{n} + \left(\frac{\frac{\Delta t}{\epsilon_{i,j,k}}}{1 + \frac{\sigma_{i,j,k}\Delta t}{2\epsilon_{i,j,k}}} \right) \left(\frac{H_y\Big|_{i+1/2,j,k}^{n+1/2} - H_y\Big|_{i-1/2,j,k}^{n+1/2}}{\Delta x} + \frac{H_x\Big|_{i,j-1/2,k}^{n+1/2} - H_x\Big|_{i,j+1/2,k}^{n+1/2}}{\Delta y} \right) \tag{1b}$$

An important observation is that all $\bar{H}$ quantities on the right-hand side of (1b) are at time step $n + 1/2$, centered in time relative to the stored electric field, $E_z\Big|_{i,j,k}^{n}$, and the newly updated electric field, $E_z\Big|_{i,j,k}^{n+1}$. Further, the bracketed coefficients are derived assuming that J_c is also evaluated at time step $n + 1/2$, taking

$$J_c\Big|_{i,j,k}^{n+1/2} = \sigma_{i,j,k} E_z\Big|_{i,j,k}^{n+1/2} = \frac{\sigma_{i,j,k}}{2}\left(E_z\Big|_{i,j,k}^{n} + E_z\Big|_{i,j,k}^{n+1}\right) \tag{2a}$$

We denote this as the "semi-implicit formulation" for the conduction current in that this current relies in part upon the updated electric field to be determined as a result of the time-stepping, and yet does not result in a system of simultaneous equations. This results in a numerically stable algorithm for arbitrary positive values of σ [1], [10].

For convenience in the discussions of this paper, we shall assume that all circuit components are located in a free-space region ($\epsilon = \epsilon_o, \sigma = 0, \bar{J}_C = 0$). Additional notational convenience is provided by denoting the Yee finite-difference analog to the curl of $\bar{H}$ observed at $E_z|_{i,j,k}$ as

$$\nabla \times H_{i,j,k}^{n+1/2} = \frac{H_y\Big|_{i+1/2,j,k}^{n+1/2} - H_y\Big|_{i-1/2,j,k}^{n+1/2}}{\Delta x} + \frac{H_x\Big|_{i,j-1/2,k}^{n+1/2} - H_x\Big|_{i,j+1/2,k}^{n+1/2}}{\Delta y} \tag{2b}$$

Then, (1b) can be rewritten very simply as

$$E_z\Big|_{i,j,k}^{n+1} = E_z\Big|_{i,j,k}^{n} + \frac{\Delta t}{\epsilon_o} \nabla \times H\Big|_{i,j,k}^{n+1/2}. \tag{2c}$$

B. Extended FD-TD Formulation

The electric field time-stepping algorithm is now modified to allow for the addition of lumped linear and nonlinear circuit elements. The basis of this formulation, reported in [9] for 2-D problems, showed that lumped circuit elements can be accounted for in Maxwell's equations by adding a lumped electric current density term, J_L, to the conduction and displacement currents on the right hand side of (1a). Equation (1a) now becomes

$$\nabla \times \overline{H} = \overline{J}_c + \frac{\partial \overline{D}}{\partial t} + \overline{J}_L. \tag{3a}$$

We assume that a lumped element is located in free-space at the field$E_z|_{i,j,k}$; is z-oriented in the grid; and provides a local current density that is related to the total element current, I_L, as

$$J_L = \frac{I_L}{\Delta x \Delta y}. \tag{3b}$$

Here, I_L is possibly a time-derivative, time-integral, scalar multiple, or nonlinear function of the electric potential,$V = E_z|_{i,j,k}\Delta z$, developed across the element. Note that the assumed positive direction of I_L is $+z$. Then, the following modified version of (2c) suffices to specify the presence of the lumped circuit element in the electromagnetic field grid

$$E_z\Big|_{i,j,k}^{n+1} = E_z\Big|_{i,j,k}^{n} + \frac{\Delta t}{\epsilon_o} \nabla \times H\Big|_{i,j,k}^{n+1/2} - \frac{\Delta t}{\epsilon_o \Delta x \Delta y} I_L^{n+1/2}. \tag{4}$$

In the key departure from the work of [9], we have defined in (4) the lumped current as being evaluated at the mid time step, $n+1/2$, just as we did in (2a) for the conduction current. Since the lumped current is a function of the electric field at the circuit element, this requires averaging the values of $E_z\Big|_{i,j,k}^{n+1}$ and$E_z\Big|_{i,j,k}^{n}$, an operation that yields a numerically-stable semi-implicit time-stepping algorithm. Generalization of all of this to x and y orientations of a lumped element is straightforward by permuting the coordinate subscripts of the field quantities.

III. Interconnect Geometries Without Lumped Loads

A. Parallel Coplanar Microstrips

Preliminary studies were performed to establish the validity of FD-TD results for characteristic impedance and delay of parallel co-planar microstrips. The first studies were done by

[1] Note added in proof. See also Tsuei *et al.* [16]. for an independent expansion of the work of [9] to 3–D.

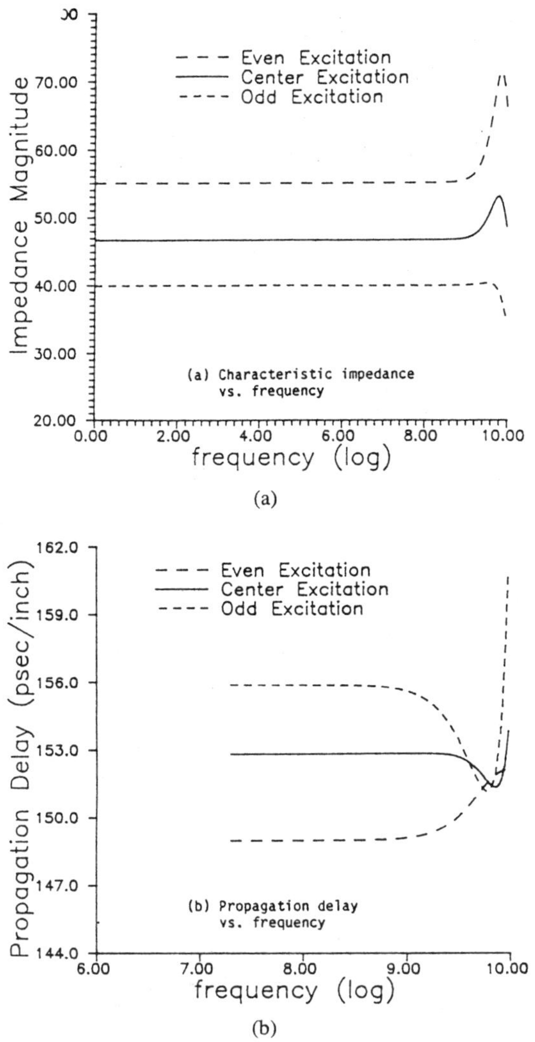

Fig. 1. FD-TD computed properties of three parallel coplanar microstrips: (a) characteristic impedance versus frequency; (b) propagation delay versus frequency.

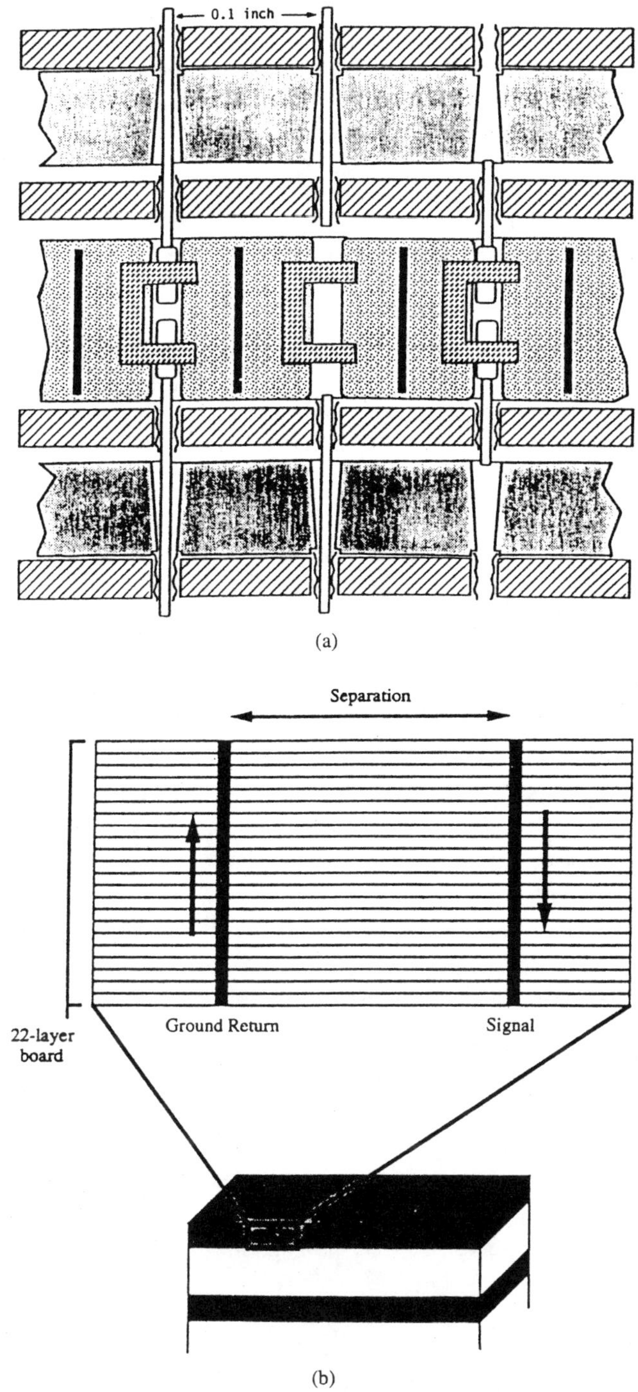

Fig. 2. 3-D connector module: (a) stack of four multilayer circuit boards perforated by scores of vertical via pins spaced on 0.1-inch centers; (b) generic sketch of digital signal and ground return paths in the top multilayer circuit board.

setting up FD-TD grids for single x-directed microstrips of negligible metalization thickness and a variety of widths over dielectric substrates in the order of 1 mil (0.001 inch) thick. Grid cell size δ for these cases was in the order of 0.1 mil with Δt in the order of 4.2 fs. In all cases, conductors were "on grid," i.e., located at planar loci of tangential E components in the FD-TD mesh that were set to zero for all time steps, and with zeroed tangential E components on the conductor edges.

Mur absorbing boundary conditions [12] were set up at the outer lattice planes. To effectively decouple the Mur boundaries from the localized fields within and near the microstrip geometries, the FD-TD space lattice extended at least 20 free-space cells beyond the microstrip structures in all directions.

Excitation of a microstrip was provided by specifying a Gaussian-pulse time history for a group of co-linear electric field components (usually E_z) bridging the gap between the ground plane and the strip conductor at the desired source location. The characteristic impedance, Z_0, of the microstrip was then found by forming the ratio of the discrete Fourier transforms of the line voltage and current, V and I, obtained from the resulting propagating E and H fields:

$$V(t, x_i) = \int_{C_V} \overline{E}(t, x_i) \cdot d\bar{l}, \qquad I(t, x_i) = \oint_{C_I} \overline{H}(t, x_i) \cdot d\bar{l}$$

$$Z_0(\omega, x_i) = \mathcal{F}[V(t, x_i)] / \mathcal{F}[I(t, x_i)]. \tag{5}$$

Here, the contour path for V extends from the ground plane to the microstrip, while the contour path for I for extends around the strip conductor at its surface. This method showed the FD-TD computed values of Z_0 to be virtually independent of frequency up to 1.0 GHz, and in the order of 1% agreement with textbook values [13].

Next, using similar FD-TD grid resolutions, we considered single microstrips with finite metalization thickness, possibly fully embedded within a dielectric layer. Here, FD-TD predictions were compared to measurements. In one example, a 1.1-mil-thick, 1.4-mil-wide metal strip was assumed to be suspended 1.1 mils above a large ground plane within a 3.1-mil-thick dielectric layer having $\varepsilon_r = 3.2$. The FD-TD numerical simulation predicted a flat characteristic impedance, Z_0, of 48 Ω up to 1.0 GHz, agreeing with the experimental results to within the measurement uncertainty (about 0.2 Ω). The computed variation of Z_0 above 1 GHz was found to be ± 2 Ω. Similar excellent agreement was found for the propagation delay, where the experimental value was 150.5 $\pm$ 1.5 psec/inch, while the FD-TD prediction was 149.5 psec/inch.

We then studied the impedance and propagation delay of three parallel, co-planar microstrips having finite metalization thickness. Each microstrip had the geometry discussed above and was separated by 3.6 mils from the adjacent line(s). Even-mode results were obtained with all three strips excited simultaneously with the same polarity, while odd-mode results were obtained with the two outer strips excited with the opposite polarity relative to the center strip. Results were also observed for the center strip excited with the two outer strips floating. Fig. 1 shows Z_0 and the propagation delay predicted by FD-TD for all three cases. These results were obtained with $\delta = 0.1$ mil, $\Delta t = 4.2$ fs, and a lattice size of $200 \times 134 \times 42$ cells. It is clear from the results that signal propagation on adjacent lines significantly influences the effective impedance of the line, and to a lesser degree affects the propagation delay. The FD-TD results were confirmed by laboratory studies which showed an approximate 7 Ω elevation of the characteristic impedance for the even-mode excitation, and an approximate 7 Ω reduction of the impedance for the odd-mode excitation (both from dc to about 1 GHz). At frequencies approaching 10 GHz, the evolution of non-transmission-line modes within the trio of parallel lines suffices to progressively disturb the observed Z_0 and propagation delay.

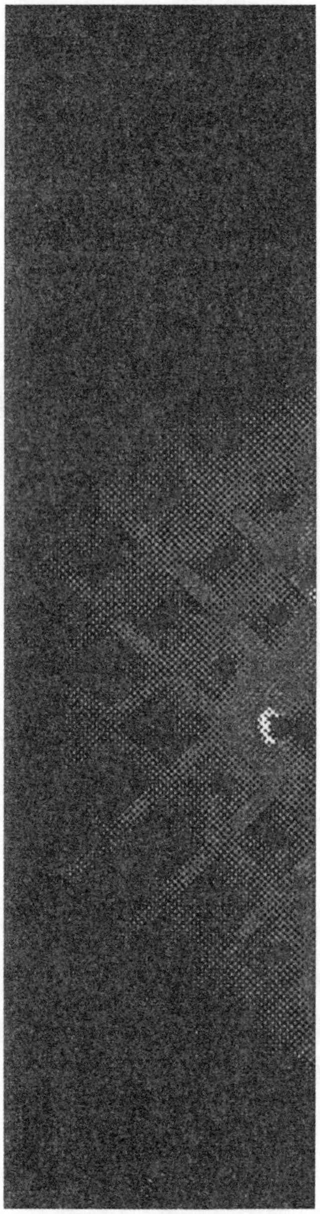

Fig. 3. Color plate showing the plan view of an outwardly propagating electromagnetic wave within the top multilayer circuit board generated by the passage of a subnanosecond pulse down a single vertical via pin. Color scale: yellow = maximum; green = moderate; dark blue = negligible.

B. Multilayered Interconnect Modeling Example

This section illustrates the power of FD-TD modeling to investigate a classical electromagnetic compatibility problem, ground loop current flow. In what may be the most complex 3-D passive circuit modeled by any means so far, we constructed a high-resolution FD-TD model of sub-nanosecond digital pulse propagation and crosstalk behavior in a real-world computer module consisting of a stack of four multilayer printed circuit boards (each with >10 metal-dielectric-metal layers) penetrated by scores of via pins forming a connector. A uniform space cell size of $\delta = 4$ mils was used with time step $\Delta t = 169$ fs. The space cell was the thickness of a single layer in the circuit boards, and 1/3 the diameter of each 12-mil-diameter circular via. In this manner, *each layer, via, and pin* of every circuit board and every connector was modeled in the final simulation, resulting in almost 60-million vector field unknowns solved. Fig. 2(a) is a diagram of the full connector module, where each multilayer circuit board is shown as a cross-hatched horizontal slab. Vertical via pins located on 0.1-inch centers penetrate the stack of boards and connectors throughout the module. Fig. 2(b) is a generic sketch of a digital signal path and a ground-return path in the top multilayer circuit board.

Single Multilayer Circuit Board, Early Time Response: The first step in the study was to model in detail the early-time response of the top multilayer circuit board to an impulse of current (90-ps Gaussian pulse, spectral width about 20 GHz) propagating down a single vertical via pin. The via pin (in the circular via hole) was excited by pulsing a vertical electric field component, E_z, in the FD-TD lattice just above the pin

Fig. 4. Color plate of the early-time coupling of magnetic fields from the excited via pin to the adjacent unexcited via pins as seen in a vertical cut through the top multilayer board and connector of the stack of Fig. 2. Color scale: red = maximum; yellow = moderate; green = low-level; dark blue = negligible.

and below a simulated ground strap connected to the outer via pins designated as ground returns. The short duration of the pulse was selected primarily to permit time resolution of layer-to-layer pulse reflection effects within the circuit board. We utilized a 300 × 92 × 52 cell FD-TD grid having 1.44 million cells (8.6 million unknowns, 19 MW of memory). Using a single processor of a Cray YMP-8, this grid size required 0.36 sec per time step. The power and signal metalizations of the circuit board were assumed to be infinitely thin.

Fig. 3 is a color plate showing the plan view of $|\overline{H}|$ of the outwardly propagating electromagnetic wave within the board generated by the passage of the pulse down the via pin. Although the relatively intense magnetic field adjacent to the excited via (shown by the yellow color) is quite localized, moderate-level magnetic fields (shown by light blue) emanate throughout the entire transverse cross section of the board and link *all* of the adjacent via pins, shown as dark dots in a diamond pattern. The complete color video of the dynamics of this phenomenon shows *repeated bursts* of outward propagating waves linking all points within transverse cross sections of the board as the pulse passes vertically through the multiple metal-dielectric-metal layers of the board. The resulting pin-to-pin cross-talk is vividly illustrated in the color plate of Fig. 4, which depicts $|\overline{H}|$ in a vertical cut through the top multilayer board and connector of the stack of Fig. 2. The coupling of magnetic fields from the excited via pin to the adjacent unexcited via pins is clearly indicated.

Complete Four-Board Connector Module, Late Time Response: The final step in the study was to model the response of the complete four-board connector module to the impulsive excitation of a single vertical via pin. Keeping the same space resolution as for the single-board model, we enlarged the FD-TD grid to 300 × 92 × 340 cells to contain the three additional multilayer circuit boards and connectors. The new grid totalled 9.4 million cells with 56 million vector field unknowns, and required 117 MW of memory. Using a single processor of a Cray YMP-8, this grid required 2.13 sec per time step, or 71 minutes for a complete run of 2000 time steps. (Using all eight of the Cray YMP processors reduced the running time by almost 8:1 to only 9 minutes.) Again, all power and signal metalizations of the circuit boards were assumed to be infinitely thin.

Color videos of pulse propagation and crosstalk were constructed to illustrate these phenomena. Fig. 5 is a color plate showing the magnitude and direction of late-time currents flowing along the vertical cross section of the complete connector module of Fig. 2 for a subnanosecond pulse assumed to excite a single vertical via pin in the top multilayer board. The currents were calculated in a post-processing step by numerically evaluating the curl of the magnetic field obtained from the 3-D FD-TD model. The color red was selected to denote downward-directed current, while the color green was selected to denote upward-directed current. At the time of this visualization, current had proceeded down the excited via through all four boards and all three connectors. But, upward-directed (green) current is seen to flow on the adjacent vias. This represents *undesired ground-loop coupling* to the circuits using these vias.

IV. Interconnect Geometries With Lumped Loads

In this section, we report 3-D FD-TD modeling of the connection of linear and nonlinear lumped loads to microstrip

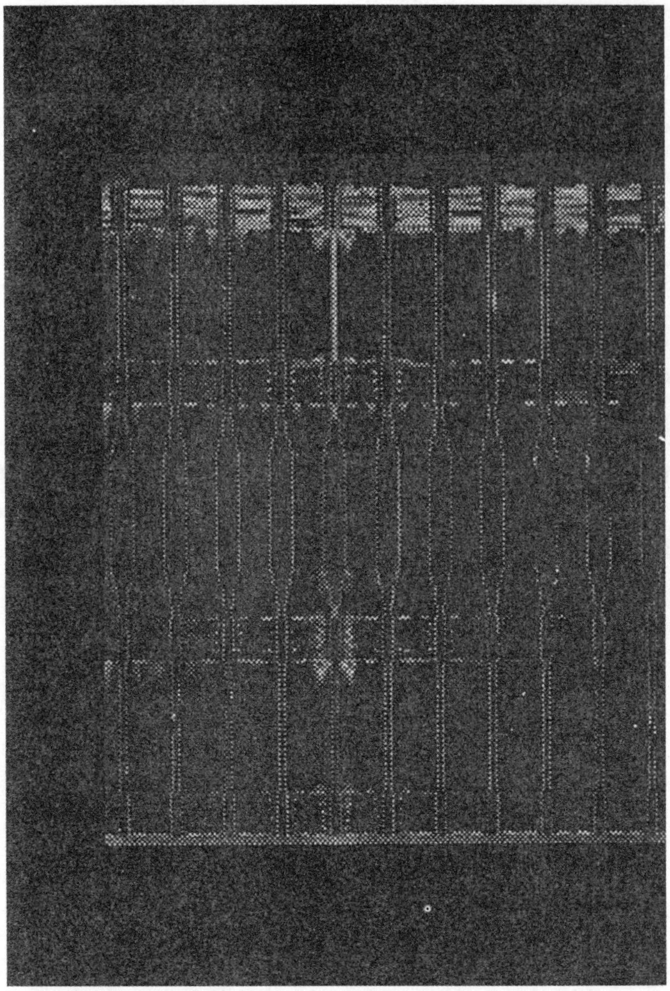

Fig. 5. Color plate showing the magnitude and direction of late-time currents flowing along the vertical cross section of the complete multiboard module of Fig. 2 for a subnanosecond digital pulse assumed to excite a single vertical via pin in the top multilayer board. Color scale: red = net downward-directed current; green = net upward-directed current; dark blue = negligible.

interconnects. Fig. 6 depicts the microstrip geometry considered. For convenience in the discussion, the microstrip is assumed to be oriented in the x-direction and the lumped load is assumed to be oriented in the z-direction. Extension of the theory to other Cartesian orientations in the FD-TD lattice is straightforward.

Lattice resolution δ in the range 1 to 8 mils was used in the simulations to study the effect of varying the number of cells spanning the strip conductor and the gap between it and the ground plane. For excitations with spectral components up to 1 GHz, the results were found to be only a very weak function of the grid space increment δ in this range. In all cases, conductors were "on grid," i.e., located at planar loci of tangential E components in the FD-TD lattice that were set to zero for all time steps, and with zeroed tangential E components on the conductor edges. Mur absorbing boundaries were used as specified in Section III.

A resistive source of the type discussed in Section B below was used in all of the FD-TD simulations of this section. The source resistance was matched to the line impedance so that retroreflections at the source were minimized.

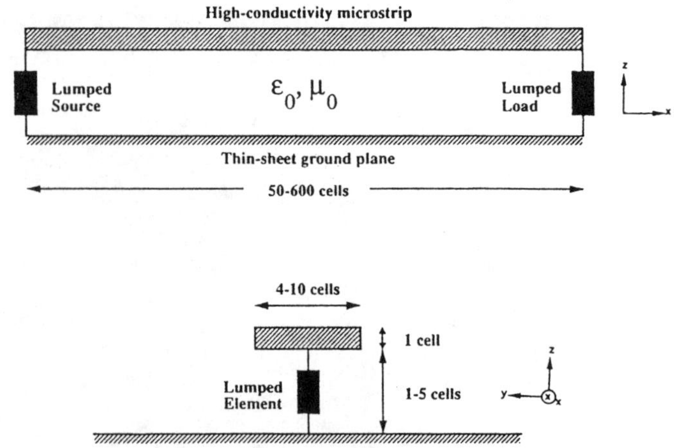

Fig. 6. Generic geometry used for 3-D FD-TD models of a linear or nonlinear lumped element terminating a stripline.

A. The Resistor

When studying microstrips it is very useful to have the capability to terminate the line in a resistive load, very likely matched to the characteristic impedance of the line. One method that we studied for terminating the line in an FD-TD grid is to insert a physical resistor block into the microstrip model using the relation $R = \rho L/A$ (where R is the desired resistance, ρ is the block's resistivity, and L and A represent the length and cross sectional area of the block) to select the parameters of the insert. The second method is to insert a numerical lumped element in an extension to 3-D of that described in [9]. Here, assuming a z-directed resistor located in free-space at the field component $E_z|_{i,j,k}$, the voltage-current characteristic that describes its behavior in a semi-implicit manner appropriate for stable operation of the FD-TD field solver is

$$I_z|_{i,j,k}^{n+1/2} = \frac{\Delta z}{2R}\left(E_z|_{i,j,k}^{n+1} + E_z|_{i,j,k}^{n}\right),$$
$$J_L = \frac{I_z|_{i,j,k}^{n+1/2}}{\Delta x \Delta y} \tag{6a}$$

where R is the value of the resistance and $\Delta x, \Delta y$ and Δz are the grid increments in the x, y and z directions. The corresponding time-stepping relation for $E_z|_{i,j,k}$ is

$$E_z|_{i,j,k}^{n+1} = \left(\frac{1 - \frac{\Delta t \Delta z}{2R\epsilon_o \Delta x \Delta y}}{1 + \frac{\Delta t \Delta z}{2R\epsilon_o \Delta x \Delta y}}\right) E_z|_{i,j,k}^{n} + \left(\frac{\frac{\Delta t}{\epsilon_o}}{1 + \frac{\Delta t \Delta z}{2R\epsilon_o \Delta x \Delta y}}\right) \nabla \times H|_{i,j,k}^{n+1/2}. \tag{6b}$$

To compare the impedance match provided by the physical and numerical resistors in the FD-TD mesh, we modeled a terminated 50-Ω transmission line similar to that of Section III-A excited by a 90-ps Gaussian pulse (spectral width of 20 GHz). After performing a discrete Fourier transform of the reflected pulse for each case, we found that both the physical and numerical resistors provided reflection coefficients of less than 1% up to 1 GHz. Fig. 7 shows the magnitude and phase of

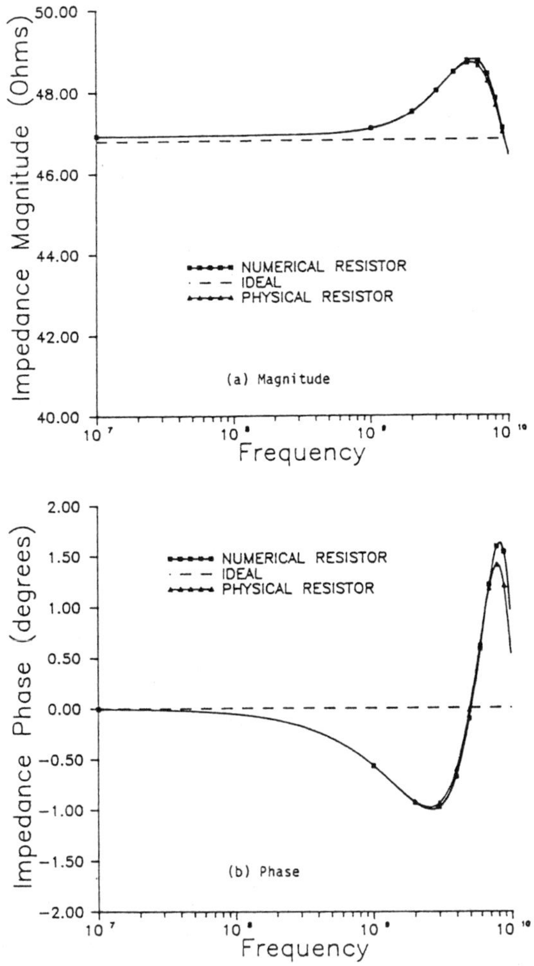

Fig. 7. Agreement of FD-TD computed effective load impedance (low frequency to 10 GHz) for the numerical resistor and the physical resistor: (a) magnitude; (b) phase.

the effective load impedance versus the ideal (matched) load impedance for the two cases. Over the span dc to 10 GHz, the maximum deviation is (positive +2Ω, –0.5Ω) and (+1.75°, –1°). The fact that the physical and numerical resistor results agree so well implies that the numerical resistor model does an excellent job of picking up the parasitic capacitance or inductance that may be present between the terminals of the physical resistor.

It is important to note that implementing (6b) at two adjacent electric field components between the microstrip signal line and ground plane simulates the presence of two parallel resistors terminating the line. The effective load resistance is thereby halved.

B. The Resistive Voltage Source

With the ability to model lumped elements in the context of FD-TD, it is a simple matter to model a nonreflecting (matched) source as a resistive voltage source in an extension to 3-D of that described in [9]. Again assuming a z-directed lumped element located in free-space at $E_z|_{i,j,k}$, the voltage-current characteristic that describes the behavior of a resistive voltage source in a semi-implicit manner is

$$I_z|_{i,j,k}^{n+1/2} = \frac{\Delta z}{2R_s}\left(E_z|_{i,j,k}^{n+1} + E_z|_{i,j,k}^{n}\right) + \frac{V_s^{n+1/2}}{R_S},$$
$$J_L = \frac{I_z|_{i,j,k}^{n+1/2}}{\Delta x \Delta y} \tag{7a}$$

where $V_s^{n+1/2}$ is the source voltage and R_s is the internal source resistance. The corresponding time-stepping relation for $E_z|_{i,j,k}$ is

$$E_z|_{i,j,k}^{n+1} = \left(\frac{1 - \frac{\Delta t \Delta z}{2R_S \epsilon_o \Delta x \Delta y}}{1 + \frac{\Delta t \Delta z}{2R_S \epsilon_o \Delta x \Delta y}}\right) E_z|_{i,j,k}^{n} + \left(\frac{\frac{\Delta t}{\epsilon_o}}{1 + \frac{\Delta t \Delta z}{2R_S \epsilon_o \Delta x \Delta y}}\right) \nabla \times H|_{i,j,k}^{n+1/2} + \left(\frac{\frac{\Delta t}{R_S \epsilon_o \Delta x \Delta y}}{1 + \frac{\Delta t \Delta z}{2R_S \epsilon_o \Delta x \Delta y}}\right) V_S^{n+1/2}. \tag{7b}$$

C. The Capacitor

We next consider the insertion of a numerical lumped capacitor into the FD-TD grid in an extension to 3-D of that described in [9]. Again assuming a z-directed lumped element located in free-space at $E_z|_{i,j,k}$, the voltage-current characteristic that describes the capacitor's behavior in a semi-implicit manner is

$$I_z|_{i,j,k}^{n+1/2} = \frac{C\Delta z}{\Delta t}\left(E_z|_{i,j,k}^{n+1} - E_z|_{i,j,k}^{n}\right),$$
$$J_L = \frac{I_z|_{i,j,k}^{n+1/2}}{\Delta x \Delta y} \tag{8a}$$

where C is the value of the capacitance. This formulation differs from that of [9] in that the electric field samples are separated here by one time step rather than two. In this manner, we are consistent with the electric field sampling used for the numerical resistor described previously. The corresponding time-stepping relation for $E_z|_{i,j,k}$ is

$$E_z|_{i,j,k}^{n+1} = E_z|_{i,j,k}^{n} + \left(\frac{\frac{\Delta t}{\epsilon_o}}{1 + \frac{C\Delta z}{\epsilon_o \Delta x \Delta y}}\right) \nabla \times H|_{i,j,k}^{n+1/2} \tag{8b}$$

For the parallel combination of a capacitor, C, and resistor, R, located at $E_z|_{i,j,k}$, the results of (6a), (6b), (8a) and (8b) can be readily combined to yield the following time-stepping relation:

$$E_z|_{i,j,k}^{n+1} = \left(\frac{1 - \frac{\Delta t \Delta z}{2R\epsilon_o \Delta x \Delta y} + \frac{C\Delta z}{\epsilon_o \Delta x \Delta y}}{1 + \frac{\Delta t \Delta z}{2R\epsilon_o \Delta x \Delta y} + \frac{C\Delta z}{\epsilon_o \Delta x \Delta y}}\right) E_z|_{i,j,k}^{n} + \left(\frac{\frac{\Delta t}{\epsilon_o}}{1 + \frac{\Delta t \Delta z}{2R\epsilon_o \Delta x \Delta y} + \frac{C\Delta z}{\epsilon_o \Delta x \Delta y}}\right) \nabla \times H|_{i,j,k}^{n+1/2}. \tag{8c}$$

To test this FD-TD modeling approach, a variety of numerical capacitive loads were modeled at the end of a long (but ≈1-mil scale in the transverse plane) 50Ω microstrip line subjected

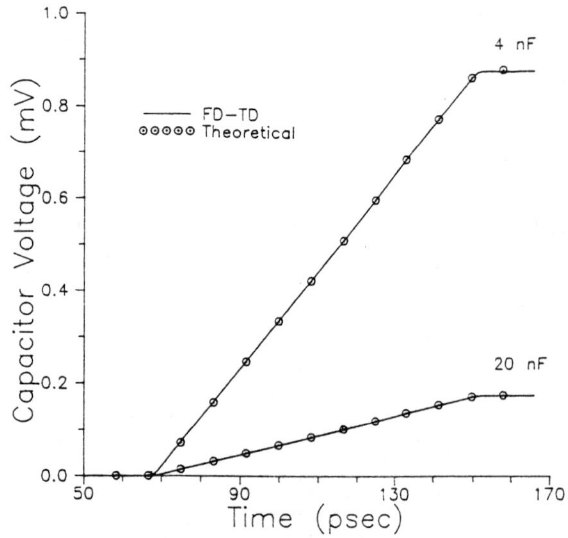

Fig. 8. Agreement of FD-TD and exact solutions for the voltage across the capacitor terminating the stripline for two values of the capacitor (stepwise incident pulse).

to a rectangular step-pulse excitation 1000 time steps long. The FD-TD computed voltage response vs. time across each capacitor was then compared to the exact theoretical response. Results are shown in Fig. 8 for microstrips terminated with 4 nF and 20 nF capacitors. The theoretical and FD-TD curves are indistinguishable.

D. The Inductor

We next consider the insertion of a numerical lumped inductor into the FD-TD grid in an extension to 3-D of that described in [9]. Again assuming a z-directed lumped element located in free-space at $E_z|_{i,j,k}$, the voltage-current characteristic that describes the inductor's behavior in a manner appropriate for stable operation of the FD-TD field solver is

$$I_z|_{i,j,k}^{n+1/2} = \frac{\Delta z \Delta t}{L} \sum_{m=1}^{n} E_z|_{i,j,k}^{m},$$
$$J_L = \frac{I_z|_{i,j,k}^{n+1/2}}{\Delta x \Delta y} \tag{9a}$$

where L is the value of the inductance. This formulation differs from that of [9] in that the electric field samples are summed only through time step n, which is consistent with the observation of the lumped current at time step $n + 1/2$ that we employ throughout this development. The corresponding time-stepping relation for $E_z|_{i,j,k}$ is

$$E_z|_{i,j,k}^{n+1} = E_z|_{i,j,k}^{n} + \frac{\Delta t}{\epsilon_o} \nabla \times H|_{i,j,k}^{n+1/2} - \frac{\Delta z (\Delta t)^2}{\epsilon_o L \Delta x \Delta y} \sum_{m=1}^{n} E_z|_{i,j,k}^{m}. \tag{9b}$$

E. The Diode

The current through a lumped-circuit diode is expressed by:

$$I_d = I_0 \left[e^{(qV_d/kT)} - 1 \right] \tag{10}$$

where q is the charge of an electron, V_d is the voltage across the diode, k is Boltzmann's constant, and T is the temperature in degrees Kelvin. According to the 2-D study of [9], if one assumes a z-directed diode located in free-space at $E_z|_{i,j,k}$, the electric field time-stepping relation is given by

$$E_z|_{i,j,k}^{n+1} = E_z|_{i,j,k}^{n} + \frac{\Delta t}{\epsilon_o} \nabla \times H|_{i,j,k}^{n+1/2} - \frac{\Delta t}{\epsilon_o \Delta x \Delta y} I_0 \left[e^{(-qE_z|_{i,j,k}^{n} \Delta z/\kappa T)} - 1 \right]. \tag{11}$$

However, it has been determined that this expression yields a numerically unstable algorithm for diode voltages larger than 0.8 volts due to its explicit formulation which employs the previously computed electric field in the exponential.

We have found that a numerically stable FD-TD algorithm for the lumped diode can be realized in 3-D by using the semi-implicit update strategy for the electric field

$$E_z|_{i,j,k}^{n+1/2} = \frac{1}{2} \left(E_z|_{i,j,k}^{n+1} + E_z|_{i,j,k}^{n} \right). \tag{12}$$

In this manner, we obtain the following transcendental equation:

$$E_z|_{i,j,k}^{n+1} = E_z|_{i,j,k}^{n} + \frac{\Delta t}{\epsilon_o} \nabla \times H|_{i,j,k}^{n+1/2} - \frac{\Delta t}{\epsilon_o \Delta x \Delta y} I_0 \left\{ e^{\left[-q\left(E_z|_{i,j,k}^{n+1} + E_z|_{i,j,k}^{n}\right)\Delta z/2\kappa T\right]} - 1 \right\}. \tag{13}$$

Upon solving (13) for the updated electric field using Newton's method, we find that the new model is numerically stable over a useful diode voltage range (up to 15 V).

To test this FD-TD modeling approach, a diode with a saturation current of 1×10^{-14} amps was modeled at the end of a 50 Ω microstrip line (of $\approx$1-mil scale in the transverse plane). The excitation was a matched resistive voltage source providing a 10-volt, 200-MHz sinusoid. This high-level source was selected both to challenge the numerical stability of the FD-TD code and to test whether FD-TD can properly simulate driving a diode into hard clipping. As shown in Fig. 9, there is excellent agreement of the diode voltage response versus time as calculated by FD-TD and SPICE. No instability of the FD-TD solver was observed.

F. The Bipolar Junction Transistor

Reference [9] presented a 2-D FD-TD model for the linearized, small-signal behavior of a transistor. While good results were obtained using this model for selected parameters, some problems remained. This model is numerically unstable for extremely high or low values of the base resistance, r_b; there is a problem with the fringing fields between the base and collector terminals; and the model lacks generality for large-signal and digital switching applications.

We have found a numerically stable FD-TD algorithm for the lumped NPN bipolar junction transistor (BJT) in 3-D that permits study of *large-signal* behavior, including digital switching. In the example considered here (Fig. 10), the BJT terminates a strip transmission line in a grounded-emitter configuration. The simulation is based upon a simple

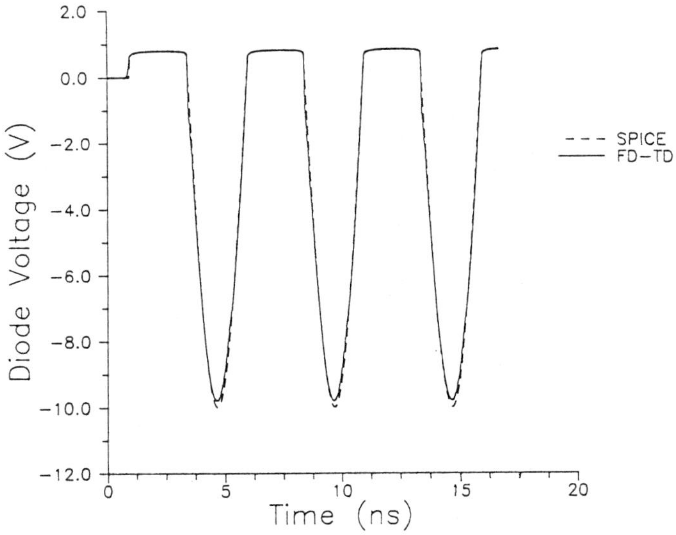

Fig. 9. Agreement of FD-TD and SPICE calculations for the voltage across the dio determinating the stripline.

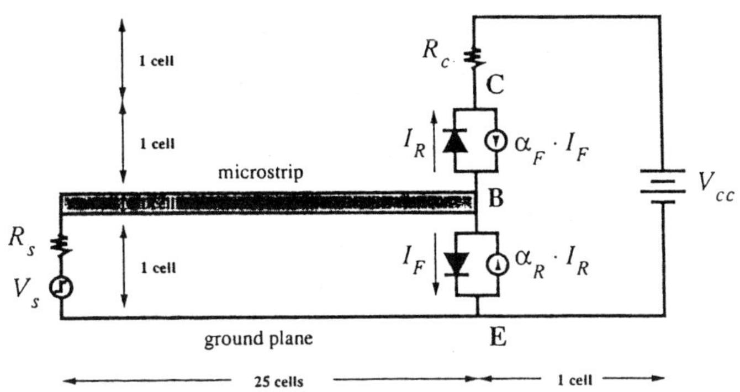

Fig. 10. Generic geometry used for 3-D FD-TD model of an NPN bipolar junction transistor (BJT) terminating a stripline in the common emitter configuration.

Ebers-Moll transistor model described by the following circuit equations [14]:

$$I_F = I_0\left[e^{(qV_{\rm BE}/kT)} - 1\right], \qquad I_R = I_0\left[e^{(qV_{\rm BC}/kT)} - 1\right], \tag{14}$$

$$I_E = \alpha_R I_R - I_F, \qquad I_C = I_R - \alpha_F I_F. \tag{15}$$

Now, assuming a transistor that is located in free-space and oriented in the z-direction in the FD-TD grid as shown in Fig. 10, we can use a semi-implicit strategy to express the base-emitter voltage, $V_{\rm BE}$, in terms of $E_Z\,|_{\rm EB}$, the FD-TD computed electric field in the one-cell gap between the ground plane and the end of the stripline

$$V_{\rm BE}^{n+1/2} = -\frac{\Delta z}{2}\left(E_Z|_{\rm EB}^{n+1} + E_z|_{\rm EB}^{n}\right). \tag{16}$$

A similar semi-implicit strategy is used to express the base-collector voltage, $V_{\rm BC}$, in terms of $E_z\,|_{\rm BC}$, the FD-TD computed electric field in the one-cell gap between the end of the stripline and the collector load:

$$V_{\rm BC}^{n+1/2} = \frac{\Delta}{2}\left(E_Z|_{\rm BC}^{n+1} + E_z|_{\rm BC}^{n}\right). \tag{17}$$

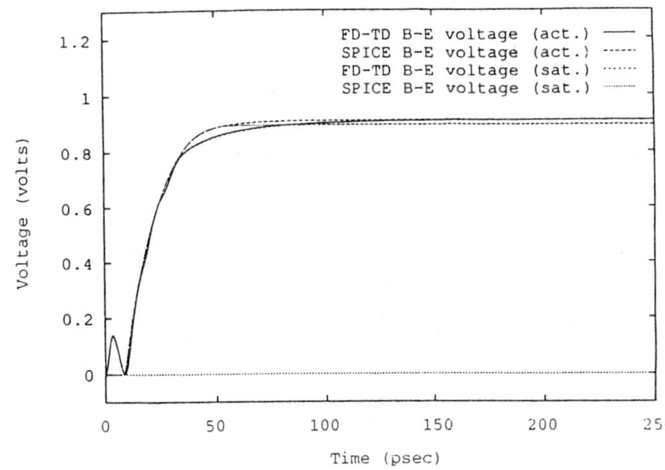

Fig. 11. Agreement of FD-TD and SPICE calculations for the transistor base-to-emitter voltage (stepwise incident pulse).

Substituting (16) and (17) into (14) and (15), we obtain:

$$\begin{aligned} I_E^{n+1/2} &= \alpha_R I_0\left\{e^{\left[\frac{q}{2}(E_z|_{\rm BC}^{n+1}+E_z|_{\rm BC}^{n})/kT\right]} - 1\right\} \\ &\quad - I_0\left\{e^{-\left[\frac{q}{2}(E_z|_{\rm EB}^{n+1}+E_z|_{\rm EB}^{n})/kT\right]} - 1\right\} \end{aligned} \tag{18}$$

$$\begin{aligned} I_C^{n+1/2} &= I_0\left\{e^{\left[\frac{q}{2}(E_z|_{\rm BC}^{n+1}+E_z|_{\rm BC}^{n})/kT\right]} - 1\right\} \\ &\quad - \alpha_F I_0\left\{e^{-\left[\frac{q}{2}(E_z|_{\rm EB}^{n+1}+E_z|_{\rm EB}^{n})/kT\right]} - 1\right\}. \end{aligned} \tag{19}$$

Now, we obtain two coupled transcendental equations for the FD-TD electric field updates at the transistor:

$$E_z|_{EB}^{n+1} = E_z|_{EB}^{n} + \frac{\Delta t}{\epsilon_o}\nabla \times H|_{i,j,k}^{n+1/2} + \frac{\Delta t}{\epsilon_o \Delta x \Delta y} I_E^{n+1/2} \tag{20}$$

$$E_z|_{BC}^{n+1} = E_z|_{BC}^{n} + \frac{\Delta t}{\epsilon_o}\nabla \times H|_{i,j,k}^{n+1/2} + \frac{\Delta t}{\epsilon_o \Delta x \Delta y} I_C^{n+1/2}. \tag{21}$$

The Newton-Raphson method may be used to solve these equations.

To test this FD-TD modeling approach, a transistor at $T = 300°$ K having $I_0 = 10^{-16}$ amp, $\alpha_R = 0.5$, and $\alpha_F = 0.9901$ was modeled at the end of a 50 Ω microstrip line (of $\approx$1-mil scale in the transverse plane) in the manner of Fig. 10. The collector dc supply was included in the electromagnetic simulation. Both the active ($R_c = 50\ \Omega$) and saturated ($R_c = 10\ \Omega$) regions of operation were observed for a step function excitation of the stripline. A typical result is shown in Fig. 11, where the FD-TD computed base-to-emitter voltage is compared to that obtained by a SPICE model. Very good agreement is observed.

V. Conclusion

Analog coupling effects for passive metallic interconnects and packaging of digital circuits operating at nanosecond clocks can be very complex. In fact, as clock speeds increase beyond 500 MHz, it may not be possible to design such systems and make them work in a timely and reliable manner without resorting to full-wave Maxwell's equations solutions in 3-D. FD-TD numerical methods appear to provide sufficient accuracy and scaling ability for large interconnection and packaging problems to be of substantial engineering importance to the digital electronics community.

An emerging possibility is that FD-TD Maxwell's equations modeling can be linked directly to SPICE [15]. This would expand full-wave electromagnetic modeling of digital interconnects to include the voltage-current characteristics of the connected logic devices. It is conceivable that eventually the logical operation of very complex, very high-speed digital electronic circuits can be directly modeled by FD-TD time-stepping of Maxwell's equations.

ACKNOWLEDGMENT

The authors wish to thank Cray Research, Inc., with special thanks to Evans Harrigan for continuous support and encouragement. The authors also acknowledge the technical contributions of Dr. Mike Jones and Dr. Vince Thomas of Los Alamos National Laboratory, and Dr. Chris Reuter of Northwestern University.

REFERENCES

[1] A. Taflove, "Review of the formulation and applications of the finite-difference time-domain method for numerical modeling of electromagnetic wave interactions with arbitrary structures," *Wave Motion*, vol. 10, pp. 547–582, Dec. 1988.

[2] K. S. Yee, "Numerical solution of initial boundary value problems involving Maxwell's equations in isotropic media," *IEEE Trans. Antennas Propagat.*, vol. 14, pp. 302–307, May 1966.

[3] G.-C. Liang, Y.-W. Liu, and K. K. Mei, "Full-wave analysis of coplanar waveguide and slotline using the time-domain finite-difference method," *IEEE Trans. Microwave Theory Tech.*, vol. 37, pp. 1949–1957, Dec. 1989.

[4] T. Shibata and E. Sano, "Characterization of MIS structure coplanar transmission lines for investigation of signal propagation in integrated circuits," *IEEE Trans. Microwave Theory Tech.*, vol. 38, pp. 881–890, July 1990.

[5] C. W. Lam, S. M. Ali, R. T. Shin, and J. A. Kong, "Radiation from discontinuities in VLSI packaging structures," *Proc. Progress in Electromagnetics Research Symp.*, Boston, MA, July 1991, p. 567.

[6] S. Maeda, T. Kashiwa, and I. Fukai, "Full wave analysis of propagation characteristics of a through hole using the finite-difference time-domain method," *IEEE Trans. Microwave Theory Tech.*, vol. 39, pp. 2154–2159, Dec. 1991.

[7] E. Sano and T. Shibata, "Fullwave analysis of picosecond photoconductive switches," *IEEE J. Quantum Electron.*, vol. 26, pp. 372–377, Feb. 1990.

[8] S. M. El-Ghazaly, R. P. Joshi and R. O. Grondin, "Electromagnetic and transport considerations in subpicosecond photoconductive switch modeling," *IEEE Trans. Microwave Theory Tech.*, vol. 38, pp. 629–637, May 1990.

[9] W. Sui, D. A. Christensen and C. H. Durney, "Extending the two-dimensional FD-TD method to hybrid electromagnetic systems with active and passive lumped elements," *IEEE Trans. Microwave Theory Tech.*, vol. 40, pp. 724–730, Apr. 1992.

[10] A. Taflove, "Basis and Application of Finite-Difference Time-Domain (FD-TD) Techniques for Modeling Electromagnetic Wave Interactions," (short course notes), *1992 IEEE Antennas and Propagation Soc. Int. Symp. and URSI Radio Sci. Meeting*, Chicago, IL, July 1992.

[11] A. Taflove and M. E. Brodwin, "Numerical solution of steady-state electromagnetic scattering problems using the time-dependent Maxwell's equations," *IEEE Trans. Microwave Theory Tech.*, vol. 23, pp. 623–630, Aug. 1975.

[12] G. Mur, "Absorbing boundary conditions for the finite-difference approximation of the time-domain electromagnetic field equations," *IEEE Trans. Electromagn. Compat.*, vol. 23, pp. 377–382, Nov. 1981.

[13] S. Ramo, J. R. Whinnery and T. Van Duzer, *Fields and Waves in Communications Electronics*, 2nd ed. New York: Wiley, 1984, p. 410.

[14] S. M. Sze, *Physics of Semiconductor Devices*. New York: Wiley, 1981, p. 152.

[15] V. A. Thomas, M. E. Jones, M. J. Piket-May, A. Taflove and E. Harrigan, "The use of SPICE lumped circuits as sub-grid models for FD-TD analysis," *IEEE Microwave Guided Wave Lett.*, vol. 4, pp. 141–143, May 1994.

[16] Y.-S. Tsuei, A. C. Cangellaris, and J. L. Prince, "Rigorous electromagnetic modeling of chip-to-package (first-level) interconnections," *IEEE Trans. Components, Hybrids, and Manufacturing Tech.*, vol. 16, pp. 876–883, Dec. 1993.

A Perfectly Matched Layer for the Absorption of Electromagnetic Waves

JEAN-PIERRE BERENGER

Centre d'Analyse de Défense, 16 bis, Avenue Prieur de la Côte d'Or, 94114 Arcueil, France

Received July 2, 1993

A new technique of free-space simulation has been developed for solving unbounded electromagnetic problems with the finite-difference time-domain method. Referred to as PML, the new technique is based on the use of an absorbing layer especially designed to absorb without reflection the electromagnetic waves. The first part of the paper presents the theory of the PML technique. The second part is devoted to numerical experiments and to numerical comparisons with the previously used techniques of free-space simulation. These comparisons show that the PML technique works better than the others in all cases; using it allows us to obtain a higher accuracy in some problems and a release of computational requirements in some others. © 1994 Academic Press, Inc.

1. INTRODUCTION

Since the initial work of K. S. Yee [1], the finite-difference time-domain technique has been widely used in electromagnetic computations. One of the inconveniences of this technique lies in the fact that the Maxwell equations have to be solved in a discretized domain whose sizes need to be restrained. Nevertheless, open problems involving theoretically boundless space extension can be solved when applying special conditions on the boundaries of the computational domain, in order to absorb the outgoing waves. Such a need of free-space simulation happens in many problems and especially in wave–structure interactions.

To absorb the outgoing waves, various techniques have been used in computer codes. The first one was the "radiating boundary" [2, 3] which seems to be left unused now. Another one was the matched layer [4–6] which consisted of surrounding the computational domain with an absorbing medium whose impedance matches that of free-space. A third technique appeared with the one-way approximation of the wave equation initially exhibited for acoustic waves by Engquist and Majda [7]. Then applied in the electromagnetic field [8] this technique has been the purpose of many works [9, 10] and seems to be the most used today. However, none of the free-space simulation techniques is faultless; a wave is absorbed without reflection in particular cases only, for instance, if it is plane and propagates perpendicularly to the boundary. These imperfections forbid treatment of some problems and impose constraints on others, as the well-known need of setting boundaries sufficiently far from the scatterer when solving interaction problems.

In this paper, we describe a new technique of free-space simulation. As in [4–6] this technique is based on the use of an absorbing layer, but the matched medium of [4–6] is now replaced by a new matched medium that we have especially designed to absorb without reflection the electromagnetic waves. With the new medium the theoretical reflection factor of a plane wave striking a vacuum–layer interface is null at any frequency and at any incidence angle, contrary to the [4–6] medium with which such a factor is null at normal incidence only. So, the layer surrounding the computational domain can theoretically absorb without reflection any kind of wave travelling towards boundaries, and it can be regarded as a perfectly matched layer. Further, we will refer to the new medium as the PML medium and to the new technique of free-space simulation as the PML technique.

The first part of the paper describes the PML technique for two-dimensional problems. The PML medium is defined, its theoretical reflectionless properties at a vacuum-layer interface are proved, and then the implementation of the PML technique in a finite-difference computational domain is adressed. The second part of the paper is devoted to numerical experiments in order to evaluate how the theoretical properties of the PML technique are preserved in practical computations. Various numerical tests are adressed: reflection of a plane wave at a vacuum–layer interface, absorption of a pulse on boundaries of a computational domain, wave-structure interaction problems, and radiation of a slot in free-space. In each case, the results computed with the PML technique are compared to those computed using the matched layer [4–6] and the one-way wave equation both in its initial form [7, 8] and in the Higdon operator form [9]. These comparisons show that the PML technique brings a real enhancement of computed results in all cases.

Reprinted with permission from *Journal of Computational Physics*, J. P. Berenger, "A Perfectly Matched Layer for the Absorption of Electromagnetic Waves," Vol. 114, pp. 185-200, October 1994. © 1994 Academic Press. Reprinted by permission of Academic Press.

2. THEORY OF THE PERFECTLY MATCHED LAYER

2.1. Definition of the PML Medium

In this paper, we will set the equations of a PML medium for two-dimensional problems, first in the TE (transverse electric) case. In Cartesian coordinates let us consider a problem that is without variation along z, with the electric field lying in the (x, y) plane (Fig. 1). The electromagnetic field involves three components only, E_x, E_y, H_z, and the Maxwell equations reduce to a set of three equations. In the most general case, which is a medium with an electric conductivity σ and a magnetic conductivity σ^*, these equations can be written as

$$\varepsilon_0 \frac{\partial E_x}{\partial t} + \sigma E_x = \frac{\partial H_z}{\partial y} \tag{1.a}$$

$$\varepsilon_0 \frac{\partial E_y}{\partial t} + \sigma E_y = -\frac{\partial H_z}{\partial x} \tag{1.b}$$

$$\mu_0 \frac{\partial H_z}{\partial t} + \sigma^* H_z = \frac{\partial E_x}{\partial y} - \frac{\partial E_y}{\partial x}. \tag{1.c}$$

Moreover, if the condition

$$\frac{\sigma}{\varepsilon_0} = \frac{\sigma^*}{\mu_0} \tag{2}$$

is satisfied, then the impedance of the medium (1) equals that of vacuum and no reflection occurs when a plane wave propagates normally across a vacuum–medium interface. Such a medium is used in the [4–6] technique in order to absorb the outgoing waves.

We will now define the PML medium in the TE case. The cornerstone of this definition is the break of the magnetic component H_z into two subcomponents which we will denote as H_{zx} and H_{zy}. In the TE case, a PML medium is defined as a medium in which the electromagnetic field has

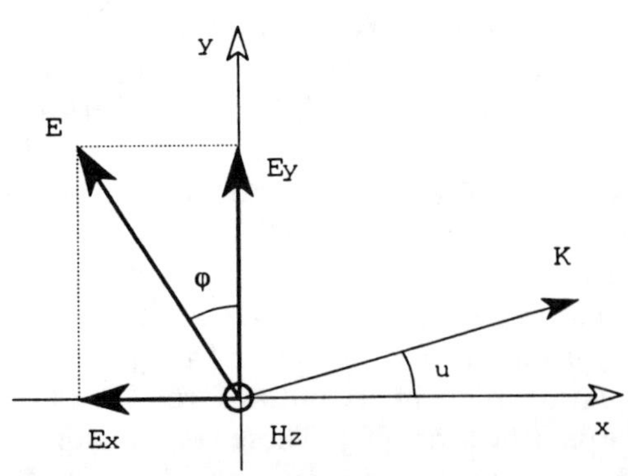

FIG. 1. The transverse electric problem.

four components, E_x, E_y, H_{zx}, H_{zy}, connected through the four following equations:

$$\varepsilon_0 \frac{\partial E_x}{\partial t} + \sigma_y E_x = \frac{\partial (H_{zx} + H_{zy})}{\partial y} \tag{3.a}$$

$$\varepsilon_0 \frac{\partial E_y}{\partial t} + \sigma_x E_y = -\frac{\partial (H_{zx} + H_{zy})}{\partial x} \tag{3.b}$$

$$\mu_0 \frac{\partial H_{zx}}{\partial t} + \sigma_x^* H_{zx} = -\frac{\partial E_y}{\partial x} \tag{3.c}$$

$$\mu_0 \frac{\partial H_{zy}}{\partial t} + \sigma_y^* H_{zy} = \frac{\partial E_x}{\partial y}, \tag{3.d}$$

where the parameters $(\sigma_x, \sigma_x^*, \sigma_y, \sigma_y^*)$ are homogeneous to electric and magnetic conductivities.

A first remark can be made when looking at system (3). If $\sigma_x^* = \sigma_y^*$, then the last two equations can merge and (3) reduces to a set of three equations involving three components E_x, E_y, and $H_z = H_{zx} + H_{zy}$. As a result, the PML medium holds as particular cases all the usual media. If $\sigma_x = \sigma_y = \sigma_x^* = \sigma_y^* = 0$, (3) reduces to the Maxwell equations of vacuum, if $\sigma_x = \sigma_y$ and $\sigma_x^* = \sigma_y^* = 0$, it reduces to the equations of a conductive medium, and, finally, if $\sigma_x = \sigma_y$ and $\sigma_x^* = \sigma_y^*$, it reduces to the equations of the absorbing medium (1).

A second remark can be made before any calculation. If $\sigma_y = \sigma_y^* = 0$, the PML medium can absorb a plane wave (E_y, H_{zx}) propagating along x, but it does not absorb a wave (E_x, H_{zy}) propagating along y, since in the first case propagation is ruled by (3.b) and (3.c), and in the second case by (3.a) and (3.d), and vice versa for waves (E_y, H_{zx}) and (E_x, H_{zy}) if $\sigma_x = \sigma_x^* = 0$. Such properties of the particular PML media $(\sigma_x, \sigma_x^*, 0, 0)$ and $(0, 0, \sigma_y, \sigma_y^*)$ are in close relationship with another one; we will prove later: if their conductivities satisfy (2), then at vacuum–medium interfaces normal respectively to x and y these two media do not activate any reflection of electromagnetic waves. Such PML media will be the basis of the PML technique.

2.2. Propagation of a Plane Wave in a PML Medium

Let us consider a wave whose electric field of magnitude E_0 forms an angle φ with the y axis (Fig. 1). We will denote as H_{zx0} and H_{zy0} the magnitudes of the magnetic components H_{zx} and H_{zy}. If a plane wave propagates in the PML medium, then the four components of the field can be expressed as

$$E_x = -E_0 \sin\varphi \, e^{i\omega(t - \alpha x - \beta y)} \tag{4.a}$$

$$E_y = E_0 \cos\varphi \, e^{i\omega(t - \alpha x - \beta y)} \tag{4.b}$$

$$H_{zx} = H_{zx0} \, e^{i\omega(t - \alpha x - \beta y)} \tag{4.c}$$

$$H_{zy} = H_{zy0} \, e^{i\omega(t - \alpha x - \beta y)}, \tag{4.d}$$

where ω is the pulsation of the wave, t is the time, and α and β are complex constants. Since the magnitude E_0 is given, the set of Eq. (4) involves four unknown quantities to be determined, α, β, H_{zx0}, H_{zy0}. Enforcing E_x, E_y, H_{zx}, H_{zy} from (4) in the PML equations (3) yields the following set of equations connecting the four unknowns:

$$\varepsilon_0 E_0 \sin\varphi - i\frac{\sigma_y}{\omega} E_0 \sin\varphi = \beta(H_{zx0} + H_{zy0}) \quad (5.a)$$

$$\varepsilon_0 E_0 \cos\varphi - i\frac{\sigma_x}{\omega} E_0 \cos\varphi = \alpha(H_{zx0} + H_{zy0}) \quad (5.b)$$

$$\mu_0 H_{zx0} - i\frac{\sigma_x^*}{\omega} H_{zx0} = \alpha E_0 \cos\varphi \quad (5.c)$$

$$\mu_0 H_{zy0} - i\frac{\sigma_y^*}{\omega} H_{zy0} = \beta E_0 \sin\varphi. \quad (5.d)$$

Obtaining H_{zx0} and H_{zy0} from (5.c) and (5.d) and bringing them respectively into (5.a) and (5.b) yields

$$\varepsilon_0\mu_0\left(1 - i\frac{\sigma_y}{\varepsilon_0\omega}\right)\sin\varphi = \beta\left[\frac{\alpha\cos\varphi}{(1 - i(\sigma_x^*/\mu_0\omega))} + \frac{\beta\sin\varphi}{(1 - i(\sigma_y^*/\mu_0\omega))}\right] \quad (6.a)$$

$$\varepsilon_0\mu_0\left(1 - i\frac{\sigma_x}{\varepsilon_0\omega}\right)\cos\varphi = \alpha\left[\frac{\alpha\cos\varphi}{(1 - i(\sigma_x^*/\mu_0\omega))} + \frac{\beta\sin\varphi}{(1 - i(\sigma_y^*/\mu_0\omega))}\right]. \quad (6.b)$$

This system of two equations connects the unknowns α and β. It may be solved by means of writing the ratio (6.a) over (6.b),

$$\frac{\beta}{\alpha} = \frac{\sin\varphi}{\cos\varphi}\,\frac{1 - i(\sigma_y/\varepsilon_0\omega)}{1 - i(\sigma_x/\varepsilon_0\omega)} \quad (7)$$

and then obtaining α^2 from (7) and (6.b) and β^2 from (7) and (6.a). That yields two sets of (α, β) of opposite signs for two opposite directions of propagation. Choosing the positive sign we have

$$\alpha = \frac{\sqrt{\varepsilon_0\mu_0}}{G}\left(1 - i\frac{\sigma_x}{\varepsilon_0\omega}\right)\cos\varphi \quad (8.a)$$

$$\beta = \frac{\sqrt{\varepsilon_0\mu_0}}{G}\left(1 - i\frac{\sigma_y}{\varepsilon_0\omega}\right)\sin\varphi, \quad (8.b)$$

where

$$G = \sqrt{w_x \cos^2\varphi + w_y \sin^2\varphi} \quad (9)$$

$$w_x = \frac{1 - i(\sigma_x/\varepsilon_0\omega)}{1 - i(\sigma_x^*/\mu_0\omega)} \quad (10.a)$$

$$w_y = \frac{1 - i(\sigma_y/\varepsilon_0\omega)}{1 - i(\sigma_y^*/\mu_0\omega)}. \quad (10.b)$$

Denoting as Ψ any component of the field, Ψ_0 its magnitude, and c the speed of light, with (4) and (8), we can write

$$\psi = \psi_0 e^{i\omega(t - (x\cos\varphi + y\sin\varphi)/cG)} e^{-(\sigma_x\cos\varphi/\varepsilon_0 cG)x} e^{-(\sigma_y\sin\varphi/\varepsilon_0 cG)y}. \quad (11)$$

The last two unknowns H_{zx0} and H_{zy0} can be found as functions of α and β from (5.c) and (5.d), and then enforcing the α and β values (8) yields

$$H_{zx0} = E_0\sqrt{(\varepsilon_0/\mu_0)}\,\frac{1}{G}\,w_x\cos^2\varphi \quad (12.a)$$

$$H_{zy0} = E_0\sqrt{(\varepsilon_0/\mu_0)}\,\frac{1}{G}\,w_y\sin^2\varphi. \quad (12.b)$$

Taking into account (9), the summation of H_{zx0} and H_{zy0} is then

$$H_0 = E_0\sqrt{(\varepsilon_0/\mu_0)}\,G \quad (13)$$

and the ratio Z of the electric magnitude over the magnetic one is

$$Z = \sqrt{(\mu_0/\varepsilon_0)}\,\frac{1}{G}. \quad (14)$$

For formulas (11) and (14), an important occurrence is when both (σ_x, σ_x^*) and (σ_y, σ_y^*) satisfy the condition (2). Then, the quantities w_x, w_y, G, equal unity at any frequency, and so the expression of the wave components (11) and of the impedance (14) become respectively

$$\psi = \psi_0 e^{i\omega(t - (x\cos\varphi + y\sin\varphi)/c)} \times e^{-(\sigma_x\cos\varphi/\varepsilon_0 c)x} e^{-(\sigma_y\sin\varphi/\varepsilon_0 c)y} \quad (15)$$

$$Z = \sqrt{\mu_0/\varepsilon_0}. \quad (16)$$

The first exponential of (15) shows that the wave phase propagates normally to the electric field (that means $u = \varphi$ on Fig. 1) with the speed of light c. The last two exponentials rule the magnitude of the wave which decreases exponentially along x and y. Formula (16) shows that the impedance of the medium equals that of vacuum. The matching impedance condition (2) of the medium (1) is a matching condition for the PML media too. Differences lie only in the fact that in the case of PML media, two couples of conductivities must satisfy (2), both (σ_x, σ_x^*) and (σ_y, σ_y^*).

Other remarks can be done when looking at (11) or (15). In the general case (11), if a wave propagates along y, that is, $\cos\varphi = 0$, and if, moreover, $\sigma_y = \sigma_y^* = 0$, then it is not absorbed. That is in accordance with the second remark achieved after the PML equations (3). In the case of matched media (15), if $\sigma_y = \sigma_y^* = 0$ the last exponential of (15) equals unity and the absorption is a function of the x coordinate only.

2.3. Transmission of a Wave through PML–PML Interfaces

In this section, we will address the problem of a plane wave moving from a PML medium to another one. We will prove that in particular cases, with adequate sets of parameters σ_x, σ_x^*, σ_y, σ_y^*, the transmission is perfect and reflectionless at any frequency and at any incidence angle. These particular cases will be the basis of the perfectly matched layer.

Interface Normal to x

We first consider the case of two PML media separated by an interface normal to the x axis (Fig. 2). Let us denote θ_1 and θ_2 the angles of the incident and transmitted electric fields E_i and E_t, with respect to the interface plane. As noted after (15) in the previous section, if the media are matched ones θ_1 and θ_2 are also the angles that the phase propagation forms with the normal to the interface. Figure 2 is drawn with that assumption. As the φ angle of a PML medium was defined with respect to the y axis (Fig. 1), in the case of Fig. 2 we have $\theta = \varphi$ in each media. Let us now assume the interface to be infinite and the incident wave to be plane. First, both the reflected and transmitted waves must be plane too. Second, the ratios of these waves over the incident one must be without variation when moving on the interface. So, for any component Ψ of the incident and transmitted fields, and for two points A and B of the interface, we can write

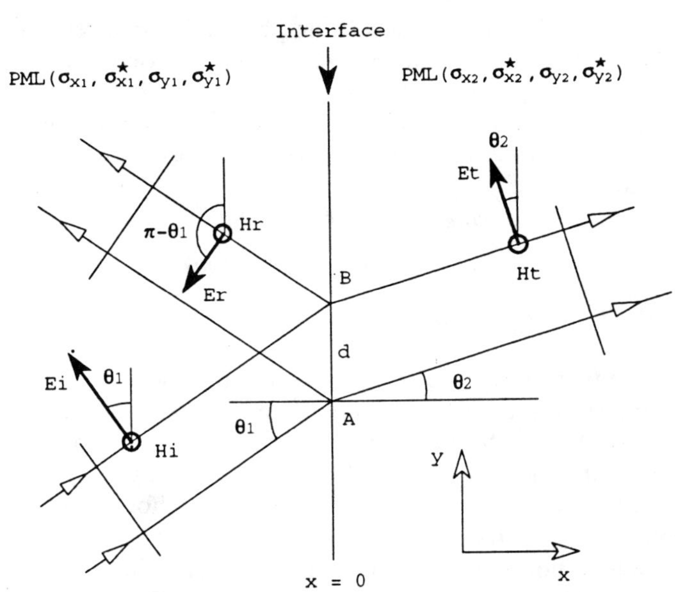

FIG. 2. Interface lying between two PML media.

$$\frac{\psi_t(B)}{\psi_i(B)} = \frac{\psi_t(A)}{\psi_i(A)}. \tag{17}$$

Denoting as d the distance from A to B, and G_1, G_2 as the quantities (9) of each media, with (11) we have

$$\psi_i(B) = \psi_i(A)\, e^{-i\omega(d \sin\theta_1/cG_1) - (\sigma_{y1} \sin\theta_1/\varepsilon_0 c G_1) d} \tag{18.a}$$

$$\psi_t(B) = \psi_t(A)\, e^{-i\omega(d \sin\theta_2/cG_2) - (\sigma_{y2} \sin\theta_2/\varepsilon_0 c G_2) d}. \tag{18.b}$$

Since (17) is true for any distance d, the exponential factors of (18.a) and (18.b) must be equal. So the relation

$$\left(1 - i\frac{\sigma_{y1}}{\varepsilon_0\omega}\right)\frac{\sin\theta_1}{G_1} = \left(1 - i\frac{\sigma_{y2}}{\varepsilon_0\omega}\right)\frac{\sin\theta_2}{G_2}, \tag{19.a}$$

must be satisfied, where

$$G_k = \sqrt{w_{xk}\cos^2\theta_k + w_{yk}\sin^2\theta_k} \qquad (\text{for } k = 1, 2). \tag{19.b}$$

This relation, which connects the incident and transmitted angles, is the Snell–Descartes law at an interface normal to x, lying between two PML $(\sigma_x, \sigma_x^*, \sigma_y, \sigma_y^*)$ media. The demonstration related to the transmitted wave could be reproduced for the reflected one, with the conclusion that the reflected angle equals $\pi - \theta_1$.

Let us now consider the incident, reflected, and transmitted electric and magnetic fields E_i, E_r, E_t, H_i, H_r, H_t (Fig. 2). First, continuity of the fields E_y and $H_{zx} + H_{zy}$ lying in the interface yields the following set of equations:

$$E_i\cos\theta_1 - E_r\cos\theta_1 = E_t\cos\theta_2 \tag{20.a}$$

$$H_i + H_r = H_t. \tag{20.b}$$

Second, denoting as E_{i0}, E_{r0}, E_{t0}, the magnitudes of E_i, E_r, E_t, and setting $x = 0$ in the interface, with (11) and (14) we can write

$$E_i = E_{i0}\, e^{-i\omega(y \sin\theta_1/cG_1)(1 - i(\sigma_{y1}/\varepsilon_0\omega))}\, e^{i\omega t} \tag{21.a}$$

$$E_r = E_{r0}\, e^{-i\omega(y \sin\theta_1/cG_1)(1 - i(\sigma_{y1}/\varepsilon_0\omega))}\, e^{i\omega t} \tag{21.b}$$

$$E_t = E_{t0}\, e^{-i\omega(y \sin\theta_2/cG_2)(1 - i(\sigma_{y2}/\varepsilon_0\omega))}\, e^{i\omega t} \tag{21.c}$$

$$H_i = E_i/Z_1, \quad H_r = E_r/Z_1, \quad H_t = E_t/Z_2. \qquad (21.d), (21.e), (21.f)$$

As a consequence of the Snell–Descartes law (19), the three exponentials on space of (21.a), (21.b), (21.c) are equal. So, reporting $E_i, \ldots, H_t$ from (21) into (20), this system (20) becomes

$$E_{i0} \cos \theta_1 - E_{r0} \cos \theta_1 = E_{t0} \cos \theta_2 \quad (22.a)$$

$$\frac{E_{i0}}{Z_1} + \frac{E_{r0}}{Z_1} = \frac{E_{t0}}{Z_2}. \quad (22.b)$$

Defining the reflection factor as the ratio of the electric components lying in the interface, that is, $-E_{r0} \cos \theta_1 / E_{i0} \cos \theta_1$, and then solving the set (22) for this ratio, the reflection factor r_p for the TE case is

$$r_p = \frac{Z_2 \cos \theta_2 - Z_1 \cos \theta_1}{Z_2 \cos \theta_2 + Z_1 \cos \theta_1}. \quad (23)$$

This factor (23) is as at an interface lying between two usual media. The formula for the transmission factor would be the same as usual, too. With (14), we can rewrite (23) as

$$r_p = \frac{G_1 \cos \theta_2 - G_2 \cos \theta_1}{G_1 \cos \theta_2 + G_2 \cos \theta_1}, \quad (24)$$

where G_1 and G_2 are functions of θ_1 and θ_2 through (19.b).

We will now address a particular case announced at the beginning of this section. Let us consider an interface lying between media having same σ_y and σ_y^* conductivities, that is, a $(\sigma_{x1}, \sigma_{x1}^*, \sigma_y, \sigma_y^*)$ and a $(\sigma_{x2}, \sigma_{x2}^*, \sigma_y, \sigma_y^*)$ media. Then the Snell–Descartes law (19.a) becomes

$$\frac{\sin \theta_1}{G_1} = \frac{\sin \theta_2}{G_2}. \quad (25)$$

If, moreover, the two media are matched ones, that is, $(\sigma_{x1}, \sigma_{x1}^*)$, $(\sigma_{x2}, \sigma_{x2}^*)$, and (σ_y, σ_y^*) satisfying (2), we have $G_1 = G_2 = 1$, so (25) reduces to

$$\theta_1 = \theta_2 \quad (26)$$

and the reflection factor (24) is then

$$r_p = 0. \quad (27)$$

So, at an interface normal to x lying between two matched PML media having the same (σ_y, σ_y^*), a plane wave is transmitted without reflection at any incidence angle and at any frequency. That is also true, of course, if the first medium is a vacuum and the second one is a $(\sigma_x, \sigma_x^*, 0, 0)$ medium, since a vacuum can be seen as a $(0, 0, 0, 0)$ medium.

In the general case of unmatched media having the same (σ_y, σ_y^*), a simple formula can be obtained for the factor (24). Bringing (25) into (24) yields

$$r_p = \frac{\sin \theta_1 \cos \theta_2 - \sin \theta_2 \cos \theta_1}{\sin \theta_1 \cos \theta_2 + \sin \theta_2 \cos \theta_1}. \quad (28)$$

Then, squaring (25), replacing G_1 and G_2 by (19.b), and taking into account that $w_{y1} = w_{y2}$, we can deduce that

$$\sqrt{w_{x2}} \sin \theta_1 \cos \theta_2 = \sqrt{w_{x1}} \sin \theta_2 \cos \theta_1 \quad (29)$$

and with (28) and (29),

$$r_p = \frac{\sqrt{w_{x1}} - \sqrt{w_{x2}}}{\sqrt{w_{x1}} + \sqrt{w_{x2}}}. \quad (30)$$

Formula (30) shows that even if the media are not matched ones, the reflection does not depend on the incidence angle θ_1. It depends only on the frequency through (10). In the case of matched media $w_{x1} = w_{x2} = 1$ and (30) reduces to (27) as expected.

Interface Normal to y

Let us consider a plane wave propagating upward and striking an interface normal to the y axis. Defining the incidence angle θ as in the previous case with respect to the interface, since φ is defined with respect to the y axis (Fig. 1) we have now $\varphi = \theta + \pi/2$. Similar treatment as in the previous case yields then the Snell–Descartes law related to an interface normal to y,

$$\left(1 - i\frac{\sigma_{x1}}{\varepsilon_0 \omega}\right) \frac{\sin \theta_1}{G_1} = \left(1 - i\frac{\sigma_{x2}}{\varepsilon_0 \omega}\right) \frac{\sin \theta_2}{G_2}, \quad (31.a)$$

where

$$G_k = \sqrt{w_{xk} \sin^2 \theta_k + w_{yk} \cos^2 \theta_k} \qquad (\text{for } k = 1, 2). \quad (31.b)$$

In the case of two media having the same σ_x and σ_x^*, (31.a) becomes (25). If, moreover, these media are matched ones, then $G_1 = G_2$ and (31.a) reduces to (26). Since the reflection factor is still (23), (27) is still true. So, at an interface normal to y lying between two matched media having the same (σ_x, σ_x^*), the reflection factor is always null.

Practical Conclusions of This Section

As a summary of this section we can now say that the reflection factor between two PML media whose conductivities satisfy (2) is null:

— at an interface normal to x in the case of the same σ_y and σ_y^*.

— at an interface normal to y in the case of the same σ_x and σ_x^*.

This result is, of course, true if some conductivities equal zero. Having in mind that a vacuum is a $(0, 0, 0, 0)$ PML medium, each of the previous cases yields two particular occurrences which will be the basis of the PML technique. These occurrences are

— the reflection factor is null at an interface normal to x lying between a vacuum and a $(\sigma_x, \sigma_x^*, 0, 0)$ matched medium or between a $(0, 0, \sigma_y, \sigma_y^*)$ and a $(\sigma_x, \sigma_x^*, \sigma_y, \sigma_y^*)$ matched media.

— the reflection factor is null at an interface normal to y lying between a vacuum and a $(0, 0, \sigma_y, \sigma_y^*)$ matched medium or between a $(\sigma_x, \sigma_x^*, 0, 0)$ and a $(\sigma_x, \sigma_x^*, \sigma_y, \sigma_y^*)$ matched media.

2.4. Perfectly Matched Layer for the Finite Difference Technique

The general frame of the PML technique is pointed out on Fig. 3. The Maxwell equations are solved by the FDTD (finite-difference time-domain) technique [1, 11] inside a computational domain in which lies a source of outgoing waves. For instance, such a source will be the scatterer in a wave–structure interaction problem. The computational domain is surrounded by an absorbing layer which is an aggregate of PML media whose properties have been predicted in previous sections. The domain is finally ended by perfectly conducting conditions.

On both the left and right sides of the computational domain, the absorbing layers are matched PML $(\sigma_x, \sigma_x^*, 0, 0)$ media. So, at interfaces vacuum–layer AB and CD normal to the x axis, the reflection factor is theoretically null, as it has been proved (Eq. (27)). Outgoing waves can propagate without reflection through the AB and CD interfaces. Similarly, matched PML $(0, 0, \sigma_y, \sigma_y^*)$ media are used on both upper and lower sides of the computational domain so that the outgoing waves can propagate without reflection through the BC and DA interfaces normal to y. At the four corners of the domain, the absorbing layers are made of PML $(\sigma_x, \sigma_x^*, \sigma_y, \sigma_y^*)$ media having conductivities equal to those of the adjacent $(\sigma_x, \sigma_x^*, 0, 0)$ and $(0, 0, \sigma_y, \sigma_y^*)$ media. As a result, there is theoretically no reflection at the interfaces lying between the side layers and the corner layers. For instance, a wave can propagate without reflection through the BB_1 and BB_2 interfaces of Fig. 3.

As physical remark, we can note that with the PML layer, the speed of propagation is the speed of light in the whole computational domain, and the Snell–Descartes law is (26) at all interfaces. So, as a wave travels from a medium to another one through interfaces, its shape is preserved. That is in close relationship with the lack of reflection at interfaces.

In the absorbing layer, the magnitude of a wave is ruled by the last two exponential factors of (15). In the side layers which are $(\sigma_x, \sigma_x^*, 0, 0)$ or $(0, 0, \sigma_y, \sigma_y^*)$ media, one factor equals unity. So, at a distance ρ of any interface, the magnitude of an outgoing plane wave can be written as

$$\psi(\rho) = \psi(0)\, e^{-(\sigma \cos\theta/\varepsilon_0 c)\rho}, \tag{32}$$

where θ is the incidence angle defined with respect to the interface, and σ is either σ_x or σ_y. After crossing the layer, a wave is reflected by the perfectly conducting conditions which ends the domain, and then, after a second crossing, it can come back into the vacuum. So, for a layer of thickness δ, an apparent reflection factor can be defined as

$$R(\theta) = e^{-2(\sigma \cos\theta/\varepsilon_0 c)\delta}. \tag{33}$$

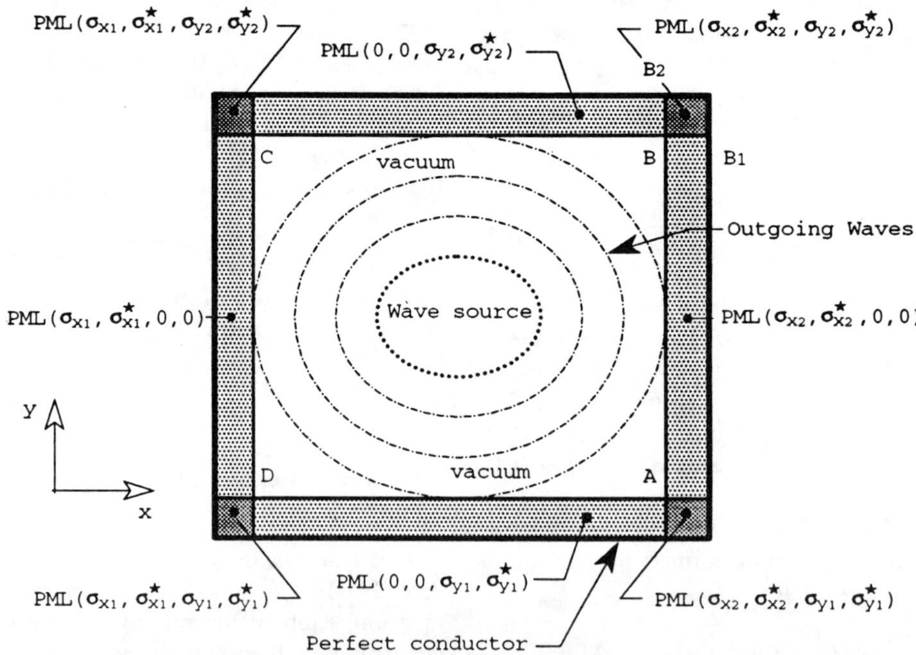

FIG. 3. The PML technique.

This formula claims a remark. If the incident wave is close to the interface ($\theta \sim \pi/2$) the factor R is close to unity, whatever σ may be. This is not troublesome in practical computations as we will see from numerical tests. It is not surprising, indeed, since for instance, a close to the interface wave on the right side of the domain will come perpendicularly to the upper (or lower) side, where it will be absorbed.

As (33) points out, the apparent reflection is a function of the product $\sigma\delta$. So, for a given layer attenuation, theoretically the layer thickness δ could be as thin as intended, reduced to one cell of the FDTD mesh, for instance. In fact, sharp variations of conductivity create numerical reflections, as with the [4–6] technique. So, in practical computations the layer has to be a few cells thick with conductivity increasing from zero at the vacuum–layer interface to a value σ_m at the outer side of the layer. For a conductivity $\sigma(\rho)$ the reflection factor is then

$$R(\theta) = e^{-2(\cos\theta/\varepsilon_0 c)\int_0^\delta \sigma(\rho)\,d\rho}. \tag{34}$$

All numerical computations reported in this paper have been performed with conductivities of the form

$$\sigma(\rho) = \sigma_{\mathrm{m}}\left(\frac{\rho}{\delta}\right)^n \tag{35}$$

for which, enforcing (35) in (34), the apparent reflection is

$$R(\theta) = e^{-(2/(n+1))(\sigma_{\mathrm{m}}\delta/\varepsilon_0 c)\cos\theta}. \tag{36}$$

Numerical implementation of the PML layer in the FDTD technique does not involve special treatment. Let us consider, for instance, the upper-right part of a gridded computational domain (Fig. 4). In the layer, the equations

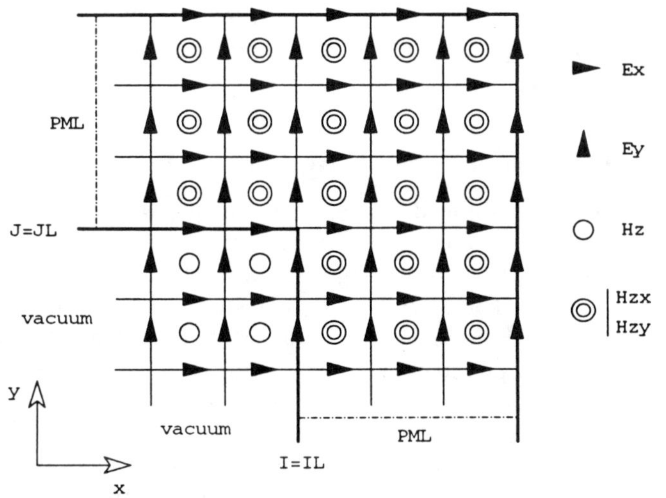

FIG. 4. Upper-right part of the FDTD grid.

to be discretized are (3). In the inner volume (vacuum and scattering structure, for instance) the set (3) could be used too, since the usual media can be seen as particular PML media. But having in mind computer storage optimization, it is more convenient to use the Maxwell equations which address only three field components, instead of four, with (3).

In the inner volume (for $\mathrm{I} < \mathrm{IL}$ and $\mathrm{J} < \mathrm{JL}$ on Fig. 4) the finite-difference equations are the usual [1, 11] discretizations of the Maxwell equations. In the PML medium, the two magnetic subcomponents are computed at the same points, in place of the magnetic component H_z. Discretization of the set (3) is straightforward. With the usual FDTD notations, (3.b) and (3.c) yield the following equations that can be applied in the whole layer, except in the interface for E_y (see below),

$$\begin{aligned} E_y^{n+1}&(i, j+1/2) \\ &= e^{-\sigma_x(i)\,\Delta t/\varepsilon_0} E_y^n(i, j+1/2) - \frac{(1-e^{-\sigma_x(i)\,\Delta t/\varepsilon_0})}{\sigma_x(i)\,\Delta x} \\ &\quad\times [H_{zx}^{n+1/2}(i+1/2, j+1/2) + H_{zy}^{n+1/2}(i+1/2, j+1/2) \\ &\quad - H_{zx}^{n+1/2}(i-1/2, j+1/2) - H_{zy}^{n+1/2}(i-1/2, j+1/2)] \end{aligned} \tag{37}$$

$$\begin{aligned} H_{zx}^{n+1/2}&(i+1/2, j+1/2) \\ &= e^{-\sigma_x^*(i+1/2)\,\Delta t/\mu_0} H_{zx}^{n-1/2}(i+1/2, j+1/2) \\ &\quad - \frac{(1-e^{-\sigma_x^*(i+1/2)\,\Delta t/\mu_0})}{\sigma_x^*(i+1/2)\,\Delta x} \\ &\quad\times [E_y^n(i+1, j+1/2) - E_y^n(i, j+1/2)], \end{aligned} \tag{38}$$

where σ_x and σ_x^* are functions of $x(\mathrm{I})$, in the left, right, and corner layers, and are equal to zero, in the upper and lower layers (for $\mathrm{I} < \mathrm{IL}$ on Fig. 4). For the E_y component lying in the interface, the magnetic field has one component H_z on one side and two subcomponents H_{zx}, H_{zy} on the other. That is a result of using the Maxwell equations in the inner volume, but it has no physical significance (H_{zx} and H_{zy} are present in the inner volume where they merge, since σ_x^* equals σ_y^*). The finite difference equations have to be modified. So, in the right side interface normal to x, (37) becomes

$$\begin{aligned} E_y^{n+1}&(il, j+1/2) \\ &= e^{-\sigma_x(il)\,\Delta t/\varepsilon_0} E_y^n(il, j+1/2) - \frac{(1-e^{-\sigma_x(il)\,\Delta t/\varepsilon_0})}{\sigma_x(il)\,\Delta x} \\ &\quad\times [H_{zx}^{n+1/2}(il+1/2, j+1/2) + H_{zy}^{n+1/2}(il+1/2, j+1/2) \\ &\quad - H_z^{n+1/2}(il-1/2, j+1/2)]. \end{aligned} \tag{39}$$

In all computations of this paper, conductivities which depend on index I (or J, too) were implemented as the average value in the cell around the index location. That can be written for $\sigma_x(\mathrm{I})$,

$$\sigma_x(i) = \frac{1}{\Delta x} \int_{x(i) - \Delta x/2}^{x(i) + \Delta x/2} \sigma_x(x')\, dx', \tag{40}$$

where $\sigma_x(x)$ was of the form (35) in our computations.

The discretized equations needed for the advance of the components E_x and H_{zy} are obtained from the remaining equations (3.a) and (3.d). That yields three equations analogous to (37), (38), (39).

2.5. PML Medium for the Transverse Magnetic Case

In TM problems, the electromagnetic field reduces to three components, E_z, H_x, H_y. In a PML medium, the component which is split into two subcomponents is the electric field E_z. The PML medium equations for the TM case are the following:

$$\varepsilon_0 \frac{\partial E_{zx}}{\partial t} + \sigma_x E_{zx} = \frac{\partial H_y}{\partial x} \tag{41.a}$$

$$\varepsilon_0 \frac{\partial E_{zy}}{\partial t} + \sigma_y E_{zy} = -\frac{\partial H_x}{\partial y} \tag{41.b}$$

$$\mu_0 \frac{\partial H_x}{\partial t} + \sigma_y^* H_x = -\frac{\partial (E_{zx} + E_{zy})}{\partial y} \tag{41.c}$$

$$\mu_0 \frac{\partial H_y}{\partial t} + \sigma_x^* H_y = \frac{\partial (E_{zx} + E_{zy})}{\partial x}. \tag{41.d}$$

Same calculations as in the TE case yield slightly changed results. In most formulas, especially in (8), (10), (19.a), (19.b), the change is only a permutation of ε_0 with μ_0 and of starred σ with unstarred ones. In (14), $1/G$ is changed to G and (23) has to be replaced by the usual r_s factor. But the most important result is left unchanged. As in the TE case, there is no reflection both at an interface normal to x between two matched media having the same (σ_y, σ_y^*) and at an interface normal to y between two matched media having the same (σ_x, σ_x^*). So, an absorbing reflectionless layer can be built as in the TE case.

3. NUMERICAL EXPERIMENTS

In order to evaluate the actual possibilities of the PML technique in practical computations, six numerical experiments are presented here. The first two are elementary tests which compare the theoretical reflection factor with its practical counterpart when a plane wave strikes either an infinite vacuum–layer interface or the corner of a computational domain. The third experiment is related to the absorption of a pulse on the boundaries of a computational domain. The last three are closer to some realistic applications. We will consider successively a wave–structure interaction problem in the TE case, a wave–structure interaction problem in the TM case, and finally the problem of a slot radiating in free space.

3.1. Techniques Used for Numerical Comparisons

In most numerical experiments, the results were computed with both the PML technique and three other techniques of free-space simulation in order to evaluate the improvement we can expect from using the PML technique. The first technique used in these comparisons was the matched layer [4–6] which we will refer to as ML on the figures. The second technique was the one-way wave equation in its Padé approximation version. Discretization was the same as in [8, 10] for the first two orders of approximation. Some results were also computed with the third order using a discretization of our own. We will refer to this technique as one-way on the figures. The third technique, referred to as operator, is that of Higdon operators defined directly in finite-difference terms. We used the following operator, which is a particular case of the general operator adressed in [9],

$$B(K, Z^{-1}) = \left[I - KZ^{-1} + \frac{\Delta x - c\,\Delta t}{\Delta x + c\,\Delta t} (K - Z^{-1}) \right]^p, \tag{42}$$

where I, K, Z^{-1} are the identity and shift operators of [9] and p is the order of the operator (2 or 3 in our computations).

For the one-way and operator techniques, the reflection factor of a plane wave is, at an incidence angle θ,

$$R(\theta) = \left(\frac{1 - \cos\theta}{1 + \cos\theta} \right)^p. \tag{43}$$

For the ML technique, with conductivities to be used in order to absorb the outgoing waves over a few cells of the FDTD mesh [5], the practical reflection factor is also (43) with $p = 1$.

3.2. Reflection of a Plane Wave Striking a Plane Boundary

As the first finite difference experiment, we have computed the reflection factor of a plane wave striking a plane boundary, in the TE case. Figure 5 shows the computational domain used. The boundary to be tested is applied on one side of the domain in which a plane wave is produced by the Huyghens surface technique [6]. Both the incident wave and the reflected one are present inside the Huyghens surface, but only the reflected wave is outside it. So, the reflected wave is obtained at a point P close to this surface but outside it. As it appears on Fig. 5, the computation is correct only during a clear time which depends on the

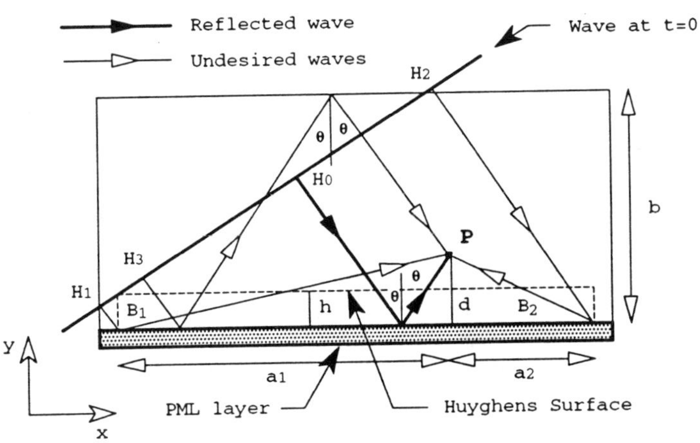

FIG. 5. FDTD domain for computing the reflection of a plane wave.

dimensions of the domain. Three undesired fields disturb the desired result at P, first, the non-physical artefacts coming from B_1 ands B_2, where the Huyghens surface is broken, and, second, the wave reflected on the boundary to be tested, and on the upper side of the domain. Setting the beginning of computation when the wave is just striking the corner of the Huyghens surface (Fig. 5), the reflected wave will arrive at a time t_0, whereas the undesired fields will arrive at times t_1, t_2, t_3. With the distances defined on Fig. 5, some calculations yield the following times:

$$ct_0 = h\cos\theta + a_1\sin\theta + d\cos\theta \tag{44.a}$$

$$ct_1 = ct_0 + \sqrt{a_1^2 + d^2} - a_1\sin\theta - d\cos\theta \tag{44.b}$$

$$ct_2 = ct_0 + \sqrt{a_2^2 + d^2} + a_2\sin\theta - d\cos\theta \tag{44.c}$$

$$ct_3 = ct_0 + 2(b - d)\cos\theta \tag{44.d}$$

In practical computations, t_1, t_2, t_3, are set in order to have a desired clear time, and then (44.b), (44.c), (44.d) yield the dimensions a_1, a_2, b of the domain.

Computations have been performed with a 5 * 5-cm cell and a 0.1-ns time step. The incident wave was a Gaussian pulse of 1 ns time constant, and the reflection factor was computed by means of a Fourier transformation of the reflected pulse observed during the clear time at P. The Huyghens surface and P were set respectively 3 and 5 cells from the boundary. Calculations have been performed from normal to 75° incidences, within domains varying from 300 * 80 to 2000 * 130 cells so that the clear times were in the interval 100–200 time steps (10–20 ns). The resolution of the Fourier transformation was then better than 100 Mhz. We used as reflection factors the values obtained from the low frequency limit of the Fourier transformation of the reflected pulse. So, these factors were actually reflection factors around the 100–MHz resolution, at frequencies only slightly disturbed by the FDTD mesh, whose cutoff is about 100 times higher.

TABLE I

Reflection Factor (%) for a Plane Wave Striking a Plane Boundary

Boundary conditions	$\theta = 0$ FDTD * theory	$\theta = 45°$ FDTD * theory	$\theta = 75°$ FDTD * theory
One-way 2	0.010 * 0.000	2.945 * 2.944	34.67 * 34.67
Operator 2	0.003 * 0.000	2.945 * 2.944	34.67 * 34.67
One-way 3	0.001 * 0.000	0.505 * 0.505	20.41 * 20.41
PML(4, C, 1.0)	3.053 * 1.000	4.953 * 3.853	30.53 * 30.36
PML(4, L, 1.0)	1.080 * 1.000	3.991 * 3.853	30.42 * 30.36
PML(4, L, 0.1)	0.059 * 0.100	0.820 * 0.756	16.84 * 16.73
PML(4, L, 0.01)	0.133 * 0.010	0.126 * 0.148	9.358 * 9.220
PML(4, P, 0.01)	0.041 * 0.010	0.234 * 0.148	9.437 * 9.220
PML(4, P, 0.001)	0.012 * 0.001	0.073 * 0.029	5.317 * 5.080
PML(8, P, 0.001)	0.0015 * 0.001	0.038 * 0.029	5.158 * 5.080
PML(8, P, 0.0001)	0.0010 * 0.0001	0.0085 * 0.0057	2.873 * 2.800
Missmatched PML	17.16 * 17.16	17.16 * 17.16	17.15 * 17.16
ML(4, L, 0.1)	0.059 * 0.100	17.10 * 17.16	58.90 * 58.88

Table I presents some results for three incidence angles and various boundary conditions. In each case two values of the reflection factor expressed as percentages are reported. The first value was computed by the FDTD technique and the second one, by the theoretical formula which is either (36) for the PML layer or (43) for the other techniques. The first three lines of the table are related to the one-way and operator techniques, with both second and third orders. One can see that the FDTD results are in very sharp agreement with the theoretical formula (43) at oblique incidences, and that the reflection is very short at normal incidence. The next eight lines are related to eight different PML layers. In Table I and always in this paper, a PML layer is defined by three parameters. The first one is the number of cells. The second one is a letter which means the variation of the conductivity, a C for constant conductivity, a L for linear conductivity, or a P for parabolic conductivity (in (35), $n = 0, 1, 2$, respectively). The last parameter is the theoretical reflection at normal incidence $R(0)$, expressed as a percentage. This factor $R(0)$ is related to the conductivity σ_m defined in (35) since (36) can be rewritten as

$$R(\theta) = [R(0)]^{\cos\theta} \tag{45}$$

with

$$R(0) = e^{-(2/(n+1))(\sigma_m \delta / \varepsilon_0 c)}. \tag{46}$$

Table I gives results for PML layers both four and eight cells thick. The first PML result is for a constant conductivity with $R(0)$ equal to 1%. In this case the reflection factor computed by FDTD is about in accordance with its theoretical value (45) at oblique incidences, but it is three times its theoretical value at normal incidence. So, a constant conductivity produces a strong numerical reflection at the vacuum–layer interface. This reflection is widely reduced when using a linear conductivity, as the next three cases of Table I show. Then, the FDTD factors are in good agreement with their theoretical values when $R(0)$ equals both 1 and 0.1 %. Numerical reflection appears only when $R(0)$ equals 0.01 %, for which the normal reflection is greater than that observed when $R(0)$ equals 0.1 %. As the last two four-cells PML layers of Table I show, a larger reduction of numerical reflection can be obtained from using a parabolic conductivity. Then the FDTD factor is reduced to about 0.01 % at normal incidence when $R(0)$ equals 0.001 %. Finally, the last two PML layers of Table I show that another way to reduce the amount of numerical reflection is to increase the thickness of the layer. With an eight-cell parabolic layer, the reflection at normal incidence is reduced to 0.001 % when using $R(0)$ equal to 0.0001 %. Such a $R(0)$ value allows very short reflection at high incidence angles, less than 3 % at 75°.

For normal and 45° incidences, three conductivity variations, and four-cell layers, Fig. 6 presents the FDTD reflection factor as a function of the theoretical factor $R(0)$. This figure confirms the results of Table I, with a constant conductivity the FDTD reflection cannot be shorter than 3 %, while with respectively linear and parabolic conductivities it can be of the order of 0.1 and 0.01 %. Moreover, we can see that in all cases, under an optimum value of $R(0)$, that reflection rises in the vicinity of the normal incidence. With a parabolic conductivity, for instance, this optimum is in the range 0.001 to 0.0001 %.

For a parabolic conductivity, Fig. 7 gives the reflection factor as a function of the incidence angle. As suggested by Fig. 6, the FDTD factor is close to its theoretical counterpart at high incidence angles. Numerical reflection appears above all in the vicinity of the normal incidence, where the theoretical reflection is short. What appears clearly on Fig. 7 is that the practical reflection can be far shorter than that of the second-order one-way technique, in most of the incidence angle range. At incidence 45° the PML reflection is then about 100 times shorter when using an $R(0)$ factor around 0.0001 %.

The next to last line of Table I gives the reflection factor computed with a PML medium which does not satisfy the matching condition (2). Computations have been performed with a highly absorbing layer in order to have an apparent reflection (45) that is negligible at the incidences of interest and, then, to observe only the reflection at the vacuum–layer interface. The results of Table I were obtained using a PML(15, L, 1.E-10) medium whose conductivity σ_m was set to its normal value 0.098 mho yielded by (46), but whose magnetic conductivity σ_m^* was set to two times the value deduced from (2). At frequencies around 100 MHz, with (10) we had $w_x = \frac{1}{2}$ and, then, the theoretical reflection factor (30) at the vacuum–layer interface was equal to 17.16 %. The results of the FDTD computation reported in Table I are in sharp agreement with that prediction and, especially, the reflection does not depend on the incidence angle.

The last line of Table I gives the reflection factor computed with the ML technique [4–6], a linear conduc-

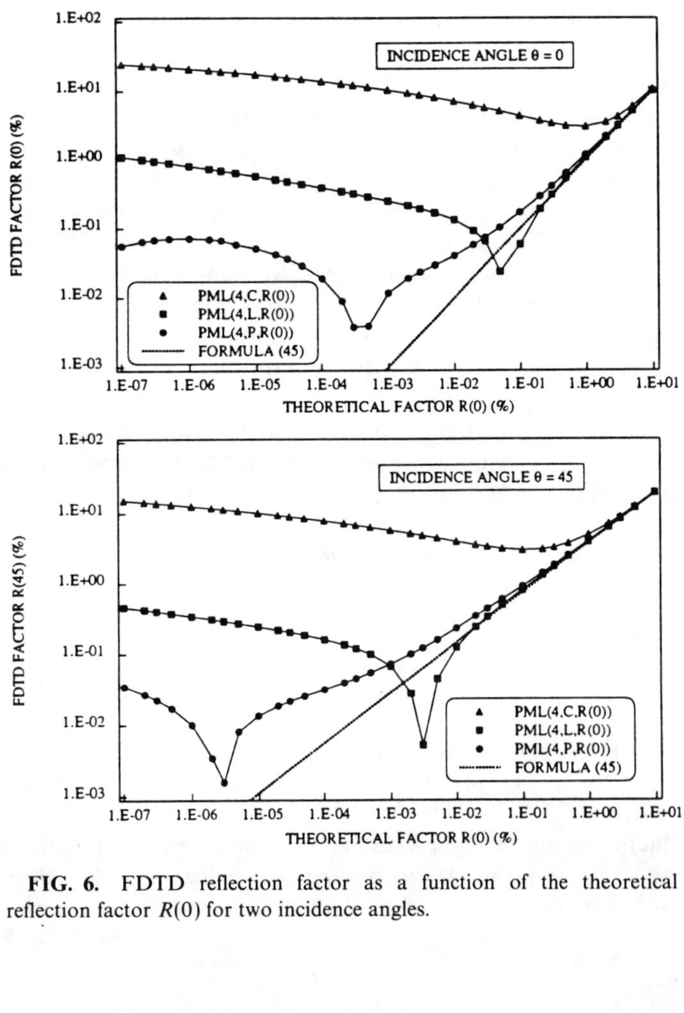

FIG. 6. FDTD reflection factor as a function of the theoretical reflection factor $R(0)$ for two incidence angles.

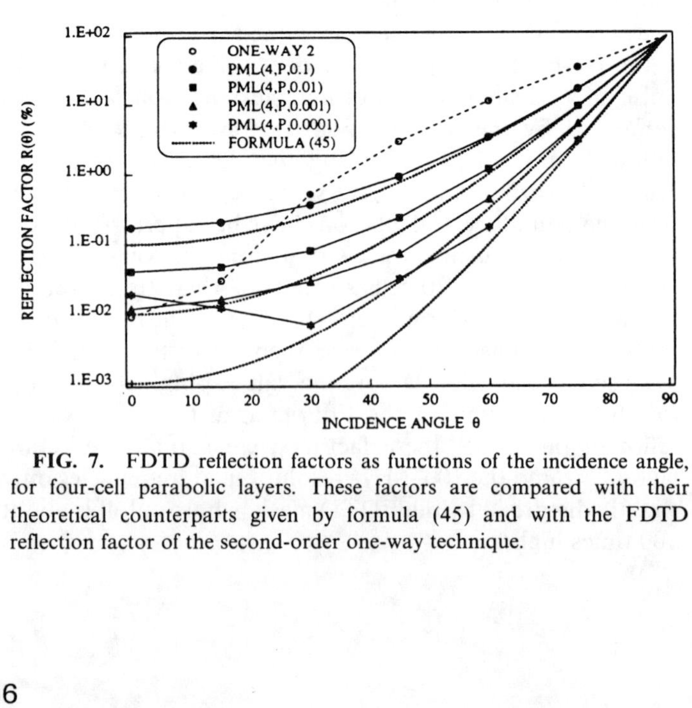

FIG. 7. FDTD reflection factors as functions of the incidence angle, for four-cell parabolic layers. These factors are compared with their theoretical counterparts given by formula (45) and with the FDTD reflection factor of the second-order one-way technique.

tivity, and a 0.1 % apparent reflection at normal incidence. As expected, at oblique incidences the reflection factor is in accordance with the first order of formula (43).

In conclusion, what appears along these first numerical experiments is that the theoretical properties of the PML layer are preserved in the FDTD computations. A small amount of numerical reflection appears at the vacuum–layer interface, but with a magnitude that can be controlled while tuning some parameters of the PML layer, mainly its conductivity profile and number of cells. So, in practical computations, very short FDTD reflection factors can be obtained using the PML technique. For instance, with a PML layer having four cells only, the actual reflection factor can be widely shorter than that of the second order one-way technique.

3.3. Reflection of a Plane Wave Striking a Computational Domain Corner

In order to check that at the corners of a computational domain the absorption of waves is in accordance with the theory of the PML technique, we have performed numericals tests with the domain of Fig. 8. Either the absorbing layer or another boundary condition is set on two sides of the domain in which a plane wave is produced by a Huyghens surface. At a point P outside this surface, the field is the summation of three waves. The first one is reflected on the boundary normal to y, the second one on the boundary normal to x, and the third one on the two boundaries. Since in PML media the shape of waves coming from vacuum is preserved, the reflected components of the field are set as if they were reflected by the conducting planes ending the domain. So, for P close to the corner, so that the phase-shift between the three reflected waves can be neglected, one can deduce the ratios of the reflected components at P over the incident ones,

$$E_x/E_{xi} = R(\pi/2-\theta) - R(\theta) - R(\theta)\,R(\pi/2-\theta) \tag{47.a}$$

$$E_y/E_{yi} = R(\theta) - R(\pi/2-\theta) - R(\theta)\,R(\pi/2-\theta) \tag{47.b}$$

$$H_z/H_{zi} = R(\theta) + R(\pi/2-\theta) + R(\theta)\,R(\pi/2-\theta), \tag{47.c}$$

where $R(\theta)$ is the magnitude of the reflection factor (45), and θ is the incidence defined on Fig. 8. The formulas (47) are also valid for both one-way and operator conditions, with the reflection factor (43) in place of (45).

As in the case of reflection on a plane boundary, the computation is exact during a clear time only. Two non-physical fields will disturb the result at P, coming from B1 and B2, where the Huygens surface is broken. Setting the beginning of computation when the incident wave is just striking the Huyghens surface (Fig. 8) and, since P is close to the corner (5 cells), the times t_0, t_1, t_2 at which the reflected wave and the undesired fields will arrive at P can be approximated as

$$ct_0 = a \sin\theta \tag{48.a}$$

$$ct_1 = ct_0 + a(1-\sin\theta) \tag{48.b}$$

$$ct_2 = ct_0 + b(1-\cos\theta). \tag{48.c}$$

Computations have been performed with 5 * 5-cm cells and a 0.1-ns time step. The results reported in Table II for 45° and 75° incidence angles were obtained within 300 * 300 and 2000 * 120 computational domains, allowing clear times greater than 10 ns.

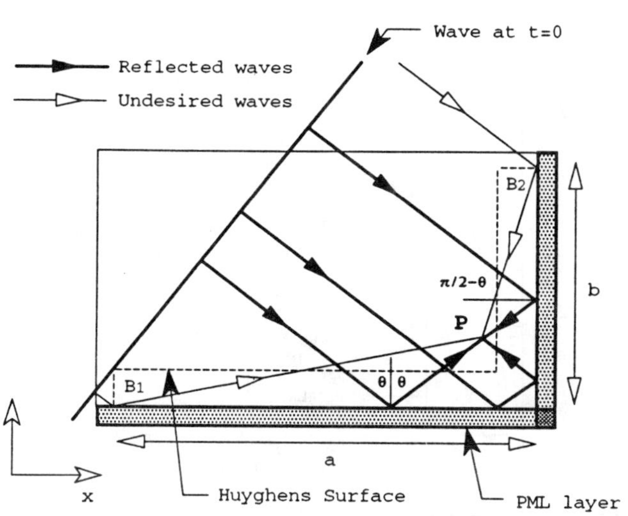

FIG. 8. Computational domain for a plane wave striking a corner.

TABLE II

Reflected Field (%) near a Computational Corner

Boundary conditions	E_x/E_{xi} FDTD * theory	E_y/E_{yi} FDTD * theory	H_z/H_{zi} FDTD * theory
	Incidence angle $\theta = 45°$		
One-way 1	2.942 * 2.944	2.942 * 2.944	37.26 * 37.26
One-way 2	16.13 * 0.087	16.13 * 0.087	18.48 * 5.974
Operator 2	0.087 * 0.087	0.087 * 0.087	5.977 * 5.974
PML(4, L, 1.0)	0.158 * 0.148	0.158 * 0.148	8.141 * 7.854
PML(4, L, 0.01)	1.7E-3 * 2.2E-4	1.7E-3 * 2.2E-4	0.251 * 0.297
PML(4, P, 0.001)	1.5E-3 * 8.5E-6	1.5E-3 * 8.9E-6	0.146 * 0.058
	Incidence angle $\theta = 75°$		
One-way 1	58.16 * 58.17	56.13 * 56.13	61.64 * 61.63
One-way 2	49.97 * 34.65	30.67 * 34.63	39.04 * 34.71
Operator 2	34.64 * 34.65	34.63 * 34.63	34.71 * 34.71
PML(4, L, 1.0)	29.54 * 29.55	28.78 * 28.84	32.06 * 31.89
PML(4, L, 0.01)	9.458 * 9.207	9.484 * 9.205	9.231 * 9.235
PML(4, P, 0.001)	5.297 * 5.079	5.300 * 5.079	5.333 * 5.082

For different techniques of free-space simulation, Table II compares the field components at P computed by both the FDTD method and the analytical formulas (47). With the first-order one-way technique, the results are in good accordance. Inversely, with the second-order one-way, FDTD results and theoretical values are strongly different. That is not surprising since the second-order discretization cannot be applied at points closest to the corner, where we applied the first-order one. With the operator technique, the second-order discretization can be applied everywhere and the FDTD fields are as expected at both incidences. With the PML technique whose three cases are shown in Table II, the FDTD results can be seen as also in good agreement with the analytical formulas. Moreover, the corner does not add any numerical reflection; the computed results are consistant with those obtained when computing the reflection factor of a wave striking a plane boundary. Indeed, using in (47) the FDTD factors as those of either Table I or Fig. 6 in place of (45) yields values close to those computed with the corner domain. For instance, in the case of the PML(4, L, 1) layer and the 45° incidence, this procedure yields 0.159, 0.159, and 8.141 for the three field components. Such values are very close to the results reported in Table II.

As a conclusion for these experiments, with the PML technique the absorption of outgoing waves at the corners is as good as on the plane boundaries. The corners do not raise any problem nor do they need any special treatment.

3.4. Absorption of a Pulse on the Boundaries of a Computational Domain

In order to evaluate the performances of the PML technique when a pulse strikes the boundaries of a computational domain, we have performed the test used in [10] to compare different one-way wave approximations. We have considered the same 100 * 50 cells domain, with the same space and time increments, respectively, 1.5 cm and 25 ps. The domain was surrounded either by some cells of PML medium or by other boundary conditions. Inversely [10], our computations have been done in the TE case, so the pulse was produced at a magnetic point. With time expressed in nanoseconds, this pulse, set at the point (50, 25) of the mesh, was

$$H_z(50, 25) = \tfrac{1}{320}(10 - 15\cos 2\pi t + 6\cos 4\pi t - \cos 6\pi t) \quad (49.a)$$

$$H_z(50, 25) = 0 \qquad \text{if} \quad t > 1 \text{ ns} \quad (49.b)$$

As in [10] a reference solution was computed using a large domain. We used a 400 * 400 domain, allowing a boundary-free solution during 500 time steps. Denoting $H_z(i, j)$ the field in the test domain and $H_{zr}(i, j)$, its counterpart in the reference domain, two kinds of results are presented below. First the reflected field along a boundary ($j = 1$), at time step 100, normalized to the peak value of the field at point (50, 1),

$$R(i) = [H_z(i, 1) - H_{zr}(i, 1)]/H_{zr}(50, 1)_{\max}, \quad (50)$$

and, second, the L^2 norm of the error on the 100 * 50 domain, as a function of time,

$$L^2 = \sum_{i=1}^{100} \sum_{j=1}^{50} [H_z(i, j) - H_{zr}(i, j)]^2. \quad (51)$$

Figure 9 shows some results computed when using the one-way and operator techniques, both with second and third orders, and the PML technique in the case of a four-cell linear layer having a 1% normal reflection $R(0)$. With this PML layer, the reflected field is about the same as that produced by the third-order techniques. Figure 10 shows the variations of the computed results when reducing the parameter $R(0)$, still with a four-cell linear layer. The reflection decreases in the first step, and then increases in the second step when $R(0)$ go below 0.01%. Such an evolution is in accordance with that of the reflection factor of a plane wave, as seen on Fig. 6. With the optimum value of $R(0)$,

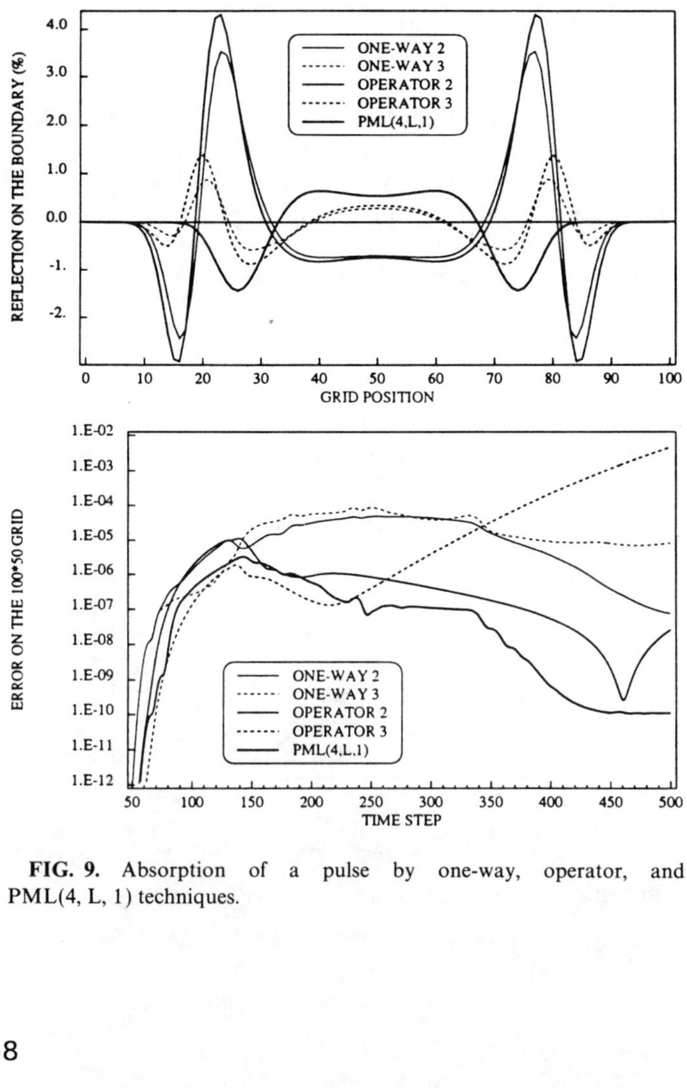

FIG. 9. Absorption of a pulse by one-way, operator, and PML(4, L, 1) techniques.

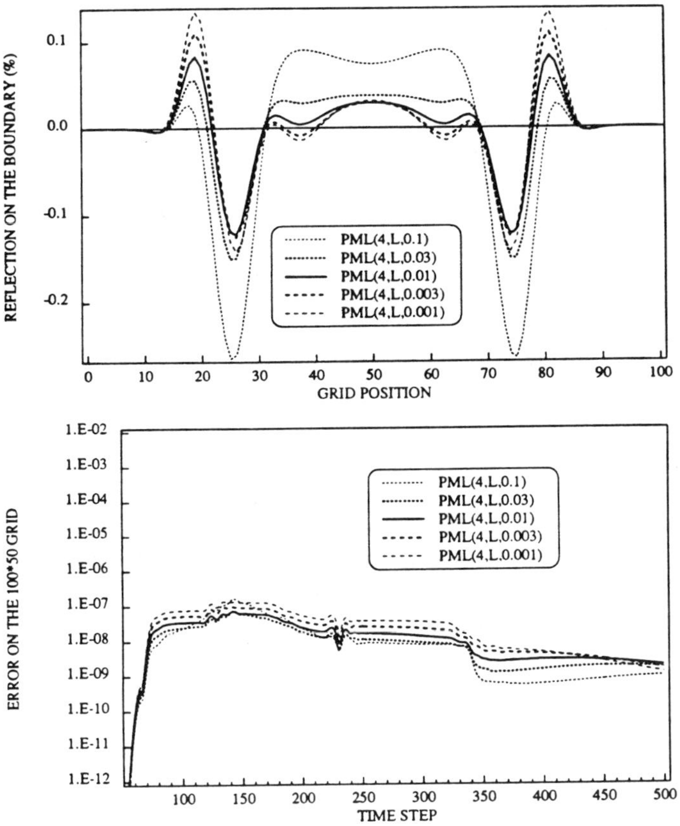

FIG. 10. Absorption of a pulse by a four-cell PML layer for various theoretical factor $R(0)$.

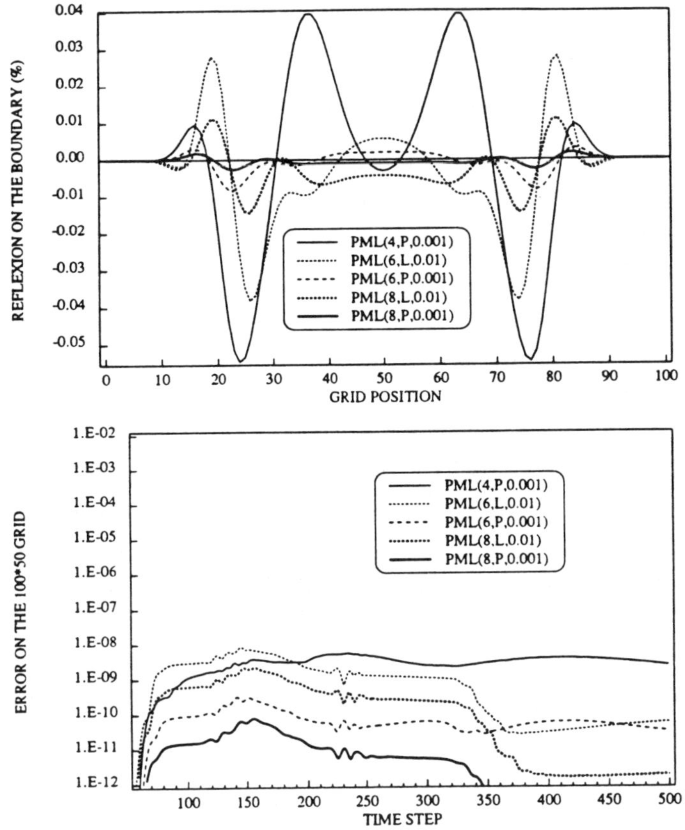

FIG. 11. Absorption of a pulse by four- to eight-cell PML layers.

the reflected field is about 10 times shorter than that of the third-order techniques.

Figure 11 is a comparison of the optimum results obtained with different PML layers, using the procedure of Fig. 10. Layers of four, six, and eight cells are addressed with both linear and parabolic conductivities. As expected, using a parabolic conductivity in place of a linear one reduces the reflection of the pulse. And a greater reduction is obtained when increasing the number of cells. Comparison of Figs. 9 and 11 shows that the PML(8, P, 0.001) layer allows the reflection along the boundary to be about 400 times shorter than that produced by the third-order techniques. Such a factor can be seen also on the average of the magnitude of the reflected field over the entire domain, since the L^2 norm, which is proportional to the energy, is reduced by a factor of the order of 100,000.

In conclusion for these tests, we have observed that the PML technique can drastically reduce the amount of reflection when outgoing waves are striking the boundaries of a computational domain.

3.5. Wave–Structure Interaction in the TE Case

Wave–structure interaction problems are the most usual application of the FDTD technique. These open problems can be satisfactorily solved using the ML or one-way techniques, on condition that the boundaries are set far enough from the scattering structure. In many applications, the structure–boundary distance must be larger than the half-length of the structure. As a result, the greatest part of the computational domain is the vacuum surrounding the scatterer. We will see below that, with the PML technique, the vacuum can be suppressed, and then the computational domain can be widely reduced.

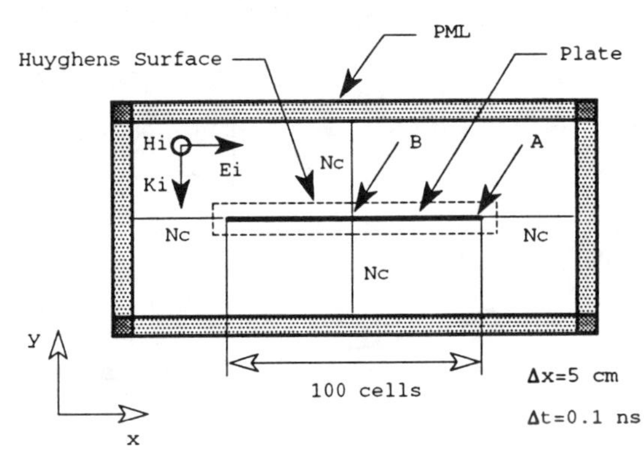

FIG. 12. Computational domain for wave–structure interaction problems.

We have considered the problem of a TE wave scattering on a conducting plate having zero thickness, an infinite width, and a 100 FDTD cells length (5 m). This problem is a two-dimensional one, the principle of computation is presented on Fig. 12. Either the PML layer or another boundary condition is set Nc cells from the plate and the incident wave is produced by a Huyghens surface one cell from the plate. The results are shown here for a wave of the form:

$$E(t) = E_{\text{inc}}[e^{-t/100} - e^{-t}] \qquad (\text{time } t \text{ in ns}). \qquad (52)$$

In order to test the results, a reference solution was computed in a large domain with 700 * 600 cells surrounded by a 10-cell PML layer. With such a domain, the scatterer–boundary distance was then about three times larger than the scatterer length. The results presented below are, first the normal electric field E_y at the end of the plate (A on Fig. 12), and second the magnetic field H_z, equal to the surface current density, at the middle of the plate (B on Fig. 12).

Figure 13 shows some results computed with various boundary conditions set a half-length from the scatterer. With the four-cell PML used, the results are close to the reference solution. Inversely, with the other techniques they are not satisfactory, a larger domain is needed, the scatterer–boundary distance must be as large as the scatterer length for solving this problem. We note that the best of the other three techniques is the operator (42) at early times, but after 2500 time steps (250 ns) there is a mild instability with this technique.

Figure 14 shows the results computed with PML layers set only two cells from the scatterer and with layer thicknesses varying from 4 to 15 cells. With the four-cell PML layer, the current density could be viewed as close to the reference, but the electric field cannot. Reducing the normal reflection factor $R(0)$ does not yield better results. Inversely, increasing the number of PML cells yields results that are closer to the reference solution. With 10 cells, the electric field is good, with 15 cells it is perfect. Considering this last case, the solution is computed within a $134 * 34 = 4556$-cell domain, involving about 18,000 field components, while computations with the other techniques of free-space simulation need domains with at least 300 * 200 cells, involving 180,000 field components. For this problem, using the PML technique can then reduce the computer

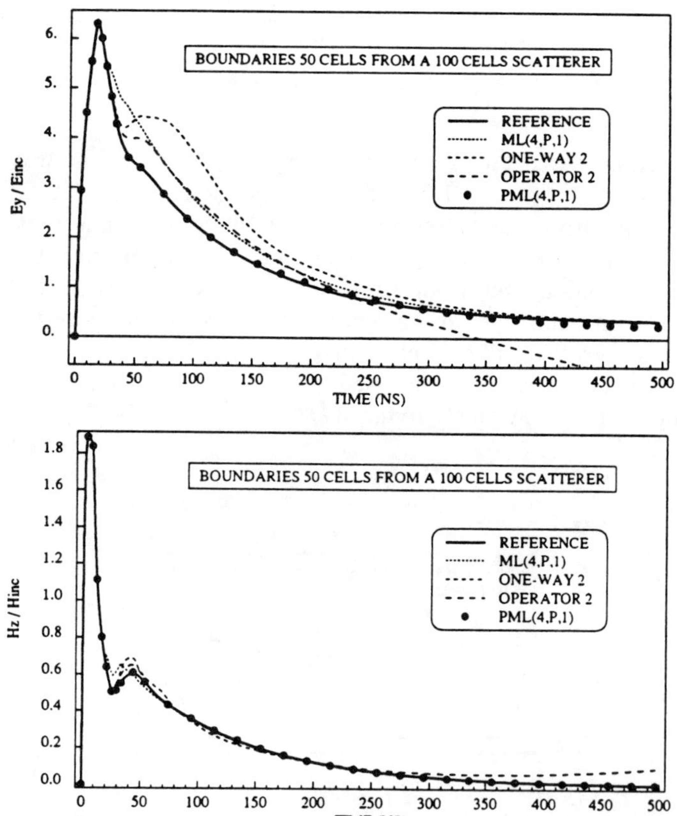

FIG. 13. Wave–structure interaction in the TE case. Electric and magnetic fields computed with boundaries positioned a half-length from the scatterer.

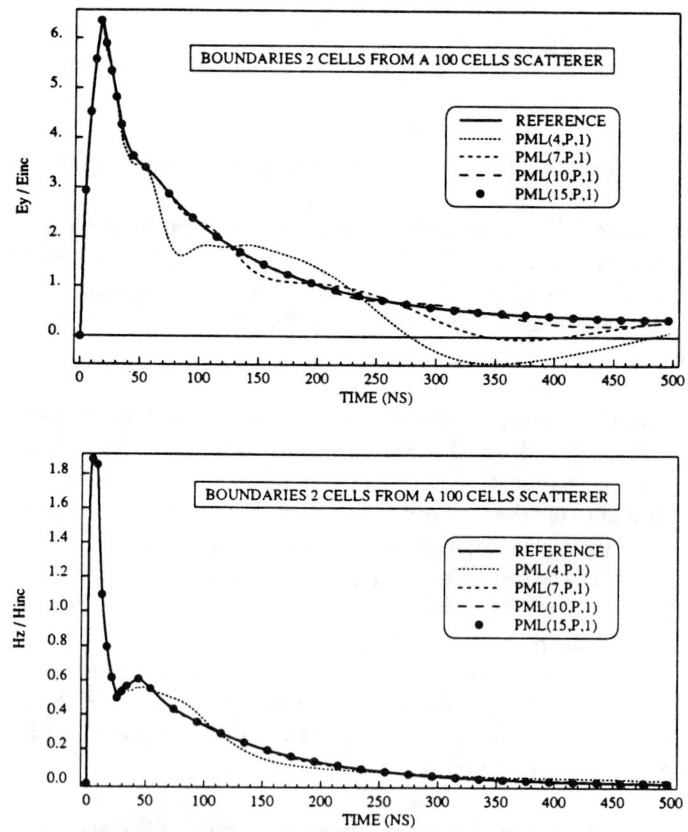

FIG. 14. Wave–structure interaction in the TE case. Electric and magnetic fields computed with the PML layer positioned two cells from the scatterer.

requirements by a factor of 10 for both memory and computational time.

So, the PML technique allows us to solve wave-structure interaction problems without vacuum around the scatterer. The layer must have a thickness of a few cells, but nevertheless the computer requirements can be widely reduced as seen with a two-dimensional problem. Since computers can now accomodate three-dimensional scatterers of 100 cells or more, drastic reductions of such requirements could be hoped for, by extending the PML technique to three-dimensional problems.

3.6. Wave–Structure Interaction in the TM Case

We have computed the current induced by a plane wave on a cylinder of a square cross section. This problem was previously solved by the FDTD method in [12] and used in [10] to compare the second and third orders of the one-way technique. Computations have been performed with the same conditions as in [10], with a cylinder of side equal to 20 cells, a 1.592-cm cell, a 100-MHz incident wave propagating normally to one side, and a six-sinusoidal cycles duration. The principle of computation is still that of Fig. 12, except that the square cylinder replaces the thin plate and that Eq. (41) are used in place of (3) in the PML layer. In order to evaluate the accuracy of the results, a reference solution was computed using a $1000 * 1000$ domain which allowed a boundary-free solution during six sinusoidal cycles.

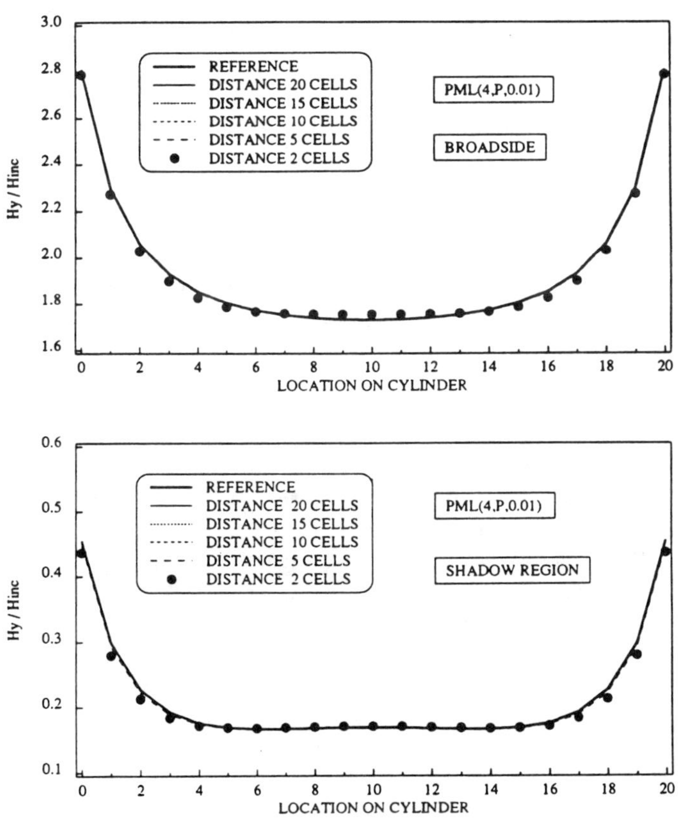

Fig. 15. Wave–structure interaction in the TM case. Magnetic field on a cylinder of square cross section for various cylinder–layer distances.

For a domain surrounded by a PML(4, P, 0.01) layer, Fig. 15 presents the magnetic field, equal to the current density, on both the broadside and the shadow side. In each case, the results have been computed with various cylinder–boundary distances, from 2 to 20 cells. We can see that these results are almost perfect for any distance. For 5 to 20 cells, the curves are not distinct from the reference ones, only a slight difference can be observed when the distance equals 2 cells. So, as in the TE case, the PML technique allows us to suppress the surrounding vacuum which is needed when using the ML, one-way, or operator techniques. Then, the computer requirements can be widely reduced too.

3.7. Free-Space Radiation of a Slot

We have computed the field radiated by an infinite slot located in a perfectly conducting plane when an electromagnetic wave propagates on one side of that plane. This problem is a two-dimensional one; the principle of computation is shown on Fig. 16. The finite-difference domain is split into two parts by the conducting plane in which a 10-cell wide slot (50 cm) is located. The incident wave above the plane is produced by a Huyghens surface, and the goal of the computation is to obtain the free-space radiated field below the plane.

The computation has been performed within a $80 * 42$-cell domain. The incident plane wave was a Gaussian pulse and the results have been translated to the frequency domain by Fourier transformation. In order to evaluate their accuracy, a reference solution was computed using a $1000 * 1000$ domain allowing a boundary-free solution during about 1500 time steps (150 ns), a duration which was also that of the main domain computations.

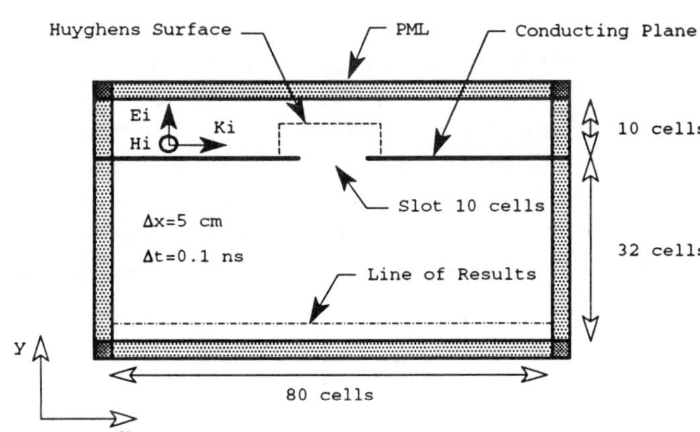

FIG. 16. FDTD domain for computing the radiation of a slot.

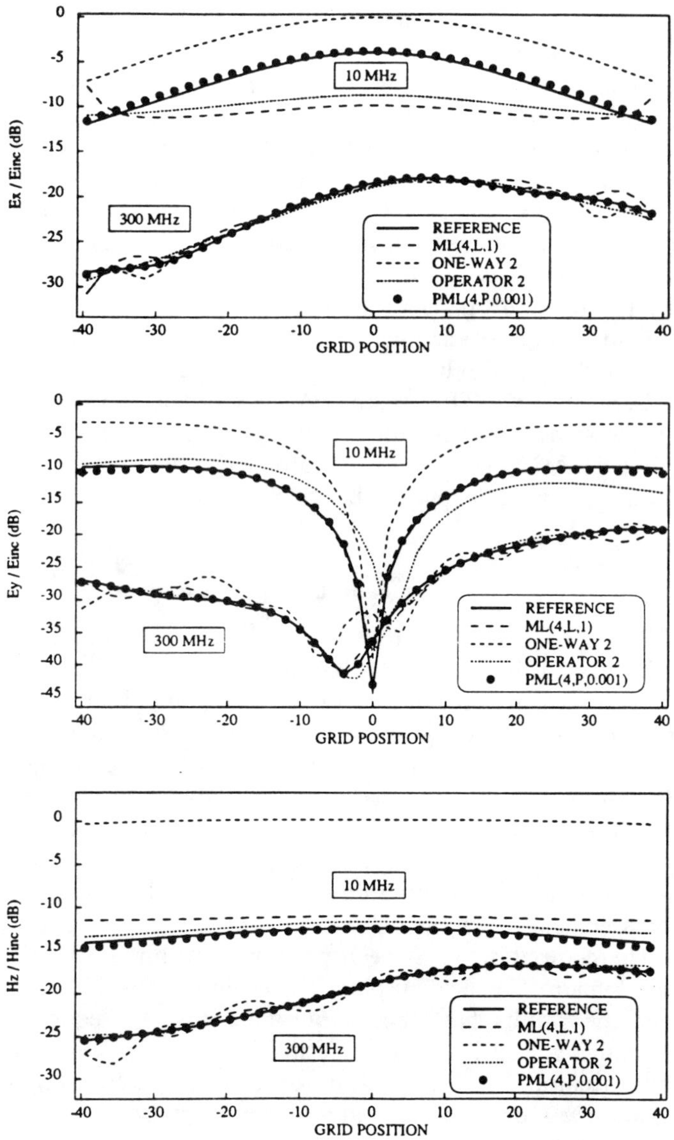

FIG. 17. Radiation of a slot. Components of the field on the line located 30 cells from the slot, computed with ML, one-way 2, operator 2, and PML techniques.

Some results are presented on Fig. 17, along a line located 30 cells below the slot, and then only two cells from the lower boundary conditions. Field components, computed when using either a four-cell PML layer or other boundary conditions, are compared with the reference solution at 10 and 300 MHz. As it appears, the results computed with the PML layer are in good accordance with the reference ones, along the whole line of Fig. 16. That is also true everywhere in the domain. Inversely, with the other techniques, reflections on boundaries produce errors whose magnitude depends on location, frequency, and component.

4. CONCLUSION

In this paper, an absorbing layer which does not activate any reflection at an interface with first vacuum has been described, and then applied to build a new technique of free-space simulation. This PML technique theoretically allows the electromagnetic waves to be absorbed with a reflection as short as needed. Finite-difference experiments have shown that a small amount of numerical reflection occurs in practical computations, but with a magnitude that can be reduced by tuning some parameters of the layer, especially its thickness. So, one can obtain actual reflection factors that are widely shorter than those of the techniques previously used for free-space simulation. As a result, using the PML technique allows us to obtain a better accuracy in some problems and a release of computational requirements in some others.

After having completed the work related above, two ways of further research were initiated: first, the generalization of the PML technique to three-dimensional problems; second, a more detailed analysis of the wave–structure interaction problem in order to understand how some numerical reflection occurs in the TE case, having in mind a reduction of the layer thickness to be used. The first step has not raised any objection; the second one is still under investigation. These topics will be addressed in further papers.

REFERENCES

1. K. S. Yee, *IEEE Trans. Antennas Propag.* **14**, 302 (1966).
2. D. Merewether, *IEEE Trans. Electromagn. Compat.* **13**, 41 (1971).
3. R. Holland, *IEEE Trans. Nucl. Sci.* **24**, 2416 (1977).
4. J. P. Bérenger, Note Technique DGA/ETCA/DET/390, 1977 (unpublished).
5. J. P. Bérenger, in *Actes du Colloque CEM, CNFRS-URSI, Trégastel, France, 1983.*
6. R. Holland and J. W. Williams, *IEEE Trans. Nucl. Sci.* **30**, 4583 (1983).
7. B. Engquist and A. Majda, *Math. Comput.* **31**, 629 (1977).
8. G. Mur, *IEEE Trans. Electromagn. Compat.* **23**, 377 (1981).
9. R. L. Higdon, *Math. Comput.* **49**, 65 (1987).
10. J. Blaschak and G. Kriegsmann, *J. Comput. Phys.* **77**, 109 (1988).
11. A. Taflove, *Wave Motion* **10**, 547 (1988).
12. K. Umashankar and A. Taflove, *IEEE Trans. Electromagn. Compat.* **24**, 397 (1982).

Using Linear and Nonlinear Predictors to Improve the Computational Efficiency of the FD-TD Algorithm

Ji Chen, *Student Member, IEEE*, Chen Wu, *Member, IEEE*, Titus K. Y. Lo, *Member, IEEE*, Ke-Li Wu, *Member, IEEE*, and John Litva, *Senior Member, IEEE*

Abstract— **It is well known that the Finite-Difference Time-Domain (FD–TD) method requires long computation times for solving electromagnetic problems, especially for high-Q structures. The reason for this is because the algorithm is based on the leap-frog formula. In this paper, both linear and nonlinear predictors, which are widely used in signal processing, are introduced to reduce the computation time of the FD-TD algorithm. A short segment of an FD-TD record is used to train the predictor. As long as the predictor is set up properly, an accurate future realization can be obtained. We demonstrate, by means of numerical results, that the efficiency of the FD-TD method can be improved by up to 90%. With this result, the FD-TD algorithm becomes a much more attractive technique for solving electromagnetic problems.**

I. Introduction

THE finite-difference time-domain (FD-TD) method has been used widely for solving electromagnetic problems [1], [2]. It gives the evolution of the fields in time, given a known excitation. This leads to a complete understanding of near fields and of transient effects. Also, it is easily applied to problems composed of complex structures, which may be difficult to solve using other numerical methods. However, a major draw back of the FD-TD method is its computation time. In order to obtain accurate frequency responses via a Fourier transform, a very long time-domain record is usually needed. For some structures, especially those with high Q-values, the computation time may require up to a few days. Premature termination in the time domain will result in inaccurate parameter extraction in the frequency domain. To enhance the FD-TD method for simulating microwave problems, one begins by combining it with signal processing techniques such as the MUSIC method [3], or the System Identification (SI) method [4]. Both of these techniques can give accurate spectra using only a short segment of the original FD-TD record. However, up to now, these methods have only been used to predict the resonant frequency or spectrum of resonant structures. Another successful example of combining the FD-TD method with a signal processing technique, in order to reduce the computational overhead, is to use the FD-TD method with Prony's technique [5], [6]. In this combination, Prony's technique is used to predict the future realization of the time domain response by training a short segment of an FD-TD record and setting up coefficient–based models, where the order is to be determined. Unfortunately, this method may require preknowledge to select the order of the coefficients for different structures [6]. Sometimes, it may be somewhat difficult to find the proper order. Even with these limitations, all of the above signal processing techniques can save computational time when using the FD-TD method to solve electromagnetic problems. It is very clear that the incorporation of signal processing techniques with the FD-TD method is an effective way to improve the efficiency of the FD-TD method. Once the FD-TD technique is formulated to solve a particular electromagnetic problem, the algorithm can be treated as a system, whose execution can be carried out using the appropriate signal processing technique. The fundamental operation of the FD-TD algorithm is based on the leap-frog formula, which means that future realizations are based on calculations that occurred in the past. In this paper, linear and nonlinear predictors are introduced, which can take advantage of the leap-frog nature of the FD-TD method to predict the later response of the system in the time domain. The following section gives detailed descriptions of the linear and nonlinear predictors used in this study. In Section III, some numerical results are given to show how these techniques are used to improve the efficiency of the FD-TD method. Conclusions and future work will be given in Section IV.

II. Methodology

A. General Description of the FD-TD Method

The FD-TD algorithm is a method in which the central difference scheme is used to discretize Maxwell's curl equations in both time and space. The central difference technique can contain the magnitude of the round-off errors so that second-order accuracy is achieved. To model the electrical and magnetic fields in space, Yee introduced the cell system in 1966 [7]. The physical basis for Yee's cell system can be easily explained using Faraday's and Ampere's laws. After using Yee's cell system to describe the computation domain, which is bounded by electric, magnetic or application specific absorbing walls, Maxwell's equations are essentially replaced by a computer which calculates the fields at the grid points associated with the cells.

In this paper, a Gaussian pulse, whose 3dB cut off frequency is designed to overlap with the frequency band of interest, is used as the excitation. The time step is given by

$$\Delta t = 0.5\Delta h/c \qquad (1)$$

Manuscript received May 19, 1993; revised November 19, 1993.

J. Chen, C. Wu, T. K. Y. Lo, and J. Litva are with the Communications Research Laboratory, McMaster University, Hamilton, Ontario, Canada.

K.-L. Wu is with COM DEV Ltd., Cambridge, Ontario, Canada.

IEEE Log Number 9404148.

Reprinted from *IEEE Transactions on Microwave Theory and Techniques*, pp. 1992-1997, October 1994.

where $\Delta h = \min\{\Delta x, \Delta y, \Delta z\}$. $\Delta x, \Delta y$ and Δz are the space steps in X, Y, and Z directions. c is the free-space velocity.

For a high dielectric constant material ($\varepsilon_r = 9.9$), the waves that propagate along the microstrip line are strongly dispersive. It is appropriate, therefore, that Litva's [8] dispersive boundary condition be used here. This boundary condition has been found to have very good performance with dispersive waves.

B. Linear Predictor

Linear predictors or autoregressive (AR) models have found applications in many fields of research [9]. It is the most popular time series modeling approach being used in modern spectral estimation [10]. This is because accurate estimation of the AR parameters can be derived by solving a set of linear equations. In contrast to Prony's method, which uses a model consisting of deterministic exponentials to fit the data, the AR model seeks a random model, to the second-order, to fit a statistical data base. Also, compared with the autoregressive moving average (ARMA) method, AR modeling does not require the solution of a set of highly nonlinear equations.

We say that the time series $x(n), x(n-1), \ldots x(n-p)$ represents the realization of an autoregressive process of order p if it satisfies the equation,

$$x(n) = -a_1 x(n-1) - a_2 x(n-2) - \cdots a_p x(n-p) + \varepsilon \quad (2)$$

where ε is an error term, and $a_1 a_2, \ldots a_p$ are constants called the AR parameters. From equation (2), we see that the present value of the process $u(n)$ is equal to a finite linear combination of past values of the process, $x(n-1), x(n-2), \ldots x(n-p)$, plus the error term $\varepsilon(n)$. Therefore, the present status can be predicted using a linear combination of previous observations. So, as long as the linear predictor is set up properly, through simple extrapolation, we can get the future realization.

There are two issues that must be addressed when setting up an AR model. One is how to choose the order of the model. The other is how to estimate the coefficients. The selection of the model order in AR is a critical problem. Using an order value which is too low results in a high rate of attenuation when extrapolating into the future, while too large an order causes general numerical instability. Two commonly used model order estimators are the final prediction error (FPE) technique and the Akaike Information Criterion (AIC)[10]. They are both based on the estimated predictor error power and are regarded as general guides for AR model selection. Although, the actual order which is selected in practice may be higher than the order given by these two techniques, at least, these two methods give a starting point for the selection of the order of the process. There are a large number of methods for estimating the AR parameters. In this paper, we choose the covariance method.

The covariance for AR parameter estimation yields the solution of the equations,

$$\begin{bmatrix} c_{xx}(1,1) & c_{xx}(1,2) & \cdots & c_{xx}(1,p) \\ c_{xx}(2,1) & c_{xx}(2,2) & \cdots & c_{xx}(2,p) \\ \vdots & \vdots & \vdots & \vdots \\ c_{xx}(p,1) & c_{xx}(p,2) & \cdots & c_{xx}(p,p) \end{bmatrix} \begin{bmatrix} \hat{a}(1) \\ \hat{a}(2) \\ \vdots \\ \hat{a}(p) \end{bmatrix} = - \begin{bmatrix} c_{xx}(1,0) \\ c_{xx}(2,0) \\ \vdots \\ c_{xx}(p,0) \end{bmatrix} \quad (3)$$

where

$$c_{xx}(j,k) = \frac{1}{N-P} \sum_{n=p}^{N-1} x(n-j)x(n-k). \quad (4)$$

The matrix is Hermitian and positive semidefinite. It can be solved by Cholesky decomposition. Under certain conditions [11], the solution of the AR parameters yields the following equations,

$$\begin{bmatrix} x(1) & x(2) & \cdots & x(p) \\ x(2) & x(3) & \cdots & x(p+1) \\ \vdots & \vdots & \vdots & \vdots \\ x(N) & x(N+1) & \cdots & x(p+N) \end{bmatrix} \begin{bmatrix} \hat{a_1} \\ \hat{a_2} \\ \vdots \\ \hat{a_p} \end{bmatrix} = - \begin{bmatrix} x(p+1) \\ x(p+2) \\ \vdots \\ x(p+N) \end{bmatrix}. \quad (5)$$

Normally the value of N is chosen to be greater than $2p$. Then the least-square algorithm is used to obtain the coefficients.

C. Nonlinear Predictor

The development of the upcoming class of nonlinear autoregressive models is motivated by the observation that many processes in nature display random variations which are essentially non-Gaussian in character [12]. In real world applications, it is found that many random processes display essentially nonlinear behavior and hence some form of nonlinear time series modeling, is required to accurately capture the process [13]. These nonlinear models are usually classified under the name of amplitude-dependent exponential autoregressive models. The basic form for an EXPAR(exponential autoregressive) model of order p is

$$x(t) = \sum_{j=1}^{p} [\alpha_j + \beta_j \exp(-\delta x^2(t-j))] x(t-j) + \varepsilon_t \qquad \delta > 0 \quad (6)$$

where ε_t is an error term. δ is a constant. α and β are the EXPAR parameters to be trained. Note that the above model contains no "constant terms" because Ozaki seems to take the view that "if the vibration process starts from the zero initiation state, it stays at zero" [14]. From (6), we can set up equations for the parameter δ as (7) shown at the bottom of the following page.

The equations are then solved using the least squares method. Also, in (7) we require N to be greater than $2p$. Experimental evidence suggests that there can be difficulties with the estimation of δ, which is a critical parameter [13]. Since δ is essentially a scaling factor, it is reasonable to look at values of δ in a range such that $e^{-\delta x_{t-i}^2}$ is different from both zero and equal to one for most values of x_{t-i} [12].

D. Discussion

There are three reasons for our introducing signal processing based models here. They are as follows:

1) As we know, the FD-TD algorithm for electromagnetic simulation is a leap-frog algorithm, which means that the present value is iterated from a previously known value. When we look at the AR and EXPAR models, we find that they also predict their future values based on past occurrences. Since they both use the past to predict the future, this similarity prompts us to incorporate the AR and EXPAR models in the FD-TD algorithm, with a considerable saving in computation time.
2) It is well known in signal processing that a signal consisting of a number of sinusoids can be modeled using AR because of the recursive nature of this algorithm. When we use FD-TD to analyze high-Q structures, we rely on the fact that the time response can be approximated by a summation of sinusoidal waves. It follows that when high-Q structures are analyzed, an AR model can be used to provide an estimate of the future realization because of the similarity between the underlying basis functions for AR and high-Q structures. However, if there are some nonlinear effects in the system, we should add nonlinear terms to the model. Thus, for some cases, an EXPAR model is needed.
3) The classical problem in AR modeling is selecting of the correct order for the process. As mentioned earlier, there are two main tests that are used to determine order. However, in practice, they can't work perfectly all of the time. Normally we will choose a value for the order which is greater than that given by either PFE or AIC in order to be sure that the AR process adequately models the problem. Complex structures normally require high order models. It is found that on occasion the choice of a high order process leads to disastrous results in that the process fails to converge. We find that this conflict can be circumvented by using the EXPAR technique. Failure of convergence does not seem to be a problem when using this technique. In practice, when EXPAR is implemented, computational difficulties disappear.

III. Numerical Results

In this section, three typical electromagnetic problems are studied. They consist of two microwave filters and a patch antenna. The usefulness of AR and EXPAR modeling for improving the computational efficiency of FD-TD will be demonstrated.

Before one uses AR or EXPAR models to analyze FD-TD records, there are two issues that must be addressed. One is the selection of the segment of the FD-TD record to be used for training the AR or EXPAR models. The segment should cover a significant fraction of a FD-TD time record except for the very beginning of the FD-TD waveform. By avoiding the very beginning of the record, we ensure that the intricacies of the structure being analyzed are captured by the data used for training our model. For complex structures, we usually choose the beginning of the training segment to coincide with three times the round–trip time required for the launched pulse to be reflected from an extremity of the structure. The other requirement is that the original FD-TD records must undergo some decimation. To meet the FD-TD stability condition, the FD-TD algorithm usually oversamples the data compared to the needs of the AR and EXPAR models. If the oversampled FD-TD records were to be modeled without decimation, a very high order AR model would be required. This would lead to divergence between predicted and the true values. If we decimate the FD-TD records by a certain factor, the divergence of the predicted result is avoided. In real applications, the FD-TD records are usually decimated by a certain factor.

As our first example, AR modeling is applied to an edge coupled bandpass filter [15] as shown in Fig. 1. In order to obtain an accurate estimation of the S parameters, over 30000 iterations are needed if the original FD-TD method is used. By using the AR approach, only 2000–3500 iterations of the FD-TD algorithm are required to generate the original data base for the AR process, denoted as $\{u\}$. Decimating the $\{u\}$ by the factor 10, we get a time series $\{x\}$. Then $\{x\}$ is used to develop an AR model of order 50. Fig. 2 gives a comparison between the result obtained by FD-TD plus AR model and the direct FD-TD computation alone. It should be added that once the AR parameters are set up, the AR model can accurately predict very long time-domain traces of transient waveforms. Fig. 3 shows the S parameter calculated by the FD-TD plus AR model, and the direct FD-TD using 3500 FD-TD time

$$\begin{bmatrix} x(1) & x(2) & \cdots & x(p) & \exp(-\delta x^2(1)) & \exp(-\delta x^2(2)) & \cdots & \exp(-\delta x^2(p)) \\ x(2) & x(3) & \cdots & x(p+1) & \exp(-\delta x(2)) & \exp(-\delta x^2(3)) & \cdots & \exp(-\delta x^2(p+1)) \\ \vdots & \vdots & \vdots & \vdots & \vdots & \vdots & \vdots & \vdots \\ x(N) & x(N+1) & \cdots & x(p+N) & \exp(-\delta x^2(N)) & \exp(-\delta x^2(N+1)) & \cdots & \exp(-\delta x^2(p+N)) \end{bmatrix} \times \begin{bmatrix} \hat{a}_1 \\ \hat{a}_2 \\ \vdots \\ \hat{a}_p \\ \hat{\beta}_1 \\ \hat{\beta}_2 \\ \vdots \\ \hat{\beta}_p \end{bmatrix} = - \begin{bmatrix} x(p+1) \\ x(p+2) \\ \vdots \\ x(p+N) \end{bmatrix} \tag{7}$$

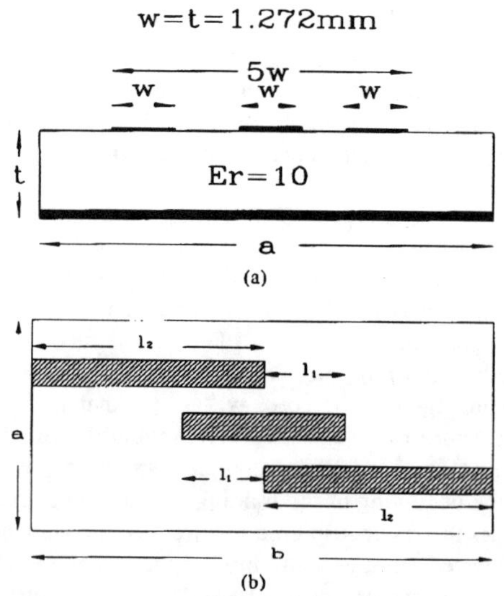

Fig. 1. (a) Plane of edged–coupled bandpass filter. (b) Cross–section of the filter. $l_1 = 6.36$ mm, $l_2 = \frac{b}{2} = 8.48$ mm, $a = 11.62$ mm, $w = t = 1.272$ mm, $\Delta x = \Delta y = \Delta z = \frac{w}{4}$.

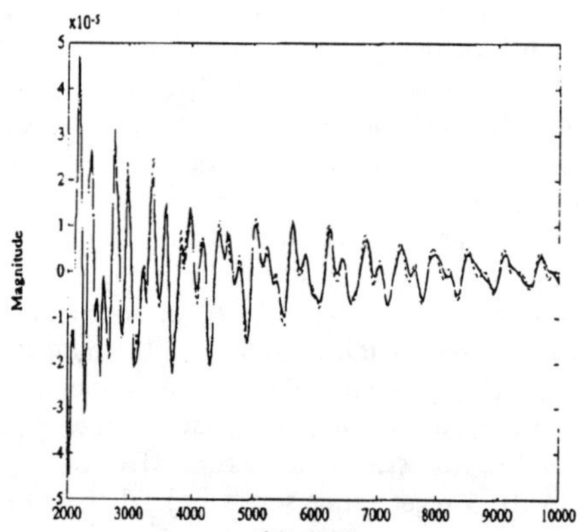

Fig. 2. AR result, composed of a short training segment, plus an extended extrapolated segment, all compared with a completed FD-TD data set.

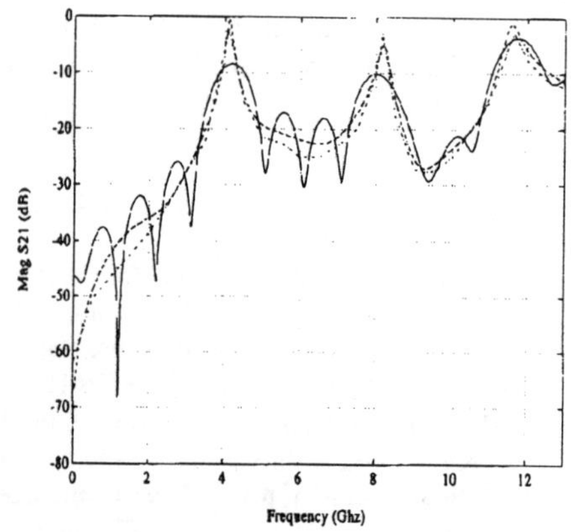

Fig. 3. Edged–coupled microstrip filter scattering parameter S_{21}.

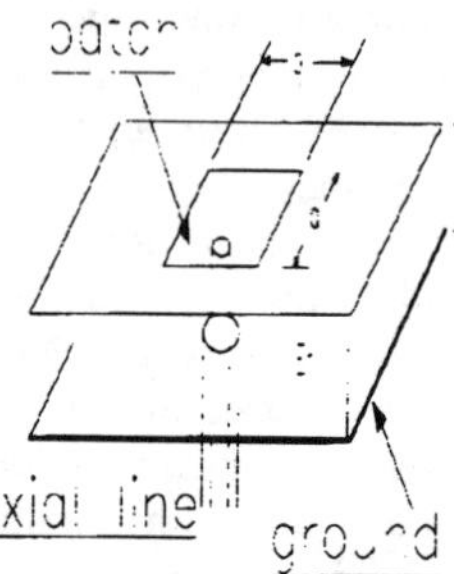

Fig. 4. Coaxial fed microstrip patch antenna,$h = 17.72$ mm, $b = 25.1$ mm, $\varepsilon_r = 2.33$, $h = 2.35$ mm, and feed point at (12.5, 4.0) mm, $\Delta x = \frac{h}{6}$, $\Delta y = \Delta z = 0.3125$ mm.

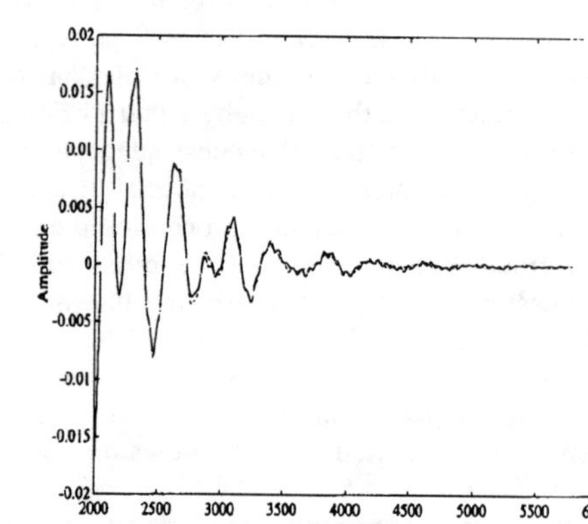

Fig. 5. AR curve fitting and extrapolation using patch antenna FD-TD records from 1200–2500.

domain iterations, as well as a measured result. It is observed that the result obtained using FD-TD plus the AR model agrees very well with the measured results.

The next example [16], is shown in Fig. 4. There we apply the proposed FD-TD plus AR model to a printed antenna. The AR model is based on a data set, which was generated during iterations 1200–2500 following the start of the FD-TD algorithm. Fig. 5 shows the AR extrapolation compared with FD-TD result. The values for the S parameter using FD-TD plus AR model are compared with measured data in Fig. 6.

The third example is the classical double-stub structure which is shown in Fig. 7 [17]. When analyzed using the FD-TD technique, this type of structure requires a large computational space, because the ratio of maximum to minimum dimensions of the structure is large. It also follows that the computational time is extremely long. So long, in fact, that it is practically impossible to solve this problem using the FD-TD algorithm and conventional computing facilities. To start, we use the FD-TD technique and generate a data set of 8900 records. The data set is then decimated by a factor of 25. After that, a 91th order AR model is applied to get the coefficients using records for iteration 3100–8900. The predictor is then used to extend the data set out to 100000 iterations. The transmission coefficient coefficient is then obtained. Thus,

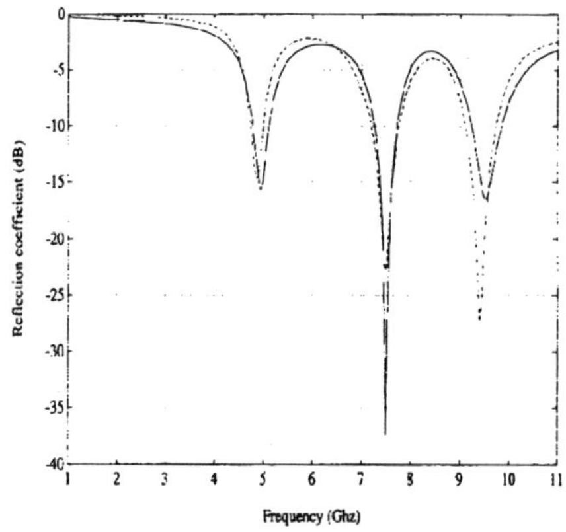

Fig. 6. Scattering parameter S_{11} of coaxial—fed patch antenna.

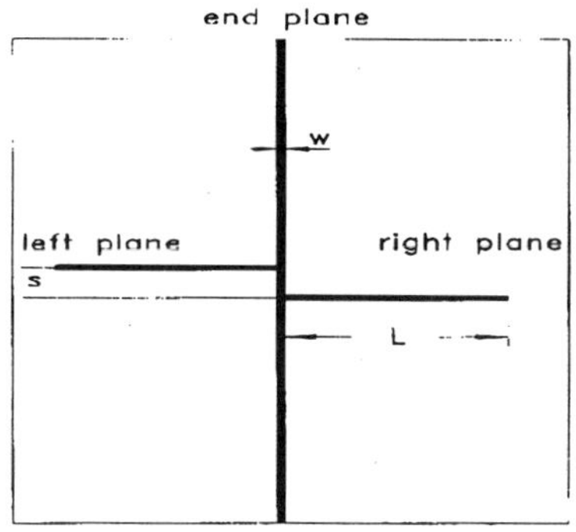

Fig. 7. Configuration of the double-stubstructure, $w = 0.1219$ mm, $L = 2.921$ mm, $s = 0.7569$ mm, $\varepsilon_r = 9.9$, $h = 12.7$ mm, $\Delta x = \frac{h}{4}$, $\Delta y = \Delta z = \frac{w}{4}$.

based on a data set consisting of 100000 records, of which only 10% were obtained using the FD-TD algorithm, we have realized very good agreement with results obtained using a frequency domain technique. However, it is not as easy as the previous example to obtain the AR order for this structure. The results of AR modeling are very sensitive to the order used for the process. When we apply EXPAR to this structure and select the order to be 80, 90 even 100, we find that there is not much change in the predicted data set. On the other hand, it is easier to determine the order of EXPAR (see Fig. 8). This suggests that in practice, if the structure is complex, we simply select an relatively large value for the order of the process, without having to determine the order for each specific problem. However, this matter is still under investigation.

IV. Conclusion

We have used linear and nonlinear predictors to improve the computational efficiency of the FD-TD algorithm. The results show that for equal levels of accuracy, great savings can be realized in CPU time. In our current approach, we use a short FD-TD data segment for setting up the AR parameters and use a predictor to get the future realization. Then we extract S parameters by means of Fourier transforms. In the future, we expect to make a significant change in this procedure. We believe that once the model is set up, that the spectral information is contained in the coefficients. Instead of using a predictor which introduces errors into the result, it is thought that the S parameters can be extracted from the model itself as well as the initial data base. This work will make the system even faster and more accurate. Thus, the EM time-domain simulation technique will become more and more powerful.

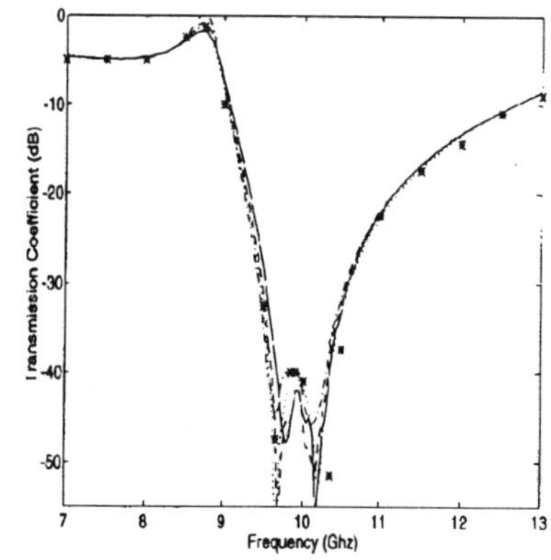

Fig. 8. Using EXPAR model to extract the transmission coefficientof the double-stub structure. $\delta = 1$.

Acknowledgment

The authors wish to thank Dr. Q. Wu, Ms. J. Li and Mr. Z. Hu at Communications Research Laboratory, McMaster University for their helpful discussion on this work. We would also wish to thank Sharon R. Aspden of Rogers Corp. for providing the dielectric material and information used in this research under the auspices of their university program. Finally, the authors acknowledge financial support provided by the Telecommunications Research Institute of Ontario.

References

[1] X. Zhang and K. K. Mei, "Time-domain finite-difference approach the calculation of the frequency-dependent characteristics of microstrip discontinuities," *IEEE Trans. Microwave Theory Tech.*, vol. MTT-36, pp. 1775–1787, Dec. 1988.

[2] C. Wu, K. L. Wu, Z. Bi, and J. Litva, "Accurate characterization of planar printed antennas using finite-difference time-domain method," *IEEE Trans. Antenna Propagat.* vol. 40, pp. 526–534, May 1992.

[3] Z. Bi, Y. Shen, K. L. Wu and J. Litva, " Fast finite-difference time-domain analysis of resonators using digital filtering and spectrum estimation technique," *IEEE Trans. Microwave Theory Tech.*, vol. 40, pp. 1611–1619, Aug. 1992.

[4] W. Kuempel and I. Wolff, "Digital signal processing of time domain field simulation results using the system identification method," *IEEE MTT-S Dig.*, 1992, pp. 793–796.

[5] W. L. Ko and R. Mittra, "A combination of FD-TD and Prony's methods for analyzing microwave integrated circuits," *IEEE Trans. Microwave Theory Tech.*, vol. 39, pp. 2176–2181, Dec. 1991.

[6] J. L. Dubard, D. Pompei, J. Le Rou,x and A. Papiernik, "Characterization of microstrip antennas using the TLM simulation associated with a Prony—Pisrenko method," *Int. J. Numer. Modelling*, vol. 3, no. 4, pp. 269–285, Dec, 1990.

[7] K. S. Yee, "Numerical solution of initial boundary value problems involving Maxwell's equation in isotropic media," *IEEE Trans. Antennas Propagat.*, vol. 14, pp. 302–307, May 1966.

[8] Z. Bi, K. L. Wu, C. Wu, and J. Litva, "A dispersive boundary condition for microstrip component analysis using the FD-TD method," *IEEE Trans. Microwave Theory Tech.* vol. 40, pp. 774–777, Apr. 1992.
[9] S. Haykin, *Nonlinear Method of Spectral Analysis.* New York: Springer-Verlag, 1979.
[10] S. M. Kay, *Modern Spectral Estimation.* Englewood Cliffs, NJ: Prentice-Hall, 1988
[11] L. Ljung, *System Identification—Theory for the User.* Englewood Cliffs, NJ: Prentice-Hall, 1987.
[12] V. Haggan and T. Ozaki, "Modelling nonlinear random vibrations using an amplitude-dependent autoregressive time series model," *Biometrika,* vol. 68, 189–196, 1981.
[13] H. Tong, *Non-linear Time Series.* Cambridge, UK: Oxford University Press1990.
[14] T. Ozaki, "nonlinear threshold autoregressive models for nonlinear random vibrations," *J. Appl. Probab.,* vol. 18, pp. 443–451, 1981.
[15] C. J. Railton and S. A. Meade, "Fast rigorous analysis of shielded planar filter," *IEEE Trans. Microwave Theory Tech.,* vol. 40, pp. 975–985, May 1992.
[16] C. Wu, K. L. Wu, Z. Bi, and J. Litva, "Modelling of coaxial-fed microstrip patch antenna by finite difference time domain method," *Electro. Lett.,* vol. 19, pp. 1691–1692, 1991.
[17] D. C. Chang and J. X. Zheng, "Electromagnetic modeling of passive circuit element in MMIC," *IEEE Trans. Microwave Theory Tech.,* vol. 40, pp. 1741–1747, 1992.

Divergence Preserving Discrete Surface Integral Methods for Maxwell's Curl Equations Using Non-orthogonal Unstructured Grids*

NIEL K. MADSEN

Engineering Research Division, Lawrence Livermore National Laboratory, Livermore, California 94550 and Research Institute for Advanced Computer Science, NASA Ames Research Center, Moffett Field, California 94035

Received March 13, 1992

Several new discrete surface integral methods for solving Maxwell's equations in the time-domain are presented. These methods, which allow the use of general non-orthogonal mixed-polyhedral unstructured grids, are direct generalizations of the canonical staggered-grid finite difference method. These methods are conservative in that they locally preserve "divergence" or charge. Employing mixed polyhedral cells (hexahedral, tetrahedral, etc.), these methods allow more accurate modeling of non-rectangular structures and objects because the traditional "stair-stepped" boundary approximations associated with the orthogonal grid based finite difference methods can be avoided. Numerical results demonstrating the accuracy of these new methods are presented. © 1995 Academic Press, Inc.

INTRODUCTION

The solution of physical problems whose behavior is governed by Maxwell's equations has been of considerable interest for many years. The propagation of electromagnetic signals such as microwaves for communication or radar pulses for the detection of aircraft are two examples of such problems. The conventional approach for numerically solving Maxwell's equations in the time domain has been the use of finite difference methods (FDTD) in conjunction with orthogonal grids [1–5]. It is well known that the use of such methods can produce very accurate results (particularly when the domain is rectangular or almost rectangular). However, many problems involve domains or objects which are sufficiently irregular in their shapes that they cannot be easily approximated with grids consisting of orthogonal cells [6].

Over the past years a number of approaches have been developed to treat problems with irregular shapes or boundaries. The simplest of these approaches has been to use the standard FDTD algorithm and to replace the curved and irregular boundaries with "stair-stepped" approximations. This approach has worked well in some cases but can give rather poor approximations in others, unless very fine discretizatons are used [6]. Another approach has been to map the original irregular domain into a rectangular one and then to apply the FDTD method in the transformed coordinates [7, 8]. This approach suffers from the facts that more complicated equations must be solved and appropriate mapping functions are not always easily obtained. These approaches have been motivated by a desire to retain the use of the basic orthogonal grid FDTD method. The FDTD method has a number of attractive numerical properties. It is charge or divergence preserving both locally and globally. It is second-order accurate and non-dissipative. It is also very computationally inexpensive.

More recently, other numerical algorithms which allow the use of unstructured and irregular grids have been studied. Finite element methods, which have been very successfully used for solving elliptic and parabolic partial differential equations, have been used with Maxwell's equations [9, 10]. These methods allow the use of boundary conforming non-orthogonal and unstructured grids which give one the ability to approximate irregular boundaries with much greater geometrical accuracy. These methods usually provide for global charge or divergence conservation but not for local element charge conservation. When applied on orthogonal structured grids, these methods do not reduce to the familiar FDTD method. In fact, some of the finite element methods, when used with orthogonal grids, reduce to methods which have known difficulties such as grid decoupled solutions [9].

In a desire to use the FDTD algorithm essentially everywhere in a problem and still be able to match irregular boundaries, Taflove [11] has developed an approximation technique that allows the use of special non-orthogonal cells immediately adjacent to the irregular boundary. Local conservation of charge for these special boundary cells is not preserved and second-order accuracy is not assured in general. The primary advantage of this approach is that of very low computational cost. Long term stability of the method is also an open question.

* Work performed under the auspices of the U.S. Department of Energy by the Lawrence Livermore Laboratory under Contract W-7405-ENG-48. The U.S. Government's right to retain a nonexclusive royalty-free license in and to the copyright covering this paper, for governmental purposes, is acknowledged.

Drawing upon experience in computational fluid dynamics (CFD), finite volume methods have recently been introduced to electromagnetics [12]. These methods use flux matching techniques across cell faces to form approximations to Maxwell's equations. In order to achieve stability, these methods require the use of time integration methods which are dissipative. Lax–Wendroff or upwind time differencing methods have typically been used. These methods preserve charge or divergence only to the truncation error level. Even though accurate short time solutions have been demonstrated, there remain accuracy questions. The longer term effects of the dissipative time integration methods and the effects of the lack of exact charge conservation are questions that remain to be answered. These methods also do not reduce to the FDTD method when orthogonal grids are used.

Not related to the above-mentioned CFD-based finite volume methods, but sharing a similar name, modified finite volume (MFV) methods have been recently presented [13]. These methods again allow the use of unstructured non-orthogonal boundary confirming grids and reduce to the traditional FDTD method when used with structured orthogonal grids. These methods are charge conservative and non-dissipative and can produce very accurate solutions. Our first impression of these methods was that they had all of the desirable numerical approximation properties (charge conservation, non-dissipative, direct generalizations of FDTD, etc.). However, for some problems using highly distorted grids, weak instabilities and subsequent non-physical solution growth have been observed.

With a continuing goal of developing algorithms without significant drawbacks for solving Maxwell's equations using unstructured non-orthogonal grids, we seek algorithms which have the following properties:

1. Allow the use of unstructured non-orthogonal grids.
2. Allow a variety of cell or element types.
3. Reduce to the FDTD method when orthogonal grids are used.
4. Preserve charge or divergence locally (and globally).
5. Conditionally stable.
6. Non-dissipative.
7. Accurate for non-orthogonal grids.

In this paper we will present three new algorithms which meet the above criteria. These methods are derived using a discrete surface integration (DSI) technique. One might refer to these methods as finite volume methods, as they are derived from an integration technique. However, since they involve only surface integrations, we prefer the terminology of discrete surface integration techniques. As formulated, the DSI techniques can be used with essentially arbitrary unstructured grids composed of convex polyhedral cells. In our implementation of the DSI algorithms, we allow for the use of unstructured grids which are composed of combinations of non-orthogonal hexahedrons, tetrahedrons, triangular prisms, and pyramids. These new methods actually reduce to the conventional FDTD method when applied on a structured orthogonal hexahedral grid. The DSI techniques are formulated so that local (and, hence, global) conservation of charge or divergence of the fields is inherent in the algorithm and easily demonstrated. They use a leapfrog time integration technique which is conditionally stable and non-dissipative. This paper extends and improves our previous unstructured grid MFV algorithms [9, 13]. The new algorithms have comparable accuracy and overcome the stability or solution growth difficulties which occurred when the MFV algorithms were applied on highly distorted grids.

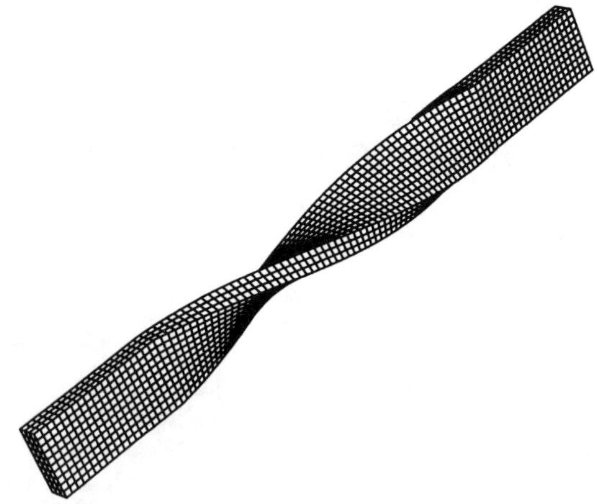

FIG. 1. Twisted waveguide discretized into hexahedral cells.

It goes without saying that algorithms based on unstructured non-orthogonal grids will be significantly more complex and costly than the orthogonal grid FDTD method. However, since the DSI methods reduce to the computationally efficient FDTD method when the grid is orthogonal, we advocate the use of grids which are orthogonal almost everywhere so that advantage may be taken of the simplifications that occur in this case. Even when non-orthogonality exists, the use of efficient data structures allow the new DSI algorithms to be quite computationally efficient.

In the next section, we will describe in detail the new DSI methods. We will then discuss issues concerning the implementation of these methods as they relate to efficiency. We will then describe the numerical experiments that have been performed to assess stability. Finally, numerical results will be presented to demonstrate the accuracy of the DSI methods.

DISCRETE SURFACE INTEGRATION METHODS

We begin by assuming that we wish to solve Maxwell's curl equations on an irregular three-dimensional domain R which has a boundary surface denoted by S. We will also assume that the domain R has been discretized into convex polyhedrons. Figure 1 shows a twisted waveguide discretized using hexahedral cells. This is an example of a problem type that we wish to consider that could not be easily solved using the conventional orthogonal grid FDTD method. Maxwell's curl equations are given by

$$\frac{\partial \mathbf{D}}{\partial t} = \nabla \times \mathbf{H} \tag{1}$$

$$\frac{\partial \mathbf{B}}{\partial t} = -\nabla \times \mathbf{E}, \tag{2}$$

where for linear isotropic materials the vectors **D**, **E**, **B**, and **H** are related by the constitutive relationships

$$\mathbf{D} = \varepsilon \mathbf{E}$$

$$\mathbf{B} = \mu \mathbf{H}.$$

The linear isotropic material properties are: ε, the permittivity, and μ, the permeability. Like the conventional FDTD method, the DSI methods can be generalized to treat more complex materials. However, for the purposes of this paper, we will assume that the linear isotropic material properties are piecewise constant over the domain R. We will also assume that S is a perfectly conducting surface, i.e., $E_{\tan} = 0$. This is sufficient to guarantee that the problem is well posed.

We will derive our new DSI algorithms for unstructured grids formed from convex polyhedral cells. We restrict the choice of cell types to convex polyhedrons whose edges are straight lines. The faces of the polyhedrons are not necessarily planar and we make the assumption that any face in the assembled grid is shared by at most two cells. These are very weak restrictions and allow a great deal of flexibility.

The DSI method requires the use of a dual grid. The dual grid and its structure are completely derivable from a knowledge of the primary grid. For each cell of the primary grid, we define its barycenter to be a node of the dual grid. The barycenter of a cell is located at the average of the coordinates of the nodes which define the cell. We construct edges of the dual grid by connecting barycenters of adjacent cells with straight lines passing through each of the interior cell faces of the primary grid. The barycenters of two cells will be connected (i.e., form a dual edge) if and only if the two cells share a common face. For primary cell faces which lie on the problem boundary, S, we form a corresponding dual edge by joining the cell barycenter to the face barycenter. There is a one-to-one correspondence between the nodes, edges, faces, and cells of the primary grid to the cells, faces, edges, and nodes of the dual grid, respectively. The dual face associated with a primary edge has as its perimeter the dual edges associated with all of the primary faces which share the given primary edge. The dual cell associated with a primary node has as its surface the dual faces which correspond to all of the primary edges which share the primary node. Although not necessary for the definition of the new algorithms, we recommend that the variations in grid sizes and

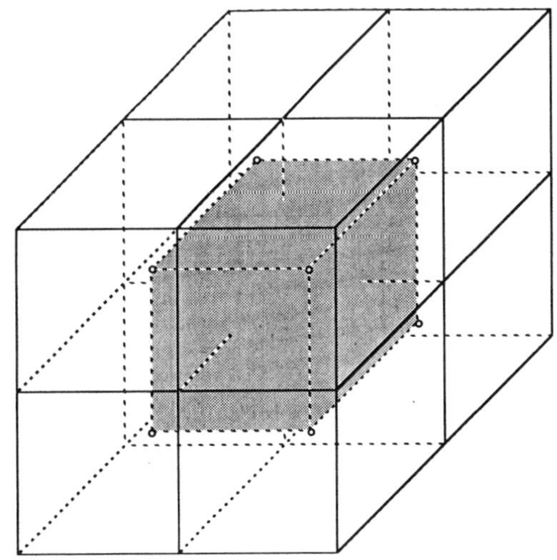

FIG. 2. Primary grid consisting of eight hexahedral cells and its one interior dual cell.

angles be sufficiently smooth so that the primary and dual edges actually intersect their corresponding dual and primary faces, respectively. Degradation of solution accuracy has been observed when this condition is not met. Figure 2 shows an eight cell hexahedral primary grid and its one interior dual cell.

Our DSI solution variables will be associated with the edges and faces of the primary grid and also with the edges and faces of the dual grid. The quantity associated with a primary cell edge is the projection of the electric field vector onto that edge, i.e., $\mathbf{E} \cdot \mathbf{s}$, where $\mathbf{s}$ is the primary cell edge vector. The magnetic field projection $\mathbf{H} \cdot \mathbf{s}^*$ is associated with a dual cell edge, where $\mathbf{s}^*$ is the dual cell edge vector. In addition, with each primary grid face we will associate a full magnetic field vector **B**, and with each dual grid face we will associate a full electric displacement vector **D**. We will denote with an asterisk, *, the geometric quantities associated with the dual grid. Figure 3 depicts these associations. We remark that these associations of field quantities with the primary and dual grid locations are entirely reciprocal and that the respective locations of the

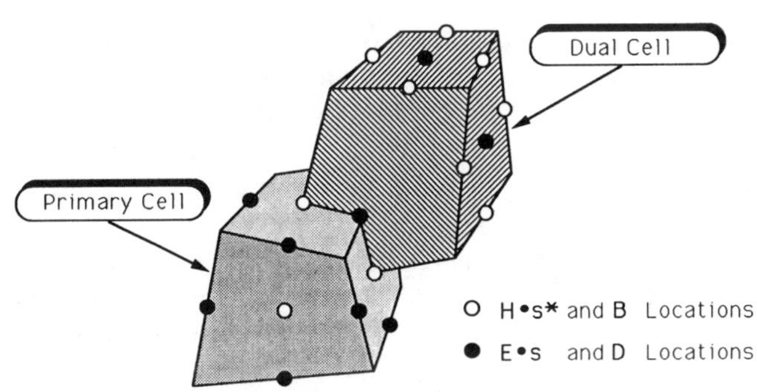

FIG. 3. Discrete electric and magnetic field variable locations relative to the primary and dual grid cells.

magnetic and electric field quantities could be interchanged. The particular choice of which field quantities to associate with each grid is best determined by deciding which field quantities one desires to have on the exterior boundary surfaces where the boundary conditions will be imposed. Since we are assuming that our domain R is surrounded by a perfect electric conductor, we will associate the electric and magnetic field quantities as described above. For open region problems, the choice for the location of the field quantities will depend on the particular radiation boundary condition algorithm used.

We will now describe the equations and algorithmic process used to advance in time the magnetic field vectors which are associated with each primary grid face. We assume that the time variable has been discretized by the choice of a timestep size, Δt. Superscripts on field quantities will denote their time state with $\mathbf{D}^k = \mathbf{D}(t^k) = \mathbf{D}(k\,\Delta t)$. As we will be using a leapfrog style time integration method, the magnetic field vectors will be associated with half-integer times, $t^{k+1/2}$, and the electric field vectors will be associated with integer times, t^k. For a particular primary cell face, we define the area-normal vector to be $\mathbf{N} = \int \mathbf{n}\, dS$, where $\mathbf{n}$ is a unit surface normal defined by the right-hand rule in relation to a specified circulation around the perimeter of the cell face. It is easily shown that $\mathbf{N}$ is uniquely determined by the perimeter of the surface and is independent of the actual interior surface shape. This fact allows for simple computations of $\mathbf{N}$ using piecewise planar approximating surfaces. Using (2) for the primary grid face, F, we define the time derivative of the normal component of the magnetic field to be

$$\frac{d\mathbf{B}^k}{dt}\cdot\mathbf{N}_F \equiv \int_F \left(\frac{\partial \mathbf{B}^k}{\partial t}\cdot\mathbf{n}\right) dS = -\int_F (\nabla\times\mathbf{E}^k)\cdot\mathbf{n}\, dS = -\oint_F \mathbf{E}^k\cdot d\mathbf{l}. \tag{3}$$

Equation (3) allows us to obtain the time derivative of the normal component of the magnetic field on a primary face, F, from the electric field projections onto the perimeter edges of that face. The last integral in (3) is easily computed numerically by summing these edge projections. This can be done for each primary cell face.

The next step in the algorithm is to use these time derivatives of the normal components of the magnetic field to compute a full vector value of $d\mathbf{B}_F^k/dt$ for the primary cell face, F. We will assume that the face F is defined by P primary edges and nodes, with the ith node being located at the intersection of the consecutive edges i and $m = (i \bmod P) + 1$. Also, we assume that the face F is shared by N_c primary grid cells, where by construction: $N_c = 1$ for boundary faces and $N_c = 2$ for interior faces. We will denote by $F_{i,j}$, the face of cell j (other than F) which shares edge i. Figure 4 depicts these associations for a dual edge associated with a primary face defined by five primary edges. At each of the P nodes of face F, we will unfold N_c vector values $d\mathbf{B}_{ij}^k/dt$ by solving the 3×3 system of equations:

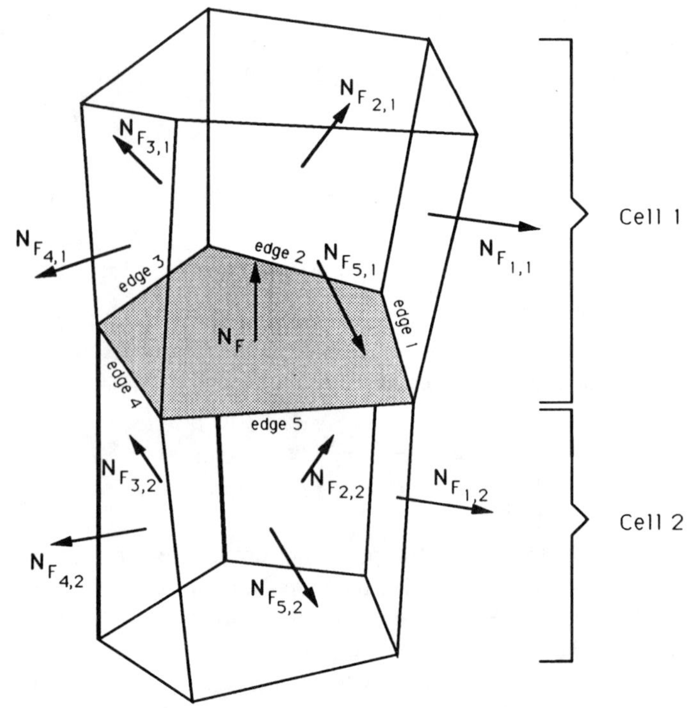

FIG. 4. Primary grid faces used to time advance a magnetic field vector.

$$\begin{aligned}
\frac{d\mathbf{B}_{ij}^k}{dt}\cdot\mathbf{N}_F &= -\oint_F \mathbf{E}^k\cdot d\mathbf{l}\\
\frac{d\mathbf{B}_{ij}^k}{dt}\cdot\mathbf{N}_{F_{i,j}} &= -\oint_{F_{i,j}} \mathbf{E}^k\cdot d\mathbf{l}\\
\frac{d\mathbf{B}_{ij}^k}{dt}\cdot\mathbf{N}_{F_{m,j}} &= -\oint_{F_{m,j}} \mathbf{E}^k\cdot d\mathbf{l},
\end{aligned} \tag{4}$$

where $i = 1, \ldots, P$, $j = 1, \ldots, N_c$, and $m = (i \bmod P) + 1$. For this primary face, F, which is shared by N_c primary cells, there are PN_c different values of $d\mathbf{B}_{ij}^k/dt$, which will now be averaged or interpolated to form a single $d\mathbf{B}_F^k/dt$ vector for the face. There are many reasonable ways to average these vectors and we will consider the three following averaging or interpolating methods:

$$\frac{d\mathbf{B}_F^k}{dt} = \left(\frac{1}{N_c P}\right)\sum_{j=1}^{N_c}\sum_{i=1}^{P}\frac{d\mathbf{B}_{ij}^k}{dt}, \tag{5a}$$

$$\frac{d\mathbf{B}_F^k}{dt} = \left(\frac{1}{N_c}\right)\sum_{j=1}^{N_c}\frac{\sum_{i=1}^{P}|w_{ij}|(d\mathbf{B}_{ij}^k/dt)}{\sum_{i=1}^{P}|w_{ij}|}, \tag{5b}$$

$$\frac{d\mathbf{B}_F^k}{dt} = \frac{\sum_{j=1}^{N_c}\sum_{i=1}^{N_c}|w_{ij}|(d\mathbf{B}_{ij}^k/dt)}{\sum_{j=1}^{N_c}\sum_{i=1}^{P}|w_{ij}|}, \tag{5c}$$

where the weight

$$w_{ij} = \mathbf{N}_F \cdot (\mathbf{N}_{F_{i,j}} \times \mathbf{N}_{F_{i+1,j}})$$

represents the volume of the jth local coordinate system at node i of face F. We note also that the weight w_{ij} is the determinant of the system of Eqs. (4). A different discretization method will be obtained for each different averaging method. We characterize (5a) as a simple vector sum average, (5b) as a partially volume weighted average, and (5c) as a fully volume weighted average. As will be subsequently discussed, numerical evidence indicates a preference for the fully volume weighted averaging scheme (5c).

The full **B** vector for each primary face, F, may now be advanced in time using the time-centered leapfrog algorithm:

$$\mathbf{B}^{k+1/2} = \mathbf{B}^{k-1/2} + \Delta t \frac{d\mathbf{B}^k}{dt}, \tag{6}$$

where Δt is the specified timestep size. Finally, a time advanced value of the projection of the magnetic field onto the dual edge, $\mathbf{s}^*$, which penetrates the primary cell face, F, is easily obtained using

$$(\mathbf{H} \cdot \mathbf{s}^*)^{k+1/2} = \frac{\mathbf{B}^{k+1/2} \cdot \mathbf{s}^*}{\mu}, \tag{7}$$

where μ is an appropriate permeability value. If the permeability is discontinuous then a value can be determined by an appropriate average as is done for FDTD algorithms.

It is important to note that the first equation in (4) (namely, the equation coming from the given face F) is common to all of the sets of 3×3 equations being used in the time advance process for the face. This implies that the averaged value (5) of the time derivative of the magnetic field vector for F also satisfies this equation. It is from this aspect of the numerical algorithm that divergence or charge conservation can be demonstrated. If we integrate the divergence of (6) over a primary cell, C, we obtain

$$\begin{aligned} \int_C (\nabla \cdot \mathbf{B}^{k+1/2})\, dV &= \int_C (\nabla \cdot \mathbf{B}^{k-1/2})\, dV \\ &\quad + \Delta t \int_C \left(\nabla \cdot \frac{d\mathbf{B}^k}{dt}\right) dV. \end{aligned} \tag{8}$$

Considering the last integral term in (8) we have

$$\begin{aligned} \int_C \left(\nabla \cdot \frac{d\mathbf{B}^k}{dt}\right) dV &= \int_{\partial C} \left(\frac{d\mathbf{B}^k}{dt} \cdot \mathbf{n}\right) dS \\ &= -\sum_i \oint_{F_i} \mathbf{E}^k \cdot d\mathbf{l} = 0, \end{aligned} \tag{9}$$

where the sum runs over all the faces F_i of the closed cell C. The last sum is zero because each edge of C will be traversed twice, once in each direction. Thus we see that the local divergence of the time derivative of the magnetic field vector is zero and so if the initial fields have zero local divergence, then (8) and (9) prove that zero divergence of the time advanced fields will be conserved.

To time advance the electric field vectors which are associated with each dual cell face, we proceed in a manner which is exactly "dual" to the magnetic field procedure described above. For any dual cell face, we define the dual area-normal vector to be $\mathbf{N}^* = \int \mathbf{n}^*\, dS^*$, where $\mathbf{n}^*$ is a unit dual surface normal defined by the right-hand rule in relation to a specified circulation around the perimeter of the dual cell face. Using (1) for a given dual grid face F^*, we define the time derivative of the normal component of the electric displacement vector to be

$$\begin{aligned} \frac{d\mathbf{D}^{k+1/2}}{dt} \cdot \mathbf{N}^*_{F^*} &\equiv \int_{F^*} \left(\frac{\partial \mathbf{D}^{k+1/2}}{\partial t} \cdot \mathbf{n}^*\right) dS^* \\ &= \int_{F^*} (\nabla \times \mathbf{H}^{k+1/2}) \cdot \mathbf{n}^*\, dS^* \\ &= \oint_{F^*} \mathbf{H}^{k+1/2} \cdot d\mathbf{l}^*. \end{aligned} \tag{10}$$

Again, the last integral in (10) is easily computed by summing the projections of the magnetic field on the edges defining the dual cell face. This can be done for each dual cell face.

The next step in the algorithm is to use these time derivatives of the normal components of the electric field to compute a full vector value of $d\mathbf{D}^k_{F^*}/dt$ for the dual cell face, F^*. We will assume that the face F^* is defined by P^* dual edges and nodes, with the ith dual node being located at the intersection of dual edges i and $m = (i \bmod P^*) + 1$. Also, we assume that the dual face F^* is shared by N^*_c dual cells. We will denote by $F^*_{i,j}$, the dual face of dual cell j (other than F^*) which shares dual edge i. Figure 4 with all quantities replaced by their appropriate duals would depict these associations. At each of the P^* dual nodes of dual face F^*, we will unfold N^*_c vector values $d\mathbf{D}^{k+1/2}_{ij}/dt$ by solving the 3×3 system of equations

$$\begin{aligned} \frac{d\mathbf{D}^{k+1/2}_{ij}}{dt} \cdot \mathbf{N}^*_{F^*} &= \oint_{F^*} \mathbf{H}^{k+1/2} \cdot d\mathbf{l}^* \\ \frac{d\mathbf{D}^{k+1/2}_{ij}}{dt} \cdot \mathbf{N}^*_{F^*_{i,j}} &= \oint_{F^*_{i,j}} \mathbf{H}^{k+1/2} \cdot d\mathbf{l}^* \\ \frac{d\mathbf{D}^{k+1/2}_{ij}}{dt} \cdot \mathbf{N}^*_{F^*_{m,j}} &= \oint_{F^*_{m,j}} \mathbf{H}^{k+1/2} \cdot d\mathbf{l}^*, \end{aligned} \tag{11}$$

where $i = 1, \ldots, P^*$, $j = 1, \ldots, N^*_c$, and $m = (i \bmod P^*) + 1$. For this dual face, F^*, which is shared by N^*_c dual cells, there are $P^*N^*_c$ different values of $d\mathbf{D}^{k+1/2}_{ij}/dt$, which will now be averaged or interpolated to form a single $d\mathbf{D}_{F*}^{k+1/2}/dt$ vector for the dual face. The averaging methods (5a)–(5c) used for the magnetic fields are also used for the electric fields.

The full **D** vector for the dual face, $F*$, may now be advanced in time using the time-centered leapfrog algorithm,

$$\mathbf{D}^{k+1} = \mathbf{D}^{k} + \Delta t \frac{d\mathbf{D}^{k+1/2}}{dt}, \tag{12}$$

where Δt is the specified timestep size. Finally, a time advanced value of the projection of the electric field onto the primary edge, **s**, which penetrates the dual cell face, $F*$, is easily obtained using

$$(\mathbf{E} \cdot \mathbf{s})^{k+1/2} = \frac{\mathbf{D}^{k+1/2} \cdot \mathbf{s}}{\varepsilon}, \tag{13}$$

where ε is an appropriate permittivity value. The local conservation of divergence for the electric field can be easily shown in a manner similar to that described above for the magnetic fields.

Equations (3)–(7) for the magnetic field quantities, Eqs. (10)–(13) for the electric field quantities, and a linear averaging method (5), constitute the new divergence conserving DSI approximation methods.

When used with structured grids composed solely of hexahedral cells, the new DSI method differs only slightly from our previous MFV method. In computing the time advanced value of an edge-projected magnetic field value, exactly the same electric field values would be used to compute the required time derivative. However, slightly different weights or coefficients would be used in the DSI method. For unstructured grids, the differences between the DSI method and the MFV method become more pronounced. In general, the DSI method requires fewer dual edge magnetic field projections to time advance a given electric field projection than does the MFV method. For the DSI method, only those dual edges defining the dual faces which are immediately adjacent to the dual face of the primary grid edge being updated are used in the update algorithm. In contrast, the MFV algorithm would use the dual edges of all of the dual faces (potentially many more) of the two dual cells which surround the primary grid edge field value that is being time advanced. Thus we see that the DSI algorithm is in general more compact than is the MFV algorithm.

We note that significant simplifications occur when a primary edge is orthogonal to its dual face, or "dually," when a dual edge is orthogonal to its primary face. When this occurs, the vectors **s** and $\mathbf{N}_{F*}^{*}$ (or $\mathbf{s}^*$ and $\mathbf{N}_F$) are aligned. The time advance of the edge-projected field value may be performed directly using (3) (or (10)) and the averaging (5) may be completely bypassed. Stated another way, when the above orthogonality conditions exist, the time advance of $\mathbf{E} \cdot \mathbf{s}$ (or $\mathbf{H} \cdot \mathbf{s}^*$) may be accomplished using a single line integral around the perimeter of the face $F*$ (or F). It is this fact that demonstrates that the DSI methods are completely equivalent to the canonical FDTD methods when orthogonal hexahedral-based grids are used. Therefore, the DSI methods are direct generalizations of the orthogonal grid FDTD methods for unstructured non-orthogonal grids. These orthogonality conditions may occur locally for only a relatively few edges or they may occur globally for the entire grid. They occur globally when structured grids composed of orthogonal hexahedral cells are used. They also occur globally when grids are used which are three-dimensional analogs of two-dimensional grids formed from Delaunay triangles and their dual Voronoi polygons. We will indicate in the next section how a properly structured code can take advantage of these simplifications when orthogonality occurs.

Exterior (perfect electric conductor) boundary conditions are very easily handled by simply setting $\mathbf{E} \cdot \mathbf{s} = E_{\tan} = 0$ for those edges **s** of the grid which form the boundary. Therefore, the above time advance procedure is not required for the electric field projections for edges which are boundary edges as these projections are determined completely from the boundary conditions. For primary cell edges which have one endpoint on the problem boundary surface S and the other in the interior of R, there will be only one complete dual cell associated with this edge (rather than two). The second dual cell is incomplete as it is clipped by the boundary surface S. For computing $\mathbf{E} \cdot \mathbf{s}$ for such edges, appropriate values for the line integrals around the incomplete dual faces can be determined from the boundary conditions using reflection principles. By choosing primary grids which are locally orthogonal near the exterior boundary, the time advance of the edge projection can be performed as mentioned above without using any of the incomplete boundary dual faces.

Local grid orthogonality at exterior boundaries also facilitates the solution of open region problems (those which are not enclosed by perfect conductors). We have not as yet developed three-dimensional radiation boundary conditions which are useable with general non-orthogonal grids. However, open region problems can be solved using the new DSI methods if the exterior boundary cells are made to be orthogonal. In this case, the radiation boundary condition methods [14] developed for the canonical FDTD methods are easily applied. Again, this is possible because of the equivalency of the DSI methods and the canonical FDTD method for orthogonal grids.

As implied in the DSI algorithm description above, interior material interfaces are also easily handled by using appropriately averaged values for the discontinuous permittivity, ε, and the permeability, μ (as is customary with the conventional orthogonal grid FDTD method).

IMPLEMENTATION OF THE DSI METHOD

In actually implementing the new DSI algorithm, there is considerable flexibility in choosing the variables to be used as the actual computational variables. If computer storage were not a significant consideration, one could keep the edge projected values, the face normal components, and the full solution vectors associated with the faces. However, storage is almost always an issue for large three-dimensional problems and in our implementation we have chosen to use the projections of the vector fields onto the primary and dual grid edges. Direct equations for the time advance of $\mathbf{H} \cdot \mathbf{s}^*$ and $\mathbf{E} \cdot \mathbf{s}$ are easily derived by combining and manipulating (3)–(7) and (10)–(13), respectively. For an implementation using edge projected field values as the variables, the total number of unknown approximate field quantities is clearly equal to the number of cell edges in the primary grid plus the number of cell edges in the dual grid. For a non-orthogonal, but structured, hexahedral grid the time advance of a typical interior $\mathbf{E} \cdot \mathbf{s}$ will involve the use of 20 surrounding magnetic field projected values, $\mathbf{H} \cdot \mathbf{s}^*$. A similar number of electric field values would be used for the time advance of a typical magnetic field value. For orthogonal hexahedral grids, this number reduces to four. These significant potential savings are the basis for our advocating the use of orthogonal grids wherever possible. These potential savings have also motivated us to design an implementation of the algorithm which can benefit from orthogonality when it occurs.

In implementing this new DSI algorithm, we have structured the code so that there first occurs a preprocessing phase. In this phase, the dual grid is constructed from the primary grid information and a dependency graph is constructed for each primary grid and dual grid variable. A dependency graph is simply a list of pointers which point to all of the other grid variables that are required to time advance a specific grid variable. In conjunction with the dependency graph, coefficients are also generated so that any given variable may be advanced in time by computing a single very simple dot product of a coefficient vector with a vector of appropriate variables determined from the dependency graph. In forming the dependency graph, checks for orthogonality are performed and variables that are not needed for the time advance due to reasons of orthogonality are never used or stored in the dependency graph. This allows us to take advantage of the greater efficiencies (both in storage and computational arithmetic) inherent in the algorithm when grid orthogonality exists, either locally or globally. Following the preprocessing phase is the execution phase where the dependency graph, coefficients, and simple dot products are repeatedly used to time advance all of the problem variables.

We have recently been solving some rather large problems using the DSI method on unstructured grids. One particular problem involved an unstructured grid consisting of 262,620 cells, resulting in 1,645,195 degrees of freedom, and required only about 17,493,000 words of memory to execute on a Cray-2. No memory saving techniques such as word packing were used. The computational time required for this problem was about 7 μs per timestep per degree of freedom. The amount of memory and computational time required for a problem can vary significantly, depending upon the structure and orthogonality present in the problem grid.

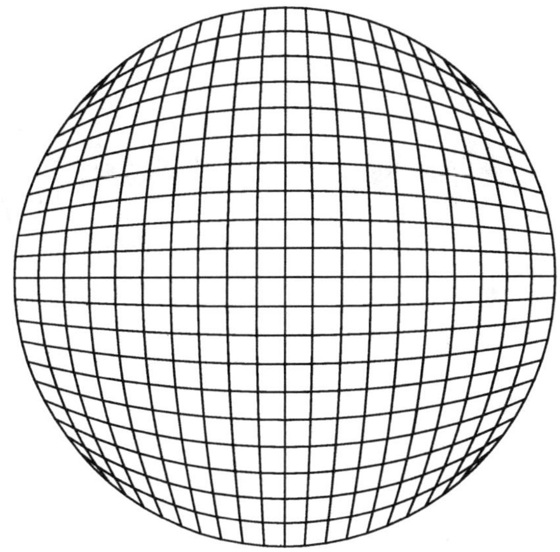

FIG. 5. Highly distorted grid obtained by mapping a 20 × 20 square grid to a circle.

STABILITY EXPERIMENTS

We have presented a description of three new DSI algorithms for solving Maxwell's equations. Each algorithm is distinguished by a different averaging method used to form the time derivative of a vector field associated with a cell face. As these methods reduce to be the canonical FDTD method on orthogonal grids, their stability properties (when used with this type of grid) are the same as for the FDTD methods. As we have been unable to mathematically analyze their stability properties for non-orthogonal unstructured grids, we have performed computational experiments to study this issue. While we recognize that these experiments do not conclusively demonstrate stability, we feel that they are useful to give general indications of stability characteristics.

For testing stability on highly distorted grids, we have found that a grid obtained from mapping a 10 × 10 × 10 cubical grid to a sphere provides a good test. The corner cells of the cubical grid become very highly distorted under the mapping. For two-dimensional testing, a square grid mapped to a circle provides a good test. Figure 5 shows a 20 × 20 two-dimensional square grid mapped to a circle and it indicates the degree of distortion that occurs in the corner cells. These problems can be randomly excited and the solution can be monitored for many thousands of time steps to see if growth occurs in the solution. We judge the stability of an algorithm by measuring the solution growth over time in terms of total energy or maximum solution magnitude.

We have experimented with the three different DSI averaging methods: (5a), a simple vector sum average of all of the time derivative vectors on both sides of the given face; (5b), a vector sum average of the two volume-weighted averages of time deriv-

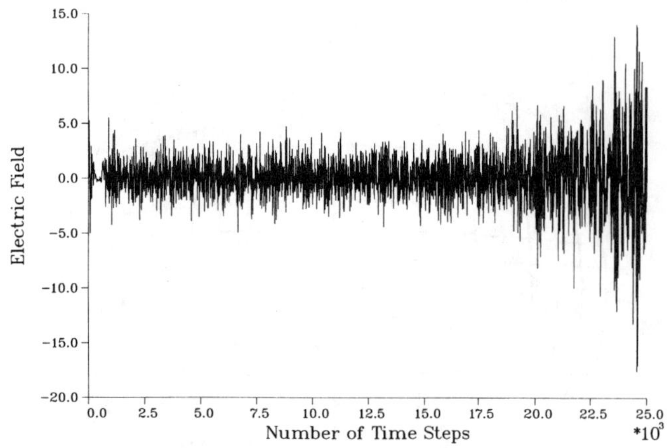

FIG. 6. Long-term solution growth plot for the weakly unstable MFV algorithm.

ative vectors on either side of the given face; and (5c), a full volume-weighted average of all of the time derivative vectors on both sides of the given face. Numerical experiments involving the use of 50,000 timesteps have indicated that all of the three DSI methods are conditionally stable. However, they differ in that they seem to have different stability limits. For the highly distorted spherical grid mentioned above, we have observed differences in the maximum allowable timestep sizes by factors as large as 5. The simple vector sum average method, (5a), always required the smallest timestep size and the full volume-weighted averaging method, (5c), allowed the largest timestep size. The mixed vector sum and volume weighted average method, (5b), always fell between the other two methods. For these stability reasons we advocate the use of the full volume weighting averaging method (5c). In contrast, our previous MFV method [13] is simply unstable for these highly distorted grids. In performing stability experiments with various algorithms, we have observed that it is essential to use a large simulation time (involving many timesteps) so as to allow enough time for signals to transit the computational domain many times. For some distorted grids, our previous MFV method would exhibit no discernible solution growth until many thousands of timesteps had occurred. Figure 6 shows an example of an MFV solved waveguide problem that showed no apparent solution growth until some 20,000 timesteps had been taken. In Fig. 6, the electric field at every tenth timestep for a single point is plotted versus the timestep number. For this same problem, the DSI algorithm does not show any spurious solution growth even to 50,000 timesteps. To determine an appropriate timestep size in solving a particular unstructured grid problem, we compute the smallest primary or dual grid edge length and divide by the local propagation velocity. We then use a timestep size which is half of this computed value. In general, this has been a reliable approach.

ACCURACY EXPERIMENTS

We feel it is important to demonstrate the accuracy of new algorithms by comparing their numerical solutions for problems which have known and computable analytic solutions. We have extensively compared the new DSI method with our previous MFV method [13]. For problems using hexahedral grids, where the MFV method does not encounter stability related solution growth, we have found that the two methods produce solutions which are almost indistinguishable. Therefore, in the results which follow, we will present results only for the full volume weighted averaging DSI method.

We first consider a simple problem which is designed to provide some assessment of the wave propagation characteristics of the new algorithm. We consider the propagation of a TE_{10} signal in a rectangular waveguide which is 1 m in width, 0.3 m in height, and 10.0 m in length. The signal is initiated in the waveguide by specifying the tangential field, E_z, at the left boundary ($y = 0.0$). The function used to drive this signal is

$$E_z(t, x, 0, z) = \sin(1.5\,\pi t)\sin(\pi x). \tag{14}$$

We assume that all of the waveguide walls are perfect electric conductors; i.e., $E_{\tan} = 0$. Normalized values for the material parameters ε and μ are used: $\varepsilon = 1.0$ and $\mu = 1.0$. The signal is allowed to propagate until $t = 8$. The analytic solution for this problem has been derived and computed so we may examine the errors precisely.

We have used a variety of grids to numerically solve this problem. Clearly one simple grid for this problem is an orthogonal hexahedral grid. However, to demonstrate the ability to use non-orthogonal grids, we deliberately skew the $10 \times 100 \times 3$ grid so that it is composed of non-orthogonal hexahedrons as shown in Fig. 7. The skewing for this grid is such that halfway down the length of the guide the cells have angles of about 57° and 123°. Figure 8 compares the time histories of the numerical solutions for the skewed and orthogonal hexahedral grids, respectively, with the analytic solution for a point in the center of the waveguide and 4.0 m down the guide. We note that the significant error occurs as the signal first reaches the observation point. Both solutions exhibit typical dispersion errors; however, the skewed grid results are somewhat better. To more precisely compare the solutions, Fig. 9 shows the maximum errors over the entire waveguide for both the skewed and

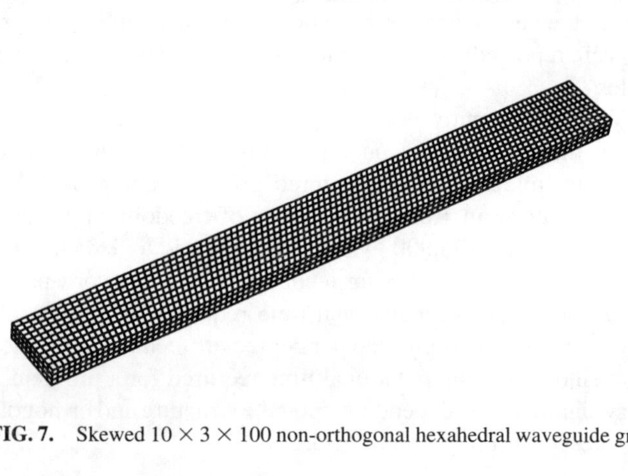

FIG. 7. Skewed $10 \times 3 \times 100$ non-orthogonal hexahedral waveguide grid.

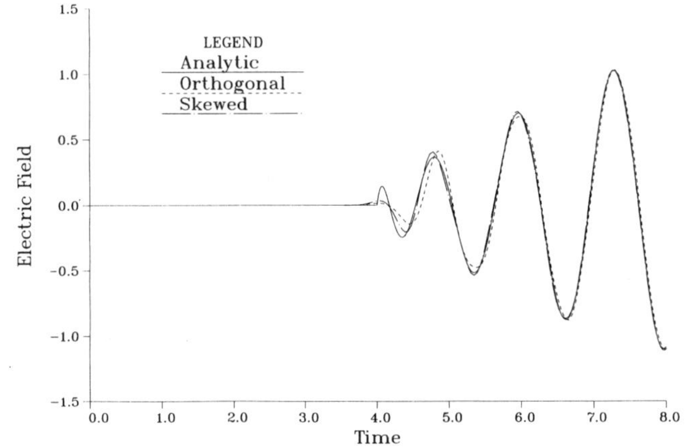

FIG. 8. Numerical TE_{10} waveguide solutions compared with the analytic solution for a point 4.0 m down the guide: (a) analytic solution; (b) orthogonal grid DSI solution; (c) skewed grid DSI solution.

orthogonal hexahedral grid solutions as a function of time. These maximum errors should be referenced to the solution magnitude which oscillates roughly between +1 and −1. Surprisingly, for all but the earliest times, the skewed grid produces a uniformly better solution. To demonstrate that other grid cell types can produce acceptable solutions Table I compares the maximum errors, run times in seconds for an IBM RISC 6000 Model 550 computer, and problem sizes for several different grids. The grids differ in the number of cells and edges. The non-hexahedral grids were all formed from the orthogonal and skewed hexahedral grids by subdividing each hexahedral cell into five tetrahedrons, six pyramids, two tetrahedrons, and two pyramids, or two triangular prisms. The subdivision of a single hexahedral cell into five tetrahedrons and the subdivision into two tetrahedrons and two pyramids is shown in Figs. 10–11.

From Table I one can see that the most accurate results were

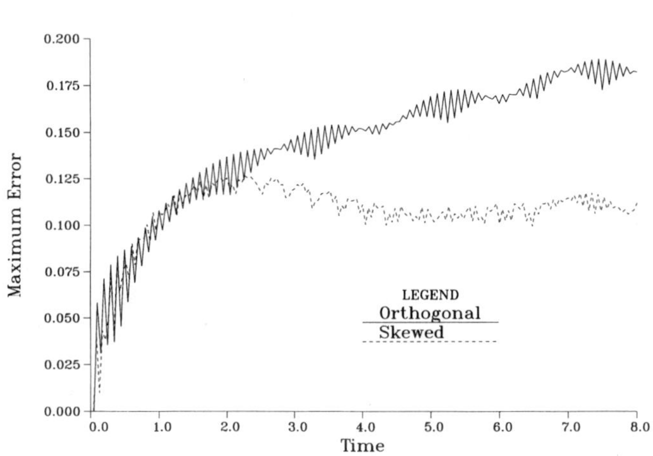

FIG. 9. Comparison of maximum errors as a function of time for the DSI orthogonal and skewed waveguide hexahedral grids.

obtained using the tetrahedral grid derived from the skewed hexahedral grid. However, these results required a great deal more computer time to obtain. The worst results resulted from the use of the skewed grid consisting of tetrahedral and pyramidal cells. The poorer results reflect the fact the grid is composed of highly distorted cells which have very large and very small internal angles. It seems apparent that this problem is most efficently solved using grids consisting of hexahedral cells.

To demonstrate the accuracy that can be gained by the better zoning of non-orthogonal surfaces than can be achieved using only orthogonal cells, we consider the problem of the scattering of a plane wave guassian pulse from a perfectly conducting sphere of radius 0.1 m. The plane-wave pulse is defined by the Gaussian-like function

$$f(t) = \begin{cases} \dfrac{\exp(-10[10^9 t - 1]^2) - \exp(-10)}{1 - \exp(-10)}, & \text{if } 0 \le t \le 2 \times 10^{-9}, \\ 0, & \text{otherwise.} \end{cases}$$

This problem again has a computable analytic solution so we may compare errors precisely. Figure 12 shows the DSI surface grid and a stair-stepped orthogonal grid approximation for the surface of a sphere of radius 0.1 m. Both surface grids are derived from a 6 × 6 × 6 cubical grid. The DSI surface grid is obtained by mapping the surface of the discretized cube to a 0.1-m radius sphere. The stair-stepped orthogonal grid is obtained by eliminating cells from the cubical grid. Figure 13 compares the orthogonal grid finite difference scattered field solution and the DSI scattered field solution with the analytic solution for a point which lies in the shadow region (behind the sphere relative to the direction of the incident pulse) two sphere radii away from the sphere surface. The maximum errors are noticeably different. Significantly more accurate results are obtained by avoiding the orthogonal ''stair-stepped'' grids. It is also noteworthy that the DSI results were obtained using a 2.2-m diameter hemispherical domain which was discretized using only 9600 radially expanding cells. In contrast, the finite difference solution used half of a 2.0-m cubical domain with a 60 × 60 × 30 uniform cubical discretization (108,000 cells).

Another problem that has a known analytic solution is that of a pulsed dipole radiating in free space. We assume that a dipole is located at the origin and is oriented in the z-direction. We will drive it for a finite amount of time with the pulsed amplitude function $f(t) = g(\alpha(t))$, where $\alpha(t) = 4 \times 10^7 - 1$ and where

$$g(x) = \begin{cases} (x^2 - 1)^4, & \text{if } -1 \le x \le 1, \\ 0, & \text{otherwise.} \end{cases}$$

As the solution of this problem depends strongly on this function

TABLE I

Comparison of TE_{10} Waveguide Solution Maximum Errors, Run Times, and Problem Sizes for Various DSI Grids

Grid type		Max error	Steps	Time	Nodes	Edges	Faces	Cells
Hexahedral	Orth	.182	160	6.7	4444	11773	10330	3000
	Skew	.112	184	8.2				
Tetrahedral	Orth	.155	370	26.6	4444	22103	32660	15000
	Skew	.069	441	67.8				
Pyramid	Orth	.188	800	85.2	7444	35773	46330	18000
	Skew	.155	919	206.3				
Tet–pyr	Orth	.181	462	72.8	4444	22130	29660	12000
	Skew	.376	552	89.8				
Tri–prism	Orth	.212	339	20.4	4444	14893	16360	6000
	Skew	.101	362	35.5				

FIG. 10. Hexahedral cell subdivided into five tetrahedrons.

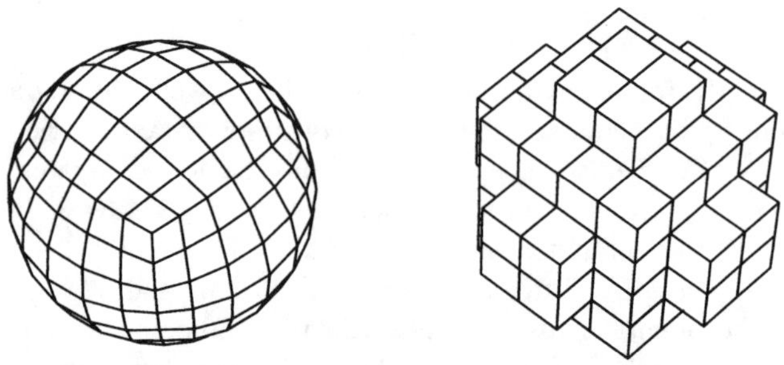

FIG. 12. Surface discretizations of a 0.1 m radius perfectly conducting sphere: (a) boundary conforming DSI surface grid (left); (b) stair-stepped orthogonal FDTD grid (right).

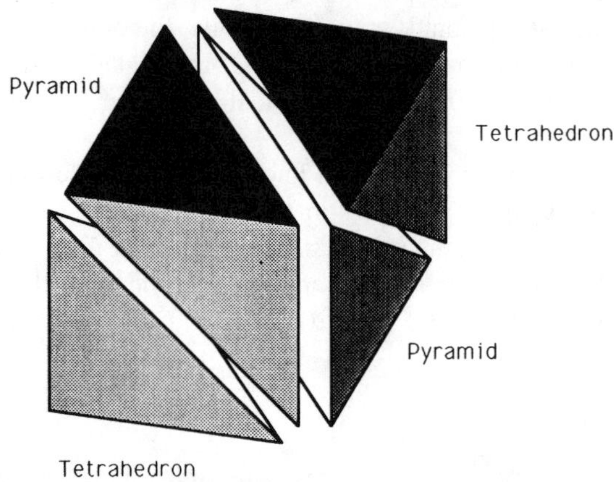

FIG. 11. Hexahedral cell subdivided into two tetrahedrons and two pyramids.

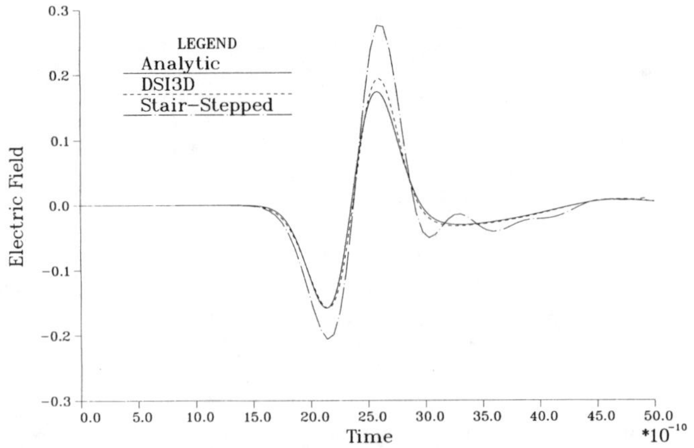

FIG. 13. Comparison of the DSI numerical solution and the stair-stepped orthogonal grid FDTD solution with the analytic solution for a point in the shadow region two sphere radii from the scatterer.

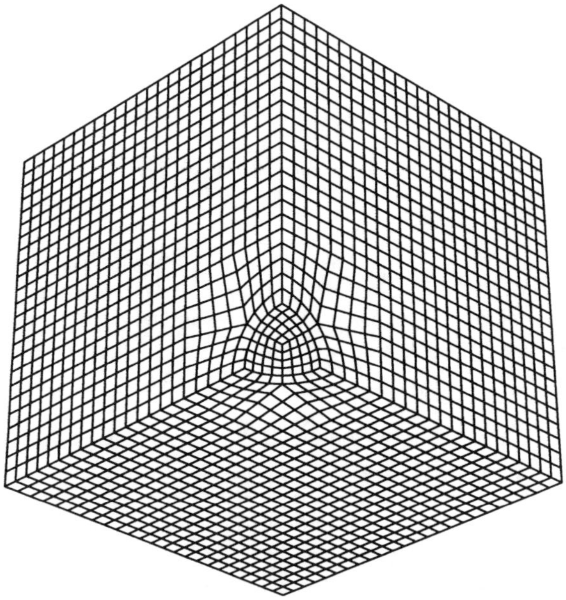

FIG. 14. Mostly orthogonal DSI grid for the dipole problem with a non-orthogonal region transitioning from a spherical surface.

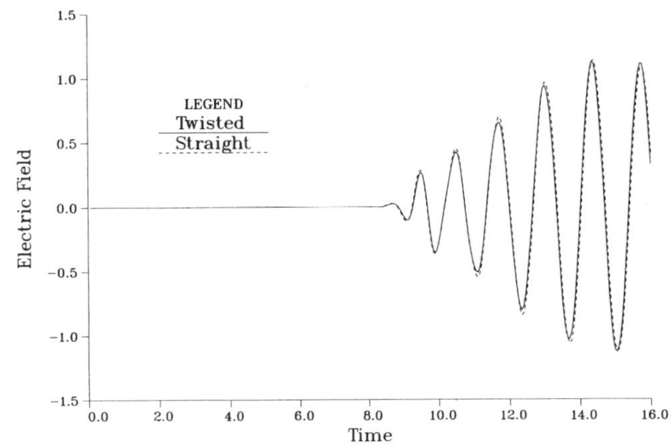

FIG. 16. Comparison of the electric field time histories for the twisted waveguide DSI numerical TE_{10} solution with a non-twisted waveguide DSI numerical solution at a point in the center of the guide immediately beyond the twisted portion.

and its first two derivatives, we have chosen the function to have smooth second derivatives. The analytic solution may be found in almost any standard text or in [15]. To solve this problem, we use a grid that is mostly orthogonal, but in the corner nearest the origin we use a small volume of non-orthogonal cells which transition from a spherical surface to the cubical grid. Figure 14 shows the grid used. We excite the problem by specifying the tangential components of the analytic solution on the spherical boundary surface and compare the numerical solution with the analytic solution at a point just outside of the non-orthogonal region. Figure 15 compares the E_z component of the DSI solution with the analytic solution for a point located slightly outside of the non-orthogonal cell region. We see that excellent results are obtained. The other solution components exhibit comparable accuracy.

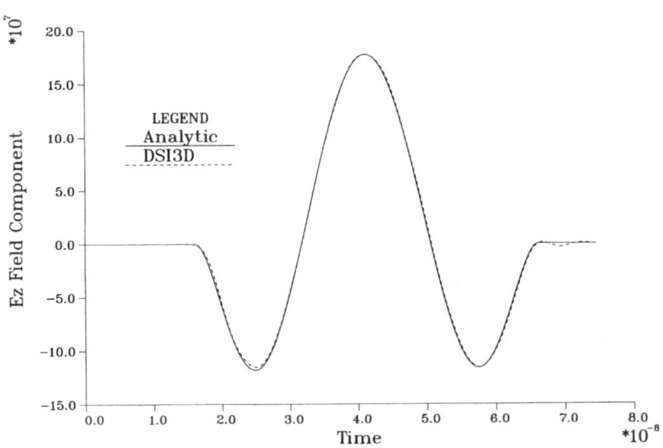

FIG. 15. Comparison of the E_z component of the DSI and analytic dipole solutions for a point just outside the non-orthogonal grid region.

As a final example we consider the problem of computing a TE_{10} waveguide solution for the rectangular waveguide (shown in Fig. 1) which goes through a 180° twist. This problem cannot be effectively solved using a stair-stepped orthogonal grid finite difference algorithm unless a tremendously large number of cells are used. The analytic solution for this problem is not known. However, from actual waveguide experiments, we know that a TE_{10} mode should propagate and be rotated 180° with little distortion by such a waveguide. Figure 1 shows the $10 \times 100 \times 3$ cell DSI grid used and Fig. 16 compares the DSI solution time history for a point in the center of the waveguide immediately after the twist with a TE_{10} solution for a similarly discretized waveguide with no twist. We note that the waves transition the twist with little distortion. Conversion of the TE_{10} mode to other modes does not occur because of the thinness of the waveguide. This problem was excited using the driving function (14).

CONCLUSIONS

The numerical results presented here lead us to believe that the DSI method presented in this paper is one with significant advantages. First, it is a direct generalization for unstructured non-orthogonal grids of the well-known and widely used orthogonal grid FDTD method for Maxwell's curl equations, so its approximation properties on regular orthogonal grids are quite well understood. Second, the DSI method is conditionally stable even for highly distorted grids. This is a distinct improvement when compared with our previous MFV algorithm. Third, the DSI method is conservative in that it conserves the divergence of the fields both locally and globally. Fourth, it is quite accurate (comparing well with our previous MFV algorithm), even when used with irregularly structured grids. Fifth, it is very flexible and can be used with various combinations of different polyhedral discretizations.

The most significant disadvantage of the DSI method is the larger amount of information (primary grid structure plus dual grid structure) that is required. This disadvantage can largely be overcome by using discrete grids which are primarily composed of orthogonal hexahedral cells, together with a relatively few number of irregularly shaped non-orthogonal cells which are used to match boundaries more accurately. Implementing the algorithm in a two-phase way, where all of the geometry information is processed first and condensed into a dependency graph and related coefficients, allows one to take advantage of grid orthogonality when it occurs and also mitigates the extra non-orthogonal grid costs.

ACKNOWLEDGMENTS

The author would like to acknowledge many fruitful conversations concerning this work with Richard Ziolkowski, Scott Ray, Dale Nielsen, Jr., Richard Sincovec, and David Mackay. This work was performed under the auspices of the U.S. Department of Energy by the Lawrence Livermore National Laboratory under Contract W-7405-ENG-48, and also supported by RIACS/USRA and NASA via Cooperative Agreement NCC 2-387.

REFERENCES

1. K. S. Yee, *IEEE Trans. Antennas Propag.* **AP-14**(3), 302 (1966).
2. A. Taflove and M. E. Brodwin, *IEEE Trans. Microwave Theory Technol.* **MTT-23,** 623 (1975).
3. R. Holland, *IEEE Trans. Nucl. Sci.* **NS-24,** 2416 (1977).
4. K. S. Kunz and K. M. Lee, *IEEE Trans. Electromagn. Compat.* **EMC-20,** 328 (1978).
5. A. Taflove and K. R. Umashankar, *J. Elec. Waves Appl.* **1**(3), 243 (1987).
6. R. Holland, "The Case Against Staircasing," in *Proceedings, 6th Annual Review of Progress in Applied Computational Electromagnetics, March 1990,* p. 89 (unpublished).
7. R. Holland, *IEEE Trans. Nucl. Sci.* **NS-30,** 4589 (1983).
8. M. Fusco, *IEEE Trans. Antennas Propag.* **38**(1), (1990).
9. N. K. Madsen and R. W. Ziolkowski, *Wave Motion* **10,** 583 (1988).
10. J. Ambrosiano, S. Brandon, and R. Lohner, "Electromagnetic Propagation on an Unstructured Finite Element Grid," in *Proceedings, 6th Annual Review of Progress in Applied Computational Electromagnetics, March 1990,* p. 155 (unpublished).
11. A. Taflove and K. R. Umashankar, *Electromagnetics* **10**(1–2), 105 (1990).
12. V. Shankar, A. H. Mohammadian, and W. F. Hall, *Electromagnetics* **10**(1–2), 127 (1990).
13. N. K. Madsen and R. W. Ziolkowski, *Electromagnetics* **10**(1–2), 147 (1990).
14. G. Mur, *IEEE Trans. Electromagn. Compat.* **EMC-23,** 1073 (1981).
15. R. W. Ziolkowski and N. K. Madsen, *J. Comput. Phys.* **50**(3), 360 (1983).

Conformal Finite-Difference Time-Domain Methods

C. H. CHAN, H. SANGANI, J. T. ELSON, AND R. F. BOWERS
ELECTROMAGNETICS AND REMOTE SENSING LABORATORY
DEPARTMENT OF ELECTRONIC ENGINEERING,
CITY UNIVERSITY OF HONG KONG
83 TAT CHEE AVE., KOWLOON, HONG KONG

I. Introduction

Finite-difference time-domain methods have been around for almost 30 years, since the monumental paper by Yee [1] in 1966. In its first fifteen years of existence, FDTD had not received much attention in the general electromagnetic community except in the electromagnetic pulse (EMP) research by Merewether [2], Kunz and Lee [3], Holland et al., [4] and others. Tavlove, on the other hand, pursued the use of FDTD for steady-state electromagnetic scattering problems [5,6]. The advent of computer technology in conjunction with the development of accurate absorbing boundary conditions [7–9] in the 1980s changed the fate of FDTD. Because of its simplicity, FDTD has become an indispensable tool in computational electromagnetics. A book devoted entirely to the discussion of FDTD was recently published by Kunz and Lubbers [10].

While FDTD enjoys increasing popularity, critique remains that it lacks the modeling flexibility of its frequency-domain counterpart, viz., the finite-element method. In the finite-element method, the computational domain is typically discretized with a triangular mesh in 2-D problems and a tetrahedral mesh in 3-D problems. These meshes can be readily generated using commercially available software such as PATRAN [11] and I-DEAS [12], among others. It is desirable to develop an FDTD algorithm to utilize these discretization models that have been previously developed for FEM, thus providing a common link between these two seemingly unrelated models.

In this chapter, we will briefly discuss various schemes developed by other researchers [13–57] to enhance the modeling capability of the finite-difference method in the time domain. A few selected reprints of these papers are included at the end of the chapter. In-depth discussions will be provided for our recent development in conformal FDTD using triangular and tetrahedral discretizations for the 2-D and 3-D problems, respectively [58–61].

II. Geometry Modeling

Local Mesh Refinement

A first step to improve the modeling accuracy of FDTD is the local mesh refinement scheme [13–16]. In 1981, Kunz and Simpson [13] proposed an expansion method to treat locally fine objects. This method entails the use of a uniform global coarse mesh and a local uniform fine mesh. A first run is performed with the coarse mesh. Spatial and temporal interpolations are employed to obtain the tangential electric field components at the boundary of the fine mesh. A second run on the fine mesh is made using the interpolated information from the coarse mesh. This local mesh refinement scheme is later improved by Kim and Hoefer [14] in which the coarse and fine mesh regions are calculated simultaneously in the same run. The tangential electric fields on the boundary of the fine mesh are obtained from linear interpolations in space and time of the surrounding electric field points on the coarse grid. For easy interpolation, the spatial and temporal discretizations in the coarse mesh are chosen as the integral multiple of the corresponding discretizations in the fine mesh. In [15], Zivanovic et al. obtained the tangential electric fields on the boundary of the fine mesh by employing a quadratic interpolation based on the homogeneous traveling wave equation which is approximated by a second-order finite-difference equation. The second-order differences are calculated from spatially interpolated field values. On the other hand, Prescott and Shuley [16] reversed the orders so that the second-order differences are first calculated. Spatial interpolation is then performed on the second-order differences. By reversing the order of calculations, they have demonstrated that less memory and CPU time are required than that in [15].

While the locally refined mesh can provide a better modeling accuracy than the original FDTD, it still suffers, to a lesser extent, from the staircasing effect in modeling curved boundaries. It is desirable to remove this modeling error completely.

Curvilinear and Non-orthogonal Grid

The original FDTD as invented by Yee discretizes Maxwell's equations in Cartesian coordinates. FDTD using other orthogonal coordinates such as cylindrical coordinates [62] and spherical coordinates [63] are simple extensions of Yee's algorithm. The first attempt in the generalization of Yee's algorithm to a non-orthogonal grid was provided by Holland in 1983 [17]. Here, the electric field is represented either by its covariant components or its contravariant components. A similar procedure is carried out for the magnetic field. The curl of the covariant components of the electric field yields the negative value of the time derivative of the contravariant components of the magnetic flux density. Similarly, the curl of the covariant components of the magnetic field yields the time derivative of the contravariant components of the electric flux density. Therefore, one needs to convert the contravariant components to its covariant counterparts through the use of a Jacobian matrix.

When compared to FDTD in rectangular coordinates, FDTD in a non-orthogonal grid requires at least twice the computer memory as both the covariant and contravariant components are stored. Despite the added memory requirements, Fusco and his colleagues [18,19] have successfully demonstrated the accuracy of Holland's method for 2-D and 3-D scattering problems. Using the similar idea of representing the field by its covariant and contravariant components, Lee and his colleagues [20–23] time-stepped the integral form of Maxwell's equations and obtained the finite-difference equations in the form similar to those of the original FDTD. The conversions between the covariant and contravariant components, however, are identical to those discussed in Holland's paper [17]. The difference equations in [20–23] are slightly more efficient than those in [17]; nonetheless, the number of operations is roughly three times that of the rectangular FDTD [22]. A microstrip line with a 45° bend, cylindrical cavities, and a cylindrical via discontinuity have been analyzed in [20], [21], and [23], respectively. A study of the numerical dispersion and stability criterion for the 2-D nonorthogonal mesh can be found in [24]. While this nonorthogonal grid FDTD provides accurate results, the extra computer memory and CPU time required are still problematic in addition to the difficulty of generating a curvilinear grid for 3-D structures.

Finite Element Time-Domain with Point Matching

A conformal time-domain finite-difference method was first proposed by Mei et al. [25] in 1984 for 2-D transverse electric (TE to x) scattering problems. In [25], a staggered grid was employed as in the conventional FDTD. The primary grid consists of the electric field grid points. The magnetic field points are located at the centers of quadrilaterals formed by four electric field points, and form the dual grid. The major difference between this method and the conventional FDTD is that the former method is nodal-based while the latter method is edge-based. In [25], both E_y and E_z are sampled in the primary grid while H_x is sampled in the dual grid. To update the magnetic field using finite-differencing in time, the curl of the electric field is evaluated at the dual grid point in which the magnetic field is sampled. The electric field at the dual grid point is evaluated by interpolation from the four surrounding electric field grid points. The interpolation functions chosen are those commonly used in the finite-element method in which the interpolation function corresponding to a particular node (grid point) has a value of one at that node and zero at the other three nodes. Using a vector identity, the curl operation on the electric field (or magnetic field) reduces to the sum of the cross products between the gradient of the interpolation functions and the corresponding nodal electric fields. A similar procedure is carried out to update the electric field. Refinements and extension of this method to 3-D are discussed in [26–32].

Special attention is required to impose the boundary condition on a perfect electric conductor for this nodal based scheme. As stated in [29], the boundary condition on a PEC defines only two of the three components of the electric field at a given node and the charge continuity equation on the conductor is required for the computation of the normal component of the electric field.

This finite-element time-domain scheme with point matching is applied to planar microstrip structures by Konrad and Lo [32]. In [32], TEM propagation is assumed and therefore, time stepping of telegrapher equations is performed. The obtained results compare well with the non-orthogonal grid method by Lee et al. [20]. For a planar geometry, hexahedral isoparametric elements employed in [29] are appropriate. The use of tetrahedral elements in this method has not been demonstrated.

Non-orthogonal Grid Using the Integral Form of Maxwell's Equations

In 1985, Yee developed an FDTD scheme based on the time-stepping of the integral form of Maxwell's equations [33]. In Cartesian coordinates, the time-stepping of the integral and differential forms of Maxwell's equations are identical. To illustrate the advantage of time-stepping the integral form over the differential form, a 2-D cavity in the form of a quarter of a circle was employed [34]. The majority of the computation domain was discretized with the well-known Yee's lattice while the region nearby the circular boundary was discretized with the well-known Yee's lattice while the region nearby the circular boundary was discretized with acute triangles. The idea of time-stepping the integral form of Maxwell's equations was also adopted in [35–40]. While most of these papers were dealing with scattering problems, Fang and Ren [38] applied this technique to a microstrip line with arbitrary shape.

In [38], fields in the whole computational domain are computed by the conventional FDTD except nearby the edges of the planar curved microstrip line in which the smooth edge is approximated by straight line segments. A local correction scheme using the time stepping of the integral form of Maxwell's equations is employed. This method retains the efficiency of the

FDTD as the locally conformed region is small. Depending on how the line segment passes through the rectangular grid, the correction scheme is divided into 16 different categories [38]. Due to the planar nature of the microstrip line, the discretization of the computational domain in the direction normal to the microstrip line is the same as the conventional FDTD. This local correction scheme becomes impractical when dealing with a general 3-D microstrip discontinuity.

Lee and his colleagues [39–40] developed a conformal scheme using a triangular discretization for 2-D scattering problems in conjunction with the integral form of Maxwell's equations. This method can be extended to analyzing the planar curved microstrip problem discussed in [38] by replacing the triangles with prisms. In [39,40], the restriction on the triangularization is that none of the interior angles is larger than 90°. For similar resolution, as we will see later, the increase in required memory and CPU time is about 33% when compared to the conventional FDTD. In [40], a rigorous stability condition is derived for the triangular mesh.

Finite-Volume Time-Domain

Finite-volume time-domain methods are discussed in [41–49]. Similar to the conventional FDTD, the Modified Finite-Volume (MFV) method by Madsen and Ziolkowski also employed a staggered grid consisting of a primary grid for the electric field and a dual grid for the magnetic field. The dual grid is defined by the barycenters of the finite-volumes bounded by the primary grid. The finite-volume algorithm is based on the time stepping of the integral form of Maxwell's equations. The circulation of the electric field along the edge on one face of the volume provides the information to update the magnetic field normal to that face. This normal may be in a different direction of the magnetic field edge on the dual grid. The MFV method calls for a projection scheme to relate the magnetic field information along the normal direction to that along the dual grid. A similar procedure is carried out when updating the electric field. Additional computations are therefore required for the projection.

As it has been pointed out by Lee [22], the MFV is similar to the nonorthogonal grid technique by Holland [17]. Unlike Holland, Lee used the nonorthogonal grid in conjunction with the integral form of Maxwell's equations and therefore, computational complexity of MFV will be the same as in [22]. The finite-volume technique by Shankar et al. in [42,43,45] is based on the techniques developed in computational fluid dynamics and is less familiar to the electromagnetic community.

Overlapping Grids

An overlapping grid technique for scattering problems has been recently proposed in [50,51]. The overlapping grid technique makes use of an orthogonal master grid and a local grid. The local grid conforms to the shape of the irregularly shaped scatterer and its exterior boundary is extended a few zones away from the scatterer. If the scatterer is perfectly conducting, the interior boundary of the local grid will be on the surface of the scatterer on which appropriate boundary conditions are enforced. On the exterior boundary of the master grid, absorbing boundary conditions and/or source excitations are imposed. To transfer information between the master and local grids, the interior boundary of the master grid is placed inside the local grid so that a region of overlapping grids exists. The field information on the exterior boundary of the local grid is obtained from the master grid via a simple bilinear interpolation. Subsequently, the field in the interior region of the local grid can be calculated. The field on the interior boundary of the master grid is then updated via the interpolation from the local grid. This method is particularly useful when the irregularly shaped region is small as the efficiency of the conventional FDTD can be maintained.

Several conformal local grids can be chosen. For simple 2-D scatterers discussed in [50], a curvilinear grid is employed. On the other hand, for 3-D PEC scatterers, a prism type element is used for the local grid [51] in which the field value is updated by the vertex-based FVTD. Similar to the point-matching finite-element time-domain method, the disadvantage of the vertex-based method is that the enforcement of boundary conditions on a material interface is more difficult than the edge-based method. Modifications needed to enforce boundary conditions are discussed in [51]. Differences between vertex-based methods and edge-based methods will be discussed shortly.

Finite-Element Time-Domain

Finite-element time-domain methods have been reported in [52–57]. These methods retain the flexibility and accuracy of their counterparts in the frequency domain. Unlike the point-matching finite-element method discussed in [25–32], a Galerkin's procedure is performed in these methods and therefore provide better accuracy. Another key feature is that these finite-element time-domain methods only involve one field quantity, i.e., the electric field rather than both the electric and magnetic fields as in the point-matching finite-element time-domain approach.

The finite-element time-domain method developed by Lee [57] begins with the two Maxwell curl equations. Taking the curl of the E-field equation and the time derivative of the curl of the H-field equation, one can eliminate the H-field in the formulation. Galerkin's procedure is then performed. Upon expanding the electric field with Whitney elements [64], one obtains a second order differential matrix equation in time in the form of $[T]d^2E/dt^2 + c^2[S]E = -F$.

Here, both $[T]$ and $[S]$ are sparse matrices while E and F are the unknown electric field column vector and known forcing column vector, respectively. With the finite-difference approximation in time for the second order derivative, one can update the electric field vector as in the conventional FDTD except that this would be an implicit scheme; that is, a solution in the form of $[T]x = y$ is required for each time step [57]. This matrix solution, however, can be solved efficiently by the pre-conditioned conjugate gradient method as $[T]$ is a positive definite matrix [57]. A similar formulation is discussed by Mur [55] and Mahadevan et al. [56], and also by Barkeshli et al. [54] with a slight difference due to the presence of a finite conductivity in the computational domain.

In [57], Lee compared the performance of the finite-element time-domain method using the central, forward and backward differences. He concluded that by proper mixing of the equations from these three approximations, an unconditionally stable algorithm is obtained. This algorithm has accuracy comparable to the algorithm based on the central difference alone while alleviating the difficulty encountered in the latter approach, namely, the small time step dictated by the smallest element size for a stable solution. While this new finite-element time-domain method has the accuracy and modeling flexibility of its frequency-domain counterpart, it requires a matrix solution in very time step.

In [52], Lynch and Paulsen employed a "lumping" scheme in which all the off-diagonal nonzero elements are set to zero with their values added on to the diagonal elements. Therefore, the inversion of the modified matrix becomes trivial. If the field values corresponding to the nonzero off-diagonal elements are comparable to the field value corresponding to the diagonal element, or if the off-diagonal elements are small compared to the diagonal element, the lumping approximate seems to be viable. While accurate results have been shown in [52], it is foreseeable that there are situations where this lumping approach may not work. One such situation is calculation of a field changing rapidly in a small region, such as near the conducting edge of a microstrip line.

III. Explicit Time-Domain Schemes

In this section, we discuss various explicit conformal schemes that we have developed [58–61]. We will start with the 2-D case using a triangular discretization and the integral form of Maxwell's equation. Figure 1 shows the backscattered radar cross section of a circular PEC cylinder with a radius of 0.25m for transverse magnetic (TM) and transverse electric (TE) polarizations. A Gaussian pulse illumination is used. The discretization is $\lambda/20$ at 3GHz. The plots are extended to 7GHz to illustrate behavior at higher frequencies. The time-marching algorithm using the triangular discretization is termed triangular element time-domain (TETD). Similarly, the conventional FDTD using the rectangular discretization is termed rectangular element time-domain (RETD) in this section. The number of sampled field points for both methods is comparable, and is about 0.5cm per linear dimension. It is obvious that a conformal scheme such as TETD yields much more accurate results than RETD when compared to analytical data. Furthermore, for the same level of discretization, TETD has similar performance for both TM and TE polarizations while RETD yields erroneous results for the TE polarization.

A Single Matrix Triangular Element Time-Domain Method

We have reported a single matrix triangular element time-domain method [59] for scattering from 2-D arbitrarily shaped cylinders. Consider the triangular discretization shown in Fig. 2 in which the shape of the triangle is controlled so that none of the interior angles of the triangles is larger than 90°. This triangular grid is known as the Delaunay tessellation. In order to minimize the amount of processing of nonuniform triangular

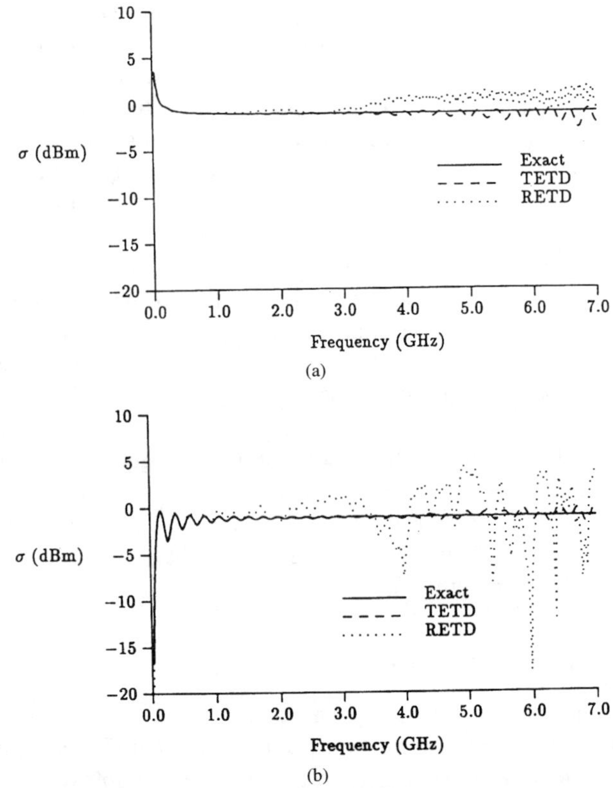

Fig. 1 Radar cross section of a circular cylinder: (a) TM case; (b) TE case.

elements for TETD, the scatterer along with its triangular mesh representation is surrounded by a rectangular grid. Employing rectangular elements on the outer portion offers an advantage in being able to change the distance between the scatterer and the radiation boundary or the grid without having to change any portion of the inner triangular grid. Furthermore, we employ the Liao absorbing boundary condition [63] in which a rectangular outer region is required.

We can take a Delaunay tessellation and locate the point of intersection of the normal bisectors of the three sides of each triangle and connect these points to form a complimentary grid

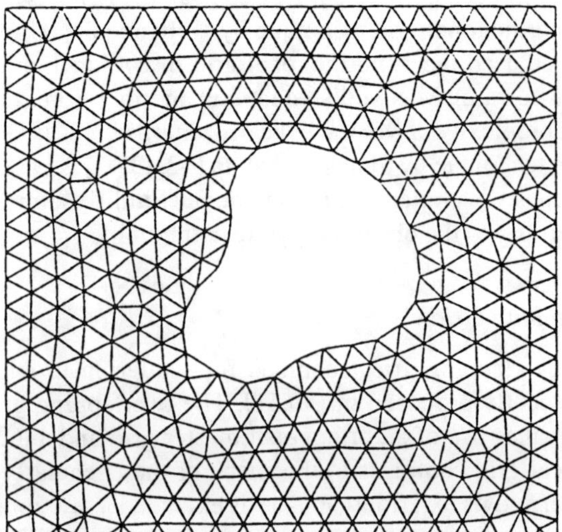

Fig. 2 Triangular grid around an arbitrarily shaped scatterer.

known as the Dirichlet tessellation. These two grids are orthogonal to each other. Let us consider the TM case in which the electric field is sampled at the vertices of the Delaunay grid while the magnetic field is sampled at the midpoint of each segment of the Delaunay grid along the same direction of the Dirichlet grid (Fig. 3). A similar procedure can be applied to the TE case except that the boundary condition is enforced differently.

From the integral form of Maxwell's equations and the discretization scheme depicted in Fig. 3, we obtain the following equations:

$$\varepsilon_0 \frac{\partial}{\partial t} \int \boldsymbol{E} \cdot d\boldsymbol{S} = \int \boldsymbol{H} \cdot d\boldsymbol{l} \tag{1}$$

$$\mu_0 \frac{\partial}{\partial t} \int \boldsymbol{H} \cdot d\boldsymbol{S} = -\int \boldsymbol{E} \cdot d\boldsymbol{l} \tag{2}$$

$$E_j^n = \frac{\delta t}{\varepsilon_0 S_j} \sum_{i=1}^{I_j} H_{k(i,j)}^{n-1/2} L'_{k(i,j)} + E_j^{n-1} \tag{3}$$

$$H_{k(i,j)}^{n+1/2} = \frac{\delta t}{\mu_0} \frac{E_j^n - E_i^n}{L_{ij}} + H_{k(i,j)}^{n-1/2} \tag{4}$$

The index n is a temporal index; j and k are spatial indices. The variable S_j is the area enclosed by the contour integral around E_j, which is indicated by the shaded area in Fig. 3a. L_{ij} is the length of the triangular edge joining E_i and E_j while $L'_{k(ij)}$ is the length of the magnetic field segment $H_k(ij)$. Since the grid is generated in a random fashion, the H_k fields adjacent to E_j are not necessarily in sequential order, hence the unusual notion of indices involving i, j, and k in the above equations. The expression $k_{(j,i)}$ means that an index k corresponds to the i-th H field surrounding E_j, and there are I_j such H fields around E_j. In Fig. 3a, I_j equals 5. The terms E_i and E_j are the two E fields on each side of H_k as seen in the local coordinate system of Fig. 3b. Equations (3) and (4) allow us to alternately update the electric and magnetic fields. Since the grid is irregular, the geometrical parameters of Eqs. (3) and (4) are not the same for a given set of (j, k). The most efficient way to handle this in terms of computer time and memory is to write the equations in matrix form. All geometric parameters are then lumped into matrix coefficients which are calculated only once. The matrix form of the linear equations, Eqs. (3) and (4), is

$$e^n = Ah^{n-1/2} + e^{n-1} \tag{5}$$

$$h^{n+1/2} = Be^n + h^{n-1/2} \tag{6}$$

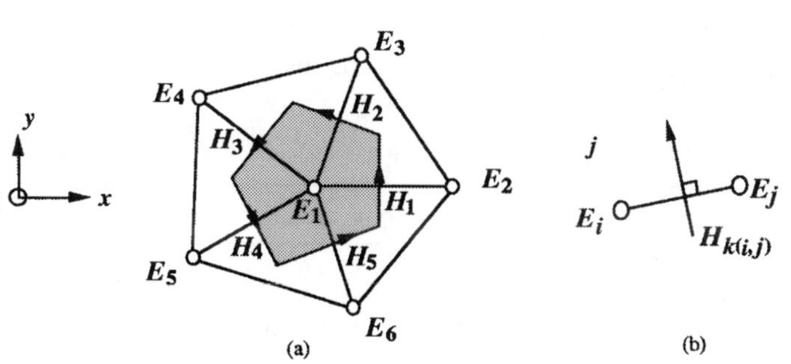

Fig. 3 A computation cell of TETD.

where

$$e = (E_1, E_2, \ldots, E_J)^T$$

$$h = (H_1, H_2, \ldots, H_K)^T$$

Here J and K are the total number of electric and magnetic fields sampled, respectively. The matrix elements of A and B are given as:

$$a_{j,k(j,i)} = \frac{\delta t L'_{k(j,i)}}{\varepsilon_0 S_j} \tag{7}$$

$$b_{k,j} = -b_{k,i} = \frac{\delta t}{\mu_0 L_{ij}} \tag{8}$$

Note that we have yet to define the size of the time step, δt. It is not difficult to see that A and B are sparse matrices and nonzero elements in A appear in the same locations of the transpose of B. As a result, the product of A and B can be easily determined which leads us to combine Eqs. (5) and (6) for the elimination of one set of equations to reduce both CPU time and computer storage requirement. If it is assumed that the initial condition is specified at $t = 0$ by e^0 and that all fields at $t < 0$ are zero, a single-matrix formulation is therefore obtained as follows:

$$e^n = AB(e^{n-1} + e^{n-2} + e^{n-3} + \cdots + e^0) + e^{n-1} \tag{9}$$

Rewrite

$$\psi_e^{n-1} = (e^{n-1} + e^{n-2} + e^{n-3} + \cdots + e^0) \tag{10}$$

Therefore,

$$e^n = C\psi_e^{n-1} + e^{n-1} \tag{11}$$

Here C is the product of A and B and its nonzero elements are given as

$$c_{j,m} = (c\delta t)^2 \frac{L'_{jm}}{L_{jm} S_j} \tag{12}$$

$$c_{j,j} = -\sum_{m=1,\, j \neq m}^{J} c_{j,m} \tag{13}$$

where c in Eq. (12) is the velocity of light in the medium. It is not difficult to identify that

$$C = SC'' \tag{14}$$

where

$$S = \mathrm{diag}(S_1^{-1}, S_2^{-1}, \ldots, S_J^{-1}) \tag{15}$$

The matrix C'' is symmetric and, therefore, only half of C'' needs to be stored in memory; however, this is not advantageous in TETD since the sparsity of the matrix leads to storing it in memory as a one dimensional array and using pointer arrays to delineate the dependence of E_j. Storing half of C'' means that another pointer array equal to half the size of C'' needs to be used to point to the correct symmetric elements, thus negating the amount of memory saved. To complete this single-matrix TETD method, we need to determine the stability condition for δt. Initially, we derive the stability condition using the Lax-Richtmyer stability condition. It is found that the derived δt is about one-half of the optimal value determined numerically. This numerically obtained value agrees with the stability condition recently derived by Lee et al. [40] which will be employed to derive the stability condition for the 2-D Whitney element time-domain method to be discussed shortly. First let us derive a stability condition from the Lax-Richtmyer stability condition [65] by rewriting Eq. (11) as:

$$e^1 = Ce^0 + e^0 \tag{16}$$

$$e^2 = C(e^1 + e^0) + e^1 = C^2e^0 + 3Ce^0 + e^0$$

$$\vdots$$

$$e^n = (C^n + a_{n-1}C^{n-1} + a_{n-2}C^{n-2} + \cdots + a_1C + I)e^0$$

The necessary and sufficient condition for the stability of the finite-difference scheme is given by the Lax-Richtmyer condition,

$$\|C^n\| \leq M \tag{17}$$

where M is a finite number and $\|C^n\|$ represents the norm of C^n. The above condition can be rewritten as:

$$\|C^n\| \leq \|CC^{n-1}\| \leq \|C\|\|C^{n-1}\| \leq \cdots \leq \|C\|^n \tag{18}$$

$$\|C\| \leq M^{1/n} \tag{19}$$

Since n is the number of time steps and is therefore rather large for a given problem, the right-hand side of (19) is approximately unity. Knowing that matrix C can be written as $\delta t^2 C'$, the stability condition is then defined as:

$$\delta t \leq 1/\sqrt{\|C'\|} \tag{20}$$

Typically, for stability analysis, the three norms that are commonly used are the 1, 2, and ∞ norms. For the matrix C' they are defined as:

$$\|C'\|_1 = \max \sum_{j=1}^{J} |c'_{j,m}|, \qquad 1 \leq m \leq J \tag{21}$$

$$\|C'\|_2 = \sqrt{\rho(C'C'^H)} \tag{22}$$

$$\|C'\|_\infty = \max \sum_{m=1}^{J} |c'_{j,m}|, \qquad 1 \leq j \leq J \tag{23}$$

In Eq. (22), ρ is the spectral radius and C^H is the Hermitian of C. The spectral radius is the largest moduli of the eigenvalues of a matrix. The evaluation of the two-norm will be difficult when the matrix size is large. On the other hand, both the one- or ∞-norm of C' can be evaluated efficiently and both will identify the same node at which instability for the problem will start (the "troublesome" node) if $\delta t > \delta t_s$, where δt_s is the optimal value. Thus, the easiest way to merely identify the "troublesome" node is to evaluate $\|C'\|_\infty$, which could be of some use to a programmer analyzing a mesh. For instance, node 1 in Fig. 4 could be the "troublesome" node. If so, the value of $\|C'\|_\infty$ would be equal to the summation of the elements in column 1 of C', with nonzero contributions coming from rows 2 through 6 corresponding to mesh connections with nodes 2 through 6. As seen in Table 1, $\|C'\|_\infty$ could also be used to obtain a quick and "safe" value of δt, as the value is much less than δt_s. On the other hand,

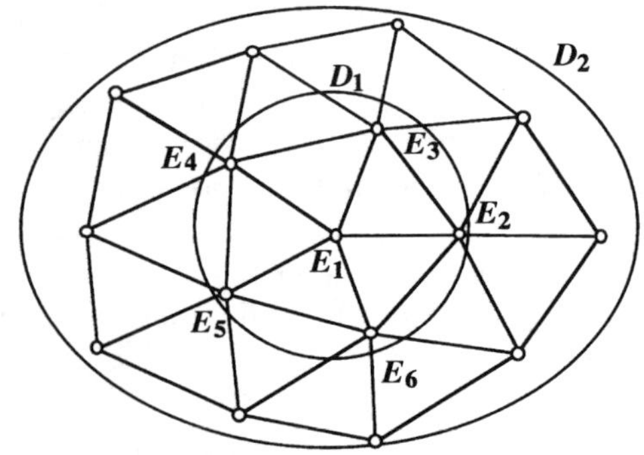

Fig. 4 First- and second-order dependents of node 1.

the two-norm can be obtained easily for D_q, where D_q is a square submatrix of C' describing the qth order dependency of node j. The submatrix of C' that includes the jth node and its immediate surrounding nodes is denoted as D_1. D_2, however, includes those nodes immediate to the surrounding nodes of the jth nodes. The nodes included in D_1 and D_2 are depicted in Fig. 4. It can be shown that $\|D_q\| \leq \|C'\|_2$. Numerical results tabulated in Table1. shows that δt obtained from $\|D_q\|_2$ is very close to $\|C'\|_2$, even for $q = 2$. We emphasize that δt obtained from the above procedure is not the optimal value. It only provides an alternate method when we cannot obtain the analytical stability condition. As we have mentioned before, the stability condition for a triangular mesh using Delaunay tessellation has been derived [40]. We will derive the stability condition for a general triangular mesh shortly.

In order to accurately represent the fields around the PEC scatterers, proper boundary conditions must be enforced on the surface of the scattering body. Figure 5 shows a portion of a triangular grid at the surface of a scatterer in which the total electric field tangential to the surface of the scatterer must be zero. In Fig. 5a, the only electric field component is E_z. All that needs to be done is to set E_z to zero on the surface of the scatterer. However, we choose to set the matrix element $c_{j,m} = 0$ from Eq. (11) for E_j on the PEC surface.

For the TM polarization the implementation of the boundary condition is just as straightforward but not as obvious. Figure 5b shows the electric and magnetic field placements on a triangular grid near a PEC scatterer for TE polarization. An important point to remember is that the E_k fields have a unit vector normal to the PEC boundary; hence, there is no need to force any E_k on the surface of the scatterer. However, there is one other consideration for enforcing the boundary conditions for TE polarization. Technically, the contour integral of the electric field needs a closer contour. Since the total field inside a perfectly conduct-

TABLE 1. Estimated and Actual Stability Conditions for Three Cases

Case #	Avg. L_{ij} (cm)	δt (ps) for type of norms used $\|C'\|_\infty$	$\|D_1\|_2$	$\|D_2\|_2$	$\|D_3\|_2$	$\|C'\|_2$	$\delta t_s(ps)$
1	10.28	108.6	143.3	136.3	136.1	134.1	268.2
2	2.58	25.1	32.8	31.4	31.4	31.2	59.0
3	0.51	4.8	6.3	6.1	6.1	—	11.8

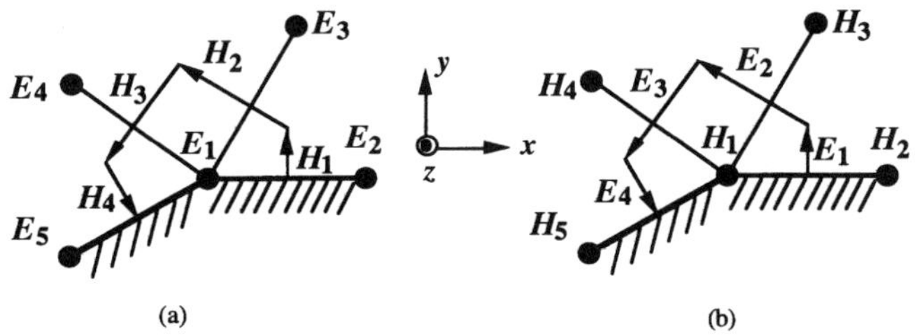

Fig. 5 Triangular grid near a PEC scatterer: (a) TM polarization and (b) TE polarization.

ing body is zero then the completion of the contour integral path, as seen in Fig. 5b, can be any path inside the scatterer. The contribution of that portion of the integral is zero and is not formally calculated. The only special consideration for TE polarization is to ensure that the nonzero part of the contour integral is truncated at the scatterer surface.

The three PEC scatterers presented for the TETD method are a circular cylinder, an ogive, and a wedge-cylinder, as shown in Fig. 6. Since there exists an exact solution for the circular cylinder, it is used for validation of TETD and RETD (see Fig. 1). The ogive is used to illustrate the difference in convergence of TETD and RETD for a scatterer that has both forward- and backward-facing sharp edges. The wedge-cylinder is a combination of the

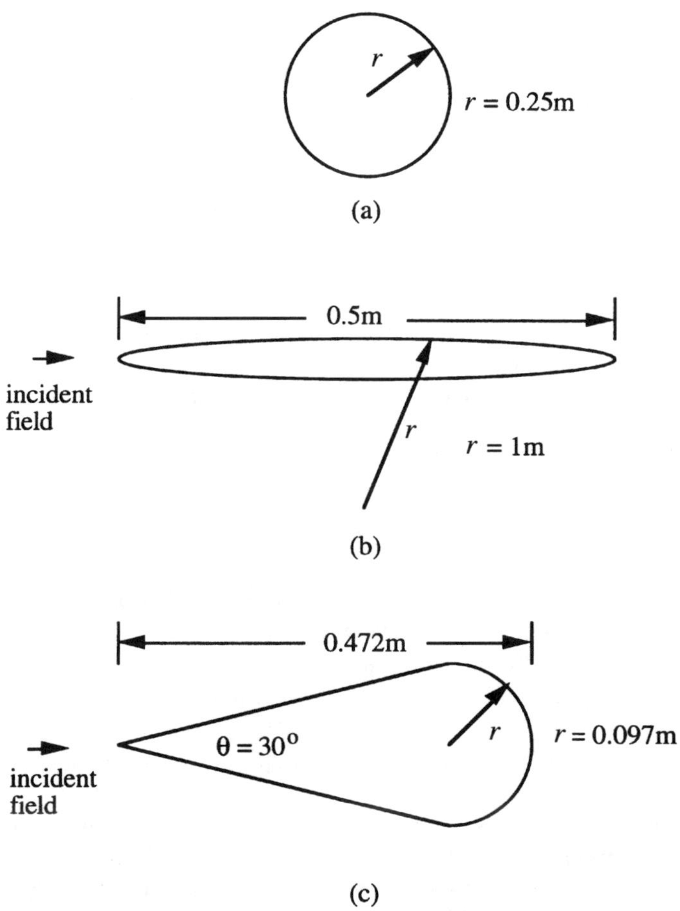

Fig. 6 Three different PEC scatterers: (a) circular cylinder, (b) ogive, and (c) wedge-cylinder.

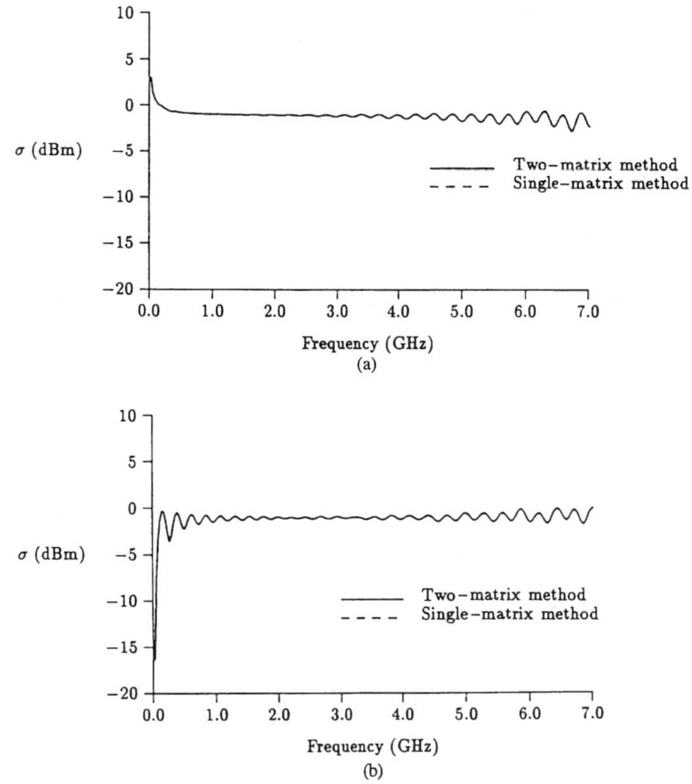

Fig. 7 Comparison of two-matrix and single-matrix methods for a circular cylinder of radius 0.25m; backscatter: (a) TM polarization and (b) TE polarization.

ogive and circular cylinder. It is used for demonstrating the accuracy of TETD for a scatterer with both sharp edge scattering and creeping wave phenomena in the same problem.

First let us compare the results obtained using the two-matrix and single-matrix methods. Figure 7 shows the comparison of the two-matrix method using Eqs. (5) and (6) and the single-matrix method using Eqs. (11) for the circular cylinder depicted in Fig. 6a at the backscattering angle for TM and TE polarizations, respectively. The two methods yield identical results. This is expected since no approximations were used to go from Eqs. (5) and (6) to Eq. (11). Table 2 compares the CPU time of these two methods normalized to the number of unknowns for a single time step. It is shown that the single-matrix method can reduce the CPU time of the conventional two-matrix method by a factor of 2.

The valid segmentation length depends on the shape of the scatterer since different shapes of scatterers have different types of scattering phenomena. Valid segmentation for rectangular and triangular grids may either converge (a rectangular cylinder) or diverge (sharp edges). Let us consider the ogive and

TABLE 2. CPU Time versus Matrix Method and Grid Types on a Sun Sparc-22 Workstation

	CPU per grid point	
Grid	*Two-matrix*	*Single-matrix*
TETD	8.0μs	4.7μs
RETD	3.6μs	1.7μs

wedge-cylinder depicted in Fig. 6. In order to get the same kind of accuracy for these scatterers as seen for the circular cylinder in Fig. 1, a smaller segmentation of 0.35cm is required. Results are compared to a 2-D method of moments (MOM) code, which used combined field integral equations with pulse basis and delta testing functions. All MOM data was run at $\lambda/20$ segmentations for all frequencies. Figure 8 shows the scattering from the ogive in TM and TE polarizations compared to the MOM code. When the MOM is run using only the electric field integral equation, the scattering at 3.8GHz has a resonance problem and the calculated cross section is higher than that shown in Fig. 8b.

The circular cylinder case shows how RETD handles the creeping wave phenomena. To show how it handles forward- and backward-facing sharp edges, the backscattering from a rectangular cell representation of the ogive of Fig. 6b was calculated. This data is also shown in Fig. 8 and uses the same segmentation size along the surface of the scatterer as TETD. It is obvious that the TE polarization case quickly diverges from the correct solution when using RETD. Even TM polarization has more difficulty than the circular cylinder case. Figure 9 shows the backscattering from the wedge-cylinder in TM and TE polarization as compared to the MOM code. Like the ogive, the agreement here is good. There are no signs of significant resonance problems in the MOM code for the wedge-cylinder configuration. Table 3 shows the discretizations needed for the conformal triangular grid and conventional rectangular grid for 1-dB and 3-dB errors when compared to the MOM. The data in Table 3 excludes frequencies at which a deep null occurred for the wedge-cylinder but TETD did not have a data point at that exact frequency. The conformal grid allows a significant reduction in memory and CPU time for the same degree of accuracy when compared to the conventional FDTD. Comparison of operations per time step between the conformal FDTD and RETD will be discussed in detail in the later part of this chapter.

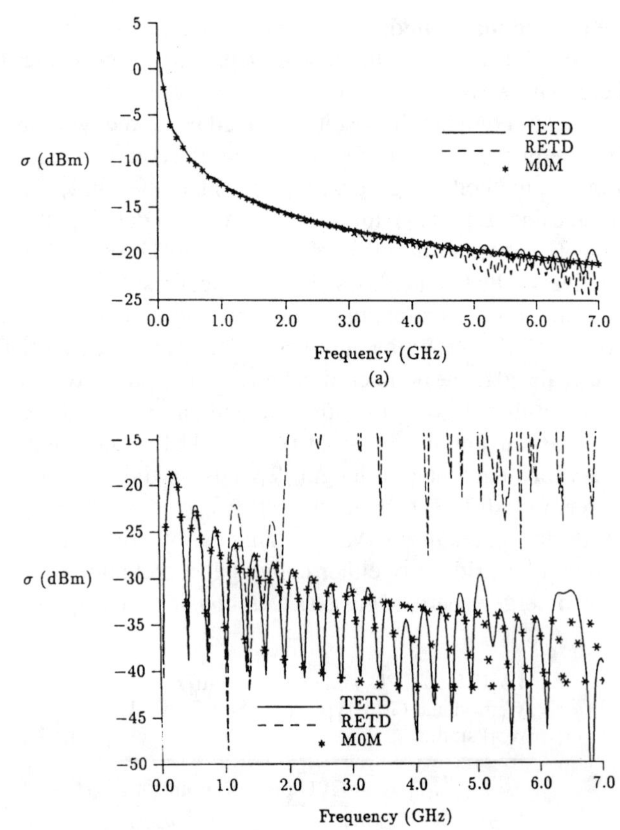

Fig. 8 Ogive backscatter cross section: (a) TM polarization and (b) TE polarization.

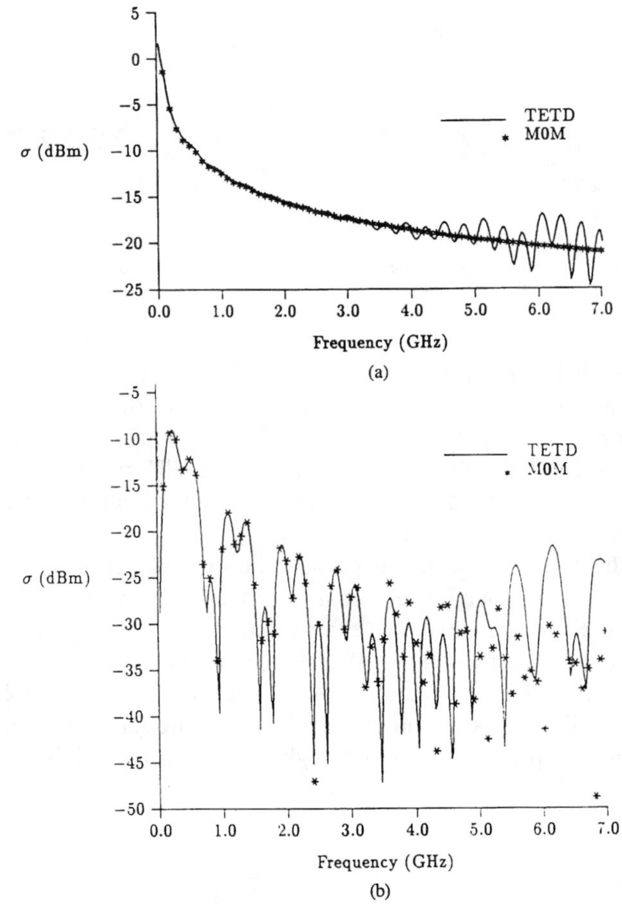

Fig. 9 Wedge-cylinder backscatter cross section: (a) TM polarization and (b) TE polarization.

A Vertex-Based Finite-Volume Time-Domain Method

If any triangle in the primary grid has an interior angle larger than 90°, the dual grid vertex associated with that triangle physically lies outside of it and causes an intolerable error for the TETD. In this section, we modify the calculation cell of the magnetic field to alleviate the constraint put on the triangular discretization. This modification results in a more flexible conformal capability.

Similar to the previous case, we sample the electric field at the vertices of the triangular grid. In contrast, the magnetic field is now sampled at the centroids of the triangles. Basic calculation cells for updating the electric and magnetic fields are given in Fig. 10. The electric and magnetic fields are assumed to be constant within the triangular area A and the polygonal area S, respectively. For each triangular edge, we define an outward

TABLE 3. Discretization Errors versus Grid Types and Polarizations

Grid type	*Polarization*	1-dB Error *Circular cylinder*	*Ogive*	*Wedge-cylinder*	3-dB Error *Circular cylinder*	*Ogive*	*Wedge-cylinder*
TETD	TM	$\lambda/9.3$	$\lambda/13.9$	$\lambda/16.9$	$\lambda/6.5$	$\lambda/10.3$	$\lambda/14.1$
	TE	$\lambda/8.4$	$\lambda/17.2$	$\lambda/23.8$	$\lambda/6.0$	$\lambda/17.1$	$\lambda/21.5$
RETD	TM	$\lambda/17.3$	$\lambda/27.7$	—	$\lambda/6.9$	$\lambda/13.9$	—
	TE	$\lambda/35.1$	$\lambda/85.7$	—	$\lambda/15.7$	$\lambda/76.9$	—

normal $\boldsymbol{n}$ and an angle θ between that edge and the x axis. Similarly, for each polygonal edge, we define an outward normal $\boldsymbol{n}'$ and an angle θ' between that edge and the x axis. The magnetic field along an edge of the polygon is assumed to be constant and is obtained from the averaged value of the fields at the two end-points of that edge. Note that we have simplified the indexing in Fig. 10 as we only concentrate on one computation cell as compared to the global indexing used in Fig. 3b and Eqs. (3) and (4).

To update the electric and magnetic fields, let us consider the following volume integral equations:

$$\int \nabla \times \boldsymbol{E}\, dv = \int (\boldsymbol{n} \times \boldsymbol{E})\, ds = -\mu_0 \frac{\partial}{\partial t} \int \boldsymbol{H}\, dv \qquad (24)$$

$$\int \nabla \times \boldsymbol{H}\, dv = \int (\boldsymbol{n} \times \boldsymbol{H})\, ds = -\varepsilon_0 \frac{\partial}{\partial t} \int \boldsymbol{E}\, dv \qquad (25)$$

For 2-D problems, dv is replaced with ds and ds with dl. For TM cases, the above equations reduce to the following scalar equations.

$$H_x^{n+1/2} = H_x^{n-1/2} + \frac{\delta t}{\mu_0 A} \sum_{i=1}^{3} \frac{E_{zi}^n + E_{zi+1}^n}{2} L_i \cos\theta_i \qquad (26)$$

$$H_y^{n+1/2} = H_y^{n-1/2} + \frac{\delta t}{\mu_0 A} \sum_{i=1}^{3} \frac{E_{zi}^n + E_{zi+1}^n}{2} L_i \sin\theta_i \qquad (27)$$

$$E_z^{n+1} = E_z^n + \frac{\delta t}{\varepsilon_0 S} \sum_{i=1}^{I_j} \left(\frac{H_{xi}^{n+1/2} + H_{xi+1}^{n+1/2}}{2} \cos\theta_i' + \frac{H_{yi}^{n+1/2} + H_{yi+1}^{n+1/2}}{2} \sin\theta_i' \right) L_i' \qquad (28)$$

L_i and L_i' denote the length between nodes i and $i + 1$ and the length between centroids i and $i + 1$, respectively. I_j is the number of nodes surrounding the E-field point. One can combine these equations to obtain a single-matrix scheme. An analytical stability condition can also be derived. Both points will be illustrated in the next section for the 2-D explicit Whitney element approach and will not be presented here. Table 4 shows the accuracy of this method when compared to the mode-matching approach for waveguide step discontinuities shown in Fig. 11, in which $a = 0.8\,\lambda_0$, $b = 0.6\,\lambda_0$, and $c = 0.2\,\lambda_0$. Here λ_0 is the free-space wavelength. Z_a (482.95Ω) and Z_b (682.02Ω) are the characteristic impedances of the waveguide in Regions I and II, respectively. A discretization of 15 points per free-space wavelength is used ($\delta x = \delta y = 0.067\,\lambda_0$) while two modes are used for the mode-matching method. Excellent agreement is obtained.

Note that the magnetic fields are sampled at the vertices of the complimentary grid as given by the shaded area in Fig. 10a and the method is appropriately described as vertex-based. In the previous section, the magnetic field is edge-based, implying that it is sampled along the edge of the complimentary polygon. The enforcement of the material boundary conditions is typically easier with an edge-based method. For example, for a PEC edge, we can simply set the edge-based electric field to be zero. On the other hand, if the electric field is vertex-based, we have to separate the electric field vector at the vertex into two components, one parallel to the edge and the other perpendicular to the edge. The parallel component can be set to zero while an additional constraint has to be imposed on the normal component. A way to impose such a constraint for 3-D problems is discussed in [51]. The following section discusses the development of an explicit edge-based time-domain method to alleviate the difficulty of imposing the boundary condition.

TABLE 4. Comparison of FVTD and Mode Matching

	$\lvert R\rvert$	$\lvert T\rvert$	Power check
Single step			$\lvert R\rvert^2 + \lvert T\rvert^2 (bZ_a)/(aZ_b)$
Mode matching	*0.156*	*1.355*	*1.000*
FVTD	*0.157*	*1.343*	*0.983*
Double step			$\lvert R\rvert^2 + \lvert T\rvert^2$
Mode matching	0.298	0.955	1.000
FVTD	0.308	0.948	0.994

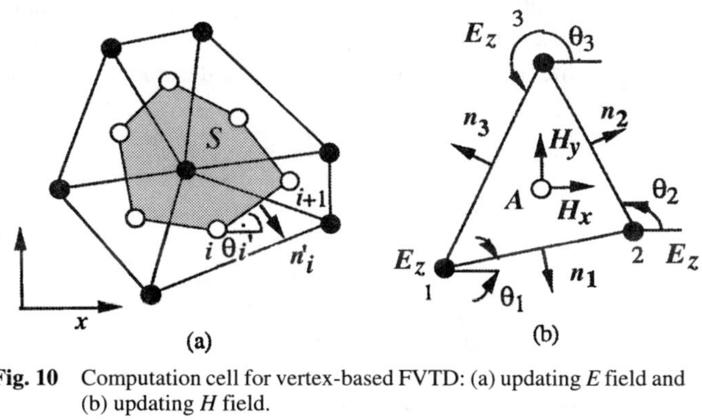

Fig. 10 Computation cell for vertex-based FVTD: (a) updating E field and (b) updating H field.

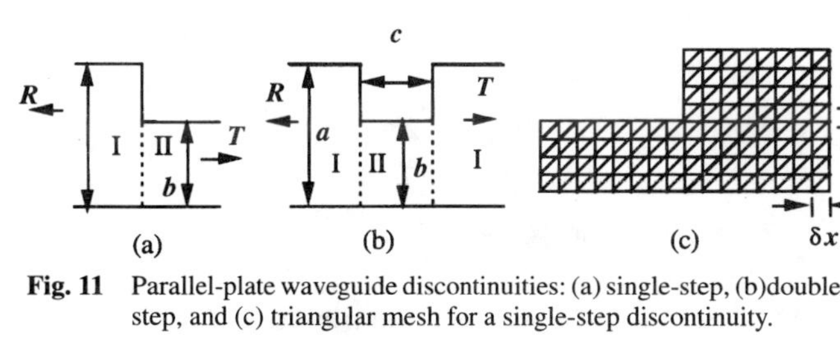

Fig. 11 Parallel-plate waveguide discontinuities: (a) single-step, (b)double-step, and (c) triangular mesh for a single-step discontinuity.

Explicit Whitney Element Time-Domain Method

Two-dimensional (2-D) and three-dimensional (3-D) explicit Whitney element time-domain methods have been recently proposed [60–61]. These methods employ the 2-D and 3-D Whitney element interpolation functions in updating the field values. The need of a sparse matrix solution for each time step in the implicit schemes proposed in [55–57] which also employ Whitney elements [64] has been alleviated in our explicit scheme.

For the 2-D algorithm, we first present an optimized explicit Whitney element time-domain approach in terms of a two-matrix approach. The algorithm is then further optimized into a scheme that uses a single matrix as discussed earlier. We compare the efficiency of the optimized Whitney element technique with the conventional FDTD. Finally, the stability criterion for choosing δt is derived. For the 3-D algorithm, a single-matrix scheme can also be developed. However, an analytical stability criterion has not been derived and we may resort to the Lax-Richtmyer stability criterion.

Two-Dimensional Explicit Whitney Element Time-Domain Method. Let us consider the 2-D parallel-plate waveguide discussed in the previous section. The electric field E_Z is sampled at the vertices while the magnetic fields H_x and H_y are sampled at the centroids of the triangles as in vertex-based FVTD. The only difference is that the magnetic field at the centroid of a triangle is interpolated from the magnetic fields normal to the edges of the triangle (see Fig. 12). 2-D Whitney two-form elements are used for the interpolation.

The electric field, E_z, is updated by the circulation of the magnetic field as shown in Fig. 10a and Eq. (28). Similar to the vertex-based FVTD, the magnetic field along the straight line joining any two centroids is assumed to be constant and is obtained by taking the average of the magnetic fields evaluated at the two controids. The magnetic field at the centroid of each triangle is updated by interpolating the magnetic fields normal to the three edges of the triangle using the Whitney two-form elements. These normal magnetic fields are computed similar to the TETD method discussed earlier. Hence the magnetic field at the centroid is updated as follows:

$$\boldsymbol{H}^{n+1/2} = \boldsymbol{H}^{n+1/2} + \sum_{i=1}^{3} h_i \boldsymbol{W}_{i,i+1}(x_c, y_c) \tag{29}$$

$$h_{1i} = \frac{\delta t}{\mu_0} \frac{(E^n_{zi+1} - E^n_{zi})}{L_i} \tag{30}$$

$$(x_c, y_c) = \left(\frac{x_1 + x_2 + x_3}{3}, \frac{y_1 + y_2 + y_3}{3}\right) \tag{31}$$

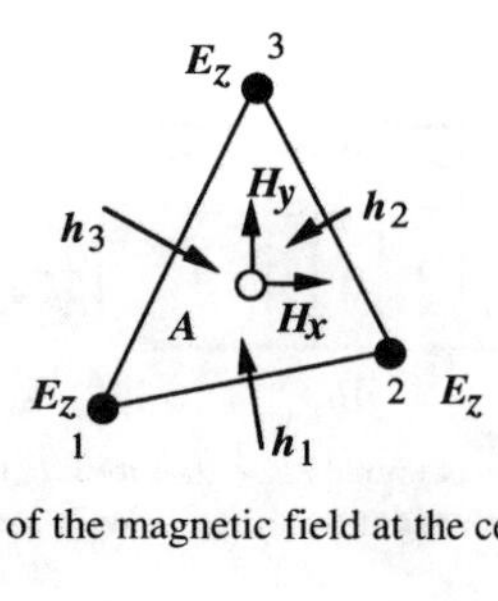

Fig. 12 Interpolation of the magnetic field at the centroid.

where (x_i, y_i) are the coordinates of the ith vertex. The Whitney two-form element interpolation functions, depicted in Fig. 13, are given by

$$\boldsymbol{W}_{i,i+1}(x, y) = L_i(z\,\xi_i \times \nabla \xi_{i+1} + \nabla \xi_i \times z\,\xi_{i+1}) \tag{32}$$

and ξ_i is the simplex coordinated defined as:

$$\xi_i(x, y) = \frac{(x_{i+1} - x)(y_{i+2} - y) - (x_{i+2} - x)(y_{i+1} - y)}{(x_{i+1} - x_i)(y_{i+2} - y_i) - (x_{i+2} - x_i)(y_{i+1} - y_i)} \tag{33}$$

The use of the Whitney two-form elements allows the magnetic field to be evaluated at the desired position, which in this case is the centroid of the triangle, from magnetic fields normal to the triangle edges. The dependence of the electric field at a node (vertex) on the magnetic field is purely linear and depends only on the geometry of the surrounding triangles. Moreover, the dependence of the magnetic field, at a centroid of a triangle, on the electric field is also purely linear and depends only on the geometry of the triangle itself. These observations lead us to conclude that a matrix formulation for the Whitney element time-domain algorithm is possible as in the TETD. The magnetic field is now separated into two components, namely, H_x and H_y, in contrast to a single component of H in TETD. As we will see later, both H_x and H_y will be eliminated in the single-matrix formulation.

Assuming that there are J triangular vertices and K triangle centroids, the electric field is expressed as a vector of length J while the magnetic field is expressed as a pair of vectors, each of length K. The electric and magnetic fields are updated as follows:

$$e_z^{n+1} = A_x h_x^{n+1/2} + A_y h_y^{n+1/2} + e_z^n \tag{34}$$

$$h_x^{n+1/2} = B_x e_z^n + h_x^{n-1/2} \tag{35}$$

$$h_y^{n+1/2} = B_y e_z^n + h_y^{n-1/2} \tag{36}$$

Here A_x, A_y, B_x, and B_y are sparse matrices. The jth rows of A_x and A_y which correspond to vertex j have only I_j nonzero elements each, where I_j is the number of triangles surrounding node j. On the other hand, the kth rows of B_x and B_y contain only three nonzero elements corresponding to the three vertices of triangle k. Similar to TETD, the transpose of B_x and B_y have their nonzero elements in the same positions as A_x and A_y. This can be explained by noting that if the magnetic field at a centroid affects the electric field at a vertex, then the electric field at that vertex must also affect the magnetic field at the centroid. Additionally, it should be noted that not only are A_x, A_y, B_x, and B_y sparse, but their nonzero elements are located at positions that depend on the particular indexing of the nodes and the centroids.

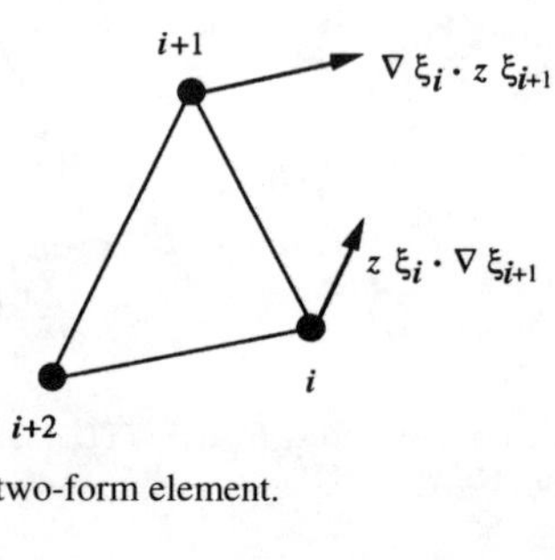

Fig. 13 Whitney two-form element.

The expression for the elements of A_x and A_y are simple after a tedious manipulation, but before expressing them, let us define global and local indexing schemes as shown in Fig. 14. In Fig. 14, both k and j are global indices. The global index j is the electric field node to be updated while k is the centroid of the triangle that contributes to this updating process. A local index i that ranges from 1 to I_j is local to vertex j. If triangle k is one of the triangles that surround vertex j and has the local index i, we have for the matrices A_x and A_y:

$$a_{xj,\,k} = \frac{\delta t}{\varepsilon_0 S_j}\left(\frac{x_{j,\,i(k)+2} - x_{j,\,i(k)} + x_{j,\,i(k)+1} - x_{j,\,i(k)-1}}{6}\right) \tag{37}$$

$$a_{yj,\,k} = \frac{\delta t}{\varepsilon_0 S_j}\left(\frac{y_{j,\,i(k)+2} - y_{j,\,i(k)} + y_{j,\,i(k)+1} - y_{j,\,i(k)-1}}{6}\right) \tag{38}$$

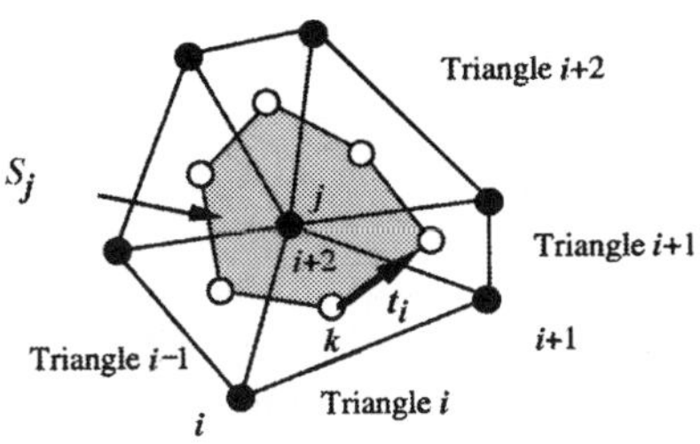

Fig. 14 Indexing surrounding node j.

S_j is the area of the I_j sided polygon obtained by connecting the I_j centroids as shown in Fig. 14. If triangle k is not one of the I_j triangles that surround node j, then $a_{xj,\,k} = a_{yj,\,k} = 0$. $(x_{j,\,i(k)}, y_{j,\,i(k)})$ is the coordinate pair for the triangle that has a local index i and a global index k. Note that i increases in a counter-clockwise manner and that if $i + 1 > I_j$, then $i + 1$ is interpreted as $i + 1 - I_j$; if $i - 1 < 1$, then $i - 1$ is interpreted as $i - 1 + I_j$. The vector $\boldsymbol{t}_i$ points from the centroid of triangle i to the centroid of triangle $i + 1$.

For the matrix elements of B_x and B_y, let us consider Fig. 15. Again, i is the local index except that it now ranges from 1 to 3 and refers to the vertices. If the node j is one of the vertices of triangle k and has the local index i, then

$$b_{xk,\,j} = \frac{\delta t}{\mu_0 A_k}\,\frac{x_{k,\,i(j)+1} - x_{k,\,i(j)+2}}{2} \tag{39}$$

$$b_{yk,\,j} = \frac{\delta t}{\mu_0 A_k}\,\frac{y_{k,\,i(j)+1} - y_{k,\,i(j)+2}}{2} \tag{40}$$

where A_k is the area of the triangle k and $(x_{k,\,i(j)}, y_{k,\,i(j)})$ is the coordinate pair of the vertex that has the local index i and the global index j. Again, $b_{xk,\,j} = b_{yk,\,j} = 0$ for all j that do not correspond to one of the vertices of triangle k.

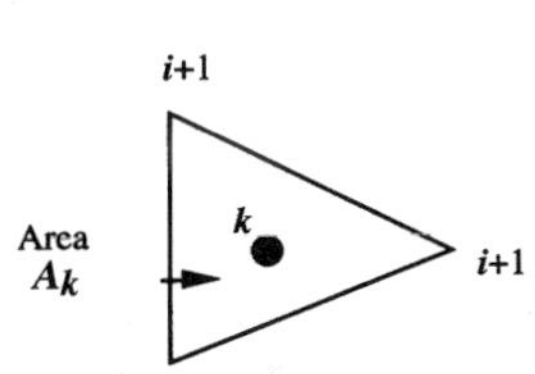

Fig. 15 Indexing for triangle k.

The boundary conditions on E_z can be either applied implicitly by assigning zeros in matrices A_x and A_y to all the elements of the rows that correspond to vertices on the boundary, or applied explicitly by zeroing the elements of E_z that lie on the conductor boundary at each time step. The implicit application results in a more optimized algorithm as in the TETD.

This two-matrix formulation of the Whitney FDTD can be reduced to a single-matrix formulation from which we can derive the stability condition.

Single-Matrix Whitney FDTD. Similar to the single-matrix formulation of the TETD, if we set both the electric and magnetic fields to zero prior to $t = 0$, then Eqs. (34), (35), and (36) can be used to eliminate h_x and h_y to give

$$e_z^{n+1} = C\psi_z^n + e_z^n \tag{41}$$

where

$$C = A_xB_x + A_yB_y \tag{42}$$

and

$$\psi_z^n = e_z^0 + e_z^1 + \cdots + e_z^n \tag{43}$$

At the price of eliminating the magnetic field entirely from the leap-frogging algorithm, we have attained a much simpler and more efficient algorithm. The matrix C, which is J by J, contains the geometrical relationships among the nodes. Note that row j of C has $I_j + 1$ nonzero elements, one for each node that shares a segment with the node j and one for node j itself. These elements can be calculated by combining A_x, A_y, B_x, and B_y in an appropriate manner. Alternately, as shown below, the elements of C can be obtained directly without first calculating A_x, A_y, B_x, and B_y.

The derivation of the elements of C is better illustrated if we concentrate on the nondiagonal ones first. Let us first look at $c_{j,\,m}$ where $j \neq m$. If the mth node is not in the immediate vicinity (not connected) of the jth node, $c_{j,\,m} = 0$. For the nonzero $c_{j,\,m}$, the expression is:

$$\begin{aligned} c_{j,\,m} = {} & \frac{(c\delta t)^2}{12S_j} \\ & \times \{[(x_j - x_{i(m)-1})(x_{i(m)+1} - x_{i(m)-1} + x_{i(m)} - x_{i(m)-2}) \\ & \quad + (y_j - y_{i(m)-1})(y_{i(m)+1} - y_{i(m)-1} + y_{i(m)} - y_{i(m)-2})]/A_{i-1} \\ & \quad + [(x_{i(m)+1} - x_j)(x_{i(m)+2} - x_{i(m)} + x_{i(m)+1} - x_{i(m)-1}) \\ & \quad + (y_{i(m)+1} - y_j)(y_{i(m)+2} - y_{i(m)} + y_{i(m)+1} - y_{i(m)-1})]/A_i\} \end{aligned} \tag{44}$$

Here A_i represents the area of triangle i, as shown in Fig. 15. The index $i(m)$ refers to the ith vertex surrounding node j, which is vertex m in the global sense, as shown in Fig. 14. The diagonal elements $c_{j,j}$ are given by

$$\begin{aligned} c_{j,\,j} = {} & \frac{(c\delta t)^2}{4S_j} \\ & \times \sum_{i=1}^{I_j} \frac{(x_i - x_{i+1})(t_{xi-1} + t_{xi}) + (y_i - y_{i+1})(t_{yi-1} + t_{yi})}{A_i} \end{aligned} \tag{45}$$

where the same conventions are used as in the case of $c_{j,\,m}$ with the additional understanding that $\boldsymbol{t}_i = \boldsymbol{x}\, t_{xi} + \boldsymbol{y}\, t_{yi}$ is the vector shown in Fig. 14. It should be noted that $c_{j,\,j} = -\sum_{m=1, m\neq j}^{J} c_{j,\,m}$.

Having derived the optimum algorithm for the Whitney FDTD, we are ready to compare its efficiency and memory requirement with the conventional FDTD. Before doing so, let us note that the single-matrix formulation is possible for the RETD as well. In the RETD, the "C matrix" would have five nonzero elements in each row for 2-D problems, and the exact expressions of these nonzero elements are easily derived. Therefore, we compare the single-matrix formulations of the Whitney FDTD and RETD for the purposes of examining the efficiency of the Whitney FDTD.

Efficiency of Whitney FDTD and RETD. Since most of the CPU time is spent in time-stepping the fields and not in the calculation of the C matrix; it is only necessary to compare the CPU time required for each time step. One can easily see that the number of floating point operations per time step in the Whitney FDTD algorithm is $2J_t + 2\sum_{j=1}^{J_t}(I_j + 1)$, where J_t denotes the total number of nodes in the triangular grid. It should be noted that we count the addition and multiplication as two separate operations here. Since the Whitney algorithm is most accurate for equilateral triangles, one can assume that $I_j = 6$. Hence, the number of floating point operations for each time step is $16\,J_t$.

For RETD, the number of floating point operations per time step is $12\,J_r$, where J_r is used to denote the total number of vertices in the rectangular grid. Note that for a fair comparison of the two techniques, J_r should be approximately equal to J_t. Therefore, it is seen that the ratio of the CPU time required by the optimized Whitney FDTD to the Yee FDTD for each time is 4:3.

The ratio of the memory required by the optimized Whitney algorithm to that required by the RETD is easily seen to be 9:7, or approximately 4:3. Therefore, the additional resources required by the Whitney technique are only a third of that required by the RETD. The superior modeling flexibility provided by the Whitney FDTD, however, outweighs the slight increases in CPU and memory requirement.

Stability Criterion for Whitney FDTD. In this section, the criterion for choosing δt is derived using the approach that is similar to the one given by Lee et al. [40]. Let us start by updating the electric field in the following form:

$$E_{zj}^{n+1} = E_{zj}^{n} + \frac{\delta t}{\varepsilon_0 S_j}\sum_{i=1}^{I_j} \boldsymbol{H}_i^{n+1/2}\cdot \boldsymbol{t}_i \tag{46}$$

Here $E_{zj}{}^n$ represents the electric field at node j and $\boldsymbol{H}_i^{n+1/2}$ is the magnetic field at the midpoint of $\boldsymbol{t}_i$. Again, the superscript on the field quantity denotes the time step, and $\boldsymbol{t}_i$ is the vector that points from centroid i to centroid $i+1$, as shown in Fig. 14. S_j is the area enclosed by the contour surrounding vertex j. $\boldsymbol{H}_i^{n+1/2}$ is obtained by taking the average of the magnetic fields at either end of $\boldsymbol{t}_i$.

Letting $c'_{j,i} = c_{j,i}/(c\delta t)^2$, where $c_{j,i}$ is the element of C as given before, Eq. (46) for updating the electric field at node j can be rewritten as:

$$E_{zj}^{n+1} = E_{zj}^{n} + \frac{\delta t}{\varepsilon_0}\left\{\left[|c'_{j,j}E_{zj}^{n} + \sum_{i=1}^{I_j} c'_{j,i}E_{zi}^{n}\right]\frac{\delta t}{\mu_0} + \frac{1}{S_j}\sum_{i=1}^{I_j}\boldsymbol{H}_i^{n-1/2}\cdot \boldsymbol{t}_i\right\} \tag{47}$$

Note that

$$E_{zj}^{n} - E_{zj}^{n-1} = \frac{\delta t}{\varepsilon_0 S_j}\sum_{i=1}^{I_j}\boldsymbol{H}_i^{n-1/2}\cdot \boldsymbol{t}_i$$

and

$$c'_{j,j} = -\sum_{i=1}^{I_j} c'_{j,i}$$

Readers should be cautious that the sum is over the I_j nodes that surround node j, and therefore $i \neq j$. We have:

$$E_{zj}^{n+1} = 2E_{zj}^{n} - E_{zj}^{n-1} + (c\delta t)^2\sum_{i=1}^{I_j} c'_{j,i}(E_{zi}^{n} - E_{zj}^{n}) \tag{48}$$

Let us assume that a plane wave is propagating in the direction of $\boldsymbol{k} = \boldsymbol{x}k_x + \boldsymbol{y}k_y$. Let us also assume that node j of Fig. 14 is located at the origin of the coordinate system with node j having a zero phase. The electric field at point i at time step n can be expressed as

$$E_{zi}^{n} = u^n e^{-j(k_x x_i + k_y y_i)} \tag{49}$$

where u is a nonzero complex constant.

Defining a complex number $z = u^{n+1}/u^n$ as the growth factor, one expects the field at node j to be stable if $|z| \leq 1$. Using Eqs. (48) and (49), we obtain the following equation for z:

$$z^2 + (a_j - 2)z + 1 = 0 \tag{50}$$

where

$$a_j = (c\delta t)^2\sum_{i=1}^{I_j} c'_{j,i}[1 - e^{-j(k_x x_i + k_y y_i)}] \tag{51}$$

The two solutions are:

$$z = \frac{(2 - a_j)}{2} + \frac{i}{2}\sqrt{a_j(4 - a_j)} \tag{52}$$

Thus $|a_j| \leq 4$ ensures stability at node j.

Depending upon the direction of the plane wave and the geometry surrounding node j, $|a_j|$ fluctuates. However, from Eq. (51), the maximum of $|a_j|$ occurs when there is constructive interference at node j from all the immediate vertices. This maximum value can be written as:

$$|a_j| = (c\delta t)^2\sum_{i=1}^{I_j}|c'_{j,i}|2 = 2\sum_{i=1}^{I_j}|c_{j,i}| = 2|c_{j,j}| \tag{53}$$

Therefore, the stability condition at node j is:

$$|c_{j,j}| \leq 2 \tag{54}$$

However, we need stability at all the nodes, so the condition $|c_{j,j}|_{\max} \leq 2$ must be met. The subscript "max" denotes that the maximum of all nodes is to be taken. Using the definition of $c'_{j,j}$, we get

$$\delta t \leq \frac{1}{c}\sqrt{\frac{2}{|c'_{j,j}|_{\max}}} \tag{55}$$

where $|c'_{j,j}|_{\max}$ refers to the maximum of $|c'_{j,j}|$ as the stability condition for the Whitney FDTD. Note that this number is approximately twice that of the δt obtained using the Lax-

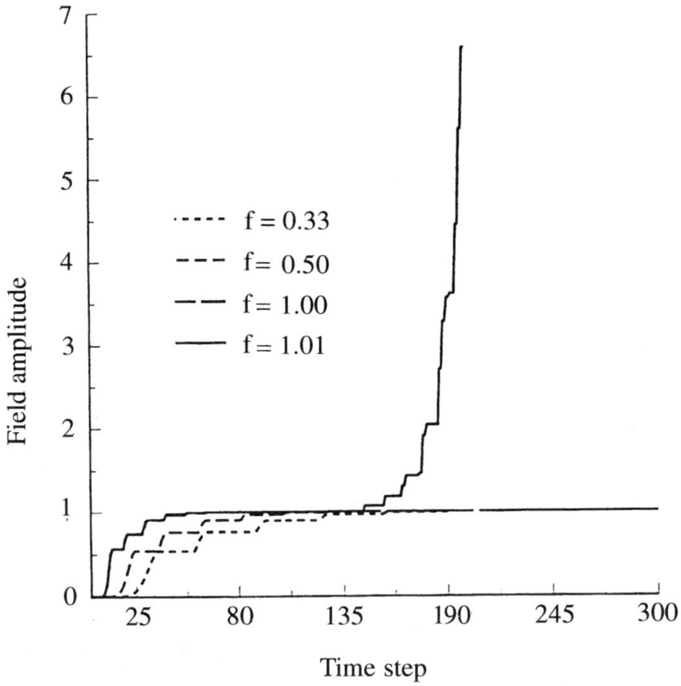

Fig. 16 Electric field amplitude as a function of time step for $\delta t = f\delta t_{\max}$.

Richtmyer condition and is consistent with the numerically determined δt shown in Table 1.

Figure 16 shows the growth of the electric field amplitude as a function of time for several different sizes of time increment δt. It illustrates that the Whitney algorithm is completely stable whenever the foregoing criterion for the time increment is met and that if the time increment is just 1% greater than the one given by Eq. (55), the Whitney algorithm becomes unstable. It should be noted that the above condition, for a triangular grid composed solely of right triangles, is equivalent to the Courant stability condition for a rectangular grid with the same discretization level.

Figure 17 shows thc contour plots of the electric field inside a parallel-plate waveguide with a discontinuity described by half a period of a sinusoidal function. A comparison between the numerical results of the Whitney FDTD and the vertex-based FVTD agree up to five significant figures. It should be noted that different coordinate systems are used here and in [58].

For 2-D problems, the performances of the Whitney FDTD and the vertex-based FVTD are about the same. However, the Whitney FDTD has a significant advantage over the vertex-based FDTD in its ease of enforcing boundary condition. In 3-D problems, the electric field will be sampled along the edges of a tetrahedron while the magnetic field can be interpolated anywhere within the volume of the tetrahedron. As a result, boundary conditions on the tangential electric field can be easily imposed.

3-D Whitney FDTD. In this section, we extend the Whitney FDTD to 3-D with the computational domain discretized into a union of tetrahedrons. The electric field vector is sampled at the midpoint of each tetrahedron edge, and the field value is assumed to be constant along the length of the edge. Magnetic fields normal to each triangular face of the tetrahedron are updated from the circulation of the electric field along the triangular edges. On the other hand, the electric field along an edge is updated from the circulation of magnetic fields on a plane perpendicular to the edge. Magnetic fields on the plane are conveniently obtained from the interpolation of the magnetic fields

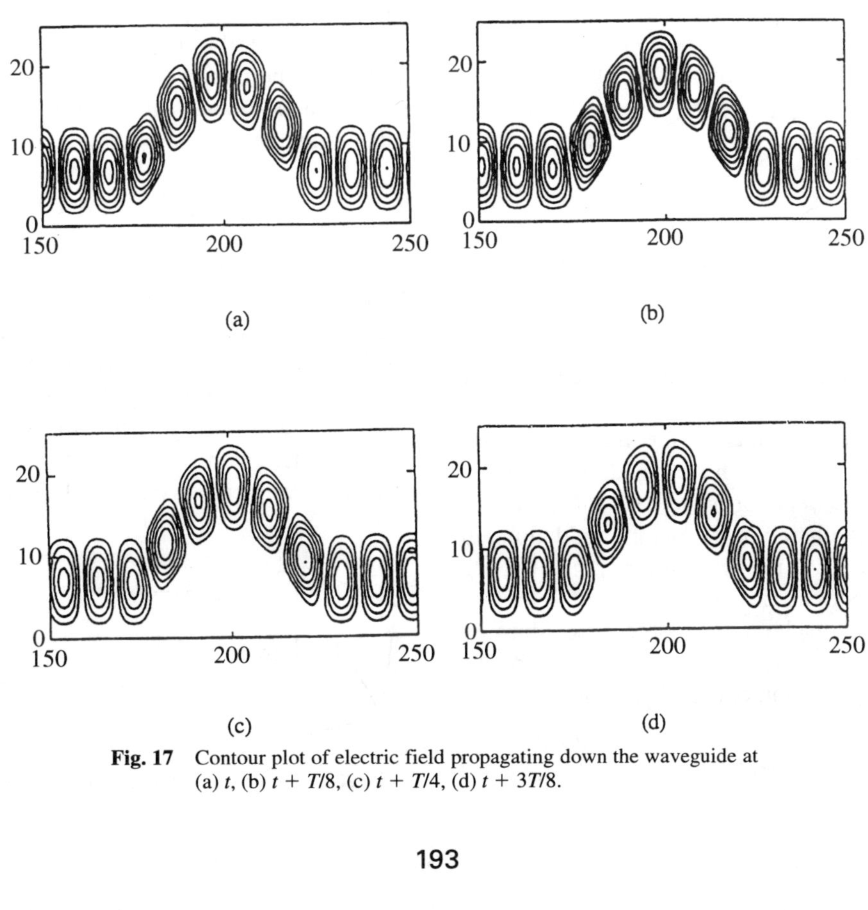

Fig. 17 Contour plot of electric field propagating down the waveguide at (a) t, (b) $t + T/8$, (c) $t + T/4$, (d) $t + 3T/8$.

normal to the triangular faces of the tetrahedrons via the use of Whitney two-form tetrahedral elements. To demonstrate the accuracy of the proposed method, we compare our numerical results with analytical ones for a rectangular resonator.

Whitney Two-Form Elements in 3-D. The Whitney two-form basis factors are defined as

$$W_i(x,y,z) = (-1)^{i+1}(\xi_{i+1}\nabla\xi_{i+2} \times \nabla\xi_{i+3} + \xi_{i+2}\nabla\xi_{i+3} \times \nabla\xi_{i+1} + \xi_{i+3}\nabla\xi_{i+1} \times \nabla\xi_{i+2}), \text{ modulus } 4 \tag{56}$$

where $i = 1, 2, 3$, and 4. ξ is a simplex coordinate defined as

$$\xi_i = \frac{d_i}{D_i} \tag{57}$$

where d_i and D_i are depicted in Fig. 18 for $i = 1$. In rectangular coordinates, ξ_i can be written as [66]:

$$\xi_i = \frac{1}{\sum_{j=1}^{4} a_j}(a_i + b_i x + c_i y + d_i z) \tag{58}$$

where

$$a_i = (-1)^{i+1}[\mathbf{x}' \cdot (\mathbf{y}' \times \mathbf{z}')],$$
$$b_i = (-1)^i(\mathbf{y}' \times \mathbf{z}') \cdot (\mathbf{x} + \mathbf{y} + \mathbf{z}),$$
$$c_i = (-1)^{i+1}(\mathbf{x}' \times \mathbf{z}') \cdot (\mathbf{x} + \mathbf{y} + \mathbf{z}),$$
$$d_i = (-1)^i(\mathbf{x}' \times \mathbf{y}') \cdot (\mathbf{x} + \mathbf{y} + \mathbf{z}),$$
$$\mathbf{x}' = \mathbf{x}\, x_{i+1} + \mathbf{y}\, x_{i+2} + \mathbf{z}\, x_{i+3}, \; \mathbf{y}' = \mathbf{x}\, y_{i+1} + \mathbf{y}\, y_{i+2} + \mathbf{z}\, y_{i+3},$$
$$\mathbf{z}' = \mathbf{x}\, z_{i+1} + \mathbf{y}\, z_{i+2} + \mathbf{z}\, z_{i+3}$$

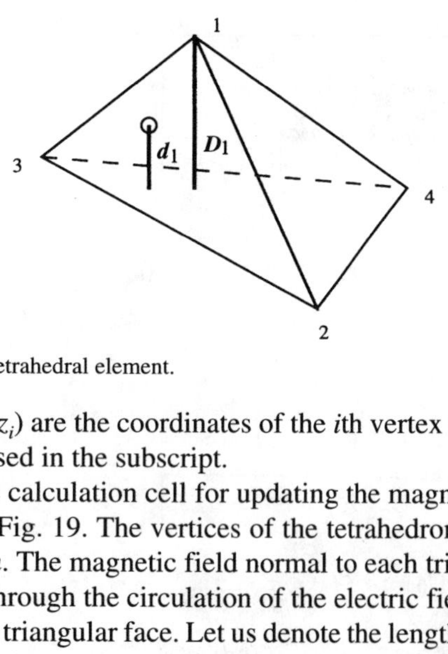

Fig. 18 A tetrahedral element.

Here (x_i, y_i, z_i) are the coordinates of the ith vertex and a modulus of 3 is used in the subscript.

The basic calculation cell for updating the magnetic fields is depicted in Fig. 19. The vertices of the tetrahedron are labeled i, j, k, and m. The magnetic field normal to each triangular face is updated through the circulation of the electric field along the edges of the triangular face. Let us denote the length of the edge joining the jth and ith nodes as L_{eji} and the electric field along the same edge as E_{ji}. The area bounded by the three edges L_{eji}, L_{eik}, and L_{ekj} is denoted by A_m and the magnetic field normal to the triangle is denoted by h_m. Mathematically, we have

$$h_m = \frac{\delta t}{A_m \mu_0}(E_{kj}L_{ekj} + E_{ji}L_{eji} + E_{ik}L_{eik}) \tag{59}$$

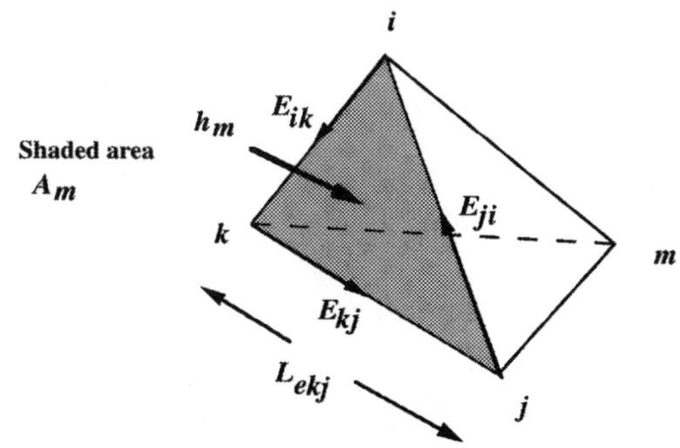

Fig. 19 Computation cell of the magnetic field.

In Fig. 20, the electric field is sampled at the midpoint of the edge common to the surrounding tetrahedrons. The electric field is updated through the circulation of the magnetic fields along the edge of the polygonal intercepting plane which is perpendicular to the edge being updated. The area of the plane is denoted by S_j and the vector joining the ith and $(i + 1)$th corners are denoted by $\mathbf{t}_i L_{hi}$. Each corner of the intercepting plane resides in each of the tetrahedrons sharing the same edge. The location of the corner is chosen such that its distance from the centroid of the corresponding tetrahedron is minimized. The magnetic field along each edge of the intercepting plane is assumed to be constant and is calculated by taking the average of the magnetic field values at both ends of the edge. The magnetic field at each corner of the intercepting plane is interpolated conveniently using the Whitney two-form elements and the known magnetic fields normal to each of the four surfaces of the tetrahedron in which the corner is residing. The expressions for updating the electric field is given as:

$$E_j^{n+1} = E_j^n + \frac{\delta t}{\varepsilon_0 S_j}\sum_{i=1}^{r}\frac{\mathbf{H}_i + \mathbf{H}_{i+1}}{2} \cdot \mathbf{t}_i L_{hi}, \text{ modulus } r \tag{60}$$

where r is the number of tetrahedrons sharing the same edge. The magnetic field $\mathbf{H}_i$ at the corner located at (x_i, y_i, z_i) which falls within the kth tetrahedron is interpolated as

$$\mathbf{H}_i(x_i, y_i, z_i) = \sum_{j=1}^{4} h_j A_j \mathbf{W}_j(x_i, y_i, z_i) \tag{61}$$

where the Whitney element interpolation functions have been defined in Eq. (56). Following the same procedure as in the 2-D case, we can also derive the single-matrix formulation for the 3-D Whitney elements. The expression is omitted here, as it has not been simplified to a form suitable for presentation.

To test the accuracy of the 3-D Whitney FDTD, let us consider a rectangular resonator. The computation domain is uniformly discretized into cubic cells. Each cell is then divided into

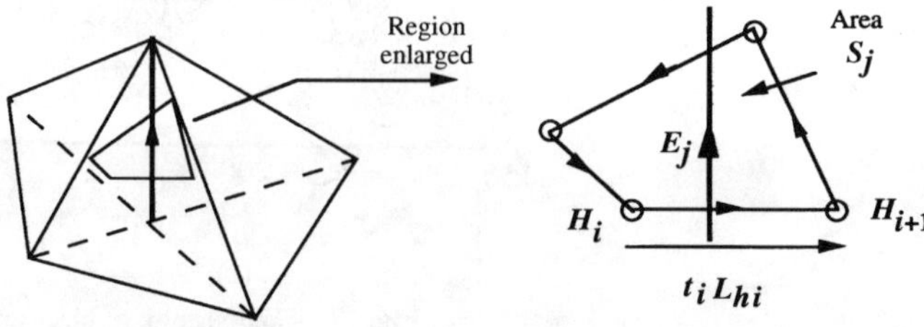

Fig. 20 Computation cell of the electric field.

five tetrahedrons. We excite an electric field edge near the center of the resonator and record the field one element from the backplane of the resonator as shown in Fig. 21. The excitation function is a modulated sine function defined as

$$f(t) = 4(\alpha t)^2 e^{-\alpha t} \sin(\omega_0 t) \tag{62}$$

where $\alpha = \omega_0/\sqrt{3}$ and $\omega_0 = 2\pi f_0$ rad/cycle. The discretization is 20 points per linear wavelength at f_0, which is set at 0.3GHz.

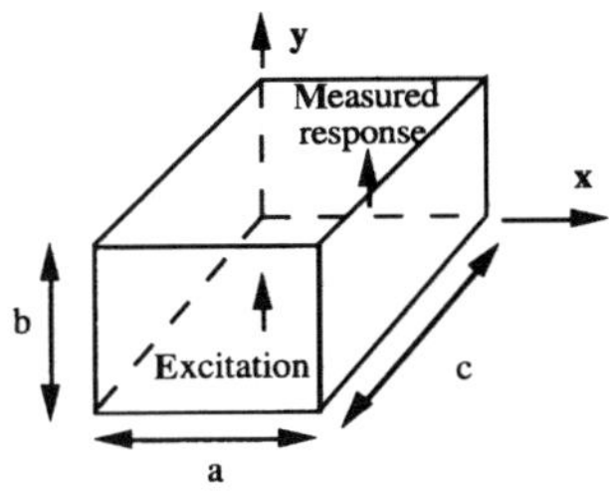

Fig. 21 A rectangular resonator.

Figure 22 shows the spectrum of the electric field measured for two different resonators while Table 5 shows the comparison of the resonance frequencies for various modes when compared to analytical values. Very good agreement is obtained. The larger error in the 101 mode in the first resonator and the 201 mode of the second resonator may be due to the smaller spectral content of the excitation field at that frequency and the coarse discretization (11.5 points per linear wavelength), respectively.

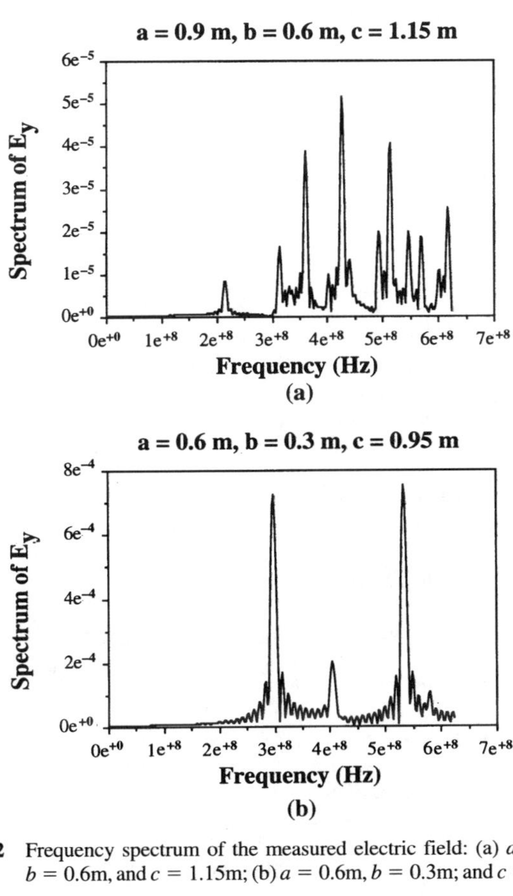

Fig. 22 Frequency spectrum of the measured electric field: (a) $a = 0.9$m, $b = 0.6$m, and $c = 1.15$m; (b) $a = 0.6$m, $b = 0.3$m; and $c = 0.95$m.

TABLE 5. Comparison of Resonance Frequencies Between Whitney FDTD and Analytical Values

$a = 0.9$m, $b = 0.6$m, $c = 1.15$m			
Mode	*WETD*	*Exact*	*% error*
f_{101}	214.8	211.6	1.51
f_{102}	312.5	309.6	0.94
f_{201}	358.9	357.9	0.28
f_{202}	424.8	423.3	0.35
$a = 0.6$m, $b = 0.3$m, $c = 0.95$m			
Mode	*WETD*	*Exact*	*% error*
f_{101}	295.7	297.9	0.74
f_{102}	402.8	402.8	0.00
f_{201}	524.3	534.7	1.95

IV. Summary

We have discussed various conformal schemes regarding the enhancing of the modeling flexibility of the finite-difference time-domain method. The ultimate goal of these conformal schemes is to attain similar modeling flexibility of the well-established finite-element method and yet preserve the computational efficiency of the FDTD. Six reprints representing six different conformal schemes are included here in a chronological fashion. All but one of these papers dealt with microwave structures such as microstrip transmission lines, microwave cavities, and waveguides and discontinuities.

The first paper by Lee [1] is one of the earliest papers to apply conformal time-domain techniques to the analysis of microwave structures. The authors formulated the non-orthogonal FDTD using the covariant and contravariant components of the electric and magnetic fields as the unknown variables. The required memory and number of operations is two and three times, respectively, that of the rectangular FDTD with similar mesh size. Several examples are used to validate the method. Results for *H*-plane waveguide junction, circular waveguides with a circular iris, and rectangular iris compared well with other methods while results for the 45° microstrip bend discontinuity are questionable [2]. Note that the curved boundary in the structures presented in [1] only constitutes a small region of the whole computation domain; a nonorthogonal or curvilinear grid only local to the curved boundary is desirable. The overlapping grids developed by Yee in the third reprint addressed this issue by overlapping a curvilinear grid local to the curved boundary with a global rectangular grid. Information is exchanged between the two grids via bilinear interpolation. This technique can be adopted to analyze a microwave structure such as a cylindrical via with circular pads connecting two uniform microstrip lines. A cylindrical grid in the vicinity of the via and circular pads can be employed as the required local grid. The majority of the computation remains in the global rectangular grid. The fourth paper by Konrad and Lo employed the point-matched time-domain finite element method to analyze planar microstrip circuits of arbitrary shape. A quasi-static approximation is assumed leading to time-stepping of finite differences of voltage and current density. Therefore, a 2-D grid is required as opposed to the

3-D mesh in the conventional FDTD. Simple and yet effective absorbing boundary conditions were also discussed. Similar to the idea of the overlapping grids with a local irregular grid, the fifth paper by Fang and Ren entailed the integral form of Maxwell's equations to correct the irregular FDTD cells locally conformed to the microstrip line. There are 32 possible irregular cells, and their corresponding magnetic and electric field components can be corrected according to one of the five groups. Examples of nonuniform line and over-crossing lines are given. Similar to the overlapping grid method, the locally conformal cells only occupy a small region of the whole computation domain and therefore, the efficiency of FDTD is maintained. Effort is needed, however, to identify this conformal cell. The sixth reprints by Chan and Elson presented a nodal-based FVTD scheme in which an irregular triangular grid is employed to model 2-D waveguide discontinuities. The advantage of this technique is that the need for the Delaunay tessellation is alleviated. The last two reprints employed Whitney elements in the time domain. The seventh reprint by Lee presented an implicit Whitney-element time-domain method in modeling microwave cavities and waveguides. A combination of central and backward difference yielded a more efficient algorithm. A rigorous stability condition based on eigenvalue analysis was also presented. This scheme, however, is implicit and therefore requires a matrix solution for each time step. The last reprint by Elson et al. presented an explicit Whitney-element time-domain scheme in which the Whitney elements were employed as interpolation functions in contrast to the finite-element formulation in [7]. The field components were updated via the integral form of Maxwell's equations. Accurate results were presented for rectangular cavity structures.

While developing new algorithms to bridge the gaps between FDTD and FEM, it is often found that tools in FEM can be used in FDTD and vice versa. The implicit Whitney element time-domain method significantly improves the modeling flexibility of previous time-domain methods. This time-domain method, however, requires solving a system of linear equations for every time step which results in additional costs in memory and CPU time. To alleviate these difficulties, we have developed an explicit 2-D and 3-D Whitney element time-domain method. The 2-D computational domain is discretized with a triangular grid while the 3-D computational domain is discretized into a union of tetrahedrons. 2-D and 3-D Whitney two-form elements are employed to interpolate magnetic fields at the required locations which are consequently used to update the electric field. These algorithms provide modeling capability similar to that of the implicit scheme while retaining the efficiency of FDTD. However, further refinement, such as the rigorous stability condition for the 3-D case, is needed to make these algorithms useful engineering tools.

Acknowledgments

The authors are indebted to Dr. K. S. Yee who has shared his insights in conformal FDTD methods and suggested to us the possibility of using Whitney elements as interpolation functions. Helpful discussions with Dr. J. -F. Lee on Whitney elements are gratefully acknowledged. This work was supported in part by the National Science Foundation (NSF-9158040).

References

[1] K. S. Yee, "Numerical solution of initial boundary value problems in isotropic media," *IEEE Trans. Antennas Propagat.*, vol. AP-14, no. 3, pp. 302–307, May 1966.

[2] D. E. Merewether, "Transient currents induced on a metallic body of revolution by an electromagnetic pulse," *IEEE Trans. Electromagn. Compat.*, vol. EMC-13, pp. 41–44, 1971.

[3] K. S. Kunz and K.-M. Lee, "A three-dimensional finite-difference solution of the external response of an aircraft to a complex transient EM environment: Part I—The method and its implementation," *IEEE Trans. Electromagn. Compat.*, vol. EMC-20, pp. 328–333, 1978.

[4] R. Holland, L. Simpson, and K. S. Kunz, "Finite-difference analysis of EMP coupling to lossy dielectric structures," *IEEE Trans. Electromagn. Compat.*, vol. EMC-22, pp. 203–209, 1980.

[5] A. Taflove and M. E. Brodwin, "Numerical solution of steady-state electromagnetic scattering problems using the time-dependent Maxwell's equations," *IEEE Trans. Microwave Theory Tech.*, vol. MTT-23, pp. 623–630, 1975.

[6] A. Taflove, "Application of the finite-difference time-domain method to sinusoidal steady-state electromagnetic penetration problems," *IEEE Trans. Electromagn. Compat.*, vol. EMC-22, pp. 191–202, 1980.

[7] G. Mur, "Absorbing boundary conditions for the finite-difference approximation of the time-domain electromagnetic field equations," *IEEE Trans. Electromagn. Compat.*, vol. EMC-23, pp. 377–382, 1981.

[8] Z. P. Liao, H. L. Wong, B. P. Yang, and Y. F. Yuan, "A transmitting boundary for transient wave analysis," *Scientia Sinica, Series A,* vol. 27, pp. 1063–1076, 1984.

[9] W. C. Chew, *Waves and Fields in Inhomogeneous Media,* New York, NY: Van Nostrand Reinhold, 1990.

[10] K. Kunz and R. Lubbers, *The Finite Difference Time-Domain Method for Electromagnetics,* Boca Raton, FL: CRC Press, 1993.

[11] PATRAN is a trademark of PDA Engineering.

[12] I-DEAS is a trademark of Structural Dynamics Research Corporation.

[13] K. S. Kunz and L. Simpson, "A technique for increasing the resolution of finite-difference solutions to the Maxwell equation," *IEEE Trans. Electromagn. Compat.*, vol. EMC-23, pp. 419–422, Nov. 1981.

[14] I. S. Kim and W. J. R. Hoefer, "A local mesh refinement algorithm for the time-domain-finite difference method using Maxwell's curl equation," *IEEE Trans. Microwave Theory Tech.*, vol. MTT-38, no. 6, pp. 812–815, June 1990.

[15] S. S. Zivanovic, K. S. Yee, and K. K. Mei, "A subgridding method for the time-domain finite-difference method to solve Maxwell's equations," *IEEE Trans. Microwave Theory Tech.*, vol. MTT-39, no. 3, pp. 471–479, March 1991.

[16] D. T. Prescott and N. V. Shuley, "A method for incorporating different sized cells into the finite-difference time-domain analysis technique," *IEEE Microwave and Guided Wave Letters,* vol. 2, no. 11, pp. 434–436, Nov. 1992.

[17] R. Holland, "Finite-difference solutions of Maxwell's equations in generalized nonorthogonal coordinates," *IEEE Trans. Nuc. Sci.*, vol. NS-30, pp. 4589–4591, Dec. 1983.

[18] M. Fusco, "FDTD algorithm in curvilinear coordinates," *IEEE Trans. Antennas Propagat.*, vol. AP-38, pp. 76–89, Jan. 1990.

[19] M. A. Fusco, M. V. Smith, and L. W. Gordon, "A three-dimensional FDTD algorithm in curvilinear coordinates," *IEEE Trans. Antennas Propagat.*, vol. 39, pp. 1463–1471, Oct. 1991.

[20] J. F. Lee, R. Palandech, and R. Mittra, "Modeling three-dimensional discontinuities in waveguides using non-orthogonal FDTD algorithm," *IEEE Trans. Microwave Theory Tech.*, vol. MTT-40, pp. 346–352, Feb. 1992.

[21] P. H. Harms, J. F. Lee, and R. Mitra, "A study of the nonorthogonal FDTD method vs. the conventional FDTD technique for computing resonant frequencies of cylindrical cavities," *IEEE Trans. Microwave Theory Tech.*, vol. 40, no. 4, pp. 741–746, April 1992.

[22] J. -F. Lee, "Finite difference time-domain algorithm for non-orthogonal grids and its application to the solution of electromagnetic scattering problems," *International Journal of Electronics and Communications,* vol. 46, no. 5, pp. 328–335, Sept. 1992.

[23] P. H. Harms, J. F. Lee, and R. Mittra, "Characterizing the cylindrical via discontinuity," *IEEE Trans. Microwave Theory Tech.,* vol. 41, no. 1, pp. 153–156, Jan. 1993.

[24] S. L. Ray, "Numerical dispersion and stability characteristics of time-domain methods on nonorthogonal meshes," *IEEE Trans. Antennas Propagat.,* vol. AP-41, no. 2, pp. 233–235, Feb. 1993.

[25] K. K. Mei, A. Cangellaris, and D. J. Angelakos, "Conformal time-domain finite difference method," *Radio Science,* vol. 19, pp. 1145–1147, Sept.-Oct. 1984.

[26] A. C. Cangellaris, C. C. Lin, and K. K. Mei, "Point-matched time-domain finite element methods for electromagnetic radiation and scattering," *IEEE Trans. Antennas Propagat.,* vol. AP-35, pp. 1160–1173, 1987.

[27] R. L. Lee and N. K. Madsen, "A Mixed finite element formulation for Maxwell's equations in the time-domain," *J. Comp. Phys.,* vol. 88, pp. 284–304, June 1990.

[28] A. C. Cangellaris and K. K. Mei, "The method of conformal boundary elements for transient electromagnetics," *Progress in Electromagnetic Research,* vol. PIER-2, M. A. Morgan ed., pp. 249–285, New York, NY: Elsevier, 1990.

[29] A. C. Cangellaris, "Time-domain finite methods for electromagnetic wave propagation and scattering," *IEEE Trans. Magnetics,* vol. MAG-27, no. 5, Sept. 1991.

[30] J. Joseph, T. J. Sober, K. J. Gohn, and A. Konrad, "Time-domain analysis by the point-matched finite element method," *IEEE Trans. Magnetics,* vol. MAG-27, no. 5, pp. 3852–3855, Sept. 1991.

[31] S. L. Ray, "Grid Decoupling in finite element solution of Maxwell's equation," *IEEE Trans. Antennas Propagat.,* vol. AP-40, no. 4, pp. 443–445, April 1992.

[32] A. Konrad and J. O. Y. Lo, "Time-domain solution of planar circuits," *J. of Electromagnetic Wave Applications,* pp. 77–92, 1993.

[33] K. S. Yee, "Numerical solution to Maxwell's equations with nonorthogonal grids," *Proc. 1st Review of Numerical Electromagnetic Code,* Mar. 1985.

[34] K. S. Yee, "Numerical solution to Maxwell's equations with non-orthogonal grids," *Lawrence Livermore National Laboratory Technical Report,* UCRL-93268, April 1987.

[35] A. Taflove, K. R. Umashankar, B. Beker, F. Harfoush, and K. S. Yee, "Detailed FD-TD analysis of Electromagnetic fields penetrating narrow slots and lapped joints in thick conducting screens," *IEEE Trans. Antennas Propagat.,* vol. AP-36, no. 2, pp. 247–257, Feb. 1988.

[36] A. Taflove and K. R. Umashankar, "The finite-difference time-domain method for numerical modeling of electromagnetic wave interactions with arbitrary structures," *Progress in Electromagnetic Research,* vol. PIER-2, M. A. Morgan, ed., pp. 287–373, New York, NY: Elsevier, 1990.

[37] T. G. Jurgens, A. Taflove, K. Umashankar, and T. G. Moore, "Finite-difference time-domain modeling of curved surfaces," *IEEE Trans. Antennas Propagat.,* vol. 40, no. 4, pp. 357–366, April 1992.

[38] J. Fang and J. Ren, "A locally conformed finite-difference time-domain algorithm of modeling arbitrary shape planar metal strips," *IEEE Trans. Microwave Theory Tech.,* vol. MTT-41, no. 5, pp. 830–838, May 1990.

[39] C. F. Lee, R. T. Shin, J. A. Kong, and B. J. McCartin, "Finite difference time-domain techniques for two dimensional triangular grids," *Proceedings of the Progress in Electromagnetics Research Symposium,* pp. 189–190, July 1989.

[40] C. F. Lee, B. J. McCartin, R. T. Shin, and J. A. Kong, "A triangular-grid finite-difference time-domain method for electromagnetic scattering problems," *Journal of Electromagnetic Waves and Applications,* vol. 8, no. 4, pp. 449–470, 1994.

[41] N. K. Madsen and R. W. Ziolkowski, "Numerical solution of Maxwell's equations in the time-domain using irregular nonorthogonal grids," *Wave Motion,* vol. 10, pp. 583–596, Dec. 1988.

[42] V. Shankar, W. F. Hall, and A. H. Mohammadian, "A time-domain differential solver for electromagnetic scattering problems," *Proceedings of the IEEE,* vol. 77, no. 5, pp. 709–721, May 1989.

[43] V. Shankar, W. Hall, and A. H. Mohammadian, "A three-dimensional Maxwell's equation solver for computation of scattering from layered media," *IEEE Trans. Magn.,* vol. 25, pp. 3098–3103, July 1989.

[44] N. K. Madsen and R. W. Ziolkowski, "A three-dimensional modified finite-volume technique for Maxwell's equations," *Electromagnetics,* vol. 10, no. 1–2, pp. 147–161, Jan.-June 1990.

[45] V. Shankar, A. H. Mohammadian, and W. F. Hall, "A time-domain finite-volume treatment for the Maxwell's equations," *Electromagnetics,* vol. 10, no. 1–2, pp. 127–146, Jan.-June 1990.

[46] R. Holland, V. P. Cable, and L. C. Wilson, "Finite-volume time-domain (FVTD) techniques for EM scattering," *IEEE Trans. Electromagnetic Compat.,* vol. 33, no. 4, pp. 281–294, Nov. 1991.

[47] N. K. Madsen, "Divergence preserving discrete surface integral methods for Maxwell's curl equations using non-orthogonal unstructured grids," *Lawrence Livermore National Laboratory Technical Report,* UCRL JC-109787, Feb. 1992.

[48] S. D. Gedney and F. Lansing, "A parallel discrete surface integral equation method for the analysis of three-dimensional microwave circuit devices with planar symmetry," *IEEE Antennas and Propagation Society International Symposium Digest,* vol. 3, pp. 1778–1781, June 1994.

[49] S. D. Gedney, "A comparison of the performance of finite difference time-domain, finite element time-domain, and discrete surface integral equation methods on high performance parallel computers," *IEEE Antennas and Propagation Society International Symposium Digest,* vol. 3, pp. 384–387, June 1994.

[50] K. S. Yee, J. S. Chen, and A. H. Chang, "Conformal finite difference time-domain (FDTD) with overlapping grids," *IEEE Trans. Antennas Propagat.,* vol. 40, no. 9, pp. 1068–1075, September 1992.

[51] K. S. Yee and J. S. Chen, "Conformal hybrid finite difference time-domain and finite-volume time-domain," *IEEE Trans. Antennas Propagat.,* vol. 42, no. 10, pp. 1450–1455, Oct. 1994.

[52] D. R. Lynch and K. D. Paulsen, "Time-domain integration of the Maxwell equations on finite elements," *IEEE Trans. Antennas Propagat.,* vol. AP-38, no. 12, pp. 1933–1942, Dec. 1990.

[53] G. Mur, "The finite-element modeling of three-dimensional time-domain electromagnetic fields in strongly inhomogeneous media," *IEEE Trans. Magnetics,* vol. 28, no. 2, pp. 1130–1133, March 1992.

[54] S. Barkeshli, H. A. Sabbagh, D. J. Radecki, and M. Melton, "A novel implicit time-domain boundary integral/finite-element algorithm for computing transient electromagnetic field coupling to a metallic enclosure," *IEEE Trans. Antennas Propagat.,* vol. AP-40, no. 10, pp. 1155–1164, October 1992.

[55] G. Mur, "The finite-element modeling of three-dimensional electromagnetic fields using edge and nodal elements," *IEEE Trans. Antennas Propagat.,* vol. 41, no. 7, pp. 948–953, July 1993.

[56] K. Mahadevan, R. Mittra, D. Rowse, and J. Murphy, "Edge-based finite element frequency and time-domain algorithms for RCS computation," *IEEE Antennas and Propagation Society International Symposium Digest,* vol. 3, pp. 1680–1683, June 1993.

[57] J. F. Lee, "WETD—A finite element time-domain approach for solving Maxwell's equations," *IEEE Microwave and Guided Wave Letters,* vol. 4, no. 1, pp. 11–13, Jan. 1994.

[58] C. H. Chan and J. T. Elson, "A vertex-based finite-volume time-domain method for analyzing waveguide discontinuities," *IEEE Microwave and Guided Wave Letters,* vol. 3, no. 10, pp. 372–374, Oct. 1993.

[59] R. F. Bower and C. H. Chan, "A unique single step algorithm for time stepping electromagnetic fields," *IEEE Microwave and Guided Wave Letters,* vol. 3, no. 11, pp. 411–413, November, 1993.

[60] J. T. Elson, H. Sangani, and C. H. Chan, "An explicit time-domain method using three-dimensional Whitney element," *Microwave and Optical Technology Letters,* vol. 7, no. 13, pp. 607–610, September 1994.

[61] C. H. Chan, H. Sangani, K. S. Yee, and J. T. Elson, "A finite-difference time-domain method using Whitney elements," *Microwave and Optical Technology Letters,* vol. 7, no. 14, pp. 673–676, October 1994.

[62] R. Holland and J. R. Hill, "Self-consistent atmospheric EMP coupling in three dimension," *IEEE Trans. Nuc. Sci.,* vol. NS-23, pp. 1283–1293, June 1976.

[63] R. Holland, "THREDS: A finite-difference time-domain EMP code in 3D spherical coordinates," *IEEE Trans. Nuc. Sci.,* vol. NS-30, pp. 4592–4595, Dec. 1983.

[64] H. Whitney, *Geometric Integration Theory,* Princeton, NJ: U.P., 1957.

[65] G. D. Smith, *Numerical Solution of Partial Differential Equations: Finite Difference Methods,* Third Edition, Oxford Applied Mathematics and Computing Science Series, pp. 47–67, 1985.

[66] S. R. H. Hoole, *Computer-Aided Analysis and Design of Electromagnetic Devices,* New York, NY: Elsevier, 1989.

Modeling Three-Dimensional Discontinuities in Waveguides Using Nonorthogonal FDTD Algorithm

Jin-Fa Lee, *Member, IEEE,* Robert Palandech, and Raj Mittra, *Fellow, IEEE*

***Abstract*—In this paper, we present a generalization of the finite difference time domain (FDTD) algorithm adapted to nonorthogonal computational grids and apply it to the investigation of three dimensional discontinuity problems. The nonorthogonal FDTD uses a body-fitted grid for meshing up the computation domain and, consequently, is able to model the problem geometry with better accuracy than is possible with the staircasing approach conventionally employed in the FDTD algorithm. In addition to extending the FDTD algorithm to nonorthogonal grids, we also derive the stability conditions for the nonorthogonal FDTD algorithm in two and three dimensions. Numerical results including an *H*-plane waveguide junction, a circular waveguide with a circular iris, a circular waveguide with a rectangular iris, and a microstrip bend discontinuity, are presented to validate the current nonorthogonal FDTD approach.**

I. Introduction

The MICROWAVE discontinuity problems in three dimensions often involve complicated shapes. In using the conventional FDTD algorithm [1] to model such problems, a staircased mesh is often employed. This approach suffers from the disadvantage that an accurate description of a non-Cartesian geometry usually requires a very fine mesh and, consequently, a very small time step in the FDTD algorithm.

To circumvent this problem, Holland [2] has generalized the FDTD algorithm to general nonorthogonal grids. Using the nonorthogonal FDTD mesh enables one to conform to the shape of a three-dimensional structure and, therefore, avoid the problem of a very small time step resulting from an overly dense mesh. In this paper, we first modify Holland's formulation in order to improve its computational efficiency, and then apply it to study some three-dimensional discontinuity problems. Additionally, we perform the stability analysis of the nonorthogonal FDTD algorithm in two and three dimensions, and derive an upper bound for the time step that guarantees stable numerical solutions. The study of several microwave junctions, viz., an *H*-plane waveguide discontinuity, a circular waveguide with a circular iris, a circular waveguide with a rectangular iris, and a microstrip bend discontinuity, are presented to validate the current approach.

This paper is organized as follows. Section II presents the formulation which extends the FDTD algorithm to nonorthogonal grids. The corresponding stability analyses for two- and three-dimensional geometries are presented in Section III. Several numerical results are included in Section IV, and brief conclusions are given in Section V.

Manuscript received April 9, 1991; revised August 1, 1991.

J.-F. Lee was with the Electromagnetic Communication Laboratory, Department of Electrical and Computer Engineering, University of Illinois 1406 N. Green St., Urbana, IL 61801-2991. He is presently with the Department of Electrical Engineering, Worcester Polytechic Institute, Worcester, MA 01609.

R. Palandech was with the Department of Electrical and Computer Engineering, University of Illinois, 1406 W. Green St., Urbana, IL 61801-2991. He is presently with Motorola, Inc., 1301 E. Algonquin Rd., Schaumburg, IL 60196.

R. Mittra is with the Electromagnetic Communication Laboratory, Department of Electrical and Computer Engineering, University of Illinois, 1406 W. Green St., Urbana, IL 61801-2991.

IEEE Log Number 9105249.

II. Three-Dimensional Nonorthogonal FDTD Algorithm

The formulation of the FDTD algorithm in nonorthogonal computational grids has been presented by Holland [2] and, in [3], Fusco has successfully used this approach to solve two-dimensional scattering problems. In this paper, we reformulate the nonorthogonal FDTD algorithm using the covariant and contravariant components of the electric and magnetic fields as the unknown variables. This modification results in a more efficient time marching procedure than that presented in [2].

A nonorthogonal coordinate system (u^1, u^2, u^3) in three dimensions may be characterized [4] by a set of vectors $\mathbf{A}_i$ (see Fig. 1) where

$$d\vec{r} = \sum_{i=1}^{3} \frac{\partial \vec{r}}{\partial u^i} du^i = \sum_{i=1}^{3} \mathbf{A}_i du^i \tag{1}$$

From Fig. 1, we see that the bases $\mathbf{A}_i$ are vectors defined by the edges of the cell. According to Fig. 1, we can also define a dual bases (reciprocal bases) $\mathbf{A}^i$ as follows

$$\mathbf{A}^1 = \frac{\mathbf{A}_2 \times \mathbf{A}_3}{\sqrt{g}}$$
$$\mathbf{A}^2 = \frac{\mathbf{A}_3 \times \mathbf{A}_1}{\sqrt{g}}$$
$$\mathbf{A}^3 = \frac{\mathbf{A}_1 \times \mathbf{A}_2}{\sqrt{g}} \tag{2}$$

Reprinted from *IEEE Transactions on Microwave Theory and Techniques,* pp. 346-352, February 1992.

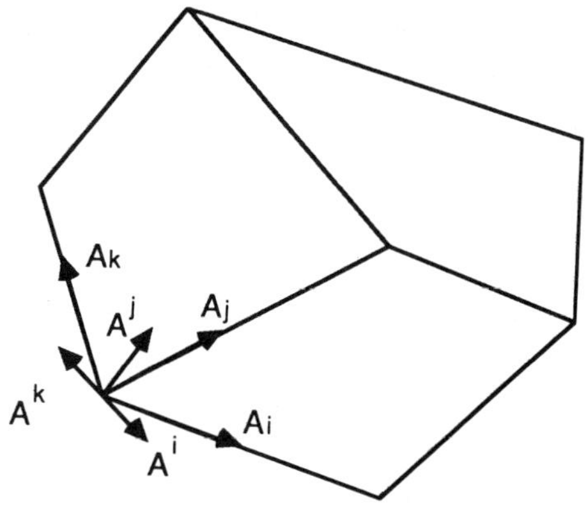

Fig. 1. Reciprocal basis vectors for the nonorthogonal coordinate system.

where $\sqrt{g}$ is the volume enclosed by the unit cell. A dual basis vector, say $\boldsymbol{A}_i$, points in the direction normal to the face i, which is a quadrilateral area spanned by $\boldsymbol{A}_j$ and $\boldsymbol{A}_k$ (i, j, k are cyclical indices). Furthermore, the dual bases $\boldsymbol{A}^i$'s and the original bases $\boldsymbol{A}_i$'s satisfy the reciprocal relationship:

$$\boldsymbol{A}^i \cdot \boldsymbol{A}_j = \delta_{ij} \tag{3}$$

where δ_{ij} is the usual Kronecker delta. An arbitrary vector can be written in two-forms expressed by these two bases. For example, the electric field $\vec{E}$, can be expressed in terms of these two bases as

$$\vec{E} = \sum_i E^i \boldsymbol{A}_i$$

$$\vec{E} = \sum_i \boldsymbol{A}^i E_i \tag{4}$$

Here, the coefficients E^i and E_i are called the contravariant and covariant components [4], respectively, of the electric field $\vec{E}$.

The physical meanings of the covariant and contravariant components will now be given. We first use the second equation in (4) to find the *flow* of the electric field along the edge i, i.e., $\vec{E} \cdot \boldsymbol{A}_i$. From the reciprocal relationship in (3), we obtain

$$\vec{E} \cdot \boldsymbol{A}_i = \left(\sum_j \boldsymbol{A}^j E_j \right) \cdot \boldsymbol{A}_i = E_i \tag{5}$$

We therefore conclude from (5) that the covariant component E_i represents the *flow* of $\vec{E}$ along the edge i. Similarly, by taking the inner product of the first equation in (4) with the dual basis $\boldsymbol{A}^i$, and using the reciprocal relationship in (3), we get

$$\vec{E} \cdot \boldsymbol{A}^i = \left(\sum_j E^j \boldsymbol{A}_j \right) \cdot \boldsymbol{A}^i = E^i \tag{6}$$

From (2) and (6), the contravariant component E^i is seen to be the total flux of the electric field passing through face i, divided by the volume $\sqrt{g}$.

To extend the FDTD algorithm to nonorthogonal grids, we need to approximate Maxwell's equations in terms of the contravariant and covariant components. To this end, we start from the source-free, integral form of Maxwell's equations, for isotropic media, given below:

$$-\frac{\partial}{\partial t} \int_\Omega \mu \vec{H} \cdot \boldsymbol{da} = \oint_{\partial\Omega} \vec{E} \cdot \boldsymbol{dl}$$

$$\frac{\partial}{\partial t} \int_\Omega \epsilon \vec{E} \cdot \boldsymbol{da} = \oint_{\partial\Omega} \vec{H} \cdot \boldsymbol{dl} \tag{7}$$

The left hand sides of (7) represent the time rates of change of the electric and the magnetic flux through a surface Ω, whereas the right hand sides of (7) give the *circulations* of the electric and magnetic fields along the corresponding closed boundary $\partial\Omega$. From above, we can obtain a finite difference approximation of the Faraday's law:

$$\begin{aligned} E^1(I, J, K)^{n+1} &= E^1(I, J, K)^n + \frac{\Delta t}{\epsilon\sqrt{g}} \left[H_3\left(I, J + \frac{1}{2}, K\right) \right. \\ &\quad - H_3\left(I, J - \frac{1}{2}, K\right) - H_2\left(I, J, K + \frac{1}{2}\right) \\ &\quad \left. + H_2\left(I, J, K - \frac{1}{2}\right)\right]^{n+1/2} \end{aligned} \tag{8}$$

The other two components, E^2 and E^3, can be derived simply by index permutations. Likewise, the finite difference forms of the Ampere's law read

$$\begin{aligned} H^1(I, J, K)^{n+1/2} &= H^1(I, J, K)^{n-1/2} - \frac{\Delta t}{\mu\sqrt{g}} \left[E_3\left(I, J + \frac{1}{2}, K\right) \right. \\ &\quad - E_3\left(I, J - \frac{1}{2}, K\right) - E_2\left(I, J, K + \frac{1}{2}\right) \\ &\quad \left. + E_2\left(I, J, K - \frac{1}{2}\right)\right]^n . \end{aligned} \tag{9}$$

Again, the index permutations are employed to obtain H^2 and H^3.

It is worthwhile mentioning here that the use of the contravariant and covariant components of the $\vec{E}$ and $\vec{H}$ fields in the finite difference forms is what distinguishes our formulation from that of Holland [2]. An important advantage of the present approach is that (8) and (9) have exactly the same form as the conventional cartesian FDTD algorithm.

As a final step in our formulation, we need to convert the contravariant components of $\vec{E}$ into its covariant components, in order to calculate the contravariant components of $\vec{H}$, and vice versa. This can be accomplished by noting that the original bases $\boldsymbol{A}_i$ are related to the dual bases through the metric tensor g_{ij} [4], according to

$$\boldsymbol{A}_i = \sum_j g_{ij} \boldsymbol{A}^j; \quad g_{ij} = \boldsymbol{A}_i \cdot \boldsymbol{A}_j . \tag{10}$$

Thus, the covariant components of $\vec{H}$, for instance, H_i, can be obtained from the contravariant components of $\vec{H}$ as follows:

$$H_i = \sum_j g_{ij} H^j \tag{11}$$

By averaging the neighboring values, we have the finite difference approximation of (11). For example, the covariant component H_1 can be written as

$$\begin{aligned} H_1(I, J, K) &= g_{11} H^1(I, J, K) + \frac{g_{12}}{4}\left[H^2\left(I + \frac{1}{2}, J - \frac{1}{2}, K\right)\right. \\ &\quad + H^2\left(I - \frac{1}{2}, J - \frac{1}{2}, K\right) \\ &\quad + H^2\left(I + \frac{1}{2}, J + \frac{1}{2}, K\right) \\ &\quad \left. + H^2\left(I - \frac{1}{2}, J + \frac{1}{2}, K\right)\right] \\ &\quad + \frac{g_{13}}{4}\left[H^3\left(I, + \frac{1}{2}, J, K - \frac{1}{2}\right)\right. \\ &\quad + H^3\left(I - \frac{1}{2}, J, K - \frac{1}{2}\right) \\ &\quad + H^3\left(I + \frac{1}{2}, J, K + \frac{1}{2}\right) \\ &\quad \left. + H^3\left(I - \frac{1}{2}, J, K + \frac{1}{2}\right)\right]. \end{aligned} \tag{12}$$

Similar expressions can be obtained for the other covariant components of either $\vec{E}$ or $\vec{H}$ fields.

Equation (12) represents the only additional computation required in the present formulation over and above that in the cartesian form of the FDTD algorithm. Therefore, the number of operations in the nonorthogonal FDTD is roughly three times that of cartesian FDTD algorithm, provided that the same number of mesh points are used. Of course, as pointed out earlier, the primary motivation for using the nonorthogonal algorithm is to reduce the number of mesh points substantially as compared to the uniform cartesian mesh.

III. Stability Analysis

For a numerical algorithm based upon the finite difference approximation to converge, the necessary and sufficient condition is that the scheme must both be consistent as well as stable [5]. The consistency condition, which is reflected in the Taylor expansions of replacing the differential operators with difference operators, requires that the local truncation error approaches zero as the cell size $h \rightarrow 0$. Consequently, to assure the convergence of the numerical procedure presented in this paper, we only need to prove that it is stable for the chosen time step. Therefore, in this section, we derive the stability conditions for the nonorthogonal FDTD algorithm in two and three dimensions. To derive the stability conditions, we appeal to the causality consideration for the two-dimensional case, and employ a more rigorous mathematical treatment for the three-dimensional case.

A. Two-Dimensional Case

Fig. 2 shows a typical non-Cartesian cell in two dimensions, where h represents the closest distance between two lattice planes in this non-Cartesian lattice. Since energy can not propagate faster than the speed of light, the law of causality requires that the time step Δt, in the non-orthogonal FDTD algorithm, be less than the shortest travel time h/c for the wave between the lattice planes. However, from Fig. 2, we see that

$$h = \frac{\sqrt{g}}{\sqrt{(\mathbf{A}_1 - \mathbf{A}_2) \cdot (\mathbf{A}_1 - \mathbf{A}_2)}} = \frac{\sqrt{g}}{\sqrt{g_{11} + g_{22} - 2g_{12}}}. \tag{13}$$

In two dimensions, $\sqrt{g}$ reduces to the area of the cell. Consequently, the necessary stability criterion is obtained as

$$\Delta t \leq \frac{1}{c}\left(\frac{\sqrt{g}}{\sqrt{g_{11} + g_{22} - 2g_{12}}}\right). \tag{14}$$

This condition was first provided by Fusco [3]. Unfortunately, however, there appears to be a typographical error in his formula in [3].

In the numerical implementation, the r.h.s. of (14) is evaluated at every grid point, the smallest value is then used as the upper bound for choosing the time step in the nonorthogonal FDTD algorithm.

B. Three-Dimensional Case

To derive the stability condition for three-dimensional nonorthogonal FDTD algorithm, we start from the vector wave equation:

$$\nabla \times \nabla \times \vec{E} = \frac{-1}{c^2}\frac{\partial^2 \vec{E}}{\partial t^2}. \tag{15}$$

As is well known [6], an arbitrary wave can be expanded in terms of a spectrum of plane waves which can also be viewed as the eigenmodes of the wave equation. Next, we argue that a time domain procedure must be stable for an arbitrary plane wave. Consequently, we let

$$\begin{aligned} \vec{E}(u^1, u^2, u^3; t) &= \vec{\mathcal{E}}(t) \exp(-j\vec{k} \cdot \vec{r}) \\ &= \vec{\mathcal{E}}(t) \exp(-j[k_1 u^1 + k_2 u^2 + k_3 u^3]) \end{aligned} \tag{16}$$

where

$$k_i = \vec{k} \cdot \mathbf{A}_i \tag{17}$$

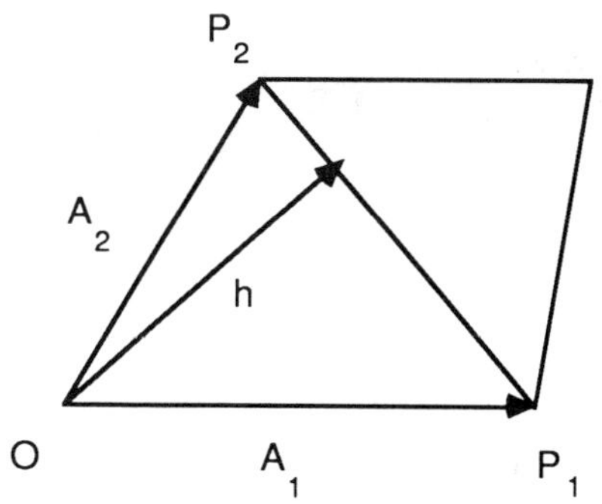

Fig. 2. Typical two-dimensional non-Cartesian cell.

In a general nonorthogonal coordinate system, the operator ∇ becomes

$$\nabla = A^1 \frac{\partial}{\partial u^1} + A^2 \frac{\partial}{\partial u^2} + A^3 \frac{\partial}{\partial u^3} \tag{18}$$

Substituting (16) into (18), performing the central differencing for each of the spatial derivative and making use of $\Delta u^i = 1$, the ∇ operator can be replaced symbolically by

$$\nabla = -2j \sum_i \left(A^i \sin \left[\frac{\Delta (k_i u^i)}{2} \right] \right) \tag{19}$$

Furthermore, any plane wave with a representation given in (16) satisfies the divergence-free condition. Therefore, (15) reduces to

$$(\nabla \cdot \nabla)\vec{E} = \frac{1}{c^2} \frac{\partial^2 \vec{E}}{\partial t^2} \tag{20}$$

Defining a solution growth factor $\lambda = \vec{E}^{n+1}/\vec{E}^n$, and substituting (19) into (20), we obtain

$$-4\left\{ \sum_l \left(A^l \sin \left[\frac{\Delta (k_l u^l)}{2} \right] \right) \cdot \sum_m \left(A^m \sin \left[\frac{\Delta (k_m u^m)}{2} \right] \right) \right\} \vec{E} = \frac{1}{c^2} \frac{\lambda^2 - 2\lambda + 1}{\lambda \Delta t^2} \vec{E}. \tag{21}$$

Solving (21) for λ we get

$$\lambda = (1 - 2s^2 \Delta t^2) \pm 2s \Delta t \sqrt{s^2 \Delta t^2 - 1}$$

where

$$s^2 = c^2 \sum_{\substack{l=1 \\ m=1}}^{3} \left(A^l \cdot A^m \sin \frac{\Delta (k_l u^l)}{2} \sin \frac{\Delta (k_m u^m)}{2} \right). \tag{22}$$

The numerical scheme would be stable if the growth factor $|\lambda| \leq 1$. From (22) we see that this condition is satisfied if and only if

$$s^2 \Delta t^2 \leq 1. \tag{23}$$

Furthermore, as pointed out earlier, the above condition must hold for an arbitrary plane wave. This can be assured by noting that for all possible waves, we have

$$s^2 \leq c^2 \sum_{\substack{l=1 \\ m=1}}^{3} (g^{lm})$$

where

$$g_{lm} = A^l \cdot A^m. \tag{24}$$

Hence, to satisfy the stability condition (23) for all the spatial modes in the lattice, we set

$$\Delta t \leq \frac{1}{c \sqrt{\sum_{\substack{l=1 \\ m=1}}^{3} g_{lm}}}. \tag{25}$$

In an attempt to extend the stability condition to a more general form, and hopefully provide more insights, we will show that the square-root term in (25) is related to the norm of the ∇ operator in the nonorthogonal FDTD algorithm. Typically the norm of an operator, for instance ∇, is defined as

$$\|\nabla\| = \sup_W \frac{\|\nabla W\|}{\|W\|} \tag{26}$$

where $\|\cdot\|$ is the usual L_2 norm, W is any admissible function in the solution process, and sup represents the upper-bound. In particular, since plane waves form a complete set of basis functions, therefore W in (26) can be limited to only plane waves, i.e.,

$$\|\nabla\| = \sup_{k_1 k_2, k_3} \|\nabla W\|,$$

$$W = \exp(-j[k_1 u^1 + k_2 u^2 + k_3 u^3]) \tag{27}$$

Substituting (19) into (27), the result is

$$\|\nabla\| = 2 \sup_{k_1 k_2 k_3} \cdot \sqrt{\sum_{\substack{l=1 \\ m=1}}^{3} A^l \cdot A^m \sin\left(\frac{\Delta[k_l u^l]}{2} \right) \sin\left(\frac{\Delta[k_m u^m]}{2} \right)} = 2 \sqrt{\sum_{\substack{l=1 \\ m=1}}^{3} g^{lm}}. \tag{28}$$

Using (28), we rewrite (25) as

$$\Delta t \leq \frac{2}{c\|\nabla\|}. \tag{29}$$

Note that (29) is based on the assumption that central-differencing is employed for both time and space derivatives. Finally, we demonstrate that (29) reduces to the familiar forms in a Cartesian system:

Two-Dimensional Case:

$$\|\nabla\| = \sup_E \frac{\|\nabla E\|}{\|E\|} = 2 \sqrt{\frac{1}{(\Delta x)^2} + \frac{1}{(\Delta y)^2}} \pi \rightarrow \Delta t \leq \frac{1}{c \sqrt{\frac{1}{(\Delta x)^2} + \frac{1}{(\Delta y)^2}}}$$

Three-Dimensional Case:

$$\|\nabla\| = \sup_{E} \frac{\|\nabla E\|}{\|E\|} = 2\sqrt{\frac{1}{(\Delta x)^2} + \frac{1}{(\Delta y)^2} + \frac{1}{(\Delta z)^2}} \rightarrow \Delta t \leq \frac{1}{c\sqrt{\frac{1}{(\Delta x)^2} + \frac{1}{(\Delta y)^2} + \frac{1}{(\Delta z)^2}}}. \quad (30)$$

IV. Numerical Results

Several numerical examples are presented in this section to demonstrate the validity of the non-orthogonal FDTD algorithm. The problems considered are: 1) *H*-plane waveguide junction; 2) circular waveguide with a circular iris; 3) circular waveguide with a rectangular iris; and, 4) 45° microstrip bend discontinuity. The numerical results of Section IV-A–D are for the steady-state responses which have been derived by using a sinusoidal signal as the excitation. However, in Section IV-D the reflection coefficient of the discontinuity is obtained from time domain results of a Gaussian pulse excitation.

A. H-Plane Waveguide Junction

Figure 3 shows the cross-section of an *H*-plane waveguide junction. The height of the junction is uniform and, therefore, this is essentially a two-dimensional problem. By using the three-dimensional non-orthogonal FDTD algorithm to model it, we can compare our results with an existing two-dimensional FEM approach [8]. Also, shown in Fig. 3 is a plane-cut of the nonorthogonal mesh that is employed for the current analysis. Note that this problem is solved without using any absorbing boundary conditions; consequently, the problem domain has to be truncated very far away from the discontinuity region as can be seen from the figure. The computed steady state field distribution along the length of the discontinuity region is plotted in Fig. 4, where the unit of the distance is arbitrary. In Fig. 4, we also show the comparison of the current results with the ones obtained from the frequency domain finite-element analysis [8]. Very close agreement between the two results is observed.

B. Circular Waveguide with a Circular Iris

Fig. 5 shows a circular waveguide with a circular iris. The waveguide is filled with an isotropic dielectric material of dielectric constant 9.7, except the thin iris region which is filled with air. This problem has been analyzed using two techniques, viz., the mode matching method and the nonorthogonal FDTD approach. In the nonorthogonal FDTD approach, the cross-section of the mesh that is used is shown in Fig. 6. Since the problem is symmetric, only a quarter of the geometry is modeled. The three-dimensional mesh is generated by extruding the two-dimensional mesh (see Fig. 6) in the third dimension. By using the first-order ABC [9], we are able to bring the artificial truncation boundaries to distances as close as 1λ away from the iris for a frequency of 6 GHz. The computed reflection and transmission coefficients derived from these two approaches are shown in Figs. 7 and 8. Good agreement between the two results confirm the validity of the nonorthogonal FDTD approach.

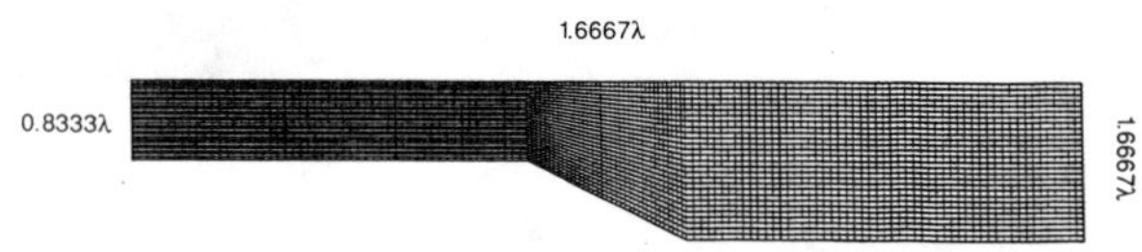

Fig. 3. Nonorthogonal mesh for an *H*-plane waveguide junction.

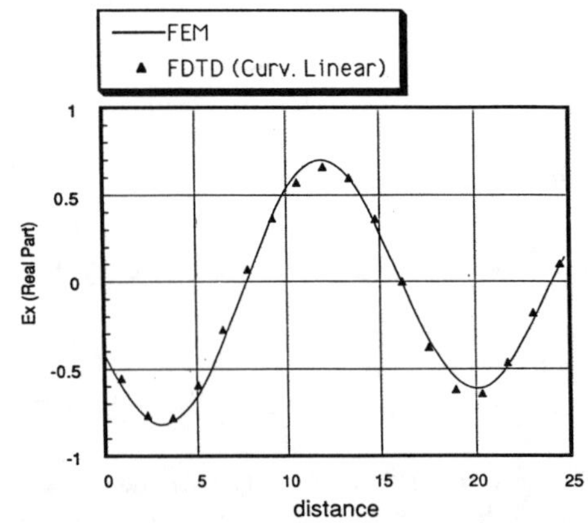

Fig. 4. Field distribution along the taper region.

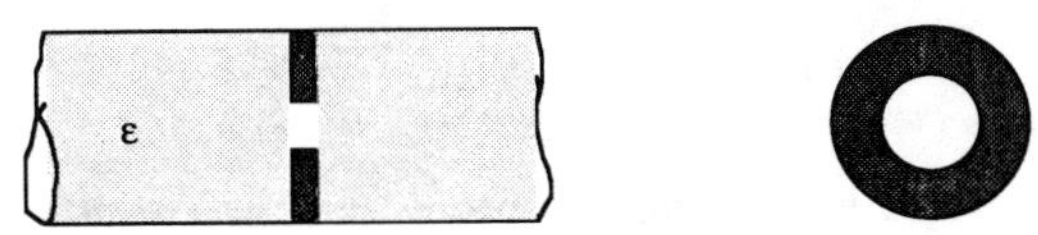

Fig. 5. Circular waveguide with a circular iris.

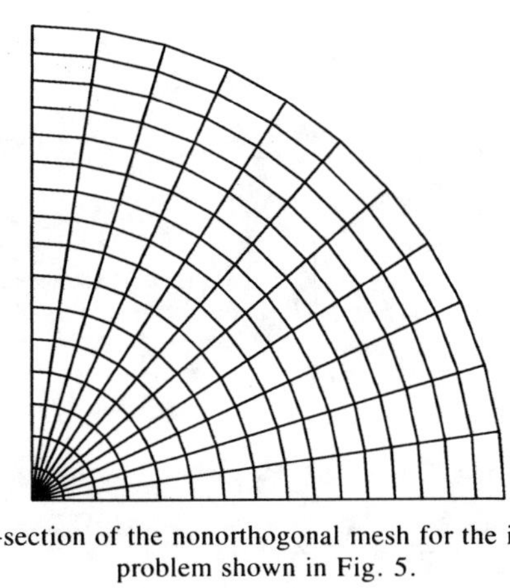

Fig. 6. Cross-section of the nonorthogonal mesh for the iris discontinuity problem shown in Fig. 5.

C. Circular Waveguide with a Rectangular Iris

The mesh used to investigate a circular dielectric-loaded waveguide with a thin rectangular iris is shown in Fig. 9.

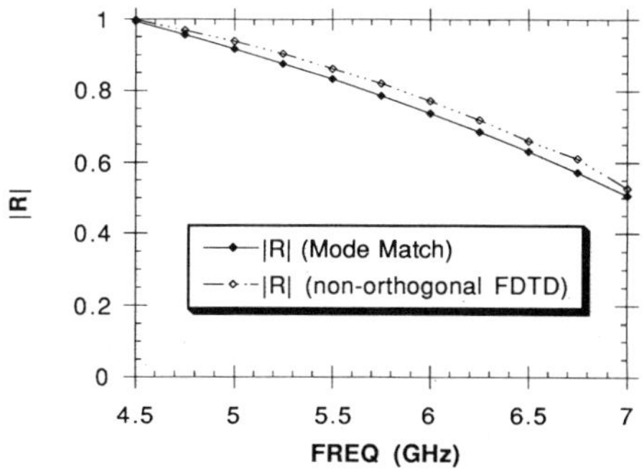

Fig. 7. Comparison of the numerical results for the reflection coefficient for the circular iris dioscontinuity computed by using the mode matching and the nonorthogonal FDTD algorithm.

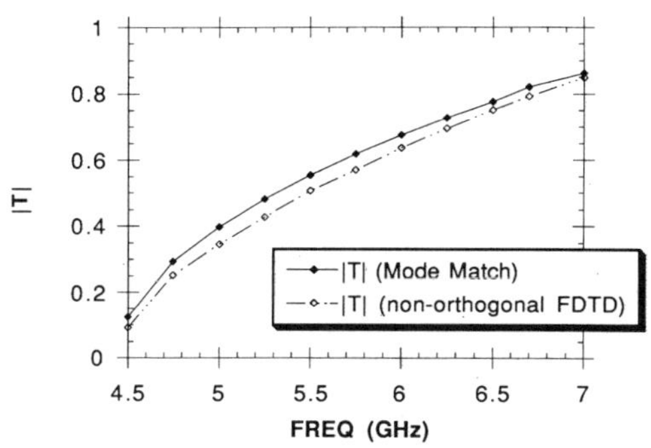

Fig. 8. Comparison of the numerical results for the transmission coefficient for the circular iris discontinuity problem.

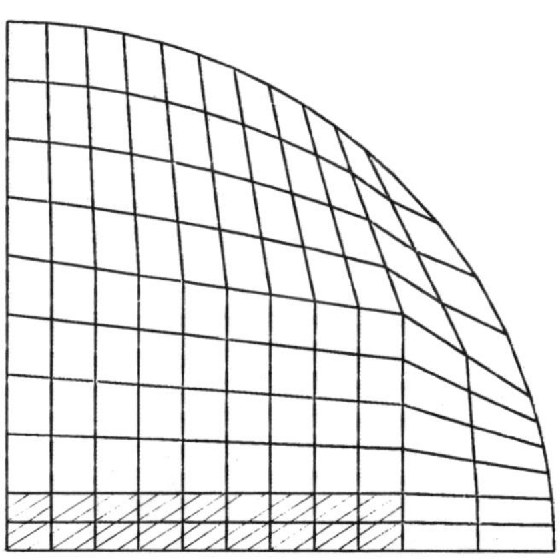

Fig. 9. Cross-sectional mesh for a circular waveguide with a rectangular iris.

The dielectric constant of the dielectric is 9.7, the radius of the waveguide is 6.35 mm, and the dimensions of the aperature is 1.3716 mm × 9.1948 mm. Again, this two-dimensional mesh is extrapolated into the third dimension to generate the required three-dimensional computational grid. The use of ABC again allows us to bring the truncation boundaries to 1λ away from the iris at 6 GHz. The computed reflection and transmission coefficients are plotted in Figs. 10 and 11, respectively. It has not been possible to compare these results with others owing to the unavailability of published results for this geometry in the literature.

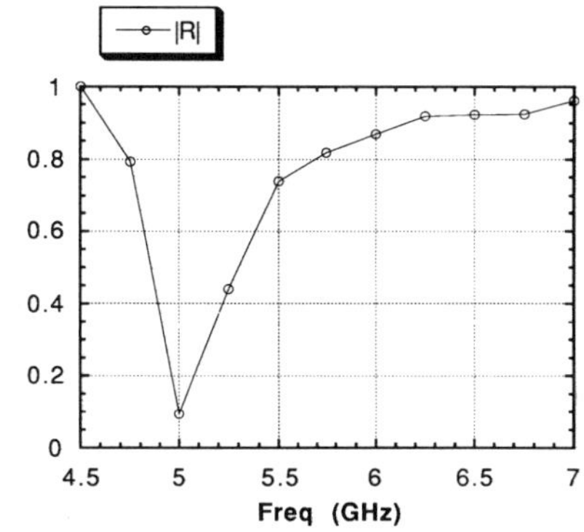

Fig. 10. Computed reflection coefficient for the rectangular iris problem using the nonorthogonal FDTD algorithm.

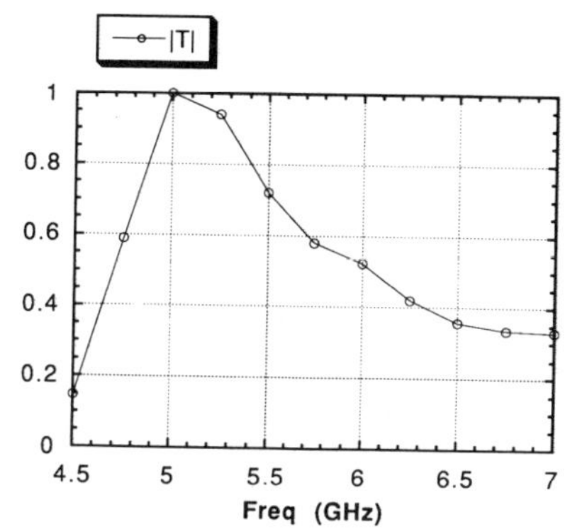

Fig. 11. Computed transmission coefficient for the rectangular iris discontinuity.

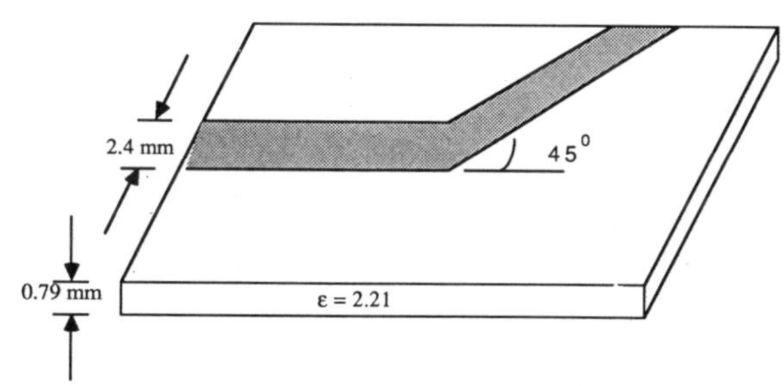

Fig. 12. A 45° microstrip bend discontinuity.

D. 45° Microstrip Bend Discontinuity

Another problem investigated by using the nonorthogonal time domain approach is the 45° microstrip bend shown in Fig. 12, which is conveniently analyzed using the nonorthogonal mesh approach that avoids staircasing. Fig. 13 displays the return loss versus frequency curves obtained by using: 1) the nonorthogonal FDTD approach; 2) the microwave computer-aided design package TOUCHSTONE; 3) a lossless T circuit model for the discontinuity; and 4) HP8510B network analyzer. We note that below 6 GHz, which is believed to be the upper limit for accurate measurements because of the connectors used, the agreement between the nonorthogonal FDTD and the measured results is good.

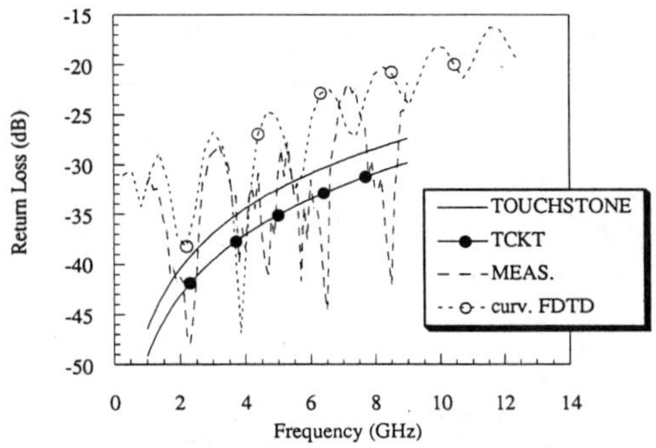

Fig. 13. Numerical results for three models versus experiment for the 45° bend.

V. Conclusion and Discussion

In this paper, we have reformulated the nonorthogonal FDTD algorithm, using the covariant and contravariant components of the electric and magnetic fields as the unknown variables, to result in a more efficient time marching procedure than that presented in [2]. Additionally, we have performed the stability analyses of the nonorthogonal FDTD algorithm in two and three dimensions, and derived an upper bound for the time step that guarantees stable numerical solutions. Finally, the modified nonorthogonal FDTD algorithm has been applied to study several waveguide discontinuities, viz., an *H*-plane waveguide junction, a circular waveguide with a circular iris, a circular waveguide with a rectangular iris, and a 45° microstrip bend. The computed results compared favorably with previously available techniques and, therefore, validated the present approach.

References

[1] K. S. Yee, "Numerical solution of initial boundary value problems involving Maxwell's equations in isotropic media," *IEEE Trans. Antennas Propagat.*, vol. AP-14, pp. 302–307, May 1966.

[2] R. Holland, "Finite-difference solution of Maxwell's equations in generalized nonorthogonal coordinates," *IEEE Trans. Nucl. Sci.*, vol. NS-30, no. 6, Dec. 1983.

[3] M. Fusco, "FDTD algorithm in curvilinear coordinates," *IEEE Trans. Antennas Propagat.* vol. 38, pp. 76–88, Jan. 1990.

[4] J. A. Stratton, *Electromagnetic Theory*. New York: McGraw-Hill, 1941.

[5] G. Strang and G. J. Fix, *An Analysis of the Finite Element Method*. Englewood Cliffs, NJ: Prentice Hall, 1973.

[6] W. C. Chew, *Waves and Fields in Inhomogeneous Media*. New York: Van Nostrand Reinhold, 1990.

[7] E. Kreyszig, Introductory Functional Analysis with Applications. New York: Wiley, 1978.

[8] Z. J. Cendes and J. F. Lee, "The transfinite element method for modeling MMIC devices," *IEEE Trans. Microwave Theory Tech.*, vol. 36, pp. 1639–1649, Dec. 1988.

[9] J. F. Lee and R. Mittra, "Analysis of microwave ferrite devices by using finite element method," *J. Appl. Phys.*, vol. 69, pp. 5032–5034, Apr. 1991.

Triangular-Domain Basis Functions for Full-Wave Analysis of Microstrip Discontinuities

Robert Kipp and Chi H. Chan

Abstract—**A numerical technique for computing the S-parameters of arbitrarily shaped microstrip discontinuities and interconnects is presented. The microstrip conducing surface is rendered in a triangular discretization. Unknown currents are expanded with triangular-domain basis functions in a method of moments solution to the mixed-potential integral equation (MPIE). Triangular basis functions offer several advantages over rectangular subdomain functions, including their ability to conform readily to arbitrary geometries without "staircase" effect. Examples comparing triangular-domain modeling with rectangular-domain modeling and experiment are given.**

I. Introduction

THE increasing switching speed and frequency of digital and analog circuits has expanded the need for characterizing board and chip-level microstrip interconnects without recourse to low-frequency or quasi-static approximations. The literature offers several full-wave approaches including mode matching [1], finite-difference time-domain FDTD [2], [3], and integral equation formulation solved with method of moments (MoM) [4], [5]. Some address specific geometries, such as microstrip step discontinuities [1] or open circuits and gaps [6], while others are intended for application to arbitrary configurations. In the latter category, the literature offers techniques based on MoM where the current on the upper conductor is expanded over rectangular subdomains in an approximate solution to the electric-field integral equation (EFIE) [7], [8] or mixed-potential integral equation (MPIE) [5]. This paper describes the application of triangular discretizations in the MoM solution of an MPIE formulation of scattering by microstrip interconnects and discontinuities.

Triangular subdomain basis functions were developed to compute scattering into a homogeneous medium by arbitrary surfaces [9]. Starting from an MPIE formulation, the unknown surface currents on a perfectly conducting object are obtained through MoM by rendering the surface in a nonuniform triangular mesh and expanding surface currents in basis functions defined over adjoining triangle pairs. More recently, this technique was extended to the inhomogeneous microstrip configuration to compute the surface current distribution and input impedances of microstrip patch antennas [10]. The efficient use of triangular basis functions in computing the S-parameters of microstrip discontinuities is detailed here.

Manuscript received May 14, 1992; revised November 2, 1992. This work was supported in part by the NSF, in part by the NSF(ECS-9158040) and NSF/Center for the design of Analog/Digital Integrated Circuits (CDADIC).

The authors are with the Department of Electrical Engineering, FT-10, University of Washington, Seattle, WA 98195.

IEEE Log Number 9209340.

Triangular discretization offers several advantages over rectangular discretization. The triangles conform exactly to any angled edges in the discontinuity, and curved edges are rendered in a line-segment rather than stair-case approximation. Transmission lines extending from the discontinuity at oblique angles are also represented without stair-case effect and handled with relative ease. Finally, triangular-domain functions offer greater flexibility in the use of nonuniform discretizations.

Section II provides a complete definition of the microstrip characterization problem considered here. The governing equation employed is the mixed-potential integral equation (MPIE) formulation, described in detail for microstrip in [11] and [12], and briefly repeated here for the sake of continuity. Section II also gives the equations for representing currents in triangular subdomains, and describes their incorporation into the MoM solution of the MPIE.

Specific implementation details for the microstrip S-parameter problem are given in Section III. The methodology is based on unimodal analysis of currents on the extended ports. In consequence, the ports are included in the scattering problem, and the number of unknowns often exceeds 1000. Hence, efficient solution requires efficient computation of the Green's functions within the MPIE. Evaluation of Green's function is accelerated with a technique developed in [13] and [14]; its important features are presented in Section III. One can also spare significant computational effort by identifying the redundant interactions among subdomain elements. These redundancies exist in abundance in the regions of the surface geometry where uniform discretization is viable. An approach for exploiting redundancy in arbitrary discretizations is described.

The MoM procedure renders the current distributions on the ports for a given excitation of the network. Scattering parameters can be extracted by computing the unimodal weighting coefficients for forward and reverse current waves on the ports. For an N-port discontinuity, no greater than N linearly independent excitations are required, with structural symmetry reducing this number. A robust procedure for obtaining scattering parameters on an N-port is described in Section IV.

Numerical results are presented in Section V, with validation against experimental data for a microstrip stub and a coupled line. The results are also compared to those from a commercial microwave CAD package and, where possible, a full-wave, rectangular MoM code. Also considered is a 45° bend, whose S_{11} reflection measurement and full-wave analysis with finite-difference time-domain (FDTD) method

Reprinted from *IEEE Transactions on Microwave Theory and Techniques,* pp. 1187-1194, June/July 1993.

are documented [3]. Finally, a comparison of the triangular and rectangular analyses of corner and miter bends is given. The corner bend can be exactly conformed by both rectangular and triangular discretization, while the miter bend can be exactly conformed only by the triangular discretization. Deficiencies in the rectangular approximation are identified.

II. MPIE Formulation and Solution

A. Problem Definition

The microstrip configuration considered here is illustrated in Fig. 1 and has arbitrary surface conductor geometry. The ground plane and substrate extend to infinity in the transverse direction, and the space above the dielectric is unbounded. The substrate is homogeneous and isotropic, but not necessarily lossless. The upper conductor and ground plane have zero resistivity, and the upper conductor is infinitely thin. The number of ports extending from the discontinuity is only limited by computation power. With some reformulation of the Green's functions, the procedure may be extended to include multiple dielectric and strip layers [12], [13].

With the goal of obtaining scattering parameters for a multiport discontinuity, the problem can be divided in two procedures: 1) obtain the surface current distribution on the discontinuity and extended ports under a number of excitation conditions; and 2) compute the scattering parameters from the currents on the ports.

B. Computing Current Distribution

For a given excitation, the surface current distribution on the discontinuity and ports can be obtained by solving a relevant integral equation through the method of moments (MoM) [15]. We choose the mixed-potential integral equation (MPIE) over the electric-field integral equation (EFIE), as it has weaker singularities in its Green's functions, rendering more quickly convergent solutions [16].

The MPIE relates the electric-field incident upon the upper conductor to the magnetic vector and electric scalar potentials generated by the unknown surface current and charge densities [5]:

$$\boldsymbol{E}^{inc} = j\omega \boldsymbol{A} + \nabla\Phi\,, \tag{1}$$

where

$$\boldsymbol{A} = \int_S dS'\boldsymbol{J}(\boldsymbol{r}')G_a(\boldsymbol{r},\boldsymbol{r}') \tag{2}$$

$$\Phi = \int_S dS' q_s G_q(\boldsymbol{r},\boldsymbol{r}')\,. \tag{3}$$

The primed and unprimed coordinates, respectively, designate the source and field points. The Green's function G_a is the horizontal component of the magnetic vector potential from an infinitesimal horizontal dipole located at the interface ($z' = 0$). G_q is the electric scalar potential for a pulsating point-charge at the interface. Both potentials are evaluated at the interface, displaced from the source by radius ρ. $\boldsymbol{J}_s$ and q_s are the surface current and surface charge densities on the upper microstrip conductor. The incident field $\boldsymbol{E}^{inc}$ provides the excitation to the system, the nature of which is addressed later in the discussion of S-parameter extraction.

As part of the MoM solution to (1), the unknown current and charge distributions are represented as a sum of vector basis functions weighted by unknown coefficients. Hence, current density $\boldsymbol{J}_s$ is approximated by

$$\boldsymbol{J}_s(\boldsymbol{r}) = \sum_{n=1}^{N} I_n \boldsymbol{f}_n\,. \tag{4}$$

Through application of the continuity condition

$$\nabla\cdot\boldsymbol{J}_s = -j\omega q_s\,, \tag{5}$$

the expansion of the charge distribution is obtained from (4), and no additional unknown coefficients are required.

Since the shape of the upper conductor is arbitrary, it is subdivided into small elements, with subdomain basis functions $\boldsymbol{f}_n$ approximating the current over each element. A possible discretization of a splitter is illustrated in Fig. 2. Here, we employ the triangular-domain basis functions developed in Rao *et al.* [9]. In this scheme, any two triangular plates sharing a side form a surface-current basis function, seen in Fig. 3. These functions overlap, and each plate can be part of up to three different basis functions. For any basis function $\boldsymbol{f}_n$, the two triangles are arbitrarily designated T_n^+ and T_n^-, with corresponding areas A_n^+ and A_n^-. The current dependence within $\boldsymbol{f}_n$ is given by

$$\boldsymbol{f}_n(\boldsymbol{r}) = \begin{cases} \frac{l_n}{2A_n^+}\,\boldsymbol{\rho}_n^+ & \text{if } \boldsymbol{r} \text{ in } T_n^+ \\ \frac{l_n}{2A_n^-}\,\boldsymbol{\rho}_n^- & \text{if } \boldsymbol{r} \text{ in } T_n^- \\ 0 & \text{otherwise} \end{cases} \tag{6}$$

where l_n is the length of the common side and $\boldsymbol{\rho}_n^+ = \boldsymbol{r} - \boldsymbol{r}_n^+, \boldsymbol{\rho}_n^- = \boldsymbol{r}_n^- - \boldsymbol{r}$. A vector mapping of this dependence is given in Fig. 3. The current flows from plate T_n^+ to T_n^-, with maximum current across the common edge, zero current at the isolated vertices V_n^+ and V_n^-, and no currents with components

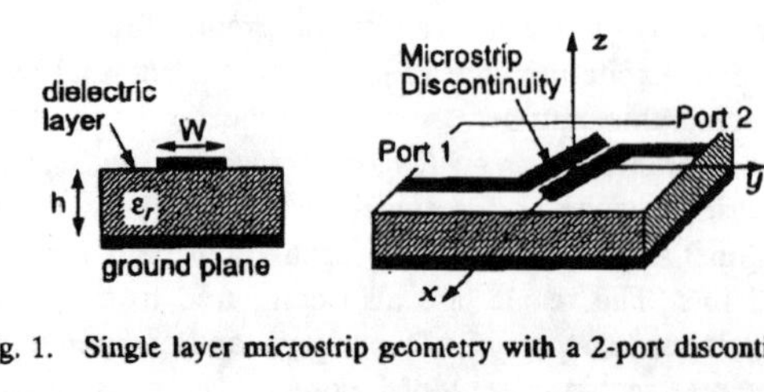

Fig. 1. Single layer microstrip geometry with a 2-port discontinuity.

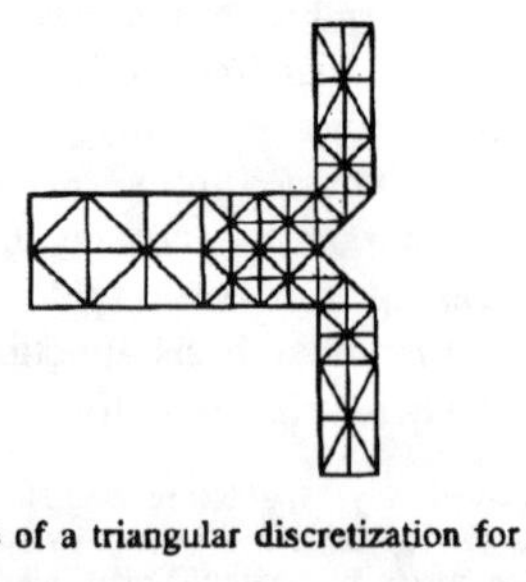

Fig. 2. Example of a triangular discretization for a splitter.

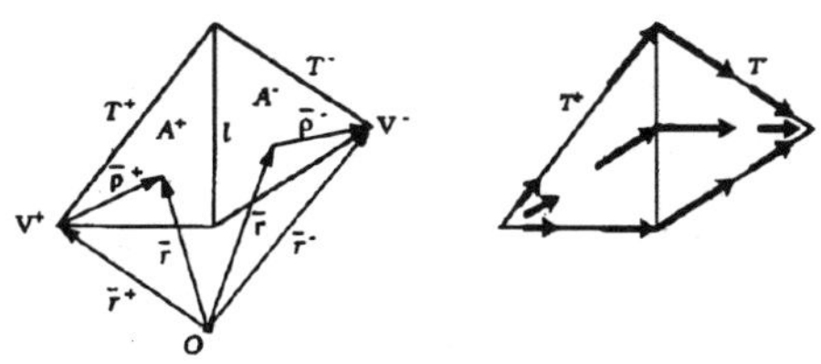

Fig. 3. Vector dependence of triangular-domain basis functions.

normal to the other four sides. The weighting in (6) is such that the current normal to the common edge is continuous across that edge and, hence, no fictitious charge singularities arise. Also, when the continuity equation is applied to (5), the charge density is a constant equal to (l_n/A_n^+) and $-(l_n/A_n^-)$ on each plate and the total charge is zero.

The N weighting coefficients in (4) are obtained through the Galerkin procedure whereby the N basis functions are adopted as testing functions [9]. The result is the familiar MoM impedance matrix given by

$$\{V\} = [Z]\{I\} \tag{7}$$

where the N rows in vector V and matrix Z correspond to the N testing functions applied, respectively, to the left-hand side (LHS) and right-hand side (RHS) of (1), and the columns of matrix Z represent the contribution of each basis function to the tested field. By solving this system of linear equations, the unknown weighting coefficients in vector I are obtained. Defining the testing procedure in terms of the inner product operator

$$\langle a, b\rangle = \int_S dS\, a \cdot b, \tag{8}$$

we have

$$\langle \boldsymbol{E}^{inc}, \boldsymbol{f}_m\rangle = j\omega\langle \boldsymbol{A}, \boldsymbol{f}_m\rangle + \langle \nabla\Phi, \boldsymbol{f}_m\rangle \tag{9}$$

where $\boldsymbol{f}_m(\boldsymbol{r})$ is the mth testing function. Evaluating (9) involves two surface integrals—one for the testing procedure and the other for the potentials within. The testing surface integral can be eliminated by assuming that potentials $\boldsymbol{A}$ and Φ and incident field $\boldsymbol{E}^{inc}$ within the domains of each triangle of $\boldsymbol{f}_m$ are constant and equal to their computed values at the triangle centroids. A further simplification is had by transferring the differential operator on Φ onto the testing function through integration by parts. Substituting (4) and (5) into (1), it can be shown that the elements of matrix Z are given by

$$Z_{mn} = -l_m\left[j\omega\left(\boldsymbol{A}_{mn}^+ \cdot \frac{\boldsymbol{\rho}_m^{c+}}{2} + \boldsymbol{A}_{mn}^- \cdot \frac{\boldsymbol{\rho}_m^{c-}}{2}\right) + \Phi_{mn}^- - \Phi_{mn}^+\right] \tag{10}$$

where

$$\boldsymbol{A}_{mn}^{\pm} = \int_S dS'\, \boldsymbol{f}_n(\boldsymbol{r}')G_a(\boldsymbol{r}_m^{c\pm}, \boldsymbol{r}') \tag{11}$$

$$\Phi_{mn}^{\pm} = \int_S dS'\, \nabla' \cdot \boldsymbol{f}_n(\boldsymbol{r}')G_q(\boldsymbol{r}_m^{c\pm}, \boldsymbol{r}'). \tag{12}$$

Here, $\boldsymbol{\rho}_m^{c+}$ and $\boldsymbol{\rho}_m^{c-}$ arise from evaluating testing function $\boldsymbol{f}_m$ at the centroids of T_m^+ and T_m^-. Observe in (12) that $\Delta \cdot \boldsymbol{f}_n$ is constant in each triangular domain and that, for a given basis/testing function pair mn, the contribution of the scalar potential in (10) depends upon the triangular plates involved but not the basis/testing function to which they are assigned. Hence, if the integral (12) is evaluated for each element Z_{mn}, many identical integrations will be performed. Considerable computational effort is saved by evaluating and storing (12) for all plate combinations and recalling these results as needed in evaluating (10).

The same cannot be said of (11), whose integrand evaluated over a particular source triangle depends on the identity of the isolated vertex and, hence, the basis function to which the triangle is assigned. However, a similar savings of effort is possible by breaking (11) into three scalar integrals

$$I_i = \int_S dS'\, g_i(\boldsymbol{r}')G_a(\boldsymbol{r}_m^{c\pm}, \boldsymbol{r}') \tag{13}$$

where $g_i(\boldsymbol{r}')$ is set alternately to x', y', and 1. The basis function dependence is removed from the integrals and reintroduced in the evaluation of Z_{mn} through a weighted sum of these three. Thus, for each plate combination, a total of four scalar integrals are evaluated and later recalled in constructing the elements in impedance matrix Z.

Finally, applying the approximate testing procedure in the LHS of (9), the elements of the excitation vector V are given by

$$v_m = -l_m\left[\boldsymbol{E}^{inc}(\boldsymbol{r}_m^{c+}) \cdot \frac{\boldsymbol{\rho}_m^{c+}}{2} + \boldsymbol{E}^{inc}(\boldsymbol{r}_m^{c-}) \cdot \frac{\boldsymbol{\rho}_m^{c-}}{2}\right]. \tag{14}$$

Most of the elements v_m will be zero, as the incident field is only applied in regions of port excitation, as discussed later.

III. Efficient Implementation

A. Efficient Evaluation of Green's Function

Generating the elements in matrix Z with (10) requires convolution of the Green's functions G_a and G_q with the triangular basis functions, so it is necessary to evaluate them efficiently if acceptable solution times are to be achieved. There are no closed-form expressions for G_a and G_q in the spatial domain, but they can be represented in closed form in the spectral domain [14]

$$\tilde{G}_a = \frac{\mu_o}{4\pi}\frac{1}{j2k_{zo}}[1 + R_{TE}(k_\rho)] \tag{15}$$

$$\tilde{G}_q = \frac{1}{4\pi\epsilon_o}\frac{1}{j2k_{zo}}\left[1 + R_{TE} + \frac{k_{zo}^2}{k_\rho^2}(R_{TE} + R_{TM})\right]. \tag{16}$$

Here, ϵ_o and μ_o are the free-space permittivity and permeability. Wave numbers k_{zo} and k_ρ are the vertical and radial components of the free-space propagation constant k_o in the cylindrical system. R_{TE} and R_{TM} are the reflection coefficients at the interface of TE and TM plane waves incident on the substrate with ground plane. The required spatial-domain Green's functions can then be expressed as an inverse Hankel transform on (15) and (16), which is commonly referred to as a Sommerfeld integral

$$G_{a,q}(\rho) = \int_{-\infty}^{+\infty} \tilde{G}_{a,q} H_0^{(2)}(k_\rho \rho) k_\rho \, dk_\rho \tag{17}$$

where ρ is the radial separation between the source and the observation point, both located on the interface, and $H_0^{(2)}$ is the zeroth-order Hankel function of the second kind. The contour of integration can be deformed to avoid surface-wave poles, which reside on the real k_ρ axis for lossless dielectrics and become complex for lossy dielectrics. Approximate analytic expressions exist for the evaluation of (17) for ρ in the near and far fields [12], but intermediate regions require quadrature of slowly convergent, oscillatory integrands. However, a technique developed in [13] (and improved in [14]) allows efficient evaluation in all regions. The Sommerfeld integral is divided into three contributions: 1) quasi-dynamic images, 2) surface waves, and 3) complex images. The first two contributions, which dominate in the near and far field, respectively, are extracted from R_{TE} and R_{TM}, and handled analytically using the Sommerfeld Identity [17]. What remains in R_{TE} and R_{TM} is relatively well behaved and exhibits exponential decay for sufficiently large values of k_ρ. This remainder can be accurately approximated with a short series of exponentials terms, which are interpreted as complex images. The function is sampled from $k_\rho = 0$ out to a truncation point where the exponential decay character dominates. The exponents of the expansion are computed using Prony's method [18], and the term weights are then obtained through a least-squares fit. The inverse Hankel transform of the exponentials can be performed analytically, again using the Sommerfeld Identity. Our experience indicates that 2–4 expansion terms are appropriate, depending on the frequency. Particular care should be taken in determining the number of expansion terms of G_q since its contribution in (10) is a second-order difference arising from the source pulse doublet and the testing procedure. Overall, the process yields a reduction in computation time on the order of 100 compared to straight quadrature, and is essential in keeping the matrix fill time comparable to the matrix solve time. For this purpose, however, it is not sufficient; a means of exploiting redundant triangular interactions is also required.

B. Efficiencies of Discretization

As explained in Section II, all elements in the impedance matrix can be computed from a linear combination of four scalar integrals evaluated for all source/test plate combinations. The number of these which must be computed can be significantly reduced by exploiting redundancy in the discretization and using approximate expressions when plates are separated by large distances. In the latter case, the surface integration over the source plate can be replaced by evaluation of the integrand at the plate centroid. Equations (11) and (12) become

$$A^{\pm}_{mn} = \frac{l_n}{2}\left[\rho_n^{c+} G_a(r^{\pm}_m, r^{c+}_n) + \rho_n^{c-} G_a(r^{\pm}_m, r^{c-}_n)\right] \tag{18}$$

$$\Phi^{\pm}_{mn} = l_n\left[G_q(r^{\pm}_m, r^{c+}_n) - G_q(r^{\pm}_m, r^{c-}_n)\right]. \tag{19}$$

The criterion for distinguishing near and far interactions is based on the approximation used for the far interactions. If the percent difference between the distances from the three vertices of the source plate to the test plate centroid is under some set maximum level, say, 20%, then that approximation is viable. Since the great majority of plate interactions will often fall into the category of far interactions, it also makes sense to evaluate the Green's functions with an interpolation table, the construction of which is greatly expedited by the above described scheme for computing Green's functions. Both G_a and G_q exhibit $(1/\rho)$ and $\log \rho$ singularities, so the table must begin at some minimum displacement ρ_0 governed by the interpolation scheme, the dominant $(1/\rho)$ singularity, and a maximum error criterion, say, 1%. We use second-order differencing with an interval equal to 1% of the substrate wavelength and ρ_0 set to 3.5 times the interval.

For near interactions, approximations (18) and (19) cannot be used, and identification of redundancy is important to keeping the computational task manageable. The interactions can be cataloged by stepping through each plate combination. Far interactions are ignored, as they are too numerous to store and can be rapidly evaluated through the G_a and G_q interpolation tables at the point of computing the impedance Z_{mn}. For near interactions, the four scalar integrals from (12) and (13) are evaluated and cataloged. Subsequent plate combinations are then checked against the stored interactions and computed only if no equivalent interaction is available. Two plate interaction integrals are equivalent if the x- and y-displacements of the test plate centroid from the source plate vertices are identical. Here, too, significant improvements in efficiency are achieved by using, where possible, the interpolation tables in evaluating the integrals.

When all the efficiencies of discretization are combined with the method for rapid evaluation of the microstrip Green's functions and the elimination of the testing surface integral, the construction time for the impedance matrix is often comparable to, if not less than, the matrix solve time when the number of unknowns exceeds 500.

IV. Extracting Scattering Parameters

Once the approximate current-density distribution is found by solving (7) for the unknown weighting coefficients, scattering parameters for an N-port discontinuity are obtained by examining the current distribution on the ports. In general, N linearly independent excitation schemes are required to evaluate an N-port network, although physical symmetries in the discontinuity will reduce this. For example, a symmetric 2-port can be fully characterized by processing the current on both ports while excitation is provided to one port.

The ports are driven by applying horizontal voltage generators at the port ends. This is achieved by setting the incident electric field to some constant over the subcells at the port end(s). The corresponding elements v_m in the excitation vector are assigned a nonzero quantity according to (14). While horizontal voltage sources are not the driving mechanism typically employed in practice, they are legitimately used here; the S-parameter extraction process only requires some form of excitation which allows examination of a sufficient number of current samples on the ports between the discontinuity and the port ends. While the open-circuited port ends do affect the amplitudes of the current waves on the ports, the relationships among the waves incident upon and reflected from the discontinuity are nonetheless uniquely defined by its S-parameters [6].

With excitation applied, the sampled currents on the ports are fitted to the unimodal wave model

$$I_{ij}(x) = \frac{1}{\sqrt{Z_{oi}}}\left[a_{ij}e^{-\gamma_i x} - b_{ij}e^{+\gamma_i x}\right]. \tag{20}$$

Here we assume a port extending in the x-direction. $I_{ij}(x)$ is the sampled current distribution on the ith port for the jth excitation, obtained by integrating J_x across the port width. If the port is discretized as shown in Fig. 4, we observe that only the highlighted basis functions have x-components at the indicated sample positions x_n, x_{n+3}. It can be shown from (5) that the current at one of these points is given by

$$I(x_n) = (i_1 + i_2)\Delta_y \tag{21}$$

where i_1 and i_2 are the weighting coefficients for the basis functions spanning the sample point. The uniform line characteristic impedance Z_{0i} in (20), which can be obtained from a rigorous empirical formula [19], is necessary to normalize the currents when the ports are not identical. The propagation constant γ_i is obtained by MoM analysis of a uniform line of the same dimensions at the port. Here, the current on an end driven uniform line is also fit to the form in (20) and γ is optimized for smallest error in the fit. With these line constants known, forward and reverse complex wave coefficients a and b are generated through least-squares fit of the current samples to (20). In doing so, samples near the port end and discontinuity should be left out owing to the presence of decaying modes.

The S-parameters are obtained directly from the computed wave coefficients using the standard S-parameter matrix definition [20]. Recasting this matrix to solve for S-parameters, one finds, for example, that the S_{11} and S_{12} for an arbitrary 2-port are obtained by solving

$$\begin{bmatrix} a_1^1 & a_2^1 \\ a_1^2 & a_2^2 \end{bmatrix}\begin{bmatrix} S_{11} \\ S_{12} \end{bmatrix} = \begin{bmatrix} b_1^1 \\ b_1^2 \end{bmatrix} \tag{22}$$

where the subscripts on wave coefficients a and b denote the port, and the superscripts denote the excitation.

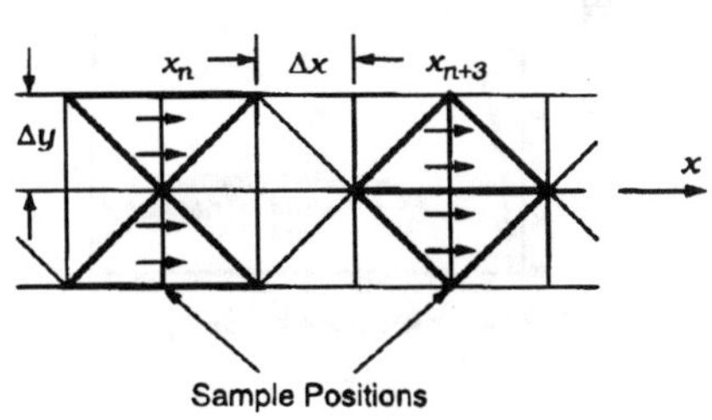

Fig. 4. Scheme for sampling the x-component of current on a uniform line.

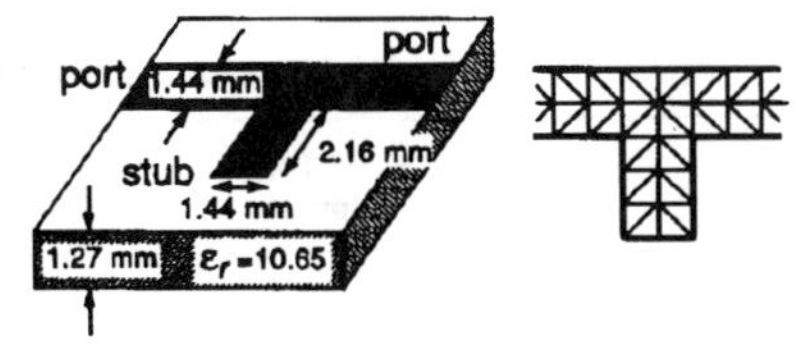

Fig. 5. Dimensions, substrate parameters, and discretization of microstrip stub.

V. Numerical Results

A. Validating Examples

In order to validate the above approach, including the approximations used to make it computationally tractable, several microstrip structures are simulated for which experimental data are available. The first of these is the microstrip stub whose geometry and discretization are given in Fig. 5. For the analysis using triangular MoM, the two ports extend 30 subcells each beyond those depicted in order to obtain a sufficient number of current samples. The bilateral symmetry in the stub is exploited in processing the port current data. A synopsis of the computational task in computing the current distribution for the stub and for the other cases which follow is provided in Table I. Itemized there are the typical CPU time (SPARC 2 workstation) per frequency point, matrix build/solve time ratio, number of unknowns (basis functions), number of triangular plates in the discretization, and the number of identified, unique plate-to-plate interactions; matrices were solved with LINPACK's complex, single-precision routine. The computed and measured responses are shown in Fig. 6; the measurement data are found in [8]. For comparison, also shown are responses generated by an MoM code using rectangular roof-top basis and razor-blade testing functions [21], and by Libra v. 3.000, the commercial microwave CAD package. The rectangular and the triangular MoM codes produce nearly identical results. Both predict the location of the measured stub resonance at 10.15 GHz to within 0.1 GHz, while the response generated by Libra shows resonance at 9 GHz. For Libra, the stub is constructed from a microstrip TEE circuit element attached to a section of open-circuited line with capacitive edge effect; these elements are operated within the specified regions of validity [22].

The second validation example is the coupled line whose geometry and discretization are seen in Fig. 7. The substrate parameters are $h = 1.524$ mm, $\varepsilon_r = 4.7$. The nonuniform discretization in the lateral direction of the couple-line pair is necessary to accurately account for the close coupling. The 45° bends at the ends of the lines are included to allow attachment of SMA connectors for measurements, conducted with a calibrated HP-8510B network analyzer and HP-8514A S-parameter test set. The measurement ends at 4 GHz due to limitations in the connectors. Results for S_{21} and S_{41} transmission coupling are given in Fig. 8(a) and (b). Simulation data are generated from the triangular MoM code and Libra. The 45° angle of the ports makes this structure cumbersome to model with rectangular basis functions, and no simulation is attempted with the rectangular MoM code. The

TABLE I
SYNOPSIS OF COMPUTATIONAL TASK FOR 3 DIFFERENT DISCONTINUITIES

Device	CPU Time (min)	Build/Solve Time Ratio	Unknowns	Plates	Stored Interact.
Stub	5.1	1.0	453	364	3210
45°-Bend (L)	5.4	1.4	436	350	935
45°-Bend (H)	75.2	0.5	1250	886	4247
Coup. Line	124.6	0.3	1536	1232	8692

test set contains two matched ports; the remaining two ports of the coupled line are terminated with match loads verified to produce less than -30 dB return over the measurement band. A separate measurement with a uniform line showed the SMA connectors to generate returns on the order of -25 to -15 dB in this measurement band. This precludes the possibility of a successful S_{11} reflection measurement. For S_{21} coupling, Libra and the MoM code are in good agreement with measurement up to 2.5 GHz, with the MoM code coming closer to predicting the magnitude of the band-pass peaks. Beyond 2.5 GHz, the measured coupling breaks out of its previous pattern. Both Libra and MoM fail to predict this, suggesting the emergence of nonideal phenomena in this frequency range, including increased reflection from the SMA connectors. For S_{41} coupling, Libra and MoM are in close agreement, and both predict the overall character of the measurement. The gap between measurement and the computed response increases with frequency, again indicating the contributions of nonideal phenomena.

A third validation example is prompted by a recent paper presenting FDTD to simulate microstrip discontinuities [3]. The discontinuity is an unmitered 45° bend with line width $W = 2.4$ mm and substrate parameters $h = 0.79$ mm, $\epsilon_r = 2.21$. The measured return loss in [3, Fig. 13] exhibits significant ringing—a result that cannot be explained by a single bend. The published FDTD results also show ringing, but they do not match well with experiment beyond 4 GHz. The simulation data using the TOUCHSTONE and TCKT commercial packages published in the same reference show a smooth increase in return loss with increasing frequency, and this agrees well with the trends predicted with our triangular MoM code under three different levels of discretization and simulation on Libra, seen in Fig. 9. Our Libra results identically match the published TOUCHSTONE results, which is to be expected, as Libra is an extension of TOUCHSTONE. We

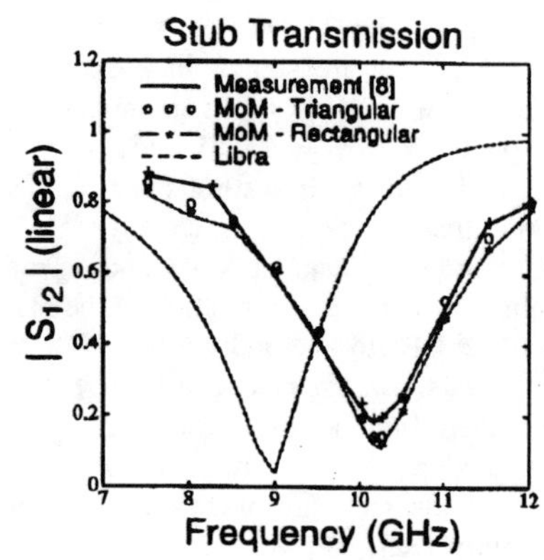

Fig. 6. Transmission response of stub; measured and computed with method-of-moments (triangular and rectangular basis functions) and Libra v 3.000.

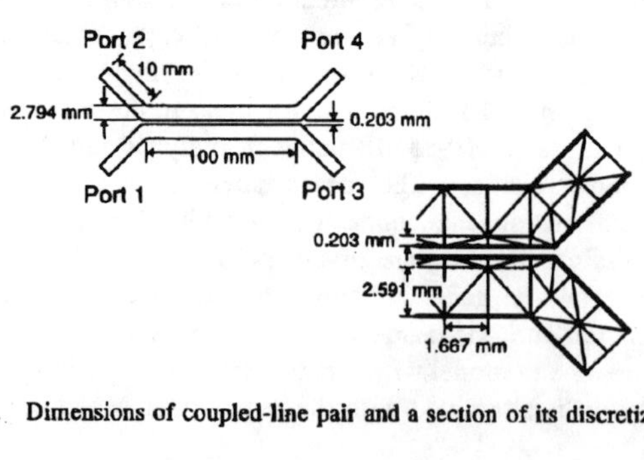

Fig. 7. Dimensions of coupled-line pair and a section of its discretization.

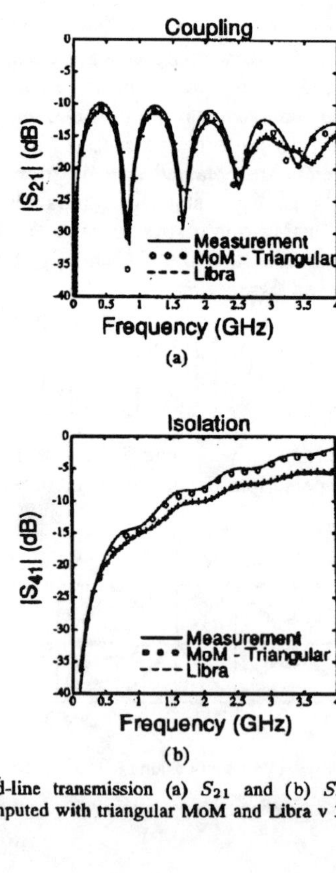

Fig. 8. Coupled-line transmission (a) S_{21} and (b) S_{41}: measured and computed with triangular MoM and Libra v 3.000.

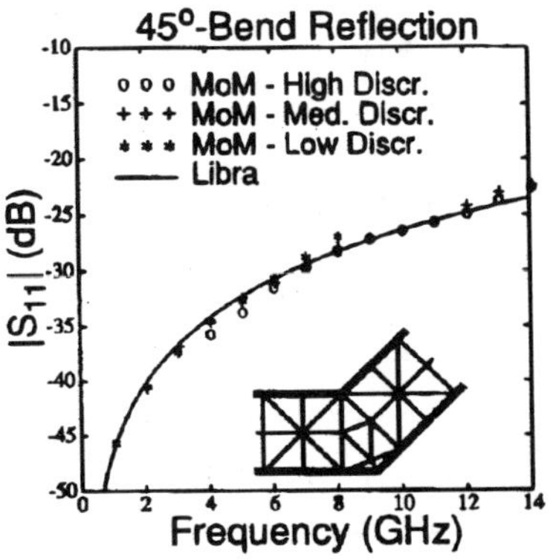

Fig. 9. Reflection response of 45°-bend computed with triangular MoM and Libra v 3.000.

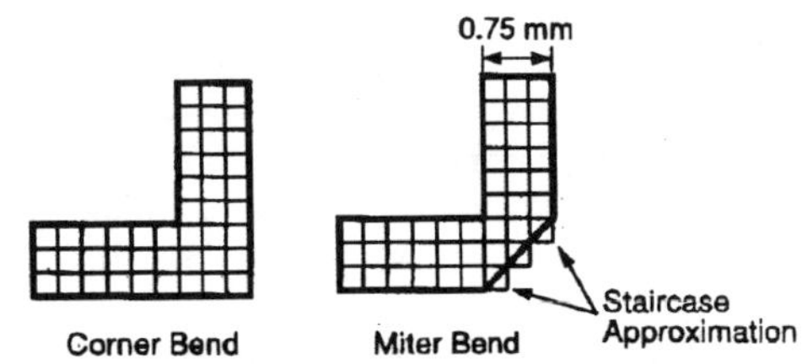

Fig. 10. Dimensions and discretization of corner and miter bends, with staircasing evident in rectangular discretization of miter bend.

believe that the measured ringing is induced by reflections at the ports, probably from connectors inadequate for this measurement. Judging from the return predicted by the Libra and TCKT packages and our MoM code, a connector with less than -45 dB return would be required for a good measurement. As mentioned earlier, our own experience using SMA coaxial-to-microstrip adaptors indicates that reflections below -20 dB cannot be reliably measured. Reference [3] does not indicate what manner of connection to the bend was used. It is our opinion that the ringing in the published FDTD response is probably due to the error introduced by the absorbing boundary condition [23].

For the MoM simulation, three different levels of discretization are used. There are 2, 4, and 6 square subcells spanning the line for the low, medium, and high levels of discretization; a portion of the low-level discretization is illustrated in Fig. 9. The two entries for the 45° bend in Table I are for the low and high discretization. All three give nearly identical results and agree with Libra in a frequency region where the reflections are all below -20 dB. No data are shown for the low discretization beyond 8 GHz, as it begins to fail at higher frequencies. Our experience indicates that there should be at least 20 subcells per wavelength to get acceptable results. This is a stiffer requirement than the 10-per-wavelength rule for MoM applied to far-field scattering problems; scattering parameters are more sensitive to errors in current than the far fields. At low frequencies, high levels of discretization are not only unnecessary but, in fact, become a liability as the electrical lengths of the ports shrink below one-quarter of a wavelength. While the port subcells can be stretched to maintain the electrical length of the port without increasing the number of unknowns, this introduces a discretization discontinuity whose effect, while small, can generate unacceptable errors in the computed reflection coefficient when the reflection from the actual discontinuity is also small, as is the case here. For this reason, the plotted results for the low, medium, and high discretizations begin at 1, 2, and 4 GHz, respectively.

B. Rectangular versus Triangular Discretization

A microstrip corner bend and miter bend were simulated with the rectangular and triangular MoM codes to demonstrate the deficiencies which can arise from approximate representation of the conductor geometry. The geometry of the bends considered is described in Fig. 10; the ports, which extend 60 subdivisions, have been truncated for clarity. The substrate parameters are $h = 1.27$ mm, $\varepsilon_r = 10$. The corner bend can be exactly duplicated with either the triangular or rectangular discretization. The miter bend can be rendered exactly with the triangular discretization, but only approximately under a rectangular discretization. The staircase approximation of the miter bend was also conducted with the triangular MoM code. The computed S_{11} reflections are shown in Fig. 11. The triangular and rectangular simulations of the corner bend and staircase approximated miter bend are in excellent agreement. The approximate miter bend has several decibels less reflection, as would be expected given a less abrupt transition. The true miter provides a still better transition, as evidenced by the simulation results under the triangular MoM code. The agreement between the triangular and rectangular code in the simulation of the corner and approximate miter bend shows that the predicted improvement in the true miter is due only to accurate rendering of the shape, and not to small numerical differences in two different computer codes.

VI. Conclusion

A frequency-domain method for characterizing arbitrarily shaped, planar microstrip discontinuities has been described. This technique is based on solving the mixed-potential integral equation with the method of moments (MoM) under a triangular discretization of the microstrip upper conductor.

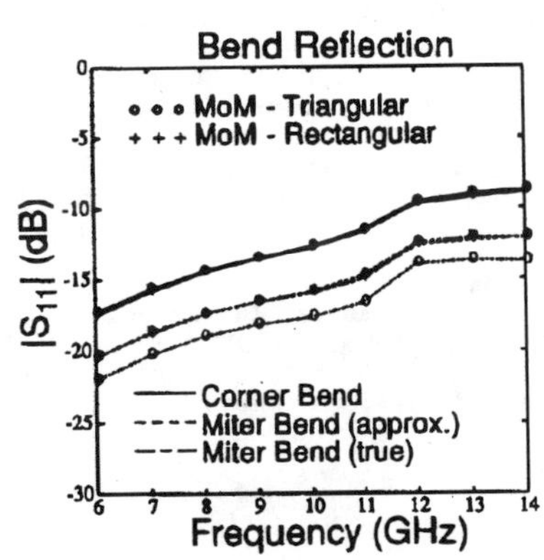

Fig. 11. Reflection response of corner bend and approximate miter bend computer with both triangular and rectangular MoM, and true miter bend computed with triangular MoM.

Triangular basis functions offer significant improvement over rectangular basis functions in their ability to conform to arbitrary shapes, including ports which extend at oblique angles. The implementation includes a number of techniques to significantly reduce the computational task including efficient evaluation of Green's function for microstrip and exploitation of redundancy in the discretization. These efficiencies allow simulations where the time spent filling the MoM impedance matrix is on the order of the time spent solving that matrix. Several validating examples confirm the overall approach. A study of a corner and miter bend shows how approximate rendering of a discontinuity in a rectangular discretization can degrade the accuracy of the simulation.

Acknowledgment

We wish to acknowledge Dr. M. Soma and G. Devarayanadurg for providing the microstrip coupled-line device and for their assistance in measuring its S-parameters.

References

[1] T.S. Chu, T. Itoh, and Y.-C. Shih, "Comparative study of mode-matching formulations for microstrip discontinuity problems," *IEEE Trans. Microwave Theory Tech.*, vol. MTT-33, pp. 1018–1023, Oct. 1985.

[2] D.M. Sheen, S.M. Ali, M.D. Abouzahra, and J.A. Kong, "Application of the three-dimensional finite-difference time-domain method to the analysis of planar microstrip circuits," *IEEE Trans. Microwave Theory Tech.*, vol. 38, pp. 849–852, July 1990.

[3] J.-F. Lee, R. Palandech, and R. Mittra, "Modeling three-dimensional discontinuities in waveguides using nonorthogonal FDTD algorithm," *IEEE Trans. Microwave Theory Tech.*, vol. 40, pp. 346–352, Feb. 1992.

[4] R.W. Jackson and D.M. Pozar, "Full-wave analysis of microstrip open-end and gap discontinuities," *IEEE Trans. Microwave Theory Tech.*, vol. MTT-33, pp. 1036–1042, Oct. 1985.

[5] J.R. Mosig, "Arbitrarily shaped microstrip structures and their analysis with a mixed potential integral equation," *IEEE Trans. Microwave Theory Tech.*, vol. MTT-36, pp. 314–323, Feb. 1988.

[6] P.B. Katehi and N.G. Alexopoulos, "Frequency-dependent characteristics of microstrip discontinuities in millimeter-wave integrated circuits," *IEEE Trans. Microwave Theory Tech.*, vol. MTT-33, pp. 1029–1035, Oct. 1985.

[7] R.W. Jackson, "Full-wave finite element analysis of irregular microstrip discontinuities," *IEEE Trans. Microwave Theory Tech.*, vol. MTT-37, pp. 81–89, Jan. 1989.

[8] W.P. Harokopus and P.B. Katehi, "Characterization of microstrip discontinuities on multilayer dielectric substrates including radiation losses," *IEEE Trans. Microwave Theory Tech.*, vol. MTT-37, pp. 2058–2065, Dec. 1989.

[9] S.M. Rao, D.R. Wilton, and A.W. Glisson, "Electromagnetic scattering by surfaces of arbitrary shape," *IEEE Trans. Antennas Propogat.*, vol. AP-30, pp. 401–418, May 1982.

[10] K.-L. Wu, J. Litva, R. Fralich, and C. Wu, "Full wave analysis of arbitrarily shaped line-fed microstrip antennas using triangular finite-element method," *IEE Proc. H*, vol. 138, no. 5, pp. 421–428, Oct. 1991.

[11] J.R. Mosig and F.E. Gardiol, "Analytical and numerical techniques in the Green's function treatment of microstrip antennas and scatterers," *IEEE Proc. H*, vol. 130, no. 2, pp. 175–182, Mar. 1983.

[12] ——, "A dynamical radiation model for microstrip structures," in *Advances in Electronics and Electron Physic*, P. Hawkes, Ed. New York: Academic, 1982, pp. 139–237.

[13] D.G. Fang, J.J. Yang, and G.Y. Delisle, "Discrete image theory for horizontal electric dipoles in a multilayered medium above a conducting ground plane," *IEE Proc. H*, vol. 135, pp. 297–303, 1988.

[14] Y.L. Chow, J.J. Yang, D.G. Fang, and G.E. Howard, "A closed-form spatial Green's function for the thick microstrip substrate," *IEEE Trans. Microwave Theory Tech.*, vol. 39, pp. 588–592, Mar. 1991.

[15] R.F. Harrington, *Field Computation by Moment Methods*. New York: Macmillan, 1968.

[16] J.R. Mosig and F.E. Gardiol, "Integral equation techniques for the dynamic analysis of microstrip discontinuities," *Alta Frequenza*, vol. 57, no. 5, pp. 171–181, June 1988.

[17] A. Sommerfeld, *Partial Differential Equations*. New York: Academic, 1949.

[18] F.B. Hildebrand, *Introduction to Numerical Analysis*. New York: McGraw-Hill, 1974.

[19] I. Bahl and P. Bhartia, *Microwave Solid State Circuit Design*. New York: Wiley, 1988.

[20] R.F. Harrington, *Times-Harmonic Electromagnetic Fields*. New York: McGraw-Hill, 1961.

[21] R. Kipp, C.H. Chan, A.T. Yang, and J.T. Yao, "Simulation of high-frequency integrated circuits incorporating full-wave analysis of microstrip discontinuities," *IEEE Trans. Microwave Theory Tech.*, accepted for publication.

[22] *EEsof Element Catalog*, EEsof Inc., CA, 1989.

[23] J.-F. Lee, R. Palandech, and R. Mittra, "Corrections to "Modeling three-dimensional discontinuities in waveguides using nonorthogonal FDTD algorithm," *IEEE Trans. Microwave Theory Tech.*, vol. 40, pp. 1736, Aug. 1992.

Conformal Finite-Difference Time-Domain (FDTD) with Overlapping Grids

Kane S. Yee, *Member, IEEE*, Jei Shuan Chen, *Member, IEEE*, and Albert H. Chang, *Member, IEEE*

Abstract—**It is shown how to use local conformal grids several zones around the scatterer and the rectangular grid away from the scatterer when we numerically solve Maxwell's equations. The locally conformal grids will allow better approximation of the boundary condition on the scattering object while the rectangular grid preserves the simplicity and accuracy of the regular finite-difference time-domain (FDTD) algorithm. These grids overlap typically about three zones. The major accomplishment here is to explain how to connect together the calculations done in different overlapping grids.**

I. Introduction

The simplicity and usefulness of the finite-difference time-domain (FDTD) [10] in solving Maxwell's equations is well known and its application is now widespread. One of the major shortcomings of the existing FDTD is that its computational volume consists of a union of cubes. The staircasing makes the approximation of a curved boundary inaccurate; therefore error is introduced in numerical computations. In some applications the error may become intolerable; and in the cases when the scatterer is covered with thin layers, the staircasing FDTD will become inapplicable. Thus a conformal FDTD is necessary and desirable. Devising a practical, conformal FDTD algorithm has occupied the attention of many researchers since the early 1980's [1]–[9].

The work thus far can be divided into two categories. One category [1], [2], [6] uses a global nonrectangular grid generated by available grid generators, and the other [4], [5], [7]–[9], [13], [14] uses a locally unstructured distorted grid blended into a predominantly rectangular grid. The common feature is that the grid used is a single distorted grid. Because of the necessity of creating a "radiation boundary," the whole computation volume will be many times the volume occupied by the scatterer. The single distorted grid generated by a grid generator has the intrinsic shortcoming that zones are vastly different in size; and in the explicit algorithm such as the FDTD, the time increment may be very small in order to satisfy the numerical stability requirement. One practical solution seems to be to keep an irregular grid around the scatterer and use the rectangular grid away from the body.

This idea has been advanced since 1985 [4], [5] and improved significantly in 1988 by Madsen and Ziolkowski [7], [8]. Madsen and Ziolkowski, and recently Holland *et al.* [9], have shown great improvement of their locally unstructured grid over the staircasing FDTD. If the zones are nearly orthogonal, the Madsen–Ziolkowski first-order method is simple and give credible results; but when the zones are too irregular one may have to use the second-order correction, which is quite complicated. Furthermore, these locally distorted grids are unstructured and may require storing the metric tensors, the various direction vectors and other auxiliary quantities necessary for an FDTD computation. Thus the bookkeeping problem is very complicated. The work of Cangellaris *et al.* [14] is based on finite elements; however, because of the use of a single contiguous grid, the comments on the bookkeeping difficulties may also apply. For these and other reasons, there exists no published practical conformal FDTD algorithm. A good conformal scheme is urgently needed. After several years of intermittent search, the present authors believe that a simple, practical scheme has been found. In this paper we propose the use of many overlapping grids in the neighborhood of the scatterer, and away from the scatterer the use of the rectangular grid. One very convenient choice consists of many tangent-normal grids (for practical reasons, the number of the curvilinear grids should be kept to a minimum) near the scattering body. The local tangent-normal grids overlap and the rectangular grid also overlap. The overlapping should be such that the necessary "boundary" data for one grid can be obtained through interpolation from the calculated data of another grid. We shall detail this algorithm. The local tangent-normal grid seems to be most convenient when a surface impedance boundary condition is imposed. Also across material boundaries the tangent-normal grids will be unusually simple. The proposed overlapping grids will attain the goal of keeping irregular grids local and using coordinate systems makes the bookkeeping simpler. Moreover the tangent-normal grid approximates more faithfully the curved boundary. The computational steps in this algorithm apply to general 3-D scatterers, but we shall explain the algorithm with mostly 2-D figures, as it is easier to draw 2-D figures. In Section II we explain the FDTD algorithm. In Section III we shall explain our overlapping grid algorithm. In Section IV we shall present some sample calculations based on this algorithm. For a 2-D circular cylinder we show the comparison between the exact solution, the staircasing FDTD solution, and the conformal FDTD solution. For 3-D examples we shall show calculations on the scattering by a finite circular

Manuscript received November 19, 1991; revised May 4, 1992.
The authors are with the Lockheed Palo Alto Research Laboratory, Palo Alto, CA 94304.
IEEE Log Number 9202227.

Reprinted from *IEEE Transactions on Antennas and Propagation*, pp. 1068-1075, September 1992.

cylinder and by a sphere. In Section V we shall explicitly list the steps of the algorithm when there is more than one bodily conformal curvilinear grids.

II. The FDTD Algorithms

We assume that the details and notations of the FDTD are known. In a rectangular grid we have the rectangular FDTD; in a curvilinear grid we have a curvilinear FDTD, and in an unstructured grid we have an irregular FDTD. All these different difference equations can be derived from the Maxwell's equations in integral forms:

$$\int_C \bar{\mathbf{H}} \cdot d\bar{\mathbf{l}} = \int_S \left(\frac{\partial \bar{\mathbf{D}}}{\partial t} + \sigma \bar{\mathbf{E}} + \bar{\mathbf{J}} \right) \cdot d\bar{\mathbf{s}} \tag{1}$$

$$-\int_C \bar{\mathbf{E}} \cdot d\bar{\mathbf{l}} = \int_S \left(\frac{\partial \bar{\mathbf{B}}}{\partial t} + \sigma_m \bar{\mathbf{E}} + \bar{\mathbf{J}}_m \right) \cdot d\bar{\mathbf{s}}. \tag{2}$$

We adopt the convention that the centers of the electric cells (the Yee cells in the case of the rectangular grid) are indexed by $(i + 1/2, j + 1/2, k + 1/2)$ and the centers of the magnetic cells are indexed by (i, j, k). The edges of the electric cells shall be called electric edges and the edges of the magnetic cells will be the magnetic edges. A face of an electric cell will be called an electric face and a face of a magnetic cell will be called a magnetic face. The centers of the electric edges are indexed by $(i + 1/2, j, k)$, $(i, j + 1/2, k)$, and $(i, j, k + 1/2)$. The centers of the magnetic edges are indexed by $(i, j + 1/2, k + 1/2)$, $(i + 1/2, j, k + 1/2)$, and $(i + 1/2, j + 1/2, k)$. The electric faces are indexed the same manner as the magnetic edges. At the center of an electric edge FDTD will advance a component of the electric field (the component will be along the edge in case of an orthogonal grid and the component will be perpendicular to a magnetic face in case of nonorthogonal grid). At the center of a magnetic edge, the FDTD will advance a component of the magnetic field. Each coordinate system will have its appropriate FDTD.

Maxwell's equations in integral forms are coordinate free; they are discretized to derive the FDTD algorithms:

$$\int_C \bar{\mathbf{H}}^{n+1/2} \cdot d\bar{\mathbf{l}} = \frac{1}{\Delta t} \left\{ \int_S (\bar{\mathbf{D}}^{n+1} - \bar{\mathbf{D}}^{n}) \cdot d\bar{\mathbf{s}} \right\} + \int_S (\sigma^{n+1/2} \bar{\mathbf{E}}^{n+1/2} + \bar{\mathbf{J}}^{n+1/2}) \cdot d\bar{\mathbf{s}} \tag{3}$$

$$-\int_C \bar{\mathbf{E}}^{n} \cdot d\bar{\mathbf{l}} = \frac{1}{\Delta t} \left\{ \int_S (\bar{\mathbf{B}}^{n+1/2} - \bar{\mathbf{B}}^{n-1/2}) \cdot d\bar{\mathbf{s}} \right\} + \int_S (\sigma_m^n \bar{\mathbf{H}}^{n} + \bar{\mathbf{J}}_m^n) \cdot d\bar{\mathbf{s}}. \tag{4}$$

Thus if the surfaces are the faces of the electric cubes and the magnetic cubes (Fig. 1(a)) we obtain the 1966 Yee algorithm; and if the surfaces are faces of nonorthogonal cubes (Fig. 1(b)) we obtain the curvilinear FDTD published by Holland [1] in 1983. Other irregular FDTD [4], [5], [7], [8] are also derived from (1) and (2).

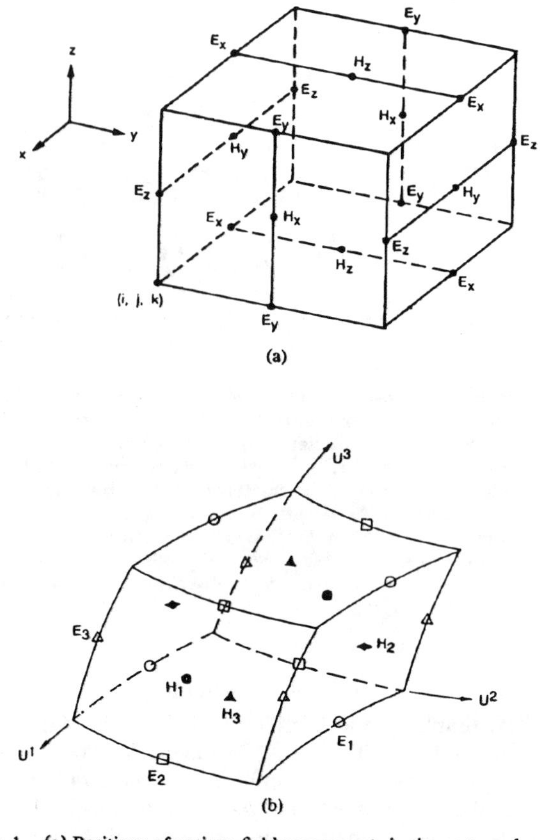

Fig. 1. (a) Positions of various field components in the rectangular cell. The E-components are in the middle of the edges and the H-components are in the centers of the faces of the electric cubes. (b) The covariant field components on a curvilinear cell in a generalized nonorthogonal coordinate system.

In an orthogonal grid, these computed components are sufficient for the evaluation of the various line integrals to obtain the circulations (the electromotive or magnetomotive forces). In a curvilinear coordinate system, however, one computes the contravariant component of the electric field along an electric edge and the contravariant component of the magnetic field along a magnetic edge. But one needs the covariant component of the electric field along an electric edge in order to calculate the electric circulation. The covariant component can be obtained by the relationship between the contravariant components and the covariant components of a vector field. Therefore, one needs the three contravariant components of the electric field at the same point unless the coordinate is orthogonal. Holland obtains the other two contravariant components at the required locations by a simple interpolation. In our overlapping grids and at a given space point we need to know the rectangular components or the three curvilinear components (covariant or contravariant). However, if two coordinate systems overlap, the various field components are not calculated at the same points. Thus an interpolation will be necessary. The interpolation we use is the simple trilinear interpolation.

III. The Conformal FDTD with Overlapping Grids

In this section we shall first show how we arrive at our proposed conformal algorithm, then give the concrete steps of our algorithm. We start with one curvilinear grid intersecting a rectangular grid as shown in Figs. 2(a) and 2(b). Each of the grids has its FDTD algorithm to advance the interior field variables. The field at the boundary of a grid cannot be advanced by the FDTD algorithm for that grid. The boundary field variables must be either prescribed from physical boundary conditions or by other means. If we give the boundary field for a period of time and prescribe the initial conditions of the field in the interior of the grid we would be able to carry out the calculations in that grid. This is the discretized FDTD version of the well-known rigorous existence and uniqueness theorem of electromagnetics. The interior electric field in each of the grids is advanced at $n\,\Delta t$ and the interior magnetic field is advanced at $(n + 1/2)\,\Delta t$.

The intersection of the two grids must be such that the boundary field of one grid can be obtained through interpolation of the calculated field in the other grid. In this way we pass on the necessary boundary data of one grid to the other. This is essentially the key idea of our conformal algorithm. In scattering problems we use a curvilinear grid that conforms with the scatterer, and the other grid will be rectangular grid. Intuitively, we believe that passing on data from one grid to the other by interpolation will be accurate if the grid sizes of the two grids in the overlapping region are comparable. To minimize the number of interpolations we would use as small a number of zones in the overlapping regions as possible. Our algorithm has attained these desirable features.

We now describe one version of our algorithm for a scattering problem.

1) Choose an appropriate ΔX.
2) Let S be the scattering surface.
   Denote by
   - $S1$ The surface consisting of all points at a distance ΔX away from S along the exterior pointing normal.
   - $S2$ The surface consisting of all points at a distance $2\,\Delta X$ away from S along the exterior pointing normal.
   - $S3$ The surface consisting of all points at a distance $3\,\Delta X$ away from S along the exterior pointing normal.

   In the volume enclosed by S and $S3$ introduced a tangent-normal grid or any other conformal grid (see Fig. 3). This will be our body conformal curvilinear grid.
3) Introduce a rectangular grid of spatial distance ΔX. Let V = union of all electric FDTD cubes whose centers are at least a distance $1.25\,\Delta X$ from S. This is the staircasing step. If the center of an electric cube is at least $1.25\,\Delta X$ from the scattering surface, none of its edges will intersect the scatterer. The inner boundary of V will be the inner boundary of our rectangular grid. The inner boundary of our curvilinear grid will be our scatterer and the outer boundary of our curvilinear grid will be approximately $S3$. We say approximately because the precise implementation of our algorithm as

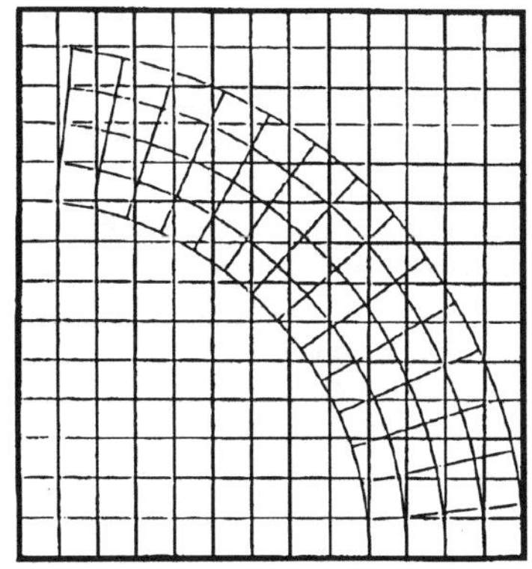

(a)

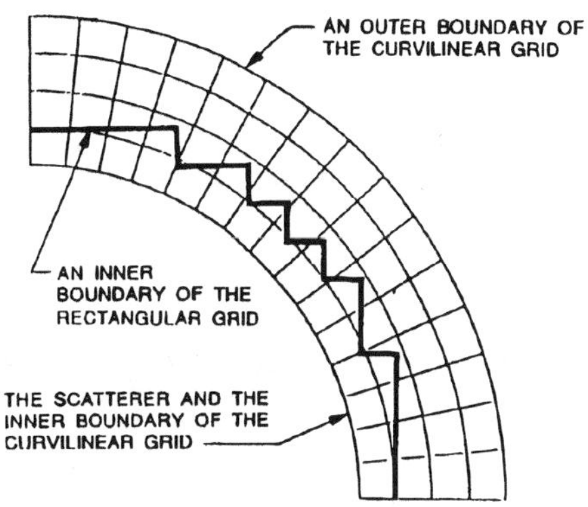

(b)

Fig. 2. (a) A rectangular grid overlapping a curvilinear grid. (b) The inner and outer boundaries of the overlapping grids.

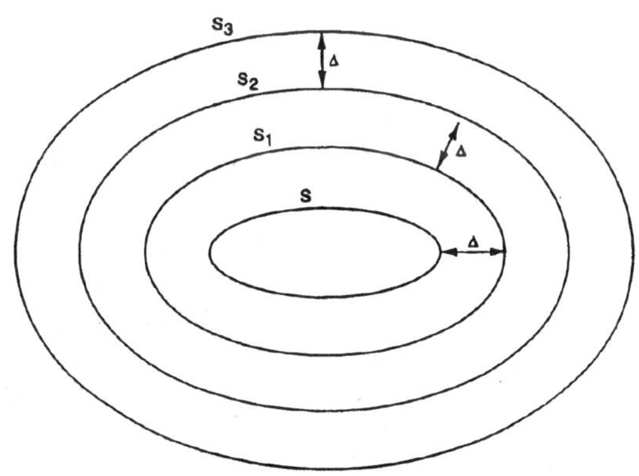

Fig. 3. The scattering surface S and its parallel surfaces $S1$, $S2$, and $S3$.

shown later reveals that the outer boundary sometimes is closer than $S3$.

4) The tangential electric field in the inner boundary of our rectangular grid is obtained through interpolation from the electric field calculated by the curvilinear FDTD algorithm and the tangential electric field at the outer boundary of the curvilinear grid are obtained through interpolation from the electric field calculated by the rectangular FDTD algorithm. The boundary field at the inner boundary of the curvilinear grid is obtained through given physical boundary conditions. The boundary field at the outer boundary of the rectangular grid are obtained through numerical radiation boundary condition truncation.

During our numerical experimentations, we notice that just prescribing the tangential electric field on the curvilinear outer boundary introduces some oscillation which die slowly. These oscillations seem to be typical of the resonant modes of the volume occupied by the curvilinear grid. The oscillation is an indication that the two grids are too loosely coupled. In order to introduce more coupling between the two grids we replace step 4 by

4′. a) interpolate for the tangential electric field at the inner boundary of the rectangular grid as in step 4.

b) Calculate the magnetic field with the rectangular FDTD algorithm at the boundary electric faces.

c) Determine the boundary field at the inner boundary of the curvilinear grid as in step 4.

d) (different from step 4) If a curvilinear field component (an electric field component or a magnetic field component) can be obtained through interpolation from the calculated (by the rectangular FDTD algorithm) rectangular field components, use this interpolated value instead of the one obtained by using the curvilinear FDTD algorithm. Otherwise, calculate with the curvilinear FDTD.

Step 4′ seems to favor the rectangular FDTD and introduces more intermixing of the two grids. We notice that the late time oscillations observed before have been eliminated significantly with step 4′ in our 2-D simulations. However, in our 3-D simulations of scattering by a finite circular cylinder we encounter some stability problem with step 4′. We modified step 4′ as follows, and found that our numerical simulation is stable.

4″. a, b, c are the same as in step 4′.

d) If a curvilinear magnetic field component can be calculated by the curvilinear algorithm and interpolated (with the rectangular field), take the average of the two.

e) Interpolate for the outer boundary curvilinear tangential electric field and calculate all other electric field either by the curvilinear algorithm or through the physical boundary condition. There are yet other variations of step 4 to be tried.

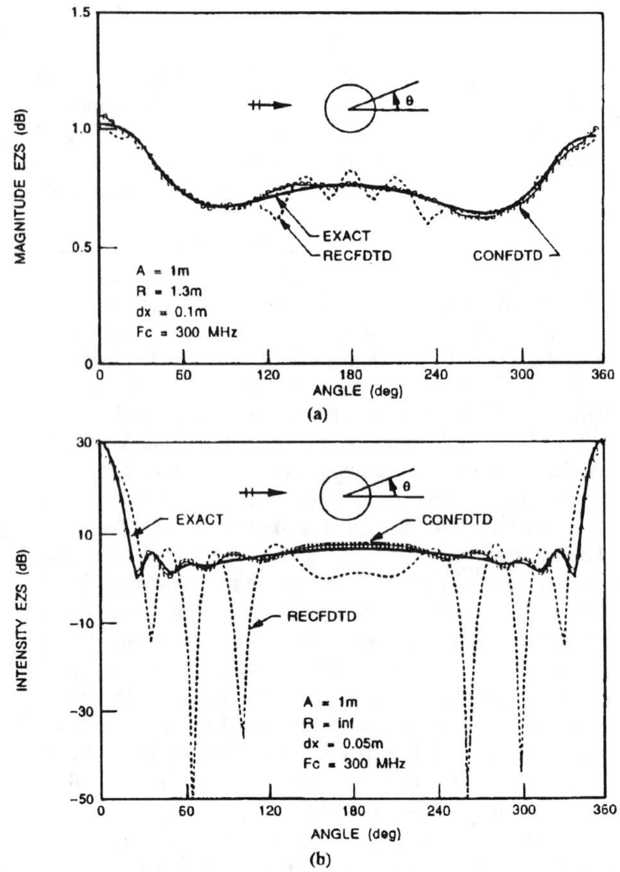

Fig. 4. (a) Near-field comparison of the amplitudes of the scattered electric field of an incident plane harmonic TM wave (PEC circular cylinder). (b) The TM bistatic RCS of a PEC circular cylinder.

It is to be emphasized again that to interpolate for field values in one grid one must use data obtained from the FDTD algorithm appropriate for another grid.

IV. Sample Calculations

We have built a 2-D code based on the steps of Section III. We have also built a 3-D code based the same algorithm. For simplicity we take the scatterers to be perfectly electric conducting.

A. Numerical Modeling Scattering by 2-D Infinite Cylinders

Our 2-D conformal FDTD algorithm code is built with a tangent-normal curvilinear grid intersecting a rectangular grid. Figs. 2(a) and 2(b) show the curvilinear grid intersecting a rectangular grid and their grid boundaries. When a plane harmonic TE or TM wave encounters a PEC infinite cylinder, there arises a TE or a TM scattered e/m wave. In Figs. 4(a) and 5(a) we show the amplitude of the scattered field due to an incident plane harmonic

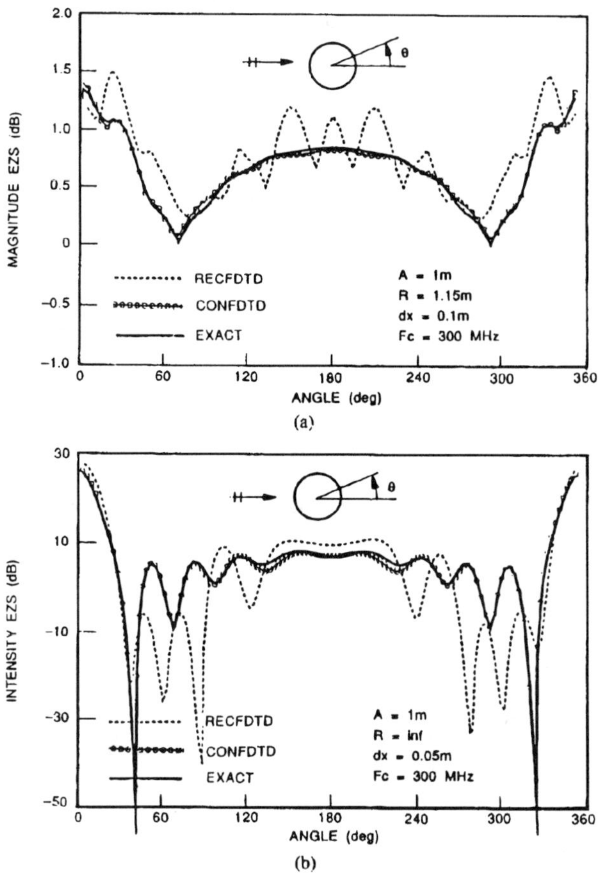

Fig. 5. (a) Near-field comparison of the amplitudes of the scattered magnetic field of an incident plane harmonic TE wave (PEC circular cylinder). (b) The TE bistatic RCS of a PEC circular cylinder.

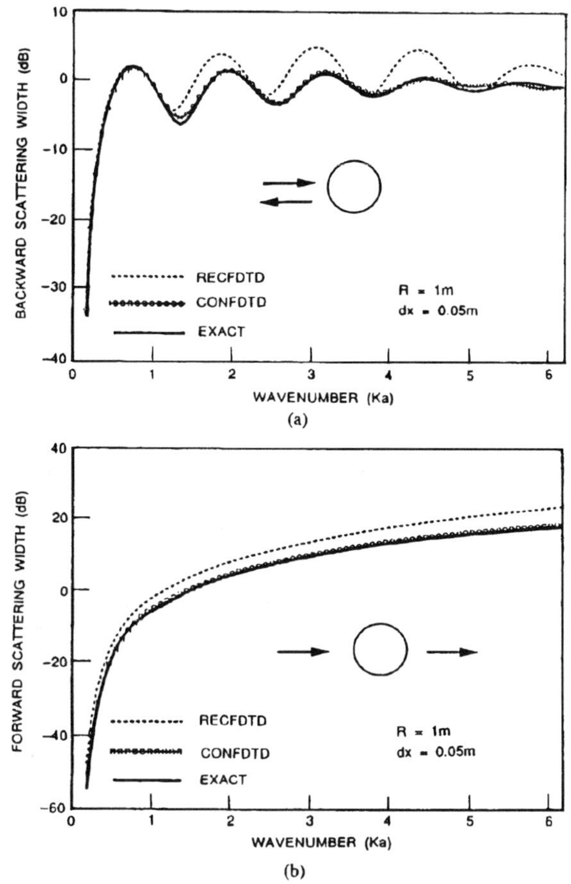

Fig. 6. (a) The TE monostatic RCS versus frequency for a PEC circular cylinder. (b) The TE forward scattering width versus frequency of a PEC circular cylinder.

wave as a function of the angle for a constant distance from the center of the scatterer. The computations with the staircasing FDTD, with the conformal FDTD, and with the exact series solution are shown. In Figs. 4(b) and 5(b) we show the bistatic radar cross section (RCS)) for the circular cylinder. In Fig. 6 we show the TM RCS of the cylinder as a function of frequency obtained by the staircasing FDTD, the conformal FDTD and the exact series. These frequency results are obtained with an incident Gaussian plane pulse and the time result are obtained through a fast Fourier transform. It is evident that the agreement between the exact solution and the conformal FDTD computations is better than that between the staircasing FDTD and the exact solution.

B. Scattering by a Finite PEC Circular Cylinder

We have applied our 3-D conformal algorithm to the scattering of a plane incident pulse by a PEC finite circular cylinder. The dimension of the scatterers are shown in the relevant figures. The overlapping grids in the finite circular cylinder case are shown in Figs. 7(a) and 7(b). One noteworthy point here is that we only use a curvilinear grid in the neighborhood of the lateral surface of the cylinder and elsewhere we use the rectangular grid. We use these overlapping grids in order to avoid the coordinate singularity near the cylindrical axis and to keep the curvilinear zones comparable to the rectangular zones. In Figs. 8(a) and 8(b) we show the comparison of RCS obtained by a method of moment code and FDTD. The results for the finite circular cylinder was obtained through step 4″. It can be seen from these graphs that the conformal FDTD agrees better with the MoM calculations although for this particular scatterer and the PEC boundary condition the staircasing FDTD RCS calculation is credible.

C. Scattering by a PEC Sphere (Near-Field Comparison)

We show in Figs. 9(a)–9(d) some near-field comparison between overlapping grids conformal FDTD calculations with the Mie series solution for a conducting sphere of radius 1 m at a frequency of 200 MHz. These graphs are the amplitudes of the various components of the scattered field. The incident field is linearly polarized and traveling along the negative z-direction. The amplitude of the inci-

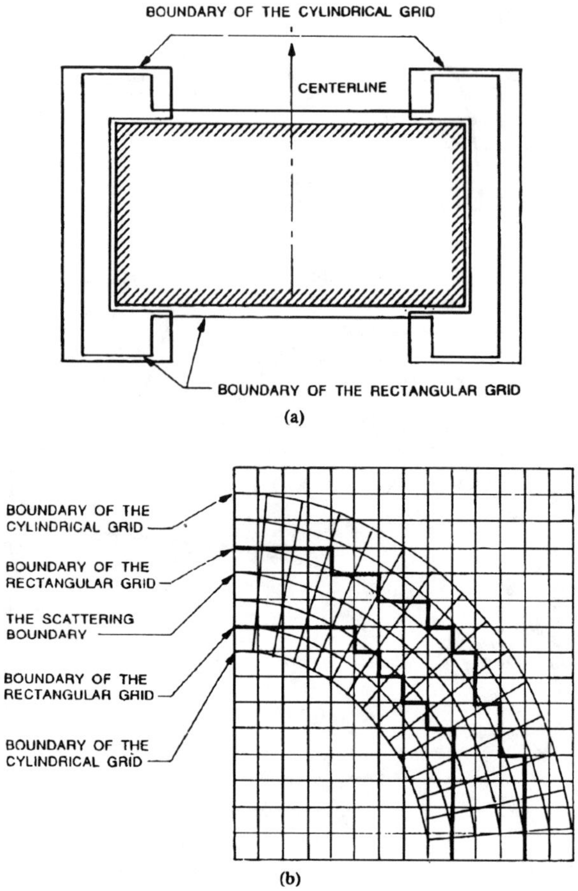

Fig. 7. (a) A finite circular cylinder scatterer and the boundaries of the overlapping grids (side view). (b) A finite circular cylinder scatterer and the boundaries of the overlapping grids (top view).

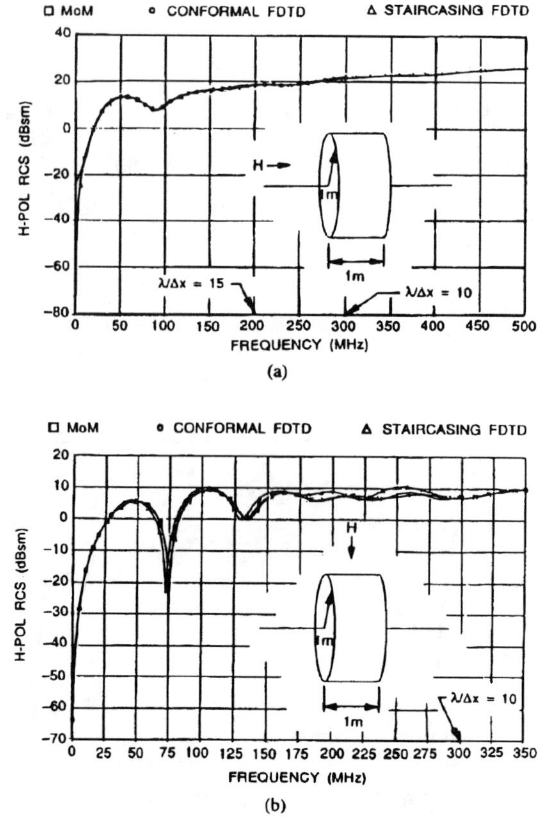

Fig. 8. (a) The H-pol RCS of a PEC finite circular cylinder (axially incidence). (b) The H-pol RCS of a PEC finite circular cylinder (90° incidence).

dent electric field is 1 V/m. The graphs show the amplitude of the scattered electric field and 377 times the scattered magnetic field. The graphs are at a constant radius and at a constant azimuthal angle. There are two bodily conformal grids:

1) a spherical coordinate system with the z-axis as the polar axis and the lattitude angle ranges from 40° to 140°;
2) another spherical coordinate system with the x-axis as the polar axis and the corresponding lattitude angle from 40° to 140°.

Both of these overlapping spherical grids extend to four zones along the radial direction and they intersect the rectangular grid whose inner boundary is within one zone from the conducting sphere. We introduced two overlapping grids in order to avoid the spherical coordinate singularities at the poles and for which the FDTD algorithm would require a smaller time increment than that otherwise required for the rectangular FDTD. These comparisons serve as further evidence that our overlapping grids conformal FDTD is credible.

V. General 3-D Conformal FDTD with More Than Two Overlapping Grids

We limit ourselves in the following to smooth, orientable surfaces. The discussion will necessarily be brief to maintain the flow of the main thought. For practical purposes we emphasize once more that we should use as few as possible the number of local conformal patches as we have done with the scattering by a sphere.

Let S be a scattering surface. Subdivide S into small disjoint sets $S = \cup A_i$. For $i \neq j$, $A_i \cap A_j$ has no interior. In other words, the sets A_i and A_j can have common boundary points only. Construct a surface coordinate patch $P_i \supset A_i$ and choose a larger surface patch $Q_i \supset P_i$ so that ∂Q_i, the boundary of Q_i, is three zones away from ∂P_i. For each j, there is at least one i_0 such that $\partial Q_j \cap P_{i_0} \neq \varnothing$ because $\partial Q_j = \partial Q_j \cap S = \cup(\partial Q_j \cap A_i)$. Let $N = \{n\colon 0 \leq n \leq 3\Delta\}$ where n is the distance from the scattering surface S along the outward pointing surface normal $\hat{\mathbf{n}}$. If (u, v) is a surface coordinate system of a coordinated patch, a point on the surface can be represented as

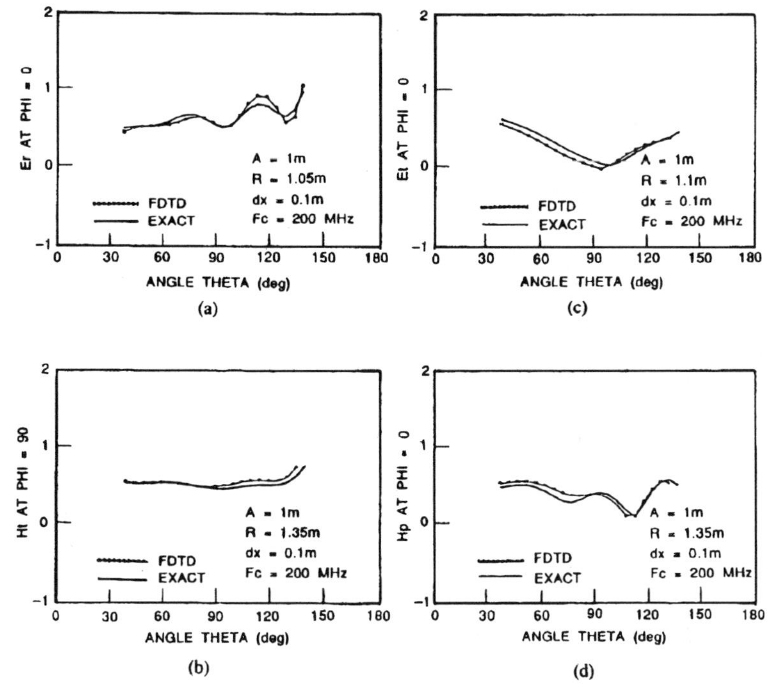

Fig. 9. (a)–(d) Near field comparison of the overlapping grids conformal FDTD with the exact solution.

$\bar{\mathbf{R}}(u, v) = (x(u, v), y(u, v), z(u, v))$ and a point in the neighborhood of the surface can be represented as $\bar{\mathbf{r}}(u, v, n) = \bar{\mathbf{R}}(u, v) + n\bar{\mathbf{n}}(u, v)$. In the volume $Q_i \otimes N$ use a curvilinear coordinate system and the corresponding curvilinear FDTD algorithm.

a) The boundary field on the interior rectangular grid boundary is obtained through interpolation from the calculated (by the appropriate FDTD algorithm) field of an appropriate curvilinear grid.
b) $\partial(Q_i \otimes N)$, where the boundary tangential electric field is obtained by interpolation, consists of the outer boundary and the lateral boundary.

On the outer boundary the tangential electric field is obtained through interpolation with the calculated fields by the rectangular FDTD. On the lateral boundary let p be a point where a boundary field is needed. We do the following:

1) find i_0 such that $\bar{\mathbf{p}} \in (\partial Q_j \cap P_{i_0}) \otimes N$;
2) obtain the field at $\bar{\mathbf{p}}$ by interpolating the appropriate calculated field in $Q_{i_0} \otimes N$.

The above discussion is appropriate if step 4 in the algorithm in Section III is used. If step 4′ is used one would have to modify the above steps slightly.

VI. Conclusion

We have detailed the steps to show how to construct a conformal FDTD algorithm. This conformal scheme is based on the basic emphasis that the various curvilinear schemes serve the rectangular scheme, although in many potential applications the local conformal schemes are absolutely necessary. We believe that the overlapping grids together with the Holland scheme of FDTD in curvilinear coordinate systems or any other curvilinear FDTD derived from the integral form of Maxwell's equations can be more easily implemented than the existing known conformal schemes. Our sample calculations show definitely that the conformal FDTD is more accurate than the staircasing FDTD. More importantly, we can always introduce local curvilinear grid without disturbing the other local grids. The modeling of scattering by a surface with thin layers of lossy materials can be done quite easily when we introduce smaller spatial discretization along the surface normal. The overlapping region does not have to be next to the scatterer; it can be a finite number zones away. This creates another convenience. This conformal scheme can be used in conjunction with local subgridding [12] in problems where more detailed local field variables are needed.

References

[1] R. Holland, "Finite difference solution of Maxwell's equations in generalized nonorthogonal coordinates," *IEEE Trans. Nucl. Sci.*, vol. NS-30, pp. 4589–4591, Dec. 1983.

[2] K. K. Mei, A. Cangellaris, and D. J. Angelakos, "Conformal time domain finite-difference method," *Radio Sci.*, vol. 19, no. 5, pp. 1145–1147, 1984.

[3] T. Weiland, "On the numerical solution of Maxwell's equations and application in the field of accelerator physics," *Particle Accelerators*, vol. 15, pp. 245–292, 1984.

[4] K. S. Yee, "Numerical solution to Maxwell's equations with non-orthogonal grids," *Proc. 1st Review of Numerical Electromagnetic Code*, Mar. 1985.
[5] T. G. Jurgens, A. Taflove, and K. R. Umashankar, "FDTD conformal modeling of smooth curved surfaces," in *Proc. 1987 URSI Radio Sci. Meet.*, Blacksburg, VA, p. 227, June 1987.
[6] M. Fusco, "FDTD algorithm in curvilinear coordinates," *IEEE Trans. Antennas Propagat.*, vol. 38, pp. 76–89, Jan. 1990.
[7] N. K. Madsen and R. W. Ziolkowski, "Numerical solution of Maxwell's equations in the time domain using irregular non-orthogonal grids," *Wave Motion*, vol. 10, pp. 583–596, Dec. 1988.
[8] ——, "A modified finite-volume technique for Maxwell's equations," *Electromagn.*, vol. 10, pp. 147–161, Jan. 1990.
[9] R. Holland, V. P. Cable, and L. Wilson, "RCS calculation by finite-volume time-domain (FVTD) techniques," Lockheed Burbank Tech. Rep., Oct. 1990.
[10] K. S. Yee, "Numerical solution of initial boundary value problems involving Maxwell's equations in isotropic media," *IEEE Trans. Antennas Propagat.*, vol. AP-14, pp. 302–307, May 1966.
[11] A. Taflove and K. R. Umashankar, "The finite-difference time-domain (FDTD) method for electromagnetic scattering and interaction problems," *J. Electromagn. Wave Appl.*, vol. 1, pp. 243–267, 1987.
[12] S. S. Zivanovic, K. S. Yee, and K. K. Mei, "A subgridding method for the time-domain finite-difference method to solve Maxwell's equations," *IEEE Trans. Microwave Theory Tech.*, vol. 39, pp. 471–479, Mar. 1991.
[13] J.-F. Lee and R. Mittra, "Solution of electromagnetic scattering problems using the non-orthogonal finite difference time domain approach," a chapter on a monograph. New York: Plenum, to appear.
[14] A. C. Cangellaris, C. C. Lin, and K. K. Mei, "Point match time domain finite element methods for electromagnetic radiation and scattering," *IEEE Trans. Antennas Propagat.*, vol. AP-35, pp. 1160–1173, Oct. 1987.

A Locally Conformed Finite-Difference Time-Domain Algorithm of Modeling Arbitrary Shape Planar Metal Strips

Jiayuan Fang, *Member, IEEE*, and Jishi Ren

Abstract—**A general algorithm to model arbitrary shape planar metal strips by the finite-difference time-domain (FDTD) method is presented in this paper. With this method, fields in the entire computation domain are computed by the regular FDTD algorithm except those near metal strips where special techniques proposed in this paper are applied. Unlike globally conformed finite-difference algorithms, the computation efficiency of the regular FDTD method is maintained while high space-resolution is obtained by this locally conformed finite-difference method. Numerical tests have verified that a higher computation accuracy is achieved by this scheme than the conventionally used staircase approximation. The modeling of electrical characteristics of two crossed strip lines is provided as an example.**

I. Introduction

THE finite-difference time-domain method has been applied to many electromagnetic problems in antennas, scattering and microwave circuit components [1]–[3]. In dealing with complex shape objects, where the boundaries of objects do not coincide with finite-difference grid lines, the staircase approximation is commonly used [3]. Due to the low space resolution of the staircase approximation, a very fine mesh has to be used to accurately represent object geometries, which can easily result in an excessive requirement of computer resources.

To overcome the difficulties in the staircase approximation, finite-difference algorithms in nonorthogonal coordinates have been proposed and applied to various problems [4]–[6]. With nonorthogonal finite-difference algorithms, finite-difference grids are conformed to surfaces of modeled objects in the way similar to that in finite-element methods. Although the resolution of surfaces of objects is significantly improved, nonorthogonal finite-difference algorithms usually require much longer computation times than the regular FDTD algorithm for the same size of problems [6]. It is also very challenging to generate grids in nonorthogonal coordinates for three-dimensional complex shape objects.

Manuscript received May 19, 1992; revised September 25, 1992. This work was supported by the National Science Foundation under contract MIP-9110203 and ECD-9108850 which supports the Integrated Electronics Engineering Center (IEEC) at the State University of New York at Binghamton.

J. Fang is with the Department of Electrical Engineering, State University of New York at Binghamton, Binghamton, NY 13902.

J. Ren was a visiting scholar at the Department of Electrical Engineering, State University of New York at Binghamton, from the Department of Electrical Engineering Northwestern Polytechnical University, Xi'an 710072, P. R. of China.

IEEE Log Number 9207412.

To maintain the computation efficiency of the original FDTD method, instead of globally deforming the grid as in nonorthogonal finite-difference algorithms, we propose a locally conformed finite-difference algorithm for modeling arbitrary shape planar metal strips. With this method, fields in the whole computation domain are first computed by the regular FDTD algorithm, fields near metal strips are then corrected according to locations of metal edges in the finite-difference mesh. The additional computation required in the present method to correct fields near irregular shape objects is only a small percentage of the overall computation still dominated by the regular FDTD operations, while the same improvement on the space-resolution is achieved by this method as by the nonorthogonal finite-difference algorithms. It will be shown that the numerical error, such as the artificially slowed propagation speed in the staircase approximation [7], can be substantially reduced by the present method.

The main idea of the locally conformed finite-difference method presented in this paper is based on the integral form of Maxwell's equations. This approach has been applied to model curved surfaces in two dimensional problems [8], and conducting wires and strips which are parallel to finite-difference grid lines in three dimensional problems [9]. In this paper we will present a method of properly applying the integral form of Maxwell's equations for modeling arbitrary shape planar metal strips in three dimensional problems.

II. Locally Conformed Finite-Difference Algorithm

The finite-difference time-domain method proposed by K. S. Yee [1] is based on the discretization of Maxwell's curl equations by the central-difference approximation in both space and time. With given initial and boundary conditions, the electric and magnetic field components are computed alternatively in the leap-frog manner for some known source excitations. When edges of metal strips are approximated by stair-stepped boundaries, all field components are computed by the regular FDTD algorithm except that the tangential electric and normal magnetic field components on metal surfaces are set to zero.

The general procedure of computation by our locally conformed finite-difference algorithm consists of the following steps for problems involving metal strips:

1. Calculate the electric field at time-step n in the entire computation domain by the regular FDTD algorithm.
2. Set tangential electric field components at time-step n to zero on surfaces of metal strips.

Reprinted from *IEEE Transactions on Microwave Theory and Techniques*, pp. 830-838, May 1990.

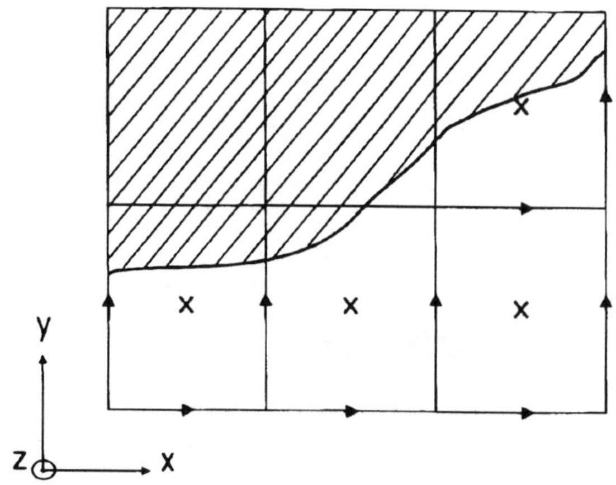

Fig. 1. Metal strip on a finite-difference mesh. × H_z nodes; → E_x nodes; ↑ E_y nodes.

3. Calculate the magnetic field at time-step $n + 1/2$ in the entire computation domain by the regular FDTD algorithm.
4. Set normal magnetic field components at time-step $n + 1/2$ to zero on surfaces of metal strips.
5. Correct the electric field components at time-step n, and the magnetic field components at time-step $n + 1/2$ around metal strips that can not be found by the regular FDTD algorithm.

Steps 1 to 4 above are straightforward regular FDTD computations. What needs to be explained is step 5 on the correction of electric and magnetic fields near metal strips.

As shown in Fig. 1, suppose a metal strip of zero thickness is placed on the x-y plane. The edge of the metal strip can pass across the finite-difference mesh in many different ways, resulting in different types of irregular elements near the metal strip. Different types of elements may require different correction procedures. When the angle between the edge of the metal strip and the x axis is between 0° to 45°, all possible cases of irregular elements are listed in Fig. 2 according to positions of various field components near the metal strip. It is generally sufficient to consider the computation of field components for cases listed in Fig. 2 only. In case the angle between the metal edge and the x axis is larger than 45°, the angle between the metal edge and the y axis is then less than 45°, and therefore the technique of dealing with elements in Fig. 2 can be applied accordingly by replacing the x axis in Fig. 2 by the y axis.

For elements near metal strips listed in Fig. 2, field components H_z, E_y, E_x, H_y and H_x need to be computed specially when they can not be computed by the regular FDTD algorithm. The sixteen types of elements in Fig. 2 can be categorized to five groups, as listed in Table I, according to their corresponding correction procedures of field components. Next, we will discuss procedures of correcting field components near metal strips for each group of elements.

Group A: The field components associated with this element, which is redrawn in Fig. 3, that can not be found by the regular FDTD algorithm are $H_z(i,j,k), H_x(i,j,k)$ and

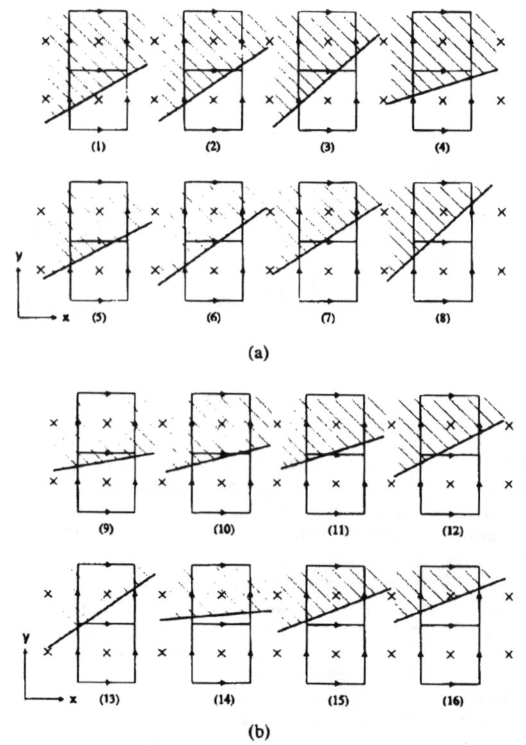

Fig. 2. (a) Finite-difference elements near metal strips. × H_z nodes; → E_x nodes; ↑ E_y nodes. (b) Finite-difference elements near metal strips. × H_z nodes; → E_x nodes; ↑ E_y nodes.

TABLE I
CATEGORIZATION OF IRREGULAR ELEMENTS LISTED IN FIG. 2

	Group A	Group B	Group C	Group D	Group E
Elements in Figure 2	(9)	(10),(11), (12),(13)	(14),(15), (16)	(1),(4)	(2),(3),(5), (6),(7),(8)

$H_x(i,j,k-1)$. $H_x(i,j,k)$ is located half space-step above $E_y(i,j,k)$ in the z direction, and $H_x(i,j,k-1)$ is located half space-step below $E_y(i,j,k)$ in the z direction. The correction procedure consists of following processes:

(a) Correcting $H_z^{n+1/2}(i,j,k)$.

From Faraday's law,

$$\oint_{c_1} \vec{E}\cdot d\vec{l} = -\mu\frac{\partial}{\partial t}\int_{s_1}\vec{H}\cdot d\vec{S}, \tag{1}$$

where c_1 is the loop ABCDA shown in Fig. 3, and s_1 is the area enclosed by loop c_1. Discretizing (1) results in the following difference equation for computing $H_z(i,j,k)$ at time-step $n + 1/2$.

$$\begin{aligned} H_z^{n+1/2}(i,j,k) &= H_z^{n-1/2}(i,j,k) - \frac{dt}{\mu s_1} \\ &\cdot [-E_y^n(i,j,k)\cdot l_{\mathrm{AB}} + E_x^n(i,j,k) \\ &\cdot dh + E_y^n(i+1,j,k)\cdot l_{\mathrm{CD}}] \end{aligned} \tag{2}$$

where dt is the time-step, dh is the space-step of the finite-difference grid, l_{AB} and l_{CD} are the lengths of sides AB and

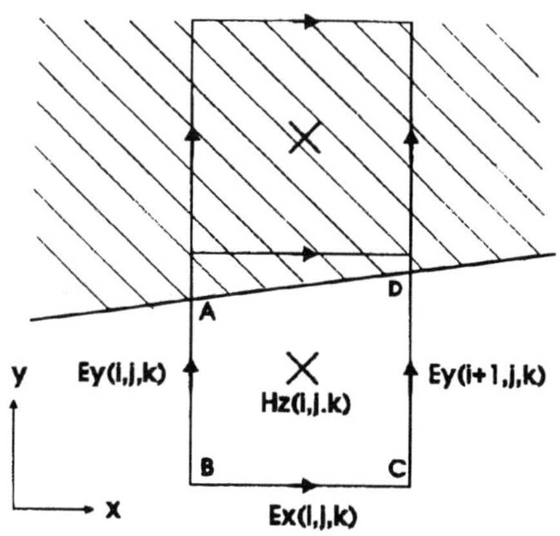

Fig. 3. An element of group A.

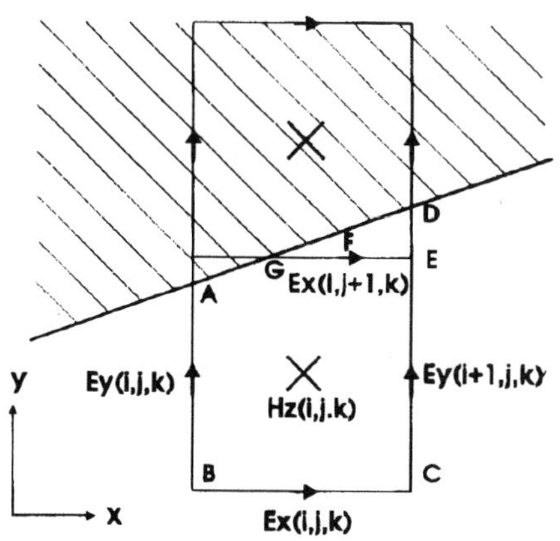

Fig. 4. An element of group B.

CD. The electric field component along the metal edge is assumed to be zero.

(b) Correcting $H_x^{n+1/2}(i,j,k)$ and $H_x^{n+1/2}(i,j,k-1)$.

The finite-difference algorithms for computing $H_x^{n+1/2}(i,j,k)$ and $H_x^{n+1/2}(i,j,k-1)$ need to be modified from the regular FDTD algorithms by taking into account that $E_y^n(i,j,k)$ is nonzero only across the length l_{AB}. The modified difference equations for computing $H_x^{n+1/2}(i,j,k)$ and $H_x^{n+1/2}(i,j,k-1)$ are:

$$\begin{aligned} H_x^{n+1/2}(i,j,k) &= H_x^{n-1/2}(i,j,k) - \frac{dt}{\mu \cdot dh}\Big[E_y^n(i,j,k)\cdot\frac{l_{\mathrm{AB}}}{dh} \\ &\quad + E_z^n(i,j+1,k) - E_y^n(i,j,k+1) - E_z^n(i,j,k)\Big], \end{aligned} \tag{3}$$

$$\begin{aligned} H_x^{n+1/2}(i,j,k-1) &= H_x^{n-1/2}(i,j,k-1) - \frac{dt}{\mu \cdot dh}\Big[E_y^n(i,j,k-1) \\ &\quad + E_z^n(i,j+1,k-1) - E_y^n(i,j,k)\cdot\frac{l_{\mathrm{AB}}}{dh} \\ &\quad - E_z^n(i,j,k-1)\Big]. \end{aligned} \tag{4}$$

In the term $l_{\mathrm{AB}}/dh = 1$, (3) and (4) will become the regular FDTD equations for computing $H_x^{n+1/2}(i,j,k)$ and $H_x^{n+1/2}(i,j,k-1)$. In dealing with the element in Fig. 3, there is no need to correct $H_x^{n+1/2}(i+1,j,k)$ and $H_x^{n+1/2}(i+1,j,k-1)$ located half space-step above and below $E_y(i+1,j,k)$ in the z direction, because these components will be taken care of when the neighboring element at the right is being processed.

Group B: A typical element of this group is drawn in Fig. 4. The field components need to be corrected are $H_z(i,j,k), E_x(i,j+1,k), H_y(i,j+1,k), H_y(i,j+1,k-1), H_x(i,j,k)$ and $H_x(i,j,k-1)$. $H_y(i,j+1,k)$ is located half space-step above $E_x(i,j+1,k)$ in the z direction, and $H_y(i,j+1,k-1)$ is located half space-step below $E_x(i,j+1,k)$ in the z direction. The correction procedure of this group of elements is described below:

(a) Correcting $H_z^{n+1/2}(i,j,k)$.

$H_z^{n+1/2}(i,j,k)$ is computed in the same way as that for group A elements. Faraday's law is applied along the loop ABCDA denoted in Fig. 4, resulting in (2) for computing $H_z^{n+1/2}(i,j,k)$.

(b) Correcting $E_x^n(i,j+1,k)$.

The location of the node of $E_x(i,j+1,k)$ is assumed to be at the point F in Fig. 4, which is at the middle of the side EG. $E_x^n(i,j+1,k)$ is calculated by making use of the corrected value of $H_z^{n+1/2}(i,j,k)$ obtained in process (a) above. From Faraday's law expressed in (1), replacing c_1 and s_1 in (1) by c_2 and s_2, where c_2 is the loop ABCEFGA and s_2 is the area enclosed by loop c_2, the difference equation for correcting $E_x^n(i,j+1,k)$ is obtained as

$$\begin{aligned} E_x^n(i,j+1,k) &= \frac{1}{l_{\mathrm{EG}}}[E_x^n(i,j,k)\cdot dh + E_y^n(i+1,j,k)\cdot dh \\ &\quad - E_y^n(i,j,k)\cdot l_{\mathrm{AB}}] + \frac{\mu s_2}{dt\cdot l_{\mathrm{EG}}}[H_z^{n+1/2}(i,j,k) \\ &\quad - H_z^{n-1/2}(i,j,k)]. \end{aligned} \tag{5}$$

(c) Correcting $H_y^{n+1/2}(i,j+1,k)$ and $H_y^{n+1/2}(i,j+1,k-1)$

With the newly corrected $E_x^n(i,j+1,k)$ in process (b), $H_y^{n+1/2}(i,j+1,k)$ and $H_y^{n+1/2}(i,j+1,k-1)$ are computed by the modified FDTD algorithms, that take into account that $E_x^n(i,j+1,k)$ is nonzero only across the length l_{EG}. This is evident in equations (6) and (7), which are shown at the bottom of the next page.

(d) Correcting $H_x^{n+1/2}(i,j,k)$ and $H_x^{n+1/2}(i,j,k-1)$

$H_x^{n+1/2}(i,j,k)$ and $H_x^{n+1/2}(i,j,k-1)$ are computed in the same way as group A elements by (3) and (4) respectively.

Group C: The field components need to be corrected for this group of elements are $H_z(i,j,k), E_x(i,j+1,k), H_y(i,j+1,k), H_y(i,j+1,k-1), H_x(i,j+1,k)$ and $H_x(i,j+1,k-1)$. A typical element of this group is drawn in Fig. 5. The correction procedure of this group of elements is:

(a) Correcting $H_z^{n+1/2}(i,j,k)$ by (2) where s_1 is the area enclosed by the loop $ABCDA$.

(b) Correcting $E_x^n(i,j+1,k)$. $E_x^n(i,j+1,k)$ is computed in the same way as that for group B elements, except that

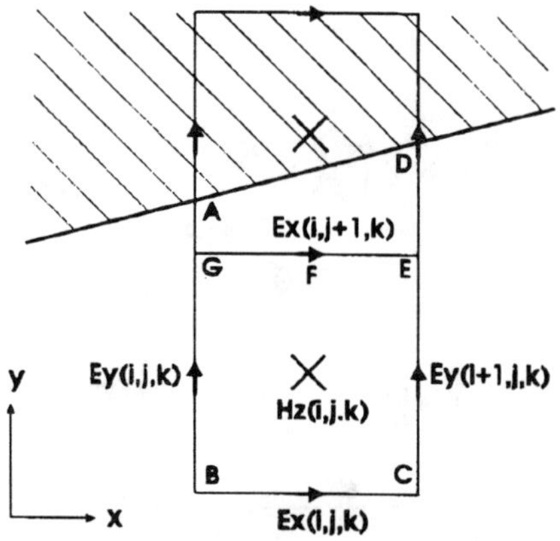

Fig. 5. An element of group C.

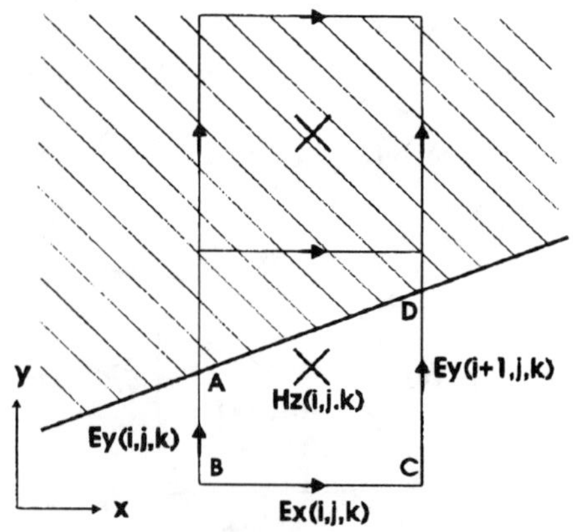

Fig. 6. An element of group D.

the loop c_2 (GBCEFG) encloses a rectangle formed by finite-difference grid lines. The difference equation for correcting $E_x^n(i,j+1,k)$ is therefore:

$$\begin{aligned} &E_x^n(i,j+1,k) \\ &\quad = E_x^n(i,j,k) + E_y^n(i+1,j,k) - E_y^n(i,j,k) \\ &\qquad + \frac{\mu \cdot dh}{dt}[H_z^{n+1/2}(i,j,k) - H_z^{n-1/2}(i,j,k)]. \end{aligned} \tag{8}$$

(c) Correcting $H_y^{n+1/2}(i,j+1,k)$ and $H_y^{n+1/2}(i,j+1,k-1)$.

$H_y^{n+1/2}(i,j+1,k)$ and $H_y^{n+1/2}(i,j+1,k-1)$ are computed with the newly corrected $E_x^n(i,j+1,k)$ in process (b). Their difference equations are:

$$\begin{aligned} &H_y^{n+1/2}(i,j+1,k) \\ &\quad = H_y^{n-1/2}(i,j+1,k) - \frac{dt}{\mu \cdot dh}[E_x^n(i,j+1,k+1) \\ &\qquad - E_z^n(i+1,j+1,k) - E_x^n(i,j+1,k) \\ &\qquad + E_z^n(i,j+1,k)], \end{aligned} \tag{9}$$

$$\begin{aligned} &H_y^{n+1/2}(i,j+1,k-1) \\ &\quad = H_y^{n-1/2}(i,j+1,k-1) - \frac{dt}{\mu \cdot dh}[E_x^n(i,j+1,k) \\ &\qquad - E_z^n(i+1,j+1,k-1) - E_x^n(i,j+1,k-1) \\ &\qquad + E_z^n(i,j+1,k-1)]. \end{aligned} \tag{10}$$

(d) Correcting $H_x^{n+1/2}(i,j+1,k)$ and $H_x^{n+1/2}(i,j+1,k-1)$

The FDTD algorithms of computing $H_x^{n+1/2}(i,j+1,k)$ and $H_x^{n+1/2}(i,j+1,k-1)$ are modified by taking into account that E_y is nonzero only along the side AG. Let the value of E_y along side AG be approximated by $E_y(i,j,k)$, and stored at the node of $E_y(i,j+1,k)$. The difference equations for computing $H_x^{n+1/2}(i,j+1,k)$ and $H_x^{n+1/2}(i,j+1,k-1)$ become (11) and (12) at the bottom of the next page.

Group D: Consider a typical element of this group shown in Fig. 6. The procedure of computing field components, $H_z(i,j,k)$, $H_x(i,j+1,k)$ and $H_x(i,j+1,k-1)$, associated with this group of elements is explained as follows:

(a) Correcting $H_z^{n+1/2}(i,j,k)$.

$H_z^{n+1/2}(i,j,k)$ is computed by (2) where $E_y^n(i,j,k)$ represents E_y along side AB in Fig. 6. $E_y^n(i,j,k)$ can not be obtained by the regular FDTD algorithm for this group of elements, and is approximated by $E_y^n(i,j-1,k)$.

(b) Correcting $H_x^{n+1/2}(i,j,k)$ and $H_x^{n+1/2}(i,j,k-1)$ by (3) and (4) respectively. Again, $E_y^n(i,j,k)$ in (3) and (4) is approximated by $E_y^n(i,j-1,k)$

Group E: Consider a typical element of this group shown in Fig. 7. The valuer of $E_y^n(i,j,k)$ located at the middle of side AB is approximated by that of $E_y^n(i,j-1,k)$. The procedure of correcting the rest of field components is:

(a) Correcting $H_z^{n+1/2}(i,j,k)$ by (2). (b) Correcting $E_x^n(i,j+1,k)$ by (5). (c) Correcting $H_y^{n+1/2}(i,j+1,k)$ and $H_y^{n+1/2}(i,j+1,k-1)$ by (6) and (7) respectively. (d) Correcting $H_x^{n+1/2}(i,j,k)$ and $H_x^{n+1/2}(i,j,k-1)$ by (3) and (4) respectively.

$$\begin{aligned} H_y^{n+1/2}(i,j+1,k) = H_y^{n-1/2}(i,j+1,k) - \frac{dt}{\mu \cdot dh}\Big[&E_x^n(i,j+1,k+1) \\ &- E_z^n(i+1,j+1,k) - E_x^n(i,j+1,k) \cdot \frac{l_{\mathrm{EG}}}{dh} + E_z^n(i,j+1,k)\Big], \end{aligned} \tag{6}$$

$$\begin{aligned} H_y^{n+1/2}(i,j+1,k-1) = H_y^{n-1/2}(i,j+1,k-1) - \frac{dt}{\mu \cdot dh}\Big[&E_x^n(i,j+1,k) \cdot \frac{l_{\mathrm{EG}}}{dh} \\ &- E_z^n(i+1,j+1,k-1) - E_x^n(i,j+1,k-1) + E_z^n(i,j+1,k-1)\Big]. \end{aligned} \tag{7}$$

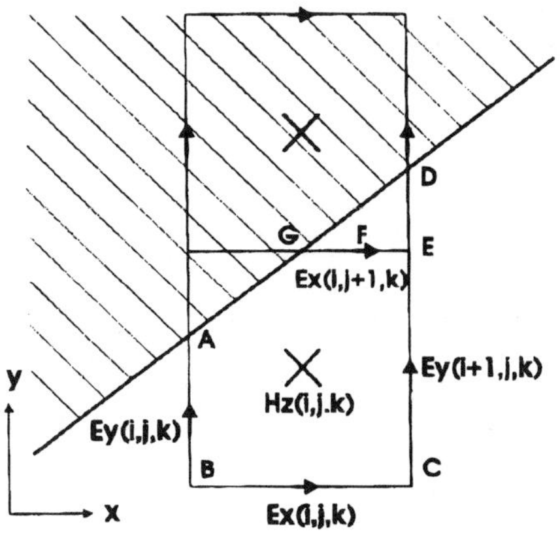

Fig. 7. An element of group E.

The procedures discussed above of correcting field components near metal strips for different groups of elements are summarized in Table II.

III. Tests on the Locally Conformed Finite- Difference Algorithm

The locally conformed finite-difference algorithm, presented in the last section, can be tested by applying it to model a pulse propagation through a strip-line shown in Fig. 8. The strip line is oriented in an angle θ with the x axis and surrounded by a dielectric medium of $\epsilon_r = 4$. The separation between two perfectly conducting reference planes is 1 mm. The metal strip is 0.4 mm above the lower reference plane. A uniform finite-difference mesh of equal space steps in the x, y and z directions is chosen for the computation. The space-step dh is 0.2 mm, and the time-step dt is 0.5 dh/v, where v is the speed of light in the medium. The width of the strip in the y direction W_y is chosen as 0.8 mm = 4 dh. The actual width of the strip is $W_y \cdot \cos\theta$.

Fig. 9 illustrates the staircase approximation of a strip line tilted from the x axis by an angle $\theta = \tan^{-1}$ (1/3) = 18.4°. It has been found that, due to the staircase approximation, a wave propagating along the conducting surface is artificially slowed down [7]. An alternative description of the numerical propagation speed along the strip line of Fig. 8 is the effective dielectric constant ϵ_{reff} which can be obtained from the transient solutions of fields along the strip line [2]. Suppose u and $u+L$ are two positions separated by a distance L along the strip line. $E(u,t)$ and $E(u+L,t)$ are numerically computed transient electric fields at the position u and $u+L$. Denote the Fourier transforms of $E(u,t)$ and $E(u+L,t)$ as $E(u,\omega)$ and $E(u+L,\omega)$ respectively. For a wave propagating in the positive u direction, $E(u,\omega)$ and $E(u+L,\omega)$ are related by

$$E(u+L,\omega) = E(u,\omega)e^{-\gamma(\omega)L}, \tag{13}$$

where

$$\gamma(\omega) = \alpha(\omega) + j\beta(\omega) \tag{14}$$

is the propagation constant. The effective dielectric constant ϵ_{reff} is defined as

$$\beta(\omega) = \omega\sqrt{\mu\epsilon_0\epsilon_{reff}}. \tag{15}$$

From (13) to (15), the effective dielectric constant ϵ_{reff} can be found from

$$\epsilon_{reff} = \frac{\beta^2}{\omega^2\mu\epsilon_0} = \frac{1}{\omega^2\mu\epsilon_0}\left[\frac{1}{L}\,\mathrm{Imag}\left[\ln\left(\frac{E(u,\omega)}{E(u+L,\omega)}\right)\right]\right]^2 \tag{16}$$

where Image is the operator of taking the imaginary part of a complex number.

Since the strip line in Fig. 8 is surrounded by the uniform dielectric medium of $\epsilon_r = 4$, the effective dielectric constant ϵ_{reff} should be the same as ϵ_r. However, the numerically computed effective dielectric constant ϵ_{reff} obtained by modeling a pulse propagation through a strip line is somewhat different from ϵ_r, due to the computation error of the numerical scheme being used. The dashed curves in Fig. 10 are the computed effective dielectric constants of the strip line modeled by the staircase approximation for $\theta = 45°, 26.6°$ (= $\tan^{-1}$ (1/2)), 18.4° (= $\tan^{-1}$ (1/3)), 14.0° (= $\tan^{-1}$ (1/4)) and 11.3° (= $\tan^{-1}$ (1/5)). The reason for choosing these particular angles of θ is for the convenience of the staircase approximation of oblique strip lines as shown in Fig. 9. As can be seen from Fig. 10, ϵ_{reff}'s obtained by the staircase approximation are all larger than the accurate solution of 4.0, representing slowed propagation speeds due to the numerical computation error, and the largest numerical error appears when $\theta = 45°$.

The effective dielectric constants of the strip line approximated by the locally conformed finite-difference algorithm of this paper are also shown in Fig. 10 by solid curves. It is seen that errors in the effective dielectric constants computed by the locally conformed finite-difference algorithm are about 60% to 70% smaller than those computed by the staircase approximation for the angles of θ chosen.

As a comparison, a locally conformed finite-difference algorithm based on a scheme in [8] is also tested. This scheme

$$H_x^{n+1/2}(i,j+1,k) = H_x^{n-1/2}(i,j+1,k) - \frac{dt}{\mu\cdot dh}[E_y^n(i,j+1,k)\cdot\frac{l_{AG}}{dh} + E_z^n(i,j+2,k) - E_y^n(i,j+1,k+1) - E_z^n(i,j+1,k)], \tag{11}$$

$$H_x^{n+1/2}(i,j+1,k-1) = H_x^{n-1/2}(i,j+1,k-1) - \frac{dt}{\mu\cdot dh}[E_y^n(i,j+1,k-1) + E_z^n(i,j+2,k-1) - E_y^n(i,j+1,k)\cdot\frac{l_{AG}}{dh} - E_z^n(i,j+1,k-1)]. \tag{12}$$

TABLE II
COMPUTATION OF FIELD COMPONENTS NEAR METAL STRIPS

Field components to be corrected	Group A	Group B	Group C	Group D	Group E
$E_y^n(i,j,k)$	—	—	—	$E_y^n(i,j-1,k)$	$E_y^n(i,j-1,k)$
$H_z^{n+1/2}(i,j,k)$	equation (2)	equation (2)	equation (2)	equation (2)	equation (2)
$E_x^n(i,j+1,k)$	—	equation (5)	equation (8)	—	equation (5)
$H_y^{n+1/2}(i,j+1,k)$ and $H_y^{n+1/2}(i,j+1,k-1)$	—	equations (6) and (7)	equations (9) and (10)	—	equations (6) and (7)
$H_x^{n+1/2}(i,j,k)$ and $H_x^{n+1/2}(i,j,k-1)$	equations (3) and (4)	equations (3) and (4)	—	equations (3) and (4)	equations (3) and (4)
$E_y^n(i,j+1,k)$	—	—	$E_y^n(i,j,k)$	—	—
$H_x^{n+1/2}(i,j+1,k)$ and $H_x^{n+1/2}(i,j+1,k-1)$	—	—	equations (11) and (12)	—	—

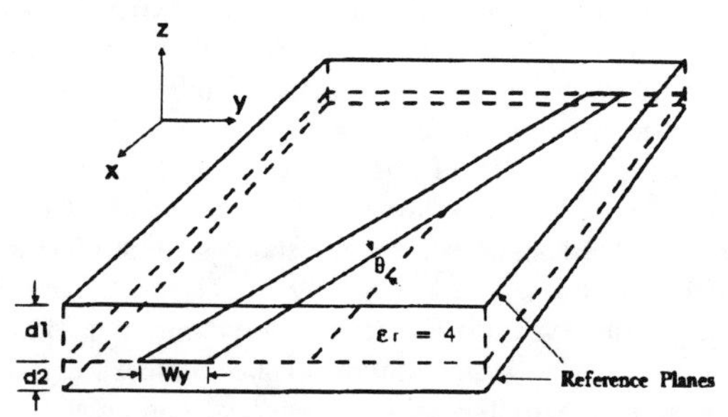

Fig. 8. Structure of a strip line tilted by an angle θ from the x axis. $d1 = 0.6$ mm, $d2 = 0.4$ mm, $Wy = 0.8$ mm.

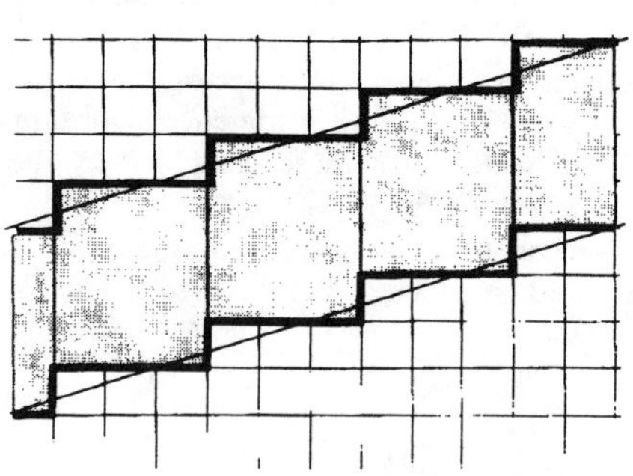

Fig. 9. The staircase approximation of a strip line.

is called the "nearest neighbor approximation" in [8]. With this method applied to the strip line in Fig. 8, the irregular electric field components around the metal strip are approximated by their nearest neighbors of the same field components, and the

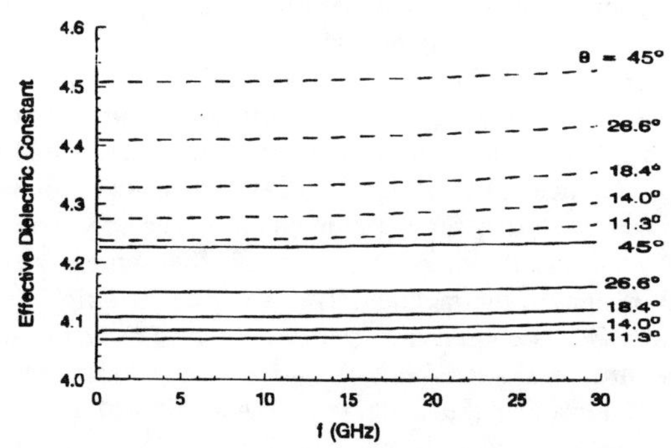

Fig. 10. Effective dielectric constants of a strip line computed by the staircase approximation (dashed curves) and the locally conformed finite-difference method of this paper (solid curves). The strip line is along the direction of an angle θ with the x axis.

irregular magnetic field components are calculated by applying the Faraday's law. The effective dielectric constant $\epsilon_{reff}(f)$ of the strip line computed by the "nearest neighbor approximation" are shown in Fig. 11 together with those obtained by the staircase approximation. It appears that an improvement on computation accuracy can also be made by the "nearest neighbor approximation," but not as much as that by the locally conformed finite-difference method of this paper.

IV. NUMERICAL EXAMPLE

An example to illustrate the application of the locally conformed finite-difference algorithm for modeling microwave circuits is shown in Fig. 12. In this example, two strip lines are

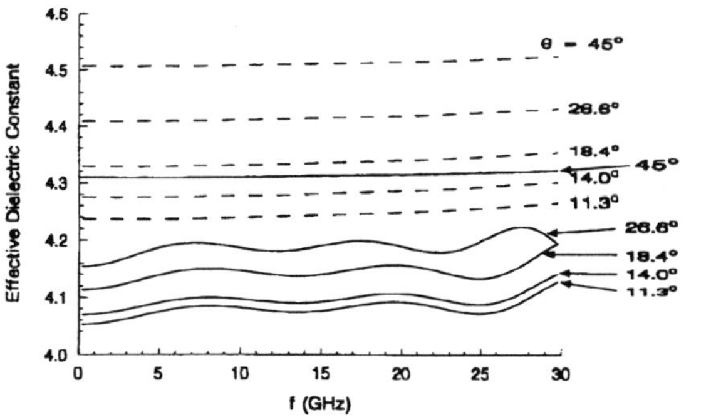

Fig. 11. Effective dielectric constants of a strip line computed by the staircase approximation (dashed curves) and the "nearest neighbor approximation" (solid curves). The strip line is along the direction of an angle θ with the x axis.

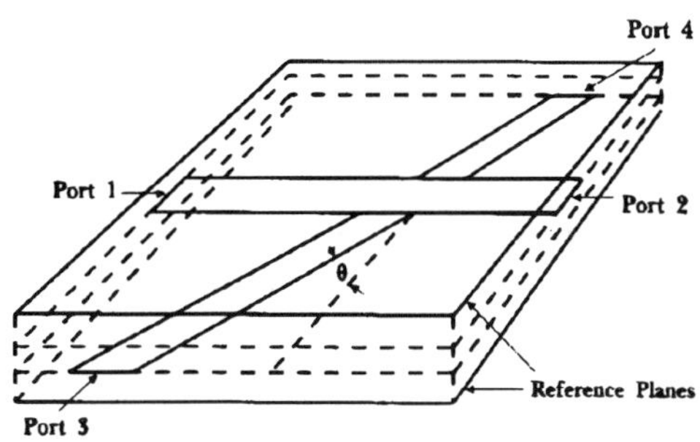

Fig. 12. Structure of two crossed strip lines.

crossed over each other with an angle θ. The distance between two ground planes is 0.9 mm. Three layers of dielectrics between ground planes are all of thickness 0.3 mm and $\epsilon_r = 4$. The space-step dh of the uniform finite-difference mesh is 0.1 mm. The time-step dt is chosen as 0.5 dh/v. Both metal strips are of width $W = 0.4$ mm.

Apparently, for such structures as the ones shown in Fig. 12, the staircase approximation results in very poor resolutions of circuit geometries. By the application of the locally conformed finite-difference method, the numerical analysis of the dependence of the characteristics of the crossed strip lines on various parameters, such as the crossing angle θ and strip-line dimensions, becomes readily feasible.

The electrical properties of the two crossed strip lines are obtained by the numerical simulation of a pulse propagation through the structure. Let a Gaussian pulse propagate into the structure from port 1. The frequency dependent S parameters of the structure are found by taking Fourier transforms of the reflected, transmitted and coupled waves. Fig. 13 shows the coupling coefficients S_{13} and S_{14} in the frequency range $f = 0$ to 15 GHz, as θ changes from 0° to 45° with a 5° interval. It is seen that, as expected, S_{13} increases with increasing θ, while S_{14} decreases with increasing θ. Both S_{13} and S_{14} increase with frequency in the frequency range of Fig. 13. An alternative presentation of S_{13} and S_{14} is shown in Fig. 14, where S_{13} and S_{14} at various frequencies are plotted as a

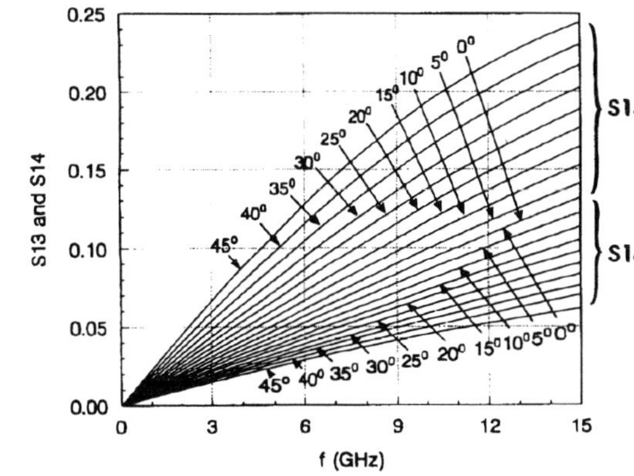

Fig. 13. Coupling coefficients between two crossed strip lines.

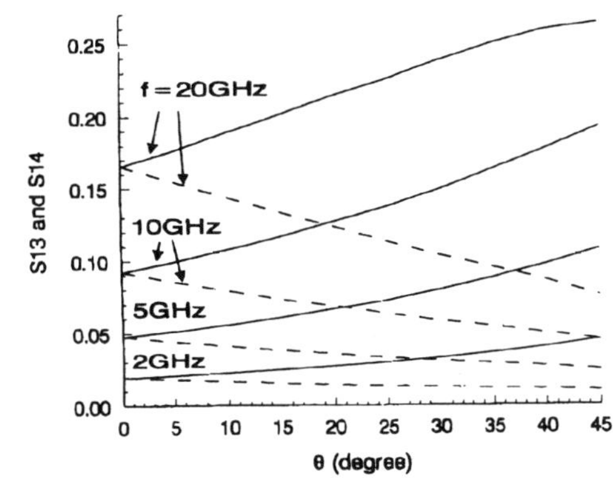

Fig. 14. Coupling coefficients (S_{13} and S_{14}) between two crossed strip lines. Solid curves: S_{13}; Dashed curves: S_{14}.

function of the crossing angle θ.

S parameters obtained above can be post-processed to extract equivalent circuit parameters of the crossed strip-lines. For the equivalent circuit model consisted of capacitors C_c, C_a and mutual inductor M shown in Fig. 15, it can be derived that

$$MG_0 = \frac{S_{13} - S_{14}}{j2\pi f(1 - S_{11} + S_{12})}, \tag{17}$$

where G_0 is the characteristic admittance of the stip line, and

$$C_a Z_0 = \frac{1 - S_{11} - S_{12} - S_{13} - S_{14}}{j2\pi f(S_{12} + S_{14})(1 + j2\pi f M G_0)}, \tag{18}$$

$$C_c Z_0 = \frac{1 - S_{11} - S_{12} - j2\pi f(S_{12} + j2\pi f M G_0 S_{14}) C_a Z_0}{j2\pi f(S_{12} - S_{14})(1 - j2\pi f M G_0)}, \tag{19}$$

where Z_0 is the characteristic impedance of the strip line. Values of $C_c Z_0, C_a Z_0$ and MG_0 as a function of frequency for different crossing angles are displayed in Figs. 16, 17, and Fig. 18 respectively. The equivalent circuit model thus obtained can be used in computer-aided design tools to analyze crosstalk between strip lines with different terminations.

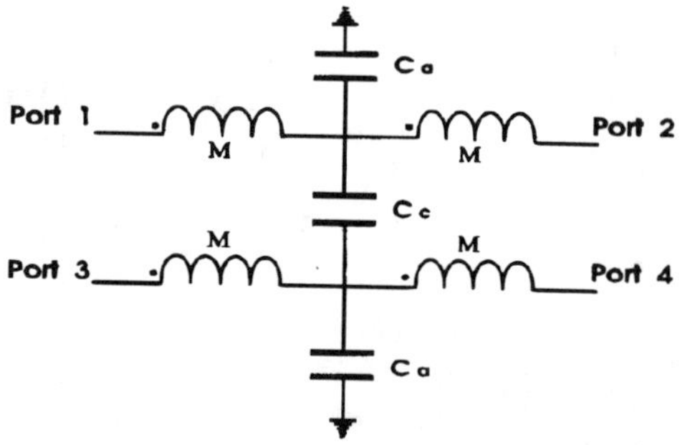

Fig. 15. The equivalent circuit of the crossed strip lines in Fig. 12.

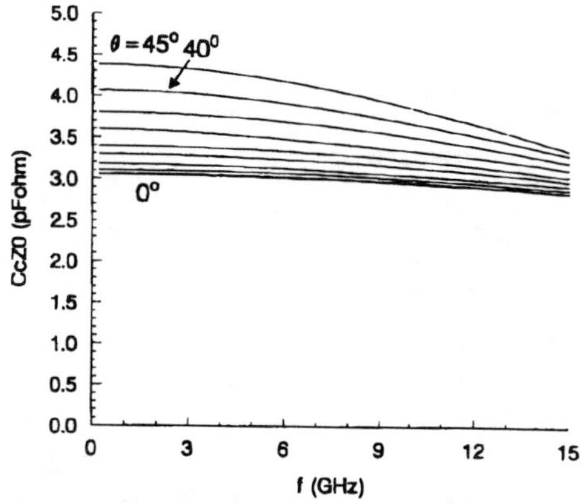

Fig. 16. Normalized equivalent capacitance $C_c Z_0$ of the crossed strip lines in Fig. 12 as a function of frequency. The crossing angle θ changes from 0° to 45° with 5° interval.

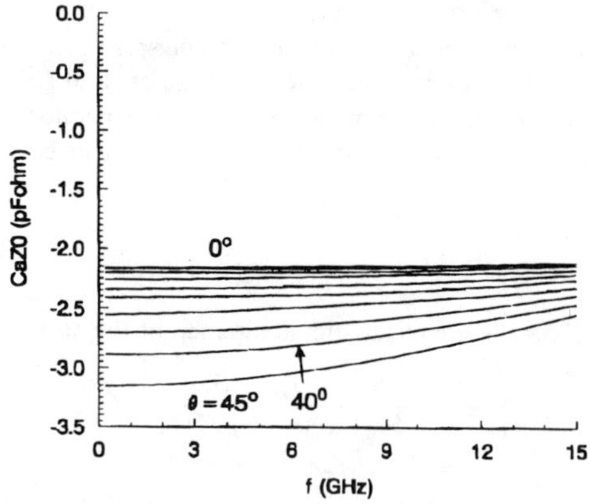

Fig. 17. Normalized equivalent capacitance $C_a Z_0$ of the crossed strip lines in Fig. 12 as a function of frequency. The crossing angle θ changes from 0° to 45° with 5° interval.

Strip-line crossings appear frequently in microwave circuits, but few results are available in present literatures on detailed analysis and modeling of strip-lines crossed with arbitrary angles. Although the computation results in Figs. 13–18 are believed to be highly reliable, they are subject to be compared

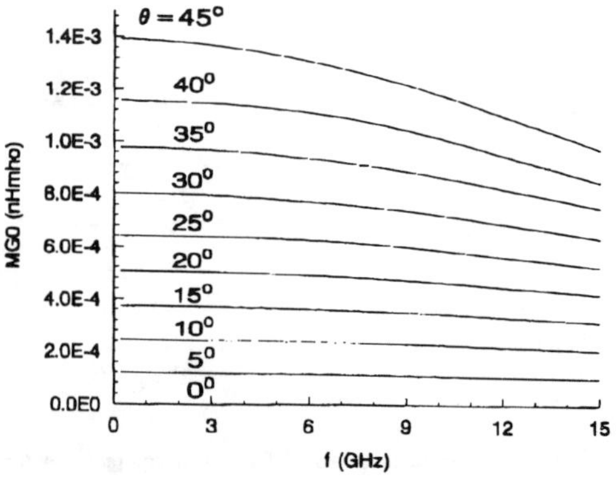

Fig. 18. Normalized equivalent inductance MG_0 of the crossed strip lines in Fig. 12 as a function of frequency. The crossing angle θ changes from 0° to 45° with 5° interval.

with the results obtained by other numerical or experimental techniques.

V. Conclusion

A locally conformed finite-difference scheme is proposed in this paper to model arbitrary shape planar metal strips. This method is computationally more efficient than the globally conformed finite-difference schemes. Besides the high space-resolution of this method, it is also shown to be much more accurate than the staircase approximation. This method can be applied to various microwave planar circuits of complex shapes.

References

[1] K. S. Yee, "Numerical solution of initial boundary value problems involving Maxwell's equations in isotropic media," *IEEE Trans. Antennas Propagat.*, vol. AP-14, pp. 302–307, May 1966.

[2] X. Zhang, J. Fang, K. K. Mei, and Y. Liu, "Calculation of the dispersive characteristics of microstrips by the time-domain finite-difference method," *IEEE Trans. Microwave Theory Tech.*, vol. 36, pp. 263–267, Feb. 1988.

[3] P. A. Tirkas and C. A. Balanis, "Finite-difference time-domain method for antenna radiation," *IEEE Trans. Antennas Propagat.*, vol. 40, pp. 334–340, Mar. 1992.

[4] R. Holland, "Finite-difference solution of Maxwell's equations in generalized nonorthogonal coordinates," *IEEE Trans. Nucl. Sci.*, vol. NS-30, pp. 4589–4591, Dec. 1983.

[5] M. A. Fusco, M. V. Smith, and L. W. Gordon, "A three-dimensional FDTD algorithm in curvilinear coordinates," *IEEE Trans. Antennas Propagat.*, vol. 39, pp. 1463–1471, Oct. 1991.

[6] J. F. Lee, R. Palandech, and R. Mittra, "Modeling three-dimensional discontinuities in waveguides using nonorthogonal FDTD algorithm," *IEEE Trans. Microwave Theory Tech.*, vol. 40, pp. 346–352, Feb. 1992.

[7] A. C. Cangellaris and D. B. Wright, "Analysis of the numerical error caused by the stair-stepped approximation of a conducting boundary in FDTD simulations of electromagnetic phenomena," *IEEE Trans. Antennas Propagat.*, vol. 39, pp. 1518–1525, Oct. 1991.

[8] T. G. Jurgens, A. Taflove, K. Umashankar, and T. G. Moore, "Finite-difference time-domain modeling of curved surfaces," *IEEE Trans. Antennas Propagat.*, vol. 40, pp. 357–366, Apr. 1992.

[9] A. Taflove, K. R. Umashankar, B. B. Beker, F. Harfoush, and K. S. Yee, "Detailed FDTD analysis of electromagnetic Fields penetrating narrow slots and lapped joints in thick conducting screens," *IEEE Trans. Antennas Propagat.*, vol. 36, pp. 247–257, Feb. 1988.

A Vertex-Based Finite-Volume Time-Domain Method for Analyzing Waveguide Discontinuities

C. H. Chan and J. T. Elson

Abstract—**A vertex based finite-volume time-domain method is proposed for analyzing parallel-plate waveguide discontinuities. An irregular grid is chosen so that it can conform linearly to any discontinuity. In this scheme, the electric field vector is located at the vertices of the grid and the magnetic vector is located at the centroids of the elements formed by the grid. Reflection and transmission coefficients are calculated for the TE case. Excellent agreement is obtained for single- and double-step discontinuities when compared with the mode matching method.**

I. Introduction

The SIMPLICITY and usefulness of the finite-difference time-domain method [1] are widely accepted. However, a significant error due to the staircasing effect may occur when a curved boundary exists in the compuational domain. Various schemes have been developed to eradicate this difficulty [2]–[7]. All these methods of discretization can be considered to be edge-based in the sense that the variables are located at the middle of the edges of the grids. In this paper, a vertex based finite-volume time-domain (FVTD) method proposed recently [8], [9] is employed to analyze parallel-plate waveguide discontinuities. The computational domain is discretized with a triangular mesh so that any curved boundary can be approximated by piece-wise linear line segments. The electric field vector is sampled at the vertices of the triangles and the magnetic field vector is sampled at the centroids of the triangles. This method is not limited by the choice of triangulation and is more general than the triangulation presented in [10]. When using the Delaunay–Voronoi property for Delaunay triangulation, the location of the magnetic field in [10] will be equivalent to the present method. This new algorithm is stable even if the triangles are moderately distorted from the equilateral situation.

II. Vertex-Based Finite-Volume Time-Domain Method

To demonstrate how this vertex-based FVTD method works, let us consider the triangular grid depicted in Fig. 1. For the TE case, E_y is sampled at the vertices and both H_x and H_z are sampled at the centroids of the triangles. The electric and magnetic fields are assumed to be constant within the triangle of area A_e and the polygon of area A_h, respectively. The fields along an edge are assumed to be constant obtained from the averaged values of the fields at the two endpoints of the edge. With these assumptions, the FVTD equations given in (1) and (2) reduce to (3) and (4), respectively, for the two-dimensional problem:

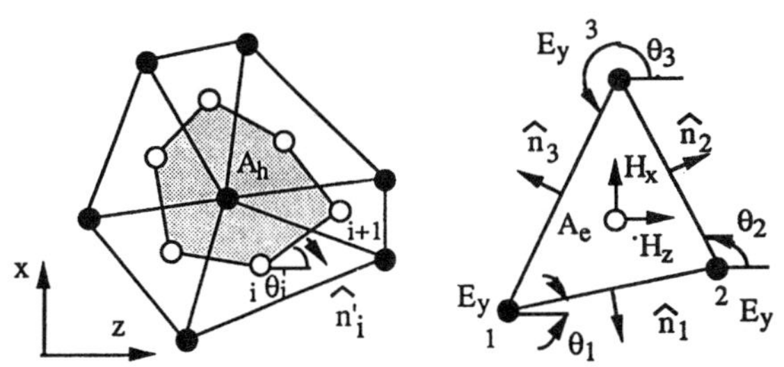

Fig. 1. A vertex-based finite-volume time-domain grid for a two-dimensional problem.

$$\int \nabla \times \vec{E} dv = \int \left(\hat{n} \times \vec{E}\right) ds = -\mu \frac{\partial}{\partial t} \int \vec{H} dv, \quad (1)$$

$$\int \nabla \times \vec{H} dv = \int \left(\hat{n} \times \vec{H}\right) ds = \frac{\partial}{\partial t} \int \vec{E} dv, \quad (2)$$

$$\vec{H}^{n+\frac{1}{2}} = \vec{H}^{n-\frac{1}{2}} - \frac{\Delta t}{\mu A_e} \sum_{i=1}^{3} \hat{n}_i \times \left(\frac{\vec{E}_i^n + \vec{E}_{i+1}^n}{2}\right) L_i, \quad i+1 = 1 \text{ when } i = 3, \quad (3)$$

$$\vec{E}^{n+1} = \vec{E}^n + \frac{\Delta t}{\mu A_h} \sum_{i=1}^{N} \hat{n}'_i \times \left(\frac{\vec{H}_i^{n+\frac{1}{2}} + \vec{H}_{i+1}^{n+\frac{1}{2}}}{2}\right) L'_i, \quad i+1 = 1 \text{ when } i = N, \quad (4)$$

where $\Delta t < 2r/c$. Here, r is the radius of the smallest circle inscribed in the triangles or polygons, c is the velocity of light, and N is the number of sides of the polygon. The lengths of the ith edge of the triangle and polygon are L_i and L'_i, respectively. While a rigorous stability condition has not been derived, the time step chosen here is stable for all the examples shown in this paper.

For the TE case, E_y, H_x, and H_y are substituted into (3) and (4) from which three scalar equations can be obtained as follows:

$$H_x^{n+\frac{1}{2}} = H_x^{n-\frac{1}{2}} + \frac{\Delta t}{\mu A_e} \sum_{i=1}^{3} \frac{E_{yi}^n + E_{yi+1}^n}{2} L_i \sin\theta_i, \quad (5)$$

Manuscript received June 16, 1993. This work was supported in part by NSF Grant ECS-9158040.

The authors are with the Electromagnetics and Remote Sensing Laboratory, Department of Electrical Engineering, University of Washington, Seattle, WA 98195.

IEEE Log Number 9212417.

Reprinted from *IEEE Microwave and Guided Letters*, pp. 372-374, October 1993.

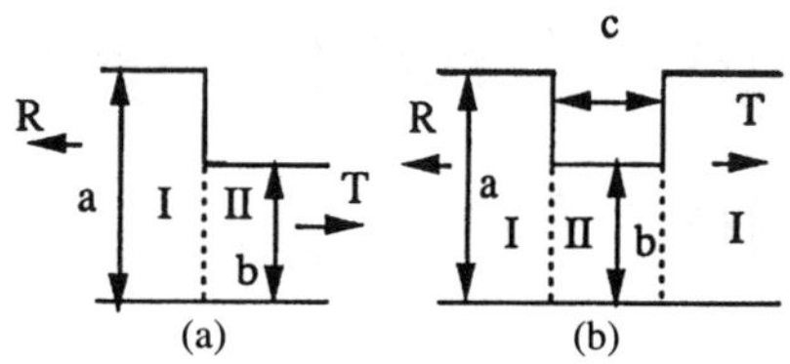

Fig. 2. Step discontinuity. (a) Single step; (b) Double step. $a = 0.8\lambda_o, b, = 0.6\lambda_o$ and $c = 0.2\lambda_o$.

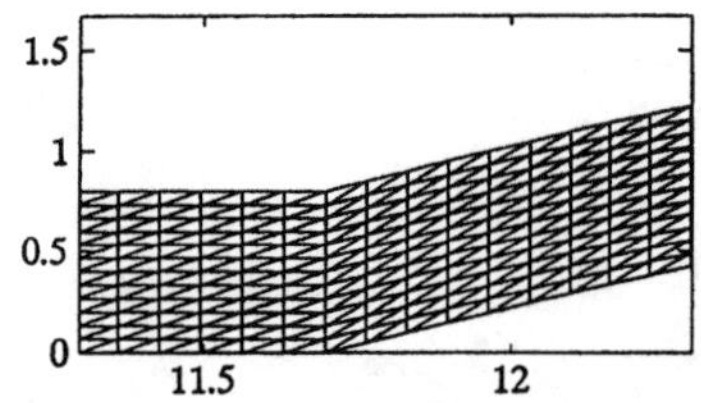

Fig. 3. Triangular mesh for the vertex based finite-volume time-domain grid.

$$H_z^{n+\frac{1}{2}} = H_z^{n-\frac{1}{2}} + \frac{\Delta t}{\mu A_e} \sum_{i=1}^{3} \frac{E_{yi}^n + E_{yi+1}^n}{2} L_i \cos\theta_i, \qquad (6)$$

$$E_y^{n+1} = E_y^n + \frac{\Delta t}{\mu A_h} \sum_{i=1}^{N} \left(\frac{H_{xi}^{n+\frac{1}{2}} + H_{xi+1}^{n+\frac{1}{2}}}{2} \sin\theta_i' + \frac{H_{zi}^{n+\frac{1}{2}} + H_{zi+1}^{n+\frac{1}{2}}}{2} \cos\theta_i' \right) L_i'. \qquad (7)$$

Time stepping these three equations yields the time-domain response of parallel-plate waveguides. The second-order Liao absorbing boundary condition [11] is enforced for the total and scattered E_y field at the terminal plane and the source plane, respectively. A detailed discussion on the application of the absorbing boundary condition on the scattered field in the total field formulation can be found in [12].

III. Numerical Results

To establish the accuracy of this new method, we calculate the reflection and transmission coefficients of single- and double-step parallel plate waveguide discontinuities (see Fig. 2) and compare the results with those obtained from the mode matching method. Excellent agreement is shown in Table I when compared to the mode matching method using 3 modes on either side of the discontinuity. A discretization of 15 points per free-space wavelength (λ_o) is chosen. In Table I, Z_a and Z_b are the characteristic impedances at Regions I and II, respectively. Next, we consider a discontinuity configuration that cannot be analyzed by the mode matching method. Fig. 3 shows a portion of the triangular discretization of the discontinuity which is described by a sine function. The scales shown in Fig. 3 are in terms of wavelengths. The contour plots of the electric field at various time steps within half a period are shown in Figures 4a to 4d. Here the scales are in terms of discretization $\Delta x = \Delta y = 0.067\lambda_o$. The magnitudes of the reflection and transmission coefficients are 0.102 and 0.993, respectively, giving a power check of 0.997. Although not implemented in this paper, this vertex based FVTD can be applied to the locally distorted grid in the vincinity of the discontinuity and blended smoothly into a rectangular grid for the conventional FDTD calculation in the region away from the arbitrary discontinuity.

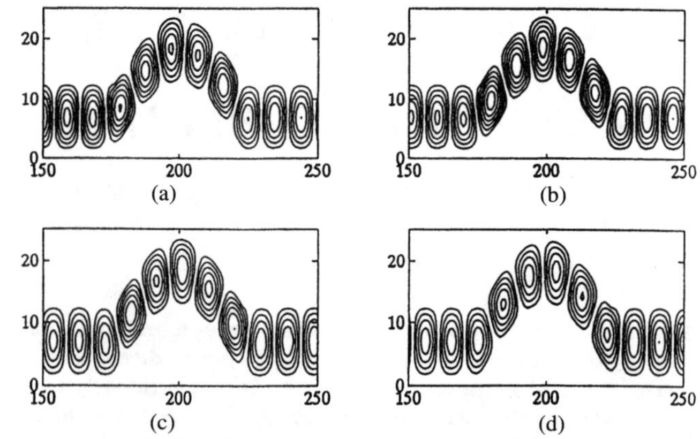

Fig. 4. Contour plots of fields propagating through a waveguide discontinuity described by a sine function. (a) At t; (b) At $t + T/8$; (c) At $t + T/4$; (d) At $t + 3T/8$.

TABLE I
Comparison of the Present Method and the Mode Matching Method
$Z_a = 482.95$ ohms and $Z_b = 682.02$ ohms

	$\|R\|$	$\|T\|$	Power Check
Fig. 2(a)			$\|R\|^2 + \|T\|^2(bZ_a)/(aZ_b)$
Mode Matching	0.156	1.355	1.000
FVTD	0.157	1.343	0.983
Fig. 2(b)			$\|R\|^2 + \|T\|^2$
Mode Matching	0.298	0.955	1.000
FVTD	0.308	0.948	0.997

IV. Conclusion

We have presented a vertex based FVTD method for analyzing arbitrary plate waveguide discontinuities. The primary grid is a triangular grid while the complementary grid consists of polygons formed by connecting the centroids of the triangles of the primary grid. Excellent agreement in the reflection and transmission coefficients is obtained when compared to the mode matching method for step discontinuities. The flexibility of the method in conforming to the shape of the waveguide discontinuity is demonstrated via the example of a sine function. A study on the extension of the method for treating three-dimensional problems is currently underway.

Acknowledgment

The authors would like to acknowlege Dr. K. S. Yee for his helpful discussions.

References

[1] K. S. Yee, "Numerical solution of initial boundary value problems involving Maxwell's equations in isotropic media," *IEEE Trans. Antennas Propagat.*, vol. AP-14, no. 3, May 1966.

[2] R. Holland, "Finite-difference solution of Maxwell's equations in generalized nonorthogonal coordinates," *IEEE Trans. Nucl. Sci.*, vol. NS-30, no. 6, Dec. 1983.

[3] K. S. Yee, "Numerical solution to Maxwell's equations with nonorthogonal grids," presented at the *Proc. 1st Review of Numerical Electromagnetic Code*, Mar. 1985.

[4] N.K. Madsen and R.W. Ziolkowski, "A three-dimensional modified finite volume technique for Maxwell's equations," *Electromagnetics*, vol. 10, pp. 147–161, Jan. 1990.

[5] M. Fusco, "FDTD algorithm in curvilinear coordinates," *IEEE Trans. Antennas Propagat.*, vol. 38, pp. 76–89, Jan. 1990.

[6] J.F. Lee, R. Palandech, and R. Mittra, "Modeling three-dimensional discontinuities in waveguides using nonorthogonal FDTD algorithm," *IEEE Trans. Microwave Theory Tech.*, vol. 40, pp. 346–352, Feb. 1992.

[7] K.S. Yee, J.S. Chen, and A.H. Chang, "Conformal finite-difference time-domain (FDTD) with overlapping grids," *IEEE Trans. Antennas Propagat.*, vol 40, pp. 1068–1075, Sept. 1992.

[8] K.S. Yee, C.H. Chan, and J.S. Chen, "Conformal hybrid finite difference time domain and finite volume time domain," in *Proc. Progress in Electromagn. Res. Symp.*, Pasadena, CA, pp. 603, June 1993.

[9] K. S. Yee and J. S. Chen, "Conformal hyrid finite difference time domain and finite volume time domain," submitted to *IEEE Trans. Antennas and Propagat.*

[10] C.F. Lee, R.T. Shin, J.A. Kong, and B. J. McCartin, "Finite difference time domain techniques for two dimensional triangular grids," in *Proc. Progress in Electromagn. Res. Symp.*, pp. 189–190, July 1989.

[11] W.C. Chew, *Waves and Fields in Inhomogeneous Media.* New York: Van Nostrand Reinhold, 1990.

[12] K. Umashankar and A. Taflove, "A novel method to analyze electromagnetic scattering of complex objects," *IEEE Trans. Electromagn. Compatibility*, vol. EMC-24, pp. 397–405, Nov. 1982.

WETD—A Finite Element Time–Domain Approach for Solving Maxwell's Equations

Jin-Fa Lee, *Member, IEEE*

***Abstract*— A family of finite element time-domain methods, WETD(Θ), is derived to solve the time-varying Maxwell's equations. The proposed methodology is based upon the application of the Faedo-Galerkin procedure and the use of the Whitney 1-forms as bases to result in an ordinary differential equation in time for the electric field. Moreover, the resultant ordinary differential equation is solved by employing central and/or backward difference approximations. Since the WETD methods presented here are used in conjunction with tetrahedral finite element meshes, it imposes no limitations on the problem geometry. Also, in this contribution, a general stability condition has been derived for the WETD(Θ) method of which the central and backward differences are special cases corresponding to Θ = 1 and Θ = 0, respectively.**

I. Introduction

The finite difference time-domain (FDTD) [1] algorithm has been used widely in solving the transient responses of electromagnetic problems. However, in its original form, it is difficult to model complex EM problems with curved surfaces using the FDTD method. Many variants have been proposed in the past with the aim to circumvent this difficulty with varying degrees of success. Almost all of these approaches are based upon, one form or the other, the use of finite difference approximation in both spatial and temporal domains. It is the purpose of this paper to show a finite element time domain formulation, Whitney element time domain (WETD) method, which uses Whitney 1-forms in the spatial domain and the finite difference in the time domain, respectively, for solving Maxwell's equations. In this way, the proposed WETD method can be used on a tetrahedral finite element mesh and consequently, it imposes no geometric limitations. Furthermore, in this contribution, we also generalize the formulation by the Θ method, will be described later, to a family of WETD methods depending on the Θ value. Therefore, the time step can always be set with compatible resolution with its spatial counterparts, i.e. $c\delta t \approx \Delta x$, by choosing a suitable Θ.

II. Faedo–Galerkin Formulation

In this paper, we consider the solution of Maxwell's equations in space-time R^4

$$\begin{aligned}\nabla \times \vec{E} &= -\mu\frac{\partial \vec{H}}{\partial t}\\ \nabla \times \vec{H} &= \vec{J} + \epsilon\frac{\partial \vec{E}}{\partial t}.\end{aligned} \tag{1}$$

Manuscript received August 24, 1993.
This work was supported by Engineering Foundation Grant RI-A-92-10.
The author is with the ECE Department, Worcester Polytechnic Institute, Worcester, MA 01609.
IEEE Log Number 9214843.

From these equations, an initial value problem can be derived in terms of the electric field $\vec{E}$ as

$$\begin{aligned}\nabla \times \frac{1}{\mu}\nabla \times \vec{E} + \epsilon\frac{\partial^2 \vec{E}}{\partial t^2} &= -\frac{\partial \vec{J}}{\partial t} \quad in\ \Omega\\ \hat{n} \times \vec{E} &= 0 \quad on\ \Gamma_e\\ \hat{n} \times \nabla \times \vec{E} &= 0 \quad on\ \Gamma_h.\end{aligned} \tag{2}$$

The weak form, Galerkin form, of (2) is just

$$\int_\Omega \left(\vec{v} \bullet \epsilon\frac{\partial^2 \vec{E}}{\partial t^2} + \vec{v} \bullet \nabla \times \frac{1}{\mu}\nabla \times \vec{E}\right) d\Omega = -\int_\Omega \vec{v} \bullet \frac{\partial \vec{J}}{\partial t} d\Omega, \tag{3}$$

where $\vec{v}$ is a test function. To achieve a greater symmetry between trial and test vector functions, we apply vector identities to result in

$$\int_\Omega \left[\vec{v} \bullet \epsilon\frac{\partial^2 \vec{E}}{\partial t^2} + \frac{1}{\mu}(\nabla \times \vec{v}) \bullet \left(\nabla \times \vec{E}\right)\right] d\Omega = -\int_\Omega \vec{v} \bullet \frac{\partial \vec{J}}{\partial t} d\Omega. \tag{4}$$

Consequently, the application of the Faedo-Galerkin procedure can be stated as following. Given an N-dimensional subspace $S^h \subset \mathcal{H}^1$, where $\mathcal{H}^1 = \{\vec{v} \mid \vec{v}, \partial\vec{v} \in (L^2)^3\}$, the Galerkin principle is to find a vector function $\vec{E}^h(\vec{r};t)$ with the following property: At each $t > 0$, $\vec{E}^h$ lies in S^h and satisfies

$$\int_\Omega \left[\vec{v}^h \bullet \epsilon\frac{\partial^2 \vec{E}^h}{\partial t^2} + \frac{1}{\mu}(\nabla \times \vec{v}^h) \bullet \left(\nabla \times \vec{E}^h\right)\right] d\Omega + \int_\Omega \vec{v}^h \bullet \frac{\partial \vec{J}^h}{\partial t} d\Omega = 0, \qquad \forall \vec{v}^h \in S^h. \tag{5}$$

Notice that the time variable is still continuous: The Faedo-Galerkin formulation is discrete in the spatial variables and yields a system of ordinary differential equations in time. To make this formulation operational, we use the Whitney 1-forms as the bases for the trial space S^h and expand the trial and test vector functions as

$$\vec{E}^h(\vec{r};t) = \vec{W}_i(\vec{r})\, e^i(t) \tag{6}$$

where $e^i, \vec{W}_i$ are the circulation of the electric field and the vector basis function associated with edge i. Also, in (6) we have used the Einstein notation for the summation over index i.

Reprinted from *IEEE Microwave and Guided Letters,* pp. 11-13, January 1994.

Finally, the optimal weights e^i are determined by the Galerkin principle

$$\int_\Omega \left(\epsilon \vec{W}^j \bullet \vec{W}_i\right) d\Omega \ \frac{d^2 e^i}{dt^2} + \int_\Omega \left(\frac{1}{\mu}\nabla\times\vec{W}^j \bullet \nabla\times\vec{W}_i\right) d\Omega \ e^i$$

$$= -\int_\Omega \left(\vec{W}^j \bullet \frac{\partial \vec{J}^h}{\partial t}\right) d\Omega, \qquad \forall j. \tag{7}$$

Now we put the Galerkin equation into vector notation, with $\mathcal{E}$ as the coefficient vector, the result is

$$[T]\frac{d^2\mathcal{E}}{dt^2} + c^2[S]\mathcal{E} = -\mathcal{F} \tag{8}$$

where

$$[T]_{ij} = \int_\Omega \left(\epsilon_r \vec{W}^j \bullet \vec{W}_i\right) d\Omega$$

$$[S]_{ij} = \int_\Omega \left(\frac{1}{\mu_r}\nabla\times\vec{W}^j \bullet \nabla\times\vec{W}_i\right) d\Omega \tag{9}$$

and the components of the right-hand-side are

$$\mathcal{F}_j = \int_\Omega \vec{W}^j \bullet \frac{\partial \vec{J}^h}{\partial t} d\Omega. \tag{10}$$

If there were inhomogeneous boundary conditions, time-dependent or not, their effects would also appear in $\mathcal{F}$.

III. Finite Differences in Time

The ordinary differential equation in (8) typically can be approximated by finite difference formulae in three ways: forward difference, backward difference, and central difference. The forward difference formula when applies to (8) becomes unconditionally unstable in the numerical implementation. Therefore, in this contribution, we will focus only on the central, backward differences and their combinations. Furthermore, to simplify the discussion here, we shall also assume that $\mathcal{F} = 0$, however, the inclusion of $\mathcal{F}$ in the final formulation is straightforward and will not affect the main concepts in this contribution.

A. Central Difference

By applying the central difference to the time derivative in (8), we obtain

$$[T]\frac{1}{\delta t^2}(\mathcal{E}^{n+1} - 2\mathcal{E}^n + \mathcal{E}^{n-1}) + c^2[S]\mathcal{E}^n = 0. \tag{11}$$

The above equation can be rearranged to result in a matrix equation which can be used to update the coefficient vector, $\mathcal{E}$, as

$$[T]\mathcal{E}^{n+1} = -[T]\mathcal{E}^{n-1} + (2[T] - c^2\delta t^2[S])\mathcal{E}^n. \tag{12}$$

As evidenced in (12), the updating of the electric field, for each time step, requires solving a matrix equation of the form:

$$[T]x = y. \tag{13}$$

The stability condition of this approach will be described later as a special case in the Θ method with $\Theta = 1$.

B. Backward Difference

When the O.D.E. (8) is approximated by backward difference, the following equation is obtained:

$$[T]\frac{1}{\delta t^2}(\mathcal{E}^{n+1} - 2\mathcal{E}^n + \mathcal{E}^{n-1}) + c^2[S]\mathcal{E}^{n+1} = 0. \tag{14}$$

Finally, the matrix equation that is used to update the coefficient vector, $\mathcal{E}^{n+1}$, is written as

$$\left([T] + c^2\delta t^2[S]\right)\mathcal{E}^{n+1} = 2[T]\mathcal{E}^n - [T]\mathcal{E}^{n-1}. \tag{15}$$

Like the central difference approach, for each time step the backward difference approach also requires solving a matrix equation of the form

$$\left([T] + c^2\delta t^2[S]\right)x = y. \tag{16}$$

However, as can be seen later, the algorithm results in an unconditional-stable time marching numerical scheme.

C. Θ Method

It is well-known that the central difference will give second order accuracy in time, and whereas, the backward difference will only be accurate upto first order. A general finite difference scheme that we proposed here is based upon a concept which is similar to the Generalized Θmethod [3] for solving parabolic equations. We multiply (12 by a real number, $0 \le \Theta \le 1$, and (15) by $1 - \Theta$, and sum the two equations together to form

$$\{[T] + (1-\Theta)c^2\delta t^2[S]\}\mathcal{E}^{n+1} =$$

$$-[T]\mathcal{E}^{n-1} + \{2[T] - \Theta c^2\delta t^2[S]\}\mathcal{E}^n. \tag{17}$$

From (17), we see that $\Theta = 0$ and $\Theta = 1$ reduce to backward and central difference schemes, respectively.

D. Stability Condition

Since the central and backward differences are special cases of the Θ method, we will only derive the stability condition for the Θ method in this section. By defining a growing factor α as

$$\alpha = \frac{\| \mathcal{E}^{n+1} \|}{\| \mathcal{E}^n \|}. \tag{18}$$

Then we say that a numerical scheme is stable if and only if

$$\lim_{n\to\infty} \alpha \le 1. \tag{19}$$

From (17), it can be shown that the Θ method will be stable if and only if the following condition holds

$$\rho\left(\{2[T] - \Theta c^2\delta t^2[S]\}^2\right) \le 4\rho\left([T]\{[T] + (1-\Theta)c^2\delta t^2[S]\}\right) \tag{20}$$

where $\rho(A)$ is the spectral radius of the matrix A. Finally, with algebraic operations, the final stability condition can be deduced as

$$c\delta t \le \frac{2}{\Theta}\sqrt{\frac{\lambda^T_{\min}}{\lambda^S_{\max}}} \tag{21}$$

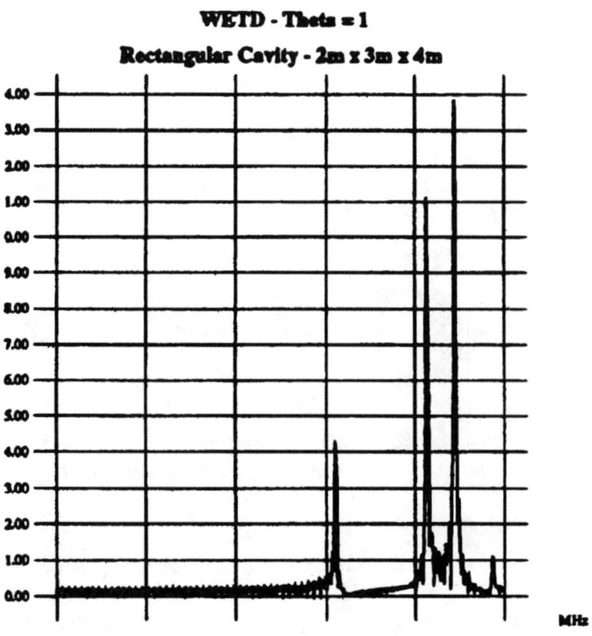

Fig. 1. The FFT of the time signal for a rectangular cavity with dimensions 2 m x 3 m x 4 m.

where $\lambda_{\min}^{T}$ and $\lambda_{\max}^{S}$ are the minimum and the maximum eigenvalues of matrices $[T]$ and $[S]$, respectively. Moreover, we have

$$\frac{\lambda_{\min}^{T}}{\lambda_{\max}^{S}} = \frac{1}{\kappa_{\max}} \tag{22}$$

where κ is the maximum eigenvalue of the following generalized eigenmatrix equation:

$$[S]x = \kappa[T]x. \tag{23}$$

In practice, (23) can be solved efficiently by using Lanczos algorithm [4].

TABLE I
RESONANT FREQUENCIES OF THE FIRST THREE MODES

Resonant frequencies (MHz)	
Exact	WETD(1.0)
62.5	62.07 407
83.853	82.59 259
90.139	88.74 074

IV. Numerical Results

A simple rectangular cavity with dimensions 2 m x 3 m x 4 m has been studied by using the WETD method presented in this contribution. The cavity is first discretized into tetrahedra with typical element size corresponds to $h = 0.25$ m. The numerical algorithm starts by assigning a divergence-free excitation with a Gaussian distribution of the form $\exp -((t-5\sigma)/\sigma)^2$ for the electric field. The constant σ is chosen as $\sigma = 1/(\pi f)$ and $f = 500$ MHz. The numerical experiment has been conducted using $\Theta = 1$, and time step is calculated as

$$c\delta t = \frac{1.9}{\Theta}\sqrt{\frac{\lambda_{\min}^{T}}{\lambda_{\max}^{S}}} = 0.06209. \tag{24}$$

It took 12 035 CPU time to run for 10 000 time steps on a HP 735 workstation. The FFT of the time signal obtained at a certain observation point is shown in Fig. 1. The resonant frequencies of the first three modes compared to the exact solutions is given in Table I. Good agreements are obtained.

References

[1] K. S. Yee, "Numerical solution of initial boundary value problems involving Maxwell's equations in isotropic media," *IEEE Trans. Antennas Propagat.*, vol. AP-14, pp. 302–307, 1966.
[2] G. Strang and G. J. Fix, *An Analysis of the Finite Element Method.* Englewood Cliffs, NJ: Prentice Hall, 1973.
[3] O. C. Zienkiewicz and K. Morgan, *Finite Elements and Approximation.* New York: Wiley, 1983.
[4] B. N. Parlett, *The Symmetric Eigenvalue Problem.* Englewood Cliffs, NJ: Prentice-Hall, 1980.

Speed-Up Methods for the FDTD Algorithm

BIJAN HOUSHMAND
JET PROPULSION LABORATORY
CALIFORNIA INSTITUTE OF TECHNOLOGY
4800 OAK GROVE DRIVE
PASADENA, CA 91109-8099

TATSUO ITOH, FELLOW, IEEE
DEPARTMENT OF ELECTRICAL ENGINEERING
UNIVERSITY OF CALIFORNIA AT LOS ANGELES
LOS ANGELES, CA 90024-1594

I. Introduction

In this chapter we will consider the methods for reducing the computational requirements of the FDTD method. The computation requirements of the FDTD algorithm is large both in terms of memory allocations and computation (CPU) time. [1] and [2] illustrate the computational requirements for a number of microwave structures such as couplers, filters, and microstrip-fed patch antennas. A moderate size structure can require a computation volume of the order $(100 \times 100 \times 100)$ grid points in the three dimensional Cartesian coordinate system. In addition, the number of time iterations can be as large as 20,000 steps. The requirement for large memory allocation arises from the need to represent the entire structure of interest in the FDTD computation volume. On the other hand, large computation time is due to the time-marching nature of the FDTD algorithm. That is, in order to account for the interaction of the electromagnetic wave with the microwave structure, the wave has to transverse the entire structure. The relationship between the spatial and temporal discretizations is governed by the stability criteria which is described in Chapter 2. The computational cost for updating the field values over all the grid points for one time step is of the order N, where N is the total number of grid points used to represent the structure. It is of interest, therefore, to reduce the number of grid points in the computation volume. The computation time also increases linearly with the number iterations in time. Therefore another venue to speed up the FDTD algorithm is to extract the required information in a shorter time interval.

Reduction of the computation requirements in space and time is carried out by applying the Diakoptics and SI concepts to the FDTD algorithm. The reduction in computation volume is achieved by representing a portion of the structure by its impulse response. The block diagrams shown in Figs. 1 and 2 illustrate these methods. The reduction of the computation volume is directly related to the issue of the ABC. The ABC essentially simulates an infinite open space, or matched condition where no reflection should occur. A number of effective ABC conditions are available for such a purpose [3–5]. Here the Diakoptics method provides an absorbing boundary condition where the interaction of the microwave structure in the FDTD computation volume with the unmatched external structure is taken into account. The reduction in the simulation time is achieved by extracting the desired information such as S-parameters and resonant frequencies from a short time interval of waveforms which are generated by the FDTD method. System identification algorithms are applied to the FDTD waveforms for this purpose. These algorithms, in effect, extrapolate the time domain signals without using the computationally expensive FDTD algorithm.

This chapter is organized as follows. Section II introduces the application of the system identification method to the FDTD algorithm. This method is used to compute the resonant frequencies and the S-parameters for microwave structures. Section III is an overview of the FDTD Diakoptics method. The application of this method to circuit structures and closed waveguides are illustrated. An efficient method to implement the Diakoptics method using the system identification methodology is also demonstrated. The numerical behavior of both methods are illustrated in Sections II and III. Concluding remarks are in Section IV.

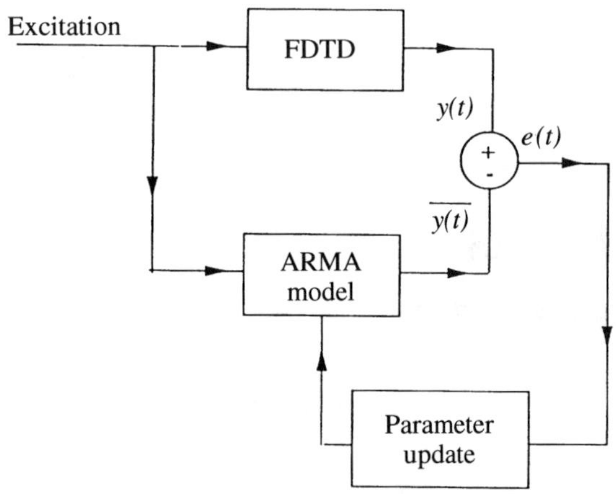

Fig. 1 Block diagram of the application of system identification (SI) to the FDTD method. The parameters of a linear system, Eq. (3), are computed such that the output of the linear system matches the FDTD-generated signal.

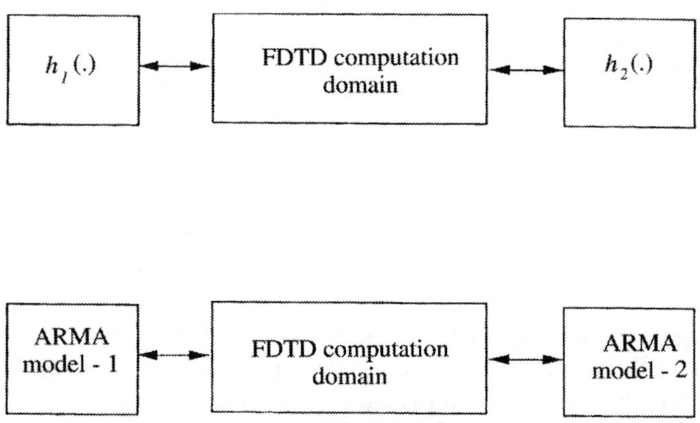

Fig. 2 Block diagram of the Diakoptics method. (a) Segments of the computation domain are represented by their impulse responses. The interaction of these segments with the FDTD computation is carried out by a convolution operation. (b) The impulse response's segment can be represented by an ARMA model. This approach results in an efficient implementation of the convolution operation.

II. Application of the System Identification to the FDTD Algorithm

Time-domain simulation of the microwave structures are used to extract wide band frequency domain information. The transformation of the time-domain information to frequency domain has traditionally been performed by the use of the Fourier Transformation. This method is computationally efficient due to the well known Fast Fourier Transformation (FFT) algorithm. Using the FFT algorithm, a frequency domain spectrum is obtained for a discrete number of frequency points. The duration of the time signal as well as the sampling frequency are important parameters for accurate time-frequency transformation. The relationship between the sampling in the time and frequency domains is well known and is repeated here for reference. For a time signal consisting of N samples, the frequency resolution is given by

$$\Delta f = f_s/N \tag{1}$$

where f_s is the sampling frequency which satisfies the Nyquist sampling theorem and is given by

$$f_s = 1/T_s \tag{2}$$

where T_s is the sampling interval and is typically larger than the sampling interval used for the FDTD updating algorithm. The duration of the time signal used for frequency transformation also affects the resulting spectrum. It is well known that the truncation of the time signal distorts the frequency spectrum of the time window used for the truncation [6]. This is an important consideration for resonant structures where the time waveform due to a short pulse interacts with the structure for a long time interval. The above relationships state that a high resolution spectrum which is free from the windowing effects requires a large number of FDTD time update iterations which is a costly CPU requirement. In order to reduce this computational requirement, the transformation of the time information to the frequency domain is carried out in an indirect way. This transformation is performed by application of the system identification techniques to the FDTD algorithm.

Theory

A number of digital signal processing methods can be used to reduce the computational requirements of the time-domain methods [7–14]. In this section, a projection system identification algorithm for a deterministic Auto-Regressive Moving Average (ARMA) model is applied to the FDTD algorithm [14]. Savings in the computation requirements are achieved by replacing the computationally intensive FDTD algorithm by the ARMA model for output signal computation, after the system parameters converge to their final values.

The computed time signal at an appropriate location in the computational volume and the corresponding input signal can be interpreted as the input and output signals of a discrete linear system. Figure 1 illustrates the application of system identification in the FDTD computations. The goal of this procedure is to match a linear system to the input/output waveforms generated by the FDTD algorithm. The linear system is

$$y(n) = -\sum_{k=1}^{K} a_k y(n-k) + \sum_{m=0}^{M} b_m x(n-m) \tag{3}$$

The output signal is completely known when the model parameters (a_k, b_m) are computed. The parameter space is taken to be large enough to allow the convergence of the model output to the FDTD simulated field values. Equation (3) can be written as a vector inner product

$$y(n) = \Phi(n-1)\Theta_0$$

where Φ is a vector containing the present and past values of the input and output, and it can be considered as data. The vector Θ_0 contains the system parameters and uniquely defines the properties of the linear system such as the resonance frequencies. Θ_0 is estimated in terms of the available data vector Φ. The estimated output is related to the estimated system parameters:

$$\hat{y}(n) = \Phi(n-1)\hat{\theta}(n-1) \tag{5}$$

The difference in Eqs. (4) and (5) is minimized with respect to the system parameters (a_k, b_m) to arrive at a parameter update law. The general parameter update law is of the form

$$\hat{\Theta}(n) = \hat{\Theta}(n-1) + M(n-1)\Phi(n-1)[y(n) - \hat{y}(n)] \quad (6)$$

where the updated system parameter vector is computed from its previous value and a term which is proportional to the difference between the estimated and the FDTD-computed output. The variable M is a gain term and can be a matrix. The specific form for the variable M is dependent on the algorithm which is used to compute the system parameters (a_k, b_m). The following is an orthogonal projection algorithm [13]:

$$\hat{\Theta}(n) = \hat{\Theta}(n-1) + \frac{P(n-1)\Phi(n-1)}{\Phi(n-1)^T P(n-1)\Phi(n-1)}[e(n)] \quad (7)$$

where

$$P(n) = P(n-1) - \frac{P(n-1)\Phi(n-1)\Phi(n-1)^T P(n-1)}{\Phi(n-1)^T P(n-1)\Phi(n-1)} \quad (8)$$

$P(n)$ provides an orthogonal projection search in the parameter space which results in rapid parameter convergence [13], where $\Theta(n)$ is the computed parameter vector, and where $e(n)$ is the discrepancy between the model output and the FDTD computed field value. The variable M for this algorithm can be identified by a comparison of Eqs. (6) and (7). Computation of Eqs. (7) and (8) requires only vector addition and multiplication, and it results in minimal additional cost to the FDTD computation. We note that the system parameters converge to their final values when the output error is sufficiently small. The frequency spectrum of the linear system described by Eq. (3) is obtained analytically using the system coefficients computed by Eqs. (7) and (8). The discrete-time Fourier transform of the linear system is

$$H(\Omega) = \frac{Y(\Omega)}{X(\Omega)} = \frac{\sum_{m=0}^{M} b_m e^{-jm\Omega}}{1 + \sum_{k=1}^{K} a_k e^{-jk\Omega}} \quad (9)$$

The discrete-time spectrum is related to the continuous-time spectrum of the following relationship:

$$f = \Omega/(2\pi T_s) \quad (10)$$

where T_s is the sampling frequency. In the above methodology the Fourier Transformation operation is replaced by the analytical expression given by Eq. (9).

Numerical Examples

The numerical behavior of this method is demonstrated by applying it to the cavity problem. An ARMA model with system parameters $K = 40$, $M = 40$ [see Eq. (2)] is used to obtain the resonance frequencies of a rectangular cavity. The cavity is excited at the center plane by imposing a TE_{10} mode distribution with impulsive temporal dependence. Figure 3 shows the parameter convergence for a number of system coefficients. The initial condition for the parameters is set to the origin of the parameter space, and the parameter values are updated at each sampling interval.

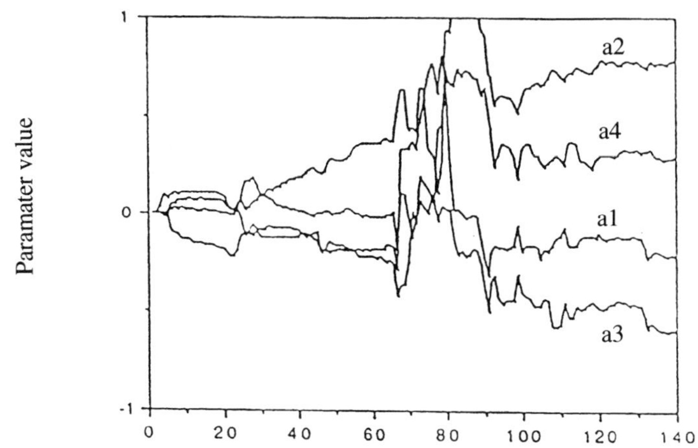

Fig. 3 Parameter convergence of the ARMA model with 80 coefficients. The evolution of the first four output coefficients is demonstrated.

The resonance frequencies of the cavity can be derived directly from the poles of the ARMA model. They can also be recovered from the spectrum of the output signal. This spectrum is computed from the Fourier Transform of the output signal, or directly by evaluation of the Z-Transform of Eq. (3) on the unit circle which is defined in terms of the system parameters. Figure 4 shows the spectrum of the output signal using the system parameters and the Fourier Transformation of the output signal. The recovered resonance frequencies of the first three odd modes are illustrated. The location of the observation point coincides with the null position of the even modes; as the result these modes cannot be recovered from this time signal. In this example, the ARMA based spectrum is computed using 140

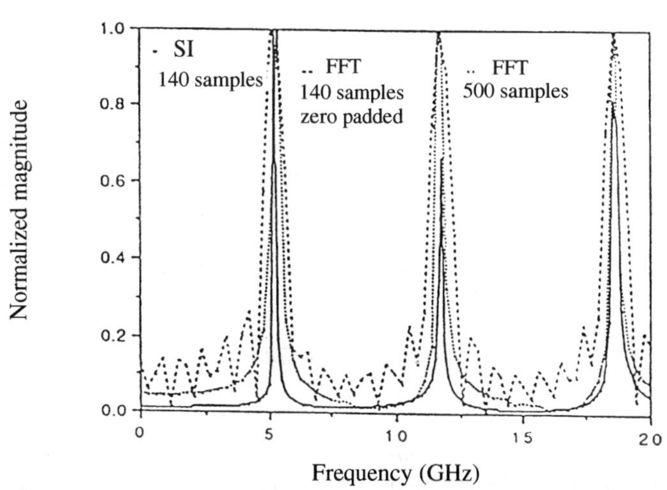

Fig. 4 FDTD generated spectrum. 140 samples (at a rate of 1 sample per 5 FDTD time steps) are used to obtain the spectrum by the system identification method. Similar spectrum is obtained by Fourier transformation of 500 samples. The Fourier transform of 140 samples with zero padding is distorted.

output samples. A similar spectrum is obtained by applying the Fourier Transformation to 500 output samples. Figure 4 also shows the Fourier Transform of the 140 output samples which are augmented with zero padding to provide sufficient spectral resolution for locating the spectrum peaks. This spectrum, while qualitatively locating the resonance frequencies, is distorted and might not provide sufficient resolution where the resonance frequency separation is small. We note that the first two resonance frequencies are predicted accurately for this example, while higher order modes are underestimated due to frequency dispersion of the FDTD spatial grid. This method is used to obtain the resonant frequencies of a partially filled rectangular cavity. Figure 5 shows the shift in the resonant frequency as the permittivity is gradually changed from $\varepsilon = 1$ to $\varepsilon = 2$.

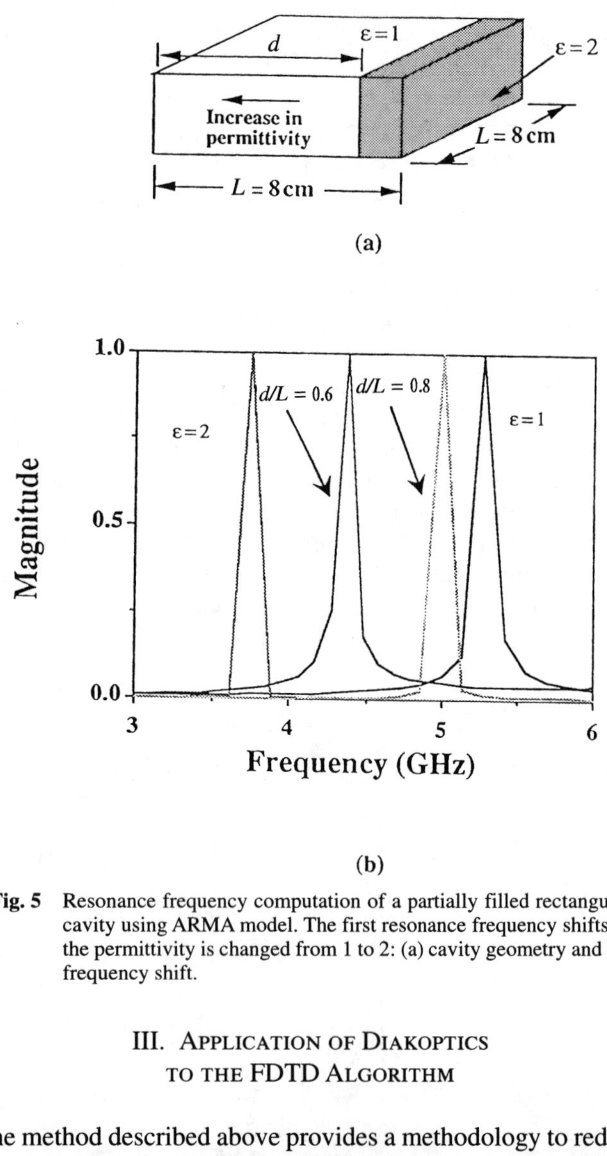

Fig. 5 Resonance frequency computation of a partially filled rectangular cavity using ARMA model. The first resonance frequency shifts as the permittivity is changed from 1 to 2: (a) cavity geometry and (b) frequency shift.

III. Application of Diakoptics to the FDTD Algorithm

The method described above provides a methodology to reduce computational requirements of the FDTD algorithm by extracting useful information from the time-domain signals in a shorter time interval than that required for a Fourier Transformation. In this section we address the reduction of the computation domain by reducing the memory requirements of the FDTD algorithm. The concept which is used here is directly related to the truncation condition which is required for the FDTD algorithm. While the ultimate goal of the usual ABCs are to simulate a one-way wave, here the FDTD Diakoptics allows the presence of waves which are propagating in both directions. This approach allows the representation of discontinuities in the absorbing boundary condition.

Theory

The concept of time-domain Diakoptics originates from the linear system theory. To illustrate this concept consider a discrete linear passive system where the output waveform $Y(n)$ is determined from the convolution of the system impulse response $h(n)$ and the system input $X(n)$. This concept has been used successfully with the transmission line matrix (TLM) method [15–20]. This relationship can be exploited in the FDTD algorithm by replacing the complete one-port linear passive structure by its impulse response. Similarly, a multi-port linear passive region in the microwave structure can be replaced by an impulse response matrix, as shown in Fig. 6. This impulse response matrix is in effect the time-domain Green's function [21]. The multi-port convolution is defined as

$$Y_m(k) = \sum_{n=1}^{N} \sum_{k'=1}^{K} g(m, n, k - k')X_n(k') \tag{11}$$

where $g(m, n, k')$ is the impulse response at port m at $t = k'$ due to the unit excitation at port n at $t = 0$. Figure 2 shows a block diagram of this operation.

The Diakoptics can be extended to structures which are composed of multiple segments. Each segment is represented by its impulse response. These impulse responses are combined to represent the impulse response of the aggregate structure. Sequential and parallel algorithms have been implemented for combining the impulse responses of a number of segments [22,23].

The convolution operation in Equation (11) can be implemented efficiently by proper representation of the impulse response [24–26]. Figure 2b illustrates a block diagram representation of

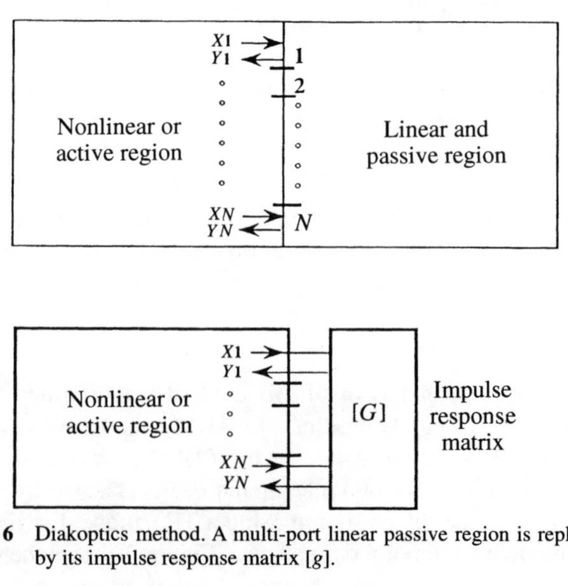

Fig. 6 Diakoptics method. A multi-port linear passive region is replaced by its impulse response matrix [g].

this concept. The impulse response of a microwave segment is obtained by exciting that structure with an impulsive source. The corresponding impulse response is represented by an ARMA model. The computation of the parameters of the ARMA model is carried out by the algorithm which is described in Section II. The ARMA model description is in effect a difference equation which relates the current field value to the previous field values and the incident field [see Eq. (3)]. As a result, the field values at the boundary points of each segment are updated using a number of the past field values and the incident field.

Numerical Examples

Figure 7 shows a one-dimensional circuit which is used to illustrate the FDTD Diakoptics. The structure is composed of an infinitely long parallel-plate transmission line loaded with lumped elements and material discontinuity. For this simulation the parameter of interest is the current in the resistive load. In order to reduce the computation domain where the FDTD algorithm is applied, the structure is decomposed into four segments. Segment 2 is the FDTD computation domain, while Segments 1, 3, and 4 are represented by their impulse responses. A sequential Diakoptics algorithm is used to combine Segments 3 and 4 [22]. Figure 8 shows the agreement between the Microwave SPICE (MWSPICE) and the FDTD algorithms augmented with the Diakoptics for the structure in Fig. 7. It is important to note that the application of the Diakoptics in Region 1 of Fig. 7 provides a mechanism to absorb a one-way wave. On the other hand, the application of the algorithm to the combined Regions 3 and 4 includes the multiple reflections at the discontinuity boundaries.

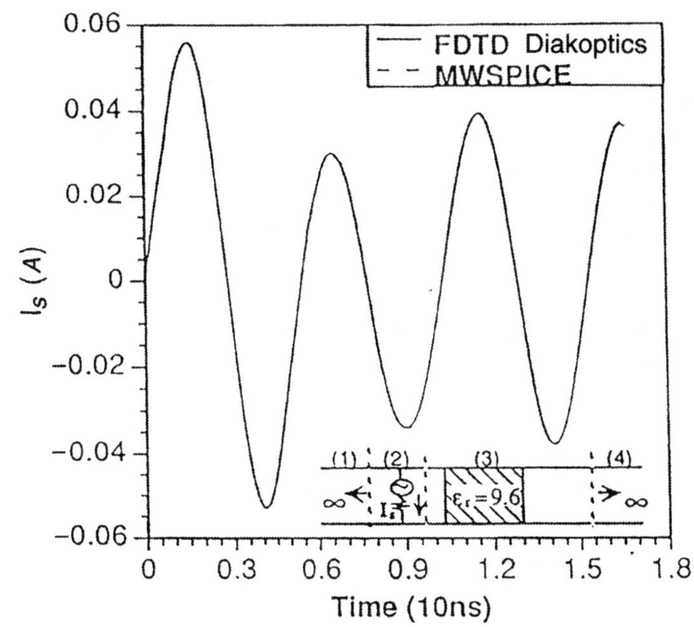

Fig. 8 Computed I_s (see also Fig. 7). FDTD Diakoptics and Microwave SPICE simulation produce identical results.

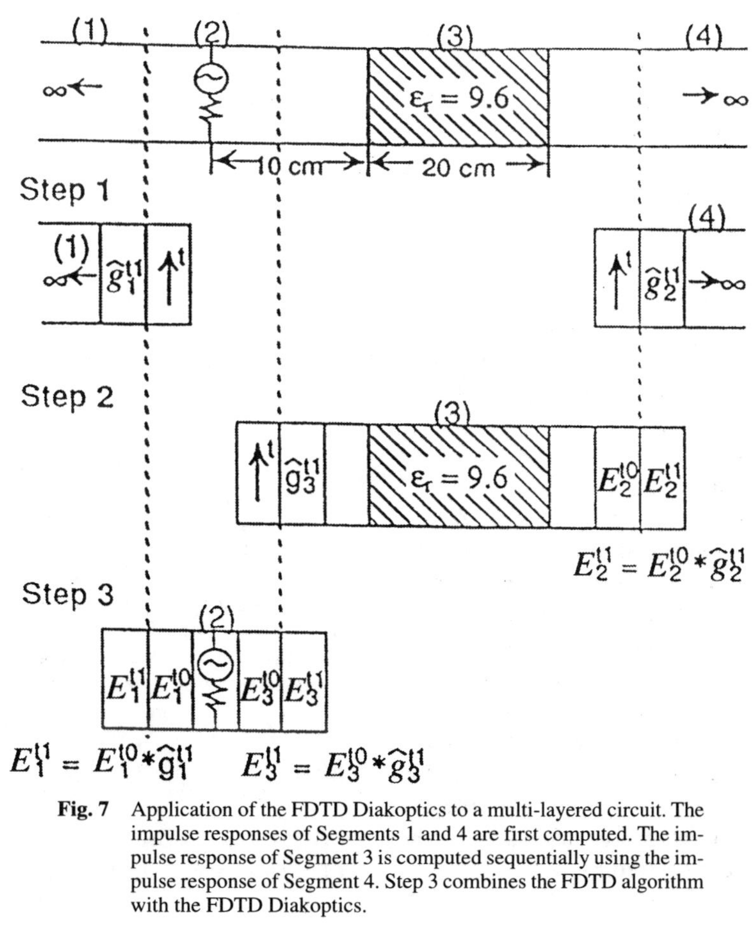

Fig. 7 Application of the FDTD Diakoptics to a multi-layered circuit. The impulse responses of Segments 1 and 4 are first computed. The impulse response of Segment 3 is computed sequentially using the impulse response of Segment 4. Step 3 combines the FDTD algorithm with the FDTD Diakoptics.

IV. Summary

In this chapter, two methods to reduce the computation requirements of the FDTD method in time and space are outlined. The reduction in the computation time is obtained by using the signal processing algorithms for parameter identification of linear systems. The SI algorithm computes the parameters of a deterministic ARMA model using the input and output signals of a FDTD simulation. The parameter convergence is rapid, and the frequency spectrum is computed directly using the system parameters. The reduction in the memory requirements is obtained by the Diakoptics method. In this method, segments of the computation volume are represented by their impulse responses. The interaction of the incident field with these segments is described in terms of a convolution operation. The convolution operation is carried out efficiently by representing the impulse response using an ARMA model.

The first four selected papers describe the application of the advance signal processing methods for extracting circuit parameters from the time domain signals. The remaining four papers describe the application of Diakoptics to the time-domain microwave circuit simulations.

The first paper is devoted to the application of the Prony-Pisarenko method (PPM) to the time-domain signals. In this formulation the physical signal is modeled as a deterministic signal plus a non-correlated noise signal. The noise is due to the numerical round off errors and the spectral limitations off the time-domain method. The deterministic signal is modeled as the sum of complex damped exponential. In the process of the solution, the resonant frequencies, damping factors, and the complex amplitude for each complex exponential is computed. The important issue of the order of the model is pointed out in this paper. This method is applied to a rectangular patch and a microstrip antenna fed by a microstrip line. The performance of the PPM method is compared with the Fast Fourier Transform for the spectral analysis.

In the second paper, the Prony method is applied to low- and band-pass filters. This paper demonstrates that accurate spectral analysis of the time domain signals for computation of the scattering parameters is carried out using a relatively short time record of the FDTD generated signal.

The third paper applied the multiple signal classification (MUSIC) method for spectral analysis of the FDTD generated signals. In this paper a cylindrically shaped dielectric resonator is used to generate the time-domain signal. It is shown that the resonant frequencies are estimated with 2,000 time iterations using the MUSIC algorithm, while a similar spectrum is computed with 20,000 time iterations using the conventional FFT.

The fourth paper describes the identification of an unknown linear system using an autoregressive moving average (ARMA) model. For our applications the unknown system is a microwave circuit structure. This paper presents the general formulation and implementation of the identification algorithm. It is also noted that the algorithm developed in this paper can be implemented on massively parallel computers.

The fifth paper describes the segmentation method for analysis of planar microwave circuits in the frequency domain. In this method the characteristics of a planar circuit is computed by combining the characteristics of the segmented elements. The interactions between segments are limited to the directly connected elements. This method is applied to the optimum design of a ladder-type 3-dB hybrid circuit. The planar circuit is divided into four segments. The S-matrices of these segments are combined to synthesize the S-matrix of the entire circuit.

The sixth paper describes the implementation of the Diakoptics in the FDTD method. The interactions at the boundaries of the segmented structure are taken into account by performing a convolution operation of the incident signal and the impulse response of each segment. Parallel and sequential formulations are presented in this paper.

The seventh paper describes the implementation of dispersive wide-band absorbing boundary conditions with diakoptics for the transmission line matrix (TLM) method. The formulation for TLM Diakoptics is presented in Section III. This section briefly reviews the origins of the frequency and time domain diakoptics and points out the potential usefulness of this method for optimization problems where only sub-sections of a microwave structure is changed and the remaining structure remains unaltered. A flow graph for the TLM method including the diakoptics module is also presented. This method is applied to an inductive waveguide iris discontinuity and an E-plane bandpass filter.

The eighth paper combines the concepts of diakoptics and the system identification methods for efficient implementation of the FDTD Diakoptics method. In this paper the impulse response of a circuit element is presented in terms of an infinite impulse response (IIR) filter. The convolution operation which is required for the diakoptics method is converted to the update of a difference equation.

References

[1] X. Zhang and K. K. Mei, "Time-domain finite-difference approach to the calculation of frequency dependent characteristics of microstrip discontinuities," *IEEE Trans. Microwave Theory Tech.,* vol. MTT-36, no. 12, pp. 1775–1787, Dec. 1988.

[2] D. M. Sheen, A. S. Ali, M. D. Abouzahra, and J. K. Kong, "Application of three-dimensional finite-difference time-domain method to the analysis of planar microstrip circuits," *IEEE Trans. Microwave Theory Tech.,* vol. 38, No. 7, July 1990.

[3] J. P. Berenger, "Perfectly matched layer for the FDTD solution of wave-structure interaction problems," *IEEE Trans. Antennas Propagat.,* vol. 44, pp. 110–117, Jan. 1996.

[4] B. Engquist and A. Majda, "Absorbing boundary conditions for numerical simulation of waves," *Math. Comput.,* vol. 31, pp. 629–651, 1977.

[5] G. Mur, "Absorbing boundary conditions for finite-difference approximation of the time-domain electromagnetic field equations," *IEEE Trans. Electromagnetic Compat.,* vol. EMC-23, pp. 377–382, Nov. 1981.

[6] A. V. Oppenheim and R. W. Schafer, *Digital Signal Processing,* Englewood Cliffs, N.J.: Prentice-Hall, 1975.

[7] G. C. Goodwin and K. S. Sin, *Adaptive Filtering Prediction and Control,* Englewood Cliffs, NJ: Prentice-Hall, NJ, 1984.

[8] P. Strobach, *Linear Prediction Theory: A Mathematical Basis for Adaptive Systems,* Berlin: Springer Verlag, 1990.

[9] W. Kuempel and I. Wolff, "Digital signal processing of time domain field simulation results using the system identification method," *IEEE MTT-S Digest,* June 1992.

[10] W. Kuempel and I. Wolff, "System identification method for transient analysis of (M)MIC-components using time iterative methods," 22nd European Microwave Conference, Finland, 24–27 August 1992.

[11] J. L. Dubard, D. Pompei, J. Le Roux, and A. Papiernik, "Characterization of microstrip antennas using the TLM simulation associated with a Prony-Pisarenko method," *Int. J. of Numerical Modelling: Electronic Networks, Devices and Fields,* vol. 3, 269–285, 1990.

[12] W. K. Ko and R. Mitra, "A Combination of FD-TD and Prony's methods for analyzing microwave integrated circuits," *Microwave Theory Tech.,* vol. 39, no. 12, pp. 2176–2181, Dec. 1991.

[13] B. Houshmand and T. Itoh, "Parameter estimation for characterization of microwave structures," presented at the National Radio Science Meeting, Jan. 5–8, 1993.

[14] C. Eswarappa and W. J. R. Hoefer, "Fast S-parameter computation of a microstrip interdigital filter using the TLM, Prony's and digital filtering techniques," *Int. J. of Numerical Modelling,* in press.

[15] P. B. Johns and K. Akhtarzad, "The use of time domain diakoptics in time discrete models of fields," *Int. J. Num. Methods Eng.,* vol. 17, pp. 1–14, 1981.

[16] P. B. Johns and K. Akhtarzad, "Time domain approximations in the solution of fields by time domain diakoptics," *Int. J. Num. Methods Eng.,* vol. 18, pp. 1861–1873, 1992.

[17] P. So, C. Eswarappa, and W. J. R. Hoefer, "A two-dimensional TLM microwave field model simulator using new concepts and procedures," *IEEE Trans. Microwave Theory Tech.,* vol. 37, pp. 1877–1884, Dec. 1989.

[18] C. Eswarappa, G. I. Costache, and W. J. R. Hoefer, "TLM modeling of dispersive wide-band absorbing boundaries with time domain diakoptics for s-parameter extraction," *IEEE Trans. Microwave Theory Tech.,* vol. 38, pp. 379–386, Apr. 1990.

[19] C. Eswarappa and W. J. R. Hoefer, "Diakoptics and wide-band dispersive absorbing boundaries in the 3-D TLM method with symmetrical condensed nodes," *IEICE Trans.,* vol. E74, pp. 1242–1250, May 1991.

[20] W. J. R. Hoefer, "The discrete time-domain Green's function or johns matrix—A new powerful concept in transmission line modeling (TLM)," *Int. J. Num. Modelling,* vol. 2, pp. 215–225, 1989.

[21] C. W. Kuo and T. Itoh, "A new time-domain transverse resonance method in solving guided wave problems," *Int. J. Num. Modelling,* vol. 3, pp. 229–234, 1990.

[22] T. W. Huang, B. Houshmand, and T. Itoh, "The implementation of time-domain diakoptics in the FDTD method," *IEEE Trans. Microwave Theory Tech.,* vol. 42, no. 11, Nov. 1994.

[23] T. W. Huang, B. Houshmand, and T. Itoh, "The FDTD wide-band absorbing boundary conditions for the excitation plane," *Proc. 1993 IEEE Antennas Propagat. Soc. Int. Symp.,* June 1993.

[24] B. Houshmand, T. W. Huang, and T. Itoh, "Microwave structure characterization by a combination of FDTD and system identification methods," *IEEE Microwave Guided Wave Lett.,* vol. 3, no. 8, pp. 262–264, Aug. 1993.

[25] T. W. Huang, B. Houshmand, and T. Itoh, "Application of system identification technique to FDTD and FDTD Diakoptics method," *23rd European Microwave Conf. Proc.,* Sept. 1993.

[26] T. W. Huang, B. Houshmand, and T. Itoh, "Fast sequential FDTD diakoptics method using the system identification technique," *IEEE Microwave Guided Wave Lett.,* vol. 3, no. 10, pp. 378–380, Oct. 1993.

Characterization of Microstrip Antennas Using the TLM Simulation Associated with a Prony-Pisarenko Method

J.L. DUBARD, D. POMPEI, AND A. PAPIERNIK
ELECTRONICS LABORATORY
UNIVERSITY OF NICE-SOPHIA ANTIPOLIS
CNRS, BAT 4, SPI, RUE A. EINSTEIN, 06560 VALBONNE, FRANCE

J. LE ROUX
COMPUTER SCIENCES SIGNALS AND SYSTEMS
UNIVERSITY OF NICE-SOPHIA ANTIPOLIS
CNRS, BAT 4, SPI, RUE A. EINSTEIN, 06560 VALBONNE, FRANCE

SUMMARY

The transmission-line matrix (TLM) method enables simulation of interior electromagnetic field propagation problems. With the use of absorbent walls, we can simulate exterior problems such as the radiaition of a microstrip antenna. The input impedance is deduced from the standing wave observed in the feedline. The radiation pattern is determined from the field over a plane located in the immediate vicinity of the antenna, using the theory of radiating apertures. However, the CPU time and memory space involved are excessive. Since the radiating structure has several resonant frequencies, it is convenient to apply classical digital signal processing techniques such as finite impulse response filtering associated with a linear prediction method. The present paper focuses on a Prony–Pisarenko method to improve on the TLM method in terms of both computation time and precision of the frequency-domain analysis of the results. In this case, overall CPU time is reduced by a factor of 2 to 3. The paper discusses the results obtained for radiation patterns. This represents a new field of application for the TLM method whose drawbacks are reduced by using appropriate signal processing methods.

INTRODUCTION: TLM AND DIGITAL SIGNAL PROCESSING TECHNIQUES

Numerous articles have appeared in the literature on microwave structures such as microstrip lines, slot-lines, and fin-lines. The techniques employed to characterize these structures are widely known. In most cases, numerical models are used, either in the spectral[1,2] or time[3] domains. Today, these techniques are being used intensively to characterize open structures such as microstrip antennas. Among the time-domain methods, the transmission-line matrix (TLM) technique[4,5] was first applied to simulate the propagation of EM waves in closed structures (cavity, waveguide, microstripline etc.). Its extension to open structures such as microwave antennas has been shown to be possible, but certain precautions must be taken.[6]

TLM simulates the propagation of EM waves by the scattering of impulses in a three-dimensional meshed network of transmission lines. We can use a variable mesh,[6,7] but the modelled space is, by necessity, bounded (Figure 1). For a given impulsive excitation, the method yields a signal which is sampled over time, which is the time-domain response of the structure at each point in the modelled space. The accuracy of the information depends on two important points:

- how finely the space is described, and this depends on the lengths Δl_i of the TLM mesh used. About this point, the variable mesh is an important improvement;
- the time over which the scattering is observed must be sufficiently long, since the resulting frequency discretization interval is inversely proportional to this duration. This aspect is the principal drawback of TLM and it is the subject of our work.

However, the observation must be limited in time; this factor has a significant impact on the accuracy of the results. On the computer, this process is a convolution of the frequency response with the spectrum of a window which is, itself, limited in duration. This is another potential source of inaccuracy. The abrupt stopping of the process corresponds to the use of a rectangular window;

Reprinted with permission from *International Journal of Numerical Modeling: Electronic Networks, Devices and Fields*, J. L. Dubard, D. Pompei, J. Le Roux and A. Papiernik, "Characterization of Microstrip Antennas Using the TLM Simulation Associated with a Prony-Pisarenko Method," Vol. 3, pp. 269-285, 1990.

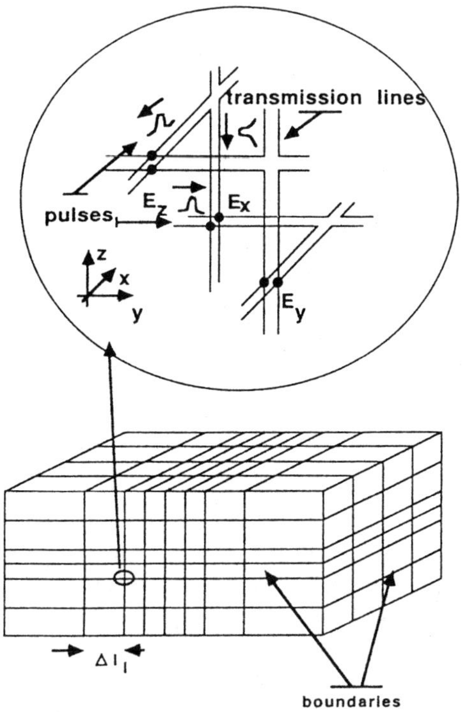

Figure 1. TLM mesh

the spectral domain response is thus strongly perturbed by the sin(F)/F functions caused by this window.

Thus the processing of the data provided by the TLM method requires a careful choice of digital signal processing techniques, in order to access the most useful data. These techniques, associated with TLM, can lead to a substantial reduction in CPU time or storage without compromising the accuracy.

Previously, the fast fourier transform (FFT) was employed to process the TLM results. Emphasis was placed on limiting inaccuracy caused by the fact that the TLM response is truncated in time. Multiple truncation windows can be used. The rectangular window's drawbacks can be reduced (−13 dB to −40 dB for the first sidelobe), but the precision about the frequency may be damaged, the width of the principal lobe increasing. Nevertheless, by concentrating on the effects of multiple truncation windows,[8,9] we were able to achieve an efficient method combining TLM and FFT for the 3-D time-domain characterization of closed structures. However, when applied to open structures such as microstrip antennas, the use of the TLM method was hampered by two predictable obstacles.

— First, if the problem is to be handled numerically, the space described must have boundaries. This difficulty was largely surmounted with the use of absorbent walls, which are currently the topic of many publications.[10,12]
— Second, the FFT requires an observation over an extended period of time, which makes the corresponding CPU time prohibitively costly. In addition, the characterization of a microstrip antenna requires the values of output signals, densely sampled over time, at a large number of points. This volume of data occupies a considerable amount of storage.

Thus, particularly for characterization of microstrip antennas, it can be concluded that the FFT is not entirely appropriate for processing TLM results for three reasons:

— it requires a large number of time samples, which increases CPU time;
— problems arise due to the window's truncation error (decreasing sidelobe but increasing width of principal lobe);
— for minimize this last drawback, considerable storage is needed to characterize an open structure.

Analysis methods for this type of sampled signal have been developed in signal-processing applications such as speech processing.[13,14] The linear prediction methods utilized appeared to be easily adaptable to the problem at hand: indeed, the TLM output is a sampled signal, often oversampled. Furthermore, this signal is in fact a superposition of different 'modes', each with its own resonant frequency, and thus corresponds to the type of signals handled with linear prediction techniques. Efforts in this direction have already been undertaken by other authors.[15,16]

One of the best-adapted methods dates from 1795, and is based on the work of Prony.[17] This method will be developed below, taking into account the various perturbations possible (which are considered as noise) with a view to applying it to radiating structures.[18] We use for this a development of the Pisarenko's method.[19] These concepts will be developed hereafter.

THE PRONY–PISARENKO METHOD (PPM)

Principle

The technique replacing the FFT for the spectral analysis associated with the TLM method consists in using the Prony model to represent the observation signal. The TLM output signal is considered to be a deterministic signal drowned in noise. The choice of an exponential representation is guided by considerations concerning the physical nature of the phenomena described by the TLM method (study of the resonant modes of a printed antenna or a cavity). Estimation of the model parameters involves minimizing the noise variance, which amounts to determining the eigenvector associated with the smallest eigenvalue of the signal covariance matrix. This method is, in fact, a general case of the Pisarenko method[20] which was applied only to stationary random processes.

Deterministic signal

The physical signal, $y(t)$, is modelled as the deterministic signal to be analysed, $s(t)$, overlayed by a non-correlated white nose, $w(t)$.

$$y(t) = s(t) + w(t)$$

Based on the technique of Prony, signal $s(t)$ is considered to be made up of superimposed complex damped exponentials whose coefficients α_k and f_k characterize the damping and frequency:

$$s(t) = \sum_{k=1}^{p} A_k\, e^{j\phi_k}\, e^{(\alpha_k + j2\pi f_k)t} \tag{1}$$

where $s(t)$ is one of the six electromagnetic field components, A_k is its amplitude for the frequency f_k, ϕ_k is the phase, α_k is its absorption coefficient and p is the order of the model (number of sine waves $= p/2$).

The analysis of signal $s(t)$ requires time-discretization, which is achieved by the TLM, and so we have

$$s(n) = \sum_{k=1}^{p} A_k\, e^{j\phi_k}\, z_k^{n-1} = \sum_{k=1}^{p} C_k z_k^{n-1} \tag{2}$$

with

$$z_k = e^{(\alpha_k + j2\pi f_k)\,\Delta t}$$

and

$$C_k = A_k\, e^{j\phi_k}$$

Introduction of the white noise

With TLM, we simulated the electromagnetic field propagation in the surrounding space of the microstrip antenna. We then obtained a sample signal $y(n)$ in one (or more) previously selected output point(s). We now suppose that this signal is the sum of a determinist signal $s(n)$ and a white noise signal $w(n)$:

$$y(n) = s(n) + w(n) \tag{3}$$

In our model the difference between the observed signal $y(n)$ and the model $s(n)$ is a noise added to the output of the antenna, conssidered here as an AR filter (Figure 2), and not an input applied to this filter as in most of Prony's model extensions. For the sake of computational simplicity, we assume that this noise is white. It is due to the rounded numbers of the calculation and to the spectral limitation of the method; even if the noise given by the TLM and the spectral limitation is not exactly white, the accuracy of the numerical results show that this assumption is reasonable. It is as follows:

$$E[w(n)\,w(n-k)] = \delta_{k0}\,\sigma^2 \tag{4}$$

$$E[s(n)\,w(n-k)] = 0 \tag{5}$$

where E is the mean value function, δ_{k0} is Kronecker's index, and σ is the white noise variance.

Determination of z_k *(resonant frequencies and attenuation coefficients)*

By use of the z-transform of $s(n)$, we obtain

$$s(z) = z \sum_{k=1}^{p} \frac{C_k}{(z-z_k)} \tag{6}$$

or

$$s(z) = \frac{B(z)}{A(z)} = \frac{b_0 + b_1 z^{-1} + \ldots + b_{p-1} z^{-p+1}}{1 + a_1 z^{-1} + \ldots + a_p z^{-p}} \tag{7}$$

The determination of a_i gives us the exact corresponding electromagnetic mode.

It is easily deduced from previous expressions that $s(n)$ is generated by recursive equations and initial conditions giving the p starting values for $s(n)$

$$s(n) = -\sum_{k=1}^{p} a_k\, s(n-k) = \sum_{k=1}^{p} C_k\, z_k^n \tag{8}$$

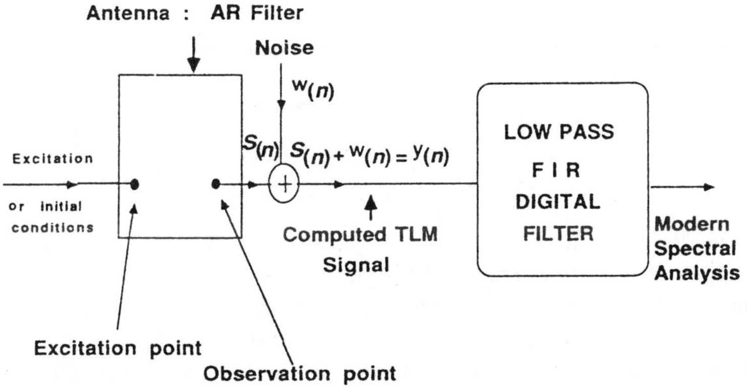

Figure 2. Model of the Prony–Pisarenko method

with initial conditions $s_0, s_1, s_2, \ldots, s_{p-1}$. The p a_k coefficients characterize this recursivity. It is easily shown that they are related to z_k by the following:

$$A(z) = 1 + \sum_{k=1}^{p} a_k z^{-k} = \prod_{k=1}^{p} \left(1 - \frac{z_k}{z}\right)$$

The a_i yield the absorption coefficients α_k and the resonant frequency f_k in equation (1). $z_1, z_2, \ldots, z_p$ are the roots of

$$1 + \sum_{k=1}^{p} a_k z^{-k} = 0$$

and give

$$f_k = \frac{1}{2\pi} \operatorname{Arg} z_k \quad \text{and} \quad \alpha_k = \ln |z_k|$$

Estimation of the autoregressive parameters a_i *of the model*

From equation (8) we have the relation:

$$\sum_{k=0}^{p} a_k s(n-k) = 0 \tag{9}$$

Equation (9) can also be written

$$\sum_{k=0}^{p} a_k [y(n-k) - w(n-k)] = 0$$

We can compute the mathematical expectation

$$E\left\{\sum_{k=0}^{p} a_k [y(n-k) - w(n-k)]\, y(n-m)\right\} = 0$$

or

$$\sum_{k=0}^{p} a_k \{E[y(n-k)y(n-m)] - E[w(n-k)y(n-m)]\} = 0$$

Noting that according to equations (4) and (5)

$$E[y(n)w(m)] = E[w(n)w(m)] = \delta_{n-m,0}\, \sigma^2$$

r_{km} being the element (k,m) of the covariance matrix **[R]** for the signal $y(n)$, the previous equations give

$$\sum_{k=1}^{p} a_k (r_{km} - \delta_{k,m}\sigma^2) = 0 \tag{10}$$

We introduce the vector **A**, whose components $(1, a_1, \ldots, a_p)$ are the values sought and the covariance matrix **[R]**, defined from the TLM signals observed. The theory shows that **A** is a solution of the equation $[\mathbf{R}]\mathbf{A} = \sigma^2\mathbf{A}$. **A** is thus an eigenvector of the covariance matrix and σ^2, the noise power, is the associated eigenvalue. We have chosen as test for the calculation of the a_i the minimization of the noise characterized by σ. We note that the a_i are the components of the eigenvector associated with the smallest eigenvalue σ of the covariance matrix for the signal

$y(n)$: it is the unique decomposition of $y(n)$ in $s(n)$ and $w(n)$ that has a physical meaning, that is where the covariance matrix of $s(n)$ is non-negative.

Note that the model allows the analysis of signals composed of damped exponentials as well as undamped sine waves (the original Pisarenko model was given for stationary signals leading necessarily to undamped sine waves).

Determination of C_k *(amplitude and phase)*

The parameters b_i give the complex amplitude of each mode by the relation

$$C_m = \frac{\sum_{k=0}^{p-1} b_k z_m^{-k}}{\prod_{i=1, i\neq m}^{p} (1 - z_i z_m^{-1})}$$

Estimation of the moving average parameters b_i *of the model*

For the determination of b_i, often called the moving average part of the model, we have

$$s(z) = \frac{B(z)}{A(z)}$$

or

$$s(z) = B(z)H(z) \qquad \text{with} \qquad H(z) = 1/A(z)$$

Having found the a_i, $H(z)$ is known. By an inverse Z-transform of $H(z)$, which gives the terms $h(n)$, we can return to the sampled time domain.

The TLM method gives us $y(n)(n = 1, \ldots, N)$. We must calculate the b_i giving the signal $s(n)$ closest to $y(n)$. The minimization of the quadratic error (Wiener) e is a good test for this approach:

$$e = \sum_{n=p+1}^{N} [y(n) - \sum_{i=0}^{p-1} b_i h(n-i)]^2 \tag{11}$$

This minimization can be translated into the equation

$$\sum_{k=0}^{p-1} b_k H_{km} = G_m \tag{12}$$

where

$$H_{km} = \sum_{n=p+1}^{N} h(n-k)h(n-m)$$

$$G_m = \sum_{n=p+1}^{N} y(n)h(n-m)$$

These last two relations can be calculated and the moving average coefficients b_i may be deduced. The flowchart of the process is given in Figure 3.

Remark. The present computation of $B(z)$ based on the minimization of the noise energy is not rigorously compatible with the criterion used for computing the autoregressive part of the model.

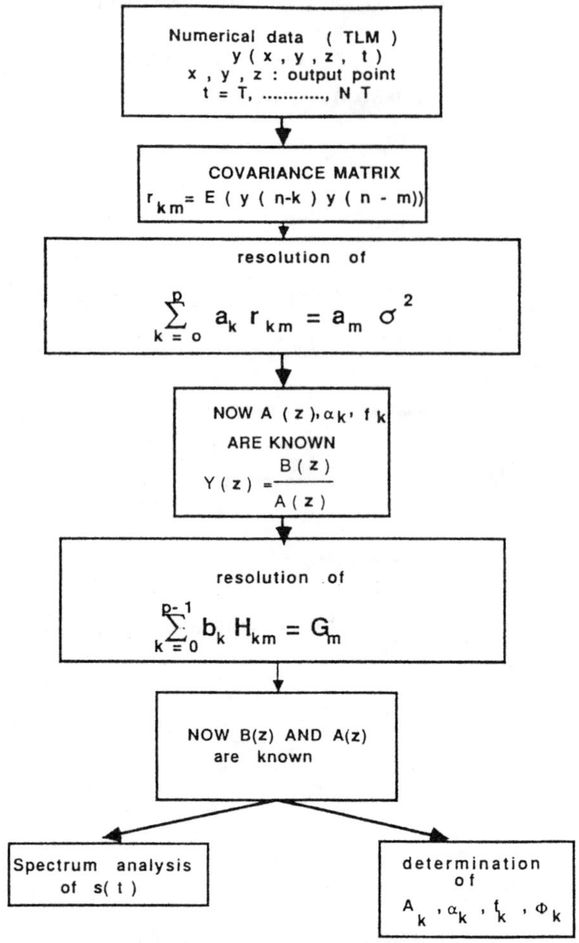

Figure 3. Flowchart of the process

Comments on the sampling, filtering and order of the model

Sampling. The accuracy of the TLM method depends on how fine the mesh is. The time response is therefore sampled at a very high frequency. However, our interest is focused on resonant electromagnetic modes whose frequencies, F, are low with respect to the sampling frequency, F_e ($F/F_e = 10^{-4}$ for a single-layer slot-fed antenna). This oversampling renders the PPM inefficient; it exhibits extremely low spectral resolution for the first modes. Actually, it is not necessary to record the TLM response at each iteration; the number of samples to be analysed can thus be reduced significantly. In order to preserve adequate spectral resolution, PPM requires a reduced number of modes, which must be appropriately distributed over the interval $[0,F_e/2]$. This involves filtering the TLM signal in a frequency band $[F_1,F_2]$ to be determined, then resampling the filtered signal such that $[F_1,F_2]$ lies in the interval $[F_e/6,F_e/3]$ where parametric spectral analysis techniques are most sensitive.

Filtering. The resampling filter is a finite impulse response[21] filter. It is always stable and exhibits linear dephasing. If the impulse response contains N coefficients, it is necessary to remove the first N non-significant samples from the filtered signal.

Order of the model. The order is the most delicate aspect of the PPM. If the order p is too small, the spectrum lacks accuracy. If it is too high, the model is unstable (the element matrix H_{kl} exhibits singularities) or causes dispersion in the results. The ideal trade-off is to use the highest order possible, yielding the most accurate estimation, while preserving the stability of the model. Useful criteria include minimization of the ratio between prediction error energy and signal energy. It is shown theoretically that, for an order p higher than the order p_0 of the model, there are $p-p_0+1$ identical eigenvalues equal to the noise variance.

The order of the model can thus be estimated by observing the number of eigenvalues identical to the lowest eigenvalue of the covariance matrix of the signal studied.

Application to test signals

The method is remarkably efficient when the signal is exactly modelled. Tests were performed by the superposition of several sine curves:

$$s(t) = \sum_{k=1}^{5} A_k \, e^{\alpha_k t} \cos(2\pi f_k t + \phi_k) \qquad (13)$$

After sampling

$$s(n) = \sum_{k=1}^{5} A_k \, e^{\alpha_k n} \cos(2\pi f_k n + \phi_k) \qquad (14)$$

Spectral analysis was performed using both the fast Fourier transform and the Prony–Pisarenko method. It was observed that:

— With PPM, only thirty samples are necessary to obtain a good spectrum. Moreover, the stability of the result is perfect since the same results is achieved with 100 samples. Only a rectangular window is used.
— The fast Fourier transform, on the other hand, gives aberrant or very rough results with 30 samples (Figure 4). In fact, 1000 samples are necessary (Figure 5) for the fast Fourier transform to show the same precision as the spectrum obtained by the Prony–Pisarenko method with 30 samples.
— We may think that, with FFT, we can use specific windows (Hanning, Hamming etc.), but to obtain the same degree of precision on frequencies as with PPM, we must increase the number of indispensable samples.

Thus, in this ideal case, in which the model corresponds exactly to the signal, PPM required 30 times fewer samples. We note also that in this test, there is no problem for determining the number of modes in Prony's method.

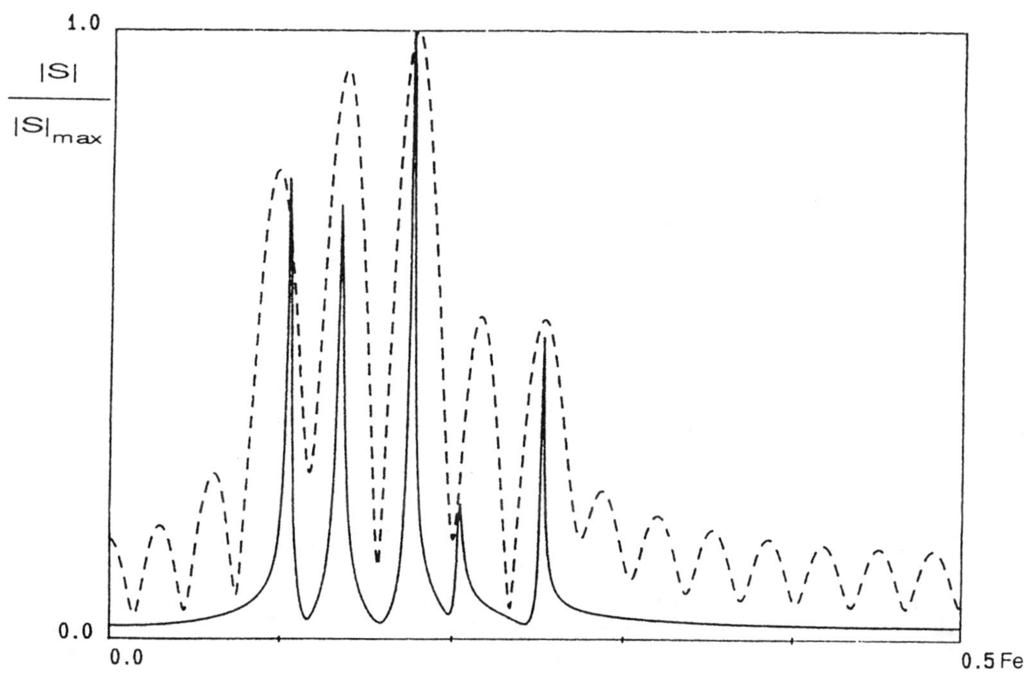

Figure 4. Spectral analysis with a test signal with 30 samples. Solid line, PPM; dashed line, FFT

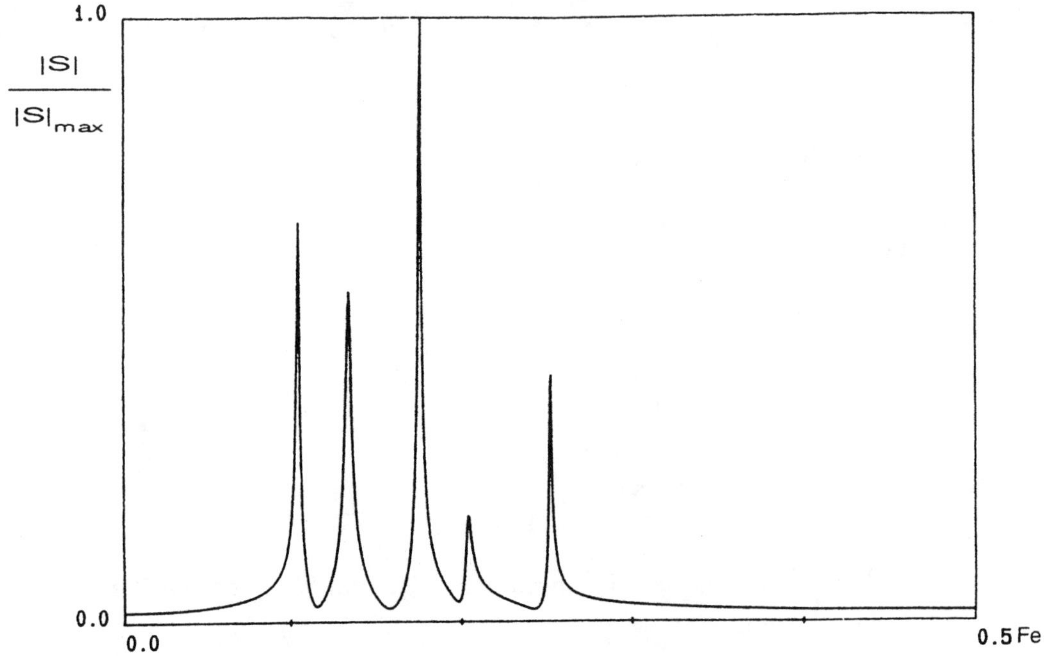

Figure 5. Spectral analysis of a test signal with FFT and 1000 samples

PROCESSING OF TLM SIGNALS FROM COMPLEX STRUCTURES

Electromagnetic cavities

Since cavities are perfectly resonating structures, they have a line frequency spectrum. Analysis combining TLM and FFT requires the convolution of the exact spectrum with that of the truncation window used; a typical result is given in Figure 6. For the same case, the Prony–Pisarenko method yields a line spectrum since the losses for each mode are null. It is pointed out that for an agreement between theory and 'experiment' of 0·3 per cent, the PPM requires an order $p = 20$ and the processing of 200 samples supplied by the TLM. To achieve the same accuracy, the FFT requires at least 800 samples.

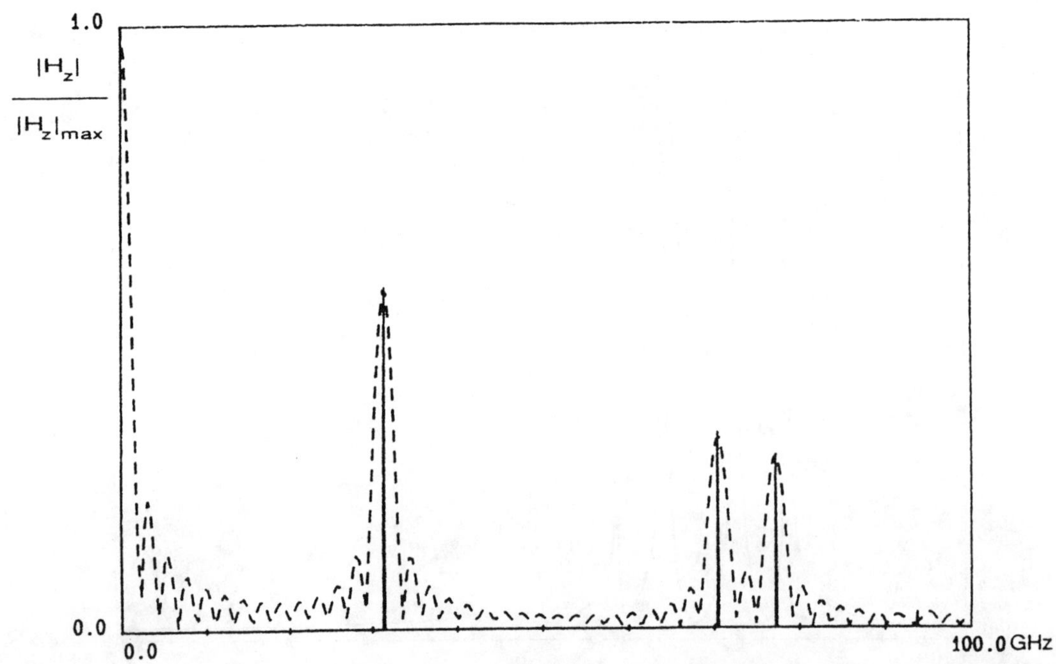

Figure 6. Spectral analysis of a resonant cavity. Solid line, PPM with 200 samples; dashed line, FFT with 800 samples

Rectangular patch antennas

The aim here is to determine the resonant frequency of a simple rectangular patch antenna (Figure 7). The spectral analysis of the E_y electromagnetic field component at a point located between the ground plane and the patch is achieved using the PPM with order p = 20 and 2500 TLM iterations (Figure 7). The FFT requires six times more iterations in order to obtain the same accuracy, both in frequency and in amplitude (Figure 8).

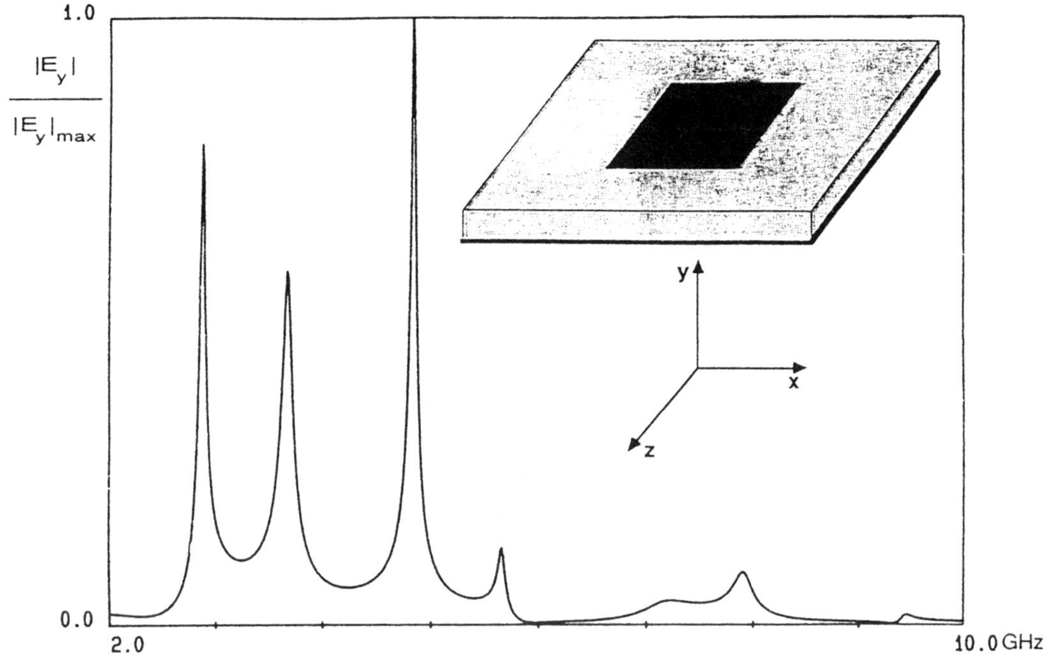

Figure 7. Spectral analysis of a rectangular patch with PPM and 2500 samples

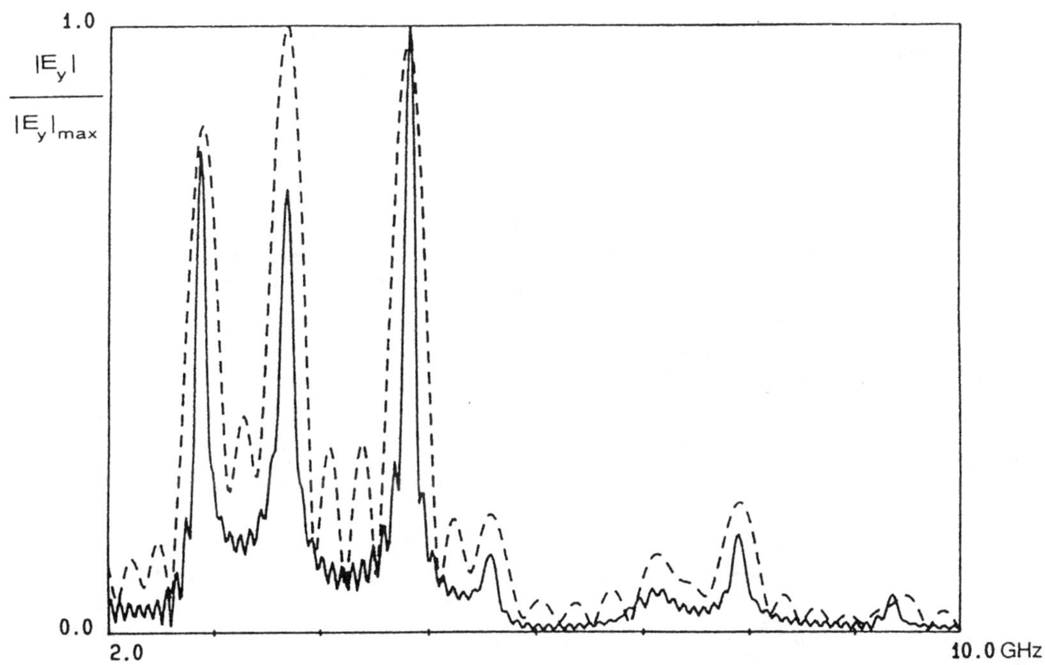

Figure 8. Spectral analysis of a rectangular patch. Solid line, FFT with 15,000 samples; dashed line, FFT with 5000 samples

The TLM method, combined with FFT on one hand and with PPM on the other, was applied to the characterization of antennas. The aim was to obtain the three principal antenna properties, i.e. resonant frequency, input impedance, and radiation pattern. Presented below are the results achieved using the TLM method with FFT, which demonstrate the feasibility of the project. However, as anticipated, the CPU time and storage required caused considerable difficulties. The results yielded by the TLM/PPM combination are also presented, bearing out the significant gains achieved through this approach.

Demonstration of feasibility using the FFT

Printed antennas are characterized by their resonant frequencies, input impedance, and radiation pattern.

Resonant frequencies. Resonant frequencies are determined in the same way as for cavities. The excitation, applied at one (or several) appropriately selected point(s), is a signal which may possibly be filtered to limit the spectrum.

The iterative process of the TLM method yields the six field components, at all points in the modelled space. The resonant frequencies are not significantly affected by the antenna feed geometry. The feed can thus be replaced by artificially exciting an appropriate component at a carefully selected point. Spectral analysis of the response gives us the various resonant frequencies. If we wish to concentrate on just a part of the spectrum, we can excite the antenna with a sinusoidal signal modulated by a window (Papoulis, Gaussian, Hamming, Blackman etc.) for a limited time. To obtain the entire spectrum, we would have to apply a Dirac pulse at time $t=0$. Depending on the point (or points) of excitation and observation, we can filter the modes. We have presented several examples concerning the simplest microstrip antenna, a rectangular patch.

Input impedance. In this case, we applied the FFT only. Work is currently under way for application of the PPM. Determining the input impedance is a delicate task in all microstrip antenna analysis methods. It depends on how the feed geometry is modelled. The TLM method theoretically makes possible an accurate description of the feed, using a tight mesh. However, the actual feedpoint and excitation signal remain artificial in nature. To overcome this problem we must integrate part of the feedline into the structure to be modelled as with coaxial feeds. With knowledge of the field in the line, we can deduce the input impedance by applying classical transmission-line theory.

We also investigated the printed dipole fed by electromagnetic coupling to a microstripline (ECM).[22] The non-perturbed region allows us to determine the input impedance. Comparison with experiment or with integral equation analysis is satisfactory (Figure 9).

The radiation pattern. The radiation pattern cannot be obtained directly by observing the antenna far field because the space studied extends no further than several wavelengths from the antenna. First, the antenna near-field is obtained by the spectral theory of radiating apertures: the radiation pattern is deduced directly by a spatial Fourier transform from the tangent electric field on a plane located in the neighbourhood of the antenna.

This theory is classical with a sinusoidal signal. We also used this method directly with an impulse excitation, and noted that the resulting radiation pattern had a very simple shape, dependent on the impulse field on a plane in the vicinity of the antenna. There are two ways to compute the radiation pattern at different frequencies, based on this field:

- by first applying a time-domain Fourier transform at each point in the reference plane, then applying a space-domain Fourier transform to obtain the pattern in all directions and at all frequencies;
- by determining the radiation pattern in selected directions (the E and H planes, for example) and applying the time-domain Fourier transform in these directions only.

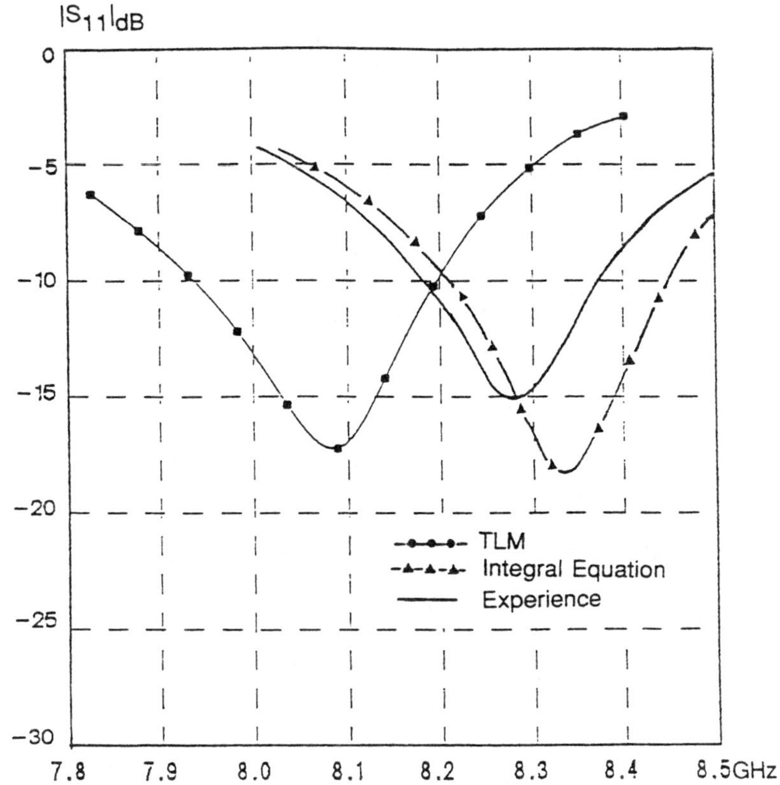

Figure 9. Reflexion coefficient versus frequency

This second method enables considerable savings in computation time and storage space. We now present a number of results obtained for the dipole described earlier (Figure 10). The results are encouraging: the computed pattern is in good agreement with measurements and with other theoretical results.

Use of advanced signal processing techniques

Before this development, microstrip antennas were to be characterized with the TLM method and FFT.[6,23] It was shown that the radiation patterns and input impedance could be obtained for a printed dipole fed by electromagnetic coupling to a microstripline. However, as mentioned, the storage space and the calculation time were prohibitive. With the TLM method and Prony–Pisarenko, we investigated a microstrip antenna fed by a microstrip line and coupling through a slot (Figure 11).

As noted earlier, a number of precautions must be taken in using signals obtained from TLM method:

— Resampling. The precision of the TLM method depends on how fine the space–time discretization is. The fact that the time-domain response is oversampled for application of Prony–Pisarenko means that the signal must be resampled, which is an easy task and involves taking 1 out of 10 or 20 TLM samples.
— Signal filtering. The TLM signal must be numerically filtered in order to limit the number of resonant modes. For this we introduce a filter with a finite impulse response.

This method was applied to a patch fed by electromagnetic coupling through the ground plane. The time-domain response at each point on a plane located immediately above the patch was sampled every 20 iterations. The model was of the order $p = 28$, which means 14 damped sinusoids. A filter was used to limit the band studied to below 10 GHz. We compared the spectra obtained by the fast Fourier transform with that of the Prony–Pisarenko method (Figures 12 and

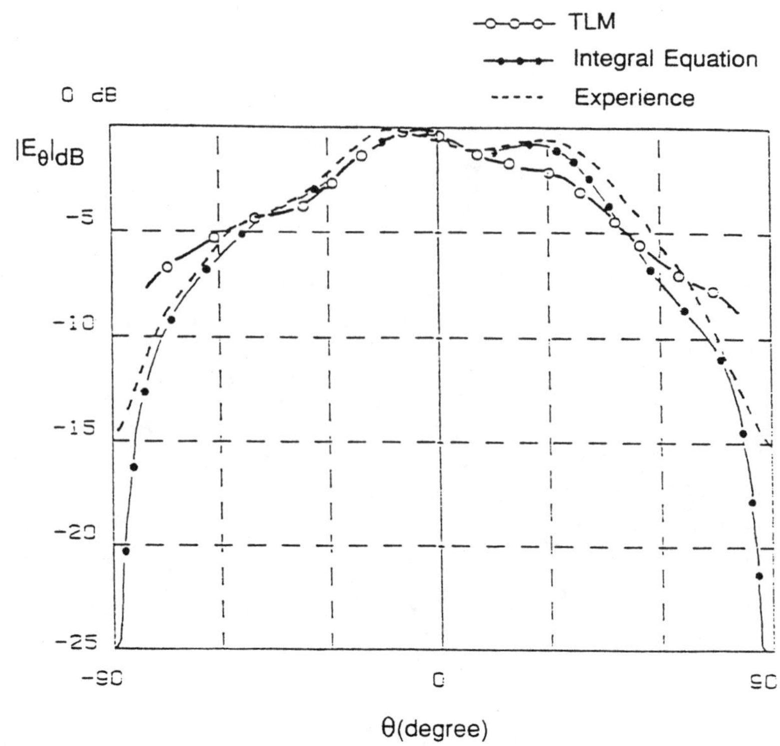

Figure 10. Radiating pattern of ECM dipole in the *E*-plane

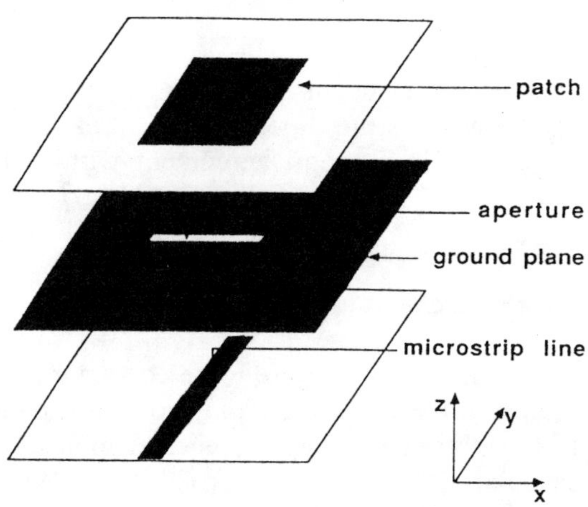

Figure 11. Microstrip antenna fed by a microstripline and coupling across a slot[24]

13). For the last, 20,000 iterations are used. With the fast Fourier transform, 60,000 iterations are needed to achieve the same precision, both in frequency and in amplitude, thus CPU time is increased by a factor of 3. We can note that the CPU time for the PPM part as for the FFT part are very short beside the CPU time of the TLM part. Thus the PPM method permits the total CPU time to be reduced.

Determination of the radiation pattern. The radiation pattern is determined in the same way as earlier, from the field over the plane located directly above the patch. Several signals are treated using Prony–Pisarenko. It is unnecessary to repeat the computations, since the frequencies f_k and damping factors α_k are the same at all points. Only the C_k are space-dependent. Initial computations yielded results similar to those obtained with the Fourier transform, but two to three times faster. Some of these results are given and compared with those of Pozar[24] (Figures 14 and 15). Input impedance results are forthcoming.

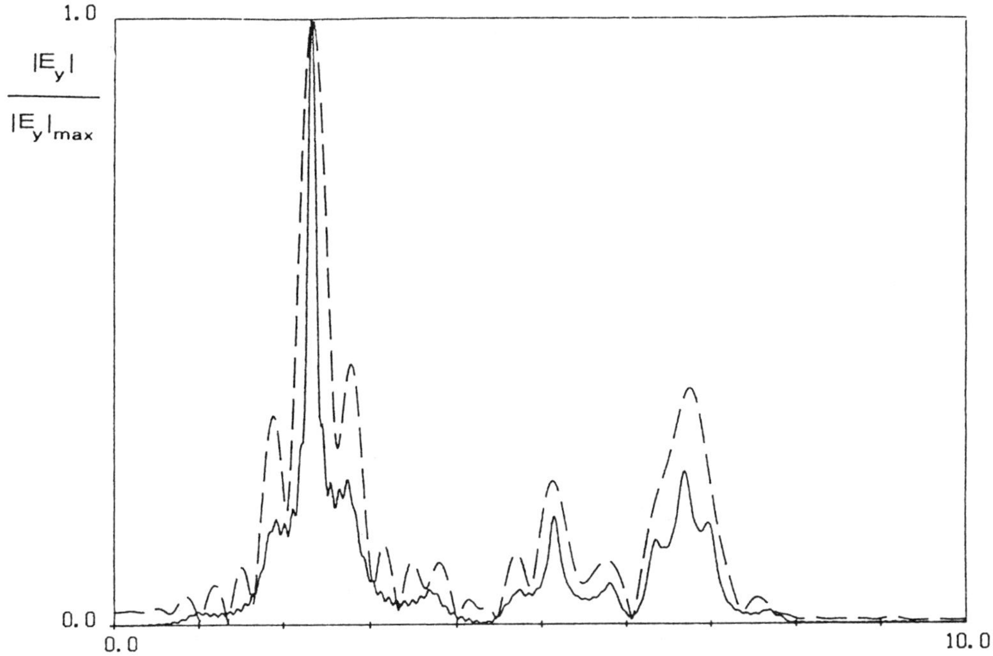

Figure 12. Spectral analysis of a microstrip antenna with FFT. Dashed line, 20,000 iterations; solid line, 60,000 iterations

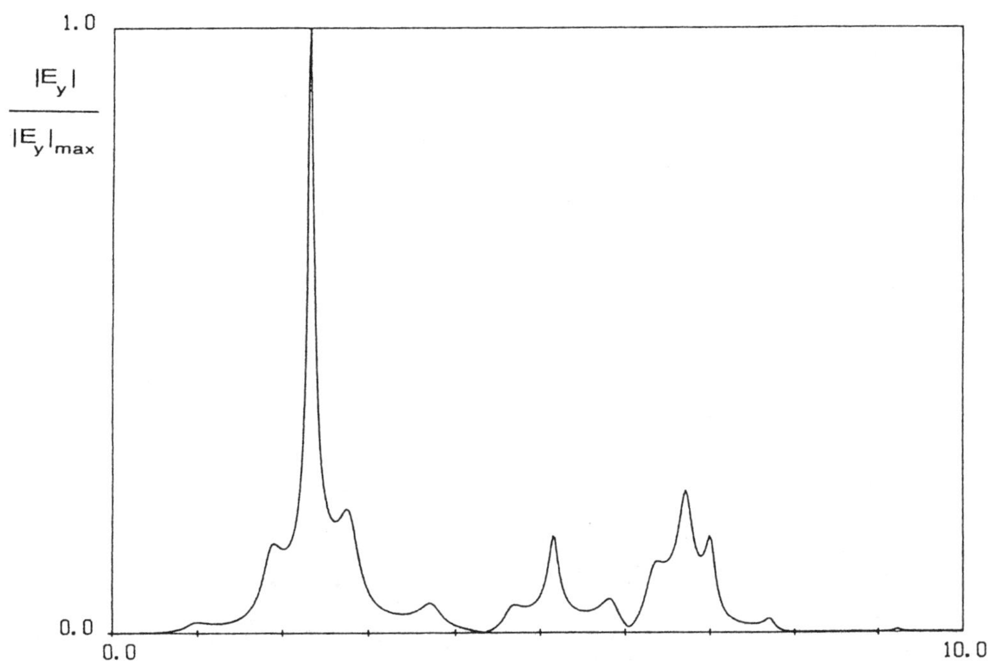

Figure 13. Spectral analysis of a microstrip antenna with PPM and 20,000 iterations

Gains in CPU time and storage space

Independently of the accuracy of the results obtained, which demonstrates the reliability of the TLM method used with PPM, this approach offers considerable gains in the CPU time and temporary storage needed. For the structure described earlier (Figure 11),[23] the TLM method associated with the FFT requires 60,000 iterations, 3000 samples, and 23 megabytess of storage. CPU time is approximately 68 hours on a Vax 8530, with, for the FFT part, 5·5 s for one spectrum. Under these circumstances the spectra can be identical with those given by TLM–PPM. But use of the PPM for the same structure, and with the same results, involves 20,000 iterations, 1000 samples, and 7·7 megabytes of storage (CPU time 23 hours with 3·2 s for the PPM part),

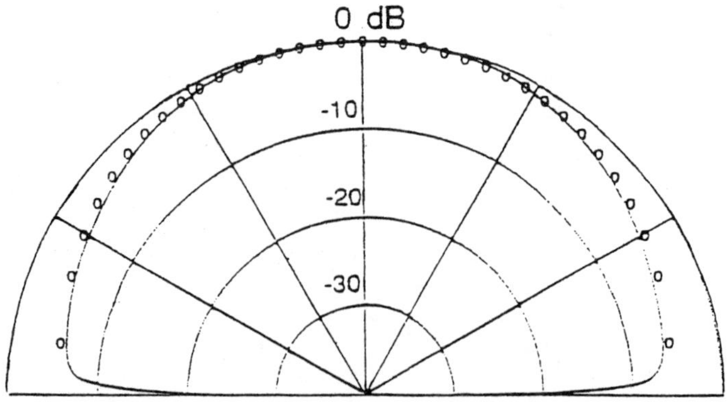

Figure 14. Radiating pattern (E-plane) of the stub-tuned aperture-coupled patch antenna determined by TLM and PPM. ○ ○ ○, by TLM and PPM: 20,000 iterations. ——— by integral equation in spectral domain

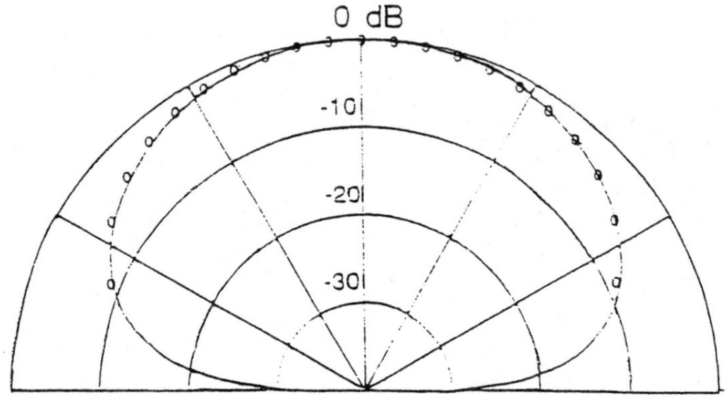

Figure 15. Radiating pattern (H-plane) of the stub-tuned aperture-coupled patch antenna determined by TLM and PPM. ○ ○ ○, by TLM and PPM: 20,000 iterations. ——— by integral equation in spectral domain

representing a threefold gain, for computing time and for storage, compared to the use of the FFT.

CONCLUSION

The TLM method is well suited to the investigation of antennas, and printed antennas in particular. However, the computational load must be reduced by adapting the method to the specific context of antennas:

— the model space is limited by introducing absorbent walls which consists in appropriately loading the TLM lines;
— the radiation pattern is determined from a field over a plane located in the immediate vicinity of the antenna, based on the theory of radiating apertures;
— input impedance is obtained by integrating a part of the feedline into the modelled structure and by applying transmission line theory to the fields within the feedline;
— lastly, the use of modern signal processing techniques offers significant savings (two to three times) in CPU time and storage.

Other techniques, such as the Johns matrix, or the use of parallel processing could enhance this method. Complex geometries could then be handled with reasonable CPU times (patches with finite ground plane, subarrays, coupled antennas, for example). Among the methods presented, the most interesting aspect, in our opinion, is the significant improvement contributed by the use of signal processing techniques. Of course, a certain amount of care must be taken in selecting the model parameters. In order to make the analysis method accessible to non-specialists, efforts should concentrate on developing an automatic procedure to evaluate the order of the model. A similar procedure could be applied to the resampling. Once these 'final touches' are completed, such techniques can be applied not only to the TLM method but also to any other time-domain method.

REFERENCES

1. T. Itoh, 'An overview on numerical techniques for modeling miniaturised passive components', *Ann. Télécommun.*, **41**, 449–462 (1986).
2. E.K. Miller, 'A selective survey of computational electromagnetics', *IEEE Trans.*, **AP-36**, 1281–1305 (1988),
3. T. Itoh, (Ed.) *Numerical Techniques for Microwave and Millimeter-Wave Passive Structures,* Wiley–Interscience, New York, 1989.
4. S. Akhtarzad, Analysis of lossy microwave structures and microstrip resonators by TLM method, Ph.D. thesis, Nottingham University, 1975.
5. P.B. Johns, 'Use of condensed and symmetrical TLM nodes in computer-aided electromagnetic design', *IEE Proc.*, **133**, 368–374 (1986).
6. O. Messaoudi, O. Benevello, D. Pompei and A. Papiernik, 'The transmission line matrix method applied to microstrip antennas', IEEE AP-S International Symposium, Syracuse, 1988.
7. D.A. Al-Mukhtar, A transmission-line matrix with irregularly graded space, Ph.D. thesis, University of Sheffield, 1980.
8. F.J. Harris, 'On the use of windows for harmonic analysis with the discrete Fourier transform', *Proc. IEEE,* **66**, 51–83 (1978).
9. D. Pompei, Caractérisation et sélection des modes de propagation dans les lignes microondes: Généralisation du formalisme quasi-TEM. Utilisation des techniques de traitement du signal dans la méthode TLM, Thèse de doctorat d'état, Université de Nice, 1987.
10. N.R.S. Simons and E. Bridges, 'Application of absorbing boundary conditions to TLM simulations', IEEE AP-S International Symposium, Dallas, 1990.
11. N.R.S. Simons, Application of the transmission-line matrix method to open region field problems, Thesis for the degree of LMaster, Winnipeg, Manitoba, 1989.
12. Eswarappa and W.J.R. Hoefer, Application of time domain diakoptics to 3-D TLM method with symmetrical condensed nodes, IEEE AP-S International Symposium, Dallas, 1990.
13. J. Makhoul, 'Linear prediction: a tutorial review, *Proc. IEEE,* **63**, 561–580 (1975).
14. A.J. Poggio, M.L. Van Blaricum, E.K. Miller and R. Mittra, 'Evaluation of a processing technique for transient data', *IEEE Trans.,* **AP-26**, 165–173 (1978).
15. M. Leroy, 'Original improvements of TLM method' *Electron. Lett.*, **17**, 684–685 (1981).
16. J.D. Wills, 'Spectral estimation for the transmission line matrix method', *IEEE Trans.*, **MTT-38**, 448–451 (1990).
17. R. Prony, 'Essai expérimental et analytique sur la loi de la dilatabilité des fluides élastiques', *J. de L'Ecole Polytechnique*, no. 1, 24–80 (1795).
18. A. Papiernik, D. Pompei and O. Benevello, 'Three dimensional TLM method for planar radiating structures', IEEE, AP-S, International Symposium, Dallas, 1990.
19. J. Le Roux and F. Giannella, 'Whiteness of the measurement noise as a criterion for ARMA modelisation of speech', IEEE International Conference on ASSP, Paris, 1982.
20. V.F. Pisarenko, 'The retrieval of harmonics from a covariance function', *Geophysics J. R. Astr. Soc.,* **33**, 347–366 (1973).
21. J.H. McClellan, T.W. Parks and L.R. Rabiner, 'A computer program for designing optimum FIR linear phase digital filters', *IEEE Trans.,* **AU-21**, 506–526 (1973).
22. P. Lepeltier, Le dipôle imprimé alimenté par couplage électromagnétique avec une ligne microruban, Thèse de doctorat, INSA, Rennes, 1986.
23. O. Messaoudi, Applications nouvelles de la méthode des lignes de transmission (TLM) à l'étude des antennes imprimées, Thèse de doctorat, Université de Nice–Sophia Antipolis, 1989.
24. D.M. Pozar, 'A reciprocity method of analysis for printed slot and slot-coupled microstrip antennas', *IEEE Trans.*, **AP-34**, 1439–1446 (1986).

A Combination of FD-TD and Prony's Methods for Analyzing Microwave Integrated Circuits

Wai Lee Ko, *Senior Member, IEEE*, and Raj Mittra, *Fellow, IEEE*

Abstract—**It is demonstrated in this paper that in applying the FD-TD technique to analyze microwave integrated circuits, the long FD-TD time record required for generating accurate frequency domain scattering parameters can be extrapolated from a relatively short FD-TD time record by using Prony's method. As shown by comparison with the direct FD-TD generated results, the new approach using the combination of FD-TD and Prony's methods achieves the same type of accuracy with a time record computed over a much shorter time.**

I. Introduction

IN APPLYING the FD-TD technique [1], [2] to analyze microwave integrated circuits, the transient time record may be required over tens of thousands of time steps in order to obtain accurate frequency domain scattering parameters via Fourier transformation. Depending on the complexity of the microwave integrated circuit being analyzed, which dictates the size of the spatial sampling grid to be used in modeling the physical structure, the computational time can run up to several hours. As shown in this paper, the accuracy of the scattering parameters is compromised if the transient time record is terminated too prematurely. However, it is also demonstrated in this paper that the required FD-TD time response can be efficiently obtained from a relatively short time record by using an extrapolation scheme based upon the Prony's method [3], [4].

II. Method of Approach

A combination of the FD-TD and Prony's methods is used to analyze microwave integrated circuits. The FD-TD technique is summarized as follows.

1) Fill computational space as shown in Fig. 1 with Yee's cells.
2) Truncate the computational space with reflectionless walls [5].
3) Place structure in the computational volume.

Manuscript received March 28, 1991; revised July 24, 1991.
W. L. Ko is with the Department of Electrical Engineering, University of South Florida, Tampa, FL 33620.
R. Mittra is with the Department of Electrical and Computer Engineering, University of Illinois at Urbana-Champaign, Urbana, IL 61801.
IEEE Log Number 9103395.

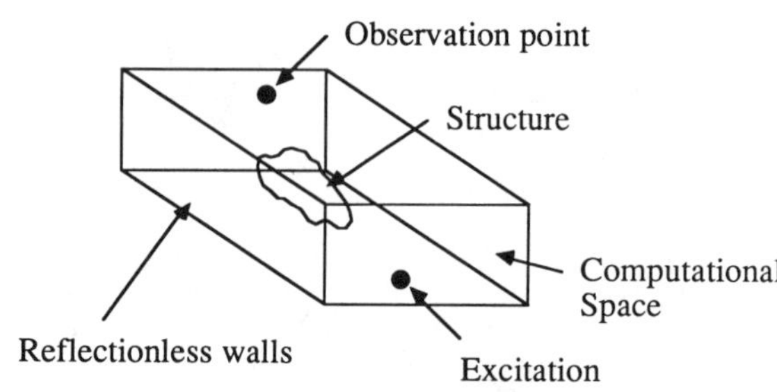

Fig. 1. FD-TD computational space.

4) Excite the structure with a Gaussian pulse whose width is chosen to cover the frequency bandwidth of interest.
5) Observe the transient wave form in the time-domain at a proper location.
6) Extract the scattering parameter in the frequency-domain by using Fourier transform.

As a first step in the extrapolation scheme, a curve fitting procedure is employed to approximate the time record, computed over a relatively short period using the conventional FD-TD scheme, in terms of complex exponential functions whose frequencies of oscillation and damping factors are determined by Prony's method. In this method, the late time transient wave form is approximated by a sum of damped complex exponentials in the following manner:

$$g_n = \sum_{k=1}^{K} c_k z_k^n, \qquad n = 1, 2, \cdots, N \tag{1}$$

where c_k and z_k are complex quantities which can be expressed as follows:

$$c_k = A_k \exp(j\phi_k) \tag{2}$$

and

$$z_k = \exp[(-\alpha_k + j2\pi f_k)\Delta t] \tag{3}$$

where

A_k is the amplitude,
ϕ_k is the phase,
α_k is the damping factor,
f_k is the frequency of the kth resonance,
Δt is the sampling interval.

Reprinted from *IEEE Transactions on Microwave Theory and Techniques*, pp. 2176-2181, December 1991.

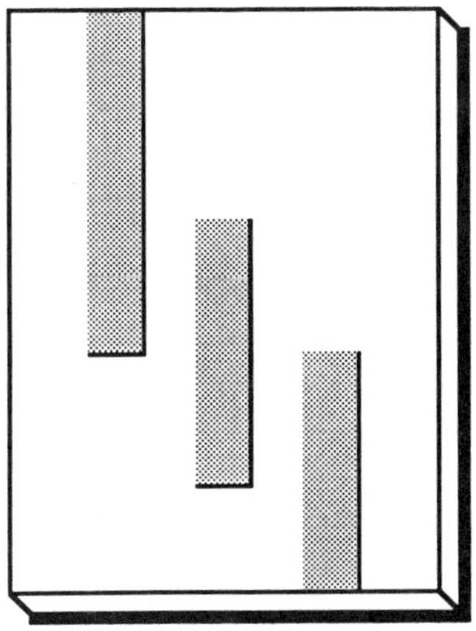

Fig. 2. Band-pass filter.

It is readily verified that g_n satisfies a difference equation

$$g_n = -\sum_{k=1}^{K} b_k g_{n-k}, \qquad n = K+1,\cdots,N \tag{4}$$

where $b_k, k=1,\cdots,K$ are the coefficients of the polynomial

$$P(z) = z^K + b_1 z^{K-1} + \cdots + b_K \tag{5}$$

which has complex roots $z_1,\cdots,z_K$ for representing g_n as given in (1).

In (4), $g_n, g_{n-1}, g_{n-2},\cdots, g_{n-K}$ are known quantities obtained from sampling the FD-TD time record. The coefficients b_k can be solved for as the least-squares solution to the over determined system of linear equations in (4). Having determined the b_k's, we can proceed to find the roots of $P(z)$ in (5) since the coefficients of that polynomial now become known. These roots are estimates of the damping factors and frequencies as shown in (3). The final step involves substituting these roots, $z_k, k=1,2,\cdots,K$, into (1) and solving the resulting least squares problem to estimate the c_k, $k=1,2,\cdots,K$. As shown in (2), these c_k's are complex quantities with amplitudes A_k and phases ϕ_k.

The results for the future time steps, i.e., the late time FD-TD record, can now be generated very efficiently using the representation given in (1) in terms of complex exponential while bypassing the time-consuming FD-TD computation altogether. The time saving resulting from the use of this extrapolation scheme can be very substantial indeed and it is not unusual to realize a reduction in computation time from hours to seconds.

III. Numerical Results

The usefulness of the scheme outlined above is illustrated in this section by considering a microstrip band-pass filter shown in Fig. 2, which is known to require a very long FD-TD time record of over 30000 iterations (See Fig. 3), in order to obtain accurate S-parameters depicted in Fig. 4, that compare well with the experimental data in [6].

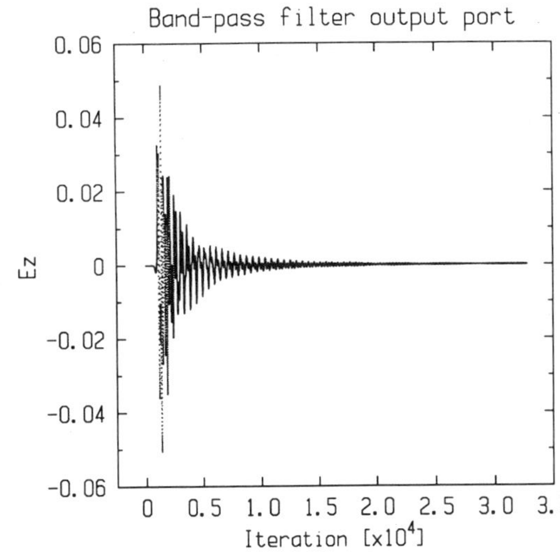

Fig. 3. FD-TD time record of 33000 iterations.

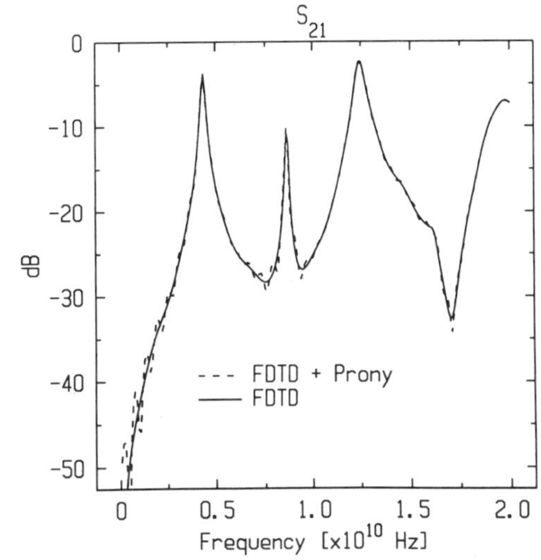

Fig. 4. S-parameter obtained by Fourier transformation of Fig. 3.

Using the new approach, a relatively short FD-TD time record of 5000 iterations is generated as shown in Fig. 5. A window of 2000 iterations is sampled at every 20 iterations to obtain 100 data points from which complex conjugate pairs of resonance frequencies and coupling coefficients are found by applying Prony's method. After experimenting a few times by shifting the window up and down the FD-TD time record, it is found that an 18-term series of complex exponential functions fits the FD-TD data satisfactorily as shown in Fig. 6. Using this 18-term series, a good approximation of the long FD-TD record of 30000 iterations can be computed in seconds, saving hours of iteration time in the process.

Fig. 7 shows the comparison between the waveform extrapolated by the new approach and the direct FD-TD waveform at late time, e.g., between the 7000 and 10000 iterations. If the truncated time record in Fig. 5 is used to compute the S-parameter, and the result is compared with the S-parameter generated by the present approach, as illustrated in Fig. 8, one sees that the peaks in the

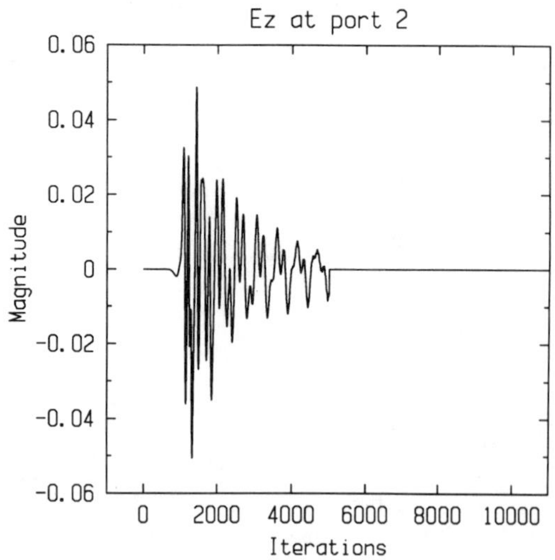

Fig. 5. FD-TD time record of 5000 iterations.

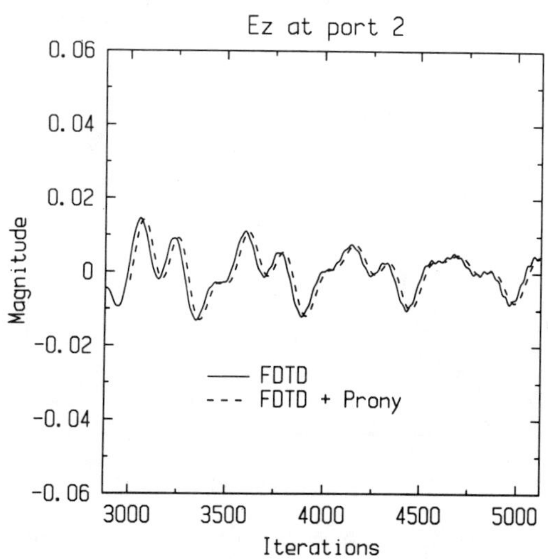

Fig. 6. Curve-fitting by Prony's method in a window of 2000 iterations.

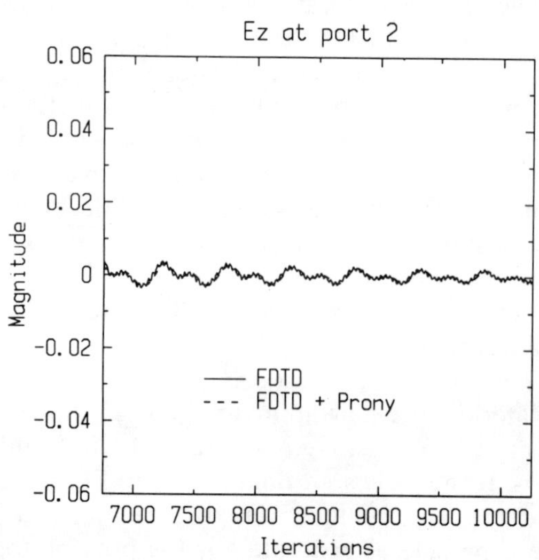

Fig. 7. Extrapolating the time transient waveform to late time.

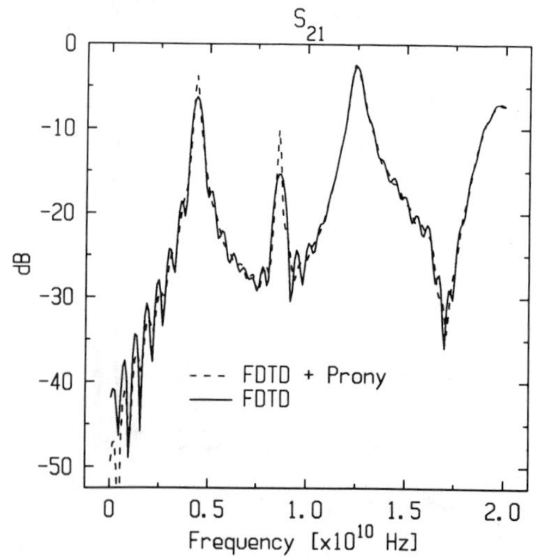

Fig. 8. Sharpening of the peaks in the S-parameter computation by including the late time transient waveform.

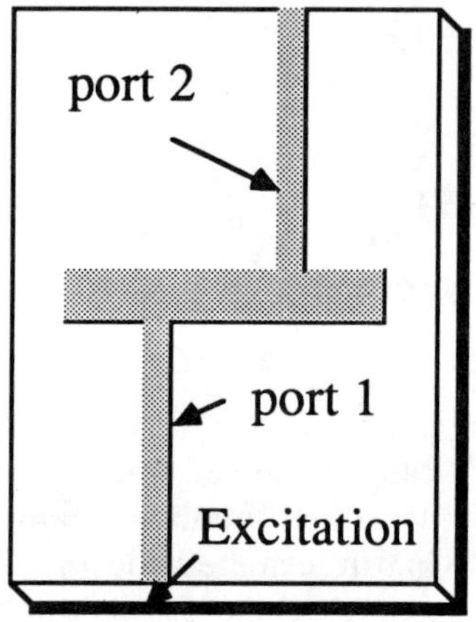

Fig. 9. Low pass filter.

S-parameter have been sharpened by including the late time transient waveform in the Fourier transform computation.

Having demonstrated the essential steps in applying the combination of FD-TD and Prony's methods to analyze the bandpass filter, we now validate the technique by applying it to analyze various structures in [7], which is chosen because experimental data are also available for comparison with the computational results. Numerical results obtained by the present approach for the lowpass filter in Fig. 9 are presented in Figs. 10 to 14, which are in good agreement with those measured as well as computed by direct FD-TD in [7]. Fig. 15 emphasizes that the same degree of accuracy in computing the frequency response in the pass band of the low pass filter is achieved by the present approach. The accuracy of the computation of the frequency response is compromised if the transient time record is terminated prematurely as shown in Fig. 16.

The actual computation time reduction realized in the implementation of the present approach to analyze microwave integrated circuits depends on the structures under investigation. The CPU times for the bandpass

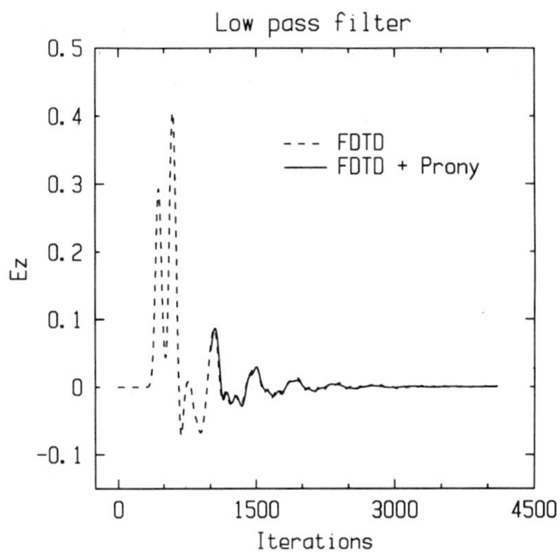

Fig. 10. Comparison of a direct 4096 iteration FD-TD time transient waveform at the output port of the low pass filter and the composite waveform generated by curve fitting a 2000 iteration FD-TD time record in a window of 1000 iterations from 1000 to 2000 iterations and extrapolating it to 4096 time steps.

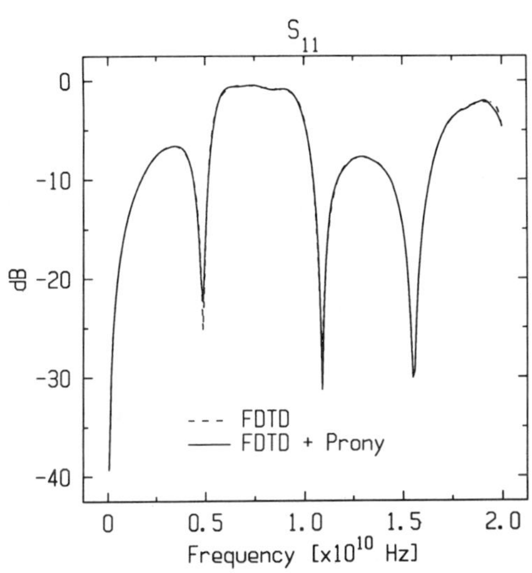

Fig. 11. Scattering parameter S_{11} of the low pass filter obtained by using a 4096 iteration FD-TD time record and a 4096 iteration time waveform extrapolated from a 2000 iteration FD-TD time record.

filter analyzed are listed in Table I and those for the low pass filter are in Table II. Generally speaking, a more significant reduction in computation time is realized for a structure that requires a longer FD-TD time record to generate accurate scattering parameters in the frequency domain.

The window, which is that portion of the short FD-TD time record sampled for curve fitting by Prony's method, is usually at the end of the short FD-TD time record. This window covers a significant fraction of the short FD-TD time record but does not include the beginning of the FD-TD waveform. The FD-TD waveform within the window is a steadily decaying oscillation. The FD-TD waveform is sampled frequently enough to allow effective curve fitting. The success of the curve fitting is checked

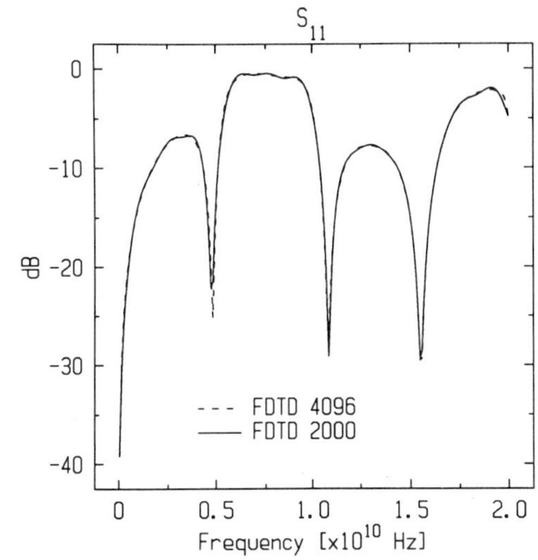

Fig. 12. Scattering parameter S_{11} of the low pass filter obtained by using a 4096- and a 2000-iteration FD-TD time record.

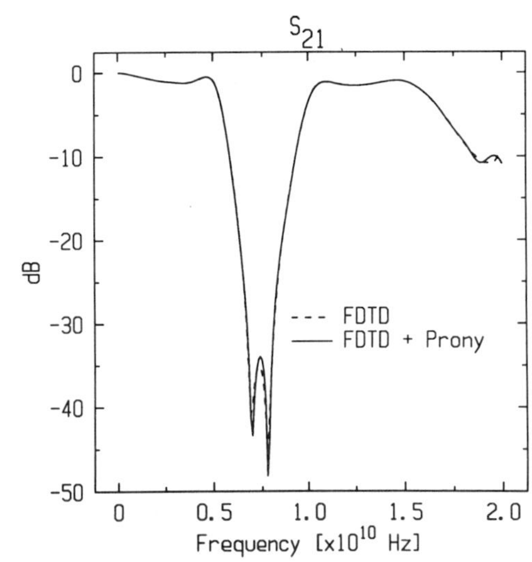

Fig. 13. Scattering parameter S_{21} of the low pass filter obtained by using a 4096 iteration FD-TD time record and a 4096 iteration time waveform extrapolated from a 2000 iteration FD-TD time record.

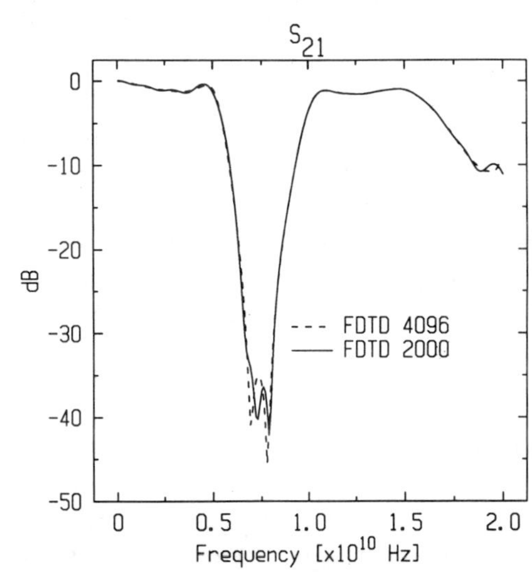

Fig. 14. Scattering parameter S_{21} of the low pass filter obtained by using a 4096- and a 2000-iteration FD-TD time record.

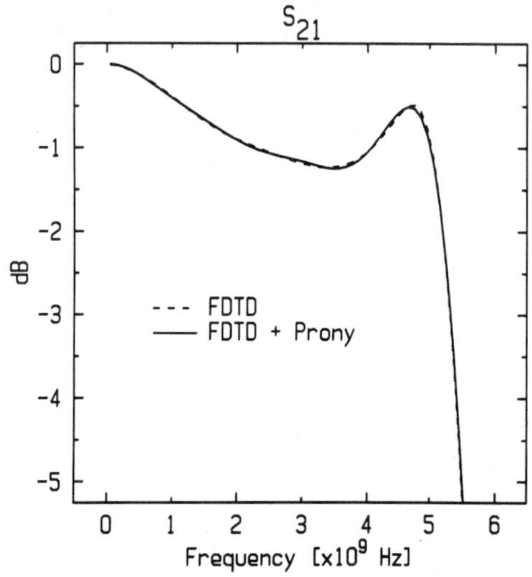

Fig. 15. Frequency response in the pass band of the low pass filter obtained by using a 4096 iteration FD-TD time record and a 4096 iteration time waveform extrapolated from a 2000 iteration FD-TD time record.

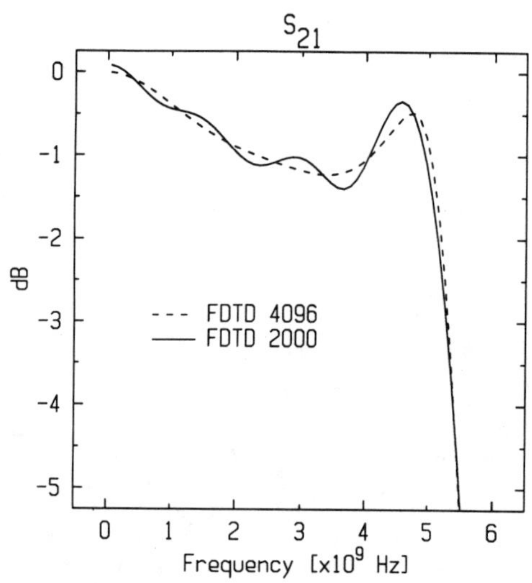

Fig. 16. Frequency response in the pass band of the low pass filter obtained by using a 4096- and a 2000-iteration FD-TD time record.

virtually as shown in Fig. 6 between the 3000 and 5000 iterations or in Fig. 10 between the 1000 and 2000 iterations. If these guidelines for choosing the window for curve fitting are followed, a relatively small number of terms in Prony's method, e.g., 18 terms for Figs. 6 and 7, and 20 terms for Fig. 10, will result in a satisfactory curve fitting and subsequently a successful extrapolation.

IV. Conclusion

As shown by comparison with the direct FD-TD generated results, the new approach using the combination of FD-TD and Prony's methods achieves the same degree of accuracy in a much shorter time. This method can be used for other potential applications that require the computation of response waveforms over a long period of time in order to obtain an accurate frequency domain result derived by Fourier transformation.

TABLE I
CPU Times for Bandpass Filter

5000 FD-TD iterations using 100×80×16 Yee's cells on a computer rated at 64 MFLOP	1.5278 h
Prony's extrapolation to 35 000 time steps, including 5 approximations to obtain 18 terms to fit the 100 points obtained by sampling at every 20 iterations in a window of 2000 FD-TD iterations between the 3000 and 5000 iterations.	74.88 s
Additional 30 000 FD-TD iterations to 35 000 time steps	9.1667 h

TABLE II
CPU Times for Low Pass Filter

2000 FD-TD iterations using 100×80×16 Yee's cells on a computer rated at 64 MFLOP	0.6112 h
Prony's extrapolation to 4096 time steps, including 6 approximations to obtain 20 terms to fit the 100 points obtained by sampling at every 10 iterations in a window of 1000 FD-TD iterations between the 1000 and 2000 iterations.	87.36 s
Additional 2096 FD-TD iterations to 4096 time steps	0.6404 h

References

[1] A. Taflove and K. R. Umashankar, "Review of FD-TD numerical modeling of electromagnetic wave scattering and radar cross section," *Proc. IEEE*, vol. 77, no. 5, pp. 682–699, May 1989.
[2] K. S. Yee, "Numerical solution of initial boundary value problems involving Maxwell's equations in isotropic media," *IEEE Trans. Antennas Propagat.*, vol. AP-14, pp. 302–307, May 1966.
[3] F. B. Hildebrand, *Introduction to Numerical Analysis.* New York: McGraw-Hill, 1974, pp. 457–465.
[4] M. L. Van Blaricum and R. Mittra, "Problems and solutions associated with Prony's method for processing transient data," *IEEE Trans. Antennas Propagat.*, vol. AP-26, pp. 174–182, Jan. 1978.
[5] G. Mur, "Absorbing boundary conditions for the finite-difference approximation of the time-domain electromagnetic field equations," *IEEE Trans. Electromagn. Compat.*, vol. EMC-23, pp. 377–382, Nov. 1981.
[6] T. Shibata, T. Hayashi, and T. Kimura, "Analysis of microstrip circuits using three-dimensional full-wave electromagnetic field analysis in the time domain," *IEEE Trans. Microwave Theory Tech.*, vol. MTT-36, no. 6, pp. 1064–1070, June 1988.
[7] D. M. Sheen *et al.*, "Application of the three-dimensional finite-difference time-domain method to the analysis of planar microstrip circuits," *IEEE Trans. Microwave Theory Tech.*, vol. MTT-38, no. 7, pp. 849–857, July 1990.

Enhancing Finite-Difference Time-Domain Analysis of Dielectric Resonators Using Spectrum Estimation Techniques

ZHIQIANG BI, YING SHEN, KELI WU, AND JOHN LITVA
COMMUNICATIONS RESEARCH LABORATORY, McMASTER UNIVERSITY
HAMILTON, ONTARIO, CANADA L8S 4K1

Abstract *In this paper we discuss the use of digital filtering and spectrum estimation techniques for improving the efficiency of the FD-TD algorithm in solving eigenvalue problems. It is demonstrated by means of numerical results that the efficiency of the FD-TD method for dielectric resonator analysis has been improved at least one order of magnitude. This new result makes it possible for the FD-TD method to be a practical designing tool for the dielectric resonator analysis.*

I. Introduction

The optimization of the performance of resonators in microwave circuits requires accurate and efficient methods for calculating the resonant frequencies and the spatial distributions of the field. Various methods have been developed to study the resonant frequencies of resonant structures. Most of them, such as; the mode matching method, integral equation method, and finite element method, are carried out in the frequency domain [1].

The finite-difference time-domain (FD-TD) method has been widely used for solving electromagnetic problems. Resently, it has been used to solve eigenvalue problems associated with resonator structures [2],[3] and to calculate critical parameters for complex microstrip antennas [4]. All of these results have shown the FD-TD method to be a very powerful tool for eigenvalue analysis, primarily because of two desirable attributes. First, it can be applied to problems exhibiting a complex structure which may be very difficult to solve using other analytical or numerical methods. Second, only one computation is required to get the frequency domain results over a large frequency spectrum. However, this method has one significant drawback, which is that it requires a very long computation time for extracting the resonant frequencies from the FD-TD results; for example, in the case of the problem discussed in [2], the time iteration N has to be as large as $N = 2^{16}$.

The main purpose of this paper is to introduce the use of digital filtering and modern spectrum estimation techniques with the FD-TD method, as a means for overcoming above limitation. By using numerical results, it will be shown that modern spectrum estimation techniques can reduce the time taken to solve a problem, such as that discussed in [2], by one order of magnitude, without any loss of accuracy in calculating the resonant frequencies.

II. FD-TD Method for Resonator Analysis

For ease of description, the method is described by referring to the generalized cylindrically shaped dielectric resonator (DR) in Fig. 1. To start the computation, the initial electric and magnetic fields are set to zero throughout the grid, except at one selected point. Here the electric field is set to 1. This unit impulse source will excite a large number of modes. Using the FD-TD algorithm [5], Fig. 2a gives the computed electric field in the time domain at the point of observation. The resonant frequencies can be obtained by taking the Fourier transform of the computed time domain response. The field distribution for any particular frequency can be obtained by performing Fourier transform at each point in the computation domain at that frequency. With the objective of getting more accurate estimates of the resonant frequency and field distribution than that have been obtained in [2], the following procedure is put forward.

The procedure to be followed is based on the signal analysis of the time domain results obtained using the FD-TD method. In this section, it is assumed that the sequence length of the FD-TD result $\{x(n)\}$ is very long, where $x(n)$ is one of the field components. The procedure that has been suggested here is to do the numerical integration of Fourier Transform of $x(t)$ directly in the interested frequency band

$$\begin{aligned} X(f) &= \int_0^\infty x(t)\exp(-j2\pi ft)dt \\ &\approx \int_0^{N\Delta t} x(t)\exp(-j2\pi ft)dt \\ &\approx \sum_{n=0}^{N-1} x(n)\exp(-j2\pi fn\Delta t)\Delta t. \end{aligned} \qquad (1)$$

The advantage of this method is that it removes ambiguities sometimes encountered with the discrete Fourier

Reprinted from *IEEE MTT-S Digest*, pp. 869-872, 1992.

Transform, due to narrowband signal components with center frequencies that lie in the gaps between the $N/2$ frequency points evaluated with the DFT. It will be shown in the following that, when FD-TD method is used for resonator analysis, the time domain results are signals which consist of many narrowband signal components. The accuracy of calculating the spectral peaks, i.e. the field distribution, is also enhanced by using (1).

The efficiency of calculating (1) can be greatly improved by using the following method. Instead of using the original sequence $\{x(n)\}$ obtained using the FD-TD result, a new sequence $\{x_1(n)\}$ is used in (1), which is obtained by desampling the $\{x(n)\}$ at a certain rate. The desampling rate is determined by the ratio of the half sampling frequency $f_s/2$ to the maximum frequency f_{max} of the long sequence $\{x(n)\}$. Because $\{x_1(n)\}$ is much shorter than the original sequence $\{x(n)\}$, the time required to analyze the new time domain sequence can be greatly reduced, with no reduction in the accuracy of the result. The theory which supports this treatment is the popular Nyquist sampling theorem. For dielectric resonator analysis, the maximum frequency of the FD-TD time domain result is equal to about $f_{max} = v/(2\Delta h)$, where v is the speed of light in the largest dielectric constant material and Δh is the space step (where a uniform grid is assumed) [5].

In Fig. 2b, the solid line gives the result obtained by applying (1) to the long sequence, and the dashed line gives the result obtained by applying (1) to the decimated sequence $\{x_1(n)\}$. The two results are exactly the same and therefore overlap in Fig. 2b.

III. Use of Digital Filtering and Modern Spectrum Estimation Techniques with FD-TD Method

The objective of this section is, based on a much shorter sequence $\{x_2(n)\}$ obtained directly from the output of FD-TD algorithm, to use digital filtering and modern spectrum estimation techniques to extract the resonant frequencies of the dielectric resonator. Suppose $\{x_2(n)\}$ is the sequence consisting of the first two thousand data points in $\{x(n)\}$. The DFT spectrum of $\{x_2(n)\}$ is shown in Fig. 3a. After desampling $\{x_2(n)\}$, using a desampling rate of $(f_s/2)/f_{max}$, which is about 10, we get a sequence $\{x_3(n)\}$ whose DFT spectrum is shown in Fig. 3b. Because we are interested in the lower frequency band, we further process the signal $\{x_3(n)\}$ by using a decimating filter to get $\{x_4(n)\}$, whose DFT spectrum is shown in Fig. 3c. In order to improve the accuracy of estimating the resonant frequencies of the first few modes, we further process $\{x_4(n)\}$ with a low-pass filter and get $\{y(n)\}$, whose DFT spectrum is shown in Fig. 3d. In the next phase of the work we carry out a search for a good high resolution spectrum estimator, with which to extract the resonant frequencies from the data $\{y(n)\}$.

From the behavior of the spectrum of $\{x(n)\}$, based on the results given in Fig. 2b, it seems reasonable to assume that $\{y(n)\}$ is composed of sinusoidal components. One of the best kinds of methods for estimating the frequencies of sinusoidal components is the multiple signal classification (MUSIC) method [6],[7]. This method belongs to the eigendecomposition-based class of super-resolution spectrum estimation methods. The term "super-resolution" refers to the fact that this class of methods have the ability to surpass the limiting behavior of classical Fourier-based methods. There are a number of reasons for our choosing the MUSIC algorithm from amongst this class of methods. These are: (i) it is easy to implement, (ii) it provides good performance, (iii) it is used as a bench mark in the field of signal processing, and (iv) it provides a good introduction to modern spectrum estimation.

IV. Numerical Results

Using the MUSIC algorithm, the signal $\{y(n)\}$ which was obtained after filtering processing in the last section is analyzed. The result is shown in Fig. 4. The dashed line was obtained by applying a Fourier transform (1) to a very long FD-TD sequence, corresponding to 20000 time iterations in the FD-TD algorithm. The dotted curve gives the result from Fourier processing (1) of the first two thousand points in the former sequence. This shortened sequence corresponds to 2000 time iterations in the FD-TD algorithm. From this curve we see that, for short data records, the resonant frequencies cannot be accurately estimated using the Fourier transform. Biases occur in the locations of the first and fourth resonant frequencies and the second and third resonant frequencies are missing altogether. The solid line gives the result of application of digital filtering and the MUSIC spectral estimation technique to the shorter data. In the MUSIC algorithm, the data length of $\{y(n)\}$, K, was equal to 100, the order of the correlation matrix, $M+1$, was determined by the relation $M = 2K/3$. The accuracy of the method increases with increasing M. However, $M+1$ should not be larger than the number of data points. The choice for the order of the signal subspace, p, is based on the eigenvalue spectrum of the estimated correlation matrix. For our example, p was equal to 21. When p was changed from 21 to a higher value, we still got accurate frequency estimates. This suggests that the method is robust. Comparing the solid and dashed lines, it is seen that the same order of accuracy is obtained in the resonant frequency estimation by applying signal processing and spectral estimation to a short data set as that obtained by applying a Fourier transform to a much longer data set.

V. Conclusions

There are two main results coming from the present study of the FD-TD method. Digital filtering and modern spectrum estimation techniques were successfully incorporated with the FD-TD method as a means of improving its efficiency for carrying out eigenvalue analysis. The efficiency and validity of the method are demonstrated using both numerical and measured results. Another relatively new spectrum estimation method, which is called Thomson's multiple-window-method (MWM) [8], was also tested with FD-TD data. The same good frequency estimates were obtained using MWM. The second main outcome of this research was the application of signal analyses to the time domain data obtained using the FD-TD algorithm. It has been shown that the FD-TD time domain signal for dielectric resonator analyses is much over sampled. The data that needs to be retained for later processing can be greatly compressed, without degrading the accuracy of the analysis. This conclusion is valid when the FD-TD method is used to analyze microstrip components and antennas. In these latter cases, the maximum frequency of time domain result, f_{max}, which sets the desampling or compressing rate, is not determined by the cutoff frequency of the FD-TD algorithm itself, but rather by the maximum frequency of the excitation gaussian pulse. According to our experience, for the analysis of microstrip antennas and components [4], the data from the FD-TD results can be compressed by at least one order of magnitude. The above two conclusions are applicable to other time domain methods, such as the Transmission Line Matrix method.

In conclusion, it should be mentioned that signal processing and spectrum estimation techniques can greatly improve both the capability and the efficiency of the time domain methods. This point has been reinforced by several papers [9-14], where the authors have to greater or less degree drawn on signal processing techniques to improve the performance of their numerical algorithms.

VI. Acknowledgement

The authors wish to thank Mr. A. Drosopoulos, Mr. C. Wu, Drs. Q. Wu, W. Chen and Mr. Y. Zhou at Communications Research Laboratory, McMaster University for their helpful discussions on this research.

References

[1] D. Kajfez and P. Guillon, "Dielectric Resonators", Artech House, 1986.

[2] A. Navarro, M. J. Nuñez and E. Martin, "Study of TE_0 and TM_0 modes in dielectric resonators by a finite difference time-domain method coupled with the Discrete Fourier transform," IEEE Trans. Microwave Theory Tech., vol. MTT-39, pp. 14-17, Jan. 1991.

[3] D. H. Choi and W. J. R. Hoefer, "The finite difference-time domain method and its application to eigenvalue problems," IEEE Trans. Microwave Theory Tech., vol. MTT-34, pp. 1464-1470, Dec. 1986.

[4] C. Wu, K. Wu, Z. Bi and J. Litva, "Accurate characterization of planar printed antennas using finite difference time domain method," to appear in IEEE Trans. Antennas Propagat..

[5] Z. Bi, Y. Shen, K. Wu and J. Litva,"Fast finite-difference time-domain analysis of resonator using digital signal processing," accepted for publication in IEEE Trans. Microwave Theory Tech..

[6] P. Stoica and T. Söderström, " Statistical analysis of MUSIC and subspace rotation estimates of sinusoidal frequencies," IEEE Trans. Acoust., Speech, Signal Processing, vol. ASSP-39, pp.1836-1847, August 1991.

[7] S. Haykin, "Adaptive filter theory," 2nd ed., Prentice-Hall, Inc., 1991.

[8] A. Drosopoulos and S. Haykin, "Adaptive radar parameter estimation with Thomson's Multiple-Window method," chapter 7, In S. Haykin and A. Steinhardt eds., Adaptive Radar Detection and Estimation, Wiley, 1991.

[9] Z. Bi, K. Wu, C. Wu and J. Litva, "A Dispersive boundary condition for microstrip component analysis using the FD-TD method," to appear in IEEE Trans. Microwave Theory Tech., April 1992.

[10] J. G. Blaschak and T. G. Moore, "Adaptive absorbing boundary condition," presented at Process in Electromagnetics Research Symposium, Cambridge, Massachusetts, USA, July 1991.

[11] W. L. Ko and R. Mittra, "A combination of FD-TD and prony's method for analyzing microwave integrated circuits," 1991 IEEE AP-S Int. Symp. Dig., pp. 1742-1745, June 1991.

[12] S. T. Chu and S. K. Chaudhuri, "Combining modal analysis and the FD-TD method in the study of dielectric waveguide problems," IEEE Trans. Microwave Theory Tech., vol. MTT-38, pp. 1755-1760, Nov. 1990.

[13] Z. Bi, J. Litva, K. Wu and C. Wu, "Using digital signal processing techniques with the FD-TD method," submitted to IEEE AP-S, 1992, Chicago, USA.

[14] Z. Bi, K. Wu and J. Litva, "Designing dispersive boundary condition (DBC) for the FD-TD method using digital filtering theory," submitted to IEEE AP-S, 1992, Chicago, USA.

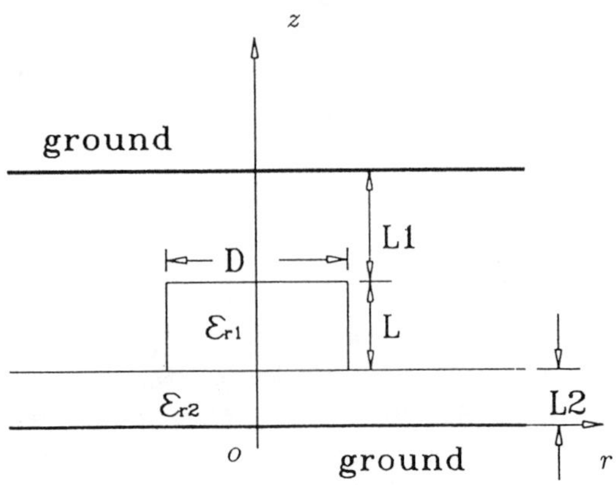

Figure 1 A generalized cylindrical shaped dielectric resonator.

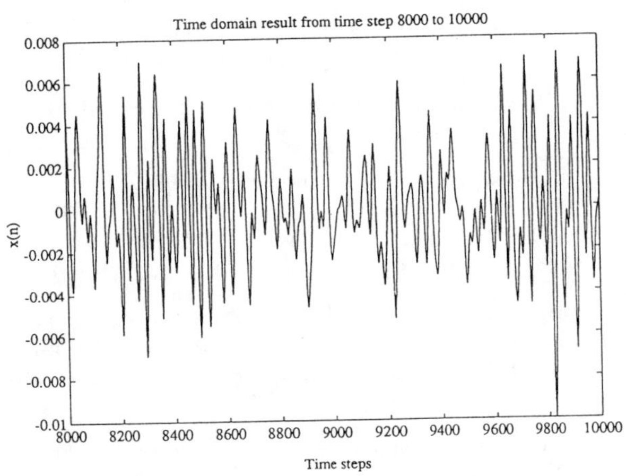

(a)

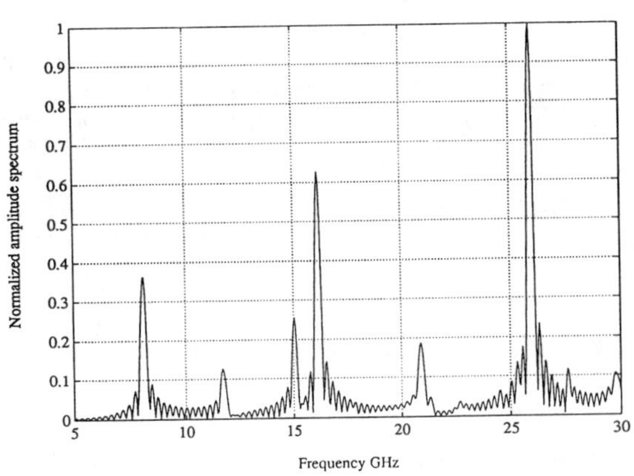

(b)

Figure 2 (a) Time domain result directly obtained from the FD-TD algorithm; (b) Normalized amplitude spectrum of Fourier transform of long sequence and desampled shorter sequence.

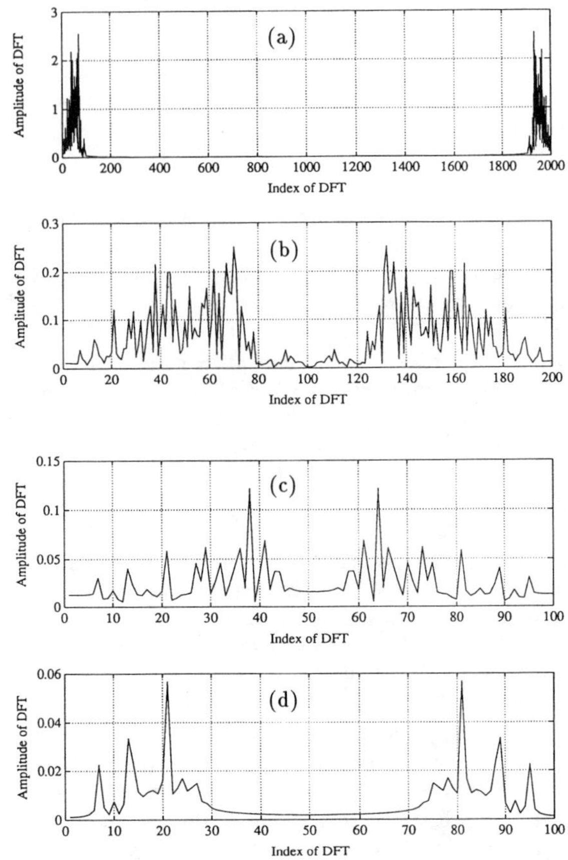

Figure 3 Digital filtering processing of the FD-TD time domain result. (a-d) DFT spectrums.

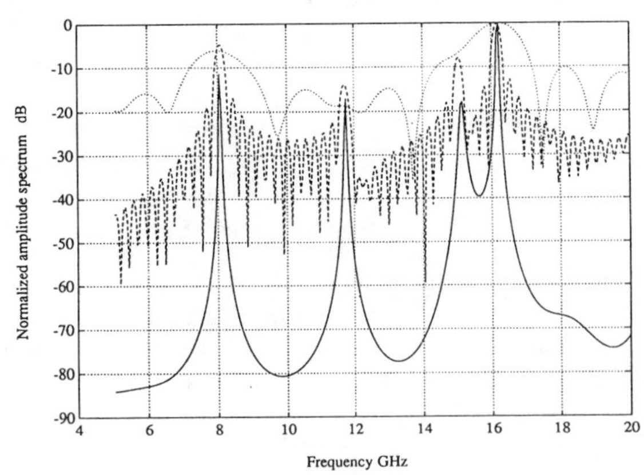

Figure 4 Resonant frequency estimation using different methods. Dashed line: 20000 iterations and using Fourier integration Eq. (1); Dotted line: 2000 time iterations and using Fourier integration Eq.(1); Solid line: Using 2000 time iterations in FD-TD algorithm and using digital filter processing and MUSIC method.

Recursive Covariance Ladder Algorithms for ARMA System Identification

PETER STROBACH, MEMBER, IEEE

Abstract—**Fast recursive-in-time procedures for ARMA system identification which realize the desired (pole-zero) transfer function in a classical transversal structure, or in the "growing-order" Ladder-Lattice form, have been available for years. Although very attractive due to their low computational costs, these procedures often suffer from the underlying ill-conditioned fast-Kalman-type algorithm structure which allows roundoff errors to accumulate in the time direction in each stage of the fast estimation algorithms. This effect causes the algorithms to degrade in an unpleasant way so that they lose their fast start-up and tracking capability and finally perform not better than simple gradient algorithms when operated in finite arithmetic environments. To cope with these problems, the recently developed method of pure order recursive ladder algorithms (PORLA) is extended in this paper so as to facilitate the identification of ARMA models. Since the time recursion in this new method is limited on the calculation of the input data covariance matrix, roundoff errors cannot propagate in time in higher stages of the pure order recursively constructed ladder form. Thus, the superior LS tracking and fast start-up capability of the new algorithms is not corrupted by roundoff error. Furthermore, the new algorithms allow the use of higher order recursive windows on the data (e.g., recursive Hanning) which again significantly improves the tracking as well as the steady-state behavior. The algorithm development presented here is backed up with a computer program, an instructive example for implementation of the new method on a massively parallel processor (MPP), and finally, we show several experimental results which confirm the superior properties of the new PORLA method over conventional techniques.**

I. Introduction

PARAMETRIC techniques for high-performance modeling of rapidly time-varying systems have been widely investigated in recent years. Especially, estimation techniques based on the autoregressive (AR) model have been studied extensively [7], [10], [16], [17], [19]. In some situations, particularly in the presence of sharp zeros in the transfer function of the unknown system, the all-pole model appears to have poor performance. This has motivated a considerable interest in least-squares (LS) system identification techniques which use the more general pole-zero (ARMA) model. In the past several years, ARMA system identification algorithms have been developed by Lee, Friedlander, and Morf [18], [22], Grenier [11], Benveniste and Chaure [7], Karlsson and Hayes [13], and others. A survey of ARMA modeling methods is given by Morf *et al.* [21], [22] and Goodwin and Payne [2]. The stationary case of ARMA system identification plays a central role in spectral estimation methods [8], [14].

The basic motivation for the development of exact LS ARMA system identification algorithms clearly was their expected superior fast LS start-up and tracking capability over conventional gradient-based techniques. Some of the most recently developed algorithms received attention since they belong to the class of "fast" LS estimation techniques, where the computational complexity is in the order of p operations per time instant, where p denotes the model order. Different algorithms of this "fast" (in terms of a small number of arithmetic operations) type which are based on the transversal structure [23], or the growing-order Ladder-Lattice form [18], have been introduced. Evaluating these algorithms, it is most important to realize that they all derive from a central time-update theorem which can be formulated using the matrix inversion lemma for partitioned matrices [1], [5] (algebraic approach), or in a more refined geometrical framework as first appeared in the basic work of Lee [16]. A more general treatment of this geometrical approach is given in [4]. Since the algorithms which have been presented there are based on a common time-update theorem, it is not surprising that they all exhibit a quite similar structure, namely, the so-called "fast-Kalman"-type algorithm structure. This algorithm structure is characterized by the property that the estimated parameters are updated via time-recursive formulas which heavily involve residuals of the inverse filter in the updating process. Such structures, which can be found in all the currently available fast LS techniques, cause several inherent difficulties when used in real-time finite arithmetic fast tracking and modeling applications. So far, fast LS ARMA estimation algorithms have only been derived for the growing (exponentially weighted) data window and the sliding rectangular window. These windows, however, can be shown to have poor performance in terms of tracking accuracy and steady-state behavior. Higher order windows, such as the Hanning window or the Barnwell window [6], could improve the tracking behavior as well as the steady-state behavior since these windows are known to have much better spectral properties than the simple rectangular and exponential windows. Unfortunately, since the time-update recursions in the fast LS algorithms prohibit the incorporation of higher order windows, the tracking performance as well as the steady-state behavior of currently available fast LS ARMA estimation algorithms is quite

Manuscript received April 7, 1986; revised October 23, 1987.

The author is with the SIEMENS Information Systems Laboratory, ZT ZTI INF 121 Otto-Hahn-Ring 6, D-8000 München 83, West Germany.

IEEE Log Number 8719019.

Reprinted from *IEEE Transactions on Acoustics, Speech, and Signal Processing,* pp. 560-580, April 1988.

limited and far from optimum. These conjunctions are discussed in detail in Section VII of this paper, where we present some experimental results in order to illustrate our claims.

But there are two more phenomena which gain considerable importance as soon as a real-time or "on-line" implementation of fast LS techniques on a real-world digital signal processor is desired. These are the problems of numerical stability and numerical accuracy. Numerical accuracy and stability are two completely different properties of adaptive algorithms caused by roundoff error. So far, these problems have been only briefly analyzed and discussed [31]–[35]. An algorithm is called numerically stable if it can be operated over infinite time with a bounded error in the estimated parameters. An algorithm is called to have a good numerical accuracy if the error in the estimated parameters is small when compared to an infinite precision realization of the same algorithm. Note that an algorithm, although very accurate on short data records, could be unstable in the long run, i.e., the error in the estimated parameters grows unbounded with time until it destroys the normal operation of the algorithm. On the other hand, an algorithm which performs very inaccurately, could exhibit the property that it runs stably over infinite time, i.e., the error may be considerably high but bounded. It is clear from these considerations that numerical stability is an indispensable requirement, as soon as "on-line" applications are an issue. Unfortunately, it can be shown that such estimation algorithms which involve sliding window computations are numerically unstable when implemented with floating-point arithmetic. The same algorithms perform numerically stable when fixed-point arithmetic is used. This very surprising conjunction is explained more in detail in Section VII of this paper and was also considered in [31]. Now, since sliding window algorithms are necessary for fast tracking, it is very important to investigate whether or not a given algorithm can be computed with a sufficient numerical accuracy when fixed-point arithmetic is used.

Under this objective, we have studied the numerical accuracy and the behavior of the most important fast LS estimation algorithms for ARMA system identification carefully. Especially, we have investigated the start-up and tracking behavior rather than the steady-state behavior. This test strategy was motivated by the fact that the investigated fast LS algorithms have been developed through the demand for fast start-up and tracking capability over existing gradient techniques. The very surprising result which we finally obtained from our experiments (some of them are reported in Section VII) can be summarized as follows: the fast LS algorithms lose their fast start-up capability and perform not significantly better than simple gradient algorithms when computed with fixed-point arithmetic. To demonstrate this, we have implemented two well-known fast LS ARMA ladder algorithms, namely, the method of Lee [16], [18], and the more refined method with a direct updating of the reflection matrices which was recently introduced by Ling *et al.* [19]. The reason for the poor results obtained by using these techniques was found to be the time-recursive update formulas which realize a certain coupling of the ladder form inverse filter and the coefficient update recursions and additionally enable roundoff errors to accumulate in the recursion variables. Until now, these important conjunctions have been commonly ignored in the modern adaptive signal processing literature. The only tentative explanation for this seems to be the fact that indeed all the currently discussed fast LS techniques have the same problems since they are all related on the fast-Kalman-type algorithm structure.

This gives rise to the question whether or not a basically different algorithm structure can be found, which prohibits roundoff errors to propagate in the time direction in higher stages of the algorithm. This demand can be satisfied by an algorithm which constructs the true LS ladder form by a pure order recursive procedure. Algorithms of this "pure order recursive" type have been introduced much earlier for scalar sequences in the stationary case. See the basic work of Makhoul [20] and Cumani [9]. These algorithms, particularly the method of Cumani, have proven superior numerical properties. However, they still exhibit the inherent drawback that they are no exact LS techniques, and mainly that they suffer from an enormous amount of computation which is proportional to the cube of the model order. From this point of view, a true LS solution of the "pure order recursive ladder algorithm" (PORLA) can be expected to provide superior numerical properties over conventional fast techniques. A closed form of this pure order recursive true LS ladder algorithm for scalar sequences was first introduced in [25]–[28]. The algorithm development presented there was based on an algebraic approach using new recursion variables, the so-called "generalized residual energies" (GRE's). The so-obtained true LS PORLA method has only a complexity proportional to the square of the model order. Most recently, several extensions of the PORLA method for scalar sequences have appeared, such as the extension to the stationary case [29] and an approach to FIR system identification [30], [24]. The numerical properties of the PORLA method in the scalar case were considered in [35], where [40] presents some interesting VLSI realization structures for the PORLA technique backed up with several instructive experimental results of the proposed structures when realized with fixed-point arithmetic.

The purpose of this paper is to extend the PORLA method to the more general ARMA (pole-zero) modeling problem. As will become apparent in this paper, the success of the PORLA method for ARMA system identification is rooted in the fact that this new technique has an algorithm structure, in which the tracking and the modeling problems are treated as independent subalgorithms as follows. First, the tracking of the nonstationary data is performed by the time-recursive "on-line" computation of the input/output data covariance block matrix. Arbitrary windowing techniques can be easily incorporated in order to control the fast start-up and tracking capability as well as the steady-state behavior effectively. Some useful recursive windowing algorithms are summarized in Appendix A. Second, the modeling problem is solved by the two-channel PORLA method in which the ARMA modeling problem is embedded. The algorithm is capable of constructing the true LS adaptive ARMA ladder form exclusively from the covariance (second-order) information. This new technique does not suffer from uncertain residual computations since these quantities are not required in the updating process of the PORLA algorithm. It is clear that the new method breaks up the loops of error propagation in time, and this was found to be the deep reason why the fast start-up and tracking capability of the PORLA algorithm is not corrupted by roundoff error. This claim is also confirmed by the simulation results which are shown in Section VII. The paper is organized as follows. In Section I we define the problem of ARMA system identification in a Normal Equations formulation. It is shown how the ARMA Normal Equations can be embedded in the more general two-channel AR system identification problem. The ARMA ladder form is introduced in Section III. In Section IV, the PORLA method for ARMA system identification is derived. In Section V, several cases of interest where the model input is white or even unknown are briefly discussed. Some examples of implementation and application of the new algorithm are discussed in Section VI. This section is backed up with a computer program which is structured such that the inherent parallelism of the new technique becomes very obvious. This suggests a massively parallel processor (MPP) structure for efficient real-time implementation of the new ARMA estimation method. Finally, Section VII presents a collection of experimental results which basically focus on the three problems of adaptive LS estimation algorithms, namely, the problem of data windowing, the problem of numerical stability, and the problem of numerical accuracy. Section VIII concludes this paper. In Appendix A we have listed the most important recursive windowing algorithms which can be combined with the new ARMA covariance ladder algorithm. Appendix B extends the PORLA technique to the most general vector autoregressive (VAR) modeling problem.

II. Problem Statement

Fig. 1 is a block diagram of the ARMA system identification model. Both the input and the output of the unknown system are accessible. No restricting preassumptions are made about the input/output processes $\boldsymbol{x}(t)$ and $\boldsymbol{u}(t)$. It is desired to model the unknown system as a time-varying ARMA (pole-zero) filter with a sufficiently large order p.

The approach used here is to determine the ARMA model parameters such that the LS cost function $E_p(t)$ is minimized, where

$$E_p(t) = \boldsymbol{e}_p^{u^T}(t)\, \boldsymbol{e}_p^u(t) \tag{2.1}$$

$$\boldsymbol{e}_p^u(t) = \left[e_{p,0}^u(t), e_{p,1}^u(t), \cdots, e_{p,L-1}^u(t)\right]^T. \tag{2.2}$$

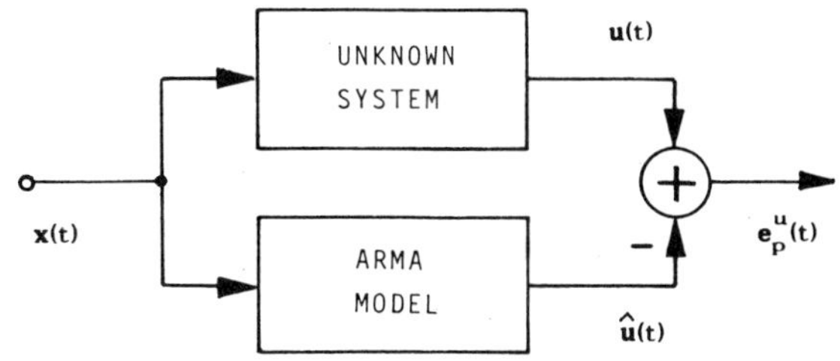

Fig. 1. ARMA system identification model.

$\boldsymbol{e}_p^u(t)$ is the residual vector of the pth-order problem and T denotes transpose. L determines a sliding rectangular window (this restriction will later be relaxed) wherein the ARMA parameters are assumed as constant. Efficient covariance ladder algorithms for estimating the ARMA parameters can be developed when the covariance matrices $\boldsymbol{W}^{xx}(t)$ and $\boldsymbol{W}^{uu}(t)$, as well as the cross covariance matrix $\boldsymbol{W}^{xu}(t)$, are known.

$$\boldsymbol{W}_{i,j}^{xx}(t) = \boldsymbol{x}^T(t-i)\, \boldsymbol{x}(t-j) \tag{2.3a}$$

$$\boldsymbol{W}_{i,j}^{uu}(t) = \boldsymbol{u}^T(t-i)\, \boldsymbol{u}(t-j) \tag{2.3b}$$

$$\boldsymbol{W}_{i,j}^{xu}(t) = \boldsymbol{x}^T(t-j)\, \boldsymbol{u}(t-j), \tag{2.3c}$$

where $\boldsymbol{x}(t)$ and $\boldsymbol{u}(t)$ are the (observable) input/output process vectors.

$$\boldsymbol{x}(t) = \left[x(t), x(t-1), \cdots, x(t-L+1)\right]^T \tag{2.4a}$$

$$\boldsymbol{u}(t) = \left[u(t), u(t-1), \cdots, u(t-L+1)\right]^T. \tag{2.4b}$$

A. The ARMA Normal Equations

Suppose now that the unknown system of Fig. 1 can be modeled by a time-varying ARMA filter of the following form (see Fig. 2), where $\boldsymbol{X}_p(t)$ and $\boldsymbol{U}_p(t)$ are the matrices of the joint input/output vectors of past observations, and $\boldsymbol{a}_p(t)$ and $\boldsymbol{b}_p(t)$ are the ARMA coefficient vectors.

$$\boldsymbol{u}(t) = \boldsymbol{x}(t) + \boldsymbol{X}_p(t)\, \boldsymbol{b}_p(t) + \boldsymbol{U}_p(t)\, \boldsymbol{a}_p(t) \tag{2.5}$$

$$\boldsymbol{X}_p(t) = \left[\boldsymbol{x}(t-1), \boldsymbol{x}(t-2), \cdots, \boldsymbol{x}(t-p)\right] \tag{2.6a}$$

$$\boldsymbol{U}_p(t) = \left[\boldsymbol{u}(t-1), \boldsymbol{u}(t-2), \cdots, \boldsymbol{u}(t-p)\right] \tag{2.6b}$$

$$\boldsymbol{b}_p(t) = \left[b_1(t), b_2(t), \cdots, b_p(t)\right]^T \tag{2.7a}$$

$$\boldsymbol{a}_p(t) = \left[a_1(t), a_2(t), \cdots, a_p(t)\right]^T. \tag{2.7b}$$

The linear LS estimate of $\boldsymbol{u}(t)$, denoted by $\hat{\boldsymbol{u}}(t)$, is then given by

$$\hat{\boldsymbol{u}}(t) = \boldsymbol{x}(t) + \boldsymbol{X}_p(t)\, \hat{\boldsymbol{b}}_p(t) + \boldsymbol{U}_p(t)\, \hat{\boldsymbol{a}}_p(t). \tag{2.8}$$

Introducing the residual vector (prediction error vector) $\boldsymbol{e}_p^u(t)$, and expressing it in terms of the estimated ARMA

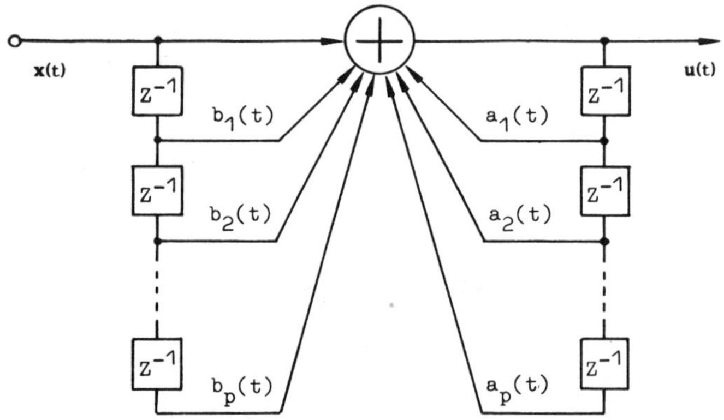

Fig. 2. ARMA representation of unknown system.

parameters, gives

$$e_p^u(t) = u(t) - \hat{u}(t) \tag{2.9}$$

$$e_p^u(t) = u(t) - x(t) - X_p(t)\,\hat{b}_p(t) - U_p(t)\,\hat{a}_p(t). \tag{2.10}$$

Utilizing (2.10), the ARMA LS cost function (2.1) can be written as a function of the estimated parameters as follows:

$$E^{uu}(t) = E^{uu}(\hat{a}, \hat{b}, \hat{a}^T, \hat{b}^T, t). \tag{2.11}$$

It is desired to minimize the prediction error energy E^{uu}. This is achieved by setting the partial derivatives of E^{uu}, with respect to $\hat{a}^T$ and $\hat{b}^T$, to zero.

$$\frac{\partial E^{uu}}{\partial \hat{a}^T} = -U^T u + U^T x + U^T X\hat{b} + U^T U\hat{a} = 0 \tag{2.12a}$$

$$\frac{\partial E^{uu}}{\partial \hat{b}^T} = -X^T u + X^T x + X^T X\hat{b} + X^T U\hat{a} = 0. \tag{2.12b}$$

The relations (2.12a, b) can be rewritten in a more convenient matrix/vector notation as follows:

$$\left[\begin{array}{c|c} U^T U & U^T X \\ \hline X^T U & X^T X \end{array}\right] \left[\begin{array}{c} \hat{a} \\ \hline \hat{b} \end{array}\right] = \left[\begin{array}{c} U^T[u - x] \\ \hline X^T[u - x] \end{array}\right]. \tag{2.13}$$

The relations (2.13) have been termed the Normal Equations of ARMA system identification. Recalling the definitions of the input/output covariance matrices (2.3a, b) and the input/output cross covariance matrix (2.3c), we can compose the ARMA Normal Equations (2.13) from elements of the following partitioning schemes:

$$\left[\begin{array}{c|c} x^T x & x^T X \\ \hline X^T x & X^T X \end{array}\right] = W^{xx} \tag{2.14a}$$

$$\left[\begin{array}{c|c} u^T u & u^T U \\ \hline U^T u & U^T U \end{array}\right] = W^{uu} \tag{2.14b}$$

$$\left[\begin{array}{c|c} x^T u & x^T U \\ \hline X^T u & X^T U \end{array}\right] = W^{xu}. \tag{2.14c}$$

B. ARMA Embedding

Instead of solving the ARMA Normal Equations directly, we introduce an embedding approach which allows the ARMA system identification model to be embedded in a two-channel AR estimation framework. Consider the order p, two-channel AR predictor of the form

$$\begin{bmatrix} e_p^{x^T}(t) \\ e_p^{y^T}(t) \end{bmatrix} = \begin{bmatrix} x^T(t) \\ y^T(t) \end{bmatrix} + \sum_{m=1}^{p} \begin{bmatrix} a_{11}^{(m)}(t) & a_{12}^{(m)}(t) \\ a_{21}^{(m)}(t) & a_{22}^{(m)}(t) \end{bmatrix} \cdot \begin{bmatrix} x^T(t-m) \\ y^T(t-m) \end{bmatrix} \tag{2.15}$$

where $x(t)$ and $y(t)$ are the predicted processes, $e_p^x(t)$ and $e_p^y(t)$ are the prediction error residuals, and $a_{**}^{(m)}(t)$ are the (time-varying) two-channel AR parameters of stage m. Rewriting the lower part of (2.15) gives

$$e_p^y(t) = y(t) + \sum_{m=1}^{p} a_{21}^{(m)}(t)\,x(t-m) + \sum_{m=1}^{p} a_{22}^{(m)}(t)\,y(t-m). \tag{2.16}$$

Now let the input process $y(t)$ of the two-channel AR predictor be the reduced output signal of the unknown system as follows:

$$y(t-m) = u(t-m) - x(t-m) \qquad 0 \le m \le p; \tag{2.17}$$

and let the input signal $x(t)$ of the two-channel AR predictor be connected with the input of the unknown system. Then, by substituting (2.17) into (2.16), we obtain

$$e_p^y(t) = u(t) - x(t) - \sum_{m=1}^{p} \left[a_{22}^{(m)}(t) - a_{21}^{(m)}(t)\right] \cdot x(t-m) - \sum_{m=1}^{p} \left[-a_{22}^{(m)}(t)\right] u(t-m). \tag{2.18}$$

Comparing (2.18) and the ARMA predictor equation (2.10), we conclude that ARMA system identification can be provided by the two-channel AR predictor as defined by (2.15). Note that the reduced output signal $y(t)$ of the unknown system, namely, the output signal $u(t)$ minus the excitation $x(t)$, must be fed into the y-input of the two-channel AR predictor, whereas the excitation signal $x(t)$ must be connected to the x-input of the two-channel AR predictor. Then, by comparing (2.18) and (2.10), it is easy to see that the ARMA parameters can be calculated from the two-channel AR parameters via the following relations:

$$a_m(t) = -a_{22}^{(m)}(t) \tag{2.19a}$$

$$b_m(t) = a_{22}^{(m)}(t) - a_{21}^{(m)}(t). \tag{2.19b}$$

III. The ARMA Ladder Form

In Section II, we have described the "fixed-order" formulation of the ARMA system identification problem based on the Normal Equations. The name "normal," however, derives from geometrical considerations. In a geometrical interpretation, the two-channel AR predictor of (2.15), which was used for ARMA system identification, provides the projection of the reduced output signal $\boldsymbol{y}(t)$ onto the joint subspace spanned by the observations $\boldsymbol{X}_p(t)$ and $\boldsymbol{Y}_p(t)$. Solving the ARMA Normal Equations (2.13) yields the parameter set $\hat{\boldsymbol{a}}_p(t)$ and $\hat{\boldsymbol{b}}_p(t)$ such that the projection is orthogonal, i.e., $\boldsymbol{y}(t)$ is decomposed in a component $\hat{\boldsymbol{y}}(t)$ which lies in the joint subspace, and a component $\boldsymbol{e}_p^y(t)$ which is orthogonal with respect to the joint subspace of past observations. The so-obtained residual vector $\boldsymbol{e}_p^y(t)$ is of least Euclidean length and therefore satisfies the LS cost function (2.1). From these considerations, it becomes obvious that an alternative way of constructing $\boldsymbol{e}_p^y(t)$ exists as follows. First, construct a set of orthogonal basis vectors of the joint subspace and, second, project the vector $\boldsymbol{y}(t)$ successively on the orthogonal basis vectors. This procedure was termed the "growing-order" or ladder formulation of the ARMA system identification problem.

Let $\boldsymbol{c}$ be the complement of the orthogonal projection of $\boldsymbol{a}$ onto a subspace $\boldsymbol{B}$ denoted by

$$\boldsymbol{c} = \boldsymbol{a}\langle \boldsymbol{B} \rangle; \quad \text{with the property} \quad \boldsymbol{c}^T\boldsymbol{b}_i = 0; \quad 1 \le i \le m$$

where

$$\boldsymbol{B} = [\boldsymbol{b}_1, \boldsymbol{b}_2, \cdots, \boldsymbol{b}_m]; \qquad \boldsymbol{c}, \boldsymbol{a}, \boldsymbol{b}_i \in \boldsymbol{R}^{m+1}.$$

Then, the observations $\boldsymbol{x}(t)$ and $\boldsymbol{y}(t)$ can be decomposed by the following nonrecursive Gram–Schmidt decomposition:

$$\boldsymbol{e}_0^y(t) = \boldsymbol{y}(t) \tag{3.1a}$$

$$\boldsymbol{e}_1^y(t) = \boldsymbol{y}(t)\langle \boldsymbol{x}(t-1); \boldsymbol{y}(t-1)\rangle \tag{3.1b}$$

$$\boldsymbol{e}_2^y(t) = \boldsymbol{y}(t)\langle \boldsymbol{x}(t-1), \boldsymbol{x}(t-2); \boldsymbol{y}(t-1), \boldsymbol{y}(t-2)\rangle \tag{3.1c}$$

$$\vdots \tag{3.1d}$$

$$\boldsymbol{e}_m^y(t) = \boldsymbol{y}(t)\langle \boldsymbol{X}_m(t); \boldsymbol{Y}_m(t)\rangle$$

$$\begin{aligned}\boldsymbol{e}_m^y(t) = {} & \boldsymbol{y}(t)\langle \boldsymbol{X}_{m-1}(t); \boldsymbol{Y}_{m-1}(t)\rangle \\ & + \boldsymbol{K}_m^{fyx}(t)\,\boldsymbol{x}(t-m)\langle \boldsymbol{X}_{m-1}(t); \boldsymbol{Y}_{m-1}(t)\rangle \\ & + \boldsymbol{K}_m^{fyy}(t)\,\boldsymbol{y}(t-m)\langle \boldsymbol{X}_{m-1}(t); \boldsymbol{Y}_{m-1}(t)\rangle.\end{aligned} \tag{3.2}$$

Similarly, we have

$$\begin{aligned}\boldsymbol{e}_m^x(t) = {} & \boldsymbol{x}(t)\langle \boldsymbol{X}_{m-1}(t); \boldsymbol{Y}_{m-1}(t)\rangle \\ & + \boldsymbol{K}_m^{fxx}(t)\,\boldsymbol{x}(t-m)\langle \boldsymbol{X}_{m-1}(t); \boldsymbol{Y}_{m-1}(t)\rangle \\ & + \boldsymbol{K}_m^{fxy}(t)\,\boldsymbol{y}(t-m)\langle \boldsymbol{X}_{m-1}(t); \boldsymbol{Y}_{m-1}(t)\rangle.\end{aligned} \tag{3.3}$$

Likewise, we can construct a set of orthogonal basis vectors of the joint input/output subspace as follows:

$$\boldsymbol{r}_0^y(t) = \boldsymbol{y}(t) \tag{3.4a}$$

$$\boldsymbol{r}_1^y(t) = \boldsymbol{y}(t-1)\langle \boldsymbol{x}(t); \boldsymbol{y}(t)\rangle \tag{3.4b}$$

$$\boldsymbol{r}_2^y(t) = \boldsymbol{y}(t-2)\langle \boldsymbol{x}(t), \boldsymbol{x}(t-1); \boldsymbol{y}(t), \boldsymbol{y}(t-1)\rangle$$

$$\vdots \tag{3.4c}$$

$$\boldsymbol{r}_m^y(t) = \boldsymbol{y}(t-m)\langle \boldsymbol{X}_m(t+1); \boldsymbol{Y}_m(t+1)\rangle \tag{3.4d}$$

$$\begin{aligned}\boldsymbol{r}_m^y(t) = {} & \boldsymbol{y}(t-m)\langle \boldsymbol{X}_{m-1}(t+1); \boldsymbol{Y}_{m-1}(t+1)\rangle \\ & + \boldsymbol{K}_m^{byx}(t)\,\boldsymbol{x}(t)\langle \boldsymbol{X}_{m-1}(t+1); \boldsymbol{Y}_{m-1}(t+1)\rangle \\ & + \boldsymbol{K}_m^{byy}(t)\,\boldsymbol{y}(t)\langle \boldsymbol{X}_{m-1}(t+1); \boldsymbol{Y}_{m-1}(t+1)\rangle\end{aligned} \tag{3.5}$$

$$\begin{aligned}\boldsymbol{r}_m^x(t) = {} & \boldsymbol{y}(t-m)\langle \boldsymbol{X}_{m-1}(t+1); \boldsymbol{Y}_{m-1}(t+1)\rangle \\ & + \boldsymbol{K}_m^{bxx}(t)\,\boldsymbol{x}(t)\langle \boldsymbol{X}_{m-1}(t+1); \boldsymbol{Y}_{m-1}(t+1)\rangle \\ & + \boldsymbol{K}_m^{bxy}(t)\,\boldsymbol{y}(t)\langle \boldsymbol{X}_{m-1}(t+1); \boldsymbol{Y}_{m-1}(t+1)\rangle.\end{aligned} \tag{3.6}$$

These nonrecursive decompositions (3.1–6) can be brought in an order recursive form as follows:

$$\begin{aligned}\boldsymbol{e}_m^x(t) = {} & \boldsymbol{e}_{m-1}^x(t) + \boldsymbol{K}_m^{fxx}(t)\,\boldsymbol{r}_{m-1}^x(t-1) \\ & + \boldsymbol{K}_m^{fxy}(t)\,\boldsymbol{r}_{m-1}^y(t-1)\end{aligned} \tag{3.7a}$$

$$\begin{aligned}\boldsymbol{e}_m^y(t) = {} & \boldsymbol{e}_{m-1}^y(t) + \boldsymbol{K}_m^{fyx}(t)\,\boldsymbol{r}_{m-1}^x(t-1) \\ & + \boldsymbol{K}_m^{fyy}(t)\,\boldsymbol{r}_{m-1}^y(t-1)\end{aligned} \tag{3.7b}$$

$$\begin{aligned}\boldsymbol{r}_m^x(t) = {} & \boldsymbol{r}_{m-1}^x(t-1) + \boldsymbol{K}_m^{bxx}(t)\,\boldsymbol{e}_{m-1}^x(t) \\ & + \boldsymbol{K}_m^{bxy}(t)\,\boldsymbol{e}_{m-1}^y(t)\end{aligned} \tag{3.8a}$$

$$\begin{aligned}\boldsymbol{r}_m^y(t) = {} & \boldsymbol{r}_{m-1}^y(t-1) + \boldsymbol{K}_m^{byx}(t)\,\boldsymbol{e}_{m-1}^x(t) \\ & + \boldsymbol{K}_m^{bxy}(t)\,\boldsymbol{e}_{m-1}^y(t)\end{aligned} \tag{3.8b}$$

where

$$\boldsymbol{e}_0^x(t) = \boldsymbol{r}_0^x(t) = \boldsymbol{x}(t) \tag{3.9a}$$

$$\boldsymbol{e}_0^y(t) = \boldsymbol{r}_0^y(t) = \boldsymbol{y}(t). \tag{3.9b}$$

The vectors $\{\boldsymbol{r}_m^x(t), \boldsymbol{r}_m^y(t); 1 \le m \le p\}$ determine the orthogonal basis of the joint subspace of past observations, whereas the recursion vectors $\boldsymbol{e}_m^x(t)$ and $\boldsymbol{e}_m^y(t)$ are known as the partial orthogonalized forward residual vectors. The scalar quantities $\boldsymbol{K}_m^{f**}(t)$ and $\boldsymbol{K}_m^{b**}(t)$ are the time-varying ARMA ladder reflection coefficients. The vector order recursions (3.7a, b) and (3.8a, b) have an important interpretation as a filter structure, termed the ARMA ladder form which is shown in Fig. 3.

Note that the ARMA ladder equations in their original formulation are fairly complex. They can be rewritten in a more convenient matrix notation, hence, (3.7a, b) and

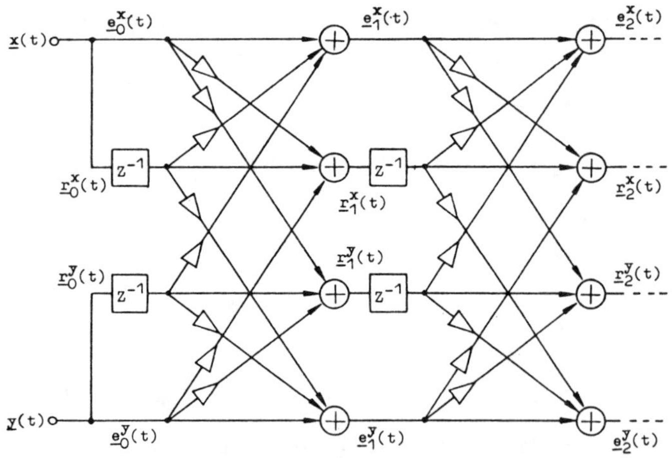

Fig. 3. Canonical ARMA ladder form.

(3.8a, b) reduce to

$$e_m(t) = e_{m-1}(t) + r_{m-1}(t-1)\, K_m^f(t) \quad (3.10a)$$

$$r_m(t) = r_{m-1}(t-1) + e_{m-1}(t)\, K_m^b(t) \quad (3.10b)$$

with the initial condition

$$e_0(t) = r_0(t) = s(t) \quad (3.11)$$

$$e_m(t) = [e_m^x(t), e_m^y(t)] \quad (3.14a)$$

$$r_m(t) = [r_m^x(t), r_m^y(t)] \quad (3.12a)$$

$$s(t) = [x(t), y(t)] \quad (3.13)$$

$$K_m^f(t) = \begin{bmatrix} K_m^{fxx}(t) & K_m^{fyx}(t) \\ K_m^{fxy}(t) & K_m^{fyy}(t) \end{bmatrix} \quad (3.14a)$$

$$K_m^b(t) = \begin{bmatrix} K_m^{bxx}(t) & K_m^{byx}(t) \\ K_m^{bxy}(t) & K_m^{byy}(t) \end{bmatrix}. \quad (3.14b)$$

We are interested in determining the ARMA ladder reflection matrices $K_m^f(t)$ and $K_m^b(t)$. The minimization of the LS cost functions

$$e_m^T(t)\, e_m(t) \overset{!}{=} \min \quad (3.15a)$$

$$r_m^T(t)\, r_m(t) \overset{!}{=} \min \quad (3.15b)$$

yields the desired rule for adjusting the ARMA ladder reflection matrices in the LS sense. Note that (3.15a, b) implies minimization of each element of the respective energy matrix. However, minimization of only the matrix traces would suffice, since it can be shown that this automatically minimizes also the cross residual energies. After some algebra, it is found that the ARMA ladder reflection matrices must be adjusted as follows:

$$K_m^f(t) = -R_{m-1}^{-1}(t-1)\, C_{m-1}^T(t) \quad (3.16a)$$

$$K_m^b(t) = -E_{m-1}^{-1}(t)\, C_{m-1}(t) \quad (3.16b)$$

where

$$E_{m-1}(t) = e_{m-1}^T(t)\, e_{m-1}(t) \quad (3.17a)$$

$$R_{m-1}(t) = r_{m-1}^T(t)\, r_{m-1}(t), \quad (3.17b)$$

$$C_{m-1}(t) = e_{m-1}^T(t)\, r_{m-1}(t-1) \quad (3.17c)$$

are the residual energy matrices of stage $m-1$.

IV. The PORLA Method for ARMA System Identification

The PORLA method is a generalized covariance ladder technique which was first introduced for scalar sequences in [25] and [26]. Note that the term "covariance" in this context derives from the property that algorithms of the PORLA type construct the adaptive true LS ladder form exclusively from the second-order information, that is, the algorithms require the input data covariance matrix (or some rows thereof) for initialization of the adaptation algorithm. Therefore, the PORLA method differs significantly from the fast estimation techniques, where also some first-order information (data and residuals) is required in the adaptation algorithm. It is the purpose of this section to extend the PORLA method to the ARMA system identification problem. This is done by utilizing the embedding approach as treated in Section II. It is then easy to extend the approach to the multichannel case which is summarized in Appendix B. Later in this section, we have also listed a Levinson-type recursion which is useful to recover the true LS ARMA parameters from the ARMA ladder reflection matrices. The key idea in the development of the PORLA method for ARMA system identification is the expansion of the residual energy matrices (3.17a–c) to block matrix forms as follows:

$$E_{m,i,j}(t) = e_m^T(t-i)\, e_m(t-j) \quad (4.1a)$$

$$R_{m,i,j}(t) = r_m^T(t-1-i)\, r_m(t-1-j), \quad (4.1b)$$

$$C_{m,i,j}(t) = e_m^T(t-i)\, r_m(t-1-j). \quad (4.1c)$$

These block matrix forms of residual energies have been termed "block generalized residual energies" (BGRE's). More precisely, we can rewrite the element (i, j) of the BGRE $E_m(t)$ as follows:

$$E_m(t) = \begin{bmatrix} E_{m,0,0} & E_{m,0,1} & \cdots & E_{m,0,J} \\ E_{m,1,0} & E_{m,1,1} & \cdots & E_{m,1,J} \\ \vdots & & \ddots & \vdots \\ E_{m,J,0} & E_{m,J,1} & \cdots & E_{m,J,J} \end{bmatrix} \quad (4.2a)$$

where $j = p - m - 1$.

$$E_{m,i,j}(t) = \begin{bmatrix} e_m^{xT}(t-i)\, e_m^x(t-j) & e_m^{xT}(t-i)\, e_m^y(t-j) \\ e_m^{yT}(t-i)\, e_m^x(t-j) & e_m^{yT}(t-i)\, e_m^y(t-j) \end{bmatrix}. \quad (4.2b)$$

The BGRE's $R_m(t)$ and $C_m(t)$ have similar structures.

A. Block Matrix Covariance Recursions

In a next step, we substitute the order recursions of the time-shifted residual matrices, namely, the ARMA vector order recursions (3.10a, b), into the definitions of the BGRE's yielding

$$E_{m,i,j}(t) = e_{m-1}^T(t-i)\, e_{m-1}(t-j) + K_m^{fT}(t-i)\, r_{m-1}^T(t-1-i)\, e_{m-1}(t-j) + e_{m-1}^T(t-i)\, r_{m-1}(t-1-j)\, K_m^f(t-j) + K_m^{fT}(t-i)\, r_{m-1}^T(t-1-i) \cdot r_{m-1}(t-1-j)\, K_m^f(t-j). \quad (4.3)$$

Similar recursions can be developed for the pure order recursive updating of $R_{m,i,j}(t)$ and $C_{m,i,j}(t)$. Obviously, the right-hand side of (4.3) can be expressed in terms of the BGRE's of ladder stage $m-1$. Following this idea, we finally obtain order recursions of BGRE's, i.e., the BGRE's of ladder stage m can be updated using only BGRE's of the previous stage and the ladder reflection matrices as follows:

$$E_{m,i,j}(t) = E_{m-1,i,j}(t) + K_m^{fT}(t-i)\, C_{m-1,j,i}^T(t) + C_{m-1,i,j}(t)\, K_m^f(t-j) + K_m^{fT}(t-i)\, R_{m-1,i,j}(t)\, K_m^f(t-j) \quad (4.4a)$$

$$R_{m,i,j}(t) = R_{m-1,i+1,j+1}(t) + K_m^{bT}(t-1-i) \cdot C_{m-1,i+1,j+1}(t) + C_{m-1,j+1,i+1}^T(t)\, K_m^b(t-1-j) + K_m^{bT}(t-1-i)\, E_{m-1,i+1,j+1}(t) \cdot K_m^b(t-1-j) \quad (4.4b)$$

$$C_{m,i,j}(t) = C_{m-1,i,j+1}(t) + K_m^{fT}(t-i)\, R_{m-1,i,j+1}(t) + E_{m-1,i,j+1}(t)\, K_m^b(t-1-j) + K_m^{fT}(t-i)\, C_{m-1,j+1,i}^T(t) \cdot K_m^b(t-1-j). \quad (4.4c)$$

The "conventional" residual energy matrices (3.17a-c) can be viewed as the upper-left elements of the BGRE's. Thus, it is easy to check that the ARMA ladder reflection matrices (3.16a, b) can be redefined in terms of the BGRE's:

$$K_m^f(t) = -R_{m-1,0,0}^{-1}(t)\, C_{m-1,0,0}^T(t) \quad (4.5a)$$

$$K_m^b(t) = -E_{m-1,0,0}^{-1}(t)\, C_{m-1,0,0}(t). \quad (4.5b)$$

This is a useful result since the order recursions of BGRE's (4.4a–c), together with the reflection matrices (4.5a, b), already constitute a closed recursion for the desired order recursive computation of the ladder reflection matrices. Exploiting the initial condition of the ARMA vector order recursions (3.11), we can immediately establish the initialization scheme of the BGRE's at stage zero of the pure order recursive ARMA ladder algorithm formed by the recursions (4.4a–c) and (4.5a, b).

$$E_{0,i,j}(t) = s^T(t-i)\, s^T(t-j) = W_{i,j}(t) \quad (4.6a)$$

$$R_{0,i,j}(t) = s^T(t-1-i)\, s(t-1-j) = W_{i+1,j+1}(t) \quad (4.6b)$$

$$C_{0,i,j}(t) = s^T(t-i)\, s(t-1-j) = W_{i,j+1}(t) \quad (4.6c)$$

where it is easy to check that $W_{i,j}(t)$ is the (i, j)th element of the covariance block matrix of input data, where we make use of the input/output covariance information as follows:

$$W_{i,j}(t) = \begin{bmatrix} W_{i,j}^{xx}(t) & W_{i,j}^{xy}(t) \\ W_{j,i}^{xy}(t) & W_{i,j}^{yy}(t) \end{bmatrix}. \quad (4.7)$$

The covariance block matrix (4.7) is updated time recursively. In Section II, we have stated the ARMA system identification problem assuming a sliding rectangular window of length L samples. However, since in our new algorithm only the updating of the covariance block matrix involves a time recursion, we are no longer restricted on simple windows. Several other sophisticated weighting functions, which incorporate a finite or infinite set of data samples, could be used. This way, the new algorithm can be tuned quite accurately for a given practical situation, concerning a desired tracking behavior or steady-state performance. Let $\boldsymbol{Q}(k)$ be an arbitrary weighting matrix, and let $W_{i,j}(t)$ be the weighted covariance block matrix.

$$\boldsymbol{Q}(k) = \begin{bmatrix} Q(k) & 0 \\ 0 & Q(k) \end{bmatrix} \quad (4.8)$$

$$W_{i,j}(t) = \sum_{k=0}^{L-1} \boldsymbol{Q}(k)\, Z_{i,j}(t-k) \quad (4.9)$$

where

$$Z_{i,j}(t) = \begin{bmatrix} x(t-i)\, x(t-j) & x(t-i)\, y(t-j) \\ y(t-i)\, x(t-j) & y(t-i)\, y(t-j) \end{bmatrix} \quad (4.10)$$

is the innovations matrix. Note that (4.9) is a nonrecursive formulation of the windowing problem. However, several well-known windows have been brought in a recursive formulation. In Appendix A, we have summarized some of the most important recursive windowing algorithms. The effect of using these windows combined with the new ARMA covariance ladder algorithm is demonstrated by the simulation results shown in Section VII.

B. Block Vector Covariance Recursions

A closer look at the order recursions of BGRE's (4.4a–c) indicates that the computation of this original version of the new ARMA covariance ladder algorithm can be troublesome since it requires, at each recursion, the computation of three block matrix polynomials with a complexity of $O(p^3)$. Therefore, in the following, we investigate two interesting properties of the BGRE's yielding a reduction of the computational complexity of the new algorithm to $O(p^2)$. First, it can be seen from the definition of the BGRE's (4.1a–c) that the following symmetric properties hold:

$$E^T_{m,i,j}(t) = E_{m,j,i}(t) \quad (4.11a)$$

$$R^T_{m,i,j}(t) = R_{m,j,i}(t) \quad (4.11b)$$

but

$$C^T_{m,i,j}(t) \neq C_{m,j,i}(t). \quad (4.11c)$$

Furthermore, it can be readily verified that the BGRE's obey a shift invariance property as follows:

$$E_{m,i+1,j+1}(t) = E_{m,i,j}(t-1) \quad (4.12a)$$

$$R_{m,i+1,j+1}(t) = R_{m,i,j}(t-1) \quad (4.12b)$$

$$C_{m,i+1,j+1}(t) = C_{m,i,j}(t-1). \quad (4.12c)$$

Substituting the symmetric property (4.11a, b) and the shift invariance property (4.12a–c) into the block matrix recursions (4.4a–c), it becomes apparent that only four recursion block vectors uniquely determine the final version of the new ARMA covariance ladder algorithm.

$$\begin{aligned} E_{m,0,j}(t) = {} & E_{m-1,0,j}(t) + K^{fT}_m(t)\, C^T_{m-1,j,0}(t) \\ & + C_{m-1,0,j}(t)\, K^f_m(t-j) \\ & + K^{fT}_m(t)\, R_{m-1,0,j}(t)\, K^f_m(t-j) \end{aligned} \quad (4.13a)$$

$$\begin{aligned} R_{m,0,j}(t) = {} & R_{m-1,0,j}(t-1) + K^{bT}_m(t-1) \\ & \cdot C_{m-1,0,j}(t-1) \\ & + C^T_{m-1,j,0}(t-1)\, K^b_m(t-1-j) \\ & + K^{bT}_m(t-1)\, E_{m-1,0,j}(t)\, K^b_m(t-1-j) \end{aligned} \quad (4.13b)$$

$$\begin{aligned} C_{m,0,j}(t) = {} & C_{m-1,0,j+1}(t) + K^{fT}_m(t)\, R_{m-1,0,j+1}(t) \\ & + E_{m-1,0,j+1}(t)\, K^b_m(t-1-j) \\ & + K^{fT}_m(t)\, C^T_{m-1,j+1,0}(t)\, K^b_m(t-1-j) \end{aligned} \quad (4.13c)$$

$$\begin{aligned} C_{m,j+1,0}(t) = {} & C_{m-1,j,0}(t-1) + K^{fT}_m(t-1-j) \\ & \cdot R^T_{m-1,0,j}(t-1) \\ & + E^T_{m-1,0,j}(t-1)\, K^b_m(t-1) \\ & + K^{fT}_m(t-1-j)\, C^T_{m-1,0,j}(t-1) \\ & \cdot K^b_m(t-1). \end{aligned} \quad (4.13d)$$

Note that the updating of the BGRE $C_{m,i,j}(t)$ requires two recursion block vectors as this BGRE is not symmetric (4.11c). In Section VI, Table III provides a complete Fortran subroutine listing of this new ARMA covariance ladder algorithm. Table II relates the formula variables to the variables used in the subroutine in order to enable the reader to relate the algorithm development to the subroutine.

C. *Computing the ARMA Parameters from the Ladder Matrices*

After constructing the true LS ARMA ladder form, it is sometimes of interest to determine the direct form ARMA parameter estimates $\hat{a}_p(t)$ and $\hat{b}_p(t)$ from the ARMA ladder reflection matrices. In the stationary case, these parameters can be easily obtained as the impulse response of the ARMA ladder form (with time-invariant parameters). In the most general nonstationary case, however, such a simple procedure does not hold. In the nonstationary case, a two-channel Levinson-type recursion is required to transform the (time-varying) ARMA ladder reflection matrices into the direct form two-channel AR parameters. This useful procedure is summarized in Table I. The direct form ARMA parameter estimates are then obtained from the two-channel AR parameters via the ARMA embedding equations (2.19a, b). In Table I, we are using a short notation for the direct form forward/backward two-channel AR parameters as follows:

$$a_m(t) = \begin{bmatrix} a^{(m)}_{11}(t) & a^{(m)}_{12}(t) \\ a^{(m)}_{21}(t) & a^{(m)}_{22}(t) \end{bmatrix} \quad (4.14a)$$

$$b_m(t) = \begin{bmatrix} b^{(m)}_{11}(t) & b^{(m)}_{12}(t) \\ b^{(m)}_{21}(t) & b^{(m)}_{21}(t) \end{bmatrix}. \quad (4.14b)$$

V. Special Cases of ARMA Ladder Recursions

A. *Stationary Case*

In many applications, where the excitation signal is nearly stationary and the ARMA model parameters change slowly and smoothly compared to the data sample rate, it can be expected that the ARMA estimation problem has a considerably simpler structure than in the most general nonstationary case. Assuming a slowly time-varying model with stationary excitation, the following two approximations can be introduced in the new ARMA covariance ladder algorithm in order to achieve a further simplification of the recursions.

Approximation 1—Time Invariant Reflection Matrices: A slowly time-varying model corresponds with nearly time-constant ARMA ladder reflection matrices. Hence, we can set

$$K^f_m(t-i) = K^f_m(t) \quad (5.1a)$$

$$K^b_m(t-j) = K^b_m(t). \quad (5.1b)$$

Approximation 1 allows us to omit the stack of reflection matrices which was required in the true LS ARMA covariance ladder algorithm. We have found, however, that the adaption behavior does not suffer very much from this approximation.

Approximation 2—Block Toeplitz Structure of BGRE's: If, additionally, the excitation is a stationary process, the joint input/output data covariance block ma-

TABLE I

LEVINSON-TYPE RECURSION FOR COMPUTING THE TWO-CHANNEL DIRECT FORM AR PARAMETERS FROM THE ARMA LADDER REFLECTION MATRICES. I IS THE IDENTITY MATRIX

FOR t = 0, 1, 2, . . . do the following:

initialize: $\mathbf{a}_0^{(1)}(t) = \mathbf{I}$ $\qquad \mathbf{a}_1^{(1)}(t) = \mathbf{K}_1^{f^T}(t)$

$\mathbf{b}_0^{(1)}(t) = \mathbf{K}_m^{b^T}(t)$ $\qquad \mathbf{b}_1^{(1)}(t) = \mathbf{I}$

FOR m = 2, 3, . . . , p

FOR j = 1, 2, . . . , m-1

$$\mathbf{b}_j^{(m)}(t) = \mathbf{b}_{j-1}^{(m-1)}(t-1) + \mathbf{K}_m^{b^T}(t)\, \mathbf{a}_j^{(m-1)}(t)$$

$$\mathbf{a}_j^{(m)}(t) = \mathbf{a}_j^{(m-1)}(t) + \mathbf{K}_m^{f^T}(t)\, \mathbf{b}_{j-1}^{(m-1)}(t-1)$$

$\mathbf{a}_0^{(m)}(t) = \mathbf{I}$ $\qquad \mathbf{a}_m^{(m)}(t) = \mathbf{K}_m^{f^T}(t)$

$\mathbf{b}_0^{(m)}(t) = \mathbf{K}_m^{b^T}(t)$ $\qquad \mathbf{b}_m^{(m)}(t) = \mathbf{I}$

FOR m = 1, 2, . . . , p

$$\mathbf{a}_m(t) = \mathbf{a}_m^{(p)}(t)$$

TABLE II

RELATIONSHIP BETWEEN FORMULA PARAMETERS AND PROGRAM PARAMETERS

ALGORITHM PARAMETERS		MEANINGS
formula	program	
p	P	fixed order of algorithm
m	M	growing order of algorithm
j	J	recursion variable
$x(t),\ e_p^x(t)$	XSAMPLE	x-data in, residual out
$y(t),\ e_p^y(t)$	YSAMPLE	y-data in, residual out
$e_m^x(t)$	EX	inverse filter x-residual
$e_m^y(t)$	EY	inverse filter y-residual
$r_m^x(t)$	RX(0,M)	inverse filter backward x-residual
$r_m^x(t-1)$	RX(1,M)	stack of backward x-residual
$r_m^y(t)$	RY(0,M)	inverse filter backward y-residual
$r_m^y(t-1)$	RY(1,M)	stack of backward y-residual
$\mathbf{W}_{0,j}(t)$	W(*,*,0,J)	1. row of data covariance block matrix
$\mathbf{W}_{1,j}(t)$	W(*,*,1,J)	2. row of data covariance block matrix
$\mathbf{E}_{m,0,j}(t)$	E(*,*,J)	forward GRE block vector
$\mathbf{E}_{m,0,j}(t-1)$	EM(*,*,J)	stack of forward GRE block vector
$\mathbf{R}_{m,0,j}(t)$	R(*,*,J)	backward GRE block vector
$\mathbf{R}_{m,0,j}(t-1)$	RM(*,*,M,J)	stack of backward GRE block vector
$\mathbf{C}_{m,0,j}(t)$	C(*,*,J)	cross GRE row block vector
$\mathbf{C}_{m,0,j}(t-1)$	CM(*,*,J)	stack of cross GRE row block vector
$\mathbf{C}_{m,j,0}(t)$	D(*,*,J)	cross GRE column block vector
$\mathbf{C}_{m,j,0}(t-1)$	DM(*,*,M,J)	stack of cross GRE column block vector
$\mathbf{K}_m^f(t-j)$	KF(*,*,M,J)	forward reflection matrix
$\mathbf{K}_m^b(t-j)$	KB(*,*,M,J)	backward reflection matrix

trix obeys a block Toeplitz structure. It can be easily verified, viewing the initialization scheme (4.6a–c), that the BGRE's will share the block Toeplitz structure. Finally, it can be shown that the block Toeplitz structure of BGRE's holds as long as Approximation 1 is satisfied.

$$\boldsymbol{E}_{m,i+1,j+1}(t) = \boldsymbol{E}_{m,i,j}(t) \tag{5.2a}$$

$$\boldsymbol{R}_{m,i+1,j+1}(t) = \boldsymbol{R}_{m,i,j}(t) \tag{5.2b}$$

$$\boldsymbol{C}_{m,i+1,j+1}(t) = \boldsymbol{R}_{m,i,j}(t). \tag{5.2c}$$

Clearly, the block Toeplitz property of BGRE's (5.2a–c) can serve as a substitute of the true LS shift invariance property (4.12a–c). Note further that the incorporation of the two approximations, as discussed above, allows the algorithm to be operated in a block processing mode, that is, we have the opportunity of updating the algorithm only each Mth time step, where M is the parameter update rate.

B. *White Input Process*

Next, the influences of a given white-noise input process on the recursions of the new ARMA covariance ladder algorithm are briefly discussed. A white-noise input process is characterized as follows:

$$\boldsymbol{x}^T(t-i)\,\boldsymbol{x}(t-j) = W^{xx}(t-i) \qquad \text{if } i = j \tag{5.3a}$$

$$\boldsymbol{x}^T(t-i)\,\boldsymbol{x}(t-j) = 0 \qquad \text{else.} \tag{5.3b}$$

Furthermore, if $\boldsymbol{x}(t)$ is a white-noise process, then

$$\boldsymbol{e}_m^x(t) = \boldsymbol{x}(t) \tag{5.4}$$

and it can be shown that the BGRE's tend to the following forms, where we distinguish two cases.

Case 1: $j = 0$

$$\boldsymbol{E}_{m,0,j} = \begin{bmatrix} * & * \\ * & * \end{bmatrix} \tag{5.5a}$$

$$\boldsymbol{R}_{m,0,j} = \begin{bmatrix} * & * \\ * & * \end{bmatrix} \tag{5.5b}$$

$$\boldsymbol{C}_{m,0,j} = \begin{bmatrix} 0 & 0 \\ * & * \end{bmatrix}. \tag{5.5c}$$

$$\boldsymbol{C}_{m,j,0} = \begin{bmatrix} 0 & 0 \\ * & * \end{bmatrix}. \tag{5.5d}$$

Case 2: $j > 0$

$$\boldsymbol{E}_{m,0,j} = \begin{bmatrix} 0 & 0 \\ * & * \end{bmatrix} \tag{5.6a}$$

$$\boldsymbol{R}_{m,0,j} = \begin{bmatrix} * & * \\ * & * \end{bmatrix} \tag{5.6b}$$

$$\boldsymbol{C}_{m,0,j} = \begin{bmatrix} 0 & 0 \\ * & * \end{bmatrix}. \tag{5.6c}$$

$$\boldsymbol{C}_{m,j,0} = \begin{bmatrix} * & * \\ * & * \end{bmatrix}. \tag{5.6d}$$

C. *Unknown Input Process*

A special problem of interest is the estimation of ARMA models when the input process is not accessible. In contrast to the all-pole (AR) model, this "unknown input problem" has no exact analytical solution in the ARMA case. However, we can use the new covariance ARMA ladder algorithm, if the unknown input is simply replaced with its best estimate obtained by using the best current parameter estimates. Assuming sufficient accurate estimates of the system parameters, then this suggests that we take $x(t) = e_p^y(t)$, and thus we can treat the residual as known input process, hence entered into the input along with the observed system output measurement $u(t)$. Such a bootstrapping has been frequently suggested in context with ARMA system identification algorithms [2], [18].

VI. Implementation and Application

A. *Fortran Program*

A Fortran program of the new least-squares ARMA covariance ladder algorithm is presented in this section. The purpose for this is not only to simplify the implementation of the new algorithm, but mainly, in our case, the Fortran coded routine provides a very clear and instructive insight in the structural properties of the presented algorithm. The program reveals the inherent parallelism of the new algorithm. Analyzing this Fortran coded routine (Table III), it becomes very obvious how the new technique can be established on a massively parallel processor (MPP).

B. *Implementation on an MPP*

It is well known that the evaluation of the computational complexity of a given algorithm, only in terms of the number of arithmetic operations per recursion, can be very misleading. Since massively parallel processors (MPP's) [36]–[40] are now available for signal processing applications, it is much more important to investigate whether or not a given algorithm can utilize the parallel processing capabilities of an MPP. Evaluating the subroutine listing (Table III), it is seen that the proposed ARMA covariance ladder algorithm fits in the class of "block algorithms" which are motivated by the desire to fully utilize given multiprocessor architectures. Indeed, all the operations in the new algorithm can be led back to the following basic operation: multiply a matrix by another and accumulate the products. No "strange" operations like square-roots appear. It is clear that such an algorithm structure suggests a very simple but massively parallel implementation. This requires a central I/O processor which directs the data streams from a correlation processor to special-purpose subprocessors which basically provide a simple matrix-by-matrix multiply-accumulator structure. These subprocessors can be viewed as the hardware realizations of the arithmetic subroutines MUL ADD KFF and KBB, as used in our Fortran coded routine. As will become apparent in Section VII, the new algorithm has a superior numerical accuracy and robust-

TABLE III
FORTRAN SUBROUTINE LISTING OF THE NEW LS ARMA COVARIANCE LADDER ALGORITHM

```
C**************************************************************
C****                                                      ****
C****    THE RECURSIVE LEAST-SQUARES COVARIANCE LADDER     ****
C****                     ALGORITHM                        ****
C****            FOR ARMA SYSTEM IDENTIFICATION            ****
C****                                                      ****
C****                                                      ****
C****                PETER STROBACH, PH.D.                 ****
C****                                                      ****
C****       SIEMENS INFORMATION SYSTEMS LABORATORY         ****
C****          ZT ZTI INF 121, OTTO-HAHN-RING 6            ****
C****          D-8000 MUNCHEN 83,  (W. GERMANY)            ****
C****                                                      ****
C****                                                      ****
C****                                     DATE: 26.NOV.86  ****
C**************************************************************
C
C***************** PARAMETER DEFINITIONS *******************
C
C (INPUT/OUTPUT) PARAMETERS:
C
C P                    (IN):   ORDER OF ALGORITHM
C W(2,2,0:1,0:P)       (IN):   TWO ROWS OF DATA COVARIANCE
C                              BLOCK MATRIX
C XSAMPLE              (I/O):  X-SIGNAL IN, X-RESIDUAL OUT
C YSAMPLE              (I/O):  Y-SIGNAL IN, Y-RESIDUAL OUT
C
C ALGORITHM PARAMETERS:
C
C E(2,2,0:P)           FORWARD RESIDUAL ENERY BLOCK VECTOR
C EM(2,2,0:P)          STACK OF E(2,2,0:P)
C R(2,2,0:P)           BACKWARD RESIDUAL ENERGY BLOCK VECTOR
C RM(2,2,0:P,0:P)      STACK OF R(2,2,0:P)
C C(2,2,0:P)           CROSS RESIDUAL ENERY ROW BLOCK VECTOR
C CM(2,2,0:P)          STACK OF C(2,2,0:P)
C D(2,2,0:P)           CROSS RESIDUAL ENERGY COLUMN BLOCK VECTOR
C DM(2,2,0:P,0:P)      STACK OF D(2,2,0:P)
C
C EX                   INVERSE FILTER FORWARD  X-RESIDUAL
C EY                   INVERSE FILTER FORWARD  Y-RESIDUAL
C RX(0:1,0:P)          INVERSE FILTER BACKWARD X-RESIDUAL
C RY(0:1,0:P)          INVERSE FILTER BACKWARD Y-RESIDUAL
C
C KF(2,2,0:P,0:P)      FORWARD REFLECTION MATRIX  (STACK)
C KB(2,2,0:P,0:P)      BACKWARD REFLECTION MATRIX (STACK)
C
C PROD1-4(2,2)         AUXILIARY VARIABLES
C
C**************************************************************
C

      SUBROUTINE ARMA(W,XSAMPLE,YSAMPLE,P)
C
      INTEGER P
C
      REAL W(2,2,0:1,0:15),XSAMPLE,YSAMPLE,EM(2,2,0:15),
     *R(2,2,0:15),CM(2,2,0:15),D(2,2,0:15),KF(2,2,0:15,0:15),
     *KB(2,2,0:15,0:15),DM(2,2,0:15,0:15),RM(2,2,0:15,0:15),
     *E(2,2,0:15),C(2,2,0:15),RX(0:1,0:15),RY(0:1,0:15),
     *PROD1(2,2),PROD2(2,2),PROD3(2,2),PROD4(2,2)

C
      IF(P.GT.15)THEN
      WRITE(6,*)' SET ORDER OF ALGORITHM .LE. 15 !'
      RETURN
      ENDIF
C
C--------------------------------------------------------------
C0 ----------- INITIALIZE BLOCK GENERALIZED RESIDUAL ENERGIES:
C
      DO J=0,P-1
      DO K=1,2
      DO L=1,2
      EM(K,L,J)=W(K,L,0,J)
      R(K,L,J)=W(K,L,1,J+1)
      CM(K,L,J)=W(K,L,0,J+1)
      D(K,L,J)=W(L,K,1,J)
      ENDDO
      ENDDO
      ENDDO
C
C--------------------------------------------------------------
C1 ----------- COMPUTE REFLECTION MATRICES AT THE FIRST STAGE:
C
      CALL KFF(CM(1,1,0),R(1,1,0),KF(1,1,1,0))
      CALL KBB(CM(1,1,0),EM(1,1,0),KB(1,1,1,0))
C
```

```
C-------------------------------------------------------------
C2 ---------------------------------------- ORDER RECURSIONS:
C
      DO M=1,P-1
C
      DO J=0,P-1-M
C
      CALL MUL(KF(1,1,M,0),D(1,1,J),PROD1,2)
      CALL MUL(CM(1,1,J),KF(1,1,M,J),PROD2,0)
      CALL MUL(KF(1,1,M,0),R(1,1,J),PROD3,1)
      CALL MUL(PROD3,KF(1,1,M,J),PROD4,0)
      CALL ADD(EM(1,1,J),PROD1,E(1,1,J))
      CALL ADD(E(1,1,J),PROD2,E(1,1,J))
      CALL ADD(E(1,1,J),PROD4,E(1,1,J))
C
      CALL MUL(KB(1,1,M,0),CM(1,1,J),PROD1,1)
      CALL MUL(D(1,1,J),KB(1,1,M,J),PROD2,1)
      CALL MUL(KB(1,1,M,0),EM(1,1,J),PROD3,1)
      CALL MUL(PROD3,KB(1,1,M,J),PROD4,0)
      CALL ADD(R(1,1,J),PROD1,RM(1,1,M-1,J))
      CALL ADD(RM(1,1,M-1,J),PROD2,RM(1,1,M-1,J))
      CALL ADD(RM(1,1,M-1,J),PROD4,RM(1,1,M-1,J))
C
      CALL MUL(KF(1,1,M,0),R(1,1,J+1),PROD1,1)
      CALL MUL(EM(1,1,J+1),KB(1,1,M,J+1),PROD2,0)
      CALL MUL(KF(1,1,M,0),D(1,1,J+1),PROD3,2)
      CALL MUL(PROD3,KB(1,1,M,J+1),PROD4,0)
      CALL ADD(CM(1,1,J+1),PROD1,C(1,1,J))
      CALL ADD(C(1,1,J),PROD2,C(1,1,J))
      CALL ADD(C(1,1,J),PROD4,C(1,1,J))
C
      ENDDO
C
      DO K=1,2
      DO L=1,2
      DM(K,L,M,0)=C(K,L,0)
      ENDDO
      ENDDO
C
      DO J=0,P-1-M
C
      CALL MUL(KF(1,1,M,J),R(1,1,J),PROD1,2)
      CALL MUL(EM(1,1,J),KB(1,1,M,0),PROD2,1)
      CALL MUL(KF(1,1,M,J),CM(1,1,J),PROD3,2)
      CALL MUL(PROD3,KB(1,1,M,0),PROD4,0)
      CALL ADD(D(1,1,J),PROD1,DM(1,1,M-1,J+1))
      CALL ADD(DM(1,1,M-1,J+1),PROD2,DM(1,1,M-1,J+1))
      CALL ADD(DM(1,1,M-1,J+1),PROD4,DM(1,1,M-1,J+1))
C
      DO K=1,2
      DO L=1,2
      R(K,L,J)=RM(K,L,M,J)
      D(K,L,J)=DM(K,L,M,J)
      EM(K,L,J)=E(K,L,J)
      CM(K,L,J)=C(K,L,J)
      ENDDO
      ENDDO
C
      ENDDO
C
      CALL KFF(CM(1,1,0),R(1,1,0),KF(1,1,M+1,0))
      CALL KBB(CM(1,1,0),EM(1,1,0),KB(1,1,M+1,0))
C
      ENDDO
C
C-------------------------------------------------------------
C3 -------------------------------- INCREMENT STACK POINTERS:
C
      DO M=P-1,1,-1
      DO J=P-1,0,-1
      DO K=1,2
      DO L=1,2
      RM(K,L,M,J)=RM(K,L,M-1,J)
      DM(K,L,M,J)=DM(K,L,M-1,J)
      KF(K,L,M,J+1)=KF(K,L,M,J)
      KB(K,L,M,J+1)=KB(K,L,M,J)
      ENDDO
      ENDDO
      ENDDO
      ENDDO
C
```

```
C------------------------------------------------------------
C5 ---------------------- COMPUTE ARMA LADDER INVERSE FILTER:
C
      EX=XSAMPLE
      EY=YSAMPLE
      RX(0,0)=XSAMPLE
      RY(0,0)=YSAMPLE
C
      DO M=1,P
      RX(0,M)=RX(1,M-1)+KB(1,1,M,0)*EX+KB(2,1,M,0)*EY
      RY(0,M)=RY(1,M-1)+KB(1,2,M,0)*EX+KB(2,2,M,0)*EY
      EX=EX+KF(1,1,M,0)*RX(1,M-1)+KF(2,1,M,0)*RY(1,M-1)
      EY=EY+KF(1,2,M,0)*RX(1,M-1)+KF(2,2,M,0)*RY(1,M-1)
      ENDDO
C
      DO M=0,P
      RX(1,M)=RX(0,M)
      RY(1,M)=RY(0,M)
      ENDDO
C
C------------------------------------------------------------
C
      XSAMPLE=EX
      YSAMPLE=EY
C
      RETURN
C
      END
```

```
      SUBROUTINE MUL(X,Y,PROD,INDX)
C
      REAL X(2,2),Y(2,2),PROD(2,2)
C-----------------------------------------------
      IF(INDX.EQ.0)THEN
C COMPUTE THE MATRIX PRODUCT:
C PROD = X * Y
      PROD(1,1)=X(1,1)*Y(1,1)+X(1,2)*Y(2,1)
      PROD(1,2)=X(1,1)*Y(1,2)+X(1,2)*Y(2,2)
      PROD(2,1)=X(2,1)*Y(1,1)+X(2,2)*Y(2,1)
      PROD(2,2)=X(2,1)*Y(1,2)+X(2,2)*Y(2,2)
      ENDIF
C-----------------------------------------------
      IF(INDX.EQ.1)THEN
C COMPUTE THE MATRIX PRODUCT:
C PROD = X(TRANSPONSE) * Y
      PROD(1,1)=X(1,1)*Y(1,1)+X(2,1)*Y(2,1)
      PROD(1,2)=X(1,1)*Y(1,2)+X(2,1)*Y(2,2)
      PROD(2,1)=X(1,2)*Y(1,1)+X(2,2)*Y(2,1)
      PROD(2,2)=X(1,2)*Y(1,2)+X(2,2)*Y(2,2)
      ENDIF
C-----------------------------------------------
      IF(INDX.EQ.2)THEN
C COMPUTE THE MATRIX PRODUCT:
C PROD = X(TRANSPONSE) * Y(TRANSPONSE)
      PROD(1,1)=X(1,1)*Y(1,1)+X(2,1)*Y(1,2)
      PROD(1,2)=X(1,1)*Y(2,1)+X(2,1)*Y(2,2)
      PROD(2,1)=X(1,2)*Y(1,1)+X(2,2)*Y(1,2)
      PROD(2,2)=X(1,2)*Y(2,1)+X(2,2)*Y(2,2)
      ENDIF
C-----------------------------------------------
      RETURN
      END
```

```
      SUBROUTINE ADD(X,Y,SUM)
C
      REAL X(2,2),Y(2,2),SUM(2,2)
C
C COMPUTE THE MATRIX SUM:
C SUM = X + Y
      DO I=1,2
      DO J=1,2
      SUM(I,J)=X(I,J)+Y(I,J)
      ENDDO
      ENDDO
C
      RETURN
      END
```

```
      SUBROUTINE KFF(X,Y,KF)
C
      REAL X(2,2),Y(2,2),KF(2,2),DET
C
      DET=Y(1,1)*Y(2,2)-Y(1,2)*Y(2,1)
C
      IF(DET.LT.0.1)THEN
C MATRIX Y IS NEGATIV DEFINITE, SET:
      KF(1,1)=0.0
      KF(1,2)=0.0
      KF(2,1)=0.0
      KF(2,2)=0.0
      ELSE
C COMPUTE FORWARD REFLECTION MATRIX:
      KF(1,1)=(X(1,1)*Y(2,2)-X(1,2)*Y(1,2))/(-DET)
      KF(1,2)=(-X(2,2)*Y(1,2)+X(2,1)*Y(2,2))/(-DET)
      KF(2,1)=(X(1,2)*Y(1,1)-X(1,1)*Y(2,1))/(-DET)
      KF(2,2)=(-X(2,1)*Y(2,1)+X(2,2)*Y(1,1))/(-DET)
      ENDIF
C
      RETURN
      END
```

```
      SUBROUTINE KBB(X,Y,KB)
C
      REAL X(2,2),Y(2,2),KB(2,2),DET
C
      DET=Y(1,1)*Y(2,2)-Y(1,2)*Y(2,1)
C
      IF(DET.LT.0.1)THEN
C MATRIX Y IS NEGATIV DEFINITE, SET:
      KB(1,1)=0.0
      KB(1,2)=0.0
      KB(2,1)=0.0
      KB(2,2)=0.0
      ELSE
C COMPUTE BACKWARD REFLECTION MATRIX:
      KB(1,1)=(X(1,1)*Y(2,2)-X(2,1)*Y(1,2))/(-DET)
      KB(1,2)=(-X(2,2)*Y(1,2)+X(1,2)*Y(2,2))/(-DET)
      KB(2,1)=(X(2,1)*Y(1,1)-X(1,1)*Y(2,1))/(-DET)
      KB(2,2)=(-X(1,2)*Y(2,1)+X(2,2)*Y(1,1))/(-DET)
      ENDIF
C
      RETURN
      END
```

ness over conventional techniques which suggests a fixed-point realization of the subprocessors.

C. Speech Recognition—A Possible Application

In the following, the application of the new ARMA method in the area of speech recognition is briefly discussed. The assumption that ARMA modeling can be a very powerful tool in speech recognition stems from two well-known observations. First, it seems that—especially for nasal sounds and consonants—the all-pole model is not realistic. Second, the conventional approaches tend to model the pitch excitation as well as the vocal tract parameters. The excitation waveform has thus a considerable influence on the estimated vocal tract parameters, which are indeed a mixed-parameter set since the modeling algorithm tends to decorrelate also the excitation waveform. However, since the pitch waveform is a very uncertain signal, the currently applied modeling techniques generally lead to very weak vocal tract parameter estimates. This is an important factor which is commonly ignored in currently available speech recognition systems. A very instructive example to this estimation problem can be observed in cases of a degraded pitch, which leads to invalid parameter estimates causing the speech recognition system to provide incorrect results when the parameter extraction routine is based on the classical all-pole model which uses only the speech signal without side information about the excitation waveform. A way to overcome these difficulties is two-channel speech analysis, where the speech signal together with some information about the excitation is processed by an ARMA system identification algorithm. The two-channel approach to speech analysis has been recently suggested by Krishnamurthy and Childers [15] who extracted the excitation signal by means of an electroglottograph (EGG). This EGG technique has previously been investigated by Hess and Indefrey [12]. A summary of pitch extraction methods is given by Hess [3].

VII. Experimental Results

We have carried out numerous simulations in order to evaluate and compare the behavior of the new ARMA covariance ladder algorithm, and we have compared the new covariance technique to some well-introduced fast time recursive ladder methods. Only a small fraction of the results, which seem to be most instructive, are presented here. We find that it is most appropriate to divide the simulation presentation in three basic categories associated with the three basic problems in LS ARMA modeling, namely, the problem of data windowing, the problem of numerical stability, and the problem of numerical accuracy. First, the problem of windowing of the observed input/output processes of the unknown system is addressed. It is well known that in all fast time-recursive techniques, the possibilities of windowing the data are quite limited. As to the author's knowledge, fast techniques have only been derived in case of the growing exponentially weighted window and the sliding rectangular window (or simple modifications thereof). In the proposed new covariance least-squares algorithm, however, we can use arbitrary windows on the data. Recall that this interesting property arises from the fact that the new algorithm limits the time recursion on the calculation of the input data covariance block matrix. A collection of useful recursive windowing algorithms is given in Appendix A, whereas the way of incorporating these routines in the covariance calculation is discussed in Section IV-A. It is expected that different windows can lead to a very different tracking performance in a nonstationary environment. To demonstrate this, we have applied the new ARMA covariance ladder algorithm on the following order $p = 4$ time-varying ARMA system.

For $0 \leq t < 150$, the poles and zeros are held fixed at the following locations:

$$\begin{aligned} \textit{poles: } r_{\infty 0} &= 0.98 & \varphi_{\infty 0} &= 30^\circ \\ r_{\infty 1} &= 0.95 & \varphi_{\infty 1} &= 65^\circ \\ \textit{zeros: } r_{00} &= 0.95 & \varphi_{00} &= 85^\circ \\ r_{01} &= 0.80 & \varphi_{01} &= 20^\circ. \end{aligned}$$

For $150 \leq t \leq 350$, one pole pair is moving with a constant speed along a trajectory in the z-plane described by the following trajectory equation:

$$r_\infty(\varphi) = r_{\infty 0}(1 - \Delta(\varphi)) \tag{7.1a}$$

$$\Delta(\varphi) = 1.895(\varphi - \varphi_{\infty 0}) \exp(-3). \tag{7.1b}$$

For $350 < t \leq 500$, the poles and zeros are held fixed at the following locations:

$$\begin{aligned} \textit{poles: } r_{\infty 0} &= 0.87 & \varphi_{\infty 0} &= 100^\circ \\ r_{\infty 1} &= 0.95 & \varphi_{\infty 1} &= 65^\circ \\ \textit{zeros: } r_{00} &= 0.95 & \varphi_{00} &= 85^\circ \\ r_{01} &= 0.80 & \varphi_{01} &= 20^\circ. \end{aligned}$$

The trajectory of the moving pole pair is marked by dashed lines in Fig. 4. This time-varying ARMA test system was excited by the nonstationary random pitch process shown in Fig. 5(a). Fig. 5(b) is the reduced output signal (output signal minus excitation) of the time-varying ARMA system. Fig. 5(c) shows the AR parameters $\boldsymbol{a}(t)$ of the time-varying system. The MA parameters $\boldsymbol{b}(t)$ are not shown as they are perfectly straight lines. Recall that the zeros are held at fixed locations. Now it is of interest to investigate how different recursive windows affect the tracking behavior when our new ARMA covariance ladder algorithm (Subroutine, Table III) is used. Three windows have been compared. First, we have used the growing exponentially weighted window with an exponential forgetting factor of $\mu = 0.96$. Second, a sliding rectangular window of length $L = 30$ samples has been used. Third, we used a recursive Hanning window of length $L = 30$ samples. The estimated AR parameter trajectories for the respective windows are then plotted in Fig. 6(a)–(c). The estimated MA parameter trajectories, which exhibit a deviation from the expected straight trajectory due to estimation errors, are shown in Fig. 7(a)–(c). To illustrate the estimation error induced by the several windowing techniques, the length of the tap error vector (including both AR and MA parameters) is displayed in Fig. 8(a)–(c). We conclude that the recursive Hanning window comes out with the best result. This technique provides both a very accurate tracking behavior and considerably smoothed parameter estimates. We have found that the recursive modified-Barnwell window, as given in Appendix A, performs almost as well as the Hanning window. However, the Barnwell window has a less complex recursive computation scheme when compared to the Hanning window.

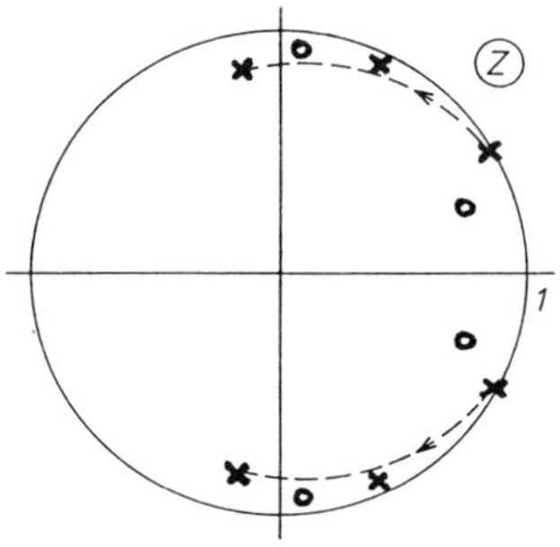

Fig. 4. z-plane of time-varying test system.

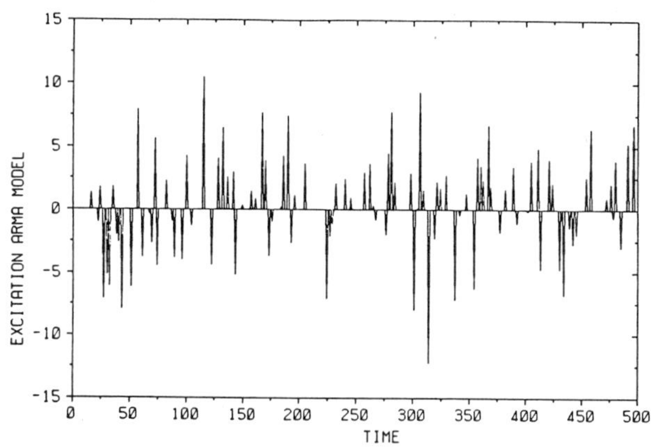

(a)

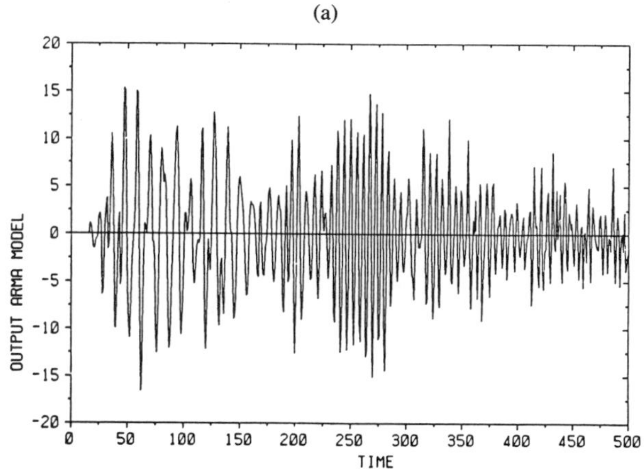

(b)

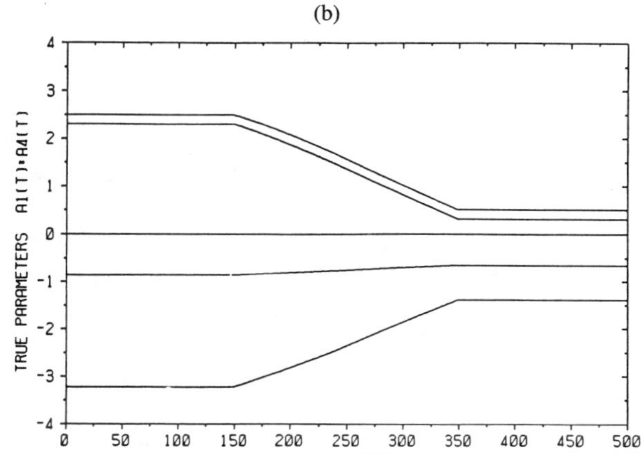

(c)

Fig. 5. (a) Random pitch excitation of ARMA test system. (b) Reduced output signal of order $p = 4$ ARMA test system. (c) Time-varying AR parameters of ARMA test system.

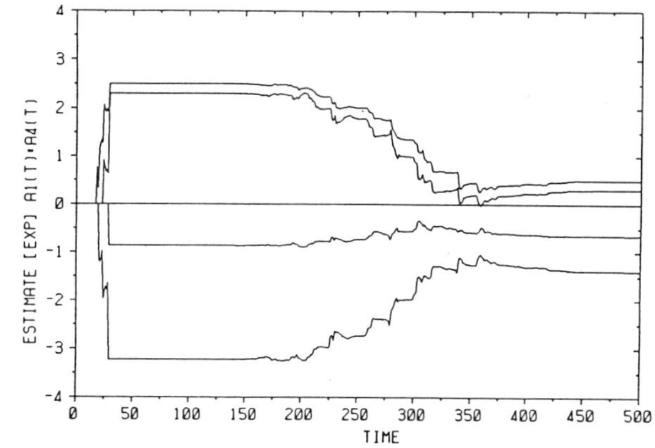

(a)

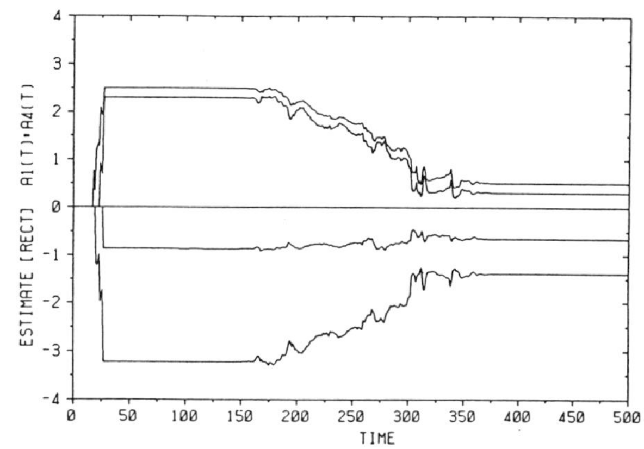

(b)

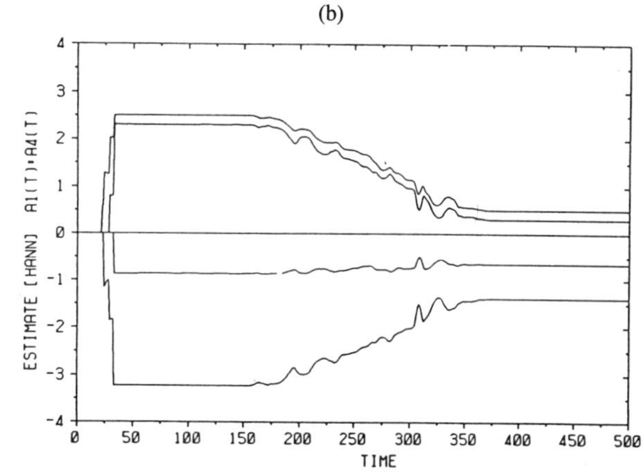

(c)

Fig. 6. Estimated AR parameters using different recursive windows. (a) Exponential growing window $\mu = 0.96$. (b) Sliding rectangular window $L = 30$. (c) Recursive Hanning window $L = 30$.

The sliding rectangular window produces very noisy estimates when only a short data record ($L = 30$) is used. However, short data records are necessary for fast tracking. The exponentially weighted growing window exhibits both a poor tracking and a poor steady-state behavior. The estimates appear to be very noisy. The parameter trajectories in Fig. 9(a) and (b) are shown in order to demonstrate the phenomenon of numerical unstability. To generate these plots, a sliding window ($L = 200$) ARMA estimation algorithm was implemented under 32-bit floating-point arithmetic. We used an order $p = 4$ ARMA test system with constant parameters as follows:

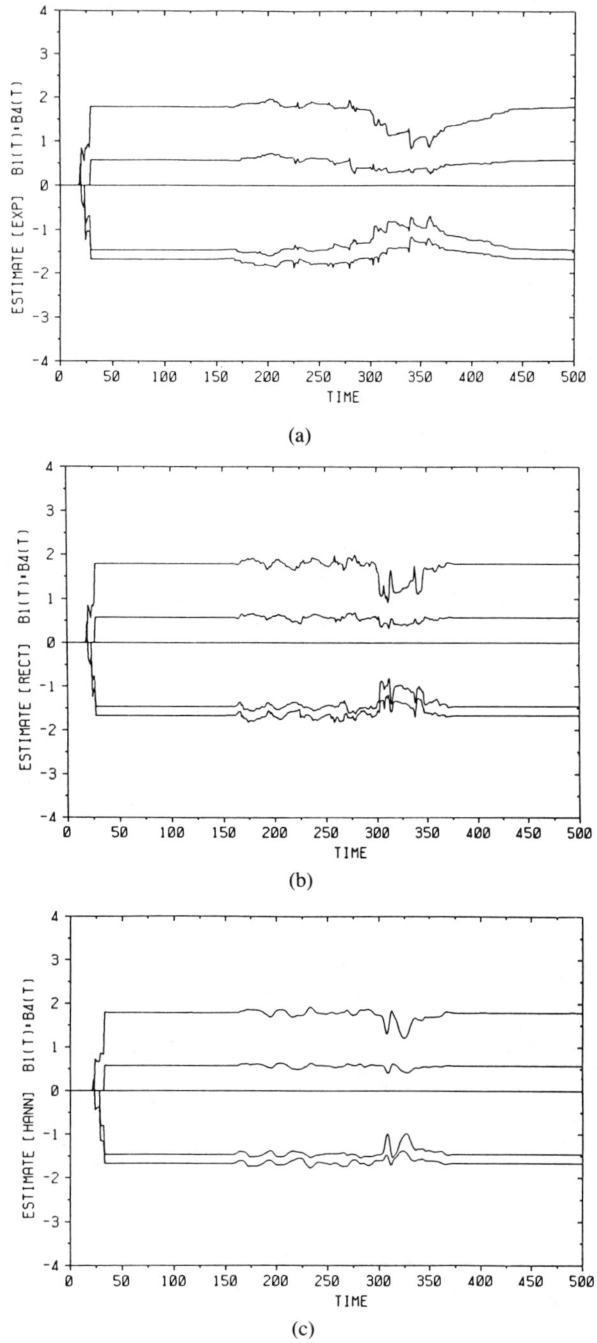

Fig. 7. Estimated MA parameters using different recursive windows. (a) Exponential growing window $\mu = 0.96$. (b) Sliding rectangular window $L = 30$. (c) Recursive Hanning window $L = 30$.

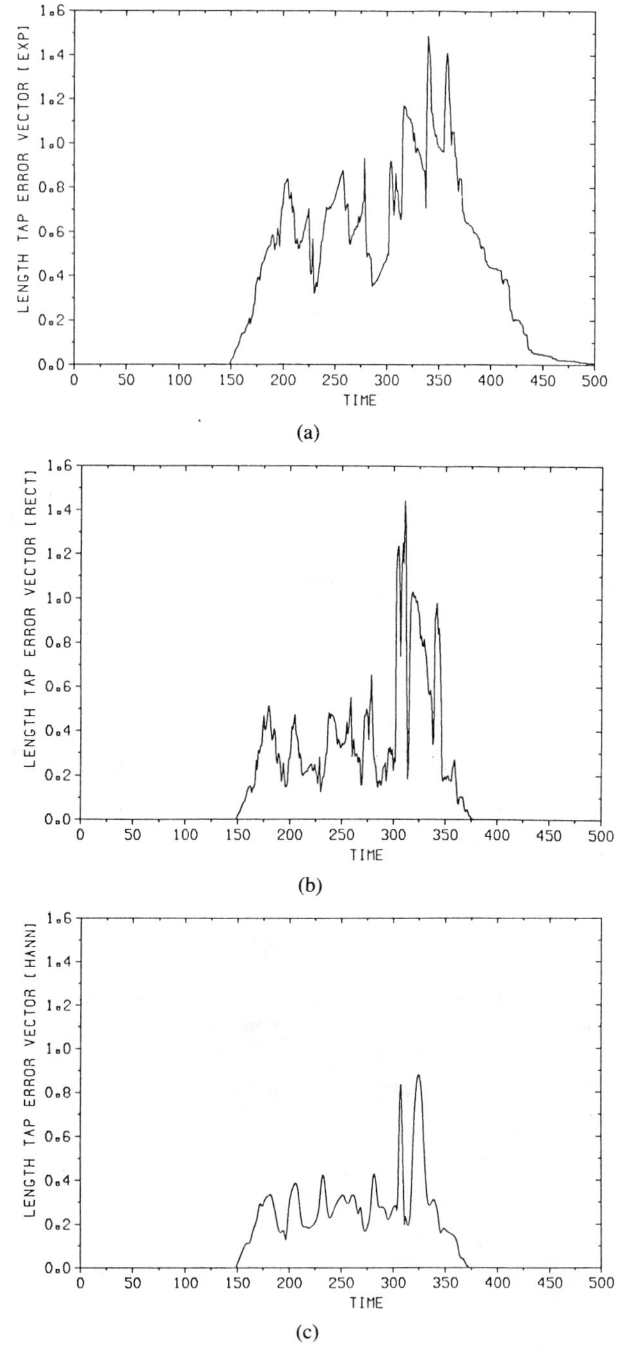

Fig. 8. Length of tap error vector of parameter estimates shown in Figs. 6 and 7. (a) Exponential growing window $\mu = 0.96$. (b) Sliding rectangular window $L = 30$. (c) Recursive Hanning window $L = 30$.

$$\begin{aligned} a_1 &= 2.2 & b_1 &= -1.4 \\ a_2 &= -2.7 & b_2 &= 1.8 \\ a_3 &= 2.0 & b_3 &= -1.0 \\ a_4 &= -0.8 & b_4 &= 0.5. \end{aligned}$$

This model was excited by a white Gaussian random process. Fig. 9(a) and (b) undoubtedly indicates that the algorithm is unstable in the long run, i.e., the error in the estimated parameters grows unbounded with time until it destroys the normal operation of the algorithm. This effect of numerical unstability has previously been reported by other authors [31], [32], [35]. The deep reason for this unpleasant effect is that floating-point computations involve rounding in both addition and multiplication. The same algorithm would run stably, over infinite time when fixed-point arithmetic is used.

This gives rise to the question whether a fast tracking algorithm can be implemented with a sufficient numerical accuracy when fixed-point arithmetic is used. Indeed, it is well known that different algorithms which perform identically over short data records when implemented with floating-point arithmetic can yield quite different results

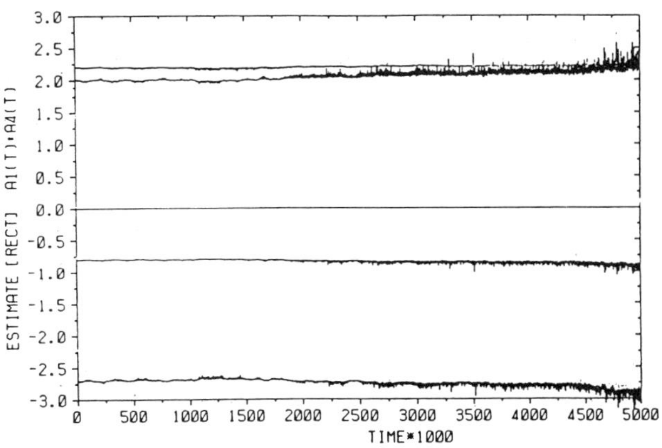

(a)

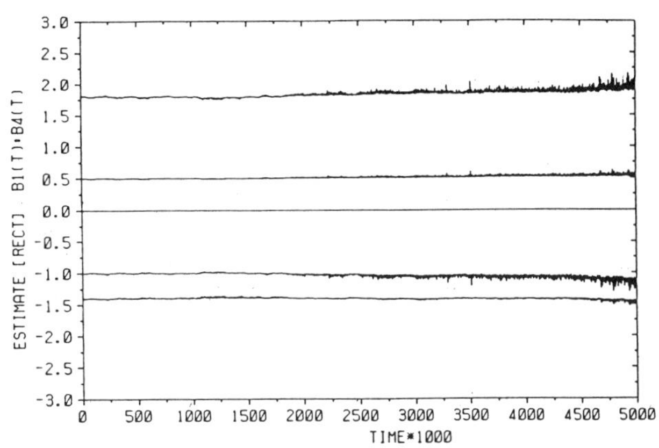

(b)

Fig. 9. Numerically unstable parameter estimates obtained from an $L = 200$ sliding window ARMA estimation algorithm implemented with 32 bits floating-point arithmetic. (a) Estimated AR parameters. (b) Estimated MA parameters.

as soon as fixed-point arithmetic is used. To demonstrate this, we have implemented the traditional time-recursive ladder algorithm of Lee [16]–[18] in the *a posteriori* formulation. A summary of this algorithm is given in [19, p. 839, Table I]. As a second time-recursive ladder algorithm, we have implemented the more refined algorithm of Ling [19] which provides the direct (time-recursive) updating of the reflection matrices. Here we have also used the *a posteriori* formulation. In [19], this algorithm has been praised for its good numerical properties. As a test system, we have used the same system as in the previous test on numerical stability. But now the system is driven by the random pitch sequence [Fig. 5(a)]. As the existence of LS techniques has been justified mainly from their fast start-up and tracking capabilities, it is of great interest to investigate how this property is affected by fixed-point roundoff error. First, the algorithms are implemented with 18 bits fixed-point arithmetic. The accumulation of products is carried out using twice the multiplier wordlength (36 bits). Only the ARMA ladder algorithms are computed under fixed-point arithmetic. The ARMA parameters are computed from the ARMA ladder reflection matrices using the Levinson-type recursion

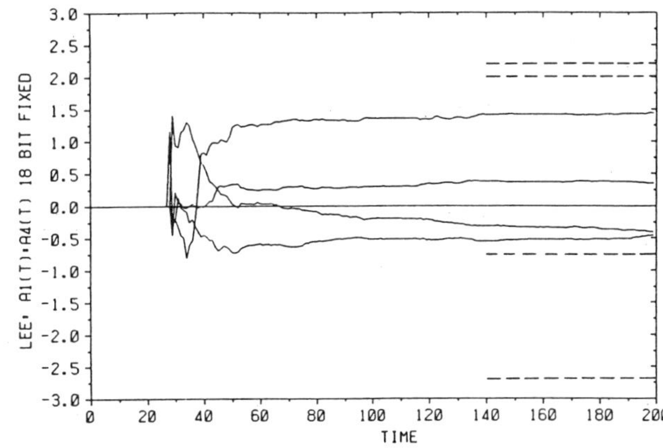

(a)

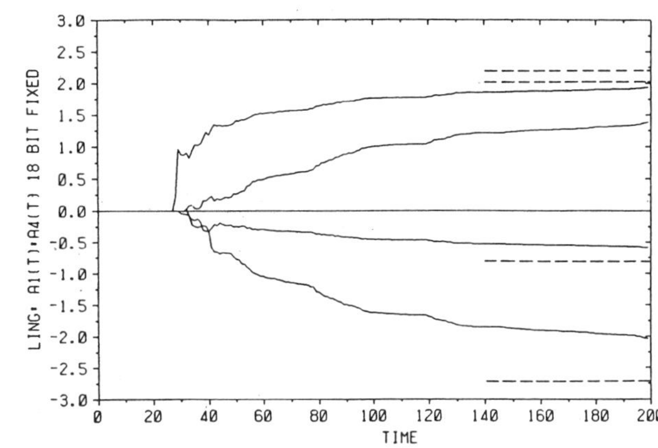

(b)

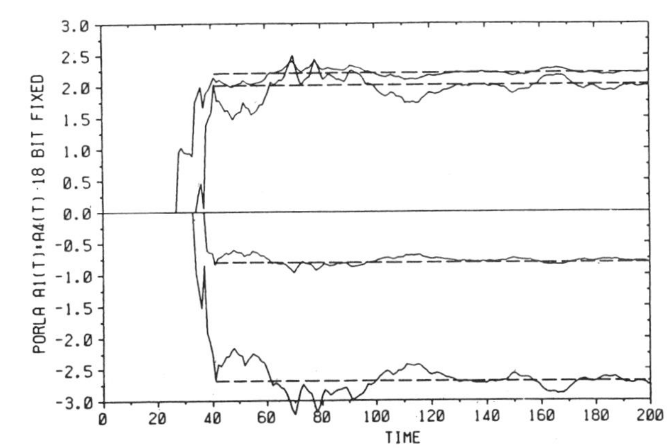

(c)

Fig. 10. Comparison of start-up and tracking capability of several ARMA ladder algorithms implemented with 18 bits fixed-point arithmetic. True parameters marked by dashed lines. (a) Method of Lee: AR parameter estimates. (b) Method of Ling: AR parameter estimates. (c) New ARMA covariance ladder algorithm: AR parameter estimates.

given in Table I. This recursion was operated under floating-point arithmetic in order to obtain correct measurements of the fixed-point roundoff effects in the ladder algorithms. The results of the 18 bits fixed-point simulations are shown in the Figs. 10(a)–(c) and 11(a)–(c). The traditional method of Lee fails to provide accurate estimates. The algorithm proposed by Ling performs better. This algorithm converges reluctantly to the true parame-

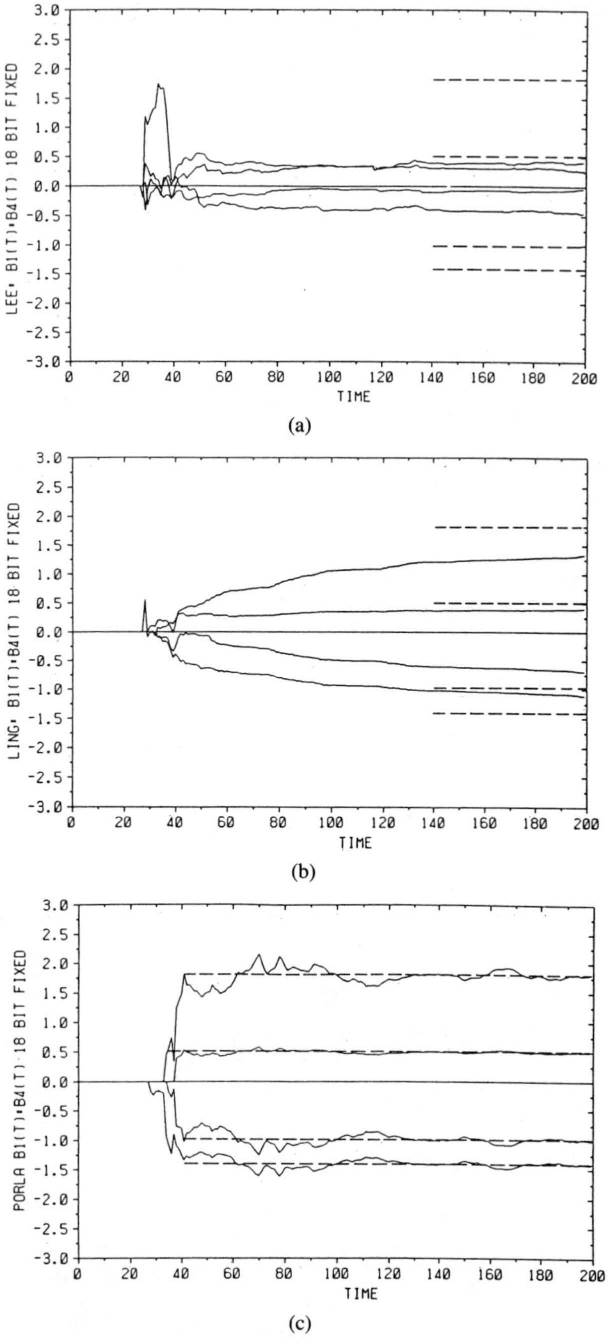

Fig. 11. Comparison of start-up and tracking capability of several ARMA ladder algorithms implemented with 18 bits fixed-point arithmetic. True parameters marked by dashed lines. (a) Method of Lee: MA parameter estimates. (b) Method of Ling: MA parameter estimates. (c) New ARMA covariance ladder algorithm: MA parameter estimates.

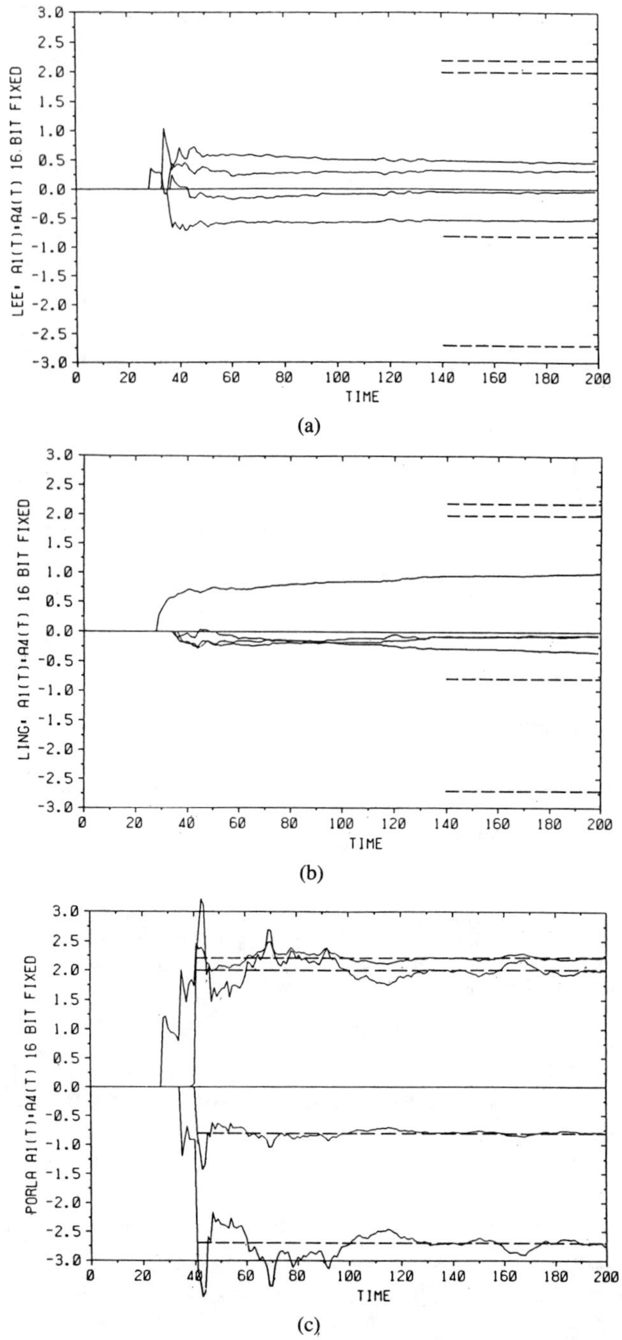

Fig. 12. Comparison of start-up and tracking capability of several ARMA ladder algorithms implemented with 16 bits fixed-point arithmetic. (a) Method of Lee: AR parameter estimates. (b) Method of Ling: AR parameter estimates. (c) New ARMA covariance ladder algorithm: AR parameter estimates.

ters (dashed lines). However, it is very obvious that the time-recursive ladder algorithms have lost their fast start-up and tracking capability due to roundoff error. Several authors, e.g., [19], have investigated only the roundoff error induced in a single (isolated) stage of the algorithm. This, however, was found to be very misleading. In practice, the problem of roundoff error is much more dramatic in higher order estimation problems. There, a small error induced at a lower stage can be significantly amplified in succeeding stages until it causes the algorithm to stop the updating of the coefficients at a somewhat higher stage. The problem is also critical for high-correlated processes and, in such cases, where the poles are located close to the unit circle. The situation changes completely as soon as the new covariance ARMA ladder algorithm is considered. In this new technique, roundoff errors cannot prop-

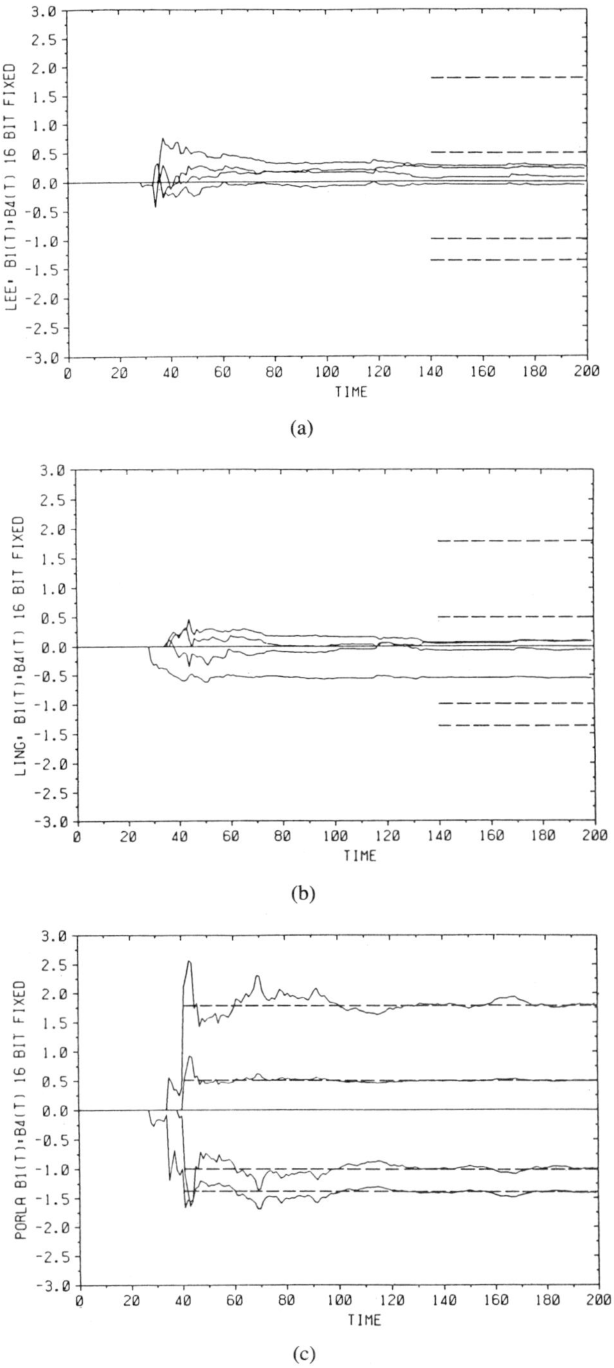

Fig. 13. Comparison of start-up and tracking capability of several ARMA ladder algoirthms implemented with 16 bits fixed-point arithmetic. (a) Method of Lee: MA parameter estimates. (b) Method of Ling: MA parameter estimates. (c) New ARMA covariance ladder algorithm: MA parameter estimates.

agate in the time direction. This is the main reason for the undegraded nice start-up and tracking capability of the new covariance ARMA ladder algorithm as demonstrated in Fig. 10(c) and Fig. 11(c). Figs. 12(a)–(c) and 13(a)–(c) exhibit the result of the experiment when the wordlength is reduced to 16 bits. It is seen that now the method of Ling (Fig. 12(b) and Fig. 13(b) there) also degrades sharply in contrast to the PORLA method [Fig. 12(c) and (13(c)] which provides a very satisfactory start-up and tracking capability.

VIII. Conclusions

The PORLA method for ARMA system identification has been presented in this paper. This new covariance ladder technique enables the separation of the tracking problem and the modeling problem. The resulting algorithm structure is completely different from the time recursive fast-Kalman-type algorithm structure associated with other existing ARMA estimation techniques. Error propagation in time cannot occur in the new method, and hence, the new algorithm has some nice features like an undegraded fast LS start-up and tracking capability when computed with fixed-point arithmetic and the possibility of arbitrary windowing of the data. Since the estimation algorithm requires only the second-order (covariance) information for the construction of the true LS adaptive ARMA ladder form, the new technique is also an appropriate tool in those modeling problems, where only the covariance information rather than the signal itself is available. The simulation results greatly confirmed the excellent behavior of the PORLA technique over conventional methods. Although the algorithm has a higher computational complexity in terms of the number of arithmetic operations, this seems to be compensated from the viewpoint of recent developments in the area of massively parallel processors for signal processing applications. The new algorithm has a very simple modular structure making it an attractive candidate for implementation on just such modern massively parallel processors.

Appendix A
Recursive Windowing Algorithms

Tables IV and V summarize the time recursive algorithms for computation of some useful higher order sliding and growing windows. These windowing algorithms can be interpreted as fixed multichannel filters with the correlation sequence $\boldsymbol{Z}_{i,j}(t)$ as input, and the respective element $\boldsymbol{W}_{i,j}(t)$ of the covariance block matrix as output. As mentioned earlier, the application of these windows is only possible in the covariance algorithm framework, where the time recursion was limited on the calculation of the covariance block matrix. Note also that the covariance block matrix obeys a symmetric property and a shift invariance property as follows:

$$\boldsymbol{W}_{i,j}(t) = \boldsymbol{W}_{j,i}^{T}(t) \tag{A.1}$$

$$\boldsymbol{W}_{i+1,j+1}(t) = \boldsymbol{W}_{i,j}(t-1). \tag{A.2}$$

We see from (A.1) and (A.2) that the recursively updated covariance block matrix is completely specified by its first row block vector. Other elements are obtained by shift-and-store operations. This property holds for any window function $Q(k)$ applied on the correlation sequence $\boldsymbol{Z}_{i,j}(t)$.

Appendix B

It is now only a small step to extend the PORLA method to the multichannel case, where a vector autoregressive (VAR) process must be fitted on a set of q times series. In the VAR case, the input signal matrix attains the form

TABLE IV
HIGHER ORDER SLIDING WINDOWS FOR TIME-RECURSIVE UPDATING OF THE COVARIANCE BLOCK MATRIX

WINDOW (FIR)	WINDOW FUNCTION		RECURSIVE WINDOWING ALGORITHM
rectangular 	$Q(k)=1$ $Q(k)=0$	$0\leqslant k\leqslant L-1$ else	$\underset{\sim}{W}_{i,j}(t)=\underset{\sim}{W}_{i,j}(t-1)+\underset{\sim}{Z}_{i,j}(t)-\underset{\sim}{Z}_{i,j}(t-L)$
triangular 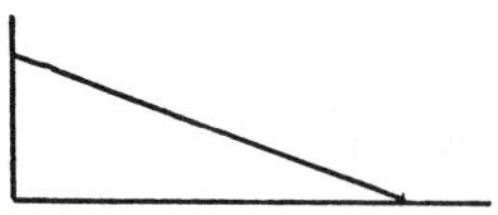	$Q(k)=1-k/L$ $Q(k)=0$	$0\leqslant k\leqslant L-1$ else	$\underset{\sim}{W}_{i,j}(t)=\underset{\sim}{W}_{i,j}(t-1)+\underset{\sim}{Z}_{i,j}(t)-\frac{1}{L}\underset{\sim}{X}_{i,j}(t-1)$ $\underset{\sim}{X}_{i,j}(t)=\underset{\sim}{X}_{i,j}(t-1)+\underset{\sim}{Z}_{i,j}(t)-\underset{\sim}{Z}_{i,j}(t-L)$
Bartlett 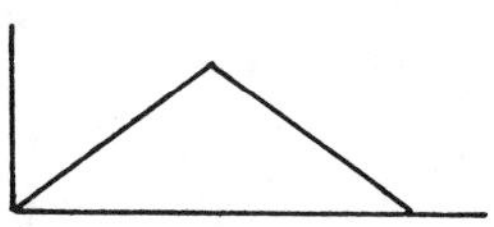	$Q(k)=2k/(L-1)$ $Q(k)=2(L-1-k)/(L-1)$ $Q(k)=0$	$0\leqslant k\leqslant (L-1)/2$ $(L-1)/2\leqslant k\leqslant L-1$ else	$\underset{\sim}{W}_{i,j}(t)=\underset{\sim}{W}_{i,j}(t-1)+\frac{2}{L-1}(\underset{\sim}{X}_{i,j}(t)-\underset{\sim}{X}_{i,j}(t-\frac{L-1}{2}))$ $\underset{\sim}{X}_{i,j}(t)=\underset{\sim}{X}_{i,j}(t-1)+\underset{\sim}{Z}_{i,j}(t)-\underset{\sim}{Z}_{i,j}(t-\frac{L-1}{2})$
Hanning 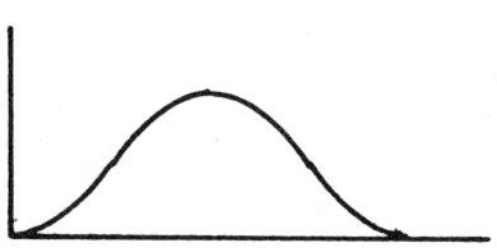	$Q(k)=0.5-0.5\cos(\frac{2\pi k}{L-1})$ $Q(k)=0$	$0\leqslant k\leqslant L-1$ else	$\underset{\sim}{W}_{i,j}(t)=0.5\underset{\sim}{X}_{i,j}(t)-0.5\underset{\sim}{Y}_{i,j}(t)$ $\underset{\sim}{X}_{i,j}(t)=\underset{\sim}{X}_{i,j}(t-1)+\underset{\sim}{Z}_{i,j}(t)-\underset{\sim}{Z}_{i,j}(t-L)$ $\underset{\sim}{Y}_{i,j}(t)=-\underset{\sim}{Y}_{i,j}(t-2)+\underset{\sim}{Z}_{i,j}(t)-\underset{\sim}{Z}_{i,j}(t-L-1)$ $+\cos(\frac{2\pi}{L-1})(2\underset{\sim}{Y}_{i,j}(t-1)-\underset{\sim}{Z}_{i,j}(t-1)-\underset{\sim}{Z}_{i,j}(t-L))$

TABLE V
HIGHER ORDER GROWING WINDOWS FOR TIME-RECURSIVE UPDATING OF THE COVARIANCE BLOCK MATRIX

WINDOW (IIR)	WINDOW FUNCTION	RECURSIVE WINDOWING ALGORITHM
exponential 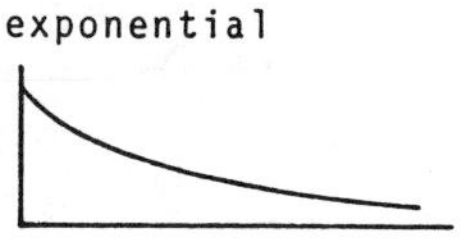	$Q(k)=\mu^k \quad \mu\leqslant 1$	$\underset{\sim}{W}_{i,j}(t)=\mu\underset{\sim}{W}_{i,j}(t-1)+\underset{\sim}{Z}_{i,j}(t)$
modified Barnwell 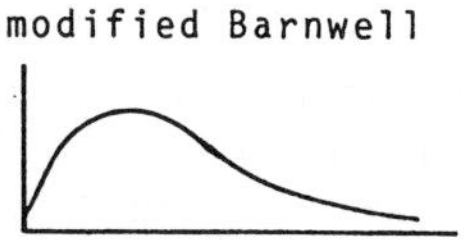	$Q(k)=(1+k)\mu^k \quad \mu\leqslant 1$	$\underset{\sim}{W}_{i,j}(t)=\mu\underset{\sim}{W}_{i,j}(t-1)+\underset{\sim}{X}_{i,j}(t)$ $\underset{\sim}{X}_{i,j}(t)=\mu\underset{\sim}{X}_{i,j}(t-1)+\underset{\sim}{Z}_{i,j}(t)$

$$s(t) = [x_1(t), x_2(t), \cdots, x_q(t)]. \tag{A.3}$$

The forward residual matrix changes to

$$e_m(t) = [e_m^1(t), e_m^2(t), \cdots, e_m^q(t)]. \tag{A.4}$$

The reflection matrices, as well as the elements of the BGRE's, will change to $q * q$ matrices. Even the Fortran coded routine (Table III) can be used for multichannel processing, if we only change the dimensions of the block vectors and the associated stacks from dimension $(2, 2, *)$ to dimension $(q, q, *)$, where q is the number of channels to be processed simultaneously.

REFERENCES

[1] G. J. Bierman, *Factorization Methods for Discrete Sequential Estimation.* New York: Academic, 1977.

[2] G. C. Goodwin and R. L. Payne, *Dynamic System Identification, Experimental Design and Data Analysis.* New York: Academic, 1977.

[3] W. Hess, *Pitch Determination of Speech Signals.* New York: Springer-Verlag, 1983.

[4] M. L. Honig and D. G. Messerschmitt, *Adaptive Filters.* New York: Academic, 1984.

[5] M. D. Srinath and P. K. Rajasekaran, *An Introduction to Statistical Signal Processing with Applications.* New York: Wiley, 1979.

[6] T. P. Barnwell, "Recursive windowing for generating autocorrelation coefficients for LPC analysis," *IEEE Trans. Acoust., Speech, Signal Processing*, vol. ASSP-29, pp. 1062–1066, 1981.

[7] A. Benveniste and C. Chaure, "AR and ARMA identification algorithms of Levinson type: An innovations approach," *IEEE Trans. Automat. Contr.*, vol. AC-26, pp. 1243–1261, 1981.

[8] J. A. Cadzow, "High performance spectral estimation—A new ARMA method," *IEEE Trans. Acoust., Speech, Signal Processing*, vol. ASSP-28, pp. 524–529, 1980.

[9] A. Cumani, "On a covariance lattice algorithm for linear prediction," in *Proc. Int. Conf. ASSP*, Paris, France, 1982, pp. 651–654.

[10] B. Friedlander, "Lattice filters for adaptive processing," *Proc. IEEE*, vol. 70, pp. 829–867, 1982.

[11] Y. Grenier, "Time-dependent ARMA modeling of nonstationary signals," *IEEE Trans. Acoust., Speech, Signal Processing*, vol. ASSP-31, pp. 899–911, 1983.

[12] W. Hess and H. Indefrey, "Accurate pitch determination of speech signals by means of a laryngograph," in *Proc. Int. Conf. ASSP*, San Diego, CA, 1984, pp. 1813.1.1–1813.1.4.

[13] E. Karlsson and M. H. Hayes, "ARMA modeling of time-varying systems with lattice filters," in *Proc. Int. Conf. ASSP*, Tokyo, Japan, 1986, pp. 2335–2338.

[14] S. M. Kay, "A new ARMA spectral estimator," *IEEE Trans. Acoust., Speech, Signal Processing*, vol. ASSP-28, pp. 585–588, 1980.

[15] A. K. Krishnamurthy and D. G. Childers, "Two-channel speech analysis," *IEEE Trans. Acoust., Speech, Signal Processing*, vol. ASSP-34, pp. 730–743, 1986.

[16] D. T. L. Lee, "Canonical ladder form realizations and fast estimation algorithms," Ph.D. dissertation, Stanford Univ., Stanford, CA, 1980.

[17] D. T. L. Lee, M. Morf, and B. Friedlander, "Recursive least-squares ladder estimation algorithms," *IEEE Trans. Acoust., Speech, Signal Processing*, vol. ASSP-29, pp. 627–641, 1981.

[18] D. T. L. Lee, B. Friedlander, and M. Morf, "Recursive ladder algorithms for ARMA modeling," *IEEE Trans. Automat. Contr.*, vol. AC-27, pp. 753–764, 1981.

[19] F. Ling, D. Manolakis, and J. Proakis, "Numerically robust least-squares lattice-ladder algorithms with direct updating of the reflection coefficients," *IEEE Trans. Acoust., Speech, Signal Processing*, vol. ASSP-34, pp. 837–845, 1986.

[20] J. Makhoul, "Stable and efficient lattice methods for linear prediction," *IEEE Trans. Acoust., Speech, Signal Processing*, vol. ASSP-25, pp. 423–428, 1975.

[21] M. Morf, T. Kailath, and L. Ljung, "Fast algorithms for recursive identification," in *Proc. IEEE Conf. Dec. Contr.*, Clearwater Beach, FL, 1976, pp. 916–921.

[22] M. Morf, D. T. L. Lee, J. Nicholls, and A. Vieira, "A classification of algorithms for ARMA models and ladder realizations," in *Proc. Int. Conf., Acoust., Speech, Signal Processing*, Hartford, CT, 1977, pp. 13–19.

[23] R. L. Moses, J. A. Cadzow, and A. A. Beex, "A recursive procedure for ARMA modeling," *IEEE Trans. Acoust., Speech, Signal Processing*, vol. ASSP-33, pp. 1188–1196, 1985.

[24] B. Porat and T. Kailath, "Normalized lattice algorithms for least-squares FIR system identification," *IEEE Trans. Acoust., Speech, Signal Processing*, vol. ASSP-31, pp. 122–128, 1983.

[25] P. Strobach, "Schnelle adaptive Algorithmen zur ordnungsrekursiven Kleinste-Quadrate-Schätzng autoregressiver Parameter," Ph.D. dissertation, Bundeswehr Univ., Munich, West Germany, 1985.

[26] —, "Pure order recursive least-squares ladder algorithms," *IEEE Trans. Acoust., Speech, Signal Processing*, vol. ASSP-34, pp. 880–897, 1986.

[27] —, "A new approach to the least-squares ladder estimation algorithm," in *Proc. IASTED Int. Symp. Appl. Signal Processing Digital Filtering*, Paris, France, 1985, pp. 37–40.

[28] —, "A new class of least-squares covariance ladder estimation algorithms," in *Proc. 19th Annu. Asilomar Conf.*, Monterey, CA, 1985, pp. 116–120.

[29] —, "Efficient and robust covariance ladder algorithms for linear prediction," in *Proc. EUSIPCO-86*, The Hague, The Netherlands, 1986, pp. 1025–1028.

[30] —, "A fast recursive approach to FIR system identification," in *Proc. EUSIPCO-86*, 1986, pp. 1055–1058.

[31] F. Ling and J. G. Proakis, "Numerical accuracy and stability: Two problems caused by round-off-error," in *Proc. Int. Conf. ASSP*, San Diego, CA, 1984, pp. 30.3.1–30.3.4.

[32] F. Ling, D. Manolakis, and J. G. Proakis, "New forms of LS lattice algorithms and an analysis of their round-off error characteristics," in *Proc. Int. Conf. ASSP*, Tampa, FL, 1985, pp. 1739–1742.

[33] C. Samson and V. U. Reddy, "Fixed-point error analysis of the normalized ladder algorithm," *IEEE Trans. Acoust., Speech, Signal Processing*, vol. ASSP-31, pp. 1177–1191, 1983.

[34] E. Satorius *et al.*, "Fixed-point implementation of adaptive digital filters," in *Proc. Int. Conf. ASSP*, Boston, MA, 1983, pp. 33–36.

[35] P. Strobach, "New forms of least-squares lattice algorithms and a comparison of their round-off error characteristics," in *Proc. Int. Conf. ASSP*, Tokyo, Japan, 1986, pp. 573–576.

[36] H. T. Kung, "Why systolic architectures," *IEEE Comput. Mag.*, vol. 15, pp. 37–46, 1982.

[37] H. T. Kung and W. M. Gentleman, "Matrix triangularization by systolic arrays," *Proc. SPIE*, vol. 298, 1981.

[38] S. Y. Kung, "VLSI array processors," *IEEE Acoust., Speech, Signal Processing Mag.*, vol. 2, pp. 4–22, 1985.

[39] P. Strobach and U. Appel, "Ein Signalprozessor mit Wirt-Gast-Kopplung über gemeinsame Speicherbereiche," in *Proc. 8 GI-NTG Fachtagung Architektur und Betrieb von Rechensystemen.* Berlin, Germany: Springer-Verlag, 1984, pp. 61–72.

[40] P. Strobach, "Efficient covariance ladder algorithms for finite arithmetic applications," *Signal Processing*, vol. 13, no. 1, pp. 29–70, July 1987.

The Segmentation Method—An Approach to the Analysis of Microwave Planar Circuits

T. OKOSHI, MEMBER, IEEE, Y. UEHARA, AND T. TAKEUCHI

Abstract—**In many practical planar circuitries, the circuit pattern can be divided into several segments which themselves have simpler shapes such as rectangles. The segmentation method proposed in this short paper is a method in which the characteristics of a planar circuit are computed by combining those of the segmented elements. It features a relatively short computer time required. The principle and computer algorithm are described. Finally, as an example, the application of the proposed method to the trial-and-error optimum design of a ladder-type 3-dB hybrid is described.**

I. Introduction

A planar circuit is a 2-dimensional circuit that should be classified between the distributed-constant (1-dimensional) circuit and waveguide (3-dimensional) circuit. It is defined as an electrical circuit having dimensions comparable to the wavelength in two directions, but much less thickness in one direction [1].

Several methods have been known for the analysis of planar circuits. When the circuit pattern is as simple as square, rectangular, circular, or annular, the impedance matrix can be obtained in a series-expansion form from the Green's function of the wave equation [1]. When the circuit pattern is more arbitrary, the numerical analysis based upon the contour-integral representation of the wave equation is most efficient [1]. Other numerical approaches applicable to arbitrary circuit patterns are the variational method, relaxation method, and the finite-element method [2].

When the circuit pattern is entirely arbitrary, we must rely upon one of those numerical analyses. However, in actual planar circuitry, an entirely arbitrary circuit pattern is not very common; in many cases the pattern consists of several "segments," which themselves have simpler shapes such as rectangles.

The "segmentation method" proposed in this short paper is a method in which the characteristics of a planar circuit are computed by combining those of the segmented elements. It features relatively short computer time required in the analysis.

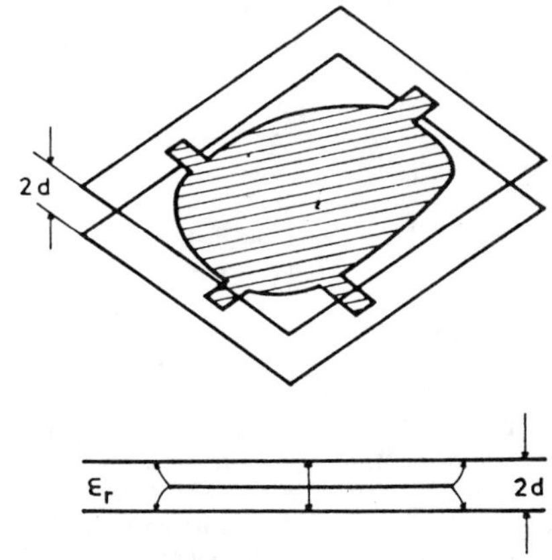

Fig. 1. A triplate-type planar circuit.

In this short paper, the principle and computer algorithm of the segmentation method are described first in Section II. In Section III, to demonstrate the feasibility of the proposed method, its application to the trial-and-error optimum design of a ladder-type 3-dB hybrid is described as an example. The obtained optimum design is verified by the experiment in Section IV.

II. Segmentation Method

A. Basic Concepts

A planar circuit consisting of two ground conductors and a circuit-conductor plate sandwiched by them as shown in Fig. 1 is called a triplate-type planar circuit. In the following only the triplate-type circuit is considered.

The basic equation governing the electromagnetic field in a triplate-type planar circuit is the 2-dimensional wave equation

$$(\nabla_T^2 + k^2)V = 0 \tag{1}$$

where

$$\nabla_T^2 = \frac{\partial^2}{\partial x^2} + \frac{\partial^2}{\partial y^2}, \qquad V = E_z d, \qquad k^2 = \omega^2 \varepsilon \mu$$

Manuscript received December 24, 1975; revised April 2, 1976.
The authors are with the Department of Electronic Engineering, University of Tokyo, Bunkyo-ku, Tokyo 113, Japan.

Reprinted from *IEEE Transactions on Microwave Theory and Techniques*, pp. 662-668, 1976.

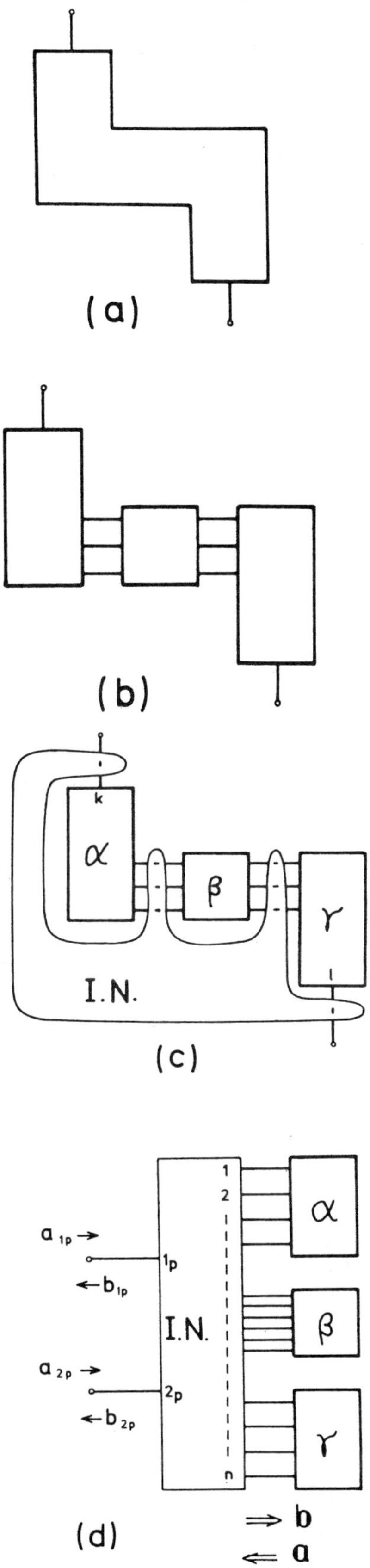

Fig. 2. Principle of the segmentation method. (a) An example of a circuit pattern to which the segmentation method is suitable. (b) Dividing into segments. (c) Introducing the interface network (I.N.). (d) A modified representation of Fig. 2(c).

ε and μ denote the permittivity and permeability of the spacing material, ω the angular frequency, d the spacing between conductors, and E_z denotes the electric-field strength normal to the conductors.

To analyze such a circuit means to solve the wave equation under given boundary conditions, and determine the circuit parameters of its equivalent n-port network as functions of the frequency. When the circuit pattern is simple (for instance, rectangular), the problem may be solved analytically [1]. The segmentation method is a method of the analysis suited to somewhat more complex patterns, as shown in Fig. 2(a). We divide such a circuit pattern into simpler segments, as shown in Fig. 2(b), compute the characteristics of each rectangular segments from their Green's functions [1], and finally obtain the characteristics of the entire circuit by synthesizing the characteristics of the segments.

B. Interface Network

To facilitate the synthesis of the segments in a unified manner, a new circuit called the "interface network" is introduced. As shown in Fig. 2(c), all connections between segments are made through this interface network. The interface network is nothing but a directly connecting network; it only connects directly those couples of ports facing each other, as shown in Fig. 2(c). However, a unified formulation of the synthesis is only possible with the aid of this fictitious network.

In the following, we consider a two-port circuit for the simplicity of the formulation. Fig. 2(c) may also be expressed as Fig. 2(d). We define, with respect to ports $1p, 2p, 1, 2, \cdots, n$, incident and reflected waves of the interface network as

$$a_{1p}, a_{2p}, a_1, a_2, \cdots, a_n$$

and

$$b_{1p}, a_{2p}, b_1, b_2, \cdots, b_n.$$

In numbering ports, we use (numerals $+ p$) for those of the interface network connected to external ports, and simple numerals for those connected to segmentary circuits.

The aforementioned incident and reflected waves are related in terms of the scattering matrix S_N of the interface network as

$$\begin{bmatrix} b_{1p} \\ b_{2p} \\ b_1 \\ b_2 \\ \vdots \\ b_n \end{bmatrix} = [S_N] \begin{bmatrix} a_{1p} \\ a_{2p} \\ a_1 \\ a_2 \\ \vdots \\ a_n \end{bmatrix} \tag{2}$$

where elements of $[S_N]$ are given by

$$[S_N]_{ij} = \begin{cases} 1, & \text{if ports } i \text{ and } j \text{ are connected} \\ 0, & \text{if ports } i \text{ and } j \text{ are not connected} \end{cases} \tag{3}$$

where $i,j = 1p, 2p, 1, 2, \cdots, n$. When $i = j$, $[S_N]_{ij} = 0$.

C. Computation of S-Matrix

If ports $1p$ and $2p$ are connected directly in the interface network to ports k and l, respectively, S_N may be expressed as

$$[S_N] = \left[\begin{array}{cc|c} 0 & 0 & S_1^{\rightarrow} \\ 0 & 0 & S_2^{\rightarrow} \\ \hline S_1^{\downarrow} & S_2^{\downarrow} & S_n \end{array}\right] \tag{4}$$

where $S_1^{\rightarrow}$ and $S_2^{\rightarrow}$ are row vectors, and $S_1^{\downarrow}$ and $S_2^{\downarrow}$ are column vectors defined by

$$S_1^{\rightarrow} = (0,\cdots,0,\underset{\hat{k}}{1},0,\cdots,0)$$

$$S_2^{\rightarrow} = (0,\cdots,0,\underset{\hat{l}}{1},0,\cdots,0) \tag{4a}$$

$$S_1^{\downarrow} = [S_1^{\rightarrow}]^t$$

$$S_2^{\downarrow} = [S_2^{\rightarrow}]^t \tag{4b}$$

where the superscript t denotes transposition. The submatrix S_n in (4) expresses connections between segments.

The S-matrix of the entire circuit should be a (2×2) matrix giving b_{1p} and b_{2p} in terms of only a_{1p} and a_{2p}. However, we should note that, because $1p$ and $2p$ are connected to k and l, respectively,

$$b_{1p} = a_k \qquad b_{2p} = a_l \tag{5}$$

hold. Therefore, in the following we compute a_k and a_l in terms of a_{1p} and a_{2p}.

From (2), taking out n rows from the bottom, we obtain

$$\begin{bmatrix} b_1 \\ b_2 \\ \vdots \\ b_n \end{bmatrix} = [S_n] \begin{bmatrix} a_1 \\ a_2 \\ \vdots \\ a_n \end{bmatrix} + S_1^{\downarrow} a_{1p} + S_2^{\downarrow} a_{2p}. \tag{6}$$

On the other hand, we may also write

$$\begin{bmatrix} a_1 \\ a_2 \\ \vdots \\ a_n \end{bmatrix} = [S_c] \begin{bmatrix} b_1 \\ b_2 \\ \vdots \\ b_n \end{bmatrix} \tag{7}$$

where S_c is the composite S-matrix of the segmentary circuits expressed as

$$[S_c] = \begin{bmatrix} S_\alpha & & 0 \\ & S_\beta & \\ 0 & & S_\gamma \end{bmatrix} \tag{8}$$

S_α, S_β, and S_γ denoting S-matrices of the segments. When the segments are rectangular these matrices are obtainable analytically [1, eq. (20)].

Eliminating the b-vector from (6) and (7), we obtain

$$\begin{bmatrix} a_1 \\ a_2 \\ \vdots \\ a_n \end{bmatrix} = T S_1^{\downarrow} a_{1p} + T S_2^{\downarrow} a_{2p} \tag{9}$$

where

$$T = [E - S_c S_n]^{-1} [S_c] \tag{9a}$$

E denoting an $(n \times n)$ unit matrix. From (4b)

$$TS_1^{\downarrow} = \text{the } k\text{th column of } T$$

$$TS_2^{\downarrow} = \text{the } l\text{th column of } T. \tag{10}$$

Putting (5) into (9), we finally obtain

$$b_{1p} = a_k = T_{kk} a_{1p} + T_{kl} a_{2p}$$

$$b_{2p} = a_l = T_{lk} a_{1p} + T_{ll} a_{2p}. \tag{11}$$

Equation (11) shows that the S-matrix of the entire circuit is a (2×2) submatrix of T; it consists of four elements of T at crosspoints between the kth, lth columns, and the kth, lth rows. Thus the circuit characteristics may be derived from S_n and S_c.

In the aforementioned derivation, the entire circuit has been assumed as a 2-port circuit. The characteristics of a multiport circuit may be obtained in a similar manner.

III. Optimum Design of a Ladder-Type 3-dB Hybrid—An Example of Application

To demonstrate the usefulness of the segmentation method in design problems, its application to the optimum design of a ladder-type 3-dB hybrid will be described as an example.

A. Computer Analysis

We consider a ladder-type 3-dB hybrid as shown in Fig. 3. At a frequency for which $l = \lambda_g/4$, where λ_g denotes the wavelength in the stripline, this circuit shows hybrid characteristics; when power is incident at port 1, it will appear at ports 2 and 3 half-and-half, and no power appears at port 4. Similar relations hold for inputs at other ports.

However, in the following two cases, the stripline width becomes comparable to the quarter wavelength as shown in Fig. 4. The first case is when the required impedance level of the hybrid is relatively low as in a mixer circuit employing Schottky-barrier diodes which have low matching impedance. The second case is when the frequency is relatively high and the length of the arms are shortened. In either case, we have to deal with the circuit as a planar circuit.

The segmentation method is most conveniently used in analyzing such a planar hybrid circuit. In the following analyses, we assume that the circuit is of triplate type, with dielectric-spacing material having ε_r (relative dielectric constant) $= 2.53$ and $d = 1.52$ mm, in accordance with the material used in the experiment.

We first divide the circuit into four segments as shown in Fig. 5. Next the S-matrices of each segment are computed by transforming the Z-matrices given in [1] into S-matrix form. Finally, we synthesize them by the method previously described to obtain the S-matrix of the entire 4-port.

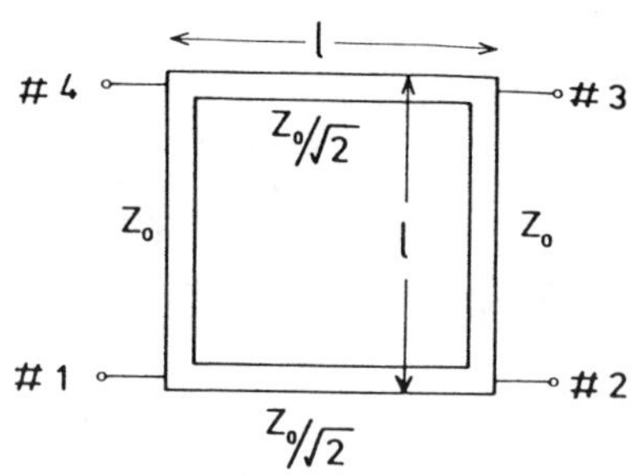

Fig. 3. Basic configuration of a ladder-type 3-dB hybrid circuit.

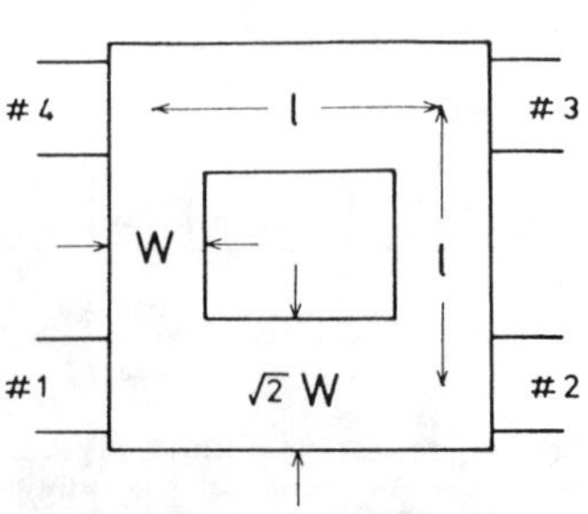

Fig. 4. A ladder-type planar circuit which requires planar-circuit approach.

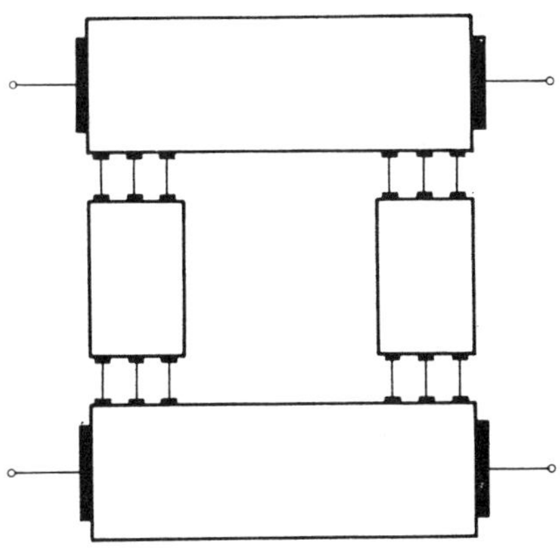

Fig. 5. Segmentation of the ladder-type 3-dB hybrid.

The method of taking into account the width of external terminals (see Fig. 5) might require some comment. When the external terminals are assumed to have certain widths, components of the impedance matrix are computed by averaging the Green's function of the wave equation over the widths of the corresponding terminals [1]. When the terminal shifts, the averaging range should be moved accordingly.

B. Starting Circuit Pattern

As a starting circuit pattern, we take one as shown in Fig. 5 and assume that l = 10.8 mm for each arm. According to the distributed-constant model (Fig. 4), this length and $\varepsilon_r = 2.53$ lead to f_0 (center frequency) = 4.4 GHz. Further, if we assume $Z_0 = 50\ \Omega$, we obtain W = 3.6 mm and $\sqrt{2}\,W$ = 5.1 mm. Comparison of l and W suggests that the planar-circuit model must be employed in exact analysis and design.

The result of the segmentation-method analysis of the aforementioned circuit pattern is shown in Fig. 6(b). These characteristics show increase of the center frequencies (approximately 20 percent above f_0) as well as asymmetries. Obviously, this starting circuit pattern requires improvement.

C. Figure of Merit

Starting at the aforementioned circuit pattern we look for better ones. As the figure of merit of a circuit pattern we use the "bandwidth" defined as follows: at any frequency within the "band," incident power at port 1 is divided and appears at ports 2 and 3 with percentages less than 55 percent and greater than 45 percent, and at ports 1 and 4 with percentages less than 5 percent. The circuit pattern is modified successively upon trial-and-error basis so that the "bandwidth" be maximized and the symmetry of the frequency response be optimized.

In modifying the circuit pattern, symmetries of the circuit pattern with respect to two axes remain. Hence, the circuit characteristics may be determined by four parameters: $|S_{11}|^2$, $|S_{12}|^2$, $|S_{13}|^2$, and $|S_{14}|^2$.

D. Variation of Characteristics by Modification of Circuit Pattern

To collect basic information needed in the optimization, we first investigate the variation of characteristics caused by modification of design parameters.

First, effects of the position of terminals were investigated. The computed characteristics are shown in Fig. 6(a)–(g). This figure shows that when terminals move from the ends of the $\sqrt{2}\,W$-width arms toward those of the W-width arms, the S-parameters vary as shown in Table I(a). The optimum pattern is found somewhere between Figs. 6(c) and Fig. 6(d). The obtained terminal position is shown in Fig. 7, for which the center frequencies of $|S_{11}|^2$, $|S_{12}|^2$, and $|S_{14}|^2$ show the best coincidence. Hereafter, we consider only the terminal positions shown in Fig. 7.

Next, for further improvement of the characteristics, variations caused by shortening the $\sqrt{2}\,W$-width arms were investigated. It was found that the variations as shown in Table I(b) took place. (The detailed data are omitted to save space.)

Finally, the variations caused by widening the $\sqrt{2}\,W$-width arms were investigated. It was found that the variations as shown in Table I(c) took place.

If we disregard a common shift of the center frequencies, the variations of the length and/or width of the W-width arms may be replaced by those of the $\sqrt{2}\,W$-width arms.

E. Optimum Circuit Pattern

Based upon the aforementioned tendencies obtained by the high-speed computer analyses using the segmentation method, we performed the final optimization as follows.

1) Determine the optimum position of ports so that center frequencies of $|S_{11}|^2$, $|S_{12}|^2$, and $|S_{14}|^2$ coincide with each other (Fig. 7).

2) Determine the optimum width of the $\sqrt{2}\,W$-width arms so that the peak value of $|S_{12}|^2$ is optimized. (At this stage, the center frequency of $|S_{13}|^2$ is somewhat deviated.)

3) Determine the length of the $\sqrt{2}\,W$-width arms so that the center frequency of $|S_{13}|^2$ coincides with the others.

The optimum circuit pattern obtained finally and the corresponding circuit characteristics are shown in Fig. 8.

IV. Comparison with Experiment

Experiment was performed using the optimum circuit pattern obtained theoretically to confirm the validity of the analysis and design procedures.

A center conductor having dimensions shown in Fig. 9 was sandwiched by two ground conductors, with spacers of Rexolite 1422 (d = 1.52 mm, $\varepsilon_r = 2.53 \pm 0.03$). The center conductor was produced by a photoetching technique. In the analysis described in Section III, the open-circuit boundary was assumed along the edges of the circuit. Actually, however, the electromagnetic field extends outward due to the edge effect. One method to take this effect into account in the analysis is to shift all the circuit boundaries outward by $0.442d$, in this case by 0.67 mm [3]. The circuit dimensions shown in Fig. 9 were chosen so that the corrected pattern agrees with Fig. 8.[1]

The computed and measured circuit characteristics are shown in Fig. 10 by solid curves and dots, respectively. They show good agreement with each other, except for a certain loss (approximately −0.7 dB) found in the measured value of $|S_{13}|^2$.

[1] However, due to error in the photoetching process, the measured arm lengths in the completed pattern (11.1 mm for all arms) were longer than the design value by 2 and 3 percent for the $\sqrt{2}W$-width and the W-width arms, respectively. The computed characteristics shown in Fig. 10 are based upon the actual circuit dimensions.

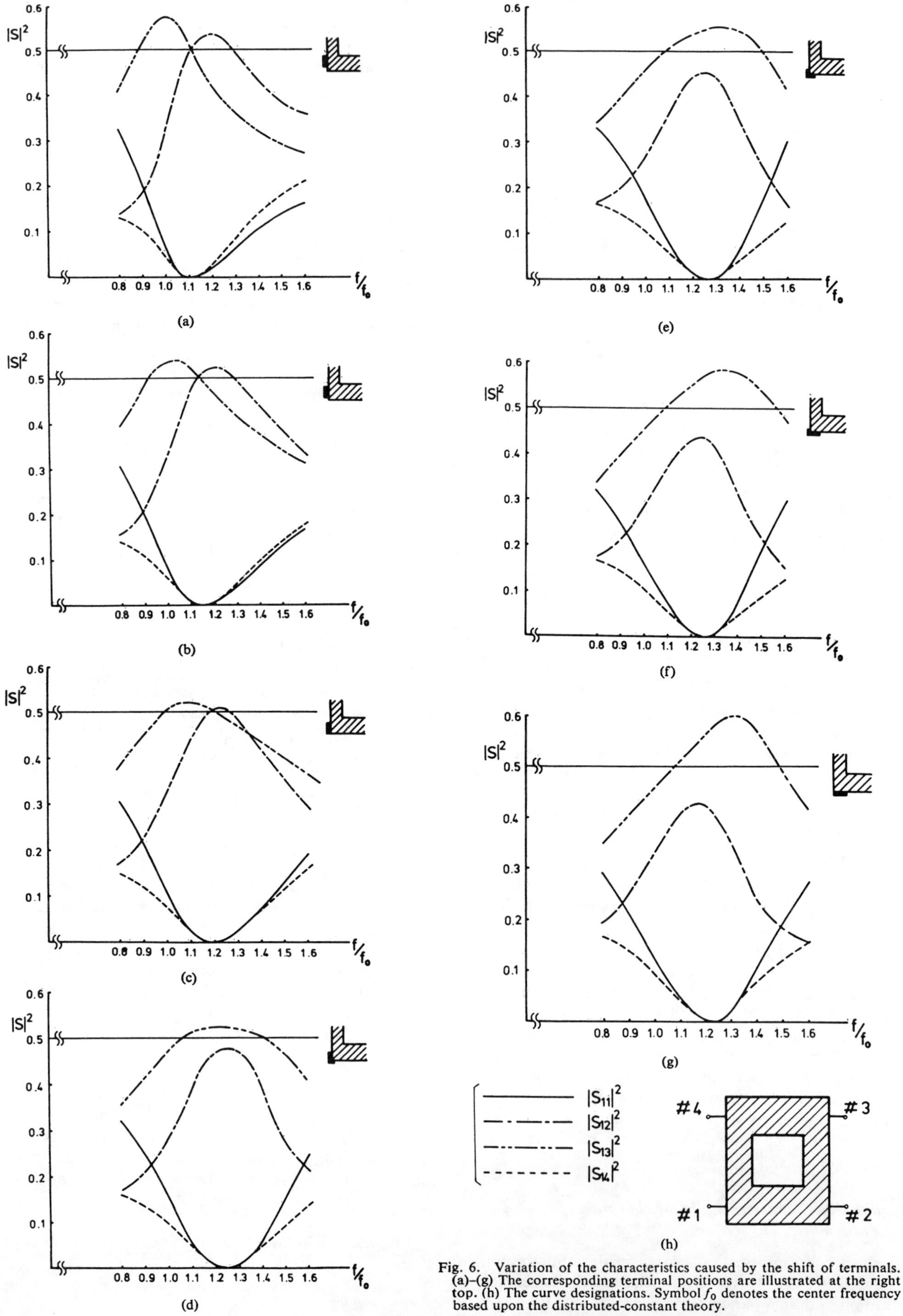

Fig. 6. Variation of the characteristics caused by the shift of terminals. (a)–(g) The corresponding terminal positions are illustrated at the right top. (h) The curve designations. Symbol f_0 denotes the center frequency based upon the distributed-constant theory.

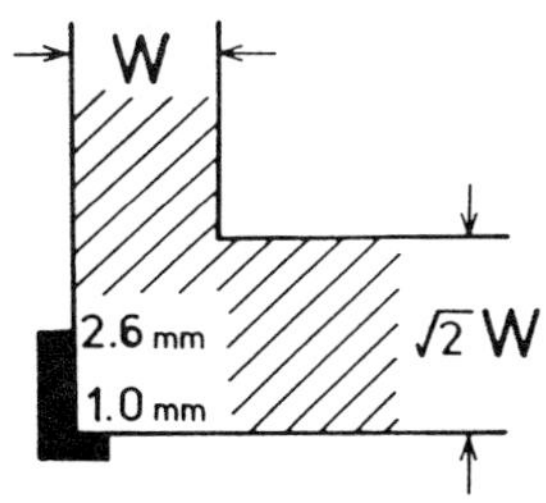

Fig. 7. The optimum terminal position.

TABLE I
VARIATIONS OF S-PARAMETERS CAUSED BY MODIFICATION OF THE CIRCUIT PATTERN

	$\|s_{11}\|^2$	$\|s_{12}\|^2$	$\|s_{13}\|^2$	$\|s_{14}\|^2$
Center frequency (p:peak, v:valley)	↗ v	× p	↗↗ p	↗ v
Peak or valley value	—	↘	↘↗	—

(a) Effects of the shift of terminal positions. (For their shift from ends the $\sqrt{2}$W-width arms toward those of the W-width arms.)

	$\|s_{11}\|^2$	$\|s_{12}\|^2$	$\|s_{13}\|^2$	$\|s_{14}\|^2$
Center frequency (p:peak, v:valley)	× v	× p	↗↗ p	× v
Peak or valley value	—	×	↘↗	—

(b) Effects of shortening the $\sqrt{2}$W-width arms.

	$\|s_{11}\|^2$	$\|s_{12}\|^2$	$\|s_{13}\|^2$	$\|s_{14}\|^2$
Center frequency (p:peak, v:valley)	× v	× p	↗ p	× v
Peak or valley value	—	↗↗	↘↘	—

(c) Effects of widening the $\sqrt{2}$W-width arms.

↗ : Increases gradually.
↗↗ : Increases abruptly.
↘↘ : Decreases abruptly.
↘↗ : Once decreases and then increases.
× : No remarkable variation.
— : These "valley" values are always almost zero.

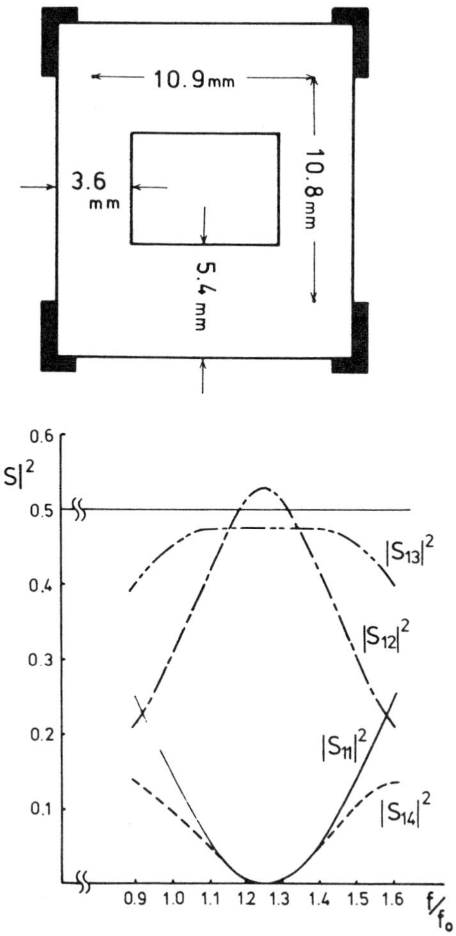

Fig. 8. The optimum circuit pattern and the corresponding characteristics.

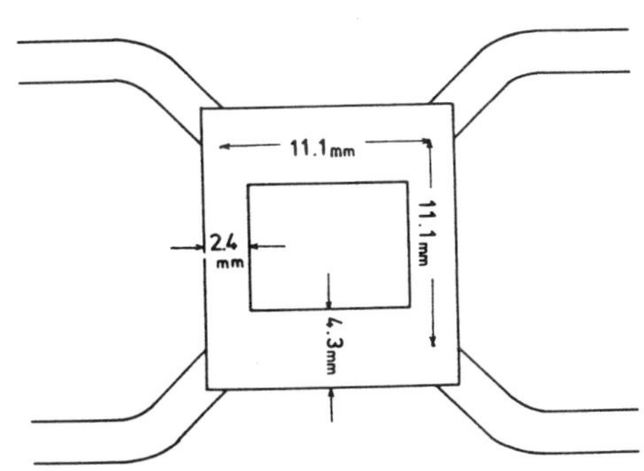

Fig. 9. The circuit pattern used in the experiment (see footnote).

V. COMPUTER TIME

A high-speed analysis technique is a premise for the trial-and-error optimum design of a microwave integrated circuit.

In the present case, the required computer time for analyzing the circuit pattern shown in Fig. 4 for a single frequency was about 2.5 s when 10 × 10 modes were considered in each segment. (The computer used was a HITAC-8800.) However, by decreasing the number of considered modes to a reasonable one (for example, 8 × 3), the required time will easily be reduced below 0.7 s. On the other hand, when we used the contour-integral method [1], the estimated time required for the corresponding accuracy was about 5 s.

Generally speaking, comparison of the previous two methods is rather difficult. However, in many cases reduction of the computer time by almost an order of magnitude was possible by the segmentation method without great loss of accuracy as compared with the contour-integral method.

VI. CONCLUSION

The principle and computation procedure of an efficient method for analyzing a planar circuit, the segmentation method, was presented. To show the usefulness of the proposed method, its application to the trial-and-error optimum design of a hybrid circuit was demonstrated.

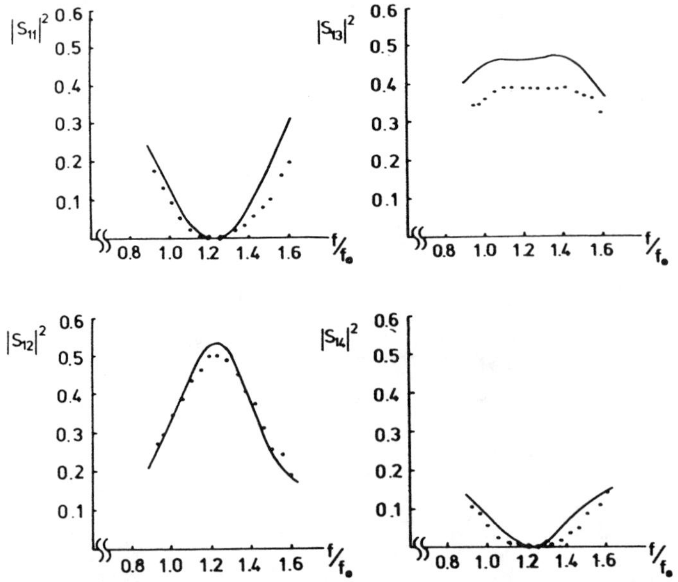

Fig. 10. Comparison of the computed and measured circuit characteristics. (Solid curves: computed; dots: measured.)

The application of the same principle to the short-boundary planar circuit, which has mainly been dealt with by contour-integral analysis [4], will be another interesting task in near future.

Acknowledgment

The authors wish to thank Prof. Y. Okabe of University of Tokyo for his helpful advice, and Prof. J. P. Hsu of Kanagawa University and his colleagues for their assistance in preparing the circuit used in the experiment.

References

[1] T. Okoshi and T. Miyoshi, "The planar circuit—An approach to microwave integrated circuitry," *IEEE Trans. Microwave Theory Tech.*, vol. MTT-20, pp. 245–252, Apr. 1972.

[2] P. Silvester, "Finite element analysis of planar microwave networks," *IEEE Trans. Microwave Theory Tech.*, vol. MTT-21, pp. 104–108, Feb. 1973.

[3] N. Marcuvitz, *Waveguide Handbook*. New York: McGraw-Hill, 1951, p. 160.

[4] T. Okoshi and S. Kitazawa, "Computer analysis of short-boundary planar circuits," *IEEE Trans. Microwave Theory Tech.*, vol. MTT-23, pp. 299–306, Mar. 1975.

The Implementation of Time-Domain Diakoptics in the FDTD Method

Tian-Wei Huang, Bijan Houshmand, and Tatsuo Itoh, *Fellow, IEEE*

Abstract—**The time-domain Diakoptics is implemented in the finite-difference time-domain (FDTD) method with two types of connecting interfaces: i) directional interface (TLM-type), and ii) total-field interface (FDTD-type). The FDTD-type interface provides a more efficient way to realize time-domain Diakoptics than TLM, especially for device optimization problems. To emulate the TLM-type interface in FDTD, two novel algorithms are developed in this paper. One is to implement an ultra-wideband absorbing boundary on the excitation plane during excitation. The other is to separate directional waves without the knowledge of incident waves. For a large circuit with cascaded modules, sequential and parallel algorithms are provided to connect them. With these connecting algorithms, time-domain Diakoptics is one candidate method to realize modular and parallel computation in FDTD simulations. The validity of these algorithms is confirmed by comparison with simulated results from Microwave SPICE.**

I. Introduction

FOR the rigorous analysis of a wide-band and nonlinear monolithic microwave integrated circuit (MMIC), it is important to include all the mutual interactions among circuit elements in the simulation. According to this requirement, several time-domain methods include the nonlinear circuit analysis directly into their time-domain field simulations. These time-domain methods are the transmission line matrix (TLM) [1], the spatial network (SNW) [2], [3], and the finite-difference time-domain (FDTD) [3], [4].

These time-domain methods also require a fine spatial discretization, to maintain numerical accuracy at high frequencies, that results in a large computer storage. For a large and complicated circuit, this memory requirement for circuit simulation is often not practical. One possible solution is the time-domain Diakoptics method. The time-domain Diakoptics is to divide one large circuit into several modules, simulated individually. Using some special interfaces to connect these modules, all the mutual interactions among them are included. Applying this modular approach in the device optimization problems, the repetitive computation of fixed large structures is avoided. Another application is the numerically exact absorbing boundary conditions, that treats the outside open space as a large uniform module. The time-domain Diakoptics is also one candidate method for the implementation of parallel computation in time-domain simulations.

Manuscript received April 23, 1993; revised December 23, 1993. This work was supported in part by the Office of Naval Research N00014-91-J-1651 and the Joint Services Electronics Program F49620-92-C-0055.

The authors are with the Department of Electrical Engineering, University of California at Los Angeles, Los Angeles, California 90024 USA.

IEEE Log Number 9404645.

The time-domain Diakoptics idea was first introduced in TLM [5]–[10]. Unlike TLM in which the waves of different directions are separately defined, the field variables in FDTD represent the total field. Due to these two different characteristics in TLM and FDTD, two types of interfaces are proposed in FDTD for the implementation of time-domain Diakoptics. One is the directional interface that tries to emulate the TLM-type interface in FDTD, and the other is the total-field interface that follows the natural properties of FDTD. Like the TLM Diakoptics, the directional interface needs absorbing boundary conditions (ABC's) on the excitation plane to absorb the reflected wave. On the contrary, using the total-field interface, no absorbing boundary conditions are required on the excitation plane. This means that the total-field interface in FDTD provides a more efficient way to realize time-domain Diakoptics than TLM.

During the development of the directional interface, techniques of directional wave separation in FDTD are needed. Although, a connecting condition is proposed in [12], [13] to separate the incident wave and the scattered wave. This connecting condition requires the knowledge of the incident wave during simulation and another absorbing boundary condition to absorb the scattered wave on a different plane. For the impulse response simulation, only the ABC's based on the time-domain Diakoptics can provide such an ultra-wideband absorption [7]–[10]. This paper develops two novel techniques in the FDTD to separate directional waves. The first one is, for the excitation plane, to separate the reflected wave from the total field and to absorb it on the same plane. This technique can move the source plane extremely close to the discontinuity and reduce the computational volume to a minimum. The second one is, for any other plane, using a recursive algorithm to divide the total field into two-directional waves. For this recursive algorithm, the knowledge of the incident wave is not required.

For a large and complicated linear passive region, a further segmentation is helpful in computational memory reduction and using the parallel computation facilities. Therefore, two algorithms are proposed to connect cascaded modules. From outer modules to inner modules, these cascaded modules are simulated sequentially, that is an "impedance transformation" type approach called the sequential algorithm. To simulate all modules independently, and then connected through some algebraic expressions, is a cascaded "matrix" type approach called the parallel algorithm.

This paper is organized as follows. The idea of time-domain Diakoptics is introduced first. Then a detailed description

Reprinted from *IEEE Transactions on Microwave Theory and Techniques*, pp. 2149-2155, November 1994.

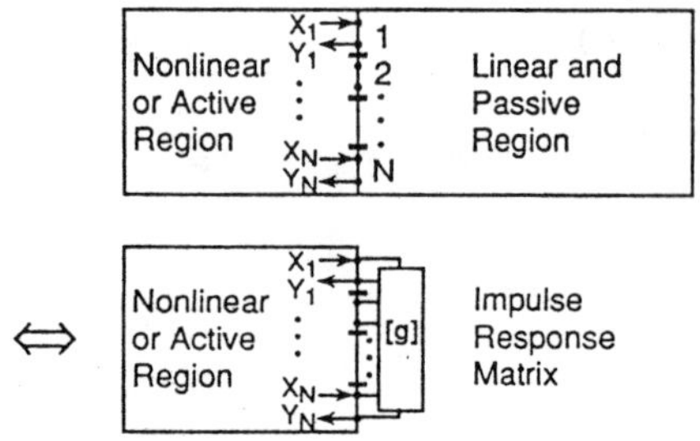

Fig. 1. A multiport linear passive region is replaced by its impulse response matrix $[g]$.

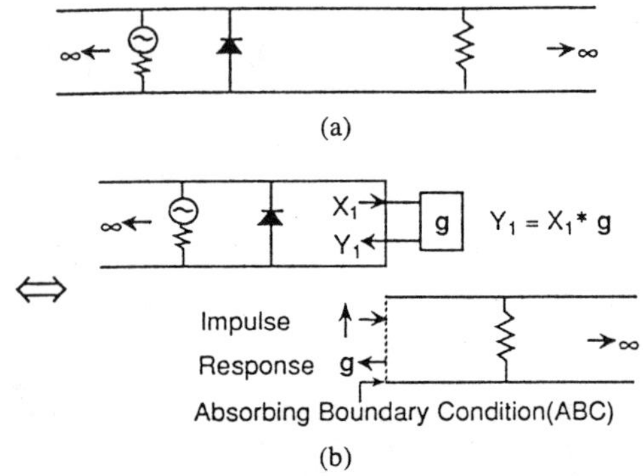

Fig. 2. An infinitely long parallel-plate transmission line loaded with lumped elements is analyzed by time-domain Diakoptics (TLM implementation).

of the implementation of time-domain Diakoptics in FDTD is given, that includes two different types of interfaces. To realize the directional interface, two types of directional wave separation in FDTD are provided. For the Diakoptics of cascaded modules, sequential and parallel algorithms are then demonstrated. Finally, several applications of the FDTD Diakoptics are illustrated.

II. The FDTD Diakoptics Method

The time-domain Diakoptics idea originates from the linear system theory. The system output $Y(n)$ of a one-port linear passive system can be determined from the convolution operation between the impulse response $h(n)$ and the system input $X(n)$. This means that the whole one-port linear passive system can be replaced by its impulse response $h(n)$. Similarly, a multiport linear passive region in the field calculation can be replaced by an impulse response matrix $[g]$ in Fig. 1, that is similar to a time-domain Green's function [14]. The multiport convolution is defined in (1).

$$Y_m(k) = \sum_{n=1}^{N} \sum_{k'=1}^{k} g(m, n, k - k') X_n(k') \qquad (1)$$

where $g(m, n, k')$ is the impulse response (or the time-domain Green's function) at port "m" at $t = k'$ due to the unit excitation at port "n" at $t = 0$.

In this paper, a one-dimensional circuit in Fig. 2(a) is selected to demonstrate the time-domain Diakoptics, although this method can be extended to two or three dimensional structures. Fig. 2(a) shows an infinitely long parallel-plate transmission line loaded with lumped elements. Assuming, only TEM wave exists along this line for two plates with a small distance. The use of one-dimensional model makes it possible that the results can be compared with those by a conventional time-domain circuit simulator, like Microwave SPICE (MWSPICE). This 1D example can be easily used to explain the complex algorithms in Sections III and IV, and be extend to 2D or 3D structures with the aid of (1).

The application of time-domain Diakoptics to the circuit in Fig. 2(a) can be explained by TLM implementation which is illustrated in Fig. 2(b). The left-hand section is a nonlinear active region and the other section is a linear passive region. First step is to compute the impulse response g of the linear passive region. After g is obtained, the active region can be simulated with a convolution on the boundary between X_1 and g to represent the load effect. Since TLM deals with directional waves, ABC on the left to the impulse is necessary for the absorption of reflected waves during excitation. On the other hand, the field variables in FDTD represent the total field only. Hence a directional interface is proposed to emulate TLM-type interface in FDTD.

In this paper, Yee's lattice [11] is utilized to show the implementation of time-domain Diakoptics in FDTD. The superscript of a field variable represents the field direction and relative positions on interface: "f" for the forward direction, "r" for the reflected direction, "t" for both directions, and "0", "1" for relative positions. The subscript indicates the output and input port number. All the field variables in this paper represent the tangential E-field to the segmentation interfaces.

A. Directional Interface

The directional interface is to emulate the TLM-type interface in FDTD. As shown in Fig. 3(a), the wave propagating to the right is denoted as E^f, the forward wave, and the one to the left is denoted by E^r, the reflected wave. There are two different cases: i) the input E^{f0} is known, or E^{f0} can be precalculated, ii) the input E^{f0} is unknown, or E^{f0} cannot be precalculated. For the case i), E^{f0} is precalculated or obtained from a pre-simulation of active region, so E^{r0} can be calculated from a convolution between the known E^{f0} and g^{r0}. Since both E^{f0} and E^{r0} are precalculated, the active region during FDTD simulation can be terminated by the known E^{t0} $(= E^{f0} + E^{r0})$. Therefore, only one cell interface is used, which is very similar to TLM Diakoptics. One example of case i) is the interface between section (2) and (3) in Fig. 9 presented later. To precalculate E^{f0}, usually one additional simulation of active region should be performed. In stead of this additional simulation, a two-cell directional interface is proposed for case ii). In Fig. 3(a), two cells are used for the interface: one is E^{t1} which is from convolution and to terminate the simulation of active region, and the other is E^{t0} which is from FDTD iterations and to generate the input E^{f0}. To separate the unknown E^{f0} from E^{t0}, a recursive algorithm is developed in the next section, without the knowledge of incident waves. This two-cell interface is similar to a connecting condition in [12], [13], but no knowledge of the incident waves is required.

Another problem for the directional interface is the ABC on the excitation plane, as in TLM Diakoptics. This ABC on

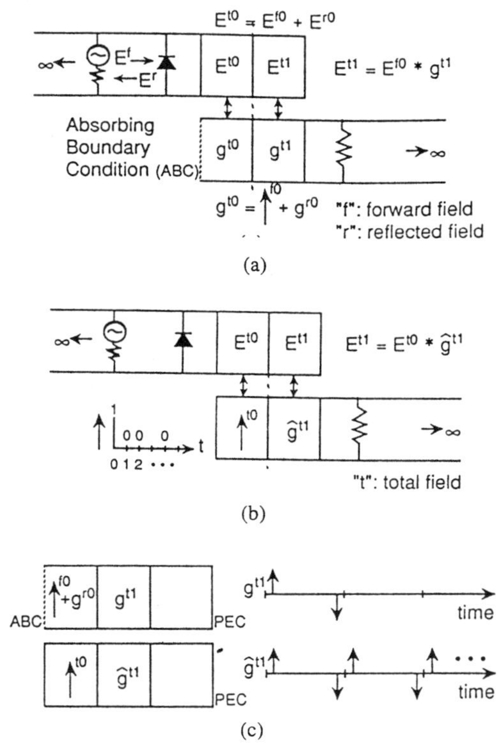

Fig. 3. The implementations of FDTD Diakoptics on a one-port connected circuit with (a) total-field interface, and (b) directional interface, (c) ideal impulse response of a shorted line with directional or total-field interface.

the excitation plane should only absorb the reflected impulse response g^r, and leave the impulse $\uparrow^{f0}$ propagating to the right freely. Hence, for directional interface, separation of directional waves in FDTD is needed on the excitation plane, and an ultra-wideband ABC is also required on the same plane to absorb the reflected impulse response g^r, just as in the case of TLM Diakoptics. The implementation of such an ABC is also explained in the next section.

B. Total-Field Interface

Directional interface tries to implement the TLM-type properties in FDTD. In such a case, some additional computations beside FDTD simulation are needed so that the separation of directional waves can be realized. In addition, a super-wideband ABC is needed as described above. If an interface can preserve the natural properties of FDTD which use the total-field variables only, then the computational efficiency is increased for FDTD Diakoptics. To this end, the total-field interface is proposed here to directly uses the total-field variable in FDTD for convolution. It is found that no separation of directional waves is required. In Fig. 3(b), E^{t0} and E^{t1} are the total tangential E-field of two adjacent cells "0" and "1" on the segmentation interface. After an impulse of total E-field $\uparrow^{t0}$ is injected into the right-hand section at the position corresponding to E^{t0}, an impulse response $\widehat{g}^{t1}$ can be found at the position corresponding to E^{t1}. The hat of the impulse response $\widehat{g}^{t1}$ in total-field interface indicates a different response from the response $\widehat{g}^{t1}$ in directional interface. From the linear system theory, an impulse needs to be injected into the input variable position, to obtain the impulse response. The total-field interface has an impulse $\uparrow^{t0}$ injected into the total-field variable position, so the total-field E^{t0} is used as the input variable in the convolution. On the other hand, for the directional interface, an impulse is injected into the forward-field variable position, so the input variable is E^{f0} instead of E^{t0} in the convolution. Hence for the total-field interface, the boundary field E^{t1} of active region can be calculated from a convolution between E^{t0} and $\hat{g}^{t1}$ which represents the right-hand section. This two-cell interface is similar to the Mur's first order absorbing boundary conditions [12] that calculate a boundary field from an inner simulated field.

The injected impulse of total E-field $\uparrow^{t0}$ is shown in the inset of Fig. 3(b) that indicates a grounding from time step 1 to infinity. This grounding from time step 1 and beyond is equivalent to have a perfect electric conductor (PEC) on the excitation plane, and no ABC's are needed on the left to this PEC. Without the requirement of ABC's, the total-field interface in FDTD is more efficient to realize time-domain Diakoptics than TLM. This simplification of excitation plane benefits from the natural properties of FDTD that use the total-field variables, and no directional waves are taken into account.

Two different ideal impulse responses g^{t1} (directional interface) and $\hat{g}^{t1}$ (total-field interface) of a shorted ideal transmission line (assuming no FDTD numerical dispersion) are shown in Fig. 3(c). The effect of an equivalent PEC on the excitation plane of total-field interface is demonstrated by $\widehat{g}^{t1}$ in Fig. 3(c) which is similar to an ideal response of a 1D cavity. Since the response $\widehat{g}^{t1}$ has interactions with the impulse $\uparrow^{t0}$, the output of the convolution E^{t1} will be also interacted with the input E^{t0} through the FDTD simulation.

For 2D or 3D problems, there are many tangential field variables on the segmentation interfaces that can be treated as a multiport system. With the aid of (1), the above algorithms can be applied to a multiport segmentation interface. For a multiport system, the impulse response is also replaced by an impulse response matrix.

III. Separation and Absorption of Directional Waves in FDTD

This section presents two algorithms to separate and to absorb the directional waves in FDTD, which can be used for the implementation of the direction interface.

A. On the Excitation Plane

Fig. 4(a) shows an impulse traveling through a uniform transmission line as g^f. If this forward excitation wave g^f is reflected by a resistor and without any second reflection, then a reflected wave g^r is generated, as shown in Fig. 4(b). Using superposition principle, the forward and the reflected wave are combined into a total wave g^t. If g^{t0} on the excitation plane is known, then the simulation can be terminated by g^{t0}. Since $\uparrow^{f0}$ for g^{t0} is already known, only the reflected impulse response g^{r0} on the excitation plane needs to be calculated.

To find g^{r0} on the excitation plane, there is an algorithm in Fig. 4. As shown in Fig. 4(b), the reflected wave g^r

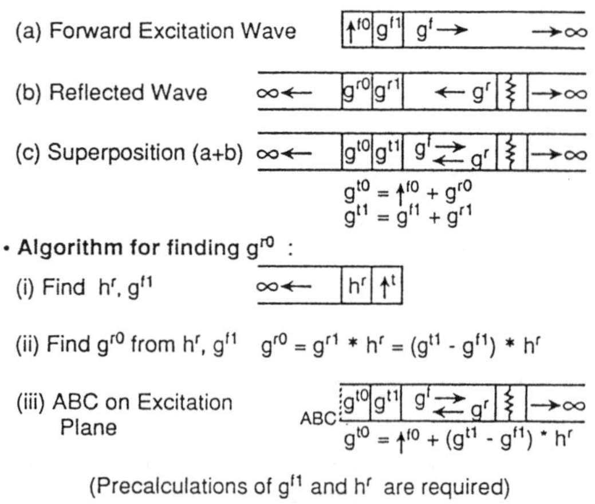

Fig. 4. Separation of directional waves on the excitation plane in FDTD in order to absorb the reflected wave during the excitation.

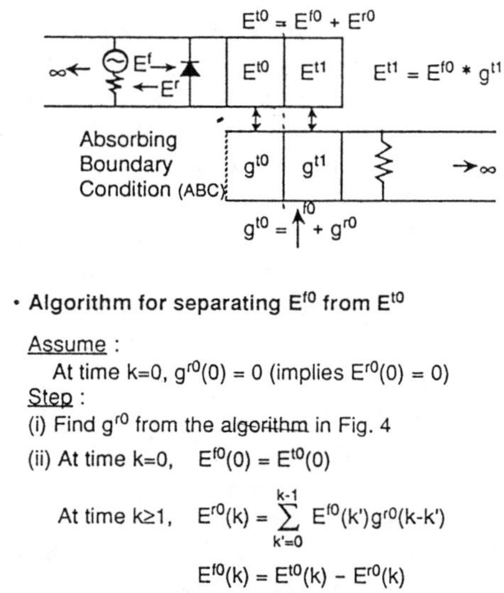

Fig. 5. Separation of directional waves from the total-field variables on the convolution plane in the FDTD Diakoptics.

will propagate to g^{r1} before arriving at the excitation plane as g^{r0}. Therefore g^{r0} can be calculated from g^{r1}, through a convolution with time domain Green's function h^r of the left-hand uniform section in Fig. 4(i). During the FDTD simulation, only the total-field variable g^{t1} is known, so g^{r1} needs to be indirectly calculated. If the forward field g^{f1} is precalculated from a simulation of an impulse through a right-hand uniform region in Fig. 4(a), then the reflected field g^{r1} can be obtained by a subtraction of g^{f1} from g^{t1}. Hence, two precalculations for h^r and g^{f1} are required before the FDTD simulation. During the simulation, one convolution between g^{r1} and h^r is needed to find g^{r0} on the excitation plane. With g^{r0} on the excitation plane, the reflected wave g^r is absorbed on the same plane. Following above algorithm, a numerically exact absorbing boundary condition [16] is implemented on the excitation plane to absorb the ultra-wideband reflected impulse response.

B. On the Convolution Plane

After the impulse response g^{r0} of the right-hand section is found, the directional wave separation can be performed on the segmentation interface, or on the convolution plane, as shown in Fig. 5. If the discontinuity of the right-hand section is not on the segmentation interface, then the instant system response $g^{r0}(0)$ of the right-hand section is zero at time $k = 0$. That means no instant reflected wave $E^{r0}(0)$ on the convolution plane.

At time $k = 0$, there is no instant reflected wave $E^{r0}(0)$, so $E^{f0}(0)$ is equal to $E^{t0}(0)$. During the FDTD simulation at time $k \geq 1$, only the total-field variable E^{t0} is available directly from the simulation. The forward field variable E^{f0} can be calculated by a subtraction of E^{r0} from E^{t0}. The reflected wave E^{r0} is computed from the convolution between the impulse response g^{r0} and the previous input E^{f0}. Hence a recursive formula in Fig. 5, ii) is obtained to compute E^{f0} from the difference between E^{r0} and E^{t0}, and E^{r0} is calculated from g^{r0} and the previous E^{f0}. This directional wave separation algorithm can be applied to any plane with source at one side and load at the other, and the knowledge of incident waves is not required.

IV. The Algorithms for Cascade Modules

From the previous sections, a circuit can be divided into a linear passive part and a nonlinear active part, and connected through an impulse response matrix. For a very large linear passive region, a further segmentation is helpful. For the FDTD Diakoptics of cascaded modules, sequential and parallel algorithms are proposed in this section.

A. Sequential Algorithm

A multilayered circuit in Fig. 6 is used to demonstrate the sequential algorithm. Sections (3) and (4) in Fig. 6 are cascaded sections. Two outer sections, (1) and (4), are calculated first in Step 1. During Step 2, when the inner section (3) is simulated by FDTD, a convolution operation is performed on the interface between (3) and (4), to establish the boundary field of section (3). That means, during the simulation of section (3), the load effect of section (4) is represented by its impulse response $\hat{g}_2^{t1}$. Following this procedure, cascaded sections (3) and (4) is represented by $\hat{g}_3^{t1}$. If there are more cascaded sections, this process is repeated until reach the interface between the linear passive part and the active part. Finally, Step 3 is solved for the active region (2).

Using the sequential FDTD Diakoptics algorithm, the transient response of the circuit in Fig. 6 is plotted in Fig. 7 and compared with simulated results from Microwave SPICE. The agreement is so good that they cannot be distinguished on the figure. If section (3) in Fig. 6 is modified, then only Steps 2 and 3 need to be recalculated. The computational efforts on sections (1) and (4) in Step 1 need not be repeated. This modified circuit is in Fig. 8, and the agreement of circuit response analyzed by the sequential Diakoptics and MWSPICE is excellent. This is one example of modular computation in the FDTD Diakoptics. Although the total-field interfaces are used in these examples, this algorithm can be also applied to the directional interfaces.

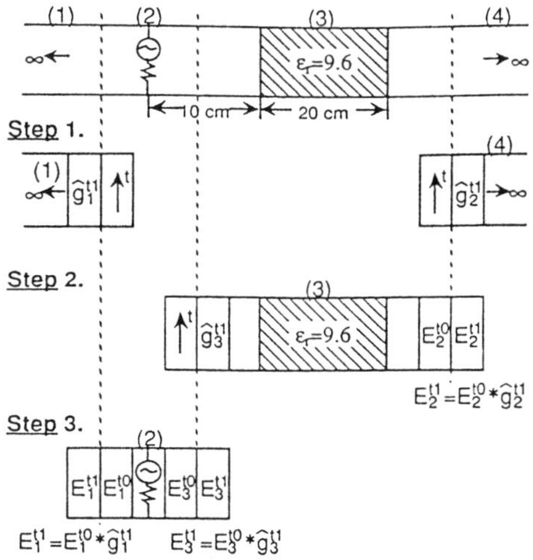

Fig. 6. A multilayered circuit is analyzed by the sequential FDTD Diakoptics algorithm with the total-field interface.

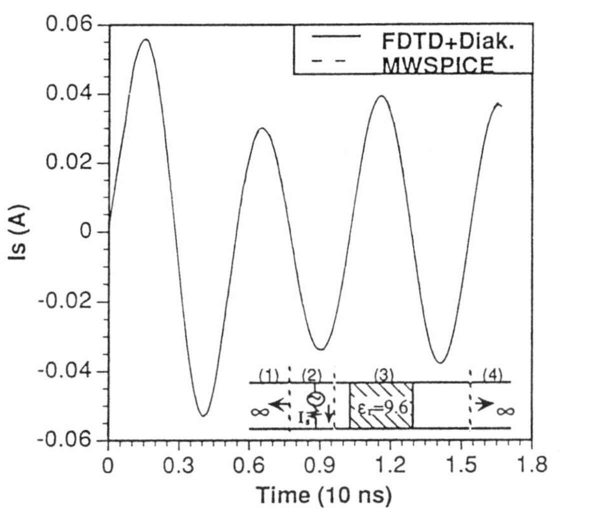

Fig. 7. The transient response of the circuit in Fig. 6 ($V_s = -0.1\sin(2\pi ft), f = 200$ MHz, $R_s = 0.5\ \Omega$, Cell Size $= 1$ cm^2).

B. Parallel Algorithm

There are two major difficulties for the application of the sequential algorithm. The first one is, due to the sequential computation from outer to inner sections, the total computational time cannot be significantly reduced by using a parallel computation machine. The second is, for the modular computation purpose, these inner sections cannot be simulated independently. A modification in one section will affect its inner section. Therefore, further improvement in computation is obtained by a parallel algorithm. The parallel algorithm [6] can simulate each section by FDTD independently. When one section is modified, no other section needs to be recalculated. For the parallel algorithm, since the directional wave is used in the connecting algorithm between cascaded modules, only the directional interface in the FDTD Diakoptics can be applied.

This algorithm is illustrated in Fig. 9. Section (2) is treated as a two-port module that needs to + be simulated twice in Step 1: one for the impulse propagating to the right denoted as "f", and the other to the left denoted by "r".

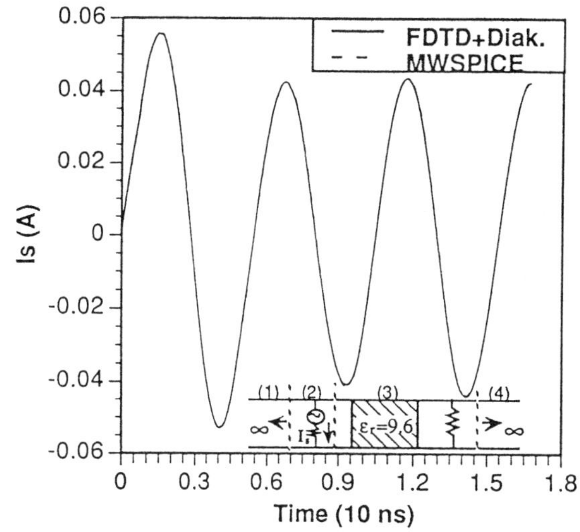

Fig. 8. The transient response of a modified circuit from Fig. 6 ($V_s = -0.1\sin(2\pi ft), f = 200$ MHz, $R_s = 0.5\ \Omega, R_L = 2.5\ \Omega$, Cell Size $= 1$ cm^2).

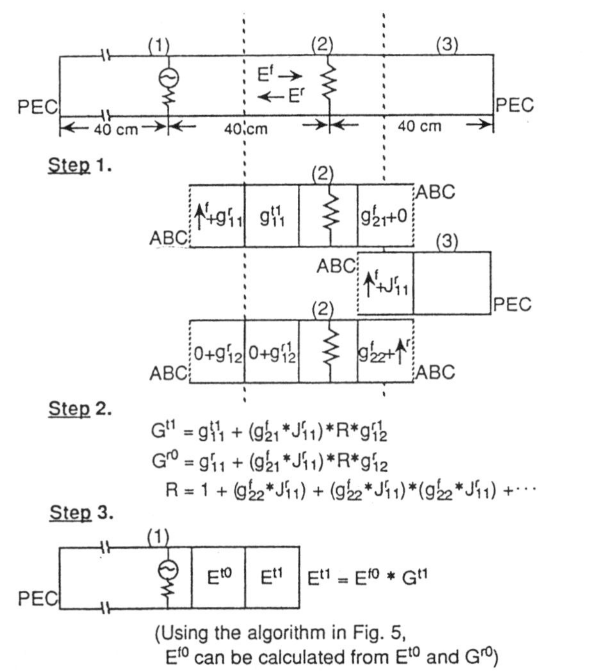

Fig. 9. Application of the parallel FDTD Diakoptics algorithm to a shorted parallel-plate transmission line circuit with the directional interface.

The superscript of the impulse response variables represents the direction of this response and the position of the cell on an interface. The subscript indicates the response port and the excitation port, individually, like the index of scattering parameters. After sections (2) and (3) are simulated in Step 1, a convolution expression, known as Bewley diagram [10], is used in Step 2 to calculated the total impulse response G^{t1} and G^{r0}. The multireflected response R in Step 2 represents the multiple reflection between sections. Choice of the number of terms in R depends on the magnitude of reflection and the ratio of the traveling time between discontinuities to the computational time of interest. For a stronger reflection and a shorter traveling time between sections, more terms in R are

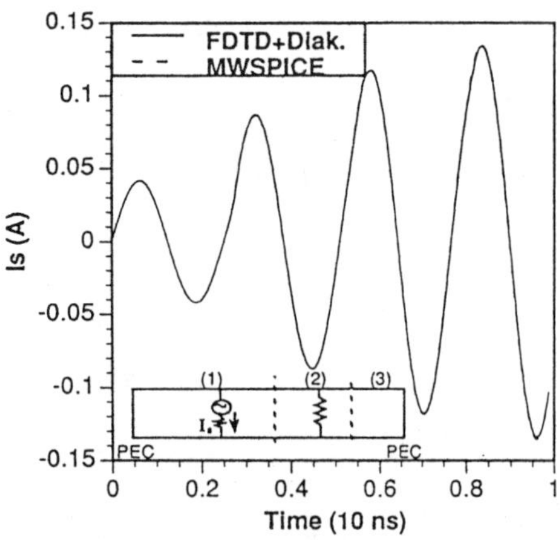

Fig. 10. The transient response of the circuit from Fig. 9 ($V_s = -0.1\sin(2\pi ft), f = 400$ MHz, $R_s = 0.5\ \Omega, R_L = 2.5\ \Omega$, Cell Size $= 1$ cm^2).

TABLE I
COMPARISON BETWEEN FDTD DIAKOPTICS AND TLM DIAKOPTICS

	Interface Field Type	Sequential Algorithm	Parallel Algorithm	ABCs on Exc. Plane	Interface Width
FDTD Diak. (i)	Total	√			2 Cells
FDTD Diak. (ii)	Directional	√	√	√	2(1) Cells
TLM Diak.	Directional	√	√	√	1 Cell

needed in Step 2. Fig. 10 shows the good agreement between the circuit response simulated by the parallel algorithm and the MWSPICE. Because no significant difference is observed from using 3 or 4 terms in R for Step 2, only 3 terms are used for this circuit which has a strong reflection from PEC.

All features of the FDTD Diakoptics are summarized in Table I and are compared with those of the TLM Diakoptics. With the directional interface, FDTD Diakoptics has similar functions as TLM Diakoptics, except that FDTD needs extra procedures to separate directional waves. Hence this feature makes FDTD Diakoptics with directional interface more complicated than TLM. On the other hand, the total-field interface in the FDTD Diakoptics which does not require ABC's on the excitation plane is more efficient for realization of the time-domain Diakoptics than TLM. No parallel algorithm is available if the total-field interface is used. However, this total-field interface of FDTD Diakoptics is especially suitable for device optimization problems in comparison to two other interfaces.

V. APPLICATIONS

Let us illustrate essential features of the proposed methods from the point of applications.

A. Device Optimization

For device optimization problems, in most cases, the optimization is required only in a small passive or active region, while the remaining large passive parts are fixed. Applying the Diakoptics method to this problem, the repetitive computation of fixed large structures is avoided. One example of using time-domain Diakoptics in the slot-line width optimization is demonstrated in [10]. In this example, to change the slot width is equivalent to cut or to add the metallization near the slot, so only a small region that contains metal length variations needs to be recalculated. For the construction of impulse response matrix at the interface made of several ports, it is necessary to simulate the entire volume as many times as the number of ports. Therefore, the reduction of computational time for this problem will happen, only when the number of optimization times is larger than the number of ports on the segmentation boundary. The FDTD Diakoptics with the total-field interface, without any ABC on the excitation plane, is more efficient than TLM Diakoptics in device optimization problems.

B. Numerically Exact ABC's

If the impulse response matrix of a uniform open volume outside circuits is found, then the convolution with this impulse response matrix is equivalent to implement a numerically exact absorbing boundary condition [16]. This numerically exact ABC is independent of incident angle and the dispersive properties of the incident wave, and includes all the numerical dispersion that results from the time and space discretization in FDTD. To calculate the impulse response, only this exact ABC can provide such an ultra-wideband absorbing. A 2D example in [16] demonstrates the capability of this ABC on the excitation plane. The TEM and TM modes of all directions in a parallel-plate waveguide can be absorbed well on the excitation plane by this ABC.

C. Modular Computation

Modular computation is the basic idea in time-domain Diakoptics that divides one large circuit into several modules, simulated individually. Through special interfaces, these modules are connected with all the mutual interactions included. Therefore, when a large circuit is modified, only a few modules need to be recalculated. After each module is characterized, a standard library of sub-circuits analyzed by a full-wave theory can be built to save the repetitive computational time. Figs. 7 and 8 are examples of modular computation that avoid the repetitive computation of outside modules. For the modular computation purpose, parallel algorithm gives more flexibility than sequential algorithm.

D. Parallel Computation

Using the parallel algorithm in modular computation, the simulation of each module is independent from others. Therefore, the parallel algorithm provides a way to utilize parallel computational facilities. Because each module has an independent time-domain simulation, no data communication is required between modules for every time step. This means no synchronization between CPUs are needed for every time step, except only one synchronization before simulating nonlinear active regions. Hence, the time-domain simulation can be fully accelerated by the combining of time-domain Diakoptics and parallel computational facilities.

VI. CONCLUSION

The time-domain Diakoptics is successfully implemented in the FDTD method with two types of connecting interfaces: i) directional interface, and ii) total-field interface. Using the total-field interface in FDTD provides a more efficient way to realize time-domain Diakoptics than TLM, especially for device optimization problems, because no ABC is required. The TLM-type interface can be emulated in FDTD by the directional interface with two novel algorithms in Section III. These novel algorithms provide a numerically exact ABC on the excitation plane during excitation, and a recursive formula to separate directional waves without the knowledge of incident waves.

For a large circuit, a further segmentation is helpful in computational memory reduction and using the parallel computational facilities. Therefore, sequential and parallel algorithms are proposed to connect cascaded modules. Using the parallel algorithm, modular computation and parallel computation are available for the time-domain simulation to avoid the repetitive computation of large structures and to reduce the total simulation time. Several numerical examples are provided to confirm the validity of these algorithms by comparison with simulated results from Microwave SPICE. These algorithms can be extended to 2D or 3D structures with the aid of the multiport convolution formula in (1). However, there is a tradeoff between memory reduction by using time-domain Diakoptics and additional memory requirement for the 3-D impulse response matrix. For future development, system identification techniques [17], [18] can be used to accelerate the time-domain convolution and to reduce the size of impulse response matrix.

REFERENCES

[1] S. A. Kosmopoulos, W. J. R. Hoefer, and A. Gagnon, "Nonlinear TLM modelling of high-frequency varactor multipliers and halvers," *Int. J. Infrared Millimeter Waves*, vol. 10, pp. 343–352, Mar. 1989.

[2] T. Shibata, "Circuit simulations combined with the electromagnetic field analysis," *IEEE Trans. Microwave Theory Tech.*, vol. 39, pp. 1862–1868, Nov. 1991.

[3] H. Kimura and N. Yoshida, "Three-dimensional full-wave analysis with nonlinearity and line characteristics of device by electromagnetic field analysis on time domain," *Electron. Communicat. Jap.*, Part 2, vol. 75, pp. 89–100, June 1991.

[4] W. Sui, D. A. Christensen, and C. H. Durney, "Extending the two-dimensional FDTD method to hybrid electromagnetic systems with active and passive lumped elements," *IEEE Trans. Microwave Theory Tech.*, vol. 40, pp. 724–730, Apr. 1992.

[5] P. B. Johns and K. Akhtarzad, "The use of time domain diakoptics in time discrete models of fields," *Int. J. Num. Methods Eng.*, vol. 17, pp. 1–14, 1981.

[6] ——, "Time domain approximations in the solution of fields by time domain diakoptics," *Int. J. Num. Methods Eng.*, vol. 18, pp. 1861–1373, 1982.

[7] P. So, C. Eswarappa, and W. J. R. Hoefer, "A two-dimensional TLM microwave field simulator using new concepts and procedures," *IEEE Trans. Microwave Theory Tech.*, vol. 37, pp. 1877–1884, Dec. 1989.

[8] C. Eswarappa, G. I. Costache and W. J. R. Hoefer, "TLM modeling of dispersive wide-band absorbing boundaries with time domain diakoptics for s-parameter extraction," *IEEE Trans. Microwave Theory Tech.*, vol. 38, pp. 379–386, Apr. 1990.

[9] C. Eswarappa and W. J. R. Hoefer, "Diakoptics and wideband dispersive absorbing boundaries in the 3-D TLM method with symmetrical condensed nodes," *IEICE Trans.*, vol. E74, pp. 1242–1250, May 1991.

[10] W. J. R. Hoefer, "The discrete time-domain Green's function or Johns matrix—A new powerful concept in transmission line modelling (TLM)," *Int. J. Num. Modelling*, vol. 2, pp. 215–225, 1989.

[11] K. S. Yee, "Numerical solution of initial boundary value problems involving Maxwell's equations in isotropic media," *IEEE Trans. Antennas Propagat.*, vol. AP-14, pp. 302–307, May 1966.

[12] G. Mur, "Absorbing boundary conditions for the finite-difference approximation of the time-domain electromagnetic-field equations," *IEEE Trans. Electromagn. Compat.*, vol. EMC-23, pp. 377–382, Nov. 1981.

[13] K. R. Umashankar and A. Taflove, "A novel method to analyze electromagnetic scattering of complex objects," *IEEE Trans. Electromagn. Compat.*, vol. EMC-24, pp. 397–405, Nov. 1982.

[14] C. W. Kuo and T. Itoh, "A new time-domain transverse resonance method in solving guided wave problems," *Int. J. Num. Modelling*, vol. 3, pp. 229–234, 1990.

[15] T. W. Huang, B. Houshmand, and T. Itoh, "The FDTD Diakoptics method," *IEEE MTT-S Int. Microwave Symp. Digest*, June 1993.

[16] ——, "The FDTD wide-band absorbing boundary conditions for the excitation plane," *Proc. 1993 IEEE Antennas Propagat. Soc. Int. Symp*, June 1993.

[17] ——, "Application of system identification technique to the FDTD Diakoptics method," submitted to *23rd Europ. Microwave Conf.*, Sept. 1993.

[18] ——, "Microwave structure characterization by a combination of FDTD and system identification methods," submitted to *IEEE Microwave Guided Wave Lett.*

Transmission Line Matrix Modeling of Dispersive Wide-Band Absorbing Boundaries with Time-Domain Diakoptics for *S*-Parameter Extraction

ESWARAPPA, STUDENT MEMBER, IEEE, GEORGE I. COSTACHE, SENIOR MEMBER, IEEE, AND WOLFGANG J. R. HOEFER, SENIOR MEMBER, IEEE

***Abstract*— A numerical modeling procedure based on Johns's time-domain diakoptics approach has been developed for efficient transmission line matrix (TLM) analysis of two-dimensional microwave circuits by introducing space interpolation techniques. Frequency dispersive boundaries are represented in the time domain by their characteristic impulse response or numerical Green's function (Johns matrix). Almost perfect wide-band absorbing boundary conditions have been obtained with this technique, permitting accurate characterization of waveguide discontinuities and components. The application of these techniques saves considerable computer run time and memory when compared with conventional TLM analysis.**

I. Introduction

The TRANSMISSION line matrix (TLM) method is a numerical time-domain technique first described by Johns and Beurle [1] in which both space and time are discretized. The details of this method and an extensive list of references on this subject can be found in a review paper [2] and a book chapter on TLM [3] by Hoefer. New concepts and procedures which were developed to speed up TLM modeling were reported in [4]. The TLM method is very general in that it can cover arbitrary geometries and account for realistic features that are often neglected in theoretical analysis. The earlier applications of this method have concentrated mainly on finding the cutoff frequencies and propagation characteristics of transmission lines and resonant frequencies of cavities.

Only a few attempts have been made so far to compute scattering parameters with this method, since wide-band absorbing boundaries cannot be modeled in the time domain, particularly in structures supporting non-TEM modes of propagation. However, in the absence of a wide-band absorbing termination, the impulse excitation capability, which is one of the main assets of the TLM method, cannot be exploited. Furthermore, the wide-band absorbing boundaries must be of high quality since the Fourier transform of time-domain results is very sensitive to imperfect boundary treatment. Small errors in the time domain may produce fairly large errors in the frequency domain. Thus, even though the time-domain results may be reasonably accurate, the frequency-domain results obtained from their Fourier transform may not be acceptable as useful data. Therefore, simulation of good absorbing boundary conditions is very important for computation of *S* parameters.

Recently, absorbing boundary algorithms for the time domain–finite difference (TD–FD) method have been proposed in the literature [5]–[7]. In [7], even though the authors claimed less than 0.02% reflections from an absorbing boundary at a particular (tuning) frequency, they obtained less than 1% reflections only over a narrow bandwidth of 7.7% (center frequency $(\Delta l/\lambda) = 0.09$, bandwidth $(\Delta l/\lambda) = 0.007$) for TE_{10} mode propagation in a standard rectangular waveguide (Δl being the TLM mesh parameter and λ the free-space wavelength). In contrast, we have achieved less than 2% of reflections over the whole operating band of the waveguide with the method presented in this paper.

In the following, two different ways of modeling wide-band absorbing boundary conditions (matching terminations) in homogeneous waveguides will be described. Efficient implementations of wide-band absorbing boundaries using Johns's time-domain diakoptics approach with space interpolation techniques are presented in Section III. Then results for certain waveguide discontinuities and components computed with these techniques are presented in Section IV.

Manuscript received August 29, 1989; revised December 6, 1989. This work was supported by the National Science Engineering Research Council (NSERC) of Canada and the Telecommunications Research Institute of Ontario (TRIO).

The authors are with the Department of Electrical Engineering, University of Ottawa, Ottawa, Ont., Canada K1N 6N5.

IEEE Log Number 8934028.

Reprinted from *IEEE Transactions on Microwave Theory and Techniques,* pp. 379-386, April 1990.

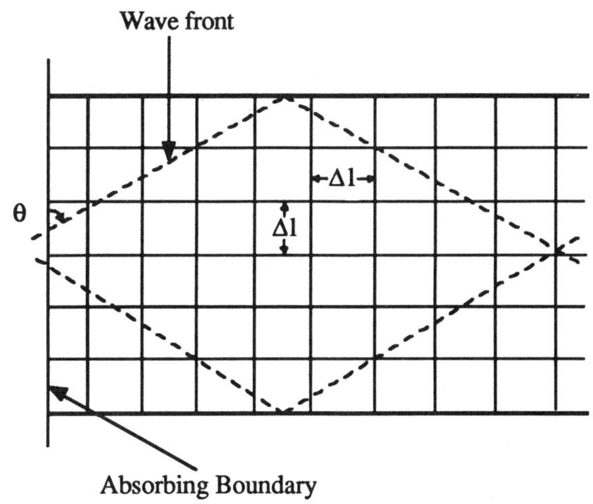

Fig. 1. Plane wave incident at an arbitrary angle θ.

II. Modeling of Wide-Band Absorbing Boundary Conditions

In the two-dimensional TLM simulation of a plane TEM wave, an absorbing boundary can be modeled by terminating the TLM mesh lines with the intrinsic impedance of the TLM mesh, $Z_0/\sqrt{2\epsilon_r}$, where Z_0 is the characteristic impedance of the mesh lines and ϵ_r is the relative permittivity of the simulated medium. This technique works well at all frequencies as long as the plane TEM wave is incident normally on the absorbing boundary and, of course, $\Delta l/\lambda$ is much smaller than unity. For a plane wave incident at an arbitrary angle θ (as in the case of waveguide modes, see Fig. 1), the mesh lines must be terminated with dispersive wave impedances $Z_0/\sqrt{2\epsilon_r(1-\sin^2\theta)}$, where ϵ_r is the relative permittivity of the medium filling the propagation space. Since this impedance is a function of the angle of incidence θ, the termination is totally absorbing only *at one frequency*.

To simulate absorbing boundaries over *a wide frequency range* in waveguides, we have employed two different approaches:

a) modeling of a waveguide termination with gradually increasing losses,
b) modeling of a very long uniform waveguide section.

A. Modeling of a Waveguide Termination with Gradually Increasing Losses

Practical waveguide terminations are made by arranging for the gradual absorption of the incident wave. A tapered resistive sheet or pyramid gradually increases the effective attenuation constant in the termination. Provided that the taper is made several wavelengths long, the reflection is very small. An alternative approach, more appropriate for modeling purposes, is to realize the termination by cascading a number of uniform lossy sections of waveguide as shown in Fig. 2(a). The loss tangent of the sections is progressively increased in such a way that reflection is minimized over a wide frequency range. We have obtained the required optimum loss profile using Touchstone™ CAD software. About nine sections of different lengths and loss tangents (the dielectric loss tangent is taken as the variable quantity) are needed to obtain a return loss of less than -40 dB over the operating band of a standard rectangular waveguide. The optimized lengths and dielectric loss tangents are given in Fig. 2(b). For a WR28 waveguide, the total length of the termination which consists of nine sections is 42.58 mm ($=3.785\lambda_g$ at the center frequency of the operating band). The return loss obtained with Touchstone is shown in Fig. 3. It is less than -40 dB throughout the operating band of the WR28 waveguide. The simulation of this lossy matched wave-

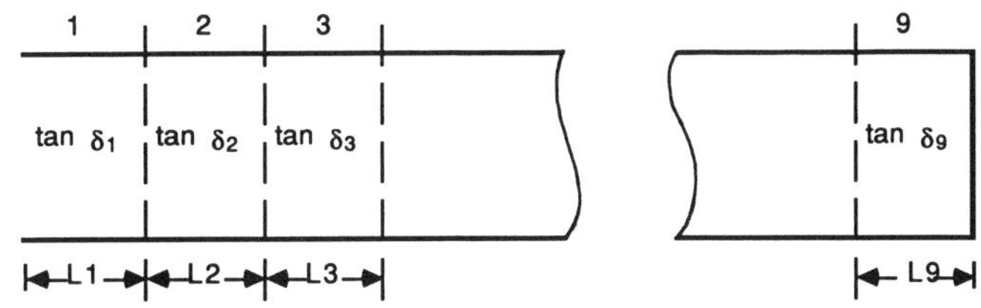

(a)

Section	Length (mm)	Dielectric Loss tangent (tan δ)
1	1.275	0.0095
2	1.739	0.0112
3	3.095	0.0499
4	3.031	0.1162
5	3.166	0.1990
6	7.980	0.2708
7	7.990	0.3686
8	7.536	0.8907
9	6.769	0.5960

(b)

Fig. 2. (a) Modeling of a wide-band absorbing waveguide termination by a cascade of nine increasingly lossy line sections. (b) Optimized lengths and dielectric loss tangents for a matched WR28 load (TE_{10} mode).

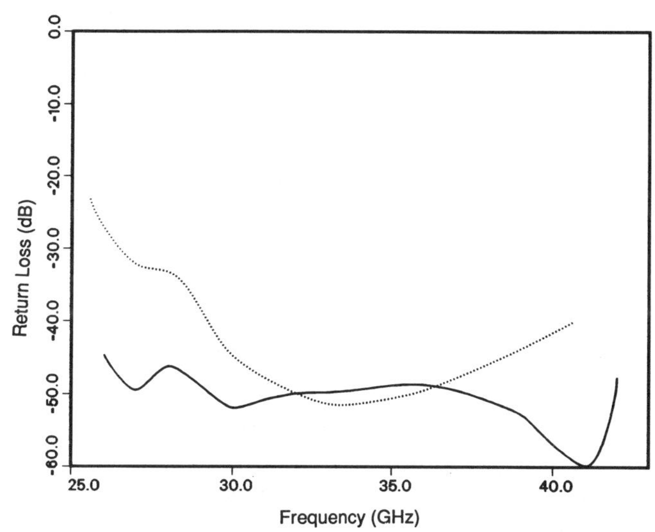

Fig. 3. Return loss of the lossy waveguide termination in Fig. 2: ——— simulation using Touchstone; · · · · · · simulation using TLM method.

guide termination with the TLM method is described below.

Simulation of the Lossy Waveguide Termination with the TLM Method: A TLM calculation usually starts by exciting the mesh at specific points by voltage or current impulses and follows the propagation of these impulses over the mesh as they are scattered by the nodes and bounce back at boundaries. Each iteration corresponds to a unit time Δt required for a pulse to travel from one node to the next. The output which is taken from a chosen point is a series of discrete impulses of varying magnitude separated by constant time intervals. This output function corresponds to a discrete sampling of the E field or H field (depending on the voltage or current) in the time domain. The frequency response (within any frequency range $\Delta l/\lambda$ much smaller than unity) can be obtained by performing the discrete Fourier transform on the output function. The TLM algorithm may be expressed as

$$_{k+1}[V]^r = S_k[V]^i + _k[V]^s \tag{1}$$

$$_{k+1}[V]^i = C_{k+1}[V]^r \tag{2}$$

where V^i and V^r are the incident and reflected impulses at the ports around the nodes at time step $k\Delta t$, S is a block diagonal scattering matrix for all nodes in the network, C is a connection matrix describing the topology of the network, and V^s is a source vector.

To account for dielectric losses, one can either consider the TLM mesh to consist of lossy mesh lines or load the nodes of a lossless mesh with so-called loss stubs [2], [3]. The latter approach has been adopted for our study, where each node is resistively loaded with a matched transmission line of appropriate characteristic admittance g_0, extracting energy from each node at every iteration. The values of g_0 are directly proportional to the local loss tangent and, for a lossy dielectric, can be derived as follows:

$$\epsilon = \epsilon_0\epsilon_r + \frac{\sigma}{j\omega} = \epsilon_0\epsilon_r(1 - j\tan\delta). \tag{3}$$

The conductivity, σ, of the medium and the equivalent attenuation constant, α, of the mesh lines can be expressed as follows:

$$\sigma \equiv \frac{g_0 c C}{\Delta l} \qquad \alpha = \frac{g_0}{4\Delta l \epsilon_r} \ \mathrm{Np/m} \tag{4}$$

where g_0 is the normalized characteristic admittance of the loss stubs (normalizing admittance being the characteristic admittance of the main mesh lines) and Δl is the distance between nodes. The attenuation constant, α_n, of the network is

$$\alpha_n = \frac{g_0}{2\sqrt{2}\,\Delta l\sqrt{\epsilon_r}} \ \mathrm{Np/m}. \tag{5}$$

The attenuation constant for dielectric losses can be written as [8]

$$\alpha_n = \frac{\omega^2\mu_0\epsilon_0\epsilon_r\lambda\tan\delta}{4\pi} \tag{6}$$

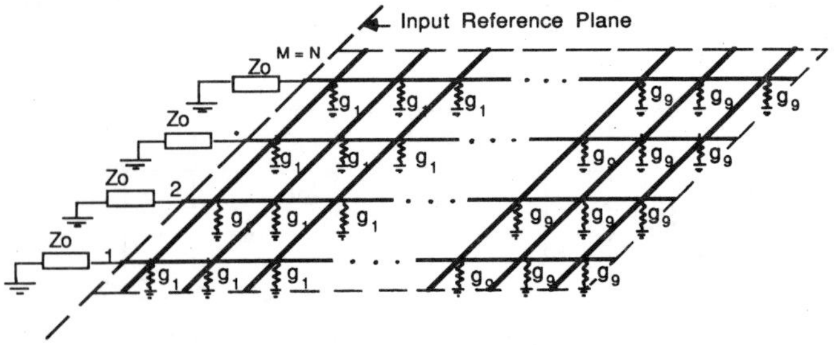

Fig. 4. Configuration for computing the discrete numerical Green's function or Johns matrix of a lossy waveguide matched termination.

where λ is the wavelength in the dielectric. From (5) and (6), g_0 follows:

$$g_0 = \frac{2\epsilon_r\sqrt{2}\,\pi f \Delta l \tan\delta}{c}. \tag{7}$$

If $\tan\delta$ and Δl are known, g_0 can be computed. The frequency can be taken as the midband frequency since g_0 does not change very much across the operating band of the waveguide. The only condition for (5) to (7) to be true is that

$$\alpha_n \Delta l \ll 1. \tag{8}$$

The TLM discretization of a matched termination (of Fig. 2) is shown in Fig. 4. Note that all boundaries are placed halfway between nodes to ensure time synchronism of impulses throughout the TLM mesh. The quantities g_1 and g_9 are the characteristic admittances of the loss stubs of sections 1 and 9, respectively. To satisfy the condition given by (8) and to keep the velocity error to a tolerable level, the width of the waveguide is discretized into $30\Delta l$ (i.e., $N = 30$ in Fig. 4), and about $180\Delta l$ are needed along the length to realize a WR28 waveguide matched termination (shown in Fig. 2(a)). The return loss obtained with a TLM simulation is given in Fig. 3. A minimum of 32 dB is obtained over the operating band of WR28 waveguide. That means the reflections off the absorbing boundary (input plane of the matched termination) are less than 2.5%. This proves the ability of the TLM method to properly account for the losses. The results can be further improved with finer discretization and more iterations.

B. Modeling of a Very Long Uniform Waveguide Section

In this approach, the wide-band termination is represented by a very long waveguide section, and computations are stopped before the reflections from the far end return to the reference plane. For example, for a computation covering 2000 iterations, we need to discretize a waveguide section which is $1000\Delta l$ long.

To compute the scattering parameters of a microwave two-port over a wide frequency band in a single TLM run, we need two absorbing boundaries, one at each port. If we wanted to include the absorbing TLM structures described above, the total additional length to discretize two absorbing boundaries using approach A would be about $400\Delta l$, while for approach B it would be about $4000\Delta l$ (for a computation requiring 4000 iterations). That means we need enormous computer run time and memory to achieve the above absorbing boundaries with the conventional TLM algorithm. However, these problems can be solved effectively with less computer resources by using diakoptics and space interpolation techniques as described in the next section.

III. Implementation of Wide-Band Absorbing Boundary Conditions with Johns's Time-Domain Diakoptics Approach

A. Diakoptics Technique

Diakoptics (or segmentation) is a method of partitioning large structures into substructures which are solved independently and later reassembled. It is very attractive for the repeated analysis of large structures in which only a small subpart is changed from one problem to another. For example, during the optimization of planar and quasi-planar circuits, only the metallization part is changed, the homogeneous dielectric regions remaining unchanged. It is wasteful to analyze the entire structure every time a small change is made.

The method was originated by Kron [9] and has since been applied extensively in conjunction with frequency-domain methods [10]–[12]. For example, complicated two-dimensional planar components can be analyzed by segmenting them into regular shapes for which the analytical Green's functions are known. However, there are only a few regular shapes, and these applications are thus limited to certain standard regular geometries. The technique was extended to the time domain for TLM modeling by Johns and Akhtarzad in 1981 [13], [14]. They showed how the substructures may be solved in the time domain using the TLM method and how the reconnection is made. Since the characteristic impulse response (which can be interpreted as a numerical Green's function) of the substructures must be computed and stored, the extra dimension of time associated with the TLM method vastly increases the computer storage (when compared with steady-state problems). To reduce the computational effort, Johns and Akhtarzad proposed space approximations along the connecting interface: they connected only a fraction of the TLM branches in the interface. Using this space interpolation technique, they computed the cutoff frequencies of simple waveguides and ridge waveguides. Even though the computed values compared reasonably well with analytical values, the frequency response curve was no longer of $\frac{\sin(x)}{x}$ shape but got distorted because of loss/gain of power during the approximate connection process.

This shows that, even though the space-approximated diakoptics may work reasonably well for computing eigenvalues, it may introduce considerable errors in the computation of scattering parameters. For this reason, and in view of the comparatively large computer resources which are required for this method, there was not much further application of this technique.

Fortunately, the diakoptics technique can be made much more efficient and accurate in the case of single-moded transmission structures. Only a single removed branch in the interface is connected, and the field values at the remaining ports are interpolated according to the transverse mode field distribution. This reduces the computer storage and run time by a few orders of magnitude. These procedures are described next.

B. Application to Wide-Band Absorbing Boundaries

We have used diakoptics to represent wide-band matched terminations (shown in Fig. 4 and described in Section II) at the input reference plane by their time-domain characteristic impulse response or numerical Green's function. We then discretize only the circuit to be characterized and convolve its time-domain response with the numerical Green's functions in the input and output reference planes of the circuit.

The branches penetrating through the input reference plane (also called removed branches) are numbered 1 through $M = N$ (see Fig. 4). A single impulse injected at any of these branches will cause impulses separated by the iteration time interval to flow in streams out of the branches of this structure. These impulse functions result from the scattering at the nodes and boundaries of the structure, and can be interpreted as a Green's function in numerical form. All removed branches are terminated in their own characteristic impedance during this procedure so as to absorb the emerging output streams. If we denote $g(m, n, k)$ as the output impulse function emerging at the mth branch at $t = k\,\Delta t$ due to a unit excitation of the nth branch at $t = 0$, the complete Green's function for the input plane of the matched load can be written in matrix form as follows:

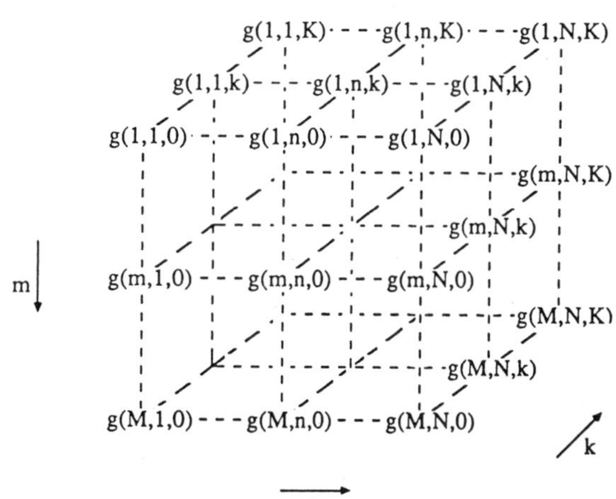

(9)

This is a three-dimensional array of dimension $(M * N * K)$, where K is the total number of iterations, and $M = N$ is the number of branches or transmission lines along the reference plane. We call this numerical Green's function a Johns matrix in honor of the late P. B. Johns, pioneer of TLM and time-domain diakoptics [4].

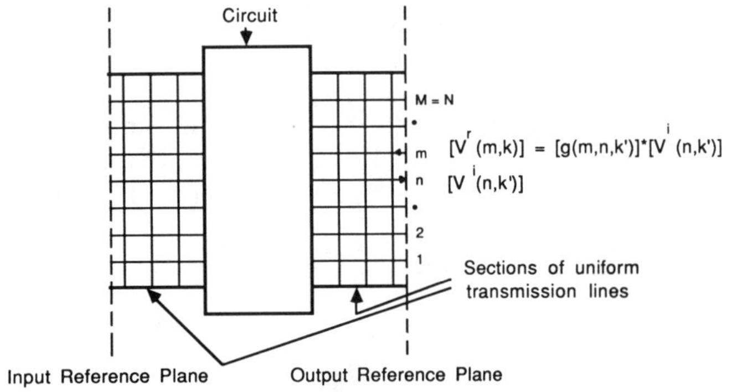

Fig. 5. Convolution of the Johns matrices of wide-band matched terminations with the impulse response of the circuit.

Note that the above matrix is computed only once and is stored. The next step is to discretize the circuit to be characterized and convolve its time-domain impulse response with these Green's functions (which simulate the wide-band absorbing boundary conditions in the time domain) along the branches of the input and output reference planes (see Fig. 5).

When impulses are injected into the circuit, they are scattered at nodes and boundaries and arrive after some time at the input and output reference planes. Any impulse which hits a reference plane will give rise to streams of impulses (characteristic of the matched termination) separated by the iteration time interval that will flow back into the structure through all branches. For example, a series of k impulses incident on the nth branch in the output reference plane will give rise to the following reflected impulse voltage on the mth branch:

$$\begin{aligned} V^r(m,k) &= V^i(n,k)*g(m,n,0) \\ &\quad + V^i(n,k-1)*g(m,n,1) \\ &\quad + \cdots + V^i(n,0)*g(m,n,k). \end{aligned} \quad (10)$$

This can be further written as follows:

$$V^r(m,k) = \sum_{k'=0}^{k} g(m,n,k')*V^i(n,k-k'). \quad (11)$$

The total reflected impulse voltage on the mth branch at time $k\Delta t$ due to the impulses incident on all N branches in previous iterations is the summation of the above term for N branches:

$$V^r(m,k) = \sum_{n=1}^{N} \sum_{k'=0}^{k} g(m,n,k')*V^i(n,k-k'). \quad (12)$$

The above equation forms the basis of the diakoptics algorithm.

The TLM algorithms with and without the diakoptics approach are shown in Fig. 6. Note the extra modules to be implemented for convolution purposes with the diakoptics approach. The computer run time and memory required with the conventional TLM algorithm (i.e., to discretize the circuit and two matched terminations together) is proportional to

$$(NX^c + 2\times NX^m)\times N\times K \quad (13)$$

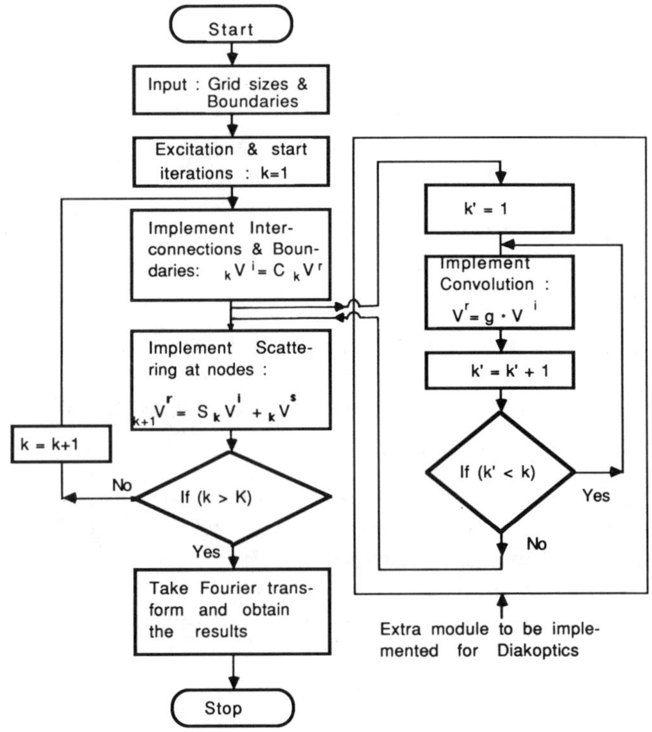

Fig. 6. TLM algorithms with and without diakoptics.

while that with the diakoptics technique is

$$(NX^c\times N\times K) + (K\times(K+1)\times N^2) \quad (14)$$

where NX^c is the number of grids along the length of the circuit and NX^m is the number of grids along the length of a matched load. In (14), the first term corresponds to the discretization of the circuit and the second part to the convolution with the Johns matrices of the matched loads.

The computer resources required for convolution can be reduced if we take the input and output reference planes far away from the circuit or discontinuity under test, so that we can assume only dominant mode propagation along the uniform guide (higher order mode effects on the transverse field distribution can be neglected). In such cases, if we excite the circuit at all the nodes along the input reference plane with impulses whose magnitudes are spatially distributed according to the dominant field distribution, the reflected impulses from these nodes at any iteration will have the same spatial distribution. Hence the impulse response of a matched load can be represented merely by storing the reflected impulse values at any one node for the required number of iterations. By knowing the transverse field distribution of the propagating mode (e.g., $\sin(\pi x/a)$ variation for TE_{10} mode propagation in waveguides), the reflected impulses at the other nodes can be calculated from these values. Hence the Johns matrix $G(M,N,K)$ becomes one-dimensional of size K, the total number of iterations. Thus the memory size required to store the Johns matrix is reduced by a factor of N^2 and the time taken to compute the Johns matrix is reduced by a factor of N, where N is the total number of branches along the reference plane.

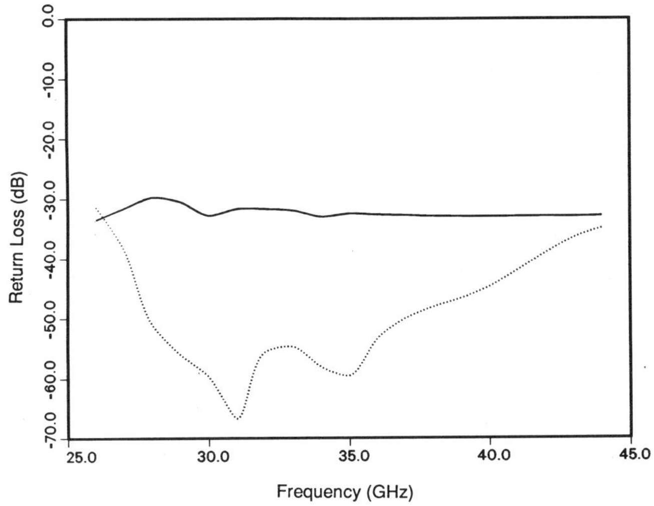

Fig. 7. Return loss of back-to-back waveguide absorbing boundaries computed with diakoptics: · · · · · · approach A; ——— approach B.

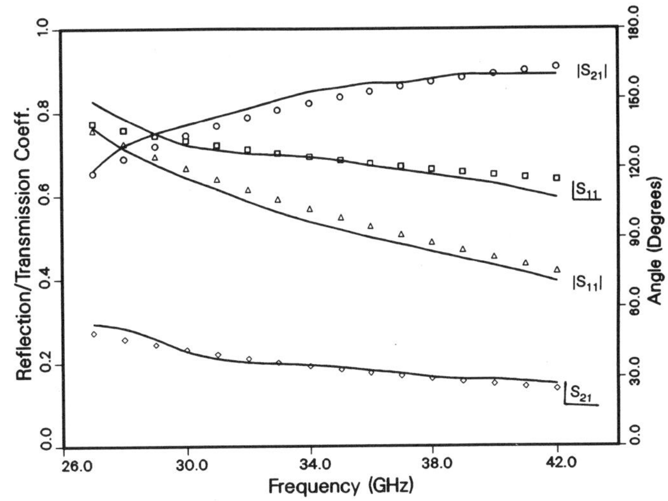

Fig. 8. S parameters of an inductive waveguide iris: ——— computed with diakoptics; O□△◇ computed by Marcuvitz [15].

Note that in the convolution algorithm, we compute the reflected impulses (given by (12)) on all N branches along the reference planes in every iteration. The number of required computational steps is given by the second term of (14). However, under the above assumption, we can perform the convolution at only one node and calculate the reflected impulses at all other branches according to the spatial distribution of the dominant mode. Hence the time and memory taken to convolve are reduced by a factor of (N^2) and (14) becomes

$$(NX^c \times N \times K) + (K \times (K+1)). \tag{15}$$

Using the above technique, we have computed the return loss of the opposing absorbing boundaries (modeled as described in subsections II-A and II-B) separated by a length of WR28 waveguide (about $50\Delta l$ long). The return loss obtained as $20\log(VSWR-1)/(VSWR+1)$ is shown in Fig. 7. It is less than -35 dB throughout the operating band of the WR28 waveguide for approach A, while for approach B it is less than -30 dB, and the response is flat as expected. The propagation constant, β, can be obtained by solving

$$e^{-j\beta(\omega)(L_2-L_1)} = \frac{E_y(\omega, z=L_1)}{E_y(\omega, z=L_2)} \tag{16}$$

where L_1 and L_2 are the distances from the origin to any two points along the waveguide, and E_y are the Fourier transforms of $E_y(t)$ at $z=L_1$ and $z=L_2$. For the uniform WR28 waveguide, these β values agree exactly with the analytical values over the whole operating frequency band. Also, the phase difference of fields between any two consecutive nodes along the length of the waveguide is the same. This demonstrates the excellent quality of the wide-band absorbing boundaries.

IV. Applications

To further check the quality of the wide-band absorbing boundary conditions and to verify the validity of the proposed space interpolation techniques, we have computed the S parameters of an inductive waveguide iris discontinuity and an E-plane band-pass filter. The S parameters over the whole operating frequency band are computed with two TLM runs—with and without the discontinuity present. The incident field is obtained from analysis of a small section of the empty waveguide terminated on both sides with simulated wide-band matched terminations. The reflected field is obtained from the difference between the total field and the incident field.

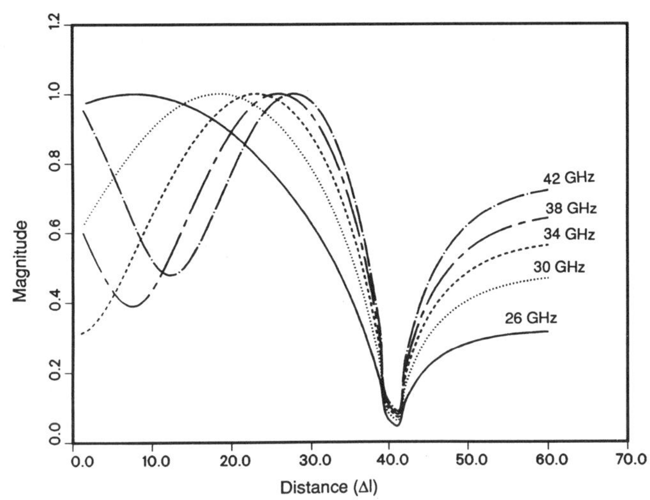

Fig. 9. Electric field variation along the length of a waveguide containing the inductive iris discontinuity at $40\Delta l$.

Fig. 8 shows computed magnitude and phase of the S parameters of a symmetrical inductive iris (of gap width equal to 3.556 mm) in a WR28 waveguide. Results compare well with those computed using empirical formulas given in [15]. The electric field variation along the center line of the waveguide around the inductive iris is shown in Fig. 9 for five different frequencies. Note a steep dip in the magnitude of the electric field at the discontinuity. The fields become almost constant for all frequencies on the right-hand side of the discontinuity, indicating the excellent quality of the matched loads. Fields vary sinusoidally

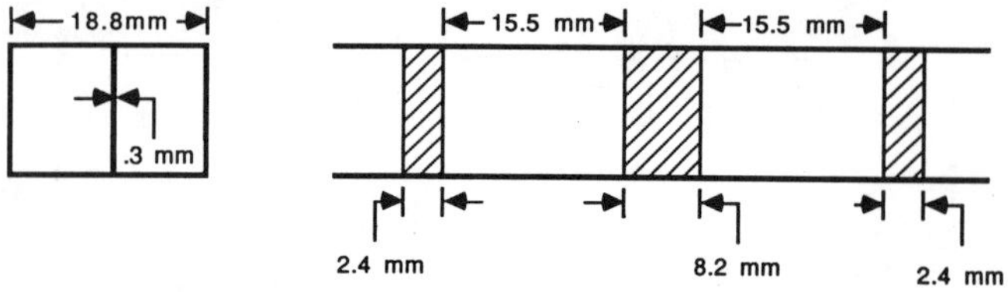

Fig. 10. The geometry of a two-section maximum flat E-plane filter.

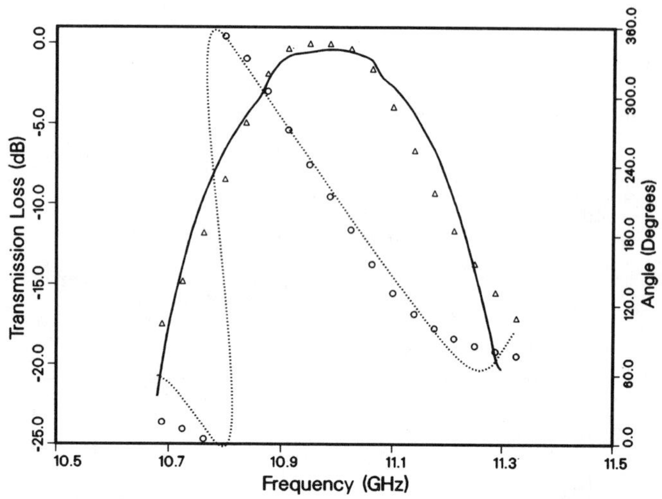

Fig. 11. Transmission characteristics of a E-plane filter: ——— $|S_{21}|$ computed with diakoptics; ······ S_{21} computed with diakoptics; △ $|S_{21}|$ computed with mode-matching technique (Uher); ○ S_{21} computed with mode-matching technique (Uher).

towards the left side of the discontinuity, as expected. Also, it can be seen that the higher order mode effect is almost negligible beyond a distance of about $20\Delta l$ on either side of the discontinuity.

Fig. 10 shows the geometry of a two-section maximum flat band-pass filter [16] with the following specifications:

center frequency:	10.95 GHz
bandwidth:	218 MHz
guide width:	18.8 mm
strip thickness:	0.3 mm

The computed transmission characteristics are given in Fig. 11. The results compare well with those computed with the mode-matching technique.

V. Conclusions

Excellent wide-band waveguide absorbing boundary conditions have been implemented using Johns's time-domain diakoptics approach. A space interpolation technique based on the dominant field distribution has been proposed for efficient S-parameter extraction. The good accuracy of this technique and the quality of wide-band absorbing boundary conditions are documented by the good agreement of the computed S parameters of waveguide components with data obtained with other methods.

Acknowledgment

The authors wish to thank J. Uher for computing the characteristics of the E-plane filter with his mode-matching program.

References

[1] P. B. Johns and R. L. Beurle, "Numerical solution of 2-dimensional scattering problems using a transmission-line matrix," *Proc. Inst. Elec. Eng.*, vol. 118, no. 9, pp. 1203–1208, Sept. 1971.

[2] W. J. R. Hoefer, "The transmission-line matrix method-theory and applications," *IEEE Trans. Microwave Theory Tech.*, vol. MTT-33, pp. 882–893, Oct. 1985.

[3] W. J. R. Hoefer, "The transmission line matrix (TLM) method," in *Numerical Techniques for Microwave and Millimeter Wave Passive Structures*, T. Itoh, Ed. New York: Wiley, 1989.

[4] P. So, Eswarappa, and W. J. R. Hoefer, "A two-dimensional TLM microwave field simulator using new concepts and procedures," *IEEE Trans. Microwave Theory Tech.*, vol. 37, pp. 1877–1884, Dec. 1989.

[5] J. Fang and K. K. Mei, "A super-absorbing boundary algorithm for solving electromagnetic problems by time domain finite difference method," in *Proc. AP-S Symp.*, June 1988, pp. 472–475.

[6] J. E. Roy and D. H. Choi, "The application of a simple absorbing boundary algorithm to cylindrical waveguide," in *Proc. AP-S Symp.* (San Jose), June 1989, pp. 58–61.

[7] J. E. Roy and D. H. Choi, "A simple absorbing boundary algorithm for the TDFD method with arbitrary incidence angle," in *Proc. AP-S Symp.* (San Jose), June 1989, pp. 54–57.

[8] R. E. Collin, *Field Theory of Guided Waves.* New York: McGraw-Hill, 1960.

[9] G. Kron, *Diakoptics.* London: MacDonald, 1963.

[10] T. Okoshi, Y. Uehara, and T. Takeuchi, "The segmentation method —An approach to the analysis of microwave planar circuits," *IEEE Trans. Microwave Theory Tech.*, vol. MTT-24, pp. 662–668, 1976.

[11] R. Chadha and K. C. Gupta, "Segmentation method using impedance matrices for analysis of planar microwave circuits," *IEEE Trans. Microwave Theory Tech.*, vol. MTT-29, pp. 71–74, 1981.

[12] R. Sorrentino, "Planar circuits, waveguide models, and segmentation method," *IEEE Trans. Microwave Theory Tech.*, vol. MTT-33, pp. 1057–1066, Oct. 1985.

[13] P. B. Johns and K. Akhtarzad, "The use of time domain diakoptics in time discrete models of fields," *Int. J. Num. Methods Eng.*, vol. 17 pp. 1–14, 1981.

[14] P. B. Johns, and K. Akhtarzad, "Time domain approximations in the solution of fields by time domain diakoptics," *Int. J. Numer. Methods Eng.*, vol. 18, pp. 1361–1373, 1982.

[15] N. Marcuvitz, *Wavegudie Handbook.* New York: Dover, 1965.

[16] Y. Konishi and K. Uenakada, "The design of a bandpass filter with inductive strip-planar circuit mounted in waveguide," *IEEE Trans. Microwave Theory Tech.*, vol. MTT-22, pp. 869–873, Oct. 1974.

Fast Sequential FDTD Diakoptics Method Using the System Identification Technique

Tian-Wei Huang, Bijan Houshmand, and Tatsuo Itoh, *Fellow, IEEE*

***Abstract*—The sequential finite-difference time-domain (FDTD) Diakoptics method is accelerated by the system identification technique. The time-domain convolution in Diakoptics is replaced by a difference equation with a small number of coefficients. Hence, the memory size is also reduced. The validity of this method is confirmed by comparison with simulated results from Microwave SPICE. The accuracy of this fast algorithm is also demonstrated by the analysis of cascaded circuit modules.**

I. Introduction

One way to simulate a large circuit is to divide this circuit into several small modules, simulated individually. According to this application, the time-domain Diakoptics method has been developed in TLM [1], [2], and in FDTD [3], [4]. The time-domain Diakoptics can include all the mutual interactions among modules through a convolution interface, so it is a full-wave analysis. Applying Diakoptics in the device optimization problems, the repetitive computation of fixed large structures is avoided. The time-domain Diakoptics is also one candidate method for the implementation of parallel computation in time-domain simulations.

The time-domain Diakoptics uses time-domain convolution for connecting modules. This convolution requires the knowledge of the entire signal history, especially for a module with an infinitely-long impulse response. Instead of using convolution approach, a linear discrete-time system can be characterized by its constant-coefficient difference equation, generated by the system identification (SI) technique [5], [6]. This difference equation depends only on a small number of past inputs and outputs. In [7], a single module is analyzed by the SI technique. This letter extended this fast algorithm to cascaded modules. The memory reduction and computational efficiency are also demonstrated.

II. Theory

There are two different approaches to characterize a discrete-time system. The first one is convolution approach that uses impulse response $h(k)$ to represent a system, like a finite-duration impulse response (FIR) filter. The second one is system identification approach that replaces a system by its difference equation with coefficients a_k, b_k, and time-delay

Manuscript received July 3, 1993. This work was supported by the Office of Naval Research N00 014-91-J-1651 and the Joint Services Electronics Program F49 620-92-C-0055. Editorial decision made by S. Maas.

The authors are with the Department of Electrical Engineering, University of California at Los Angeles, 405 Hilgard Avenue, Los Angeles, CA 90024–1594.

IEEE Log Number 9212416.

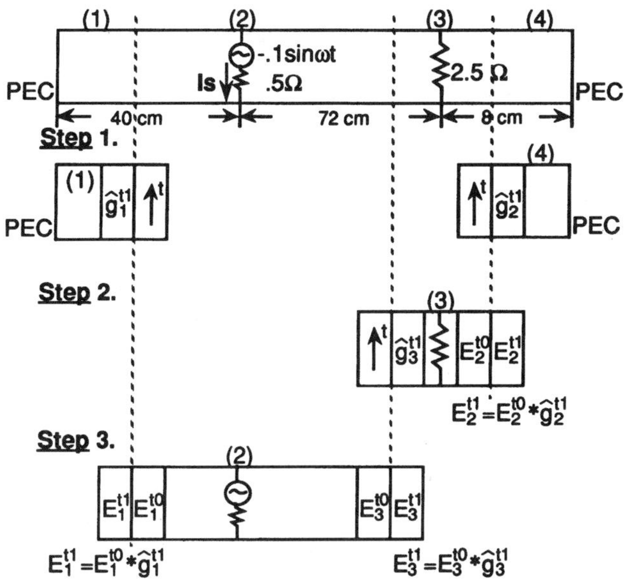

Fig. 1. A shorted parallel plate waveguide is analyzed by the sequential FDTD Diakoptics method.

t_d. Because the SI approach has feedback, it is an infinite-duration impulse response (IIR) filters. A detailed description of SI technique is available in [8].

1) Convolution Approach (Nonrecursive):

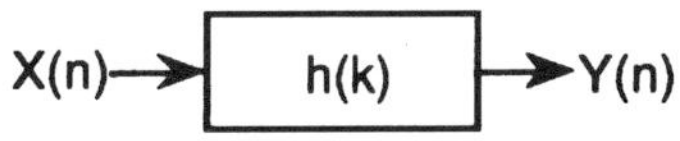

$$Y(n) = \sum_{k=0}^{n} h(k)X(n-k). \quad (1)$$

2) System Identification Approach (Recursive):

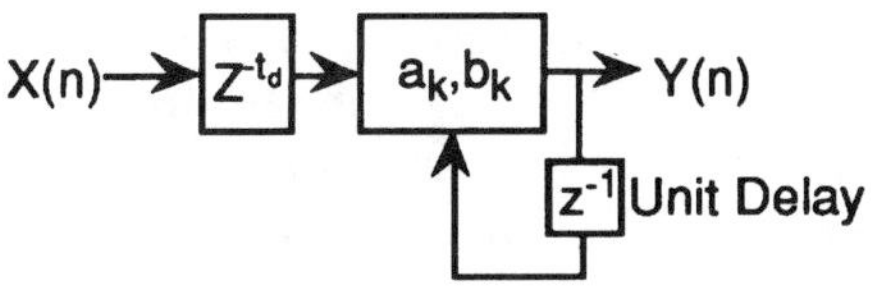

$$Y(n) = -\sum_{k=1}^{N} a_k Y(n-k) + \sum_{k=0}^{M} b_k X(n-k-t_d). \quad (2)$$

Where, $X(n)$ and $Y(n)$ are the input and output of this system. For FDTD Diakoptics, $X(n)$ and $Y(n)$ are the total

Reprinted from *IEEE Microwave and Guided Letters,* pp. 378-380, October 1993.

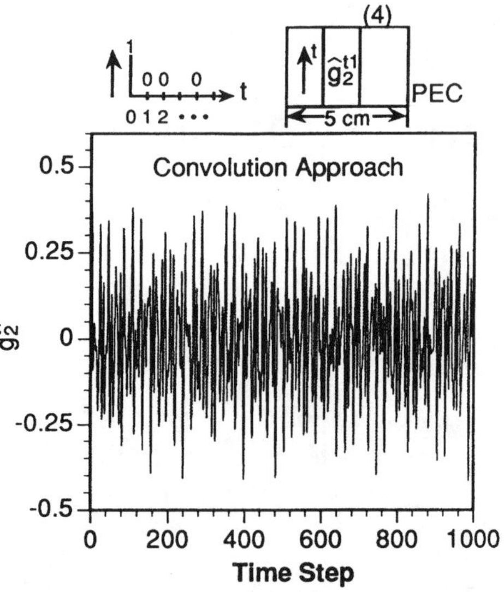

Fig. 2. Impulse response of module (4) in Fig. 1, which can be replaced by a difference equation with 80 coefficients obtained from SI approach using 90 time steps. This difference equation can generate the indistinguishable impulse response to the above.

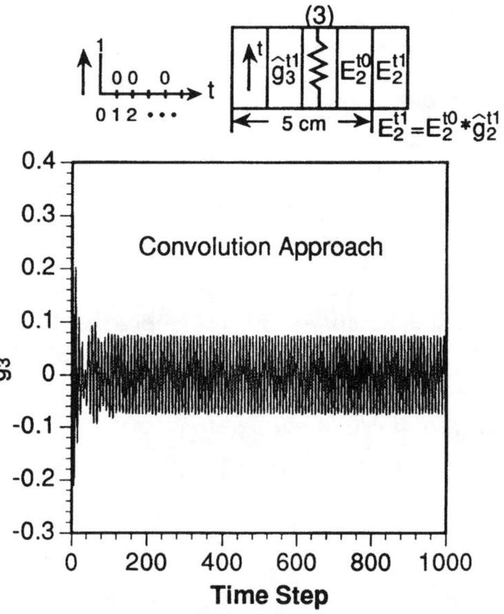

Fig. 3. Impulse response of cascaded module (3) and (4) in Fig. 1, which can be replaced by a difference equation with 80 coefficients obtained from SI approach using 90 time steps. This difference equation can generate the indistinguishable impulse response to the above.

tangential E-field, defined in [3]. Since the TLM and FDTD discretize both time and space, there is a stop-band for high frequencies. For a low-pass system, an IIR filter can provide a smaller order realization than a FIR filter. Hence it is more efficient to use SI approach in Diakoptics than convolution.

III. Results

Fig. 1 shows a parallel-plate transmission line with TEM mode analyzed by the sequential FDTD Diakoptics [3], [4]. During Step 1 the impulse responses of two outer modules are calculated, as shown in Fig. 2. This impulse response has infinite-duration, that is easily characterized by SI approach within 90 time steps. To simulate the inner module (3), one

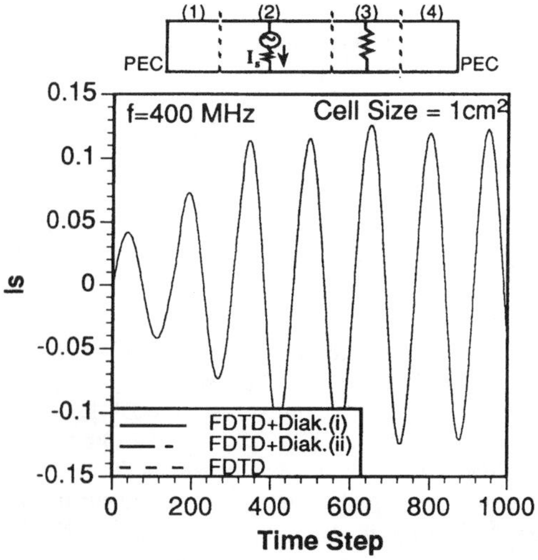

Fig. 4. Transient response simulated by FDTD method and sequential FDTD Diakoptics with (i) convolution approach, and (ii) SI approach.

TABLE I
Comparison of Simulation Time (in Time Steps) Between (i) Convolution Approach, and (ii) System Identification Approach

	Impulse Duration (i)	No. Coeff. in Diff. (ii)	Simulation Time (i)	Simulation Time (ii)
Uniform Line [7]	300 (Approx.)	80	300	60
Shorted Line (Fig. 2)	Infinite	80	Time of Interest	90
Cascaded (Fig. 3)	Infinite	80	Time of Interest	90

convolution with the result from Fig. 2 is needed at the interface between (3) and (4). Fig. 3 is the impulse response of cascaded modules (3) and (4), that can be also obtained by applying SI technique on module (3) terminated with a difference equation of module (4). Fig. 4 shows excellent agreement between the circuit responses simulated by the FDTD Diakoptics with SI technique and the FDTD. This example also demonstrates the accuracy of SI approach on cascaded modules.

Table I compares approaches (i) and (ii) for three types of impulse responses to show the efficiency of SI technique. For a infinite-duration impulse response, the SI approach is necessary to make its computational time finite. For an uniform line, its impulse response can be approximated by 300 time steps. Even for a finite-duration response, SI technique can also reduce the computational time and memory.

IV. Conclusion

The convolution operation in time-domain Diakoptics is accelerated by the system identification technique. In this letter, this technique is applied to cascaded circuit modules to verify its accuracy. The SI technique has particular efficiency for an infinitely-long impulse response system, that can be characterized with a small number of time steps and coefficients. For the multimode problems in [9], the SI technique can be used to accelerate the simulation of each element in an impulse response matrix. This technique can also be applied to the time-domain Diakoptics in TLM.

REFERENCES

[1] P. B. Johns, and K. Akhtarzad, "The use of time domain diakoptics in time discrete models of fields," *Int. J. Num. Methods Eng.*, vol. 17, pp. 1–14, 1981.

[2] W. J. R. Hoefer, "The discrete time-domain Green's function or Johns matrix—A new powerful concept in transmission line modeling (TLM)," *Int. J. Num. Modeling*, vol. 2, pp. 215–225, 1989.

[3] T. W. Huang, B. Houshmand, and T. Itoh, "The FDTD Diakoptics method," in *IEEE MTT-S Int. Microwave Symp. Dig.*, June 1993, pp. 1435–1438.

[4] ——, "The sequential FDTD Diakoptics method for modular computation," to be published in *Asia–Pacific Microwave Conf. Proc.*, Oct. 1993.

[5] B. Houshmand, T. W. Huang, and T. Itoh, "Microwave structure characterization by a combination of FDTD and system identification methods," to be published in *IEEE Microwave Guided Wave Lett.*

[6] W. Kuempel and I. Wolff, "Digital signal processing of time domain field simulation results using the system identification method," in *IEEE MTT-S Int. Microwave Symp. Dig.*, vol. 2, June 1992, pp. 793–796.

[7] T. W. Huang, B. Houshmand, and T. Itoh, "Application of system identification technique to FDTD and FDTD Diakoptics method" to be published in *23rd European Microwave Conf. Proc.*, Sept. 1993.

[8] G. C. Goodwin, and K. S. Sin, *Adaptive Filtering Prediction and Control.* Englewood Cliffs, NJ: Prentice-Hall, 1984.

[9] T. W. Huang, B. Houshmand, and T. Itoh, "The FDTD wide-band absorbing boundary conditions for the excitation plane," in *Proc. 1993 IEEE Antennas Propagat. Soc. Int. Symp.*, June 1993, pp. 2–5.

Efficient Implementation of the FDTD Algorithm on High-Performance Computers

STEPHEN D. GEDNEY

UNIVERSITY OF KENTUCKY

DEPARTMENT OF ELECTRICAL ENGINEERING

LEXINGTON, KY 40506-0046

I. Introduction

Over the last decade, the FDTD method has become one of the most widely used computational techniques for the full wave analysis of electromagnetic phenomena. Its popularity can be accredited to the simplicity of the algorithm, while providing a robust and accurate analysis of electromagnetic fields [1]. The FDTD algorithm has a number of inherent characteristics that render it ideal for analyzing microwave circuit devices. Since it is based on a volume discretization, 3-D material inhomogeneities are inherently modeled, allowing for the analysis of planar stratified media, non-planar devices, devices on finite dielectric substrates and ground planes, and lossy conductors and dielectrics. Lumped or hybrid elements including passive loads and active elements [2] can also be incorporated into the discrete Maxwell's equations. Other recent advances in absorbing boundary conditions have provided for the accurate broad-band analysis of printed microwave circuits and antennas in unbounded media [1]. Finally, both broad-band analysis and narrow-band characteristics of microwave devices can be obtained with a single FDTD simulation.

One principal factor that has delayed the extensive application of the FDTD method is that substantial computational resources are required for the analysis of practical three-dimensional problems. An orthogonal lattice with dimensions $N_x \times N_y \times N_z$ requires $N = 6\ N_xN_yN_z$ degrees of freedom to represent the three vector components of the electric and magnetic fields throughout the 3-D space. The total memory required to store the fields and the update coefficients thus increases as $O(N)$. The number of floating point operations per time iteration will also increase as $O(N)$, while the total number of time iterations roughly increases as $O(N^{1/3})$. Therefore, the number of floating point operations of the FDTD algorithm will increase as $O(N^{4/3})$. While this appears ideal, N is based on a 3-D volume discretization and the computational requirements can grow quickly.

In the last decade, the introduction of inexpensive high-speed memory, powerful vector processors, high-speed RISC processors, and the maturation of high-performance parallel computers has provided the economical and powerful computational resources that are necessary to perform the analysis of practical engineering applications via the FDTD method. Correspondingly, the FDTD method has received extraordinary popularity over the last five years. Consider the low-pass filter modeled in [3]. This simple structure was modeled by the FDTD method using an $80 \times 110 \times 16$ lattice. A simulation of this structure requires 3.4MB to store the fields (using 32-bit precision) and took eight hours to perform 4,000 time iterations on a vintage VAX station 3,500 [3]. However, on a 20 MFLOPS (millions of floating point operations per second) single-processor workstation such as a 120-MHz Pentium-based PC or a Sparc 20 this same simulation can be performed in less than 15 minutes. A somewhat larger circuit employing a $320 \times 440 \times 16$ grid will require 54MB of memory, and assuming the simulation requires 8,000 time iterations, will take eight hours to complete on the 20-MFLOPS processor. If the simulation was performed on a high-end multiprocessor workstation such as a 16-processor SGI Power Challenge realizing 1.0 GFLOPS, this larger simulation would be performed in only five minutes.

If the FDTD algorithm is to be a truly viable design tool in industry, it will be necessary to obtain reasonable computational speeds on inexpensive workstations. This will require advanced algorithms and acceleration techniques, such as those presented in Chapter 5, as well as affordable high-performance computing. The good news is that the next generation of RISC-based processors are reportedly operating at peak rates upwards to 400 MFLOPS and can host large, fast-memory capacities. Affordable multiprocessor workstations employing high-speed RISC processors by Silicon Graphics Inc. (SGI), Hewlett Packard (HP), and Digital Equipment Corp. (DEC) have also been introduced. Furthermore, workstations are typically installed on wideband networks providing file sharing and communication. A cluster of workstations can be utilized as a parallel computer, or a *workstation cluster,* over such a network. This economical loosely coupled distributed memory computer can execute programs in parallel using message-passing protocols such as PVM [4,5] or MPI-achieving supercomputer speeds at a fraction of the cost.

The intent of this chapter is to present efficient implementations of the FDTD method. It is realized that the optimal FDTD algorithm will be dependent on the computer's architecture as well as the language and compiler being employed. Thus, specific attention is given in Section II to the various architectures of high-performance computers. In Section III, the implementation of the FDTD algorithm on pipelined RISC and vector processors is discussed. In Section IV, programming paradigms for the FDTD algorithm on DM-MIMD computers using explicit message-passing are discussed. Finally, Section V provides a brief summary of the reprints exemplifying the development of efficient FDTD algorithms on high-performance computers.

II. High-Performance Computing

Super computers have historically utilized a small number of powerful processors. Most renowned are the Cray supercomputers, which employed expensive vector processors. Computational speeds nearing 20 GFLOPS combined with gigabytes of memory have provided substantial computing power over the last decade. However, such machines have cost tens of millions of dollars and are only available to a small number of research centers, large corporations, and academic institutions. More recently, pipelined RISC processors have provided computational speeds that are approaching those of expensive vector processors at a fraction of the cost. Because RISC processors are low cost, low power, and air cooled, it has been possible to develop multiprocessor computers equipped with hundreds and even thousands of RISC processors. While these multiprocessor computers are still in their infancy, the technology is maturing such that they are providing true supercomputer performance, and the cost per MFLOPS has dramatically decreased so that they are becoming affordable on more of a wide-scale basis.

It is difficult to classify high-performance computers due to the great diversity of architectures. Nevertheless, the taxonomy introduced by Flynn [6] is still the most commonly used and can still provide a global classification of high-performance computers. Specifically, they are classified as being SISD (single instruction, single data), SIMD (single instruction, multiple data), or MIMD (multiple instruction, multiple data) computers. SISD computers are typical uniprocessor systems running a single program with its own addressable memory. SIMD computers are multiprocessor computers that typically have a larger number of processing units that execute the same instruction on different data in lock-step. Examples of SIMD computers are the vintage Thinking Machine's CM-2, and MasPar's MP-2. On the other hand, MIMD computers execute independent instruction streams on each processor and operate on different data. Table 1 references a number of popular high-performance computers and their classification [7].

The second means of classifying high-performance computers is the manner by which memory is accessed. Shared memory (SM) systems have multiple CPUs which all share the same address space. Thus, the location of the memory is transparent to the user. Some SM-SIMD and SM-MIMD multiprocessors are listed in Table 1. The second class includes distributed memory (DM) systems. In a distributed memory architecture, each processor has its own local addressable memory. Addressing memory that is owned by a remote processor requires communication over a high speed network interconnecting the processors. Typically, this is done explicitly by the user by calling message-passing subroutines. Alternatively, many DM-MIMD or DM-SIMD computers have programming languages or compiler directives that provide virtual shared memory. Even though the data is distributed, it is transparent to the user and the compiler is responsible for distributing the data or addressing data on remote processors.

DM-MIMD is undoubtedly the most widely used architecture for high-performance computing since it provides the greatest potential for economical massive parallelism. Unfortunately, this architecture does have some drawbacks as compared to SM-MIMD computers. Initially, the communication between processors on DM-MIMD computers is typically much slower. Secondly, DM-MIMD computers typically require user-defined data distribution and explicit message-passing to exchange data between processors. The most portable message-passing protocol is PVM (parallel virtual machine) [4]; although this is not supported by all vendors, most support PVM as well as their own native message-passing protocol. Unfortunately, the use of explicit message-passing is extremely cumbersome and has been a limiting factor for DM-MIMD computers. Most vendors

Table 1. Statistics of Select High-Performance Computers

Computer	Class	Max number of processors	Peak performance (theoretical)	Memory	Int. comm. speed
MASPar MP2216	SM-SIMD	16,384	34 GFLOPS	1GB	20GB/s
Cray YMP-C90	SM-MIMD	16	16 GFLOPS	16GB	4GB/s
SGI Power Chal.	SM-MIMD	36	5.4 GFLOPS	16GB	1.2GB/s
Cray T3D	DM-MIMD	2,048	307 GFLOPS	64MB/node	300MB/s
Convex Exemplar	DM-MIMD	128	25.6 GFLOPS	256MB/node	4GB/s
IBM SP2	DM-MIMD	128	8 GFLOPS	64–512MB/n	20MB/s
Intel Paragon XP	DM-MIMD	4,000	300 GFLOPS	128MB/node	200MB/s
T.M. CM-5	DM-MIMD	16,384	2 TFLOPS	32MB/node	20MB/s

offer their own version of a high-performance FORTRAN (HPF) compiler, offering a virtual shared memory paradigm that uses in-line compiler directives for data distribution and compiler-derived message-passing. Unfortunately, a standard for HPF compilers has not yet been established, though efforts in that direction are being pursued [8], and for the meantime these codes are typically not portable. Furthermore, before HPF compilers mature, codes written in HPF will scale poorly over large numbers of processors as compared to codes invoking user-defined message-passing. As with early vector processing, efficient compilers will eventually be developed to overcome these obstacles.

III. RISC-Based Workstations and Vector Processors

Most of the microprocessors being used in commercially available workstations are RISC (reduced-instruction-set-computer) based processors. The one prominent underlying factor defining a RISC processor is the relative simplicity of the instructions and address modes embedded in the processor hardware. Using single word length (32-bit or 64-bit) instructions and employing simplified address modes facilitates a number of enhanced features such as pipelining and superscalar operation. Furthermore, this allows for one or more instructions to be issued per clock cycle, accelerating the performances of today's microprocessors. Previous generations' computers contained floating point units and fast memory cache on separate chips called coprocessors. The current generation of microprocessors achieve even greater computational speeds by incorporating floating point units, integer units, and cache on the same chip as the microprocessor. This allows for reduced latencies and increased clock speeds, yielding substantially faster microprocessors.

Through the advancement of RISC microprocessor architectures and compiler technology, the ratio of peak MFLOPS and MIPS (millions of instructions per second) rates per dollar has drastically increased for workstations in the last five years, and will continue to increase at similar rates in the years to come. However, the actual performance one can achieve when programming using a high-level language such as FORTRAN or C, is very much dependent on the compiler and the program structure. That is, the code must be written in a manner that the compiler can execute most efficiently. Otherwise the performance achieved can be less than adequate.

The kernel of the FDTD algorithm consists of six triple loops updating each component of the vector fields intensities in a 3-D space [1]. The triple loops are readily vectorizable, and can be executed with high efficiency on a vector processor. Furthermore, FORTRAN or C code written for a vector processor will also yield optimal performance on a RISC-based microprocessor since vector operations are ideally pipelined and also result in high cache-hit rates. In fact, the longer the vector lengths, the easier it is to keep the pipelines full and to minimize the ratio of cache misses to cache hits. Thus, loops can be performed with high floating-point operation speeds on most RISC-based and vector-based processors if: 1) the inner loops are void of branches, namely, they are void of function calls, logical statements, or indirect addressing; 2) in FORTRAN, the inner most column index of multi-dimensional arrays corresponds to the index of the inner loop, or in C or C++ , the outer row index of the multi-dimensional array corresponds to the inner loop index; and 3) the length of the inner loop has the largest dimension. It is discussed in [9] that violating any of these principle concepts can result in a significant degradation of the processors' MFLOPS performance. In fact it is more expedient to take special cases out of the main loops, such as absorbing boundary conditions, impedance boundary conditions, active loads, and place them in separate loops. One can also incorporate other special cases, such as PEC boundaries, material discontinuities, and passive loads, and incorporate them directly into the main loop by appropriately pre-constructing the update coefficients.

Figure 1 illustrates a nested loop structure for the FDTD time-domain algorithm for the update of E_x and H_x throughout the entire space which yields excellent performance on a RISC or vector processor. The coefficients *dex* and *cex* for the electric field update are general coefficients which include the effects of discontinuities in the dielectric profile, including lossy dielectrics. These coefficients can also include other special conditions. For example, if the electric field is on the surface of a PEC (or within a PEC), these coefficients are simply set to zero. They can also be used to update an electric field on an impedance boundary, or the field on an edge loaded with a passive load [2]. Additional memory is required to store these constants. However, it is suggested by Gedney and Barnard [10] that these coefficients can be stored in a packed form. Subsequently, one integer array is constructed for each of the electric and magnetic field updates which is used as a pointer to the packed coefficients, greatly reducing the memory requirements. Furthermore, for printed circuit applications, the coefficients change infrequently, and the memory reference is already in cache, speeding up the application. It is noted that this type of indirect addressing will do very poorly on a vector processor such as the CRAY YMP; however, it benefits the pipelined RISC processor.

Note that the update for E_x in Fig. 1 excludes the indices $j = 1, j = ny, k = 1$, and $k = nz$. At these indices, E_x lies on the exterior boundary in unbounded media and is updated using an absorbing boundary condition, or simply remains zero in bounded media. The former operation is removed from the primary loop to avoid any additional branch statements, which would reduce computational efficiency, and the updates of these fields are simply performed in a separate loop.

In [9], it is illustrated that the floating point operation speed that a RISC or vector processor can perform the nested loop is a function of the length of the inner loop, e.g., *nx*. The larger *nx* is, up to an optimal vector length, the better the performance. The optimal length varies from processor to processor, but is typically somewhere between 64 and 128. Unfortunately, for smaller problems, the computational speed will be much slower than larger problems due to the shorter vector lengths. This was also found to result in inefficiencies for the parallel FDTD algorithm when a small problem is distributed across a large number of processors [9].

```
      do 10 k = 1,nz-1
         do 10 j = 1,ny-1
            do 10 i = 1,nx
               hx(i,j,k) = hx(i,j,k)+chx(i,j,k)*(ey(i,j,k+1)-ey(i,j,k)
     1                                      -ez(i,j+1,k)+ez(i,j,k))
10    continue
```

```
      do k = 2,nz-1
         do j = 2,ny-1
            do i = 1,nx-1
               ex(i,j,k) = dex(i,j,k)*ex(i,j,k)+cex(i,j,k)*
     1                  (hz(i,j,k)-hz(i,j-1,k)-hy(i,j,k)+hy(i,j,k-1))
10    continue
```

Fig. 1 Vectorized explicit updates of h_x and e_x where nx, ny, and nz are the vector lengths on each processor along the x-, y-, and z-directions, respectively, and bnx and cnx are the update coefficient arrays.

An alternative way of expressing the field updates is using an un-nested loop structure, as suggested in [11], as presented in Fig. 2. One composite index m is used here, resulting in one long vector loop rather than three nested loops. The only difference between the un-nested loop and the nested loop structure is that meaningless calculations are being performed for the boundary points. This results in a small number of additional floating point operations. Furthermore, if the exterior boundary is an absorbing boundary, an additional memory location is needed to store the previous field values which will be overwritten by the un-nested loop.

Figure 3 illustrates the MFLOPS rates realized on a single i860 RISC processor and a single Cray-YMP processor by the FDTD algorithm using both the nested loop and the un-nested loop structures versus the length of the innermost nested loop, i.e., nx. In this study $nx = ny$, i.e., ny is scaled in the same manner, and $nz = 20$. As expected the un-nested loop structure results in faster floating point speeds which are nearly independent of the problem size. The overall CPU seconds required per FDTD time iteration on the i860 and Cray-YMP of the two loops are presented in Fig. 4. It is observed that as the problem size increases, the CPU times required by both of the methods converge (the un-nested loop structure remains slightly faster despite the additional FLOPs) as nx and ny become large. However, for smaller problems, the un-nested loop structure is going to yield improved performance.

```
      cons = co*dt
      cy = cons/dz
      cz = cons/dy
      m2 = nx+(ny-2)*nx+(nz-2)*nx*ny
      do 10 m = 1,m2
         hx(m) = hx(m)+cy*(ey(m+nx*ny)-ey(m))
     1            -cz*(ez(m+nx)-ez(m))
10    continue
```

```
      cy = 1.0/dz
      cz = 1.0/dy
      m1 = 1+nx+nx*ny
      m2 = nx-1+(ny-2)*nx+(nz-2)*nx*ny
      do 10 m = m1,m2
         ex(m) = bnx(m)*ex(m)+cnx(m)*(cz*(hz(m)-hz(m-nx))
     1                           -cy*(hy(m)-hy(m-nx*ny)))
10    continue
```

Fig. 2 Explicit updates of h_x and e_x using an un-nested loop.

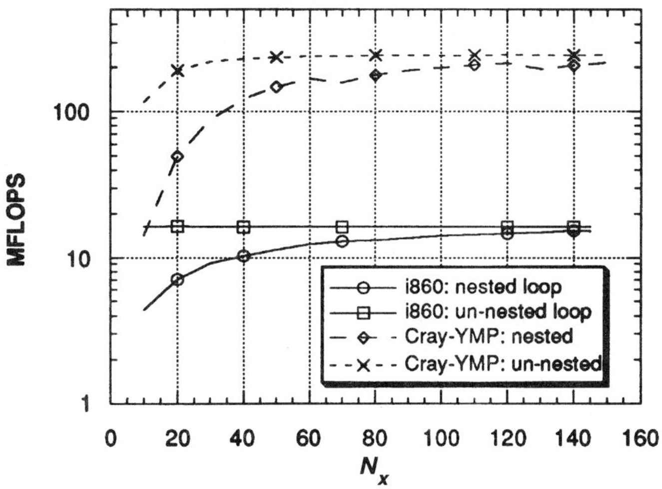

Fig. 3 MFLOPS rates recorded for the FDTD method using the nested loop and the un-nested loop updates versus the *nx* (where *nx* = *ny*) on single i860 RISC and Cray-YMP vector processors.

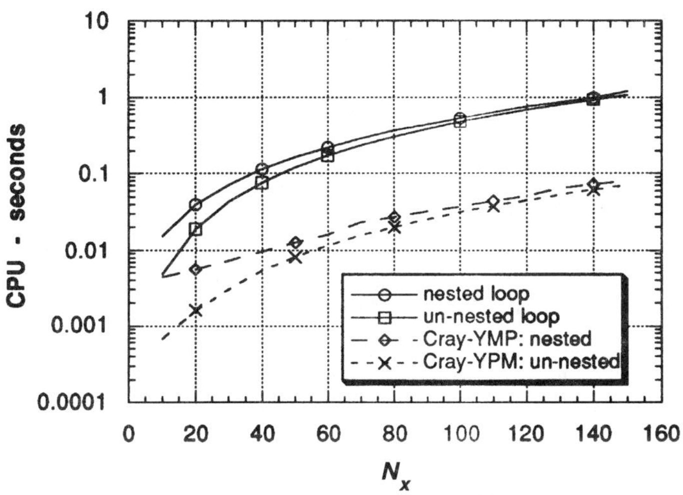

Fig. 4 CPU-time per time iteration for the FDTD method using the nested loop and the un-nested loop updates versus the *nx* (where *nx* = *ny*) on single i860 RISC and Cray-YMP vector processors.

IV. High-Performance Parallel Computers

The FDTD algorithm is inherently a scalable algorithm. Due to the explicit nature of the FDTD solution, the update of each vector field component is only dependent on its nearest neighbor. As a result, a high degree of parallelism can be realized. The FDTD has achieved excellent speedups on a plethora of parallel platforms, including coarse-grained SM-MIMD computers [12], coarse-grained DM-MIMD transputer arrays [13–15], massively parallel DM-SIMD multiprocessor computers [16–19], tightly coupled distributed MIMD multiprocessor computers [9, 20–22], and workstation clusters [23–25]. The common element of these parallel FDTD algorithms is that the field update tasks are distributed via a spatial decomposition, viz., the lattice grid is evenly distributed among all the processors by decomposing it into non-overlapping contiguous sub-lattices. Due to the regularity of the grid, excellent load balances can easily be achieved. Subsequently, each processor concurrently updates the fields within its local subregion. If the spatial decomposition is performed along the primary grid edges, the magnetic field can be updated independently on all processors without any interprocessor communication (this is at the expense of redundant updates of the normal magnetic fields lying on shared interfaces) [9]. Interprocessor communication is necessary to update the tangential electric fields on the shared boundary, since this is dependent on the central difference approximation of the normal derivative of the tangential magnetic field across the shared boundary. On a SM-MIMD computer, this is done by direct memory addressing. On a DM-MIMD or DM-SIMD computer, this will be done using explicit message-passing. The principal focus of this section will be the development of a parallel FDTD algorithm for a DM-MIMD computer.

DM-MIMD computers typically support two types of programming formats: 1) explicit message-passing and 2) virtual shared memory. The latter requires the support of a compiler such as HPF. However, HPF compilers are still in their infancy, and have yet to compete with explicit message-passing for performance and scalability over a large number of processors. On the other hand, explicit message-passing typically requires in-line subroutine calls to control the exchange of data between processors. While the actual data communication is transparent to the user, this still requires significant fore thought to plan the distributing of data and to coordinate the processors sending and receiving data.

The parallel FDTD algorithm will be developed using a single program multiple data (SPMD) paradigm. Specifically, every processor executes the same program while simultaneously operating on its local data, and it addresses remote memory via message-passing. Note that this is differentiated from SIMD operation in which every processor performs the same operation simultaneously.

The parallel FDTD program has the following structure [9]:

```
Initialize Fields and Update Coefficients
loop{
   update Magnetic Fields
   communicate Magnetic Fields
   update Electric Fields
   update Electric Field Absorbing Boundary Conditions
}
```

To ensure efficient parallelism the data must be distributed across all the processors in a manner such that the work load is balanced. Assume that the $N_x \times N_y \times N_z$ FDTD lattice is distributed across P processors. The lattice can be spatially decomposed into P subdomains that are contiguous and non-overlapping and are arranged as a $P_x \times P_y \times P_z$ grid. Each subdomain is assigned a lattice of dimension $N_x/P_x \times N_y/P_y \times N_z/P_z$, with the possibility of having a slight imbalance of $N_i/P_i - 1$ along each dimension due to non-integer factors. The fields within each processor and the associated update coefficients are stored in arrays local to the processor with this dimension. Each processor then independently updates the fields within its local lattice using the update loops in Fig. 1 (or Fig. 2).

It is assumed that the boundaries of the partitions are cut along the primary edges of the FDTD lattice. Subsequently, on a shared boundary the discrete electric fields are tangential and the discrete magnetic fields are normal to the shared boundary face (Fig. 5). The update of the tangential electric fields on a shared boundary cannot be updated locally without the values of the tangential magnetic fields resident on the neighboring processors. Subsequently, the electric field can be updated in one of two ways: 1) by partially updating the electric field on each processor and then concatenating the total field via interprocessor communication, or 2) by communicating the tangential magnetic fields needed to perform the electric field updates from the neighboring processor. The latter approach requires additional storage for the magnetic fields. In practice, however, it is much easier to implement since it greatly reduces the logic needed to perform the parallel updates. In both cases, each processor must exchange data with neighboring processors which share a lattice face.

To avoid networking conflicts, it is often desirable to maintain nearest neighbor communication. Thus, the topology of the DM-MIMD's communication network will determine the dimension of the spatial decomposition. For example, the high-speed communication network of the Intel Paragon has a 2-D mesh topology. As a result, it is natural to use a 2-D decomposition of the mesh (e.g., $P_x \times P_y \times 1$). Transputer arrays typically have a 1-D topology, and a 1-D decomposition is warranted. Hypercube topologies, employed by the Intel iPSC/860, the NCUBE, or the CM-2 can provide nearest neighbor communication for 1-, 2-, or 3-D decompositions. The Cray T-3D's network has a 3-D toroid topology, which again is well suited for 3-D decompositions. For a 3-D decomposition of the lattice, each processor communicates with 8 processors; for a 2-D decomposition, each processor communicates with 4 processors; and for a 1-D decomposition, each processor communicates with only 2 processors.

It is noted that while higher-order decompositions require communication with more processors, the lower-dimensional decompositions can result in much larger shared boundaries which require more data to be communicated. The time to send a data packet of n bytes via interprocessor communication can be modeled by the following expression:

$$t_{comm}(n) = t_{startup} + (n/1024)t_{pack} + nt_{trans}$$

where t_{comm} is the time required to initiate the communication, commonly referred to as the *latency* time; t_{pack} is the time per kilobyte required to pack the communication packets; and t_{trans} is the time per byte to transmit the data and is governed by the bandwidth of the network (see Table 1 for some typical bandwidths). The largest incremental time of any DM-MIMD computer is the latency time. Tightly coupled DM-MIMD computers, such as the Intel Paragon or the Cray T3D have latencies on the order of 1 to 10μs. Loosely coupled DM-MIMD computers, such as a workstation cluster, will have latencies on the order of 1 to 10ms. On the other hand, t_{pack} and t_{trans} are typically on the order of nanoseconds (these parameters can be determined experimentally for a given architecture using the formula above). Therefore, it is more efficient to send one large data packet rather than smaller data packets for each field component. Finally, the total communication time can then be computed based on the problem size, the dimension of the partitioning, and the architecture dependent values for $t_{startup}$, t_{pack}, and t_{trans} [26].

The exchange of the magnetic fields via message-passing is typically done by calls to subroutine libraries supplied by the DM-MIMD vendor. For example, the following subroutine performs the exchange of tangential magnetic fields across the shared boundaries written for the Intel Paragon:

```
subroutine Intel_Communicate_Magnetic_Fields()
include 'global_dat.inc'
do is = 1, n_share
   iproc = ishare(is)
   call buf_pack(is,hx,hy,hz,bufout,nbyte)
   call csend(itype(myp),bufout,nbyte,iproc,0)
enddo
```

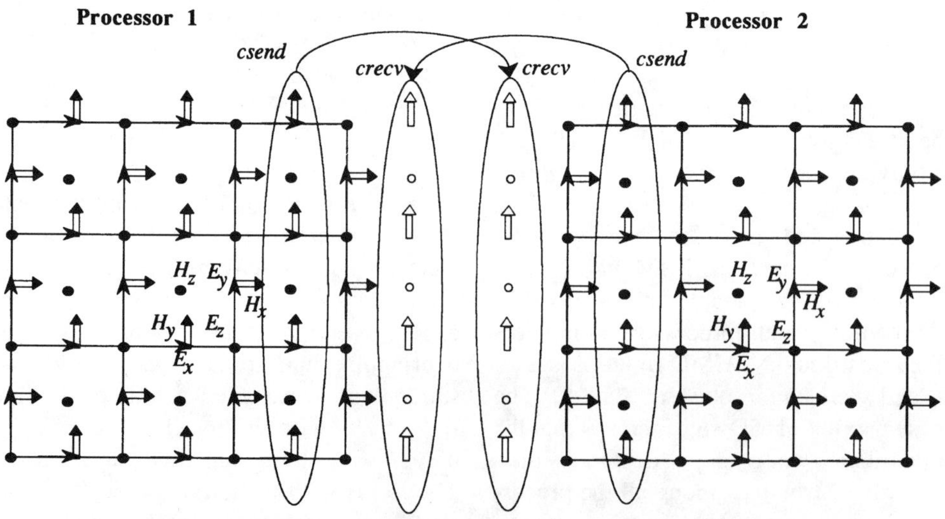

Fig. 5 Interprocessor communication of magnetic fields tangential to the boundary shared between two adjacent processors.

```
do is = 1, n_share
   iproc = ishare(is)
   nbyte = leng(is)
   call crecv(itype(iproc),bufin,nbyte)
   call buf_unpack(is,hx,hy,hz,bufin,nbyte)
enddo
return
end
```

where, n_share is the number of processors the local processor exchanges data with, bufout and bufin are vectors that are used to temporarily store the communicated magnetic fields and are packed by subroutine buf_pack (this is written by the user), nbyte is the length of the data in bytes, iproc is the processor ID (an integer) of the adjacent processor which will receive the data, myp is the local processor ID, and itype is an integer array providing an identification for the data packet referred to as the message type. The subroutines csend and crecv are Intel specific subroutines that transmit and receive the data asynchronously. This can similarly be written with the PVM message-passing library[1][27]:

```
subroutine PVM_Communicate_Magnetic_Fields()
include 'global_dat.inc'
do is = 1, n_share
   iproc = ishare(is)
   call buf_pack(is,hx,hy,hz,bufout,nbyte)
   call pvmfinitsend(PVMDEFAULT, ibufid)
   call pvmfpack(BYTE1,bufout,nbyte, 1, info)
   call pvmfsend(iproc,bufout,info)
enddo

do is = 1, n_share
   iproc = ishare(is)
   nbyte = leng(is)
   call pvmfrecv(-1,bufin,info)
   call pvmfunpack(BYTE1,bufin,nbyte,1,info)
   call buf_unpack(is,hx,hy,hz,bufin,nbyte)
enddo
return
end
```

where pvmfinitsend, pvmfpack, pvmfsend, pvmfrecv, and pvmfunpack are subroutines provided by the PVM library. The variable info is returned by pvmfpack and serves as a message identifier as well as an error flag. It is noted that the overall format of the programs using the Intel or PVM libraries is identical. The only difference is in the syntax and usage of the message-passing subroutines.

The user-defined subroutine buf_pack copies the tangential magnetic fields associated with the current boundary into the buffer bufout. This is then sent asynchronously to the adjacent processor iproc by the subroutines csend (or pvmfsend). The message packet sits in a buffer space until processor iproc posts a crecv (or pvmfrecv) with the same message type. The user-defined subroutine buf_unpack then unpacks the received buffer bufin into the local magnetic field space. This essentially requires the storage of one extra layer of tangential magnetic fields just outside the shared boundary, as illustrated in Fig. 5. Finally, the electric fields within and on the shared boundary can be updated using the update in Fig. 1 or Fig. 2 without any special logic to identify if the fields lie on the shared boundary.

The exterior boundaries of the local lattice are either on a shared boundary, or an exterior boundary. If a PML boundary condition is implemented (see Chapter 2), the exterior boundary is simply a PEC, and the lossy material medium is represented within the update coefficients. If a pseudo-differential boundary operator, such as Liao's or Higdon's boundary operators, is employed, then a separate update expression is needed to update the tangential electric fields on the exterior boundary. This requires a separate update loop and results in additional computation; however, since the processor does not perform an interprocessor communication across this boundary, the load remains balanced.

Benchmark Comparisons

The parallel FDTD algorithm has been successfully implemented on a number of SM-MIMD and DM-MIMD computers. On coarse-grained SM-MIMD parallel computers, such as the Cray YMP C90, or an SGI Power Challenge, near linear speedups have been realized by simply using the appropriate compiler options. On DM-MIMD computers, the parallel FDTD algorithm described earlier in this section was implemented. As a benchmark comparison, the low-pass filter described in [3] was simulated on a number of parallel platforms. The model consists of a $80 \times 110 \times 16$ grid and requires 4,000 time iterations. The resultant speedups recorded on the 512 processor Intel Touchstone Delta at the California Institute of Technology (the prototype for the Paragon), a 4 processor SGI Power Challenge, and a cluster of Sun Sparc 20 workstations are recorded in Figs. 6 and 7.

Figure 6 illustrates the speedup of the parallel FDTD algorithm measured over 128 processors of the Intel Delta. Two

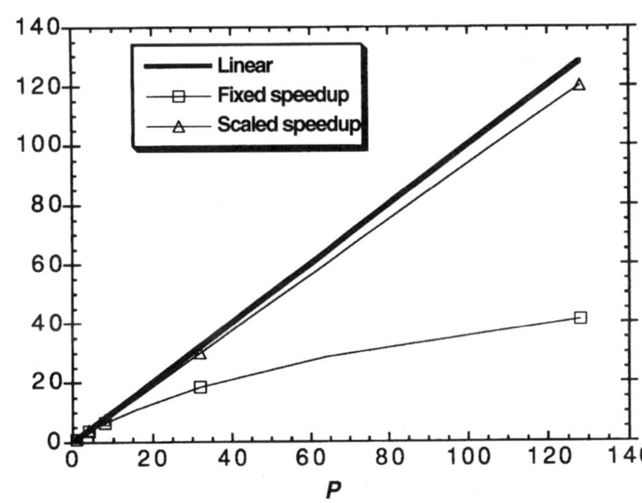

Fig. 6 Fixed and scaled speedup over 128 processors on the Intel Delta.

[1]PVM is a public domain library and can be obtained through anonymous FTP from netlib.ornl.com or through the world wide web at www.netlib.no/netlib/pvm3/index.html.

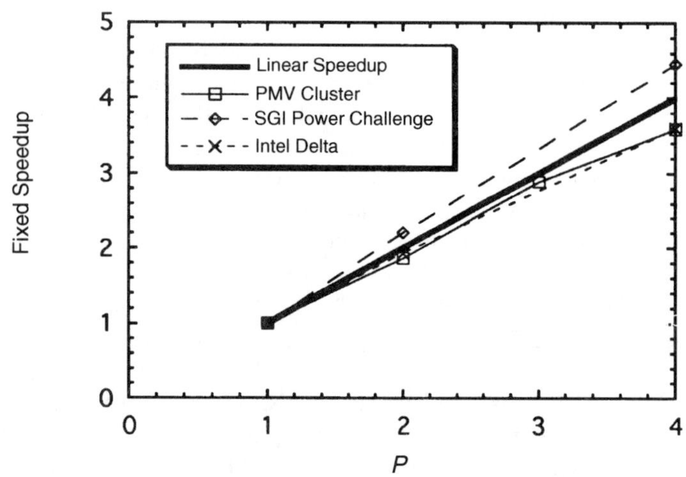

Fig. 7 Fixed speedup over four Sun Sparc 20 processors and a four processor SGI Power Challenge using the PVM message-passing library.

types of speedups are recorded here, *fixed speedup* and *scaled speedup*. Fixed speedup is the speedup of the overall wall clock time as the number of processors are increased when operating on a fixed problem size, and is calculated as

$$S_f(P) = \frac{\text{Time to Perform Problem of size } N \text{ on 1 Processor}}{\text{Time to Perform Problem of size } N \text{ on } P \text{ Processors}}$$

From Fig. 6 it is observed that a fixed speedup of about 42 was realized over 128 processors for this small problem. Larger problems will realize higher speedups [9]. The observed loss of parallel efficiency as the number of processors is increased has multiple causes. Initially, as the original 80 × 110 × 16 lattice is distributed over 128 processors, it reduces to sub-assembly of 10 × 7 × 16 lattices on each of 128 processors. As a result, the ratio of computation to interprocessor communication greatly decreases. Slight load imbalances also become more predominant as the problem size per processor becomes much smaller. Furthermore, from Fig. 3 it is observed that for the nested loop algorithm, as the inner loop length is reduced from 80 to 10, the floating-point speed of the processor reduces by 60%, which reduces the expected speedup.

Scaled speedup assumes that the problem size scales with the number of processors, and is computed as [28]:

$$S_s(P) = \frac{P \times (\text{Time to Perform Problem of size } N \text{ on 1 Processor})}{\text{Time to Perform Problem of size } (N \times P) \text{ on } P \text{ Processors}}$$

The scaled speedup recorded in Fig. 6 is much closer to the optimal linear speedup. This is due to the fact that the percentage of serialism to parallelism is held constant as the number of processors is increased. Specifically, the amount of communication does not continually increase with P, since the interprocessor communication is limited to nearest neighbor communication. Therefore, the ratio of communication to computation remains small, and a high level of scalability can be maintained. This implies that if a problem of size N takes S seconds per time iteration on one processor, then we can expect that a problem of size $N \times P$ will require roughly S seconds per time iteration on P processors.

The previous example was recorded on a tightly coupled DM-MIMD computer. Specifically, each processor and the high-speed data network interconnecting the processors are dedicated to one user at a time. A second class of DM-MIMD computers described in Section II is a loosely coupled multiprocessor computer composed of a cluster of workstations interconnected through a local area network. This can be facilitated through the use of a message-passing library such as PVM. To this end, the same problem studied in Fig. 6 was studied on a cluster of four Sun Sparc 20 workstations interconnected through an ATM switch. To port the program, a simple library was written to convert the Intel message-passing protocol to a PVM protocol. The fixed speedup for simulating the low-pass filter device is illustrated in Fig. 7, where a speedup of 3.6 was achieved. The speedup is also compared with that realized by the Intel Delta. Over four processors, it is seen they are roughly the same. However, as the number of processors is increased, the communication bandwidth and latency delays will become more significant and the fixed speedup of the workstation cluster will level off much more rapidly.

It is noted that the PVM simulation was performed under ideal conditions. Namely, there was no additional load on any of the processors and the FDTD program was allotted 95% of each CPU. However, if any one of the processors received an external load from another application, then the total wall clock time would be governed by the slowest processor. This is due to the recursive nature of the FDTD algorithm such that the algorithm cannot advance before receiving the necessary data from the neighboring processors. A dynamic load-balancing scheme would be needed to compensate for this. Secondly, on a heterogeneous cluster, the load balance would also need to be altered to account for the various speeds of the platforms.

The PVM version was also executed on a four-processor SGI Power Challenge. While this computer is a SM-MIMD computer, it was run using the DM-MIMD paradigm. The speedups running PVM were almost identical to those realized using the parallel compiler options. The results are also illustrated in Fig. 7. It is observed here that the speedup is supralinear on this platform! This is explained by the fact that each processor of the SGI has 4Mb of cache. The original 80 × 110 × 16 problem is too big to fit in cache on one processor. However, the two and four processor runs, which reduced to per-processor lattices of dimension 80 x 55 x 16 and roughly 80 x 28 x 16, respectively, fit entirely in cache. As a result, the floating point speeds were greatly increased since there were no memory fetches to cache.

V. Summary

It has been the purpose of this chapter to demonstrate that the FDTD algorithm can be designed to be highly computationally efficient on RISC-based workstations, high-performance vector-processor computers, and high-performance parallel computers. This is further illustrated in the reprinted articles that appear at the end of this chapter. Specifically,

1) R. Larson, T. Rudolph, and P. Ng, "Special purpose computers for the time domain advance of Maxwell's equations," *IEEE Transactions on Magnetics,* vol. 25, pp. 2913–2915, July 1989.
2) A. Perlik, T. Opsahl, and A. Taflove, "Predicting scattering of electromagnetic fields using the FD-TD on a connection machine," *IEEE Transactions on Magnetics,* vol. 25, pp. 2910–2912, July 1989.
3) D. B. Davidson and R. W. Ziolkowski, "A connection machine (CM-2) implementation of a three-dimensional parallel finite difference time domain code for electromagnetic field simulation," *International Journal of Numerical Modelling: Electronic Networks, Devices and Fields,* vol. 8, pp. 221–232, May-Aug. 1995.
4) S. Gedney, "Finite-difference time-domain analysis of microwave circuit devices on high-performance vector/parallel computers," *IEEE Transactions on Microwave Theory and Techniques,* vol. 34, pp. Oct. 1995.
5) U. Effing, W. Kümpel, and I. Wolff, "Parallel FDTD simulator for MIMD computers," *International Journal of Numerical Modelling: Electronic Networks, Devices and Fields,* vol. 8, pp. 277–281, May-Aug. 1995.
6) W. D. Gropp and E. E. Smith, "Computational fluid dynamics on parallel processors," *Computers and Fluids,* vol. 18, no. 3, pp. 289–304, 1990.
7) S. Gedney and F. Lansing, "A parallel planar generalized Yee-algorithm for the analysis of microwave circuit devices," *International Journal for Numerical Modelling: Electronic Networks, Devices and Fields,* vol. 8, pp. 249–264, May-Aug. 1995.

The focus of the first article by Larson et al. is on the design of a FDTD computer. The authors propose a pipelined computer architecture dedicated to performing FDTD simulations. The contribution of this article is not just an illustration of pipelined computer architectures, but it also provides a treatise on how one can develop an efficient pipelined FDTD algorithm that can exploit today's RISC processors.

The second article by Perlik et al. is one of the earliest papers on the development of a parallel FDTD algorithm, specifically for the massively parallel CM2 (SIMD). The reader is also referred to [17] for additional results obtained from this work. The CM2 provided good scaled speedups. However, due to the lack of memory, it was restricted in the size of the problems which one could solve. The third article by Davidson and Ziolkowski also considers the implementation of the FDTD method on the CM2. This article provides some basic time-complexity models for predicting the CM2 performance, as well as to demonstrate a FORTRAN 90 implementation of the FDTD algorithm.

The fourth article by Gedney provides an efficient FDTD algorithm for RISC processors and vector processors, as well as DM-MIMD multiprocessor computers. To this end, scaled and fixed speedups are presented for the 512-processor Intel Touchstone Delta computer. It is demonstrated that significantly large problems can be solved on such an architecture using the FDTD algorithm in reasonable times. The fifth article by Effing et al. demonstrates the application of the FDTD algorithm on a transputer array, classified as a SM-MIMD multiprocessor computer. Transputer arrays have been very successful in providing economical high-performance computing.

The sixth article by Gropp and Smith presents the development of a parallel algorithm for the explicit time-domain solution of Navier-Stokes equations. For this algorithm, detailed time-complexity models for a variety of DM-MIMD and SM-MIMD multiprocessor computer architectures are provided. While this work is based on the solution of Navier-Stokes equations, the computational methods used have significant overlap with the FDTD solution of Maxwell's equations, and similar formulations can be applied to developing efficient parallel FDTD algorithms as well as obtaining performance predictions for specific multiprocessor computer architectures.

The final paper by Gedney and Lansing presents a parallel algorithm for the planar generalized-Yee (PGY) algorithm. The PGY algorithm, like the FDTD method, is based on a direct solution of Maxwell's equations. The difference is that the PGY algorithm assumes a general unstructured dual grid, rather than a regular orthogonal grid. This presents a new set of challenges for mapping the algorithm onto a parallel platform. This paper deals exclusively with mapping the PGY algorithm onto DM-MIMD multiprocessor computers. Similar to the FDTD method, the parallel PGY algorithm is based on a spatial decomposition of the lattice grid; however, the decomposition is based on a much more general algorithm to decompose the unstructured grid. Similar to the FDTD algorithm, the parallel PGY algorithm is found to be highly scalable.

Acknowledgements

This research was supported in part by the Jet Propulsion Laboratory, California Institute of Technology under JPL/NASA Contract 959534 with the University of Kentucky, by the National Science Foundation under award ECS-9206014, and by the Army Research Office, Research Triangle Park, NC under Grant DAAH04-94-G-0243. Access to the Touchstone Delta at the California Institute of Technology was provided by NASA.

References

[1] A. Taflove, *Computational Electrodynamics, The Finite-Difference Time-Domain Method.* Boston, MA: Artech House, 1995.

[2] M. Piket-May, A. Taflove, and J. Baron, "FD-TD modeling of digital signal propagation in 3-D circuits with passive and active loads," *IEEE Transactions on Microwave Theory and Techniques,* vol. 42, pp. 1514–1523, August 1994.

[3] D. Sheen, S. Ali, M. Abouzahra, and J. A. Kong, "Application of the three-dimensional finite-difference time-domain method to the analysis of planar microstrip circuits," *IEEE Transactions on Microwave Theory and Techniques,* vol. 38, pp. 849–856, July 1990.

[4] J. J. Dongerra, G. A. Geist, R. Manchek, and V. S. Sunderam, "Integrated PVM framework supports heterogeneous network computing," *Computers in Physics,* April 1993.

[5] C. H. Cap and V. Strumpen, "Efficient parallel computing in distributed workstation environments," *Parallel Computing,* vol. 19, pp. 1221–1234, 1993.

[6] M. J. Flynn, "Some computer organizations and their effectiveness," *IEEE Transactions on Computing,* vol. C-21, pp. 948–960, 1972.

[7] A. J. Van der Steen, "Overview of recent supercomputers," *Technical Report,* Stichting Nationale Computer Faciliteiten, The Netherlands, September 1994.

[8] High Performance Fortran Forum, "Draft high performance Fortran language specification, Version 0.4, *Technical Report,* Houston, TX: Rice University, November 1992.

[9] S. Gedney, "Finite-difference time-domain analysis of microwave circuit devices on high performance vector/parallel computers," *IEEE Transactions on Microwave Theory and Techniques,* vol. 43, pp. 2510–2514, October 1995.

[10] S. D. Gedney and S. Barnard, "Efficient FD-TD algorithms for vector and multiprocessor computers," in *Computational Electrodynamics, The Finite-Difference Time-Domain Method,* A. Taflove, ed. Boston, MA: Artech House, 1995.

[11] M. E. Braaten, "Development of a parallel computational fluid dynamics algorithm on a hypercube computer," *International Journal for Numerical Methods in Fluids,* vol. 12, pp. 947–963, 1991.

[12] J. P. Brooks, K. K. Ghosh, E. Harrigan, D. S. Katz, and A. Taflove, "Progress in Cray-based algorithms for computational electromagnetics," in *Computational Electromagnetics and Supercomputer Architecture,* vol. 7, T. Cwik and J. Patterson, eds., in *Progress in Electromagnetics Research,* vol. 7, J. A. Kong, ed. Cambridge, MA: EMW Publishing, 1993, pp. 23–55.

[13] W. Kümpel, U. Effing, M. Rittweger, and I. Wolff, "Parallel FDTD simulator for microwave structures," in *23rd European Microwave Conference,* Palacio de Congresos, Madrid, Spain, 1993.

[14] U. Effing, W. Kümpel, and I. Wolff, "Parallel FDTD simulator for MIMD computers," *International Journal of Numerical Modelling: Electronic Networks, Devices and Fields,* vol. 8, pp. 277–281, May-Aug. 1995.

[15] K. C. Chew and V. F. Fusco, "A parallel implementation of the finite-difference time-domain algorithm," *International Journal of Numerical Modelling: Electronic Networks, Devices and Fields,* vol. 8, pp. 293–299, May-Aug. 1995.

[16] A. T. Perlik, T. Opsahl, and A. Taflove, "Predicting scattering of electromagnetic fields using FD-TD on a connection machine," *IEEE Transactions on Magnetics,* vol. 25, pp. 2910–2912, July 1989.

[17] A. Perlik and S. Moraites, "Electromagnetic wave analysis using FD-TD and its implementation on the connection machine," in *Computational Electromagnetics and Supercomputer Architecture,* vol. 7, M. Morgan, ed., in *Progress in Electromagnetics Research,* vol. 7, J. A. Kong, ed., Cambridge, MA: EMW Publishing, pp. 266–308, 1993.

[18] R. Guerrieri, K. Tadros, J. Gamelin, and A. Neureuther, "Massively parallel algorithms for scattering in optical lithography," *IEEE Transactions on Computer-Aided Design,* vol. 10, pp. 1091–1100, September 1991.

[19] D. B. Davidson and R. W. Ziolkowski, "A connection machine (CM-2) implementation of a three-dimensional parallel finite difference time domain code for electromagnetic field simulation," *International Journal of Numerical Modelling: Electronic Networks, Devices and Fields,* vol. 8, pp. 221–232, May-Aug. 1995.

[20] R. Calalo, T. Cwik, W. Imbriale, N. Jacobi, P. Liewer, T. Lockhart, G. Lyzenga, and J. Patterson, "Hypercube parallel architecture applied to electromagnetic scattering analysis," *IEEE Transactions on Magnetics,* vol. 25, pp. 2898–2900, July 1989.

[21] R. D. Ferraro, "Solving partial differential equations for electromagnetic scattering problems on coarse-grained concurrent computers," in *Computational Electromagnetics and Supercomputer Architecture,* vol. 7, T. Cwik and J. Patterson, ed., in *Progress in Electromagnetics Research,* vol. 7, J. A. Kong, ed., Cambridge, MA: EMW Publishing, pp. 111–154, 1993.

[22] J. Patterson, T. Cwik, R. Ferraro, N. Jacobi, P. Liewer, T. Lockhart, G. Lyzenga, J. Parker, and D. Simoni, "Parallel computation applied to electromagnetic scattering and radiation analysis," *Electromagnetics,* vol. 10, pp. 21–40, Jan.-June 1990.

[23] V. Varadarajan and R. Mittra, "Finite-difference time-domain analysis using distributed computing," *IEEE Microwave and Guided Wave Letters,* vol. 4, pp. 144–145, May 1994.

[24] D. P. Rodohan and S. R. Saunders, "Rapid solutions of the finite difference time domain method using parallel associative techniques," in *Eighth International Conference on Antennas and Propagation,* London, UK: IEE Publications, 1993.

[25] D. P. Rodohan and S. R. Saunders, "Parallel implementation of the finite difference time domain method," in *Second International Conference on Computation in Electromagnetics,* London, UK: IEE Press, 1994.

[26] W. D. Gropp and E. E. Smith, "Computational fluid dynamics on parallel processors," *Computers and Fluids,* vol. 18, no. 3, pp. 289–304, 1990.

[27] A. Geist, A. Beguelin, J. Dongarra, W. Jiang, R. Manchek, and V. Sunderam, *PVM: Parallel Virtual Machine. A User's Guide and Tutorial for Networked Parallel Computing,* Boston MA: MIT Press, 1994.

[28] J. L. Gustafson, G. R. Montry, and R. E. Benner, "Development of parallel methods for a 1024–processor hypercube," *SIAM Journal of Scientific and Statistical Computing,* vol. 9, pp. 609–638, July 1988.

Special Purpose Computers for the Time Domain Advance of Maxwell's Equations

RONAL W. LARSON, TERENCE RUDOLPH, AND POH H. NG
ELECTRO MAGNETIC APPLICATIONS, INC.
P.O. BOX 26263
DENVER, CO 80226

1. ABSTRACT

Several special purpose computer designs will be presented for the efficient time advance of Maxwell's Equations. These involve both paralleling and pipelining techniques with specialized hardware that takes advantage of simplicity and symmetry of Maxwell's Equations. A time speedup of a factor of 1000 over operation on a VAX at reduced cost seems immediately achievable. There is a direct applicability to algorithms for general purpose computers as well.

An apparently new normalization for the electric and magnetic fields will also be discussed. This normalization simplifies the hardware design, but also saves approximately 30% computation time in computations on a general purpose computer. A simplification will be described for the first order radiating boundary conditions. This gives slightly different, but apparently equally accurate results with approximately a 30% time or hardware savings. A term is added for use of the condition in conducting media. A discussion will be given of the preferred use of E and H conditions at different boundaries.

An alternative method of computing Huygens' sources will be discussed. This new method has speeded up Huygens' computations on a general purpose computer by a factor of three, and also greatly simplified the hardware design. New results will be discussed on modifications of standard thin wire formulations by changing the properties of the cell in which the thin wire is embedded.

The work to be reported is the result of a recently concluded, first phase Small Business Innovation Research (SBIR) contract from the Harry Diamond Laboratories (under Prime Contract # DAAL02-87-C-0091).

2. INTRODUCTION

The reason for proposing this work initially was the long delay sometimes experienced at EMA in obtaining Time Domain Finite Difference (TDFD) results. Even with several VAX-size computers, the queue experienced by approximately twenty analysts was often greater than 24 hours, mostly due to a few large problems that might take that long. A review of this computing load showed that the vast majority of the computing time was spent in the time advance of Maxwell equations.

For most universities and many companies, a serial or parallel supercomputer would probably be a cost-effective option for reducing the queue. However, their high cost and/or remote location made this option unattractive. A larger number of small machines also looks unattractive because of the high cost of memory, which can far exceed the cost of computing power. This left the option of developing a pipelined machine that could more effectively handle the set of equations that comprised such a large fraction of EMA's computing time. This thinking preceded announcement of the SBIR opportunity.

Our initial proposal to the Harry Diamond Laboratories did not offer a specific time-savings or cost-savings goal. However, our TDFD computations on the VAX take approximately 1000 clock steps per spatial cell. We soon came to believe that pipelining could accomplish the same in one time step, and this became the project time-savings goal. Our cost goal is to be less than a VAX.

3. BASIC EQUATIONS

We begin with the six Maxwell curl equations, which in vector form are:

$$\mu \frac{\partial \mathbf{H}}{\partial t} + \nabla \times \mathbf{E} = \mathbf{M} \tag{3.1a}$$

$$\varepsilon \frac{\partial \mathbf{E}}{\partial t} + \sigma \mathbf{E} - \nabla \times \mathbf{H} = -\mathbf{J} \tag{3.1b}$$

Here, ε is the permittivity, μ is the permeability, and σ is the conductivity of the medium. **M** and **J** represent magnetic and electric current sources, respectively.

The finite difference solution of these equations in the time domain may be accomplished by making use of a centered difference approximation of the time and space derivatives in the curl equations [2,3] or their integral forms. We also employ an apparently new normalization that puts both the E and H fields into the units of volts, using:

$$\Delta x \; E_x \rightarrow E_x \tag{3.2a}$$

$$\frac{\mu \, \Delta y \, \Delta z}{\Delta t} H_x \rightarrow H_x \tag{3.2b}$$

where the left hand side of the replacement ($\rightarrow$) operation refers to equations (3.1), but the right hand nomenclature will be used hereafter.

After this substitution, we obtain for two of the six difference equations:

$$H_x^{n+1}(i,j,k) = H_x^n(i,j,k) + \left[E_z^n(i,j,k) - E_z^n(i,j+1,k)\right] + \left[E_y^n(i,j,k+1) - E_y^n(i,j,k)\right] + \dot{M}_x(i,j,k) \tag{3.3a}$$

$$E_x^{n+1}(i,j,k) = A_x' E_x^n(i,j,k) + C_1\left[H_z^{n+1}(i,j,k) - H_z^{n+1}(i,j-1,k)\right] + C_2\left[H_y^{n+1}(i,j,k-1) - H_y^{n+1}(i,j,k)\right] - J_x'(i,j,k) \tag{3.3b}$$

with

$$A_x' = \frac{1-g_x}{1+g_x}, \quad g_x = \frac{\sigma_x(i,j,k)\,\Delta t}{2\varepsilon_x(i,j,k)}, \quad C_1 = \frac{f_y^2}{1+g_x}, \quad C_2 = \frac{f_z^2}{1+g_x},$$

$$f_y = \frac{v\Delta t}{\Delta y}, \quad f_z = \frac{v\Delta t}{\Delta z}, \quad v = \text{propagation velocity} \tag{3.3c}$$

Reprinted from *IEEE Transactions on Magnetics,* pp. 2913-2915, July 1989.

The remaining four Maxwell difference equations can be obtained by permutation from these two. Note that the three H advance equations contain no multiplications. Other normalizations in the literature achieve this by assuming that the three spatial increments are equal. The three constants, A', reduce to unity for lossless media, while the six constants, C, reduce to a single constant if the spatial increments are equal. We make neither of these assumptions, although we have assumed free space permeabilities throughout the problem space.

The five values needed to compute one of the three H-fields are itself (previous iteration) and the four E-fields resident on the edges of "its" face (the face used in the integral form of the equation). Figure 3.1 shows that only 12 (rather than 3x5=15) values of E and H are required to compute the three H-fields. However, because we compute along an x-pencil, two of the values are repeated in the x-direction, so that there are only 10 unique values: 1-6 are internal to the i,j,k call, while 7-10 come from outside as E-values.

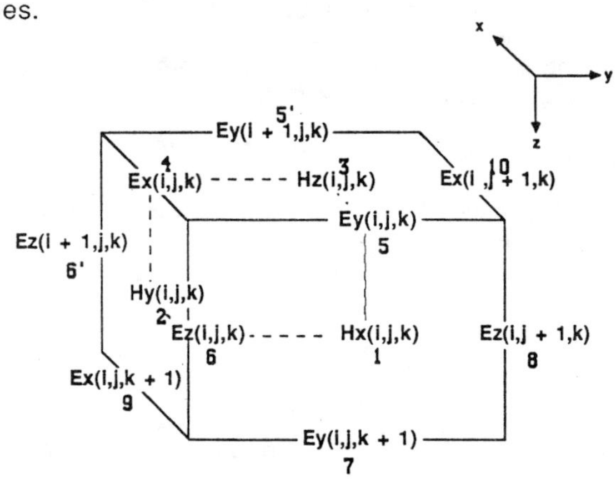

Figure 3.1 3 Faces Used with an H Computation

For computational purposes it is convenient to think of E values as not only residing on edges of cubes, but also at the center of another set of faces which are offset from the H cubes by a negative 1/2 of a cube dimension in each of the three axis directions as shown in Figure 3.2 Again, one can think of 10 values as being required for the complete determination of the E-fields, but since six of these can be carried over from the H-field computations, only four new H-fields are being added: fields 11-14.

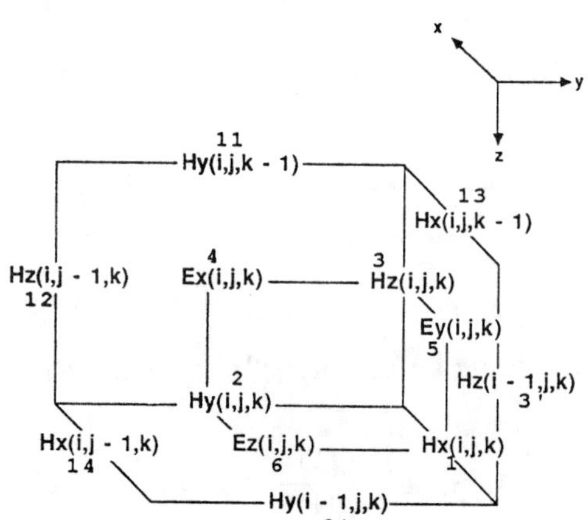

Figure 3.2 3 Faces Used with an E Computation

4. IMPLEMENTATION IN PIPELINING HARDWARE

There are several ways in which pipelining computation can traverse through space:

a. Coarse: one pencil at a time,
b. Medium: a plane of pencils simultaneously,
c. Fine: a volume simultaneously

Each of these must be moved to fill up the total space. They differ primarily in their degree of computational element complexity, with attendant reliability considerations. In the medium and fine grain approaches, message passing techniques must be employed over a nearest neighbor grid, with storage in local memory. The coarse grain approach uses only a global shared memory.The same amount of total memory is required for each of these approaches.

Another choice must be made in the number of equations to be advanced simultaneously; logical alternatives are one, two, three, or six. A third choice must be made between calculating the boundary conditions, Huygens' current, and thin wire computations as the main Maxwell Equation routine is being run, or coming back to calculate them later. It is imperative that these subsidiary computations not be made using an auxiliary computer, since this could negate the speedup of the pipelined computer.

There is a tradeoff between cost (for additional processing elements) and speed (by avoiding calling the fields from memory twice) in calculating six fields simultaneously. In small memory applications (and therefore very rapid computation), it may be preferable to minimize the computational hardware, thereby leading to three (or possibly even one or two) processors, rather than the six shown in Figure 4.1. Figure 4.1 is obtained by replacing the symbols of Equation 3.3 by the numbers used in Figure 3.1-3.2. The left hand side shows the nine steps in the computation flow with the first step being the reading (R) of E-fields 4-10 from memory.

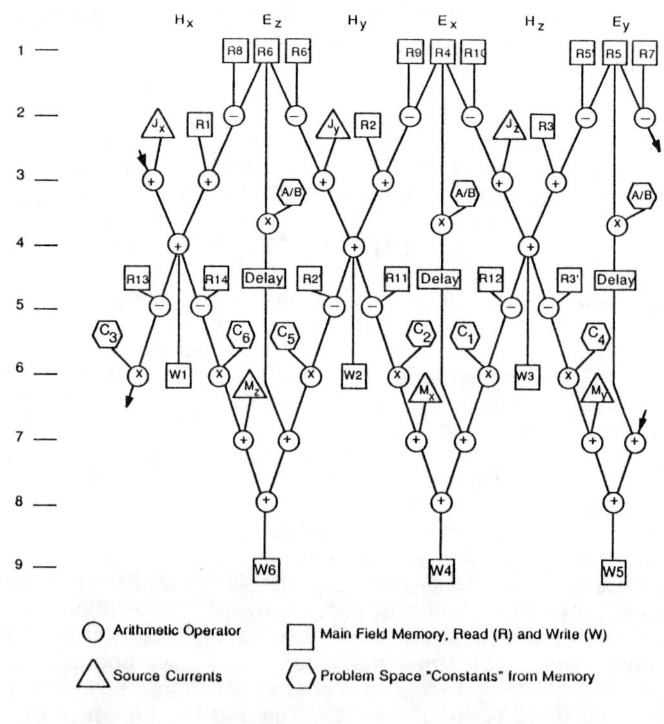

Figure 4.1 Flow Chart for 39 Operations

In the second row (or clock step) are six subtractions and the collection of additional H-field data (variables 1-3) and current data. In the third row or time step, six additions are performed. Three more additions in the fourth time step complete the computation of the three magnetic fields. These are shown (for clarity) as being written (W) back to memory in the 6th time step, but it could be done as early as the 5th time step, if the memory were ready to receive it.

Computing the E-fields requires several subtractions of H-Fields in step 5 and multiplications that are shown in step 4 (using the three values of A/B=A') and step 4 (using the six C's) . After additions in steps 7 and 8 to bring in prior E's, then the final E fields are ready for writing back to memory by the ninth time step.

Each operation such as a subtraction or an addition requires latches and control wires in addition to the signal lines. Each signal wire shown is also in fact a cable of (probably) 32 wires. There are 39 operations shown in Figure 4.1: 12 subtractions, 9 multiplies, and 18 additions. Each is performed every clock step.

5. MEMORY MANAGEMENT

Memory has turned out to be the most expensive part of the problem (for an assumed size of $(215)^3$ = 10 million cells). Speed is also a critical problem, for which the present solution is interleaving adjacent cells. It is then possible to read three cells (for H-fields, the present and to the left and above; for E-fields, present, right, and below) and write one all in the same time step. This will require a four-to-four switch arrangement for each of the six field memories, as cells and pencils become doubly interleaved.

6. BOUNDARY CONDITIONS

The standard first order Mur [4] condition in the normalized form becomes:

$$E_y^{n+1}(1,j,k) = E_y^n(2,j,k) - \frac{1 - f_x}{1 + f_x}\left[E_y^{n+1}(2,j,k) - E_y^n\,(1,j,k)\right]$$

This has the large disadvantage of requiring back-storage of a field. After some review of the Mur condition, it became clear that he was averaging two first order extrapolations, although this was not the way it was derived. No increase in accuracy results from the averaging, so the easier (for pipelining) of the two extrapolations was retained. After including a term to account for attenuation, we obtain:

$$E_y^{n+1}(1,j,k) = \frac{1}{1 + f_x}\, E_y^n(1,j,k)\exp(-g_y) + \frac{f_x}{1 + f_x}\, E_y^n(2,j,k)$$

We have not previously seen either the three term or lossy forms in the TDFD literature. Using it saved approximately 30% computation time on a small sample problem.

7. HUYGENS' SOURCES

It presently appears that it will be possible to add a small (add and multiply) computation that will compute the Huygens' current sources during the normal operations. The terms will then be added through the electric and magnetic current source terms, **J** and **M.** Space limitations preclude showing the new approach, which will possibly calculate the geometric spatial delays, rather than store these in memory.

8. THIN WIRES

Many problems of interest often contain such small structures in the scattering object as communication wires, thin slots or struts, booms and antenna elements. It is usually not possible to choose a grid size small enough to resolve these small structures and to include the whole scattering object in the same problem space. The main constraints are computer memory size and the finite difference time step which has to satisfy the Courant condition. Special techniques have been developed to model objects whose sizes are smaller than the finite difference cell size.

Consider a wire in the z-direction along the E_z grid points. As a conductor, the electric field inside the wire is zero. Therefore, all E_z nodes in the wire are set to zero. Modification of the magnetic components adjacent to the wire can be calculated by applying Stokes' theorem (before normalization):

$$H_x^{n+1}(i_w,j,k) = H_x^n(i_w,j,k) - \frac{\Delta t}{\mu}\frac{E_z^n(i_w,j+1,k) - E_z^n(i_w,j,k)}{\Delta y} + \frac{\Delta t}{\mu}\left(1 - \frac{r}{\Delta y}\right)\frac{E_y^n(i_w,j,k+1) - E_y^n(i_w,j,k)}{\Delta z}$$

where $j = j_w, j_w-1$, with similar terms for the other two components of H.

The effect of the wire is to modify the constant factor multiplying the relevant E field terms. These factors can either be absorbed into the normalized E fields in the advance equations or be inserted as "currents". Preliminary results from this technique agree with the standard method.

9. SUMMARY

The SBIR phase 1 study has been encouraging, with the project speed and cost goals apparently still attainable. The Phase 2 demonstration, recently initiated, will ultimately determine the value of a pipelining approach to computational economy.

REFERENCES

[1] Larson, Ronal W. et al, "Numerical Approaches to the Solution of Electromagnetic/Coupling Problems", EMA-88-R-25, March 1988, SBIR Final Report to HDL.

[2] Yee, Kane S., "Numerical Solution of Initial Boundary Value Problems Involving Maxwell's Equations in Isotropic Media," IEEE Trans. Antennas & Propag.,Vol. AP-14, No. 3, May 1966, pp. 302-307.

[3] Merewether, David, and Robert Fisher, "Finite Difference Solution of Maxwell's Equation for EMP Applications," DNA 5301F (Contract No. DNA 001-78-C-0231) Electro Magnetic Applications, Inc. April 22, 1980.

[4] Mur, G., "Absorbing Boundary Conditions for the Finite Difference Approximations of the Time-Domain Electromagnetic-Field Equations," IEEE Trans. EMC, Vol. EMC-23, No. 4, November, 1981.

Predicting Scattering of Electromagnetic Fields Using FD-TD on a Connection Machine

ANDREW T. PERLIK, MEMBER, IEEE
MRJ, INC.
10455 WHITE GRANITE DRIVE, OAKTON, VIRGINIA 22124

TORSTEIN OPSAHL
CENTER FOR INNOVATIVE TECHNOLOGY
13873 PARK CENTER ROAD, HERNDON, VIRGINIA 22071

ALLEN TAFLOVE, SENIOR MEMBER, IEEE
DEPARTMENT OF ELECTRICAL ENGINEERING AND COMPUTER SCIENCE
TECHNOLOGICAL INSTITUTE, NORTHWESTERN UNIVERSITY
EVANSTON, ILLINOIS 60201

The Finite Difference-Time Domain (FD-TD) numerical technique for solving Maxwell's equations is mapped onto a massively parallel Single Instruction Multiple Data (SIMD) architecture. A Connection Machine™ was chosen over other contemporary SIMD machines as the most promising candidate. The fundamental FD-TD algorithm developed by Taflove is decomposed into its serial and parallel segments. Connection Machine implementation is discussed in detail including processor assignment, processor utilization, run time, problem size, and future directions.

Introduction

FD-TD is an explicit time stepping finite difference formulation of Maxwell's curl equations that was developed in the mid-1980's by Allen Taflove [1]. It has broad applicability to the study of electromagnetic scattering by three dimensional objects because the scatterers can be closed or open, conducting, dielectric, inhomogeneous, or anisotropic. The algorithm has been validated against experimental data on objects that exhibit scattering physics at edges, corners, and cavity penetration. Agreements to within 1 dB of experimental data have been reported for scatterers that physically span 9 wavelengths and have a bistatic scatter dynamic range of 40 dB [1]. Consequently, FD-TD is certainly among the most robust scattering predictive algorithms that are currently available. If a parallel computer algorithm could be developed that could handle larger scatterers than serial code and/or run faster than serial codes, then this new development could have broad applicability to the electromagnetics community.

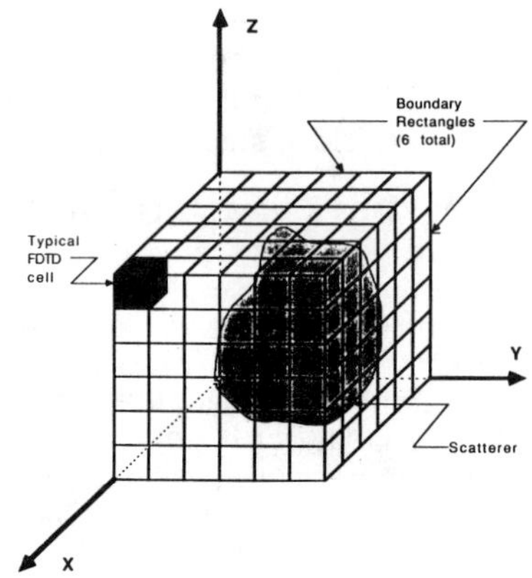

Figure 1. FD-TD Volumetric Grid Enclosing a Scatterer

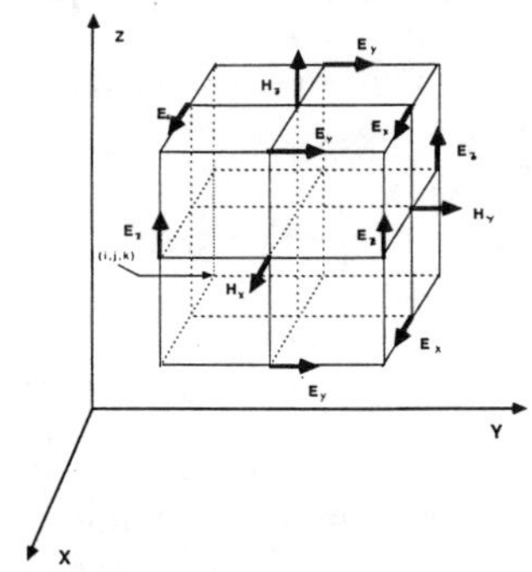

Figure 2. Yee FD-TD Lattice

FD-TD Algorithm

FD-TD models the propagation of a plane wave in a finite volume of space containing the electromagnetic scatterer, Figure 1. A cubic cell spatial lattice grids the volume under study. Although simplistic, attempts to generalize the gridding procedure to a non-uniform, scatterer surface conformal one have not been totally satisfactory because both the computational dynamic range and algorithm run time degrade severely. Alternative gridding techniques still remain an area for continued research. An exploded view of a typical FD-TD cell appears in Figure 2, the well known Yee lattice. Note that the E and H field components are computed on a half cell staggered grid. This formulation guarantees second order accuracy in the difference equations that are being solved. Two computational regions are defined, a total field region and a scattered field region. The total field region completely encloses the scatterer. In it, the total field gets updated computationally in order to preserve a large computational dynamic range. Scattered fields are needed to obtain far field information. The traveling wave incident fields get added at the interface of these two computational regions respecting causality.

A problem solution is obtained by time stepping the difference equations. As time evolves the incident wave travels through space exciting the scatterer. Time stepping continues until steady state is achieved for all field components in the grid. At steady state all field components are sinusoidal in time with converged magnitudes and converged relative phases.

Update equations for the volumetric grid decompose into four categories: discretized Maxwell's equations for grid interior points, Mur update equations for E components at the interior points of the boundary rectangles, interpolation and extrapolation equations for the edges of the boundary rectangles, and equations for the incident fields. Mur update equations [2] are

Reprinted from *IEEE Transactions on Magnetics,* pp. 2910-2912, July 1989.

used to computationally terminate the volumetric grid. Basically, they are discrete versions of a one-way wave equation and are designed to "absorb" outgoing plane waves. In order to update a Mur boundary point to a new time, all nearest neighbor spatial values are needed. Consequently, the Mur updates do not apply for fields along the edges of the volumetric grid. Edges are updated heuristically using an interpolation and extrapolation scheme on nearest neighbor data.

The Connection Machine

The Connection Machine (CM) is a massively parallel SIMD computer manufactured by Thinking Machines Corporation in Cambridge, Massachusetts. CMs are hosted by serial computers that broadcast instructions to it, and the same instruction is executed by each processor on data in its own memory. A commercially available CM-2 contains a maximum of 65,536 processors, each having 65,536 bits of dedicated memory (63,995 bits are user addressable). Instructions are bit oriented giving the programmer the unusual, but extremely useful flexibility to match word length to suit desired dynamic range or to match interprocessor message lengths. Floating point coprocessors are available, but at the present time they support only 32 bit single precision arithmetic. An elapsed time for a complete cycle of arithmetic (i.e., retrieve operands, perform the arithmetic operation, and store the result) of 40 μsec is easily achievable, using non-optimized non-pipelined code. Provided that all 65k processors are productively computing, 1.6 gigaflops (65k/40 μs) establishes a lower limit on the machine capability. Typical performance for algorithms developed at MRJ, Inc. is equivalent to a serial machine performance of 5 gigaflops, and peak performance rates equivalent to 28 gigaflops on a serial machine have been demonstrated.

General processor-to-processor communication is available; however, for the algorithm capabilities incorporated into the present model, only nearest neighbor processor communication is needed. All processors can get data from their nearest neighbors simultaneously. For a 32 bit message the elapsed time ranges from 30 μsec to 140 μsec depending on algorithm design implying that a net memory transfer rate of 70 gigabits is possible on a full machine.

On-line disk storage (data vault) is available up to 80 gigabytes for a 65k CM-2, but at present it is not compatible with all hosts. Software will be available in the near future to remedy the situation.

The host computer plays an integral part in executing algorithms. Not only does it broadcast instructions to the CM, but it can also read data out, perform computations, and write data back to selected processors. This affords the user with flexibility to use both serial and parallel computational capability, as needed.

Programs can be written on small (8k) machines and can be run on larger machines without changing the code. This is accomplished by using the good coding practice of exploiting the operating software supplied machine constants. The concept of virtual processors is also supported. Each physical processor can partition its memory by factors of 2 and assign to each partition a virtual processor ID. An 8k machine can, therefore, act like a 65k machine provided that each processor needs only 1/8 of a processor's physical memory. Run time degrades with virtualization.

Mapping FD-TD onto the CM

FD-TD was mapped onto the CM with two guiding principles: maximize the size of the volumetric grid that can be analyzed and minimize run time. Development took place under the constraints that no on-line disk storage was available and that all floating point arithmetic would be single precision (32 bit) so that the floating point coprocessors could be used. The first constraint arose because both development risk and time were high when the programming was started. Data vault software was new, unproven, and was available under an operating system that was less flexible than the one desired. Floating point arithmetic using the coprocessors reduced run time by a factor of 10--sufficiently attractive to warrant their use. Contrary to what might be expected, however, no compromise in desired accuracy is incurred. Single precision arithmetic was demonstrated to support a computational dynamic range of 40 dB and is believed to support a range of over 60 dB. The real trade was against volumetric grid size because using less than 32 bits would free memory for more grid points.

Step 1

The FD-TD algorithm is examined and broken into its algorithmic serial and parallel computational steps. Start by identifying the computational chronology at each time step:

update incident fields
update interior H fields
update interior E fields
update rectangular boundary edges
update rectangular boundary interiors
test for steady state

Each computational step above depends on the result of the previous one. Unless the basic algorithm is changed, these steps are sequential on both serial and parallel computers. All spatial points can theoretically be updated simultaneously in each computational step. In the second step, for example, all three components of H can be updated simultaneously as well.

Step 2

Identify the role(s) to be played by each processor for each parallel step. This step is critical for robust parallel code development. Performance both in run time and problem size could vary by orders of magnitude among implementations. Returning to the second computational step above, minimum execution time would result if each H field component and each E field component at a grid cell were assigned to individual processors. Such an assignment would greatly sacrifice problem size since each FD-TD cell would require six processors. Instead, the decision was made to assign each processor the role of updating the three H components and the three E components assigned to each FD-TD cell, Figure 3. The trade here was in favor of problem size at the expense of run time. Total storage requirements needed to support computations for each FD-TD cell sums to 1,536 bits, significantly less than the 63,995 user addressable bits. Each processor can, therefore, by virtualized by a factor of 39, the remaining storage being reserved for stack space.

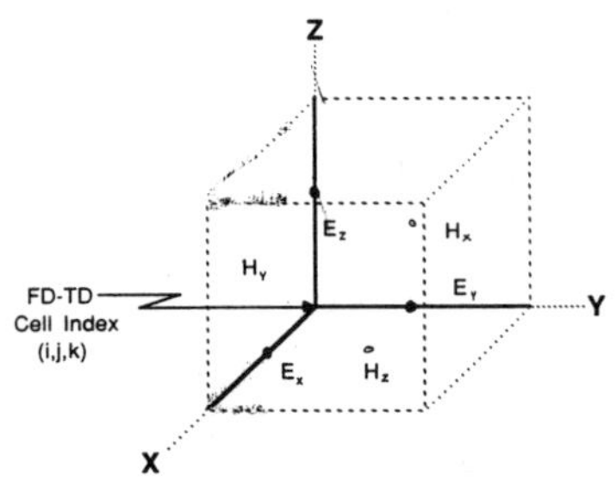

Figure 3. Fields Assigned to a CM Processor

Figure 4 summarizes the essential breakdown between the serial and parallel processing steps. Some steps have been left out for clarity, i.e., update of incident fields, convergence testing, and computation of peak and phase for scattered fields, but the breakdown in Figure 4 captures the philosophy.

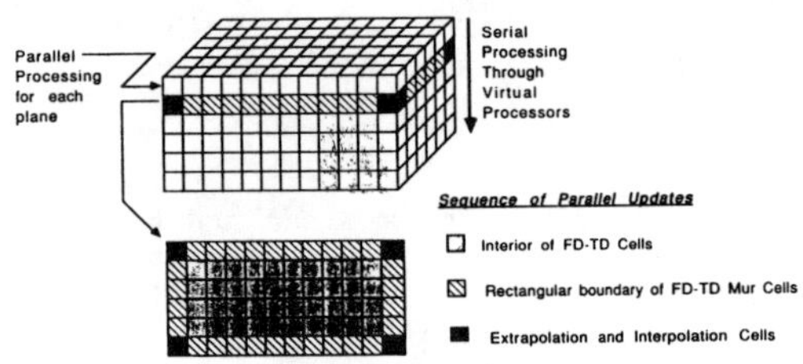

Figure 4. Overview of Essential Parallel and Serial Steps

Assessment of Parallel Implementation

Each FD-TD cell needs only 1,536 bits for data storage. Creating virtual processors is clearly a good idea; otherwise, the excess memory per processor would not get used. Now consider processing for an individual plane. Let each plane be n cells by m cells. Then (n-2)(m-2) cells are interior ones, all getting updated simultaneously. Edge processors update next. Only four processors are active independent of grid size. Clearly, this is poor utilization of the processors since only a small number are being used. The true measure of its impact on overall performance is the relative amount of time consumed compared to the other grid update functions for each plane. Table 1 provides the data needed. Edge updates require nearly half as much time to compute as the interior points do. Updating the edges on the CM instead of the host was a poor choice, one that will be remedied in the next version of the software. Face updates are computed using 2(n+m)-8 processors, on the order of the square root of the number of available processors. Again referring to Table 1, the conclusion is drawn that a significant amount of time is required to update the face processors relative to the interior ones. This does not imply that face updates should be performed in the host, however, because the I/O between the host and the CM for the root of CM processors needed to service the faces may be too expensive. Algorithms for face update run time reductions are currently being examined.

Table 1. Normalized Elapsed Time for Major Parallel Program Segments

PROGRAM SEGMENT	NORMALIZED RUN TIME
interior update	1 unit
edge update	.496 units
face update	.402 units

The three program segments discussed above constitute the core code that gets executed for each plane in the FD-TD grid. The execution time required to service those segments that employ only a fraction of the number of available processors, the edge and face updates, is approximately 47 percent of the total time required to service a plane. This suggests that the run time improvement for the current algorithm is at best a factor of 2. Run time improvements by factors of 10 or more will require a fundamental algorithm redesign.

Assessment of Algorithm Performance

Performance assessment will touch on the following topics: problem size as a function of machine size, run time as a function of machine size, and comparison between serial and parallel codes. The most succinct method for conveying results is to express them in terms of the number of FD-TD cells that fit into the CM's random access memory (RAM). Otherwise, results would be coupled to scatterer size and the desired solution accuracy, since for fixed scatterer size solution accuracy is a function of the thickness of the scattered field computational region. With the present code an 8k processor CM can evaluate 38 planes of 62x126 cells giving a total FD-TD volume of 296,856 cells. A 65k processor CM can evaluate 38 planes of 254x254 cells for a total of 2,451,608 FD-TD cells. The same code will run on both machines at the same execution time of 0.045 seconds/time step/virtual plane. Total run time per time step is 38 times larger, 1.7 seconds/time step. Run time is flat with the increase in machine size because the architecture is SIMD. Run time does change as the number of virtual planes change because the physical processors serially service the virtual processors that they represent. Benchmarks were run to show that reducing the number of virtual planes by a factor of two also reduces the run time by a factor of two. The largest scatterer studied and reported in the literature as far as we are aware is the 9λ cross tee plate reported by Taflove [1] using a Cray based code. It was embedded in a 258,048 cell lattice. By comparison the CM code can evaluate a computational lattice 9.5 times larger in volume on a full CM-2 in RAM. Furthermore, the CM code as currently written is somewhat more general because the lattice cells can be rectangular parallelepipeds instead of just cubes. If the CM code is restructured in the same way, lattice size would double by another factor of two in volume and execution time per virtual level would fall by roughly 10 percent.

Summary

The advent of massively parallel processors is still in its infancy; the CM, in particular, being commercially available for only two years and being based on technology that is 10 years old. Already, we are compiling data that suggests parallel codes are capable of exceeding lattice sizes and executing faster than even Cray based codes for lattice optimally matched to the CM's memory. The applicability of parallel processing to grid based techniques appears to have a bright future.

References

[1] A. Taflove and K. R. Umashankar, "Analytical Models for Electromagnetic Scattering," Final Technical Report RADC-TR-85-87 by ITT Research Institute, Chicago, IL, to Rome Air Development Center, Hanscom AFB, MA 01731 on Procurement Number F19628-82-C-0140, May 1985.

[2] G. Mur, "Absorbing Boundary Conditions for the Finite-Difference Approximation of the Time-Domain Electromagnetic-Field Equations," IEEE Trans. Electromag. Compat., Vol. EMC-23, Nov. 1981, pp. 377-382.

A. Taflove was supported in part for this research by NSF Grant No. ASC-8811273.

A Connection Machine (CM-2) Implementation of a Three-Dimensional Parallel Finite Difference Time-Domain Code for Electromagnetic Field Simulation

DAVID B. DAVIDSON
DEPARTMENT OF ELECTRICAL AND ELECTRONIC ENGINEERING
UNIVERSITY OF STELLENBOSCH
STELLENBOSCH 7600, SOUTH AFRICA

RICHARD W. ZIOLKOWSKI
ELECTROMAGNETICS LABORATORY
DEPARTMENT OF ELECTRICAL AND COMPUTER ENGINEERING
UNIVERSITY OF ARIZONA
TUCSON, ARIZONA 85721 USA

SUMMARY

This paper describes a parallel three-dimensional finite difference time-domain (FDTD) code for electromagnetic field simulation that has been developed for the Connection Machine (CM-2). The CM-2 is briefly discussed. Then the FDTD method is reviewed using a one-dimensional example, and the extensions required for the 3-D case are outlined. The parallelization of the FDTD method is considered, and a simple analytical timing model is derived. This model predicts the efficiency of the parallelized algorithm as a function of grain size. Some specific points relating to the implementation of the parallel FDTD algorithm in Fortran-90 on the CM-2 are discussed. Timing data for the parallel 3-D FDTD code measured on a CM-2 is presented and compared qualitatively with the theoretical model. These results are then put into perspective for a particular computational electromagnetics problem, viz. the development of software tools for full-wave modelling of 3-D optical devices. Finally, we draw some conclusions about this work.

1. INTRODUCTION

The finite difference time-domain (FDTD) method is now widely used in computational electromagnetics and was first introduced by Yee[1] almost three decades ago. The method discretizes the Maxwell equations using a second-order central differencing scheme, with electric and magnetic fields interleaved in time and space on staggered grids. The electric and magnetic field components are successively updated by repeatedly implementing the finite difference equivalents of the curl equations; this process is widely known as leapfrog time-marching. The scheme is an explicit scheme; no matrix inversion is required. Yee's initial work was directed at electromagnetically small problems with closed boundaries, but great strides were made in the following decades. These developments included: absorbing boundary conditions to simulate open boundaries; incident waves of arbitrary time duration and spatial variation; and the availability of very fast and capable computers with very large memory capacities. Subsequently, by the end of the 1980s, the FDTD approach had been applied to a number of large and complex problems. Kunz and Luebbers recently published a comprehensive textbook on the FDTD method and its applications;[2] other recent references on the subject are Chew's textbook[3] which presents a very readable introduction to the method and Taflove's research monograph[4] which provides an overview of more complex applications. Merewether and Fisher's report[5] is a very comprehensive reference, although it may prove difficult to obtain.

Because it discretizes space directly, the FDTD method is very convenient for handling problems with material inhomogeneities, and is especially suitable when wide bandwidth data is required, since it can generate all required frequency data with one run. The drawback with the method is that it requires prodigious computer resources for problems which involve structures that are a large number of wavelengths in size. The computational cost grows at least as the fourth power

Reprinted with permission from *International Journal of Numerical Modeling: Electronic Networks, Devices and Fields,* D. B. Davidson and R. W. Ziolkowski, "A Connection Machine (CM-2) Implementation of a Three-Dimensional Parallel Finite Difference Time Domain Code for Electromagnetic Field Simulation," Vol. 8, pp. 221-232, May-August 1995.

of the number of unknowns along any one dimension, assuming a box-like volume of interest. It will be seen that the method has a very simple 'stencil' update scheme, and the operations are *local* (i.e. restricted to grid points in the vicinity of the point being updated). Hence the FDTD method should be an ideal candidate for parallel processing, in particular using a fairly fine grain. In this paper, we discuss a parallel FDTD code that we have implemented for the Connection Machine CM-2.

The outline of this paper is the following. Firstly, the CM-2 is briefly discussed. Then the FDTD method is reviewed using a one-dimensional example, and the extensions required for the 3-D case are outlined. The parallelization of the FDTD method is considered, and a simple analytical timing model is derived. Then some specific points relating to the implementation of the parallel FDTD algorithm on the CM-2 are discussed, followed by measured timing data. These results are then put into perspective for a particular computational electromagnetics modelling problem, namely the full-wave simulation of optical devices. Finally, we draw some conclusions about this work.

2. PARALLEL PROCESSING AND THE CM-2

The 1980s were a time of intensive experimentation with parallel processing, resulting in the marketing of a number of innovative machines and microprocessors. The major architectural issue of the period was the debate between the exponents of the single instruction, muliple data (SIMD) and multiple instruction, multiple data (MIMD) paradigms. (In essence, the former applies one operation concurrently to a number of items of data; in the latter the operations performed concurrently may be different.) The CM-2 epitomizes the former, and INMOS transputer arrays the latter. The first author has had experience on transputer arrays for parallel moment-method codes; details may be found in References 6–8. The first of these references contains much tutorial material on parallel processing, which will not be repeated here.

The Connection Machine CM-2 was introduced in 1987. (At the time of writing, the machines were starting to be withdrawn from service, demonstrating the rapid pace of technological change in high-speed computation.) The CM-2 is a massively parallel bit-serial processor, with a maximum of 65,536 1-bit processors. Floating-point hardware is also present on a CM-2. The basic architecture is as follows. The 1-bit processors are grouped 16 per chip. Two of these chips, plus memory and a floating-point accelerator chip, make up a block, 16 blocks make a board and 16 boards make an octant with 8k (8192) 1-bit processors (or 256 32-bit floating-point processors). Eight octants make up a full CM-2. For most scientific work, the floating-point mode is the most useful.

The processors are connected via a *hypercube* topology; an example of a three-dimensional hyper-cube is given in Figure 1. The CM-2 has a twelve-dimensional hypercube, with one chip at each node. Physically, a CM-2 is a cube about 1·5 m on a side. Overviews of the CM-2 may be found in a number of recent texts on computing, an example being Reference 9, pp. 340–346. For a more comprehensive discussion of the CM-2, with a fairly extensive list of references, the reader is referred to Reference 10.

Users can access octants in a 'space-shared' mode—i.e. a user has *exclusive* access to one octant (or more)—although time-shared modes on particular octants are normally provided for testing and debugging. A full CM-2 has a peak rated throughput of 14–28 Gflop/s (different sources

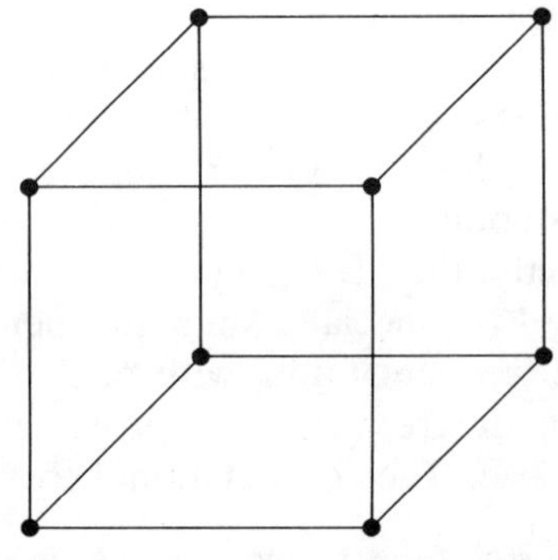

Figure 1. A hypercube of dimension three

give different figures), and has a maximum memory capacity of 2 gigabytes. Throughput as used in this paper is simply the floating-point computation rate in million floating-point operations per second (Mflop/s) or billion (10^9) floating-point operations per second (Gflop/s).

The CM-2 should be ideal for applications emphasizing local referencing, such as the FDTD algorithm. In the next section, this algorithm is reviewed prior to considering parallelizing it.

3. THE FDTD METHOD

To illustrate the basic operation of the FDTD code, a one-dimensional example will be presented; although simple, it conveys the flavour of the Yee algorithm. We consider a wave with only E_x and H_y components (such as a z propagating TEM wave), and write the relevant components of the Maxwell equations, assuming a non-dispersive medium:

$$\frac{\partial E_x(z,t)}{\partial t} = -\frac{1}{\epsilon(z)}\frac{\partial H_y(z,t)}{\partial z} \tag{1}$$

$$\frac{\partial H_y(z,t)}{\partial t} = -\frac{1}{\mu(z)}\frac{\partial E_x(z,t)}{\partial z} \tag{2}$$

The non-dispersive (i.e. frequency-independent) assumption is necessary to permit writing the permitivity ϵ and permeability μ in the simple time-independent form shown above. (The FDTD formulation has been extended fairly recently to include: linear dispersive media (see Reference 2, Chapter 8 for details); and nonlinear dispersive materials (see, for instance, Reference 11).)

Now the differential operators are approximated using central differencing, viz.

$$\frac{\partial E_x(z,t)}{\partial z} \approx \frac{E_x(z + \Delta z/2,t) - E_x(z - \Delta z/2,t)}{\Delta z} \tag{3}$$

for the spatial differential and a similar expression for the temporal differential. This now introduces a complication of the Yee scheme: the central differencing implies that the fields are evaluated at locations and times offset in both space and time. This can be easily understood by referring to equation (1); if E_x is evaluated at $k\ \Delta z$, then the values of H_y (required for the partial derivative in z) must be taken at $(k - \frac{1}{2})\ \Delta z$ and $(k + \frac{1}{2})\ \Delta z$, and the same applies to t.

The result is a set of update expressions for the discretized field components:

$$E_x(k,n) = E_x(k,n-1) - \frac{\Delta t}{\epsilon(k)} \times \left[\frac{H_y\left(k+\frac{1}{2},n-\frac{1}{2}\right) - H_y\left(k-\frac{1}{2},n-\frac{1}{2}\right)}{\Delta z}\right] \tag{4}$$

$$H_y\left(k+\frac{1}{2},n+\frac{1}{2}\right) = H_y\left(k+\frac{1}{2},n-\frac{1}{2}\right) - \frac{\Delta t}{\mu(k)} \times \left[\frac{E_x(k+1,n) - E_x(k,n)}{\Delta z}\right] \tag{5}$$

where field component $E_x(k,n)$ is evaluated at $z = k\ \Delta z$, $t = n\ \Delta t$ etc. Equations (4) and (5) are often referred to as the 'stencil'. Note the time and space offsets already discussed above. The algorithm is an initial value procedure; a source is introduced into the domain and **E** and **H** are repetitively updated, first E_x and then H_y, for all the discretized locations on the z-axis, for sufficient time samples to capture the response of interest. For computer coding, the time and space offsets are understood implicitly and $H_y(k + \frac{1}{2}, n + \frac{1}{2})$ is written simply as HY(K,N). Although the arrays EX and HY may each be stored in a two-dimensional array, it is customary to rather use two one-dimensional arrays for each field, one for the 'future' values (index n in equation 4, $n + \frac{1}{2}$ in equation (5)), and one for the 'present' values (index $n - 1$ in equation (4), $n - \frac{1}{2}$ in equation (5)). (The use of 'future' and 'present' here is rather arbitrary; 'present' and 'past' respectively would also be quite correct.) These are then overwritten at the end of each timestep with the new updates. This reduces the storage requirements by a dimension, and the savings become progressively more important as the physical dimensionality of the problem grows.

The timestep size cannot be made arbitrarily large since the time-marching becomes unstable. The upper limit on the temporal step-size is the well-known Courant–Friedrichs–Levy stability criterion,[3,2] namely

$$\Delta t < \Delta s/(v_{\max}\sqrt{n}) \tag{6}$$

where n is the dimensionality of the problem, Δs is the grid size (assumed uniform) and $v_{\max}$ the maximum speed of the wave in the mesh.

This development generates an *explicit* scheme, so called because the future update depends only upon the present (and possibly also past) values. (*Implicit* schemes couple the future values, requiring the solution of a system of linear equations at each timestep. However, the stability requirements are very different and implicit schemes can use much larger timesteps).

For the above one-dimensional formulation, the computational cost is easily calculated. Assume that the z-axis is sampled at N points, and the calculations are repeated for M timesteps. Two arrays each of length N are required and the computational cost is $O(NM)$. Since the stability requirement couples M and N linearly, the cost can also be expressed as $O(N^2)$.

The full Yee scheme is based on precisely the above ideas, but applied in three dimensions. The full set of equations becomes lengthy to write, although they are inherently simple. The details may be found in References 2, Chapter 2, and 5. An example of the Fortran code will be given later for one of the components.

The mesh size Δs fundamentally determines the accuracy of the method. It has been widely accepted that a minimum of around eight to ten samples per wavelength is necessary for adequate representation of the field amplitude and phase,[2] and to control the mesh dispersion. (The FDTD mesh is inherently dispersive, viz. different frequency components move at different frequencies.) This widely used heuristic—rule of thumb—in the high-frequency computational electromagnetic community was established when problems spanned only a few wavelengths. Using elegant reasoning based on the solution of different equations, Cangellaris and Lee have recently shown it to be valid only for electromagnetically small structures;[12] as the electromagnetic size of the structure increases, so does the required number of samples per wavelength. This is required to contain the grid dispersion errors, which become significant for large structures.

4. PARALLELIZING THE FDTD ALGORITHM

The natural parallelism inherent in this algorithm is ready apparent and does not require the tools of formal methodologies to extract it. By inspection of equations (4) and (5), it can be seen that all elements of array $E_x(k,n)$ may be updated in parallel (i.e. the operations for spatial index k may be performed in any sequence), although each overall update must still of course be performed sequentially (i.e. temporal index n must be computed in the ordered sequence $1,2,\ldots,n_{\max}$, which corresponds simply to increasing time). So assuming sufficient processors P, the lower bound on the cost of the parallel algorithm is $O(M)$.

This is of course an over-simplified analysis; it ignores the *communication* cost associated with the algorithm. Referring again to equation (4), we note that for the next update, the processors need the information from adjacent processors to proceed. Assume that $H_y(k + \frac{1}{2},n - \frac{1}{2})$, $H_y(k + \frac{1}{2},n + \frac{1}{2})$, $E_x(k,n - 1)$ and $E_x(k,n)$ are stored on processor k. (This is not the only possible mapping.) To compute $E_x(k,n)$ from equation (4), processor k will need information from processor $k - 1$ on the 'present' value of $H_y(k - \frac{1}{2},n - \frac{1}{2})$, and similarly to compute $H_y(k + \frac{1}{2},n + \frac{1}{2})$ from equation (5), processor k will need information from processor $k + 1$ on the 'present' value of $E_y(k + 1,n)$.

Counting the floating-point operations, each update of equation (4) requires two additions/

subtractions, one multiplication and two divisions. The sign reversal is ignored since this is very fast operation. (The operation count can be reduced by computing the term $(\Delta t/(\epsilon(k)\Delta z)$ beforehand, since it is a constant for all n; if this is done, then the operation count is two additions/subtractions and one multiplication). The communication count per update is one floating-point communication from the nearest neighbour (assuming that the processor interconnection topology is connected such that nearest neighbours on a Cartesian grid remain nearest neighbours on the interconnection topology). On most purpose-built parallel systems, the times for a basic floating-point operation and a nearest-neighbour communication are of the same order. (As an example, on a typical T800 transputer array, the ratio of communication to computation is about 1·6 for real single-precision data.)

From this it can be seen that this parallel FDTD algorithm utilizing the maximum possible number of processors has comparable amounts of computation and communication. Hence the efficiency cannot be expected to be very high for this case of maximum parallelism. (Efficiency will be defined formally shortly.) However, for most real applications, the number of spatial samples is likely to exceed the number of processors, and this becomes more pronounced as the dimensionality of the problem increases. Hence the data must be distributed across the processor array. This process is often known as domain decomposition.[13] Only the values of **E** and **H** on the processor boundaries now need be communicated to the neighbouring processors, so the dimensionality of the communication tends to one less than the dimensionality of the problem as the grain of the problem (the number of sample points per processor) increases. One expects that such a decomposition should be quite efficient; this will be confirmed in the following analysis.

In three dimensions, one way of doing the domain decomposition is by dividing the computational volume equally into smaller cubic volumes. With this decomposition, the communication required is across the *surface* demarcating data sets on the different processors, whereas the computation is a function of the *volume* of each data set. To formalize this, consider a cubic grid of size $N \times N \times N = N^3$, running on a processor array with $\sqrt[3]{P} \times \sqrt[3]{P} \times \sqrt[3]{P} = P$ processors. (For simplicity, it is assumed here that P is some convenient integral cube.) The amount of computation per time update on a serial machine is

$$T_s = \alpha_1 N^3 t_{calc} \tag{7}$$

where α_1 is a constant for the FDTD algorithm (about 50 for the 3-D code as implemented) and t_{calc} is the time for a real floating-point operation. (This is of course approximate since an addition/subtraction and a multiply take similar amounts of time, whereas a division can take somewhat longer.)

The amount of computation per time update on a parallel machine is

$$T_p = \frac{\alpha_1 N^3}{P} t_{calc} \tag{8}$$

to which must be added the communication time

$$T_c = \frac{\alpha_2 N^2}{\sqrt[2/3]{P}} t_{comm} \tag{9}$$

where α_2 is a constant for the FDTD algorithm (12 for a 3-D code) and t_{comm} is the time for a nearest-neighbour communication of a simple real floating-point number. This result follows very simply by noting that the surface of each small cubic volume is $(N/\sqrt[3]{P})^2$; the factor of 12 arises from communicating one field component of both **E** and **H** over each of the six faces.

Hence an expression for the efficiency ϵ (the ratio of serial to parallel execution times, normalized by the number of processors; see Reference 6 for further discussion), can be expected to have the following form:

$$\epsilon = \frac{T_s}{P(T_p + T_c)} = \left[1 + \beta \frac{\alpha_2}{\alpha_1} \frac{\sqrt[3]{P}}{N}\right]^{-1} \tag{10}$$

where β is the ratio $t_{\text{comm}}/t_{\text{calc}}$. Recognizing N^3/P as the grain size n, the above can be rewritten as

$$\epsilon = \left[1 + \beta \frac{\alpha_2}{\alpha_1} \frac{1}{\sqrt[3]{n}}\right]^{-1} \tag{11}$$

The term in n is also known as the fractional communication overhead. Note the asymptotic behaviour as $n \to \infty$; $\epsilon \to 1$. This is as expected, given that the amount of computation increases as the volume whereas the amount of communication increases as the surface.

In general, one can expect a synchronization overhead to appear in an expression for ϵ, but with a SIMD system and the parallel algorithm outlined here, the algorithm has no synchronization problems.

Another possibility is a $\sqrt{P} \times \sqrt{P} \times 1$ decomposition. This is effectively a two-dimensional decomposition, with all the relevant z elements of the 3-D arrays held locally on each processor. Instead of the previous cubes of size $(N/\sqrt[3]{P}) \times (N/\sqrt[3]{P}) \times (N/\sqrt[3]{P})$, we now have 'bricks' of size $(N/\sqrt{P}) \times (N/\sqrt{P}) \times (N/1)$. (The volumes of the cubes and bricks are of course the same.) This decomposition might be expected to improve the speed of operations involving the z indices, since no communication is now required for such operations, although the amount of communication in x and y can be expected to increase. This decomposition is considered here theoretically since it was found it practice to result in a higher throughput when the entire code, and not just the FDTD updates, was considered—this will be discussed in sections 5 and 6. All we need change from our previous analysis is to modify equation (9), which becomes

$$T_c = \alpha_2' N^2/\sqrt{P} t_{\text{comm}} \tag{12}$$

since the communication is now across the surfaces of size $N \times N/\sqrt{P}$; α_2' is reduced to 8 since all z indices of each 'block' of **E** and **H** are now contained entirely on a processor. The efficiency of this algorithm is

$$\epsilon = \left[1 + \beta \frac{\alpha_2'}{\alpha_1} \frac{\sqrt{P}}{N}\right]^{-1} \tag{13}$$

and rewriting as a function of the grain n yields:

$$\epsilon = \left[1 + \beta \frac{\alpha_2'}{\alpha_1} \frac{\sqrt[6]{P}}{\sqrt[3]{n}}\right]^{-1} \tag{14}$$

The efficiencies predicted by equations (11) and (14) can be expected to be very similar; the $\sqrt[6]{P}$ term is a very weak function of P. If all the penalties imposed by the parallelization have been accounted for, the throughput of the system should be simply the efficiency multiplied by the maximum rated throughput, and previous work on transputer arrays showed good agreement between predicted and measured results.[6–8] Low-level technical data for computation and communication speed was available for the transputer. However, the necessary data on communication speed was not available for the CM-2. The basic thrust of this section has been to show mathematically the fundamental suitability of the FDTD algorithm for parallelization, rather than make specific numerical predictions. We will return to this point in section 6.

An analysis of a parallelized two-dimensional finite difference solution of Laplace's equation may be found in Reference 13, Chapter 7. The basic approach is similar to that presented here.

This completes the exposition and analysis of the parallel FDTD algorithm. In the next section details specific to the implementation on the CM-2 will be discussed.

```
FORTRAN 77 code:
      DO 10 I =1,1000
        A(I) = B(I) + C(I)
10    CONTINUE

FORTRAN 90 equivalent:
      A = B + C
```

Figure 2. A comparison of the Fortran-77 and -90 operations to add two arrays, demonstrating the array operations in Fortran 90

5. IMPLEMENTING THE FDTD ALGORITHM ON A CM-2

For floating-point operations, the CM-2 supports a subset of the full Fortran-90 array operations. Amongst the many significant features of Fortran-90 is explicit support for an array type and hence array operations. An example is shown in Figure 2. The simplest method to implement finite differencing is using the Fortran-90 shift operations, shown in Figure 3.

It has already been mentioned that the basic interconnection topology of the machine is a twelve-dimensional hypercube. It may be shown that lower-order Cartesian grids can be embedded therein in a nearest-neighbour preserving fashion (Reference 13, pp. 260–261). This is important for a finite difference code.

The parallel arrays must be explicitly allocated to the CM-2. The CM-2 is accessed via a front-end processor (a Sun 4-47 workstation for the specific CM-2 used), and data not explicitly assigned to the CM-2 resides on the front-end. The data allocation is managed via the LAYOUT command. The default distribution is called NEWS, and a 3-D array is laid out distributed evenly along each axis (the $\sqrt[3]{P} \times \sqrt[3]{P} \times \sqrt[3]{P}$ decomposition). Some experimentation revealed that this was *not* the optimal distribution for the overall code, and better results were obtained by laying the 3-D arrays out using LAYOUT HX(:NEWS, :NEWS, :SERIAL) instead. This is the $\sqrt{P} \times \sqrt{P} \times 1$ distribution discussed in the preceding section. One index is thus retained as local on a processor. Timing results will be presented and discussed in the next section.

The Connection Machine supports some specialized subroutines to implement the typical stencils required by finite differencing. These are the POLYSHIFT routines, and they implement a concurrent multidimensional shift of data. Referring to Figure 3, it will be noted that the HX update requires a shift of EZ along dimension 2 by 1 and similarly of EY along dimension 3 by 1. An example of the use of these routines is shown in Figure 4.

Array operations can require substantial overheads in temporary storage: for instance, the example shown in Figure 3 requires temporary storage for at least the shifted versions of EZ and EY. The strategy for the allocation of temporary storage is not very clear from the CM-2 documentation. This is an unavoidable additional memory overhead incurred by the use of the array constructs.

```
FORTRAN 77 code for Hx update:
      DO 10 I=1,ICAN
        DO 20 J=1,JCAN
          DO 30 K=1,KCAN
**          Hx
            TEMP1=(EZ(I,J+1,K)-EZ(I,J,K)) / DELTAY
            TEMP2=(EY(I,J,K+1)-EY(I,J,K)) / DELTAZ
            HX(I,J,K) = HX(I,J,K) + DELTAT/MU*(-TEMP1+TEMP2)
30        CONTINUE
20      CONTINUE
10    CONTINUE

FORTRAN 90 equivalent:
      HX = HX + DELTAT/MU * (
     &      - (EOSHIFT(EZ,2,1)-EZ) / DELTAY
     &      + (EOSHIFT(EY,3,1)-EY) / DELTAZ)
```

Figure 3. Fortran-77 and Fortran-90 code for one of the field component updates of a 3-D code

```
FORTRAN 90 code for HX update:
      INVDELTAY = 1./DELTAY
      INVDELTAZ = 1./DELTAZ

      CALL PSHIFT(2,SHIFTHX,IER,CMSSL_EOSHIFT_0,TMP1,EZ,2,1,
     & CMSSL_EOSHIFT_0,TMP2,EY,3,1)
      HX = HX + DELTAT/MU * (
     &      - (TMP1-EZ) * INVDELTAY
     &      + (TMP2-EY) * INVDELTAZ)
```

Figure 4. The use of POLYSHIFT routines. In this case, the scalars $1/\Delta y$ and $1/\Delta z$ are computed *before* the 'stencils' are applied, to replace the slower division operation by a multiplication

6. TIMING RESULTS

Timing data reported in this section was computed on one octant of the CM-2 at the Pittsburgh Supercomputer Centre. (It was possible to reserve time for up to four octants on the system, but the tests reported here were run on only one octant owing to a limited CPU time allocation.) The actual physical problem addressed will be described in section 7. The peak throughput is expected to be about 1·75 Gflop/s. Results are shown in Table I. The timing data is for processor time, not elapsed (wall-clock) time. The initial data distribution scheme referred to in this table uses LAYOUT(:NEWS, :NEWS, :NEWS) (the default layout) to generate approximately the $\sqrt[3]{P} \times \sqrt[3]{P} \times \sqrt[3]{P}$ decomposition ($\sqrt[3]{256}$ is not an integer, hence the approximation). The 'improved' data distribution scheme uses LAYOUT(:NEWS, :NEWS, :SERIAL) to generate the $\sqrt{P} \times \sqrt{P} \times 1$ decomposition. (Note that :NEWS is *not* a dimension, it is an instruction regarding the mapping of the 3-D array to the hypercube.) For Table I, the problem size was $120 \times 120 \times 240$; the corresponding grain size was 13,500. The amount of storage explicitly required in single precision for the three components of both **E** and **H** was approximately 83 Mbyte. The memory available per octant was specified as 256 Mbyte, but we were unable to run larger problems. (We speculate that this may have been due to temporary arrays which are not explicitly dimensioned in the code, as discussed in section 5, or alternatively due to limitations imposed by the queueing software.)

In section 4, the theoretical efficiencies for the different data decompositions have been presented, and it has been shown theoretically that one should expect similar values of ϵ for the two cases shown. For the 256 floating-point processors available on the octant, $\sqrt[6]{256} \approx 2{\cdot}5$, so as far as the fractional communication overhead term in equations (11) and (14) is concerned, the decrease in α_2 from 12 to 8 is more than outweighed by the increase due the $\sqrt[6]{P}$ term. In comparing the theoretical timing model to the results in Table I one should remember that the theoretical model is *only* applicable to the actual FDTD updates; these correspond to the 'peak' results in Table I. The theoretical model does indeed correctly predict that the $\sqrt[3]{P} \times \sqrt[3]{P} \times \sqrt[3]{P}$ decomposition will be superior to the $\sqrt{P} \times \sqrt{P} \times 1$ approach for the FDTD 'core' of the code. There are other requirements on the parallel code, including handling absorbing boundaries and injecting the source (driving) function; these aspects of the code are not included in the theoretical model and account for the different trends regarding the average through-puts.

Table I. Timing data on one octant of a CM-2. The problem size was $120 \times 120 \times 240$; grain size 13,500. Refer to text for further details

	Initial code		Improved data distribution	
	Mflop/s	Percentage of peak specs	Mflop/s	Percentage of peak specs
Peak	85	4·9	77	4·4
Average	24	1·4	51	2·9

Timing data for the multidimensional shift operations are given in Table II. The columns headed 'Normal shift ops' use a number of one-dimensional shift calls, as in Figure 3; the columns headed 'POLYSHIFTS' use the multidimensional concurrent shifts shown in Figure 4. Both use the $\sqrt{P} \times \sqrt{P} \times 1$ decomposition. The supposedly 'improved' code is now actually running more slowly than the original code which simply calls a one-dimensional shift as needed. Note that the problem size is slightly smaller than that of Table I; as predicted by the theoretical model, the peak throughput increases as the grain size increases.

Aside from the issue of the data decomposition and 'optimization' procedures, there is the fundamental, rather disappointing, result of the percentage of peak throughput obtained. It is of course quite usual to obtain only some fraction of the peak throughput rated by the manufacturer, but these values are only a few per cent of the peak rate. Some users compute a throughput based on counting floating-point operations and clock cycles in the CM-2 assembler, but quite apart from the difficulty involved in this, our approach seems more honest in that it is based on the speed that the user actually sees. Our timing results do include a substantial amount of non-floating-point operations, such as those required to perform the `SHIFT` operations, but it seems reasonable to include these in the timings since they are absent in a comparable serial code.

It is interesting to compare the general timing results with previous work by one of us (DBD) using transputer arrays for boundary element frequency-domain calculations, where efficiencies of 70 to 80 per cent were obtained.[6–8] Efficiency as used in section 4 and in References 6–8 relates speed on one processor to speed on many processors, and does not take into account what one might term *scalar* efficiency, namely the percentage of peak rated throughput obtainable on *one* processor. For the transputer arrays, scalar efficiency was around 33 per cent, so overall these arrays ran at around 25 per cent of their claimed peak throughput. However, the transputer systems involved at most 32 processors, and it is generally easier to get good efficiencies from smaller parallel arrays. Furthermore, the transputer results were obtained with hand-crafted Occam code, permitting very low-level and hence efficient use of the transputer's parallel features. (The performance degradation in the case of the transputers was primarily due to the peak rate being based on the program and the data being stored entirely on the fast on-chip 4kB RAM. This scheme would not possible with any finite difference/element scheme with a non-trivial grid size.)

As was mentioned in section 4, low-level technical data was not available on the CM-2, and furthermore, it is not possible to measure the scalar efficiency, since the smallest computational unit of a CM-2 that can be accessed by a user is the octant. Hence, the results for the CM-2 are given in terms of percentage of peak claimed throughput. It is interesting to try to estimate the scalar efficiency using the analysis of section 4. Consider the results shown in Table I. If we assume that $\beta \approx 1$, then with $\alpha_1 = 50$, $\alpha_2 = 12$, and $n = 13{,}500$, equation (11) predicts $\epsilon \approx 99$ per cent. However, the measured overall efficiency is between 4 and 5 per cent. One could conclude that the scalar efficiency is very poor, but the actual reason is more likely to lie in the timing analysis of section 4; the actual communication regime on the CM-2 is rather more complex than that considered here. (See, for example, Johnsson's work[14] for a more detailed communication model.) The disagreement between the present analysis and measurements show clearly the problems posed by massively parallel processors as regards timing predictions. (We note in passing that a similar 3-D FDTD code running on a workstation has a throughput of around 12·8 Mflop/s, close to the peak rated throughput claimed the manufacturer—a Hewlett-Packard 720 workstation in this case.)

Table II. Timing data on one octant of a CM-2. The problem size was 128 × 128 × 128; grain size 8192. Refer to text for further details

	Normal shift ops		POLYSHIFTS	
	Mflop/s	Percentage of peak specs	Mflop/s	Percentage of peak specs
Peak	47	2·7	36	2·1
Average	35	2·0	25	1·4

7. APPLICATIONS TO FULL-WAVE 3-D OPTICAL DEVICE SIMULATIONS

We had intended to use the CM-2 FDTD code for full-wave three-dimensional modelling of optical devices. Following the work reported in the paper, we realized that this aim was premature; the computational cost is simply too high for parallel systems of the CM-2 generation. The code that generated the timing results presented in the previous section was a free-space-only prototype code; the actual problems run are discussed below. We eventually used a different formulation for the actual optical device simulations—the reasons for this will be discussed shortly—hence this paper concentrates on the parallelization aspects rather than the eventual application.

Consider the following. Typical optical devices that might be encountered in fibre-optic systems have dimensions on a side of several micrometres, and typical sources in the near-infrared have wavelengths on the order of 1 μm. This translates into a requirement of approximately a 100 nm *maximum* sample spacing. This means the problem size will involve many millions of sample points, and the numerical complexity can easily approach a trillion (10^{12}) floating-point operations. The largest problem that we have run on the CM-2 used around one-third of a million sample points to discretize a volume of about $20{,}000\lambda^3$ at $\lambda/10$ sampling, using two planes of symmetry to reduce the problem to effectively $5000\lambda^3$. The problem was a Gaussian beam propagating in free space, similar to the case presented in Reference 15, section 5. This problem took about one-half hour on the CM-2 octant.

Whilst moderately satisfactory results were obtained from this $5000\lambda^3$ case, increasing the sample density by only two exceeds the capacity of the computer. Thus even simple convergence tests cannot be performed. Non-linear materials frequently require *much* higher sampling densities.[11] Hence we conclude that given the current state of available massively parallel processors, modelling a realistic optical device in 3-D is on the borderline of practicality: this may well change once new MPPs become available for modelling work.

However, by exploiting rotational symmetry, we were able to develop a full-wave body of revolution (BOR) FDTD formulation; this approach assumes rotational symmetry, and expands the azimuthal dependence of the fields as a Fourier series. Each azimuthal mode in the series is orthogonal, and a two-dimensional FDTD problem is solved for each mode. For a significant class of actual optical device/beam interactions, such as a linearly polarized Gaussian beam interacting with a lens, only one mode is required. This makes the computational and memory requirements of the BOR-FDTD code essentially two-dimensional, and such problems may be run on a variety of more conventional systems. (We used a CONVEX C-240.) This research has been reported in detail in Reference 16 and some novel applications for intensity enhancement using ultra-wideband pulsed waveforms are presented in Reference 17.

8. CONCLUSIONS

The theoretical analysis presented in this paper has demonstrated that FDTD codes can be expected to map well into massively parallel environments. However, the results presented have shown that the reality of a real parallel code running on an actual massively parallel processor unfortunately falls far short of expectations, at least as far as throughput goes. We should state that the issue of benchmarking machines is a very controversial one, and the approach we have used in this paper may well be criticized. However, for engineering purposes, the actual speed of a real problem is the issue of fundamental interest, which we have addressed. It should also be stated that although we have worked on improving the throughput of the code, a skilled CM-2 specialist may well be able to further improve the speed of the code substantially.

This paper has not addressed some of the software tools available on the CM-2 for optimization purposes. The PRISM performance monitoring tool can provide useful data on the computational 'hot-spots' in the code. Especially compared to the very crude tools available for earlier work on transputer arrays, PRISM is an impressive software tool. However, limited use was made of PRISM by the authors, since it became apparent that our original aim of a 3-D FDTD code for full-wave simulation of optical devices was simply too ambitious for the CM-2. This application falls into the class of 'Grand Challenge' applications owing to the multiwavelength optical device sizes; for non-linear simulations, which is where most of the interesting possibilities lie, the computational task is even more challenging.

It is appropriate to end this paper by reviewing recent trends in the supercomputer industry.

The CM-5 has recently been released, and differs substantially from the CM-2. It combines both SIMD and MIMD paradigms, and uses a 'fat tree' interconnection topology rather than a hybercube. (On a fat tree, the inter-processor communications bandwidth increases as one traverses up the tree, as opposed to a conventional tree where the bandwidth remains constant. The upper levels of a conventional tree often become communication bottlenecks.) It would be very interesting to port our CM-2 code to the CM-5, but at the time this research was performed, no CM-5 systems were generally available outside a few national laboratories in the U.S.A. Unfortunately, Thinking Machines Corporation (the manufacturers of the CM-2 and CM-5) sought bankruptcy protection in August 1994, so such tests are unlikely to ever be performed. Another high-performance parallel computer manufacturer, Kendall Square Research, also failed in 1994, fanning speculation about the viability of this specialist industry.

ACKNOWLEDGEMENTS

This paper is based partly on research pursued whilst one of us (DBD) was a visiting scholar at the University of Arizona, on sabbatical leave from the University of Stellenbosch: this leave, and additional funding provided by the latter institution, are gratefully acknowledged. This work was also performed in part by the Lawrence Livermore National Laboratory under the auspices of the U.S. Department of Energy under contract W-7405-ENG-48. The provision of CM-2 time on the (N.S.F.) Pittsburgh Supercomputer Center is gratefully acknowledged.

REFERENCES

1. K. Yee, 'Numerical solution of initial boundary value problems involving Maxwell's equation in isotropic media', *IEEE Trans. Antennas Propagat.*, **AP-14**, 302–307 (1966).
2. K. S. Kunz and R. J. Luebbers, *The Finite Difference Time Domain Method for Electromagnetics*, CRC Press, Boca Raton, FL, 1993.
3. W. C. Chew, *Waves and Fields in Inhomogeneous Media*, Van Nostrand Reinhold, New York, 1990.
4. A. Taflove and K. R. Umashankar, 'The finite-difference time-domain method for numerical modelling of electromagnetic wave interactions with arbitrary structures', in *Finite Element and Finite Difference Methods in Electromagnetic Scattering*, M. A. Morgan (Ed.), Progress in Electromagnetic Research Series Number 2, Elsevier, New York, 1990.
5. D. E. Merewether and R. Fisher, 'Finite difference solution of Maxwell's equation for EMP applications', Defense Nuclear Agency, Washington, DC 20305, Report no. EMA-79-R-4, April 1980.
6. D. B. Davidson, 'A parallel processing tutorial', *IEEE Antennas Propagat. Magazine,* **32**, 6–19 (1990).
7. D. B. Davidson, 'Parallel processing revisited: a second tutorial', *IEEE Antennas Propagat. Magazine,* **34**, 9–21 (1992).
8. D. B. Davidson, 'Parallel matrix solvers for moment method codes for MIMD computers', *Applied Computational Electromagnetics Society Journal,* **8**, 2, 144–175 (1993).
9. H. M. Deitel, *Operating Systems*, 2nd edn, Addison-Wesley, Reading, MA, 1990.
10. B. Boghosian, 'Computational physics on the Connection Machine', *Computers in Physics*, 14–33 (1990).
11. R. W. Ziolkowski and J. B. Judkins, 'Full-wave vector Maxwell equation modelling of the self-focusing of ultrashort optical pulses in a nonlinear Kerr medium exhibiting a finite response time', *J. Opt. Soc. Am. B.,* **10**, 186–198 (1993).
12. A. C. Cangellaris and R. Lee, 'On the accuracy of numerical wave simulations based on finite methods', *Journal of Electromagnetic Waves and Applications,* **6**, 12, 1635–1653 (1992).
13. G. C. Fox, M. A. Johnson, G. A. Lyzenga, S. W. Otto, J. K. Salmon and D. W. Walker, *Solving Problems on Concurrent Processors*, Vol. I: *General Techniques and Regular Problems*, Prentice Hall, Englewood Cliffs, New Jersey, 1988.
14. S. L. Johnsson, 'Performance modelling of distributed memory architectures', *Journal of Parallel and Distributed Computing,* **12**, 300–312 (1991).
15. R. W. Ziolkowski and J. B. Judkins, 'Propagation characteristics of ultrawide-bandwidth pulsed Gaussian beams', *J. Opt. Soc. Am. A,* **9**, 2021–2030 (1992).
16. D. Davidson and R. W. Ziolkowski, 'Body-of-revolution finite-difference time-domain modelling of space-time focusing by a three-dimensional lens', *J. Opt. Soc. Am. A,* **11**, 1471–1490 (1994) (Special issue on 3-D scattering.)
17. R. W. Ziolkowski and D. B. Davidson, 'Designer pulsed-beams for enhanced space-time focusing', *Optics Letters,* **19**, 284–286 (1994).

Finite-Difference Time-Domain Analysis of Microwave Circuit Devices on High Performance Vector/Parallel Computers

STEPHEN D. GEDNEY

Abstract— **In this paper, an efficient finite-difference time-domain algorithm for high performance distributed memory vector/parallel computers is presented. The algorithm is developed in a manner which requires only one interprocessor communication per time step. Illustrated examples based on the analysis of microwave circuit devices are presented demonstrating the efficiency and scalability of the finite-difference time-domain algorithm.**

I. Introduction

The finite-difference time-domain (FDTD) algorithm is an ideal algorithm for both vector processors and distributed memory multiprocessor computers. The kernel of the algorithm is a vector update operation that is readily vectorizable. Employment on distributed memory multiprocessor computers requires limited amounts of message passing between neighboring processors, leading to a highly scalable algorithm. By exploiting the architecture's of high performance vector/parallel computers, the problem sizes that can be currently solved using the FDTD are orders of magnitudes larger than problems that could have been treated only two or three years ago. As a result, the FDTD method has been highly effective for numerous applications, including the modeling of microwave circuit devices [1], [2]. As memory densities and CPU speeds continue to increase at extraordinary rates, so will the potential of this technique.

In this paper a parallel FDTD algorithm is presented. The algorithm is based on a spatial decomposition of the orthogonal lattice structure. It is shown that only one interprocessor communication per time iteration is necessary, leading to higher parallel efficiencies than previous algorithms. Test cases based on a simple 3 dB Wilkinson power divider, and an 8-way power divider are presented. It is thus demonstrated that the parallel algorithm presented herein is scalable, and extremely efficient for parallel platforms.

II. The FDTD Algorithm

The finite-difference time-domain algorithm is based on central difference approximations of the spatial and time derivatives of Maxwell's curl equations. This is achieved by projecting orthogonal components of the electric and magnetic field intensity vectors onto the edges of a dual, staggered, orthogonal grid [3], [4]. By staggering the vector fields both in space and time, a second-order accurate explicit time-marching solution is obtained. This leads to the explicit time-domain solution for the discrete electric and magnetic field vectors [3], [4], e.g., (1) and (2) as shown at the bottom of the next page, where, ε_r, μ_r, and σ represent the averaged relative permittivity and permeability, and the averaged absolute conductivity, respectively, along the grid edges, c_o is the speed of light in free space, and the remaining field components are similarly updated. Note that the electric and magnetic field intensities have been normalized as

$$\vec{e} = \frac{\vec{E}}{\sqrt{\eta_o}}, \qquad \vec{h} = \vec{H}\sqrt{\eta o} \tag{3}$$

where η_o is the free space characteristic wave impedance. The result of this normalization is that the magnitudes of $\vec{e}$ and $\vec{h}$ are of the same order.

The most computationally intensive portion of the FDTD algorithm is the explicit updates of the six vector field components, as expressed in (1) and (2). Each vector update can be implemented using a triple nested loop. A naive implementation of (1) and (2) would require 87 N floating-point operations per time iteration to update the six field components throughout the entire space, where $N = N_x N_y N_z$, and N_x, N_y, and N_z are the grid dimensions. A more traditional approach, is to precompute the material related multipliers, which thus reduces the number of floating-point operations to $45N$ [4]. If the space is assumed to be inhomogeneous in three-dimensions the material related constants are also stored in three-dimensional arrays. Thus, the total memory required to store the six field components and the material constants associated with each of the field components will be $15N$ floating point real numbers. In homogeneous media, or one or two-dimensional inhomogeneities, this memory requirement is relaxed.

The computational time can be further reduced by multiplying the electric and magnetic field intensity vectors by their respective edge lengths, e.g

$$e'_{x_{i+(1/2),j,k}} = e_{x_{i+(1/2),j,k}} \Delta x. \tag{4}$$

Subsequently, (1) and (2) are rewritten as

$$\begin{aligned} e'^{n+(1/2)}_{x_{i+(1/2),j,k}} = {} & D'_{x_{i+(1/2),j,k}} e'^{n-(1/2)}_{x_{i+(1/2),j,k}} + C'_{x_{i+(1/2),j,k}} \\ & \cdot \Big\{ \Big[h'^{n}_{z_{i+(1/2),j+(1/2),k}} - h'^{n}_{z_{i+(1/2),j-(1/2),k}} \Big] \\ & - \Big[h'^{n}_{y_{i+(1/2),j,k+(1/2)}} - h'^{n}_{y_{i+(1/2),j,k-(1/2)}} \Big] \Big\} \end{aligned} \tag{5}$$

and

$$\begin{aligned} h'^{n+1}_{x_{i,j+(1/2),k+(1/2)}} = {} & h'^{n}_{x_{i,j+(1/2),k+(1/2)}} - B'_{x_{i,j+(1/2),k+(1/2)}} \\ & \cdot \Big[e'^{n+(1/2)}_{z_{i,j+1,k+(1/2)}} - e'^{n+(1/2)}_{z_{i,j,k+(1/2)}} \\ & + e'^{n+(1/2)}_{y_{i,j+(1/2),k+1}} - e'^{n+(1/2)}_{y_{i,j+(1/2),k}} \Big] \end{aligned} \tag{6}$$

where

$$\begin{aligned} B'_{x_{i,j+(1/2),k+(1/2)}} &= \frac{c_o \Delta t \, \Delta x}{\Delta y \, \Delta z \, \mu_{i,j+(1/2),k+(1/2)}} \\ D'_{x_{i+(1/2),j,k}} &= \frac{2\varepsilon_{r_{i+(1/2),j,k}} - c_o \Delta t \, \sigma_{i+(1/2),j,k} \eta_o}{2\varepsilon_{r_{i+(1/2),j,k}} + c_o \Delta t \, \sigma_{i+(1/2),j,k} \eta_o}, \\ C'_{x_{i+(1/2),j,k}} &= 2 \Delta x \, c_o \Delta t / \Delta y \, \Delta z [2\varepsilon_{r_{i+(1/2),j,k}} \\ & \quad + c_o \Delta t \, \sigma_{i+(1/2),j,k} \eta_o]. \end{aligned} \tag{7}$$

Similar expressions can be derived for the remaining field components by permuting the indices in a right-handed manner. Based on these expressions, the field updates will require $33N$ floating point operations per time iteration. This results in more than a 60% reduction as compared to a naive approach and a 25% reduction compared to the traditional approach. Furthermore, there are no additional memory

Manuscript received February 7, 1995; revised June 29, 1995.

This work was supported in part by Grant DAAH04-94-G-0243 from the Army Research Office, Research Triangle Park, NC, and by NSF Award ECS-9309179. It was performed in part on the University of Kentucky's 32-node iPSC/860, which was supported by Grants from the National Science Foundation (EMS-9206014), Intel Corp. (iSC022492), and the Army Research Office (DAAH04-93-G-0453).

The author is with the Department of Electrical Engineering, University of Kentucky, Lexington, KY 40506-0046 USA.

IEEE Log Number 9414244.

Reprinted from *IEEE Transactions on Microwave Theory and Techniques*, pp. 2510-2513, October 1995.

constraints as compared to the traditional approach. This technique also has the advantage that special cases can be included directly in the update expressions without the need for additional programming logic in the update loops. For example, within a perfect electrical conductor (PEC), the constants C' within the PEC or on a PEC surface can be set to zero. Subsequently, the electric and magnetic fields within the PEC will remain zero throughout the computation without any additional logic. In addition, passive lumped loads [5] can also be modeled directly within the field updates in (5) and (6) by correctly constructing the coefficients C' and D'. This leads to a reduction in the overall CPU time of the FD-TD solution due to improved vectorization and reduced operations, and also simplifies the programming task.

III. Parallel Implementation on Distributed Memory Parallel Computers

It has been demonstrated that the finite-difference time-domain algorithm is ideally suited for implementation on multiple-instruction multiple data (MIMD) [6], [7] and single-instruction multiple-data (SIMD) [8], [9] high performance parallel computers. This is principally due to the regularity of the dual grid and the even distribution of effort in time and space. The algorithm presented in this section is primarily focused on MIMD type-architectures, such as the Intel iPSC/860, Intel Paragon, or a PVM cluster, but can certainly be implemented on SIMD architectures as well. Unlike the previous algorithms presented in [6]–[9], it will be shown that only one interprocessor communication needs be done per time iteration, which increases the parallel efficiency of the algorithm. It will also be demonstrated that it is extremely important to vectorize the core of the FDTD algorithm to achieve optimal performance on today's high performance computers.

The parallel algorithm is based on a spatial decomposition of the regular grid structure. To this end, the original domain is spatially decomposed into contiguous subdomains [6]–[9]. The subdomains are rectangular in shape, nonoverlapping, sharing common surfaces only, and of equal size. The boundaries, or surfaces, shared by subdomains are chosen by taking slices along edges of the primary grid in the x, y, and z-directions. Thus, the discrete electric fields are tangential and the discrete magnetic fields are normal to the shared boundary surfaces. Each sub domain is then mapped onto independent processors of the parallel computer. The electric fields are updated using (5). Each edge is updated using the magnetic field vectors normal to the four faces sharing this edge. However, on a shared boundary, each processor has in local memory at most three of the four faces needed to update the electric field, and it becomes necessary to retrieve the needed data via interprocessor communication. On the other hand, the magnetic field vectors normal to the shared interface are updated using (6). Since each processor has the updated value of the tangential electric fields on the shared interface, the normal magnetic field can be updated independently on each processor, and interprocessor communication is not needed. Rather, it is more expedient to simply update the normal magnetic fields redundantly on each processor sharing the face.

Tangential electric field vectors on exterior absorbing boundaries must be updated using an absorbing boundary condition (ABC). The choice of an absorbing boundary condition can affect both the accuracy as well as the efficiency of the parallel FDTD algorithm. The widely used Mur's second-order ABC [4] requires transverse derivatives, which will result in additional interprocessor communication on shared boundaries. On the other hand, Liao's ABC [10] or the second-order dispersive boundary condition (DBC) [11], are based on normal derivatives only. Based on the decomposition described above, these operators will not require any additional interprocessor communication, since all the necessary information needed to compute the update is resident on each processor. The only consequence, is that each processor sharing a field value will perform the update redundantly.

IV. Vectorization

Almost all of today's distributed memory parallel computers utilize RISC processors as central processing units (CPU's). Most RISC processors rely on pipelining to achieve maximum floating point operation speeds and high speed cache to reduce memory access time. The optimization of many RISC processors can be achieved in the same manner as one would vectorize a program. Vectorization is realized on the innermost loops of any multi-dimensional loop structure and can be achieved in an optimal manner when: 1) the inner loops are truly vector operations and are not corrupted by function calls, or branch statements, 2) the inner-most column index corresponds to the index of the inner loop, and 3) the length of the inner loop is equal to or greater than the optimal vector length (typically determined by the vector length of a vector processor, or the cache size of a pipelined processor).

The triple loops updating e_x and h_x based on (5) and (6), are implemented as

```
      do 10 k = 2, nz - 1
         do 10 j = 2, ny - 1
            do 10 i = 1, nx - 1
               ex(i, j, k) = dx(i, j, k)*ex(i, j, k) + cx(i, j, k)*
               ·[hz(i, j, k) - hz(i, j - 1, k) - hy(i, j, k)
               + hy(i, j, k - 1)]
10          continue
```

$$e^{n+(1/2)}_{x_{i+(1/2),j,k}} = \frac{\dfrac{\varepsilon_{r_{i+(1/2),j,k}}}{c_o\,\Delta t} - \dfrac{\eta_o\sigma_{i+(1/2),j,k}}{2}}{\dfrac{\varepsilon_{r_{i+(1/2),j,k}}}{c_o\,\Delta t} + \dfrac{\eta_o\sigma_{i+(1/2),j,k}}{2}}\, e^{n-(1/2)}_{x_{i+(1/2),j,k}} + \frac{1}{\dfrac{\varepsilon_{r_{i+(1/2),j,k}}}{c_o\,\Delta t} + \dfrac{\eta_o\sigma_{i+(1/2),j,k}}{2}} \cdot \left\{ \frac{h^n_{z_{i+(1/2),j+(1/2),k}} - h^n_{z_{i+(1/2),j-(1/2),k}}}{\Delta y} - \frac{h^n_{y_{i+(1/2),j,k+(1/2)}} - h^n_{y_{i+(1/2),j,k-(1/2)}}}{\Delta z} \right\} \tag{1}$$

$$h^{n+1}_{x_{i,j+(1/2),k+(1/2)}} = h^n_{x_{i,j+(1/2),k+(1/2)}} - \frac{c_o\,\Delta t}{\mu_{r_{i,j+(1/2),k+(1/2)}}} \cdot \left\{ \frac{e^{n+(1/2)}_{z_{i,j+1,k+(1/2)}} - e^{n+(1/2)}_{z_{i,j,k+(1/2)}}}{\Delta y} - \frac{e^{n+(1/2)}_{y_{i,j+(1/2),k+1}} - e^{n+(1/2)}_{y_{i,j+(1/2),k}}}{\Delta z} \right\} \tag{2}$$

TABLE I
MFLOPS PER PROCESSOR VERSUS INNER LOOP LENGTH ($Ny = 110$, $Nz = 40$)

Nx	i860 (MFLOPS)	Cray-YMP(MFLOPS)
200	20.0	208
128	16.9	213
100	14.9	200
50	12.7	165
25	11.0	100
10	8.1	53.4

```
do 20 k = 1, nz - 1
   do 20 j = 1, ny - 1
      do 20 i = 1, nx
         hx(i, j, k) = hx(i, j, k) + bx(i, j, k)*
         ·[ey(i, j, k + 1) - ey(i, j, k) - ez(i, j + 1, k)
         + ez(i, j, k)]
20    continue.                                          (8)
```

Note that there is no extraneous logic embedded in either of these triple loops. Rather, special conditions such as perfect conductors, impedance boundaries, or lumped loads are modeled directly within the coefficients bx, cx, and dx, preserving the inner vector loops.

The j and k indices of the e_x-update in (8) are looped over the ranges $(2, ny - 1)$ and $(2, nz - 1)$, respectively. These indices are chosen in this manner such that at $j = 1$ or $j = ny$, e_x is either on an absorbing boundary, or a shared boundary if it is an interior partition. There is no reason to introduce additional logic that will upset the vectorization of the inner loop, and these fields are simply updated within a separate subroutine. Also note that h_x is looped over the entire range of i, j, and k. This is due to the fact that all the necessary information is available on each processor to completely update h_x whether it is on a shared boundary or on an absorbing boundary.

Table I illustrates the effects of the inner vector length on the floating point speed. These results were generated using the loops in (8) (case #1) on a single i860 processor and a single Cray-YMP processor. As expected, longer vector lengths result in faster floating point speeds, up to the optimal vector length (which was found to be ~128 for the Cray-YMP, and ~200 for the i860). This leads to the conclusion that better floating point performance is realized by maximizing the length of the inner loop. Unfortunately, as the problem size is decreased the floating point speeds realized will decrease. It will be shown in the following section, that this leads to a degradation in the measured speedups when doing a fixed speedup study.

V. NUMERICAL RESULTS

The parallel FDTD algorithms described in the previous section were implemented on a number of platforms. The programs were initially developed on the 32-node iPSC/860 at the University of Kentucky and then ported directly to the Intel Delta at the California Institute of Technology, as well as a number of sequential computers, including the JPL Cray-YMP. The program has been fully validated through a large number of numerical examples [12].

The purpose of this section is to present the computational efficiency of the vector/parallel algorithm. The numerical results have been obtained using the 512-node Intel Delta and a Cray-YMP, for a benchmark comparison. Each processing node of the Intel Delta hosts an i860 RISC processor with 12 Mb of RAM. The Intel Delta has a high-speed network interconnecting nodes with a two-dimensional mesh topology. The primary advantage of the mesh topology is that the routing hardware is greatly simplified resulting in much larger communication bandwidths (200 Mb/sec). The efficiency of the parallel code will be measured by the *speedup* of the parallel

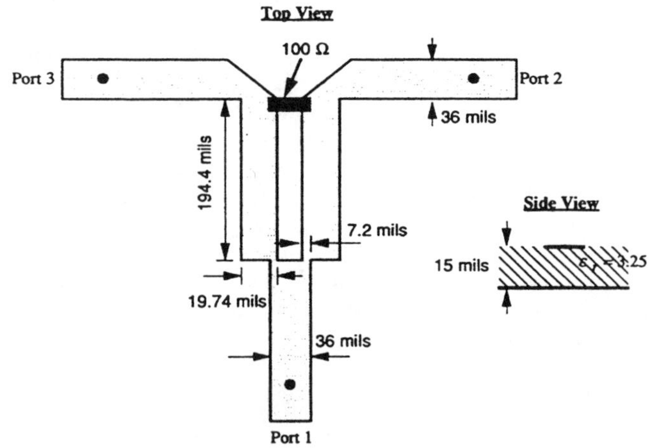

Fig. 1. 3 dB, in-phase Wilkinson power divider printed on a 15 mil TMM substrate ($\varepsilon_r = 3.25$).

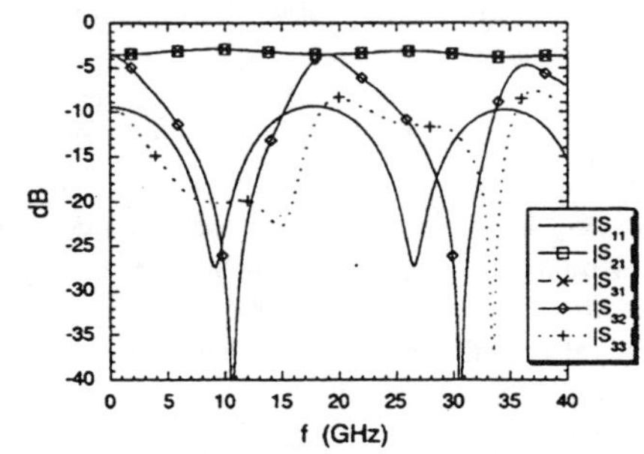

Fig. 2. Magnitude of the S-parameters of the Wilkinson power divider illustrated in Fig. 1.

algorithm. To this end, two types of speedups are referenced: fixed speedup and scaled speedup. Fixed speed up (S_f) is a measure of the speedup of a fixed sized problem as the number of processors is increased, whereas, scaled speedup (S_s) is a measure of the speedup of a problem whose size scales with the number of processors. Finally, the parallel efficiency is expressed as $\eta(P) = S(P)/P \times 100\%$. Speedup is typically used to measure the degree of parallelism in a code, or more specifically, the relationship between parallelism and the serialism in a computer program. Unfortunately, fixed speedups are asymptotically bound by the algorithm serialism, thus the amount of serialism in an algorithm must be very small to obtain reasonable fixed speedups over large numbers of processors. Scaled speedups have the advantage that the problem size can be chosen such that it maximizes the capacity of each processor. Large speedups are expected from this perspective since the percentage of serialism is constant as the problem is scaled.

As an example, the microstrip-line Wilkinson power divider illustrated in Fig. 1 was simulated using the FDTD method. The power divider is printed on a 15 mil TMM substrate ($\varepsilon_r = 3.25$), and is designed for 3 dB, equal phase power division at 9 GHz. For isolation, a 100 Ω chip resistor was placed across the apex of the power divider, which was modeled as a distributed lumped load [5]. The mitered bends were modeled via a staircase approximation of the regular orthogonal grid. Fig. 2 illustrates the S-parameters computed using the parallel FDTD program. This model was based on a $99 \times 151 \times 20$ mesh, where $dx = dy = 3.6$ mils, and $dz = 3.75$ mils. The simulation required 2500 time iterations ($dt = 0.15$ ps).

Table II presents the time required to complete 2500 time-iterations on the Intel Delta versus the number of processors using the proposed

TABLE II
FIXED SPEEDUP STUDY FOR WILKINSON POWER DIVIDER ON THE INTEL DELTA

P	$nx \times ny$	CPU s	$S_f(P)$	$\eta(P)$
1	99 x 151	2327.	1	100 %
2	99 x 76	1187.	1.93	98 %
4	99 x 38	629.3	3.69	92 %
8	99 x 19	337.1	6.90	86 %
16	50 x 19	190.1	12.23	76 %
32	50 x 10	110.8	21.00	65 %
64	25 x 10	69.5	33.48	52 %
128	13 x 10	48.0	48.47	38 %

TABLE III
SCALED SPEEDUP STUDY ON THE INTEL DELTA
(99 ×151× 20 LATTICE PER PROCESSOR)

P	FDTD	$S_s(P)$	$\eta(P)$
1	.930 s	1	–
8	.978 s	7.6	95 %
16	.997 s	14.9	93 %
32	.997 s	29.8	93 %
64	.997 s	59.5	93 %
128	.997 s	119.1	93 %
256	.997 s	237.7	93 %

FDTD algorithm. The benchmarked times are recorded in CPU-seconds, which is identical to the wall clock time. The last two columns of Table II are the fixed speedup and the parallel efficiency, respectively. The speedup recorded for this small problem realizes 38% efficiency over 128 processors. The observed loss of parallel efficiency as the number of processors is increased has multiple causes. Initially, as the original problem is distributed over 128 processors, it reduces from 99 × 151 × 20 on a single processor to roughly 13 × 10 × 20 on each of 128 processors. As a result, the ratio of floating point operations to interprocessor communication greatly decreases. Slight load imbalances also become more predominant for smaller problem sizes since the CPU time is based on the *slowest* processor. Furthermore, as discussed in Section IV, as the problem size decreases, the FLOPS rate per processor decreases. From Table I, the 99 × 151 × 20 sized problem is being performed at 14.9 MFLOPS on the single processor partition, whereas the 12 × 10 × 20 sized problem is being performed at approximately 8.1 MFLOPS per processor on the 128-processor partition. This is a 55% reduction in CPU speed. For benchmark comparison this problem was also executed on a Cray-YMP. This simulation required 413 CPU's on a single processor, which is equivalent to roughly ten i860 processors.

The above example was also simulated using a parallel FDTD method which utilizes the interprocessor communication scheme suggested in [6]–[9]. The field updates were based on (5), (6), and (8) so that the degradation in efficiency as compared to the method described in this paper is only due to additional communication. A speedup of 37 over 128 processors was recorded on the Intel Delta (η = 29%) using this approach, as compared to 48.5 (η = 38%) recorded in Table II. Since the Intel Delta is a tightly coupled distributed computing environment and has relatively small latencies and high network bandwidths, the speedup is only reduced by 25%. However, in a loosely coupled distributed computing environment, such as a workstation cluster, the loss of efficiency will be more catastrophic due to large latencies and low network bandwidths.

A scaled speedup study was also performed on the Intel Delta. This was done by scaling the problem size with the number of processors. Maintaining a 99 × 151 × 20 grid on each processor, the number of processors was scaled from 1–256 (the partitions were kept as square as possible). The CPU times per time iteration and the scaled speedups are recorded in Table III. Excellent speedups are realized. Note that there is an initial jump in CPU time from one to two processors, which is principally due to interprocessor communication. This communication time is again increased for 16 processors, when the central processors in the mesh partition are communicating about all four boundaries. Beyond 16 processors, the scaled speedup remained linear.

A final example is the simulation of a larger device, specifically the 8-way power divider illustrated in Fig. 3. The 8-way power divider consisted of three stages of the Wilkinson power dividers in Fig. 1. The FDTD model was based on a 475 × 343 × 20 mesh, where $dx = dy$ = 3.6 mils, and dz = 3.75 mils. The simulation

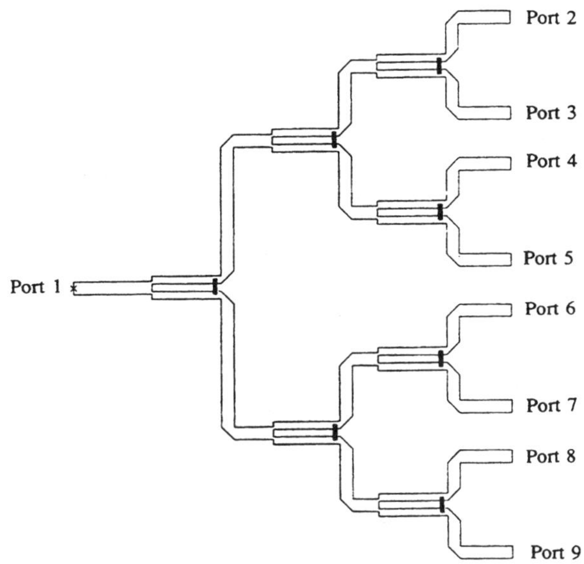

Fig. 3. 8-way Wilkinson power divider network printed on a 15 mil substrate (ε_r = 3.25).

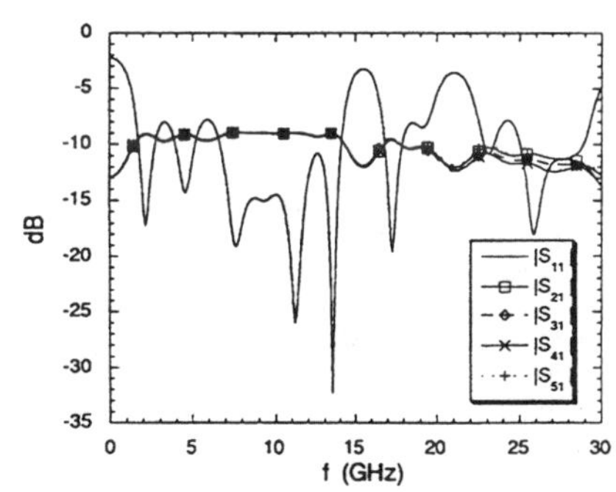

Fig. 4. S-parameters of the power divider illustrated in Fig. 3. (a) Magnitude of the S-parameters. [By symmetry $S_{21} = S_{91}$, $S_{31} = S_{81}$, $S_{41} = S_{71}$, and $S_{51} = S_{61}$.]

required 6000 time iterations (dt = 0.15 ps), which required 3700 CPU's on 16 processors of the Delta (1950 CPU's on 32 processors). The amplitudes of the S-parameters computed for this device are illustrated in Fig. 4. It is demonstrated that −9 dB power division is achieved at each output port at resonance. The ports were also of equal phase over the spectrum, although this is not illustrated here.

VI. CONCLUSION

In this paper, a parallel FDTD algorithm was presented for the analysis of microwave circuits and devices. The algorithm, which requires only one interprocessor communication per time iteration, results in high parallel efficiencies. Benchmarked analyses presenting the fixed and scaled speedups of the parallel FDTD algorithm were provided, demonstrating the efficiency and the scalability of the algorithm on the highly parallel Intel Delta. The existing parallel FDTD algorithm is capable of supporting upwards of one billion degrees of freedom on the Intel Delta, and is capable of solving a problem of this magnitude in quite reasonable amounts of time. With the rate of advances of RISC processors and dynamic random access memory, high performance computers will be capable of handling 10's of billions of degrees of freedom in a fraction of the time using highly scalable algorithms such as the FDTD in the near future.

ACKNOWLEDGMENT

This research was performed in part using the Intel Touchstone Delta System operated by the California Institute of Technology on behalf of the Concurrent Supercomputing Consortium. Access to this facility as well as to the Cray-YMP was provided by the Jet Propulsion Laboratory, Pasadena, CA.

REFERENCES

[1] D. Sheen, S. Ali, M. Abouzahra, and J. A. Kong, "Application of the three-dimensional finite-difference time-domain method to the analysis of planar microstrip circuits," *IEEE Trans. Microwave Theory Tech.*, vol. 38, pp. 849–856, July 1990.

[2] X. Zhang, J. Fang, K. Mei, and W. Lin, "Calculation of the dispersive characteristics of microstrips by a time-domain finite-difference method," *IEEE Trans. Microwave Theory Tech.*, vol. 36, pp. 263–267, Jan. 1988.

[3] A. Taflove and M. E. Brodwin, "Numerical solution of steady-state electromagnetic scattering problems using the time-dependent Maxwell's equations," *IEEE Trans. Microwave Theory Tech.*, vol. 23, pp. 623–630, Aug. 1975.

[4] K. Kunz and R. Luebbers, *The Finite-Difference Time-Domain Method for Electromagnetics.* Boca Raton, FL: CRC Press., 1993.

[5] M. Piket-May, A. Taflove, and J. Baron, "FD-TD modeling of digital signal propagation in 3-D circuits with passive and active loads," *IEEE Trans. Microwave Theory Tech.*, vol. 42, pp. 1514–1523, Aug. 1994.

[6] J. Patterson, T. Cwik, R. Ferraro, N. Jacobi, P. Liewer, T. Lockhart, G. Lyzenga, J. Parker, and D. Simoni, "Parallel computation applied to electromagnetic scattering and radiation analysis," *Electromagnetics,* vol. 10, pp. 21–40, Jan.–June 1990.

[7] R. D. Ferraro, "Solving partial differential equations for electromagnetic scattering problems on coarse-grained concurrent computers," in *Computational Electromagnetics and Supercomputer Architecture,* T. Cwik and J. Patterson, Eds. vol. 7; also in *Progress in Electromagnetics Research,* J. A. Kong, Ed. Cambridge, MA: EMW Publishing, 1993, vol. 7, pp. 111–154.

[8] A. Perlik, T. Opsahl, and A. Taflove, "Predicting scattering of electromagnetic fields using the FD-TD on a connection machine," *IEEE Trans. Magn.*, vol. 25, pp. 2910–2912, July 1989.

[9] A. Perlik and S. Moraites, "Electromagnetic wave analysis using FD-TD and its implementation on the connection machine," in *Computational Electromagnetics and Supercomputer Architecture,* M. Morgan, Ed., vol. 7; also in *Progress in Electromagnetics Research,* J. A. Kong, Ed. Cambridge, MA: EMW Publishing, 1993, vol. 7, pp. 266–308.

[10] W. C. Chew, *Waves and Fields in Inhomogeneous Media.* New York: Van Nostrand, 1990.

[11] V. Betz and R. Mittra, "Comparison and evaluation of boundary conditions for the absorption of guided waves in an FDTD simulation," *IEEE Microwave and Guided Wave Lett.*, vol. 2, pp. 499–401, Dec. 1992.

[12] Stephen D. Gedney, "Finite-difference time-domain analysis of microwave circuit devices on high performance parallel computers," Univ. of Kentucky, Lexington, KY, Elec. Eng. Tech. Rep. EE-1-94, Feb. 1994.

Parallel FDTD Simulator for MIMD Computers

U. EFFING, W. KUMPEL AND I. WOLFF
DEPARTMENT OF ELECTRICAL ENGINEERING (FB9/ATE)
GERHARD-MERCATOR-UNIVERSITY—GH DUISBURG, BISMARCKSTRASSE 81
D-47048 DUISBURG, GERMANY

SUMMARY

A parallel implementation of the finite difference time-domain algorithm for the analysis of planar microwave circuits is presented in this paper. For this the FDTD mesh is divided in equal-sized subspaces. Every processor of the parallel computer calculates the field in one subspace. Furthermore detailed efficiency investigations are made to predict the timing performance of the parallel FDTD simulator.

INTRODUCTION

The finite difference time-domain method (FDTD)[1,2] is a very flexible and powerful approach for three-dimensional electromagnetic field analysis with a large numerical but a low analytical expense. Owing to the large amount of CPU time and memory required, only small subcircuits or single line discontinuities can be simulated in an acceptable time at this time. The coupling of several (M)MIC components via the electromagnetic field is not neglectable in every case. Thus a full-wave analysis of a large coherent circuit is desirable to get accurate simulation results.

The use of large parallel computers seems to be very promising because the scalable properties of these systems allow to achieve potentially any performance level by adding more processors.[3–5] For the development of the parallel FDTD simulator a transputer system consisting of eight T800 processors was used in conjunction with the operating system Helios™ and the programming language C. Large field problems should be solved by using a modern parallel computer based on PowerPC or i860 processors. The concept of the parallelization and the efficiency investigations of this work are also usable for other modern parallel computers based on the MIMD (multiple instruction, multiple data) principle.

PARALLELIZATION CONCEPT

To distribute the problem as evenly as possible over all N available processors, the FDTD mesh is divided by space dividing planes in subspaces which are nearly equal in size (Figure 1). The

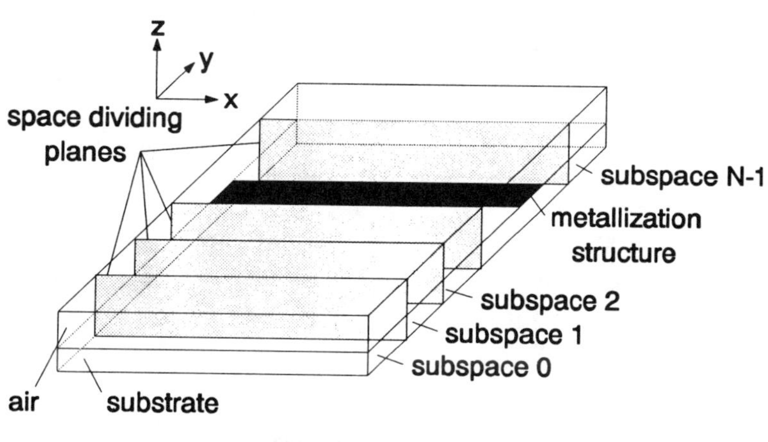

Figure 1. Dividing of the field into subspaces

Reprinted with permission from *International Journal of Numerical Modeling: Electronic Networks, Devices and Fields,* U. Effing, W. Kumpel and I. Wolff, "Parallel FDTD Simulator for MIMD Computers," Vol. 8, pp. 277-281, May-August 1995.

y-axis of the co-ordinate system is perpendicular to all space-dividing planes. The subspaces are rectangular prisms.

The organization of the program tasks is shown in Figure 2.[3] The program consists of one manager-task and N worker-tasks in which on every available processor exactly one worker-task is run. The communication channels (pipes) are represented by arrows in Figure 2. The task force manager of Helios places the pipes on the corresponding transputer links.

The manager task does all central jobs such as reading the information describing the microwave structure from the fileserver, determination of the field dividing and initialization of the worker-tasks. After successful initialization the manager starts the calculation and waits for the terminating signals of the workers. Thus the manager can run together with a worker on one processor without slowing down the calculation of the worker-task.

The worker-tasks include the FDTD algorithm. Every worker calculates the field distribution in one subspace without external control of the manager, completely independent like a serial program. Thus all subfields are simulated at the same time on different processors. The worker gets all necessary information about the microwave structure in the subfield during the initialization from the manager. At every timestep the workers have to communicate with their neighbour worker to exchange the field in the space dividing planes. For the calculation of the magnetic field in subspace $\nu + 1$ worker ν sends the E_x and E_z field components in the boundary to worker $\nu + 1$ and for the calculation of the electric field in subspace ν worker $\nu + 1$ sends the H_x and H_z field components in the boundary to worker ν.

EFFICIENCY INVESTIGATIONS

In this section speed-up and efficiency are investigated in dependence on the number of available processors N and the number of FDTD grid elements in x-, y-, and z-directions, X_{nge}, Y_{nge} and Z_{nge}, respectively. The speed-up factor, S, is a measure of how much faster a parallel program will run on N processors when compared to only one available processor.[6,7] S can be calculated from the ratio of the execution times: $S = T_{ser}/T_{par}$. The efficiency ϵ is simply the speed-up factor normalized by N: $\epsilon = S/N \leq 1$. Because the parallel FDTD simulator can not run on one

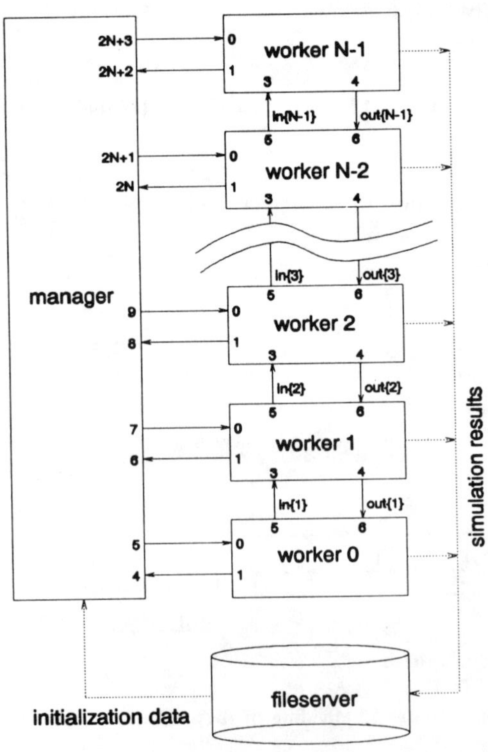

Figure 2. Task organization of the parallel FDTD simulator

processor, in this work the speed-up factor is normalized to the execution time for one FDTD timestep with two processors $t_{\text{step},2}$ as follows:

$$S = 2\frac{t_{\text{step},2}}{t_{\text{step},N}} \tag{1}$$

The execution time of one step is the sum of the communication time t_{com} and the calculation time t_{cal} where

$$t_{\text{com}} = c_1 X_{\text{nge}} Z_{nge} \tag{2}$$

and

$$t_{\text{cal}} = c_2 X_{nge} Z_{nge} \frac{Y_{\text{nge}}}{N} \tag{3}$$

Thus the execution time of one step can be approximated by

$$t_{\text{step}} = t_{\text{com}} + t_{\text{cal}} = X_{\text{nge}} Z_{\text{nge}} \left(c_1 + c_2 \frac{Y_{\text{nge}}}{N} \right) \tag{4}$$

The parameters c_1 and c_2 are determined experimentally with several FDTD simulation time measurements to $c_1 = 196\ \mu s$ and $c_2 = 35\ \mu s$. The values are valid for a single precision FDTD-code[2] running on T800 transputers with the operating system Helios™. Thus all interesting performance parameters are given by the following formulas:

$$t_{\text{step}} = 35 \mu s X_{\text{nge}} Z_{\text{nge}} \left(5{\cdot}6 + \frac{Y_{\text{nge}}}{N} \right) \tag{5}$$

$$S = \frac{Y_{\text{nge}} + 11{\cdot}2}{\dfrac{Y_{\text{nge}}}{N} + 5{\cdot}6} \tag{6}$$

$$\epsilon = \frac{Y_{\text{nge}} + 11{\cdot}2}{Y_{\text{nge}} + 5{\cdot}6N} \tag{7}$$

Equation (5) shows that the execution time for one timestep is constant for a constant ratio Y_{nge}/N. Furthermore the speed-up factor and the efficiency are independent of X_{nge} and Z_{nge} in the above equations.

Figure 3 shows the speed-up factor and Figure 4 the efficiency in dependence on the number

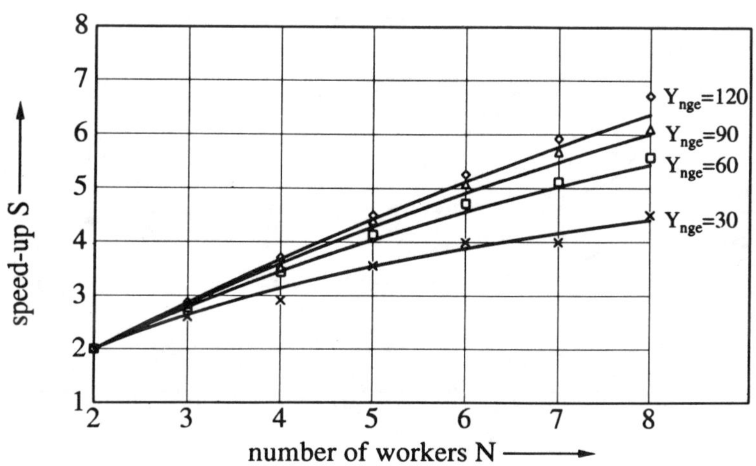

Figure 3. Speed-up factor of the parallel FDTD simulator: —— calculated, ×,□,△,◇ measured results

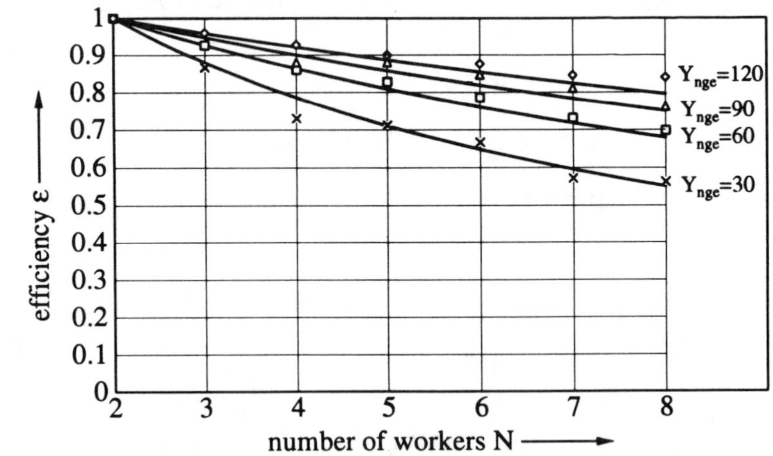

Figure 4. Efficiency of the parallel FDTD simulator: —— calculated; ×,□,△,◇ measured results

of workers and on the number of grid elements in y-direction Y_{nge} with constant $X_{nge} = 30$ and $Z_{nge} = 10$. The agreement of the calculated results obtained by using equation (6) and (7) (solid lines in Figures 3 and 4) and the measured ones (marks in Figures 3 and 4) is very good. A more detailed consideration of effects like communication offset time will only lead to slight accuracy improvements but increase the expenditure significantly. A speed-up factor of 6·72 by using eight workers and with $Y_{nge} = 120$ is an excellent result which indicates that the parallel FDTD simulator works very efficient ($\epsilon = 0{\cdot}84$). Figure 4 also shows that ϵ decreases for smaller Y_{nge}. A good efficiency of $\epsilon \geqslant 0{\cdot}73$ is obtained if the ratio Y_{nge}/N is greater or equal than 15 which is indicated in Figure 5. Thus it is not useful to simulate small FDTD meshes on many processors. In case of very small Y_{nge} and large X_{nge} and Z_{nge} the proposed parallelization becomes less efficient and a division of the mesh with planes perpendicular to the x- and z-axis is necessary.

The speed-up factor (6) and the efficiency equation (7) are very useful to determine the number of required processors for a given application especially in the case of procurement of a new transputer system.

The algorithms used for the parallel FDTD simulator are described in References 2 and 8 in detail. The simulation results of both programs are exactly equal.

CONCLUSION

The parallelization concept presented in this work is a significant improvement of the FDTD approach. Owing to the fact that computation power and memory of transputer systems are nearly

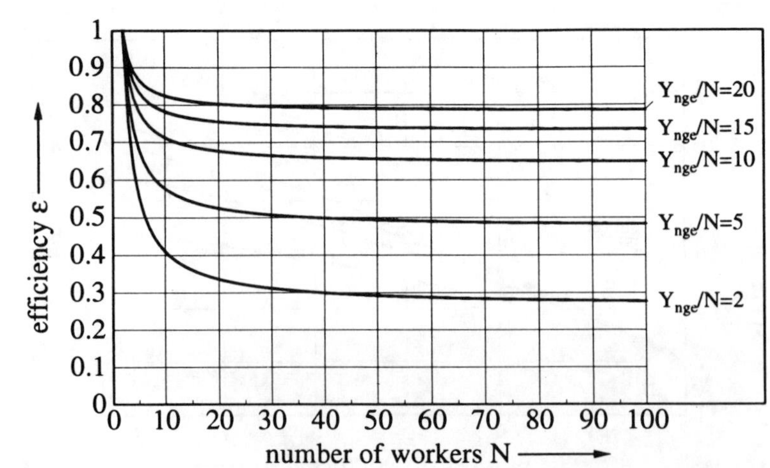

Figure 5. Efficiency calculated by using equation (7) for different constant ratios of Y_{nge}/N

arbitrarily scalable, it seems to be possible to simulate the electromagnetic field of very complex problems like large microwave circuits, superconductive structures and microwave packages in an acceptable time. A very promising application of the parallel FDTD simulator will be the investigation of biomedical effects in human body caused by electromagnetic waves radiated by mobile communication systems. In combination with parallel computers based on modern processors and with other improvements like local mesh refinements and modern signal processing techniques[9] the FDTD approach will be a very powerful and flexible design tool for microwave circuits.

REFERENCES

1. X. Zhang and K. K. Mei, 'Time-domain finite difference approach to the calculation of the frequency-dependent characteristics of microstrip discontinuities', *IEEE Trans. Microwave Theory Techniques*, **MTT-36**, 1775–1787 (1988).
2. I. Wolff and M. Rittweger, 'Finite difference time-domain analysis of planar microwave circuits', *Archiv für Elektrotechnik*, **74**, 189–201 (1991).
3. U. Effing, Implementation eines Verfahrens zur Berechnung elektromagnetischer Felder planarer Mikrowellengrundschaltungen mit der FDTD-Methode auf einam Parallelrechnersystem, diploma-thesis, Duisburg University, 1993.
4. W. J. Buchanan, 'Simulation of radiation from a microstrip antenna using three-dimensional finite-difference time-domain (FDTD) method', 8th Int. Conf. on Antennas and Propagation, IEE Conf. Pub. No. 370, Part 2, 1993, pp. 639–642.
5. D. B. Davidson, 'FDTD modelling of 3D optical pulse propagation in linear and non-linear materials', 2nd Int. Conf. on Computation in Electromagnetics, IEE Conf. Pub. No. 384, 1994, pp. 166–169.
6. D. B. Davidson, 'A parallel processing tutorial', *IEEE Antennas and Propagation Society Magazine*, **32**, 2, 6–19 (1990).
7. D. B. Davidson, 'Large parallel processing revisited: a second tutorial', *IEEE Antennas and Propagation Society Magazine*, **34**, 5, 9–21 (1992).
8. M. Rittweger, Simulation transienter elektrodynamischer Ausbreitungsphänomene zur Analyse der Übertragungseigenschaften von Systemen der Mikro- und Millimeterwellentechnik, Ph.D. thesis, Duisburg University, 1992.
9. W. Kümpel and I. Wolff, 'Digital signal processing of time domain field simulation results using the system identification method', *IEEE Trans. Microwave Theory Techniques*, **MTT-42**, 667–671 (1994).

COMPUTATIONAL FLUID DYNAMICS ON PARALLEL PROCESSORS

William D. Gropp[1] and Edward B. Smith[2]

[1]Department of Computer Science, Yale University, New Haven, CT 06520, U.S.A. and
[2]United Technologies Research Center, East Hartford, CT 06108, U.S.A.

(*Received* 1 *May* 1989; *in revised form* 17 *January* 1990)

Abstract—Greater computational power is needed for solving Computational Fluid Dynamics (CFD) problems of interest in engineering design. Parallel architecture computers offer the promise of providing orders of magnitude greater computational power. In this paper we quantify that promise by considering an explicit CFD method and analyze the potential parallelism for three different parallel computer architectures. The use of an explicit method gives us a "best case" analysis from the point of view of parallelism, and allows us to uncover potential problems in exploiting significant parallelism. The analysis is validated against experiments on three representative parallel computers. The results allow us to predict the performance of different parallel architectures. In particular, our results show that distributed memory parallel processors offer greater potential speedup. We discuss the importance of our model for the development of parallel CFD, algorithms and parallel computers. We also discuss our experiences in converting our model code to run on the three different parallel computers.

1. NOTATION

$\overline{(\ )}$	value of ( ) after first step
c_v	specific heat
e	energy
F	group of terms defined by eqn (2)
h_T	total enthalpy
H	group of terms defined by eqn (3)
i	radial index
j	axial index
k	number of processors
L	message startup time
L_2	message startup time for two hop messages
L_3	message startup time for three hop messages
m	number of mesh points
n	time index
p	pressure
p_0	upstream pressure
Q	heat source term
r	radial distance
r^-	radial distance to $-r$ edge of cell
r^+	radial distance to $+r$ edge of cell
r_{ave}	average radial distance across cell
R	pipe radius
R_{gas}	real gas constant
s	speedup
t	time
T	temperature
T_0	constant temperature for Poiseuille flow
$\mathbf{v}$	velocity vector
v_r	radial velocity
v_z	axial velocity
v_θ	azimuthal velocity
V_0	centerline velocity
w	time to send on word
W	time to compute one time step
z	axial distance
α	thermal conductivity
Δr	radial step size
Δt	time step size
Δz	axial step size
κ	bulk viscosity
μ	viscosity
ρ	density

2. INTRODUCTION

Computational Fluid Dynamics (CFD) problems are among the most demanding scientific computing problems in terms of the computational resources they require. Currently, CFD problems are extensively solved on vector supercomputers, primarily Cray and Cyber. While current supercomputers have adequate computational power to solve CFD problems which are 3-D or 2-D and unsteady or viscous, extending CFD to problems which are 3-D and unsteady and viscous requires vastly more computational power.

One example of a computationally intensive CFD problem is modeling the air flow through an airplane's gas turbine engine. The flow through a gas turbine engine is 3-D, unsteady and viscous. Further the geometry is extremely complex including several turbine stages, several compressor stages and a combustor. Gas turbine engineers must design the entire engine and would simulate the flow through the entire engine if the capability were available. As an example of the computational intensity of CFD calculations, Rai recently took over 100 h on a Cray X-MP to run a 3-D viscous unsteady model for the air flow past two stator blades and a single rotor blade of an axial flow gas turbine [1]. Although this run time is far too excessive for the use of such a code in engineering design, only part of the gas turbine was modeled. As the Cray is one of the fastest machines in the world and near the physical limits on the speed of a uni-processor, no existing computer is fast enough for these calculations.

To speed needed calculations up several orders of magnitude faster supercomputers are needed in gas turbine engine design. Computer processors, however, are reaching their physical speed limits. Processor designs appear to be within a single order of magnitude of their speed limits due to physical limitations such as the speed of light and gate switching limits. Therefore needed throughput cannot be achieved with single processor computers.

Parallel processor computers offer the possibility of achieving the throughput needed in CFD. Researchers are now investigating the effectiveness of using parallel computers for the solution of CFD problems [2–4]. The need to understand and exploit the architecture of parallel computers, however, makes it unclear whether we can design parallel algorithms which will achieve the needed throughput for CFD problems. In particular, the results in Ref. [4] on a variety of CFD kernals gave examples where both hardware and software problems reduced the effectiveness of a parallel CFD code. Our goal is to provide a model that can help understand some of these effects.

To test the promise of parallel computers, we consider a model CFD problem and apply it to representatives of several types of parallel computers. This model is based on an explicit difference technique and thus has the greatest opportunities for parallelism (as compared to implicit difference techniques). For each class of parallel computer, we develop a time complexity model and validate it against our model problem. These complexity estimates are analytical estimates of the computer time needed to solve a CFD problem. The complexity estimates are expressed in terms of basic machine and problem parameters, such as floating point and communication speeds and number of grid points. From these estimates, we can estimate the performance of CFD codes on supercomputers utilizing significant parallelism.

Finally, we discuss our experiences in using parallel computers for solving CFD problems.

Our model problem is relatively simple and does not address some features of advanced CFD codes, such as implicit schemes or unstructured grids. Methods for implicit problems on parallel computers are an area of active research [5, 6]. Current results are promising, but more needs to be done here before implicit methods can be efficiently solved on massively parallel computers, and we have restricted our study to explicit methods. Similarly, algorithms using unstructured grids that are efficient for vector and parallel computers and do not rely on fine-grained scatter-gather operations are also under study [7, 8]. We have not considered these approaches here. One point that should be made clear is that the algorithms that are appropriate for a uniprocessor may not be appropriate for a massively parallel machine, and that as an algorithm becomes more complicated (and more carefully tuned to the features of the computer it is running on), it may be very hard to implement it efficiently on a very different kind of computer. Instead, a different algorithm may need to be chosen. The complexity models developed in this paper, and those for other parallel algorithms, should serve as a guide for the development of such algorithms.

3. ARCHITECTURES

A number of different types of parallel architecture computers are available today. These run from very fine grain machines such as vector processors and very long instruction word machines to large numbers of almost independent processors. In this paper, we will consider three architectures which represent an important part of the spectrum of possible parallel computers. These are multiple vector processors, tightly coupled MIMD and loosely coupled MIMD. Of course, these are artificial distinctions and any real machine may fall between several of these categories.

Multiple vector processors consist of several vector computers operating in parallel. Exchanges of information and synchronization between the processors is facilitated by special high-speed hardware. Examples are the Cray X-MP, Cray 2, ETA 10, and the Alliant FX/8. The Alliant was used in this study.

Tightly and loosely coupled MIMD computers work on essentially independent data in parallel, with differing kinds of access to each other's data.

Tightly coupled MIMD computers are collections of independent processors which are closely connected, usually by shared memory. A parallel program running on this type of machine consists of independent threads of control which may access each other's data and can tightly control each other's operation by manipulating this shared data. These are similar to Multiple vector processors, but usually have only scalar floating point processors and little hardware for synchronizing operations beyond shared memory and perhaps a mechanism for managing data locks.

The advantage of the tightly coupled MIMD approach is that there is no single thread of control, and hence no *a priori* serial execution. There have been a number of machines of this type built by recent startup companies, such as the Encore Multimax 120, used in this study and the Sequent Balance series.

Loosely coupled MIMD machines are collections of independent processors which communicate through some reliable mechanism, but which don't directly share any data. Instead, all interprocessor sharing of data is done by I/O operations, typically the sending and receiving of message packets. This provides a measure of programming safety and reproducibility of results often absent in shared memory (tightly coupled MIMD) machines, since all modifications to "shared" data structures is handled explicitly by the programmer, rather than implicitly through a shared memory access. However, most systems currently on the market have a high overhead associated with interprocessor communication.

Example machines of this type include the various hypercubes from Intel, NCUBE and others, and some research machines, such as the LCAP system of IBM. The systems themselves vary from a few, fast processors (LCAP) to many slow processors (Intel Hypercube). The Intel Hypercube was used in this study.

4. FINITE DIFFERENCE ALGORITHMS

Finite difference algorithms used in Computational Fluid Dynamics for time accurate calculations can be divided into two major types: explicit and implicit.

With explicit finite difference algorithms the value of a variable at the new time is determined from the values of the variables at the old time directly (that is explicitly) without dependence on the values at the new time. Practically speaking this means that the values of the variables can be determined without solving a system of equations. The time step size that can be taken is limited by a time step on the order of the Courant–Freidrichs–Lewy (CFL) criterion (the smallest value of the time for a particle traveling at the local fluid velocity to cross a finite difference grid interval). In many applications the CFL time step is much smaller (often orders of magnitude smaller) than the time step required for accurate calculation of the time variation of the variables.

The model problem studied here used MacCormack's algorithm [9] which was chosen for its simplicity and wide familiarity to the CFD community. Major features of MacCormack's algorithm are:

- An explicit difference scheme which is second order accurate in space and first order accurate in time.
- A two step scheme in which the present time values are used to calculate the intermediate values in the first step, and the present time and intermediate values are used to calculate the values at the new time.

Implicit finite difference algorithms differ from explicit finite difference algorithms in that the determination of the values of the variables at the new time depends (implicitly) on the values of the variables at the new time. The advantage of a well-chosen implicit scheme is that numerical stability can be achieved with a time step size much greater than given by the explicit CFL limit [10].

The price to be paid by most implicit schemes is that the solution of a system of equations is required at each step. Except for the solution of a tightly banded linear system it is often faster to solve the linear system iteratively.

We left implicit schemes for later study in order to address the basic question of whether parallel architecture computers were applicable to solving CFD problems. We intend to study implicit schemes and linear solution techniques in future work. The use of sparse matrix methods in CFD problems is an area of current research on serial computers [11] as well as on parallel computers.

5. MODEL PROBLEM

Descriptions of the physical model problem, its geometry, its mathematical formulation, and the numerical solution method are described below.

The physical model problem we consider is unsteady compressible viscous flow through an insulated duct. We further assume the flow to be axisymmetric with swirl. A time dependent model problem was chosen for this study since time marching schemes are generally used in CFD for both steady and unsteady problems. A two dimensional problem was chosen both to provide a start at physically reasonable models for combustor flow and to introduce the complexities of multi-dimensional problems. For the model problem studied here the fluid velocities and pressure are known so that the temperature can be obtained by integrating in time the energy equation which is linear in temperature. The density was obtained from the perfect gas law.

The case studied was obtained from the model problem by assuming "Poiseuille flow" (parabolic velocity profile and linear pressure gradient) and then adding in "source" terms so that the computed flow remained steady. The fictitious source terms enabled us to simply check whether the calculations were being performed correctly on the different computers while retaining all the computations needed for a compressible transient flow. Poiseuille flow is an exact solution for steady, incompressible, isothermal, viscous pipe flow. We wished to retain the compressible effects in case they affected the parallelization of the code. Poiseuille flow does not exactly satisfy the compressible flow equations. There are residual terms which are collectively treated as a heat source. Thus the solution to the case studied was steady Poiseuille flow and, in particular, constant temperature. This facilitated debugging.

The geometry of the model problem is a rectangular domain with a tensor product mesh. The left side is the inflow and the right side is the outflow. The top is the insulated wall, and the bottom is the centerline of the cylinder.

The mathematical formulation of the energy equation in cylindrical coordinates is given by eqns (1)–(6).

$$\frac{\partial \rho e}{\partial t} + \frac{1}{r}\frac{\partial rF}{\partial r} + \frac{\partial H}{\partial z} = Q \tag{1}$$

where

$$F = \rho v_r h_T - \alpha\frac{\partial T}{\partial r} + (\tfrac{2}{3}\mu - \kappa)(\nabla\cdot\mathbf{v})v_r - \mu\left[2v_r\frac{\partial v_r}{\partial r} + v_\theta\left(\frac{\partial v_\theta}{\partial r} - \frac{v_\theta}{r}\right) + v_z\left(\frac{\partial v_r}{\partial z} + \frac{\partial v_z}{\partial r}\right)\right] \tag{2}$$

and

$$H = \rho v_z h_T - \alpha\frac{\partial T}{\partial z} + (\tfrac{2}{3}\mu - \kappa)(\nabla\cdot\mathbf{v})v_z - \mu\left(2v_z\frac{\partial v_z}{\partial z} + v_r\left(\frac{\partial v_z}{\partial r} + \frac{\partial v_r}{\partial z}\right) + v_\theta\frac{\partial v_\theta}{\partial z}\right) \tag{3}$$

where

$$\begin{aligned} e &= \rho(\tfrac{1}{2}\mathbf{v}\cdot\mathbf{v} + c_v T) \\ h_T &= e + \frac{p}{\rho} \\ \mathbf{v} &= (v_r, v_\theta, v_z) \\ \nabla\cdot\mathbf{v} &= \frac{1}{r}\frac{\partial r v_r}{\partial r} + \frac{\partial v_z}{\partial z}. \end{aligned} \tag{4}$$

Q is the heat source term. In the case studied Q was set equal to the right hand side of eqn (1) evaluated for Poiseuille flow

$$\begin{aligned} \mathbf{v} &= [0, 0, V_0(1 - r^2/R^2)] \\ p &= p_0 - \frac{4\mu V_0 z}{R}. \end{aligned} \tag{5}$$

Density is determined from the equation of state (perfect gas law)

$$p = \rho R_{\text{gas}} T, \tag{6}$$

and T is the unknown to be solved for. To simplify debugging the solution $T = T_0$ (T_0 a constant) was used.

The numerical solution method for the above partial differential equations is by MacCormack's explicit algorithm [9]. The finite difference form of these equations using MacCormack's algorithm is given by eqns (7) and (8).

In the first step of MacCormack's two step difference scheme the derivatives of F and H are calculated using backward differences with central differences for the terms involving second derivatives.

$$\overline{(\rho e)}_{ij}^{n+1} = (\rho e)_{ij}^{n} - \frac{\Delta t^n}{r_{\text{ave}_{ij}}} \frac{1}{\Delta r_{ij}} (r_{ij}^{+} F_{ij}^{n} - r_{ij}^{-} F_{i-1j}^{n}) - \frac{\Delta t^n}{\frac{1}{2}(\Delta z_j + \Delta z_{j-1})} (H_{ij}^{n} - H_{ij-1}^{n}). \tag{7}$$

In the second step of MacCormack's scheme the derivatives of F and H are calculated using forward differences with central differences for the terms involving second derivatives.

$$(\rho e)_{ij}^{n+1} = \frac{1}{2} \left\{ \begin{array}{ll} (\rho e)_{ij}^{n} + \overline{(\rho e)}_{ij}^{n+1} - & \dfrac{\Delta t^n}{r_{\text{ave}_{ij}}} \dfrac{1}{\Delta r_{ij}} (r_{ij}^{+} \bar{F}_{i+1j}^{n+1} - r_{ij}^{-} \bar{F}_{ij}^{n+1}) \\ - & \dfrac{\Delta t^n}{\frac{1}{2}(\Delta z_j + \Delta z_{j-1})} (\bar{H}_{ij+1}^{n+1} - \bar{H}_{ij}^{n+1}) \end{array} \right\}. \tag{8}$$

In CFD studies of duct flow typical grid sizes are order of 50–100 (radial) by 100–1000 (axial), and the CFL limited time steps generally number in the tens of thousands. In this study we used grids from 50 × 50 to 250 × 250 in size and took from 10 to 50 time steps. The matrix sizes were chosen to test effects of memory size and the number of time steps was chosen large enough to get accurate timings. The choice of a square simplifies the analysis and the comparison with the time complexity estimate developed below by eliminating a parameter. This choice does not affect the actual arrangement of the code or significantly affect the performance. It should be noted, however, that on a hypercube parallel processor, the most efficient grid dimensions are powers of 2 (though there can be a different power in each coordinate direction); this effect is not considered in our analysis (but see Fig. 4 and the discussion of it).

The boundary conditions are specified inflow and axially smooth outflow with insulated cylinder walls. The geometry was chosen to keep the code simple for transporting from one parallel machine to another.

6. TIME COMPLEXITY MODEL

In order to understand the potential of parallel programming for CFD problems, we need to develop models which will allow us to predict the performance of a parallel computer. Time complexity models are constructed for each of the three classes of parallel computers used in this study. Since the time complexity model is an analytic expression for the computer time needed to solve a problem, based on fundamental parameters of the model problem and the computer, we can use the complexity estimates to predict the performance of new parallel computers and suggest which architectures offer the most potential for CFD problems.

In this section we will first give some basic definitions. Then we will give time complexity estimates for simple stencils and apply and extend these estimates to our model problem.

We will denote the time complexity by $W(k, m)$, where k is the number of processors and the problem is on an $m \times m$ mesh. Another useful measure is the speedup, s, defined by $s(k, m) = W(1, m)/W(k, m)$. In a perfectly parallel algorithm with perfect hardware and software, $W(k, m) = W(1, m)/k$. Algorithms, hardware, and software are not, however, perfect and this section is concerned with modeling the "imperfections." We therefore define the computational efficiency as $W(1, m)/[kW(k, m)]$. A perfect parallel computer and algorithm has a computational efficiency of 1.

The nature of these imperfections depends on the type of parallel computer. However, they are related to the access to memory. In a loosely-coupled MIMD parallel processor, it is the interprocessor communication time, which can be thought of as access to remote memory, which is important. In a tightly-coupled MIMD parallel processor, it is the access to shared memory, and the establishment of critical regions (access control to memory) which is important. In a multiple vector parallel processor, it is the access to shared memory again and cache contention among the processors which is important. In addition, various "non-parallel" operations may reduce the possible speedup.

Time complexity estimates for simple stencils

In order to motivate our choice of complexity model, we first develop a model for finite difference schemes with 5 and 9 point stencils for the three parallel architectures used in this study. Using these results, we can construct a model which predicts the time complexity of our model problem. We also discuss some global operations which are present in our model algorithm, and how they are handled on different computer architectures.

We first establish some basic ideas concerning two main classes of parallel programming styles: message passing and shared memory. We first describe these two approaches and then discuss the similarities in terms of both expressiveness and of efficiency.

In a message passing computer, memory access for data on different processors is via inter-processor communication which is handled by explicitly sending a message from one processor to another. There are two major parameters of interest. These are the message startup time L, and the transfer rate w. To make L and w independent of particular hardware implementations, L and w have been nondimensionalized by dividing by the floating point speed in flops/s. Thus L is in ops/startup and w is in ops/word. We will denote the floating point computation rate by f flops/s.

For the tightly coupled computers, memory access is from shared memory to processor and the three parameters of interest are the memory access time, the speed of a floating point operation and the number of simultaneous memory references. The programming constructs used with shared memory machines include barriers and critical regions. A barrier is a synchronization point which all processors must reach before any of them can proceed. A critical region allows only one process to access and modify data at a time. In a parallel program, barriers may be used, for example, to wait for all processors to reach the end of a time step. A critical region may be used to provide a unique do-loop index to each parallel processor. These are discussed in more detail in, e.g. Ref. [12].

Message passing computers such as the Intel Hypercube and shared memory machines such as the Encore Multimax are often considered two completely different types of parallel computer. In fact, when viewed in the right way, they are very similar. What distinguishes them is the relative cost of referencing remote or shared information. The following table shows the correspondence between the costs of interprocessor communication for the two types of machines.

Message Passing	Shared Memory
I/O startup time	Cache miss overhead and time to establish critical regions and barriers
Transfer rate	Memory transfer rate

Each of these machines have different strengths. The shared memory machines have faster access to shared information, but pay a penalty in terms of higher overheads in the forms of barriers and critical regions. In general, these barriers and critical regions are intrinsic to the multiprocess computation and are therefore unavoidable. Message passing machines have simpler access control but at a higher cost in sharing data. Barriers and critical regions are sometimes used with message passing computers; however, the style of programming normally used with message passing makes them less important.

Consider the cost of computing a step of an explicit PDE on an $m \times m$ (in 2-d) and an $m \times m \times m$ (in 3-d) grid. This step will usually consist of an estimate of the time step size to use, based on CFL estimates, followed by the application of a local stencil. There may be additional steps (such as smoothing steps) or some special non-local computation (such as integrating along a boundary). For simplicity we do not consider these costs here; many of them have a similar ratio

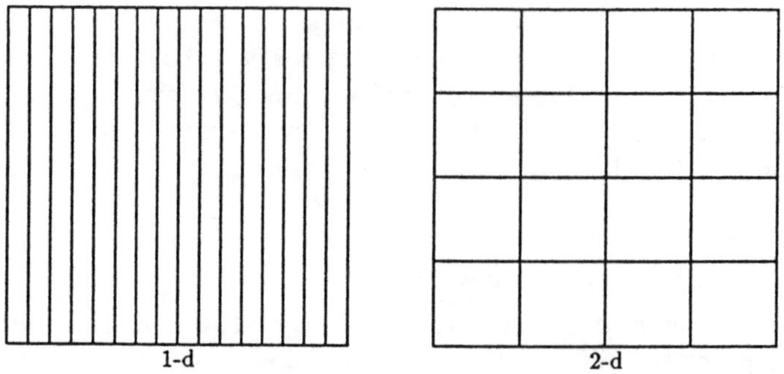

1-d 2-d

Fig. 1. Two different decompositions of a 2-d domain. Each outlined area is given to a different processor.

of computation to communication cost as the application of the stencil. Our full complexity model developed below does include a sample cost for a non-local computation step.

The estimate of the time step requires global information (the solution everywhere); the application of the stencil just local or neighbor information. We describe first the local communication and then the global communication. Since these effects are most noticable in the context of message passing, we describe them in those terms. However, similar considerations apply to shared memory.

The first step in any parallel program is the division of work among the processors. Divide this grid among the processors by slicing the problem domain by planes in 1, 2 or all 3 cartesian directions. Call the resulting decomposition 1-d, 2-d or 3-d slices. An example of 1-d and 2-d slices for a 2-d domain is shown in Fig. 1.

In each case, it is the length of the boundary of the slice which determines the communication cost. For example, in 2-d where the domain is sliced in one direction, the boundaries between processors are m mesh-points long, so that the effort for one step is

$$\frac{m^2}{k} + 2(L + wm).$$

(All times are expressed in nondimensional units of floating point operations.) The "2" comes from sending the left internal boundary to the left neighbor and the right internal boundary to the right neighbor. While in principle these operations could go on simultaneously, in practice they require essentially exclusive use of the processor and memory bus. For slices in two directions, there are now either 4 neighbors (for a 5-point stencil) or 8 neighbors (for a 9-point stencil), and the cost becomes

$$\frac{m^2}{k} + 4\left(L + w\frac{m}{\sqrt{k}}\right) + 4(L_2 + w).$$

Here L_2 to denotes the cost of a two hop link and is only present in the 9-point case. Similar analysis can be carried out for other cases; they are summarized in Table 1.

Table 1. Times for various decompositions of the domain, for both 2 and 3-d domains

Slicing	Dim	Time estimate
1-d	2-d	$\frac{m^2}{k} + 2(L + wm)$
2-d	2-d	$\frac{m^2}{k} + 4\left(L + w\frac{m}{\sqrt{k}}\right) + 4(L_2 + w)$
1-d	3-d	$\frac{m^3}{k} + 2(L + wm^2)$
2-d	3-d	$\frac{m^3}{k} + 4\left(L + w\frac{m^2}{\sqrt{k}}\right) + 4(L_2 + wm)$
3-d	3-d	$\frac{m^3}{k} + 6\left(L + w\frac{m^2}{\sqrt[3]{k^2}}\right) + 12\left(L_2 + w\frac{m}{\sqrt[3]{k}}\right) + 8(L_3 + w)$

L_2 is the cost for a two-hop link and L_3 the cost for a three hop link. The terms containing them are not present for 5 or 7 point stencils.

From these formulas it is clear that once the problem is large enough (m large), the communication costs become negligible. However, in practice the problems are neither enormous nor the constants the same size. Thus, the communication terms, though asymptotically negligible, can be of dominant importance. For example, take

$$k = 64$$

$$m = 256$$

$$f = 1\ \mu\text{s/mesh-point}$$

$$fL = 3\ \text{ms}$$

$$fw = 1\ \mu\text{s/word.}$$

(Recall that L and w are expressed in terms of the floating point computation rate f.) This represents a fast node (over 1 megaflop) with moderate communication speeds and has at least 2 megabytes of memory. An existing machine with similar parameters is the Intel VX hypercube. In this case, in 3-d with 1-d slicing and a 5-point stencil and using the time estimates in Table 1, the communication time is roughly 0.12 s and the computation time is roughly 0.24 s. The communication is taking one third of the total time. Further, this is assuming that the message requires only one startup, despite its large size (m^2 words). If we also require all messages to be less than 16384 bytes, this adds another 0.1 s of startup time, making communication almost half of the total time.

The analysis so far has been adequate for centered single step schemes such as Lax–Wendroff or leapfrog. However, many schemes use combinations of one-sided stencils. The MacCormack scheme used in our model problem is one such scheme. In this case, we simply consider each access separately. Specifically, sharing information at a boundary with another processor takes time

$$L + wm. \tag{9}$$

One problem with considering each access separately is that there may be synchronization delays caused by different processors finishing at different times. We will ignore these effects for now; however, they are likely to become more important in large scale MIMD parallelism.

A certain amount of global communication is necessary in these algorithms as well. For example, the choice of time step size requires the global value of the maximum velocity. In addition, the values of interest in the computation include integrals along the boundary, which may need to be computed across processors. These operations fall into the general category of reductions: taking data from many places and combining it to produce a single result.

Reductions are often done using a fast distribution algorithm such as those in Ref. [13], with the operation (maximum or sum) inserted into the distribution. For example, a tree-based distribution algorithm such as those described in Ref. [13] is fast on both shared memory and hypercube/message passing computers. In this case, the time to reduce k items is

$$(L + w + f)\log k$$

assuming simultaneous send/receives. One potential problem here is possible round-off error; to avoid that, we reduce upward using the tree to a single node, then distribute that single value downward to all nodes, again using the tree. The time is then

$$(2(L + w) + f)\log k. \tag{10}$$

When using a small number of processors and a shared-memory or complete interconnect, it is often easier for a single processor to compute the reduction, then make that value available to all processors. In this case, the time is pf plus the time to establish two barriers, one before the reduction to insure that the data is ready, and one after it to insure that the result is ready.

Time complexity estimates for the model problem

Our model problem is more complicated than a simple 5-point explicit scheme because it is multistep and each step is not centered. However, we can break it down into

- floating point work
- local data exchanges
- global data exchanges.

This form is suitable to a program where the parallelism is expressed explicitly. In our particular implementation, we can write the total computation time per step as

$$T = F_1 \frac{m^2}{k} + F_2 \frac{m}{k_y} + 8\text{“global reduce”} + 6(\text{“send in } y\text{”} + \text{“send in } x\text{”}) \tag{11}$$

where F_1 is the floating point time per mesh point for operations on the whole grid, F_2 is the floating point time per mesh point for operations along one boundary, the grid is $m \times m$, k is the number of processors, and k_y is the number of processors in the y direction.

Using eqns (9) and (10) as the communication model, eqn (11) becomes

$$T = F_1 \frac{m^2}{k} + F_2 \frac{m}{k_y} + 8(L + w)\log k + 6\left(L + w\frac{m}{k_x}\right) + 6\left(L + w\frac{m}{k_y}\right). \tag{12}$$

The memory access terms this equation are the direct communication terms involving L and w.

For this local communication model to be applicable, it must be possible to map a grid onto the parallel processor in such a way as to make adjacent slices adjacent in the parallel processor. We will assume that any 2 or 3-d grid of interest can be efficiently embedded in the parallel processor. Such embeddings are easy on hypercubes; alternatively, everything we say here applies equally well to a mesh connected processor of the correct dimension.

For a shared memory machine with sufficient memory bandwidth, the figures are similar, except the global reduction is done differently, and the overhead of data sharing is slightly different. In this case, the results are

$$T = F_1 \frac{m^2}{k} + F_2 \frac{m}{k_y} + 16\text{“barriers for reduce”} + 6\text{“barriers for data”}. \tag{13}$$

Here, the “barriers” are synchronization points in the code. Depending on the implementation, these can be order k or $\log k$. Further, it is possible to reduce the number of these by using barriers with values [14].

On a tightly coupled multiple vector processor such as the Alliant FX/8 the work estimates are slightly different. In these machines, “close coupling” of the processors allows the rapid transfer of information from one processor to another, and extremely fast synchronization. On the negative side, since there is no explicit parallelism, the programmer may be dependent on the Fortran compiler to generate parallel code and it is possible for there to be substantial amounts of non-parallel code generated. Further, any shared memory machine suffers from a potential bottleneck in memory access; depending on cache or register utilization and the exact pattern of operations chosen by the compiler, the performance may exhibit anything from superlinear speedup to a plateau in performance.

We can model this bottleneck in shared resources such as memory as follows. Let the computation consist of two parts: one which is arbitrarily parallel, and one which uses the shared resource. For example, consider a machine with k processors in which k floating point operations may be done in parallel. However, only K processors may use the shared memory at any time. Then the time to do a computation has the form

$$W(k) = \frac{a}{k} + \frac{b}{\min(k, K)}. \tag{14}$$

For $K = 1$, this is the well-known Amdahl's law. The speedup predicted for such a machine as k gets large ($k > K$) is

$$s(k) = \frac{a + b}{(a/k) + \{b/[\min(k, K)]\}} \to \frac{(a + b)Kk}{aK + bk}. \tag{15}$$

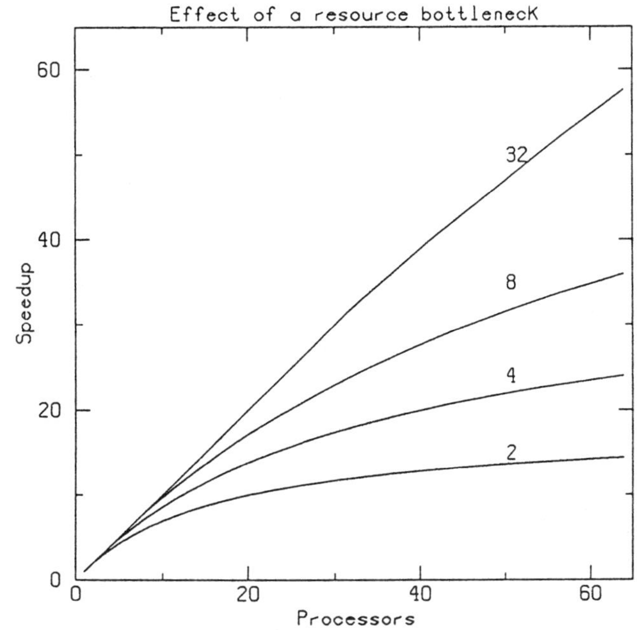

Fig. 2. Graph of eqn (15) for $a = 8$ and $b = 1$, for various values of K (K is labeled on the lines). The ratio a/b is similar to the ratios we observed in our experiments.

As $k \to \infty$, the speedup tends to $(a + b)K/b$ or

$$s \to \left(1 + \frac{a}{b}\right) K. \tag{16}$$

Figure 2 shows the behavior of s for different values of K. Note the approach to the asymptotic value of s, which for the example values of a and b is 9 K.

This "bottleneck" effect is present in both multiple vector processors and in tightly-coupled shared memory computers. Jordan [15] discusses these effects in more detail.

We have not considered all possible effects. One important effect is load balancing. On distributed memory machines, the current approach is to divide the problem statically among the processors. This is a large grain division of work. If the computational work is not equally divided among the processors, there will be additional costs caused by some processors going idle while others continue to work. This effect diminishes, however, as the size of the problem relative to the number of processors increases, and in fact this behavior can be seen in our results on the Intel Hypercube. For more general algorithms (such as adaptive mesh refinement), load balancing such as that described in Ref. [16] must be considered.

7. EXPERIMENTAL RESULTS

In this section we show the results of computing our model problem on several different types of architectures.

We ran our model problem on three different parallel computers. An Intel iPSC/1 Hypercube, an Encore Multimax 120, and an Alliant FX/8. The Intel Hypercube is a loosely coupled MIMD, the Encore is a tightly coupled MIMD, and the Alliant is a multiple vector processor. Each of these represents a different class of parallel computer.

Figures 3–6 show the results of our experiments. We have computed fits to these experiments using the models developed in Section 4. These fits were obtained by a non-linear least-squares fit to the timing data, after scaling the timing data by m^2/k.

Intel iPSC Hypercube

The Intel iPSC Hypercube is an example of a loosely coupled MIMD computer. The processors are connected in a hypercube architecture, and communicate by sending messages over a dedicated ethernet link (one per processor-pair). Each processor has 0.5 megabytes of memory, and on our machine, there is an additional 4.0 megabytes of memory on an attached board.

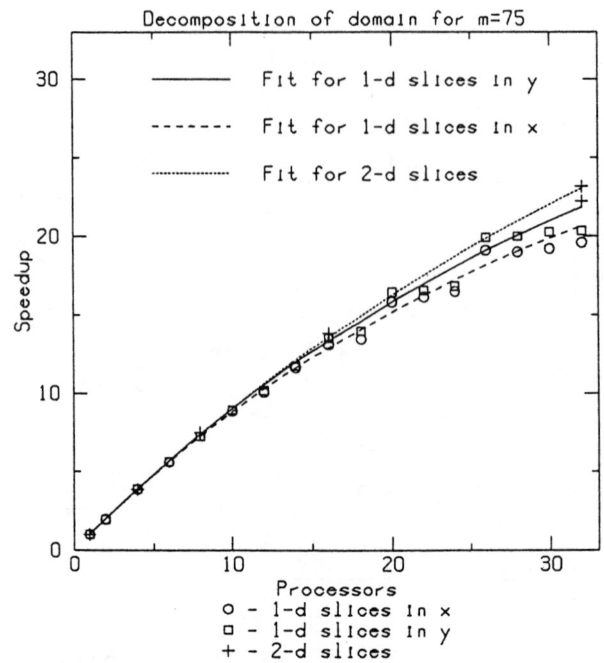

Fig. 3. Speed up figures for the Intel iPSC hypercube. The graph shows fits as a function of decomposition at a fixed problem size. The same fit parameters were used in all graphs (this figure and Fig. 4), and are drawn as curves.

Figures 3 and 4 present the results of these experiments. Each computation represents 10 time steps over the grid. A fit to the data using eqn (12) gives

$$W = 0.066\frac{m^2}{k} + 0.014\frac{m}{k_y} + 0.861 \log k + 0.0302(m_x + m_y) + 0.868, \tag{17}$$

where $m_x = m/k_x$ and $m_y = m/k_y$.

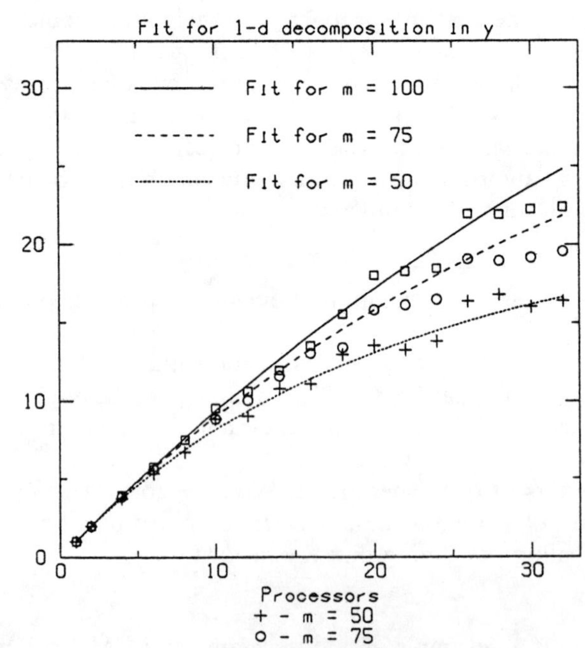

Fig. 4. Speed up figures for the Intel iPSC hypercube. The graph shows fits as a functions of problem size for fixed decomposition. The same fit parameters were used in all graphs (this figure and Fig. 3), and are drawn as curves.

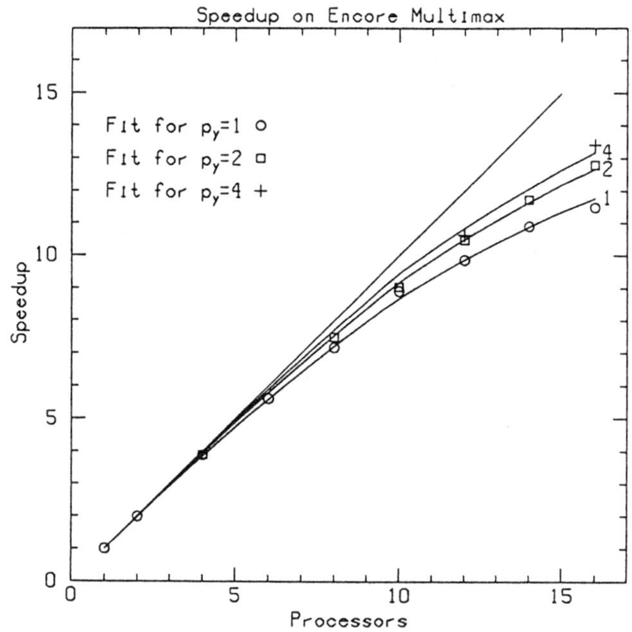

Fig. 5. Speedup figures for the Encore Multimax. The fit uses $W = 589.5/k + 79.5/\min(k, 10) + 8.22/k_y + 0.29k$. The problem size is $m = 100$. The line labels are the values of k_y used for that curve. The solid line represents a perfect speedup. Note the sharper deviation from perfect speedup at roughly 10 processors.

In Figs 3 and 4, we see the effects of the uneven distribution of work among the processors in the stepped increase in the speedup figures. Each processor has some fixed number of "slices", each of which has roughly $m/k_x \times m/k_y$ mesh points. If k_x or k_y do not divide m, then some processors have $\lceil m/k_x \rceil \times \lceil m/k_y \rceil$ and some have $\lfloor m/k_x \rfloor \times \lfloor m/k_y \rfloor$. The relative difference is approximately (assuming both k_x and k_y don't divide m)

$$\frac{k_x}{m} + \frac{k_y}{m}.$$

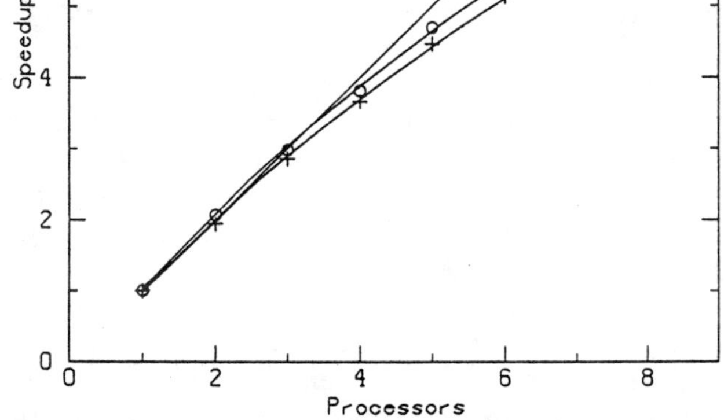

Fig. 6. Results for the Alliant FX/8. The fit uses $W = 0.00645\,\mathrm{m}^2/k + 0.00085\,\mathrm{m}^2/\min(k, 2.45)$. The speedup is almost independent of the problem size. The solid line represents a perfect speedup; note the deviation from this line by both the experimental results and the theory at around three processors.

As the number of processors approaches m, the difference can be very large. The points in k where there are jumps in the speedup for the 1-d decomposition are just those values of k which divide m, as can be seen in Figs 3 and 4. Note that the hypercube interconnections are not of major importance; a 2-d mesh interconnection would give almost the same results.

Encore Multimax

The Encore Multimax is an example of a shared memory MIMD computer. There are two processors per board, along with a cache memory. These boards are connected to a very fast bus, on which the shared memory resides. The computer runs Unix™ and provides a time-sharing environment. One problem when doing timing measurements in a multi-user environment is that any parallel job is competing for processors with other users, mailer daemons, etc. Timings are thus somewhat erratic, though we found ours to be reproducible to within 5%. Further, they don't reflect synchronization problems well, since idle processes won't use CPU time, and "real time" measurements are even more inaccurate on a time-sharing system. Thus, our results shown in Fig. 5 are at best approximate. However, they do match our speedup estimate very well with parameters in eqn (18), which is a combination of eqn (13) and (14).

$$W = 0.059\frac{m^2}{k} + 0.0080\frac{m^2}{\min(k, 10)} + 0.08\frac{m}{k_y} + 0.29k. \tag{18}$$

These parameters give a very good fit to our experimental data as shown in Fig. 5.

Note that since the onset of the "bottleneck" term is so high, this may be due more to the availability of processors than to limitations in the hardware. In fact, the Encore bus is very fast relative to the speed of the processors, and we do not expect to see a significant degradation for the number of processors available.

Alliant FX/8

The Alliant FX/8 is an example of a multiple vector processor computer. It has up to eight "computational elements", each of which is a vector processor. These processors share two high-speed caches, which in turn have a high speed channel to memory. It is possible for the processors to exceed speed of the cache.

Computations were done for 51 time steps and mesh sizes of $m = 50$, 100, 150, 200 and 250 on an eight processor Alliant FX/8. Figure 6 shows the speedups observed for various problem sizes compared to a fit against eqn (14). The fit used is

$$W = 0.00645\frac{m^2}{k} + 0.00085\frac{m^2}{\min(k, K)},$$

where $K = 2.45$. The fit is insensitive to K in the range $2 \leqslant K \leqslant 3$.

The apparent super-linear speedup for two processors shown by the curves in Fig. 6 is an artifact of the curve fit, caused by the rapid turn-down in speedup at three processors.

The data in Fig. 6 show a dropoff below linear speedup as the number of processors increases. This dropoff is due in part to a memory bottleneck. The memory bottleneck occurs when many processors require the same data from memory. In the Alliant FX/8 each of the four quadrants of the fast cache memory can be accessed by any two computational elements (processors) via a crossbar interconnect. When more than two computational elements try to access the same data there is "cache contention" which is a form of memory bottleneck. In addition cache and main memory are updated via hardware to maintain memory system consistency. Thus if more cache memories were added there would be a memory bottleneck as cache and main memories are updated to maintain consistency. Another part is due to some of the program running in serial (using only a single processor); this is due in part to the Fortran compiler on the Alliant, in a deliberate trade-off between ease of use and maximum parallelism.

Memory bottleneck will occur in any shared memory parallel computer. Since memory bottleneck will cause a falloff in the speedup curve below linear speedup, it is a very significant limitation on shared memory parallel computers for massive parallelism.

Summary of experimental results

The agreement between the experiments and the complexity estimates is very good. The complexity estimates capture the important contributions to the computation time. Thus we can use the complexity results to predict performance for differing values of the parameters, including the number of processors.

In the case of a distributed memory parallel computer, such as the Intel Hypercube, the time estimate for the computation has the form shown in eqn (12). For simplicity, we will consider the case $k_x = k_y = \sqrt{k}$. Then the estimate can be simplified to

$$W = F_1 \frac{m^2}{k} + F_2 \frac{m}{\sqrt{k}} + 8(L + w)\log k + 12\left(L + w\frac{m}{\sqrt{k}}\right).$$

Because of the $\log k$ term there is a value of k beyond which adding additional processors actually slows the computation down. The maximum speed up is achieved when $\mathrm{d}W/\mathrm{d}k = 0$, and is at

$$\sqrt{k} = m\frac{\frac{1}{2}F_2 + 6w + \sqrt{(\frac{1}{2}F_2 + 6w)^2 + 32F_1(L + w)}}{16(L + w)}.$$

From this we can see that the optimal number of processors depends on m. It can be shown that, as $m \to \infty$ and with this optimal k, the speedup tends to

$$\frac{Am^2}{\log m}.$$

Thus, the speedup for the distributed memory parallel computer (e.g. Intel Hypercube) increases without bound as the problem size and number of processors grow.

For a shared memory parallel computer, such as the Alliant FX/8 or Encore Multimax, the time estimate for the computation has the form shown in eqn (14). (Note that the shape of the Encore and Alliant results are very similar; the terms corresponding to eqn (14) in both cases are the dominant terms.) The predicted speedup is independent of the problem size, as we observed experimentally. The parameters determined by our experiments give a maximum possible speedup with an infinite number of processors of roughly 20 for the Alliant and roughly 80 for the Encore. Of course, the number of processors available on both the Alliant and the Encore were picked to match the available memory bandwidths. Other technologies and designs would allow larger limits. The limits themselves are, however, intrinsic in any shared resource design. Thus, because of the bottlenecks in the shared memory, even with an infinite number of processors, the maximum speedup of a shared memory parallel computer is limited.

Comparing these results, we see that the distributed memory computers offer more potential speedup than shared memory computers as the number of available processors increases.

8. PREDICTIONS AND OBSERVATIONS

As an example of the predictive use of our complexity estimates, we can estimate the performance of a distributed memory parallel processor, starting with our results for the Intel Hypercube. The relevant formula is eqn (17), written as

$$W = f_1 \frac{m^2}{k} + f_2 \frac{m}{k_y} + c_1 \log k + c_3(m_x + m_y) + c_2.$$

Terms f_1 and f_2 depend only on the floating point speed. c_1 and c_2 depend on the communication startup times, and c_3 on the communication transfer rate (c_1 also depends on floating point speed and transfer rate; however, startup times will usually dominate c_1).

Table 2 shows possible parameters for some distributed memory hypercubes. All of these estimates are drawn from machines which are either available today or planned for the near future. Machine (a) is the Intel hypercube used in our tests. Machine (b) represents a machine with 4 megaflop vector boards. Machine (c) is a more balanced machine, with both floating point and communication startup times reduced through a faster processor (such as the Intel 80386). Machine (d) is (c), but with four times as fast floating point. This might come from using a non-vector/pipelined fast floating point chip. Machine (e) is (c), using the vector boards from (b).

Table 2. Parameters for some possible distributed memory parallel processors

Machine	f_1	f_2	c_1	c_2	c_3
(a)	0.066	0.014	0.861	0.868	0.0302
(b)	0.00066	0.00014	0.861	0.868	0.0302
(c)	0.0132	0.0028	0.215	0.217	0.0302
(d)	0.0033	0.0007	0.215	0.217	0.0302
(e)	0.00066	0.00014	0.215	0.217	0.0302
(f)	0.00132	0.00028	0.0172	0.0176	0.00755

Finally, Machine (f) uses high speed combined communications and computation hardware, such as in the INMOS Transputer chips.

In Table 3 we show the predicted performance of a number of possible distributed memory parallel computers described in Table 2.

The relatively poor showing of machines (b) and (e) is due to the large communication terms, and suggests changes to both the algorithm (to reduce data exchanges) and the hardware (to make communication speeds comparable to floating point speeds). The figures for Machine (f) show that if communication speeds (particularly startups, the "L" term) are increased, very respectable speedups can be obtained, even for large number of processors. Similar calculations can be performed for proposed shared memory machines.

It is also interesting to compare the amount of programming work needed to run our experiments. In no case was much effort required to modify the original serial code to run on any of the parallel machines. The easiest to run was the Alliant, since the Fortran compiler on the Alliant automatically makes reasonably good use of the parallelism. Neither the Encore nor the Intel machines have such compilers, and each required a small amount of special coding. In addition, the Alliant required some special coding to improve the performance. This included some simple loop-rewriting to remove boundary condition testing from the insides of some loops, and the replacement of some loops with Fortran 8X-like intrinsic functions (in particular, sum) to speed up non-vector operations. The overall efforts were similar.

9. CONCLUSIONS

We have shown that significant parallelism is available at least for explicit methods for CFD. Our complexity estimates, validated by our experiments, show speedups limited only by the problem size for distributed memory architectures. For shared memory architectures, bottlenecks at the shared resource constrain the available speedups.

In actual practice, our results show that the communications costs associated with distributed memory parallel computers can seriously degrade the performance of these systems. Commercial shared memory computers such as the Alliant FX/8 have achieved a better balance of system components, at the (recognized) cost of limited parallelism. Since only massive parallelism has the promise of orders of magnitude increases in speed, we hope that our results and ones like them will stimulate both algorithm and hardware designers to overcome these practical problems.

In this paper we have considered shared memory and distributed memory parallel computers. In reality, actual computers are rarely so easily categorized. For example, a computer presenting the programmer with a shared memory model but implemented in hardware with several levels of distributed memory (a hierarchical memory) offers an intermediate form of parallel computer. The significance of our results is that, for significant parallelism, use of any shared resource that presents a bottleneck must be extremely limited. The success of the distributed memory code shows that explicit CFD codes can be easily written in such a form which avoids these bottlenecks.

Table 3. Predicted times and speedups for various distributed memory parallel processors. Times are in the column W and speedups are in the column s

		a		b		c		d		e		f	
k	m	W	s	W	s	W	s	W	s	W	s	W	s
32	250	137	30	9.1	4.5	30	28	10	20	5.3	7.8	3.4	25
64	250	73	57	8.6	4.8	16	50	6.6	31	4	10	1.9	44
128	250	41	101	8.6	4.8	9.6	86	4.7	44	3.4	12	1.1	74
32	500	527	31	16	11	110	30	32	25	12	14	12	28
64	500	268	61	12	13	57	58	18	45	7.9	21	6.2	53
128	500	139	119	11	15	30	109	11	76	5.7	30	3.4	97
256	500	75	221	10	16	17	196	7	117	4.5	37	1.9	171

Acknowledgements—The authors wish to thank James D. Wilson of AFOSR, Michael J. Werle and Joseph R. Caspar of UTRC, and Martin H. Schultz of Yale University, for their comments and suggestions.

This work was supported in part by the Air Force Office of Scientific Research under contract AFOSR-86-0098 and in part by United Technologies Research Center.

REFERENCES

1. M. M. Rai, Unsteady Three-dimensional Navier–Stokes simulations of turbine rotorstator interaction. *23rd Joint AIAA/ASME Propulsion Conf.*, San Diego, California, AIAA paper 87-2058 (1987).
2. A. Jameson, Successes and challenges in computational aerodynamics. *AIAA 8th Computational Fluid Dynamics Conf.*, Honolulu, Hawaii, AIAA paper 87-1184 (1987).
3. G. M. Johnson, Parallel processing in fluid dynamics. *ASME Fluids Engineering Conference*, Cincinnati, Ohio, ISC Technical Report 87003, Institute for Scientific Computing. Fort Collins (1987).
4. E. M. Murman, D. Modiano, R. Haimes and M. Giles, Performance of several CFD loops on parallel processors. MIT Department of Aeronautics and Astronautics report CFDL-TR-88-3 (1988).
5. W. Gropp and D. E. Keyes, Domain decomposition on parallel computers. In: *Domain Decomposition Methods* (Edited by T. Chan *et al.*). SIAM, Philadelphia (1989).
6. T. Chan, Y. Saad and M. Schultz, Solving elliptic partial differential equations on hypercubes. In: *Hypercube Multiprocessors 1986* (Edited by M. Heath). SIAM, Philadelphia (1986).
7. M. Berger and A. Jameson, Automatic adaptive grid refinement for the Euler equations. Courant Mathematics and Computing Laboratory, New York University, Technical Report DOE/ER/03077-202 (1983).
8. W. Gropp and D. Keyes, Domain decomposition with local mesh refinement. Yale University Department of Computer Science Research Report Yale/DCS/RR-726 (1989).
9. R. W. MacCormack, The effect of viscosity in hypervelocity impact cratering. *AIAA Hypervelocity Impact Conference*, Cincinnati, Ohio, AIAA Paper No. 69-354 (1969).
10. A. Jameson, Transonic flow calculations. Department of Mechanical and Aerospace Engineering, Princeton University, MAE Report 1651 (1983).
11. L. B. Wigton, N. J. Yu and D. P. Young, GMRES acceleration of computational fluid dynamics codes. *AIAA 7th Computational Fluid Dynamics Conf.*, Cincinnati, Ohio, AIAA paper 85-1494 (1985).
12. F. André, D. Herman and J.-P. Verjus, *Synchronization of Parallel Programs*. The MIT Press, Cambridge (1985).
13. Y. Saad and M. H. Schultz, Topological properties of hypercubes. Yale University Department of Computer Science Research Report Yale/DCS/RR-389 (1985).
14. S. Eisenstat, private communication.
15. H. F. Jordan, Interpreting parallel processor performance measurements. *Siam J. sci. stat. Compt.* **8,** s220–s226.
16. D. Nicol and J. Saltz, Optimal pre-scheduling of problem remappings. ICASE Report 87-52 (1987).

A PARALLEL PLANAR GENERALIZED YEE ALGORITHM FOR THE ANALYSIS OF MICROWAVE CIRCUIT DEVICES

STEPHEN D. GEDNEY

Department of Electrical Engineering, University of Kentucky, Lexington, KY 40506–0046, U.S.A.

AND

FAIZA LANSING

Spacecraft Telecommunications Equipment Section, Jet Propulsion Laboratory, Pasadena, CA 91109, U.S.A.

SUMMARY

The planar generalized Yee (PGY) algorithm is an extension of the generalized Yee algorithm and the discrete surface integral (DSI) methods, which are based on explicit time-marching solutions of Maxwell's equations. Specifically, the PGY algorithm exploits the planar symmetries of printed microwave circuit devices, achieving great savings in both CPU time and memory. Since the PGY algorithm is an explicit method, it has a high degree of parallelism. To this end, a highly scalable parallel algorithm based on a spatial decomposition of the general unstructured mesh is presented. Two spatial decompositions are compared, the recursive inertia partitioning (RIP) algorithm and the Greedy algorithm. The Greedy algorithm provides optimal load balance, whereas the RIP algorithm more effectively minimizes shared boundary interface lengths. Through numerical examples, it is demonstrated that the Greedy algorithm provides superior speed-ups. It is also demonstrated that the parallel PGY algorithm is a highly scalable algorithm.

1. INTRODUCTION

The generalized Yee algorithm,[1,2] and the discrete surface integral (DSI) algorithm[3] have proven to be highly robust and accurate techniques for the analysis of microwave circuit devices. These methods are explicit time-marching schemes derived from the discretization of Ampère's and Faraday's laws in their integral form.[1–3] To this end, the vector fields are projected onto the edges of a dual, staggered grid which is assumed to be unstructured and irregular. This is in contrast to the traditional finite difference time-domain (FDTD) method[4] which is based on regular and orthogonal grids, or the non-orthogonal FDTD method,[5–8] which is based on irregular structured grids. The advantage of using unstructured grids, is that highly complex geometries can be accurately modelled with the aid of automatic grid generation techniques. Unfortunately, a disadvantage of generalized Yee and DSI algorithms is that the numerical grid must be stored. Moreover, the sparse matrices associated with them must also be stored,[1,2] greatly limiting the size of the problem which can be solved.

The memory requirements of the generalized Yee algorithm and the DSI algorithm can be greatly relaxed by exploiting symmetries in the model. A large class of microwave circuits can be said to have *planar symmetry*, which is recognized for three-dimensional geometries that can be uniquely described by a projection onto a two-dimensional plane. It was shown in Reference 9 that by exploiting this symmetry, the entire three-dimensional circuit geometry can then be described uniquely by a two-dimensional grid. Subsequently, the grid used to analyse the three-dimensional problem can be described by an unstructured two-dimensional grid in a transverse plane and as a regular grid in the third dimension. Thus only the two-dimensional grid need be stored. Furthermore, it is shown below that the sparse update matrices of this planar generalized Yee (PGY) algorithm need only be constructed for the two-dimensional grid. This greatly reduces the memory requirements of the algorithm to the extent that it is actually as memory efficient as the FDTD algorithm.

The PGY algorithm is an efficient computational method. However, since it is based on a

Reprinted with permission from *International Journal of Numerical Modeling: Electronic Networks, Devices and Fields*, S. Gedney and F. Lansing, "A Parallel Planar Generalized Yee-Algorithm for the Analysis of Microwave Circuit Devices," Vol. 8, pp. 249-264, May-August 1995.

volume discretization, the computational demand can increase substantially with the size of the problem geometry. For example, the number of floating-point operations per time iteration will increase as $O(N)$, where N is the number of unknowns in the discrete volume. However, the number of time iterations required to reach a steady state will also increase as $O(\sqrt{N})$ (the number of time iterations will be inversely proportional to the minimum edge length in the model as well). The analysis of single component circuits can typically be modelled on conventional workstations or sequential computers in reasonable amounts of time. On the other hand, the analysis of multicomponent circuits requires much greater resources.

It is becoming much more evident that distributed parallel computing is a highly cost-effective means of achieving supercomputing performance. With the rapid increase in computational power of RISC microprocessors, as well as the increase in speeds of local area networks, highly cost-effective supercomputing can be achieved through loosely coupled distributed parallel systems using message-passing protocols such as PVM (parallel virtual machine). Tightly coupled distributed memory multiprocessor computers, such as the Intel Paragon, the Cray T3D, the IBM SP2, or the Convex Exemplar, will provide higher performance, principally because they have faster dedicated networks interconnecting processors with greatly reduced latency times. These architectures have already demonstrated the potential of 100 Gflops performance for practical scientific applications.

The PGY algorithm is extremely well suited for high performance distributed memory computing. Since the method is explicit, the kernel of the algorithm consists of a series of matrix–vector products. These operations are conveniently parallelized using a spatial decomposition of the unstructured mesh. It is shown that by treating the matrix as a subassembly of matrices, where each submatrix is associated with each spatial subdomain, interprocessor communication can be minimized, resulting in a highly scalable parallel algorithm. Furthermore, owing to the regularity of the grid along the vertical direction, vector/pipelining can be exploited to further increase the floating-point speed of the algorithm.

The focus of this paper is on the development of an efficient implementation of the PGY algorithm on high-performance parallel computers. Section 2 presents the PGY algorithm. Section 3 discusses the parallelism inherent within the algorithm, and presents an efficient parallel algorithm based on a spatial decomposition of the three-dimensional mesh. Section 4 presents the inherent vectorism in the PGY algorithm, and the advantages of exploiting the vector/pipelining of RISC-based parallel computers. Finally, some numerical examples are presented in section 5, which illustrate the efficiency and the scalability of the parallel PGY algorithm.

2. THE PLANAR GENERALIZED YEE ALGORITHM

The planar generalized Yee algorithm is based on a direct solution of the time-dependent Maxwell's equations in their integral form. The electric and magnetic field intensities are initially normalized as

$$\mathbf{e} = \mathbf{E}/\sqrt{\eta_o}$$
$$\mathbf{h} = \mathbf{H}\sqrt{\eta_o} \qquad (1)$$

where η_o is the characteristic wave impedance in free space. Faraday's law and Ampère's law are then expressed in their integral form as

$$\oint_c \mathbf{e} \cdot d\mathbf{l} = -\frac{1}{c_o}\frac{\partial}{\partial t} \oiint_s \mu_r \mathbf{h} \cdot d\mathbf{S} \qquad (2)$$

$$\oint_c \mathbf{h} \cdot d\mathbf{l} = \frac{1}{c_o}\frac{\partial}{\partial t} \oiint_s \epsilon_r \mathbf{e} \cdot d\mathbf{S} + \eta_o \oiint_s \sigma \mathbf{e} \cdot d\mathbf{S} \qquad (3)$$

where c_o is the free space velocity of light, μ_r and ϵ_r are the relative permeability and permittivity, respectively, and σ is the absolute conductivity. The principal advantage of this normalization is that the magnitudes of **e** and **h** will be of the same order, reducing rounding error. Furthermore, it is much more convenient to work with the relative permittivity and permeabilities rather than their absolute values.

Faraday's and Ampère's laws are expressed in a discrete form by mapping **e** and **h** into a discrete three-dimensional space. The mapping consists of projecting the vector fields onto the edges of a dual grid, composed of two staggered grids, referred to as the primary and secondary grids. Each grid is a three-dimensional grid that is described as being regular along the vertical direction (assumed to be the z-direction), and is unstructured in the horizontal direction.

Conceptually, this grid can be generated by extruding a two-dimensional unstructured grid in the vertical direction, and segmenting it at discrete heights, as illustrated in Figure 1. The secondary grid is staggered within the primary grid such that its vertices lie at the centroids of the primary grid cells, and the edges of the secondary grid connect the centroids by passing through the faces of the primary grid.

The electric and magnetic fields are then decomposed into orthogonal components

$$\mathbf{e} = \mathbf{e}_t + \hat{\mathbf{z}}e_z$$
$$\mathbf{h} = \mathbf{h}_t + \hat{\mathbf{z}}h_z \tag{4}$$

Subsequently, the transverse electric and magnetic fields are mapped onto the horizontal edges of the primary and secondary grids, respectively. Likewise, the vertical electric and magnetic fields are mapped onto the vertical edges of the primary and secondary grids, respectively. The vector fields are assumed to be constant along their respective edge lengths, as well as over the dual face through which they pass.

Based on the above discretization, Faraday's and Ampère's laws are then mapped into the discrete space. The time derivative is then approximated using a central difference expression, which is second-order accurate if the fields are staggered in time. This leads to[9]

$$h_{z_j}^{n+1}(k) = h_{z_j}^{n}(k) - \frac{c_o}{A_p}\frac{\Delta t}{\mu_r(k)}\left[\sum_{i=1}^{N_{pj}} e_{t_i}^{n+(1/2)}(k)l_i\right] \tag{5}$$

$$b_{t_i}^{n+1}\left(k+\frac{1}{2}\right) = b_{t_i}^{n}\left(k+\frac{1}{2}\right) - \frac{c_o\Delta t}{\Delta z|l_i|}$$
$$\left[\left(e_{z_m}^{n+(1/2)}\left(k+\frac{1}{2}\right) - e_{z_{m+1}}^{n+(1/2)}\left(k+\frac{1}{2}\right)\right)\Delta z + \left(e_{t_i}^{n+(1/2)}(k+1) - e_{t_i}^{n+(1/2)}(k)\right)l_i\right] \tag{6}$$

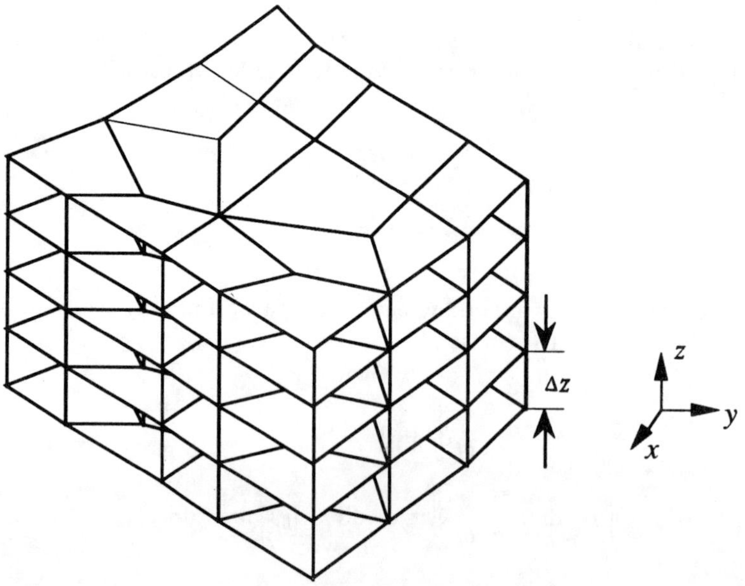

Figure 1. An example of the primary grid described by similar two-dimensional unstructured grids cascaded in the vertical z-direction in a regular sense

$$e_{z_m}^{n+(3/2)}\left(k+\frac{1}{2}\right) = \left[\left(\frac{\epsilon_{r_m}\left(k+\frac{1}{2}\right)}{c_o\,\Delta t} - \frac{\sigma_m\left(k+\frac{1}{2}\right)\eta_o}{2}\right)e_{z_m}^{n+(1/2)}\left(k+\frac{1}{2}\right)\right.$$

$$\left. = -\frac{1}{A_s}\left[\sum_{i=1}^{N_{s_m}} h_{t_i}^{n+1}\left(k+\frac{1}{2}\right)l_i\right]\right]\frac{1}{\left(\frac{\epsilon_{r_m}\left(k+\frac{1}{2}\right)}{c_o\,\Delta t} + \frac{\sigma_m\left(k+\frac{1}{2}\right)\eta_o}{2}\right)} \tag{7}$$

$$d_{t_i}^{n+(3/2)}(k) = \frac{1}{\left(\frac{1}{c_o\,\Delta t} + \frac{\sigma_i(k)\eta_o}{2\epsilon_{r_i}(k)}\right)}\left[\left(\frac{1}{c_o\,\Delta t} - \frac{\sigma_i(k)\eta_o}{2\epsilon_{r_i}(k)}\right)d_{t_i}^{n+(1/2)}(k) - \frac{1}{\Delta z/l_i}\right.$$

$$\left.\left[\left(h_{z_j}(k) - h_{z_{j+1}}(k)\right)\Delta z + \left(h_{t_i}^{n+1}\left(k-\frac{1}{2}\right) - h_{t_i}^{n+1}\left(k+\frac{1}{2}\right)\right)l_i\right]\right] \tag{8}$$

where d_t and b_t are the flux densities in the transverse plane, Δt is the time increment, n is the time index, k is the index along z, A_p and A_s are the areas of the primary and secondary grid faces, respectively, N_{p_i} and N_{s_j} are the number of edges bounding the ith primary and the jth secondary grid faces, respectively, and the l_i are the length vectors of the primary or secondary grid edges. The material parameters ϵ_r, μ_r and σ are assumed to be piecewise homogeneous in both the z-direction and the transverse direction. At the interface of two unlike media, the parameters are assigned an average value, as described in Appendix A of Reference 1.

Based on (2) and (3), it is recognized that the flux densities updated in (6) and (8) are normal to the faces. However, the corresponding field intensities on the dual edges passing through these faces are not necessarily normal to the faces. As a result, the flux densities must be projected onto the edges before the dual fields can be updated. Since only one component of the field is locally known, an auxiliary operator must be introduced to perform the projection. To this end, the projection operators implemented in [1] are used to project the fields onto the dual edge passing through the face.

Let $\mathbf{N}_P$ be the normal area vector to a primary grid transverse face, and $\hat{\mathbf{s}}$ be the unit vector along the dual grid edge passing through the face (Figure 2). Using (6) the magnetic flux densities projected onto the normals of all the primary grid transverse faces are updated. Subsequently, for each face, a general flux density vector $\mathbf{b}$ is introduced. From (6), $\mathbf{b}\cdot\mathbf{N}_P$ is known at each edge. In Figure 2, the edge identified is bound by vertices *1* and *2*, which is identified by the index $i = 1, 2$. Each vertex is also shared by two additional edges which share a common cell. Let j represent one of these edges, where $j = 1, 2$. The normal area vector to the jth edge associated with the ith vertex is $\mathbf{N}_{D_{i,j}}$. Subsequently, we define a general flux density vector $\mathbf{b}_{i,j}$

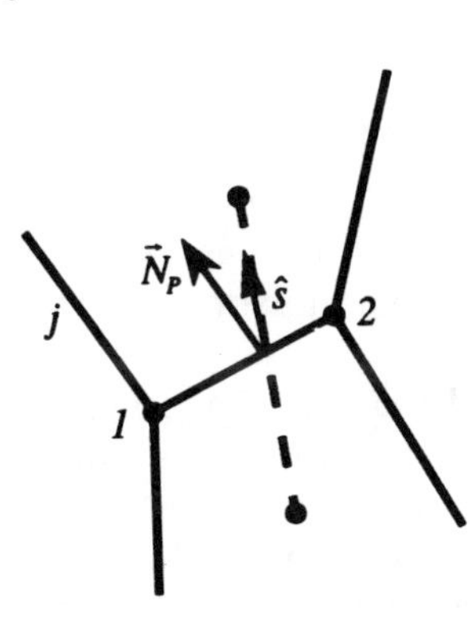

Figure 2. Normal to a transverse primary face, and a dual edge passing through the face

to be the local estimate of the magnetic flux vector associated with the ith vertex and the jth edge, where $\mathbf{b}_{i,j}$ is computed by solving the two-dimensional system of equations

$$\begin{aligned}\mathbf{b}_{i,j} \cdot \mathbf{N}_D &= \mathbf{b} \cdot \mathbf{N}_D \\ \mathbf{b}_{i,j} \cdot \mathbf{N}_{D_{i,j}} &= \mathbf{b} \cdot \mathbf{N}_{D_{i,j}}\end{aligned} \tag{9}$$

where the right-hand side is known from (6). Subsequently, introducing the weighting coefficient $w_{i,j} = |\hat{\mathbf{z}} \cdot (\mathbf{N}_D \times \mathbf{N}_{D_{i,j}})|$, the flux density projected onto the dual edge is expressed as

$$\mathbf{b} \cdot \hat{\mathbf{s}} = \frac{\sum_{i=1}^{2} \sum_{j=1}^{2} w_{i,j}(\mathbf{b}_{i,j} \cdot \hat{\mathbf{s}})}{\sum_{i=1}^{2} \sum_{j=1}^{2} w_{i,j}} \tag{10}$$

The field updates are then computed using (6)–(10). However, it is realized that computing the parameters for these equations requires a significant number of floating-point operations, leading to a highly inefficient algorithm. However, by employing standard finite element type techniques, the computational efficiency can be greatly enhanced by treating these linear operators as sparse matrices. To this end, (6)–(10) can be expressed in reduced form as

$$[H_z^{n+1}]_k = [H_z^n]_k + \overline{\overline{A}}_{h_z} [E_t^{n+(1/2)}]_k \tag{11}$$

$$[B_t^{n+1}]_{k+(1/2)} = [B_t^n]_{k+(1/2)} + \overline{\overline{A}}_{h_t} \begin{bmatrix} E_{z_{k+(1/2)}}^{n+(1/2)} \\ E_{t_{k,k+1}}^{n+(1/2)} \end{bmatrix} \tag{12}$$

$$[H_t^{n+1}]_{k+(1/2)} = \overline{\overline{A}}_{h_p} [B_t^{n+1}]_{k+(1/2)} \tag{13}$$

$$[E_z^{n+(3/2)}]_{k+(1/2)} = \overline{\overline{D}}_{e_z} [E_z^{n+(1/2)}]_{k+(1/2)} + \overline{\overline{A}}_{e_z} [H_t^{n+1}]_{k+(1/2)} \tag{14}$$

$$[D_t^{n+(3/2)}]_k = \overline{\overline{D}}_{e_t} [D_t^{n+(1/2)}]_k + \overline{\overline{A}}_{e_t} \begin{bmatrix} H_{z_{k,k+1}}^{n+1} \\ H_{t_{k+(1/2)}}^{n+1} \end{bmatrix} \tag{15}$$

$$[E_t^{n+(3/2)}]_k = \overline{\overline{A}}_{e_p} [D_t^{n+(3/2)}]_k \tag{16}$$

where the subscript k refers to the discrete height along the z-direction, D_t and B_t are the flux densities, the $\overline{\overline{D}}$s are diagonal matrices, and the $\overline{\overline{A}}$s are sparse matrices. Note that these matrices are only associated with the two-dimensional grid since they are the same for all values of k (inhomogeneities in material parameters are easily built into these expressions). As a result, the additional memory required to store these matrices is nominal.

Finally, the field updates are performed using (11)–(16), providing an extremely efficient computational technique for the simulation of the time-varying fields that is second-order accurate. Furthermore, the solution is stable, providing the timestep satisfies the stability criterion[9]

$$\Delta t < \frac{1}{c_o \sqrt{\left(\frac{1}{\Delta z}\right)^2 + 2\left(\frac{1}{l_{\min}}\right)^2}} \tag{17}$$

where $l_{\min}$ is the minimum edge length in the model. This stability criterion is approximate and has been derived through numerical experimentation. An exact expression will depend on the eigenspectrum of the operators in (11)–(16). However, this approximation closely estimates the exact stability criterion. It is noticed that the numerical stability is dependent upon the smallest edge length in the model, or the *finest* portion of the mesh. On the other hand, the accuracy of the method will be dependent on the *coarsest* portion of the mesh. To this end, it is recommended that cell sizes are smaller than one-fifteenth of a wavelength (corresponding to the maximum frequency component of signal). Furthermore, care should be taken that the automatic mesh generator does not unnecessarily generate extremely small cells as this will lead to small timesteps.

3. THE PARALLEL ALGORITHM

In this paper, we are principally concerned with the development of a parallel PGY algorithm that exploits distributed-memory multiprocessor computers. Specifically, these are computers that host independent processors, each with its own local memory, which communicate over a network via message passing. This applies to both a loosely coupled architecture, such as a cluster of workstations, or a tightly coupled architecture, such as the Intel Paragon or the Cray T3D. It is the purpose of this section to present the development of a highly efficient parallel PGY algorithm.

Amdahl's law[10] implies that if a computational process is distributed over P processors of a multiprocessor computer, the maximal *speed-up* one can achieve over performing the same process on a single processor of the computer is P. In fact, owing to *serialism*, the speed-up will always be less than P. Thus the principal objective in designing a parallel algorithm is to minimize the amount of serialism in the parallel algorithm.

Serialism can be defined as computation that would be better done on a uniprocessor system than a parallel system. It also includes additional computation that must be performed by the parallel algorithm, but is not required by a sequential algorithm. Some examples of tasks that lead to serialism are: (1) load imbalances, (2) interprocessor communication, (3) latency, (4) synchronization, and (5) redundant computation. In a parallel system, the total computational time to complete a global task is equal to the time required by the *slowest* processor to complete its local task. Therefore, it is important to evenly distribute the work effort among all the processors, namely, the work load must be balanced. In contrast, load imbalances lead to processor idle time and reduce parallel efficiency.

Interprocessor communication, i.e., message passing, introduces additional effort that is required by the parallel algorithm which is not performed by the sequential algorithm, and it leads to the degradation of the parallel efficiency. The computational time required to perform an interprocessor communication can be divided into two parts, (1) the time it takes to initiate the communication, which is referred to as the *latency time*, and (2) the time/byte required to transmit a data packet between two remote processors. In most tightly coupled distributed-memory multiprocessor computers, the time required to transmit a byte of data is of the order of nanoseconds. However, the latency time is typically of the order of tens of microseconds. Therefore, it is much more profitable to send a small number of large data packets rather than a large number of small data packets. Furthermore, since single floating-point operations are performed on the orders of nanoseconds, it is extremely important to maximize the ratio of the time a processor spends performing floating-point operations to the amount of time spent performing interprocessor communication. Unfortunately, as the number of processors increases for a fixed problem size, this ratio inevitably decreases, leading to the degradation of the parallel efficiency.

Synchronization can also lead to serialism in a parallel algorithm. This leads to processors which are left idle while waiting for information from a remote processor. A good example of this is a recursive algorithm, where a process on one processor is dependent on data from another processor, whereas, another processor is dependent on the data still to be processed on the first processor. Even though the workload may be balanced, the parallel efficiency can be extremely poor due to the required synchronization.

Finally, redundant computation is computation done on a number of processors concurrently that could have been done by a single processor. Often, if the time spent doing redundant computation is small, it is cheaper in terms of overall CPU time to do it redundantly rather than to have to perform an additional interprocessor communication. Again, the reason being that the latency time to initiate the communication will be of the order of microseconds, compared to the nanoseconds required to perform the floating-point operations. However, a substantial amount of redundant computation can lead to degradation in parallel efficiency.

It is therefore important to design a parallel algorithm which is load balanced, has a minimal number of interprocessor communications, minimizes the size of the data packets being communicated, is asynchronous in operation, and minimizes the amount of redundant or serial computation. To this end, the parallel PGY algorithm is based on a spatial decomposition of the three-dimensional grid into contiguous, non-overlapping subdomains. The partitioning of the two-dimensional unstructured mesh is being performed using one of two different techniques: (1) the recursive inertia partitioning (RIP) algorithm,[11] which is a power-of-two algorithm that is ideal for hypercube computers, and (2) the Greedy algorithm,[12] which a non-power-of-two algorithm providing a more general decomposition. Both algorithms are quite simple to implement, and are computationally efficient even for large meshes. The RIP algorithm has the advantage that it minimizes the number of grid edges on shared boundaries, however, it typically has 10 per cent load imbalances in the partition. The Greedy algorithm has the advantage that it partitions the mesh in a manner that is ideally load balanced; however, the number of edges on shared boundaries is typically greater than that yielded by the RIP algorithm. Furthermore, the Greedy algorithm can sometimes result in disjoint subdomains.[12] Nevertheless, in section 5, it is illustrated that the load balancing is more important than minimizing shared boundary lengths, and the Greedy algorithm leads to improved parallel efficiencies. The spatial decomposition along the third, regular dimension is done using a trivial partitioning scheme of the regular grid.

Once the mesh is decomposed into subdomains, one subdomain is assigned to each processor. The matrices in (11)–(16) are then expressed as a subassembly of matrices, where each submatrix represents the updates of the fields within each subdomain. Subsequently, the matrix vector products are simply expressed as

$$\overline{\overline{A}}x = \sum_{i=1}^{P} \overline{\overline{A}}_i x_i \tag{18}$$

where P is the total number of processors.

This approach has a number of advantages. The spatial decomposition is done in a manner such that each processor performs roughly the same number of floating-point operations each time iteration, leading to a balanced parallel algorithm. Secondly, only the local matrices and field vectors need be stored on each processor. This algorithm maximizes the ratio of computation to communication leading to a highly scalable algorithm. This can be seen by further decomposing $\overline{\overline{A}}_i$ as

$$\overline{\overline{A}}_i = \overline{\overline{A}}_i^{\text{int}} + \sum_{j=1}^{N_{\text{shared}}} \overline{\overline{A}}_{i,j}^{\text{shared}} \tag{19}$$

where $\overline{\overline{A}}_i^{\text{int}}$ are the rows of $\overline{\overline{A}}_i$ associated with all field vectors internal to the ith processors' subdomain, $\overline{\overline{A}}_{i,j}^{\text{shared}}$ are the rows of $\overline{\overline{A}}_i$ associated with all field vectors in the ith processors' subdomain that lie on the boundary shared with the jth domain, and N_{shared} is the number of processors that share boundaries with the ith processor. Subsequently, local to each processor, (18) is actually performed as

$$\overline{\overline{A}}_i x_i = \overline{\overline{A}}_i^{\text{int}} x_i + \sum_{j=1}^{N_{\text{shared}}} (\overline{\overline{A}}_{i,j}^{\text{shared}} x_i) + \sum_{j=1}^{N_{\text{shared}}} R_x (\overline{\overline{A}}_{j,i}^{\text{shared}} x_j) \tag{20}$$

where R_x is the receive operator, receiving the vector of data from the jth processor. The first two expressions on the right-hand side of (20) are done completely in parallel on each processor, and the final term requires interprocessor communication. Since the RIP and the Greedy algorithms both attempt to minimize the lengths of the shared boundaries, the ratio of computation to communication is high, leading to a highly scalable algorithm, as demonstrated in section V.

4. THE VECTOR ALGORITHM

Almost all of today's distributed memory parallel computers utilize RISC processors as central processing units (CPUs). Many of the RISC processors rely on pipelining to achieve maximum floating-point operation speeds. They also rely on high-speed cache to reduce memory access time. As a result, to optimize the processor floating-point operation speeds, dominant computational tasks must be vectorized. Vectorization is realized on the innermost loops of any multidimensional loop structure and can be achieved in an optimal manner when: (1) the inner loops are truly vector operations and are not corrupted by function calls, logical statements, or indirect addressing, (2) the innermost column index of multidimensional arrays corresponds to the index of the inner loop, (3) the length of the inner loop is equal to or greater than the optimal vector length (typically determined by the vector length of a vector processor, or the cache size of a pipelined or a superscalar processor).

The matrices in (11)–(16) are assumed to be stored in a compressed format, namely, only the non-zero entries are actually stored and pointers are used to identify the row and column number of each entry. Thus, the matrix vector product performed in (20) is performed using indirect addressing, i.e., the effective index of the vector x is determined by either a pointer, or an integer array, and x is addressed in a random fashion. As a result, poor floating-point operation will result when performing the linear operations in (20) on RISC processors that have architectures employing either vector units, pipelined floating-point units, or even superscalar floating-point pipelined units. This appears to be predominately due to the inability to stream the vectors into the local high-speed cache, and subsequently through the floating-point control unit.

Vectorization can be achieved by exploiting the regularity of the sparse update matrices along the vertical direction. Specifically, the operations dependent on the discrete index k corresponding to the vertical z-direction can be placed on the inner loop of the matrix–vector product operation. As a result, the indirectly addressed variables become constants within the inner loop, and are addressed only in the outer loop. This leads to a very efficient operation that has increased floating-point speeds on a RISC processor. As an example, Figure 3 illustrates the Fortran loop that is used to perform the update of the interior vertical electric-field intensities, specifically from (14), assuming $\sigma = 0$. It is assumed that the secondary grid cells in the transverse plane are arbitrary polygons, and each row has a random number of non-zero elements. To this end, the pointer *iez* points to the first non-zero entry of the ith row in the vectors *aez* and *jez, jez* is the column indices of each entry of *aez*, and *aez* contains the non-zero entries of the matrix. Also *epsz* is the inverse of the relative permittivity local to the vertical edge. The inner loop, loops through the vertical index k, and within this loop *jez* and *aez* are constants.

By exploiting vectorization on the RISC processor, a substantial speed-up can be achieved. As an example, the CPU speed versus the vector loop lengths were measured on an Intel i860 microprocessor. The i860 is a 32-bit RISC processor with a quoted peak performance of 80 Mflops (millions of floating-point operations per second) single precision and 60 Mflops double-precision at a 40-MHz clock cycle. The i860 has a 4-Kbyte instruction cache organized as a two-way set-associative memory with 32 bytes per cache block. It also has a separate data cache which is a two-way set-associative memory of 8 Kbytes. There are also two floating-point units within the processor, namely, a multiplier unit and an adder unit, which can be used separately or simultaneously under the co-ordination of the floating-point control unit. This design supports dual operation floating-point instructions such as 'add-and-multiply' and 'subtract-and-multiply' by using both the adder and multiplier units in parallel. Vector operations can be pipelined through the floating-point units yielding maximal performance.

The benchmarked performance of the Fortran loop in Figure 3 versus the vector length *nz* is illustrated in Figure 4 (Case 1). This was performed using single-precision floating-point arithmetic. This is compared to the case in which the two inner loops in Figure 3 are reversed (Case 2). Interestingly, as *nz* is increased, the performance of this loop decreases. Overall, this illustrates that loops with indirect addressing cannot take advantage of pipelining. This is a problem that plagues many finite element codes. An additional benchmark was performed to vary the value of *nnod* (in Figure 4, *nnod* = 1000). In both cases, the performance is effectively independent of the length of *nnod*. These results are not illustrated here.

5. NUMERICAL RESULTS

A Fortran program based on the planar generalized Yee-algorithm has been developed on a 32-node Intel iPSC/860 hypercube. This same program has been directly ported to the 512-node Intel Delta Supercomputer, a two-processor Cray-YMP, and an HP 720 workstation. The program

```
        do 10 i = 1,nnod
           do 10 j = iez(i),iez(i+1)-1
              j1 = jez(j)
              aj = aez(j)*codt
              do 10 k = 1,nz-1
                 ez(k,i) = ez(k,i)+aj*ht(k,j1)*epsz(k,i)
10         continue
```

Figure 3. Fortran loop performing the update of the vertical electric field using (14)

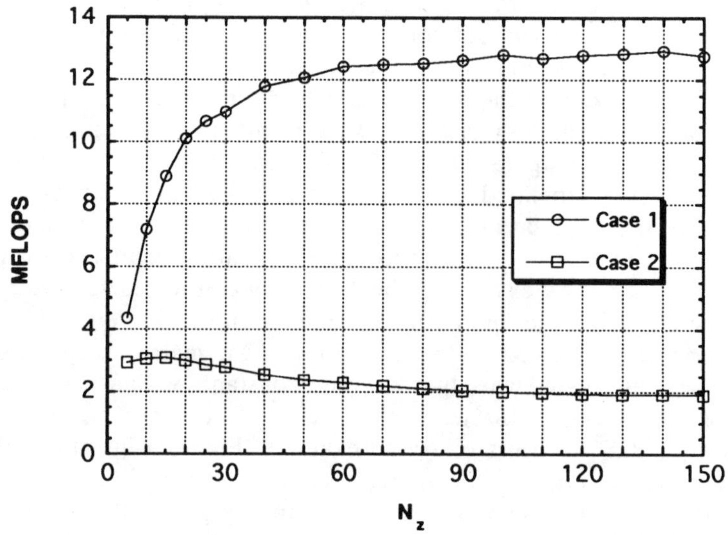

Figure 4. Mflops versus number of cells in the vertical direction recorded on a single 40-MHz i860 RISC processor. Case 1 refers to the loop in Figure 3, and Case 2 is the same update with the two inner loops switched

is interfaced with a commercial CAD software package (SDRC I-DEAS) running on an HP 720 workstation. The CAD software is used to design and build the circuit models. It is also used to generate the two-dimensional unstructured mesh via automatic grid generation techniques. The node-based two-dimensional mesh is subsequently partitioned on the workstaion using either the Greedy or the RIP algorithms. Since the generation of the meshes of very large models can be extremely time-consuming and memory-intensive, an automatic mesh refinement technique has been implemented within the parallel algorithm. The refinement is done in a fairly trivial manner, as illustrated in Figure 5, and can be done completely in parallel. Furthermore, since it is a gobal refinement, a mesh partitioning does not need to be repeated.

A second-order absorbing boundary condition (ABC) known as the dispersive boundary condition[13] is used to update the fields on truncation boundary walls, minimizing any non-physical reflections. To maintain the second-order accuracy of the boundary condition, the two-dimensional mesh is padded with two layers of rectangular cells (this is done prior to the partitioning to maintain load balance of the parallel algorithm).

A number of numerical simulations have been run to validate the code and to demonstrate its robustness.[9] As an example, consider a circular cylindrical via through a ground plane, as illustrated in Figure 6. The via connects two 50-Ω microstrip lines using a cylindrical post passing through a circular hole in a ground plane. The two-dimensional mesh representing this geometry is illustrated in Figure 7. The mesh models the three-dimensional geometry as it is projected onto a two-dimensional plane. Since there are conductors and dielectrics at different heights in the

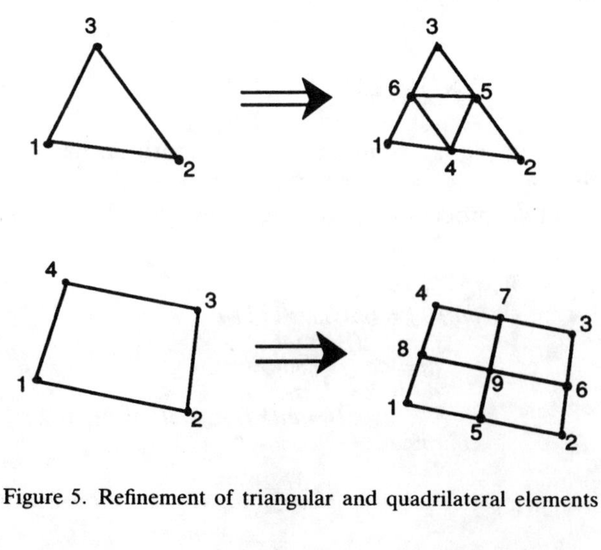

Figure 5. Refinement of triangular and quadrilateral elements

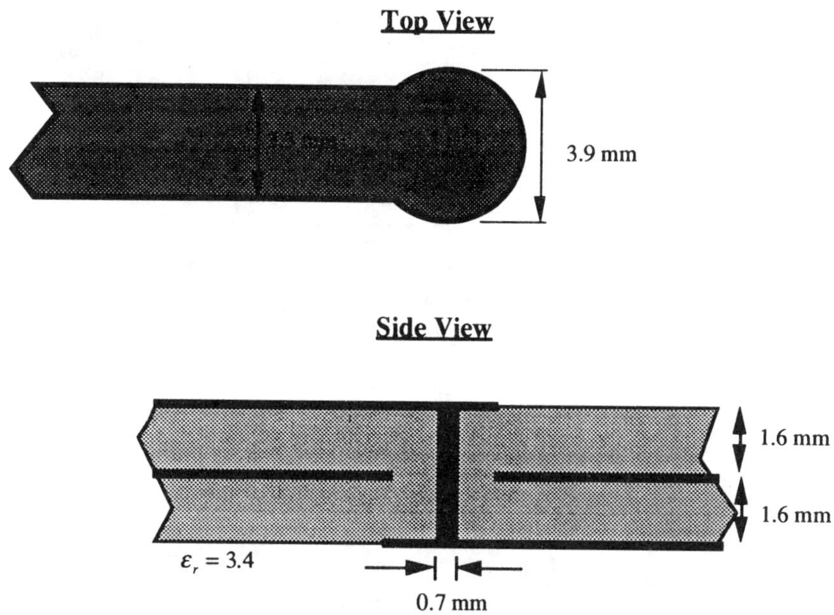

Figure 6. Geometry of a cylindrical via through a PEC ground plane

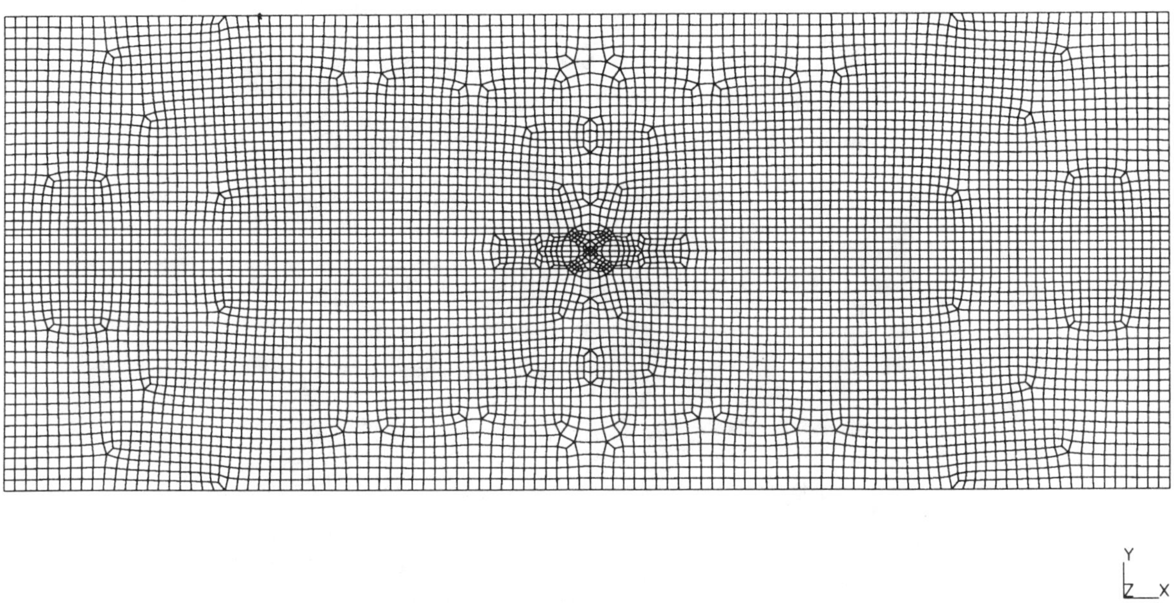

Figure 7. Two-dimensional mesh representing the cylindrical via

three-dimensional model, each cell is assigned a material identification number. Correspondingly, a table is built which identifies the discrete heights, thickness and properties of conductor strips and material slabs.

Both the RIP algorithm[11] and the Greedy algorithm[12] were used to perform the spatial decomposition of the mesh. As described in section 3, the Greedy algorithm provides a better load balance, while the RIP algorithm better minimizes the lengths of the shared boundaries. Figures 8 and 9 illustrate the spatial decomposition of the two-dimensional quadrilateral mesh illustrated in Figure 7 into 32 subdomains using the RIP and Greedy algorithms, respectively. A means of comparing the load balancing of the two algorithms, is to compare the maximum number of cells and nodes in a subdomain to the minimum number. This is illustrated in Table I. It is seen that the Greedy algorithm yields a more optimal load balancing overall. However, the number of edges on shared boundaries are greater in the mesh decomposed by the Greedy algorithm than that decomposed by the RIP algorithm.

The via was analysed on a 32-node iPSC/860. Each node of the iPSC/860 hosts a 40-MHz i860

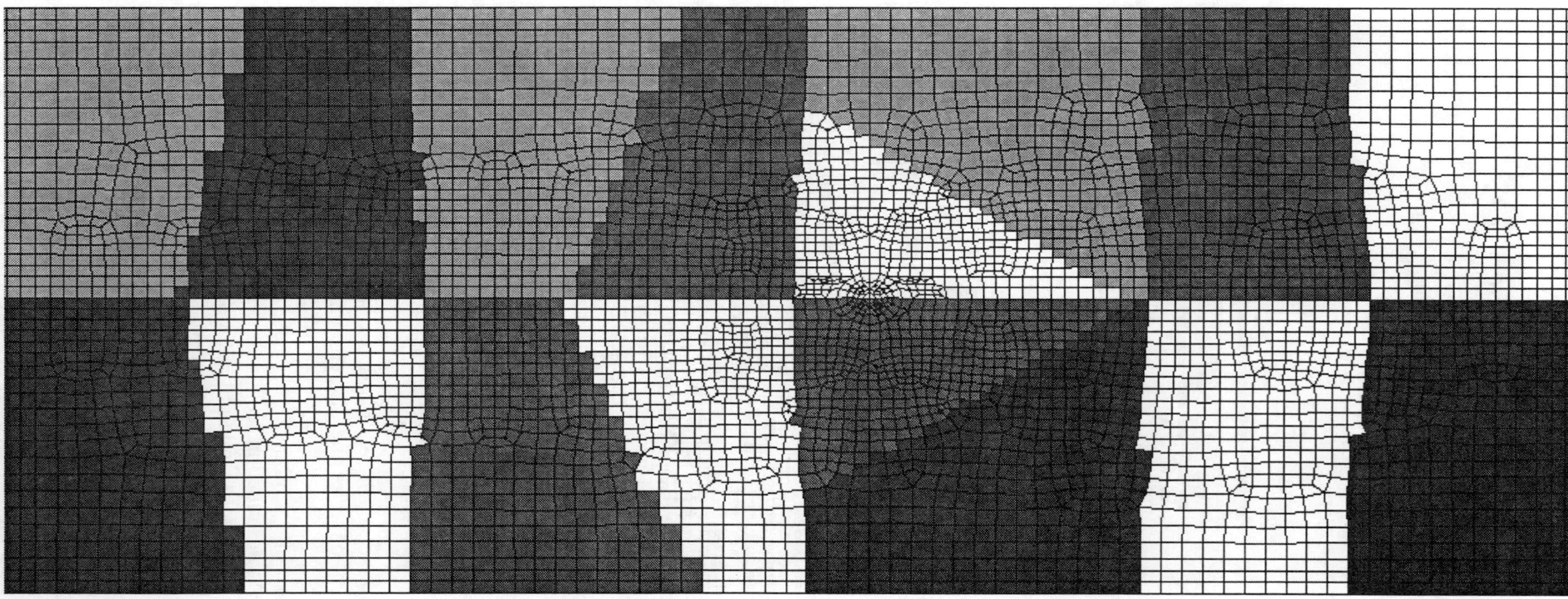

Figure 8. Spatial decomposition of the two-dimensional mesh using the RIP algorithm (16 subdomains)

Figure 9. Spatial decomposition of the two-dimensional mesh using the Greedy algorithm (16 subdomains)

Table I. Load balance of the RIP and Greedy algorithms

P	RIP		Greedy	
	$N_{c_{max}}:N_{c_{min}}$	$N_{n_{max}}:N_{n_{min}}$	$N_{c_{max}}:N_{c_{min}}$	$N_{n_{max}}:N_{n_{min}}$
1	4867:4867	4988:4988	4867:4867	4988:4988
2	2542:2325	2629:2414	2434:2433	2541:2541
4	1353:1142	1421:1210	1217:1216	1313:1289
8	661:571	743:621	609:608	670:662
16	374:284	421:322	319:289	384:341
32	203:131	236:151	153:151	190:177

Table II. CPU times recorded versus number of processors (P) on iPSC/860 for the microstrip via (4000 time iterations)

P	RIP	Greedy
1	7936·0	7936·0
2	4224·0	4144·0
4	2406·0	2219·0
8	1301·0	1216·0
16	815·0	769·0
32	520·0	384·0

RISC processor and 16 Mb of memory. The two-dimensional mesh in Figure 7 was used, and consists of 4867 quadrilateral cells. The three-dimensional mesh was 40 cells high along the vertical direction. The full simulation required 4000 time iterations. The CPU times required to perform the simulation versus the number of processors are illustrated in Table II, comparing the times that resulted from using the RIP and the Greedy algorithms. Clearly, the Greedy algorithm results in improved CPU times. Figure 10 illustrates the speed-ups of the parallel algorithm, again based on the RIP and Greedy decompositions. The speed-up here is defined as being the ratio of the CPU time required to execute the problem on P processors to that required by a single processor. This is also compared to the ideal case of a linear speed-up. Excellent speed-ups are observed

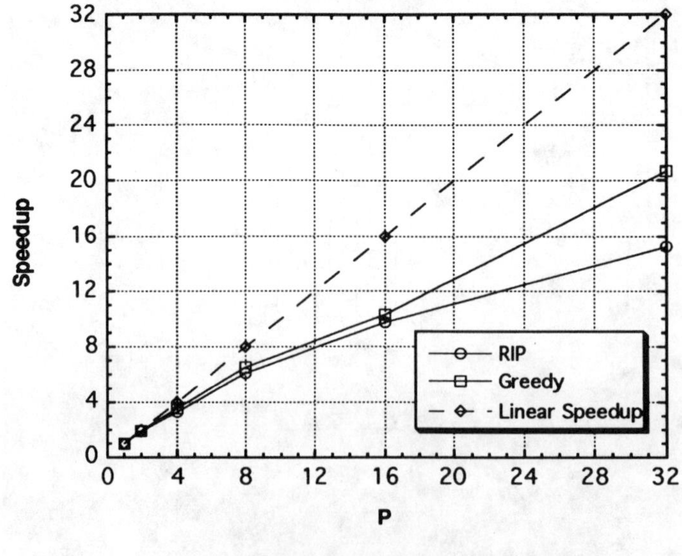

Figure 10. Speed-up of the parallel PGY algorithm over 32 processors of an iPSC/860 using the RIP and the Greedy spatial decomposition methods

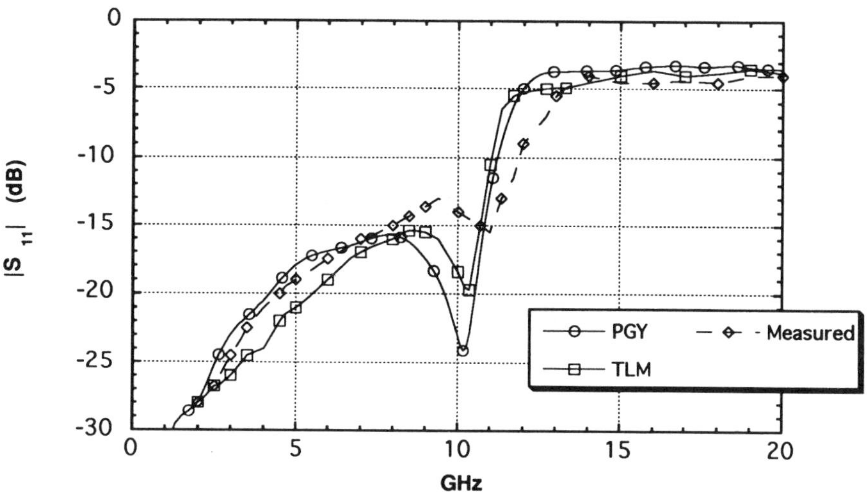

Figure 11. Magnitude of the S-parameters for the cylindrical via through a ground plane computed using the planar generalized Yee algorithm

over the 32 processors. Finally, the magnitude of the S-parameters are illustrated in Figure 11. These results are compared with the measured results presented in Reference 8 and those computed using an orthogonal grid TLM method.[14]

6. SUMMARY

In this paper, the parallel planar generalized Yee (PGY) algorithm has been presented. Initially, it was shown that by exploiting the planar symmetries of printed microwave circuit devices, great savings in both CPU time and memory can be achieved. It was also shown that significant speed-ups in floating-point operation speeds can be achieved by exploiting inherent vectorism in the PGY algorithm owing to the regularity of the grid along one dimension.

The parallel PGY algorithm presented was based on a spatial decomposition of the general unstructured mesh. By treating the update matrices as subassemblies of matrices, a very efficient parallel scheme was obtained. Two spatial decompositions were compared, the recursive inertia partitioning (RIP) algorithm and the Greedy algorithm. The Greedy algorithm provides optimal load balance, whereas the RIP algorithm more effectively minimizes shared boundary interface lengths. Through a numerical example, it was demonstrated that the Greedy algorithm provides superior speed-ups. From this, it can be concluded that load balancing is extremely important, and will lead to scalable parallel algorithms. Finally, it is demonstrated that the parallel PGY algorithm has a high level of parallel efficiency and provides the means to efficiently and accurately solve practical engineering problems.

ACKNOWLEDGEMENTS

This research was supported in part by the Jet Propulsion Laboratory, California Institute of Technology under JPL/NASA Contract 959534 with the University of Kentucky and by the National Science Foundation under award ECS-9309179. This research was also supported by a 32-node Intel iPSC/860 funded under NSF grant EMS-9206014, Army Research Office Grant # DAAHO4-93-G-0453, and the CEM grant awarded through the Intel Supercomputer Systems Division. This research was also performed in part using the Intel Touchstone Delta System operated by the California Institute of Technology on behalf of the Concurrent Supercomputing Consortium. Access to this facility was provided by the Jet Propulsion Laboratory.

REFERENCES

1. S. Gedney, F. Lansing and D. Rascoe, 'A generalized Yee-algorithm for the analysis of MMIC devices', *IEEE Transactions on Microwave Theory and Techniques*, submitted.
2. S. Gedney and F. Lansing, 'Full wave analysis of printed microstrip devices using a generalized Yee-algorithm', in *IEEE Antennas and Propagation Symposium Digest*, Ann Arbor, MI, 1993.

3. N. Madsen, 'Divergence preserving discrete surface integral methods for Maxwell's equations using nonorthogonal unstructured grids', UCRL-JC-109787, Technical Report, LLNL, February 1992.
4. A. Taflove and K. Umashankar, 'Finite-difference time-domain (FDTD) modeling of electromagnetic wave scattering and interaction problems', *IEEE Antennas and Propagation Magazine,* **30**, 5–20 (1988).
5. J.-F. Lee, R. Palendech and R. Mittra, 'Modeling three-dimensional discontinuities in waveguides using nonorthogonal FDTD algorithm', *IEEE Transactions on Microwave Theory and Techniques,* **40**, 346–352 (1992).
6. J. F. Lee, 'Finite difference time domain algorithm for non-orthogonal grids and its application to the solution of electromagnetic scattering problems', *Archiv fur Elektronik und Uebertragungstechnik,* **46**, 328–335 (1992).
7. P. Harms, J.-F. Lee and R. Mittra, 'A study of the non-orthogonal FDTD method versus the conventional FDTD technique for computing resonant frequencies of cylindrical cavities', *IEEE Transactions on Microwave Theory and Techniques,* **40**, 741–746 (1992).
8. P. Harms, J.-F. Lee and R. Mittra, 'Characterizing the cylindrical via discontinuity', *IEEE Transactions on Microwave Theory and Techniques,* **41**, 153–156 (1993).
9. S. Gedney and F. Lansing, 'A generalized Yee-algorithm for the analysis of three-dimensional microwave circuit devices with planar symmetry', *IEEE Transactions on Microwave Theory and Techniques*, submitted.
10. S. Gedney, 'Finite-difference time-domain analysis of microwave circuit devices on high performance vector/parallel computers', *IEEE Transactions on Microwave Theory and Techniques*, submitted.
11. B. Nour-Omid, A. Raefsky and G. Lyzenga, 'Solving finite element equations on concurrent computers', in *Symposium on Parallel Computation and their Impact on Mechanics*, Boston, MA, 1987; ASME, 1988.
12. C.Farhat and M. Lesoinne, 'Automatic partitioning of unstructured meshes for the parallel solution of problems in computational mechanics', *International Journal on Numerical Methods in Engineering,* **36**, 745–764 (1993).
13. V. Betz and R. Mittra, 'Comparison and evaluation of boundary conditions for the absorption of guided waves in an FDTD simulation', *IEEE Microwave and Guided Wave Letters,* **2**, 499–401 (1992).
14. C. Eswarappa and W. J. R. Hoefer, 'One-way equation absorbing boundary conditions for 3-D TLM analysis of planar and quasi-planar structures', *IEEE Transactions on Microwave Theory and Techniques,* **42**, 1169–1175 (1994).

Applications of Finite-Difference Time-Domain Technique to Planar Microwave Circuit Design

BY INGO WOLFF
DEPARTMENT OF ELECTRICAL ENGINEERING
GERHARD-MERCATOR-UNIVERSITY DUISBURG
BISMARCKSTR. 81
D-47057 DUISBURG, GERMANY

***Abstract*—This chapter gives an overview of the application of FDTD analysis techniques to the simulation of electromagnetic fields and the design of microwave integrated circuits. In particular the application to the analysis of 2-D and 3-D discontinuities in microstrip and coplanar microwave circuits is described. The numerical results are compared to measurement results and the applicability of the technique in CAD of microwave circuits is discussed. Special problems like the simulation of planar lines with very thin conducting layers where the internal field of the layers have to be taken into account (e.g., superconducting circuits) will be addressed, using a 2-D formulation of the FDTD technique and special solution techniques. Techniques for accelerating the computation like the signal identification method, the time series method and the parallel processing are also a subject of this chapter.**

The design of microwave circuits containing nonlinear active components using the FDTD technique in combination with a hamonic balance method (HBM) is shortly mentioned and the feasibility of total RF and microwave package simulation including nonlinear components and circuits will be possible in the future as first results show. For all applications the state of the art will be discussed using references from the international literature and the special selected papers which are reprinted in this chapter.

I. INTRODUCTION

The FDTD technique is a numerical method for the solution of electromagnetic field problems which has a large numerical but only a low analytical expense. Despite the large numerical expense, it is believed to be one of the most efficient techniques because it stores only the field distribution at one moment in memory instead of working with a large equation system matrix. Additionally, it is a very flexible technique because it can address problems with arbitrarily inhomogeneous, anisotropic, and also nonlinear material properties. The field solution for each time is determined directly from Maxwell's equations and is calculated using a time-stepping procedure based on the finite difference formulation of Maxwell's equations in time and space as it has been described in detail in chapter 2. The "leapfrog algorithm" as it has been first introduced by Yee [1] fits well on modern computer architectures, so that the data required to describe a 3-D field distribution can be handled in a reasonable time. Therefore the method can be efficiently be implemented on vector or on parallel computers. Sufficiently accurate results can be obtained by using a single precision floating-point expression requiring only four bytes, as has been shown by Railton and McGeehan [2].

For a long time the FDTD technique was only used to qualitatively demonstrate electromagnetic field solutions in the time domain. Only the introduction of absorbing walls [3–6] made this technique into a powerful tool for application, e.g., in real microwave circuit design. This was soon realized by several authors who presented first results for the time-domain analyses of eigenvalue problems [7], planar circuits [8,9], microstrip lines, and discontinuities [2,10–17]. A combination of time-domain analysis techniques with the method of lines was presented in [14].

In this chapter, as already mentioned in the summary, an overview shall be given on FDTD simulations of electromagnetic fields as applied to planar microwave circuit design. It will be shown that this technique makes it possible to model 2-D and 3-D circuit structures in microstrip and coplanar technique and adjoint problems, e.g., of electromagnetic compatibility. It will be also demonstrated that this technique is a powerful analysis technique for nonlinear microwave circuit design by combining semiconductor modeling with the FDTD analysis of passive circuit elements or even total packages if, for reasons to be discussed later, a harmonic balance method (HBM) is used for this combination.

II. SOLVING PLANAR MICROWAVE CIRCUIT DESIGN PROBLEMS WITH FDTD

In this chapter the direct application of Yee's FDTD scheme to planar microwave circuit design problems shall be discussed without repeating the fundamentals of the "leapfrog algorithm," and of the absorbing boundary wall, as both have been presented in Chapter 2. Only special problems directly connected to the mentioned application shall be addressed here. These are:

- the excitation of the electromagnetic field in FDTD simulations,
- the correct choice of the absorbing boundary wall algorithm,
- the application of different grid sizes in the simulation,
- the use of a 2-D version of FDTD for planar line parameter analysis,
- the use of different methods to speed up the calculations, and
- the demonstration of interesting results and the discussion of possible errors.

A. *The Excitation of the Electromagnetic Field in FDTD Simulations*

The electromagnetic field in the structure which is to be analyzed must be "excited," i.e., at the beginning of the calculation a certain field structure must be defined in the grid of the FDTD scheme. There are three possible methods:

a) At the time t_0 (starting time of the calculation), the electromagnetic field is defined within the total space of the considered structure. This method has been used, for example, in [7] for the analysis of microwave resonators. The field distribution used at t_0 depends on the experience of the program user who, for example, may define the field distribution similar to that of a special resonant mode in a resonator for which eigenvalues shall be determined.

b) The second possibility is to define a harmonic field oscillation on a boundary of the structure. Again the transversal distribution of the electromagnetic field in the boundary may depend on the experience of the programmer. This method is only applicable in cases where no reflections occur in the circuit which is to be analyzed, because by using a harmonic excitation, incident and reflected waves cannot be separated on the structure and, for example, scattering parameters cannot be estimated under these conditions.

c) The third form of excitation is to use an electromagnetic field pulse with a finite length in time and space. Using this kind of excitation, in principle arbitrary *n*-ports and their transmission properties can be analyzed, if the incident, transmitted, and reflected pulses can be separated. Because such a pulse contains a broad spectrum of frequencies (dependent on its time length and pulse form), the transmission properties in a desired frequency range can be determined in one analysis cycle.

The spatial transversal distribution of the pulse again must be defined by the user, but very crude assumptions are good enough, if only enough space is left between the excitation plane and the reference plane of the structure which is to be analyzed, so that the electromagnetic fields, which are real solutions of Maxwell's equations in the considered structure, can be established in the space between the source and the structure. For example, in the case of a microstrip line, the assumption of a constant electric field across the strip width and zero field outside the strip leads to good results if the microstrip line is long enough to build up the typical electromagnetic wave on the microstrip line with, for example, large electric field values at the edges of the strip while propagating along the line section (e.g., Fig. 3)

The pulse length in time must be so short that the spectrum contains all desired frequencies. The length in space must be so short that incident and reflected pulses can be exactly separated. On the other hand, the used grid size must be smaller than the space length of the pulse to guarantee a proper resolution of the spatial field distribution.

Version c) of the discussed excitations is suitable for the analysis of planar microwave circuits. An exciting Gaussian pulse in the time domain described by $f(t) = \exp\{-[(t - t_0)/T]^2\}$ is used [11]. Its length in time is determined by T. For example, because the microstrip line is dispersive, the pulse form is changed when propagating along the line.

From the calculated electromagnetic field, time-dependent voltages and currents can be calculated at the ports of the circuits. Of course the same problems exist as in the case of the frequency-domain calculations that, for example, the voltage is no longer defined exactly in a line structure with two different dielectric materials (in frequency domain, modes are no longer TEM-modes but instead hybrid modes), but these problems will not be discussed here. From the time-dependent signals the equivalent transmission parameters in the frequency domain may can be determined using Fourier transform so that the scattering parameters at the defined ports or reference planes can then be calculated.

B. *The Correct Choice of the Absorbing Boundary Wall Algorithm*

The principle of the absorbing boundary walls has already been described in Chapter 2 and must not be repeated here. Only some very simple aspects essential for the application of the absorbing boundaries for simulating planar microwave components and circuits shall be discussed. If the transient analysis is to be applied, magnetic or electric walls cannot be used in the boundaries of a problem which shall be solved using FDTD technique. This is because reflections from these boundaries would be superimposed upon the reflected and transmitted pulses in the circuit and therefore would disturb the information on the circuit properties. To allow the separation of the pulses which are scattered at the component or circuit without disturbance by other signals, absorbing boundary conditions enclose the structure to be analyzed have to be used.

An ideal absorbing boundary can be defined by an infinite space behind this boundary in which a wave penetrates without exciting a reflected wave at the boundary. Such an ideal absorbing boundary wall leads to a very high computational expense unacceptable in circuit design techniques. But an approximate solution for an absorbing wall can be found easily [4–6,11] if the field in the boundary always has the same value as at a distance Δh from the boundary at a time $n_p\Delta t$ before the actual time. The expression $n_p\Delta t$ is the time the wave needs to propagate along the distance Δh. Using the stability condition for the FDTD algorithm with an equidistant mesh size Δh (see Chapter 2), the following condition must hold:

$$\frac{\Delta h}{v\Delta t} = 1 + (\sqrt{3\varepsilon_{\text{reff}}}) \tag{1}$$

The paranthetical term defines the integer value, $\varepsilon_{\text{reff}}$ is the effective dielectric constant of the considered planar line structure, and v is the maximum phase velocity in the analyzed structure. If $E^n_{\tan}(I, j, k)$ is the tangential component of the electric field in the absorbing wall at a time step $n\Delta t$ and $E^{n-n_p}_{\tan}(I-1, j, k)$ is the tangential electric field in the distance Δh perpendicular before the wall at a time $(n - n_p)\Delta t$, the condition for the absorbing wall can be formulated as follows:

$$E^n_{\tan}(I, j, k) = E^{n-n_p}_{\tan}(I-1, j, k) \tag{2}$$

Equation (2) is only a good approximation for a real absorbing boundary if the phase velocity of the wave is dispersionless and if the wave is purely transversal and propagates perpendicular to the boundary. It is well known that the most planar lines are dispersive. Therefore, Eq. (2) leads to a wall with a small reflection coefficient (3 to 5%). But using a similar simple model for the tangential magnetic fields, a description of the wall can be found with an error of similar magnitude but with a changed sign. The weighted superposition of both processes leads to a good approximation of the absorbing wall [5,6,11], as long as the wave propagates perpendicular to the wall. The same model can also be used for walls parallel to the line structure if the wall is far enough from the line, so that the incident fields are to a first approximation transversal field. In the case of a wall above the line structure, Eq. (2) must be used without any error compensation.

It can be shown that the described simple boundary algorithm which fulfils the wave equation exactly in the case of a non-dispersive and single direction wave propagation also allows a wave excitation in the opposite direction for a frequency f_{osc} corresponding to the following period:

$$T_{\text{osc}} = \frac{\Delta h}{v_{ph}} + \frac{\Delta h}{c} \tag{3}$$

This expression describes the time needed to travel one space step on the line structure (phase velocity v_p) plus the time needed to travel one space step in free space (phase velocity c). So, the stability of the absorbing boundary is not given for f_{osc} because an oscillation can be built up in the boundary from the reflections and the two possible propagation directions.

Figure 1 shows a typical result of such a boundary oscillation for an example of a radial microstrip stub. The figure shows the time-domain signals at the input and the output ports of the radial stub (defined at the input and the output microstrip line far away from the stub structure). After more than 3,000 time-iterations, an instability arises in the time-domain solution as a result of the absorbing wall oscillation described above.

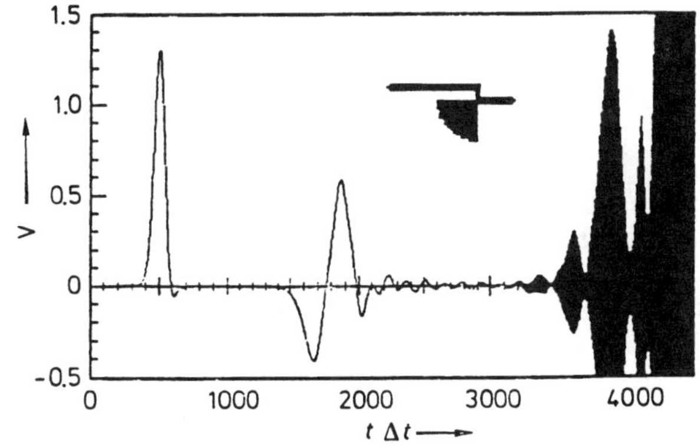

Fig. 1 Unstable time-domain solution from an absorbing boundary oscillation.

Alternative absorbing boundary conditions without this disadvantage normally have higher reflection coefficients or they are (as mentioned above) numerically too expensive. But owing to the fact, that the absorbing boundary algorithm described above offers a number of field samples over the actual moment of time, together with buffered samples of passed moments, digital filter techniques even with non-causal systems can be used to suppress the oscillation of the absorbing boundary. The simplest method is to use an average of some neighboring samples in the time domain—which means nothing else but the convolution of a rectangular time function with a width T_{osc}—or to multiply the boundary field with a real low-pass filter function in the frequency domain. This will not influence the signals at microwave frequencies in the time domain very much, but it improves the stability of the absorbing boundary drastically. The introduction of such a boundary finally reduces the computation domain in which the electromagnetic field must be calculated. Hence, the storage requirements and the computation time are reduced.

C. *Matched Sources*

As already mentioned, the excitation of the pulse inside the circuit is not a critical process, but a certain space must be available so that, for example, a constant electric field distribution across the microstrip line (the assumed excitation field) can change to a real electromagnetic field solution of the microstrip line. Such an exact field solution is needed at the input port of the component or circuit to be analyzed. To avoid long lines at the input of the circuit (which need additional computation space), other techniques must be developed which provide an electromagnetic field distribution in the form of a valid solution on the input line. Such a solution shall be called a matched source. Its principle is shown in Fig. 2.

In [11], the problem of matched sources was solved by separating the stimulated pulse and the reflected pulse by using a transmission line connected to the component under test, which is long compared to the pulse width. After the pulse has left the excitation plane, the reflected pulse propagates through an absorbing boundary which is switched on at this time. There must be a certain distance between the excitation plane and the

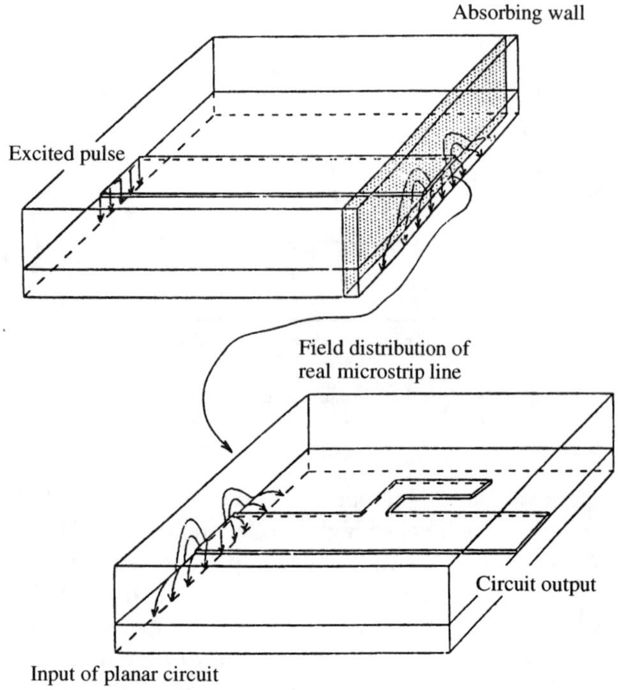

Fig. 2 Principle of a "matched source" for exciting field pulses in microstrip circuits.

absorbing wall because the homogeneous pulse excitation induces a dc-offset in the tangential H-field because of the violation of Maxwell's equations.

Accuracy suggests the application of an excitation pulse with a duration in space which is sufficiently long. Arbitrary long pulse duration can be realized if "matched sources" are available. They can be built up by separating the calculation domain into two parts. The first part contains only a transmission line which is excited in the conventional way. The transmission line is terminated with an absorbing boundary (Fig. 2) at the other end, and it should be long enough to stabilize the typical pulse form of the microstrip line after excitation and furthermore to let the mentioned dc-offset of the H-field vanish.

The field close to the termination of this line, which is a real physical field distribution on the microstrip line, is used to excite the second part of the computation domain containing the component to be analyzed and short connecting lines at the input and the output. In this area, superposition of pulse propagation in both directions is allowed. The absorbing boundary algorithm can be used for the reflected part as before because separation of the transmitted and reflected part of the pulse is possible using the information of the first computation domain.

D. Applications in Planar and 3-D Component Simulation

Various first applications showing the principle and the possibilities of FDTD simulation in planar microwave component and circuit design shall be discussed. In the available computer code, the metallization structure of the planar or 3-D structures on a dielectric carrier substrate can be defined using a graphical editor. Input and output ports of the structures are uniform planar lines, e.g., microstrip lines or coplanar lines (such as one with a 50Ω impedance for comparison with measurements).

The structure to be analyzed is enclosed in a "shielding" of absorbing walls. The following data are typical for an analysis of a simple microstrip line. The total space is discretized into 40 grid elements of length Δh parallel to the line length, 100 elements parallel to the width of the line, and 30 elements perpendicular to the substrate plane. The time step Δt is [considering the stability criteria in Eq. (1)] typically chosen so that the wave propagates along the grid length in five time steps. This time is calculated using a static approximation of the phase velocity. The spatial width of the exciting field pulse is about $40\Delta h$. From these data all the other values are determined. If the pulse width is defined for the time or space where it has decreased to 0.05% of its maximum value, the maximum frequency which is considered in the simulation, using the above mentioned pulse data, is about 50GHz.

If in the description of the applied Gaussian pulse $f(t) = \exp\{-[(t - t_0)/T]^2\}$ the starting time t_0 is chosen to be $300\Delta t$, the pulse is "switched on" at a moment when its value is still very small (2×10^{-12} of the maximum value). The fields are evaluated at the absorbing boundaries for the electric fields and 0, $5\Delta h$ in front of the boundaries for the magnetic field. About 1,000 to 3,000 time steps or more (dependent on the complexity of the structure) are used for the analysis; however 2^{13} time points are used for the Fourier transform to get accurate results in the frequency domain. The values for the time steps which have not been calculated then are filled with zeros.

The application of the method to different planar line discontinuities shall be discussed. Figure 3 (see [18–20]), as a first example, shows the electromagnetic field distribution, the time-domain signal at the input port (a) and the phase of the reflection coefficient in dependence on the frequency (b) compared to measurements and a spectral domain analysis solution for a microstrip open end. Figure 3a shows in detail the input pulse on the microstrip line excited by a matched source at the beginning of the line, the pulse which has reached the open end of the line, and the reflected pulse together with some surface wave signals which can be detected behind the open end of the microstrip line, especially if a moving picture of the pulse propagation is produced.

Finally, Fig. 3a shows the voltage at the input port in dependence on the time. The incident pulse at $500\Delta t$ and the reflected pulse at $1{,}500\Delta t$ can be clearly recognized. By Fourier transform the S-parameters (reflection coefficients) at the input port in the frequency domain can be calculated (Fig. 3b). As can be seen from the comparison between measurement and calculation, the agreement is excellent. It is much better in the case of a spectral domain solution which is also shown in Fig. 3b. The reason for the discrepancies between analysis and measurement in the case of the spectral domain solution may be that the mea-

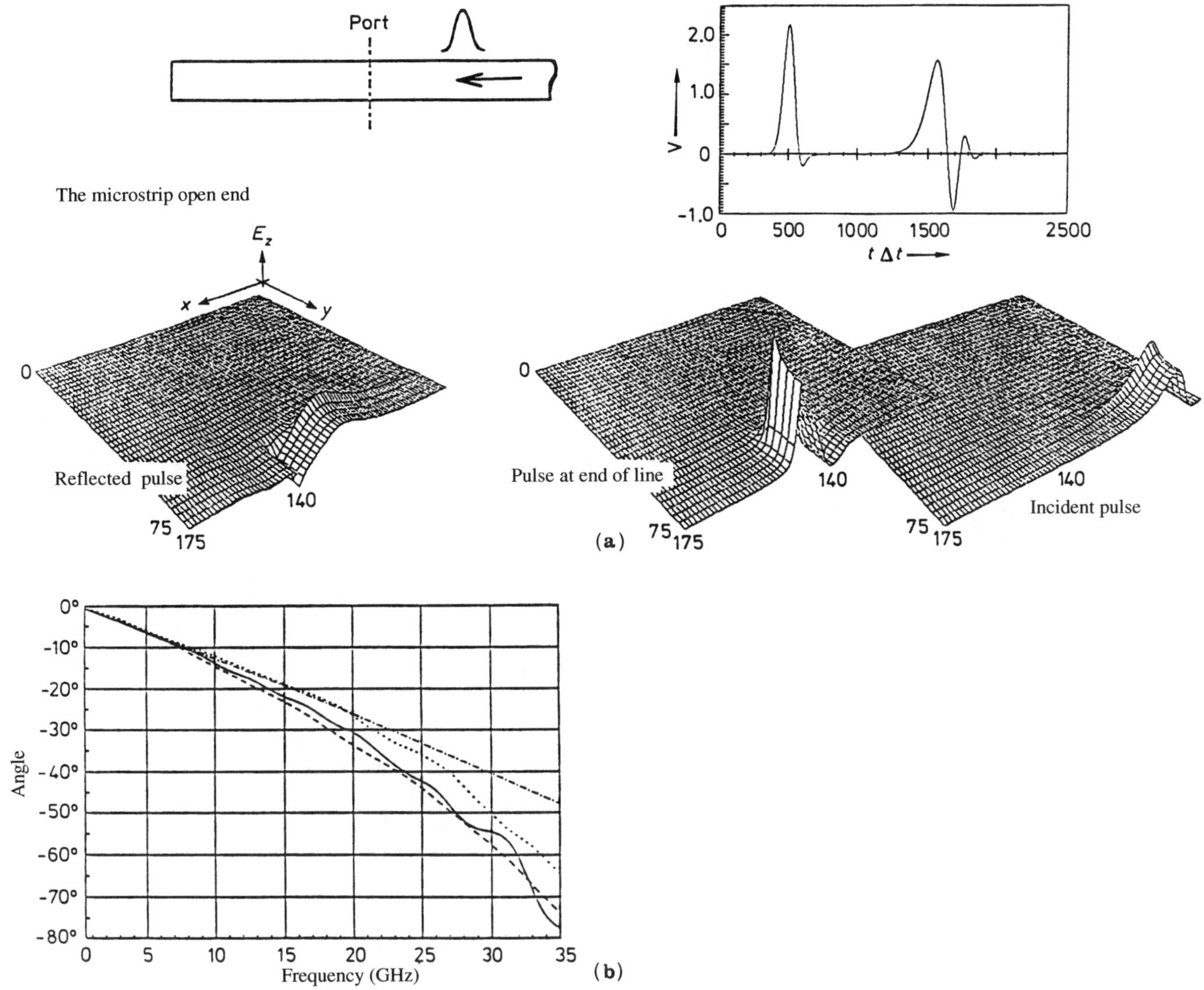

Fig. 3 (a) The electric field component perpendicular to the substrate plane on a microstrip open end, shortly after the time of pulse excitation (incident pulse), for the time, the pulse arrives at the open end, and after reflection at the open end as well as the time domain signals at the defined port. Only one half of the field distribution on the symmetrical microstrip line is calculated and shown, to save computation space. (b) Comparison of calculated and measured results for the reflection coefficient phase of a microstrip open end: — · — · — Kirschnings formula [18], — — — FDTD method, · · · · · spectral domain analysis [19], and ——— measurements [20].

surements have been performed using an open structure. The spectral analysis technique referenced here [19] used a shielded structure for analysis. This is not the case for the analysis where radiation effects as well as surface waves are considered in the analysis procedure.

Figure 4 shows similar results in the case of an impedance step twoport. At the discontinuity, one pulse is transmitted and another one reflected. The figure shows the typical time-domain signals and the electric field at the two ports. At port 1 the incident and the reflected pulses can clearly be recognized; at port 2 the transmitted pulse arrives after about 600 time steps. From the time domain signals the scattering parameters again can be calculated using Fourier transform.

In planar microwave integrated circuits problems often have to be solved where nearby placed transmission lines couple with each other and closely spaced discontinuities lead to multiple reflections and excitation of higher order modes within a circuit or component. In this case the usually applied segmentation technique cannot be used for analysis because the line segments between the discontinuities are not long enough to ensure the decrease of the higher order modes, and the component or circuit must be analyzed in total.

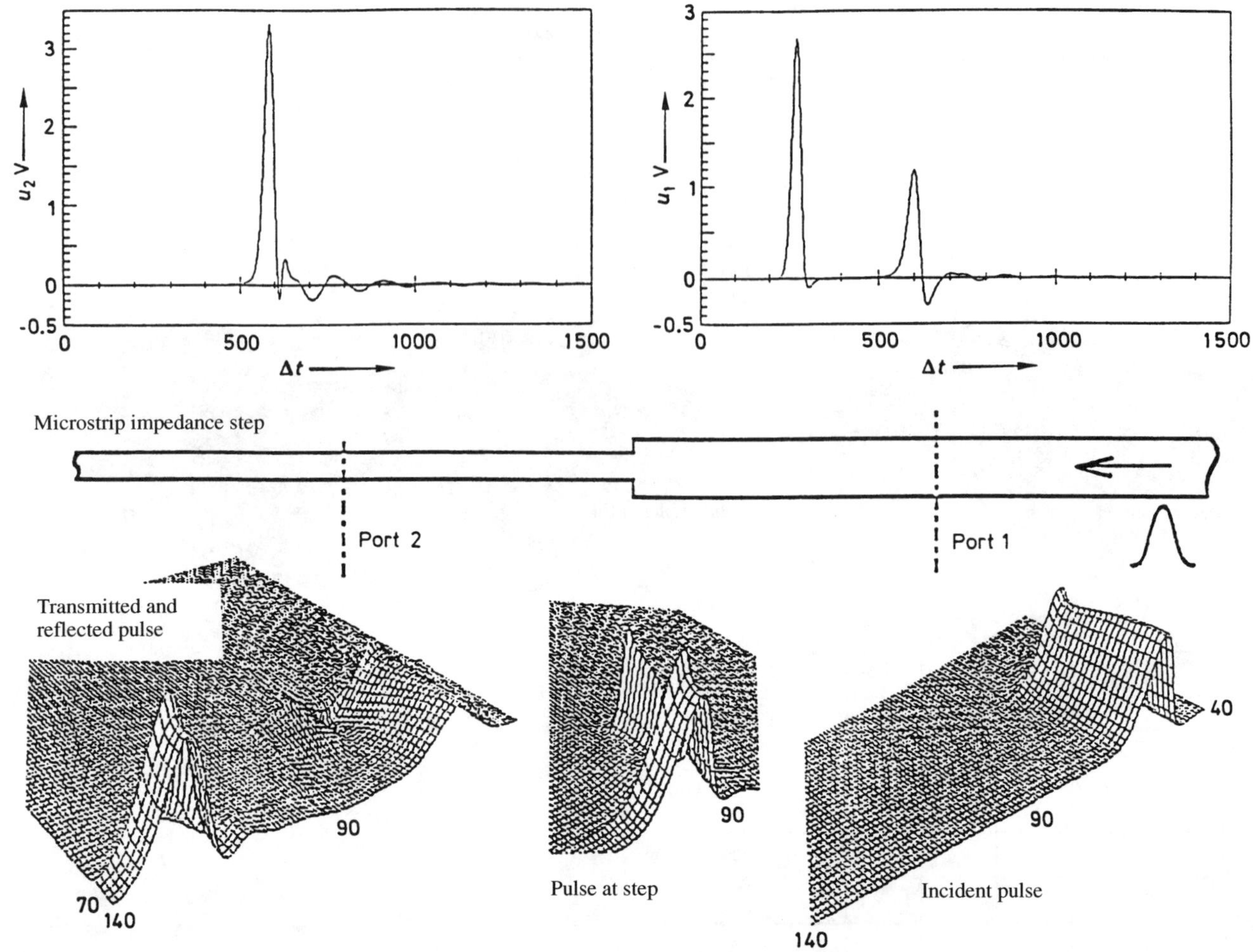

Fig. 4 Electric field of the incident pulse, pulse at the discontinuity, and transmitted pulse (from right to left) as well as the time domain signals at port 1 and port 2 in a FDTD analysis of a microstrip impedance step.

In Figure 5, such an example in the form of a meander line which consists of a number of coupled line sections is shown. This structure has been discretized into 212 elements in the x-direction, 120 elements in the y-direction, and 16 elements in the z-direction. Only four grid elements are used across the line width and along the height of the substrate material, respectively. The theoretical frequency range of the analysis is 35GHz. Figure 5a shows the line structure and the applied discretization in the x-y-plane, Fig. 5b shows the propagation of the wave along the structure at different times, and Fig. 5c compares the calculated scattering parameters to measurement results. Again the agreement between theory and experiment is excellent. The good agreement of the analyzed and measured scattering parameter phases must be especially mentioned.

As an example for an arbitrarily shaped planar line discontinuity with closely coupled lines and discontinuities, a radial stub which first was investigated by Wertgen [21] is analyzed using the FDTD technique. The circular shape of the radial stub (Fig. 6a) can be approximated in the rectangular mesh by a step function. Even very rough mesh sizes lead to acceptable accuracy of the results. As an example for this thesis, the circular radial stub which is closely connected to two microstrip bends has been discretized with only three spatial steps over the width of the microstrip line and the substrate height. Nevertheless, as a comparison of the calculated S-parameters (Fig. 6c) with accurate measurements shows, the amplitude and the phase response of this complicated structure is described very accurately, only at higher frequencies a small frequency offset may be recognized. This offset is a typical error of the large mesh size, which is not able to resolve the spectral components of the higher frequencies, i.e., short wavelengths. It can be avoided by reducing the mesh size.

A final example shows that the FDTD analysis technique can be used to analyze real 3-D structures. The spiral inductor shown in Fig. 7 includes an airbridge above the surface of the substrate material. The structure is analyzed using the FDTD method. As the simulated and measured S-parameters of the quasi-planar component show, the agreement between theory and measurement is good even for frequencies higher than first resonant frequency of the inductor, for the magnitude as well as for the phases. Figure 7c shows that because of the complicated structure of the component, a large number of

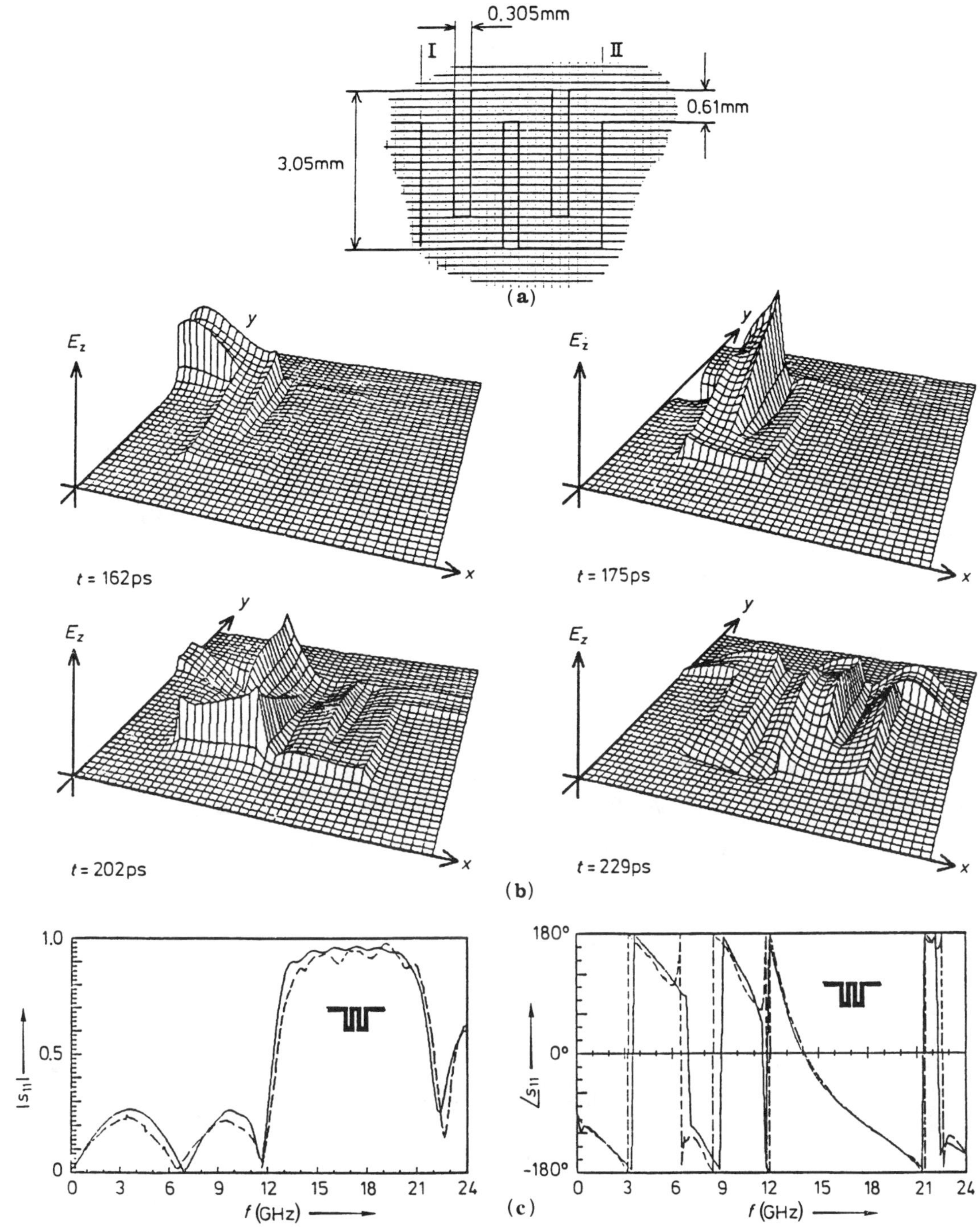

Fig. 5 (a) Microstrip meander line structure; substrate: Al_2O_3, ε_r = 9.98 (measured), height = 635μm. (b) Wave propagation along the meander line at different times. (c) Scattering parameters of the meander line structure, ——— simulations, — — — measurements.

time iterations are needed to come to a final solution in the time domain. Multiple reflections of the electromagnetic pulses inside the structure require up to 6,000 time steps for a good convergence. At the same time it is recognized from Fig. 7c that the time domain solution is always stable and no instabilities from the absorbing boundaries occur even for such high iteration times.

The big geometrical dimensions of the inductor, as indicated in Fig. 7a, have been chosen because it was intended to have all the resonant phenomena of the component in a frequency range, that allows to verify the simulation by measurements. Smaller dimensions lead to higher resonant frequencies, but they do not influence the simulation and its accuracy.

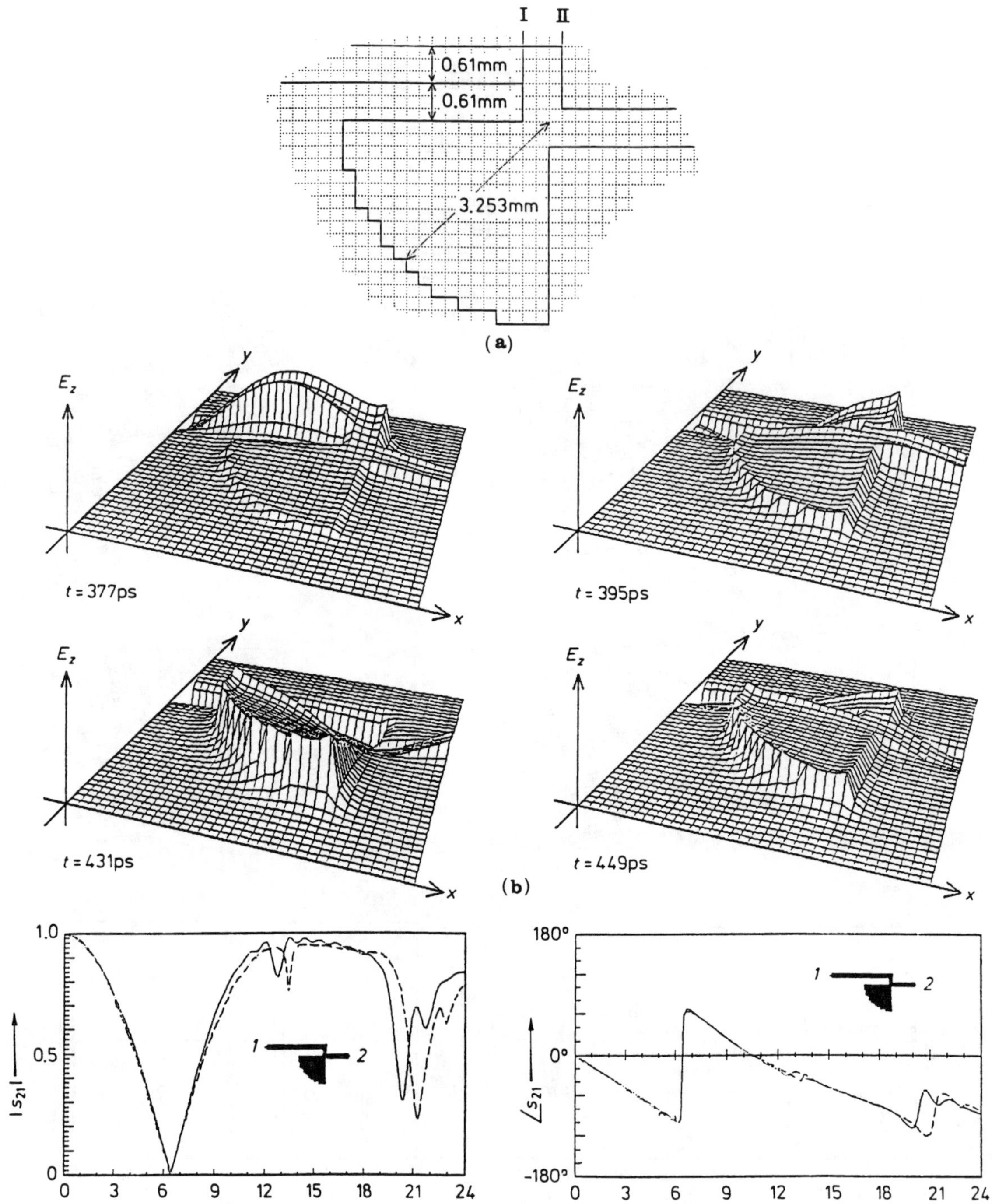

Fig. 6 (a) The structure of the analyzed radial stub, (b) the electric field of the wave propagating along the structure at different times, and (c) the calculated (———) and the measured transmission parameter S_{21} (— — —) in dependence on the frequency. Parameters: $w = 610\mu m$, $s = 610\mu m$, $r = 3253\mu m$. Substrate: $\varepsilon_r = 9.8$, $h = 635\mu m$.

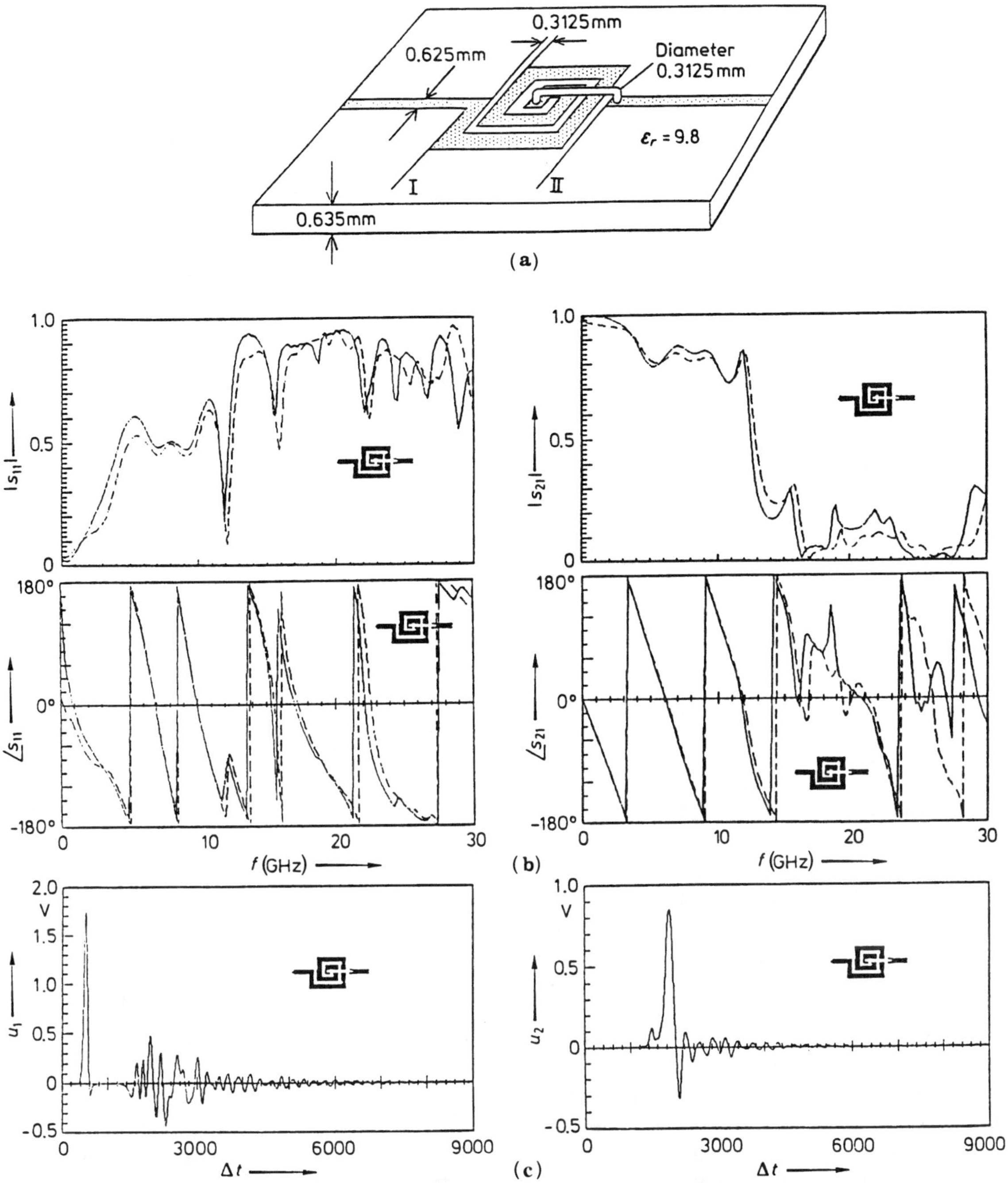

Fig. 7 (a) A planar spiral inductor with airbridge, (b) comparison of calculated (———) and measured (— — —) *S*-parameters, and (c) convergence of the time-domain signal during the analysis. Substrate as in Fig. 6.

E. The Application of Non-equidistant Grid Sizes

Before discussing the phenomena which occur in connection with a non-equidistant grid size, the fundamental definitions shall be cleared.

Yee's [1] original scheme is defined as an equidistant discretization scheme. In this scheme, the coordinates of each node can be calculated by multiplication of the space coordinate indices (i, j, k) with the grid size elements Δx, Δy, Δz. A non-equidistant discretization is defined by a variation of the grid size length in dependence on the space coordinate. The most simple variant could be the direct dependence of the grid size length from its adjoint coordinate, e.g., a dependence of Δx on the coordinate x. An implementation of such a kind of non-equidistant discretization into a software is very simple and therefore this technique has been described by several authors in the literature [22,23]. Connecting this method, the following comments have to be given:

- Space regions without large changes of the electromagnetic field or without field singularities will also be discretized using the fine grid size.
- The discretization of time will be dependent on the smallest grid size and therefore is not well adjusted in the regions with a larger grid size.

Unphysical reflections between regions with different grid sizes will always occur if a non-equidistant discretization is applied. Different propagation coefficients of electromagnetic waves in regions with different discretizations do not inevitably lead to such reflections, but on the one hand no central differences can be defined in the boundary between the two regions, or the boundary conditions for the electrical and the magnetic field cannot be fullfilled in the same boundary plane.

Reflections may heavily disturb a transient analysis because they easily lead to a misinterpretation in the assignment of the signals. As a consequence, an erroneous evaluation of time-windows may result in a nonlinear distortion and significant errors. An optimum numerical efficiency in the application of a non-equidistant discretization in conjunction with maximum accuracy can only be achieved if a local grid size refinement is used where also the stability criteria are considered. It means that the discretization of time in the refined local mesh must be changed adequate to the reduction of the grid size length in space.

Examples of such local mesh refinements are described in the literature [24,25]. Here another variant of such a local mesh refinement shall be described [26] which has the special advantage of a simple implementation into an available software because the fundamental structure of a FDTD program must not be changed.

Independent of the fact that each integer factor between the original and the refined mesh size can be used with the below described technique, a reduction factor of three is suggested. This is the lowest nontrivial odd factor which enables the magnetic field to run continuously from the larger mesh into the refined mesh. Instead of using a larger odd reduction factor, it is advisable, to apply a recursive mesh refinement because the electromagnetic field normally changes gradually. An abrupt change of the mesh size is therefore not adequate.

Figure 8 shows how such a refined mesh can be embedded into one with larger mesh size. For reasons of demonstration, a plane is shown which separately enables the calculation of TE_{m0}-fields in waveguide structures. The shadowed region is the overlapping area, as it is needed for the synchronization of both meshes. In the adjustment of the meshes the following steps are needed:

- Step from E_z^n to E_z^{n+1} in the total coarse mesh.
- Step from E_z^n to $E_z^{n+1/3}$ in the total refined mesh.
- Determination of E_z^{n+1} on the boundary of the refined mesh by interpolation in space and time from the adjoint values of the coarse mesh (prolongation, Fig. 9).

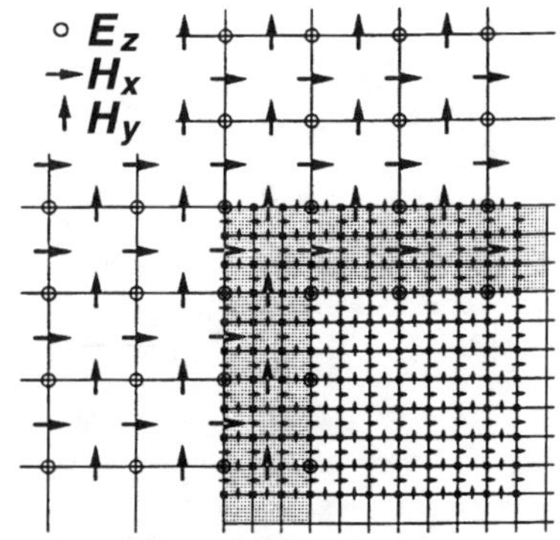

Fig. 8 Local mesh refinement.

- Step from $E_z^{n+1/3}$ to $E_z^{n+2/3}$ in the total refined mesh.
- Determination of $E_z^{n+2/3}$ on the boundary of the refined mesh by interpolation in space and time from the adjoint values of the coarse mesh (prolongation).
- Step from $E_z^{n+2/3}$ to E_z^{n+1} in the total refined mesh.
- Determine E_z^{n+1} on the boundary of the refined mesh by interpolation in space and time from the adjoint values of the coarse mesh (prolongation).
- Determine E_z^{n+1} on the boundary of the coarse mesh from a medium value of the adjoint values of the refined mesh (restriction, Fig. 9).

The value E_z is a representative for the tangential electric field components in the 3-D simulation.

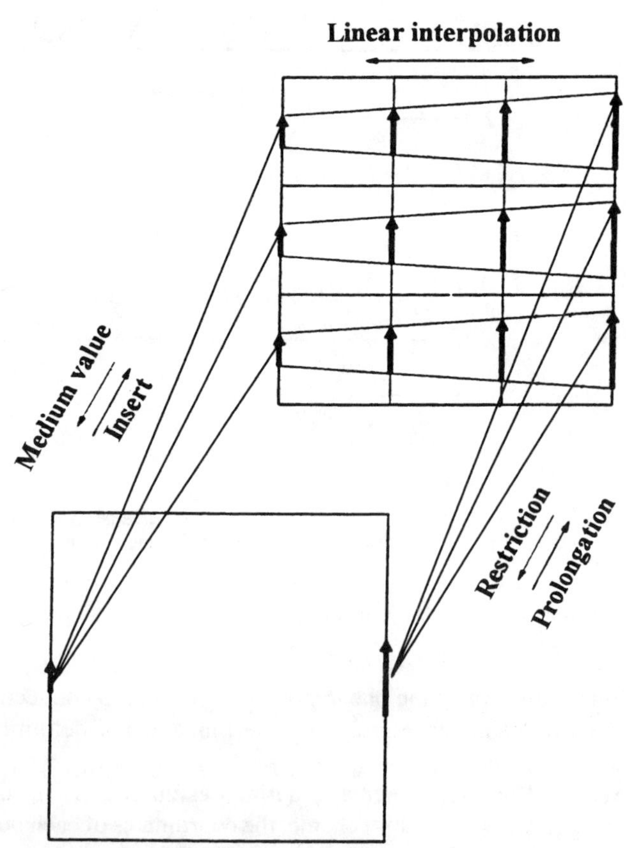

Fig. 9 Matching procedure in the boundary between the coarse mesh and the refined mesh.

Figure 9 demonstrates the essential steps of the matching procedure in the boundary between the coarse mesh and the refined mesh. Only the nodes for one field component of the electric field and one cell of the coarse mesh as well as nine cells of the refined mesh are shown. When switching the field from the coarse grid to the refined grid, a step approximation is chosen in the direction of the field component (prolongation). From this process, three equal values for this field component result in the refined mesh. If a line integral is considered along the electrical field strength in the boundary, then by this technique it is ensured that this line integral always has the same value in the coarse and the refined mesh. Should some kind of interpolation in this field direction been applied under these conditions, problems will arise because of steps in the electrical field at dielectric boundaries or because of missing grid nodes at the boundary of the FDTD grid. In the remaining direction perpendicular to these field components, no steps of the field occur (continuity of the field is guaranteed) and also grid nodes are available up to the boundary of the grid. Therefore a linear interpolation is applied.

While transmitting the fields from the refined mesh into the coarse mesh (restriction), the line integral of the electric field is again preserved. In this process the medium value of the three field components of the considered direction is calculated and replaced for the field value in the coarse mesh.

If the condition for the line integral over the electric field strength is not considered in both meshes, e.g., by simple replacement of the fields instead of the medium value calculation in the restriction process, an electrical conductivity which is not available in reality is simulated in the boundary between the two meshes.

The method as it is described above, will not be able to simulate electromagnetic fields which change strongly near the boundary of the two meshes. Therefore near field singularities the space must be totally discretized using a small grid size. No singularities must be located near a plane between the coarse and the refined mesh.

The reflection properties of the described boundaries depend heavily on the field structure of the electromagnetic field. No general statement can be given on this subject.

Figure 10 shows a first application of the mesh refinement: a microstrip line of width *w* with a gap of length *s* separating the two line sections as shown in the figure on a ceramic substrate ($\varepsilon_r = 9.8$, $h = 635\mu$m). The discretization of the coarse mesh is using five grid elements across the line width and the substrate height is already finer than that used in the examples discussed above. Nevertheless, for this example there is a significant influence when refining the mesh further. The absorbing walls are a relative large distance from the gap. This is needed so that the absorbing boundary does not deal with the large-sized reactive near-field of the discontinuity as well as not with radiated electromagnetic waves.

The influence of the refined discretization is demonstrated in Fig. 11. The electric field directly under the metallization plane of the microstrip line is shown for a time in which the incident pulse is just maximum inside this area. The field distribution can only be resolved with difficulties near the field singularities despite the mesh refinement. The field contour at the boundary between the coarse and the refined mesh is the typical result of the applied prolongation process, as described above.

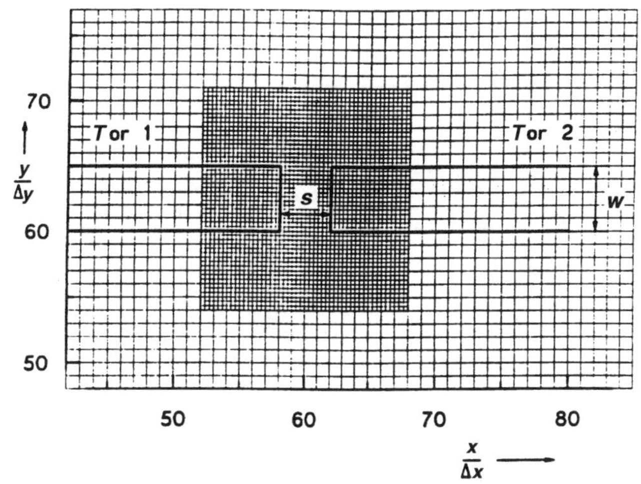

Fig. 10 Microstrip gap with local mesh refinement in the gap area: $w = 635\mu$m, $s = 508\mu$m.

To test the influence of the refined mesh on the scattering parameters, the microstrip gap has been analyzed once only with the fundamental coarse grid and a second time using the mesh refinement near the gap as shown in Fig. 10. Under the assumption of a steady convergence, the scattering parameters are extrapolated to the case which would result for an arbitrarily small grid size in the refinement zone. Additionally a comparison with measurements is shown together with the other results for the magnitude of the scattering parameters in Fig. 12. Similar results are valid for the phases [26], not shown here.

As has been shown in the example of a radial stub (Fig. 6, Section II-D), a coarse mesh size results in a frequency offset of the calculated scattering parameters at higher frequencies whereas the magnitude of the parameters was well simulated. Using the technique of mesh refinement such a structure can be analyzed with an acceptable computation time and better accuracy. If the grid size is reduced by a factor of three in the total

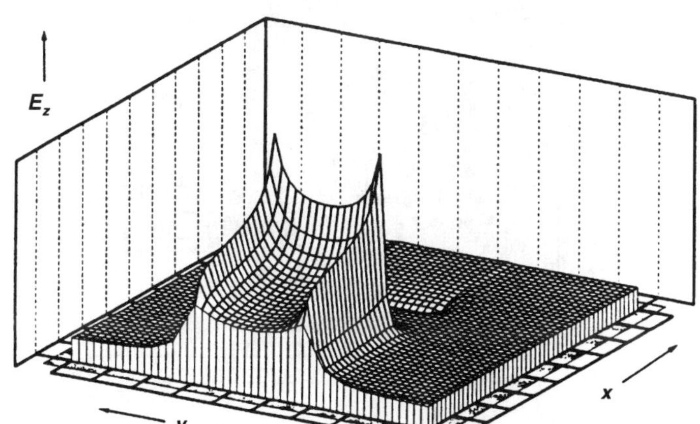

Fig. 11 Distribution of the *z*-component of the electric field in the area of the refined mesh, directly beneath the metallization of the microstrip line.

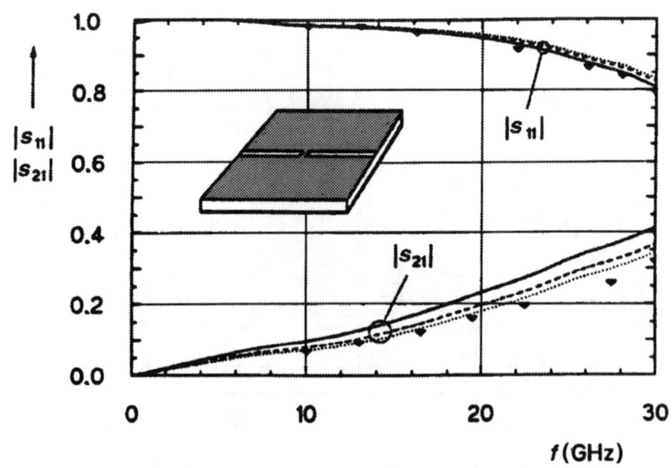

Fig. 12 Magnitude of the scattering parameters of the microstrip gap, calculated ——— with the coarse grid, — — — with the local grid refinement, and · · · · · by extrapolation to an arbitrarily fine mesh size (∇ measurements).

computation space, the storage requirements increase by a factor of 27, and the computation time is multiplied by a factor of 81. Using only a local grid refinement as shown in Fig. 13, spares computation time as well as computer storage. The same mesh refinement reduces the frequency discretization error nearly by the same factor as it used for the grid size reduction. If again an extrapolation technique is used, the scattering parameters in Fig. 14 result.

In principle it must be mentioned that all mesh refinement techniques are connected to finite reflections at the plane between the two meshes of different grid size even in the case of the technique described here. This may lead to stability problems in cases where long FDTD iterations are needed. This subject, using the example of the radial stub, will be taken up again at the end of the next section and in Section III.

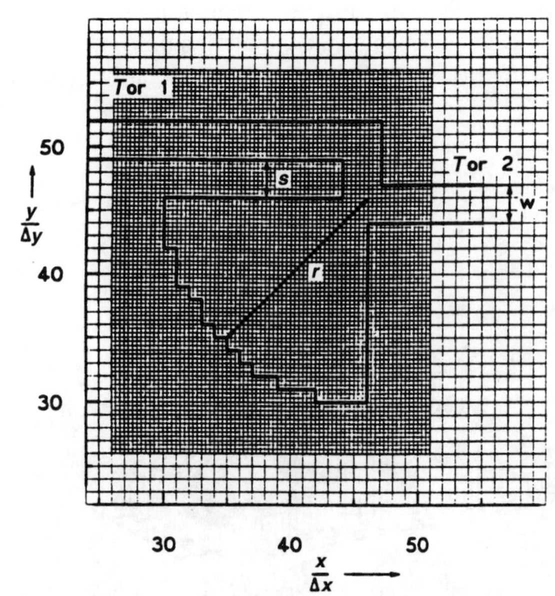

Fig. 13 Radial stub in microstrip technique and applied mesh refinement: $w = 610 \mu m$, $s = 610 \mu m$; substrate: $\varepsilon_r = 9.8$, $h = 635 \mu m$.

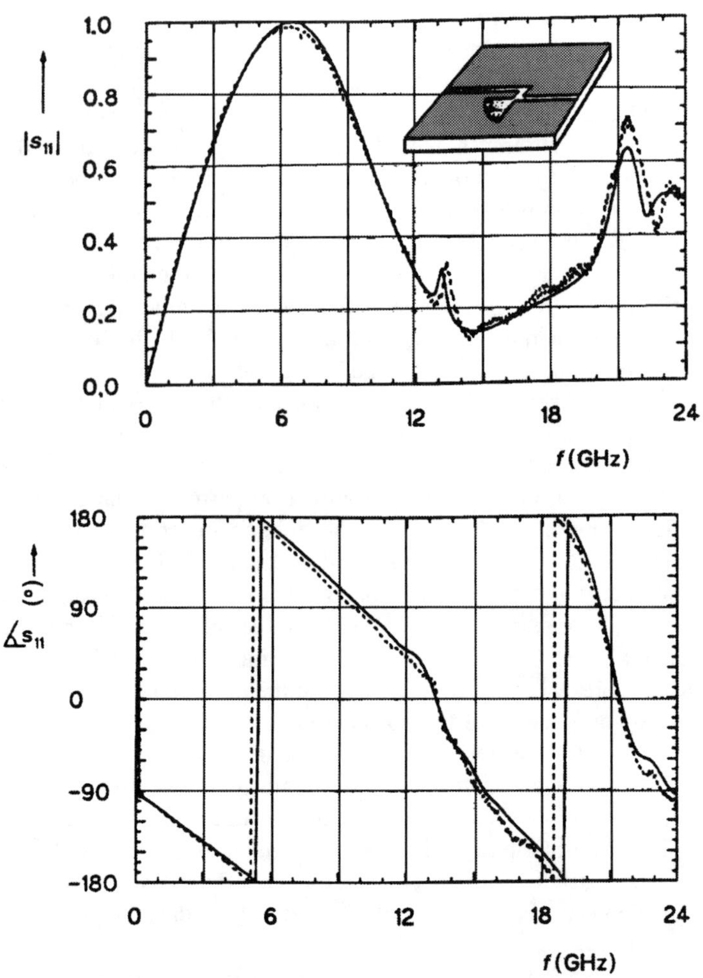

Fig. 14 Magnitude and phase of the scattering parameters for the radial stub in dependence on the frequency; ——— extrapolated FDTD simulation, and — — — measurements.

F. Digital Signal Processing of FDTD Results Using the SI Method

The computation time needed to analyze complex structures using FDTD technique becomes large, and methods have to be developed to reduce this. In the literature, Prony's method [27–30] and the MUSIC method [31] are known to reduce the required number of time steps of time-domain simulations. Here the SI method [32–34] shall be briefly described. The SI method is able to determine a model of the microwave structure from a short observation interval of the input and output data of the time-domain simulation. Long output signals can be computed in a very short time using this simple digital filter model. Using an analytical formula, the scattering parameters of the investigated microwave structure can be calculated directly from the filter coefficients of the model.

The FDTD mesh can be interpreted as a high-order digital filter with accessible input and output signals $[x_0(n), y_0(n)]$, where n is the *discretized time, normalized to the corresponding sampling interval* Δt: $n = t/\Delta t$. The excitation pulse again is Gaussian in shape with a maximum notable frequency, f_{max}. Due to the stability condition of the FDTD algorithm [Eq. (3)], the sampling interval Δt is much smaller than is required by the Nyquist formula [$\Delta t \ll 1/(2f_{max})$]. For application in the SI method it is useful to increase the sampling interval Δt_{SI} by the factor $k_{\Delta t}$:

$$k_{\Delta t} = 1/(2f_{max}\Delta t) \tag{4}$$

$$\Delta t_{SI} = k_{\Delta t} \cdot \Delta t \tag{5}$$

$k_{\Delta t}$ must be truncated to the next smaller integer value. The original time series $x_0(n)$ and $y_0(n)$ are sampled at every $k_{\Delta t}$ iteration to obtain $x(n)$ and $y(n)$, which are the input time series of the SI method. Note that the sampling interval Δt_{SI} used for the SI method differs from that used for the FDTD simulation.

The normalized time delay n_0 of the original system is represented by a delay line of the order $(n_0 - 1)$ in the model (Fig. 15). The remaining part is approximated by a digital filter of the order N, which is determined with the SI method. The input signals of this approach are $x_d(n) = x(n - n_0 + 1)$ and $y(n)$.

Without going into further details of the technique which may be found in [32–34], the application of the SI method to a time-domain field simulation is described as follows:

- First the time-domain method is started to calculate $y_0(n)$ as a reaction of $x_0(n)$.
- $x_0(n)$ and $y_0(n)$ are sampled at every $k_{\Delta t}$ iteration to obtain $x(n)$ and $y(n)$.
- The normalized delay time n_0 is determined automatically.
- Now the SI method is started which operates in synchronism with the field simulator. The filter order N can be determined using the final prediction-error criterion (FPE) [35].
- The field simulator and the SI method are stopped at time-step $n = n_1$ when the variation of the coefficients [$ci(n)$, $b_i(n)$] of the filter is negligible and the error $e(n)$ is small enough.
- Using the prediction error filter (Fig. 15) with the coefficients [$c_i(n_l)$, $bi(n_l)$] at the time step n_l, the model shown in Fig. 16 is determined.
- The frequency response $\hat{H}(\omega)$ can be calculated from the filter coefficients with the following formula at arbitrary frequencies lower than f_{max}:

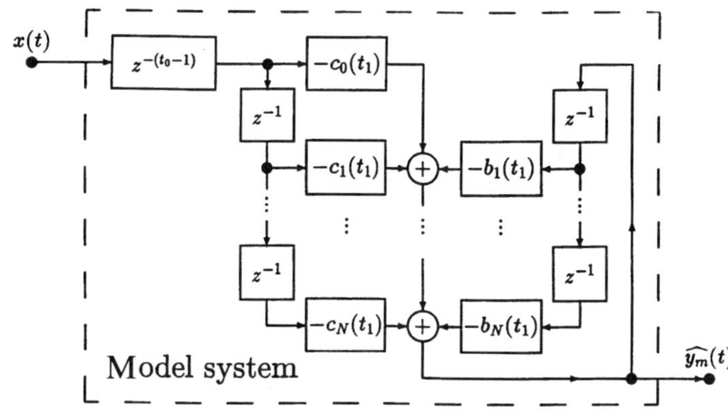

Fig. 16 Model of the microwave structure.

$$\hat{H}(\omega) = \frac{-\sum_{i=0}^{N} c_i(n_1)\cdot e^{-j\omega(i+n_0-1)\Delta t_{SI}}}{1+\sum_{i=1}^{N} b_i(n_1)\cdot e^{-j\omega i\Delta t_{SI}}} \tag{6}$$

As an example, the method is applied to a coplanar band-reject filter which is described in [34]. The FDTD mesh of the investigated structure is excited with a Gaussian pulse to calculate $y_0(n)$ over 30,000 time steps. The frequency range of the excited pulse $x_0(n)$ is from dc up to about f_{max} = 140 GHz. The sampling interval of the original signals $x_0(n)$ and $y_0(n)$ is $\Delta t = 5.9511 \times 10^{-14}$s. Using (4) and (5), a window of 3,300 iterations of the original signals (Fig. 17) is sampled at every $k_{\Delta t} = 60$ iterations to obtain 55 samples of $x(n)$ and $y(n)$. Thus the sampling interval of the SI-method is $\Delta t_{SI} = 3.5707 \times 10^{-12}$s. The delay time of $y(n)$ is $n_0 = 3$, and the filter order of the model is determined using the FPE criterion to $N = 19$.

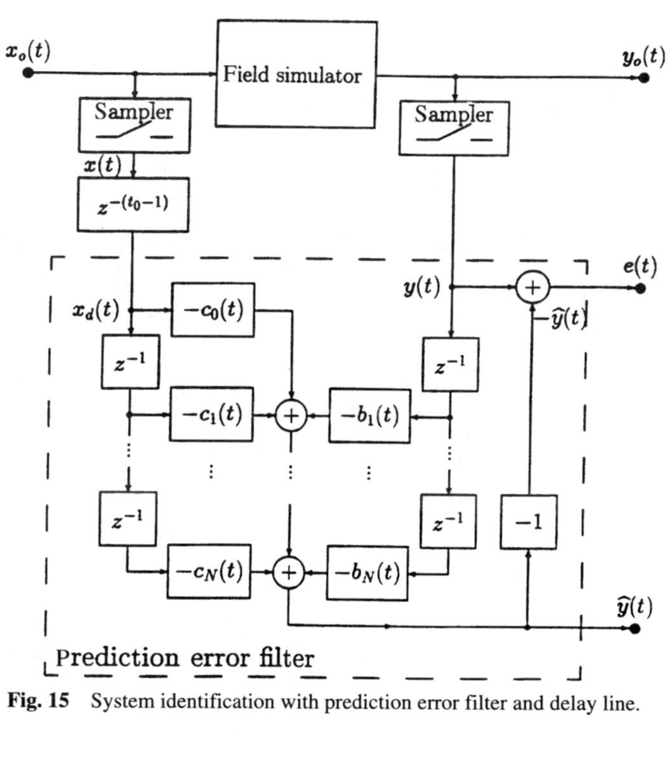

Fig. 15 System identification with prediction error filter and delay line.

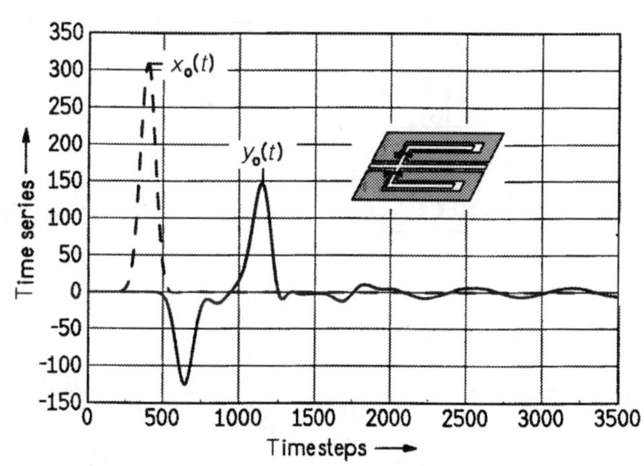

Fig. 17 Time series $x_0(n)$ (— — —) and $y_0(n)$ (———) of the FDTD band-reject filter simulation.

The reflection coefficient of the band-reject filter calculated with the FDTD/SI method is compared to the original one obtained by Fourier transform of the long time series from the FDTD simulations (Fig. 18). The agreement of both results is excellent. The deviation between the original S-parameter $|S_{11}|$ and the estimated S-parameter $|\hat{S}_{11}|:||S_{11}|-|\hat{S}_{11}||$ is smaller than 0.01, and therefore the solid lines in Fig. 18 enter the cross marks. Also the resonant phenomenon at 26GHz is determined exactly.

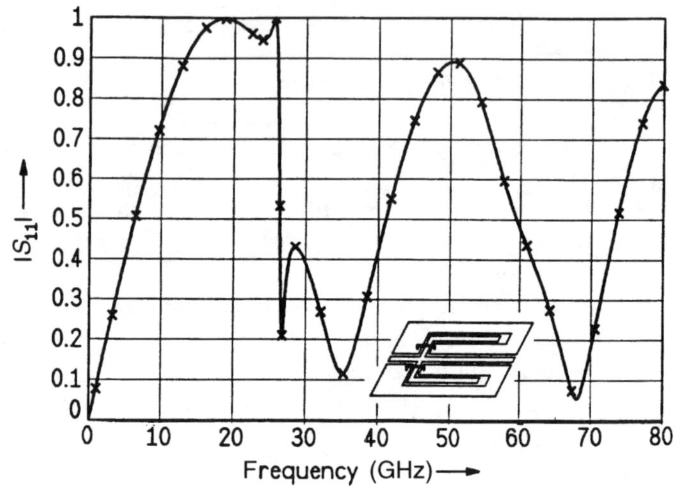

Fig. 18 Input reflection coefficient of the coplanar band reject-filter calculated using the FDTD/SI method (———) and calculated using a FDTD/FFT technique (x x x x x): magnitude.

In some cases, stability of FDTD simulations cannot be achieved, since critical absorbing boundary or special local mesh refinement techniques are used (see Sections II-B and II-E). Figure 19 shows the output signal $y_0(n)$ resulting from a FDTD of the radial stub with a mesh refinement (Section II-E). After more than 3,500 iterations, an instability of the solution occurs which comes from the small reflection at the boundary of the refined mesh. If only the solution for about 3,500 time steps and a simple fast Fourier transform (FFT) is used (Fig. 20),

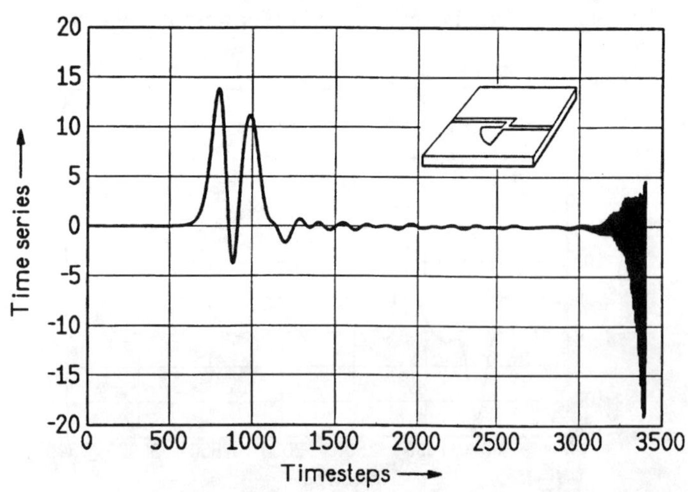

Fig. 19 Time series $y_0(n)$ of a FDTD simulation using a local mesh refinement surrounding the radial stub.

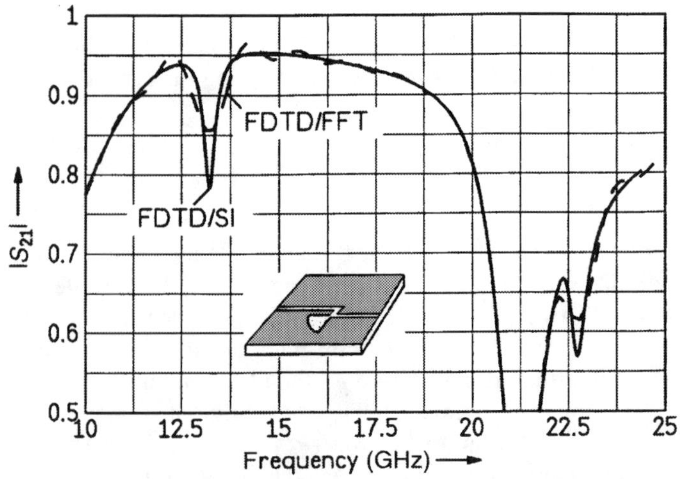

Fig. 20 Transmission coefficient $|S_{21}|$ of the radial stub as calculated with the FDTD/FFT technique (— — —) and using the FDTD/SI method (———).

the accuracy of the solution due to the truncation error in the FFT is not very high. Under application of the FDTD/SI method, the achieved accuracy is excellent (Fig. 21) and the truncation error is eliminated by the SI method.

In summary the SI method not only leads to a reduction of the computation time by a factor of 10 if it is applied to FDTD transient analyses, but also the accuracy of the time-domain simulations can be increased in cases where a restricted usable time window is only available. In this case, only a short sequence of the time series is necessary to obtain an acceptable accuracy of the calculated S-parameters.

III. Two-Dimensional Finite-Difference Time-Domain Analysis of Planar Microwave Waveguides

There are several problems in the analysis of modern planar microwave integrated circuits which can be solved only with large

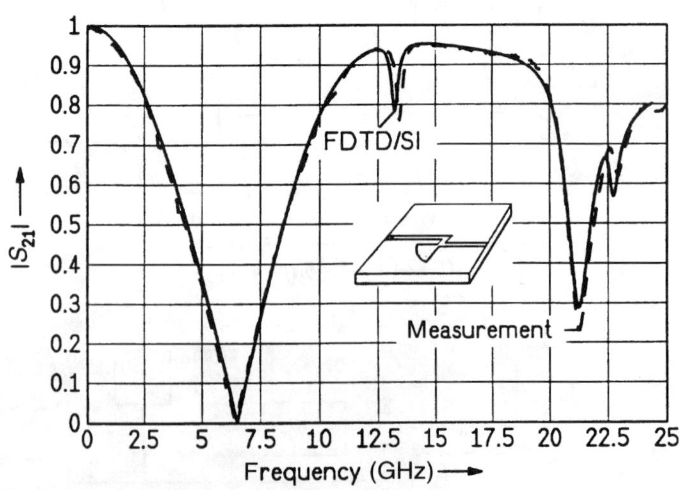

Fig. 21 Transmission coefficient $|S_{21}|$ of the radial stub as calculated with the FDTD/SI method (———) and compared to measured results (— — —).

expense using numerical analysis techniques. One of these problems is the correct determination of the phase velocity and the characteristic impedance of a planar, superconducting microstrip or coplanar waveguide, where the conducting layer is only about 200 to 300nm thick. Using the FDTD technique, the electromagnetic field distribution inside a high-Tc superconducting microstrip line has been analyzed recently [26,36], but the conductor thickness was still 1μm in this case. For very thin conductors as they are used in real microwave circuits, a very fine spatial discretization is required. Due to the stability condition, the fine spatial discretization causes a very fine time discretization, too. Therefore, a large number of iterations is necessary to get a time domain signal which is suitable for a Fourier transform, containing at least a few periods at the desired frequency.

If at least the characteristic parameters of such superconducting lines shall be analyzed for a subsequent independent circuit analysis, a 2-D FDTD, as it has been introduced recently by [37,38] can be helpful, even the direct application of this technique still involves many problems, for the mentioned reasons. Reference [37] is reprinted in this chapter, so that no further introduction is given to this 2-D FDTD analysis technique. Only the specific aspects of analyzing microwave planar lines with very thin conductors or superconductors shall be addressed here.

Normally, as has been described in the above section, the simulation results in frequency domain are extracted from the calculated time series of the transient FDTD analysis using the Fourier transform. For very small structures and thin conductors, as mentioned above, a very fine spatial discretization and subsequently a fine time discretization is required. This needs a large number of iterations to derive at a time signal suitable for the Fourier transform. Even applying the SI method (Section II-F) would not solve the problem. Therefore a time series analysis technique using a Fletcher-Powel optimization technique to extract dispersion and loss characteristics of planar waveguiding structures from only a part of a single period is introduced here. Especially if the fields inside the conductors have to be calculated to analyze the waveguide losses and the internal inductance, this method reduces the required computation time up to a factor of 25. Therefore, it is possible, to simulate planar and coplanar waveguides with a metallization thickness of only 2μm or a superconducting film thickness of 300nm, for example, which cannot be simulated with the conventional FDTD or 2-D FDTD technique due to the enormous CPU-time needed.

A. *Theoretical Background*

The 2-D FDTD algorithm is well suited to analyze the propagation characteristics of waveguiding structures with arbitrarily shaped cross sections. Therefore, based on the original formulation of the FDTD algorithm, using Yee's 3-D mesh [1], the spatial derivatives for the transverse fields in one direction, e.g., the x-direction, which shall be the propagation direction of the guided wave, are replaced using the analytical formulas [37]

$$\underline{\vec{E}}_t(x \pm \Delta l) = \underline{\vec{E}}_t(x) \cdot e^{\mp j\beta\Delta l}$$

and

$$\underline{\vec{H}}_t(x \pm \Delta l) = \underline{H}_t(x) \cdot e^{\mp j\beta\Delta l} \tag{6}$$

Thus a 2-D FDTD mesh formulation can be derived [37]. Since losses inside the conductor materials shall be taken into account here, all field components may be defined as complex or using the method described in [39] as real values. During the simulation, a propagation constant β and a proper excitation are chosen. After a certain number of iterations, the corresponding modal frequencies can be extracted using Fourier transform of the time series obtained. In the conventional approach, repeating the Fourier transform after some iterations makes it possible to calculate the losses of the waveguiding structure [37]. For the 2-D FDTD method, a new stability condition based on the calculations of Taflove and Brodwin [40] can be derived [41]:

$$c_0\Delta t = \left[\frac{1}{(\Delta y)^2} + \frac{1}{(\Delta z)^2} + \left(\frac{\beta}{2}\right)^2\right]^{-\frac{1}{2}} \tag{7}$$

It can be seen that the timestep Δt depends on the maximum wave phase velocity c_0 in the air-filled region above the substrate and mainly on the spatial discretizations Δy and Δz.

The discretizations inside the conductor material should be smaller than the skin depths in the case of a normal conductor or smaller than the London penetration depth in the case of a superconductor for the given material and the considered frequency [40]. Due to the extreme short corresponding timestep Δt, a tremendous number of iterations is necessary to obtain the time series of voltage and current on the waveguide for at least one period. The simulations demonstrated in the following would require at least 100,000 iteration steps for one period at 8GHz using a discretization of 0.5μm for example. Keeping in mind that several periods are necessary for an exact Fourier transform, the simulation of such a structure becomes nearly impossible using the conventional or even the FDTD/SI method due to the large amount of required CPU-time on commonly used workstations.

The major idea to reduce the required computation time is the application of a locally refined mesh similar to that described in Section II-E and a specialized method to extract the desired information for the planar waveguides after only a small number of timesteps.

B. *The Local Mesh Refinement in 2-D FDTD*

As already discussed in Section II-A, if only a very simple excitation field is used as the starting condition for the simulation, wave components are enforced on the investigated waveguide that are not part of the normal guided wave solution. In 3-D FDTD therefore, a special line section is needed at the input to generate the wanted field solution on the waveguide, or a matched source as described in Section II-C must be used. For the 2-D FDTD, the unwanted, excited waveforms are initiated as a kind of cylindrical wave or pulse, propagating from the excitation point to the outer boundaries of the computational domain. Figure 22a shows the magnitude of the electrical field $|\vec{E}|$ in the cross section of a coplanar waveguide (CPW) 10

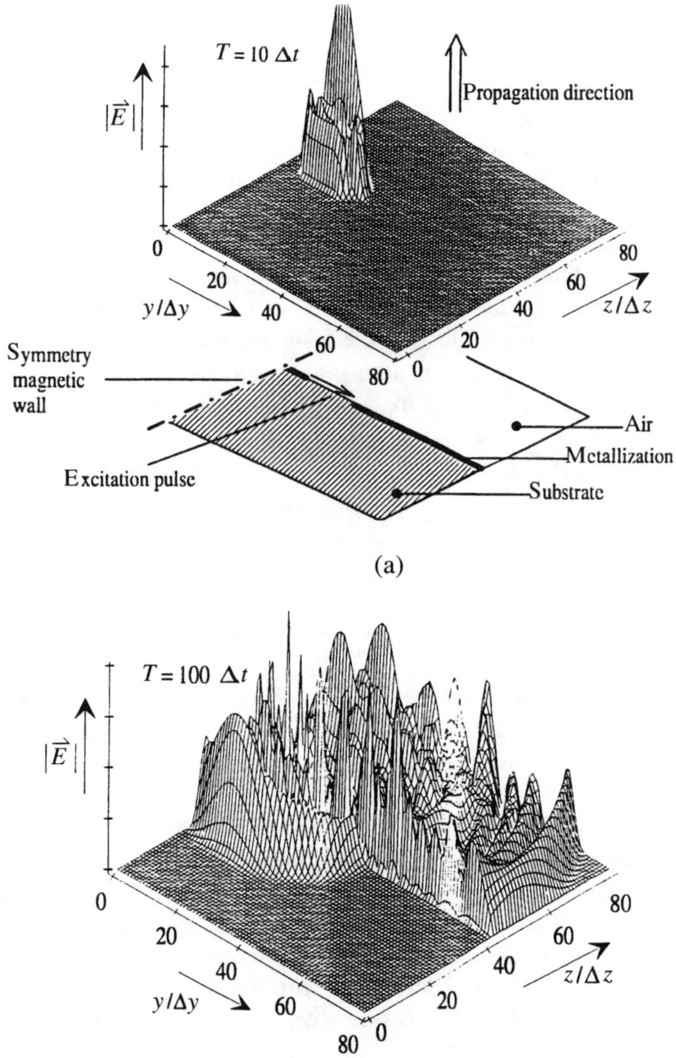

Fig. 22 Electric field (in arbitrary units) in the cross section of a (a) CPW 10 and (b) 100 timesteps after the excitation inside the gap between inner and outer conductor in the case of an equidistant discretization. In (b) the cylindrical waves reach the absorbing boundaries.

timesteps after the excitation pulse has been launched inside the gap between the inner and the outer conductor. This excitation pulse is a kind of Dirac excitation in the spatial domain, and the initiated waves and pulses are only band limited by the mesh resolution. As can be seen from Fig. 22b which shows $|\vec{E}|$ after 100 timesteps in the case of an equidistant discretization, the cylindrical waves and pulses reach the absorbing boundaries (which are similar to those described in Section II-B) and are absorbed. Therefore, after a certain simulation time, only the field components corresponding to the fundamental mode of the CPW can be observed inside the computational domain.

A problem occurs if a local mesh refinement (Section II-E) is used perpendicular to the propagation direction x. The cylindrical waves and pulses traveling towards the outer absorbing boundaries are multiply reflected at the mesh discontinuity. Additionally, the waves and pulses initiated in the fine mesh regions cannot propagate into the regions with the coarse mesh and are therefore strongly reflected. In Fig. 23 the corresponding field distribution after 100 timesteps is illustrated. In contrast to Fig. 22b, a graded mesh has been used starting at $y = 40\Delta y$, $z = 64\Delta z$, and $z = 23\Delta z$ (see Fig. 23). The mesh cell size in this case is increased towards the outer boundaries from cell to cell by a factor of 1.1. As it can be seen from Fig. 23 only a small part of the initiated parasitic wave components are propagating through the graded mesh region to the outer absorbing boundaries. For that reason, many field components which are not part of the fundamental mode may be recognized for a long time as a field disturbance enclosed in a virtual box inside the computational domain which makes it impossible to analyze currents or voltages on the waveguide.

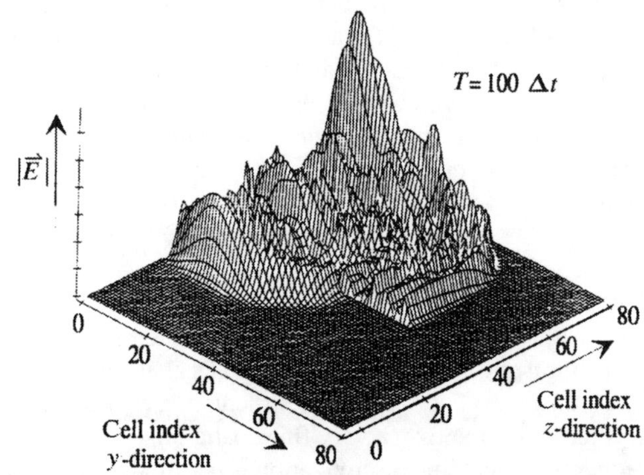

Fig. 23 Electric field in the cross section of a CPW 100 timesteps after the excitation in the case of a graded mesh. The cylindrical waves and pulses are multiply reflected at the mesh discontinuities.

These disturbances can be suppressed if a field distribution calculated from a quasi-static finite-difference method (FDM) [42] is used as the excitation condition for the 2-D FDTD which is demonstrated in Fig. 24. The small disturbances due to the

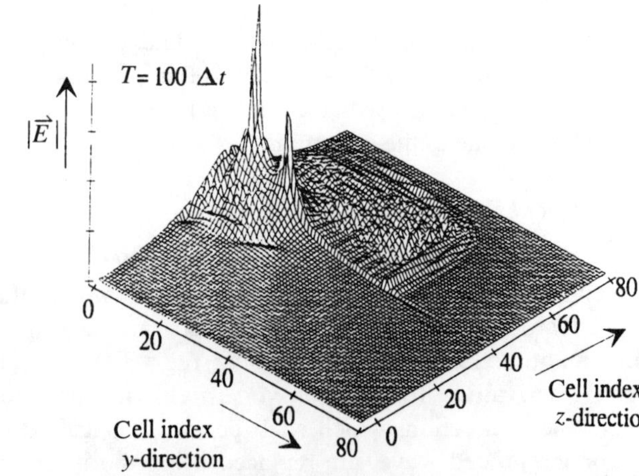

Fig. 24 Electric field in the cross section of a CPW 100 timesteps after the excitation in the case of a graded mesh, using an excitation with a quasistatic FDM field solution.

transition from the quasistatic FDM to the dynamic FDTD analysis will fade away after a short simulation time and will not hinder the calculation of currents and voltages. Thus, a graded mesh can be used without introducing additional computational steps as proposed in [39].

Figure 25 shows that it is important to use the graded mesh. As an example the effective dielectric constant of a CPW on GaAs with a center strip width of 30μm (which will be discussed

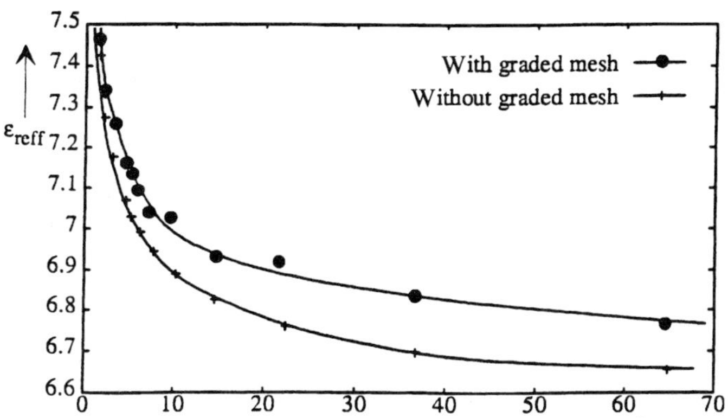

Fig. 25 Comparison of the simulated effective dielectric constants of a CPW on GaAs (ε_r = 12.9) using an equidistant and a graded mesh discretization with 70 × 80 mesh points, respectively.

in detail later) has been calculated with a discretization of 70 × 80 mesh points, which was a practical limit concerning the computation time on the used workstation. To provide an accurate loss calculation, a spatial resolution of $\Delta y = 1.0\mu$m and $\Delta z = 0.5\mu$m is required inside and near the conductors. Therefore a physical size of only 70μm × 40μm for one-half the symmetric CPW structure can be achieved using an equidistant discretization. As it can be clearly seen from Fig. 25, neither the magnitude of the effective dielectric constant nor the increase at low frequencies is comparable with the results for the graded mesh, which are in good agreement with measurements. Using the graded mesh together with the introduced excitation technique, a physical size of 214μm × 1,000μm can be realized for the computational domain. This guarantees accurate results which are not significantly influenced by the absorbing boundaries. To achieve the same physical size without a graded mesh, a discretization of 214 × 2,000 grid points would be needed. This would cause an increase of the memory consumption and the computation time by a factor of 76.

C. *Time Series Analysis*

If a CPW would be simulated, a magnetic wall is introduced in the symmetry plane of the structure, whereby the odd mode is suppressed. Thus, a nearly monomodal response to the excitation can be achieved. In the case that the line is lossy, it is known *a priori* that the time series resulting from the simulation will be a damped sinusoidal wave. Hence, the theoretical time signal can be fitted to the simulation results after only an iteration time of 1/15 up to 1/2 of one cycle duration which leads to a drastically reduced CPU-time. Figure 26 shows the fitting result. The

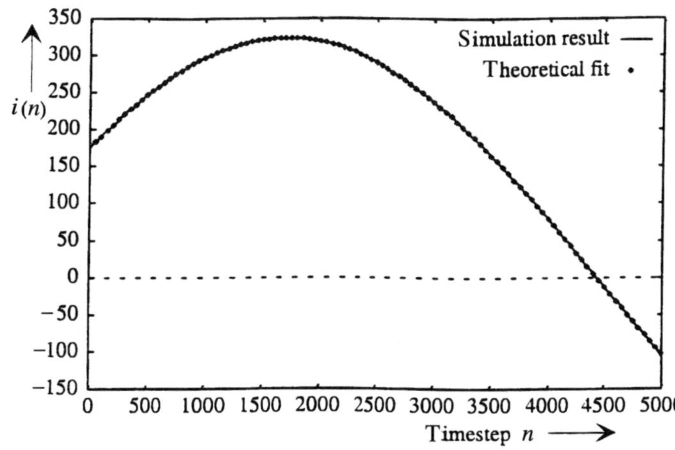

Fig. 26 Simulated time series and theoretical fit for the current i (in arbitrary units) on the center strip of a CPW.

current on the center strip of the CPW has been evaluated and stored for each time step.

The theoretical time signal

$$i(n\Delta t) = x_1 e^{-x_4 n} \cos = \left[\frac{2\pi}{x_2}(n + x_3)\right] \tag{8}$$

has been fitted to the simulated data, using Fletcher-Powel optimization technique [43] with the optimization variables x_1 to x_4. The relative error for each data point is less than 10^{-4} which shows that the response is really a single-frequency signal. The same procedure is applied to the voltage between the inner and outer conductor.

From the variables x_1, x_2, x_4 and the frequency $f = 1/(x_2\Delta t)$, the magnitude of the characteristic impedance $|Z_c|$, the effective dielectric constant ε_{reff} corresponding to the chosen β and the attenuation constant α can be calculated by the following equations:

$$\sqrt{\varepsilon_{reff}} = \frac{c_0 t}{2\pi}\beta x_2 \tag{9}$$

$$|Z_c| = \frac{\hat{u}}{\hat{\imath}} \tag{10}$$

$$\alpha = \frac{x_4\sqrt{\varepsilon_{reff}}}{c_0\Delta t}. \tag{11}$$

Here, $\hat{u}$ and $\hat{\imath}$ are the amplitudes of the voltage and current, respectively, on the waveguide which can be evaluated from the corresponding value x_1 for u and i, respectively.

Figure 27 shows a structure which has been used as a test structure for this work. Several CPWs have been fabricated depositing gold on GaAs substrates in a MMIC process. The spatial resolution of the applied 2-D FDTD analysis has been chosen to be 1/4 or 1/12 of the metallization thickness t.

In Fig. 28 the results for this CPW are shown. The parameters of the line are: the first metallization layer is of thickness 0.5μm, its conductivity is only 70% of the second upper layer which is $\chi = 35.4 \times 10^6 \Omega^{m-1}$. All technology data have directly been taken from the technology process description. The agreement between simulation and experiment especially concerning ε_{reff}

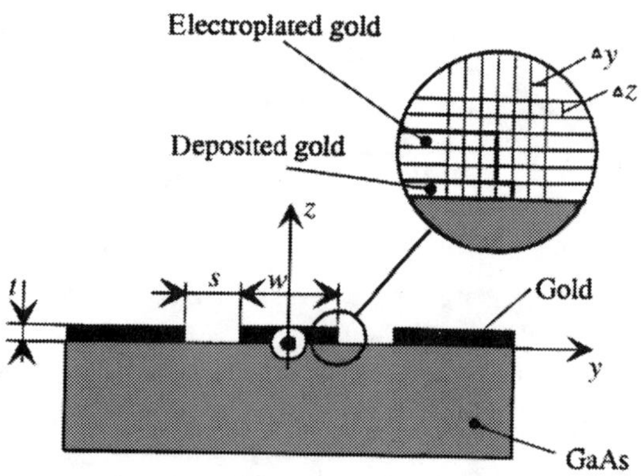

Fig. 27 The used test structure.

(Fig. 28a) is excellent and the occurring deviations are comparable to the errors in the fabrication and measurement process. From Fig. 28b it can be seen that the attenuation coefficient calculated using a quasi-static finite difference method (FDM) is only valid below 20GHz. The deviation of the results, obtained by the 2-D FDTD technique, from the measurement near and above 38GHz can be explained by the fact that at these frequencies the skin depth is nearly equal to the minimum discretization length Δz. Therefore, no further exact loss calculation is possible for higher frequencies [44]. This means that an even better spatial resolution is required at these high frequencies.

Due to the strong influence of the kinetic inductance in superconducting planar waveguides, it is very important to take the field penetration inside a superconducting field into account if the effective dielectric constant and the losses of such a CPW shall be calculated accurately. In this case the requirements for a simulation method are even higher than for normal conductors because the conductor film has only a thickness of about 300nm. Additionally, the London penetration depth is only about 150nm at low temperatures. Therefore, a discretization of only 75nm inside the conductor was chosen for this example. The effective dielectric constant $\varepsilon_{\text{reff}}$ of a superconducting CPW fabricated from YBaCuO on a $LaAlO_3$ substrate has been measured using a CPW resonator technique [43]. The center strip was 10μm and the gap 20μm wide to obtain a characteristic impedance of $Z_c = 50\Omega$. The critical temperature after the fabrication of the CPW was 86K and the London penetration depth $\lambda_L(T = 0\text{K})$ was 150nm. For the simulation of the superconductor, the London theory [45] has been introduced into the 2-D FDTD following the approach of [36].

Figure 29 shows the simulation and measuring results in dependence on the temperature T. They are in good agreement and the field penetration seems to be well modeled over the total temperature range. $\varepsilon_{\text{reff}}$ changes about 15% in this temperature range due to the strong increase of the London penetration depth and the resultant kinetic inductance. Therefore, a prediction of this behavior is very important to guarantee an accurate circuit design. The variation of the simulation results around the fitted line is caused by the observation window chosen for the time series analysis. Due to the discretization of 75nm, the timestep is even smaller for this example than for the above discussed example of the normal conducting CPW, but the observation window has been left constant. Thus, only a very small part of a single period has been used for the parameter extraction. Using this method, it should be possible to develop and test different models for the microwave and nonlinear behavior of superconductors, based on measuring results [46].

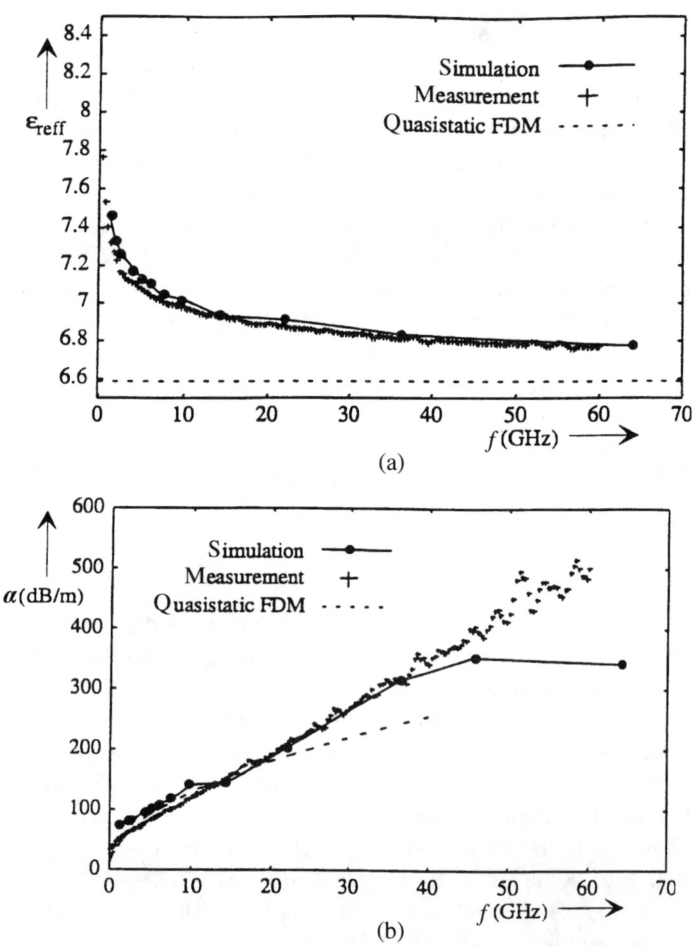

Fig. 28 (a) Effective dielectric constant $\varepsilon_{\text{reff}}$, and (b) attentuation coefficient α of a CPW on GaAs ($\varepsilon_{\text{reff}} = 12.9$) as a function of frequency. Parameters: $w = 30\mu$m, $s = 25\mu$m, $t = 2\mu$m, $\Delta z = 0.5\mu$m, $\Delta y = 1.0\mu$m.

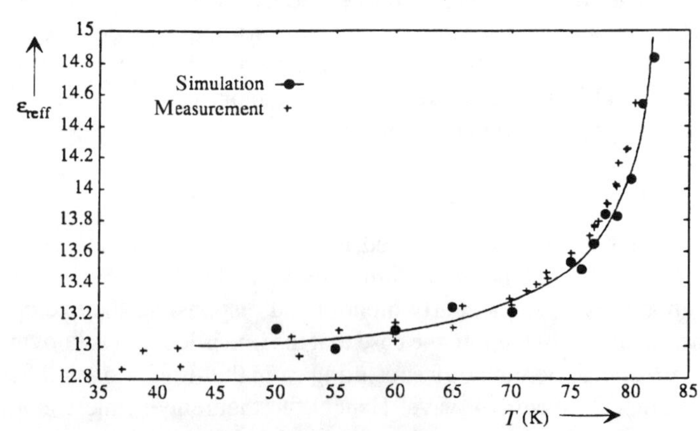

Fig. 29 Simulated and measured effective dielectric constant $\varepsilon_{\text{reff}}$ of a superconducting CPW on $LaAlO_3$ ($\varepsilon_r = 24.5$) as a function of temperature. Parameter: $w = 10\mu$m, $s = 20\mu$m, $\Delta y = 660$nm, $\Delta z = 75$nm.

IV. Analysis of Microwave Distributed Circuits and Packages Containing Nonlinear Active Elements Using FDTD Technique

The standard approach to the analysis of microwave circuits containing non-linear components consists of dividing the circuit into several segments which may be modeled independently of each other, e.g., transmission lines, discontinuities, FETs, etc. These models are then used for the analysis of the complete circuit with some circuit analysis software. The method which has found widest acceptance for the analysis of nonlinear microwave circuits is the harmonic balance method (HBM) as described, for example, in [47,48]. A major drawback of the segmentation approach is its inability to account for couplings in complex, narrow structures, except for the rare cases where a model for these structures is known. Moreover, the choice of the correct segmentation, if it exists, is a matter of estimation and therefore a potential source of error. Very similar to this problem is the question of analyzing a complete microwave package containing active, nonlinear elements or circuits, a simulation task which finally goes to the edges of the available numerical analysis techniques.

To overcome these problems, time-domain methods have gained in interest, but these methods, as has been described in the previous sections, up to now have been primarily employed for the simulation of linear, passive components. As already has been shown [49,50] they also may be enhanced to account for active, nonlinear devices as well. However, time-domain methods with nonlinear enhancement are subject to specific restrictions as well. In the presence of nonlinear elements, a characterization of the circuit as in the linear case is impossible. Therefore, if excitations of the circuits change even slightly, the analysis has to be performed entirely again, including the computationally expensive linear space domain. Hence the effort for *n* analyzes is as high as *n* times the effort for a single analysis, because with time-domain methods there is no way of utilizing the results of the former analyses as a solution estimate, as is possible in the HBM case. Multitone excitations (e.g., those of mixer circuits) cannot easily be handled with time-domain methods.

A. *The Compression Approach*

To circumvent these restrictions, a method is proposed which combines time-domain methods for the simulation of linear structures with the HBM for nonlinear analyses [51,52], resulting in several advantages:

- Nonlinear components embedded in complex structures may be considered.
- All field-theoretical aspects which, for example, time-domain methods are able to account for are fully integrated into the nonlinear analysis.
- All advantages of the HBM with respect to multiple analyses are exploited.
- All analysis options of the HBM, especially the possibility to analyze multitone excitations, are preserved.
- The characterization of the linear part has to be done only once and is independent of the actual circuit analysis.
- In contrast to other approaches for the characterization of linear parts as, for example, the Green's function-based John's Matrix concept [49], the compression approach yields as a result the minimum of required information in a highly condensed form.

Basically, the circuit is considered to be a superposition of a single linear space domain governed by Maxwell's equations, and a number of nonlinear regions made up by nonlinear elements. An interface between both domains is introduced by defining ports in the space domain to which the nonlinear elements are connected. The nonlinear elements are assumed to be described by standard CAD models. As characterization of the linear part, the entire passive, linear space domain, including the regions where the physical nonlinear devices are located, is described by a single matrix S_c of scattering parameters (which is termed the *compression matrix*) plus appropriate reference resistances.

It is important to note that this is an essential difference to the segmentation method where the linear part is divided into several components which are considered to be independent—with respect to the coupling fields—to each other. This prevents segmentation methods from utilizing the accuracy benefits of, for example, time-domain methods.

The resulting equivalent circuit of the nonlinear elements connected to the ports of the linear part (Fig. 30), is then analyzed with an HBM simulator. A crucial point of this approach is the necessity to consider *inner ports* (i.e., ports) which are located within complex structures, being connected to nonlinear elements but not to any reference line. Special attention has to be paid to these ports with respect to the definition of port variables and reference resistances.

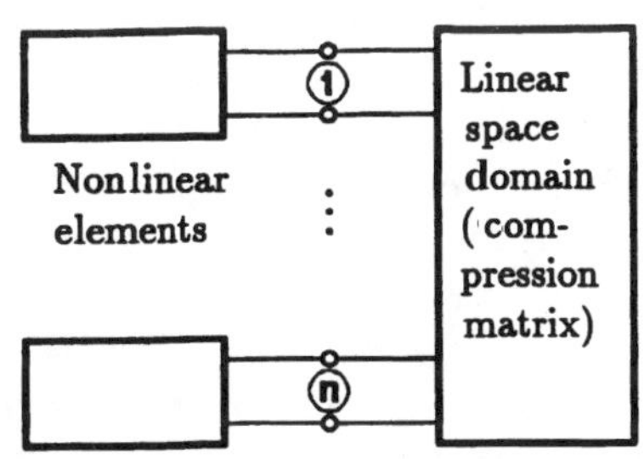

Fig. 30 Interface between nonlinear elements and the surrounding linear space domain. Nonlinear elements interact by means of the compression matrix.

The way to define the inner ports and to determine the compression matrix has been described in detail in [51,52], a paper which is reprinted within this chapter. The application of the technique to the simulation of packages with enclosed nonlinear active circuits is described in [53–56]. The papers [53–55] again are reprinted within this chapter.

V. Conclusions

The short description of the state of the art in application of FDTD to planar microwave integrated circuits shows the great progress in this field during the last 10 years. It was only in 1985 when the first real application of the FDTD method to planar circuit analysis was published [8]. Wojciech K. Gwarek described a time-domain approach for the analysis of arbitrarily shaped planar circuits using a 2-D FDTD. He formulated a differential equation for the Hertzian potential describing an E-mode electromagnetic field and for the surface current density flowing in the metallization of a planar circuit. This differential equation finally was solved by Gwarek using a time-domain approach which is quite similar to the FDTD method of Yee, though he apparently was not aware of Yee's work and the following publications. Already in this first publication Gwarek could show that his method very flexibly can be applied to the analysis of complicated planar components and that the accuracy of the theoretical results as compared to measurements is very high. Gwarek in his paper already stated the following:

- The method is very versatile. A universal program may be worked out for an analysis of a planar circuit of any shape defined by the user.
- The accuracy of the calculation can be made sufficient to assure a very good agreement between the theory and the experiment.
- There are possibilities for further development of the method (e.g., including losses, linear, and nonlinear lumped elements).
- The method simulates a natural process of the wave propagation, and the actual field distribution may be displayed during the calculation.

These advantages of the FDTD method are used today in many applications, as has been shown in this chapter. In 1988 Gwarek demonstrated the applicability of his method to analyze real arbitrarily shaped planar circuits [10].

The first application of the FDTD method as *i*st has been defined by Yee to planar component and circuit design has been described by Xiaolei Zhang, Jiayuan Fang, Kenneth K. Mei, and Yaowu Liu [9] who applied the method to the analysis of the dispersive characteristics of microstrip lines. Here for the first time, the characteristic field distribution of pulses traveling along a microstrip line have been shown and the dispersion characteristics in dependence on the frequency have been drawn from the time transfer functions. The authors of this paper already recognized the problems applying simple boundary conditions to solve the fields in microstrip structures. To minimize the reflection errors, the authors used an open and a short-circuited boundary condition in parallel and averaged the solutions of these two differently defined problems.

Two of the above mentioned authors, X. Zhang and K. K. Mei, in 1988 published their analyses of fundamental microstrip discontinuities like microstrip open ends, microstrip T-junctions, and crossings using the FDTD method [11]. This opened the large field of new applications for the FDTD. It was also this author group which recognized the need for better absorbing boundaries if the results in planar circuit design should be good enough for technical applications. J. Fang as a consequence developed the so called superabsorbing boundary [4],[1] which indicated great progress for the FDTD analysis technique and which has been used for many years in advanced computer programs.

Also in 1988 a Japanese publication debuted on the application of the FDTD technique to the analysis of microstrip circuits [13]. Its authors—T. Shibata, T. Hayashi, and T. Kimura—derived an equivalent circuit for the microstrip structure and defined node voltages of the equivalent circuit which correspond to the field components of the electric and the magnetic field strength, respectively. By application of the FDTD technique to this equivalent circuit, the authors were already able to analyze quite complex circuit structures like coupled microstrip filters.

More advanced structures in microstrip planar circuits have been analyzed by John Moore and Hao Ling [57] who demonstrated the influence of a miter in a 90° microstrip bend by using the FDTD technique, using an electric wall instead of an absorbing boundary condition for closing the computational space. Also in 1990, a paper of D. M. Sheen, A. M. Ali, M. D. Abouzahra, and J. A. Kong appeared [15] in which the authors describe the analysis of microstrip patch antennas, low-pass filters, and branch line couplers using the FDTD method and Mur's absorbing boundary [3].

First applications of the FDTD method the analysis of coplanar components and circuits have been described by Rittweger and Wolff [58] who also introduced the so called "matched source" for exciting the circuits with a Gaussian pulse which has the correct field distribution on the considered planar line basis (e.g., microstrip or coplanar waveguide) as to reduce the computational space. In these applications, Fang's superabsorbing boundaries have also been used, leading to a high accuracy of the analyzed results as compared to intensive measurements.

N. M. Pothecary and C. J. Railton showed in 1991 [59] that the FDTD with advantage also can be applied to the problem of digital signal circuits where the problem of nearby line coupling is often essential. They demonstrated how this coupling effectively can be analyzed using the FDTD method.

As has been demonstrated in this chapter, many problems in electromagnetic field analysis are so complicated that a full 3-D analysis cannot be provided because the computational space or the computation time become too large. In such cases often a 2-D analysis can help solve the problems. Effective applications of the FDTD method to 2-D problems for the first time have been demonstrated in a paper of V. J. Brankovic, D. V. Krupezevic, and F. Arndt [37] who applied the method to the analysis of shielded and open waveguides. Other authors, as described in this chapter, have overtaken this method and have analyzed more sophisticated problems like superconducting planar waveguides with a modified version of this method.

[1] See also: J. Fang, "Time domain finite difference computation for Maxwell's equations," doctoral thesis, Departm. Electrical Engineering and Computer Science, University of California at Berkeley, 1989.

One of the interesting aspects of the FDTD method is that it is able to analyze 3-D electromagnetic field problems, including nonlinear active or passive components. A very effective method for this application has been developed by J. Kunisch et al. [45,51] who combined the 3-D field analysis in the time domain with the possibilities of the Hamonic-Balance technique in the frequency domain to analyze complicated circuits including nonlinear components (e.g., diodes or transistors) and total RF packages with enclosed nonlinear components or circuits. This technique very soon was adopted by T. Shibata [53] for solving similar problems and in the mean time has been developed as a professional tool for the analysis of microwave packages and interconnect problems [56].

References

[1] K. S. Yee, "Numerical solution of initial boundary value problems involving Maxwell's equations in anisotropic media," *IEEE Trans. Antennas Propagat.,* vol. AP-14, pp. 302–307, 1996.

[2] C. J. Railton and J. P. McGeehan, "Analysis of microstrip discontinuities using the finite difference time domain technique," in *1989 IEEE MTT-S Internat. Microwave Symposium Digest,* Long Beach, CA, pp. 1009–1012, June 1989.

[3] A. Taflove and M. E. Brodwin, "Numerical solution of steady-state electromagnetic scattering problems using the time-dependent Maxwell's equations," *IEEE Trans. Microwave Theory Tech.,* vol. MTT-23, pp. 623–630, 1975.

[4] G. Mur, "Absorbing boundary conditions for the finite-difference approximation of the time-domain electromagnetic field equations," *IEEE Trans. Electromagn. Compat.,* vol. EMC-23, pp. 377–382, 1981.

[5] J. Fang and K. K. Mei, "A super-absorbing boundary algorithm for solving electromagnetic problems by time-domain finite-difference method," in *1988 IEEE AP-S International Symposium Digest,* Syracuse, NY, pp. 472–475, June 1988.

[6] K. K. Mei and J. Fang, "Superabsorption—a method to improve absorbing boundary conditions," *IEEE Trans. Antennas Propagat.,* vol. 40, pp. 1001–1010, 1992.

[7] D. H. Choi and W. J. R. Hoefer, "The finite-difference-time-domain method and its application to eigenvalue problems," *IEEE Trans. Microwave Theory Tech.,* vol. MTT-34, pp. 1464–1470, 1986.

[8] W. K. Gwarek, "Analysis of an arbitrarily-shaped planar circuit—a time domain approach," *IEEE Trans. Microwave Theory Tech.,* vol. MTT-33, pp. 1067–1072, 1985.

[9] X. Zhang, J. Fang, K. K. Mei, and Y. Liu, "Calculation of the dispersive characteristics of microstrips by the time-domain finite difference method," *IEEE Trans. Microwave Theory Tech.,* vol. MTT-36, pp. 263-267, 1988.

[10] W. K. Gwarek, "Analysis of arbitrarily shaped two-dimensional microwave circuits by finite-difference time-domain method," *IEEE Trans. Microwave Theory Tech.,* vol. MTT-36, pp. 738–744, 1988.

[11] X. Zhang and K. K. Mei, "Time-domain finite difference approach to the calculation of the frequency-dependent characteristics of microstrip discontinuities," *IEEE Trans. Microwave Theory Tech.,* vol. MTT-36, pp. 1775–1787, 1988.

[12] S. Koike, N. Yoshida, and I. Fukai, "Transient analysis of microstrip line on anisotropic substrate in three-dimensional space," *IEEE Trans. Microwave Theory Tech.,* vol. MTT-36, pp. 34–43, 1988.

[13] T. Shibata, T. Hayashi, and T. Kimura, "Analysis of microstrip circuits using three-dimensional full-wave analysis in the time domain," *IEEE Trans. Microwave Theory Tech.,* vol. MTT-36, pp. 1064–1070, 1988.

[14] S. Nam, H. Ling, and T. Itoh, "Time-domain method of lines applied to the uniform microstrip line and its step discontinuity," in *1989 IEEE MTT-S Internat. Microwave Symposium Digest,* Long Beach, CA, pp. 997–1000, June 1989.

[15] D. M. Sheen, S. M. Ali, M. D. Abouzahra, and J. A. Kong, "Application of the three-dimensional finite-difference time-domain method to the analysis of planar microstrip circuits," *IEEE Microwave Theory Tech.,* vol. 38, pp. 849–857, 1990.

[16] I. Wolff and M. Rittweger, "Finite difference time-domain analysis of planar microwave circuits," *Arch. Elektrotechn.,* vol. 74, pp. 189–201, 1991.

[17] I. Wolff, "Finite difference time-domain simulation of electromagnetic fields and microwave circuits," *Internat. J. Numerical Modelling,* vol. 5, pp. 163–182, 1992.

[18] M.Kirschning, R. H. Jansen, and N. H. L. Koster, "Accurate model for the open end effect of microstrip line," *Electron. Lett.,* vol. 17, pp. 123–125, 1981.

[19] R. H. Jansen, "Hybrid mode analysis of end effects of planar microwave and millimeterwave transmission lines," *IEEE Proc.,* vol. 128, pt. H, pp. 77–86, 1981.

[20] G. Gronau and I. Wolff, "A simple broad-band device deembedding method using an automatic network analyzer with time-domain option," *IEEE Trans. Microwave Theory Tech.,* vol. MTT-37, pp. 479–483, 1989.

[21] W. Wertgen, *Elektrodynamische Analyse geometrisch komplexer (M)(MIC-Strukturen mit effizienten numerischen Strategien (Electrodynamical analysis of geometrically complex (M)MIC structures with efficient numerical strategies),* PhD thesis, Duisburg University, Duisburg, FRG, 1989.

[22] D. Choi and W. J. R. Hoefer, "A graded mesh FD-TD algorithm for eigenvalue problems," in *Proceedings 17th European Microwave Conference,* pp. 413–317, 1987.

[23] C. J. Railton and J. P. McGeehan, "Analysis of MMIC components including the effects of finite metallization thickness," in *Proceedings 19th European Microwave Conference,* pp. 187–192, 1989.

[24] I. S. Kim and W. J. R. Hoefer, "A local mesh refinement algorithm for the time domain-finite difference method using Maxwell's curl equations," *IEEE Trans. Microwave Theory Tech.,* vol. MTT-38, pp. 812–815, 1990.

[25] S. S. Zivanovic, K. S. Yee, and K. K. Mei, "A subgridding method for the time-domain finite-difference method to solve Maxwell's equations," *IEEE Trans. Microwave Theory Tech.,* vol. 39, pp. 471–479, 1991.

[26] M. Rittweger, *Simulation transienter elektrodynamischer Ausbreitungsphänomene zur Analyse der Übertragungseigenschaften von Systemen der Mikrowellen—und Millimeterwellentechnik (Simulation of transient electrodynamic propagation phenomena for the analysis of transmission properties of microwave and millimeterwave systems)* Doctoral Thesis, Duisburg University, Duisburg, 1992.

[27] J. L. Dubard, D. Pompei, J. Le Roux, and A. Papiernik, "Characterization of microstrip antennas using the TLM simulation associated with a Prony-Pisarenko method," *Int. J. Numerical Modeling,* vol. 3, pp. 269–285, 1990.

[28] M. L. van Blaricum and R. Mittra, "Problems and solution associated with Prony's method for processing transient data," *IEEE Trans. Antennas Propagat.,* vol. AP-26, pp. 174–182, 1978.

[29] J. D. Wills, "Spectral estimation for the transmission line matrix method," *IEEE Trans. Microwave Theory Tech.,* vol. 38, pp. 448–451, 1990.

[30] W. L. Ko and R. Mittra, "A combination of FDTD and Prony's methods for analyzing microwave integrated circuits," *IEEE Trans. Microwave Theory Tech.,* vol. 39, pp. 2176–2181, 1991.

[31] Z. Bi, Y. Shen, K. Wu, and J. Litva, "Fast finite-difference time-domain analysis of resonators using digital filtering and spectrum estimation techniques," *IEEE Trans. Microwave Theory Tech.,* vol. 40, pp. 1611–1619, 1992.

[32] W. Kuempel and I. Wolff, "Digital signal processing of time domain field simulation results using the system identification method," in *IEEE MTT-S Int. Microwave Symp. Digest,* Albuquerque, NM, pp. 793–796, June 1992.

[33] W. Kuempel and I. Wolff, "System identification method for transient analysis of (M)MIC-components using time iterative methods," in *22nd European Microwave Conference Proceedings,* Espoo, Finland, pp. 345–349, Aug. 1992.

[34] W. Kuempel and I. Wolff, "Digital signal processing of time domain field simulation results using the system identification method," *IEEE Trans. Microwave Theory Tech.,* vol. 42, pp. 415–423, 1994.

[35] T. Sönderström and P. Stoica, *System Identification,* Englewood Cliffs, NJ: Prentice Hall, 1989.

[36] M. Rittweger and I. Wolff, "Finite difference time-domain formulation of transient propagation in superconductors," in *IEEE Int. Antennas Propagation Symposium Digest,* Chicago, IL, pp.1960–1963, July 1992.

[37] V. J. Brankovic, D. V. Krupezevic, and F. Arndt, "An efficient two-dimensional graded mesh finite-difference time-domain algorithm for shielded or open waveguide structures," *IEEE Trans. Microwave Theory Tech.,* vol. 40, pp. 2272–2277, 1992.

[38] S. Xiao, R. Vahldieck, and H. Jin, "A fast two dimensional FDTD fullwave analyser with adaprive mesh size," in *1992 IEEE MTT-S Int. Microwave Symp. Digest,* Albuquerque, NM, pp. 783–786, 1992.

[39] S. Xiao and R. Vahldieck, "An improved 2-D-FDTD algorithm for hybrid mode analysis of quasi-planar transmission lines," in *1993 IEEE MTT-S Int. Microwave Symp. Digest,* Atlanta, GA, pp. 421–424, 1993.

[40] A. Taflove and M. E. Brodwin, "Numerical solution of steady-state electromagnetic scattering problems using the time-dependent Maxwell's equations," *IEEE Trans. Microwave Theory Tech.,* vol. MTT-23, pp. 623–630, 1975.

[41] A. C. Cangellaris, "Numerical stability and numerical dispersion of a compact 2-D/FDTD method used for the dispersion analysis of waveguides," *IEEE Microwave Guided Wave Letters,* vol. 3, pp. 3–5, 1993.

[42] M. Naghed and I. Wolff, "A three-dimensional finite difference calculation of equivalent capacitances of coplanar waveguide discontinuities," in *IEEE MTT-S Int. Microwave Symp. Digest,* pp. 1143–1146, May 1990.

[43] S. Hofschen and U. Salz, "Determination of propagation and attenuation constants of superconducting coplanar waveguides for the design of MIC using stripline resonator measurements, in *Proc. 1993 European Conference on Applied Superconductivity,* Göttingen, Germany, pp. 1065–1068, Oct. 1993.

[44] J. G. Maloney and G. S. Smith, "The use of surface impedance concepts in the finite-difference time-domain method," *IEEE Trans. Antenna Propagat.,* vol. AP-40, pp. 38–48, 1992.

[45] S. M. El-Ghazaly, R. B. Hammond, and T. Itoh, "Analysis of superconducting microwave structures: Application to microstrip lines," *IEEE Trans. Microwave Theory Tech.,* vol. 40, pp. 499–508, 1992.

[46] J. G. Ma and I. Wolff, "Modelling the microwave properties of superconductors," *IEEE Trans. Microwave Theory Tech.,* vol. 43, pp. 1053–1059, May 1995.

[47] V. Rizzoli and A. Neri, "State of the art and present trends in nonlinear microwave CAD techniques," *IEEE Trans. Microwave Theory Tech.,* vol. MTT-36, pp. 343–365, 1988.

[48] K. S. Kundert and A. S. Vincentelli, "Finding the steady-state response of analog and microwave circuits," *Alta Frequenza,* vol. LVII, pp. 379–387, 1988.

[49] R. H. Voelker and R. J. Lomax, "A finite-difference transmission line matrix method incorporating a nonlinear device model," *IEEE Trans. Microwave Theory Tech.,* vol. 38, pp. 302–312, 1990.

[50] P. Russer, M. Schwab, and F. W. Kaertner, "Numerical analysis of microwave oscillators," in *1990 Internat. Workshop of German IEEE MTT/AP Joint Chapter on Integrated Nonlinear Microwave and Millimeterwave Circuits (INMMC 1990) Digest,* Duisburg, Germany, pp. 155–175, Oct. 1990.

[51] J. Kunisch, M. Rittweger, S. Heinen, and I. Wolff, "The compression approach: a new technique for the analysis of distributed circuits containing nonlinear elements," in *Proc. 21st European Microwave Conf.,* Stuttgart, Germany, pp. 1296–1302, 1991.z

[52] J. Kunisch and I. Wolff, "Steady-state analysis of nonlinear forced and autonomous microwave circuits using the compression approach," *Int. J. Microwave Millimeter-Wave Computer-Aided Engineering,* vol. 4, pp. 241–255, 1995.

[53] T. Shibata and H. Kimura, "Computer-aided engineering for microwave and millimeter-wave circuits using the FD-TD technique of field simulations," *Internat. J. Microwave Millimeter-Wave Computer-Aided Engineering,* vol. 2, pp. 238–250, 1993.

[54] M. Rittweger, M. Werthen, and I. Wolff, "3D FDTD analysis applied to the investigation of the resonant behaviour of ceramic feedthrus," in *1994 IEEE MTT-S Internat. Microwave Symp. Digest,* San Diego, CA, pp. 1719–1722, May 1994.

[55] M. Rittweger, M. Werthen, and I. Wolff, "FDTD simulation of microwave packages and interconnects," in *Proc. of Workshop WSFC: EM Modeling of Microwave Packages and Interconnects, 1993 IEEE MTT-S Internat. Microwave Symp.,* Atlanta, GA, pp. 78–91, May 1993.

[56] M. Rittweger, M. Werthen, J. Kunisch, I. Wolff, P. Chall, B. Bahm, and P. Lok, "3D FDTD analysis of a SOT353 package containing a bipolar wideband cascode transistor using the compression approach," in *1995 IEEE MTT-S Internat. Microwave Symp. Digest,* Orlando, FL, pp. 1587–1590, May 1995.

[57] J. Moore and H. Ling, "Characterization of a 90° microstrip bend with arbitrary miter via the time-domain finite difference method," *IEEE Trans. Microwave Theory Tech.,* vol. 38, pp. 405–410.

[58] M. Rittweger, M. Abdo, and I. Wolff, "Full-wave analysis of coplanar discontinuities considering three-dimensional bond wires," in *1991 IEEE MTT-S Int. Microwave Symp. Digest,* Boston, pp. 465–468.

[59] N. M. Pothecary and C. J. Railton, "Analysis of cross-talk on high-speed digital circuits using the finite difference time-domain method," *Int. J. Numerical Modelling,* vol. 4, pp. 225–240, 1991.

Analysis of an Arbitrarily-Shaped Planar Circuit—A Time-Domain Approach

WOJCIECH K. GWAREK

(Invited Paper)

Abstract —**A study of an arbitrarily-shaped planar circuit is reported. The theoretical background is presented and then a numerical method of analysis is introduced. Experiments show good agreement between the theoretical calculations and the measurements. The examples of applications concern stripline circuits but the method may be also applied to waveguides.**

I. Introduction

THE GENERAL CONCEPT of a planar circuit was introduced by Okoshi [1] as a circuit having one dimension very small in comparison with the wavelength and an arbitrary shape in two other dimensions. The idea proved to be important from both the theoretical and the practical points of view. In microwave theory, it initiates a two-dimensional circuit theory which is an extension of the 0-dimensional lumped-element theory and one-dimensional transmission-line theory. In microwave techniques, there are a wide range of applications of the theory including the use for microwave integrated circuits.

The idea of a planar circuit leads to a mathematical problem of solving a two-dimensional Helmholtz equation for a potential V, with proper boundary conditions. To solve the problem, Okoshi and Miyoshi [1] used Weber's solution for cylindrical waves to obtain relations between the potentials on different parts of the boundary. This led to an approximate linear matrix relation which had to be solved numerically. Hsu Jui-Pang and Anada [2] have developed a method based on a set of eigenfunctions obtained numerically for a particular circuit. There were also some efforts to characterize planar elements of particular shape and to analyze a circuit composed of such elements [3], [4].

Much progress has been made and many practical problems solved; however, none of the mentioned methods seems to lead to a computer program capable of solving an arbitrarily-shaped planar circuit without an extensive theoretical and programming effort by the user. This limits their applications and justifies the search for alternative methods more practical for engineers.

All the mentioned methods, while being very different, are based on the same idea of considering only a steady-state solution for a complicated boundary problem. The idea presented in this paper is to simulate numerically the wave propagation in the structure starting from zero initial conditions. The wave propagating and reflecting from the boundaries sets up automatically the proper boundary conditions. When the steady state is reached, all the circuit parameters can be obtained.

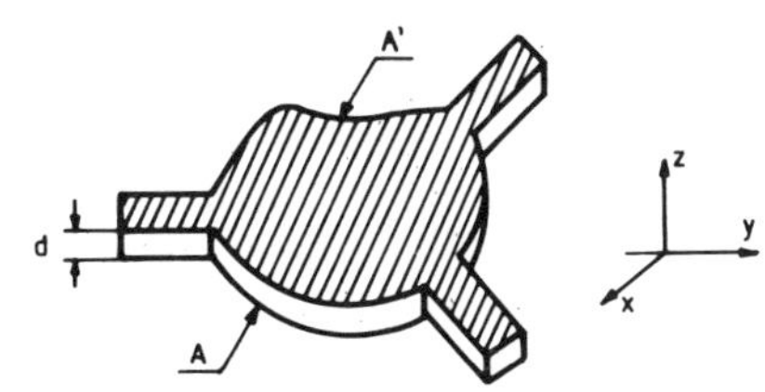

Fig. 1. A planar circuit.

II. Basic Definitions and Relations for the Considered Circuit

Hsu Jui-Pang and T. Anada [2] have presented a general circuit theory of a planar circuit extending its definition beyond the circuit of a very small height. They have based the theory on a set of E and H height modes. Their idea is followed in this paper but the formal description of the circuit is based on the Hertz potentials. This approach gives mathematical precision for the definitions and is convenient in practical use.

Let us consider the planar circuit shown in Fig. 1. The space in which the wave is transmitted is limited by two electric walls A and A' situated in the planes $z=0$ and $z=d$. We consider two sets of modes E_n and H_n.

Definition: An E_n mode is a mode described by an electric Hertz potential π_e [5] of the form

$$\pi_e^n = a_z \psi_e^n(x,y)\cos\left(\frac{n\pi}{d}z\right)e^{j\omega t} \tag{1}$$

where a_z is a unit vector parallel to the z-axis, and $n=0,1,2,\cdots$.

An H_n mode is a mode described by a magnetic Hertz potential π_h^n of the form

$$\pi_h^n = a_z \psi_h^n(x,y)\sin\left(\frac{n\pi}{d}z\right)e^{j\omega t} \tag{2}$$

where $n=1,2,\cdots$.

Manuscript received February 5, 1985; revised June 3, 1985.
The author is with the Instytut Radioelektronikci, Politechnika Warszawska, Warszawa, Nowowiejska 15/19 Poland.

Reprinted from *IEEE Transactions on Microwave Theory and Techniques*, pp. 1067-1072, October 1985.

From the general properties of the Hertz potentials [5], we obtain for an E_n mode

$$E_z = \psi_e^n(x, y)\cos\left(\frac{n\pi}{d}z\right)\beta_t^2 e^{j\omega t}$$

$$\boldsymbol{E}_t = -\frac{n\pi}{d}\sin\left(\frac{n\pi}{d}z\right)\nabla\psi_e^n(x, y)e^{j\omega t}$$

$$H_z = 0$$

$$\boldsymbol{H}_t = -j\omega\epsilon\boldsymbol{a}_z\times\nabla\psi_e^n(x, y)\cos\left(\frac{n\pi}{d}z\right)e^{j\omega t} \tag{3}$$

where

$$\beta_t^2 = \omega^2\mu\epsilon - \left(\frac{n\pi}{d}\right)^2.$$

The surface current which flows in the upper plane A' is equal to

$$\boldsymbol{J}_e = -\boldsymbol{a}_z\times\boldsymbol{H}_t|_{z=d} = j\omega\epsilon\nabla\psi_e^n(x, y)e^{j\omega t}. \tag{4}$$

Let us define a potential

$$V_e = -d\psi_e^n(x, y)\beta_t^2 e^{j\omega t}. \tag{5}$$

Using the above relations and taking into consideration that the potential (1) satisfies the Helmholtz equation, we obtain the relations between $\boldsymbol{J}_e$ and V_e

$$\nabla V_e = -j\frac{\beta_t^2}{\omega\epsilon}d\boldsymbol{J}_e \tag{6}$$

$$\nabla\cdot\boldsymbol{J}_e = -j\frac{\omega\epsilon}{d}V_e. \tag{7}$$

In the case of an H_n mode, we obtain the formulas dual to (3), (4), and (5). These formulas include a magnetic current $\boldsymbol{J}_h$ and a magnetic potential V_h, which are related by the equations

$$\nabla V_h = -j\frac{\beta_t^2}{\omega\mu}d\boldsymbol{J}_h \tag{8}$$

$$\nabla\cdot\boldsymbol{J}_h = -j\frac{\omega\mu}{d}V_h. \tag{9}$$

There is a following correspondence between the defined modes and the popular transmission lines.

1) Stripline may be considered as a composition of two planar structures with E_0 modes. The fringing fields may be included in the calculations by extending the real dimensions of the circuit in the xy-plane by a value of [1]

$$\delta = 2d(\log_e 2)/\pi. \tag{10}$$

2) Microstrip is basically a planar structure with an E_0 mode, but the complicated nature of its fringing fields makes its analysis a more difficult problem than in the case of stripline.

3) H_{01} rectangular waveguide situated with the E field parallel to the z-axis of Fig. 1 may be considered as a planar circuit of E_0 mode with the short-circuit boundary conditions on the sides of the guide. In this case, the planar circuit methods solve the problem of a guide with changing width. When the guide is turned by $\pi/2$, it becomes a planar circuit of an H_1 mode with open-boundary conditions for the V_h potential on the sides. In this case, the planar circuit methods may be applied to a guide with changing height.

As shown, all the cases for planar circuits lead to a set of equations (6) and (7) or (8) and (9) which may be written in a standard form

$$\nabla V(x, y) = -j\omega L_s\boldsymbol{J}(x, y) \tag{11}$$

$$\nabla\cdot\boldsymbol{J}(x, y) = -j\omega C_s V(x, y). \tag{12}$$

We shall now concentrate on the E_0 mode. It does not mean that the formulas presented in the next section are invalid for the higher modes. They are valid, but their physical interpretations are more complicated and will not be considered here.

For the mode E_0, the value C_s is the capacitance of a unitary square of the circuit, and the value of L_s is the inductance of an arbitrary square of the circuit.

III. The Proposed Method of Analysis

The proposed method of analysis is based on the finite-difference method [6], [7]. Let us consider (11) and (12) in the time-dependent form

$$\nabla V(x, y, t) = -L_s\frac{\partial\boldsymbol{J}(x, y, t)}{\partial t} \tag{13}$$

$$\nabla\cdot\boldsymbol{J}(x, y, t) = -C_s\frac{\partial V(x, y, t)}{\partial t}. \tag{14}$$

The xy-plane (the domain of the functions V and $\boldsymbol{J}$) is divided into a set of square meshes of the size a. The coordinates of the middle of a mesh in the kth row and the lth column are denoted by x_l and y_k. Let us assume that (13) and (14) describe propagation of a wave of the frequency ω and the wavelength λ. If we assume that $a \ll \lambda$ and $\Delta t \ll \frac{2\pi}{\omega}$, we may replace the differentials in (13) and (14) by finite differences Δt and a. After algebraic transformations, we obtain the relations for V and two components of the current J_x and J_y inside the circuit

$$J_x\left(x_l+\frac{a}{2}, y_k, t_0+\frac{\Delta t}{2}\right) = J_x\left(x_l+\frac{a}{2}, y_k, t_0-\frac{\Delta t}{2}\right) - \left(V(x_l+a, y_k, t_0) - V(x_l, y_k, t_0)\right)\frac{\Delta t}{L_s a} \tag{15}$$

$$J_y\left(x_l, y_k+\frac{a}{2}, t_0+\frac{\Delta t}{2}\right) = J_y\left(x_l, y_k+\frac{a}{2}, t_0-\frac{\Delta t}{2}\right) - \left(V(x_l, y_k+a, t_0) - V(x_l, y_k, t_0)\right)\frac{\Delta t}{L_s a} \tag{16}$$

$$V(x_l, y_k, t_0+\Delta t) = V(x_l, y_k, t_0) + -\left(J_x\left(x_l+\frac{a}{2}, y_k, t_0+\frac{\Delta t}{2}\right) - J_x\left(x_l-\frac{a}{2}, y_k, t_0+\frac{\Delta t}{2}\right) + J_y\left(x_l, y_k+\frac{a}{2}, t_0+\frac{\Delta t}{2}\right) - J_y\left(x_l, y_k-\frac{a}{2}, t_0+\frac{\Delta t}{2}\right)\right)\frac{\Delta t}{C_s a}. \tag{17}$$

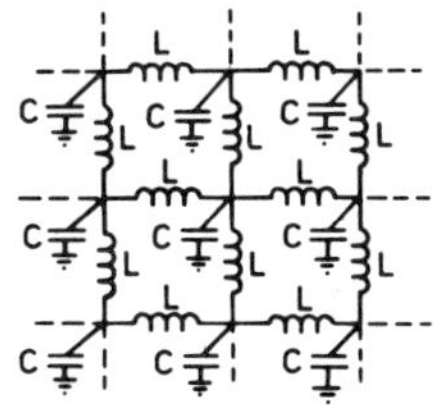

Fig. 2. Lumped circuit equivalent to a segment of a planar circuit.

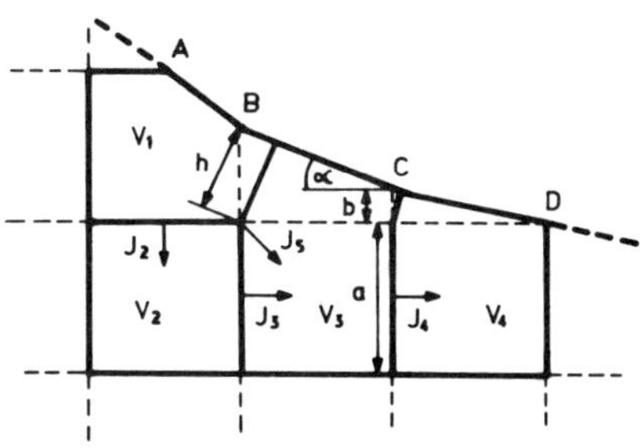

Fig. 3. Example of an open boundary approximation.

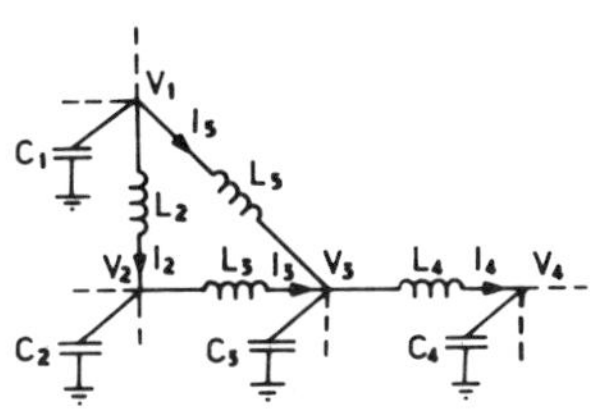

Fig. 4. Lumped circuit equivalent to the circuit considered in Fig. 3.

Equations (15)–(17) describing the circuit have an interesting interpretation. The circuit may be represented as a set of lumped capacitors C and inductances L, as presented in Fig. 2. The potential V has the meaning of the voltage. The current flowing in the inductances may be calculated as $I_x = J_x a$ and $I_y = J_y a$. The elements values are $C = C_s a^2$, $L = L_s$.

A. Boundary Conditions

If a short-circuited border line passes through the centers of the meshes in one column or one row, the boundary conditions are satisfied by assuming $V = 0$ at those points. If an open-circuited border line follows a side line of a row or a column of meshes, the boundary conditions are satisfied by putting $J_n = 0$, where J_n is the current density component perpendicular to the border line.

Those simple rules are sufficient to calculate a circuit of a rectangular shape, but if we want to analyze an arbitrarily-shaped circuit without significant loss of accuracy, a procedure conforming the boundary shape is needed. That kind of procedure for an open boundary will be introduced using an example of a segment of a circuit shown in Fig. 3, and its equivalent LC circuit shown in Fig. 4. The border line was approximated by a broken line $ABCD$. Four meshes close to the border are considered and the potentials in the centers of them are denoted by $V_1 \cdots V_4$.

To match the border line, the meshes 1, 3, and 4 are deformed as shown in Fig. 3. In the neighborhood of the border, they are limited by the line $ABCD$ or by perpendiculars dropped on it from the nodal points. Since (14) may be written in an integral form

$$\oint_{L_m} \boldsymbol{J} \cdot \boldsymbol{a}_n \, dl = -C_s \int_{S_m} \frac{\partial V}{\partial t} ds \simeq -C_s S_m \frac{dV_m}{dt} \tag{18}$$

where S_m is the surface of the mth mesh, a_n is the unit vector perpendicular to L_m, L_m-border line of the mth mesh, and the capacitances C_1, C_2, and C_4 in the equivalent circuit in Fig. 4 are proportional to the surfaces of the corresponding meshes.

The inductances L_2 and L_3 are the standard inductances between meshes $L_2 = L_3 = L_s$.

To calculate L_4, we note that

$$\frac{dJ_4}{dt} \simeq -\frac{V_4 - V_3}{aL_s}$$

and that the current passing between the meshes 3 and 4 is equal to

$$I_4 \simeq J_4(a+b). \tag{19}$$

Thus

$$-\frac{V_4 - V_3}{L_4} = \frac{dI_4}{dt} \simeq -\frac{(V_4 - V_3)}{L_s} \frac{(a+b)}{a}$$

or

$$L_4 \simeq L_s \frac{a}{a+b}. \tag{20}$$

To calculate L_5, we put

$$\frac{dJ_5}{dt} \simeq \frac{V_3 - V_1}{L_s a\sqrt{2}}.$$

Since near the open boundary there exist only a tangential component of the current

$$I_5 \simeq \frac{J_5 h}{\cos\left(\frac{\pi}{4} - \alpha\right)} \tag{21}$$

which gives

$$-\frac{V_3 - V_1}{L_5} = \frac{dI_5}{dt} \simeq -\frac{h}{\cos\left(\frac{\pi}{4} - \alpha\right)} \frac{V_3 - V_1}{L_s a\sqrt{2}}$$

and

$$L_5 \simeq L_s \frac{a\sqrt{2}\cos\left(\frac{\pi}{4} - \alpha\right)}{h}. \tag{22}$$

B. Lumped Elements in the Circuit

Direct correspondence between the algorithm used for calculation and the equivalent LC circuit makes it easy to consider lumped elements branched into the circuit. These elements may be passive (R, L, C) or active (voltage or current sources), and, since a time-domain method is used, they may be nonlinear. It should be noted, however, that (15)–(17) describe the voltages at the points of time and space different from those where and when the currents are described. The differences are $\Delta t/2$ in time and $a/2$ in space and they should be considered to avoid errors in impedance calculations.

C. Algorithm Stability

A computer algorithm based on (15)–(17), with excitation by a set of sinusoidal sources of a frequency ω, has an important condition of stability. This condition will be introduced through its physical interpretation.

In the circuit considered, any perturbation of V or $\boldsymbol{J}$ propagates with the velocity of

$$V_g = \frac{d\omega}{d\beta} = \frac{\lambda}{T} \tag{23}$$

where λ is the wavelength of the wave in the planar circuit, and

$$T = \frac{2\pi}{\omega}.$$

In the algorithm after a time Δt, the perturbation reaches the adjacent mesh. It requires $2\Delta t$ to transmit the perturbation from the mesh k, l to the mesh $k+1, l+1$. Thus, the velocity of propagation in the algorithm is not smaller than

$$V_a = \frac{a\sqrt{2}}{2\Delta t}. \tag{24}$$

To assure the proper functioning of the algorithm we must have

$$V_a > V_g$$

or

$$\frac{a}{\sqrt{2}\lambda} > \frac{\Delta t}{T}. \tag{25}$$

The same result may be obtained from the theory of finite-difference methods (see [6, sec. 26.2]).

D. Accuracy of the Calculations

There are three principal causes of errors in the proposed numerical method of planar-circuit analysis.

1) replacing the differential equations (13) and (14) by (15), (16), and (17),
2) approximation of the circuit border line, and
3) assumption of a steady state in the circuit after a finite period of time.

To analyze the first cause of errors, let us consider, for example, (15). To obtain it from (13), we have replaced the differentials by finite differences. Since some error of this replacement exists, we may write

$$\frac{\partial J_x\left(x_l + \frac{a}{2}, y_k, t_0\right)}{t} = \frac{J_x\left(x_l + \frac{a}{2}, y_k, t_0 + \frac{\Delta t}{2}\right) - J_x\left(x_l + \frac{a}{2}, y_k, t_0 - \frac{\Delta t}{2}\right)}{\Delta t} + \delta_{Jx} \tag{26}$$

$$\frac{\partial V\left(x_l + \frac{a}{2}, y_k, t_0\right)}{\partial x} = \frac{V(x_l + a, y_k, t_0) - V(x_l, y_k, t_0)}{a} + \delta_V. \tag{27}$$

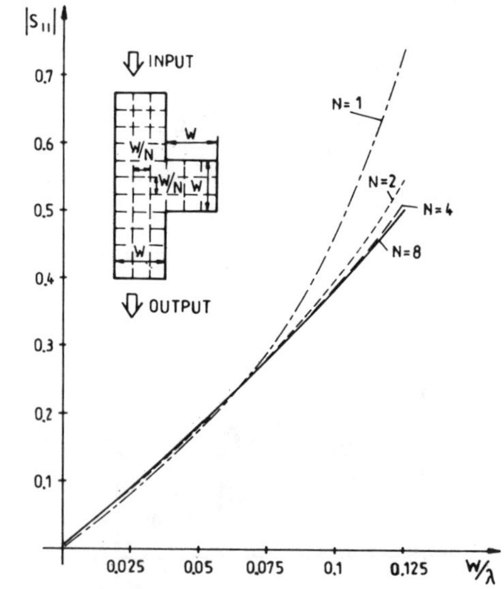

Fig. 5. Comparison of the results of calculation for different mesh size $a = \frac{W}{N}$.

Development of the functions V and J_x in Taylor series gives

$$\delta_{Jx} = \frac{\partial^3 J_x\left(x_l + \frac{a}{2}, y_k, t_0\right)}{\partial t^3} \frac{\Delta t^2}{24} + \cdots \tag{28}$$

$$\delta_V = \frac{\partial^3 V\left(x_l + \frac{a}{2}, y_k, t_0\right)}{\partial x^3} \frac{a^2}{24} + \cdots. \tag{29}$$

From (28) and (29), it is seen that the errors are proportional to Δt^2 or a^2. This means that the errors decrease rapidly with a decrease in the mesh size and a decrease in the size of a step in time.

Since the values of Δt and a are bound by the condition (25), the error caused by the approximation of the function of time has a minor effect. The value of δ_V and similar values of errors found in the analysis of (16) and (17) depend on the field distribution in the circuit. Typically, the functions $V(x, y)$ and $\boldsymbol{J}(x, y)$ change slowly in space following a sinusoidal-type distribution over a wavelength. Thus, the third derivatives of these functions are not very big and the errors are small even for relatively high values of a. An example of this is shown in Fig. 5, where the results are displayed for different sizes of the meshes assumed for calculations. Nevertheless, if the boundary conditions force rapid changes of $V(x, y)$ and $\boldsymbol{J}(x, y)$, the third derivatives of these functions may cause some increase in the errors.

It is difficult to obtain mathematical approximations for the other two causes of errors. We may formulate only some general remarks. The second cause of errors obviously depends on the possibility of reasonable approximation of the boundary line by a straight line within one mesh. The experiments have confirmed that if such an approximation is possible, the error introduced by approximating the shape of the boundary does not have a dominating role in the total error of calculations.

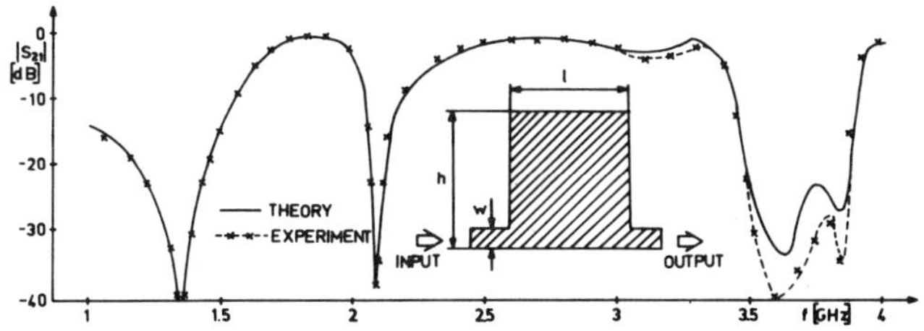

Fig. 6. $|S_{21}|$ of a rectangular stripline circuit calculated and measured in the 1–4-GHz band.

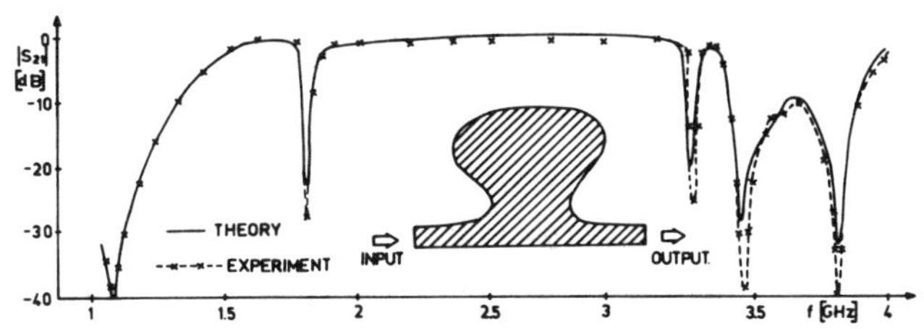

Fig. 7. $|S_{21}|$ of a stripline circuit of irregular shape calculated and measured in the 1–4-GHz band.

The time needed to stabilize the oscillations in the circuit may be assumed by the user taking into account the circuit dimensions and its loaded Q-factor. Another possibility is to leave the time of terminating the calculations to be set up automatically in the algorithm on the basis of the shape of the $V(t)$ function monitored in several chosen points of the circuit. It is always possible to reduce the error caused by a transient response to a negligible value, but it is very time consuming for the circuits of a high loaded Q-factor.

E. Examples of Application

Figs. 6 and 7 show the comparison of the calculated and measured values of $|S_{21}|$ as a function of the frequency in 1–4-GHz band for two stripline circuits. The shape of the inner strip of each circuit is also presented. Both circuits were built on a stripline of the height $2d = 6.94$ mm filled with a dielectric of $\epsilon_r = 2.5$.

The dimensions of the circuit in Fig. 6 are $w = 5.2$ mm, $l = 46.5$ mm, and $h = 54.8$ mm. The fringing fields were included in the calculations by increasing these dimensions according to (10) to the values $w' = 8.27$ mm, $l' = 49.6$ mm, and $h' = 57.9$ mm. This circuit was designed to be matched exactly by a set of squares of the size $a = \frac{1}{3}w'$. The results of the calculations are in good agreement with those of the measurements. The difference between them visible for higher frequencies may be attributed to higher value of the ratio a/λ. For example, let us assume the frequency $f = 4$ GHz ($a/\lambda = 1/17$) and calculate the error described by (29) assuming a sinusoidal distribution of $V(x, y)$ along the x-axis. We obtain the error at the level of about 45 dB, and, since several similar errors accumulate, the error level of about 30 dB is what we may expect.

Fig. 7 shows the results similar to those of Fig. 6 but obtained for a circuit of a shape drawn at random (which is also shown in Fig. 7). The mesh size assumed for calculations is the same as in the previous example. There is no visible loss of accuracy of the calculations due to an irregular boundary.

F. Computer Memory and Time Needed for Calculation

The method may be implemented on a wide range of computers, including small personal computers. 64 kB of RAM should be sufficient to calculate circuits with 50×50 rectangular elements. The time depends strongly on the number of meshes and also on the period of time required to stabilize the oscillations in the circuit. For example, analyzing the circuit of Fig. 7 for a fixed frequency takes about 1 min in Fortran on a CDC computer of the Cyber 70 series or about 80 min in Pascal on a standard personal computer.

IV. Conclusion

The method presented in this paper shows many advantages.

1) It is very versatile. A universal program may be worked out for an analysis of a planar circuit of any shape defined by the user.

2) The accuracy of calculations can be made sufficient to assure a very good agreement between the theory and the experiment.

3) There are possibilities for the further development of the method. For example, the circuit losses may be easily included in the calculations. It is also possible to analyze a circuit with lumped elements (both linear and nonlinear) branched into it.

4) The method simulates a natural process of the wave propagation, and the actual field distribution may be displayed during the calculations. This may help in debugging the program.

It may also give some valuable information to the circuit designer trying to understand better the physical properties of his circuit. It may also be useful in the teaching of microwaves.

A disadvantage of the method is the long computing time. This is especially important for the circuits of high loaded Q-factors. We may assume, however, that this is the price paid for the mentioned advantages, which may have crucial significance in many applications. We may also assume that due to constant improvement in the speed of computers, the future user will be more willing to pay that price.

References

[1] T. Okoshi and T. Miyoshi, "The planar circuit—An approach to microwave integrated circuitry," *IEEE Trans. Microwave Theory Tech.*, vol. MTT-20, pp. 245–252, Apr. 1972.

[2] Hsu Jui-Pang and T. Anada, "Planar circuit equation and its practical application to planar-type transmission-line circuit," in *IEEE MTT Int. Symp. Dig.* (Boston) 1983, pp. 574–576.

[3] R. Chadha and K. C. Gupta, "Green's functions for triangular segments in planar microwave circuits," *IEEE Trans. Microwave Theory Tech.*, vol. MTT-28 pp. 1139–1143, Oct. 1980.

[4] T. Okoshi, Y. Uehara, and T. Takeuchi, "The segmentation method —An approach to the analysis of microwave planar circuits," *IEEE Trans. Microwave Theory Tech.*, vol. MTT-24, pp. 662–668, Oct. 1976.

[5] R. E. Collin, *Field Theory of Guided Waves*. McGraw-Hill, 1960, chs. 1 and 5.

[6] G. E. Forsythe and W. R. Wasow, *Finite-Difference Methods for Partial Differential Equations.* New York: Wiley, 1960.
[7] G. D. Smith, *Numerical Solution of Partial Differential Equations.* London: Oxford University Press, 1975.

Analysis of Arbitrarily Shaped Two-Dimensional Microwave Circuits by Finite-Difference Time-Domain Method

WOJCIECH K. GWAREK

Abstract—The paper presents a version of the finite-difference time-domain method adapted to the needs of S matrix calculations of microwave two-dimensional circuits. The analysis is conducted by simulating the wave propagation in the circuit terminated by matched loads and excited by a matched pulse source. Various aspects of the method's accuracy are investigated. Practical computer implementation of the method is discussed and an example of its application to an arbitrarily shaped microstrip circuit is presented. It is shown that the method in the proposed form is an effective tool of circuit analysis in engineering applications. The method is compared to two other methods used for a similar purpose, namely the contour integral method and the transmission-line matrix method.

I. Introduction

A TWO-DIMENSIONAL circuit as understood here is a circuit in which the fields may be characterized by a scalar function $V(x, y)$ which obeys a two-dimensional wave equation

$$\nabla_{xy}^2 V(x,y,t) - \beta^2 \frac{\partial^2 V(x,y,t)}{\partial t^2} = 0 \tag{1}$$

with proper boundary conditions.

There are many microwave circuits characterized accurately or approximately by (1). As examples we suggest stripline or microstrip junctions or resonators, and also some types of waveguide discontinuities. That is why solving this equation was investigated by many researchers. If the boundary is of complicated shape, only numerical methods can be used. These methods can be divided into two groups. The methods of the first group require a certain amount of analytical preprocessing before the particular problem may be solved by a computer. As an example we propose the Green's function method supported by segmentation techniques [1]. Another group of methods assumes that the analytical preprocessing should be practically nonexistent and that it is up to the computer to do the entire job. In this group we find the contour integral method [2], the transmission-line matrix method [3], and the finite-difference method, which will be the subject of this paper.

The finite-difference time-domain method (FD-TD) was first introduced by Yee [4] and since then has been applied by many authors. These applications, however, were concentrated in the domain of scattering [5], wave absorption [17], and accelerator physics [6]. Application of the FD-TD method to microwave circuit analysis has so far attracted little attention.

The aim of this paper is to show that the FD-TD method adapted to the needs of 2-D circuit analysis can be a strong competitor to the other mentioned methods. It can lead to a universal and effective computer program capable of solving a wide range of practical problems. The paper extends the ideas presented in [7] and [8]. It presents an FD-TD method of 2-D circuit analysis based on pulse excitation by a matched source. It discusses various aspects of the method's accuracy and presents the author's experience with its implementation on an IBM PC AT computer.

II. Outline of the FD-TD Method

In the FD-TD method instead of solving the second-order equation (1) a pair of first-order equations is solved:

$$\nabla V(x,y,t) = -L_s \frac{\partial \boldsymbol{J}(x,y,t)}{\partial t} \tag{2}$$

$$\nabla \cdot \boldsymbol{J}(x,y,t) = -C_s \frac{\partial V(x,y,t)}{\partial t}. \tag{3}$$

In microwave planar circuits, the variables and constants in (2) and (3) have the following interpretations: V—voltage, $\boldsymbol{J}$—surface current density, C_s—capacitance of a unitary square of the circuit, L_s—inductance of an arbitrary square of the circuit. The xy plane is divided into a set of meshes which are basically square but may have their shapes modified to match the boundary line. The coordinates of the middle of a mesh in the kth row and the lth column are denoted by x_l and y_k. Replacing the differentials in (2) and (3) by finite differences Δt and a

Manuscript received June 29, 1987; revised November 16, 1987.
The author is with the Institute of Radioelectronics, Warsaw University of Technology, 00-665 Warsaw, Nowowiejska 15/19, Poland.
IEEE Log Number 8719199.

Reprinted from *IEEE Transactions on Microwave Theory and Techniques*, pp. 738-744, April 1988.

yields

$$J_x\left(x_l+\frac{a}{2}, y_k, t_0+\frac{\Delta t}{2}\right) = J_x\left(x_l+\frac{a}{2}, y_k, t_0-\frac{\Delta t}{2}\right) - (V(x_l+a, y_k, t_0) - V(x_l, y_k, t_0))\frac{\Delta t}{L_s a f_1(l,k)} \quad (4)$$

$$J_y\left(x_l, y_k+\frac{a}{2}, t_0+\frac{\Delta t}{2}\right) = J_y\left(x_l, y_k+\frac{a}{2}, t_0-\frac{\Delta t}{2}\right) - (V(x_l, y_k+a, t_0) - V(x_l, y_k, t_0))\frac{\Delta t}{L_s a f_2(l,k)} \quad (5)$$

$$V(x_l, y_k, t_0+\Delta t) = V(x_l, y_k, t_0) - \left(J_x\left(x_l+\frac{a}{2}, y_k, t_0+\frac{\Delta t}{2}\right) - J_x\left(x_l-\frac{a}{2}, y_k, t_0+\frac{\Delta t}{2}\right) + J_y\left(x_l, y_k+\frac{a}{2}, t_0+\frac{\Delta t}{2}\right) - J_y\left(x_l, y_k-\frac{a}{2}, t_0+\frac{\Delta t}{2}\right)\right)\frac{\Delta t}{C_s a f_3(l,k)} \quad (6)$$

where $f_1(l,k)$, $f_2(l,k)$, and $f_3(l,k)$ are mesh shape functions which are equal to unity for all square meshes (that is, those inside the circuit) but adopt different values (calculated by a boundary matching procedure like that of [7]) for the meshes modified to match boundary lines.

Consecutive calculations of (4), (5), and (6) simulate the process of the wave propagation in the circuit.

III. Problems of FD-TD Analysis of Two-Dimensional Circuits

A. Modeling of Matched Loads and Sources

The most convenient way of describing a microwave linear circuit is by its S matrix. Since the FD-TD method models the energy flow in the circuit it may be used to compute the S matrix directly, provided that the matched loads of the output lines are properly modeled in the algorithm. Matching the source, although not absolutely necessary, is highly desirable, since any reflections from the input would prolong the transient response of the circuit, thus prolonging the computing time.

The absorbing boundary conditions have been investigated in [14] and [15], but the procedures developed there are not useful in our case. They cannot handle the situation of matched source and also, being designed for general absorbing conditions, they are unnecessarily com-

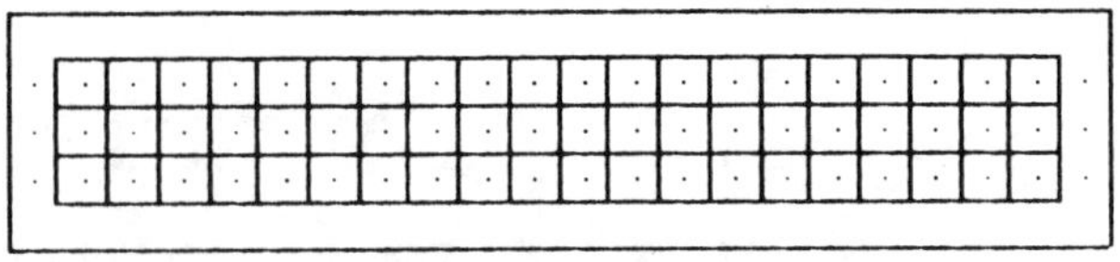

Fig. 1. A uniform transmission line as a grid of meshes.

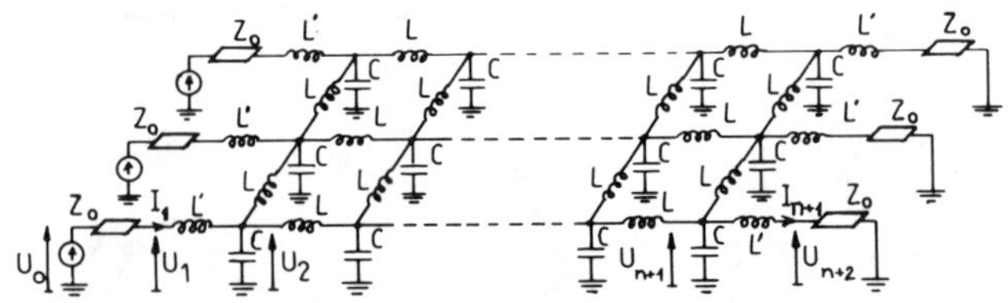

Fig. 2. Lumped circuit model of the line of Fig. 1.

plicated when applied to the case of normal incidence. That is why a different approach will be used here.

Let us consider a model of a uniform transmission line of length $n=20a$ and width $w=3a$, where a is the mesh size (Fig. 1). The lumped circuit model of this line corresponding to the FD approximation is presented in Fig. 2. Propagation inside the circuit is described by (4), (5), and (6). The input and output matching is obtained by introducing in each of the rows of meshes at the input and output the following operations:

$$I_1\left(t_0+\frac{\Delta t}{2}\right) = I_1\left(t_0-\frac{\Delta t}{2}\right) - (V_2(t_0)-V_1(t_0))\frac{\Delta t}{L'} \quad (7)$$

$$V_1(t_0+\Delta t) = V_0(t_0+\Delta t) - I_1\left(t_0+\frac{\Delta t}{2}\right)Z_0 \quad (8)$$

$$I_{n+1}\left(t_0+\frac{\Delta t}{2}\right) = I_{n+1}\left(t_0-\frac{\Delta t}{2}\right) + (V_{n+1}(t_0) - V_{n+2}(t_0))\frac{\Delta t}{L'} \quad (9)$$

$$V_{n+2}(t_0+\Delta t) = I_{n+1}\left(t_0+\frac{\Delta t}{2}\right)Z_0 \quad (10)$$

where

$$Z_0=\sqrt{\frac{L}{C}} \qquad L'=\frac{L}{2}+L''=\frac{L}{2}+\frac{\Delta t\omega}{2a\beta}L \qquad C=C_s a^2 \qquad L=L_s.$$

The meaning of the voltages and the currents used in (7)–(10) is explained in Fig. 2. The additional inductance L'' was introduced as a correction element to lower the matching errors caused by the fact that in the FD-TD algorithm the voltage is defined at points of time and space which are different from the points where the current is defined. Because of this it is also necessary to introduce some correction terms in the formulas for the S matrix elements. Let us assume that $V_0(\omega), V_1(\omega), \cdots$ are the complex amplitudes calculated by the Fourier transformation of $V_0(t), V_1(t), \cdots$ and $I_1(\omega), I_2(\omega), \cdots$ are the complex amplitudes calculated by the Fourier transforma-

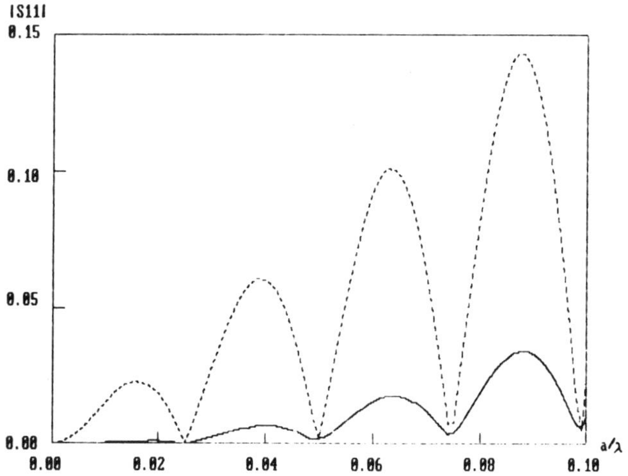

Fig. 3. Results of calculations of $|S_{11}|$ versus a/λ for the line of Fig. 1 with the special correction (continuous line) and without it (dashed line).

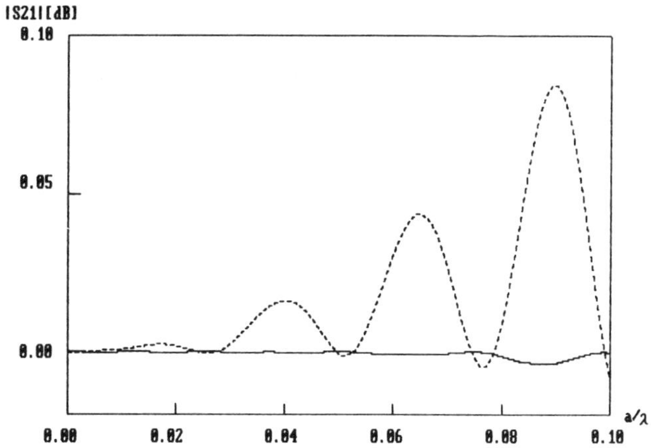

Fig. 4. Results of calculations of $|S_{21}|$ versus a/λ for the circuit of Fig. 1 with the special correction (continuous line) and without it (dashed line).

tion of $I_1(t-\Delta t/2), I_2(t-\Delta t/2), \cdots$. Resolving the circuit equations at the input and output we obtain

$$S_{21} = \frac{2V_{n+2}(\omega)\cos(\phi)}{V_0(\omega)} e^{j\phi} \tag{11}$$

$$S_{11} = \frac{Z_{\text{in}} - Z_o}{Z_{\text{in}} + Z_o} \qquad \text{with } Z_{\text{in}} = \frac{V_1(\omega)}{I_1(\omega)} e^{-j\phi} - j\omega L'' \tag{12}$$

where $\phi = \omega \Delta t/2$.

Figs. 3 and 4 present the results of calculations of $|S_{11}|$ and $|S_{21}|$ versus frequency for the line of Fig. 1 with the correction terms containing ϕ and L'' (continuous line) and without them (dashed line). The frequency is characterized by the mesh size to wavelength ratio a/λ. It is seen that when the corrections are applied, the errors of matching drop to negligible values even for relatively high a/λ ratios.

Modeling of a matched source and a matched load was introduced for a uniform transmission line. It can be applied to an arbitrarily shaped circuit provided that at the input and output the circuit includes segments of uniform transmission lines long enough to ensure effective attenuation of all the width modes except the dominant one.

B. Pulse Excitation

In the approach used in [7] the circuit with zero initial values of $V(x,y,t)$ and $\boldsymbol{J}(x,y,t)$ was excited by a sinusoidal source of frequency ω. The circuit's parameters at that frequency were obtained from the steady state achieved in the circuit after a sufficiently long time. This caused the time for computing a wide-band frequency response to be very long.

Equations (4), (5), and (6) show that all the operations in the FD-TD algorithm are linear. Thus it is possible to use Fourier analysis to obtain the circuit's frequency response from the transient response. However, we need a proof that this approach will not sacrifice the accuracy of the method.

Let us consider a relation between a voltage at the input, V_0, and a voltage at the output, V_l. We may write

$$V_l(t) = \boldsymbol{L}(V_0(t)) = \boldsymbol{L}'(V_0(t)) + \Delta\boldsymbol{L}(V_0(t)) \tag{13}$$

where $\boldsymbol{L}$ is a linear operator of the FD-TD algorithm. $\boldsymbol{L}'$ is an operator of the transformation between input and output of the original circuit, which is linear due to the linearity of the Maxwell's equations; and $\Delta\boldsymbol{L}$ is an operator describing the error of the FD-TD method, which has to be linear due to the linearity of the two former operators.

Let us write the Fourier transform of (13):

$$V_l(\omega) = T(\omega)V_0(\omega) = (T'(\omega) + \Delta T(\omega))V_0(\omega) \tag{14}$$

where $V_0(\omega), V_l(\omega)$ are the Fourier transforms of $V_0(t)$, $V_l(t)$ and $T(\omega), T'(\omega)$, and $\Delta T(\omega)$ describe the operators $\boldsymbol{L}$, $\boldsymbol{L}'$, and $\Delta\boldsymbol{L}$ in the frequency domain. This yields

$$\frac{V_l(\omega)}{V_0(\omega)} = T(\omega) = T'(\omega) + \Delta T(\omega). \tag{15}$$

Equation (15) suggests that for any particular frequency ω, the error of the FD-TD method does not depend on the shape of the input signal and that the function $V_0(t)$ may be chosen to be the most convenient for use in the computer algorithm. However, there is one aspect of the pulse excitation which needs additional checking. The Fourier transforms in (15) are assumed to be calculated in an infinite period of time. The limited period of time assumed in any computer calculation causes additional error. Thus we must determine how much this error depends on the shape of the source.

Let us consider two types of pulses. The first is a δ pulse which in the algorithm adopts the form

$$V_0(t) = \begin{cases} 0 & \text{for } t = 0 \\ 1 & \text{for } t = n\Delta t;\ n = \pm 1, \pm 2, \cdots. \end{cases} \tag{16}$$

The second is a pulse of limited spectrum approximating a δ pulse after passing through a bandpass filter of cutoff

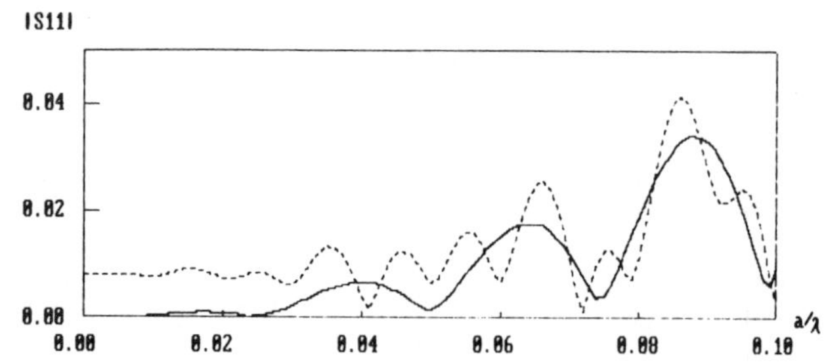

Fig. 5. $|S_{11}|$ of the circuit of Fig. 1 calculated after 200 iterations; with δ-type excitation (dashed line) and with excitation of limited spectrum (continuous line).

frequency ω_c:

$$V_0(t) = \begin{cases} 2\omega_c \Delta t/\pi & \text{for } t=0 \\ \dfrac{2\,\Delta t \sin(\omega_c t)}{\pi t} & \text{for } t = m\,\Delta t;\ m = \pm 1, \pm 2, \pm 3, \cdots, \pm M \\ 0 & \text{for } t = m\,\Delta t;\ |m| > M \end{cases} \tag{17}$$

Fig. 5 shows the results of calculations of $|S_{11}|$ of the line of Fig. 1 after 200 iterations ($t = 200\Delta t$) with the two types of excitation. The dashed line was obtained with δ type excitation, while the continuous line was obtained with the excitation of limited spectrum (with ω_c corresponding to $a/\lambda = 0.2$ and $M = 50$). When the computation is prolonged, the results obtained with the second type of pulse change very little while the ripples on the curve obtained with the δ type of excitation decrease slowly, and after about 1500 iterations the shape of the curve approaches the result obtained before with the second type of pulse.

The reason for this effect is that the FD approximation produces some resonances above the investigated frequency band. A δ pulse excites the circuit at these frequencies and produces a ringing-type response. Cutting off the high frequencies from the exciting pulse eliminates the effect.

However, it should be noted that when the resonances of the investigated circuit lie within the band of interest their effect cannot be eliminated by changing the exciting pulse spectrum. An example of such a circuit will be shown later in this paper.

We may conclude that since using a δ type of pulse simplifies the algorithm and speeds it up (because the Fourier transform of the source does not have to be computed), it is a reasonable choice in many cases. However if we consider a relatively resonance free band and the resonances are grouped outside that band, the time of computing can be brought substantially down by applying a pulse of the spectrum limited in such a way that the unwanted resonances are not excited.

C. Microstrip Circuits Analysis

As was already shown in [7], the FD-TD method is effective in arbitrarily shaped stripline circuit analysis. The fringing fields were included in the calculations by assuming that the circuit is bounded by a magnetic wall shifted by some distance from the real edge of the circuit. For a microstrip circuit a more complicated model is needed to describe the complicated nature of the fringing fields. Here is a proposal of such a model.

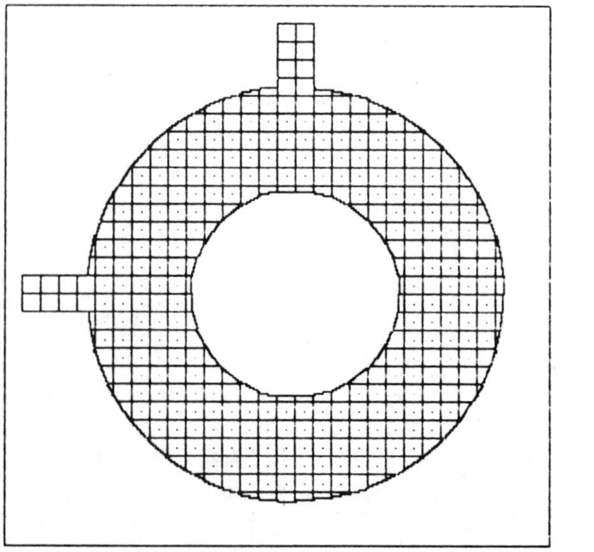

Fig. 6. A microstrip ring circuit as a grid of meshes.

Let us consider a microstrip line of width w. This line is characterized by its unit capacitance $C(w)$, unit inductance $L(w)$, and effective permittivity $\epsilon_{\mathrm{eff}}(w)$ (with permeability $\mu = \mu_0$). Let us now imagine a line filled with a uniform dielectric characterized by $\epsilon' = \text{const}$ and $\mu' = \text{const}$ and having such properties that its unit capacitance C' and unit inductance L' obey the relations

$$C'(w + \Delta w) = C(w) \tag{18}$$

$$L'(w + \Delta w + \Delta w') = L(w). \tag{19}$$

In many practical cases it is possible to find such values of Δw and $\Delta w'$ that (18) and (19) are obeyed with good accuracy over a wide range of w. For example, for a duroid substrate of $\epsilon = 10\epsilon_0$ and height $h = 0.635$ mm, drawing the functions $C(w)$ and $L(w)$ from the closed-form expressions for Z_o and ϵ_{eff} after [9] we find that (18) and (19) are obeyed with 1 percent accuracy in the range $2h < w < 10h$, assuming that $\Delta w = 1.05h$, $\Delta w' = h$, $\epsilon' = 10\epsilon_0$, and $\mu' = 0.93\mu_0$. For wider strips the error of this approximation rises only slightly. For very narrow strips it is bigger but can be corrected by adopting for the edges of these strips different values of Δw and $\Delta w'$.

We will now apply the FD-TD method to the analysis of a ring circuit already analyzed by another method by D'Inzeo *et al.* [10]. The circuit was built on the duroid substrate, had the dimensions of $r_{\mathrm{out}} = 7$ mm, $r_{\mathrm{in}} = 4$ mm, and was connected to two 50 Ω lines making a 90° angle. For FD-TD calculations we assumed the circuit as presented in Fig. 6. Its dimensions were obtained by shifting the circuit's boundary by $\Delta w/2 = 0.525h$ and it was assumed that $\epsilon' = 10\epsilon_0$ and $\mu' = 0.93\mu_0$. We applied the boundary matching procedure described in [7] but took into account additional inductance distributed along the

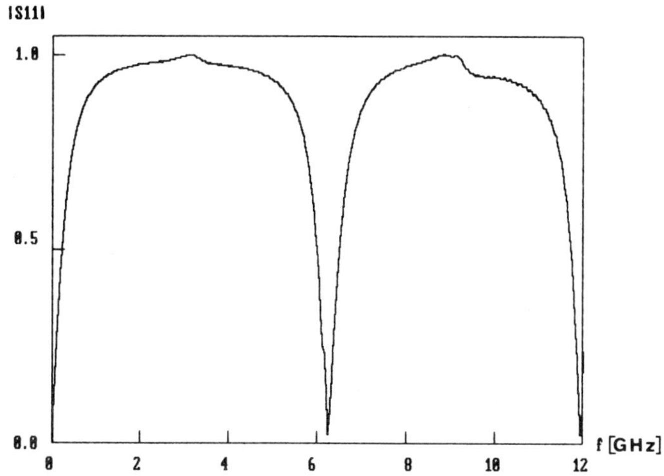

Fig. 7. $|S_{11}|$ versus frequency of the circuit of Fig. 6.

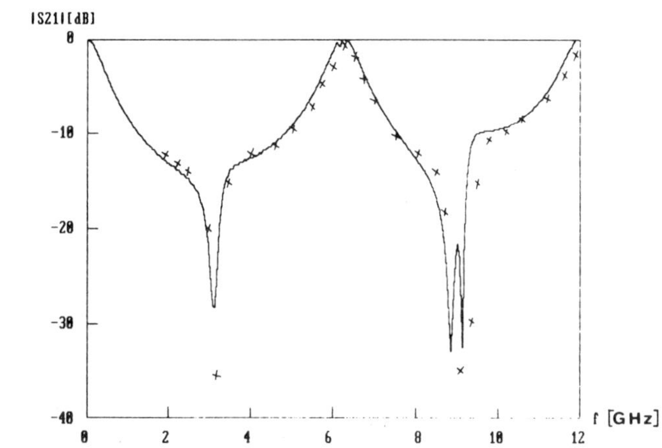

Fig. 8. $|S_{21}|$ versus frequency of the circuit of Fig. 6. —— calculations, ××××× measurements after [10].

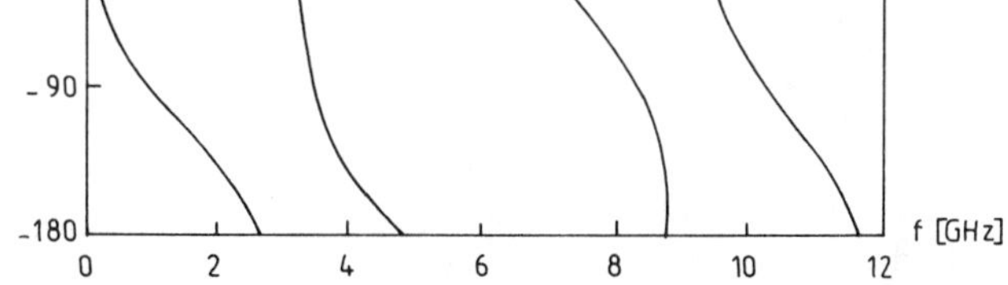

Fig. 9. Arg(S_{21}) versus frequency of the circuit of Fig. 6.

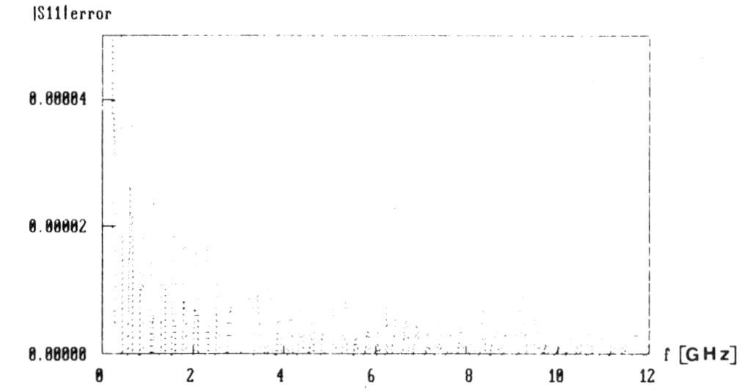

Fig. 10. Error in calculations of $|S_{11}|$ in Fig. 7 caused by a simulated low accuracy of the computer arithmetics equal to 10^{-6}.

border. This corresponded to the inductance of a line of width $\Delta w'/2 = 0.5h$. The results of calculations of $|S_{11}|$, $|S_{21}|$, and Arg(S_{21}) are presented in Figs. 7, 8, and 9. In the case of $|S_{21}|$ they are compared with the measurements after [10]. The agreement is quite good. It has to be stressed that the method presented here does not assume any regularities of the circuit's shape. Thus it is a method for truly arbitrarily shaped circuits.

We may conclude that the presented example of the FD-TD method application to an arbitrarily shaped microstrip circuit is encouraging, but more work has to be done to check its value for various circuits, especially in the higher frequency band, when dispersion becomes important.

D. Roundoff Errors

In many programming languages there is a choice of the precision of the floating point arithmetic and it is important to know which precision to choose to keep the computer roundoff errors negligible while not boosting the memory requirements and the computing time. To check the level of the roundoff errors the calculations of the circuit of Fig. 6 were repeated with a simulated low computer precision of 10^{-6}. The difference between the results of $|S_{11}|$ obtained with full and with limited precision with δ-type excitation are presented in Fig. 10. It is seen that computer precision even as low as 10^{-6} introduces negligible errors into FD-TD calculations. This and other numerical experiments have shown that the error level increases with decreasing a/λ ratio (as seen in Fig. 10) and that it is slightly smaller for other than δ-type excitations but these dependencies have negligible effect due to generally low level of the roundoff errors. We may draw the conclusion that the FD-TD method is very resistant to roundoff errors and in most cases can be applied even with the lowest (4 bytes) precision of the floating point arithmetics used in personal computers.

E. Algorithm Implementation

The described problems of the FD-TD analysis were checked with a Pascal program prepared for an IBM PC AT computer. The program was written in a "user friendly" manner under the assumption that it will serve a variety of users.

As an example, the time needed in computing the S matrix of the circuit of Fig. 6 was about 20 min. for 1000 iterations. The results presented in Figs. 7–9 were obtained after 3000 iterations, but reasonably accurate results can be obtained even after 1000 iterations. In this case the

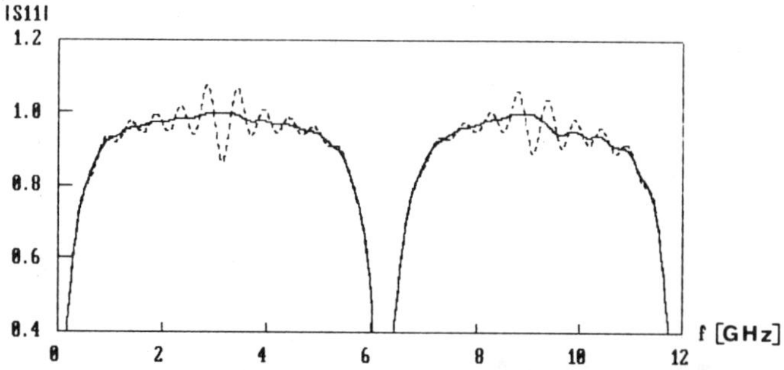

Fig. 11. $|S_{21}|$ of the circuit of Fig. 6 after 1000 iterations (dashed line) and the same result after smoothing procedure using the criterion of power conservation (continuous line).

functions obtained have some ripples (see, for example, the dotted line in Fig. 11) but these ripples can be eliminated using the criterion $|S_{11}|^2+|S_{21}|^2=1$, valid for lossless circuits (see the continuous line in Fig. 11, which is practically the same as that of Fig. 7).

F. Comparison to Other Methods

As was mentioned in the introduction, the methods which can be assigned to the same class as possible competitors to the FD-TD method are the contour integral (CI) method [2], [13] and the transmission-line matrix (TLM) method [3], [11], [12].

In the contour integral method [2], the circuit's boundary is divided into a set of N elements of finite width. Using the properties of cylindrical waves for a particular frequency, we can obtain the relations between the voltages at the elements. This leads to a set of N linear equations to be solved.

When comparing the FD-TD and the CI methods we have to note first that their domains of application are not the same. The FD-TD method can be applied to circuits filled with nonuniform media and also to circuits with nonlinear elements (when restricted to sinusoidal excitation). These two classes of circuits cannot be treated by the CI method.

It is difficult to say which of the two methods needs less computer time. There are three reasons for this.

1) It depends on the type of circuit. Many factors contribute, for example,

- the computer time needed by the CI method depends on the length of the boundary line while the time of the FD-TD calculations depends on the surface of the circuit;
- for some circuits greater advantage may be taken from the flexibility of the CI method in applying varying boundary element length.
- the time of computing by the FD-TD method depends on the presence of high-Q resonances in the circuit.

2) Specialized hardware and software can bring a dramatic drop in computing time for either of the two methods, but the gain obtained with particular computer resources may be quite different for each of them.

3) In the CI method calculations are conducted for a particular frequency and have to be repeated from the very beginning when the circuit's parameters are to be calculated for another frequency. On the contrary, in the FD-TD method calculating the circuit's parameters for several hundred frequencies brings only a fractional increase in the computing time with respect to one-frequency calculations.

As an example, the computing time by both methods was compared for the circuit of Fig. 6 (but with no additional inductance along the border, a situation which cannot be handled by the CI method) on an IBM PC AT computer using standard library routines. In the time needed for calculating the wide-band frequency characteristics by the FD-TD method only a few (about five) frequency points could be calculated by the CI method.

An important advantage of the FD-TD method is that all algorithm operations have clear physical interpretations. That is why if a computing error occurs its cause can be quite easily spotted. This is in contrast to calculations by the CI method involving large-scale matrix operations which are very difficult to control.

The FD-TD method and the transmission-line matrix (TLM) method [3], [11] are similar in many respects. They both use a net of square meshes and both consider a pulse propagation in the net (although the algorithm of simulating the pulse propagation is different). Comparison of the FD-TD and the TLM methods has been discussed in recent publications [12], [16]. The general conclusion is that those methods are developed in a parallel way. Progress in one of them triggers development of the other. For example, to the author's knowledge the TLM method was not yet applied to S matrix calculations by modeling the circuit with matched input and output, but taking into account the discussion of this paper such an approach seems straightforward. Although the author found that in the considered application his version of the FD-TD method is faster and more convenient than the versions of the TLM known to him, he thinks of his work as a step forward in developing the entire group of wave-simulating methods, which includes the FD-TD and the TLM.

III. Conclusions

The paper has presented a version of the finite-difference time-domain method adapted to the needs of the S matrix calculations of two-dimensional microwave circuits. To allow direct S matrix calculations, matched loads and matched sources were modeled in the algorithm. This modeling gave good results over a wide frequency range (covering mesh size to wavelength ratios $0<a/\lambda<0.1$).

Pulse excitation was introduced in the method for fast calculations of frequency-dependent circuit characteristics. It was shown that the pulse excitation does not degrade the FD-TD accuracy. Different types of pulses were studied. In some cases a δ-type pulse is the most practical choice but in the cases when the circuit's main resonances are grouped outside the band of interest a pulse of limited spectrum can bring a substantial drop in the time of computing.

It was shown that the FD-TD method is not sensitive to the computer roundoff errors. Low precision of the floating point numbers may be chosen in computer programs to keep down the computer time and memory.

An analysis of an arbitrarily shaped microstrip circuit was presented as an example. Good agreement between the results of calculations and measurements was obtained but more work has to be done to check the applied fringing fields models for various circuits.

Implementation of the described method on an IBM PC AT computer shows that it can become a practical tool in engineering applications. In many cases it performs better than other widely used methods such as the contour integral method and the transmission-line matrix method.

References

[1] R. Sorrentino, "Planar circuits, waveguide models, and segmentation method," *IEEE Trans. Microwave Theory Tech.*, vol. MTT-33, pp. 1057–1066, Oct. 1985.

[2] T. Okoshi and T. Miyoshi, "The planar circuit—An approach to microwave integrated circuitry," *IEEE Trans. Microwave Theory Tech.*, vol. MTT-20, pp. 245–252, Apr. 1972.

[3] P. B. Johns and R. L. Beurle, "Numerical solution of 2-dimensional scattering problems using a transmission-line matrix," *Proc. Inst. Elec. Eng.*, vol. 118, no. 9, pp. 1203–1208, Sept. 1971.

[4] K. S. Yee, "Numerical solution of initial boundary value problems involving Maxwell's equations in isotropic media," *IEEE Trans. Antennas Propagat.*, vol. AP-14, pp. 302–307, May 1966.

[5] K. Umashankar and A. Taflove, "A novel method to analyze electromagnetic scattering of complex objects," *IEEE Trans. Electromagn. Compat.*, vol. EMC-24, pp. 397–405, Nov. 1982.

[6] T. Weiland, "Numerical solution of Maxwell's equations for static, resonant and transient problems," in *Proc. U.R.S.I. Int. Symp. Electromagn. Theory* (Budapest), 1986, pp. 537–542.

[7] W. K. Gwarek, "Analysis of an arbitrarily-shaped planar circuit—A time domain approach," *IEEE Trans. Microwave Theory Tech.*, vol. MTT-33, pp. 1067–1072, Oct. 1985.

[8] W. K. Gwarek, "Two-dimensional circuit analysis by wave simulation," in *Proc. U.R.S.I. Int. Symp. Electromagn. Theory* (Budapest), 1986, pp. 580–582.

[9] K. C. Gupta, R. Garg, and R. Chadha, *Computer-Aided Design of Microwave Circuits*. Dedham, MA: Artech House, 1981, ch. 3.

[10] G. D'Inzeo *et al.*, "Wide-band equivalent circuits of microwave planar networks," *IEEE Trans. Microwave Theory Tech.*, vol MTT-28, pp. 1107–1113, Oct. 1980.

[11] W. J. R. Hoefer, "The transmission-line matrix method—Theory and applications," *IEEE Trans. Microwave Theory Tech.*, vol. MTT-33, pp. 882–893, Oct. 1985.

[12] P. B. Johns, "On the relationship between TLM and finite-difference methods for Maxwell's equations," *IEEE Trans. Microwave Theory Tech.*, vol. MTT-35, pp. 60–61, Jan. 1987.

[13] T. Okoshi, *Planar Circuits for Microwaves and Lightwaves*. Berlin; Heidelberg: Springer-Verlag, 1985.

[14] B. Engquist and A. Majda, "Absorbing boundary conditions for the numerical simulation of waves," *Math. Comput.*, vol. 31, pp. 629–651, July 1977.

[15] G. Mur, "Absorbing boundary conditions for the finite-difference approximation of the time-domain electromagnetic-field equations," *IEEE Trans. Electromagn. Compat.*, vol. EMC-23, pp. 377–382, Nov. 1981.

[16] W. K. Gwarek and P. B. Johns, "Comments on 'On the relationship between TLM and finite difference methods for Maxwell's equations'," *IEEE Trans. Microwave Theory Tech.*, vol. MTT-35, pp. 872–873, Sept. 1987.

[17] D. T. Borup, D. M. Sullivan, and O. P. Gandhi, "Comparison of the FFT conjugate gradient method and the finite-difference time-domain method for the 2-D absorbtion problem," *IEEE Trans. Microwave Theory Tech.*, vol. MTT-35, pp. 383–395, Apr. 1987.

Calculations of the Dispersive Characteristics of Microstrips by the Time-Domain Finite Difference Method

XIAOLEI ZHANG, JIAYUAN FANG, KENNETH K. MEI, FELLOW, IEEE, AND YAOWU LIU

Abstract —The dispersive characteristics of microstrips have been investigated by many authors [1]–[4] using various numerical and empirical methods. Those results showed a lack of agreement with each other, and the true dispersive characteristics of microstrips still need to be identified. In this paper, a direct time-domain finite difference method is used to recharacterize the microstrip. Maxwell's equations are discretized both in time and space and a Gaussian pulse is used to excite the microstrip. The frequency-domain design data are obtained from the Fourier transform of the calculated time-domain field values. Since this method is completely independent of all the above-mentioned investigations, the new results can be considered as an impartial verification of the published results.

The comparison of the time-domain results and those from the frequency-domain methods has shown the integrity of the time-domain computations. Since this method is very general and can be applied to model many other microwave components, its success in the microstrip problem is an important step toward its general application.

I. Introduction

The application of the time-domain method to model microstrip problems is not new [7]–[11]. However, it is more frequently used to obtain qualitative results that graphically illustrate the field propagation and the transport of power than to obtain design data. It is interesting to note that when the time-domain method is used to obtain frequency-domain data, the investigators often go back to sinusoidal excitations [7]–[10]. This compromises the advantage of the time-domain approach, which is believed to be capable of producing wide-band frequency-domain results with a single computation process.

In the process of this investigation, it is found that the Fourier transform of the time-domain results are very sensitive to numerical errors, notably those resulting from the imperfect treatment of the absorbing boundary conditions used to truncate the numerical computations of that of an open structure. Thus, even though the time-domain results may be reasonably accurate, the frequency-domain results obtained from their Fourier transform may not be acceptable as useful data. That may explain, at least in part, why many investigators mysteriously reverted to sinusoidal signals while it seems obvious that a pulse in the time domain could have been used to advantage, and others turned to solve only the closed structures [11].

To improve the absorbing boundary conditions, we have used an open-circuit, short-circuit technique to cancel the reflections from the terminations of the finite difference meshes. The improvements in the resulting computations enabled us to get reliable frequency-domain data, i.e., the effective dielectric constant and the characteristic impedance, through Fourier transforms, and to compare with various numerical results of the microstrip, presently available in the open literature.

II. Finite Difference Method and Absorbing Boundary Conditions

The finite difference method was first introduced by K. S. Yee [12] to solve electromagnetic scattering problems. In using this method, Maxwell's equations are discretized both in time and space. Knowing the initial, boundary and excitation conditions, the fields on the nodal points of the space–time mesh can be calculated in a leapfrog time marching manner. For more details of the method, [11]–[15] could be consulted.

Since the finite difference equations cannot be applied to the boundary nodes, an absorbing type of boundary condition must be used at the mesh boundary where an open space is to be simulated. There are several publications discussing the absorbing boundary conditions [16]–[17]. Most of the methods are quite good to terminate the finite difference equations, and they have been applied to the scattering problems [13]–[15]. But, they are not good enough for the microstrip problems, since the Fourier transform of the time-domain results is very sensitive to the imperfect boundary treatment. Minute errors in the time domain may produce fairly large errors in the frequency domain. To minimize the reflection error, we have chosen to use the open- and short-circuit boundary conditions [16]. In this method, the problem is solved twice for each absorbing boundary, once with vanishing tangential electric field (termination with electrically conducting wall) and once with vanishing tangential magnetic field (termination with magnetically conducting wall). The results of the two solutions are averaged to give the desired solution. The rationale of this method is quite simple. Because the reflection coefficient of a short-circuit boundary is -1, while that of an open circuit is $+1$, the

Manuscript received April 15, 1987; revised September 21, 1987. This work was supported in part by the Joint Services Electronics Program under Contract F49620-84-C-0057, the Office of Naval Research under Contract N00014-86-K-0420, and the National Science Foundation under Grant ECS-8311340.

The authors are with the Department of Electrical Engineering and Computer Sciences and the Electronics Research Laboratory, University of California, Berkeley, CA 94720.

IEEE Log Number 8717986.

Reprinted from *IEEE Transactions on Microwave Theory and Techniques*, pp. 263-267, February 1988.

sum of the two results should have the reflected fields completely canceled out. However, caution must be exercised for computations of long duration, because the multiple reflections may not be canceled out. So, it is necessary to consider the size of the finite difference mesh domain to accommodate the anticipated duration of computations.

Using the open-circuit, short-circuit boundary conditions, we are able to reduce the reflection error to an acceptable level such that the corresponding frequency domain results are smooth and accurate.

III. Numerical Results

Using the time-domain method discussed above, a typical microstrip on a gallium arsenide substrate has been investigated. Unlike other investigations, which look for the propagation velocity of the field only [1], [3], the time-domain approach actually finds the space–time distributions of the fields everywhere within the finite difference space–time mesh. In many cases this provides clear pictures and illuminating details of the field variations. At the same time, the frequency-domain design data can also be easily obtained through Fourier transform of the time-domain fields.

A. Parameters Used in Computations

The microstrip structure used in our calculations is shown in Fig. 1. (Because of the symmetry of the problem, only half of the structure is considered.) The parameters used for most of our computations are as follows:

thickness of the substrate:	$H = 0.1$ mm
width of the metal strip:	1) $W = 0.075$ mm ($W/H = 0.75$)
	2) $W = 0.15$ mm ($W/H = 1.5$)
dielectric constant of the substrate:	$\varepsilon_r = 13.0$
thickness of the metal strip:	0.

To accommodate the structural details of the microstrip, we have used the mesh parameters as follows:

space interval	$dh = 0.1\ \text{mm}/16 = 6.25 \cdot 10^{-3}$ mm (for $W/H = 0.75$)
	$dh = 0.1\ \text{mm}/8 = 1.25 \cdot 10^{-2}$ mm (for $W/H = 1.5$)
	$n_1 = 30,\ n_2 = 55,\ n_3 = 160$
time step	$dt = k \cdot dh/c$ sec.

where c is the velocity of light in air and k is the constant restricted by stability criterion [13] ($k = 0.515$ in our calculation). The most often used values for r_1, r_2, r_3 are $r_1 = r_2 = r_3 = 1.0$. But sometimes other choices of r_1, r_2, r_3 (with $r_1, r_2, r_3 \geqslant 1$) are used to ease the problem with a particular W/H.

A Gaussian pulse excitation is used at the front surface, uniform under the strip (in plane "abcd" of Fig. 1) with only E_x component:

$$E_x(t) = \exp\left[-\frac{(t-t_0)^2}{T^2}\right] \tag{1}$$

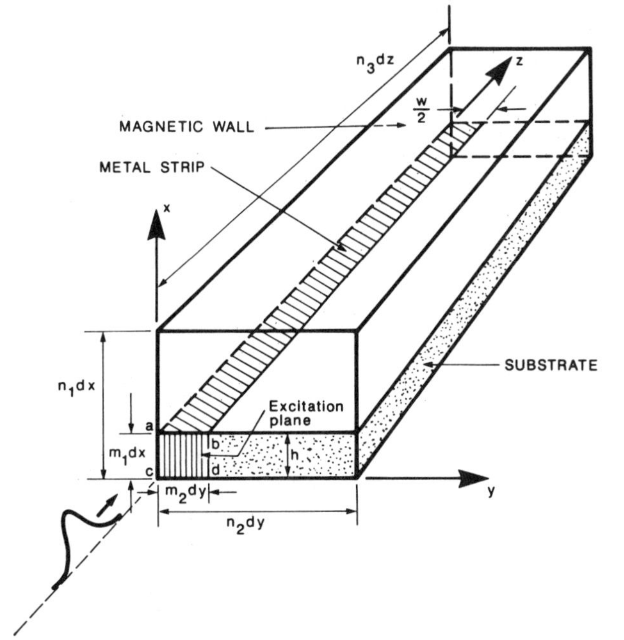

Fig. 1. Microstrip structure.

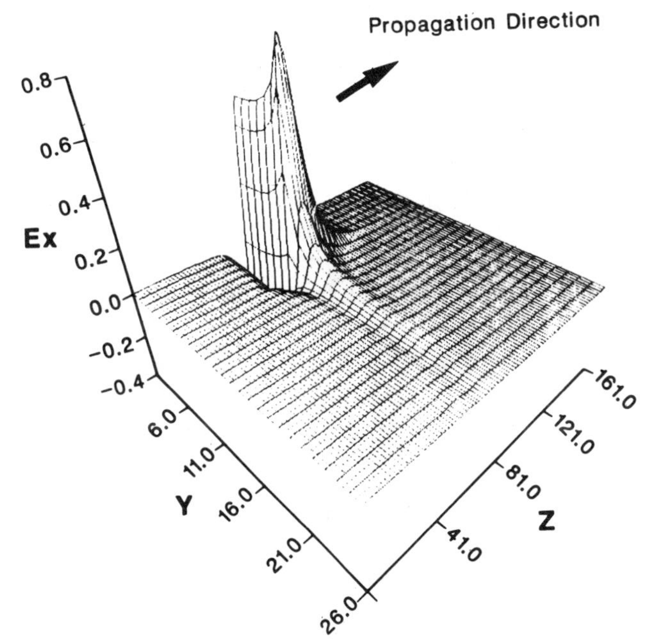

Fig. 2. Gaussian pulse propagation in microstrip (E_x component just underneath the air dielectric interface).

where $t_0 = 140dt$ and $T = 140dt$; elsewhere on the front surface, we set $E_x = E_y = 0$. The pulse width in space is about $20dh$, which is wide enough to obtain good resolution. The frequency spectrum of the pulse is from dc to about 700 GHz.

B. Time-Domain Fields

Fig. 2 shows the propagation of a Gaussian pulse along the microstrip. This is a plot of the space distributions of the vertical electric field on a surface just underneath the microstrip. Fig. 3 shows the time variations of vertical electric field at different positions along the propagation direction. The dispersive properties of the microstrip are

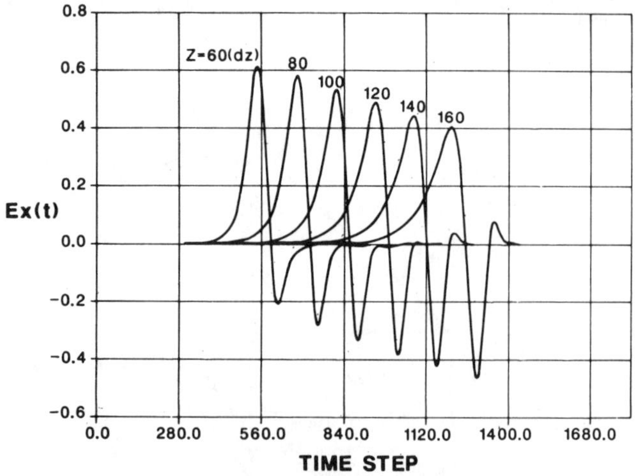

Fig. 3. Time variation of E_x at different positions along the direction of propagation.

quite obvious from the distortion of the pulse as it travels away from the feeding point. Indeed, the result (Fig. 3) is very similar to that calculated by Veghte and Balanis [6], who reconstructed the time-domain waveforms using the frequency-domain empirical formulas of effective dielectric constant.

C. Effective Dielectric Constant

The effective dielectric constant $\varepsilon_{reff}(\omega)$ can be used to account for the dispersive characteristics of the microstrip. It is calculated as follows:

Take the Fourier transforms of $E_x(t)$ at two different positions (underneath the center of the strip), with a separation of L, along the propagation direction:

$$E_x(\omega, z=0) = \int_{-\infty}^{\infty} E_x(t, z=0) e^{-j\omega t}\, dt \tag{2}$$

$$E_x(\omega, z=L) = \int_{-\infty}^{\infty} E_x(t, z=L) e^{-j\omega t}\, dt. \tag{3}$$

Taking the ratio of (2) and (3), we can get the transfer function of this section of microstrip, which is

$$e^{-\gamma(\omega)L} = \frac{Ex(\omega, z=L)}{Ex(\omega, z=0)} \tag{4}$$

where

$$\gamma(\omega) = \alpha(\omega) + j\beta(\omega). \tag{5}$$

The constant $\varepsilon_{reff}(\omega)$ is defined through $\beta(\omega)$ as

$$\beta(\omega) = \omega\sqrt{u_0 \varepsilon_0 \varepsilon_{reff}(\omega)} \tag{6}$$

or

$$\varepsilon_{reff}(\omega) = \frac{\beta^2(\omega)}{\omega^2 \varepsilon_0 \mu_0}. \tag{7}$$

The theoretical study of the effective dielectric constant has been done by many authors. Fig. 4 shows a set of results of different investigators which were gathered together by Kuester and Chang [1]. These results were obtained by using different numerical techniques and it can

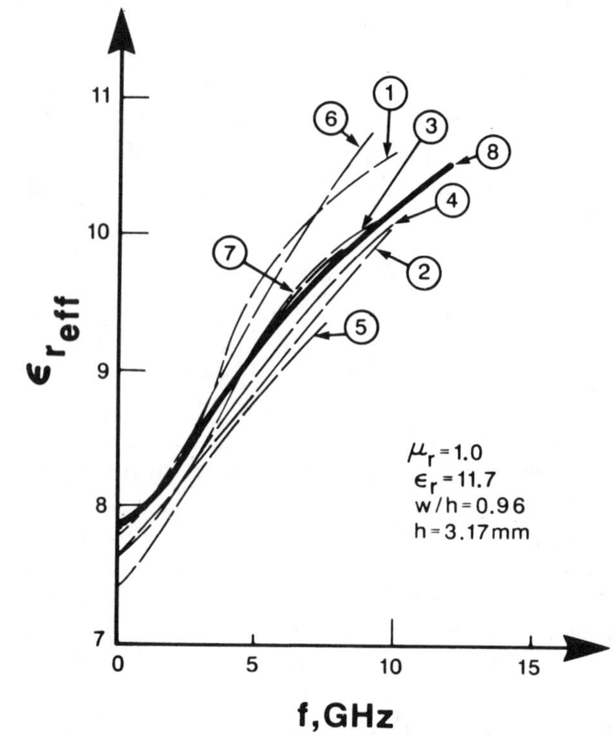

Fig. 4. Comparison of effective dielectric constant ε_{reff} as computed by different authors [1) Farrar and Adam, 2) Itoh and Mittra, 3) Van de Capelle and Luypaert, 4) Denlinger, 5) Schmitt and Sarges ($\varepsilon_r = 11.2$), 6) Chang and Kuester, 7) Pregla and Kowalaki, 8) time-domain result in this paper]. All of the results except the last one are from Kuester and Chang's paper [1].

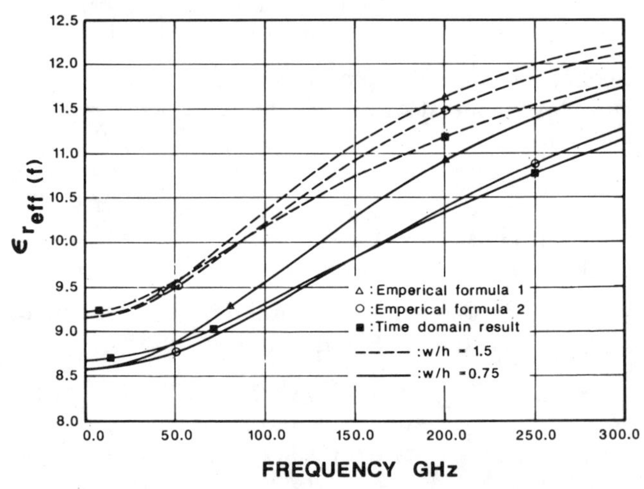

Fig. 5. Effective dielectric constant ($\varepsilon_r = 13.0$). Comparison of the results from time-domain method with two empirical formulas: 1) Edward and Owen [5], 2) Pramanick and Bhartia [6], [4].

be observed that they do not agree with one another very well. Our result is plotted on the same figure.

Fig. 5 shows the comparisons of our result with the empirical formulas of Edward and Owen [5] and that of Pramanick and Bhartia ([4], [6]). For $W/H < 1$, our result is closer to that of Pramanick and Bhartia than to Edward and Owen's. For $W/H > 1$, both empirical formulas are off from our result in the frequency range shown. It is noticed that for both $W/H < 1$ and $W/H > 1$, the time-domain results provide fairly accurate results at dc (the dc value

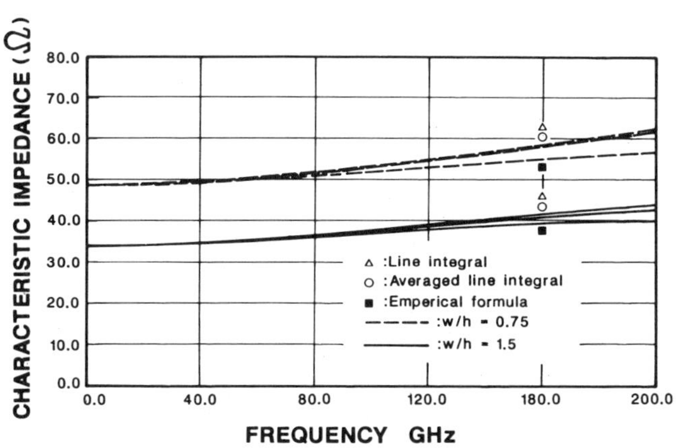

Fig. 6. Characteristic impedance ($\varepsilon_r = 13.0$). Comparison of the results from time-domain method with empirical formula (Bianco *et al.* [5]).

used in these empirical formulas has 1 to 2 percent error, as claimed in [5]), and that the accuracy of the finite difference method, unlike the empirical formulas, is not affected by the ratio of W/H. In view of our agreement with Pramanick and Bhartia at $W/H<1$, it is not an unrealistic assumption to say that the time-domain results are more credible than those of the empirical formulas in the range $W/H>1$.

D. Characteristic Impedance

The variation of the characteristic impedance with frequency is obtained through the ratio of $V(\omega)/I(\omega)$. Here $I(\omega)$ is the Fourier transform of the current defined as the loop integral of the magnetic field around the metal strip. For $V(\omega)$, two kinds of definitions are used, one is the line integral of vertical electric field under the center of the strip, the other is the averaged line integral of vertical electric field under the whole strip.

Fig. 6 shows the comparison of our results with the result from the empirical formulas of Bianco *et al.* [5]. For a fairly large frequency range starting from dc, it turns out that the results of center line integral and averaged line integral are very close to each other, indicating that the voltage uniqueness is well satisfied in that frequency range.

IV. Conclusions

It has been demonstrated that the time-domain finite difference method is a credible numerical method capable of producing frequency-domain design data through Fourier transform. It has been a frequent practice to use a sinusoidal source in the time-domain approach to obtain frequency-domain results. Our experience leads us to believe that investigators use monochromatic sources because the Fourier transform of a time-domain result is very sensitive to time-domain errors. But the monochromatic approach is expensive, especially at low frequencies. This is because, using the explicit time-domain finite difference method to model the microstrip problems, will lead to an "oversampled" time mesh. That is, we are using a much smaller time step than that required by the Nyquist criterion. This situation is brought about by the Courant condition requiring $c \cdot dt < dh/\sqrt{3}$, (dh = smallest dimension of the mesh), and dh is kept small by the requirement of representing the small dimension of the microstrip. The above sequence of conditions limits the size of dt in the computation and results in a seemingly paradoxical situation in that it requires a longer time to calculate the response of a broad pulse than that of a sharp pulse. Consequently, it is much more economic to use a sharp Gaussian pulse source to obtain wide-band frequency results than to do it one frequency at a time. Indeed, it takes much more time to caluclate the response of a single sinusoidal wave than to calculate that of a sharp Gaussian pulse.

Many of the theoretical analyses of the microstrip are solely for microstrips, but the time-domain finite difference method is very general. With minor modifications this method has already been used by the authors to investigate many interesting microstrip-related microwave components and the results are to be published. This investigation paves the way for accurate modeling of passive microwave components at high frequencies and in mm-MMIC's.

References

[1] E. F. Kuester and D. C. Chang, "An appraisal of methods for computation of the dispersion characteristics of open microstrip," *IEEE Trans. Microwave Theory Tech.*, vol. MTT-27, pp. 691–694, July 1979.

[2] T. Itoh and R. Mittra, "Spectral-domain approach for calculating the dispersion characteristics of microstrip lines," *IEEE Trans. Microwave Theory Tech.*, vol. MTT-21, pp. 496–499, July 1973.

[3] M. Kobayashi and F. Ando, "Dispersion characteristics of open microstrip lines," *IEEE Trans. Microwave Theory Tech.*, vol. MTT-35, pp. 101–105, Feb. 1987.

[4] P. Pramanick and P. Bhartia, "An accurate description of dispersion in microstrip," *Microwave J.*, pp. 89–96, Dec. 1983.

[5] K. C. Gupta, R. Grag, and R. Chadha, *Computer-Aided Design of Microwave Circuits*, Dedham, MA: Artech House, 1981.

[6] R. L. Veghte and C. A. Balanis, "Dispersion of transient signals in microstrip transmission lines," *IEEE Trans. Microwave Theory Tech.*, vol. MTT-34, pp. 1427–1436, 1986.

[7] N. Yoshida and I. Fukai, "Transient analysis of a stripline having a corner in three-dimensional space," *IEEE Trans. Microwave Theory Tech.*, vol. MTT-32, pp. 491–498, May 1984.

[8] S. Koike, N. Yoshida, and I. Fukai, "Transient analysis of microstrip side-coupled filter in three-dimensional space," *Trans. IECE Japan*, vol. E69-B, pp. 1199–1205, Nov. 1986.

[9] S. Koike, N. Yoshida, and I. Fukai, "Transient analysis of microstrip gap in three-dimensional space," *IEEE Trans. Microwave Theory Tech.*, vol. MTT-33, pp. 726–730, Aug. 1985.

[10] S. Koike, N. Yoshida, and I. Fukai, "Transient analysis of coupling between crossing lines in three-dimensional space," *IEEE Trans. Microwave Theory Tech.*, vol. MTT-35, pp. 67–71, Jan. 1987.

[11] D. H. Choi and W. J. R. Hoefer, "The finite-difference time-domain method and its application to eigenvalue problems," *IEEE Trans. Microwave Theory Tech.*, vol. MTT-34, pp. 1464–1470, Dec. 1986.

[12] K. S. Yee, "Numerical solution of initial boundary value problems involving Maxwell's equations in isotropic media," *IEEE Trans. Antennas Propagat.*, vol. AP-14, pp. 302–307, May 1966.

[13] A. Taflove and M. E. Brodwin, "Numerical solution of steady-state electromangetic scattering problems using the time-dependent Maxwell's equations," *IEEE Trans. Microwave Theory Tech.*, vol. MTT-23, pp. 623–630, 1975.

[14] K. K. Mei, A. C. Cangellaris, and D. J. Angelakos, "Conformal time domain finite difference method," *Radio Sci.*, vol. 19, pp. 1145–1147, Sept.–Oct. 1984.

[15] A. C. Cangellaris, C. C. Lin, and K. K. Mei, "Point-matched time domain finite element methods for electromagnetic radiation and scatterings," University of California, Berkeley, Electronics Research Laboratory Memorandum No. UCB/ERL M85/25, Apr. 1985 (submitted to *IEEE Trans. Antennas Propagat.*).

[16] W. D. Smith, "A nonreflecting plane boundary for wave propagation problems," *J. Comp. Phys.*, vol. 15, pp. 492–503, 1974.
[17] G. Mur, "Absorbing boundary conditions for finite-difference approximation of time-domain electromagnetic field equations," *IEEE Trans. Electromagn. Compat.*, vol. EMC-23, pp. 1073–1077, 1981.

Time-Domain Finite Difference Approach to the Calculation of the Frequency-Dependent Characteristics of Microstrip Discontinuities

XIAOLEI ZHANG AND KENNETH K. MEI, FELLOW, IEEE

Abstract —**The frequency-dependent characteristics of the microstrip discontinuities have previously been analyzed using several full-wave approaches. The time-domain finite different (TD–FD) method presented in this paper is another independent approach and is relatively new in its application for obtaining the frequency-domain results for microwave components [26]. The purpose of this paper is to establish the validity of the TD–FD method in modeling circuit components for MMIC CAD applications.**

I. Introduction

MICROSTRIP discontinuities (Fig. 1) are the basic constituent elements of microstrip integrated circuits. The accurate modeling of these discontinuities using different numerical approaches is one of the most important topics in microwave CAD. The current cut-and-try cycles in the design of microstrip integrated circuits will be greatly reduced if the frequency-dependent characteristics of the discontinuities can be obtained with certainty. Using network concepts, various microstrip resonators, couplers, and filters can be directly analyzed from the interconnection of microstrip discontinuities and microstrip line segments.

The study of microstrip discontinuities started in the early 1960's. For nearly a decade, the analyses were mostly quasi-static in nature [1]–[12]. The first accurate full-wave frequency-dependent analysis appeared around 1975 [13]–[15]. This approach began with the use of a waveguide model with electric-wall top and bottom planes and magnetic-wall sides planes to characterize the microstrip. The effective dielectric constant of the filling and the width of the guide are assumed to be frequency dependent and are determined in such a way that the model and the actual microstrip line have the same frequency-dependent propagation constant and characteristic impedance. Using the waveguide model to represent the original microstrip, the fields at the region of the discontinuities are expanded into waveguide modes, and the modes of different regions are matched at intersection planes. From the matching coefficients the S matrix for different propagation modes can thus be calculated. The waveguide model approach is efficient and has reasonable accuracy for calculating the magnitude of the S parameters in the lower frequency range, but it is not able to take into account the radiation effect (since the model is a closed one) and the surface wave generation. Besides, the mode-matching step will also introduce error due to the fact that the actual modes excited in the microstrip discontinuities are not the same as those used in the model and accordingly will not match in exactly the same way. There is also an obvious limitation on the kinds of structures this method can be applied to. It cannot, for example, be used to analyze the microstrip open-end structure where one side of the discontinuity is not connected to a microstrip and where the radiation and surface waves are present.

A full-wave approach to the microstrip open end problem was first proposed by James and Henderson in 1979 [16]. The analysis on the far end of the microstrip open end, where the surface wave and the radiation wave are the constitutents of the fields, is carried out using an analytic mode-expansion technique. On the microstrip side, a TEM wave is taken as the dominant mode incident field, and the semiempirical results for the propagation constant and the characteristic impedance are used for this incident wave. The fields at both sides are matched at the interface and a variational step is taken to reduce the error introduced by the assumption of the TEM field pattern where the electric field has a vertical constant value under the strip and is zero elsewhere in the transverse plane. Mainly due to the roughness of the field pattern assumed, the results of this method are not very accurate, but the analysis did provide valuable physical insight.

Another important method which has been used by several investigators to model the microstrip discontinuities is the spectral–domain approach [17]–[19]. In using this method to analyze the shielded or covered structures, the fields and currents involved are Fourier transformed (with respect to the space variables) into the so-called spectral domain. The shape of the current on the microstrip is assumed to be close to the actual current distribution and is easily Fourier-transformable. The spectral-domain components of the fields and currents are

Manuscript received May 3, 1988; revised July 29, 1988. This work was supported by the Office of Naval Research Under Contract N00014-86-K-0420.

The authors are with the Department of Electrical Engineering and Computer Sciences, University of California at Berkeley, Berkeley, CA 94720.

IEEE Log Number 8823766.

Reprinted from *IEEE Transactions on Microwave Theory and Techniques*, pp. 1775-1787, December 1988.

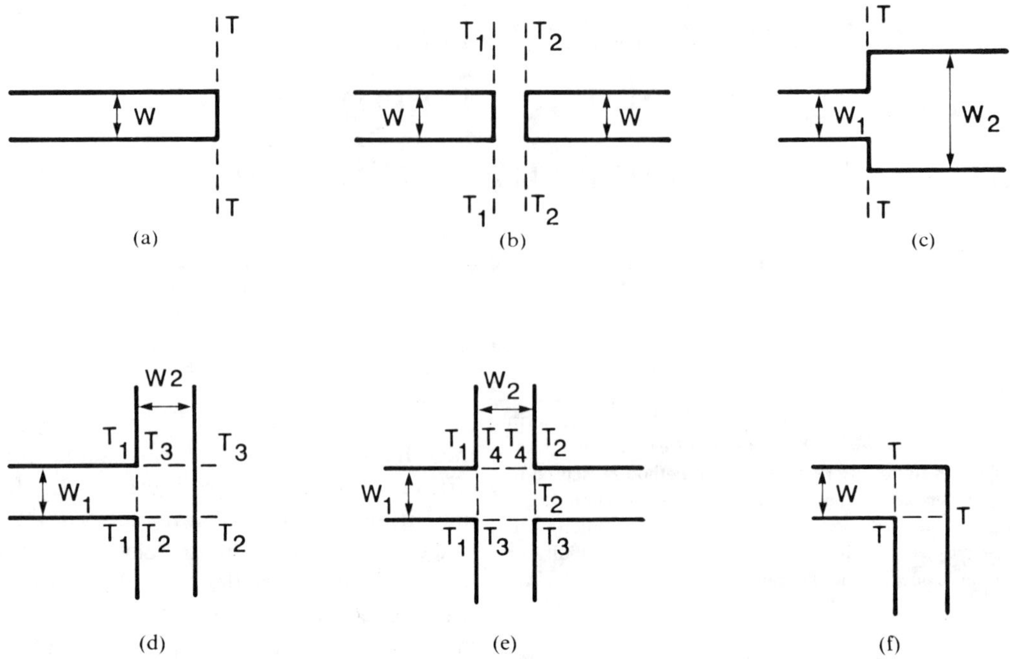

Fig. 1. Microstrip discontinuities. (a) Open-end. (b) Gap. (c) Step-in-width. (d) T junction. (e) Cross-junction. (f) Bend.

related according to the field continuity and boundary conditions and thus set up a system of equations for the variables. The inverse-transformed field solutions are used to calculate the S parameters.

Although a relatively accurate method for the type of components it is capable of calculating, the spectral-domain approach depends strongly on the current distributions assumed, which in many cases are hard to specify with high accuracy; thus it is limited in its applications. Besides, the frequency range which can be dealt with by this method is limited due to the difficulties which arise near the cutoff frequency of the higher order mode of the microstrip.

In recent years, the moment method has also been used by several investigators [20], [21] on the discontinuity problems. This method could in principle be an accurate one with wide applications, but due to the complexity of the Green's functions for the microstrip configurations it is not economical to make very fine numerical divisions to the microstrip for accurate results. In fact in many cases only a rational function form is used to represent the transverse current distribution on the microstrip, which may not correspond to the actual current distribution up to a certain frequency.

All the above-mentioned investigations are done in frequency domain; that is, the data for the whole frequency range are calculated one frequency at a time. It is an expensive task when the results of a wide frequency range are sought. This led us to seek an alternative way of calculating the frequency-domain data. Since a pulse response contains all the information of a system for the whole frequency range, it is a natural approach to use a pulse in the time domain to excite the microstrip structures, and from the time-domain pulse response to extract the frequency-domain characteristics of the system via the Fourier transform [26].

One numerical scheme which can be used to calculate the time-domain fields is the time-domain finite difference (TD–FD) method. It was first proposed by K. S. Yee in 1966 [22] and has been used by many investigators to solve electromagnetic scattering problems. Other numerical methods which can be used to solve this type of initial boundary value problem include the TLM method and Bergeron's method. Among these methods the TD–FD method is the most direct from a mathematical point of view, and is especially suitable for the accurate calculation of the microstrip fields, the reasons for which will be explained below.

Early investigators used the time-domain methods as a tool to obtain qualitative results that graphically illustrate the field propagation rather than to obtain design data via the Fourier transform of time domain results [24], [25]. In the process of our investigation, it has been found [26] that the Fourier transform of the time-domain results is very sensitive to numerical errors, notably those resulting from the imperfect treatment of the absorbing boundary conditions used to truncate the numerical computations of an open structure. Thus, even though the time-domain results may be reasonably accurate, the frequency-domain results obtained from their Fourier transform may not be acceptable as useful data.

The present available absorbing boundary conditions for the discretized wave equations are either not good enough in quality or require impractically large computer memories. Recently, a new type of absorbing boundary algorithm has been developed [29], [30] which can greatly improve the quality of the local absorbing boundary conditions. Using this new boundary treatment, together with a

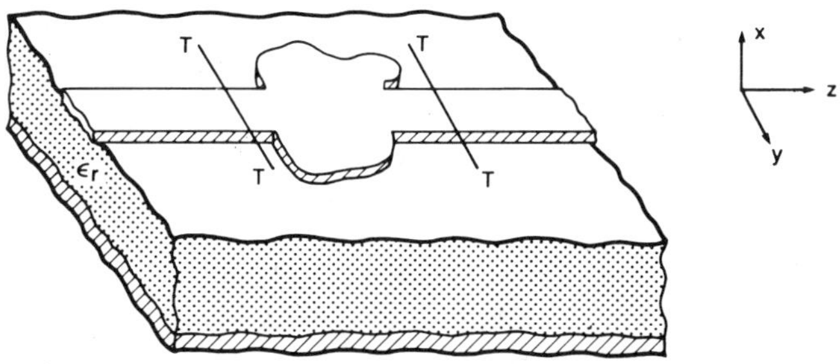

Fig. 2. A generalized microstrip discontinuity.

large enough computation domain, an accurate time-domain field can be obtained which can be used in the following Fourier transform step.

The calculated frequency-domain design data were compared with the available published results. These comparisons further demonstrate that the TD–FD approach is a viable method for modeling microstrip components.

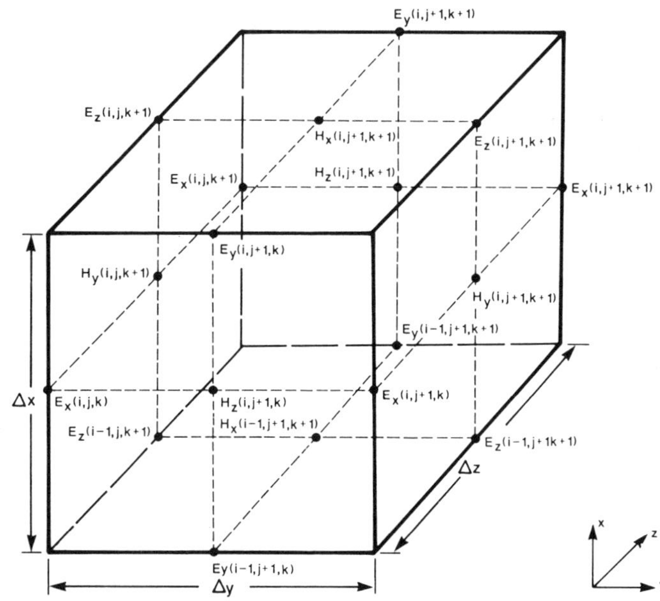

Fig. 3. Yee's mesh.

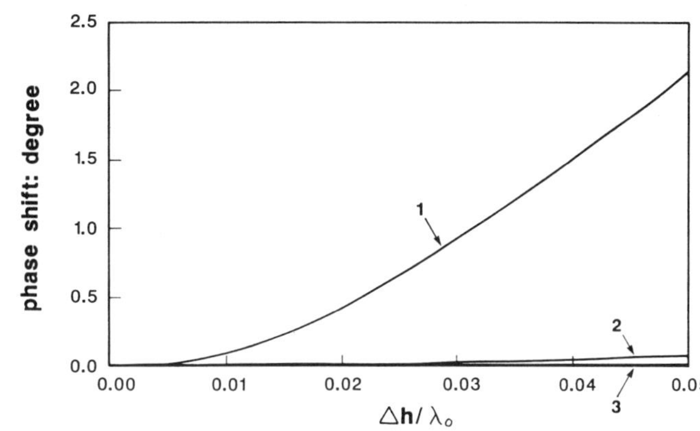

Fig. 4. Comparison of the numerical and microstrip dispersions (phase shift per time step): 1: microstrip dispersion ($w/h = 1.5$, $\epsilon_r = 13$); 2: leapfrog scheme, $c\Delta t/\Delta h = 1/7$; 3: a fourth-order scheme, $c\Delta t/\Delta h = 1/7$ [27].

II. Formulation of the Problem and the Numerical Method

A. General Formulation of the Problem

The generalized microstrip discontinuity under investigation is shown in Fig. 2 (a more general one will be an N-port structure instead of a two-port), where the strip and the bottom plane are made of a perfect conductor ($\sigma = \infty$) and the substrate has a relative dielectric constant of ϵ_r. The structure is assumed to be in an open environment, that is, above the dielectric and the metal strip surface, free space is assumed to extend to infinity; in the horizontal direction, apart from the discontinuity region, the substrate-ground structure also extends uniformly into infinity.

The Maxwell equations governing the solution of this problem are

$$\frac{\partial \vec{E}}{\partial t} = \frac{1}{\epsilon_i} \nabla \times \vec{H}$$

$$\frac{\partial \vec{H}}{\partial t} = -\frac{1}{\mu_0} \nabla \times \vec{E} \tag{1}$$

where $i = 1,2$ represents the substrate and the free-space region, respectively. At the interface of the two regions, the field continuity conditions are enforced.

For the uniqueness of the solution of these Maxwell equations, the following conditions must be satisfied:

a) The initial condition for the fields must be specified on the whole domain of interest; that is, $\vec{E}(\vec{r}, t = 0)$ and $\vec{H}(\vec{r}, t = 0)$ must be given everywhere inside the computation domain.

b) The tangential components of $\vec{E}$ and $\vec{H}$ on the boundary of the domain of interest must be given for all $t > 0$. For the boundary at infinity, Sommerfeld's radiation condition must be satisfied; that is, the wave at infinity must be of an outgoing type.

B. The Time-Domain Finite Difference Algorithm

There are many ways to solve the system of Maxwell equations in (1) numerically. The TD–FD algorithm is one of the most suitable schemes for the purpose of this investigation.

To simulate the wave propagation in three dimensions, Yee [22] arranged the spatial nodal points, where different components of $\vec{E}$ and $\vec{H}$ are to be calculated, as in Fig. 3. The repetitive arrangement of the cells of Fig. 3 fills the computation domain with a finite difference mesh. Every component of $\vec{H}$ can be obtained by the loop integral of $\vec{E}$ using the four surrounding $\vec{E}$ nodal values according to Maxwell's curl equation for $\vec{E}$. A similar approach holds for the calculation of $\vec{H}$.

In this algorithm, not only the placement of the $\vec{E}$ and $\vec{H}$ nodes are off in space by half a space step, but the time instants when the $\vec{E}$ or $\vec{H}$ fields are calculated are also off by half a time step. To be more specific, if the components of $\vec{E}$ are calculated at $n\Delta t$, where Δt is the discretization unit in time, or the time step, and n is any nonnegative integer, the components of $\vec{H}$ are calculated at $(n+1/2)\Delta t$. For this reason, this algorithm is also called the leapfrog method.

To summarize, for a homogeneous region of space, the discretization of the Maxwell curl equations (1) leads to the following:

$$E_x^{n+1}(i,j,k) = E_x^n(i,j,k) + \frac{\Delta t}{\epsilon} \cdot \left[\frac{H_z^{n+1/2}(i,j+1,k) - H_z^{n+1/2}(i,j,k)}{\Delta y} - \frac{H_y^{n+1/2}(i,j,k+1) - H_y^{n+1/2}(i,j,k)}{\Delta z} \right]$$

$$E_y^{n+1}(i,j,k) = E_y^n(i,j,k) + \frac{\Delta t}{\epsilon} \cdot \left[\frac{H_x^{n+1/2}(i,j,k+1) - H_x^{n+1/2}(i,j,k)}{\Delta z} - \frac{H_z^{n+1/2}(i+1,j,k) - H_z^{n+1/2}(i,j,k)}{\Delta x} \right]$$

$$\cdot E_z^{n+1}(i,j,k) = E_z^n(i,j,k) + \frac{\Delta t}{\epsilon} \cdot \left[\frac{H_y^{n+1/2}(i+1,j,k) - H_y^{n+1/2}(i,j,k)}{\Delta x} - \frac{H_x^{n+1/2}(i,j+1,k) - H_x^{n+1/2}(i,j,k)}{\Delta y} \right]$$

$$H_x^{n+1/2}(i,j,k) = H_x^{n-1/2}(i,j,k) - \frac{\Delta t}{\mu} \cdot \left[\frac{E_z^n(i,j,k) - E_z^n(i,j-1,k)}{\Delta y} - \frac{E_y^n(i,j,k) - E_y^n(i,j,k-1)}{\Delta z} \right]$$

$$H_y^{n+1/2}(i,j,k) = H_y^{n-1/2}(i,j,k) - \frac{\Delta t}{\mu} \cdot \left[\frac{E_x^n(i,j,k) - E_x^n(i,j,k-1)}{\Delta z} - \frac{E_z^n(i,j,k) - E_z^n(i-1,j,k)}{\Delta x} \right]$$

$$H_z^{n+1/2}(i,j,k) = H_z^{n-1/2}(i,j,k) - \frac{\Delta t}{\mu} \cdot \left[\frac{E_y^n(i,j,k) - E_y^n(i-1,j,k)}{\Delta x} - \frac{E_x^n(i,j,k) - E_x^n(i,j-1,k)}{\Delta y} \right] \quad (2)$$

where Δx, Δy, and Δz are the space discretization units in the x, y, and z directions, respectively, and Δt is the time discretization interval. The numbering of the different $\vec{E}$ and $\vec{H}$ components is illustrated in Fig. 3, which is different from that in Yee's original paper due to programming considerations.

The TD–FD algorithm has several advantages over other schemes for the calculation of microstrip time-domain fields. First, the central difference nature of the leapfrog method makes it a relatively accurate method (second-order accuracy in both time and space), compared to other first-order schemes. Second, there is no need for special treatment of the edge of the microstrip if the tangential $\vec{E}$ and vertical $\vec{H}$ components are arranged on the metal strip and only the parallel components of the electric field are arranged on the edge of the strip. Finally, the leapfrog algorithm has the unique characteristics that the numerical scheme has no dissipation (amplitude increase or decrease for any frequency component) and only a small amount of dispersion [27]. It has been shown [28], [29] that the numerical dispersion is negligible compared to the physical dispersion of the microstrip structure, as seen in Fig. 4 for comparing the microstrip dispersion and numerical dispersions for the leapfrog scheme and for a fourth-order finite difference scheme used to solve the one-dimensional wave equation. (The actual frequency range of interest for the microstrip discontinuity problems corresponds to $\Delta h/\lambda_0 \approx 0$–$0.03$.) Thus, no higher order finite difference is needed for the accurate modeling of the microstrip structures.

For any finite difference scheme, a stability condition must be found which guarantees that the numerical error generated in one step of the calculation does not accumulate and grow. The stability criterion of Yee's algorithm is the Courant condition [23]:

$$v_{\max} \cdot \Delta t \leqslant \frac{1}{\sqrt{\dfrac{1}{\Delta x^2} + \dfrac{1}{\Delta y^2} + \dfrac{1}{\Delta z^2}}}. \quad (3)$$

For the special case of $\Delta x = \Delta y = \Delta z = \Delta h$, (3) becomes

$$v_{\max} \cdot \Delta t \leqslant \frac{1}{\sqrt{3}} \cdot \Delta h. \quad (3a)$$

where $v_{\max}$ is the maximum signal phase velocity in the configuration being considered.

The stability of the absorbing boundary condition cannot be achieved exactly due to the imperfection of nearly all the presently available absorbing boundary conditions for the numerical solution of wave equations (there will always be some unrealistic reflection wave going back to the computation domain due to the boundary treatment). But since the computation lasts for only a limited time, one can always minimize the influence of the unrealistic reflections by making the computation domain sufficiently large and stopping the computation after the useful information has been obtained.

C. Choice of the Excitation Pulse

The excitation pulse used in this investigation has been chosen to be Gaussian in shape. A Gaussian pulse has a smooth waveform in time, and its Fourier transform (spectrum) is also a Gaussian pulse centered at zero frequency. These unique properties make it a perfect choice for investigating the frequency-dependent characteristics of the microstrip discontinuities via the Fourier transform of the pulse response.

An ideal Gaussian pulse which propagate in the $+z$ direction will have the following expression:

$$g(t,z) = \exp\left[-\frac{\left(t - t_0 - \dfrac{z - z_0}{v}\right)^2}{T^2}\right] \tag{4}$$

where v is the velocity of the pulse in the specific medium, and the pulse has its maximum at $z = z_0$ when $t = t_0$.

The Fourier transform of the above Gaussian pulse has the form

$$G(f) \propto \exp\left[-\pi^2 T^2 f^2\right]. \tag{5}$$

The choices of the parameters T, t_0, and z_0 are subject to two requirements. The first is that after the space discretization interval Δz has been chosen fine enough to represent the smallest dimension of the structure and the time discretization interval Δt has been chosen small enough to meet the stability criterion (3), the Gaussian pulse must be wide enough to contain enough space divisions for a good resolution. And at the same time, the spectrum of the pulse must be wide enough (or the pulse must still be narrow enough) to maintain a substantial value within the frequency range of interest. If these last two conditions cannot be satisfied simultaneously, Δz has to be rechosen to be even smaller.

The pulse width W chosen in this work is approximately 20 space steps. We define the pulse width to be the width between the two symmetric points which have 5 percent of the maximum value of the pulse. Therefore, T is determined from

$$\exp\left[-\frac{\left(\dfrac{W}{2}\right)^2}{(vT)^2}\right] = \exp(-3)(\approx 5\%) \tag{6}$$

or

$$T = \frac{1}{\sqrt{3}} \cdot \frac{10\Delta z}{v}. \tag{7}$$

By making this choice of T, the maximum frequency which can be calculated is

$$f_{\max} = \frac{1}{2T} \qquad \left(G\left(\frac{1}{2T}\right) \approx 0.1\right) \tag{8}$$

$$= \frac{1}{2} \cdot \frac{\sqrt{3}\,v}{10\Delta z} \tag{9}$$

which, with the specific Δz chosen, is high enough to cover the entire frequency range of interest, as will be shown below in the discussion of the numerical results.

The second requirement is that the choice of z_0 and t_0 be made such that initial "turn on" of the excitation will be small and smooth.

Another consideration in excitation is the specification of the spatial distribution of the field on the excitation plane. Ideally, the use of the dominant mode distribution is preferred. But this distribution is generally not known with high enough accuracy. By the use of our knowledge of the modes in the microstrip structure, a very simple field distribution can be specified at the excitation plane which serves our purposes almost as well.

The tangential electric field to be specified on the excitation plane is assumed to have only the E_x component, which is distributed uniformly under the strip and is zero elsewhere. This is not the exact dominant mode field distribution, although in the latter case the energy is also concentrated under the strip. But for frequencies under the cutoff frequency of the first waveguide type higher order mode and below the strong coupling frequency of the substrate modes (the lowest one of them will be referred to below as the inflection frequency), the only mode which can propagate down the microstrip is the dominant mode. The substrate modes which do not cut off at those frequencies have imaginary propagation (phase) constants with respect to the major wave propagation direction due to the presence of the metal strip and thus can only propagate sidewards. After the inflection frequency, there will be modes other than the dominant mode which can propagate down the line. But those modes will not contaminate our lower frequency results after the Fourier transform. This is due to the fact that the physical model and the numerical method we used are both linear; thus the modes at different frequencies will not couple energy from each other.

Therefore, as long as we allow a certain distance for the Gaussian pulse to propagate out of the excitation plane, and thus allow the unwanted substrate modes at lower frequencies to leave the central region of the microstrip, the low-frequency component of the pulse will consist of the dominant mode only. Graphically, the pulse pattern in the transverse direction gradually becomes stabilized (to its actual physical form) and a prominent edge effect manifests itself as the calculation goes on and the pulse is seen to propagate down the line.

Also, for the reasons mentioned above, caution must be taken when interpreting the Fourier transformed results after the inflection frequency, as the time-domain method does not have the capability of distinguishing between modes. These results do not correspond to the exact dominant mode results in general, although how great an influence the higher order modes have on the dominant mode results is still left to be determined.

D. Dielectric–Air Interface Treatment and the Artificial Absorbing Boundary Conditions

Fig. 5 shows the finite difference computation domain used for the discontinuity problems (in this case, a mi-

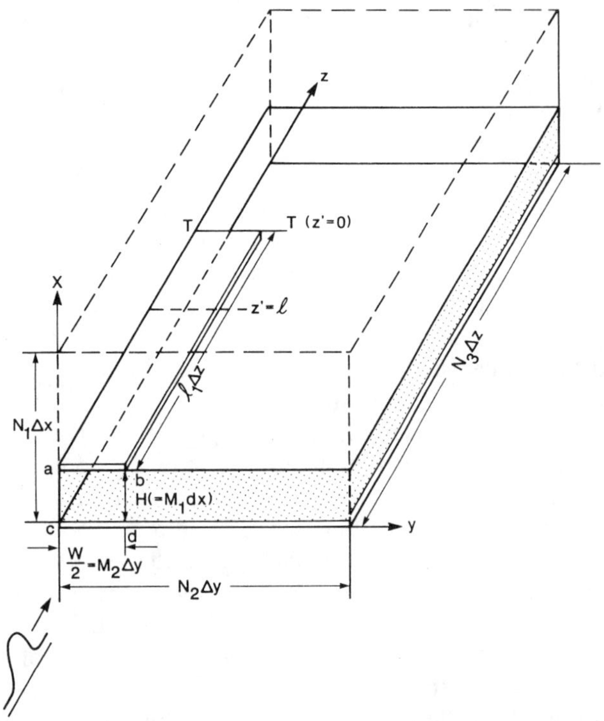

Fig. 5. Finite difference computation domain for microstrip open-end.

crostrip open end is shown as an example). Due to symmetry, only half of the structure is placed in the mesh domain with a magnetic wall at the plane of symmetry.

The finite difference form (2) of Maxwell's equations is derived in the uniform region of the medium and therefore cannot be applied to the nodal points on the dielectric–air interface or on the boundary planes of the finite difference mesh. All these points require special treatment.

The field components which lie on the dielectric–air interface are the tangential components of $\vec{E}$ (E_y and E_z) and the vertical component of $\vec{H}$ (H_x). In calculating H_x, (2) can still be used because the value of μ does not change across the boundary, and the E_y and E_z components used to calculate H_x are the tangential components with respect to the interface and are thus continuous across the boundary. To calculate E_y and E_z, however, a finite difference formulation other than (2) must be derived from the field continuity conditions across the boundary. The derivation is given in the Appendix, which is similar to that of Lin [31] for a two-dimensional finite element scheme, and the result is that E_y can be obtained by the discretization form of

$$\frac{\epsilon_1+\epsilon_2}{2}\cdot\frac{\partial E_y}{\partial t}=\frac{\partial H_x}{\partial z}-\frac{\Delta H_z}{\Delta x} \tag{10}$$

and E_z can be obtained through

$$\frac{\epsilon_1+\epsilon_2}{2}\cdot\frac{\partial E_z}{\partial t}=\frac{\Delta H_y}{\Delta x}-\frac{\partial H_x}{\partial y}. \tag{11}$$

In other words, the average value of ϵ is used in (2) for the calculation of the interface E_y and E_z nodes.

The values of E_y, E_z, and H_x vanish on the metal strip because of the assumption of a perfectly conducting surface. This also holds for the ground plane.

Since the computation domain cannot include the whole space, the finite difference mesh must be truncated to accommodate the finite computer memories. In solving our problems, the truncation planes are the side, top, and end surfaces (Fig. 5). The numerical algorithm on the truncation planes must simulate the propagation of the outgoing waves; this is known as the artificial absorbing (or radiation) boundary condition.

The perfect absorbing boundary conditions are usually global in nature, which makes them quite expensive to implement and requires excessively large computer memories. The local absorbing boundary conditions, which make use of only the fields at the neighboring space and time nodes, are relatively inexpensive to implement.

There are quite a few local absorbing boundary conditions available, but most of them are not "absorbing" enough for the purpose of this investigation. As mentioned in the Introduction, the Fourier transform of the time-domain results is very sensitive to the reflection errors. A small amount of reflection may not visibly influence the time-domain fields, but the transformed results could be far off.

To improve the local absorbing boundary conditions, a new approach based on the "local cancellation of the leading order errors" has been developed and shown to provide substantial improvement of absorbing qualities. The boundary treatment used in this investigation will be discussed below.

Consider the end boundary ($z=N_3\Delta z$) first. In most of the cases under consideration, the end surface will contain one end of the microstrip with its other end connected to the microstrip discontinuity. For most high-dielectric-constant substrates, owing to the guiding nature of the metal strip, the major direction of the power flow is in the $+z$ direction, or nearly normal incident. The sideways leakage and radiation are small. This is quite similar to a one-dimensional propagation case, and a natural choice of the boundary condition is to use the field values a few Δz before and a few Δt earlier for the present boundary fields.

In a microstrip structure, the existence of the dielectric substrate makes the wave velocity in the main propagation direction $+z$ be less than the velocity of light c in free space. Denote this velocity as v (usually v is some fraction of c and is also weakly dependent on frequency due to the dispersive nature of the structure). Assume $a_1\cdot v=c$, where a_1 is a constant. (Rigorously speaking, it will be a weak function of frequency. Here for the purpose of the boundary treatment only its low-frequency value is adapted). For the stability criterion (3a) to be satisfied, we must have (choose $\Delta x=\Delta y=\Delta z=\Delta h$)

$$v_{\max}\Delta t=c\,\Delta t=a_1 v\,\Delta t\equiv k\,\Delta h \tag{12}$$

where k is a certain constant satisfing

$$k\leqslant\frac{1}{\sqrt{3}}$$

and therefore

$$v \cdot \Delta t = \frac{k}{a_1} \cdot \Delta h. \tag{13}$$

From (13) it is clear that we can always make the wave travel a certain integral number of space steps in some integral number of time steps by choosing a specific k (better to be less than but close to $1/\sqrt{3}$ for high accuracy), thus avoiding the need for interpolation, and the stability condition (3a) or (13) is still satisfied.

Take the 50 Ω line on the alumina substrate ($\epsilon_r = 9.6$, $W/H = 1.0$) as an example. Here $\epsilon_{reff}(f=0) \approx 6.6$; then $v = 1/\sqrt{6.6} \cdot c$, or $a_1 = \sqrt{6.6}$. If we choose $k = 0.514$, then k/a_1 will be approximately 1/5; that is, the wave will travel one space step in approximately five time steps.

After choosing the parameters as above, the boundary value of the fields can now be specified as the value of the inner nodes at several time steps earlier. Again, for the alumina case, the boundary condition of the fields will be

$$E_i(N_3 \Delta z, n\Delta t) = E_i[(N_3 - 1)\Delta z, (n-5)\Delta t] \tag{14}$$

where $E_i(N_3 \Delta z, n\Delta t)$ is any electric field component which lies on the boundary of the computation domain (actually only the tangential components of the fields are needed for later calculations). By using this treatment, a storage of the next-to-boundary nodal fields for several time steps is needed. Since this is a storage of two-dimensional data, it will not increase the memory requirement significantly, compared to the major three-dimensional storage.

Usually, merely applying the boundary operation as above will still leave a visible (3–5 percent) amount of reflection. This is partly due to the fact that the true wave propagation is not one-dimensional, and partly because the velocity of the wave is not a constant but rather a function of frequency.

To improve the earlier boundary treatment, it is found that if we apply the same kind of boundary condition on the tangential $\vec{H}$ field next to the boundary, i.e. (for the alumina case),

$$H_i[(N_3 - 1/2)\Delta z, (n+1/2)\Delta t] = H_i[(N_3 - 3/2)\Delta z, (n + 1/2 - 5)\Delta t] \tag{15}$$

and compare it with those $\vec{H}$ values calculated from the loop integration of $\vec{E}$ fields (here the boundary $\vec{E}$'s used are obtained in the previous computation step by using boundary condition (14)), these two $\vec{H}$ fields will always have the property that the errors contained in them due to the imperfect treatment of the boundary condition will have opposite signs and the magnitudes of these errors will maintain a known ratio. Therefore by a weighted average (1 : 5 in this case) of these two $\vec{H}$ fields, we can get an error cancellation effect and the resulting boundary operation will have a much improved quality. This kind of error cancellation approach can actually be applied to any kind of linear local boundary conditions. A general discussion of it for the one-dimensional case can be found in [30].

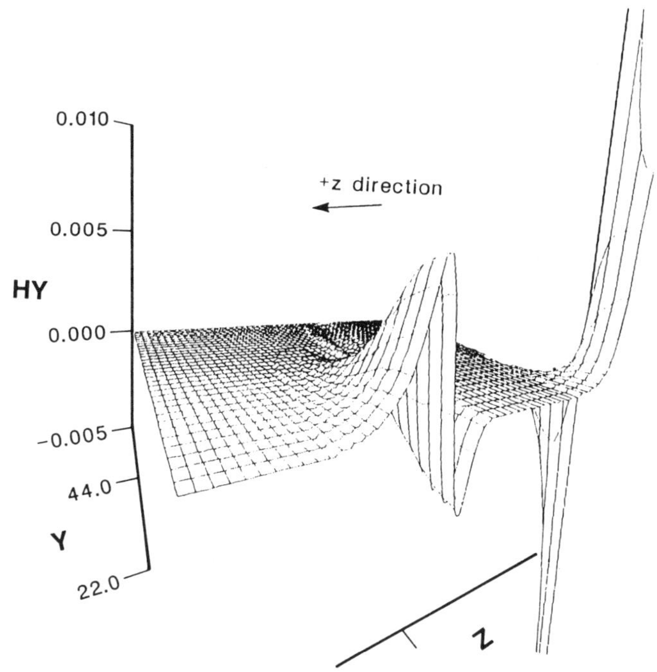

Fig. 6. dc magnetic field on the front surface due to the electric wall boundary treatment.

The treatment of the side boundary is similar to that of the end. But in this case the normal incident approximation is no longer valid. The direction of incidence of the wave changes with both time and position. An accurate and economic treatment of the side boundary condition has not been found so far. As a compromise, the boundary tangential $\vec{E}$ fields will always be given the value of their inner neighbors one time step earlier. A similar operation is done for the $\vec{H}$ nodes 1/2 space step from the boundary, and the error cancellation algorithm still works in this case.

On the top plane, only the boundary condition for $\vec{E}$ fields is applied in the same way as for the side plane. The cancellation scheme does not work for the top plane, since the field is of the evanescent type in that direction.

The front surface needs some special treatment. During the time when the Gaussian pulse is excited, under the strip on plane *abcd* of Fig. 5, the vertical field is given the value of the Gaussian pulse. Elsewhere on the front surface the electric fields are fixed to be zero. This is equivalent to an electric wall boundary condition (the magnetic wall or symmetric boundary condition turned out not to be able to give a clear tail of the pulse). Following the passing of the pulse with part of it reflected back from the discontinuities, the front surface should now behave in a "transparent" way, as in the real case. This means that from the moment the reflected wave reaches the front surface a radiation type of boundary condition must be "switched on." It is found that the early enforced electric wall boundary condition induced a dc current or tangential magnetic field on the front surface and nearby (Fig. 6). This local dc field, although it has no influence on the traveling pulse, does cause trouble in the boundary treatment. That is, if we switch on the radiation condition on this wall, the numerical errors will accumulate very rapidly and the solutions soon "blow up."

To solve this problem, what actually has been done is that after the pulse leaves the source plane and before it is reflected back from the discontinuities, the radiation boundary condition is switched on at a surface which is parallel to the source plane but a few space steps into the computation domain, and this is sufficient to avoid the trouble caused by the dc current.

At this stage, after all the boundary conditions have been properly treated, the numerical solution of the discontinuity problems is quite direct.

III. Numerical Results

In this investigation, five kinds of symmetric microstrip discontinuities on alumina substrate ($\epsilon_r = 9.6$) for a 50 Ω transmission system ($W/H = 1.0$, $W = 0.6$ mm) have been studied. They are (refer to Fig. 1) microstrip open-end, cross and T junctions, step-in-width and gap. Among these discontinuities, the microstrip open-end case has been given the most detailed discussion and compared with all the available published results.

A. Microstrip Open-End Terminations

A microstrip open-end on alumina substrate ($\epsilon_r = 9.6$) as shown in Fig. 5 is studied first. The parameters of the structure are as follows:

thickness of substrate: $H = 0.6$ mm
width of metal strip: $W = 0.6$ mm
thickness of metal strip: $t = 0.0$.

To accommodate the structural details of the microstrip, the mesh parameters have been chosen to be

space interval: $\Delta h = H/10 = 0.06$ mm;
$\Delta x = \Delta y = \Delta z = \Delta h$;
$N_1 = 40$, $N_2 = 120$, $N_3 = 190$;
$l_1 = 120$, $M_1 = 10$, $M_2 = 5$;
$l = 20, 30, 40, 50$ (Δh) have all been used to calculate the results for comparison;
time step $\Delta t = k \cdot \Delta h / c$ s, where c is the velocity of light in air and k is a constant restricted by the stability criterion (3);
$k = 0.514$ in this calculation.

A Gaussian pulse excitation is used at the front surface. It is uniform under the strip (in plane *abcd* of Fig. 4) and has only the E_x component with the following specified value:

$$E_x(t) = \exp\left[-\frac{(t - t_0)^2}{T^2}\right] \tag{16}$$

where $t_0 = 350\,dt$ and $T = 40\,dt$; elsewhere on the front surface, set $E_x = E_y = 0$. The pulse width in space is about $20dh$, which is wide enough to obtain good resolution. The

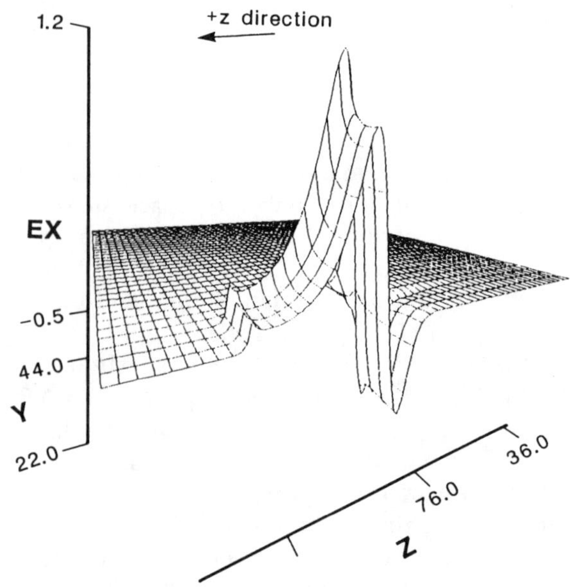

(a)

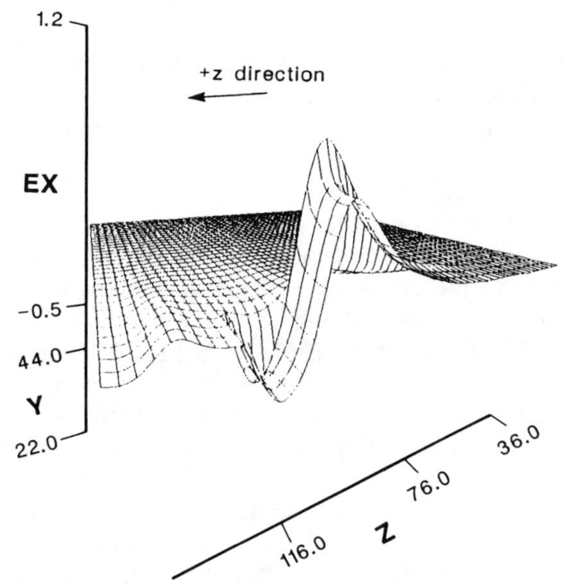

(b)

Fig. 7. Gaussian pulse propagation and reflection in the microstrip open-end structure: E_x component just underneath the strip. (a) Incident pulse just reaching the open end. (b) Pulse being reflected back and the surface wave being generated.

frequency spectrum of this pulse is from dc to about 100 GHz.

Fig. 7 shows the calculated time-domain field (E_x component) for a microstrip open end. The plane where the plot is drawn is just underneath the metal strip. Part (a) is the field distribution at the moment when the Gaussian pulse just reaches the open end and is reflected. The reflected wave is seen to have the same sign as the incident wave and is added to the incident wave. Part (b) shows the reflected wave and a small amount of traveling surface wave.

The microstrip open end structure is a one-port network (Fig. 5). Its scattering matrix has only one element, that is,

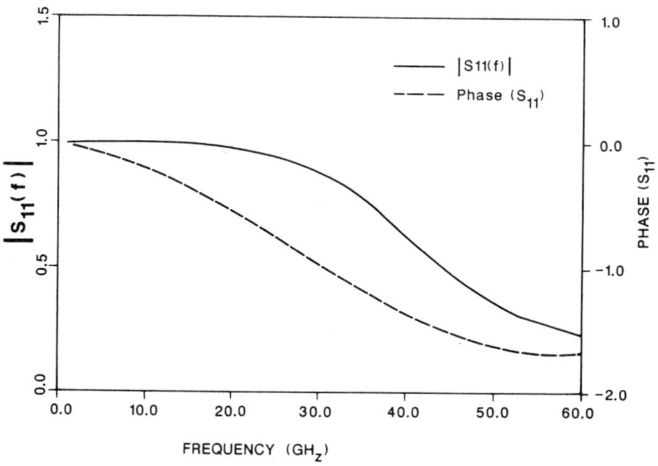

Fig. 8. Frequency-dependent S parameter of open end: magnitude and phase.

S_{11} or the reflection coefficient. S_{11} is defined as

$$S_{11}(f)=\frac{V_{\text{ref}}(f)}{V_{\text{inc}}(f)} \tag{17}$$

where $V_{\text{ref}}(f)$ is the transformed reflection voltage at the input plane (i.e., the reference plane $T-T$ in Fig. 5) of the one-port, and V_{inc} is the transformed incident voltage at the same position. In this calculation, the incident field is obtained from that of an infinitely long microstrip, and the reflected field from the open end is obtained from the difference between the total open-end field and the incident field.

It is common practice in microwave network calculation for S_{11} to be calculated away from the discontinuity through the transmission line formula (refer to Fig. 5.)

$$S_{11}(f)=\frac{V_{\text{ref}}(f,z'=l)\cdot e^{\gamma(f)l}}{V_{\text{inc}}(f,z'=l)\cdot e^{-\gamma(f)l}}$$

$$=\frac{V_{\text{ref}}(f,z'=l)}{V_{\text{inc}}(f,z'=l)}\cdot e^{2\gamma(f)l}. \tag{18}$$

This is done here to allow the higher order, evanescent modes which were generated near the discontinuity to die out. They are not included in the calculation.

Fig. 8 shows the calculated results of the magnitude and phase of $S_{11}(f)$ for the open end under consideration. The uniqueness of the solution to the use of either the field at one point under the metal strip or the voltage between the strip and the ground, when substituted into (18) for the V's there, has been checked and found to be well satisfied. This is expected to be true once the mode distribution is well established on the line.

The equivalent circuit as shown in Fig. 9 with frequency-dependent circuit parameters is used to model the microstrip open end. Here

$$Y(f)=G(f)+j2\pi fC(f) \tag{19}$$

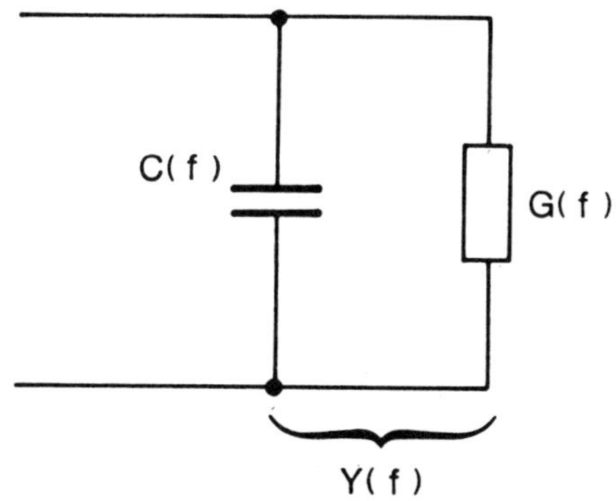

Fig. 9. Equivalent circuit of the microstrip open end.

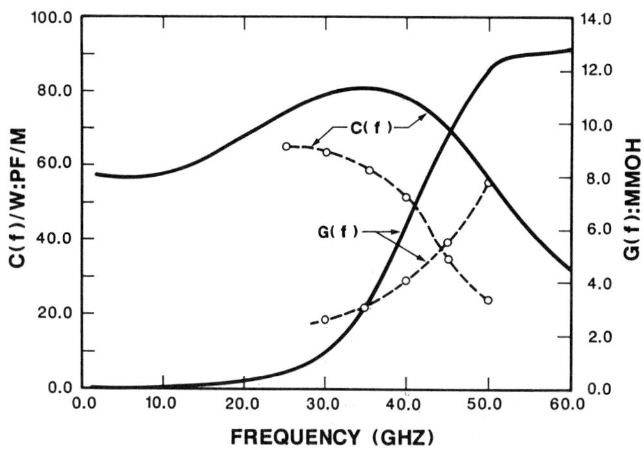

Fig. 10. Frequency-dependent equivalent circuit parameter $C(f)/W$ and $G(f)$ for open end. Solid lines denote time-domain results; dashed lines denote the results of Katehi and Alexopoulos [21].

and $Y(f)$ is related to $S_{11}(f)$ through

$$Y(f)=\frac{1-S_{11}(f)}{1+S_{11}(f)}\cdot\frac{1}{Z_0(f)}. \tag{20}$$

Here $Z_0(f)$ is the characteristic impedance of the microstrip, which is calculated using the ratio of voltage and current [26].

The calculated $C(f)$ and $G(f)$ are plotted in Fig. 10 together with the results presented by Katehi and Alexopoulos [21] for the same structure. The comparison shows quite an amount of discrepancy, especially for higher frequencies. Although both models should be questioned, there is an obvious question in the result of [21] for $G(f)$ in that it does not exhibit a trend to go smoothly down to zero as the frequency goes to dc. For $C(f)$, an equivalent parameter $\Delta l(f)$ can be derived from it and has been compared with extensive published results below.

The parameter which can also be used to account for the capacitive characteristic of the open end is the effective increase in length Δl. It is related to C through [32]

$$\frac{C(f)}{W}=\frac{\Delta l(f)}{h}\sqrt{\epsilon_{\text{eff}}}\,\frac{1}{Z_0(f)}\,\frac{h}{W} \tag{21}$$

where ϵ_{eff} is the effective dielectric constant of the microstrip [32].

The calculated $\Delta l(f)$, together with the comparison with several other published results, is given in Fig. 11.

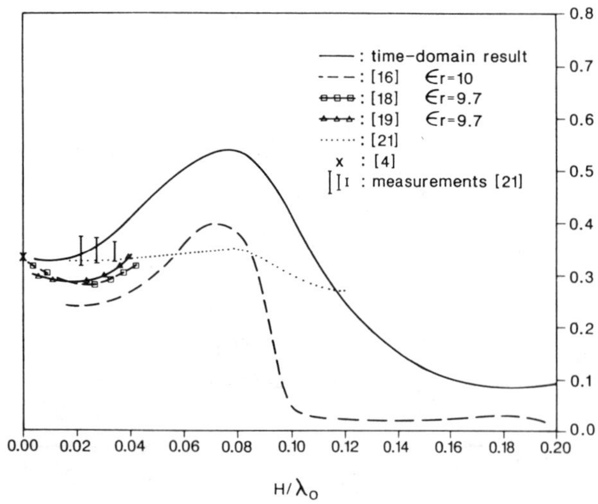

Fig. 11. Effective length increase $\Delta l(f)/H$.

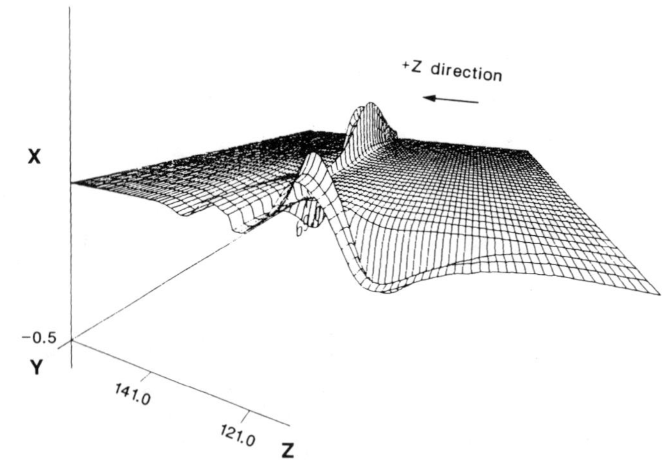

Fig. 13. Gaussian pulse propagation and reflection in the microstrip T junction: E_x component just underneath the strip.

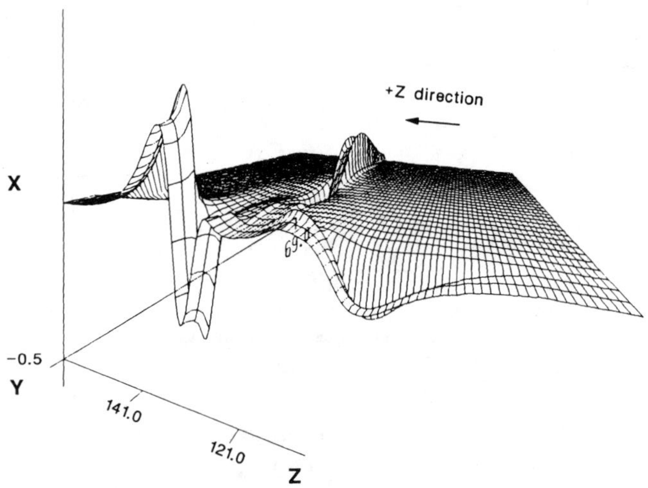

Fig. 12. Gaussian pulse propagation and reflection in the microstrip cross-junction: E_x component just underneath the strip.

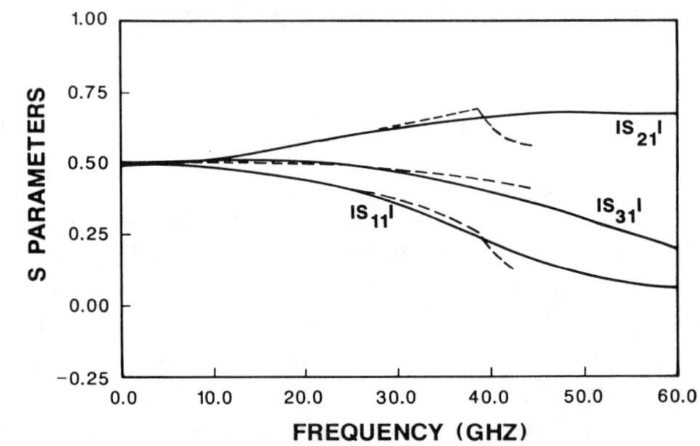

Fig. 14. Frequency-dependent S parameters of the microstrip cross-junction. Solid: time-domain result, dash: Mehran, $\epsilon_r = 9.7$, $h = 0.635$ mm, $W_1 = W_2 = 0.56$ mm [14].

The normalized frequency range used here corresponds to the frequencies from zero to 100 GHz for the structure concerned. The dc result of the time-domain calculation is very close to the quasi-static result of [4], the low frequency part is close to the experimental result of Edward ($W/H = 0.9$) [21], and the shape of the curve is similar to that of James and Henderson [16], which has too low a dc value though. The peak in the curve which appeared near the cutoff frequency of the first TE-type higher order mode of the microstrip has been predicted in [3].

B. Microstrip Cross-Junctions and T Junctions

Using a computation mesh domain similar to that for the open end (this time with $N_1 = 30$, $N_2 = 65$, $N_3 = 180$, and $\Delta h = H/8$), the symmetric microstrip cross-junctions and T junctions ($W_1 = W_2 = W = 0.6$ mm) under symmetric excitations are studied. Here the side boundary is given the same type of boundary treatment as for the end, since in this case sideways flow becomes another major direction of power flow.

Figs. 12 and 13 show the time-domain field distributions for the cross-junction and T junction. The Gaussian pulse which travels into the cross-junction is seen to split four ways after it hits the cross-junction. For the case of the T junction, a small amount of surface wave is observed to travel past the junction, as most of the energy is either reflected backward or transmitted sideways.

Figs. 14 and 15 plot the calculated S parameters for the cross- and T junctions (magnitudes only), together with the results of Mehran [14] as comparison. These results are calculated directly from the definition of each S parameter using the transformed time-domain fields in a way similar to the open-end case. The reference planes for the networks which represent the discontinuities are indicated in Fig. 1.

The independent S parameters for the cross-junction are S_{11}, S_{21}, and S_{31}. From Fig. 14, it is seen that these three S parameters all acquire the value 0.5 at dc and very low frequencies, indicating that the four branches of the cross-junction each get an equal $1/4$ share of the total field energy; thus the U condition for the S matrix is well satisfied, and no (detectable) radiation occurs at those frequency ranges. The same is not true for higher frequency ranges.

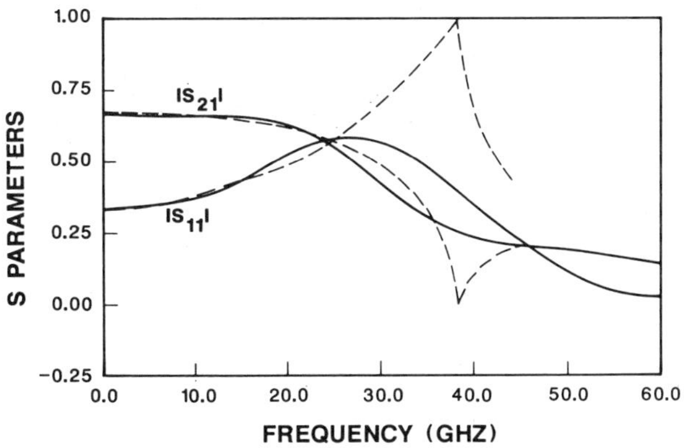

Fig. 15. Frequency-dependent S parameters of the microstrip T junction. Solid: time-domain result, dash: Mehran, $\epsilon_r = 9.7$, $h = 0.635$ mm, $W_1 = W_2 = 0.56$ mm [14].

The independent S parameters S_{11} and S_{21} for the T junction under symmetric excitation are plotted in Fig. 15. Again the U condition is checked to be satisfied in the very low frequency range.

From Figs. 14 and 15, it is seen that the time-domain results and the results of Mehran have very good agreement at dc and the lower frequency range. The discrepancies which occurred at very high frequencies are believed to be partly due to the fact that the waveguide model approach used in [14] is not able to take into account the radiation and surface wave generation effects which happened at higher frequency, and partly due to the fact that the time-domain results will not represent the exact dominant mode parameters after the inflection frequency, which is around 35 GHz for this configuration. The fact that the T junction results have earlier and larger discrepancy further confirmed this point, because here the surface wave becomes another reason for the waveguide model to fail, adding to the radiation loss.

C. Microstrip Step-in-Width and Gaps

Figs. 16 and 17 show the calculated time-domain fields and S parameters for the microstrip step-in-width ($W_1/H = 1.0$, $W_2/W_1 = 2.0$) using a mesh with size similar to that of T and cross-junction calculations. The S parameters for the step-in-width are defined as

$$S_{11} = \frac{V_{1\,\mathrm{ref}}(f)}{V_{1\,\mathrm{inc}}(f)} \tag{22}$$

$$S_{21} = \frac{\dfrac{V_{2\,\mathrm{trans}}(f)}{\sqrt{Z_{02}(f)}}}{\dfrac{V_{1\,\mathrm{inc}}(f)}{\sqrt{Z_{01}(f)}}} \tag{23}$$

$$S_{22} = \frac{V_{2\,\mathrm{ref}}(f)}{V_{2\,\mathrm{inc}}(f)} \tag{24}$$

$$S_{12} = S_{21} \tag{25}$$

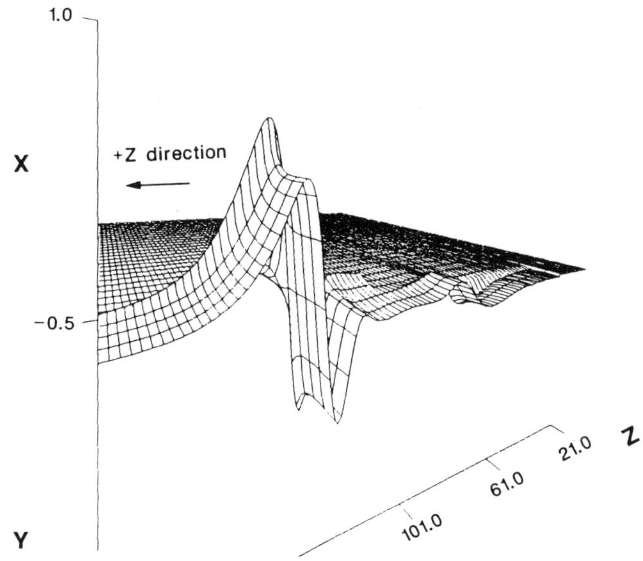

Fig. 16. Gaussian pulse propagation and reflection in the microstrip step-in-width structure: E_x component just underneath the strip.

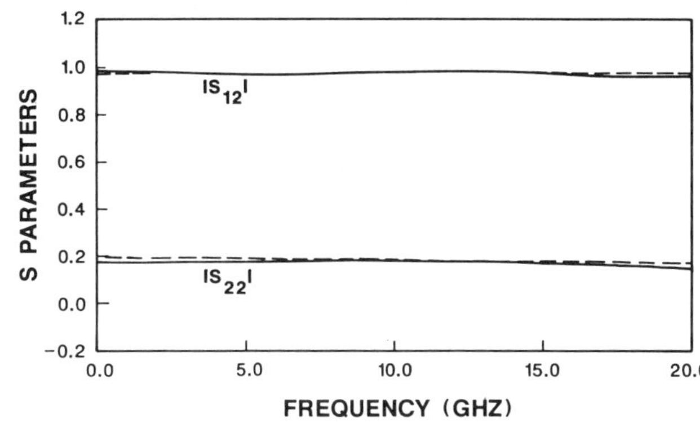

(a)

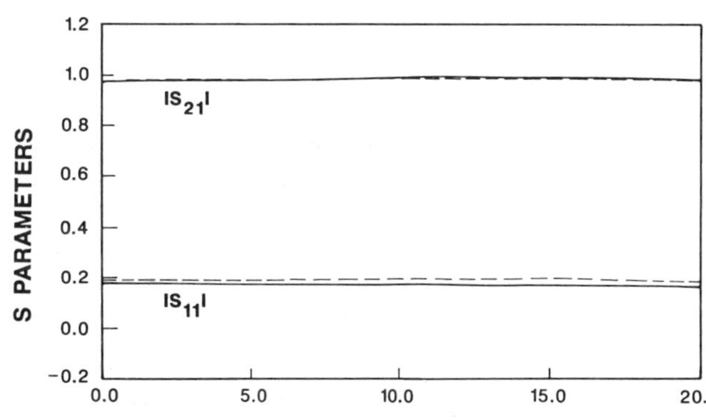

(b)

Fig. 17. Frequency-dependent S parameters of the microstrip step-in-width. Solid: time-domain result, dash: Koster and Jansen, $\epsilon_r = 10$ [33].

where $Z_{01}(f)$ and $Z_{02}(f)$ are the characteristic impedances of the microstrip lines connected to port 1 and 2 of the step, respectively, and are calculated in the same way as shown in [26].

The calculated frequency dependence of the S parameters of the step-in-width is quite flat over a large frequency

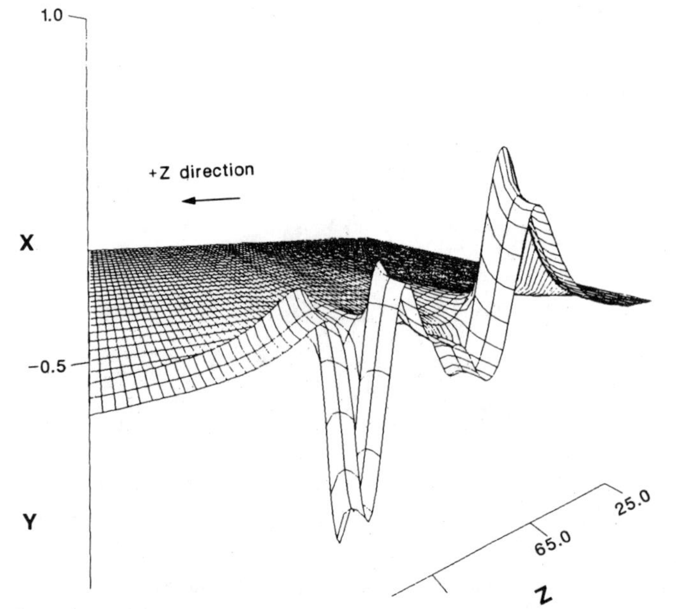

Fig. 18. Gaussian pulse propagation and reflection in the microstrip gap: E_x component just underneath the strip.

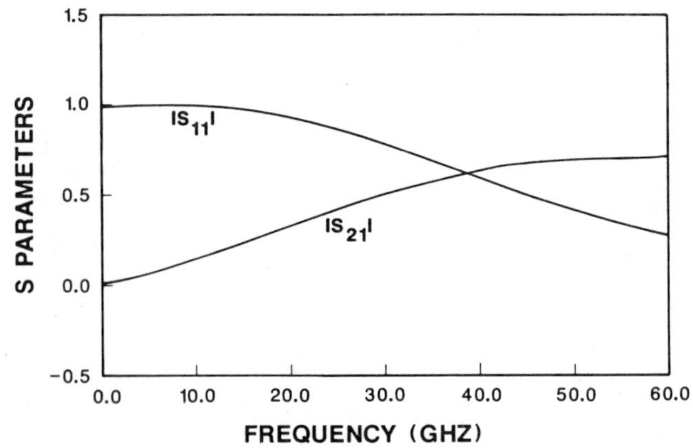

Fig. 19. Frequency-dependent S parameters of the microstrip gap.

range. It is in good agreement with the result of Koster and Jansen [33].

Figs. 18 and 19 are the calculated time-domain fields and S parameters for the microstrip gap discontinuity ($S/H = 0.5$). The dc block characteristics of the gap are exhibited in the S parameter results.

IV. Conclusion

It has been shown that the time-domain finite difference approach is capable of calculating the dispersion characteristics of the microstrip discontinuities over a large frequency range. It is a very general method and can find wide applications in modeling various microwave components. Further investigations on different computer-memory-saving schemes will make this method more suitable for CAD purposes.

Appendix

To calculate the E_y and E_z components on the dielectric–air interface, we start with the Maxwell equation

$$\frac{\partial \vec{E}}{\partial t} = \frac{1}{\epsilon_i} \nabla \times \vec{H} \tag{A1}$$

where $i = 1, 2$ denotes the dielectric constant in the substrate and the air region, respectively. Taking the calculation of E_y as an example, we have, from (A1),

$$\frac{\partial E_y}{\partial t} = \frac{1}{\epsilon_i}\left(\frac{\partial H_x}{\partial z} - \frac{\partial H_z}{\partial x}\right). \tag{A2}$$

Since E_y, H_x and $\partial H_x/\partial z$ are continuous across the interface, it is very obvious that $\partial H_z/\partial x$ is discontinuous across the boundary; i.e., we can get from (A2)

$$\frac{1}{\epsilon_2}\left(\frac{\partial H_z}{\partial x}\right)_2 - \frac{1}{\epsilon_1}\left(\frac{\partial H_z}{\partial x}\right)_1 = \left(\frac{1}{\epsilon_2} - \frac{1}{\epsilon_1}\right)\frac{\partial H_x}{\partial z} \neq 0. \tag{A3}$$

Therefore on each side of the interface and at positions very close to the boundary, we can write

$$\epsilon_1 \frac{\partial E_y}{\partial t} = \frac{\partial H_x}{\partial z} - \left(\frac{\partial H_z}{\partial x}\right)_1$$
$$\epsilon_2 \frac{\partial E_y}{\partial t} = \frac{\partial H_x}{\partial z} - \left(\frac{\partial H_z}{\partial x}\right)_2. \tag{A4}$$

Approximate $(\partial H_z/\partial_x)_1$ and $(\partial H_z/\partial_x)_2$ by

$$\left(\frac{\partial H_z}{\partial x}\right)_1 \approx \frac{H_z(m) - H_z(m-1/2)}{\frac{\Delta x}{2}}$$
$$\left(\frac{\partial H_z}{\partial x}\right)_2 \approx \frac{H_z(m+1/2) - H_z(m)}{\frac{\Delta x}{2}} \tag{A5}$$

where m is assumed to be the position of the interface, and $m+1/2$ and $m-1/2$ denote the positions a half step above and below the interface, respectively.

Substituting (A5) into (A4), we get

$$H_z(m) \approx \frac{\epsilon_1}{\epsilon_1 + \epsilon_2} H_z(m+1/2) + \frac{\epsilon_2}{\epsilon_1 + \epsilon_2} H_z(m-1/2) + \frac{\epsilon_2 - \epsilon_1}{\epsilon_1 + \epsilon_2}\frac{\partial H_x}{\partial z}\Delta x/2. \tag{A6}$$

Substituting the $H_z(m)$ value of (A6) back into (A5), substituting the resulting (A5) into the two expressions in (A4), and adding the two together, we get

$$\frac{\epsilon_1 + \epsilon_2}{2} \cdot \frac{\partial E_y}{\partial t} = \frac{\partial H_x}{\partial z} - \frac{H_z(m+1/2) - H_z(m-1/2)}{\Delta x} = \frac{\partial H_x}{\partial z} - \frac{\Delta H_z}{\Delta x}. \tag{A7}$$

Similarly, we get

$$\frac{\epsilon_1 + \epsilon_2}{2} \cdot \frac{\partial E_z}{\partial t} = \frac{\Delta H_y}{\Delta x} - \frac{\partial H_x}{\partial y}. \tag{A8}$$

Thus we can further discretize (A7) and (A8) to calculate the tangential $\vec{E}$ components on the dielectric–air interface.

It is straightforward to show that this boundary approximation is of first order in space and second order in time.

REFERENCES

[1] J. J. Campbell, "Application of the solutions of certain boundary value problems to the symmetrical four-port junction and specially truncated bends in parallel-plate waveguides and balanced strip-transmission lines," *IEEE Trans. Microwave Theory Tech.*, vol. MTT-16, pp. 165–176, Mar. 1968.

[2] L. S. Napoli and J. J. Hughes, Foreshortening of microstrip open circuits on alumina substrate," *IEEE Trans. Microwave Theory Tech.*, vol. MTT-19, pp. 559–561, June 1971.

[3] D. S. James and S. H. Tse, "Microstrip end effects," *Electron. Lett.*, vol. 8, pp. 46–47, Jan. 27, 1972.

[4] P. Silvester and P. Benedek, "Equivalent capacitances of microstrip open circuits," *IEEE Trans. Microwave Theory Tech.*, vol. MTT-20, pp. 511–516, Aug. 1972.

[5] T. Itoh, R. Mittra, and R. D. Ward, "A method for computing edge capacitance of finite and semi-infinite microstrip lines," *IEEE Trans. Microwave Theory Tech.*, pp. 847–849, Dec. 1972.

[6] C. Gupta and A. Gopinath, "Equivalent circuit capacitance of microstrip step change in width," *IEEE Trans. Microwave Theory Tech.*, vol. MTT-25, pp. 819–822, Oct. 1977.

[7] A. A. F. Thomson and A. A. Gopinath, "Calculation of microstrip discontinuity inductances," *IEEE Trans. Microwave Theory Tech.*, vol. MTT-23, pp. 648–655, Aug. 1975.

[8] P. Stouten, "Equivalent capacitances of T junctions," *Electron. Lett.*, vol. 9, pp. 552–553, Nov. 1973.

[9] I. Wolff, A. G. Kompa, and R. Mehran, "Calculation method for microstrip discontinuities and T-junctions," *Electron. Lett.*, vol. 8, pp. 177, Apr. 1972.

[10] P. Silvester and P. Benedek, "Microstrip discontinuity capacitances for right-angle bends, T junctions, and crossings," *IEEE Trans. Microwave Theory Tech.*, vol. MTT-21, pp. 341–346, May 1973.

[11] M. Maeda, "An analysis of gap in microstrip transmission lines," *IEEE Trans. Microwave Theory Tech.*, vol. MTT-20, pp. 390–396, June 1972.

[12] A. Farrar and A. T. Adams, "Matrix method for microstrip three-dimensional problems," *IEEE Trans. Microwave Theory Tech.*, vol. MTT-20, pp. 497–504, Aug. 1972.

[13] G. Kompa and R. Mehran, "Planar waveguide model for calculating microstrip components," *Electron. Lett.*, vol. 11, pp. 459–460, Dept. 1975.

[14] R. Mehran, "The frequency-dependent scattering matrix of microstrip right-angle bends, T-junctions and crossings," *Arch. Elek. Übertragung.*, vol. 29, pp. 454–460, Nov. 1975.

[15] W. Menzel and I. Wolff, "A method for calculating the frequency dependent properties of microstrip discontinuities," *IEEE Trans. Microwave Theory Tech.*, vol. MTT-25, pp. 107–112, Feb. 1977.

[16] J. R. James and A. Henderson, "High-frequency behavior of microstrip open-circuit terminations," *IEE J. Microwave Opt. Acoust.*, vol. 3, pp. 205–211, Sept. 1979.

[17] R. H. Jansen, "The spectral-domain approach for microwave integrated circuits," *IEEE Trans. Microwave Theory Tech.*, vol. MTT-33, pp. 1043–1056, Oct. 1985.

[18] R. H. Jansen, "Hybrid mode analysis of end effects of planar microwave and millimeterwave transmission lines," *Proc. Inst. Elec. Eng.*, vol. 128, pt. H, no. 2, pp. 77–86, Apr. 1981.

[19] J. S. Hornsby, "Full-wave analysis of microstrip resonator and open-circuit end effect," *Proc. Inst. Elec. Eng.*, vol. 129, pt. H, no. 6, pp. 338–341, Dec. 1982.

[20] R. W. Jackson and D. M. Pozar, "Full-wave analysis of microstrip open-end and gap discontinuities," *IEEE Trans. Microwave Theory Tech.*, vol. MTT-33, pp. 1036–1042, Oct. 1985.

[21] P. B. Katehi and N. G. Alexopoulos, "Frequency-dependent characteristics of microstrip discontinuities in millimeter-wave integrated circuit," *IEEE Trans. Microwave Theory Tech.*, vol. MTT-33, pp. 1029–1035, Oct. 1985.

[22] K. S. Yee, "Numerical solution of initial boundary value problems involving Maxwell's equations in isotropic media," *IEEE Trans. Antennas Propagat.*, vol. AP-14, pp. 302–307, May 1966.

[23] A. Taflove and M. E. Brodwin, "Numerical solution of steady-state electromagnetic scattering problems using the time-dependent Maxwell's equations," *IEEE Trans. Microwave Theory Tech.*, vol. MTT-23, pp. 623–630, 1975.

[24] S. Koike, N. Yoshida, and I. Fukai, "Transient analysis of microstrip gap in three-dimensional space," *IEEE Trans. Microwave Theory Tech.*, vol. MTT-33, pp. 726–730, Aug. 1985.

[25] S. Koike, N. Yoshida, and I. Fukai, "Transient analysis of coupling between crossing lines in three-dimensional space," *IEEE Trans. Microwave Theory Tech.*, vol. MTT-35, pp. 67–71, Jan. 1987.

[26] X. Zhang, J. Fang, K. K. Mei, and Y. Liu, "Calculation of the dispersive characteristics of microstrips by the time-domain finite difference method," *IEEE Trans. Microwave Theory Tech.*, vol. MTT-36, pp. 263–267, Feb. 1988.

[27] L. Lapidus and G. H. Pinder, *Numerical Solution of Partial Differential Equations in Science and Engineering.* New York: Wiley, 1982.

[28] J. Fang, X. Zhang, K. K. Mei, and Y. Liu, "Time domain computation and numerical dispersion," presented at XXIInd General Assembly of the International Union of Radio Science, Tel Aviv, Israel, Aug. 24–Sept. 2, 1987.

[29] X. Zhang, "Time-domain finite difference calculation of the frequency-dependent characteristics of the microstrip discontinuities," master's report, Dept. EECS, University of California, Berkeley, Dec. 1987.

[30] K. K. Mei and J. Fang, "A super-absorption boundary algorithm for one dimensional wave equations," submitted to *J. Comput. Phys.*

[31] C. C. Lin, "Numerical modeling of two-dimensional time domain electromagnetic scattering by underground inhomogeneities," Ph.D dissertation, Dept. EECS, University of California, Berkeley, 1985.

[32] K. C. Gupta, R. Garg, and I. J. Bahl, *Microstrip Lines and Slotlines.* Dedham, MA: Artech House, 1979.

[33] N. H. L. Koster and R. H. Jansen, "The microstrip step discontinuity: A revised description," *IEEE Trans. Microwave Theory Tech.*, vol. MTT-34, pp. 213–223, Feb. 1986.

Analysis of Microstrip Circuits Using Three-Dimensional Full-Wave Electromagnetic Field Analysis in the Time Domain

TSUGUMICHI SHIBATA, MEMBER, IEEE, TOSHIO HAYASHI, AND TADAKATSU KIMURA, MEMBER, IEEE

Abstract —**Calculation of the frequency characteristics for microstrip circuits based on a three-dimensional full-wave electromagnetic field analysis in the time domain is proposed. In this method, the circuit is excited by a pulse which includes broadened frequency components. The frequency characteristics are then computed at once from the Fourier transform of the output transient responses. To evaluate the validity and capability of the method, a side-coupled microstrip filter is analyzed and the frequency characteristics are calculated. A quasi-static analysis of this filter is also presented and the results compared with measurements. The frequency characteristics calculated with the full-wave analysis in the time domain show excellent agreement with the measured values, thus demonstrating the validity and the power of the analytical method.**

I. Introduction

With the development of high-speed device technology, it is becoming increasingly important to develop high-performance MMIC's which will serve as key components in compact and low-cost communication systems of the future. In MMIC's, microstrip circuits such as transmission lines (i.e., interconnect), filters, and couplers are used as circuit elements. Circuit designers are turning their attention to techniques for precisely analyzing their electrical properties. Some simple circuits have already been analyzed using quasi-static analysis. In most practical situations, however, because of the complicated electromagnetic fields that result from their three-dimensional shapes and locations, precise analysis cannot be expected from the more common analytical methods, or the so-called quasi-static analysis method. For greater accuracy, computer-aided numerical analysis of full-wave three-dimensional electromagnetic fields is required. This numerical analysis technique is thus an important key for the development of MMIC designs.

Two approaches have been taken with respect to full-wave electromagnetic field analysis. One is the time-domain analysis approach which solves the initial boundary value problem of the wave or Maxwell's equations. The other approach is a frequency-domain analysis method based on the eigenvalue problem of the vector Helmholtz equation. One of the advantages of the time-domain analysis, which is the main reason we have employed it here, is that the analysis is stable and yields a unique result, while the frequency-domain analysis suffers from spurious solutions and considerable care must be taken in the calculation process.

The numerical formulation of the time-domain analysis is given by Yee (the FD-TD method) [1], which uses finite difference equations of Maxwell's equations. An alternative method is based on the equivalent circuits of electromagnetic fields [2]. Johns *et al.* constructed an equivalent circuit with finite length transmission lines and developed a method of analysis based on the scattering matrix of each line (the T.L.M. method) [3]–[7]. These methods have been substantially modified in a number of reports [8]–[11]. Especially, Yoshida, Fukai and their coworkers have refined the T.L.M. method by expressing the medium constants by additional lumped elements to the equivalent circuit. They call their approach the Bergeron method. They also applied this method to the analysis of microstrip circuits and successfully showed the time variation of electromagnetic fields and power flow around the circuits [12]–[16]. But, for microstrip circuit design, it is necessary to calculate the frequency characteristics of the circuit. Since their analysis uses sinusoidal sources to excite the circuits, it is insufficient for this purpose.

The aim of this paper is to demonstrate calculation of the frequency characteristics for microstrip circuits and to establish the frequency-domain MMIC design method based on time-domain electromagnetic field analysis. The frequency characteristics are calculated using a raised cosine pulse source and applying the Fourier transform to the transient response of the circuit. In this way the frequency characteristics can be obtained at once over the full range of frequencies, whereas only one point of the frequency characteristic can be given using a sinusoidal source. By comparing the computed results and measured

Manuscript received October 20, 1987; revised January 19, 1988.

The authors are with NTT LSI Laboratories, 3-1 Morinosato Wakamiya, Atsugi-shi, Kanagawa, 243-01, Japan.

IEEE Log Number 8821075.

Reprinted from *IEEE Transactions on Microwave Theory and Techniques*, pp. 1064-1070, June 1988.

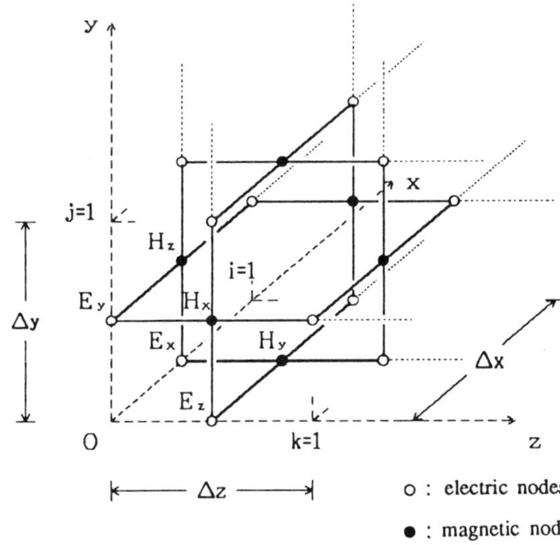

Fig. 1. The net of an equivalent circuit. Electric nodes,

$$E_x\left(i+\frac{1}{2}, j, k\right), \quad i=0,1,\cdots,L-1;\ j=0,1,\cdots,M;\ k=0,1,\cdots,N$$

$$E_y\left(i, j+\frac{1}{2}, k\right), \quad i=0,1,\cdots,L;\ j=0,1,\cdots,M-1;\ k=0,1,\cdots,N$$

$$E_z\left(i, j, k+\frac{1}{2}\right), \quad i=0,1,\cdots,L;\ j=0,1,\cdots,M;\ k=0,1,\cdots,N-1$$

and magnetic nodes,

$$H_x\left(i, j+\frac{1}{2}, k+\frac{1}{2}\right), \quad i=0,1,\cdots,L;\ j=0,1,\cdots,M-1;\ k=0,1,\cdots,N-1$$

$$H_y\left(i+\frac{1}{2}, j, k+\frac{1}{2}\right), \quad i=0,1,\cdots,L-1;\ j=0,1,\cdots,M;\ k=0,1,\cdots,N-1$$

$$H_z\left(i+\frac{1}{2}, j+\frac{1}{2}, k\right), \quad i=0,1,\cdots,L-1;\ j=0,1,\cdots,M-1;\ k=0,1,\cdots,N$$

are indicated by open circles and closed circles, respectively.

data, the time-domain analysis scheme can be verified. A side-coupled microstrip filter will be analyzed as an example to demonstrate the practicality of the scheme.

In the next section (Section II), we will briefly review the Bergeron method. The analysis of a single microstrip line is described in Section III as a preliminary to the analysis of a side-coupled microstrip filter in Section IV. In Section V, the results of the quasi-static analysis are presented and the necessity of the full-wave analysis is emphasized. In Section VI, we will discuss the accuracy of the full-wave, time-domain analysis. Finally, a few concluding remarks will be offered in Section VII.

II. The Equivalent Circuit of the Electromagnetic Field and Numerical Scheme of Analysis

In this section, we review the Bergeron method [12], which is used to solve for three-dimensional electromagnetic fields in the time domain using the equivalent circuits of the fields. Fig. 1 shows an equivalent circuit disposed in three-dimensional space. There are six kinds of nodes in the equivalent circuit: three electric nodes, E_x, E_y, and E_z, which are indicated by open circles in the figure, and three magnetic nodes H_x, H_y, and H_z, shown as closed circles. Branches consist of $\Delta l/2$ length transmission lines and gyrators, as shown in detail in Fig. 2. Values of the dielectric constant ϵ and permeability μ are expressed in terms of lumped capacitances connected to the corresponding electric and magnetic nodes, respectively. Electromagnetic field components E_x, E_y, E_z and H_x, H_y, H_z correspond to the node voltages of the equivalent circuit V_x, V_y, V_z and V_x^*, V_y^*, V_z^*, respectively, at each point (the * denoting the variables at the magnetic nodes). Electromagnetic field analysis results in the calculation of this equivalent circuit response. Fig. 3 shows the equivalent circuit at the E_x node. The following equations describe the node voltages and branch currents at the E_x node.

$$V_x(i,j,k,t)+Z_0I_{yi}(i,j,k,t) = I_{yo}^*\left(i, j-\frac{1}{2}, k, t-\Delta t\right) + Z_0V_z^*\left(i, j-\frac{1}{2}, k, t-\Delta t\right) \tag{1}$$

$$V_x(i,j,k,t)-Z_0I_{yo}(i,j,k,t) = I_{yi}^*\left(i, j+\frac{1}{2}, k, t-\Delta t\right) - Z_0V_z^*\left(i, j+\frac{1}{2}, k, t-\Delta t\right) \tag{2}$$

$$V_x(i,j,k,t)+Z_0I_{zi}(i,j,k,t) = I_{zo}^*\left(i, j, k-\frac{1}{2}, t-\Delta t\right) + Z_0V_y^*\left(i, j, k-\frac{1}{2}, t-\Delta t\right) \tag{3}$$

$$V_x(i,j,k,t)-Z_0I_{zo}(i,j,k,t) = I_{zi}^*\left(i, j, k-\frac{1}{2}, t-\Delta t\right) - Z_0V_y^*\left(i, j, k-\frac{1}{2}, t-\Delta t\right) \tag{4}$$

$$V_x(i,j,k,t)-\frac{\Delta t}{4C(i,j,k)}I_c(i,j,k,t) = V_x(i,j,k,t-\Delta t) + \frac{\Delta t}{4C(i,j,k)}I_c(i,j,k,t-\Delta t) \tag{5}$$

$$I_{yi}(i,j,k,t)-I_{yo}(i,j,k,t) + I_{zi}(i,j,k,t)-I_{zo}(i,j,k,t)-I_c(i,j,k,t)=0 \tag{6}$$

where Z_0 is the characteristic impedance of the transmission line and C is given by

$$C=(\epsilon_s-1)\epsilon_0\Delta \tag{7}$$

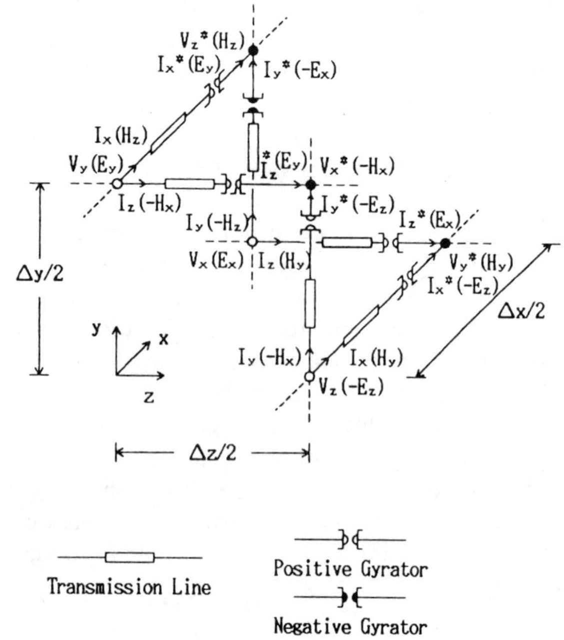

Fig. 2. Equivalent circuit elements. The equivalent circuit consists of finite length transmission lines and gyrators. $V-I$ characteristics of gyrators are

$$\begin{bmatrix} V \\ I \end{bmatrix} = \begin{bmatrix} 0 & 1 \\ 1 & 0 \end{bmatrix} \begin{bmatrix} V^* \\ I^* \end{bmatrix}$$

for the positive gyrator and

$$\begin{bmatrix} V \\ I \end{bmatrix} = \begin{bmatrix} 0 & -1 \\ -1 & 0 \end{bmatrix} \begin{bmatrix} V^* \\ I^* \end{bmatrix}$$

for the negative gyrator. Correspondence of circuit variables to electromagnetic field components is also shown in this figure.

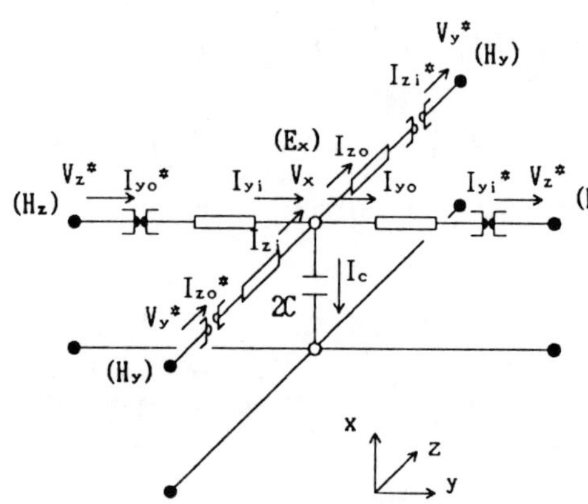

Fig. 3. Equivalent circuit at the E_x node.

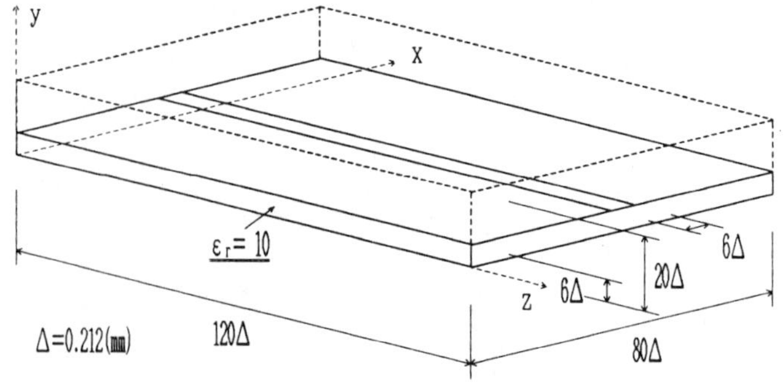

Fig. 4. Configuration of the single microstrip line.

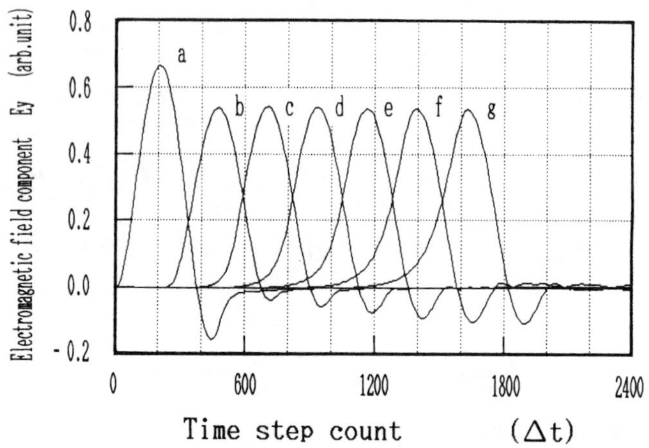

Fig. 5. The time variations of the electromagnetic field component E_y. The waves (a)–(g) are at the point $z = 0$, 20Δ, 40Δ, 60Δ, 80Δ, 100Δ, and 120Δ along the center line of the microstrip conductor in the substrate, respectively.

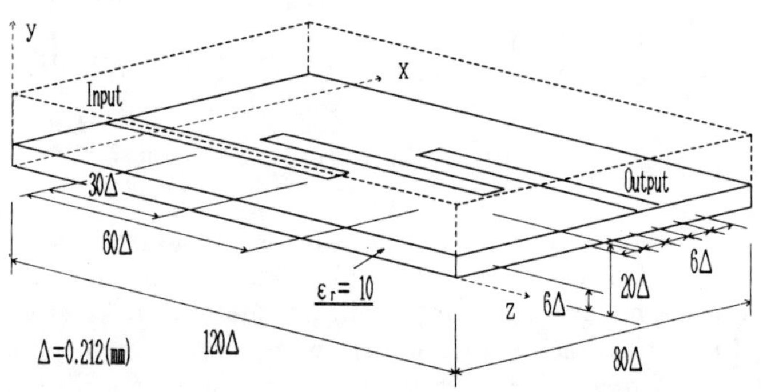

Fig. 6. Configuration of the side-coupled microstrip filter.

at the electric nodes and

$$C = (\mu_s - 1)\mu_0 \Delta \tag{8}$$

at the magnetic nodes, where $\epsilon = \epsilon_s \epsilon_0$ and $\mu = \mu_s \mu_0$. The simulation time step Δt is related to the space distance Δ as

$$\Delta t = \frac{\Delta}{4c}. \tag{9}$$

We denoted the variables at the space point $x = i$, $y = j$, $z = k$ and at the time t as $V(i, j, k, t)$. Similar equations can be written relating to the E_y, E_z, H_x, H_y and H_z nodes. From these equations, we can calculate the node voltages at t from $t - \Delta t$.

III. Analysis of a Single Microstrip Line

Prior to the description of a side-coupled microstrip filter, we discuss the results of a single microstrip line analysis. The configuration of the analyzed microstrip line is shown in Fig. 4. The net size of the equivalent circuit used here is $80\Delta x \times 20\Delta y \times 120\Delta z$ ($\Delta x = \Delta y = \Delta z = \Delta$). At the top and side faces of the equivalent circuit, irregular nodes are ended with matching impedances Z_0; thus, the infinite boundary condition is almost realized. The micro-

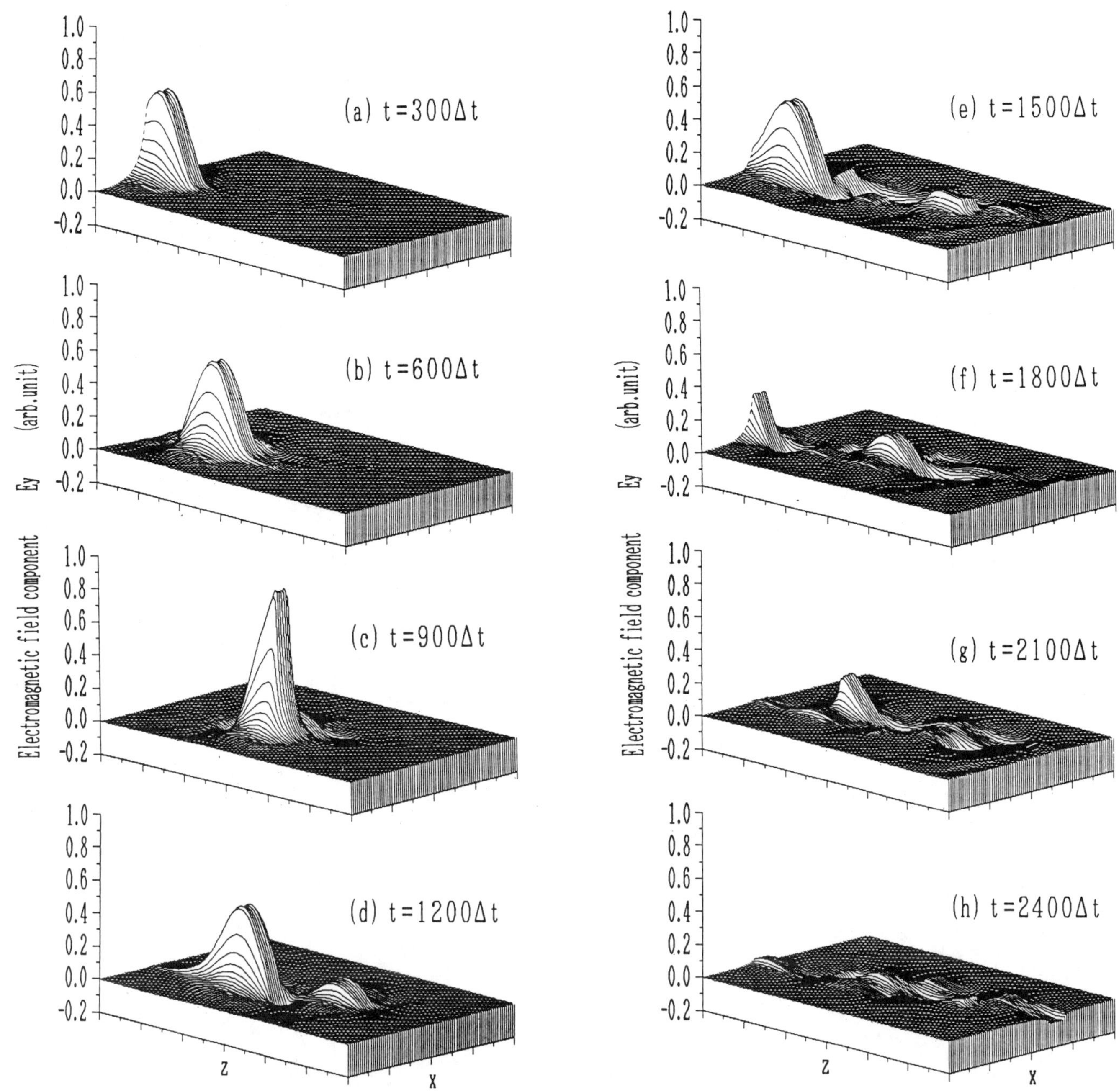

Fig. 7. The time variation of electromagnetic field component E_y at the $z-x$ plane in the substrate.

strip line parameters are as follows:

width of the microstrip conductor: $W=6\Delta$;
thickness of the substrate: $H=6\Delta$;
dielectric constant of the substrate: $\epsilon_r=10.0$;
length of each transmission line in the equivalent circuit: $\Delta/2=0.106$ (mm).

Then, the simulation time step Δt becomes 0.176 (ps) from (9). The boundary conditions of the microstrip conductor and the bottom plane of the substrate are realized by short-circuiting the E_x and E_z nodes and open-circuiting the H_y node on the plane. As an input signal, a raised cosine pulse which equals a band-limited wave was adopted of the form

$$\begin{aligned} E_y(t) &= 1-\cos(2\pi f_{\text{band}}t) \qquad \text{for } 0 \leqslant t < 1/f_{\text{band}} \\ &= 0 \qquad \text{for } 1/f_{\text{band}} \leqslant t \\ f_{\text{band}} &= 11.8\ (\text{GHz}). \end{aligned} \tag{10}$$

Here, an internal impedance of Z_0 is applied to the E_y nodes under the microstrip conductor at the $z=0$ plane, and initial values of the other field components are made zero. We applied only the E_y component under the microstrip line conductor to simplify the procedure instead of the total external electromagnetic field. The time variations of the electromagnetic field component E_y at the points under the center line of the microstrip conductor are plotted in Fig. 5.

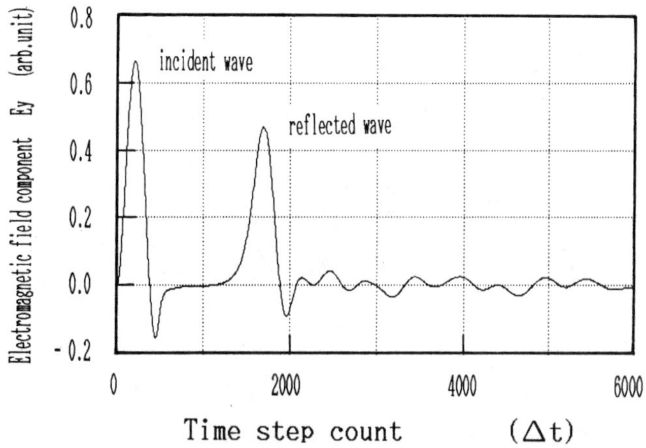

Fig. 8. The time variation of E_y at the input end.

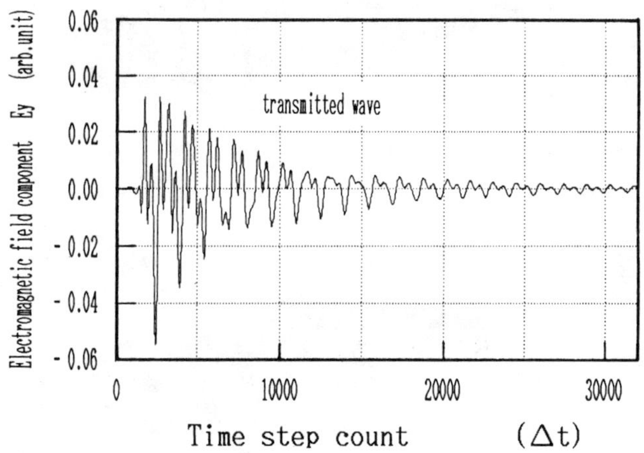

Fig. 9. The time variation of E_y at the output end.

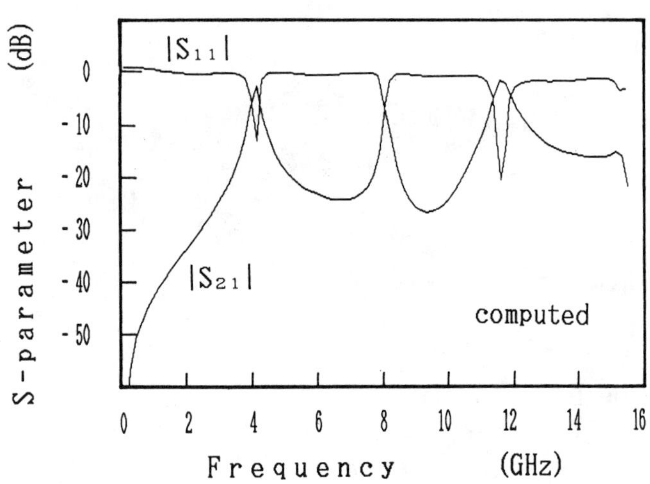

Fig. 10. Computed frequency characteristics of $|S_{21}|$ and $|S_{11}|$.

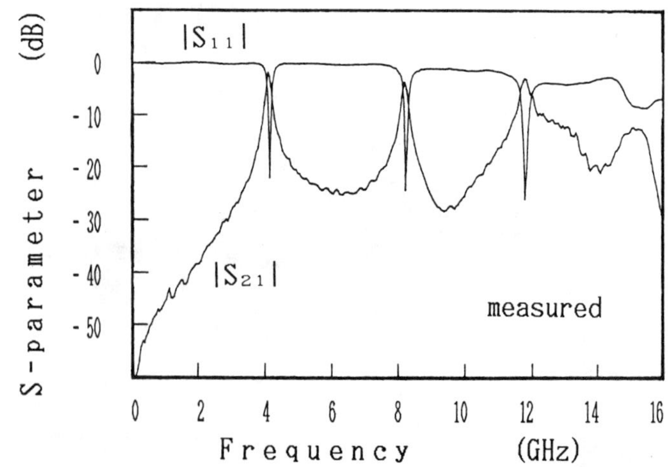

Fig. 11. Measured frequency characteristics of $|S_{21}|$ and $|S_{11}|$.

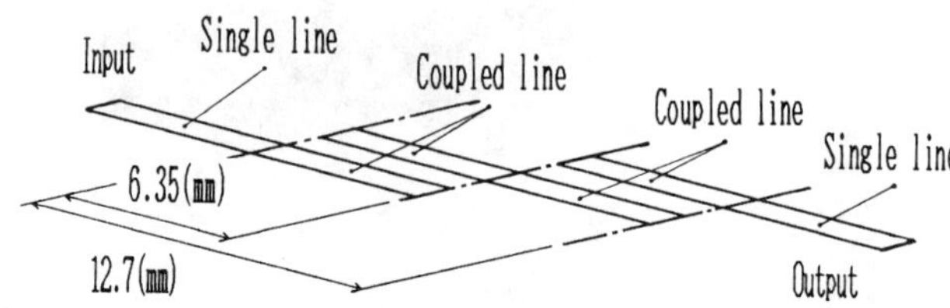

Fig. 12. Quasi-static analysis of the side-coupled microstrip filter. The filter is divided into four parts consisting of single and coupled lines.

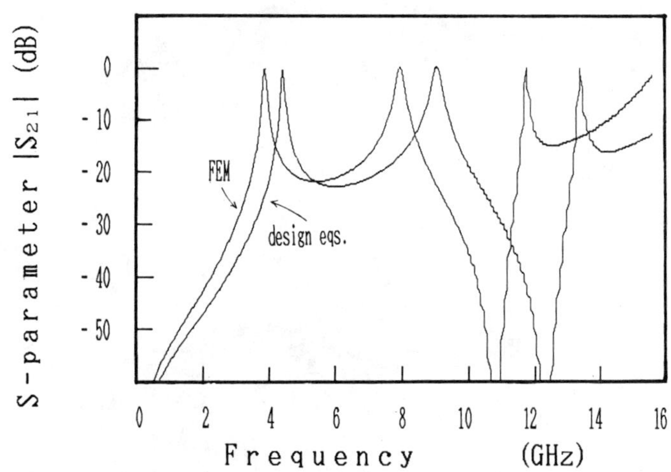

Fig. 13. The frequency characteristics of $|S_{21}|$ obtained by quasi-static analysis.

As a result of simplification of the input signal, the abrupt pulse height change from wave (a) to (b) has been reduced. The initial field component E_y was spread in space, thus creating a fringing field. The rest of the component was scattered and lost as radiation. The pulse height decreases rapidly near the input end and settles after propagating some distance (a few Δ lengths) along the microstrip in the z direction. At that point, the electromagnetic field distribution of the propagating mode is realized. We will take account of this pulse height reduction to calculate the frequency characteristics of the circuit in the next section.

Notice also that the dispersion of the microstrip line can be simulated, which smoothes the slope of the pulse front and ripples the pulse tail because low-frequency components propagate faster than high-frequency components. Compare wave (g) with wave (b) in Fig. 5, for example.

IV. Analysis of a Side-Coupled Microstrip Filter

Here we describe the results of a side-coupled microstrip filter analysis. The configuration of the side-coupled microstrip filter is illustrated in Fig. 6. The width of the

TABLE I
PARAMETERS USED IN THE QUASI-STATIC ANALYSIS

	Single line		Coupled line			
	capacitance [pF/m]	inductance [nH/m]	self capacitance [pF/m]	mutual capacitance [pF/m]	self inductance [nH/m]	mutual inductance [nH/m]
Design equations	178.	422.	173.	11.0	420.	80.9
FEM	176.	469.	163.	11.8	572.	114.

microstrip line and the thickness and dielectric constant of the substrate are the same as the single microstrip line in the previous section. Other important size parameters are

space between microstrip conductors: $S = 6\Delta$;
length of the center microstrip conductor: $L = 60\Delta$.

The input signal given by (10) is applied at the input end and the electromagnetic field is computed step by step in time. The time variation of the electromagnetic field component E_y at the $z-x$ plane in the substrate is drawn in Fig. 7, in which (a)–(c) show that the incident wave propagates along the first microstrip conductor and is reflected at the end of the line. A part of the wave transfers to the center microstrip, while the reflected wave returns to the input end, in (d)–(e). The reflected wave fades away from view and the remainder resonates around the center microstrip in (f)–(g). The resonant component transfers to the third microstrip line and is transmitted to the output end in (h).

Figs. 8 and 9 are plots of E_y at input and output ends, respectively. From these waves, the frequency characteristics of the filter can be calculated. S parameters can be calculated from the Fourier transform of the reflected wave in Fig. 8 and the transmitted wave in Fig. 9, normalizing by the frequency spectrum of the incident wave (b) in Fig. 5 to correct the effect of the pulse height reduction discussed previously. Computed S parameters $|S_{11}|$ and $|S_{21}|$ are shown in Fig. 10.

In order to verify this computational method, we fabricated the side-coupled microstrip filter and measured $|S_{11}|$ and $|S_{21}|$ using the network analyzer. The results are shown in Fig. 11. The computed results show excellent agreement with the measured data.

V. THE RESULTS OF QUASI-STATIC ANALYSIS

In this section we describe the quasi-static analysis of the side-coupled microstrip filter which was previously analyzed using the time-domain, full-wave analysis. Then the findings based on the two different methods will be compared. Here the side-coupled microstrip filter is divided into four parts, consisting of single and coupled microstrip lines, as shown in Fig. 12. Single and coupled line parameters for a unit length can be estimated from static line capacitances on the assumption of TEM mode propagation. We calculated two sets of line capacitances from the design equations in [17] and the static field analysis using the finite element method (FEM) in [18]. The line inductance of the single line is calculated by

$$L = \frac{1}{c^2 C_{\text{air}}}. \tag{11}$$

The self- and mutual inductances of the coupled line are given by

$$L_{\text{self}} = \frac{1}{2c^2}\left(\frac{1}{C_{\text{self, air}} - C_{\text{mutual, air}}} + \frac{1}{C_{\text{self, air}} + C_{\text{mutual, air}}}\right) \tag{12}$$

$$L_{\text{mutual}} = \frac{1}{2c^2}\left(\frac{1}{C_{\text{self, air}} - C_{\text{mutual, air}}} - \frac{1}{C_{\text{self, air}} + C_{\text{mutual, air}}}\right) \tag{13}$$

where c is the velocity of an electromagnetic wave in free space, and "air" means the capacitance of the line configuration with the dielectric substrate replaced by air. Calculated parameters are summarized in Table I. We obtained the $|S_{21}|$ of the side-coupled microstrip filter by circuit simulation using SPICE. The results are shown in Fig. 13. The differences between the inductances calculated by design equations and those by our numerical calculation using FEM range from 10 to 30 percent. However, both of the $|S_{21}|$ in Fig. 13 show the same tendency except for the scale of frequencies. Also important to note is that the results of quasi-static analysis are in disagreement with measured data, especially in the frequency domain over 8 GHz. These disagreements stem from the assumption of TEM mode propagation and/or from the separation of the filter into single and coupled lines. Thus, three-dimensional, full-wave analysis is needed for more precise computation.

VI. DISCUSSION

We turn now to a consideration of the accuracy of frequency characteristic calculations using time-domain, full-wave analysis. Comparing the computed $|S_{11}|$ in Fig. 10 to the measured $|S_{11}|$ in Fig. 11, some difference in the peak level at the resonance point can be observed, resulting from the lack of frequency resolution. According to Fourier transform theory, frequency resolution Δf is given by the inverse of the observed time period T. On the other hand the time step Δt depends on the branch length $\Delta/2$, as shown in (9). Moreover, the length of Δ should be made small enough to express not only the shapes and locations of the microstrips but also the electromagnetic field distribution using the network variables. Our simulations using other discretizations indicated that the accuracy of $|S_{21}|$, especially the resonance peak level around 8 GHz, depends on how many Δ's there are between microstrip lines. We need at least 5Δ lengths between isolated conductors. For this reason, it is important to make Δ sufficiently small, although it seems wasteful or oversampled to make the computational time step Δt too small and to require many iterations in time merely to obtain the frequency characteristics up to 16 GHz. To obtain the frequency characteristics shown in Fig. 10, we have computed the electromagnetic field distribution up to the $32768\Delta t$ time step. The resulting frequency resolution of Δf becomes 173 MHz.

VII. Conclusions

We propose an analysis method using pulse excitation in the time-domain, equivalent to a full-wave electromagnetic field analysis. The characteristics of a side-coupled microstrip filter were computed and excellent agreement with the measured values was obtained, thus confirming the validity of this analytical scheme. This example essentially requires the three-dimensional electromagnetic field treatment since no other method can provide the same level of precision. Moreover, the Bergeron method is effective for various media and is expected to produce various kinds of analysis, such as full-wave analysis in lossy conductors and dielectrics. Thus, the techniques described here are also promising for microstrip circuit analysis in the case of low conductivity of lines and/or semiconducting substrate.

Acknowledgment

The authors would like to thank Professor I. Fukai and Associate Professor N. Yoshida of Hokkaido University for their helpful discussions and A. Iwata, NTT LSI Laboratories, for his encouragement throughout this work.

References

[1] K. S. Yee, "Numerical solution of initial boundary value problems involving Maxwell's equations in isotropic media," *IEEE Trans. Antennas Propagat.*, vol. AP-14, pp. 302–307, May 1966.

[2] G. Kron, "Equivalent circuit of the field equations of Maxwell-I," *Proc. I.R.E.*, pp. 289–299, May 1944.

[3] P. B. Johns and R. L. Beurle, "Numerical solution of 2-dimensional scattering problems using a transmission-line matrix," *Proc. Inst. Elec. Eng.*, vol. 118, pp. 1203–1208, Sept. 1971.

[4] P. B. Johns, "The solution of inhomogeneous waveguide problems using a transmission-line matrix," *IEEE Trans. Microwave Theory Tech.*, vol. MTT-22, pp. 209–215, Mar. 1974.

[5] S. Akhtarzad and P. B. Johns, "Solution of Maxwell's equations in three space dimensions and time by t.l.m. method of numerical analysis," *Proc. Inst. Elec. Eng.*, vol. 122, pp. 1344–1348, Dec. 1975.

[6] S. Akhtarzad and P. B. Johns, "Generalised elements for t.l.m. method of numerical analysis," *Proc. Inst. Elec. Eng.*, vol. 122, pp. 1349–1352, Dec. 1975.

[7] P. B. Johns and G. Butler, "The consistency and accuracy of the TLM method for diffusion and its relationship to existing methods," *Int. J. Numer. Methods Eng.*, vol. 19, pp. 1549–1554, 1893.

[8] Y. Shih and W. J. R. Hoefer, "The accuracy of TLM analysis of finned rectangular waveguides," *IEEE Trans. Microwave Theory Tech.*, vol. MTT-28, pp. 743–746, July 1980.

[9] D. A. Al-Mukhtar and J. E. Sitch, "Transmission-line matrix method with irregularly graded space," *Proc. Inst. Elec. Eng.*, vol. 128, pp. 299–305, Dec. 1981.

[10] G. E. Mariki and C. Yeh, "Dynamic three-dimensional TLM analysis of microstrip- lines on anisotropic substrate," *IEEE Trans. Microwave Theory Tech.*, vol. MTT-33, pp. 789–799, Sept. 1985.

[11] W. J. R. Hoefer, "The transmission-line matrix method—Theory and applications," *IEEE Trans. Microwave Theory Tech.*, vol. MTT-33, pp. 882–893, Oct. 1985.

[12] N. Yoshida and I. Fukai, "Transient analysis of a stripline having a corner in three-dimensional space," *IEEE Trans. Microwave Theory Tech.*, vol. MTT-32, pp. 491–498, May 1984.

[13] S. Koike, N. Yoshida, and I. Fukai, "Transient analysis of microstrip gap in three-dimensional space," *IEEE Trans. Microwave Theory Tech.*, vol. MTT-33, pp. 726–730, Aug. 1985.

[14] S. Koike, N. Yoshida, and I. Fukai, "Transient analysis of a directional coupler using a coupled microstrip slot line in three-dimensional space," *IEEE Trans. Microwave Theory Tech.*, vol. MTT-34, pp. 353–357, Mar. 1986.

[15] S. Koike, N. Yoshida, and I. Fukai, "Transient analysis of microstrip side-coupled filter in three-dimensional space," *Trans. IECE Japan*, vol. E69, no. 11, pp. 1199–1205, Nov. 1986.

[16] S. Koike, N. Yoshida, and I. Fukai, "Transient analysis of coupling between crossing lines in three-dimensional space," *IEEE Trans. Microwave Theory Tech.*, vol. MTT-35, pp. 67–71, Jan. 1987.

[17] K. C. Gupta, R. Garg, and I. J. Bahl, *Microstrip Lines and Slotlines*. Dedham, MA: Artech House, 1979, pp. 87–89 and 337–340.

[18] P. Silvester, "High-order polynomial triangular finite elements for potential problems," *Int. J. Eng. Sci.*, vol. 7, pp. 849–861, 1969.

Characterization of a 90° Microstrip Bend with Arbitrary Miter via the Time-Domain Finite Difference Method

JOHN MOORE, STUDENT MEMBER, IEEE, AND HAO LING, MEMBER, IEEE

Abstract —A 90° microstrip bend with an arbitrary miter is characterized using the finite difference time-domain (FDTD) method. To simplify computations, the microstrip structure is enclosed by four electric walls; thus radiation effects are neglected. Time histories generated by FDTD are Fourier-transformed to yield broad-band S parameters of the microstrip bend. A miter is introduced to improve the transmission characteristics of the bend, and an optimal miter length is found such that the reflection from the microstrip bend over a broad frequency range is minimized.

I. Introduction

The finite difference time-domain (FDTD) method [1], [2] has become recognized as a useful tool for analyzing complicated wave interactions. The versatility of the FDTD method allows it to analyze microstrip line and microstrip discontinuities. Gwarek has developed a method for two-dimensional planar circuit analysis with the FDTD method [3], [4]. Zhang and Mei have computed the dispersive characteristics of microstrip [5] as well as the scattering parameters for the microstrip cross junction, tee junction, step in width, and gap via the FDTD method [6]. This paper focuses on the full-wave analysis of the microstrip right-angle bend. The microstrip bend has been studied extensively in the literature by such approximate methods as quasi-static analysis [7] and the planar waveguide model [8]–[11]. Recently, a qualitative study of wave propagation through the microstrip bend was reported by Yoshida and Fukai [12] using the transmission-line matrix method. In this paper, the time history of a short pulse as it propagates through a microstrip bend is simulated using the FDTD algorithm. Quantitative frequency-domain information contained in the time-domain data is extracted via the Fourier transform.

Computer memory constraints dictate that the computation domain of the FDTD be of finite size. In order to simulate open structures, various localized absorbing boundary conditions have been employed to truncate the FDTD grid [13], [14]. However, for dispersive structures such as the microstrip line, noise from the imperfect absorbing condition can give rise to significant degradation in the frequency-domain data. The FDTD method used in this paper does not employ absorbing conditions. Instead, the microstrip structure is enclosed by electric walls located far enough away from the strip to not seriously affect the field distributions near the strip. This electric wall assumption neglects radiation effects which are reported to be small for the open microstrip bend [15]. The FDTD grid is kept large enough in the direction of propagation so that an expanding wave never reflects off the edge of the grid.

Cutting a miter into a microstrip or waveguide bend is a well-known method for reducing reflections caused by sharp corners in waveguide bends [16]. Campbell and Jones [17] studied the 90° parallel-plate waveguide bend with a 45° miter of arbitrary length and showed that there exists an optimal cut size that minimizes reflections. Later, Douville and James [15] performed experiments on microstrip and constructed an empirical formula that relates the optimal miter to the width-to-height ratio of the strip. Results obtained by the FDTD method for the optimal cut size of the enclosed microstrip bend are also presented in this paper for a range of width-to-height ratios. They confirm Douville and James's earlier experimental results.

II. Time-Domain Simulation

Fig. 1(a) shows the top view of the 90° microstrip bend with a 45° miter to be analyzed. The distance from the inner corner of the bend to the outer corner, assuming no cut, is denoted d, and x is the distance from the inner corner of the bend to the midpoint of the cut. The amount of cut is the quantity $m = x/d$, where $m = 0$ indicates a bend with no cut and $m = 1$ indicates a cut large enough to barely separate the two sections of microstrip from the bend. The microstrip is enclosed on four sides by electric walls as shown in Fig. 1(b). The width, a, and height, b, of this box must be large enough to not disturb the field distributions near the strip.

Manuscript received May 25, 1989; revised October 2, 1989. This work was supported by the National Science Foundation under Grant ECS-8657524 and by the Texas Advanced Technology Program.
The authors are with the Department of Electrical and Computer Engineering, University of Texas at Austin, Austin, TX 78712.
IEEE Log Number 8933714.

Reprinted from *IEEE Transactions on Microwave Theory and Techniques*, pp. 405-410, April 1990.

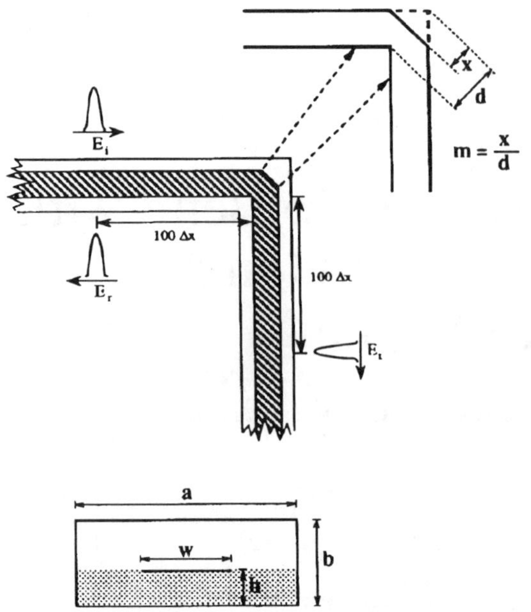

Fig. 1. The enclosed microstrip with right-angle bend. (a) Top view and detail of the mitered bend. (b) Cross-sectional view.

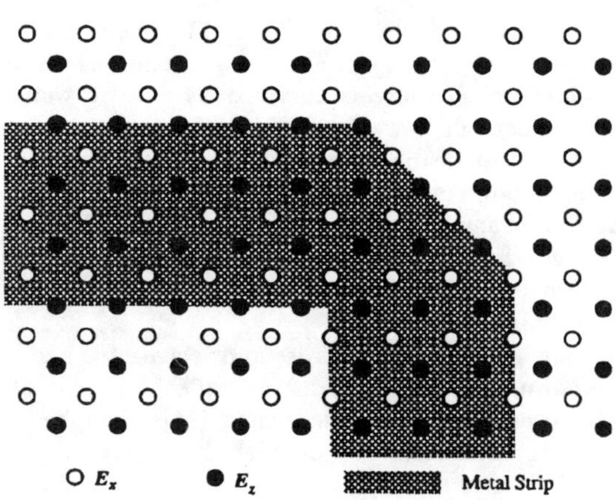

Fig. 2. Detail of the E-field nodes defining the metal strip.

In order to adequately define the bend and miter for the computer algorithm, this structure is first discretized into a three-dimensional finite difference mesh. Fig. 2 shows how electric wall boundary conditions are applied to individual nodes in order to define the metal strip. The fineness of the mesh, Δx, is chosen such that it is one tenth the length of the shortest wavelength of interest. Once the fineness of the mesh is given, the time step for iteration, Δt, is chosen to satisfy the Courant stability criterion [2].

Due to the dispersive nature of the microstrip and the complicated dielectric/metal/air boundary, conventional absorbing boundary conditions are avoided. Instead, a simple solution to this problem which saves some computer time and memory is to begin calculations on a short structure and to adaptively increase the structure length as a wave expands. The driving condition is enforced at what starts as the left edge of the FDTD grid, so that only an incident wave traveling to the right is produced. If the driving condition were left turned on while the grid expanded, then an incident wave traveling to the left would also be produced. This problem is avoided first by using an incident pulse of short duration and then by removing the driving condition when the grid begins to expand.

The incident pulse is a Gaussian pulse of short enough duration such that the driving condition is imposed on the structure for only the first few time steps. The time history of the reflected and transmitted waves is recorded. These data are then fast-Fourier-transformed and normalized with respect to the incident wave spectrum to obtain S_{11} and S_{21} as a function of frequency. The S_{11} and S_{21} phase references are chosen to be at the inner corner of the bend.

III. Results and Discussions

The microstrip line to be modeled was designed to be 50 Ω on a 3M substrate ($\epsilon_r = 2.5$). This design resulted in the following parameter values:

$$w = 2.31 \text{ mm} \qquad h = 0.84 \text{ mm}$$
$$\Delta x = 0.2096 \text{ mm} \qquad \Delta t = 0.4 \text{ ps}.$$

The cross section of the microstrip structure was first discretized into a 23×9 finite difference mesh. The length of the microstrip would begin at 2 cm and grow to about 17 cm by the end of the iterating process. The metal strip was 11 nodes wide, and this width enables the relative size of the cut, m, to be varied from 0 to 1 by increments of 1/22.

The FDTD method was first run for the no-cut case. Fig. 3(a) shows the incident pulse which then propagated towards the right-angle bend and resulted in a reflected pulse (Fig. 3(b)) and a transmitted pulse (Fig. 3(c)). In order to calculate the proper phases of the S parameters, the propagation constant was obtained by observing the phase difference between two incident pulses separated by a known distance. This propagation constant is plotted against frequency in Fig. 4(a) along with the free-space propagation constant and the dielectric propagation constant. The dip in the propagation constant at about 27 GHz corresponds approximately to the first hybrid cutoff mode of the metal box enclosing the strip. The effective permittivity was computed directly from the propagation constant and is shown in Fig. 4(b) for three different values of box height ($b = 2h$, $3h$, and $4h$) and compared to a curve-fitted formula derived for the open microstrip case [18]. A change from $b = 2h$ to $b = 3h$ resulted in a sharp increase in effective permittivity, but the change then from $b = 3h$ to $b = 4h$ resulted in little change in the effective permittivity. This suggests that the box height at $b = 3h$ is large enough to not significantly affect the dispersion characteristics of the microstrip. The width of the box, a, appears to affect the propagation constant less than the height of the box, b. The effective permittivity is not significantly altered when the width of the box is changed from $a = 2w$ to $3w$, except for the shift in the cutoff frequency of the first hybrid mode. The characteristic

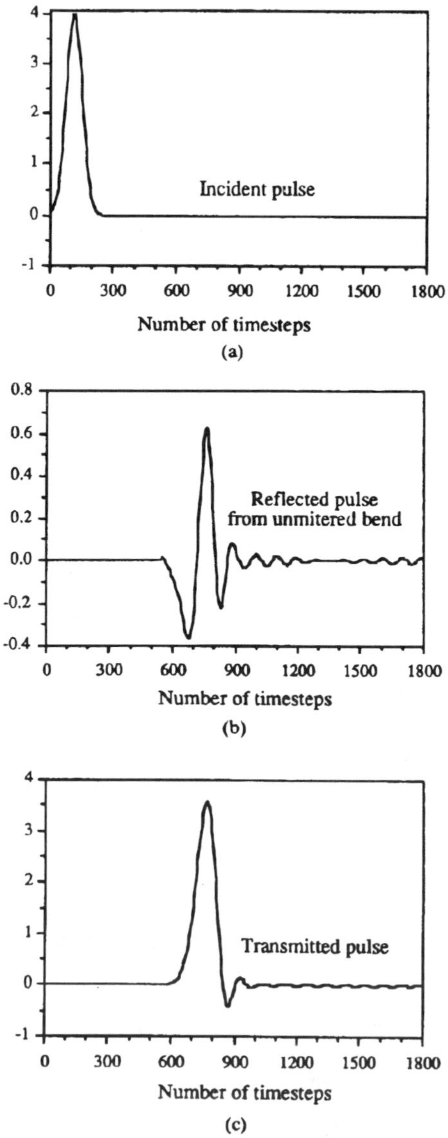

Fig. 3. Time histories for (a) the incident pulse, (b) the reflected pulse from the unmitered bend, and (c) the transmitted pulse for the unmitered bend.

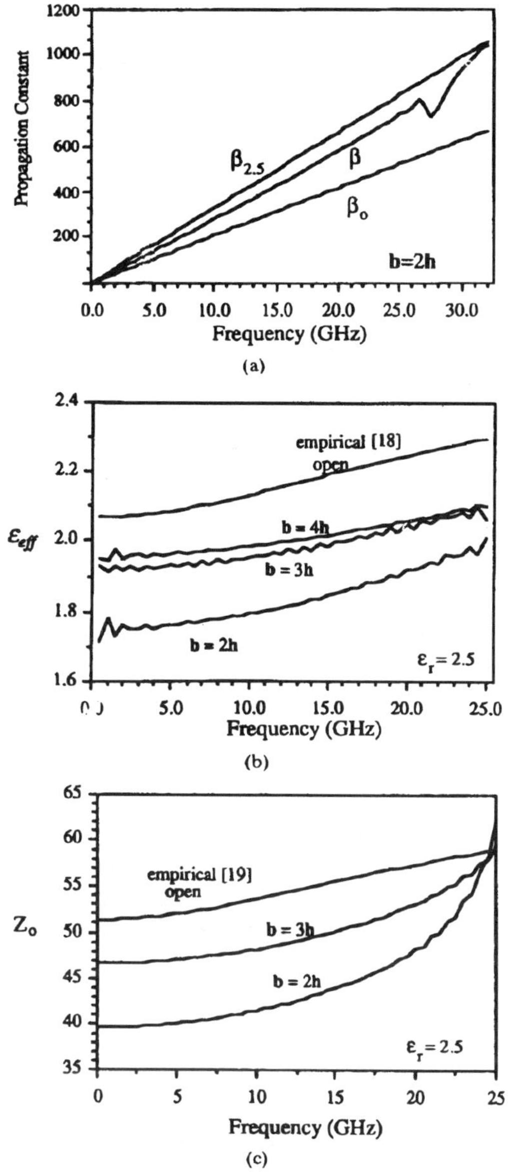

Fig. 4. Uniform microstrip characteristics calculated by the FDTD as compared with curve-fitted data for $\epsilon_r = 2.5$: (a) propagation constant; (b) effective dielectric constant; and (c) characteristic impedance.

impedances of the microstrip calculated by the FDTD method for two different box heights are compared in Fig. 4(c) with a curve-fitted formula for the open microstrip case [19]. An increase in b appears to move the characteristic impedance as well as the propagation constant closer to the open microstrip characteristic impedance. Bahl [20] reported similar effects of the enclosure upon the effective dielectric constant and the characteristic impedance of microstrip.

With the propagation constant in hand, the S parameters for the right-angle bend are completely determined. Both the magnitude and the phase of S_{11} and S_{21} for the no-cut case are shown in Fig. 5. Since the microstrip is enclosed, no radiation can occur, so the energy conservation criterion $|S_{11}|^2 + |S_{21}|^2 = 1$ was checked and found to be satisfied at all frequencies up to the cutoff frequency of the first higher order hybrid mode.

Mehran computed, via a waveguide model in conjunction with a two-dimensional mode-matching technique [10], a value of $|S_{11}| \approx 0.41$ which, when scaled in frequency to relate to the size of the microstrip modeled by the FDTD method, corresponds to $|S_{11}| \approx 0.30$ at 13 GHz. This discrepancy is probably due to the ambiguous definition of the "effective width" within the discontinuity region when analyzing the bend through the waveguide model.

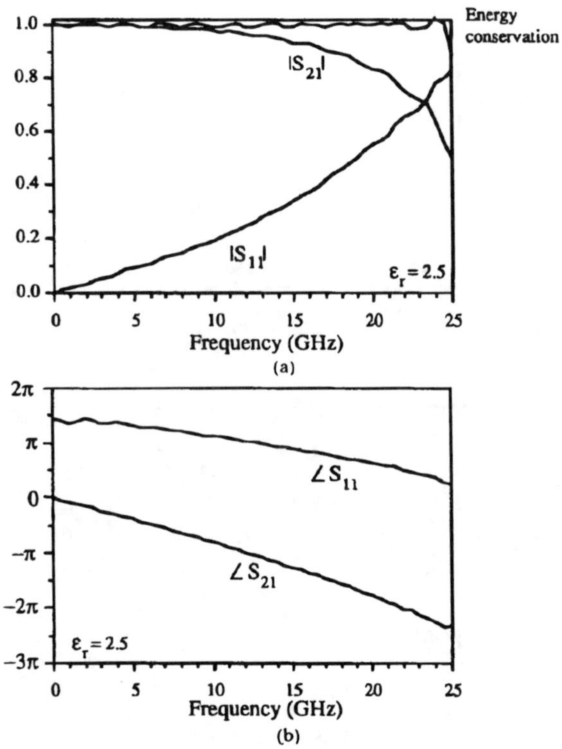

Fig. 5. S parameters for the no-cut case: (a) magnitude; (b) phase.

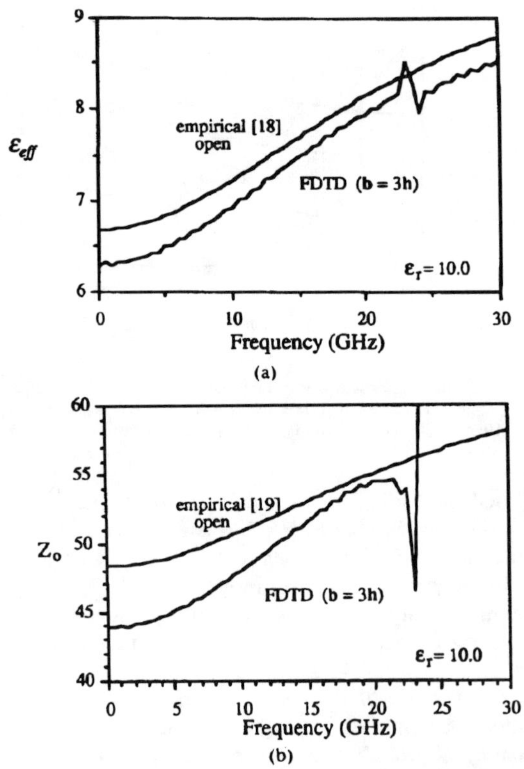

Fig. 6. Uniform microstrip characteristics calculated by the FDTD as compared with curve-fitted data for $\epsilon_r = 10.0$: (a) effective dielectric constant; (b) characteristic impedance.

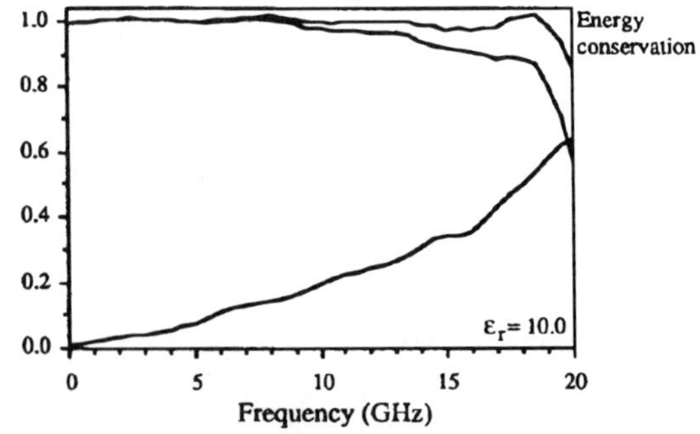

Fig. 7. Magnitude of the S parameter for $\epsilon_r = 10.0$ without a miter.

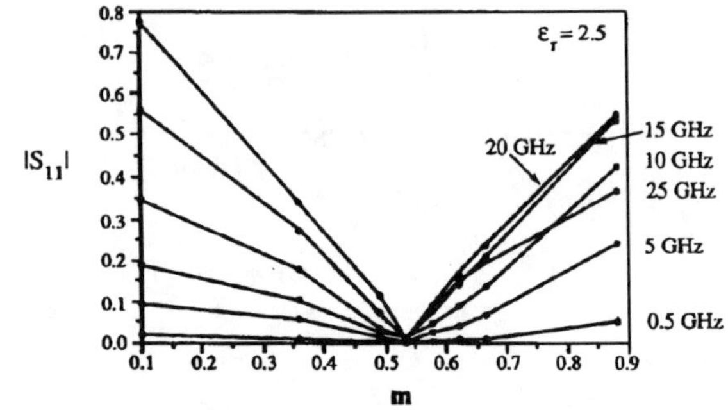

Fig. 8. Magnitude of S_{11} versus cut size for $\epsilon_r = 2.5$.

A second high-dielectric microstrip line to be modeled was designed to be 50 Ω on alumina substrate ($\epsilon_r = 10.0$) using the same values of h, Δx, and Δt as for the $\epsilon_r = 2.5$ case. This design resulted in $w = h = 0.84$ mm. The effective dielectric constant and the characteristic impedance were calculated and compared to curve-fitted data in Fig. 6. The effect of the enclosure of the box upon the effective dielectric constant appears less noticeable in the high-dielectric-substrate case than in the $\epsilon_r = 2.5$ case. Fig. 7 is a plot of the magnitudes of S_{11} and S_{21} for the high-dielectric case.

The cut size, m, was then varied to determine the optimal cut for the $\epsilon_r = 2.5$ case. Fig. 8 is a plot of S_{11} versus m. A broad-band optimal cut exists at $m = 0.53$. The time history of the reflected pulse for the optimal cut is shown in Fig. 9. The reflected pulse is barely noticeable for the optimal miter geometry. This optimal cut and six others for different width-to-height ratios and for substrates with high and low dielectric constants are compared with an empirical formula derived from measurement of open microstrip bends [15] in Fig. 10. The agreement is very good.

Finally, $|S_{11}|$ was checked against two different values of box height. $|S_{11}|$ for the no-cut case is shown in Fig. 11 for $b = 2h$, $3h$, and $4h$. The effect of the increase in b is noticeable only at higher frequencies, and there is almost no change in $|S_{11}|$ for the increase from $b = 3h$ to $b = 4h$.

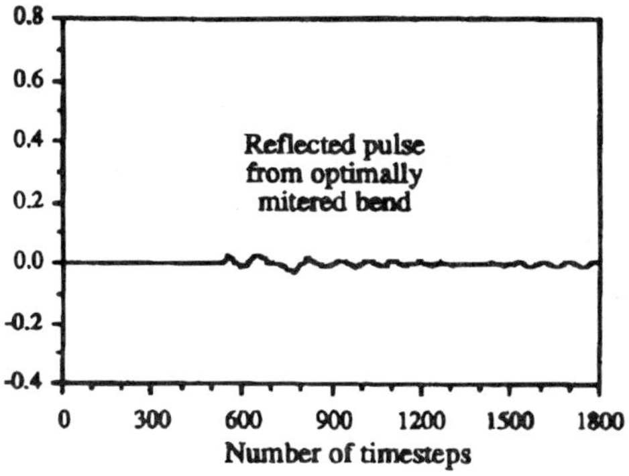

Fig. 9. Time history of the reflected pulse for the optimally mitered bend.

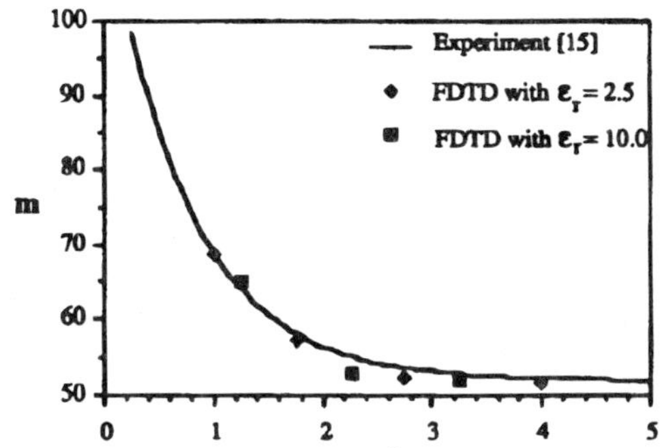

Fig. 10. FDTD optimal cut versus curve-fitting formula.

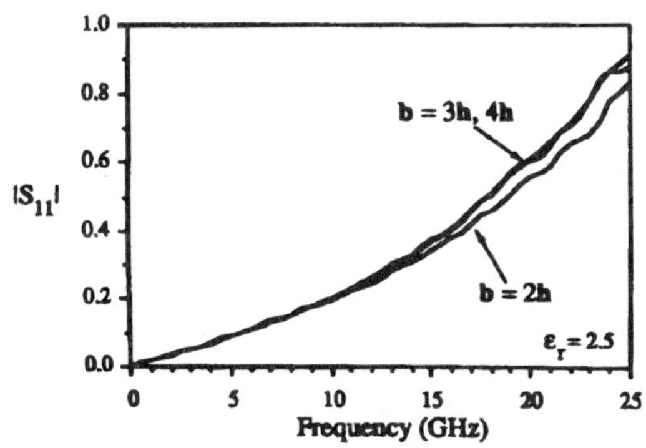

Fig. 11. Effects of change in box height on magnitude of S_{11}.

As a final check, the optimal cut for $b = 3h$ was computed and is the same size as the optimal cut for $b = 2h$.

IV. Conclusions

A microstrip with a right-angle bend has been simulated via the time-domain finite difference method. Electric walls were used to enclose the microstrip structure. Numerical results for the enclosed bend appear to agree well with experimental results for the open bend. Douville and James [15] noted that radiation associated with the bend is not important. Furthermore, the slight change in S parameters of the enclosed microstrip bend when the enclosing box is increased in height from $b = 3h$ to $b = 4h$ suggests that the FDTD results are close to results expected for the open microstrip bend. Finally, the FDTD method had equal success for both high- and low-dielectric materials.

References

[1] K. S. Yee, "Numerical solution of initial boundary value problems involving Maxwell's equations in isotropic media," *IEEE Trans. Antennas Propagat.*, vol. AP-14, pp. 302–307, May 1966.

[2] A. Taflove and M. E. Brodwin, "Numerical solution of steady-state electromagnetic scattering problems using the time-dependent Maxwell's equations," *IEEE Trans. Microwave Theory Tech.*, vol. MTT-23, pp. 623–630, Aug. 1975.

[3] W. K. Gwarek, "Analysis of an arbitrarily-shaped planar circuit—A time-domain approach," *IEEE Trans. Microwave Theory Tech.*, vol. MTT-33, pp. 1067–1072, Oct. 1985.

[4] W. K. Gwarek, "Analysis of arbitrarily shaped two-dimensional microwave circuits by finite-difference time-domain method," *IEEE Trans. Microwave Theory Tech.*, vol. 36, pp. 738–744, Apr. 1988.

[5] X. Zhang, J. Fang, K. K. Mei, and Y. Liu, "Calculations of the dispersive characteristics of microstrips by the time-domain finite difference method," *IEEE Trans. Microwave Theory Tech.*, vol. 36, pp. 263–267, Feb. 1988.

[6] X. Zhang and K. K. Mei, "Time-domain finite difference approach to the calculation of the frequency-dependent characteristics of microstrip discontinuities," *IEEE Trans. Microwave Theory Tech.*, vol. 36, pp. 1775–1787, Dec. 1988.

[7] P. Silvester and P. Benedek, "Microstrip discontinuity capacitances for right-angle bends, T-junctions, and crossings," *IEEE Trans. Microwave Theory Tech.*, vol. MTT-21, pp. 341–346, May 1973.

[8] G. Kompa, "S-matrix computation of microstrip discontinuities with a planar waveguide model," *Arch. Elek. Übertragung.*, vol. 30, pp. 58–64, 1975.

[9] W. Menzel and I. Wolff, "A method for calculating the frequency-dependent properties of microstrip discontinuities," *IEEE Trans. Microwave Theory Tech.*, vol. MTT-25, pp. 107–112, Feb. 1977.

[10] R. Mehran, "Calculation of microstrip bends and Y-junctions with arbitrary angle," *IEEE Trans. Microwave Theory Tech.*, vol. MTT-26, pp. 400–405, June 1978.

[11] R. Chadha and K. C. Gupta, "Compensation of discontinuities in planar transmission lines," *IEEE Trans. Microwave Theory Tech.*, vol. MTT-30, pp. 2151–2156, Dec. 1982.

[12] N. Yoshida and I. Fukai, "Transient analysis of a stripline having a corner in three-dimensional space," *IEEE Trans. Microwave Theory Tech.*, vol. MTT-32, pp. 491–498, May 1984.

[13] G. Mur, "Absorbing boundary conditions for the finite-difference approximation of the time-domain electromagnetic-field equations," *IEEE Trans. Electromagn. Compat.*, vol. EMC-23, pp. 377–382, Nov. 1981.

[14] J. Fang and K. K. Mei, "A super-absorbing boundary algorithm for solving electromagnetic problems by time-domain finite-difference method," in *IEEE AP-S Int. Symp. Dig.* (Syracuse, NY), June 1988, pp. 472–475.

[15] R. J. P. Douville and D. S. James, "Experimental study of symmetric microstrip bends and their compensation," *IEEE Trans. Microwave Theory Tech.*, vol. MTT-26, pp. 175–182, Mar. 1978.

[16] N. Marcuvitz, *Waveguide Handbook* (Radiation Laboratory Series, vol. 10). New York: McGraw-Hill, 1951.

[17] J. J. Campbell and W. R. Jones, "Symmetrically truncated right-angle corners in parallel-plate and rectangular waveguide," *IEEE Trans. Microwave Theory Tech.*, vol. MTT-16, pp. 517–529, Aug. 1968.

[18] R. P. Owens, "Accurate analytical determination of quasi-static microstrip line parameters," *Radio and Electron. Eng.*, vol. 46, pp. 360–364, 1976.

[19] B. Bianco *et al.*, "Frequency dependence of microstrip parameters," *Alta Frequenza*, vol. 43, pp. 413–416, 1974.

[20] I. J. Bahl, "Easy and exact methods for shielded microstrip design," *Microwaves*, vol. 17, Dec. 1974.

Application of the Three-Dimensional Finite-Difference Time-Domain Method to the Analysis of Planar Microstrip Circuits

DAVID M. SHEEN, SAMI M. ALI, SENIOR MEMBER, IEEE, MOHAMED D. ABOUZAHRA, SENIOR MEMBER, IEEE, AND JIN AU KONG, FELLOW, IEEE

Abstract —**A direct three-dimensional finite-difference time-domain (FDTD) method is applied to the full-wave analysis of various microstrip structures. The method is shown to be an efficient tool for modeling complicated microstrip circuit components as well as microstrip antennas. From the time-domain results, the input impedance of a line-fed rectangular patch antenna and the frequency-dependent scattering parameters of a low-pass filter and a branch line coupler are calculated. These circuits are fabricated and the measurements are compared with the FDTD results and shown to be in good agreement.**

I. Introduction

FREQUENCY-domain analytical work with complicated microstrip circuits has generally been done using planar circuit concepts in which the substrate is assumed to be thin enough that propagation can be considered in two dimensions by surrounding the microstrip with magnetic walls [1]–[6]. Fringing fields are accounted for by using either static or dynamic effective dimensions and permittivities. Limitations of these methods are that fringing, coupling, and radiation must all be handled empirically since they are not allowed for in the model. Also, the accuracy is questionable when the substrate becomes thick relative to the width of the microstrip. To fully account for these effects, it is necessary to use a full-wave solution.

Full-wave frequency-domain methods have been used to solve some of the simpler discontinuity problems [7], [8]. However, these methods are difficult to apply to a typical printed microstrip circuit.

Modeling of microstrip circuits has also been performed using Bergeron's method [9], [10]. This method is a modification of the transmission line matrix (TLM) method, and has limitations similar to the finite-difference time-domain (FDTD) method due to the discrete modeling of space and time [11], [12]. A unique problem with this method is that the dielectric interface and the perfectly conducting strip are misaligned by half a space step [12].

The FDTD method has been used extensively for the solution of two- and three-dimensional scattering problems [13]–[17]. Recently, FDTD methods have been used to effectively calculate the frequency-dependent characteristics of microstrip discontinuities [18]–[21]. Analysis of the fundamental discontinuities is of great importance since more complicated circuits can be realized by interconnecting microstrip lines with these discontinuities and using transmission line and network theory. Some circuits, however, such as patch antennas, may not be realized in this way. Additionally, if the discontinuities are too close to each other the use of network concepts will not be accurate due to the interaction of evanescent waves. To accurately analyze these types of structures it is necessary to simulate the entire structure in one computation. The FDTD method shows great promise in its flexibility in handling a variety of circuit configurations. An additional benefit of the time-domain analysis is that a broad-band pulse may be used as the excitation and the frequency-domain parameters may be calculated over the entire frequency range of interest by Fourier transform of the transient results.

In this paper, the frequency-dependent scattering parameters have been calculated for several printed microstrip circuits, specifically a line-fed rectangular patch antenna, a low-pass filter, and a rectangular branch line coupler. These circuits represent resonant microstrip structures on an open substrate; hence, radiation effects can be significant, especially for the microstrip antenna. Calculated results are presented and compared with experimental measurements.

The FDTD method has been chosen over the other discrete methods (TLM or Bergeron's) because it is extremely efficient, its implementation is quite straightforward, and it may be derived directly from Maxwell's equations. Many of the techniques used to implement this

Manuscript received September 29, 1989; revised March 21, 1990. This work was supported by NSF Grant 8620029-ECS, the Joint Services Electronics Program (Contract DAAL03-89-C-0001), RADC Contract F19628-88-K-0013, ARO Contract DAAL03-88-J-0057, ONR Contract N00014-89-J-1019, and the Department of the Air Force.

D. M. Sheen, S. M. Ali, and J. A. Kong are with the Department of Electrical Engineering and Computer Science and the Research Laboratory of Electronics, Massachusetts Institute of Technology, Cambridge, MA 02139.

M. D. Abouzahra is with MIT Lincoln Laboratory, Lexington, MA 02173.

IEEE Log Number 9036153.

Reprinted from *IEEE Transactions on Microwave Theory and Techniques*, pp. 849-857, July 1990.

method have been demonstrated previously [18]–[20]; however, simplification of the method has been achieved by using a simpler absorbing boundary condition [22]. This simpler absorbing boundary condition yields good results for the broad class of microstrip circuits considered by this paper. Additionally, the source treatment has been enhanced to reduce the source effects documented in [18]–[20].

II. Problem Formulation

The FDTD method is formulated by discretizing Maxwell's curl equations over a finite volume and approximating the derivatives with centered difference approximations. Conducting surfaces are treated by setting tangential electric field components to 0. The walls of the mesh, however, require special treatment to prevent reflections from the mesh termination.

A. Governing Equations

Formulation of the FDTD method begins by considering the differential form of Maxwell's two curl equations which govern the propagation of fields in the structures. For simplicity, the media are assumed to be piecewise uniform, isotropic, and homogeneous. The structure is assumed to be lossless (i.e., no volume currents or finite conductivity). With these assumptions, Maxwell's curl equations may be written as

$$\mu \frac{\partial \boldsymbol{H}}{\partial t} = -\nabla \times \boldsymbol{E} \tag{1}$$

$$\epsilon \frac{\partial \boldsymbol{E}}{\partial t} = \nabla \times \boldsymbol{H}. \tag{2}$$

In order to find an approximate solution to this set of equations, the problem is discretized over a finite three-dimensional computational domain with appropriate boundary conditions enforced on the source, conductors, and mesh walls.

B. Finite-Difference Equations

To obtain discrete approximations to these continuous partial differential equations the centered difference approximation is used on both the time and space first-order partial differentiations. For convenience, the six field locations are considered to be interleaved in space as shown in Fig. 1, which is a drawing of the FDTD unit cell [13]. The entire computational domain is obtained by stacking these rectangular cubes into a larger rectangular volume. The $\hat{x}$, $\hat{y}$, and $\hat{z}$ dimensions of the unit cell are Δx, Δy, and Δz, respectively. The advantages of this field arrangement are that centered differences are realized in the calculation of each field component and that continuity of tangential field components is automatically satisfied. Because there are only six unique field components within the unit cell, the six field components touching the shaded upper eighth of the unit cell in Fig. 1 are considered to be a unit node with subscript indices i, j, and k corresponding to the node numbers in the $\hat{x}$, $\hat{y}$, and $\hat{z}$

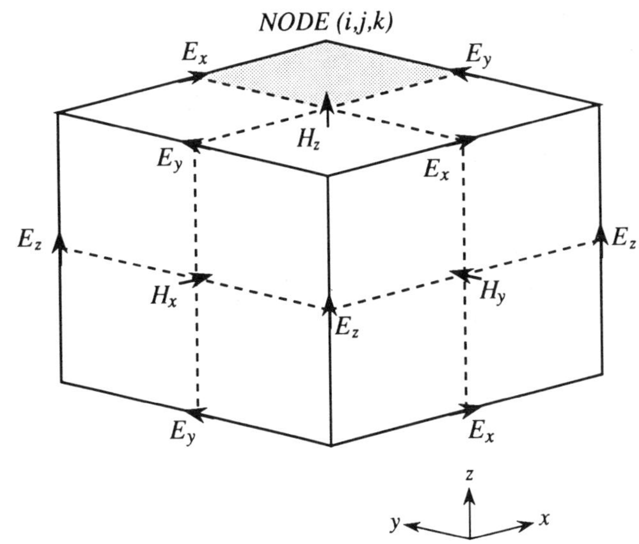

Fig. 1. Field component placement in the FDTD unit cell.

directions. This notation implicitly assumes the $\pm 1/2$ space indices and thus simplifies the notation, rendering the formulas directly implementable on the computer. The time steps are indicated with the superscript n. Using this field component arrangement, the above notation, and the centered difference approximation, the explicit finite difference approximations to (1) and (2) are

$$H_{x\,i,j,k}^{n+1/2} = H_{x\,i,j,k}^{n-1/2} + \frac{\Delta t}{\mu \Delta z}\left(E_{y\,i,j,k}^{n} - E_{y\,i,j,k-1}^{n}\right) - \frac{\Delta t}{\mu \Delta y}\left(E_{z\,i,j,k}^{n} - E_{z\,i,j-1,k}^{n}\right) \tag{3}$$

$$H_{y\,i,j,k}^{n+1/2} = H_{y\,i,j,k}^{n-1/2} + \frac{\Delta t}{\mu \Delta x}\left(E_{z\,i,j,k}^{n} - E_{z\,i-1,j,k}^{n}\right) - \frac{\Delta t}{\mu \Delta z}\left(E_{x\,i,j,k}^{n} - E_{x\,i,j,k-1}^{n}\right) \tag{4}$$

$$H_{z\,i,j,k}^{n+1/2} = H_{z\,i,j,k}^{n-1/2} + \frac{\Delta t}{\mu \Delta y}\left(E_{x\,i,j,k}^{n} - E_{x\,i,j-1,k}^{n}\right) - \frac{\Delta t}{\mu \Delta x}\left(E_{y\,i,j,k}^{n} - E_{y\,i-1,j,k}^{n}\right) \tag{5}$$

$$E_{x\,i,j,k}^{n+1} = E_{x\,i,j,k}^{n} + \frac{\Delta t}{\epsilon \Delta y}\left(H_{z\,i,j+1,k}^{n+1/2} - H_{z\,i,j,k}^{n+1/2}\right) - \frac{\Delta t}{\epsilon \Delta z}\left(H_{y\,i,j,k+1}^{n+1/2} - H_{y\,i,j,k}^{n+1/2}\right) \tag{6}$$

$$E_{y\,i,j,k}^{n+1} = E_{y\,i,j,k}^{n} + \frac{\Delta t}{\epsilon \Delta z}\left(H_{x\,i,j,k+1}^{n+1/2} - H_{x\,i,j,k}^{n+1/2}\right) - \frac{\Delta t}{\epsilon \Delta x}\left(H_{z\,i+1,j,k}^{n+1/2} - H_{z\,i,j,k}^{n+1/2}\right) \tag{7}$$

$$E_{z\,i,j,k}^{n+1} = E_{z\,i,j,k}^{n} + \frac{\Delta t}{\epsilon \Delta x}\left(H_{y\,i+1,j,k}^{n+1/2} - H_{y\,i,j,k}^{n+1/2}\right) - \frac{\Delta t}{\epsilon \Delta y}\left(H_{x\,i,j+1,k}^{n+1/2} - H_{x\,i,j,k}^{n+1/2}\right). \tag{8}$$

The half time steps indicate that $\boldsymbol{E}$ and $\boldsymbol{H}$ are alternately calculated in order to achieve centered differences for the time derivatives. In these equations, the permittivity and the permeability are set to the appropriate values depending on the location of each field component. For the electric field components on the dielectric–air interface the average of the two permittivities, $(\epsilon_0+\epsilon_1)/2$, is used. The validity of this treatment is explained in [20].

Due to the use of centered differences in these approximations, the error is second order in both the space and time steps; i.e., if Δx, Δy, Δz, and Δt are proportional to Δl, then the global error is $O(\Delta l^2)$. The maximum time step that may be used is limited by the stability restriction of the finite difference equations,

$$\Delta t \leqslant \frac{1}{v_{\max}}\left(\frac{1}{\Delta x^2}+\frac{1}{\Delta y^2}+\frac{1}{\Delta z^2}\right)^{-1/2} \tag{9}$$

where $v_{\max}$ is the maximum velocity of light in the computational volume. Typically, $v_{\max}$ will be the velocity of light in free space unless the entire volume is filled with dielectric. These equations will allow the approximate solution of $\boldsymbol{E}(\boldsymbol{r},t)$ and $\boldsymbol{H}(\boldsymbol{r},t)$ in the volume of the computational domain or mesh; however, special consideration is required for the source, the conductors, and the mesh walls.

C. Source Considerations

The volume in which the microstrip circuit simulation is to be performed is shown schematically in Fig. 2. At $t=0$ the fields are assumed to be identically 0 throughout the computational domain. A Gaussian pulse is desirable as the excitation because its frequency spectrum is also Gaussian and will therefore provide frequency-domain information from dc to the desired cutoff frequency by adjusting the width of the pulse.

In order to simulate a voltage source excitation it is necessary to impose the vertical electric field, E_z, in a rectangular region underneath port 1 as shown in Fig. 2. The remaining electric field components on the source plane must be specified or calculated. In [18]–[20] an electric wall source is used; i.e., the remaining electric field components on the source wall of the mesh are set to 0. An unwanted side effect of this type of excitation is that a sharp magnetic field is induced tangential to the source wall. This results in some distortion of the launched pulse. Specifically, the pulse is reduced in magnitude due to the energy stored in the induced magnetic field and a negative tail to the pulse is immediately evident. An alternative excitation scheme is to simulate a magnetic wall at the source plane. The source plane consists only of E_x and E_z components, with the tangential magnetic field components offset $\pm\Delta y/2$. If the magnetic wall is enforced by setting the tangential magnetic field components to zero just behind the source plane, then significant distortion of the pulse still occurs. If the magnetic wall is enforced directly on the source plane by using image theory (i.e., $\boldsymbol{H}_{\tan}$ outside the magnetic wall is equal

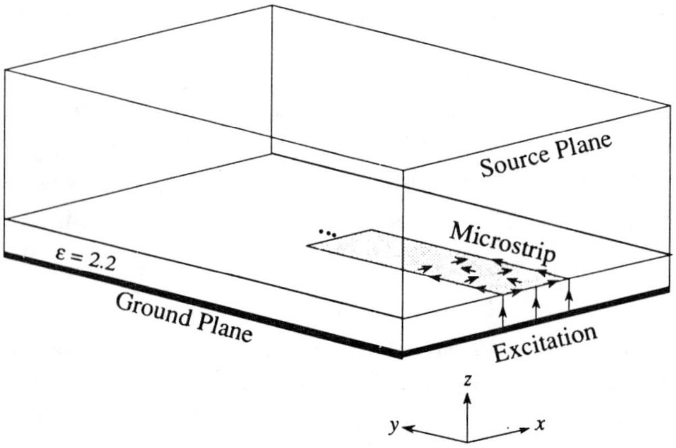

Fig. 2. Computational domain.

to $-\boldsymbol{H}_{\tan}$ inside the magnetic wall), then the remaining electric field components on the source plane may be readily calculated using the finite-difference equations. Using this excitation, only a minimal amount of source distortion is apparent. The launched wave has nearly unit amplitude and is Gaussian in time and in the $\hat{y}$ direction:

$$E_z = f_s(t) = e^{-(t-t_0)^2/T^2}. \tag{10}$$

It is assumed that excitation specified in this way will result in the fundamental mode only propagating down the microstrip in the frequency range of interest.

The finite-difference formulas are not perfect in their representation of the propagation of electromagnetic waves. One effect of this is numerical dispersion; i.e., the velocity of propagation is slightly frequency dependent even for uniform plane waves. In order to minimize the effects of numerical dispersion and truncation errors, the width of the Gaussian pulse is chosen for at least 20 points per wavelength at the highest frequency represented significantly in the pulse.

D. Conductor Treatment

The circuits considered in this paper have a conducting ground plane and a single dielectric substrate with metallization on top of this substrate in the ordinary microstrip configuration. These electric conductors are assumed to be perfectly conducting and have zero thickness and are treated by setting the electric field components that lie on the conductors to zero. The edge of the conductor should be modeled with electric field components tangential to the edge lying exactly on the edge of the microstrip as shown in Fig. 2.

E. Absorbing Boundary Treatment

Due to the finite capabilities of the computers used to implement the finite-difference equations, the mesh must be limited in the $\hat{x}$, $\hat{y}$, and $\hat{z}$ directions. The difference equations cannot be used to evaluate the field components tangential to the outer boundaries since they would require the values of field components outside of the mesh. One of the six mesh boundaries is a ground plane and its tangential electric field values are forced to be 0.

The tangential electric field components on the other five mesh walls must be specified in such a way that outgoing waves are not reflected using the absorbing boundary condition [22], [23]. For the structures considered in this paper, the pulses on the microstrip lines will be normally incident to the mesh walls. This leads to a simple approximate continuous absorbing boundary condition, which is that the tangential fields on the outer boundaries will obey the one-dimensional wave equation in the direction normal to the mesh wall. For the $\hat{y}$ normal wall the one-dimensional wave equation may be written

$$\left(\frac{\partial}{\partial y} - \frac{1}{v} \frac{\partial}{\partial t} \right) \boldsymbol{E}_{\tan} = 0. \tag{11}$$

This equation is Mur's first approximate absorbing boundary condition and it may be easily discretized using only field components on or just inside the mesh wall, yielding an explicit finite difference equation [22],

$$E_0^{n+1} = E_1^n + \frac{v\Delta t - \Delta y}{v\Delta t + \Delta y} \left(E_1^{n+1} - E_0^n \right) \tag{12}$$

where E_0 represents the tangential electric field components on the mesh wall and E_1 represents the tangential electric field components one node inside of the mesh wall. Similar expressions are immediately obtained for the other absorbing boundaries by using the corresponding normal directions for each wall. It should be noted that the normal incidence assumption is not valid for the fringing fields which are propagating tangential to the walls; therefore the sidewalls should be far enough away that the fringing fields are negligible at the walls. Additionally, radiation will not be exactly normal to the mesh walls. Second-order absorbing boundary conditions [22], [23] which account for oblique incidence will not work on the mesh walls where the microstrip is incident because these absorbing boundary conditions are derived in uniform space.

The results presented show that the first-order absorbing boundary treatment is sufficiently accurate and particularly well suited to the microstrip geometry. It is possible to obtain more accurate normal incidence absorbing boundary conditions [24], which have been used in the FDTD calculation of microstrip discontinuities [20]. Due to the more dynamic, resonant behavior of the circuits considered, Mur's first approximate absorbing boundary condition is used and this allows for accurate simulation of the various microstrip circuits.

F. Time Marching Solution

The finite difference equations, (3)–(8), are used with the above boundary and source conditions to simulate the propagation of a broad-band Gaussian pulse on the microstrip structure. The essential aspects of the time-domain algorithm are as follows:

- Initially (at $t = n = 0$) all fields are 0.
- The following are repeated until the response is ≈ 0:
  - Gaussian excitation is imposed on port 1.
  - $\boldsymbol{H}^{n+1/2}$ is calculated from FD equations.
  - $\boldsymbol{E}^{n+1}$ is calculated from FD equations.
  - Tangential $\boldsymbol{E}$ is set to 0 on conductors.
  - Save desired field quantities.
  - $n \to n+1$.
- Compute scattering matrix coefficients from time-domain results.

One additional consideration is that the reflections from the circuit will be reflected again by the source wall. To eliminate this, the circuit is placed a sufficient distance from the source, and after the Gaussian pulse has been fully launched, the absorbing boundary condition is switched on at the source wall.

G. Frequency-Dependent Parameters

In addition to the transient results obtained naturally by the FDTD method, the frequency-dependent scattering matrix coefficients are easily calculated.

$$[V]^r = [S][V]^i \tag{13}$$

where $[V]^r$ and $[V]^i$ are the reflected and incident voltage vectors, respectively, and $[S]$ is the scattering matrix. To accomplish this, the vertical electric field underneath the center of each microstrip port is recorded at every time step. As in [20], it is assumed that this field value is proportional to the voltage (which could be easily obtained by numerically integrating the vertical electric field) when considering propagation of the fundamental mode. To obtain the scattering parameter $S_{11}(\omega)$, the incident and reflected waveforms must be known. The FDTD simulation calculates the sum of incident and reflected waveforms. To obtain the incident waveform, the calculation is performed using only the port 1 microstrip line, which will now be of infinite extent (i.e., from source to far absorbing wall), and the incident waveform is recorded. This incident waveform may now be subtracted from the incident plus reflected waveform to yield the reflected waveform for port 1. The other ports will register only transmitted waveforms and will not need this computation. The scattering parameters, S_{jk}, may then be obtained by simple Fourier transform of these transient waveforms as

$$S_{jk}(\omega) = \frac{\mathcal{F}\mathcal{T}\{V_j(t)\}}{\mathcal{F}\mathcal{T}\{V_k(t)\}}. \tag{14}$$

Note that the reference planes are chosen with enough distance from the circuit discontinuities to eliminate evanescent waves. These distances are included in the definition of the circuit so that no phase correction is performed for the scattering coefficients. For all of the circuits considered, the only unique coefficients are in

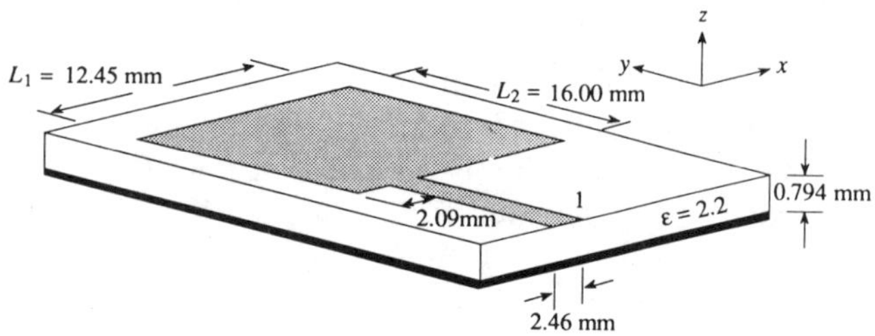

Fig. 3. Line-fed rectangular microstrip antenna detail.

the first column of the scattering matrix (i.e., $S_{11}(\omega), S_{21}(\omega), S_{31}(\omega), \cdots$).

III. Numerical Results

Numerical results have been computed for three configurations, a line-fed rectangular patch antenna, a low-pass filter, and a branch line coupler. These circuits have dimensions on the order of 1 cm, and the frequency range of interest is from dc to 20 GHz. The operating regions of all of these circuits are less than 10 GHz; however, the accuracy of the computed results at higher frequencies is examined. These circuits were constructed on Duroid substrates with $\epsilon = 2.2$ and thickness of 1/32 inch (0.794 mm). Scattering matrix coefficients were measured using an HP 8510 network analyzer, which is calibrated to 18 GHz, but provides measurement to 20 GHz.

A. Line-Fed Rectangular Microstrip Antenna

The actual dimensions of the microstrip antenna analyzed are shown in Fig. 3. The operating resonance approximately corresponds to the frequency where $L_1 = 1.245$ mm $= \lambda/2$. Simulation of this circuit involves the straightforward application of the finite-difference equations, source, and boundary conditions. To model the thickness of the substrate correctly, Δz is chosen so that three nodes exactly match the thickness. An additional 13 nodes in the $\hat{z}$ direction are used to model the free space above the substrate. In order to correctly model the dimensions of the antenna, Δx and Δy have been chosen so that an integral number of nodes will exactly fit the rectangular patch. Unfortunately, this means the port width and placement will be off by a fraction of the space step. The sizes of the space steps are carefully chosen to minimize the effect of this error.

The space steps used are $\Delta x = 0.389$ mm, $\Delta y = 0.400$ mm, and $\Delta z = 0.265$ mm, and the total mesh dimensions are $60 \times 100 \times 16$ in the $\hat{x}$, $\hat{y}$, and $\hat{z}$ directions respectively. The rectangular antenna patch is thus $32\,\Delta x \times 40\,\Delta y$. The length of the microstrip line from the source plane to the edge of the antenna is $50\,\Delta y$, and the reference plane for port 1 is $10\,\Delta y$ from the edge of the patch. The microstrip line width is modeled as $6\,\Delta x$.

The time step used is $\Delta t = 0.441$ ps. The Gaussian half-width is $T = 15$ ps, and the time delay t_0 is set to be $3T$ so the Gaussian will start at approximately 0. The simulation is performed for 8000 time steps, somewhat longer than for other circuits, due to the highly resonant behavior of the antenna. The computation time for this circuit is approximately 12 h on a VAXstation 3500 workstation.

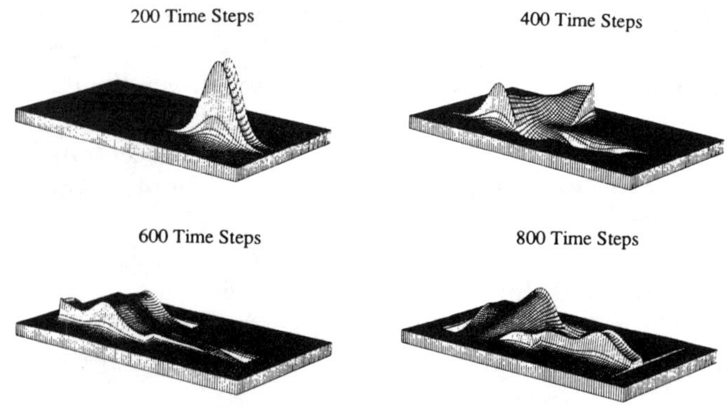

Fig. 4. Rectangular microstrip antenna distribution of $E_z(x, y, t)$ just underneath the dielectric interface at 200, 400, 600, and 800 time steps.

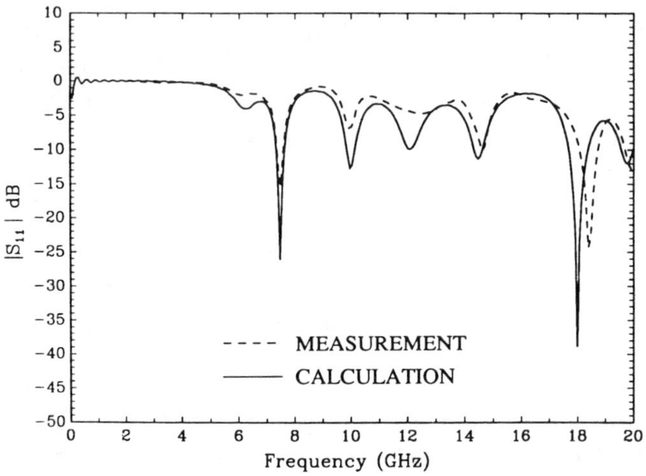

Fig. 5. Return loss of the rectangular antenna.

The spatial distribution of $E_z(x, y, t)$ just beneath the microstrip at 200, 400, 600, and 800 time steps is shown in Fig. 4, where the source Gaussian pulse and subsequent propagation on the antenna are observed. Notice that the absorbing boundary condition for the source has been implemented several nodes from the source plane to eliminate any undesirable effects of switching from source to absorbing boundary conditions. Three-dimensional properties of the propagation are observed including enhancement of the field near the edges of the microstrip.

The scattering coefficient results, shown in Fig. 5, show good agreement with the measured data. The operating resonance at 7.5 GHz is almost exactly shown by both theory and measurement. This result is a significant advancement over planar circuit techniques, which without empirical treatment will allow only $|S_{11}| = 1.0$ because they do not allow for radiation. Also, the resonance frequency calculated using planar circuit concepts will be sensitive to errors in the effective dimensions of the patch. Additional resonances are also in good ageement with experiment, except for the highest resonance near 18 GHz, which is somewhat shifted.

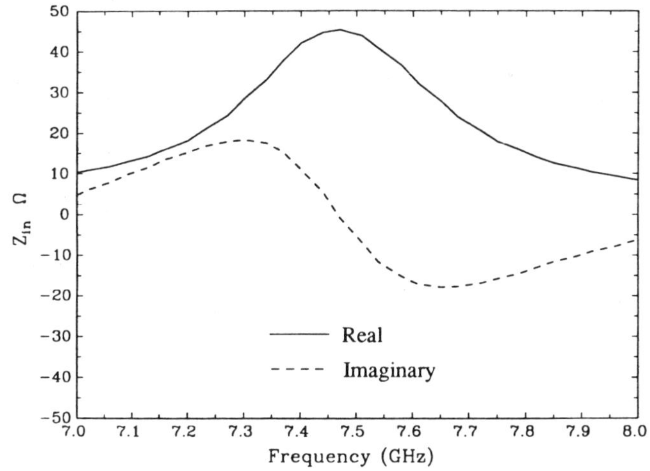

Fig. 6. Input impedance of the rectangular antenna near the operating resonance of 7.5 GHz as calculated from the FDTD results.

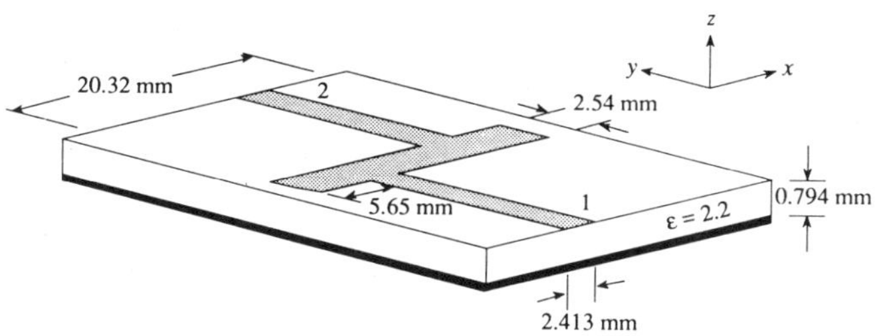

Fig. 7. Low-pass filter detail.

Input impedance for the antenna may be calculated from the $S_{11}(\omega)$ calculation by transforming the reference plane to the edge of the microstrip antenna,

$$Z_{\mathrm{in}} = Z_0 \frac{1 + S_{11} e^{j2kL}}{1 - S_{11} e^{j2kL}} \tag{15}$$

where k is the wavenumber on the microstrip, L is the length from the edge of the antenna to the reference plane ($10\Delta y$), and Z_0 is the characteristic impedance of the microstrip line. For simplicity of the Z_{in} calculation, the microstrip is assumed to have a constant characteristic impedance of 50 Ω, and an effective permittivity of 1.9 is used to calculate the wavenumber. Results for the input impedance calculation near the operating resonance of 7.5 GHz are shown in Fig. 6.

B. Microstrip Low-Pass Filter

The low-pass filter analyzed is one designed according to the criteria established by [2] and is shown in Fig. 7. As in the rectangular antenna, Δx, Δy, and Δz are carefully chosen to fit the dimensions of the circuit. The space steps Δx and Δy are chosen to exactly match the dimensions of the rectangular patch; however, the locations and widths of the ports will be modeled with some error.

The space steps used are $\Delta x = 0.4064$ mm, $\Delta y = 0.4233$ mm, and $\Delta z = 0.265$ mm, and the total mesh dimensions are $80 \times 100 \times 16$ in the $\hat{x}$, $\hat{y}$, and $\hat{z}$ directions respectively. The long rectangular patch is thus $50\Delta x \times 6\Delta y$. The distance from the source plane to the edge of the long patch is $50\Delta y$, and the reference planes for ports 1 and 2 are $10\Delta y$ from the edges of the patch. The strip widths of ports 1 and 2 are modeled as $6\Delta x$.

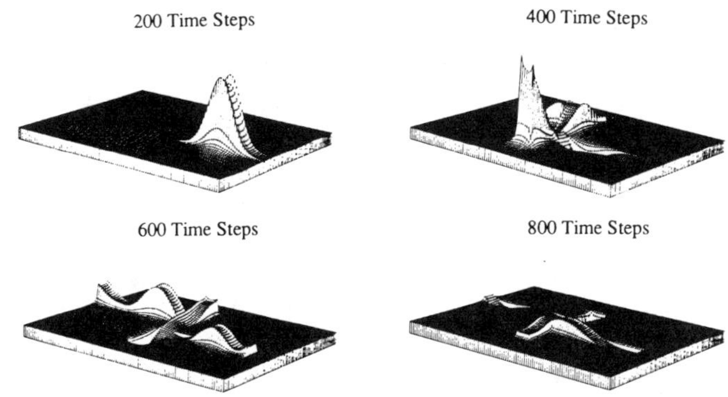

Fig. 8. Low-pass filter distribution of $E_z(x,y,t)$ just underneath the dielectric interface at 200, 400, 600, and 800 time steps.

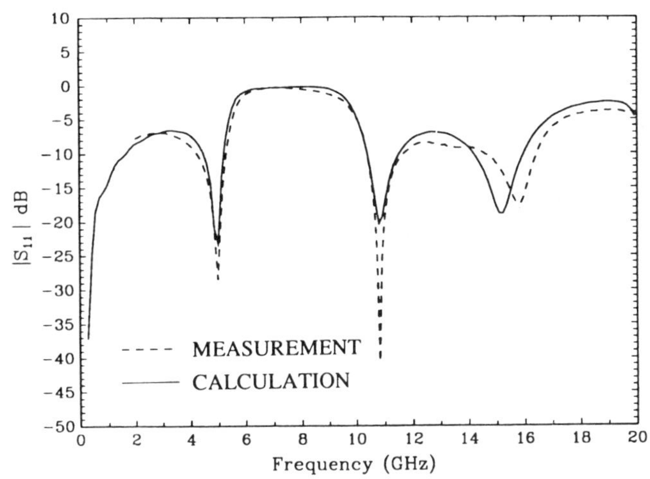

Fig. 9. Return loss of the low-pass filter.

The time step used is $\Delta t = 0.441$ ps. The Gaussian half-width is $T = 15$ ps and the time delay t_0 is set to be $3T$. The simulation is performed for 4000 time steps to allow the response on both ports to become nearly 0. The computation time for this circuit is approximately 8 h on a VAXstation 3500 workstation.

The spatial distribution of $E_z(x,y,t)$ just beneath the microstrip at 200, 400, 600, and 800 time steps is shown in Fig. 8. The scattering coefficient results, shown in Figs. 9 and 10, again show good agreement in the location of the response nulls. The desired low-pass filter performance is seen in the sharp S_{21} roll-off beginning at approximately 5 GHz. There is again some shift near the high end of the frequency range. In the S_{21} results, the stopband for the calculated curve is somewhat narrower than the measured results. Some experimentation with planar circuit techniques has led to the conclusion that this narrowing is caused predominantly by the slight misplacement of the ports inherent in the choice of Δx and Δy.

C. Microstrip Branch Line Coupler

The branch line coupler, shown in Fig. 11, is used to divide power equally between ports 3 and 4 from ports 1 or 2. This occurs at the frequency where the center-to-center distance between the four lines is a quarter wave-

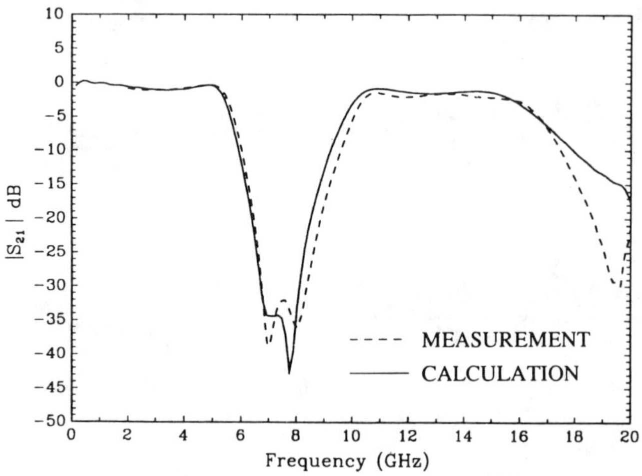

Fig. 10. Insertion loss of the low-pass filter.

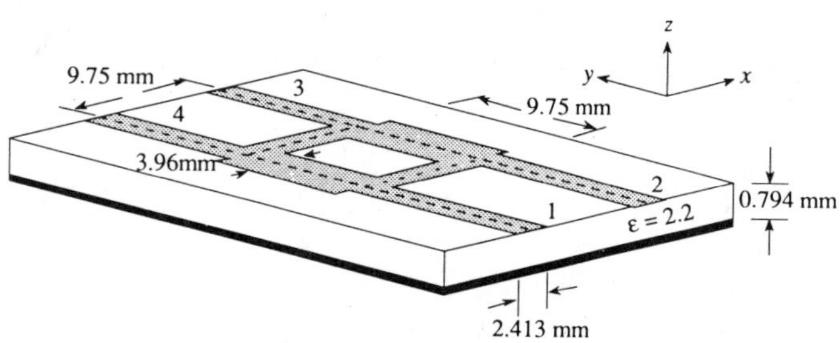

Fig. 11. Branch line coupler detail.

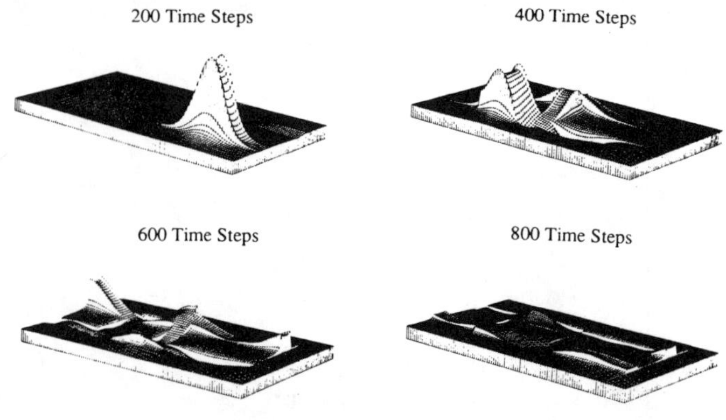

Fig. 12. Branch line coupler distribution of $E_z(x, y, t)$ just underneath the dielectric interface at 200, 400, 600, and 800 time steps.

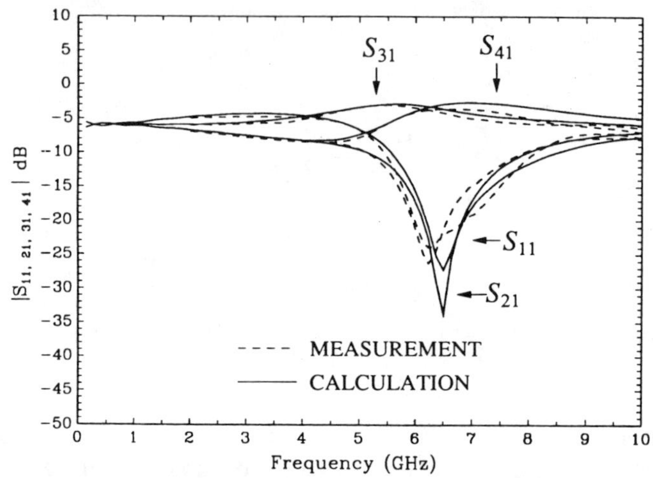

Fig. 13. Scattering parameters of the branch line coupler.

length. Also, the phase difference between ports 3 and 4 is 90° at this frequency. To model this circuit, Δx, Δy, and Δz are chosen to match the dimensions of the circuit as effectively as possible. The space step, Δz, is chosen to match the substrate thickness exactly. The space steps, Δx and Δy, are chosen to match the center-to-center distance (9.75 mm) exactly. Again, small errors in the other $\hat{x}$ and $\hat{y}$ dimensions occur.

The space steps used are $\Delta x = 0.406$ mm, $\Delta y = 0.406$ mm, and $\Delta z = 0.265$ mm and the total mesh dimensions are $60 \times 100 \times 16$ in the $\hat{x}$, $\hat{y}$, and $\hat{z}$ directions, respectively. The center-to-center distances are $24\Delta x$ and $24\Delta y$. The distance from the source plane to the edge of the coupler is $50\Delta y$, and the reference planes for ports 1 through 4 are $10\Delta y$ from the edges of the coupler. The strip widths of ports 1 through 4 are modeled as $6\Delta x$. The wide strips in the coupler are modeled as $10\Delta x$ wide.

The time step used is $\Delta t = 0.441$ ps. The Gaussian half-width is $T = 15$ ps and the time delay t_0 is set to be $3T$. The simulation is performed for 4000 time steps to allow the response on all four ports to become nearly zero. The computation time for this circuit is approximately 6 h on a VAXstation 3500 workstation.

The spatial distribution of $E_z(x, y, t)$ just beneath the microstrip at 200, 400, 600, and 800 time steps is shown in Fig. 12. The scattering coefficient results, shown in Fig. 13, again show good agreement in the location of the response nulls and crossover point. The desired branch line coupler performance is seen in the sharp S_{11} and S_{21} nulls which occur at approximately the same point (6.5 GHz) as the crossover in S_{31} and S_{41}. At this crossover point S_{31} and S_{41} are both approximately -3 dB, indicating that the power is being evenly divided between ports 3 and 4. The nulls in S_{11} and S_{21} at the operating point indicate that little power is being transmitted by ports 1 and 2. The phase difference between S_{31} and S_{41} is verified to be approximately 90° at the operating point (≈ 6.5 GHz) in both the calculated and measured coefficients, as shown in Fig. 14. Some shift in the location of the nulls is again observed and is likely due to the slight modeling errors in the widths of the lines.

D. Error Discussion

In general, agreement between measured and calculated results has been good; however, there are several reasons to explain the small discrepancies in the results. The modeling error occurs primarily in the inability to match all of the circuit dimensions. The space steps, Δx, Δy, and Δz, may be freely chosen; this allows exact matching of only one circuit dimension (or two edges) in each of the $\hat{x}$, $\hat{y}$, and $\hat{z}$ directions. Another small source of error is the exclusion of dielectric and conductor loss in the FDTD calculation. This causes the calculated S parameters to be shifted up in amplitude from the measured data at the higher frequencies. Measurement errors occur because of the microstrip-to-coaxial transitions, which are

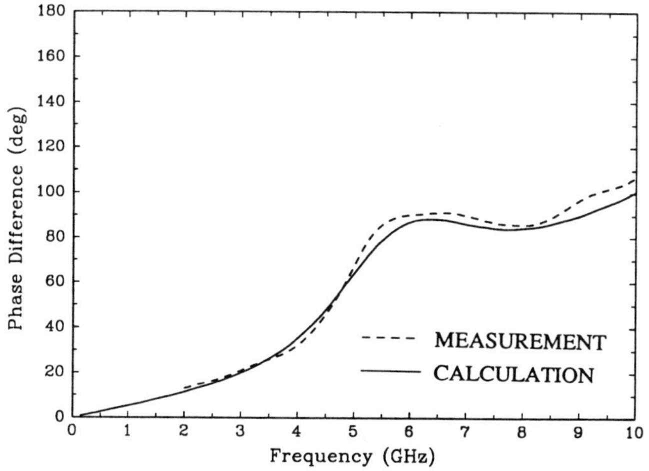

Fig. 14. Phase difference of port 3 relative to port 4 of the branch line coupler.

not de-embedded in the measurement, and because the network analyzer and its connectors are rated only to 18 GHz.

IV. Conclusions

The finite-difference time-domain method has been used to perform time-domain simulations of pulse propagation in several printed microstrip circuits. In addition to the transient results, frequency-dependent scattering parameters and the input impedance of a line-fed rectangular patch antenna have been calculated by Fourier transform of the time-domain results. These results have been verified by comparison with measured data. The versatility of the FDTD method allows easy calculation of many complicated microstrip structures. With the computational power of computers increasing rapidly, this method is very promising for the computer-aided design of many types of microstrip circuit components.

References

[1] G. D'Inzeo, F. Giannini, C. Sodi, and R. Sorrentino, "Method of analysis and filtering properties of microwave planar networks," *IEEE Trans. Microwave Theory Tech.*, vol. MTT-26, pp. 462–471, July 1978.

[2] G. D'Inzeo, F. Giannini, and R. Sorrentino, "Novel microwave integrated low-pass filters," *Electron. Lett.*, vol. 15, no. 9, pp. 258–260, Apr. 26, 1979.

[3] K. C. Gupta, R. Garg, and R. Chadha, *Computer-Aided Design of Microwave Circuits*. Dedham, MA: Artech House, 1981.

[4] T. Okoshi, *Planar Circuits for Microwaves and Lightwaves*. Berlin: Springer-Verlag, 1985.

[5] W. K. Gwarek, "Analysis of arbitrarily shaped two-dimensional microwave circuits by finite difference time domain method," *IEEE Trans. Microwave Theory Tech.*, vol. 36, pp. 738–744, Apr. 1988.

[6] K. C. Gupta and M. D. Abouzahra, "Planar circuit analysis," in *Numerical Techniques for Microwave and Millimeter-Wave Passive Structures*, T. Itoh, Ed. New York: Wiley, 1989, pp. 214–333.

[7] P. B. Katehi and N. C. Alexopoulos, "Frequency dependent characteristics of microstrip discontinuities in millimeter wave integrated circuits," *IEEE Trans. Microwave Theory Tech.*, vol. MTT-33, pp. 1029–1035, Oct. 1985.

[8] R. W. Jackson, "Full-wave, finite element analysis of irregular microstrip discontinuities," *IEEE Trans. Microwave Theory Tech.*, vol. 37, pp. 81–89, Jan. 1989.

[9] S. Koike, N. Yoshida, and I. Fukai, "Transient analysis of microstrip line on anisotropic substrate in three-dimensional space," *IEEE Trans. Microwave Theory Tech.*, vol. 36, pp. 34–43, Jan. 1988.

[10] T. Shibata, T. Hayashi, and T. Kimura, "Analysis of microstrip circuits using three-dimensional full-wave electromagnetic field analysis in the time domain," *IEEE Trans. Microwave Theory Tech.*, vol. 36, pp. 1064–1070, June 1988.

[11] S. Akhtarzad and P. B. Johns, "Solution of Maxwell's equations in three space dimensions and time by the T. L. M. method of numerical analysis," *Proc. Inst. Elec. Eng.*, vol. 122, no. 12, pp. 1344–1348, Dec. 1975.

[12] W. J. R. Hoefer, "The transmission-line matrix method—Theory and applications," *IEEE Trans. Microwave Theory Tech.*, vol. MTT-33, pp. 882–893, Oct. 1985.

[13] K. S. Yee, "Numerical solution of initial boundary value problems involving Maxwell's equations in isotropic media," *IEEE Trans. Antennas Propagat.*, vol. AP-14, pp. 302–307, May 1966.

[14] A. Taflove and M. E. Brodwin, "Numerical solution of steady state electromagnetic scattering problems using the time dependent Maxwell's equations," *IEEE Trans. Microwave Theory Tech.*, vol. MTT-23, pp. 623–630, Aug. 1975.

[15] A. Taflove, "Application of the finite difference time domain method to sinusoidal steady state electromagnetic penetration problems," *IEEE Trans. Electromagn. Compat.* vol. EMC-22, pp. 191–202, Aug. 1980.

[16] A. Taflove and K. R. Umashankar, "The finite difference time domain (FD–TD) method for electromagnetic scattering and interaction problems," *J. Electromagn. Waves and Appl.*, vol. 1, no. 3, pp. 243–267, 1987.

[17] A. C. Cangellaris, C. C. Lin, and K. K. Mei, "Point matched time domain finite element methods for electromagnetic radiation and scattering," *IEEE Trans. Antennas Propagat.*, vol. AP-35, pp. 1160–1173, Oct. 1987.

[18] X. Zhang, J. Fang, K. K. Mei, and Y. Liu, "Calculations of the dispersive characteristics of microstrips by the time-domain finite difference method," *IEEE Trans. Microwave Theory Tech.*, vol. 36, pp. 263–267, Feb. 1988.

[19] X. Zhang and K. K. Mei, "Time domain finite difference approach for the calculation of microstrip open-circuit end effect," *IEEE Trans. Microwave Theory Tech.*, in *IEEE MTT-S Int. Microwave Symp. Dig.*, 1988, pp. 363–366.

[20] X. Zhang and K. K. Mei, "Time domain finite difference approach to the calculation of the frequency dependent characteristics of microstrip discontinuities," *IEEE Trans. Microwave Theory Tech.*, vol. 36, pp. 1775–1787, Dec. 1988.

[21] D. H. Choi and W. J. R. Hoefer, "The finite-difference time-domain method and its application to eigenvalue problems," *IEEE Trans. Microwave Theory Tech.*, vol. MTT-34, pp. 1464–1470, Dec. 1986.

[22] G. Mur, "Absorbing boundary conditions for the finite difference approximation of the time domain electromagnetic field equations," *IEEE Trans. Electromagn. Compat.*, vol. EMC-23, pp. 377–382, Nov. 1981.

[23] B. Enquist and A. Majda, "Absorbing boundary conditions for the numerical simulation of waves," *Math. Comput.*, vol. 31, no. 139, pp. 629–651, July 1977.

[24] K. K. Mei and J. Fang, "A super-absorption boundary algorithm for one dimensional wave equations," submitted to *J. Comput. Phys.*

Full-Wave Analysis of Coplanar Discontinuities Considering Three-Dimensional Bond Wires

M. Rittweger, M. Abdo, and I. Wolff, FELLOW, IEEE

Department of Electrical Engineering and Sonderforschungsbereich 254
Duisburg University, Bismarckstr. 69, D-4100 Duisburg, FRG

Abstract

The analysis of a simple coplanar band-reject filter based on open ended stubs using the finite difference time-domain (FDTD) method is presented. The influence of bond wires for grounding the side conductors of the coplanar lines is investigated. A comparison of the obtained results with accurate measurements is shown. Results are used for the discussion of the effect on coplanar circuit design.

Introduction

The coplanar line has some advantages in monolithic microwave integrated circuit design. However, it has not achieved high dissemination yet because of missing design tools. With the exception of single or coupled line structures the knowledge of coplanar elements is not comparable with that of microstrip components. There are some facts which make circuit design more easy in coplanar technique. These are essentially the small dispersion of the coplanar line, the simple realization of short circuited ends and the simple integration of lumped elements or active components. Parasitic effects are also much smaller for coupled structures where the ground capacitance is unwanted, as in spiral inductors. On the other hand, the grounding of the side-conductors of the coplanar line sometimes causes trouble, a problem that is hardly considered up to now. If grounding is assured the coplanar structures can be analyzed with quasi-static methods [1]. The effects due to mode conversion can be analyzed with nearly all planar analysis techniques such as the spectral domain analysis or the method of lines. However, the influence of bond wires or air bridges on grounding the side conductors can only be analyzed using a true three-dimensional full-wave analysis technique.

The FDTD Method

The three-dimensional FDTD method applied to planar microwave circuits is well introduced in [2]. A transient analysis in the time-domain is used to determine the S-parameters of the investigated structure. Therefore a single computation run is sufficient to deliver the complete frequency response. First applications of the FDTD method to coplanar lines have been reported in [3]. In this publication the main emphasis is put on the accurate description of the fields in the coplanar line in order to investigate the current distribution, the corresponding characteristic impedance and the propagation constant. Therefore a fine mesh for spatial discretization is used. But with the application of the FDTD method to coplanar discontinuities a coarse mesh must be used to receive an acceptable computational expense. Despite the rough discretization the dynamics of the coplanar structures can be analyzed with good accuracy.

Pulse Excitation

A retarded gaussian pulse of the form

$$g(t) = e^{-(\frac{t-t_0}{T})^2} \quad (1)$$

is excited at one port of the investigated structure using the matched source technique introduced in [4]. The field distribution in the excitation plane is only the electric field at the dielectric interface between the center electrode and the side electrodes. A more detailed field distribution as e.g. the quasi-static solution only leads to faster stabilization of the physical pulse form, but does not change the result. The pulse duration has been set with the time constant T in such a way that the pulse width is about 60 space elements whereby the threshold is 5 percent of the maximum. The time delay turns on the pulse with a distance of

Reprinted from *IEEE MTT-S International Microwave Symposium Digest*, pp. 465-468,

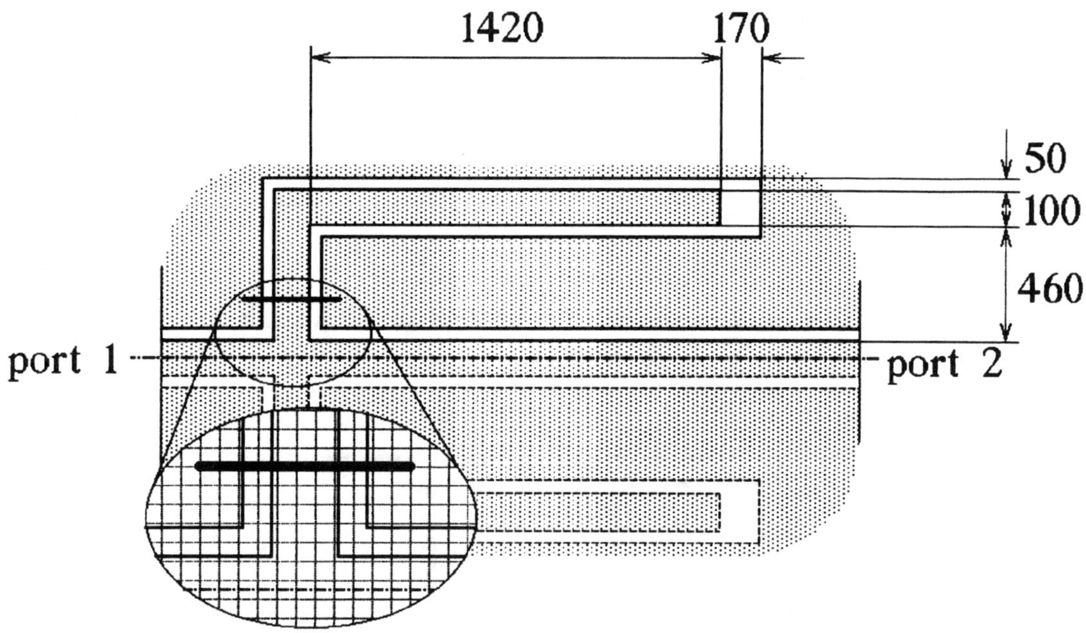

Figure 1: The coplanar band-reject filter.

also 60 space elements from the center of the pulse. Therefore the frequency range of the excited pulse is from dc up to about 130 GHz with the used discretization.

Boundary Treatment

The computation domain is truncated with artificial absorbing boundaries based on Sommerfeld's radiation condition. The used algorithm is mainly based on the super absorption method [5] and is described in [4]. The symmetry plane in the structure (Fig. 1) is replaced by a magnetic wall which reduces the computer memory need and therefore the computation time by half.

The Band-Reject Filter

The coplanar band-reject filter has been designed for application in a frequency multiplier. It should have no transmission for 18 GHz and good transmission for 36 GHz. With the assumption of negligible dispersion and open end capacitance the effective length of the bended stub has been selected to 1780 μm on a 635 μm Al_2O_3-substrate with a permittivity of $\varepsilon_r = 9.8$. The size of the structure was reduced using a bend in the stub. Furthermore the band-reject filter has been realized using a symmetric stub to avoid additional bond wires. The S-parameters of the band-reject filter with bond wires (Fig. 1) are compared with measurement results (dashed lines) performed with a HP 8510 and a CASCADE prober station (Fig. 2).

The essential difference between the measured and calculated results is the absence of attenuation in the calculated results and therefore the sharper resonant phenomenon at about 26 GHz. The measured results also show this effect which must be explained with the interaction of the connected discontinuities together with higher order modes different from the symmetric quasi-TEM mode on the stubs. Each used discontinuity in the structure analyzed separately shows negligible or only a weak frequency dependence in the investigated frequency range.

The position and length of the bond wires take big influence on the resonant frequency of the stubs. They must be positioned as close as possible to the cross junction. Otherwise the resonant frequency of the stubs increases as several simulation results showed. This means that theoretical field approaches that are not able to consider the bond wires have strong restrictions in coplanar circuit design.

The S-parameters of the investigated structure without bond wires are shown in Fig. 3. There is no similarity between the transfer functions of the structure with and without the bond wires. Apparently the structure can now be interpreted as two symmetrically coupled winding slotlines.

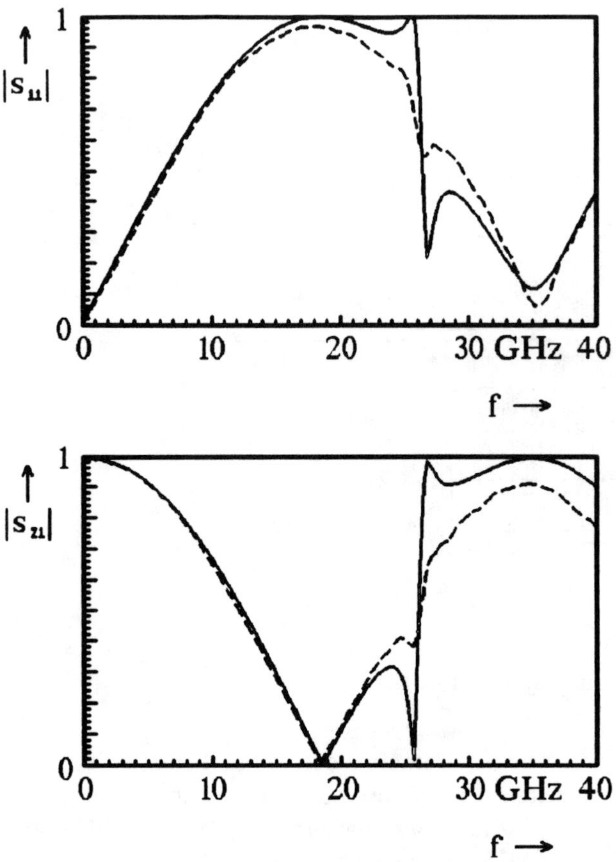

Figure 2: S-parameters of the coplanar band-reject filter with bond wires, theory (solid lines), measurement (dashed lines).

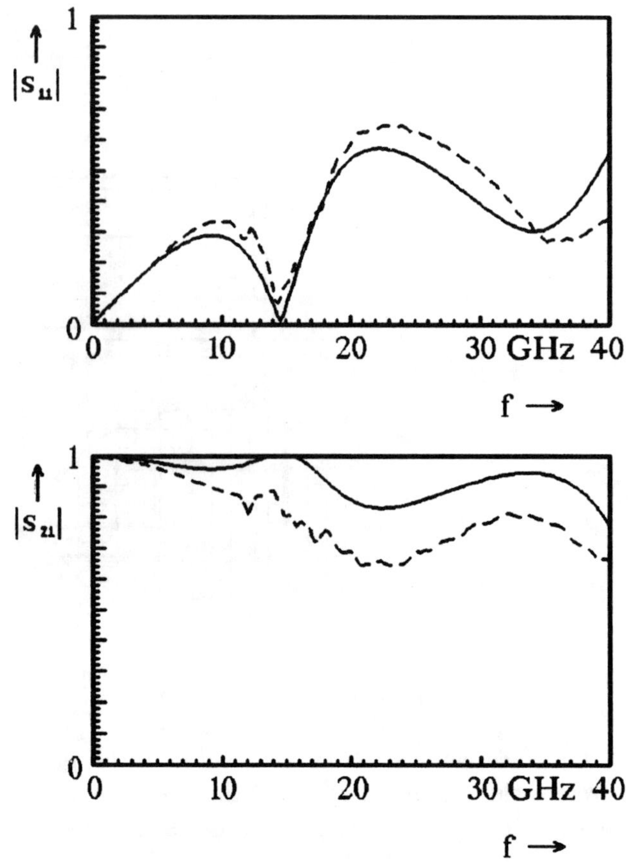

Figure 3: S-parameters of the coplanar band-reject filter without bond wires, theory (solid lines), measurement (dashed lines).

The excitation of the odd mode in the coplanar cross junction for different positions of the bond wires is shown in Fig. 4 in a simulation. u_1^{even} is the voltage of the coplanar incident wave at port 1. u_3^{odd} is the voltage of the unsymmetric outgoing wave at port 3. All the lines connected to the coplanar cross junction are infinitely long. The amplitude of the excited asymmetrical fundamental mode (which is unwanted) increases with increasing distance of the bond wires from the crossing. This means, that best succes in suppressing the asymmetrical mode is obtained by placing the bond wires as close as possible to the discontinuity.

In [6] a T-junction with an air bridge center conductor (Fig. 5) is introduced. This technique is also applicable for cross junctions. The odd mode suppression of the air bridge T-junction is significantly better than that of the conventional one as it is shown in Fig. 6.

Conclusion

Coplanar circuit design requires the use of different air bridges to supress asymmetrical modes excited at several discontinuities. True three-dimensional full-wave simulators are necessary to analyze the influence of such elements. This has been shown in the case of a coplanar band-reject filter using the FDTD method. Measured and calculated results are in good agreement.

Acknowledgement

The authors wish to thank Mr. M. Naghed for performing the measurements.

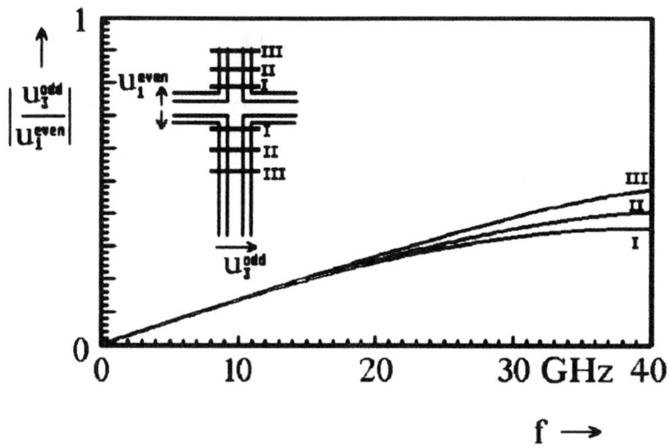

Figure 4: Mode conversion with different bond wire positions.

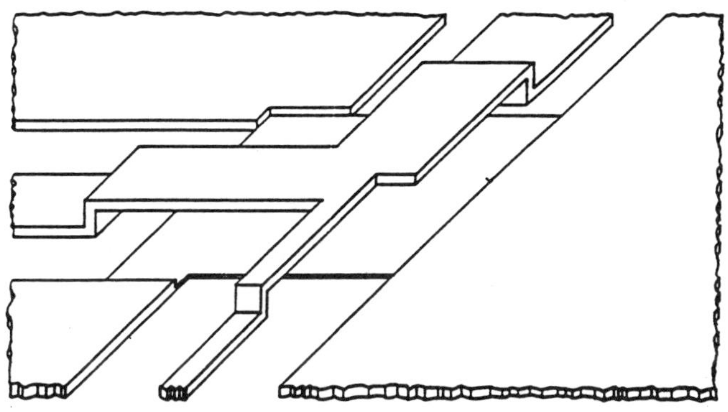

Figure 5: The air bridge T-junction from [6].

References

[1] M. Naghed and I. Wolff, "A three-dimensional finite-difference calculation of equivalent capacitances of coplanar waveguide discontinuities," IEEE MTT-S Digest, pp. 1143-1146, May 1990.

[2] X. Zhang and K. K. Mei, "Time-domain finite difference approach to the calculation of the frequency-dependent characteristics of microstrip discontinuities," *IEEE Trans. Microwave Theory and Tech.*, vol. MTT-36, pp. 1775-1787, Dec. 1988.

[3] G. C. Liang, Y. W. Liu, and K. K. Mei, "Analysis of coplanar waveguide by the time-domain finite-difference method," IEEE MTT-S Digest, pp. 1005-1008, June 1989.

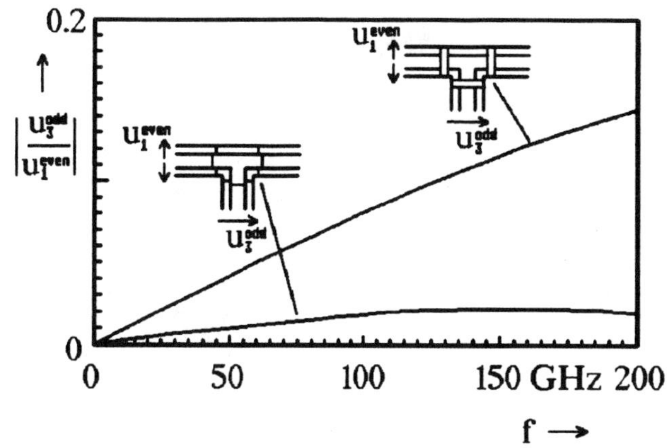

Figure 6: Mode conversion for different T-junctions, $w = 15\mu m$, $s = 10\mu m$, $\varepsilon_r = 12.9$.

[4] M. Rittweger and I. Wolff, "Analysis of complex passive (M)MIC-components using the finite difference time-domain approach," IEEE MTT-S Digest, pp. 1147-1150, May 1990.

[5] K. K. Mei and J. Fang, "Super-absorption: a method to improve local absorbing boundary conditions," submitted for publication in *J. Comput. Physics*.

[6] N. H. L. Koster, S. Koßlowski, R. Bertenburg, S. Heinen, and I. Wolff, "Investigations on air bridges used for MMICs in CPW technique," 19th EuMC Proceedings, London, Sept. 1989.

Analysis of Cross-Talk on High-Speed Digital Circuits Using the Finite Difference Time-Domain Method

N.M. POTHECARY AND C.J. RAILTON
CENTRE FOR COMMUNICATIONS RESEARCH, FACULTY OF ENGINEERING
UNIVERSTIY OF BRISTOL, BRISTOL, BS8 1TR, U.K.

SUMMARY

Cross-talk and pulse distortion in high-speed digital circuits have long been major design considerations. With the present trend towards increased speed and packing density the commonly used TEM approximation is no longer capable of producing sufficiently accurate results. In this contribution, the finite-difference time-domain technique incorporating a novel means of launching the pulses is used to provide a rigorous three-dimensional simulation of cross-talk on high-speed digital circuits. This includes the often-neglected effects of higher-order modes and surface waves. Results are presented for pulses propagating along a single track and for coupling between two and three tracks using square, gaussian and raised cosine pulse excitation functions. The results are shown to be in good agreement with other rigorous analyses.

INTRODUCTION

The current trend in high-speed digital circuit design is towards increased circuit density and faster switching speeds. Device switching speeds of 30 ps are already obtainable in FETs and this is likely to improve in the future. The interconnections in such circuits require careful consideration and the designer must take into account effects such as delays, reflections, dispersion and cross-talk. To model the behaviour of circuit interconnections such as metal tracks on a PCB or in digital integrated circuits, the dispersive nature and close proximity of other transmission lines must be considered. These factors can result in significant distortion of pulses that travel even a short distance. The effect of cross-talk is particularly insidious since it can cause false switching only under certain conditions which may occur only infrequently. A rigorous model of the circuit can identify problems of this nature at the design stage and thus save much time and effort on fault diagnosis which would otherwise be particularly difficult.

The causes of pulse distortion can be separated into three categories, namely dispersion, attenuation and coupling.

Pulse propagation on single- and multi-conductor transmission lines has received much attention in the literature and various methods have been used to obtain the characteristics of microstrip-like structures. Many use a TEM approximation,[1–5] but this ignores dispersion and the existence of higher-order modes. With picosecond pulses these effects can no longer be neglected if good accuracy is to be obtained.

Of the published rigorous analyses many use frequency-domain methods such as the method of moments (MoM),[6, 7] but these necessitate the use of Fourier Transforms when used with non-sinusoidal signals which increase the complexity of the analysis. Moreover, the MoM requires computer storage and computation time of the order $(3N)^2$ and $(3N)^3$ respectively, where N is the number of cells. Numerically efficient algorithms have been developed but the very long computation time and storage requirements prohibit its application to problems involving complex geometries requiring a large number of unknowns.

In contrast to the MoM the finite-difference time-domain, (FDTD), method has storage and time requirements proportional to N which allows more complex geometries to be handled. Although for an equivalent accuracy N may be larger for the FDTD method than for the MoM, it still provides a good alternative. In addition a time-domain method such as FDTD can be used with any desired excitation function and will provide information on the transient effects of cross-

Reprinted with permission from *International Journal of Numerical Modeling: Electronic Networks, Devices and Fields*, N. M. Pothecary, and C. J. Railton, "Analysis of Cross-Talk on High-Speed Digital Circuits Using the Finite Difference Time-Domain Method," Vol. 4, pp. 225-240, 1991.

talk anywhere in the circuit without the need to use Fourier transforms. It is also possible to predict the level of radiation emanating from the circuit and therefore the likelihood of interference with neighbouring equipment. As will be shown later, a time-domain analysis can give much valuable information concerning the mechanism by which the cross-talk is caused. With this information, not normally available from traditional analyses, the finding of a remedy is facilitated.

Few detailed results of coupling between several conductors on a PCB are available and the effect of pulse shaping on cross-talk between multiple transmission lines has received little attention. The FDTD method is used in this paper to investigate these problems and, where available, the results are compared to those of other authors. The extension of the FDTD method to include effects such as conduction and dielectric losses[8] and radiation losses[9] is straightforward and is the subject of future work.

THE FINITE DIFFERENCE TIME-DOMAIN METHOD

The finite difference time-domain method was first proposed by Yee[10] and has subsequently been used for a wide variety of applications including the analysis of various planar transmission line structures.[10, 11] The implementation of the basic method is well described in the literature[10–14] and need not be detailed here.

The implementation of the Yee Algorithm is necessarily based upon a finite number of cells, but by using a non-uniform mesh a greater number of nodes can be placed in regions of large field variations (e.g. track edges) to achieve good computational efficiency and accuracy.[11, 13] In this contribution a non-uniform grid has been used in conjunction with absorbing boundary conditions.[19]

Using a time-domain method enables the propagation of a pulse to be seen clearly without the need for any frequency–time transformations. In addition the ability to be able to use any desired pulse function means that it is easy to see the effects of pulse shaping.

Complex structures with features that would cause difficulties with other rigorous formulations can be accommodated with little penalty in computation time. For example, the finite thickness and non-uniform cross-section of tracks as a result of under-etching can become significant when circuit dimensions are small.[13] Using the analysis method described here these features can be taken into account with little or no increase in computer time.

Another important and useful advantage of using a time-domain technique is that non-linear components can be accommodated. Using frequency-domain methods to analyse non-linear circuits creates major difficulties.

COMPARISON WITH OTHER METHODS

In order to prove the effectiveness of the approach described here the structure of Figure 1 was analysed using the FDTD method and the results compared to those obtained using the quasi-static method[3] and the frequency-domain method of Gilb and Balanis.[7] The results for the FDTD

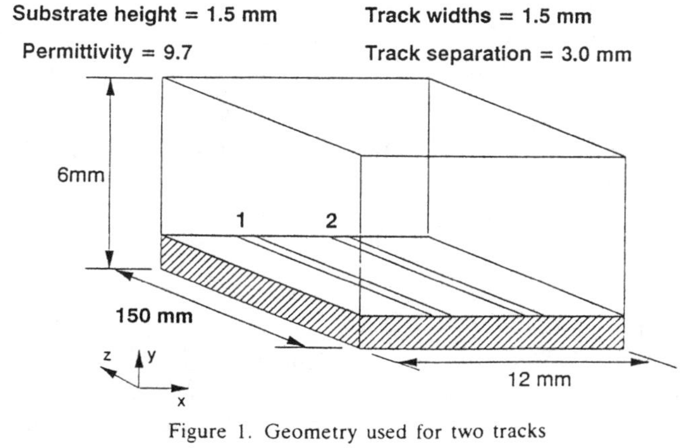

Figure 1. Geometry used for two tracks

method were obtained using the mode template technique (see later) and the geometry used is the same as that described by Gilb and Balanis.[7] To facilitate comparison we have used the same Gaussian excitation as Gilb and Balanis.[7] This has the following form:

$$E(t) = A \exp\left\{-\ln(2)\left(\frac{t}{\tau}\right)^2\right\} \tag{1}$$

Here A is the amplitude of the pulse, τ is the half-maximum, half-width time, and the pulse is centred about $t = 0$.

Figure 2(a) shows the signal pulse at 40 mm using the three different methods and Figure 2(b) shows the sense line response at this distance. There is good agreement between the FDTD method and that of Gilb and Balanis whilst the quasi-static method is less accurate in this case. There is further evidence of this in Figures 2(c) and 2(d) showing the response at 80 mm. The rigorous approaches are still in good agreement, whereas the results from the quasi-static method show considerable discrepancy.

SURFACE WAVES FROM A SINGLE TRACK

The propagation of a narrow pulse along a single isolated track was investigated (Figure 3) in order to show the presence of surface waves on the dielectric interface. These waves which have zero cutoff frequency will interact with other conductors which are in close proximity (e.g. neighbouring tracks) and give rise to cross-talk. In this example the pulse propagates along an infinitely long metallic track placed on a GaAs substrate with permittivity 12.9 and a first-order Mur boundary[14] is placed in the x–y plane at $z = L$. Since the purpose of this simulation is to study the cross-talk between tracks, the effects of the terminations are not considered.

For this test a voltage source is simulated by exciting a vertical cross-section between the track and ground plane, at $z = 0$, by a uniform electric field which can vary with time in any desired manner. This enables the effect of using different excitation functions, e.g. square, truncated gaussian, or raised cosine pulses, to be determined.

The pulse used in this test is a raised cosine pulse of duration, T, 30 ps, and peak amplitude, A, 1·5. The excitation function $E(t)$ is given by (2):

$$\begin{aligned} E(t) &= \frac{A}{2}\left(1 - \cos\frac{2\pi}{T}t\right) \qquad && t \leqslant T \\ &= 0 && t > T \end{aligned} \tag{2}$$

The amplitude of the pulse is zero at $t = 0$ and $t = T$; therefore no truncation is required.

Figure 4 shows the pulse at four different positions along the track. As the pulse propagates it undergoes dispersion, hence the amplitude falls and the pulse width increases. More information can be seen using a three-dimensional plot of field strength at a particular instant in time. Figure 5 shows the pulse after 120 ps. This illustrates two important details that cannot be seen clearly using only two-dimensional plots.

1. A surface wave can be seen spreading out from the main pulse.
2. At the track edges there are peaks in the electric field strength.

The peaks at the track edges are expected because of the well-known asymptotic behaviour of fields at a sharp edge. The ability to see surface waves enables sources and mechanisms of interference in a circuit to be more easily found and cured.

COUPLING BETWEEN THREE TRACKS

Figure 6 shows the geometry used to investigate coupling between three tracks. The combinations of excitation functions given in Table I were used. Initially only the centre track was excited (Test 1) and tracks 1 & 3 were used as sense lines. Figure 7(a) is a snapshot of electric field strength taken after 60 ps. The presence of signals on the sense lines is clearly shown. On tracks 1 & 3

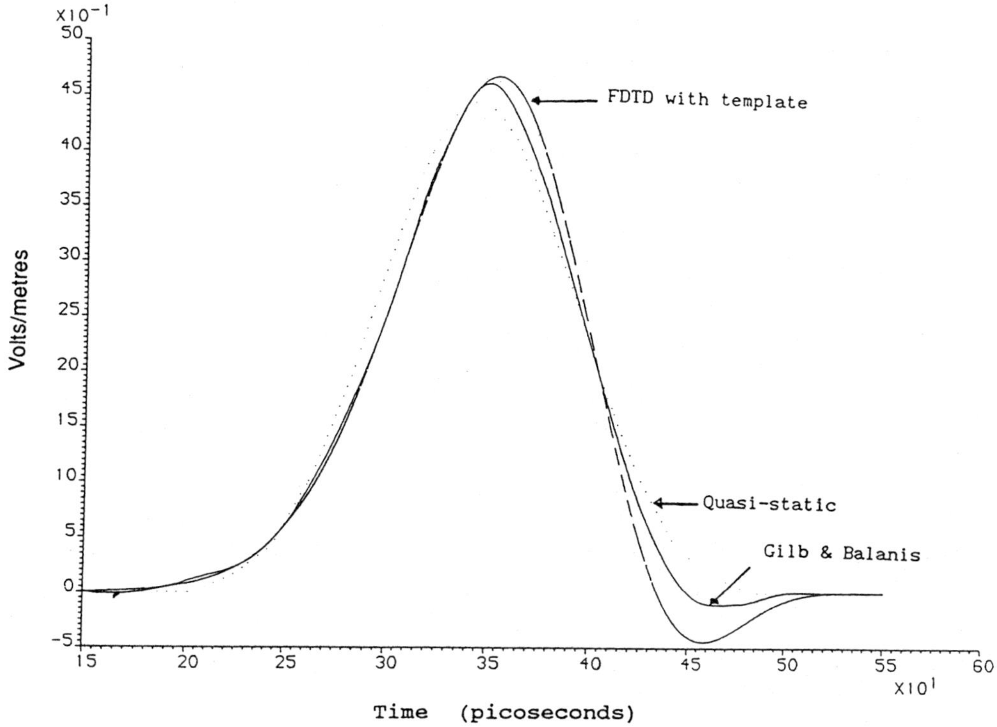

(a)

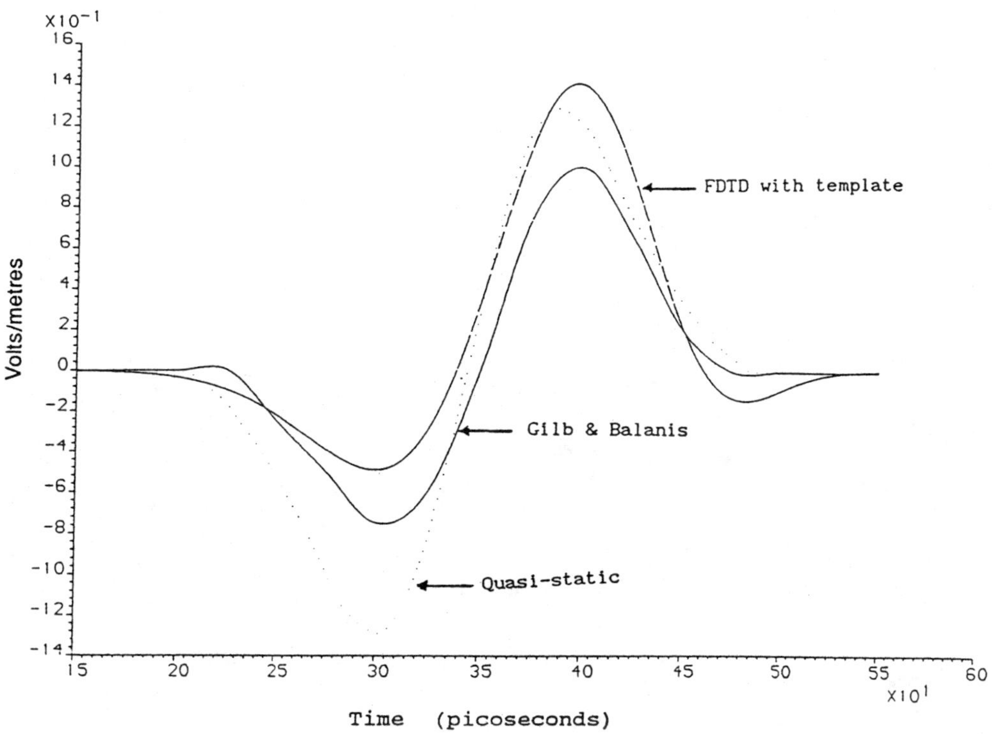

(b)

Figure 2. Comparison with other methods. Response on (a) signal line at 40 mm, (b) sense line at 40 mm, (c) signal line at 80 mm, and (d) sense line at 80 mm for the geometry of Figure 1

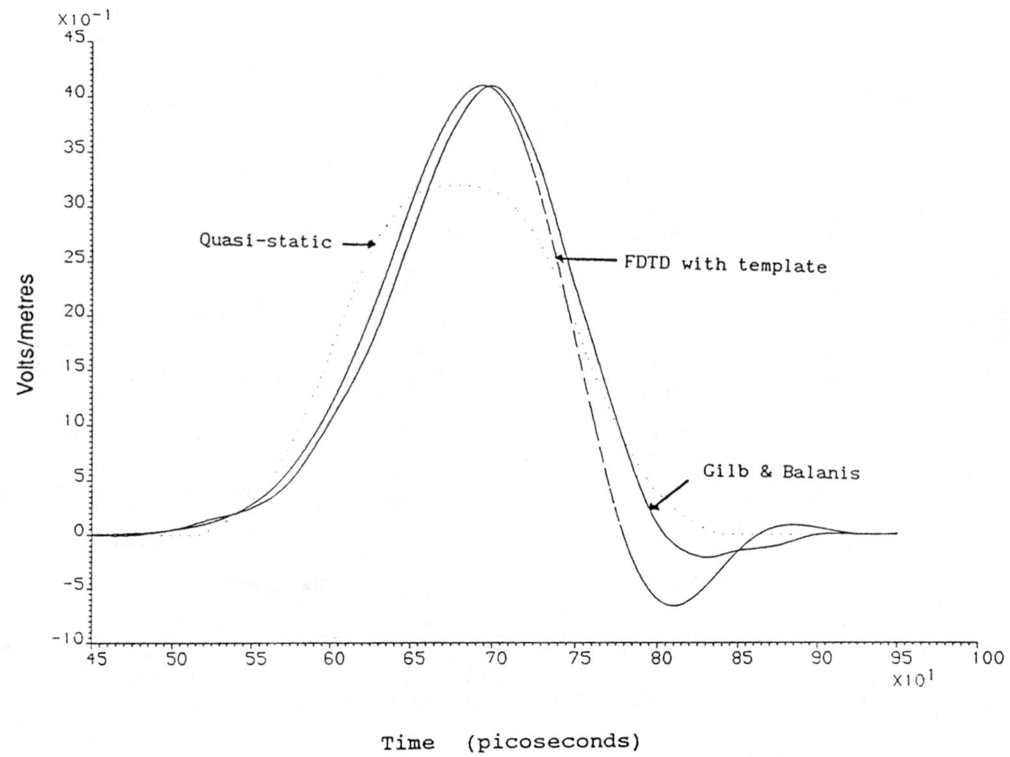

(c)

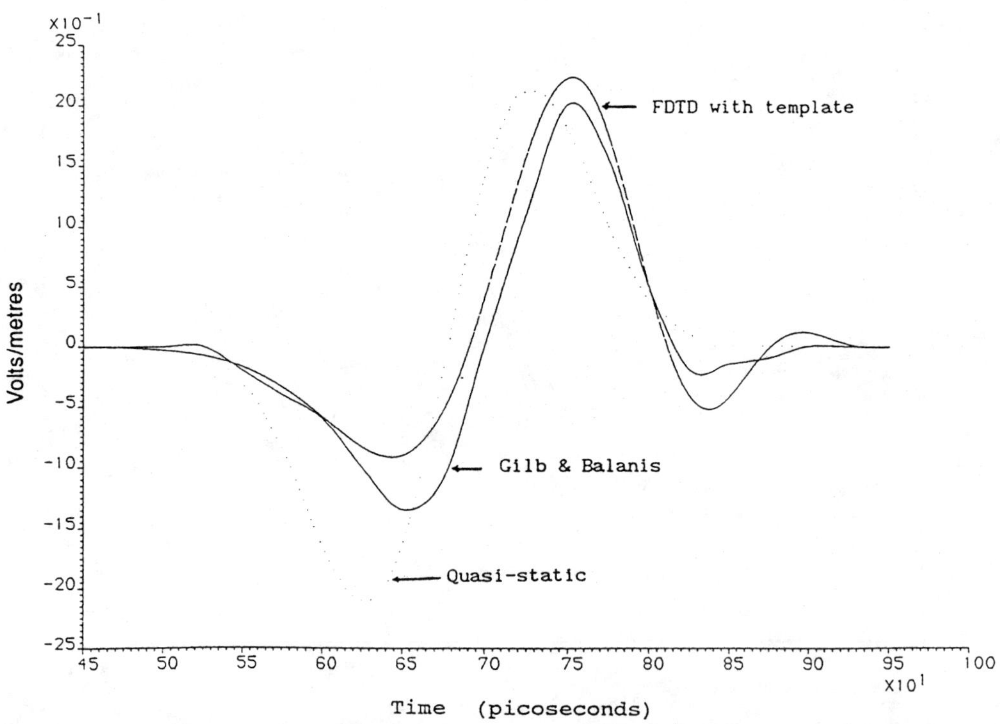

(d)

Figure 2. (Continued)

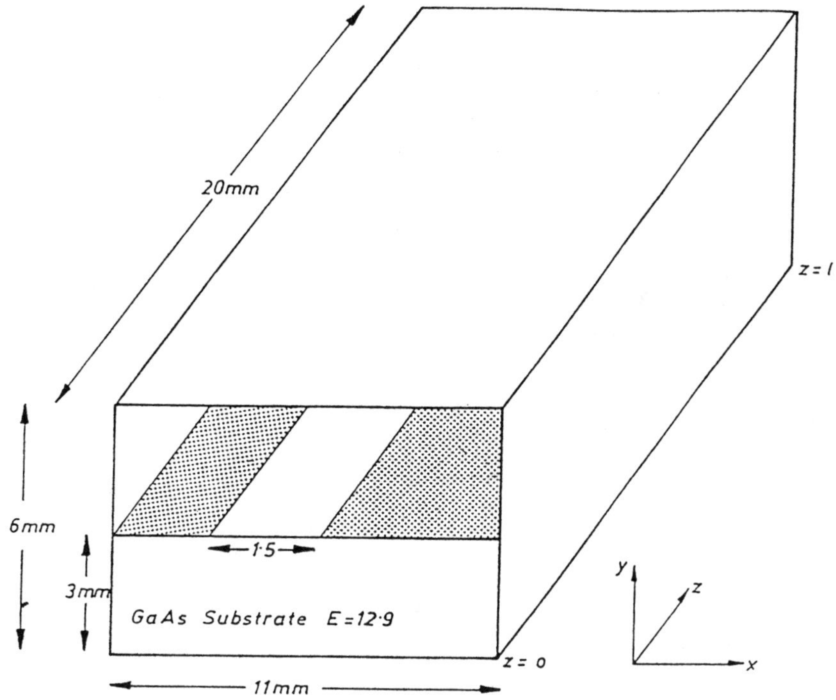

Figure 3. Geometry used for single-track test

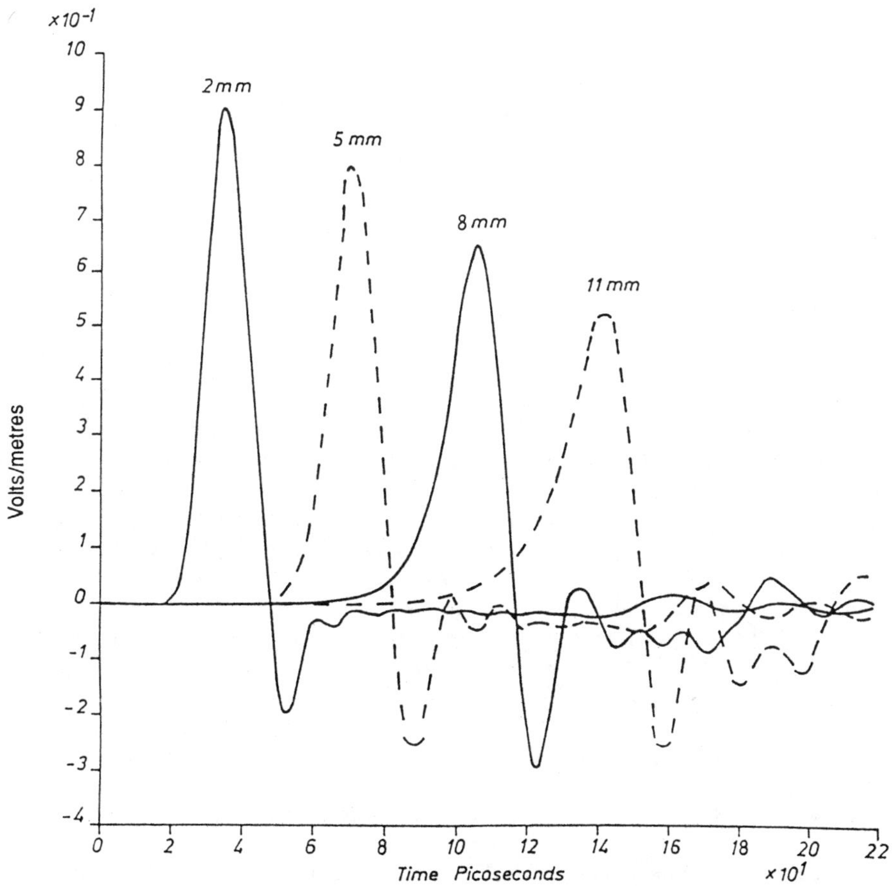

Figure 4. Propagating pulse at four positions along the track for the geometry of Figure 3

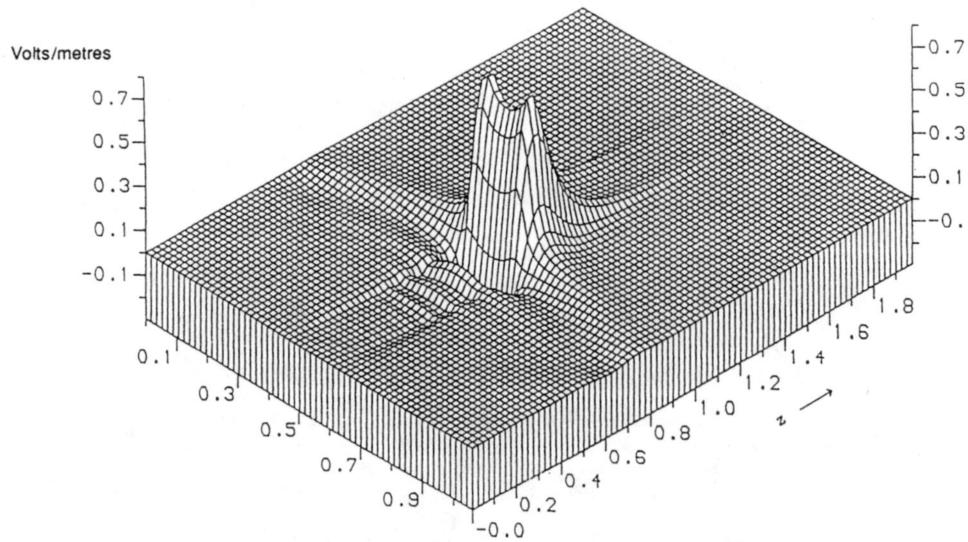

Figure 5. E field distribution at 120 ps for single-track test

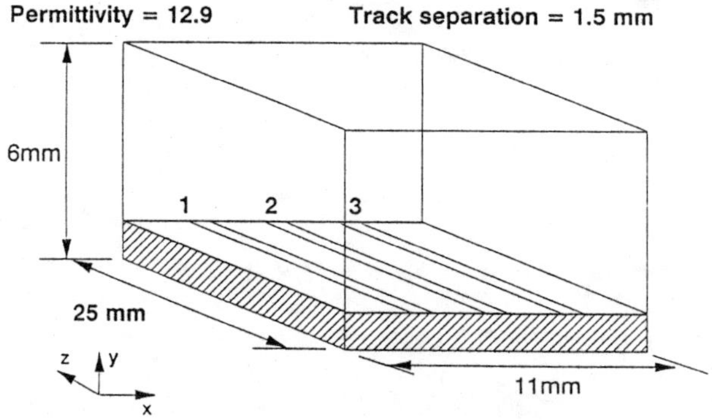

Figure 6. Geometry used for coupling between three tracks

the leading edges of the pulses have the same shape as the leading edge of the surface wave in Figure 5. The surface wave is spreading outwards from the pulse on track 2 and the wavefronts on tracks 1 & 3 are at an angle to the direction of the tracks. The coupled pulses on tracks 1 & 3 do not have peaks in the field strength at the track edges unlike the pulse on track 2. The peaks result from a build-up of charge at the track edges, but this takes a finite amount of time. In figure 7(a) sufficient time has not elapsed for this effect to be manifested.

Figure 7(b) shows the pulses after 120 ps. The wavefront across tracks 1 & 3 is no longer angled and a peak in the field strength is appearing at their outer edges. The magnitude of the pulses on tracks 1 & 3 has also increased whilst that of the pulse on track 2 has decreased. Figure 5 shows a single pulse, also after 120 ps. The amplitude of the excited pulse is decreasing more rapidly when three tracks are present than in the case of a single track. This is because both dispersion and coupling are present when three tracks are used.

After 180 ps (Figure 7(c)) the pulses on tracks 1 & 3 are now greater in magnitude than the pulse on track 2. Owing to dispersion and coupling the pulse on track 2 is spreading out and decreasing in amplitude. The pulses on tracks 1 & 3 are taking the same form as the pulse on track 2 before it changed shape, i.e the pulses on tracks 1 & 3 are delayed versions of the pulse on track 2.

A common situation in digital circuits is for several tracks to act as switching lines, e.g. bus lines in a computer. One possible situation is for the two outer tracks in a set of three to be

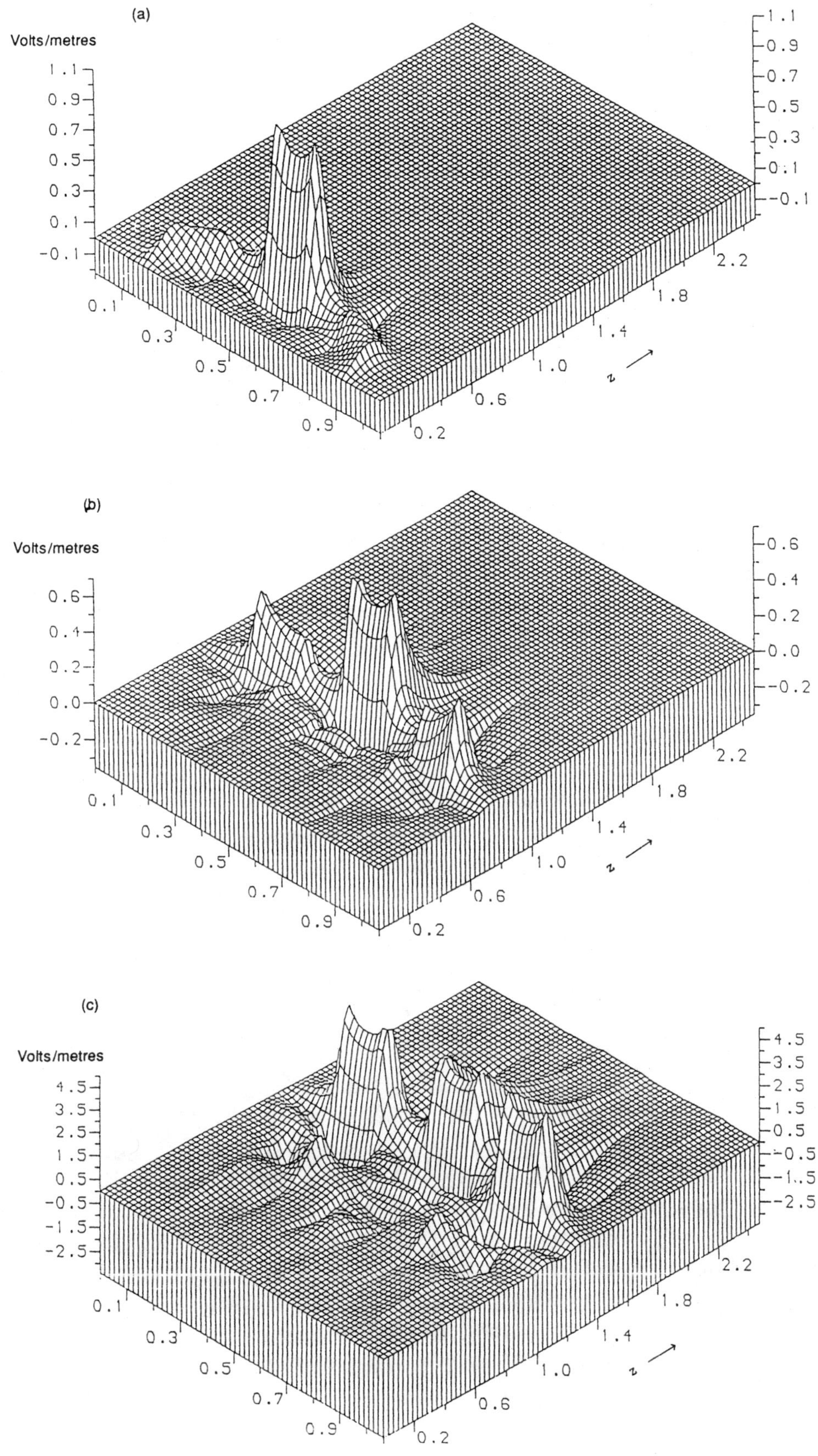

Figure 7. *E* field distribution at (a) 60 ps, (b) 120 ps, and (c) 180 ps for Test 1

excited but no signal is applied to the centre one. A signal appearing on this centre track due to cross-talk may cause spurious results which is clearly an undesirable feature of a circuit.

In Test 2 tracks 1 & 3 are excited using raised cosine pulses of 30 ps duration whilst track 2 acts as a sense line. Figure 8(a) which is a snapshot of electric field strength after 90 ps shows the presence of a signal on the sense line. As in Test 1, the coupled pulse lags slightly behind the main pulse and initially has no peaks in the field strength at the track edges. After 150 ps (Figure 8(b)) the pulse on track 2 has increased in magnitude above that on tracks 1 & 3. The rate of increase in magnitude of the coupled pulse in Test 2 is quicker than in Test 1 since there are now two tracks from which coupling can occur and only a single track to carry the coupled pulse. Figure 8(b) also shows the pulses on tracks 1 & 3 becoming broader as a result of dispersion.

In Test 3 a square pulse with a very short rise-time (0·3 ps) is used. The short rise-time means that the pulse contains more high-frequency components than the raised cosine pulse. Figures 9(a) (60 ps) and 9(b) (120 ps) show results for 30 ps square pulses propagating along tracks 1 & 3 and the resulting signal which appears on track 2. An interesting property of the pulse on track 2 is that it takes the same form (slightly delayed in time) as the pulses on tracks 1 & 3. This same fact was noted in Tests 1 and 2 but is more obvious here.

An initially square pulse changes shape as it propagates, becoming smoother as the higher-frequency components (associated with the short rise-time) separate from the main pulse. In

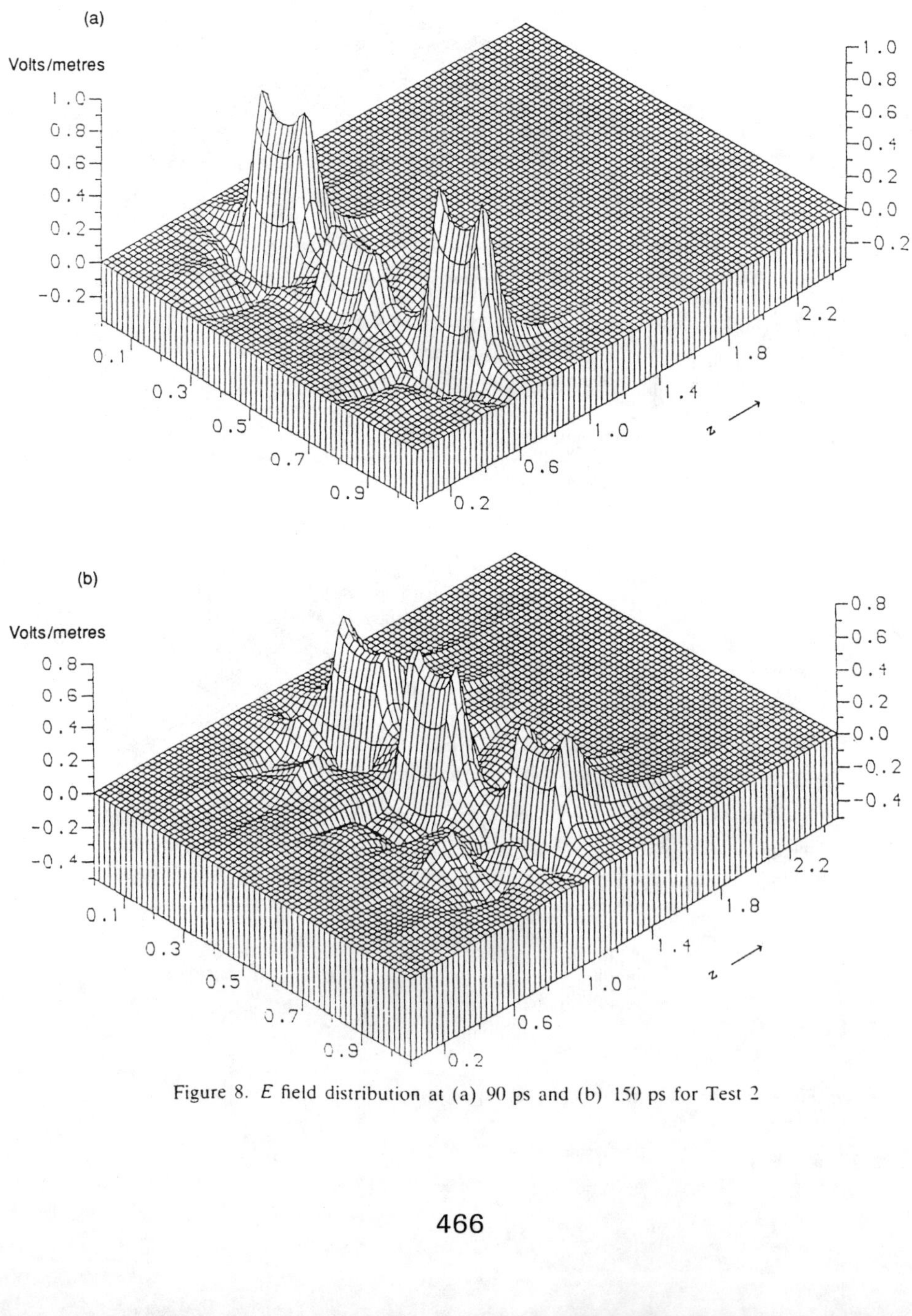

Figure 8. E field distribution at (a) 90 ps and (b) 150 ps for Test 2

Table I

Test number	Form of excitation	Duration of pulse (ps)	Tracks excited
1	raised cosine	30	track 2
2	raised cosine	30	tracks 1 & 3
3	square pulse	30	tracks 1 & 3

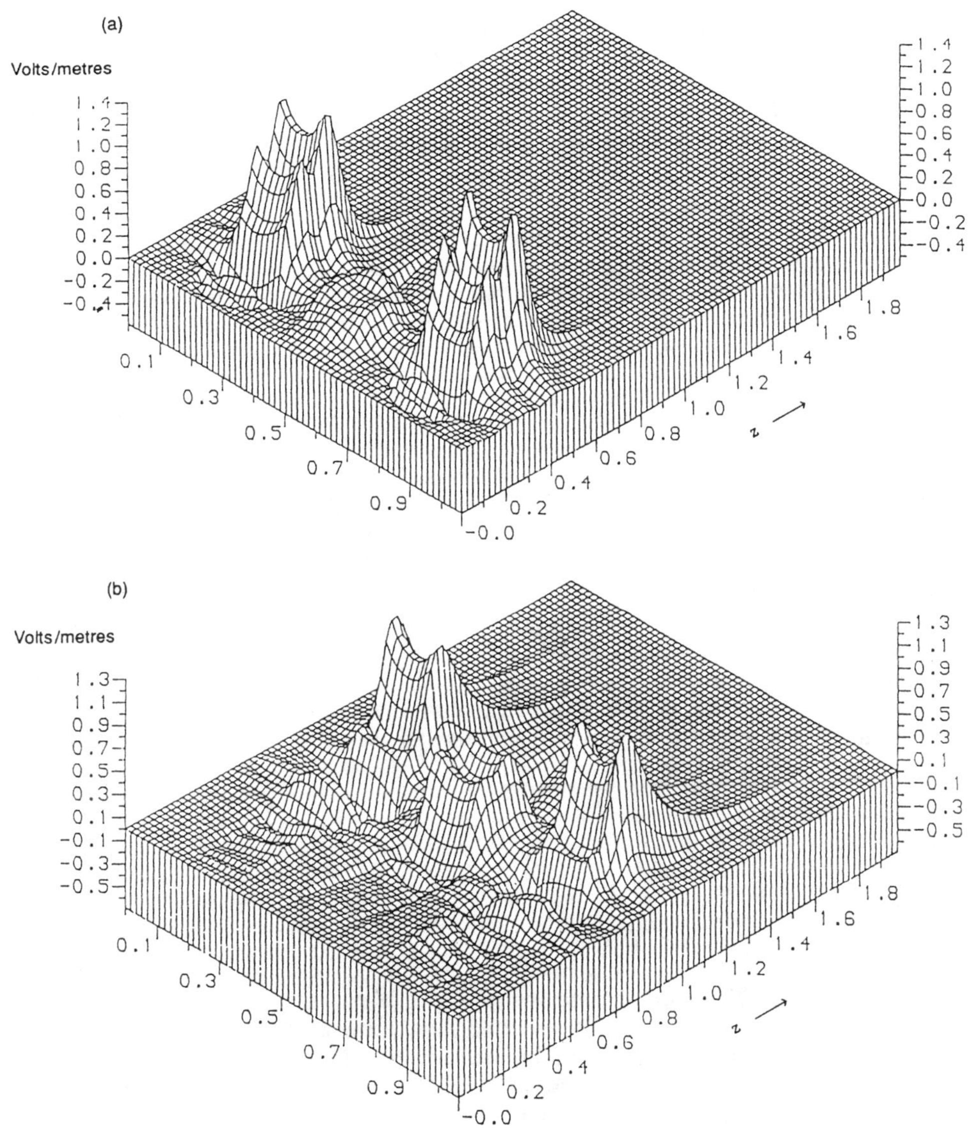

Figure 9. E field distribution at (a) 60 ps and (b) 120 ps for Test 3

Figure 9(a) the signal on track 2 is approximately square in shape and has no peaks in field strength at the track edges. However, in Figure 9(b) small peaks are forming at the four corners of the pulse which now has a very similar shape to the original pulses applied to tracks 1 & 3. The time delay between the pulses on tracks 1 & 3 and the pulse on track 2 is also evident. As the pulse on track 2 continues to propagate it loses its square shape becoming smoother in the same manner as the pulses on tracks 1 & 3.

USE OF MODE TEMPLATES

One of the biggest advantages of using a time-domain method, such as the FDTD technique, is that pulses of any required shape can be placed on the structure under test and the behaviour of the system can be seen directly. In order for this to be achieved, however, care must be taken when launching the wave onto the track. If the launching is not correctly modelled then pulse distortion will take place, owing to the mismatch at the source, which will make it difficult or impossible to ascertain the distortion introduced by the lines. This is analogous to the requirement for a good transition when connecting microstrip to another type of transmission line such as coaxial line. In the literature this aspect has been given very little attention. In Reference 17 it is pointed out that if a simple voltage source is used to launch the wave, then a spurious tail appears on the propagating pulse and the magnitude of the pulse is reduced. It is, however, not clear exactly how the authors of Reference 17 have implemented their suggested solution, although the importance of exciting the *H* field as well as the *E* field is stressed. In Reference 18 the importance of matching the excitation to the field pattern of the dominant mode is stated. In Reference 18 this has been achieved by using an excitation pulse of the following form:

$$\mathbf{E}(x,y,t) = \psi(x,y)G(t) \tag{3}$$

where $\psi(x,y)$ is a template which has been derived from a separate quasi-static analysis of the excitation plane and $G(t)$ has a gaussian time-dependence. Only the *E* field is excited. In order to make this possible for an arbitrary feed line, the existence of the results of an appropriate quasi-static analysis is presupposed.

The results presented in Figures 2(a–d) and 10(a–b) make use of the existing FDTD algorithm to provide a template for the transverse *E* and *H* fields which ensures that, as far as possible, only the dominant mode of the structure is excited and that pulse distortion at the excitation source is kept to a minimum. This technique is immediately applicable to any type of feed line. Having discretized the structure under analysis, which we will refer to as the *original* structure, into a suitable mesh, we derive a second structure which we refer to as the *derived* structure. This derived structure has the same cross-section as the plane of excitation of the original structure and is uniform in the z-direction. An absorbing boundary is then placed at $z = L$. This procedure is illustrated by considering the structure of Figure 1. From this we produce the derived geometry which only differs from Figure 1 by having a box length of 18 mm instead of 150 mm.

In order to produce the required template we analyse the response of the derived geometry to an excitation having the form of an incident y-directed *E* field immediately under the strip and with a time-dependence as follows:

$$E_y(t) = \frac{(1 - \cos \frac{\pi t}{T})}{2} \qquad t \leqslant T$$

$$E_y(t) = 1 \qquad t \geqslant T \tag{4}$$

The value of the constant, T, is chosen large enough to avoid excessive ringing in the mesh without unnecessarily increasing the time required to reach a steady state. Because the absorbing boundary acts as a matched load, we are simulating an infinite section of transmission line with a DC voltage source, V, at the point $z = 0$. Consequently, after the transient effects have decayed, a constant current of V/Z_0 will flow, where Z_0 is the characteristic impedance of the transmission line. Away from the source, the field pattern will correspond to the mode pattern for the dominant quasi-TEM mode of the structure.

After the field has settled, the transverse components of the field at the plane $z = L/2$ are recorded for use as a template. An example of such a template is shown by Figures 10(a–d), clearly illustrating the mode pattern of the structure under analysis. The original structure is then excited at the input plane with the function:

$$\mathbf{E}(x,y,t) = P(t)\ \psi_e(x,y)$$

$$\mathbf{H}(x,y,t) = P(t)\ \psi_h(x,y) \tag{5}$$

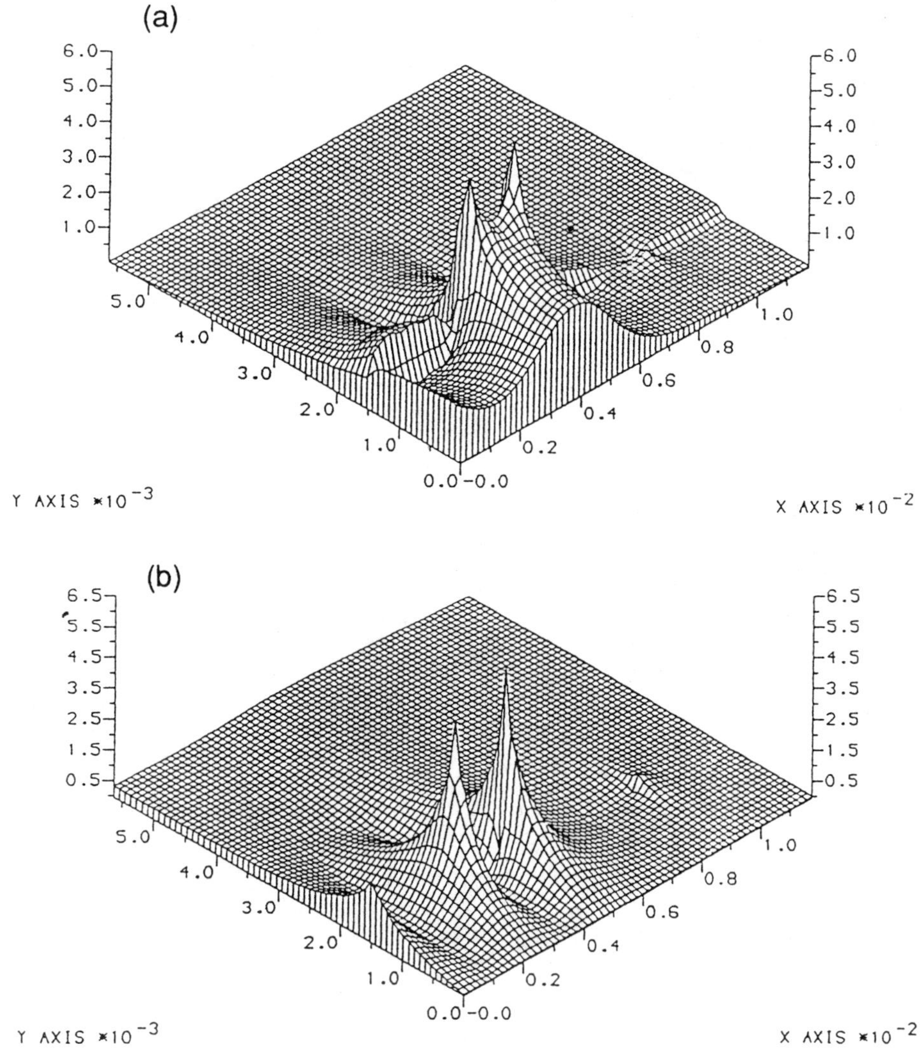

Figure 10. Snapshot of template. (a) *Ey*; (b) *Ex*; (c) *Hy*; (d) *Hx*

where $P(t)$ is the required shape of the pulse, and $\psi_e(x,y)$ and $\psi_h(x,y)$ are the derived E and H field templates.

The effectiveness of this technique is illustrated in Figures 11(a) and 11(b) which show the pulses appearing on the track at 40 mm and 80 mm respectively from the plane of excitation (a gaussian source of time constant 50 ps was used). The pulse launched using the mode template was compared to those launched using a uniform electric field and to the published results from Reference 7. It can be seen that the use of the mode template results in a pulse of the correct amplitude being launched. There is some disagreement between our results and those given in Reference 7 for the magnitude of the tail of the pulse.

The use of a mode template to launch the wave onto the track requires only a small computational overhead. Because the derived geometry is much smaller than the original geometry, the creation of the template requires only about 3 per cent of the time required for the analysis of the original structure. The overhead during the analysis of the original structure is negligible. The advantages of creating the template in this way is that any type of feed can be accommodated, the form of the template is exactly matched to the FDTD mesh used in the original structure and no *a priori* knowledge of the mode patterns is needed.

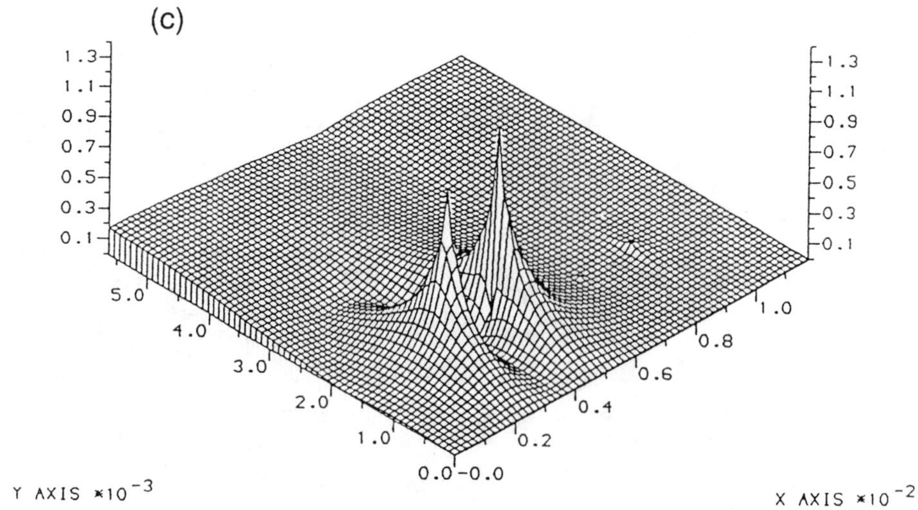

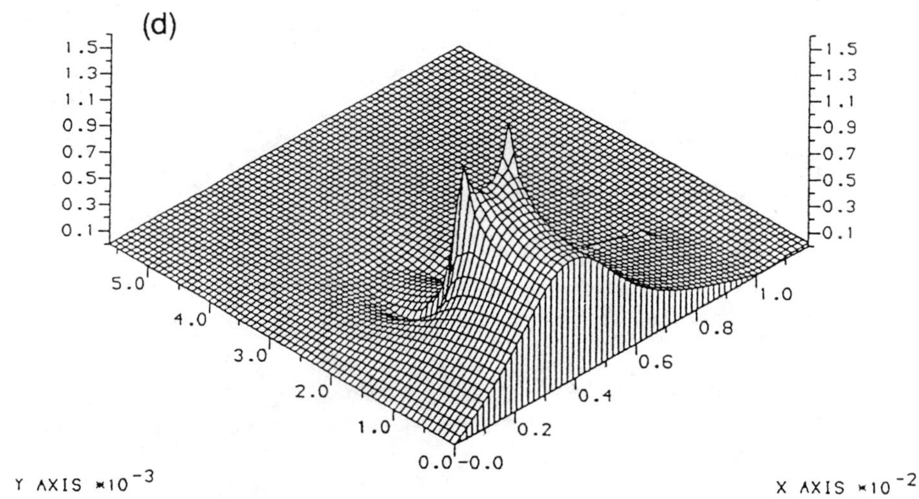

Figure 10. (Continued)

CONCLUSION

We have shown that with the introduction of fast-switching digital devices the often used quasi-static approximation is no longer adequate for accurate prediction of cross-talk and pulse distortion on interconnects. Consequently there is a need for a versatile, rigorous technique with the capability of accurately predicting the level of cross-talk and pulse distortion.

The finite-difference time-domain method has been shown to be very suitable for this purpose. In addition to including the often-ignored phenomena of higher-order modes and surface waves, it can readily take account of non-linear components, under-etching, finite track thickness and thin passivation layers, all of which occur in practice.

In this contribution a number of improvements to the basic FDTD method have been incorporated in order to improve efficiency. These include the use of a non-uniform grid in conjunction with absorbing boundary conditions and the use of a mode template to launch the pulses. This simulation provides a clear pictorial representation of the mechanisms by which cross-talk is caused thereby increasing understanding and facilitating a remedy. The ability to see and take account of not only the pulse amplitude on a line but also the current distribution across the line gives

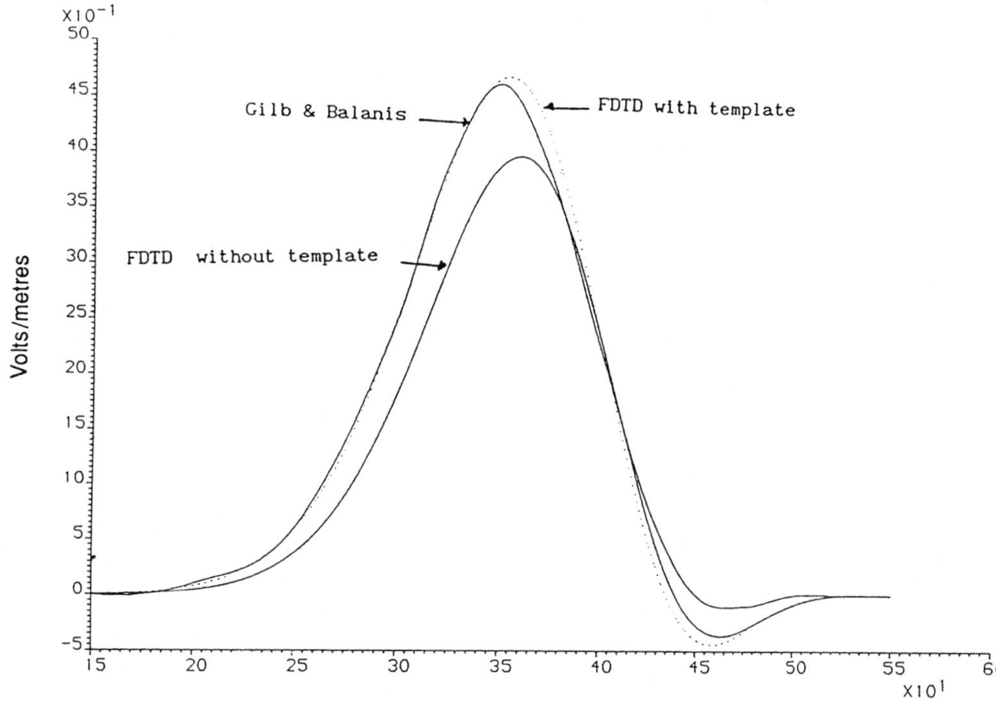

(a)

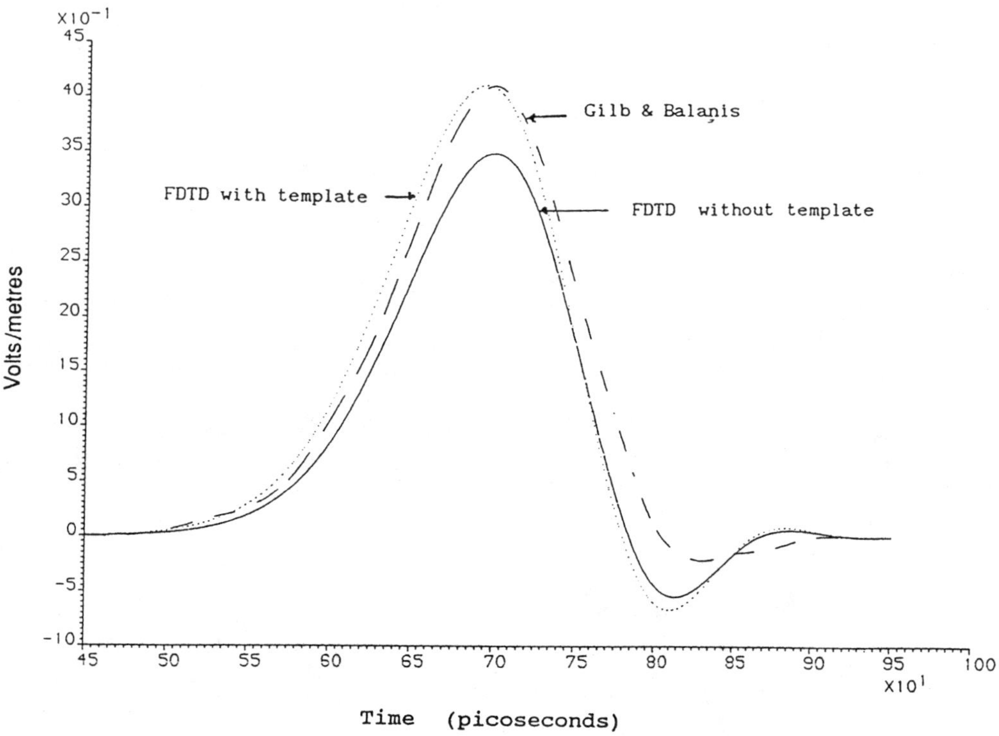

(b)

Figure 11. Comparison of excitation methods. Response (a) at 40 mm on signal line, and (b) at 80 mm on sense line, for the geometry of Figure 1 greater insight into the situation and more accurate results. The distortion of signal pulses and cross-talk has been demonstrated for three parallel tracks and the surface wave which actually causes cross-talk has been shown. Three different pulse-excitation functions have been used, although many others are possible. It has been established that a coupled pulse will take the same shape as the excited pulse and is then also subject to dispersion. Further investigation into the effects of pulse shaping may provide an alternative way of reducing cross-talk (other than changing the geometry of the circuit).

Future work is aimed at applying the technique to more complicated geometries which include discontinuities[15] and crossovers, and in reducing the computation time by incorporating into the algorithm analytical descriptions of the asymptotic field behaviour near the metal tracks.[16]

REFERENCES

1. C. Wei, R. F. Harrington, J. R. Mautz and T. K. Sarkar, 'Multiconductor transmission lines in multilayered dielectric media', *IEEE Trans. Microwave Theory and Techniques*, **MTT-32**, 439–450 (1984).
2. F. Romeo and M. Santomauro, 'Time-domain simulation of *n* coupled transmission lines', *IEEE Trans. Microwave Theory and Techniques*, **MTT-35**, 131–137 (1987).
3. A. R. Djordjevic, T. K. Sarkar and R. F. Harrington, 'Time domain response of multiconductor transmission lines', *Proc. IEEE,* **75**, 111–132 (1987).
4. G. Ghione, I. Maio and G. Vecchi, 'Modelling of multiconductor busses and analysis of crosstalk, propagation delay, and pulse distortion in high speed GaAs logic circuits', *IEEE Trans. Microwave Theory and Techniques*, **MTT-37**, 445–456 (1989).
5. K. W. Goossen and R. B. Hammond, 'Modelling of picosecond pulse propagation in microstrip interconnections on integrated circuits', *IEEE Trans. Microwave Theory and Techniques*, **MTT-37**, 469–478 (1989).
6. V. K. Tripathi and H. Lee, 'Spectral-domain computation of characteristic impedances and multiport parameters of multiple coupled microstrip lines', *IEEE Trans. Microwave Theory and Techniques*, **MTT-37**, 215–221 (1989).
7. J. P. Gilb and C. A. Balanis,'Pulse distortion on multilayer coupled microstrip lines', *IEEE Trans. Microwave Theory and Techniques*, **MTT-37**, 1620–1628 (1989).
8. T. Leung and C. A. Balanis, 'Attenuation distortion of transient signals in microstrip', *IEEE Trans. Microwave Theory and Techniques*, **MTT-36**, 765–769 (1988).
9. E. Belohoubek and E. Delinger, 'Loss considerations for microstrip resonators', *IEEE Trans. Microwave Theory and Techniques*, **MTT-23**, 522–526 (1975).
10. Kane S. Yee, 'Numerical solution of initial boundary value problems involving Maxwell's equations in isotropic media', *IEEE Trans. on Antennas and Propagation*, **AP-14**, 302–307 (1966).
11. D. H. Choi and W. J. R. Hoefer, 'A graded mesh FD-TD algorithm for eigenvalue problems', in *Proc. 17th European Microwave Conference*, Rome, 1987, pp. 413–418.
12. X. Zhang and K. K. Mei, 'Time-domain finite-difference approach to the calculation of the frequency-dependent characteristics of microstrip discontinuities', *IEEE Trans. Microwave Theory and Techniques*, **MTT-36**, 1775–1787 (1988).
13. C. J. Railton and J. P. McGeehan, 'The analysis of microstrip with rectangular and trapezoidal conductor cross section', *IEEE Trans. Microwave Theory and Techniques*, **MTT-38**, 1017–1022 (1990).
14. Gerrit Mur, 'Absorbing boundary conditions for the finite difference approximation of the time-domain electromagnetic-field equations', *IEEE Trans.*, **EMC-23**, 377–382 (1981).
15. C. J. Railton and J. P. McGeehan, 'Characterisation of discontinuities in microstrip with rectangular and trapezoidal cross-section', *Proc. 20th European Microwave Conference*, Budapest, 1990, pp. 549–554.
16. D. B. Shorthouse and C. J. Railton, 'Incorporation of static singularities into the finite difference time domain technique with application to microstrip', *Proc. 20th European Microwave Conference*, Budapest, 1990, pp. 531–536.
17. D. Sheen, S. Ali, M. Abouzahra and J. Kong, 'Application of the three dimensional finite difference time domain method to the analysis of planar microstrip circuits', *IEEE Trans. Microwave Theory and Techniques*, **MTT-38**, 849–857 (1990).
18. G. Liang, Y. Liu and K. K. Mei, 'Full wave analysis of coplanar waveguide and slotline using the time-domain finite-difference method', *IEEE Trans. Microwave Theory and Techniques*, **MTT-37**, 1949–1957 (1989).
19. E. M. Daniel and C. J. Railton, 'Fast finite-difference time-domain analysis of microstrip patch antennas', to be published in Int. IEEE/AP-S symposium, Ontario, Canada, June, 1991.

An Efficient Two-Dimensional Graded Mesh Finite-Difference Time-Domain Algorithm for Shielded or Open Waveguide Structures

Veselin J. Brankovic, Dragan V. Krupezevic, and Fritz Arndt, *Senior Member, IEEE*

Abstract—**A finite-difference time-domain (FD-TD) algorithm is described for the efficient full-wave analysis of a comprehensive class of millimeter-wave and optical waveguide structures. The FD-TD algorithm is based on a two-dimensional graded mesh combined with adequately formulated absorbing boundary conditions. This allows the inclusion of nearly arbitrarily shaped, fully or partially lateral open or shielded guiding structures with or without layers of finite metallization thickness. Moreover, lossy dielectrics and/or lossy conductors are included in the theory. The algorithm leads to a significant reduction in cpu time and storage requirements as compared with the conventional three-dimensional eigenvalue FD-TD mesh formulation. Dispersion characteristic examples are calculated for structures suitable for usual integrated circuits, such as insulated image guides, ridge guides, dielectric waveguides, trapped image guides, coplanar-lines and microstrip lines. The theory is verified by comparison with results obtained by other methods.**

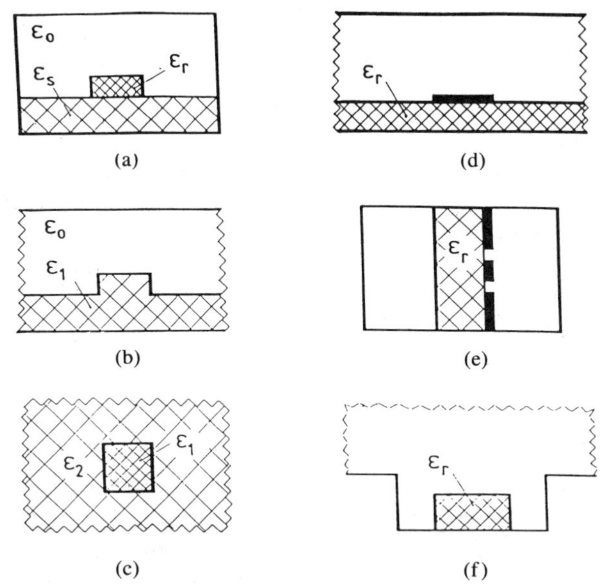

Fig. 1. The class of shielded, partially or fully open millimeter-wave and optical waveguide structures investigated with the new 2-D FD-TD method. (a) Shielded insulated image guide. (b) Lateral open ridge guide. (c) Open dielectric waveguide. (d) Lateral open microstrip line. (e) Coplanar line. (f) Open trapped image guide.

I. Introduction

SHIELDED, open or partially open waveguide structures of the class shown in Fig. 1 have found increasing interest for integrated circuit applications in the millimeter-wave and optical frequency range [1]–[17]. As this class includes a wide variety of specially shaped waveguiding structures used, in the design of integrated circuits, it is highly desirable to dispose of a reliable computer analysis which is sufficiently general and flexible to allow dominant and higher-order mode solutions of all desired cases including open or partially open structures.

Various methods of analyzing one or several of the structures of Fig. 1 have been the subject of many papers, including different kinds of mode-matching techniques [1]–[10], [13], the finite-element method [11], the finite-difference frequency-domain method [12], [14], [15], and, more recently, the finite-difference time-domain (FD-TD) method [18]–[28]. As for the criteria of flexibility, accuracy and computational efficiency, the FD-TD method is considered to be a very appropriate candidate for solving the waveguide eigenvalue problems with particular cross-sectional shapes, such as those shown in Fig. 1. The conventional FD-TD approach, however, utilizes a three-dimensional Yee's mesh for appropriate resonating sections which may be obtained by placing shorting planes along the axis of propagation of the structure under investigation. By repeating the calculation of the resonant frequency of these resonators for different distances of the shorting planes, the dispersion characteristic is then determined step by step. Consequently, the conventional approach requires considerable cpu time, needs a relatively large memory size, and tends to inaccuracies in the near of the cutoff frequencies.

A new two-dimensional FD-TD formulation has been introduced very recently [30] which helps to alleviate the above mentioned shortcomings of the conventional FD-TD approach, similar to [29], [33] where a 2-D TLM analysis has been described. Although many similarities exist between the TLM and the FD-TD method, cf. e.g.

Manuscript received March 31, 1992; revised July 27, 1992.
The authors are with the Microwave Department, University of Bremen Kufsteiner Str., NW1, D-2800 Bremen, Germany.
IEEE Log Number 9203677.

Reprinted from *IEEE Transactions on Microwave Theory and Techniques*, pp. 2272-2277, December 1992.

[17], [31], the new 2-D FD-TD approach has the advantage of being nearly an order of magnitude faster than the 2-D TLM method. Moreover, the numerical difficulties in the field simulation in the vicinity of conductive edges reported in [32] for the classical TLM condensed node formulation (which is used also in [29], [33]) are inherently avoided by the 2-D FD-TD formulation presented in this paper.

The purpose of this paper is to extend the two-dimensional FD-TD method introduced in [30] by involving adequate absorbing boundary conditions, a graded mesh algorithm and an efficient calculation procedure for lossy structures. The inclusion of absorbing boundary conditions allows a simple, flexible, and versatile FD-TD solution for analyzing the complete class of shielded, partially or fully open waveguide structures of Fig. 1. The graded mesh permits the investigation of structures with realistic dimensions by making the mesh finer in regions of particular interest. Losses are an important factor in modern circuit design. Numerical results will be presented to elucidate the usefulness of the method. The theory is verified by comparison with results available from other methods.

II. Theory

A. Two-Dimensional FD-TD Mesh Formulation

The usual FD-TD method is formulated by discretizing Maxwell's curl equations over a finite volume and approximating the derivatives with centered difference approximations. This leads to the known three-dimensional Yee's mesh [18] in various modifications [19]–[28].

In order to derive the new two-dimensional FD-TD mesh formulation, we utilize the relationships between the transverse fields $\vec{E}_t$ and $\vec{H}_t$ of a guided wave travelling in $+z$ direction (Fig. 2)

$$\vec{E}_t(z \pm \Delta l) = \vec{E}_t(z) \cdot e^{\mp j\beta\Delta l}$$
$$\vec{H}_t(z \pm \Delta l) = \vec{H}_t(z) \cdot e^{\mp j\beta\Delta l}. \quad (1)$$

By applying Yee's formulations, e.g. in the form of [21] for anisotropic structures, a set of FD-TD equations for the six components $\vec{H}$ and $\vec{E}$ is then derived. The result is shown for H_x as an example,

$$\begin{aligned} H_x^{n+1/2}(i, j+1/2) &= H_x^{n-1/2}(i, j+1/2) \\ &\quad + s/[\mu_{xx}(i, j+1/2) \cdot Z_{F0}] \\ &\quad \cdot \{E_y^n(i, j+1/2) \cdot [e^{-j\beta\Delta l} - 1] \\ &\quad + E_z^n(i, j) - E_z^n(i, j+1)\}, \quad (2) \end{aligned}$$

where the stability factor is $s = c\Delta t/\Delta w$ with $\Delta w = \Delta x = \Delta y = \Delta l$, c is the velocity of light, Z_{F0} is the characteristic impedance of free space, and μ_{xx} is the diagonal element of the relative permeability tensor. The condition for stability in free space for a uniform mesh is $s \leq 1/\sqrt{3}$ [21]. The remaining finite difference equations related to the other five field components can be similarly calculated.

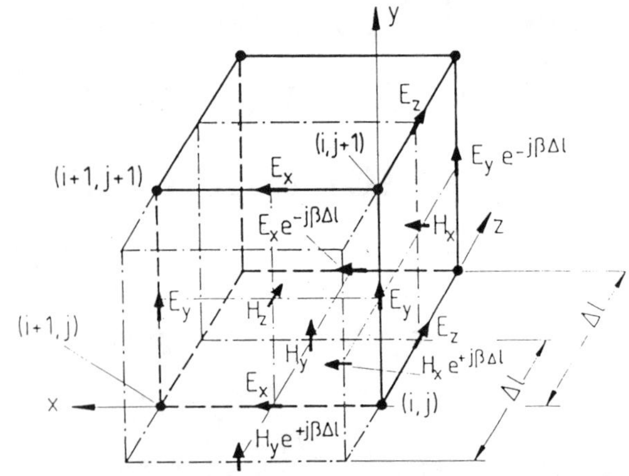

Fig. 2. New two-dimensional FD-TD mesh (i, j) for solving 2-D eigenvalue problems of waveguiding structures.

B. FD-TD Formulation of the Absorbing Boundary Conditions

For simplification, the equations of the boundary conditions are given only for the x-direction. In y-direction, the equations may be found analogously. Following Mur [20], we assume that the mesh is located in the region $x \geq 0$, that the boundary conditions are given for the plane $x = 0$, and that the outgoing wave has normal incidence. We then obtain the approximation for the boundary condition which would determine each field component W on the outer surface. This is consistent with an outgoing wave (i.e. this is absorbed), in the following form [20]:

$$\left(\partial_x - \frac{\partial_t}{v_{px}}\right) W \bigg|_{x=0} = 0, \quad (3)$$

where v_{px} is the phase velocity in x-direction.

The finite-difference formulation of (3) may be written in terms of the tangential electrical field components in the subregions 1) (outer region), and 2) (mesh region) as

$$E_{1t}^{n+1} = E_{2t}^n + \frac{v_{px}\Delta t - \Delta x}{v_{px}\Delta t + \Delta x}(E_{2t}^{n+1} - E_{1t}^n). \quad (4)$$

For the graded mesh, the following approximation of equation (4) is used

$$E_{1t}^{n+1} = E_{2t}^n + \frac{s_{xloc} - \sqrt{\epsilon_{xloc}}}{s_{xloc} + \sqrt{\epsilon_{xloc}}}(E_{2t}^{n+1} - E_{1t}^n), \quad (5)$$

where s_{xloc} is the local stability factor $s_{xloc} = c\Delta t/(\Delta w \cdot p_{loc})$, with the local graded mesh scaling factor p_{loc}, and where ϵ_{xloc} is the local relative premittivity in x-direction.

C. Graded Mesh Algorithm

In order to improve the computational efficiency, a graded mesh has turned out to be very advantageous [26]. This is particularly true for open structures where, if necessary, the enclosing absorbing boundary box is made sufficiently large so that the conditions mentioned above may be adequately satisfied. The basic structure of the FD-TD equations is similar to that of the uniform mesh with the exception that the uniform mesh parameters $\Delta w = \Delta x = \Delta y = \Delta l$ are replaced by $\Delta x = p\Delta w$, $\Delta y = q\Delta w$, $\Delta l = r\Delta w$, with the graded mesh scaling factors p, q, r.

With these relations, the set of FD-TD equations for the six components $\vec{H}$ and $\vec{E}$ is given in the form (e.g. for H_x) of

$$H_x^{n+1/2}(i, j + 1/2) = H_x^{n-1/2}(i, j + 1/2) + s/[\mu_{xx}(i, j + 1/2) \cdot Z_{F0}] \cdot \left\{ \frac{1}{r} E_y^n(i, j + 1/2) \cdot [e^{-j\beta\Delta l} - 1] + \frac{1}{q} [E_z^n(i, j) - E_z^n(i, j + 1)] \right\}. \quad (6)$$

In order to maintain numerical convergence, the stability factor has to be chosen on the basis of the smallest mesh parameter. To ensure continuity of fields across the interfaces between coarse and fine meshes, space averaging [28] is applied.

D. FD-TD Formulation for Lossy Structures

For deriving the FD-TD equations for lossy structures, two approaches are possible: The first approach differs between a) lossy dielectric regions (with the conductivity $\sigma_D << \omega\epsilon$) and b) lossy metal regions (with the finite conductivity $\sigma_M >> \omega\epsilon$). 2) The second, more general approach includes both cases simultaneously. In our calculations we used the general approach which has the advantage that only one formulation is necessary for the whole lossy structure.

By applying the Maxwell equations and Yee's formulations, we obtain for the component E_y, as an example:
a) In the case of a lossy dielectric region ($\sigma_D << \omega\epsilon$):

$$E_y^{n+1}(i, j + 1/2) = \left[1 - \frac{\Delta t}{\epsilon_0} \frac{\sigma_D(i, j + 1/2)}{\epsilon_{ry}(i, j + 1/2)}\right] \cdot E_y^n(i, j + 1/2) + \frac{s}{\epsilon_{ry}(i, j + 1/2)} \cdot Z_{F0} \cdot DH, \quad (7)$$

where

$$DH = \frac{H_x^{n+1/2}(i, j + 1/2)[1 - e^{j\beta\Delta l}]}{r} - \frac{H_z^{n+1/2}(i + 1/2, j + 1/2) - H_z^{n+1/2}(i - 1/2, j + 1/2)}{p},$$

and $\Delta x = p\Delta w$, $\Delta y = q\Delta w$, $\Delta l = r\Delta w$, s: stability factor, $\epsilon = \epsilon_r \epsilon_0$.
b) In the case of a lossy metallic region ($\sigma_M >> \omega\epsilon$):

$$E_y^{n+1}(i, j + 1/2) = -E_y^n(i, j + 1/2) + \frac{2}{\Delta w \sigma_M(i, j + 1/2)} \cdot DH. \quad (8)$$

The more general approach is obtained, assuming

$$E_y^{n+1/2}(i, j + 1/2) = \tfrac{1}{2}[E_y^{n+1}(i, j + 1/2) + E_y^n(i, j + 1/2)],$$

by the following equation where σ denotes the finite conductivity for lossy dielectric or metall regions

$$E_y^{n+1}(i, j + 1/2) = \frac{1 - \dfrac{\Delta t}{2\epsilon_0} \cdot \dfrac{\sigma(i, j + 1/2)}{\epsilon_{ry}(i, j + 1/2)}}{1 + \dfrac{\Delta t}{2\epsilon_0} \cdot \dfrac{\sigma(i, j + 1/2)}{\epsilon_{ry}(i, j + 1/2)}} \cdot E_y^n(i, j + 1/2) + \frac{\dfrac{sZ_{F0}}{\epsilon_{ry}(i, j + 1/2)}}{1 + \dfrac{\Delta t}{2\epsilon_0} \cdot \dfrac{\sigma(i, j + 1/2)}{\epsilon_{ry}(i, j + 1/2)}} \cdot DH. \quad (9)$$

The attenuation factor α of lossy structures is then calculated via the quality factor by using the equations already outlined in [33]. For the determination of the quality factor, however, an exponential regression procedure is used, i.e., for the corresponding resonance frequency, the envelope of the signal curve is evaluated by integration over several time periods. This yields high accuracy.

The principal numerical calculation steps in the two-dimensional FD-TD algorithm are similar to those in the conventional FD-TD approach with the exception that a propagation factor β has to be selected first (the choices of ß are indicated in the figure legends). After launching an excitation pulse, waiting until the distribution of the pulse is stable and performing the Fourier transformation, the modal frequencies related to the selected propagation factor are obtained. The mesh for the pulse propagation iteration needs only to be built up in the cross section x, y dimension, the number of nodes in z direction is reduced to ± 1. Also it should be emphasized that the amplitudes of all field components in the new formulation are complex quantities.

III. Results

Good agreement between our 2-D FD-TD dispersion results and those of the FD-frequency domain method [14] may be observed in Fig. 3 for the shielded insulated image guide. Fig. 4 shows the dispersion curves for different permittivities for the lateral open dielectric ridge guide. The results of the FD-TD method for a relatively low distance of the absorbing boundaries $a = 5$ h (i.e. only 2 h distance from the ridge) compare well with those obtained by the FD frequency domain method [14] for the shielded

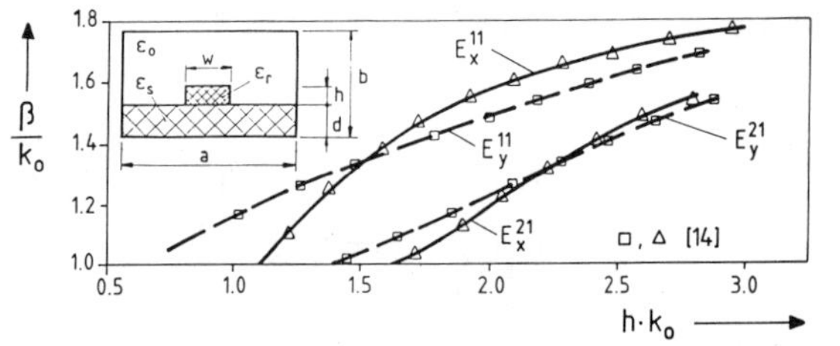

Fig. 3. Shielding insulated image guide. $w/h = 2.25$, $d/h = 0.5$, $a/h = 13.5$, $b/h = 8$, $\epsilon_r = 3.8$, $\epsilon_s = 1.5$. discretization: 54 × 64. Number of iterations $N_i = 1000$. Comparison with the FD frequency domain method [14]. (Choice of the propagation constant: $\beta = 0.5/h \cdots 5/h$, in e.g. 20 intermediate steps.)

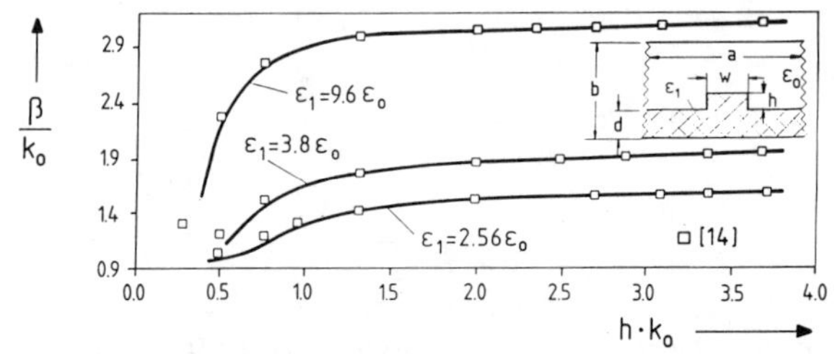

Fig. 4. Lateral open dielectric ridge guide. $w/h = 2$, $d/h = 1$, $b/h = 6$. Discretization: 60 × 72. Number of iterations $N_i = 2000$. Distance of the absorbing boundaries $a = 5$ h. Comparison with the FD frequency domain method [14] for the shielded dielectric ridge guide, lateral shield distance $a = 100$ h. (Choice of the propagation constant: $\beta = (0.3\sqrt{\epsilon_r})/\text{h} \cdots (3.8\sqrt{\epsilon_r})/\text{h}$, in e.g. 20 intermediate steps.)

dielectric ridge guide where the lateral shield distance is $a = 100$ h. These curves are considered to verify the absorbing boundary formulations made for the 2-D FD-TD method.

As an example for open structures, Fig. 5 presents the E_{11}^y-mode dispersion curves for a dielectric guide. A graded mesh was used with a mesh ratio for the dielectric kernel (ϵ_{r1}) to the outer region of 1 to 5. The normalized dispersion curves, $B = [(\beta/k_0)^2 - 1]/(\epsilon_{r1} - 1)$ versus $V = k_0 a\sqrt{\epsilon_{r1} - 1}$, are plotted for different discretizations: 1 (8 × 16), 2 (16 × 32), 3 (40 × 80). The absorbing boundary is located in a distance of $5a/2$. The comparison with the FD frequency domain method [14] for the shielded dielectric waveguide (shield distance $100a$) demonstrates good agreement with the discretization case number 3.

Fig. 6 shows the EH_0- and HE_1-mode normalized propagation factor as a function of frequency for a realistic mm-wave shielded coplanar line with finite metallization thickness. Due to the finite thickness included in the 2-D FD-TD method, there are slight deviations in comparison with own calculations by the spectral domain approach (SDA) for $t = 0$. Two slot widths w_1 and w_2 are included in the calculations in order to demonstrate the flexibility of the method also for varying typical structure dimensions.

In principle, the limitations of the applicability of the method—like for all time domain methods—are given by the limited storage and cpu time capabilities of the computer and by the fact of decreasing accuracy concerning the propagation factors of the higher-order modes with

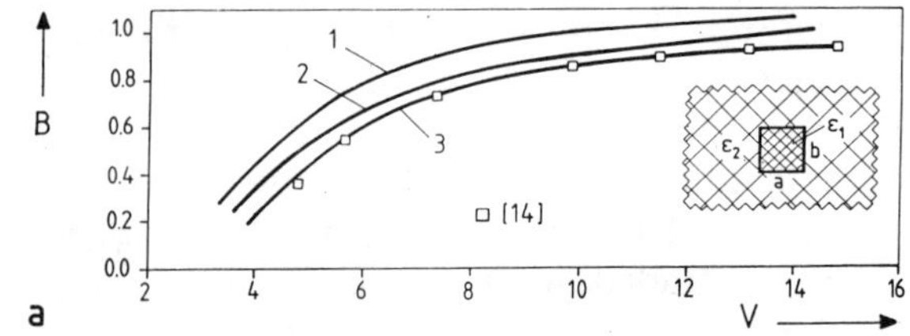

Fig. 5. Open dielectric guide $a = b$, $\epsilon_1 = 13.1\ \epsilon_0$, $\epsilon_2 = \epsilon_0$. Normalized dispersion curves of the E_{11}^y-mode, $B = [(\beta/k_0)^2 - 1]/(\epsilon_{r1} - 1)$ versus $V = k_0 a\sqrt{\epsilon_{r1} - 1}$, of the single dielectric waveguide for different discretizations: 1 (8 × 16), 2 (16 × 32), 3 (40 × 80). Comparison with the FD frequency domain method [14] for the shielded dielectric waveguide, shield distance 100a. Absorbing boundary in distance $5a/2$. Number of iterations $N_i = 2000$. Graded mesh ratio for the dielectric kernel: 1:5. (Choice of the propagation constant: $\beta = 2/a \cdots 15/a$, in e.g. 20 intermediate steps.)

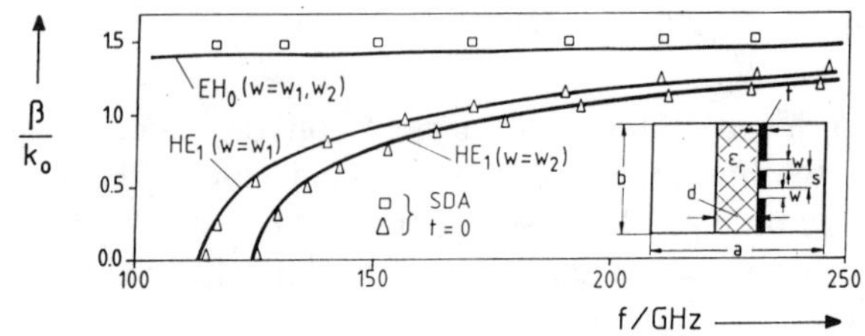

Fig. 6. EH_0- and HE_1-mode normalized propagation factor for the shielded coplanar line with finite metallization thickness t for different slot widths $w = w_1$, w_2. (For EH_0, the two curves are identical within the drawing accuracy of this figure.) Dimensions: $a = 0.8636$ mm, $b = 0.4318$ mm, $s = 0.072$ mm, $d = 0.126$ mm, $t = 0.018$ mm, $\epsilon_r = 3.7$, $w_1 = 0.09$ mm, $w_2 = 0.144$ mm. Discretization: 96 × 24. Number of iterations $N_i = 2000$. Comparison with own calculations with the Spectral Domain Approach (SDA) for $t = 0$. (Choice of the propagation constant: $\beta = 0 \cdots 7800$ 1/m, in e.g. 20 intermediate steps.)

increasing order. The computer requirements, however, are significantly reduced by utilizing the 2-D FD-TD method described in this paper and by introducing a graded mesh.

A shielded and a lateral open microstrip line is calculated in Fig. 7. The influence of the lateral shield on the expanded normalized propagation factor ratio $B\ [\%] = 100 \cdot (\sqrt{\epsilon_r} - (\beta/k_0))/\sqrt{\epsilon_r}$ as a function of frequency is demonstrated for typical dimensions for mm-wave applications. For a distance c with more than about ten times the half strip width, the influence of the lateral shield becomes negligible and tends to values obtained with the lateral absorbing boundary (curves 4, 5).

The 2-D FD-TD method presented in this paper may be utilized to calculate the characteristic impedance. After performing the Fourier transformation at the corresponding modal frequency, in each point of the mesh, the complete set of electric and magnetic field components is obtained. Dependent from the chosen definition of the characteristic impedance, the related integration can be carried out numerically. This procedure has already successfully been tested for the shielded microstrip line by using the voltage-current definition. Because this topic would be beyond the scope of this paper, results will be presented in a further paper.

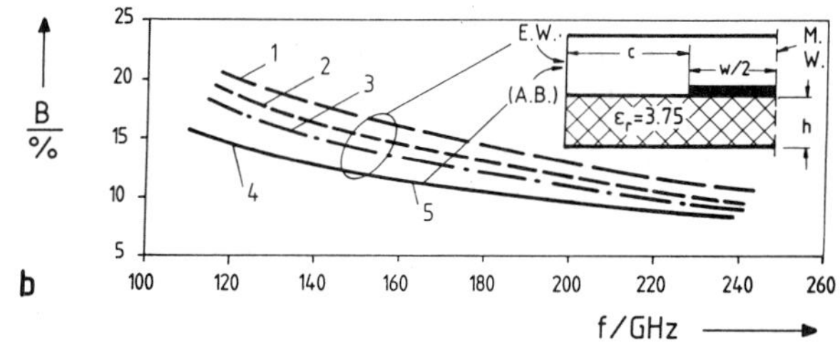

Fig. 7. Shielded and lateral open microstrip line. Influence of the lateral shield on the expanded normalized propagation factor ratio B [%] = $100 \cdot (\sqrt{\epsilon_r} - (\beta/k_0))/\sqrt{\epsilon_r}$ as a function of frequency for a single microstrip line. Dimensions: w = 0.16 mm, height b = 0.192 mm, ϵ_r = 3.7, h = 0.128 mm. Shield distances c: 1 (c = 0.248 mm ≙ 31Δl), 2 (c = 0.328 mm ≙ 41Δl), 3 (c = 0.408 mm ≙ 51Δl), 4 (c = 1.28 mm ≙ 160Δl). 5: Lateral open structure (curves 4 and 5 are identical within the drawing accuracy of Fig. 7.) (Choice of the propagation constant: β = 3800 · · · 9800 1/m, in e.g. 20 intermediate steps.)

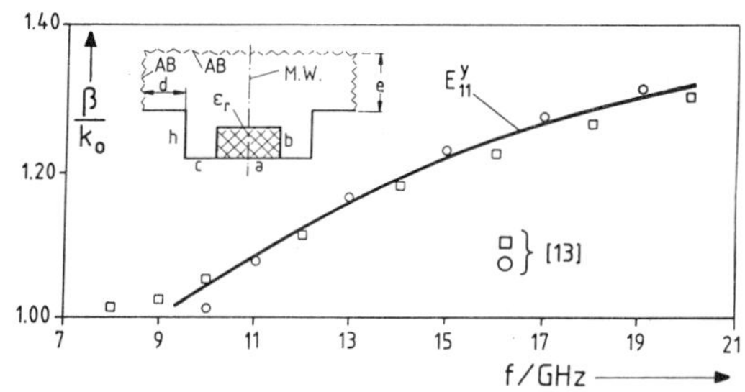

Fig. 8. Open trapped image guide. E^y_{11}-mode normalized propagation factor as a function of frequency. Dimensions: $a = b = c$ = 6 mm, h = 7 mm, ϵ_r = 2.23. Discretization: 54 × 42. Distance of the absorbing boundaries d = 6 mm, e = 7 mm. Number of iterations N_i = 1200. Comparison with [13]: □ □ □ combined effective dielectric constant transverse resonance method, ○ ○ ○ effective dielectric constant method. (Choice of the propagation constant: β = 200 · · · 550 1/m, in e.g. 20 intermediate steps.)

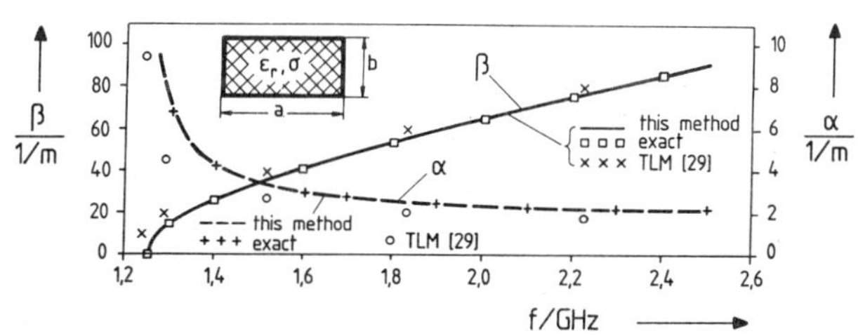

Fig. 9. Waveguide filled with a lossy dielectric. Attenuation factor α and propagation factor β as a function of frequency. Comparison with the TLM method results (○ ○ ○, × × ×) of [33] and with the exact results (+ + +, □ □ □) obtained analytically. Parameters: a = 60 mm, b = 40 mm, ϵ_r = 4, σ = 0.02 S/m. Discretization: 30 × 40. Number of iterations N_i = 8000. (Choice of the propagation constant: β = 0 · · · 90 1/m, in e.g. 20 intermediate steps.)

Fig. 8 shows the E^y_{11}-mode normalized propagation factor as a function of frequency for the open trapped image guide. The results compare well with those calculated in [13] with the transverse resonance effective dielectric constant techniques.

As an example where exact results are available, a waveguide filled with a lossy dielectric is investigated in Fig. 9. The attenuation factor α and propagation factor β calculated with our method as a function of frequency are compared with the TLM method results of [33] and with the exact results obtained analytically. The curves in Fig. 9 verify the presented FD-TD method also for lossy structures by excellent agreement with the exact results.

Conclusion

A very efficient finite-difference time-domain formulation is presented for the full-wave analysis of shielded or open waveguiding structures. Lossy dielectrics and/or lossy conductors are included in the theory. This method utilizes advantageously a two-dimensional instead of the original three-dimensional Yee's mesh combined with adequately formulated absorbing boundary conditions and a graded mesh structure. The algorithm leads to a significant reduction in cpu time and storage requirements as compared with the conventional three-dimensional eigenvalue FD-TD mesh formulation and allows the inclusion of nearly arbitrarily shaped, fully or partially lateral open or shielded guiding structures with or without layers of finite metallization thickness. Dispersion characteristic examples for structures suitable for usual integrated circuits, such as insulated image guides, ridge guides, dielectric waveguides, trapped image guides, coplanar-lines, and microstrip lines elucidate the usefulness of the method. The theory is verified by comparison with results available from other methods.

References

[1] W. V. McLevidge, T. Itoh, and R. Mittra, "New waveguide structures for millimeter-wave and optical integrated circuits," *IEEE Trans. Microwave Theory Tech.*, vol. MTT-23, pp. 788–794, Oct. 1975.

[2] R. M. Knox "Dielectric waveguide microwave integrated circuits—An overview," *IEEE Trans. Microwave Tech.*, vol. MTT-24, pp. 806–814, Nov. 1976.

[3] J. A. Paul and Y.-W. Chang, "Millimeter wave image-guide integrated passive devices," *IEEE Trans. Microwave Theory Tech.*, vol. MTT-26, pp. 751–754, Oct. 1978.

[4] T. Itoh, "Open guiding structures for mmW integrated circuits," *Microwave J.*, pp. 113–126, Sept. 1982.

[5] S. E. Miller, "Integrated optics: An introduction," *Bell Syst. Tech. J.*, vol. 48, pp. 2059–2069, Sept. 1969.

[6] R. M. Knox and P. D. Toulios, "Integrated circuits for the millimeter through optical frequency range," in *Proc. Symp. Submillimeter Waves*, New York, May 1970, pp. 497–516.

[7] R. M. Knox, "Dielectric waveguides: A low-cost option for IC's," *Microwaves*, pp. 56–64, Mar. 1976.

[8] M. Ikeuchi, H. Swami, and H. Niki, "Analysis of open-type dielectric waveguides by the finite element iterative method," *IEEE Trans. Microwave Theory Tech.*, vol. MTT-29, pp. 234–239, Mar. 1981.

[9] N. Deo and R. Mittra, "A technique for analyzing planar dielectric waveguides for millimeter wave integrated circuits," *Arch. Elek. Übertragung*, vol. 37, pp. 236–244, July/Aug. 1983.

[10] U. Crombach, "Analysis of single and coupled rectangular dielectric waveguides," *IEEE Trans. Microwave Theory Tech.*, vol. MTT-29, pp. 870–874, Sept. 1981.

[11] B. M. A. Rahman and J. B. Davies, "Finite-element solution of integrated optical waveguides," *J. Lightwave Technol.*, vol. LT-2, pp. 682–687, Oct. 1984.

[12] S. M. Saad, "Review of numerical methods for the analysis of arbitrarily-shaped microwave and optical dielectric waveguides," *IEEE Trans. Microwave Tech.*, vol. MTT-33, pp. 894–899, Oct. 1985.

[13] W.-B. Zhou, and T. Itoh, "Analysis of trapped image guides using effective dielectric constant and surface impedances," *IEEE Trans. Microwave Theory Tech.*, vol. MTT-30, pp. 2163–2166, Dec. 1982.

[14] K. Bierwirth, N. Schulz, and F. Arndt, "Finite-difference analysis of rectangular dielectric waveguide structures," *IEEE Trans. Microwave Theory Tech.*, MTT-34, pp. 1104–1114, Nov. 1986.

[15] T. K. Sarkar, M. Manela, V. Narayanan, and A. R. Djordjevic, "Finite difference frequency-domain treatment of open transmission structures," *IEEE Trans. Microwave Theory Tech.*, vol. 38, pp. 1609–1616, Nov. 1990.

[16] T. Itoh, "Overview of quasi-planar transmission lines," *IEEE Trans. Microwave Theory Tech.*, vol. 37, pp. 275–280, Feb. 1989.

[17] M. Celuch-Marcysiak, W. K. Gwarek, "Formal equivalence and efficiency comparison of the FD-TD, TLM and SN methods in application to microwave CAD programs," in *Proc. European Microwave Conf.*, Stuttgart, Sept. 1991, pp. 199–204.

[18] K. S. Yee, "Numerical solution of initial boundary value problems involving Maxwell's equations on isotropic media," *IEEE Trans. Antennas Propagat.*, vol. AP-14, pp. 302–307, May 1966.

[19] A. Taflove, and M. E. Brodwin, "Numerical solution of steady state electromagnetic scattering problems using the time dependent Maxwell's equations," *IEEE Trans. Microwave Theory Tech.*, vol. MTT-23, pp. 623–630, Aug. 1975.

[20] G. Mur, "Absorbing boundary conditions for the finite-difference approximation of the time-domain electromagnetic-field equations," *IEEE Trans. on Elect. Compatibility*, vol. EMC-23, pp. 377–382, Nov. 1981.

[21] D. H. Choi, and W. J. R. Hoefer, "The finite-difference time-domain method and its applications to eigenvalue problems," *IEEE Trans. Microwave Theory Tech.* vol. MTT-34, pp. 1464–1470, Dec. 1986.

[22] W. Gwarek, "Analysis of arbitrarily shaped two-dimensional microwave circuits by finite-difference time-domain method," *IEEE Trans. Microwave Theory Tech.*, vol. 36, pp. 738–744, Apr. 1988.

[23] G.-C. Liang, Y.-W. Liu, and K. K. Mei, "Full-wave analysis of coplanar waveguide and slotline using the time-domain finite-difference method," *IEEE Trans. Microwave Theory Tech.*, vol. 37, pp. 1949–1957, Dec. 1989.

[24] D. M. Sheen, S. A. Ali, M. D. Abouzahra, and J. A. Kong, "Application of the three-dimensional time-domain method to the analysis of planar microstrip circuits," *IEEE Trans. Microwave Theory Tech.*, vol. 38, pp. 849–857, July 1990.

[25] M. Rittweger, M. Abdo, and I. Wolff, "Full-wave analysis of coplanar discontinuities considering three-dimensional bond wires," in *IEEE MTT-S Int. Symp. Dig.*, June 1991, pp. 465–468.

[26] D. H. Choi and W. J. R. Hoefer, "A graded mesh FD-TD algorithm for eigenvalue problems," in *Proc. European Microwave Conf.*, Rome, Sept. 1987, 413–417.

[27] Z. Chen, M. M. Ney, and W. J. R. Hoefer, "A new finite-difference time-domain formulation and its equivalence with the TLM symmetrical condensed node," *IEEE Trans. Microwave Theory Tech.*, vol. 39, Dec. 1991, pp. 2160–2169.

[28] I. S. Kim, and W. J. R. Hoefer, "A local mesh refinement algorithm for the time domain-finite difference method using Maxwell's curl equations," *IEEE Trans. Microwave Theory Tech.*, vol. 38, pp. 812–815, June 1990.

[29] H. Jin, R. Vahldieck, and S. Xiao, "An improved TLM full-wave analysis using a two dimensional mesh," in *IEEE MTT-S Int. Microwave Symp. Dig.*, July 1991, pp. 675–677.

[30] F. Arndt, V. J. Brankovic, and D. V. Krupezevic, "An improved FD-TD full wave analysis for arbitrary guiding structures using a two-dimensional mesh," in *IEEE MTT-S Int. Microwave Symp. Dig.*, 1992.

[31] N. R. S. Simons and E. Bridges, "Equivalence of propagation characteristics for the transmission-line matrix and finite-difference time-domain methods in two dimensions," *IEEE Trans. Microwave Theory Tech.*, vol. 39, pp. 354–357, Feb. 1991.

[32] J. S. Nielsen, and W. J. R. Hoefer, "Modification of the condensed 3-D TLM node to improve modeling of conductor edges," *IEEE Microwave Guided Wave Lett.*, vol. 2, pp. 105–110, Mar. 1992.

[33] H. Jin, R. Vahldieck, and S. Xiao, "A full-wave analysis of arbitrary guiding structures using a two-dimensional TLM mesh," in *Proc. European Microwave Conf.*, Stuttgart, pp. 205–210, Sept. 1991.

Steady-State Analysis of Nonlinear Forced and Autonomous Microwave Circuits Using the Compression Approach

Jürgen Kunisch and Ingo Wolff

Institute of Mobile and Satellite Communication Techniques, Carl-Friedrich-Gauss-Str. 2, D-47475 Kamp-Lintfort, Germany

Received January 25, 1995; revised January 25, 1995.

ABSTRACT

This article discusses techniques for the analysis of nonlinear microwave circuits that contain lumped nonlinear elements. Emphasis is put on specific harmonic balance–based technique which allows us to retain the field theoretical accuracy in cases where lumped nonlinear elements are embedded closely within complex metallization structures. © 1995 John Wiley & Sons, Inc.

I. INTRODUCTION

The analysis of nonlinear microwave circuits may usually be viewed from two different perspectives. First, the field theoretical properties of the circuit have to be known. This is mainly a matter of applying Maxwell's theory and in most cases results in a description of the circuit or parts of it in terms of an n-port scattering matrix (either by numerical simulations or by measurements) which relates wave amplitudes at the ports of the corresponding part. Obviously, as scattering matrices are a linear concept, this is suited to describe only linear aspects of the circuit.

Second, models for the nonlinear devices are required. Many CAD-suited models for nonlinear devices are based on an intrinsic model which in turn is given in the form of analytical or semianalytical formulas which determine the behavior of characteristic quantities of the nonlinear device. For FETs, this might, for example, be formulas for I_{DS} or Q_{GS} as a function of V_{DS} and V_{GS}. Such formulas may be derived directly from semiconductor physics, with or without the use of fitting parameters, or may consist of empirical expressions fitted to measurements, or anything in between. As every nonlinear device is embedded in some metallization, which is required for connections, every model contains an additional extrinsic part to take these into account. Eventually, an equivalent circuit made up of a linear extrinsic model with the intrinsic nonlinear device model results.

To simulate the overall circuit, an equivalent circuit of the total circuit is established by properly connecting the nonlinear devices' equivalent circuits and the linear n-ports given as scattering matrices. However, the determination of scattering parameters and wave amplitudes within this context is usually related to the expansion of electromagnetic fields into eigenmodes, where the expansions are to be evaluated at particular reference planes that constitute the linear devices' ports. This is a description which is matched to the distributed nature of most microwave circuits.

On the other hand, the equivalent circuit of a nonlinear device is given in terms of circuit nodes, voltages, and currents, and therefore assumes the

This article belongs to the previous special issue (Vol. 5, No. 3) on "CAD of Nonlinear Microwave Circuits," but it could not be included therein because of delays in transit.—Editor applicability of the Kirchhoff laws. This is a concept suited for lumped devices.

In this article, we will focus on the problems associated with this slight incompatibility, which is an inherent property of the interfacing procedure as described above. We will present an alternative approach to remove this incompatibility to yield an enhanced performance of the overall simulation. As will be shown in section III, this mainly requires the introduction of a proper interface between the intrinsic nonlinear devices and the embedding linear parts of the circuit.

The article is organized as follows. The next section gives a short overview of some of the techniques for analysis of nonlinear microwave circuits. Two commonly used methods resulting from combining different approaches for the analysis of the linear and the nonlinear part of the circuit are discussed with particular attention to their benefits and merits as well as their disadvantages. In section III, a technique is presented which is intended to combine the advantages of both of the methods while circumventing their disadvantages. The application of the method to a model problem, as well as a comparison with the concurrent methods, is given in section IV.

II. ANALYSIS OF NONLINEAR MICROWAVE CIRCUITS

General Aspects

The analysis of microwave circuits may be approached from two starting points. The classical circuit analysis techniques are based on the Kirchhoff laws and are therefore formulated in terms of nodal or port voltages and branch currents. The circuit under investigation is viewed as being made up of several devices which are connected by circuit nodes. Obviously, this approach requires that the interaction between any two adjacent devices can be described completely by the influence the devices exert on their connecting nodes.

The numerical evaluation of the circuit equation may be done either in time domain or in frequency domain. Which one of these is to be preferred depends on particular aspects of the analysis to be performed. For steady-state analyses a frequency domain approach may be preferable, whereas in the case of transient analyses a time domain integration procedure is required. A well-known example of the latter case is the SPICE analysis program. The time domain formulation of the circuit equations results in a set of algebraic and differential equations. In a case in which there are only resistive devices connected to a node, the nodal analysis yields a purely algebraic equation for that node. As a consequence, it is usually not possible to rearrange circuit equations such that for every unknown quantity like voltages and currents one state equation exists which relates the first time derivative of that quantity to the actual values of all quantities. Circuit analysis software which operates in time domain therefore usually has to carry out the solution of a linear system of equations for every time step to determine the incremental changes of all circuit quantities. Many researchers have devised elaborate techniques to reduce the corresponding numerical effort and to cope with related problems, for instance, the stiffness properties of the circuit equations. However, problems are still encountered when devices having delay times or frequency dependent properties are to be considered.

When it comes to microwave (monolithically integrated) circuits, new aspects have to be considered. Now the fundamental assumptions of the Kirchhoff laws do not hold any longer. However, the invention of scattering parameters allows us to proceed in a manner similar to the classical approach by a simple "change of coordinates." Instead of port or nodal voltages and branch currents now wave amplitudes of back- and forth-going waves are to be considered. By way of a Heaviside transformation, wave amplitudes are transformed into voltages and currents which actually behave similarly to those in the classical approach. Thus circuits built out of devices which are described by scattering matrices are again amenable to the classic analysis methods and tools.

However, it clearly has to be kept in mind that the similarity in the behavior of voltages and currents determined from a Heaviside transformation is only due to a formal equivalence to the voltages and currents in the lumped circuit case. Actually, suppose in the vicinity of a reference plane R the transversal electromagnetic field of a forth-going wave on a waveguide is given as a superposition of several modes by:

$$\mathbf{E}_{\mathrm{tr}}^{+}(x, y, z) = \sum_{i} V_i^{+} \mathbf{t}_i(x, y) \exp(-\gamma_i z),$$

$$\mathbf{H}_{\mathrm{tr}}^{+}(x, y, z) = \sum_{i} I_i^{+} \mathbf{e}_z \times \mathbf{t}_i(x, y) \exp(-\gamma_i z)$$

with $I_i^{+} = V_i^{+}/Z_{g,i}$.

For every mode i, the characteristic function $\mathbf{t}_i(x, y)$ characterizes the distribution of the transversal (and thus the corresponding longitudinal) electromagnetic field components, while the factors V_i^+ and I_i^+ determine the weight of every mode in the superposition. If only a single mode with a known characteristic function $\mathbf{t}(x, y)$, characteristic impedance Z_g, and propagation constant γ is considered, the electromagnetic field is completely determined by giving $V = V^+ + V^-$ and $I = I^+ - I^-$, where V^- and I^- with $I^- = V^-/Z_g$ describe the back-going wave, respectively. The quantities V and I are to be determined from surface integrations by:

$$V = \int_A \mathbf{t}(x, y) \cdot \mathbf{E}(x, y)\, \mathrm{d}A, \tag{1}$$

$$I = \int_A (\mathbf{e}_z \times \mathbf{t}(x, y)) \cdot \mathbf{H}(x, y)\, \mathrm{d}A \tag{2}$$

where A is the waveguide cross-section and the dot "·" denotes the inner vector product. This is opposed to the theory of lumped devices where, for example, voltages V are determined by a line integration

$$V = \int_c \mathbf{E}(x, y) \cdot \mathbf{ds} \tag{3}$$

with the assumption $\mathbf{rot}\,\mathbf{E} = \mathbf{0}$, thus requiring TEM waves for a unique result if waveguides are considered.

As a consequence, interconnections of lumped element subcircuits and n-ports defined by scattering matrices which refer to wave amplitudes require a clear distinction between definitions. As scattering parameters relate to wave amplitudes, voltages and currents that occur at the ports of such an n-port primarily correspond to the definitions of eqs. (1) and (2) rather than line integrals like eq. (3). To ensure that V and I completely determine the electromagnetic field at the reference plane, it is therefore crucial to ensure that the single mode assumption holds. Otherwise, a pair (V_i, I_i) for every actually appearing mode i and generalized scattering matrices [1] have to be used. Moreover, if a mode mixture is present, there is usually no simple relationship between the voltages as defined by eqs. (1) and (3), because higher order modes usually are anything but TEM.

These properties will have to be remembered when it comes to the analysis of circuits with active lumped devices that are embedded within complex metallization structures.

Harmonic Balance and Segmentation

If steady-state analyses are required, a commonly used practice consists of dividing the total circuit into several segments each of which is modeled independently from the others. Linear segments are usually characterized by scattering parameters, whereas active devices are modeled by some equivalent circuit. For the analysis the well-known harmonic balance method is employed. Obviously, this approach has a number of attractive features. The harmonic balance method has reached a high state of maturity and offers a large number of analysis options, including signal, noise, stability, sensitivity, and multitone analysis in forced or autonomous circuit regimes [2–7]. An important feature of the harmonic balance method is the ability to use the results of former analyses as a solution estimate, thus reducing computation times significantly. Because linear parts are described in frequency domain, circuits which contain high-Q components may easily be considered.

A disadvantage with the segmentation approach is the requirement for the single mode assumption at the reference planes as discussed above. This is not a very severe restriction if interconnections of passive components are considered, which are connected by transmission lines. Usually there are precautions taken anyway to ensure that only the (fundamental) mode is able to propagate. However, the segmentation has to be done such that every higher order mode which may be excited at a discontinuity is sufficiently attenuated at the reference planes. This implies that every active device model has to contain and extrinsic part which accounts for a sufficiently long part of the device feeding lines and the discontinuity which is formed by the device metallization structure. In cases in which there are several active devices located closely together it may be impossible to find a correct segmentation. This may lead to a deterioration of the overall field theoretical accuracy.

As an example, consider the schematic LUFET (line unified field effect transistor) structure shown in Figure 1a and the corresponding simple, segmented equivalent circuit (Figure 1b). It is quite difficult to ensure that only a single mode is present at the inner reference planes. Moreover, there is a clear need for a field theoretically accurate model of the inner T-junction part, which in turn would require the introduction of new, dubious reference planes.

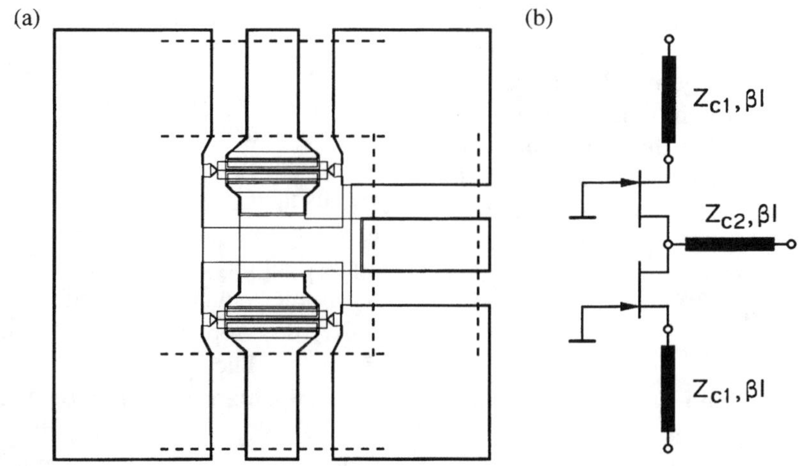

Figure 1. LUFET combiner with coplanar lines. (a) Schematic structure with reference planes (- - -). (b) Simple segmented equivalent circuit.

Time Domain Methods with Nonlinear Elements

A more direct approach to microwave circuit analysis is based on the integration of Maxwell's equations in time domain. A number of different integration schemes like finite differences in time domain (FDTD), the transmission line matrix (TLM) method, the spatial network (SN) method, etc., have been used in various modifications [8–10]. Unlike the time domain analysis methods for lumped circuits (like SPICE), these methods are, in principle, not required to solve a linear system of equations in every time step, because for every node there is an equation involving the first time derivative of the corresponding field component.

Lumped active devices which are described, for example, by an i–v-characteristic may be considered within these methods quite naturally. Time domain methods are not subject to mode considerations, as far as the determination of the electromagnetic field is concerned. The above-described situation in which there are several active devices located in close vicinity does not pose any specific problems similar to those encountered with the segmentation approach. If a device is described by an i–v-characteristic, the voltage may conceptually be determined from eq. (3), where c is a path that corresponds to the path of current flow within the device. This does not require the existence of an expansion of the field into modes or even the existence of any modes at all. The resulting currents yield constraints which are used in the determination of fields in the next time step.[1] Therefore, time domain methods are capable of analyzing circuits with active devices within complex metallization structures with excellent field theoretical accuracy.

Unfortunately, many kinds of analyses to be performed advantageously with the segmentation approach now turn out to be problematic. The analysis of multitone regimes with very close incommensurable frequencies (e.g., as can be found in mixers) may require huge numbers of iteration steps to resolve low frequency mixing products. If many analyses have to be performed, which is usually the case for nonlinear circuits, it is not possible to reuse former results as starting values due to the nature of integrating methods. This is even worse, because almost all of the computation time is spent in simulating the linear part of the circuit, which principally could be characterized by a few runs. Similar considerations are valid with respect to noise and sensitivity analyses.

III. THE COMPRESSION APPROACH

A comparison of the two commonly used methods discussed above shows that the merits and drawbacks of the one method are almost interchanged for the other. The approach to be described in this section is intended to combine the advantages of both methods to yield an analysis proce-

[1] In the case of implicit formulations, this may in turn require the solution of a small nonlinear system of equations.

dure that has a superior performance compared to both of them when lumped active devices within complex metallization structures are to be considered.

Inner Ports

The basic approach is to introduce an interface between the active device and the embedding metallization structure which is tailored to this specific situation. To see how this can be achieved, it is instructive to recall how lumped active (nonlinear) devices can be considered in field theoretical methods which solve the Maxwell equations by a time domain integration process (like the FDTD, TLM, or SN method). Lumped active devices which are characterized in terms of port voltages and currents may be incorporated into the time domain analysis by replacing the constitutive relations of the linear material by the active device model. Every time step of the time domain integration may usually—at least from a conceptual point of view— be split into two half-steps such that in the first half-step the electric field, say, is updated from the magnetic field, while in the second half-step in turn the magnetic field is updated. Active devices may now be incorporated by determining, after the first half-step, the active device port voltages by spatial integration of the electric field along a path which lies within the active element. If the active element is described by i–v-characteristics, the port current may be determined, which gives rise to an additional effective current density J to be used in the update of the magnetic field in the next half-step (cf., e.g., refs. 11 and 12 for details and a more comprehensive description).

The essence of this procedure is that the only aspect of the overall electromagnetic field which the active device model "feels" is the port voltage as determined from a spatial integration of the electrical field along a path which is somehow related to the geometric shape of the device. On the other hand, the effect of the active device on the electromagnetic field is exerted exclusively by means of an additional current, usually flowing in the direction of the integration path (cf. Fig. 2).

It is intuitively clear that this approach is a quite direct model of the actual physical situation. Therefore, this approach can yield very accurate results. Moreover, there are no specific restrictions as to the permissible locations of the active devices implied. In particular, no modes or reference planes as required by a segmentation approach are involved here.

This suggests that the proper interface between the active device and the embedding electromagnetic field is—from a geometrical point of view—given as a reference path which is to be used for the spatial integration of the electric field as well as a current path for the active devices port currents. Therefore, the lumped element is in this approximation viewed as being at most one-dimensional; that is, as having a finite length, but no thickness. Using this approximation, consideration of the lumped element in a time domain integration method is accomplished by demanding:

$$f[v(t), i(t)] = 0 \tag{4}$$

where:

$$v(t) = \int_c \mathbf{E}(x, y, t) \cdot \mathbf{ds} \tag{5}$$

and:

$$i(t) = \oint_{c'} \mathbf{H}(x, y, t) \cdot \mathbf{ds} \tag{6}$$

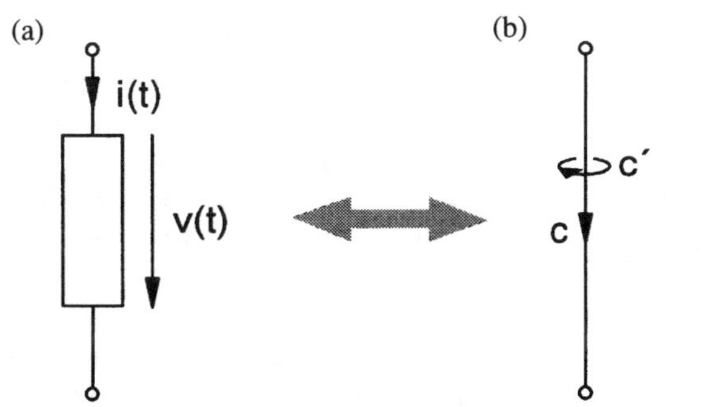

$$v(t) = \int_c \mathbf{E}(x, y, t) \cdot \mathbf{ds}$$

$$i(t) = \oint_{c'} \mathbf{H}(x, y, t) \cdot \mathbf{ds}$$

$$f[v, i] = 0$$

Figure 2. Interface to lumped active element. (a) One-port lumped element with definition of port voltage and current. (b) Interface to the embedding electromagnetic field (integration path). (c) Constitutive relations.

are the port voltage and currents, respectively, of the active device; c is the reference path corresponding to the shape of the element; c' is a path which is supposed to surround c closely; and f is the characteristic function of the device (Fig. 2).

Due to the linear part of the Maxwell equations, there is a linear relationship between $v(t)$ and $i(t)$ implied which has the form

$$\alpha(t) * v(t) - \beta(t) * i(t) = \gamma(t) \quad (7)$$

where $*$ denotes the temporal convolution and α, β, and γ are time-variant functions, which characterize the behavior of the linear part of the circuit as described by the Maxwell equations. γ represents the contributions of remote sources.

In the time domain integration paradigm, the evaluation of eqs. (4) and (7) is done simultaneously; that is, at every time instant a new sample of the electromagnetic field is determined based on the previous field state and the device characteristic eq. (4).

However, there is no need to do so. Fourier transforming eq. (7) and formally introducing normalized wave amplitudes $a(f)$ and $b(f)$ with:

$$v(f) = Z(f)^{1/2}(a(f) + b(f)) \quad (8)$$

and:

$$i(f) = Z(f)^{-1/2}(a(f) - b(f)) \quad (9)$$

with a reference impedance $Z(f)$ into eq. (7), yields

$$b(f) = \frac{\gamma(f)Z(f)^{1/2} + a(f)(\beta(f) - \alpha(f)Z(f))}{\beta(f) + \alpha(f)Z(f)} \quad (10)$$

where $\alpha(f)$, $\beta(f)$, and $\gamma(f)$ are the Fourier spectra of $\alpha(t)$, $\beta(t)$, and $\gamma(t)$, respectively, or

$$b(f) = q(f) + r(f)a(f) \quad (11)$$

The electromagnetic field, which is described by eq. (7) (w.r.t. the port voltage and current of the active device), therefore, acts as a wave source of amplitude $q(f)$ with:

$$q(f) = \frac{\gamma(f)Z(f)^{1/2}}{\beta(f) + \alpha(f)Z(f)} \quad (12)$$

and a reflection coefficient:

$$r(f) = \frac{\beta(f) - \alpha(f)Z(f)}{\beta(f) + \alpha(f)Z(f)}. \quad (13)$$

Eq. (11) is essentially the same as that for any wave source which is connected to the input line of a circuit. However, both cases are significantly different from a conceptual point of view. For the line connected source, $a(f)$ and $b(f)$ are the frequency dependent amplitudes of a forward and a backward propagating wave corresponding to one particular mode (usually the fundamental one). In the case of the electromagnetic field acting upon a lumped active device, $a(f)$ and $b(f)$ are the frequency dependent amplitudes of forth- and back-going formal waves, which, however, are related to the total port voltage $v(t)$ and current $i(t)$. There is no expansion of $v(t)$ or $i(t)$ into any modes involved or required here. Due to this, the prefix formal is required, because the resulting wave amplitudes $a(f)$ and $b(f)$ are determined in a manner which is only formally identical to the case of eigenmode related wave amplitudes.

Evidently, a reference path, c, together with eqs. (5) and (6) forms a kind of port in its own right in the same manner as a reference plane on an input or output line together with eqs. (1) and (2) forms a port. A port defined by a reference path c is obviously completely embedded (spatially) in the linear system it belongs to. The consequence is that only lumped elements can be connected to such a port, because lumped elements may be regarded as spatially one-dimensional[2] as far as their effect on the electromagnetic field is concerned. Due to this, such a port will be called an "inner port," subsequently, in contrast to "outer" ports, which are represented by reference planes.

Prior to the further discussion, a simple example shall illustrate these concepts. Figure 3 shows a part of a CPW which is loaded by two shunt diodes. The black areas represent the active areas of the diodes. Both diodes are connected to the CPW by connection pads. Evidently, here there are two active devices—the diodes—located in the middle of a discontinuity, namely a double step-in-width consisting of the connection pads. To analyze such a structure by the standard seg-

[2]The length of the path has to be short enough to allow for the lumped element assumption.

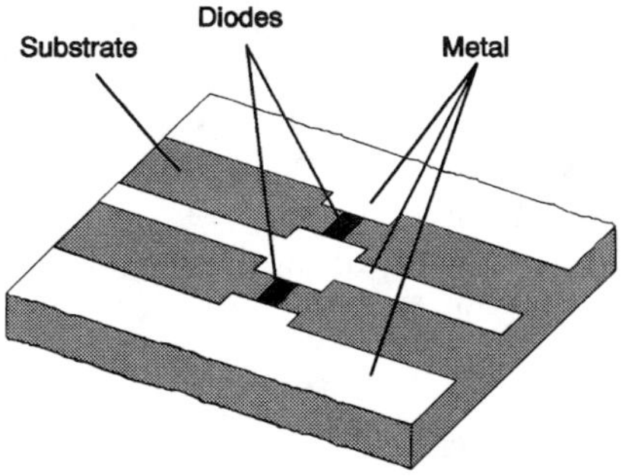

Figure 3. Part of a coplanar line with shunt diodes.

mentation approach, reference planes have to be introduced (Fig. 4a).

The problem here is to find the proper position of reference planes R_3 and R_4 (cf. Fig. 4a). Approaching the diodes too much results in a violation of the single mode assumption, because higher order modes will be present in the close vicinity of the discontinuity. On the other hand, removing the reference planes too much requires the diode models to account for a piece of the CPW including the connection pads,[3] which in turn requires coupling of a (nonlinear) diode model based on, for example, semiconductor physics and a (linear) field theoretical model of the pads plus part of the CPW. So this does not solve the problem, but only shifts it.

In any case, an equivalent circuit, like that shown in Figure 4b results,[4] where $\mathbf{S}_1$ and $\mathbf{S}_2$ are scattering matrices for the segments bounded by the reference plane pair (R_1, R_3) and (R_4, R_2), respectively.

The resulting equivalent circuit can be analyzed for linear or nonlinear regimes with standard circuit analysis software. However, the results may be subject to errors due to the problem of modeling the connection pads correctly.

To overcome that problem, the diodes, which are assumed to be lumped, may be replaced by a reference path (i.e., an inner port) each, whereas reference planes may be introduced remote from the discontinuity, giving the situation as depicted in Figure 5a. The linear part of the CPW, including the connection pads, may now be characterized by means of field theoretical analysis software to yield a four-port scattering matrix $\mathbf{S}_c$ with

[3]The CPW plus connection pad part of the diode model is symbolized by the transmission line symbols in Figure 4a.

[4]Due to the symmetry of the structure and the single (even) mode assumption, both diodes and the two CPW grounds can be represented by one diode and one ground, respectively, that is, every segment effectively forms a two port here.

(a)

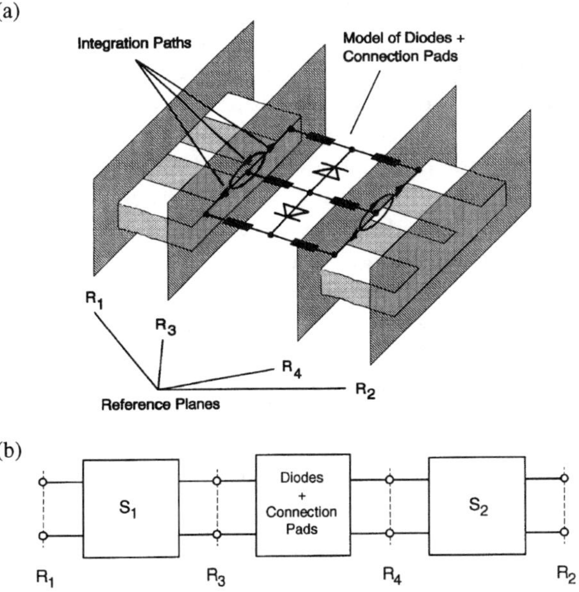

(b)

Figure 4. Segmentation. (a) Segments and reference planes. (b) Equivalent circuit assuming single even mode.

(a)

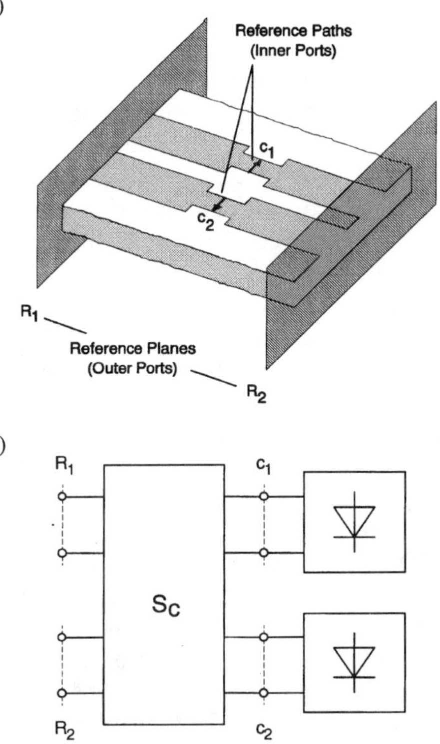

(b)

Figure 5. Analysis with inner ports. (a) Linear part to be described by $\mathbf{S}_c$. (b) Equivalent circuit.

respect to the outer ports (represented by R_1 and R_2) and the inner ports (represented by c_1 and c_2) (cf. Figure 5a). The resulting equivalent circuit consists of the single scattering matrix, $\mathbf{S}_c$, which describes all of the CPW and the connection pads, with diode models being connected to the inner ports (Fig. 5b). It should be recognized that the diode models now only have to account for the intrinsic diode without any external metallization. This is exactly the scope of a diode model which is based on semiconductor physics. There is no need for an extrinsic part of the diode model. Again, the overall circuit may then be analyzed for linear or nonlinear regimes with standard tools.

The matrix $\mathbf{S}_c$ will be called the "compression matrix" subsequently, where the index c shall indicate that some of the elements of $\mathbf{S}_c$ refer to ports defined by a reference path c.

Circuit Analysis

The above-described technique, which is known as the "compression approach" [13–15] may be summarized as follows.

1. All m inputs and outputs of the circuit which are located on a transmission line where higher order modes may be assumed to be sufficiently attenuated, form conventional, reference plane–based "outer" ports.
2. The n active devices are replaced by respective reference paths, which form "inner" ports.
3. By means of a suited field theoretical method of analysis, the remaining linear passive part of the circuit is characterized by an $(n + m)$-port scattering matrix $\mathbf{S}_c(f)$ (see "Determination of Compression Matrix" subsection below for details) with respect to both the m outer and n inner ports.
4. The overall equivalent circuit is established by connecting $\mathbf{S}_c$ to the intrinsic models of the active devices via inner ports and to sources, loads, and other parts of the circuit (which are connected by transmission lines) via outer ports. The analysis of the total circuit is performed with a standard linear circuit analysis or harmonic balance tool.

The major difference between this technique and the standard segmentation approach is the use of a matrix of scattering parameters, $\mathbf{S}_c$, to provide for an explicit conversion between the reference plane-based, mode-dependent ports on transmission lines and the reference path–oriented, mode-independent interface which is suited for lumped elements.

The comparison of both approaches is discussed in section IV.

Determination of Compression Matrix

Basically, the compression matrix may be determined by performing p linear field theoretical analyses where $p = m + n$ is the total number of considered ports. Using p linear-independent excitations $\mathbf{a}_i = (a_i^1, a_i^2, \ldots, a_i^p)^\top$ yields p responses $\mathbf{b}_i = (b_i^1, b_i^2, \ldots, b_i^p)^\top$, where a_i^j and b_i^j denote the amplitudes of the incident and reflected wave, respectively, at port j in analysis i. The compression matrix is then given by:

$$\mathbf{S}_c = \mathbf{B}\mathbf{A}^{-1} \tag{14}$$

with $\mathbf{A} = (\mathbf{a}_1, \mathbf{a}_2, \ldots, \mathbf{a}_p)$ and $\mathbf{B} = (\mathbf{b}_1, \mathbf{b}_2, \ldots, \mathbf{b}_p)$.

Due to their definition, at inner ports, a_i^j and b_i^j are to be determined from eqs. (8) and (9), together with eqs. (5) and (6).

This requires the specification of reference impedances Z_j at the inner ports. However, because the wave amplitudes at the inner ports are not related to any modes or characteristic wave impedances, the Z_j may be chosen for convenience. This is because the only purpose of wave amplitudes and scattering parameters corresponding to inner ports is to finally reproduce relation eq. (7) (which is independent of any reference impedances) between the port voltage and current (either in time or in frequency domain) in a subsequent analysis of the overall circuit. Therefore, reference impedances at inner ports simply serve as a kind of temporary scaling factor.

To evaluate eq. (14), the matrix of excitations $\mathbf{A}$ has to be regular. The test of linear independence of the columns of $\mathbf{A}$ may be reformulated in terms of the terminations and excitations used for performing the analyses. If the analysis of $s - 1$ excitations with linear-independent $\mathbf{a}_i$ has been performed, a set of terminations and excitations has to be found which maintains the linear independence, that is:

$$\sum_{i=1}^{s} \lambda_i \mathbf{a}_i = \mathbf{0} \Rightarrow \lambda_i = 0, \qquad i = 1, \ldots, s \tag{15}$$

which yields:

$$\begin{pmatrix} \Delta r_1^1 b_1^1 + q_1^1 & \cdots & \Delta r_{s-1}^1 b_{s-1}^1 + q_{s-1}^1 & \lambda_1 \\ \vdots & \ddots & \vdots & \vdots \\ \Delta r_1^p b_1^p + q_1^p & \cdots & \Delta r_{s-1}^p b_{s-1}^p + q_{s-1}^p & q_s^p \end{pmatrix} \times \begin{pmatrix} \lambda_1 \\ \vdots \\ \lambda_s \end{pmatrix} = \mathbf{0} \quad (16)$$

where $\Delta r_i^m = r_i^m - r_s^m$ is the change in reflection coefficient at port m between the ith and the sth analysis (cf. Fig. 6). Because, normally, only a few terminations are changed during the p analyses, the above shown matrix (which is not quadratic unless $s = p$) will have only a few nonzero entries. Therefore, in practice, it is fairly easy to find a set of new terminations for the sth experiment, if $s - 1$ linear-independent ones have already been performed, such that eq. (16) yields $\lambda_i = 0$. Usually it is sufficient for every port to either excite it or to change its termination at least once during all p analyses to achieve the regularity of **A**.

Inner ports are used in situations where there is no simple means to decompose the electromagnetic field into eigenmodes in the sense as it is done at reference planes defined on homogeneous line sections. However, many field simulation methods are restricted to a circuit excitation via reference lines. This is mainly due to the fact that a physically meaningful field distribution must be established by the excitation on the input and output lines. But such a field distribution is unknown a priori at those locations where the inner ports can be found, that is, in the inner area of the complex metallization structure.

Therefore, an indirect excitation of the inner ports can be used to determine the compression matrix by providing varying terminations to these ports. In this case, opens and shorts may be used, yielding a reflection coefficient of $+1$ or -1, respectively, for any reference impedance.

Local Groups

In the general case it is not possible to define a unique voltage between any two points because of $\mathbf{rot\,E} \neq \mathbf{0}$. However, occasionally, a complex active device will cover such a small area that within that area the Kirchhoff laws will be approximately applicable; many CAD FET models are based on such an assumption. Therefore, the inner ports corresponding to that device may be grouped into a "local group." For every local group of inner ports it is assumed that the Kirchhoff laws are valid in that group. There is one ground for each group which is shared by all inner ports of that group. Many groups will be made up of only one two-terminal nonlinearity port (Fig. 7).

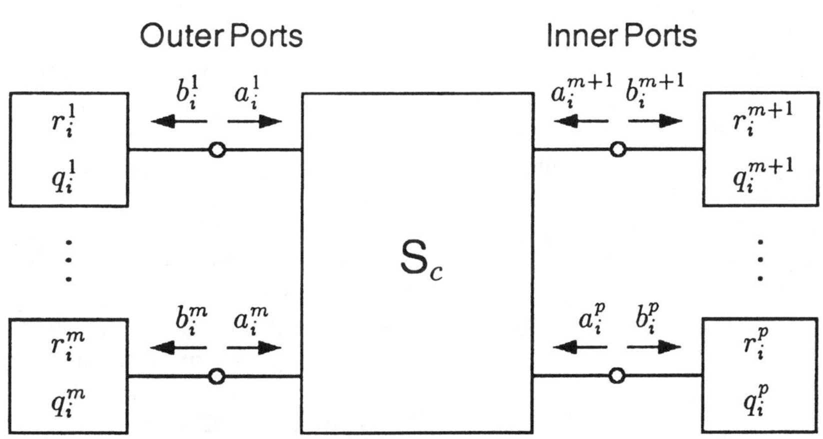

$$\mathbf{a}_i = \mathbf{R}_i \mathbf{b}_i + \mathbf{q}_i \qquad \mathbf{R}_i = \mathbf{diag}(r_i^j)$$

$$\mathbf{b}_i = \mathbf{S}_c \mathbf{a}_i \qquad i = 1, \ldots, p$$

Figure 6. Excitation of the circuit to determine the compression matrix. Superscripts and subscripts denote port and analysis number, respectively. If inner ports are not available for excitation by external sources, then $q_i^j = 0$ at inner ports. Usually, there is only one outer port excited by a source; ideally all outer ports are matched, so $r_i^j = 0$ at outer ports.

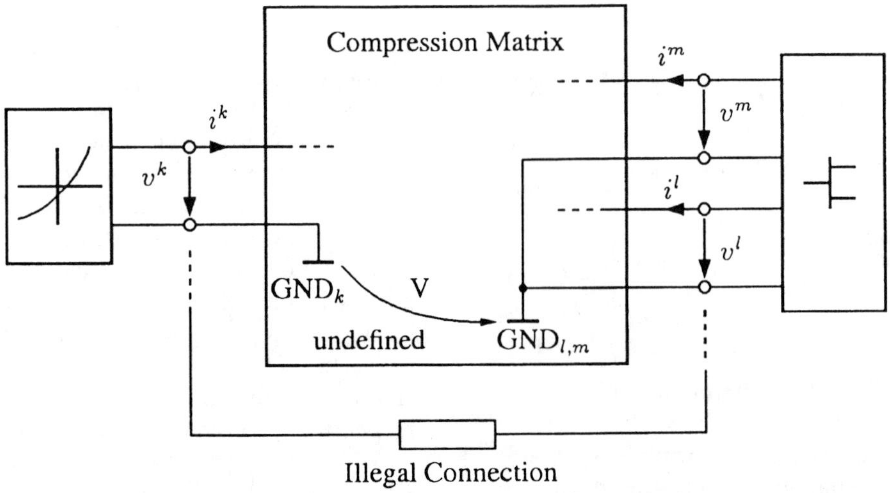

Figure 7. Definition of local groups of inner ports. Simple two-terminal nonlinearities form a local group on their own (port k). Many FET CAD models assume the applicability of the Kirchhoff laws within their vicinity and thus corresponding ports have to be grouped (ports l, m). Voltage between nodes of different local groups is undefined, thus there are no additional connections between nodes of different groups permissible.

There is no voltage defined between nodes belonging to different local groups. However, for the purpose of a nodal based circuit simulation it is required to have an absolute potential value for each ground. These may be fixed to arbitrary values, because the interaction between different local groups will be described completely by the compression matrix and depends only on port voltages and currents. It does not depend on any absolute potential values. Therefore, ground potentials as chosen for circuit simulation purposes actually do not have any influence.

This implies that it is not permissible to connect nodes of different local groups by any additional devices. Though this could be done easily during a simulation, the physical meaning would be undefined due to the lack of a properly defined voltage between nodes of different local groups (cf. Fig. 7).

IV. RESULTS

Nonlinear Transmission Line

To demonstrate the concepts discussed so far a nonlinear transmission line (NLTL) has been analyzed by a pure time domain full-wave analysis approach; namely, the FDTD method, a simple segmentation approach (SA), and the compression approach (CA). The NLTL is made up of coplanar strips which are loaded by 16 equidistant shunt diodes. It is a modified version of the example line used in ref. 10 in connection with the spatial network method. The structure is shown in Figure 8.

The numerical simulation results represent (approximate) solutions of the same underlying equations. However, it is important to recognize the fundamental differences in how the results are achieved. The methods operating in frequency domain are limited a priori to steady-state solutions, that is, all signals are at most quasiperiodic. Time domain integrating methods are, in principle, not limited (within the limits of stability criteria and the sampling theorem) with regard to time domain waveforms. For a proper comparison of both kinds of methods it is therefore required that the time domain results have reached a steady state. In particular, if the main interest is on steady-state results, which is often the case in the context of microwave circuits, the number of iteration steps which an integrating method needs to achieve steady-state conditions is a figure of merit for this method with respect to microwave CAD environments.

The compression matrix has been determined with the FDTD method [16].[5] A comparison of the time domain waveforms of port voltages of the NLTL between the FDTD (after $2\frac{1}{2}$ periods

[5]It has to be stressed that the compression matrix may be determined by any field theoretical method operating in either time or frequency domain.

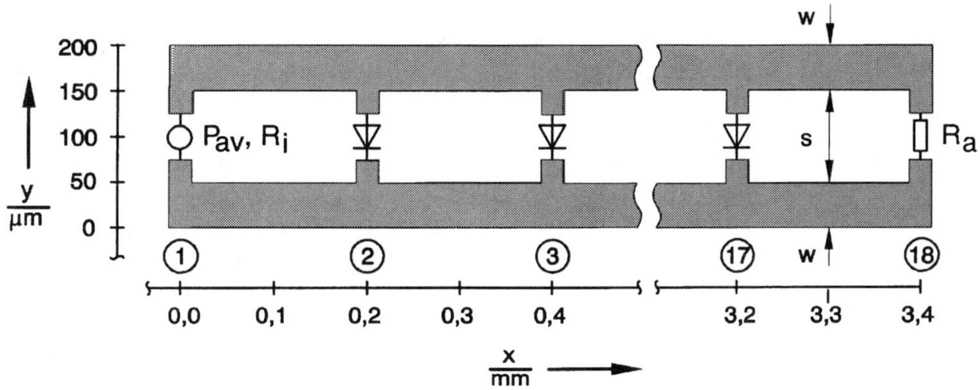

Figure 8. Nonlinear transmission line (after ref. 10, modified). Length: $N = 16$ diodes, diode spacing = 200 μm, total length = 3.4 mm. Diode: $i(v) = I_0(\exp(v/\eta V_G) - 1)$, $c(v) = C_{j0}/(1 - v/v_0)^{\nu}$, with $I_0 = 18.4$ pA, $\eta V_T = 38.775$ mV, $C_{j0} = 62.6$ fF, $v_0 = 0.65$ V, $\nu = 0.5$. Line: coplanar strips on GaAs, $w = 50$ μm, $s = 100 \mu$m. Excitation: $P_{av} = 2.5$ mW at 10 GHz, $R_i = R_a = 50$ Ω.

of the excitation frequency integration time) and CA results is shown in Figure 9a and b (extending over one period of the excitation frequency). There are still considerable visible differences between the results of both methods due to the presence of transients in the time domain integration results. After an FDTD integration time of $4\frac{1}{2}$ periods of the excitation frequency a much better agreement between both methods has been achieved, indicating that transients have almost

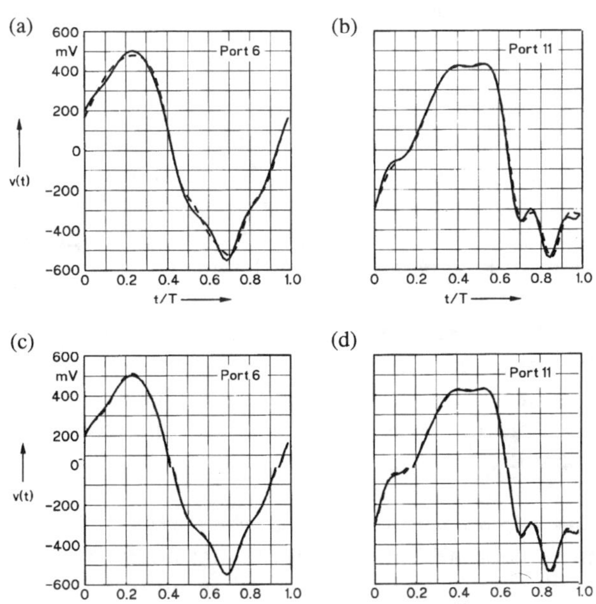

Figure 9. Voltage waveform at port 6 (a, c) and 11 (b, d) of the NLTL. (- - -) FDTD after $2\frac{1}{4}$ (a, b) and $4\frac{1}{4}$ (c, d) periods of the fundamental frequency. (——) Compression approach.

vanished (Figure 9c and d). The results of both methods coincide. Within the accuracy of the FDTD, therefore, both methods yield a field theoretically accurate description including the effects of the diode connection pads.

For an analysis by segmentation, the coplanar strips were replaced by suitable ideal transmission lines, however, without any model for the connection pads. The resulting time domain waveforms exhibit clear differences to those achieved with the compression approach (which coincide with the FDTD results) (Figure 10a and b). The differences are mainly due to a higher ripple frequency of the SA results, whereas the overall waveform matches quite well. It is well known that a periodic structure like the NLTL exhibits a small signal cut-off frequency f_{per}, which is given by:

$$f_{per} = \frac{1}{\pi(L'(C' + C'_D))^{1/2}} \tag{17}$$

where L', C', and C'_D are the inductance and capacitance of the line and the diode capacitance, respectively, per unit length (cf. ref. 17). The lower ripple frequency that is found in the FDTD/CA results (which account for the diode connection pads) indicates that the connection pads at least result in an additional capacitive load to the line. Adding a parallel capacitance of $C_p = 20$ fF to every diode in the segmented equivalent circuit to model that effect results in a decrease of the ripple frequency. However, due to the additional capacitive load, the overall line

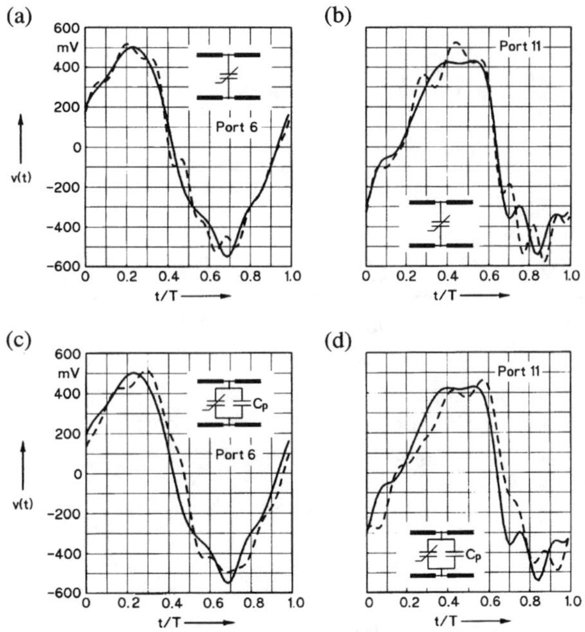

Figure 10. Voltage waveform at port 6 (a, c) and 11 (b, d) of the NLTL. (- - -) Segmentation without (a, b) and with (c, d) correction capacitance $C_p = 20$ fF. (——) Compression approach.

tends to get slower, now resulting in a slight delay of the SA results (Fig. 10c and d).

Moreover, there is still a poor coincidence. Changing the value of C_p does not yield any significant improvement, thus indicating that a simple parallel capacitance is not a sufficiently accurate model for the connection pads.

Obviously, replacing the coplanar strips by ideal transmission lines is a quite coarse approximation. However, because differences in the dispersive behavior of the linear lines tend to result in differences of the overall waveform rather than in the ripple fine structure,[6] and because the overall waveforms match quite well, the choice of ideal transmission lines seems to be still an acceptable approximation. The differences in the CA and SA results clearly show that even a small discontinuity may lead to significant effects in the circuit behavior and underline the need to precisely account for the metallization structure in the vicinity of active elements.

Discussion

The compression approach has some properties which result in enhanced analysis speed and increased accuracy compared to other techniques like FDTD and segmentation, respectively, in particular if nonlinear circuit regimes are to be considered.

Accuracy. The introduction of inner ports for active devices which are embedded in complex metallization patterns allows for the preservation of the whole accuracy of field theoretical analysis methods like FDTD, TLM, and SDA[7] to be used for the determination of the compression matrix. In contrast, the conventional segmentation approach suffers from the need of reference planes for the connection of every lumped device.

In fact, if a completely time domain–based circuit analysis, including nonlinear elements, is replaced by the compression approach utilizing inner ports and harmonic balance analysis, this does not change the underlying system of equations. Only the evaluation sequence is changed, which principally does not influence the result, as demonstrated by the above example.[8]

This is an essential advantage of the compression approach over the segmentation. It makes the compression approach suited especially for the analysis of coplanar components which, due to their nature, are prone to excitation and/or conversion of modes with different symmetries. Using inner ports to connect active devices allows one to pick exactly that part of the electromagnetic field which is physically acting upon the device regardless of any symmetry or mode considerations.

Flexibility. If p ports are given, the determination of $\mathbf{S}_c$ requires at most p linear analyses, which may be done with any suited field theoretical method. After the characterization is finished, the overall circuit simulation for nonlinear regimes may be done with harmonic balance, which offers a large number of different types of analyses, which can be performed efficiently by state-of-the-art harmonic balance tools. The compression approach has also been used for the characterization of microwave packages and interconnects [12, 20].

Speed Improvement for Nonlinear Regimes. One of the major advantages of the harmonic balance method is the ability to use the result of a former

[6]For the fundamental properties of nonlinear transmission lines compare refs. 18 and 19.

[7]Spectral domain approach.

[8]Due to the nature of the harmonic balance method this requires the confinement to periodic or quasiperiodic circuit states.

analysis as a solution estimate for the next analysis. In nonlinear regimes, every change of excitations or active device parameters requires a new analysis. Quite often parameters are swept with small increments, so the last solution provides a good solution estimate. However, sweeping parameters leads to very large cpu times in the case of time domain field theoretical methods which include nonlinear elements. This is due to the fact that virtually all of the cpu time is required to resimulate the linear part of the system again for each new analysis. On the other hand, with the compression approach, the linear part has to be characterized only once.

If hundreds or thousands of analyses have to be performed, this amounts in an enormous savings of cpu time, because a single harmonic balance analysis is much faster than a single time domain FDTD or TLM run including nonlinear elements. As an example, Figures 11 and 12 show the influence of parameter variations on the waveform at the end of the NLTL, and Figure 13 shows a comparison of the required cpu times for the compression approach versus those of the FDTD. It is evident from Figure 13 that the required analysis time is proportional to the number of analyses in the time domain analysis case. On the other hand, the compression approach requires an initial effort to determine the compression matrix $\mathbf{S}_c$. Once this has been done, the following harmonic balance simulations need much less time in comparison to the corresponding time domain simulation. However, due to the usage of inner ports at the active devices, the overall accuracy is approximately equal in both cases. It can be seen from Figure 13 that if more than $N_{\text{Break-Even}}$ (= 5 for the case of Figure 13) analyses have to be performed, a gain of approximately 2 cpu hours per analysis has been achieved

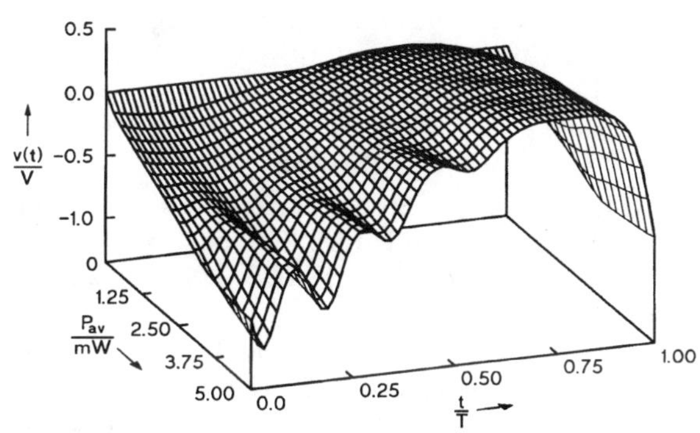

Figure 11. Voltage waveform at port 18 of the NLTL as function of the available input power P_{av} ($Z = 50\ \Omega$).

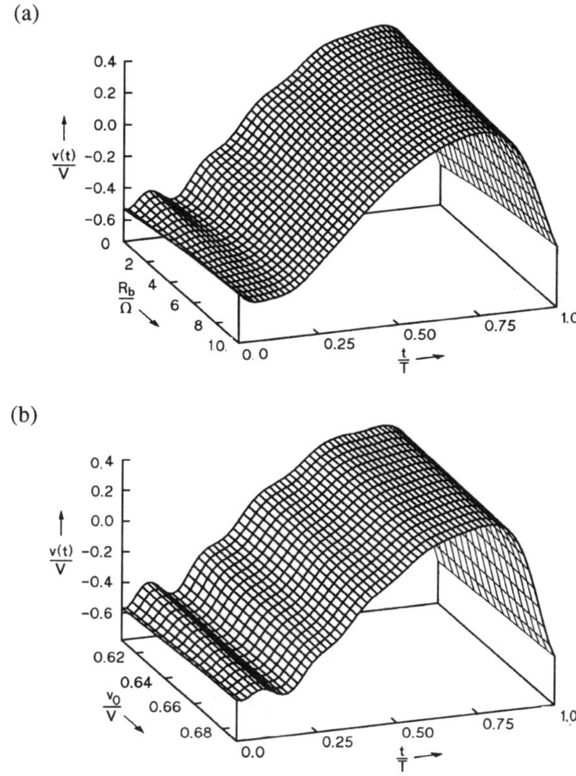

Figure 12. Voltage waveform at port 18 of the NLTL as function of (a) the diode series resistance R_b; and (b) the diode capacitance built-in voltage v_0.

for that particular example by using the compression approach. Even for the small number of analyses used for Figures 11 and 12 (a total of 59 analyses), this amounts in an approximate cpu-time reduction by a factor of 10 for this example.

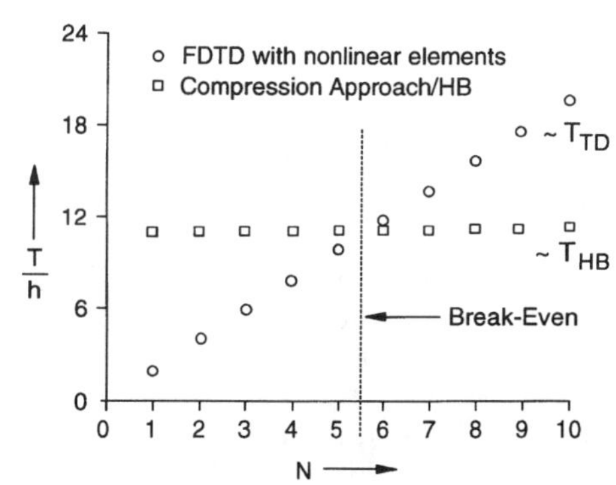

Figure 13. Comparison of cpu times for FDTD with nonlinear elements and compression approach with harmonic balance (HB) for the NLTL. The slopes are $\sim T_{TD}$ for the FDTD and $\sim T_{HB}$ for the compression approach with $T_{TD} \gg T_{HB}$. The compression approach saves computation time if more than $N_{\text{Break-Even}}$ analyses have to be performed (in this example $N_{\text{Break-Even}} = 5$) while maintaining FDTD accuracy.

V. CONCLUSION

Different approaches for the analysis of microwave circuits with lumped active devices have been considered. The compression approach, an alternative procedure which provides a tailored interface for lumped active devices, was discussed in detail. For a model problem it was shown that the compression approach may yield considerable savings in cpu time in comparison to the FDTD method while retaining both the full field theoretical accuracy of that method and all analysis options of the harmonic balance. In comparison to a standard segmentation approach, the compression approach offers an increased field theoretical accuracy.

REFERENCES

1. T. Itoh, "Generalized Scattering Matrix Technique," in T. Itoh (ed.), *Numerical Techniques for Microwave and Millimeter-Wave Passive Structures*. Wiley, New York, 1989, pp. 622–636.
2. M. S. Nakhla and J. Vlach, "A Piecewise Harmonic-Balance Technique for Determination of Periodic Response of Nonlinear Systems," *IEEE Trans. Circ. Syst.*, Vol. CAS-23, Feb. 1976, p. 85–91.
3. V. Rizzoli and A. Lipparini, "General Stability Analysis of Periodic Steady-State Regimes in Nonlinear Microwave Circuits," *IEEE Trans. Microwave Theory Tech.*, Vol. MTT-33, Jan. 1985, pp. 30–37.
4. R. J. Gilmore and M. B. Steer, "Nonlinear Circuit Analysis Using the Method of Harmonic Balance—a Review of the Art. II. Advanced Concepts," *Int. J. MIMICAE*, Vol. 1, Feb. 1991, pp. 159–180.
5. M. B. Steer, "Simulation of Nonlinear Microwave Circuits—an Historical Perspective and Comparison," *1991 IEEE MTT-S Int. Microwave Symp. Dig.* (Boston), June 1991, pp. 559–562.
6. V. Rizzoli, A. Lipparini, A. Costanzo, F. Mastri, C. Cecchetti, A. Neri, and D. Masotti, "State-of-the-Art Harmonic-Balance Simulation of Forced Nonlinear Microwave Circuits by the Piecewise Technique," *IEEE Trans. Microwave Theory Tech.*, Vol. MTT-40, Jan. 1992, pp. 12–27.
7. A. Ushida, T. Adachi, and L. O. Chua, "Steady-State Analysis of Nonlinear Circuits Based on Hybrid Methods," *IEEE Trans. Circ. Syst.—I*, Vol. CAS-39, Aug. 1992, pp. 649–661.
8. K. S. Yee, "Numerical Solution of Initial Boundary Value Problems Involving Maxwell's Equations in Isotropic Media," *IEEE Trans. Ant. Propagat.*, Vol. AP-14, May 1966, pp. 302–307.
9. S. Akhtarzad and P. Johns, "Solution of Maxwell's Equations in Three Space Dimensions and Time by the T.L.M. Method of Numerical Analysis," *Proc. Inst. Elec. Eng.*, Vol. 122, Dec. 1975, pp. 1344–1348.
10. T. Shibata, "Circuit Simulations Combined with the Electromagnetic Field Analysis," *IEEE Trans. Microwave Theory Tech.*, Vol. MTT-39, April 1991, pp. 1862–1868.
11. R. H. Voelker and R. J. Lomax, "A Finite-Difference Transmission Line Matrix Method Incorporating a Nonlinear Device Model," *IEEE Trans. Microwave Theory Tech.*, Vol. MTT-38, March 1990, pp. 302–312.
12. T. Shibata and H. Kimura, "Computer-Aided Engineering for Microwave and Millimeter-Wave Circuits Using the FD–TD Technique of Field Simulations," *Int. J. MIMICAE*, Vol. 3, July 1993, pp. 238–250.
13. J. Kunisch, M. Rittweger, S. Heinen, and I. Wolff, "The Compression Approach: A New Technique for the Analysis of Distributed Circuits Containing Nonlinear Elements," in *Proceedings of the 21st European Microwave Conference* (Stuttgart), Sept. 1991, pp. 1296–1301.
14. J. Kunisch, A. Bahr, M. Rittweger, S. Heinen, and I. Wolff, "Analysis of Nonlinear Forced and Autonomous Microwave Circuits Using the Compression Approach: Theory and Application to a Nonlinear Transmission Line," *International Workshop on Integrated Nonlinear Microwave and Millimeter-wave Circuits Digest* (Duisburg), Oct. 1992, pp. 289–299.
15. J. Kunisch, A. Bahr, M. Rittweger, and I. Wolff, "Analysis of Nonlinear Microwave Circuits Using the Compression Approach," in *1993 IEEE MTT-S Int. Microwave Symp. Dig.* (Atlanta), June 1993, pp. 637–640.
16. A. Bahr, "Untersuchung eines Verfahrens zur Analyse nichtlinearer Mikrowellenschaltungen," Master's thesis, Universität Duisburg, 1992.
17. M. J. W. Rodwell, M. Case, R. Yu, S. Allen, Y. Konishi, M. Kamegawa, E. Carman, and J. Pusl, "GaAs Nonlinear Transmission Lines for Picosecond and MM-Wave Applications," in *International Workshop on Integrated Nonlinear Microwave and Millimeterwave Circuits Digest* (Duisburg), Oct. 1992, pp. 249–263.
18. R. Landauer, "Parametric Amplification Along Nonlinear Transmission Lines," *J. Appl. Phys.*, Vol. 31, March 1960, pp. 479–484.
19. R. Hirota and K. Suzuki, "Theoretical and Experimental studies of Lattice Solitons in Nonlinear Lumped Networks," *Proc. IEEE*, Vol. 61, Oct. 1973, pp. 1483–1491.
20. M. Rittweger, M.Werthen, and I. Wolff, "FDTD Simulation for Microwave Packages and Interconnects," in *1992 IEEE MTT-S International Microwave Symposium Workshop on EM Modeling of Microwave Packages and Interconnects Digest* (Atlanta), June 1993.

Computer-Aided Engineering for Microwave and Millimeter-Wave Circuits Using the FD-TD Technique of Field Simulations (Invited Article)

Tsugumichi Shibata and Hideaki Kimura

NTT LSI Laboratories, 3-1 Morinosato Wakamiya, Atsugi, Kanagawa 243-01, Japan

Received December 3, 1992; revised January 31, 1993.

ABSTRACT

This article describes our techniques for the three-dimensional full-wave field simulation of the computer-aided design of broad-band circuits operating in the microwave and millimeter-wave frequency region. After reviewing the numerical algorithm of the finite-difference time-domain approach, the concept of the lumped device model is introduced into the simulation. We then propose an efficient way of calculating the scattering parameters of passive elements that uses the lumped device model as a standard termination of the circuit port. The analysis of a nonlinear circuit directly combined with the field simulation is also presented, and finally, a method of simulating integrated circuits mounted in a package is discussed. © 1993 John Wiley & Sons, Inc.

I. INTRODUCTION

The design of microwave and millimeter-wave circuits nowadays trends toward achieving first-pass success. This is because reducing additional iterations in the design and fabrication cycles is the most effective way to reduce costs and turnaround time. One of the key technologies required for this purpose is powerful electromagnetic field simulation, which, when done prior to fabrication, provides a clear picture of circuit performance. Up until a few years ago, the bottleneck had been in the availability of sufficient computational power, which inevitably lead to the use of expensive supercomputers because three-dimensional full-wave field simulations usually require large computational ability and memory size. However, with the advent of recent high-performance workstation-class computers, full-wave simulations have become more and more feasible in CAD of microwave and millimeter-wave circuits. In addition, new effective techniques are continuously being devised through the research of simulation methods.

This article describes our recent studies on the finite-difference time-domain (FD-TD) method and their applications to the design of microwave circuits [1]. Among the many other numerical techniques available for field simulations, the majority yield harmonic solutions of Maxwell's equations in the space or spectral domain. In contrast, the FD-TD method represents true computer simulation that shows field evolution directly in the time domain, and offers the following advantages.

- It employs a simple finite difference formulation of Maxwell's equations and does not rely on a specific assumption on boundary and material configurations, thus it has remarkable generality in terms of the structures that can be dealt with. Microstrip and coplanar circuits as well as many other complex waveguides and discontinuity structures have been analyzed. Of course, the structure can be inhomogeneous and metal and dielectric losses can be involved as well [2,3,5].

- Transient behavior of circuit response is immediately known with this method. Consequently, it is especially suited for applications in which the waveform itself is of interest, for instance, in analyses of signal reflection and crosstalk in high-speed digital circuits and pulse generation and deformation in photoconductive switching circuits [6,7].
- Frequency characteristics can be calculated by applying the Fourier transform to the time domain response of the circuit. If only the characteristic at a specific frequency point is required, other methods such as the finite element method are recommended. However, when a wide range of frequency characteristics need to be calculated, the time domain method is effective because all of the characteristics can be obtained from one pulse response at a time [8,9].
- Various lumped device elements can be easily modeled and incorporated in the analyzed structure yielding flexibility in representing structures and functions of the circuits [6,10–12].
- Because the analysis is in the time domain, nonlinearity can be included in the lumped device models as well as the medium constants, which means that we can directly combine the simulation of nonlinear circuits with field simulations using this method [10–12].

In the next section, the principal algorithm of the FD-TD method is described. Calculation of the scattering parameter is the main theme of section III. Section IV focuses on nonlinear circuit simulations, and section V states our conclusions.

II. FD-TD ALGORITHM

Maxwell's equations are the basis of electromagnetic field simulations. The time-dependent curl equations can be approximated into second-order accurate central-difference equations [2,3]. The system equations for updating the field components at each time step are given in an explicit form, therefore, the field behaviors can be calculated in the discretized computational space without relying on any complex numerical procedures such as solving large matrix equations.

II.A. Elemental Routine

Let us assume a rectangular parallelepiped analysis region of $L\Delta x \times M\Delta y \times N\Delta z$, where Δx, Δy, and Δz are the discretization lengths in each direction of the space, and L, M, and N are positive integers that provide the scale of mesh size in the analysis. In this region, Maxwell's curl equations:

$$\frac{\partial \mathbf{B}}{\partial t} = -\nabla \times \mathbf{E} \tag{1}$$

$$\frac{\partial \mathbf{D}}{\partial t} = \nabla \times \mathbf{H} - \mathbf{J} \tag{2}$$

can be discretized into the following difference equations. For instance, regarding the x component of $\mathbf{B}$, we can get

$$\begin{aligned}&\frac{B_x^{n+1/2}(i, j+\frac{1}{2}, k+\frac{1}{2}) - B_x^{n-1/2}(i, j+\frac{1}{2}, k+\frac{1}{2})}{\Delta t}\\&= \frac{E_y^n(i, j+\frac{1}{2}, k+1) - E_y^n(i, j+\frac{1}{2}, k)}{\Delta z}\\&- \frac{E_z^n(i, j+1, k+\frac{1}{2}) - E_z^n(i, j, k+\frac{1}{2})}{\Delta y},\end{aligned} \tag{3a}$$

and for the x component of $\mathbf{D}$,

$$\begin{aligned}&\frac{D_x^n(i+\frac{1}{2}, j, k) - D_x^{n-1}(i+\frac{1}{2}, j, k)}{\Delta t}\\&= \frac{H_z^{n-1/2}(i+\frac{1}{2}, j+\frac{1}{2}, k) - H_z^{n-1/2}(i+\frac{1}{2}, j-\frac{1}{2}, k)}{\Delta y}\\&- \frac{H_y^{n-1/2}(i+\frac{1}{2}, j, k+\frac{1}{2}) - H_y^{n-1/2}(i+\frac{1}{2}, j, k-\frac{1}{2})}{\Delta z}\\&- J_x^{n-1/2}(i+\frac{1}{2}, j, k).\end{aligned} \tag{4a}$$

Similar equations for the y and z components of $\mathbf{B}$ and $\mathbf{D}$ can be easily represented in the same way. Noticed here is that six field components should be defined at different positions shifted by half of the discretization length. That is, the electric field components are defined at the following points:

$$E_x(D_x)\{(i + \tfrac{1}{2})\Delta x, j\Delta y, k\Delta z\}:$$
$$i = 0, 1, \ldots, L - 1; j = 0, 1, \ldots, M;$$
$$k = 0, 1, \ldots, N,$$

$$E_y(D_y)\{i\Delta x, (j + \tfrac{1}{2})\Delta y, k\Delta z\}:$$
$$i = 0, 1, \ldots, L; j = 0, 1, \ldots, M - 1;$$
$$k = 0, 1, \ldots, N,$$

$$E_z(D_z)\{i\Delta x, j\Delta y, (k + \tfrac{1}{2})\Delta z\}:$$
$$i = 0, 1, \ldots, L; j = 0, 1, \ldots, M;$$
$$k = 0, 1, \ldots, N - 1.$$

These are called "electric nodes." The magnetic field components are defined at

$$H_x(B_x)\{i\Delta x, (j + \tfrac{1}{2})\Delta y, (k + \tfrac{1}{2})\Delta z\}:$$
$$i = 0, 1, \ldots, L; j = 0, 1, \ldots, M - 1;$$
$$k = 0, 1, \ldots, N - 1,$$

$$H_y(B_y)\{(i + \tfrac{1}{2})\Delta x, j\Delta y, (k + \tfrac{1}{2})\Delta z\}:$$
$$i = 0, 1, \ldots, L - 1; j = 0, 1, \ldots, M;$$
$$k = 0, 1, \ldots, N - 1,$$

$$H_z(B_z)\{(i + \tfrac{1}{2})\Delta x, (j + \tfrac{1}{2})\Delta y, k\Delta z\}:$$
$$i = 0, 1, \ldots, L - 1;$$
$$j = 0, 1, \ldots, M - 1;$$
$$k = 0, 1, \ldots, N,$$

which are called "magnetic nodes." The superscript n in eqs. (3a) and (4a) is an integer that indicates the time step in the unit of Δt. The electric field components and the magnetic field components are also defined alternately in time.

At the electric nodes, the components of $\mathbf{E}$ are calculated from that of $\mathbf{D}$ for each time step as

$$E_u = \frac{1}{\varepsilon_0 \varepsilon_{ru}} D_u \tag{5}$$

and the components of $\mathbf{H}$ are calculated from that of $\mathbf{B}$ at the magnetic nodes using

$$H_u = \frac{1}{\mu_0 \mu_{ru}} B_u, \tag{6}$$

where the suffix $u = x, y$, or z, and ε_{ru} and μ_{ru} are relative permittivity and permeability defined for each electric node and magnetic node, respectively. The last term of eqs. (4), the conductive current $\mathbf{J}$, should be calculated from

$$J_u = \sigma_u E_u, \tag{7}$$

where σ_u is conductivity defined for each electric node. However, because the components of $\mathbf{E}$ are defined at the time $n\Delta t$, we need to take an average between $\mathbf{E}$ at the time of $n\Delta t$ and that of $(n - 1)\Delta t$ to estimate $\mathbf{J}$ at the time of $(n - \frac{1}{2})\Delta t$. This average is obtained in regard to, for example, the x component as follows:

$$J_x^{n-1/2}(i + \tfrac{1}{2}, j, k) = \frac{\sigma_x(i + \frac{1}{2}, j, k)}{2} \times \{E_x^n(i + \tfrac{1}{2}, j, k) + E_x^{n-1}(i + \tfrac{1}{2}, j, k)\}. \tag{8a}$$

Substituting eqs. (5) and (8) into eq. (4), we can get an expression of $\mathbf{E}^n$ which can be calculated from $\mathbf{E}^{n-1}$ and $\mathbf{H}^{n-1/2}$. Similarly, eqs. (3) and (6) offer us a way of calculating $\mathbf{H}^{n+1/2}$ from $\mathbf{H}^{n-1/2}$ and $\mathbf{E}^n$. Consequently, all of the field components defined in the analysis region can be integrated step by step in time starting from the initial field distribution. Circuit structures are constructed in this analysis region by assigning values ε_{ru}, μ_{ru}, and σ_u for all nodes to the appropriate medium constants. We can choose the discretization lengths Δx, Δy, and Δz arbitrarily, but the time interval Δt must satisfy the following Courant relation to ensure the numerical stability of the analysis.

$$c\Delta t \leq \frac{1}{\sqrt{\dfrac{1}{\Delta x^2} + \dfrac{1}{\Delta y^2} + \dfrac{1}{\Delta z^2}}}, \tag{9}$$

where c is the wave velocity in free space.

II.B. Boundary Conditions

There are six boundary faces around the analysis region, and additional conditions are needed to calculate field variables defined on these boundaries. A combination of the following three kinds of conditions are often used.

1. Electric wall condition. When the analysis region is surrounded by a perfect conductor wall, the electric wall condition can be applied to the boundary. This boundary condition is also useful when the circuit structure has a symmetrical plane with this condition. Examples of this case are shown in Figures 2 and 6 and will be discussed later. This boundary can be obtained by simply assigning all field variables on the boundary, i.e., the components of the electric field parallel to and the component of the magnetic field perpendicular to the face, constantly to zero.
2. Magnetic wall condition. When the circuit structure has a symmetrical plane with the magnetic wall condition, this type of boundary is desired. There are two possible ways to form this type of boundary. One is using the plane shifted to the inside of the analysis region by a half of the discretization length. For instance, the plane $x = \Delta x/2$ contains nodes on which the parallel components of the magnetic field and the perpendicular component of the electric field are defined. Thus, by assigning all field variables on this plane to zero, we can obtain the magnetic wall condition of this plane. In this case, the variables on the plane $x = 0$ are ignored. The other way is applying eqs. (3) and (4) even for calculating field variables on the boundary. In doing this, the required values outside the analysis region should be the same except for the sign with the field variable at the symmetrical point against the boundary.
3. Absorbing boundary condition. When an open space is analyzed, the analysis region is usually truncated by the absorbing boundary. This condition is derived from the one-way wave equation that must be satisfied by a plane wave passing through the boundary from the inside of the analysis region at an arbitrary angle [4]. As a first-order approximation in regard to, for example, the y and z components of $\mathbf{E}$ on the boundary $x = 0$, we use

$$\begin{aligned} E_y^n(0, j + \tfrac{1}{2}, k) &= E_y^{n-1}(1, j + \tfrac{1}{2}, k) \\ &\quad + \frac{c\Delta t - \Delta x}{c\Delta t + \Delta x}\{E_y^n(1, j + \tfrac{1}{2}, k) \\ &\quad - E_y^{n-1}(0, j + \tfrac{1}{2}, k)\} \end{aligned} \tag{10a}$$

and

$$\begin{aligned} E_z^n(0, j, k + \tfrac{1}{2}) &= E_z^{n-1}(1, j, k + \tfrac{1}{2}) \\ &\quad + \frac{c\Delta t - \Delta x}{c\Delta t + \Delta x}\{E_z^n(1, j, k + \tfrac{1}{2}) \\ &\quad - E_z^{n-1}(0, j, k + \tfrac{1}{2})\}. \end{aligned} \tag{10b}$$

II.C. Lumped Device Models

As stated previously, we can assign medium constants ε_{ru}, μ_{ru}, and σ_u to each node in the analysis region. Hence, the building blocks of the circuit structure can be resistors and dielectric materials as well as metal conductors. Voltage and current sources can be also formulated in the analysis region.

1. Resistor model. Let us consider a parallelepiped volume of $a\Delta x \times b\Delta y \times c\Delta z$ in the analysis region. If we assign a constant value σ to all electric nodes in this portion of the region, the resistance between the opposite faces $b\Delta y \times c\Delta z$ separated by $a\Delta x$ becomes $a\Delta x/b\Delta y\, c\Delta z\, \sigma$. Therefore, if we choose an appropriate value of σ, this cube will act as a resistor of the intended resistance value. We can, of course, allocate a number of nodes for this resistor model to take the distributed effect of the device itself into account. However, the distributed effects usually appear on a circuit level and many devices can be regarded as a lumped device depending on their size. In such cases, it is enough to represent a model using one node in the analysis region. Figure 1 shows an arbitrary E_x-node for this model. We suppose here that each node in the discretized space is accompanied by a fictitious volume of $\Delta x \times \Delta y \times \Delta z$ as shown in the figure. This supposition is natural because each

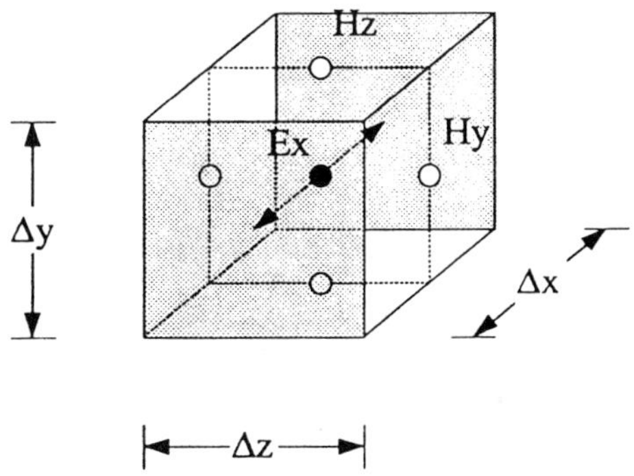

Figure 1. An E_x-node and accompanying fictitious volume.

kind of node appears at the rate of one per $\Delta x \times \Delta y \times \Delta z$ in the sapce. Then, it can be considered that the resistance between the opposite faces $\Delta y \times \Delta z$ of this cube becomes the desired value R_{lump} if we assign to this node the following value of σ_x:

$$\sigma_x = \frac{\Delta x}{\Delta y \Delta z R_{\text{lump}}}. \tag{11}$$

An arbitrary E_y- and E_z-node is also capable of representing a lumped resistor model.

2. Capacitor model. Similarly, a capacitor model can be represented using an electric node too. Referring to Figure 1 again, it can be considered that the capacitance between the opposite faces $\Delta y \times \Delta z$ of this cube becomes the desired value C_{lump} if we assign to this node the following value of ε_{rx}:

$$\varepsilon_{rx} = \frac{\Delta x C_{\text{lump}}}{\Delta y \Delta z \varepsilon_0}. \tag{12}$$

3. Voltage and current source models. For the lumped voltage or current source model, a slight modification of eq. (8) is needed. If we add a term for the internal electrical motive force E_{int}, eq. (8a) is rewritten as

$$J_x^{n-1/2} = \sigma_x \left(\frac{E_x^n + E_x^{n-1}}{2} + E_{\text{int}} \right). \tag{13a}$$

As a result, this node forms a lumped voltage source model of the voltage V_{int} with internal resistance R_{int} or the equivalent lumped current source model of the current I_{int} with the internal conductance $1/R_{\text{int}}$, where

$$R_{\text{int}} = \frac{\Delta x}{\Delta y \Delta z \sigma_x}, \tag{14}$$

$$V_{\text{int}} = -\Delta x E_{\text{int}}, \text{ and} \tag{15}$$

$$I_{\text{int}} = \Delta y \Delta z \sigma_x E_{\text{int}}. \tag{16}$$

Figure 2 illustrates a numerical experiment conducted to verify the above-mentioned formulations. We analyzed a structure of coplanar strips (CPS) driven by a lumped voltage source model with internal resistance equal to 50 Ω at one end and terminated by a lumped resistor model of 50 Ω at the other end. The size of the analysis region was $L = 15$, $M = 30$, and $N = 250$. Because of the symmetry, half of the structure was constructed in the analysis region adapting the electric wall condition on the boundary $i = 15$. The absorbing boundary con-

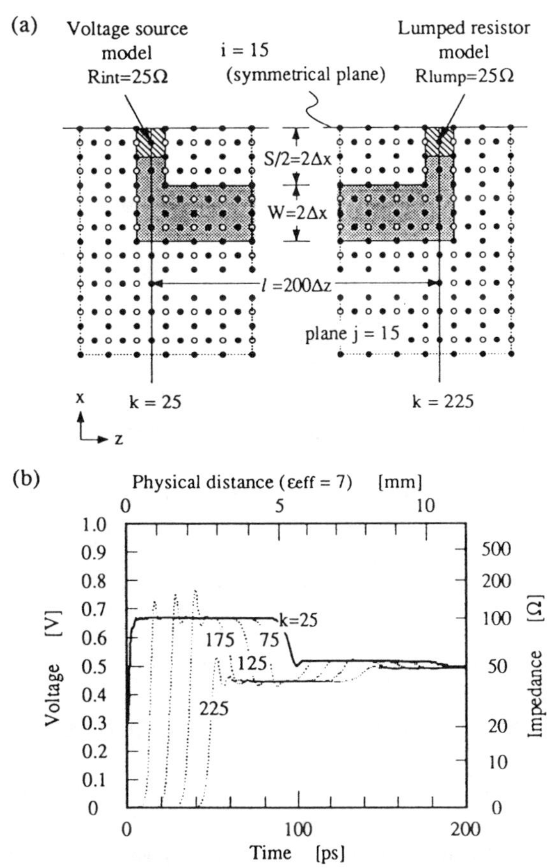

Figure 2. Analysis of coplanar strips with drive and termination models: (a) part of the analyzed structure, and (b) step response of the line voltage (TDR test).

dition was used on the other boundaries. The CPS dimensions are:

width of the striplines: $W = 2\Delta x = 50\ \mu\text{m}$;
spacing between striplines: $S = 4\Delta x = 100\ \mu\text{m}$;
length of the striplines: $l = 200\Delta z = 5$ mm,

where $\Delta x = \Delta y = \Delta z = 25\ \mu\text{m}$ is assumed. The permittivity of the substrate is assumed to be 13. Therefore, according to the result of quasistatic analysis based on the conformal mapping calculation, the characteristic impedance and the effective dielectric constant of this line should be about 100 Ω and 7, respectively. The voltage source and the terminating resistor are modeled in each E_x-node in the hatched regions in the Figure 2a. A step-function-like voltage waveform was generated by the lumped voltage source and the field behavior was calculated in the analysis region. Figure 2b shows calculated waveforms of the line voltage at different positions along the propagation direction. Particularly, the solid line shows the waveform at the driving point. One can see that the TDR waveform is in the 100-Ω level of the CPS characteristic impedance at first, and that it finally settles to the 50-Ω level of the terminating resistor, which demonstrates that the above-mentioned formulations successfully acted as intended.

III. *S*-PARAMETER CALCULATION

The main interest in the field simulation of passive microwave and millimeter-wave circuit components should be the calculation of scattering parameters at the circuit ports. The procedure for doing this using the FD-TD method can be summarized like this:

- Construct the circuit structure in the analysis region by assigning the medium constants to each node.
- Set all field variables in the analysis region to zero.
- Start time stepping along with excitation of the field at the input port of the circuit.
- Monitor the voltage and current responses at the input and output ports of the circuit.
- Extract the scattering parameters from the circuit responses.

Characteristics of the scattering parameters over a wide range of frequencies can be obtained from a pulse response using the Fourier transform, therefore, the pulse exciting the circuit should have a broad-band spectrum in the frequency range of interest.

III.A. Using a Lumped Device Model as a Standard Termination

In the early stages of our work, we demonstrated a calculation of scattering parameters for microstrip circuits using the time domain method of field simulations. In these reports, ingoing and outgoing waveforms at the reference planes on the input and output lines were used in calculations. Namely, for a two-port circuit, $S_{11}(\omega)$ and $S_{21}(\omega)$ were defined as

$$S_{11}(\omega) = \frac{V_r(\omega)}{V_i(\omega)} \tag{17}$$

and

$$S_{21}(\omega) = \frac{V_t(\omega)}{V_i(\omega)}, \tag{18}$$

where $V_i(\omega)$ and $V_r(\omega)$ are the Fourier transforms of the incident and reflected waves at the input reference plane and $V_t(\omega)$ is that of the transmitted wave at the output reference plane. The input and output lines were truncated at the boundary of the analysis region by the absorbing boundary condition. This approach has two main drawbacks. First, the precision of the result is affected by spurious reflections at the absorbing boundary of the analysis region, indicating a need for an improved absorbing boundary algorithm. Separation of incident and reflected waves at the input is also required. This requires an additional simulation for evaluating the incident waveform without the return from the circuit. Taking these drawbacks into consideration, we now describe an alternate method that uses the lumped device model as the standard termination. As already shown in Figure 2, the lumped voltage source and resistor model can be successfully incorporated in the circuit structure. Therefore, we can terminate input and output lines with these models and, for example, when port 1 of a two-port circuit is driven by a lumped voltage source $V_{\text{int}}(t)$ with an internal 50-Ω resistor and port 2 is terminated with a lumped resistor model of 50 Ω, the scattering parameters can be calculated as

$$S_{11}(\omega) = \frac{V_1(\omega)}{V_{int}(\omega)} \tag{19}$$

and

$$S_{21}(\omega) = \frac{V_2(\omega)}{2V_{int}(\omega)}, \tag{20}$$

where $V_{int}(\omega)$ is the Fourier transform of $V_{int}(t)$ and $V_1(\omega)$ and $V_2(\omega)$ are the transforms of the simulated port voltages at the input and output termination, respectively. Noticed here is that we do not need to deal with calculated waveforms for separating ingoing and outgoing components at all. Of course, because the terminations are included inside the analysis region, the quality of the absorbing boundary algorithm does not affect the results directly. In a strict sense, the above definition is a little bit ill-defined because the physical shapes of the terminating structures have been immaterialized. However, one can see the practical validity of the results obtained from this way of calculating in the following examples.

III.B. Examples of Microstrip Circuits

Figures 3 and 4 illustrate calculated circuit structures. The figures show microstrip lines having two gaps and a stub, respectively. In the simulations, $\Delta x = \Delta y = \Delta z = 0.424$ mm and resulting circuit dimensions are:

width of the microstrip conductor: 1.272 mm;
thickness of the substrate: 1.272 mm;
gap length in Figure 3a: 1.272 mm;
center microstrip line length in Fig. 3a: 12.72 mm;
stub line length in Figure 4a: 11.448 mm.

In order to make a comparison with measurements of samples fabricated on alumina substrate, the relative permittivity of the substrate was assumed to be 9.8. Each circuit was excited by the voltage source model with an internal 50-Ω resis-

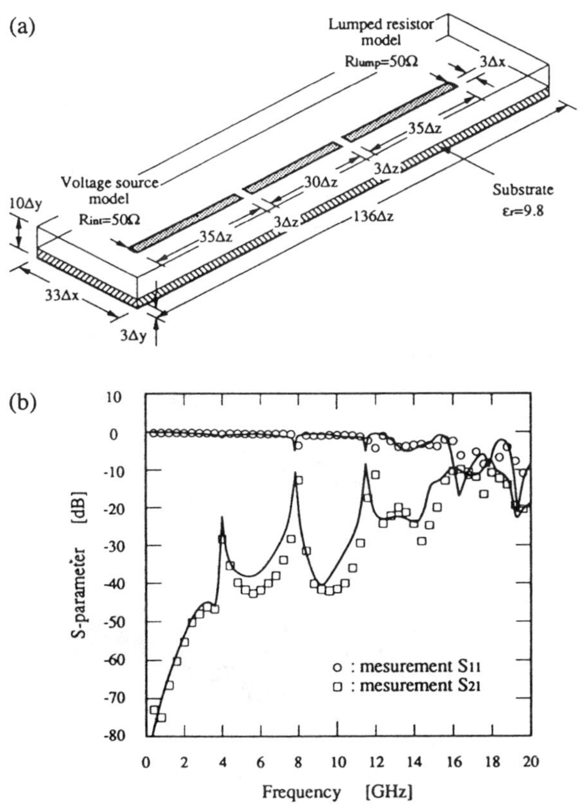

Figure 3. Microstrip circuit #1: (a) analyzed microstrip line having two gaps, and (b) the frequency characteristics of calculated and measured scattering parameters.

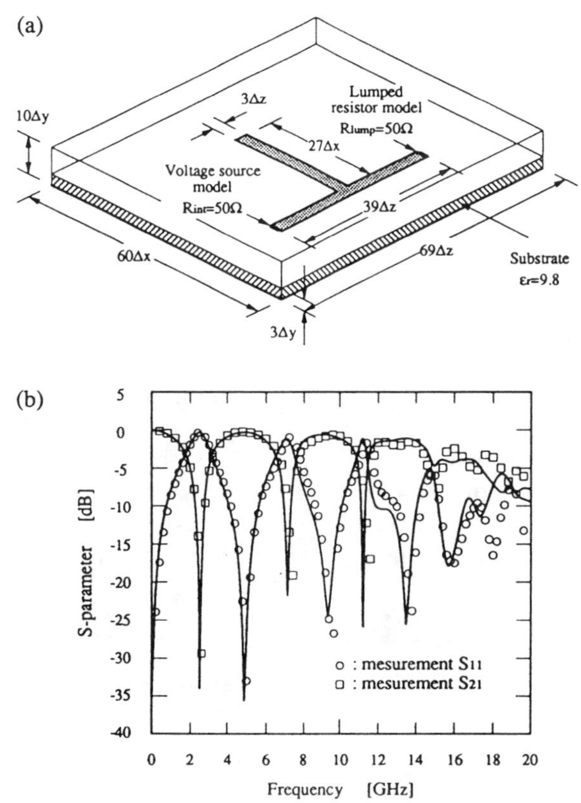

Figure 4. Microstrip circuit #2: (a) analyzed microstrip line having a stub, and (b) the frequency characteristics of calculated and measured scattering parameters.

tor at one end which generates an internal voltage waveform of

$$V_{\text{int}}(t) = 1 - \cos(2\pi f_{\text{band}} t)$$

for $0 \le t \le 1/f_{\text{band}}$ and

$$V_{\text{int}}(t) = 0$$

for $1/f_{\text{band}} \le t$.

Here $f_{\text{band}} = 11.8$ GHz. The other end was terminated by the lumped 50-Ω resistor model. Solid lines in Figures 3b and 4b present calculated results based on eqs. (19) and (20). Measured data are also plotted in these figures and excellent agreement is found between the calculation and the measurement.

IV. NONLINEAR CIRCUIT SIMULATIONS

In the FD-TD algorithm, we are able to vary ε_{ru}, μ_{ru}, and σ_u of the lumped device models at each time step in the calculation. This means that the devices can be nonlinear. If the resistance and the capacitance of the lumped device model shown in Figure 1 are changed at every time step to satisfy the equations

$$C_{\text{lump}} \equiv C_D = \frac{C_{j0}}{\sqrt{1 - V_D/V_B}} \tag{21}$$

and

$$R_{\text{lump}} \equiv \frac{1}{G_D} = \frac{V_D}{I_D};\ I_D = I_S\left(e\frac{V_D}{NV_T} - 1\right), \tag{22}$$

then this model represents simple diode characteristics which take the voltage dependence of the junction capacitance as well as the V_D-I_D characteristics of the diode into account. If needed, the effect of the ohmic series resistance can also be added using a neighboring electric node. Similarly, a more complex model can be incorporated in the circuit structure, as shown in Figure 5 for a MESFET model. This model uses five nodes to represent the voltage dependence of gate capacitances, drain, and source resistances as well as V-I characteristics and, of course, the drain-source current can be a complex function of the port voltages of these nodes [1].

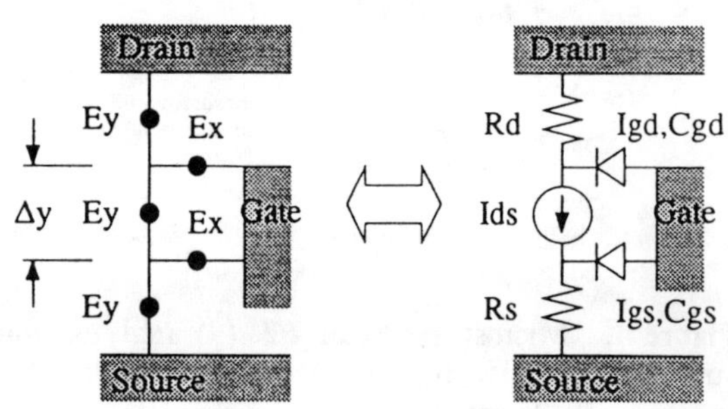

Figure 5. A MESFET model.

IV.A. Simulation of an NLTL Circuit

Here, we describe a simulation of a nonlinear transmission line (NLTL) circuit. The NLTL is a transmission line periodically loaded with Schottky diodes and implemented in GaAs IC form. The signal velocity of the NLTL is voltage dependent because of the loaded diodes. This feature makes applications such as pulse generation possible [14]. When the NLTL is driven by a large-signal sinusoidal source, the negative-oriented transition is sharpened as the wave propagates and the NLTL can generate picosecond transient pulses. This is because the negative half cycle of the waveform propagates faster than the positive half cycle. The simulated circuit structure is illustrated in Figure 6. This is a CPS line 5 mm in

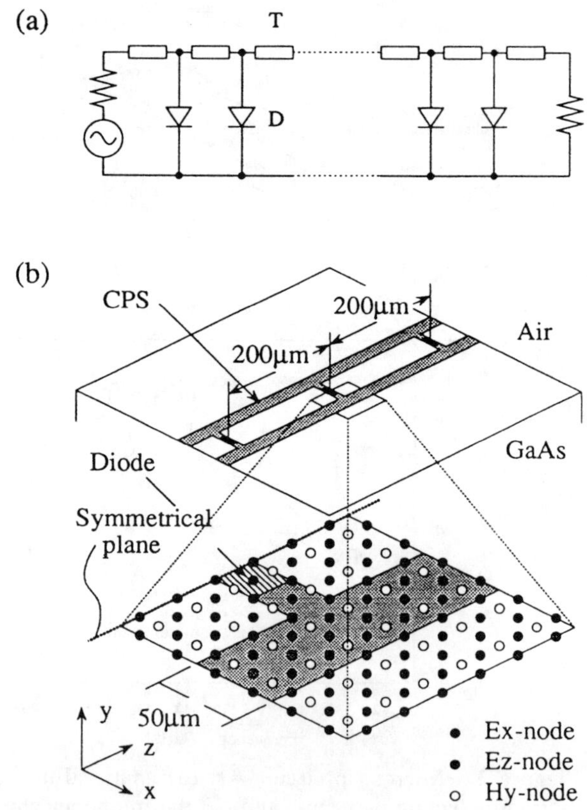

Figure 6. Circuit diagram and the structure of the NLTL circuit.

length that loads 24 diodes every 200 μm. The employed mesh size is $L = 15$, $M = 30$, and $N = 250$, assuming the discretization length $\Delta x = \Delta y = \Delta z = 25$ μm. The CPS dimensions are:

width of the striplines: $W = 2\Delta x = 50$ μm;
spacing between striplines: $S = 4\Delta x = 100$ μm.

The line was driven by a sinusoidal voltage source with internal 50-Ω resistance at one end and terminated by a 50-Ω resistor at the other end. The diodes were represented using each E_x-node in the hatched region in Figure 6b based on eqs. (21) and (22), where the following device parameters were used:

zero-bias junction capacitance: $C_{j0} = 62.6$ fF;
junction contact potential: $V_B = 0.65$ V;
reverse saturation current: $I_S = 18.4$ pA;
emission coefficient: $N = 1.5$;
thermal voltage: $V_T = 25.85$ mV.

Figure 7 presents simulated waveforms of line voltage at different positions from the driving point (denoted as $z = 0$) to the terminating point ($z = 5$ mm) when the NLTL is driven at 10 GHz. It is clear that the negative-oriented transition is sharpened as the wave propagates and the fall time reaches the picosecond domain. These pulses involve a wide frequency spectrum over 100 GHz, thus the parasitic effects caused from the circuit

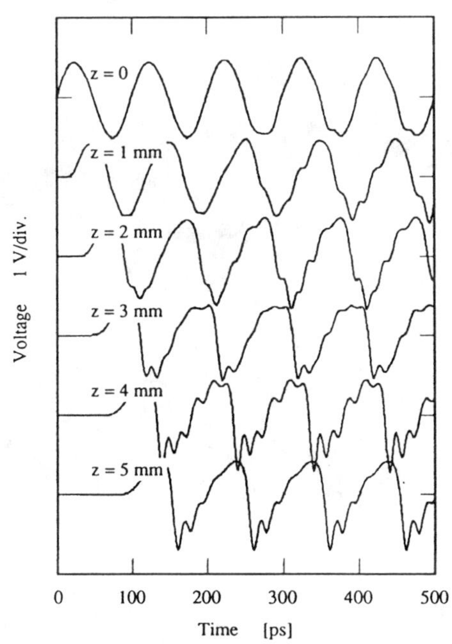

Figure 7. Simulated waveforms of the NLTL circuit at different positions alone the line.

structure significantly affect the generated waveform. These effects can hardly be taken into account without field simulations, but the above simulation is capable of involving all these effects, making it quite promising for the accurate prediction of circuit performance.

IV.B. A Package Design

The last topic of this article is the package design of broad-band, high-gain amplifier ICs. Circuit designers are faced with the problem of devising packages with sufficient electrical properties for mounting ICs that perform at higher and higher speeds. In package development, there are usually large initial costs for making the mold. Nevertheless, the accurate prediction of the electrical performance prior to the fabrication had been difficult because of its fully three-dimensional structure. Therefore, computer simulations are strongly expected to aid in designing the packages efficiently.

Here, we present a simulation method of ICs mounted in the package, starting with a description of the analysis of isolation characteristics. Figure 8 illustrates an analyzed metal package and the results. This package was developed with amplifier ICs for 10-Gbps optical communication systems [15]. In the design of such packages, the isolation characteristics between input and output ports are critical parameters. The isolation must be high enough over the operating frequency range to ensure the stability of circuit operations. This package is composed of a metal frame and built-in alumina blocks containing two signal lines (input and output) and four power supplying lines as shown in Figure 9. The size of the inner cavity in which the IC chip and additional components are mounted is 2×4 mm^2. After mounting chips, the package is sealed with a metal cap. Figure 8b shows the predicted isolation characteristics for the empty package with the cap and corresponding measured results. The simulation was performed in the same manner as when the microstrip circuits were analyzed, using the mesh size of $L = 140$, $M = 28$, and $N = 100$ with the discretization length of $\Delta x = \Delta y = \Delta z = 100$ μm. The agreement between the simulation and the measurement is fairly good up to 25 GHz and the package performance, especially the fact that the isolation characteristics should degrade at the resonance peaks around 20 GHz in this case, is predictable.

One way of obtaining the performance of the IC mounted in this package would be to do a field

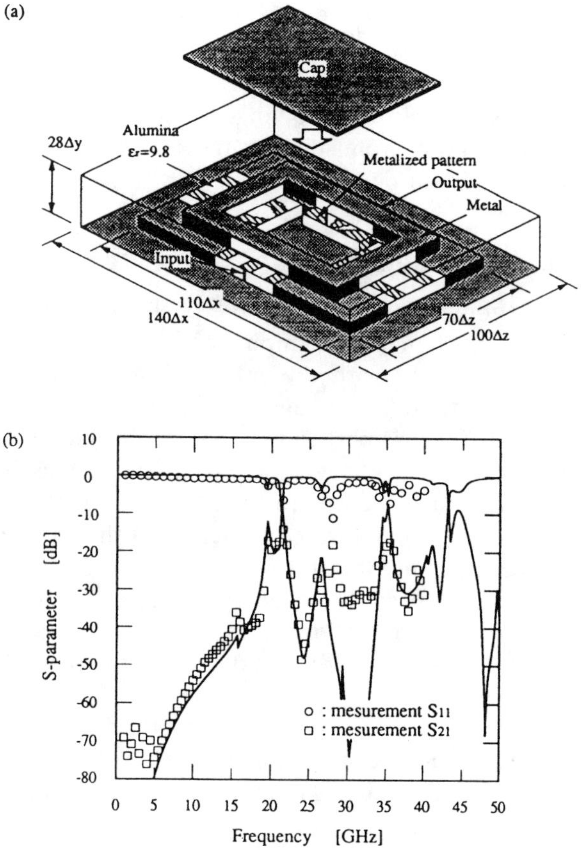

Figure 8. Package structure and isolation characteristics between the input and output ports.

simulation for the whole structure, like that described for the NLTL circuit, using details of the IC as well as the package. However, this might require more fine mesh for representing the IC patterns and does not seem feasible with presently available computational power. Consequently, we took another approach [13] where the structures of the IC chip, shunting chip capacitors for power supplies, and connecting bonding wires are added to the empty package as shown in Figure 10a. However, the IC chip is a dummy which has only six pads and a ground plane on its surface instead of the actual IC. Furthermore, 50-Ω resistor model is installed between each pad and the ground plane. This resistor defines additional internal ports on the IC chip. Then each component of a twelve-port scattering matrix for the six external and the corresponding six internal ports was calculated base on the field simulations. Figure 10b shows the frequency characteristics of some of the components in this matrix. These data were compiled in the *S*-parameter data file form of a conventional circuit simulator and combined with the circuit description of the actual IC. Figure 11 illustrates results of ac analyses of such simulations. In this figure, we compare two cases: a par-

Figure 9. Photograph of the package.

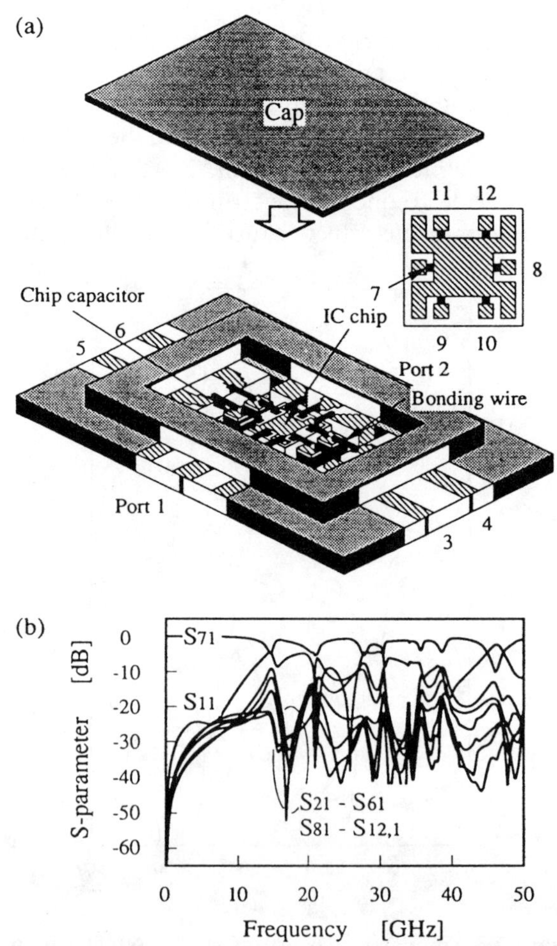

Figure 10. Package and mounted chip components.

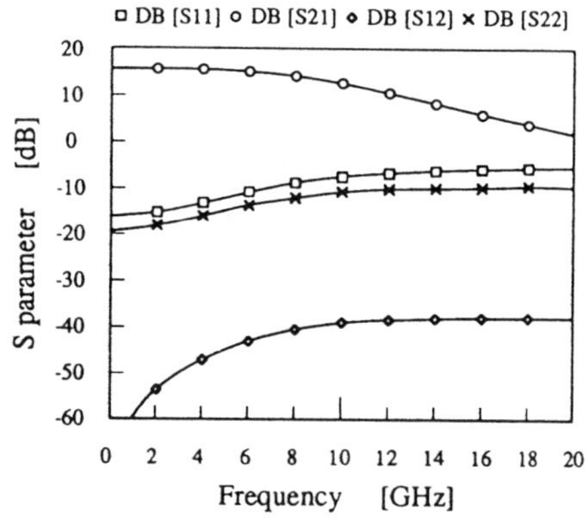

(a) Before mounting in the package

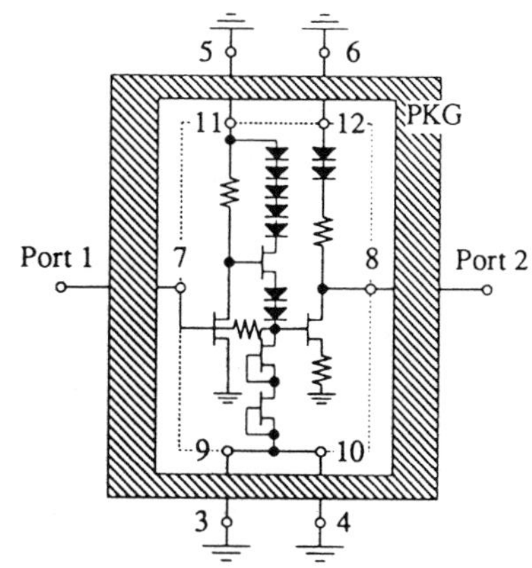

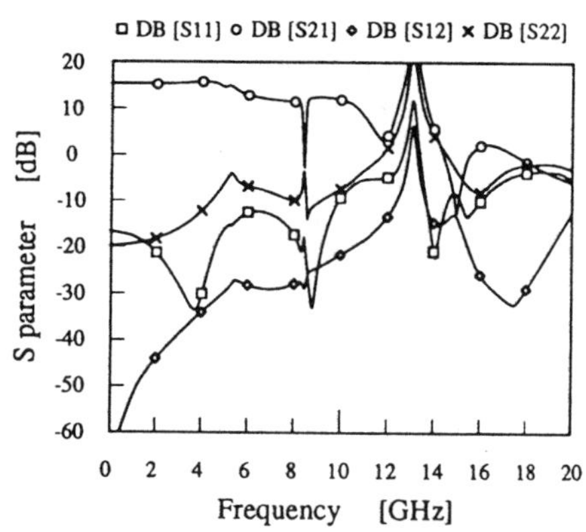

(b) After mounting in the package without chip capacitors

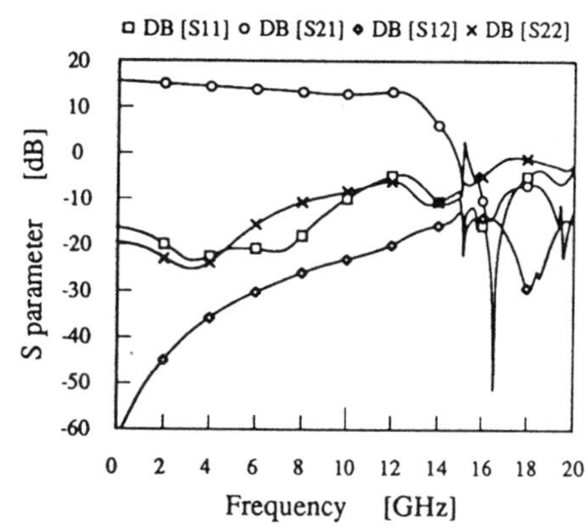

(c) After mounting in the package with chip capacitors

Figure 11. Simulated results of a parallel feedback amplifier mounted in a package.

allel feedback amplifier mounted in the package without shunting chip capacitors and with shunting chip capacitors. These results are presented in Figure 11b and c, respectively, which show that the circuit becomes unstable around 13 GHz when the IC chip is mounted alone. This instability can be settled by using shunting chip capacitors.

V. CONCLUSIONS

We have presented some ways of applying the FD-TD field calculation to microwave and millimeter-wave circuits. An efficient technique for calculating scattering parameters of passive components that uses the lumped device model as the standard termination of the circuit port has been proposed. Calculations of microstrip circuits show the practical validity of this technique. In addition, two approaches for nonlinear circuit simulations have been introduced that enable us to design a circuit taking the packaging effect as well as the parasitic effects in the IC chip into account. All simulations described in this paper were carried out using an HP-750 workstation-class computer. The required CPU time ranged from 1 hour to 40 or 50 hours, depending on the mesh size and the computed time step count. Such simulations are now much more feasible than they were a few years ago. We expect the techniques presented here to become powerful CAD tools for first-pass success in circuit design.

ACKNOWLEDGMENT

The authors wish to thank Dr. A. Iwata, Y. Akazawa, E. Sano, Dr. S. Horiguchi, NTT LSI Laboratories, and Associate Professor N. Yoshida of Hokkaido University for their continuous discussions and encouragement throughout this work.

REFERENCES

1. T. Shibata, "Circuit Simulations Combined with Electromagnetic Field Analysis—Time-Domain Electromagnetic Field Analyses and Their Application to Nonlinear Microwave Circuit Simulations," *IEEE MTT-S Workshop Dig.*, WS-K, June 1992, pp. 74–89.
2. K. S. Yee, "Numerical Solution of Initial Boundary Value Problems Involving Maxwell's Equations in Isotropic Media," *IEEE Trans. Ant. Propagat.*, Vol. 14, May 1966, pp. 302–307.
3. A. Taflove and K. R. Umashankar, "Finite-Difference Time-Domain (FD-TD) Modeling of Electromagnetic Wave Scattering and Interaction Problems," *IEEE Ant. Propagat. Newsletter*, April 1988, pp. 5–19.
4. G. Mur, "Absorbing Boundary Conditions for the Finite-Difference Approximation of the Time-Domain Electromagnetic-Field Equations," *IEEE Trans. Electromagn. Compat.*, Vol. 23, Nov. 1981, pp. 377–382.
5. T. Shibata and E. Sano, "Characterization of MIS Structure Coplanar Transmission Lines for Investigation of Signal Propagation in Integrated Circuits," *IEEE Trans. Microwave Theory Tech.*, Vol. 38, July 1990, pp. 881–890.
6. N. Fukutomi, H. Nakamura, N. Yoshida, and I. Fukai, "Cross-Talk Analysis by the Spatial Network Method," *Trans. IEICE*, Vol. J72-C-I, Oct. 1989, pp. 642–649.
7. E. Sano and T. Shibata, "Fullwave Analysis of Picosecond Photoconductive Switches," *IEEE J. Quantum Electron.*, Vol. 26, Feb. 1990, pp. 372–377.
8. T. Shibata, T. Hayashi, and T. Kimura, "Frequency Response Characteristics Analysis of Planar Circuit by T. L. M. Method," *1987 IEICE National Convention Record*, Part 3, March 1987, p. 795; and "Analysis of Microstrip Circuits using Three-Dimensional Full-Wave Analysis in the Time Domain, *IEEE Trans. Microwave Theory Tech.*, Vol. 36, June 1988, pp. 1064–1070.
9. J. Fang, X. Zhang, and K. K. Mei, "Dispersion Characteristics of Microstrip Lines in the Vicinity of a Coplanar Ground," *Electron. Lett.*, Vol. 23, Oct. 1987, pp. 1142–1143. X. Zhang and K. K. Mei, "Time-Domain Finite Difference Approach to the Calculation of the Frequency-Dependent Characteristics of Microstrip Discontinuities," *IEEE Trans. Microwave Theory Tech.*, Vol. 36, Dec. 1988, pp. 1775–1787.
10. R. H. Voelker and R. J. Lomax, "A Finite-Difference Transmission Line Matrix Method Incorporating a Nonlinear Device Model," *IEEE Trans. Microwave Theory Tech.*, Vol. 38, March 1990, pp. 302–312.
11. T. Shibata, H. Kimura, and Y. Akazawa, "Circuit Simulations Combined with the Electromagnetic Field Analysis on a Nonlinear Transmission Line Circuit Generating Picosecond Pulses," *IEE Japan Technical Report ECT-90-19*, Dec. 1990, pp. 17–26. T. Shibata, "Circuit Simulations Combined with the Electromagnetic Field Analysis," *IEEE Trans. Microwave Theory Tech.*, Vol. 39, Nov. 1991, pp. 1862–1868.
12. H. Kimura and N. Yoshida, "Three-Dimensional Full Wave Analysis with Nonlinearity and Line Characteristics of Device by Electromagnetic Field Analysis on Time Domain," *Trans. IEICE*, Vol. J75-C-I, Jan. 1992, pp. 42–51.
13. J. Kunisch and I. Wolff, "The Compression Approach: A New Technique for the Analysis of Distributed Circuits Containing Nonlinear Elements," *IEEE MTT-S Workshop Dig.*, WS-K, June 1992, pp. 15–31.
14. C. J. Madden, M. J. W. Rodwell, R. A. Marsland, D. M. Bloom, and Y. C. Pao, "Generation of 3.5 ps Fall-Time Shock Waves on a Monolithic GaAs Nonlinear Transmission Line," *IEEE Electron Device Lett.*, Vol. 9, June 1988, pp. 303–305.
15. N. Ishihara, E. Sano, Y. Imai, H. Kikuchi, and Y. Yamane, "A Design Technique for a High-Gain, 10-GHz Class-Bandwidth GaAs MESFET Amplifier IC Module," *IEEE J. Solid-State Circuits*, Vol. SC-27, No. 4, 1992, pp. 554–562.

FDTD Simulation for Microwave Packages and Interconnects

MATTHIAS RITTWEGER, MARKUS WERTHEN, AND INGO WOLFF, FELLOW, IEEE
INSTITUTE FOR MOBILE AND SATELLITE COMMUNICATION TECHNIQUES
MOERSER STR. 316, D-47475 KAMP-LINTFORT, FRG
TEL. ++49-2842-9110, FAX ++49-2842-911100

Abstract

Some aspects concerning with packaging of monolithic microwave and millimeterwave circuits are discussed by use of the finite difference time-domain (FDTD) method. After characterizing a class of problems generally arising in this field a collection of specific reasons for the application of the FDTD method is briefly presented. Special consideration is given to an efficient implementation of the simulation technique described. Selected examples of fundamentals in packaging and interconnection are examined. These are in particular chip connection and feedthrus considering different waveguide concepts. Beside the determination of disturbing resonant effects of complete packages an alternative approach for embedding chips in a real environment (package mounted on a substrate) is described. This method originally belonged to the analysis of nonlinear elements in distributed circuitries is also applicable for embedding multiple chips in connection with complex matching network elements.

Introduction

EM modelling of packaging has not really always been that important objective in the microwave field that it should have been. A survey of the investigations made for circuit design compared to housing problems indicates that there is a lot of work to make up in this topic. Especially entering consumer electronics applications a discrepancy in the costs of circuits and high performance packages seems to arise. Therefore a considerable interest can be registered in recent times.

Beside packaging of discrete devices and single MMICs nowadays attention is focused on a common integration technique using multichip packages. This seems to be the most promising way to increase system performance in connection with attractive costs. Possible applications for such multichip packages are systems with considerable complexity for high frequencies where parasitic effects due to interconnects have to be reduced. Highest reliability can be achieved using this kind of compact design. Even for consumer electronics such a technique possibly provides mass production in a low cost range. A schematic configuration of a multichip package is shown in Fig. 1.

The essential requirements for multichip packages are extended to those from single chip packages with respect to interaction between single chips from electrical coupling and additional integration of e.g. impedance matching networks. The requirements listed above are identical for single and multichip packages:

Reprinted from *Proceedings of Workshop WSFC: EM Modeling of Microwave Packages and Interconnects, 1993 IEEE MTT-S International Microwave Symposium,* pp. 78-91, May 1993.

- **mechanical**
  - protection (shock, vibration ...)
  - hermetic enclosure

- **electrical**
  - controlled line impedances
  - low transition parasitics and coupling
  - low-inductance grounding

- **thermal**
  - low thermal impedance

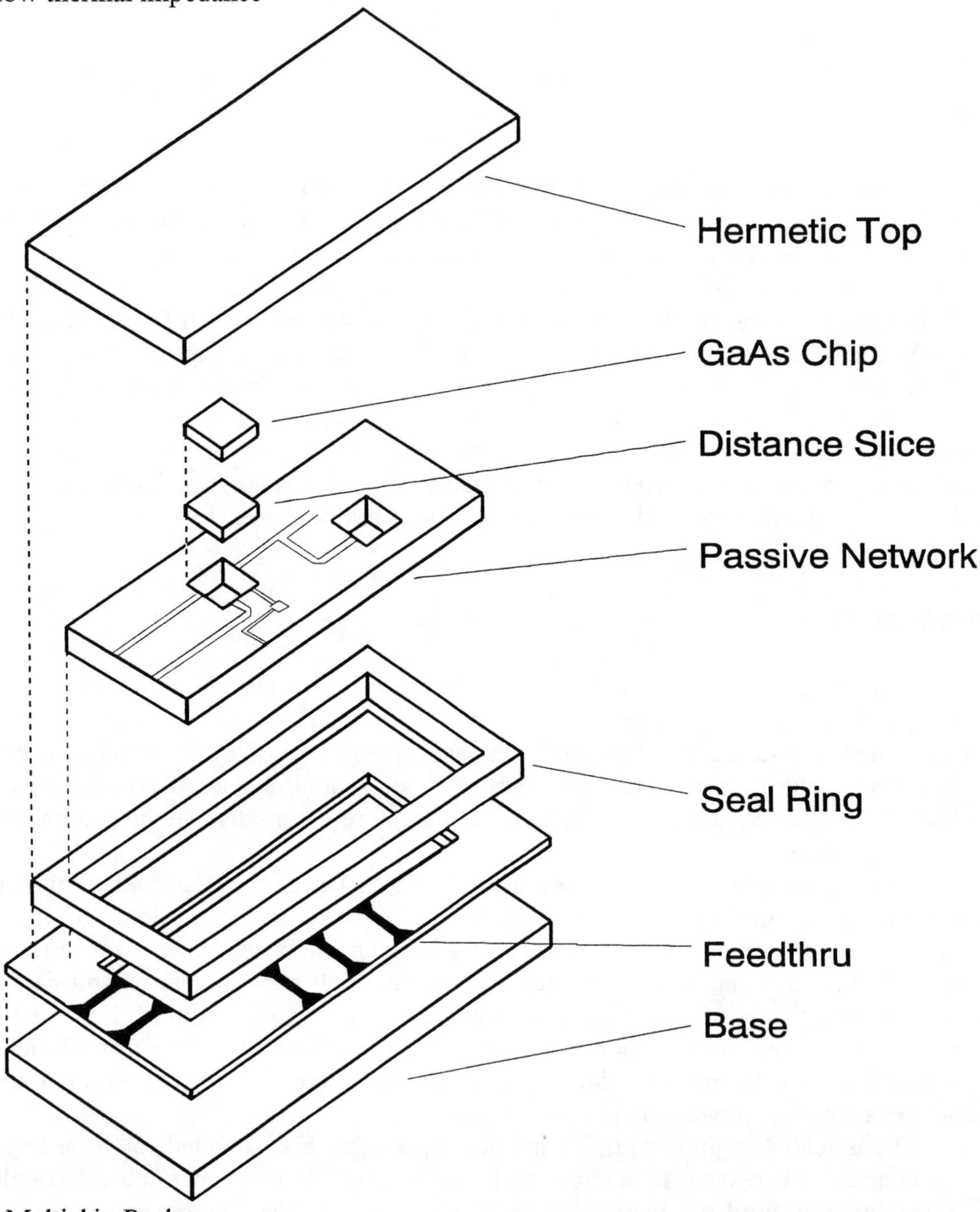

Fig. 1: Multichip Package

From the point of view of electromagnetic modelling a microwave package is not only a considerable complex structure. There are different operational levels, that have to be separated whereby interaction is described using a restricted set of parameters. The first level is the function of the embedded chips. Talking about package simulation we need not to go into further detail of the circuit. The chip is treated as a n-port whereby its global interaction with the surrounding package is considered. A current flow from one port of the chip to another e.g. leads to a global magnetic coupling to the environment. There is no need to register the local current distribution due to the internal circuitry as long as the chips are small compared to the package. This agrees well to measurement concepts using so-called surrogate chips either to contact probe tips or to make other specific connections [1].

The second level therefore is the package itself containing the carrier, sealing, feedthrus, s.o.. If the network on the carrier contains elements exceeding a specific level of complexity this elements can be treated in the same manner as the chips. This offers the possibility to keep large capacitors or inductances as they are used e.g. for biasing or impedance matching out from the chip.

The next level is the environment of the package. Generally it can be expected that there is a significant influence from the surrounding the package is mounted in. This aspect has to be integrated in the analysis procedure of the package. The number of simulation levels is to exceed whenever further system integration steps (package in a complex network, coupling between packages) have to be simulated using electromagnetic approaches.

As for circuit design not only the complete analysis of a microwave package is of interest. Moreover, specific use of EM simulators are restricted problems concerning with interconnection. This is in particular simulation of different mounted chips (Wire-Bonding, TAB, Flip-Chip). Optimization of feedthrus is another important question that is also concerned with different connecting techniques inside and outside the package. Crosstalk becomes important with increasing number of feedthrus or small dimensions of the package.

An investigation of package resonances requires the simulation of the entire package. This can be done in a simple eigenvalue analysis that is able to find the resonant frequencies and to estimate the damping characteristics.

For to know how the performance of the microwave chips is affected by the package the designer needs a characterization of the package in a real environment in form of any network description (s-parameters). This can be done using the compression approach [2] in connection with any suitable 3D field simulator.

The Finite Difference Time-Domain Method

A common property of the problems mentioned above is that they are all leading to a full three-dimensional field analysis. There is hardly any restriction with respect to the flexibility a field simulation technique needs. Therefore strongly numerical oriented methods are advantageous for investigation of packaging. Even for planar structures the three-dimensional FDTD method developed to a serious competitor against established simulation techniques working in the frequency-domain. In 1988 Zhang et al. [3] published their work in computing frequency dependent characteristics of microstrip discontinuities using the FDTD method. The transient analysis technique described in this paper using absorbing boundaries to extend the computation domain to infinity has been taken up from several investigators up to now. There is practically no task in microwave engineering where the FDTD technique has not been applied since that time. As a result suitable extensions and improvements have been developed in recent works. The simulation of packages and interconnects using the FDTD method is not so much a question of new theories for field simulation. Most of them have been developed anyway for simulation of planar based structures. It is more a problem of convenient structure handling, efficient implementation and furthermore suitable analysis strategy. These points cannot be considered separately, because there are some dependencies between each other.

Obviously unlimited 3D data base is need for the simulation of packages and interconnects. The FDTD method is a solution of time dependent Maxwell's equations discretized in both, space and time. So the computation domain consists of field nodes representing samples of the electrical and magnetical field components. A set of electrical parameters can be assigned to each of the field nodes. If there is unlimited variation in the electrical properties of the computational space a direct assignment delivers most efficient access when running the algorithm. Memory requirements for problem description therefore are about as high as for field description. For package simulation purposes there exist piecewise homogenous subdomains with equal electrical parameters. Therefore indirect assignment of electrical parameters to the field nodes limits computer memory requirements at the expense of CPU time. This is a simplified description because irregular grids are treated in a similar manner. As a result of this reflection the implementation of the algorithm has to be carefully matched to the class of problems to handle with.

Our in-house FDTD simulator is planar based. This means that there is simple and fast access for a planar base structure consisting of a substrate, metallization layer and air. Data handling is optimized in a way that optimum CPU performance can be achieved. The layered structure is predefined whereas the planar metallization is described using a discrete structure matrix. All three dimensional objects leaving the planar base are handled using a node oriented list containing all other node based data too. The list controlled concept offers the possibility to have easy access on every single node. This way characteristic port voltages, currents or field amplitudes based on space integration can be defined for output in a very flexible manner. The same is for excitation based on arbitrary field distribution. Arbitrarily shaped boundary conditions to describe three-dimensional objects are considered as well as different material parameters for every single field node. So maximum flexibility is provided. Objects based on a list oriented description are computed slower than field nodes for the planar structure. Moreover such field nodes are computed twice whereby this normally takes less time than to differ between planar based and special objects even for complex structures as e.g. packages.

Structure editing is done using a conventional object oriented drawing software. Single graphical objects and lines are interpreted as planar metallization pads and discretization lines, respectively. Compound objects, i.e. graphical objects plus text, are used to describe excitation, port definitions and three-dimensional structures. The text contains key words and suitable parameters. The graphical output is used to generate the input for the simulator.

Chip Mounting

As a first application of the FDTD simulator a problem concerning with chip connection shall be investigated. There are a lot of possibilities to mount a chip in a package or on a substrate. Only a few of them where analyzed using electromagnetic simulators as e.g. surface-to-surface transition from Jackson et al. [4] or chip-bonding from Christ et al. [5]. Fig. 2 shows a TAB-connected microstrip chip on a coplanar substrate that is expected to lead to lower inductance than a wire-bonded connection.

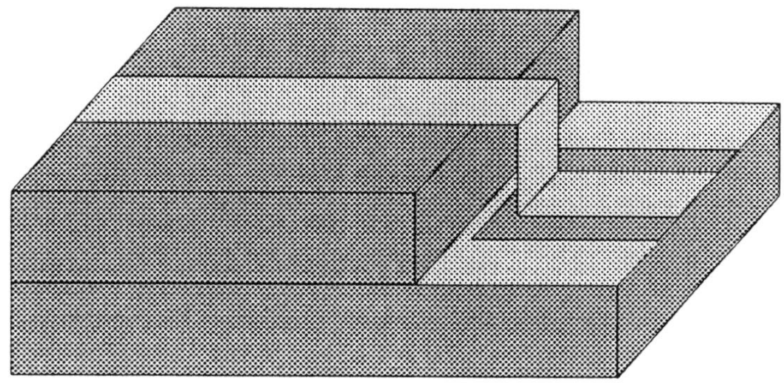

Fig. 2: Microstrip chip on a coplanar substrate.

For testability reasons the structure is build up of standard alumina substrates with 5 mil and 25 mil thickness for the chip and the carrier, respectively. The line width is 0.14 mm for both lines whereas the slots of the coplanar waveguide are 0.035 mm. These dimensions lead to the characteristic impedances seen in Fig. 3.

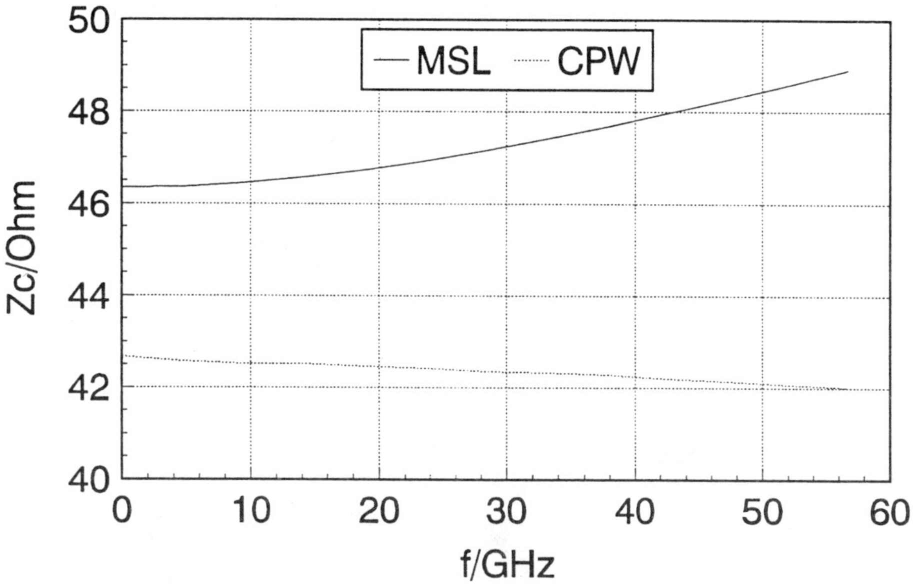

Fig. 3: Characteristic impedances for the connected waveguides.

No effort has been made to match the static values of the impedances for both transmission lines. Moreover the dispersion characteristic of the microstrip is reversed compared to that of the coplanar waveguide. As it can be seen in Fig. 4 the reflection characteristic is mainly influenced from the discontinuity itself. The different characteristic impedances would lead to the curve named ideal in the diagram.

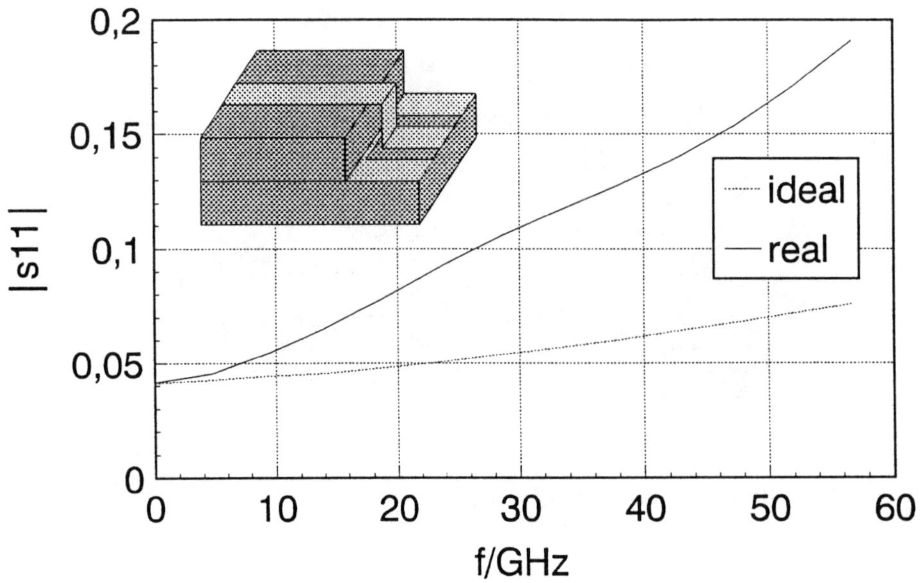

Fig. 4: Reflection coefficient for the TAB transition.

The reflection coefficient corresponding to this type of chip bonding is considerable low for the simulated structure. In practice there is some influence of gaps located between grounds as well as under the interconnect lead. This may cause problems with reproducibility of the structure. On the other hand the used die-bonding technique is suitable for power applications because excellent heat spread is expected.

Feedthru

Usually design of feedthrus begin from classical equations for transmission lines. This approach neglects basic influences from interaction with the entire package concept. This problem is taken into account for the next example. The microstrip feedthru as an example for planar feedthrus is analyzed in different realization steps shown in Fig. 5.

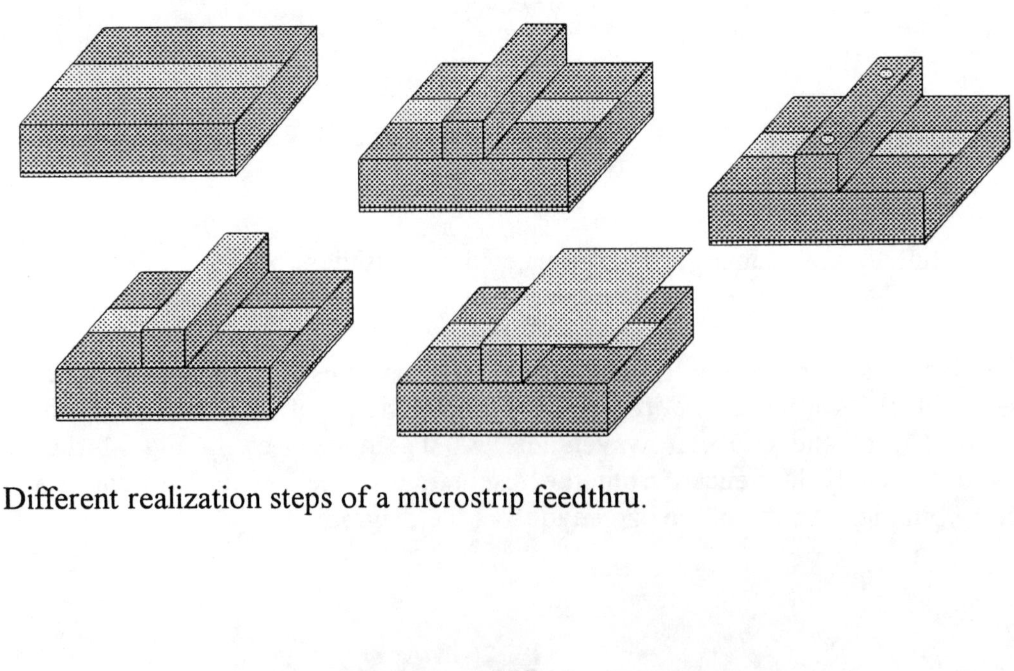

Fig. 5: Different realization steps of a microstrip feedthru.

Again for testability reasons the structure is designed using a 25 mil standard alumina substrate. First the single 50 ohms microstrip has been analyzed for to check the simulation environment. The next step was to place a ceramic stick onto the microstrip. As expected the wave is hardly affected from this disturbance since the field is concentrated under the strip. The wall can be interpreted as a small sized impedance variational element. Due to the small dimensions (thickness of the wall equal to the width of the line) only a small reflection is registered (Fig. 6).

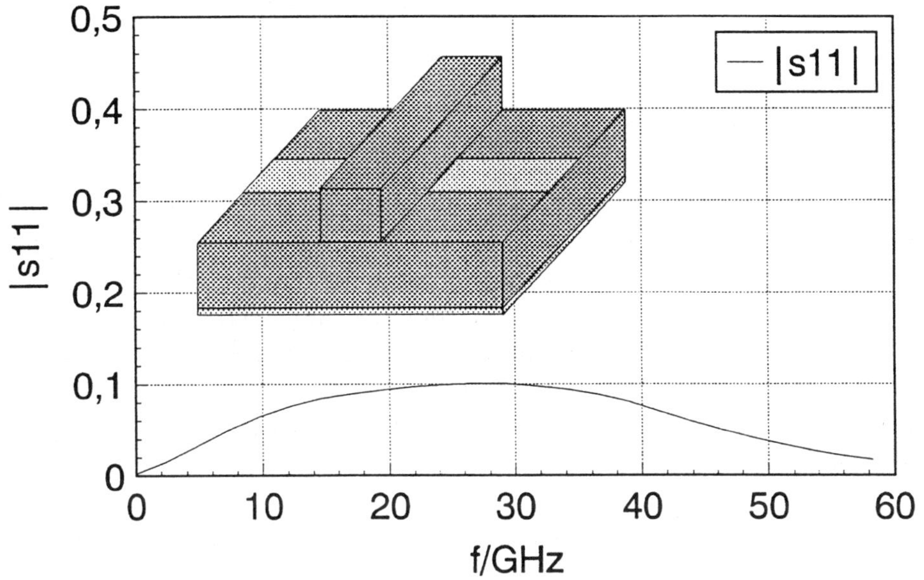

Fig. 6: Reflection coefficient of the microstrip through the ceramic wall.

Principally the ceramic stick is a waveguide itself. For the registered frequency range corresponding to the dimensions of the test structure this effect is negligible. For the simulation the crossection of the structure and therefore the length of the stick is unlimited, because the computation domain is extended to infinity using absorbing boundaries. So the structure is not influenced from transversal resonances.

The next step is to consider via holes beside the feedthru that are used either to suppress ring resonances of the seal ring or to contact a closed shielding. These vias can lead to bad resonance effects as it can be seen in Fig. 7.

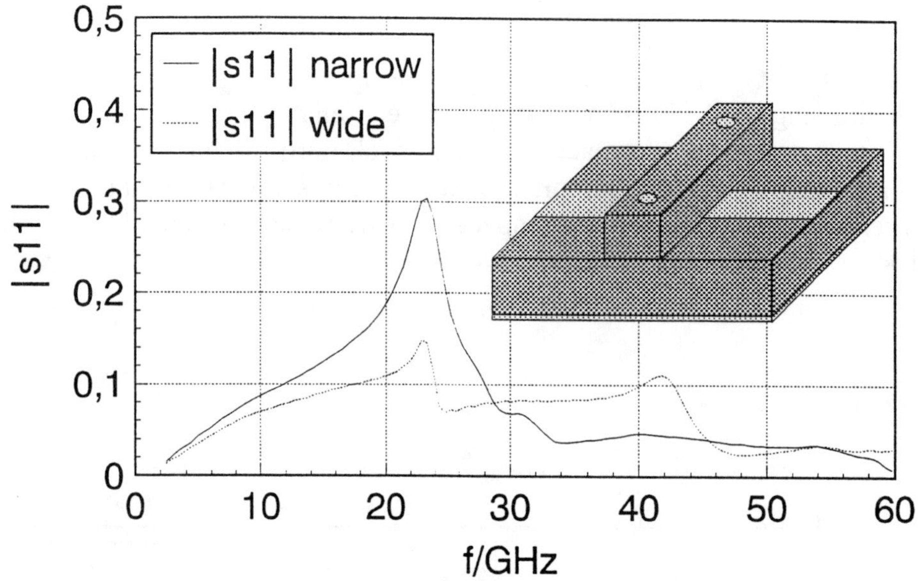

Fig. 7: Influence of different positioned via holes.

The curve named "narrow" represents the reflection coefficient for the case that the distance between the center of the via hole and the microstrip is 1.5 mm. The magnitude decreases when the distance reaches 2.2 mm as for the curve named "wide" but on the other hand a second resonance is found in the investigated frequency range. Despite the different positions of the vias for both cases the first resonance is located at nearly the same frequency which leads to conclude that it is not easy to estimate such effects.

The magnitude of the reflection coefficient increases when a conducting strip is placed on the wall as it can be seen in Fig. 8.

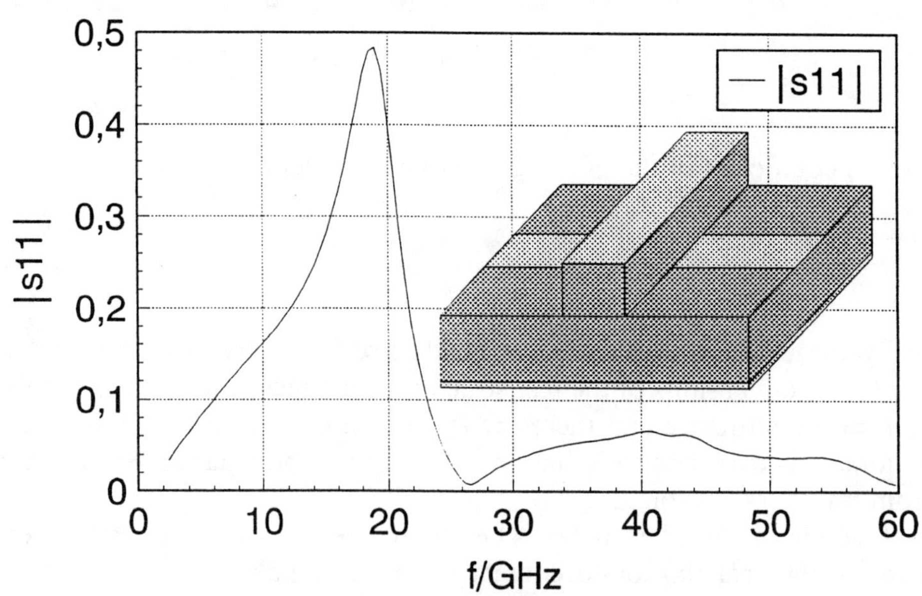

Fig. 8: Reflection coefficient with conducting metallization.

Now there is no cutoff for transversal wave propagation in a quasi TEM mode. So for the investigated frequency range more energy is coupled in the ceramic stick. The magnitude of the reflection coefficient decreases again when the conducting top is placed on the package as simulated in Fig. 9.

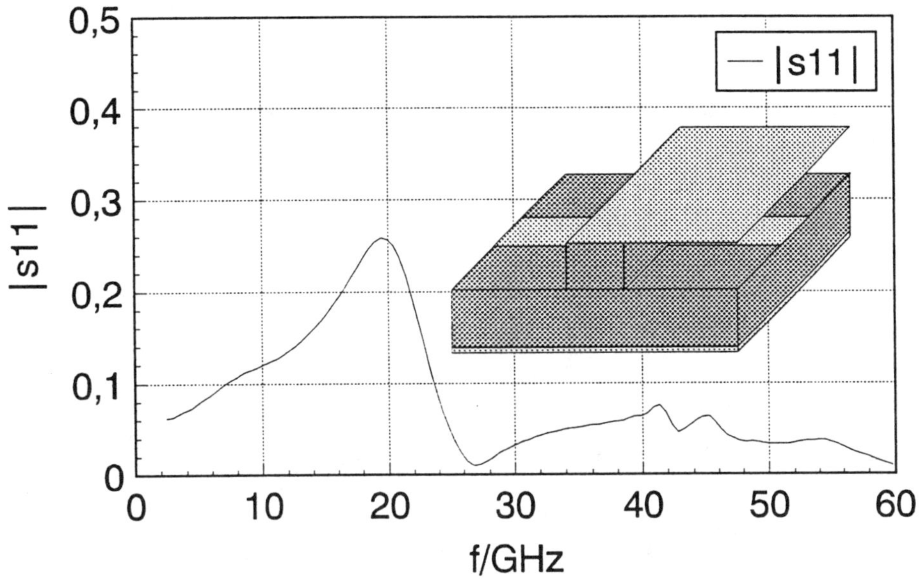

Fig. 9: Influence of a metallized top.

It can not be expected that a circuit mounted in a package using any feedthrus behaves in a way as a circuit simulator predicts when connecting the intrinsic circuit with the corresponding feedthru networks.

Package Simulation

The performance of a circuit is mainly affected from the entire package with its coupling and resonance effects. The FDTD method has been used for solving eigenvalue problems for shielded structures [6]. Resonating structures can generally be estimated using such a formulation [7]. The structure is excited using any simple pulse whereby the time-domain response is investigated using a DFT, Prony's method [8] or the SI-method [9]. The eigenvalues of this time-domain response can be used as a simple estimation where strong resonances can be expected for the package under test. In practice it is really hard to find out these resonances because there is hardly any information indicating the package resonances. Other eigenfrequencies are caused from the environment, i.e. there were resonances on peripheral transmission lines. The eigenvalue analysis technique shall not be consolidated in the present paper.

The designer needs a network description of the package that allows to predict the electrical properties of the packaged circuit from the base of the known intrinsic circuit (chip). The analysis can not be separated from the environment the package is mounted in. The analysis technique described in this paper offers the possibility to achieve such a description.

As an example a surface mounted s-mini package containing a bipolar cascode is analyzed using the compression approach. The package is shown in Fig. 10. The dimensions of the package are 1.25 mm and 2 mm for the width and the length, respectively. The leadframe is 0.7

mm from the bottom whereby the height of the pack is 0.9 mm. The relative permittivity of the plastic material is assumed to be 3.8. The package is solded on a standard alumina substrate (h = 25 mil, ε_r = 9.8). Port 1 is the coplanar input connected to the base of T1 (port 6). The emitter of T1 (port 5) is grounded. Port 5 and port 6 were both directed to the collector of T1. Port 3 and port 4 are directed to the collector of T2 which is connected to the coplanar output (port 2). Port 3 as the base of T2 is grounded and port 4 as the emitter of T2 is connected to the collector of T1. Both transistors T1 and T2 are mounted with their collectors on the lead-frame. Base and emitter are connected using bond-wires.

Port 1 and port 2 are conventional ports, i.e. reference planes on transmission lines. Port 3 to port 6 are used to connect the intrinsic transistors to the package. Therefore they are called inner ports.

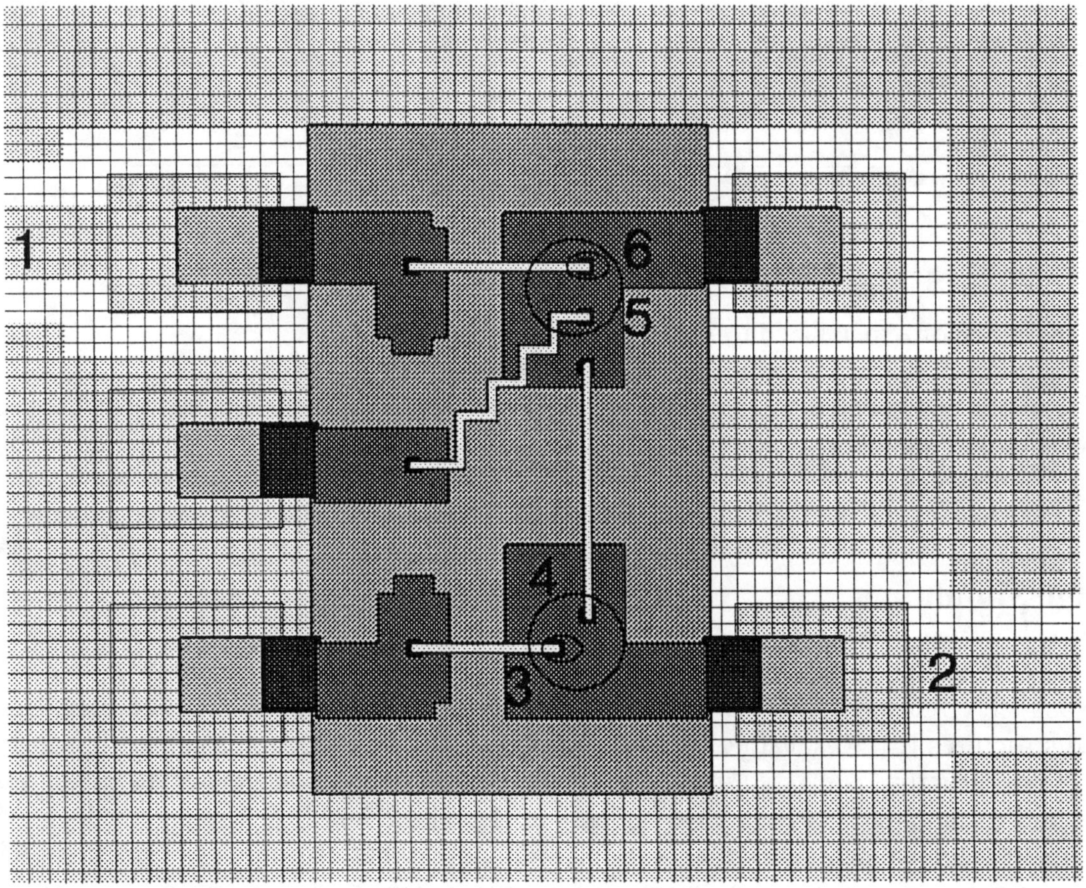

Fig. 10: Surface mount s-mini package with inner ports (3,4 and 5,6) to connect two bipolar transistors and input (1) and output (2).

The analysis procedure is to simulate n linear independent excitation cases for a linear n-port. The structure is excited by using incident waves on the transmission lines whereas the inner ports are connected to different terminations. The reflection coefficients for each port and the excitation with an incident wave is described in Table 1 for each excitation case.

number	port 1	port 2	port 3	port 4	port 5	port 6
1	0, a11	0	-1	-1	-1	-1
2	0, a12	0	1	-1	1	-1
3	0, a13	0	1	1	1	-1
4	0	0, a24	1	-1	1	1
5	0	0, a25	-1	1	1	1
6	0	0, a26	1	-1	-1	1

Tab. 1: Description of the termination conditions and incident waves for the six linear independent cases.

To achieve linear independence each transmission line has to be excited at least one time. Furthermore each inner port has to be connected to different terminations whereby the corresponding waves at this port must not vanish.

For each excitation case the wave quantities at each port are determined. For the inner ports this is a formal procedure using currents and voltages in connection with Heaviside transformation whereby the normal impedance can be chosen anyhow. For this example all ports are normalized to 50 ohms which is also the static characteristic impedance for the coplanar lines. The dispersion of the coplanar feed lines is negligible for the investigated frequency range. So each excitation case delivers one column of the excitation matrix **A** and the corresponding response matrix **B**. For the case of linear independent excitation cases the scattering matrix S can than be determined using the following relation:

$$\mathbf{S} = \mathbf{B} \cdot \mathbf{A}^{-1}$$

Fig. 11 shows selected s-parameters for the resulting 6-port.

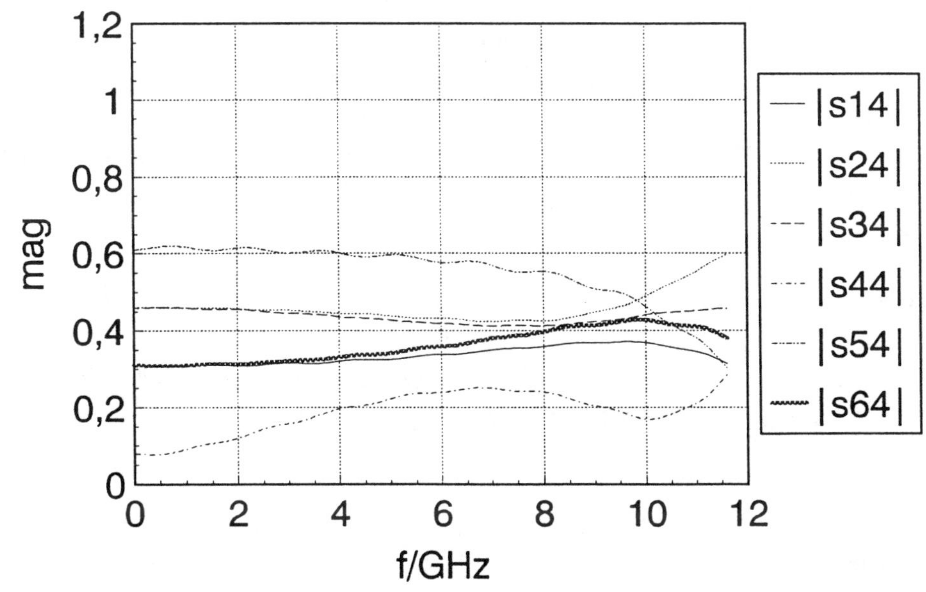

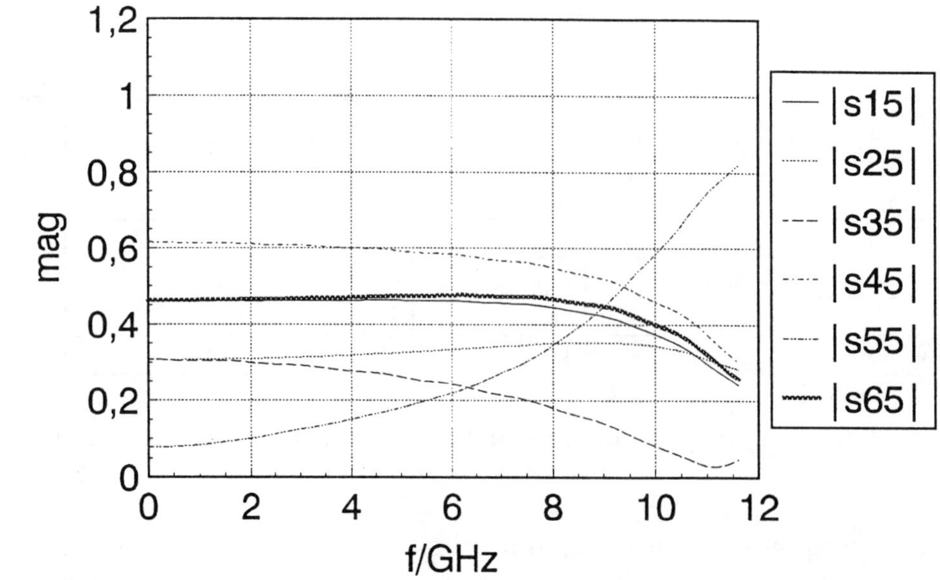

Fig. 11: Two columns of the scattering matrix of the s-mini package.

There is significant frequency dependence even for frequencies up to 3 GHz which is about the range the package is constructed for. The intrinsic transistors can be connected to the passive linear network using a conventional circuit simulator. Large signal analysis as well as noise analysis can be done based on the available linear description of the package, which is an exact description in a field-theoretical sense.

Conclusions

Analysis of packages and interconnects for microwave purposes using the FDTD technique was presented. Implementation issues for efficient analysis were discussed and a selection of examples has been shown to demonstrate the performance achieved. This was in particular a microstrip chip mounted on a coplanar carrier and a microstrip feedthru considering shielding aspects and therefore the required via holes.

The application of the compression approach for entire package simulations has been presented based on the example of a surface mounted plastic s-mini package containing a bipolar cascode.

References

[1] J. Williams, "Improved techniques for package characterization," *IEEE MTT-S Workshop on New Packaging Techniques for MMICs and Discrete Devices Proceedings*, Boston, June 1991.

[2] J. Kunisch, M. Rittweger, S. Heinen, and I. Wolff, "The compression approach: a new technique for the analysis of distributed circuits containing nonlinear elements," *21st EuMC Proceedings*, Stuttgart, pp. 1296-1301, Sept. 1991.

[3] X. Zhang and K. K. Mei, "Time-domain finite difference approach to the calculation of the frequency-dependent characteristics of microstrip discontinuities," *IEEE Trans. Microwave Theory Tech.*, vol. MTT-36, pp. 1775-1787, Dec. 1988.

[4] R. W. Jackson and D. W. Matolak, "Surface-to-surface transition via electromagnetic coupling of coplanar waveguides," *IEEE Trans. Microwave Theory Tech.*, vol. MTT-35, pp. 1027-1032, Nov. 1987.

[5] A. Christ and H. L. Hartnagel, "Three-dimensional finite-difference method for the analysis of microwave device-embedding," *IEEE Trans. Microwave Theory Tech.*, vol. MTT-35, pp. 688-696, 1987.

[6] D. Choi and W. J. R. Hoefer, "The finite-difference time-domain method and its application to eigenvalue problems," *IEEE Trans. Microwave Theory Tech.*, vol. MTT-36, pp. 1775-1787, Dec. 1988.

[7] J. A. Pereda, L. A. Vielva, and A. Prieto, "Computation of resonant frequencies and quality factors of open dielectric resonators by a combination of the finite-difference time-domain (FDTD) and Prony`s methods," *IEEE Microwave and Guided Wave Letters*, vol. 2, pp. 431-433, Nov. 1992.

[8] J. D. Wills, "Spectral estimation for the transmission line matrix method," *IEEE Trans. Microwave Theory Tech.*, vol. MTT-38, pp. 448-451, April 1990.

[9] W. Kuempel and I. Wolff, "Digital signal processing of time domain field simulation results using the system identification method," *IEEE MTT-S International Microwave Symposium Digest*, Albuquerque, Juni 1992.

Author Index

Index

About the Editors

Tatsuo Itoh received the Ph.D. in electrical engineering from the University of Illinois, Urbana, in 1969. From September 1966 to April 1976, he was with the Electrical Engineering Department, University of Illinois. From April 1976 to August 1977, he was a senior research engineer in the Radio Physics Laboratory, SRI International, Menlo Park, CA. From August 1977 to June 1978, he was an associate professor at the University of Kentucky, Lexington. In July 1978, he joined the faculty at The University of Texas at Austin, where he became a professor of electrical engineering in 1981 and Director of the Electrical Engineering Research Laboratory in 1984. During the summer of 1979, he was a guest researcher at AEG-Telefunken, Ulm, West Germany. In September 1983, he was selected to hold the Hayden Head Centennial Professorship of Engineering at The University of Texas. In September 1984, he was appointed Associate Chairman for Research and Planning of the Electrical and Computer Engineering Department at The University of Texas. In January 1991, he joined the University of California, Los Angeles, as Professor of Electrical Engineering and holder of the TRW Endowed Chair in Microwave and Millimeter-Wave Electronics. He is currently director of the Joint Services Electronics Program (JSEP) and also director of the Multidisciplinary University Research Initiative (MURI) program at UCLA.

He was an honorary visiting professor at Nanjin Institute of Technology, China, and at the Japan Defense Academy. In April 1994, he was appointed as Adjunct Research Officer for the Communications Research Laboratory, Ministry of Post and Telecommunication, Japan. He currently holds a visiting professorship at the University of Leeds, United Kingdom, and is an external examiner of the Graduate Program of the City University of Hong Kong.

Dr. Itoh is a Fellow of the IEEE, a member of the Institute of Electronics and Communication Engineers of Japan, and a member of Commissions B and D of the USNC/URSI. He served as the editor of *IEEE Transactions on Microwave Theory and Techniques* for 1983–1985. He serves on the administrative committee of the IEEE Microwave Theory and Techniques Society. He was vice president of the Microwave Theory and Techniques Society in 1989 and president in 1990. He was the editor-in-chief of *IEEE Microwave and Guided Wave Letters* from 1991 through 1994. He was elected as an Honorary Life Member of the MTT Society in 1994. He was the chairman of USNC/URSI Commission D from 1988 to 1990, the vice chairman of Commission D of the International URSI for 1991–93, and chairman of the same commission for 1993–1996. He is on the Long-Range Planning Committee of URSI. He serves on the advisory boards and committees of a number of organizations, including the National Research Council and the Institute of Mobile and Satellite Communication, Germany.

He has more than 240 journal publications, 450 refereed conference presentations, and has written more than 20 books/book chapters in the area of microwaves, millimeter-waves, antennas, and numerical electromagnetics. He has generated 37 Ph.D. students.

Bijan Houshmand received his B.S. (with Honors), M.S., and Ph.D. degrees from the Department of Electrical and Computer Engineering at the University of Illinois at Urbana-Champaign, in 1985, 1986, and 1990, respectively. In the 1990–1991 academic year, he was a visiting assistant professor at the University of Illinois at Urbana-Champaign, where he conducted research for development of an Expert System for reflector antenna design and algorithm development for scattering from multiple scatterers. He was a visiting assistant professor at the Electrical Engineering Department of the University of California at Los Angeles, where he conducted research on the application of the finite-difference time-domain (FDTD) to microwave structures. He applied system identification and diakoptics methods to the FDTD algorithm and researched the possibility of simulating microwave structures that include passive, nonlinear, and active lumped elements.

Currently, he is at the Jet Propulsion Laboratory, California Institute of Technology, where he has been involved in Interferometric Synthetic Radar research and remote-sensing data fusion. He has applied high-frequency and time-domain methods to the modeling of radar signal interactions with realistic environments. Dr. Houshmand is a member of URSI Commission B.